THE SIDDHANTAS AND THE INDIAN CALENDAR

D1878645

Owing to the antiquity and rarity of this book, certain pages are regrettably absent, and readability is compromised due to the blurring of the original text. If you possess any insights regarding the availability of these missing sections, kindly inform us. Your assistance is crucial in ensuring the completion of this significant work for the benefit of future generations.

THE SIDDHANTAS

AND

THE INDIAN CALENDAR

BEING A CONTINUATION OF THE AUTHOR'S

INDIAN "CHRONOLOGRAPHY"

WITH AN ARTICLE BY THE LATE DR. J. F, FLEET

ON THE MEAN PLACE OF THE PLANET SATURN

ROBERT SEWELL

GYAN

Published by

Gyan Publishing House
5, Ansari Road
Daryaganj, New Delhi-110002
Phone: 011-47034999, 9811692060
E-mail: books@gyanbooks.com

Distribution Network
gyanbooks.com
India, USA, Canada, UK, Australia

© Publisher

All rights reserved. No part of this work may be reproduced, stored, adapted, or transmitted in any form or by any means, electronic, mechanical, photocopying, micro-filming, recording or otherwise, or translated in any language, without the prior written permission of the copyright owner and the publisher.

ISBN: 978-93-6433-387-0 (PB)

First Published, 1924

2nd Impression 2025

Printed at: Gyan Press, Delhi.

The book is sold subject to the condition that it shall not, by way of trade or otherwise, be lent, resold, hired out, or otherwise circulated without the prior publisher's written consent.

THE SIDDHANTAS AND THE INDIAN CALENDAR
Author: ROBERT SEWELL

THE SIDDHANTAS AND THE INDIAN CALENDAR

THE SIDDHANTAS AND THE INDIAN CALENDAR

BEING A CONTINUATION OF THE AUTHOR'S

"INDIAN CHRONOGRAPHY"

WITH AN ARTICLE BY THE LATE DR. J.F. FLEET
ON THE MEAN PLACE OF
THE PLANET SATURN

ROBERT SEWELL

ASIAN EDUCATIONAL SERVICES
NEW DELHI ★ MADRAS ★ 1989

The Siddhantas and the Indian Calendar

BEING A CONTINUATION OF THE AUTHOR'S

"INDIAN CHRONOGRAPHY"

WITH AN ARTICLE BY THE LATE DR. J. F. FLEET

ON THE MEAN PLACE OF THE PLANET SATURN

BY ROBERT SEWELL, M.R.A.S.,

LATE OF H. M.'S INDIAN CIVIL SERVICE; JOINT AUTHOR OF "THE INDIAN CALENDAR", AUTHOR OF "THE AMARAVATI TOPE", "ANTIQUITIES OF MADRAS", "INDIAN CHRONOGRAPHY", "ECLIPSES OF THE MOON IN INDIA", "A FORGOTTEN EMPIRE", ETC.; HONORARY CORRESPONDENT OF THE ARCHÆOLOGICAL DEPARTMENT, GOVERNMENT OF INDIA.

[Reprinted by Order of the Government of India from a series of articles contributed to the "Epigraphia Indica".

CALCUTTA: GOVERNMENT OF INDIA
CENTRAL PUBLICATION BRANCH
1924

WORKS ON INDIAN ASTRONOMY, ETC.

BY THE SAME AUTHOR.

THE INDIAN CALENDAR, jointly with the late Sankara Balkrishna Dikshit, containing Tables for verification of ancient dates by the *First Ārya-Siddhānta* and the *Present Sūrya-Siddhānta*, and Tables of Eclipses of the sun visible in India by Dr. Robert Schram (1896).

ECLIPSES OF THE MOON IN INDIA (1898).

INDIAN CHRONOGRAPHY (1912).

Messrs. GEORGE ALLEN AND UNWIN, Ruskin House, Museum Street, Bloomsbury, London, W.C. 1

CONTENTS

a

ERRATA PAGE.

Indian Chronography, p. 62. For the calculation at top of page substitute the following :—

	Date	(Day)	Week-day	*h.*	*m.*	*s.*
True Mēsha-saṁkrānti by *Ārya-Siddh.* (*Table 1*), A.D. 1899	12 April	(102)	4 Wed.	6	7	30
Add śōdhya by *Ārya-Siddh.*	2	2	2	3	32	30
Mean Mēsha-saṁkrānti by *Ārya-Siddh.* . .	14 April	(104)	6 Fri.	9	40	0
For 5,000 years deduct (*Table on p. 61*) .	−1	−1	−1	−5	10	0
Mean Mēsha-saṁkrānti by *B.-S.* and *S.-Śirō.* .	13 April	(103)	5 Thur.	4	30	0
Deduct *Siddh.-Śirōmaṇi* śōdhya	−2	−2	−2	−4	20	56·8
True Mēsha-saṁkrānti by *Siddh.-Śirōmaṇi* .	11 April	(101)	3 Tues.	0	9	3·2

True Mēsha-saṁkrānti, then, by the *Siddhānta-Śirōmaṇi*, occurred on Tuesday, April 11th, A.D. 1899, at $0^h\ 9^m\ 3^s{\cdot}2$ after mean sunrise.

Line 18 from top.—For $101^d\ 0^h\ 22^m$ read $101^d\ 0^h\ 9^m$.

Line 19 from top.—For $217^d\ 17^h\ 11^m$ read $217^d\ 16^h\ 58^m$.

Line 21 from top.—For $578^d\ 17^h\ 40^m$ read $578^d\ 17^h\ 27^m$.

Line 24 from top.—For $17^h\ 11^m$ read $16^h\ 58^m$, and for $17^h\ 40^m$ read $17^h\ 27^m$.

PREFACE.

The present volume contains a number of articles separately published from time to time in the pages of the *Epigraphia Indica* and forming a continuation of my former work on the same subject—*Indian Chronography*,[1] —which itself was supplementary to *The Indian Calendar*[2] (Sewell and S. B. Dikshit) issued in 1896. At the end is reproduced, by the kind permission of the Council of the Royal Asiatic Society, a treatise with Tables by the late Dr. J. F. Fleet dealing with the planet Saturn.

The Tables in *Indian Chronography* having been numbered in continuation of those in *The Indian Calendar*, and the Tables contained in the *Epigraphia Indica* as well as the paragraphs of the texts having been similarly numbered in continuation of those in *Indian Chronography*, it is considered advisable, rather than start afresh here with new numbers, to adhere to the original design ; and so to prevent confusion and to avoid giving trouble to those workers who may have become habituated to the use of the older books and of the sets of Tables as originally published.

There appears to be no necessity to describe over again in this volume the whole chronological and calendrical system of the Hindus, nor the particular method adopted in this and in the former works. Full explanation has been given in those volumes. Our method is the method called the *a, b, c* system of Largeteau, with which Professor Jacobi of Bonn made us familiar, and with which students of the subject must by now have become well acquainted. It is based on measurement by division of the great circle into ten-thousand parts, and has the great advantage of being applicable to both time and space. It is described in *Indian Chronography* (*§§ 19A-26, pp. 7-9*).

Results of greater accuracy than heretofore can be obtained by the use of the Tables here presented, since the figures are given with four decimal places instead of as previously in whole numbers, and so give us planetary positions correct to a quarter of a second whether of space or time. The time-unit of the *Indian Calendar* is $4\frac{1}{4}$ minutes ; that of Rao Bahadur L. D. Swamikannu Pillai's *Indian Chronology* is about 14 minutes. Very correct results can also be obtained by Professor Jacobi's *Special Tables* published in Vol. I of the *Epigraphia Indica*, but as these are stated in degrees, minutes and seconds they are a little troublesome to convert into time-reckoning.

The processes to be followed in computing the details of a date by the Tables are in each case explained in the Examples given at the end of the several articles. It is only necessary to work by these and to be careful to use the proper Tables. The most detailed set of examples is that which is included in the article on "*The First Ārya-Siddhānta—true system*" ; and any student of the subject who is not thoroughly acquainted with our method of calculation (when using the apparent motion of sun and moon) is recommended to go through these carefully before he embarks on computation by the other astronomical authorities of India. The manner of fixing the *mean* places of the sun and moon at any moment is described in the articles devoted to *The First Ārya-Siddhānta* and *Brahma-Siddhānta* mean systems.

Several General Tables applicable to all the *Siddhāntas* have been taken from *The Indian Calendar* and *Indian Chronography*. These are required in order to fix the day of the month and week-day according to the European calendar, as well as for other purposes. Most of them

---

[1] Messrs. George Allen & Unwin, Ruskin House, Museum Street, Bloomsbury, London, W. C. (1912).

[2] Messrs. Swan Sonnenschein & Co. The *Indian Calendar* (1896) was followed by *Eclipse of the Moon in India* (1898) published by the same Firm ; now Messrs. George Allen & Unwin.

are included amongst the Tables which deal with the *First Ārya-Siddhānta*—"*true*" *system*. Such are Tables LXII, LXVIII, LXIX, LXX (to which a supplement has now been added by Tables XCIVA-F at the end of the volume), and LXXI. This assists the worker to complete all necessary calculations without having to refer to any other volume.

The Tables now published enable dates to be verified according to the requirements of the *First Ārya-Siddhānta* (mean motions of sun and moon) from A.D. 500 to 1400, and (true or apparent motions) from A.D. 900 to 1900; by the *Brahma-Siddhānta* (mean motions) from A.D. 600 to 1400, and (true motions) from A.D. 600 to 1200; and by the *Siddhānta-Śirōmaṇi* (true motions) from A.D. 1100 to 1900.

These Tables, coupled with those for the *Sūrya-Siddhānta* given in the *Indian Calendar* and in Rao Bahadur L. D. Swamikannu Pillai's *Indian Chronology*, cover the whole ground as yet possible to explore.

The Indian Astronomical authorities.

The earliest available information as to the study of astronomy in India is obtained from the *Vedāṅga Jyōtisha*, the character of which is, however, mostly astrological. Here, as well as in the *Brāhmaṇas* mention is made of the most ancient division of the year into three natural seasons, evidently, like those of the Egyptians, agricultural in origin and therefore essentially solar. The Egyptian division was into the three seasons of sowing, growing, and harvest. The three early Indian divisions, each of four months, were *Grīshma*, *Varshā*, and *Hēmanta*. This division, being one seemingly of natural origin, and therefore popular, lasted for many centuries. An inscription of a Pallava king[1] in South-India at the close of the 5th century A.D. records the date as in the third fortnight of *Hēmanta* and the 13th day; and similarly with other records of about the same period issued by Kadamba kings[2] and the Guptas.[3]

Lunar motions were, of course, carefully observed from the earliest times, and the twelve lunar months were adapted to the solar seasons by the periodical interpolation of a lunar month.

A later solar division of the year was into six double-months, viz. *Vasanta* (spring), *Grīshma* (summer), *Varshā* (rains), *Śarad* (autumn), *Hēmanta* and *Śiśira* (the cool season).

Later still, when the knowledge of solar astronomy had considerably developed, came the modern division into twelve solar months, with the lunar months adapted by interpolation.

Anciently the lunar months had seasonal names, a list of which is given in the *Indian Calendar*, p. 24. The modern names of the lunar months are stellar, being derived from the *nakshatras*.

The 27 *nakshatras*, or divisions of the ecliptic circle, otherwise "asterisms" or "lunar mansions," are mentioned in the *Vedāṅga*, but were not commonly used for recording dates or as essential parts of the daily calendar till about the 10th century A.D.

From about B.C. 300 onwards there was constant communication and traffic, both by sea and land, between India, Persia, and Greece, and the Hindus became acquainted with the principles of Greek, and later on of Roman, astrology and astronomy. Attracted at first by the astrology of the Westerns they were eventually led, after several centuries, to adopt their astronomy also

Professor Jacobi has called attention to the fact that the twelve signs of the zodiac were not heard of till the time of Firmicus Maternus (A.D. 336); and it was near about A.D. 400 before these were finally accepted as essential parts of the Indian astronomical system, which was based on the astrology of Firmicus and of Paulus Alexandrinus (A.D. 378). Thus it is

[1] The Omgodu (? Ongole), Nellore District, inscription of Vijaya Skandavarman Pallava. *Ep. Ind.*, XV, 246

[2] E.g. *Indian Antiquary*, VII, 37.

[3] Kumāragupta, A.D. 431 (*Ep. Ind.*, II, 262 f.).

probable that all the known astronomical works earlier than the *First Ārya-Siddhānta* (A.D. 499), with the exception of the very ancient *Vedāṅga*, were composed between (about) A.D. 350 and 500.

Four such works are mentioned in the *Pañcha Siddhāntikā* of Varāhamihira (*c.* A.D. 550). They are the *Paitāmaha-Siddhānta* and the *Rōmaka*, *Pauliśa*,[1] and "Original"[2] *Sūrya-Siddhāntas*. Fleet considered that the *Paitāmaha-Siddhānta* was merely the *Jyōtisha Vedāṅga* under another name.[3] The elements of none of these four authorities are known and therefore no reliable Tables can be drawn up for calculation according to their requirements. We only come to firm ground at the end of the 5th century A.D.

In A.D. 499 was produced the *Āryabhatīya*, or *First Ārya-Siddhānta*, of the astronomer Āryabhaṭa. The elements of his system are well known and are fully dealt with in the section of this volume devoted to that work, so far as they affect the preparation of the almanac.

About a century later was composed the *Brahma-Siddhānta* of Brahmagupta (A.D. 628), which introduced certain new principles into the Hindu astronomical system, notably the slight but constant shift of the points of the sun's apsis (Hindu astronomy always treats the sun as a planet).

In A.D. 638 or thereabouts Lalla introduced a *bīja*, or correction, into three of the elements of the *Ārya-Siddhānta*.

About A.D. 950 appeared the *Mahā Ārya-Siddhānta*, called in these volumes "the Second Ārya." S. B. Dikshit thought that it was nowhere in use for a long time; and for that reason it has not been thought necessary to provide general working Tables based on its requirements. Allusion is made in it to another work, the *Parāśara Siddhānta*, which is not now extant.

Fifty years or so later—the exact date has not been discovered—was composed the "*Present*" *Sūrya-Siddhānta* by an author whose name is lost. It has become the most important authority for the preparation of almanacs in large parts of India, and its contents have been made available from several manuscript copies. It is supposed to have come into general use about the beginning of the 12th century A.D., superseding the "Original" *Sūrya-Siddhānta* in the tracts where the latter had been used.

From about this period therefore there have existed three distinct schools of astronomy in India, namely the Ārya, Brahma, and Saura schools.

The *Rājamrigāṅka* (A.D. 1042) was the next important work to appear. It followed the *Brahma-Siddhānta*, but with certain corrections. No complete copy of it is known to exist, but S. B. Dikshit was convinced that, so far as regards the preparation of the almanac, its results were the same as those obtained by the use of the later and better known *Siddhānta-Śirōmaṇi*.

The *Karaṇa-prakāśa*, a commentary and guide based on the *Ārya-Siddhānta* of Āryabhaṭa as corrected by Lalla, was composed in A.D. 1092. It is an authority still used in Central India by the framers of *pañchāṅgs*.

In A.D. 1150 Bhāskarāchārya produced his *Siddhānta-Śirōmaṇi*. It followed the Brahma school and was adopted as a standard in succession to the *Brahma-Siddhānta*, whose elements as corrected by the *Rājamrigāṅka*, it generally accepted. It differed however in certain respects, and amongst others in its estimate of the rate of shift of the sun's apsidal points; and

[1] Two other *Pauliśa-Siddhāntas* are mentioned by a writer of A.D. 966. The name is derived, so Albērūnī tells us, from "Paulus the Greek," otherwise Paulus Alexandrinus.

[2] So called to distinguish it from the "Present" *Sūrya-Siddhānta* of about the early 11th century A.D.

[3] See note 2, p. 157, *Indian Chronography*.

in consequence of this it differed in the fixture of the exact moment at the beginning of each solar year when the true sun reached long. 0°, or the moment known as "true *Mēsha-saṁkrānti*," marking the true sun's entry into the first zodiacal sign Mēsha.

Following one or other of these schools there have been prepared at different times a number of *Karaṇas*, or treatises, for the guidance of those whose duty it has been to prepare local almanacs in all parts of India. And since these authorities differ slightly in their estimates of the laws that govern planetary phenomena it follows that there must be differences between them in the results obtained. There will often, for example, be a difference of one in the number of the *tithi* associated with (because ending within the limits of) a certain civil day. In intercalary years there will often be a difference of one in the lunar month intercalated or suppressed; and there are cases where by one authority a lunar month was intercalated and another suppressed, while by another there was neither intercalation nor suppression in the year concerned. There are also a number of cases in which the cycle-name of the Jovian cycle of sixty saṁvatsaras given to a year by one authority is different from that given by another.

Hence it is obvious that if anyone attempts to verify a date, whether for historical or judicial purposes, solely by one of these authorities to the exclusion of others he is liable to arrive at an erroneous conclusion. No one set of Tables, still less any ephemeris, based on the principles of a single authority can be safely used for the settlement of dates of all times and places. The correct course is to test the date by the authority generally believed to have been in use in the tract and at the period to which it apparently belongs, and, if such examination does not yield satisfactory results, then to try it by other possible authorities and systems.

Verification of dates of different periods.

It is of the utmost importance to remember that prior to the middle of the 11th century A.D. dates were, probably in all parts of India, calculated by the mean motions of the sun and moon; and that the same system may have lasted for many years later in some parts. It is only since that time that it has become the custom to use "true" or apparent solar and lunar motions.[1]

Dates earlier than A.D. 500.

It has been stated above that the earliest Hindu astronomical authority on which we can depend for the formation of reliable Tables is the *First Ārya-Siddhānta*, composed in A.D. 499-500. It is almost certain that no mention will be made in dates earlier than this of any other detail than the year and the lunar month and *tithi*, and for that reason such dates cannot be verified. An approximation, however, is possible, and as a guide to the attainment of this a note of some length will be found at the end of this Preface.

Dates between A.D. 500 and 628.

Dates between A.D. 500 and 628 should be computed by the Tables herein given for the *First Ārya-Siddhānta—mean system*, and with the use of *pūrṇimānta* lunar months, i.e. months beginning with full, and not with new, moon. The Tables are numbered LXXVI to LXXXI.

Dates between A.D. 628 and 1000.

Dates between A.D. 628 and 1000 must be tested by both the *Ārya-Siddhānta—mean system* and the *Brahma-Siddhānta—mean system* Tables. These are respectively Tables LXXVI to LXXXI and XC to XCIII. The lunar month system may have been *pūrṇimānta*, beginning with full moon, or *amānta*, beginning with the next following new moon. The earlier the date is

[1] Some Western Chālukyan records in the Bellary District of the Madras Presidency seem to prove that the *Brahma-Siddhānta mean system* was used till late in the 11th century; certainly one of them carries the practice down to A.D. 1141. This is an inscription of the fourth year of Jagadēkamalla II at Sindigēri, Bellary Taluq.

the less likely is it that the *amānta* system was used. Moreover the *pūrṇimānta* system is more common in the north than in the south of India.

Dates between A.D. 1000 and 1150.

The *Present Sūrya-Siddhānta* was composed about A.D. 1000, and, as already mentioned, there were since that time three distinct schools of astronomy in India—*Ārya*, *Brahma*, and *Saura*—each of which had its devotees. Dates between A.D. 1000 and 1150 should be examined in turn by the Tables given below for the *Ārya* and *Brahma-Siddhāntas* as well as by the Tables for the *Sūrya-Siddhānta* contained in the *Indian Calendar* (Tables I to X); testing them first by the mean system and afterwards by the "true" or apparent system.

Dates since A. D. 1150.

The *Siddhānta-Śirōmaṇi* supplanted the *Brahma-Siddhānta* at some period subsequent to A. D. 1150, the date of its composition. Dates therefore subsequent to A.D. 1150 should be examined by the *Ārya-* and *Sūrya-Siddhāntas* and the *Siddhānta-Śirōmaṇi*; in later times solely by "true" solar and lunar motions, but in earlier times by their mean motions also. The *Siddhānta-Śirōmaṇi* Tables are LIV, A and B, to LX. They have been calculated solely by true or apparent planetary motion; but since Bhāskarāchārya, the author of that work, was a follower of the Brahma school the Table prepared for the *Brahma-Siddhānta* mean system may be used for *Siddhānta-Śirōmaṇi* mean system calculation once the year is known. It is not probable that the mean system was anywhere in use after A.D. 1400. Since A.D. 1150 it may be taken for granted that the lunar month system in Southern India has been *amānta* and in Northern India *pūrṇimānta.*

Some hints.

The mistake generally made by those who, whether for pseudo-historical purposes or in perpetration of a fraud, take upon themselves to invent the details of an imaginary past date, is to enlarge too much. They state not merely a lunar month and *tithi*, but add a week-day, the number and name of a solar month and day, the name of a *nakshatra* or *yoga* and so on, with the idea of creating an impression of great accuracy. And here they trip themselves up. For, the almanacs of years long past having of course disappeared, it would be little less than a miracle if all these details, depending as they do on the exact positions of the sun and moon at a particular moment of time, could be guessed correctly.[1]

But there are other ways by which, sometimes, a stated date may *primâ facie* be judged and condemned, and it will be well to call attention to some of these. A forged date often mentions details which were not in use at the time pointed at, or states the year of an era belonging to a time when that era was never quoted. The following points should be noted and borne in mind by those concerned in arriving at the truth.

The planetary names of the days of the week—the day of the sun, of the moon, etc.—were introduced from Greek astronomy into India not long prior to A.D. 400, the Romans having adopted them for general use from about the year A.D. 200. Fleet treats of this matter in an article in the *Journal of the Royal Asiatic Society* for 1912 (pp. 1039 ff.), explaining the order of these names from the rules of Paulus Alexandrinus. The earliest known

[1] I have lately published in the *Journal of the Royal Asiatic Society* a paper containing a critical examination of the dates quoted by the author Merutunga, in his *Prabandha Chintāmaṇi*, a work of professedly historical character, in which the dates—many of them nominally belonging to a time long past—contain a number of the details referred to. The result of the examination goes to shew that at least many of these details were inserted at random, and therefore that no date can be depended upon as genuine. If some parts of a date are manifestly the outcome of the author's imagination, no trust can be put upon any part of it. In every date quoted in the work the name of the *nakshatra*, which gives the position of the moon in the heavens, is totally wrong, and quite incompatible with the moon's place on the day intended as set forth in the oher details of the date. The author was evidently in no sense an astronomer. He entered details at random and trusted that none of his readers would discover the truth

genuine instance of the use in India of these planetary names is in a Gupta inscription of A.D 484. The next is a record of date just earlier than A.D. 578. Kielhorn noted two, one from the Nellore District on the east coast of the Madras Presidency, and one from Banavâśi in North Kanara, respectively in A.D. 664 and 692. The practice only became more common after A.D. 900. So that a date professedly earlier than that should, if it mentions the day of the week, be looked upon with suspicion; and, if it should profess to belong to a year earlier than A.D. 400, should be treated as almost certainly fabricated.

The *pūrnimānta system of naming the lunar months* as beginning astronomically with the moment of full moon prevailed over all India in early years; and still does so in the north; while the *amānta* system, by which the month begins with the succeeding new moon, has succeeded it in the south. The earliest genuine inscription-date known to Kielhorn which was in *amānta* reckoning belongs to the year A.D. 794, and is contained in the Paiṭhān plates of the Rāshṭrakūṭa king, Govinda III.

The solar saṁkrānti—the entrance of the sun into one of the signs of the zodiac—is not known to have been definitely mentioned in any inscription earlier than the 10th century A.D. It is found, however, in a record of one of the Western Gaṅga kings of the peninsula in A.D. 975. But setting aside the actual mention of a *saṁkrānti* as such, we know for a fact that the solar months, as divisions of time, were used in the Tamil country of the south, in preference to the lunar months, from about A.D. 900 onwards. A record in South Arcot of the Chōḷa king Parāntaka I,[1] dated in a year corresponding to A.D. 943, mentions the *nakshatra*, solar month and week-day—"Rēvati, Saturday in Makara." In more modern times the lunar tithi is also stated, but not the lunar month. In the Telugu country after about A.D. 950 the solar months were often named, but they were ancillary to the lunar months which took first place.

The nakshatras, or stellar divisions of the ecliptic, were known in late Vedic times and were used for astrological purposes; but they were not commonly mentioned in dates till about the 10th century, after which their employment became common. The Singhalese *Dīpavaṁsa*, however, the compilation of which ceased about the middle of the 4th century, mentions the *nakshatra* in which the moon stood at the time of the anointing of one of the kings of Ceylon. Only one of the Gupta records mentions a *nakshatra*; this was in A.D. 705, in the reign of Mānadēva.[2]

The *yoga* is a purely astrological fixture, and is seldom mentioned in the dates of inscriptions, though doubtless it was held to be of great importance in the matter of ceremonial observances, rites and sacrifices.

The saṁvatsaras of the sixty-year and twelve-year cycles of Jupiter. Dr. Burgess was of opinion that the years of the Jupiter cycle with their individual names were first introduced into the Indian calendar about A.D. 350. Judging from discovered records it would appear that the cycle more commonly used in early years was that consisting of twelve years, named after the twelve lunar months with the prefix *Mahā* (e.g. *Mahā Chaitra, Mahā Vaiśākha*), the cycle of sixty *saṁvatsaras* being contained in five 12-year cycles. A table shewing the working of this arrangement is given in *The Indian Calendar* (*Table XII, p. cxxx*) and in *Indian Chronography* (*Table XXXII, p. 152*). Three Gupta inscriptions of A.D. 475, 482 and 510 fix the dates by the number of the year of the Gupta era and by the 12-year cycle-names "*Mahā Vaiśākha,*" "*Mahā Āśvina,*" and "*Mahā Chaitra*" respectively.[3] From about

---

[1] No. 559 of Mr. Rangāchāri's List, Vol. I, South-Arcot *Epig. Reports*, No. 735 of 1905. *Epig. Ind.*, VIII, 261. This is the earliest Chōḷa date that, according to the late Dr. Kielhorn, is capable of verification.

[2] No. 494 of Professor Kielhorn's List of inscriptions in Northern India. (*Epig. Ind.*, V, *Appendix*, p. 69.)

[3] Kielhorn's *Inscriptions of Northern India* (*Epig. Ind.*, V), Nos. 451, 453, 456.

A.D. 550 onwards the sixty saṁvatsara-names were more generally used. Varāhamihira, who died in A.D. 587, mentions them all. No instance, however, has been as yet met with in a record of date earlier than A.D. 602, and doubt has been expressed whether the name in that case was really intended to be read as being the saṁvatsara-name of the year. If this is set aside the earliest instance is in the Alās plates of the Rāshṭrakūṭa king Govinda II, A.D. 770.

The lagna, or the rising on the horizon of a sign of the zodiac, is sometimes noted on a record. Its function is to fix the time of day of the action commemorated to within a space of two hours. Kielhorn states that the earliest instance of its use with which he was acquainted is in an Eastern Chālukyan inscription of King Amma II in the Telugu country, the date of which is A.D. 945. But it is said to have been used in Cambodia at an even earlier date.

Eras.

It is advisable to take careful note also of the mention of an era in dates of professedly very early times; for it sometimes happens that a document (perhaps a copper-plate title-deed) can be readily recognised as a forgery by reason of the quoted date stating the year of an era belonging to a period when that era had not come into use in the preparation of almanacs. In such cases the following notes will be found useful.

The *Māḷava-Vikrama era.* Up to the present no date has been found which definitely mentions this era earlier than A.D. 436; though one has been brought to light at Bijayagaḍh in Rajputāna, which has been held to be *possibly* a genuine date and belonging to this era, and which is as old as A.D. 372.

The *Kalachuri-Chēdi era.* The oldest known inscription in this era, dated in the year "207," is engraved on the Pārḍī (Surat) plates of Dahrasena, the corresponding year being A.D. 456 or 457.

The Śaka era. The earliest known date in this era is "Śaka 500 expired," or A.D. 578. This is at Bādāmi. In the north the earliest known is dated "Śaka 784 expired" or A.D. 862. It was found at Dēōgaḍh in the Central Provinces.

The Kaliyuga era. The earliest known record which mentions this era is a Chālukyan inscription of King Pulakēśin II found at Aihole, the corresponding year A.D. being A.D. 634-35. The next belongs to the year A.D. 770, and the next to A.D. 866. These are all in the peninsula. In Northern India the earliest known is one of date A.D. 1169, or 1170.

Variation in Hindu practices.

Variation in Hindu practices.

The Tables in this volume are designed for the purpose of enabling workers to obtain the desired result *scientifically*—that is to say, a result following from calculation based on the elements and postulates of each of the Siddhāntas dealt with. Whether these elements and postulates were on all occasions fully and accurately adhered to by the framers of local almanacs is another matter altogether. And again it must never be forgotten that whereas the Tables deal always with the moment of *mean sunrise* on the civil day concerned, the almanac employed at the time of the composition of the record may have been prepared for the moment of *true sunrise* at the principal town in the locality. *True time* also may have been used instead of *mean time*; and whole numbers alone may have been employed for the necessary calculations, all fractions being omitted. Any one of these things may, in close cases, make a difference of one in the number of the *tithi* that gave its name to the day, and sometimes also a difference in the name of the lunar month.

An instance of the difference of practice referred to will be found in the following notes made by a scientific writer a hundred and thirty years ago. Henry Cavendish, F.R.S. read a paper in A.D. 1792 before one of the learned societies of London on the Hindu calendar. It was published in *Philosophical Transactions (Vol. 82, p. 383 ff.)* and has lately been reproduced with his other essays by the Cambridge University Press in a volume entitled "*Scientific papers.*" The author had been carrying on a correspondence with Mr. Charles Wilkins in India, and had obtained from him three *patras* (*pañchāngs*, almanacs), one from Benares, one from Thānā in the island of Salsette near Bombay, and one from Nadiya, north of Calcutta. As to the second he writes:—"It appears to be a copy of a Benares patra, as it is disposed in the same form as the first, and *is adapted to the same latitude and longitude.*" We learn therefore that the Pañchāng-Brahmans of Thānā did not make any changes in the Benares almanac so as to suit the precise geographical requirements of their own country. They were content, at Bombay, to calculate for sunrise as it befell at Benares.

But another of Cavendish's correspondents, Samuel Davis of Bhāgalpūr, who was in possession of a copy of the *Sūrya-Siddhānta* and had translated part of it, informed him that, whereas in the north of India almanacs were prepared by specially trained men at three centres, Benares, Nadiya and Tirhut, they (the almanacs) were subject to alteration when scattered over the country to different places. These patras, he says, "are annually dispensed throughout the adjacent country. Every Brahmin in charge of a temple, or whose duty it is to announce the time for the observance of religious ceremonies, is furnished with one of these almanacs *and, if he be an astronomer, he makes such corrections in it as the difference of latitude and longitude render necessary.*" Here then is evidence that at least in some parts of India, if not in all, the local almanac of one tract may have differed slightly from that used in another even in the same year.

Tables F and G in my "*Eclipses of the Moon in India*" *(pp. li to lv)* state the correction from mean to apparent time for every day in the year and for 1700 years past, and also give the apparent ("true") time of the rising and setting of the sun in different latitudes at all seasons of the year. Rao Bahadur L. D. Swamikannu Pillai has given a very elaborate Table of sunrises in his *Indian Chronology* (Table XIII), occupying 36 pages.

These differences must of course be allowed for before condemning a date as unsound.

Four distinct rules governing beginning of solar months.

When examining a date which states the number of a day of a solar month, as, for instance, "the 12th day of Kanyā," it must not be forgotten that there are four distinct rules, observed respectively in Bengal, Orissa, in the Tamil country and in Malabar, for fixing the first civil day of the solar month. These rules are clearly given in the *Indian Calendar* (p. 12) and in *Indian Chronography (§ 43, pp. 18, 19)*. The operation of these rules depends upon the hour of the day on which the solar *saṁkrānti*, that is the entrance of the sun into the zodiacal sign, takes place. If, to take our example as an instance, the Kanyā *saṁkrānti* in the given year was found, in the ordinary course of calculation by any of the Tables, to have occurred more than 18 hours after sunrise on a certain day, then by the Bengal rule the civil day called "1st Kanyā" was the third day later; whereas by the Orissa rule, when the Amli or Vilāyati era was in use, the "1st Kanyā" was the same day as that on which the *saṁkrānti* took place; and by the Tamil rule it was the next day. Hence the day called "12th Kanyā" was in one tract two days later than the day so called by the people in another tract. The difference, however, can never be more than two days.

Rules for intercalation of lunar months.

Lastly a word about the intercalation of lunar months when the *pūrṇimānta* system of naming the months was in force, i.e. the system whereby the month begins at the full moon next previous to the new moon which marks the beginning of the *amānta* lunar month. It will be seen from the *Indian Calendar (§§ 45-49, and Table, p. 26)* that there has existed more than one system of naming the halves, or fortnights, of intercalated, *pūrṇimānta* months. It is not necessary to reproduce here all the articles and Table relating to the subject, but merely to call attention to it.

Note on calculation in N. India in A.D. 1792.

It may be as well to note one or two interesting points in the essay by Henry Cavendish referred to above and written in 1792. He makes it clear that the almanacs of that day at Benares were prepared by the *Sūrya-Siddhānta,* while, so it may be inferred, those framed at Pondicherry followed the *Ārya-Siddhānta*. This of course was to be expected.

Analyzing a Benares patra of 1792 Cavendish states that the true solar year "began, according to the principles delivered in the *Sūrya-Siddhānta*, on April 9 at 22^h 14^m after midnight of their first meridian, which is about 41^m of time west of Calcutta"[1]; and adds: "But according to Mr. Gentil's account of the Indian astronomy it began 3^h 24^m earlier."

M. Le Gentil went to Poudichēri in 1769 to study the transit of Venus and stayed there nearly two years, employing his time in acquiring a general knowledge of Hindu astronomy.

By the *Sūrya-Siddhānta* (*Indian Calendar, Table I, p. xcv, col. 17a*) the moment of beginning of the true solar year on "the first meridian," i.e. on the longitude of Ujjain, was, in A.D. 1792, at 16^h 12^m after mean sunrise on 9 April, i.e. at 22^h 12^m after the previous midnight. Mr. Swamikannu Pillai (*Indian Chronology*, Table X, p. 120) quotes the moment as "9 April ·6747," or 16^h 11^m 34^s 08. Thus the difference between us and the Benares patra is only 2 minutes.

Now M. Le Gentil's account made the year begin, so says Cavendish, 3^h 24^m earlier. I suspect that "3^h" is a mistake, either by Le Gentil or Cavendish or the printers, for 2^h. For as a fact according to the *Ārya-Siddhānta*—the authority generally used in South-India —the solar year corresponding to A.D. 1792-93 began 13^h 50^m after mean sunrise (*Table I, Indian Calendar, or Table LXI below*); or 2^h 24^m earlier than it did by the *Sūrya-Siddhānta* if we accept Cavendish's figure for the latter as 16^h 14^m.

Cavendish proceeds to describe the divisions of the year solar and lunar, the tithi, the lunar months, and their intercalations; and he notes a difference of practice between Benares and Nadiya. As to the former he writes:—"The civil day begins at sunrise . . . . The civil year is luni-solar, consisting of 12 lunar months with an intercalary month inserted between them occasionally. It [the luni-solar year] begins the day after the new moon next before the beginning of the solar year . . . Moreover, in the years which have an intercalary month, this [intercalary] month begins at the day after the new moon; but notwithstanding this the ordinary civil month begins at the day after the full moon. To make their method more intelligible I will call the time from new moon to new moon the natural month. The civil month, Visākha begins at the day after the full moon of that natural month which commences at the beginning of the civil year, or, in other words, at the day after the full moon of that natural month during which the sun enters the first Hindoo sign . . . . A consequence of this way of counting the months is that the first half of Chitra falls in one year, and the latter half in the following year . . . . . *In these almanacs no notice is taken of solar months* . . . . which seems to shew that in the countries which use the Benares patra it is not customary to date by the solar month.

"In those parts of India which use the Nadeea patra the case is quite different. *This almanac contains the names of the solar and lunar month* . . . . The lunar months begin, not at the full, as in the Benares patra, but at the new moon, and are called by the name of that solar month which ends during the course of them; for example the lunar month during which the solar month Visākha ends, is called Chandra (or lunar) Visākha, so that each month begins a fort-

[1] The meridian of Ujjain is 12° 38′ west of Calcutta, the time-difference being actually 50^m 32^s.

night later than by the Benares patra. Mr. Wilkins informs me that the Hindoos of Bengal, in all their common transactions, date according to solar time . . . . . and use what is commonly called the Bengal era, but in the correspondence of the Brahmins, dating books, and regulating feasts and fasts they generally use the teethee [tithi]."

It appears therefore that the *pūrṇimānta* system of lunar months obtained in A.D. 1792 at Benares, while at Nadiya in the same year the system was *amānta*. This should not be forgotten when dealing with the old dates of these countries.

The computation of dates earlier than A.D. 500.

It has been stated above that prior to the appearance of the *Āryabhatiya* or *First Ārya-Siddhānta* of Āryabhaṭa (A.D. 499), though it is known that several astronomical treatises had been composed, their leading principles and postulates have not been brought to light, and therefore that no reliable Tables can be prepared for the purpose of calculation of a date by any of them. How then are we to proceed when desirous of examining a date belonging to such an early period?

It seems useless to attempt more than an approximation for two reasons. The first is that—since it is almost certain that no detail will, if the date be genuine, be mentioned other than the year of one of the eras and the lunar month and tithi,[1] – the actual day cannot be verified; and the second is that, even if it could be verified, there is no historical or other reason why any particular trouble should be taken in that direction. The information will enable us to state the year A.D. and the time of year within, probably, a month. That will surely suffice. If a number of other details are given the document must be looked on with suspicion, as before remarked.

But the following hints may be found of use to those engaged in the decipherment of such records.

If no era is mentioned all mere guessing is useless, and the period when the inscription or document was engraved or written can only be learned from the characters. Such a date must be entrusted to a skilled palæographist.

When the year of an era is definitely stated it can be converted into the corresponding European year by aid of the notes, *a-f*, which follow, but with the reservation that it cannot, perhaps, be definitely stated whether the quoted year was a solar year, or a luni-solar year, and if the latter whether it began with the month Chaitra or some previous month such as Kārttika or Āśvina.

(*a*) *The Kaliyuga era.* It is most unlikely that the year of the Kaliyuga will be found quoted in a date earlier than A.D. 500, but should it be so it is necessary to remember that, by reason of the length of one solar year being differently estimated by different authorities, the same year may not always have borne the same Kaliyuga number. According to the *Vedāṅga Jyōtisha* and the *Paitāmaha-Siddhānta* the solar year consisted of 366 days; the *Rōmaka* made it $365^d\ 5^h\ 55^m\ 12^s$; the *Paulisa* $365^d\ 6^h\ 12^m\ 0^s$; while the *Original Sūrya-Siddhānta* and the other two *Pauliśa-Siddhāntas* mentioned by Varāhamihira estimated it at $365^d\ 6^h\ 12^m\ 36^s$. Thus by the year A.D. 500 the number of the year of the Kaliyuga according to the *Jyōtisha* would have fallen seven years earlier than the same year calculated by the rules of Āryabhaṭa. "K.Y. 3600" by the Ārya would be K.Y. 3593 or thereabouts by the *Jyōtisha* rule. The same year, K.Y. 3600 began by the *Romaka* 42 days earlier than it did by the *Ārya*; by the *Pauliśa* it began 30 hours earlier; and by the *Original Sūrya* and the other two *Pauliśas* it began 6 hours later.

(*b*) *The Mālava-Vikrama era.* To convert a year of this era into a year A.D., deduct 57 from the number quoted. Chaitrādi Vikrama 428 expired=A.D. 371-72. For years B.C., or

[1] Only one record is at present known to exist of earlier date than A.D. 500 which mentions more than the month and tithi. This is the Eran pillar inscription of Budhagupta, and it includes the name of a week-day; enabling Prof. Kielhorn to fix the date as 21 June A.D. 484 (*Epig. Ind.*, V, App., p. 64, No. 454).

a Vikrama year of number less than 58, refer to Table XXXVIIIA, *Indian Chronography*, p. 160. In Kielhorn's List in *Epigraphia Indica*, Appendix, Vol. V,[1] there are only three records earlier than A.D. 500. The Vikrama year generally began with the month Āshādha or Kārttika.

(*c*) *The Śaka era.* To obtain the year A.D. add 78 to the number of the quoted year. Śaka 223 expired = A.D. 301-2. All records known to Kielhorn bearing Śaka dates earlier than A.D. 500 were found, on careful examination, to be spurious.

(*d*) *The Kalachuri-Chēdi era.* To obtain the year A.D. add 247 to the given number of the year. Kal. Ch. 252 expired = A.D. 499-500. Note that the *Kalachuri-Chēdi* year begins with the beginning of the lunar month Āśvina preceding the month Chaitra which marks the beginning of the Chaitrādi year. Kielhorn notes eight such records earlier than A.D. 500.

(*e*) *The Gupta era.* To obtain the A.D. year add 319 to the number of the year quoted. Chaitrādi Gupta 129 = A.D. 448-49. Kielhorn's List contains 21 inscriptions dated in this era earlier than A.D. 500.

(*f*) *The Valabhi era.* This was a continuation of the Gupta era. Its years begin, not with Chaitra, but with the preceding Kārttika.

The epochs of the other eras are subsequent to A.D. 500.

For a Table of correspondence of all eras refer to Table II, Part III, *Indian Calendar*.

R. SEWELL.

THE CYCLE OF JUPITER,

AND

THE NAMES OF THE SAMVATSARAS APPLIED TO HINDU SOLAR YEARS

(Previously published in Epigraphia Indica, Vol. XIII, pp. 61—103.)

Introductory.

199. In my "*Indian Chronography*" (pp. 46-65 and Tables XXVII to XXXI A) I have shewn how the exact beginning and ending of a Jovian year can be ascertained, according to the various astronomical authorities in use in India, from K. Y. 3117 (A.D. 16-17) to 5133 (A.D. 2032-33). These calculations were made, as regards the motion of Jupiter, by the mean sign system, that is to say, by conceiving the length of each saṁvatsara as being the time occupied by the planet in passing by his mean motion through one sign, or 30°, of the Hindu zodiac; and they were made as regards the solar year by determining the number of days and decimals of a day by which each saṁvatsara began after apparent Mēsha-saṁkrānti[1] in each solar year. In the single case of the *Original Sūrya-Siddhānta*, however, (Tables XXX and XXX A) the computation was made with reference to the moment of mean Mēsha-saṁkrānti; for the reason that it is almost certain that during the whole period of its use the Hindu calculators worked entirely on the mean system.

200. Since the publication of the *Indian Chronography* I have examined a large number of dates of Indian inscriptions, and have come across many cases where the name of the given saṁvatsara does not exactly accord with the solar year with which it should be connected according to rule framed with apparent Mēsha-saṁkrānti as the guiding-point. Sometimes this may be due to mere accident; sometimes it may arise from the use of the name of the saṁvatsara current at the moment of the action commemorated by the record instead of that of the saṁvatsara current at Mēsha-saṁkrānti of the current year. But it is certain that at least up to the time of Śrīpati (about A.D. 1040) and probably for a long time afterwards the Hindu calculators based their determination of the Jovian saṁvatsara current at Mēsha-saṁkrānti (and, therefore, according to custom giving its name to the entire solar year) not with reference to the *apparent* but to the *mean* Mēsha-saṁkrānti; and this would often cause the solar year to be called by a different Jovian cycle-name. The late Sankara Balkrishna Dikshit hinted (*Indian Calendar*, p. 28) that possibly this practice lasted till as late as the 15th century.

201. My tables in the *Indian Chronography* were intended to enable the beginning and ending time of a saṁvatsara to be calculated by time measured from a known point, and since Table I of the *Indian Calendar* stated that point (apparent Mēsha-saṁkrānti) in each year it was obviously most simple to use that point. The tables were not framed to serve as a guide to the Jovian name to be correctly applied to each solar year, though that could be gathered from them with a little trouble and care.

202. It is evident, however, that we can only be secure in our acceptance of, or rejection as irregular of, an inscription-date, if, besides the tables calculated by the apparent Mēsha-saṁkrānti, we have others calculated by the mean Mēsha-saṁkrānti; and furthermore have at hand a table containing the Jovian cycle-name properly (*i.e.* by Hindu rule) connected with each solar year with reference to both apparent and mean Mēsha-saṁkrānti, and by all the Hindu Siddhāntas, *i.e.* such a table as will shew at a glance whether a cycle-name is properly applicable to a particular solar year by *any* system or by *any* known Hindu authority. This then is the work partly done in the present paper.

[1] The Mēsha-saṁkrānti point marks the first moment, or beginning, of each solar year.

203. Before explaining the method of preparation and the use of the tables which follow a few remarks may not be considered out of place.

204. As mentioned below, the late Mr. S. Balkrishna Dikshit expressed the opinion that the *Second Ārya-Siddhānta*, whose date is believed to be about A.D. 950, was in no part of India in use for a long time. The Siddhānta which has obtained most general acceptance, except in the south, is the *Present Sūrya-Siddhānta*, which dates perhaps from about A.D. 1000, and which in parts was corrected by the author of the *Makaranda* in A.D. 1478. My table XLII (below) shews all the years in which suppressions of Jovian samvatsaras took place according to each authority. These suppressions are marked with asterisks. Now it will be apparent to anyone using that table that in this respect the results afforded by calculation from the elements of the *Second Ārya-Siddhānta* are much nearer to those of the *Present Sūrya-Siddhānta* with the correction (*bīja*) than to results obtained by the use of any other authority. The position of Jupiter, that is, as calculated by the *Second Ārya* differed considerably from that calculated by the *Sūrya-Siddhānta* until the Hindu astronomer in the 15th century introduced the correction to the latter's elements; after which the two come much closer together. If, therefore, the corrected *Sūrya-Siddhānta* is really the most accurate authority, we must hold that at least in the matter of the motion of Jupiter the *Second Ārya-Siddhānta* was unworthily dealt with and received scant justice.

205. Although the *Second Ārya-Siddhānta* seems to have been in use for a very short time I was induced to continue the calculations according to its elements through the whole period of over 1,400 years embraced in the general Table XLII below, partly in order to call attention to this peculiarity.

206. In ordinary cases it would suffice, when once the moment of beginning of a saṁvatsara had been calculated with reference to apparent Mēsha-saṁkrānti, merely to add to it the time-difference or śōdhya, between apparent and mean Mēsha-saṁkrānti in order to arrive at the moment of its beginning with reference to mean Mēsha-saṁkrānti; and in ordinary cases the four decimal points given in my tables would suffice. But in order that there may be no mistake in very close cases I have worked the whole of these tables by nine places of decimals. One instance, and that a very interesting and instructive one, will shew how important it is that this should be done, especially with reference to the information afforded by Table XLII.

207. Note the year K. Y. 3710, A.D. 609-10, in which No. 1 Prabhava of a cycle began, according to the *First Ārya-Siddhānta* and as tabulated for four decimals of a day, 169·440 days after mean Mēsha-saṁkrānti (Table XXIX B below). We see that during that cycle 41 Plavaṅga was suppressed because it both began and ended within the limits of the solar year A.D. 649-50. Turning to the complementary Table XXIX A of the *Indian Chronography* we see that 41 Plavaṅga began in its year 169·4400 days prior to the time when No. 1 Prabhava began in its year which means that in A.D. 649 it began precisely at the moment of mean Mēsha saṁkrānti. Was it or was it not suppressed? Did it begin after or before that moment? If before, it was current at that moment and gave its name to the year; if later, it both began and ended within the limits of the solar year, and did not give its name to the year. Calculation by nine decimals settles the question. 1 Prabhava in A.D. 649-50 really began 169·439979088 days after mean Mēsha-saṁkrānti and 41 Plavaṅga began 169·439978320 days earlier than No. 1 Prabhava. So 41 Plavaṅga actually began 0·000000768 or ·066 of a second after the moment of mean Mēsha-saṁkrānti. Consequently it began and ended within the solar year; it was not current at mean Mēsha-saṁkrānti, and on that basis did not give its name to the year; it was suppressed. But if it had begun a tenth of a second earlier it would have been current at the critical instant and the solar year would have been named after it. I am confident that the Hindu framers of pañchāṅgs would have insisted on the year A.D. 649-50

being named after 40 Parābhava even though that saṁvatsara expired less than a tenth of a second after the beginning of the year and 41 Plavaṅga was current from that instant till shortly before its close. The rule was strict as to the naming of the year according to *actual currency at Mēsha-saṁkrānti*, and it would have been adhered to.

208. We have yet to learn, and our knowledge can only come from careful and painstaking research and study of a large number of inscription-dates, how far the practice of naming a *solar* year after a Jovian saṁvatsara was extended to the *luni-solar* year in these parts of India where such reckoning was used, and when such extension took place. In the *Indian Calendar* (§ 57, p. 33) it was noted that evidence exists to shew that such a practice was followed, at least for a time in some tracts; and the system adopted would doubtless be similar to that obtaining in the case of the solar year, but applied to the luni-solar year; that is to say, the year would be called after the name of the saṁvatsara current at the moment of beginning of the luni-solar year, or at the exact moment when, at the time of the new moon at the end of the lunar month Phālguna, the longitude of the moon's centre coincided with that of the sun. This moment always takes place earlier than the moment of the solar Mēsha-saṁkrānti, and of course the Jovian name thus given to the luni-solar year might be one different from that given to the solar year with which it was mostly connected. Careful calculation as to the arc travelled by Jupiter between the moment of beginnings of the luni-solar and solar year would have to be made by the framers of luni-solar pañchāṅgs for each year separately, in order to find the appropriate saṁvatsara whose name the luni-solar year was to bear. This cannot be determined by any general table. In such a system no expunction of a saṁvatsara can take place except in a luni-solar year which has an intercalary month, since the luni-solar common year is in length roughly seven days less than the saṁvatsara.

209. I begin Table XLII from the year A.D. 490 when a cycle began, and not from an earlier date, because at present the earliest certain date yet found in India which contains the saṁvatsara-name of a year belongs to the 8th century A.D. Scholars are not quite clear about the Chalukya inscription of A.D. 602 (see *Indian Chronography*, p. 3). It seems useless to begin from an earlier date.

---

210. The present Tables XXVII B to XXXI E supplement the work of Tables XXVII to XXXI A published in *Indian Chronography*, and enable the beginning and ending time of a Jovian saṁvatsara to be ascertained by any of the principal Indian Siddhāntas, when calculation is made on the basis of *mean* Mēsha-saṁkrānti.

211. The present Table XXVII B follows the *Present Sūrya-Siddhānta* without the bīja (or correction introduced in A.D. 1478) on the basis of mean Mēsha-saṁkrānti, Table XXVII of *Indian Chronography* being calculated by *apparent* Mēsha-saṁkrānti; and Table XXVII B is to be used with Table XXVII A just as is Table XXVII. The rule is given in § 146, p. 51, and examples in § 147, and (pp. 117-120) "Examples" 48 to 52.

The present Table XXVIII B is calculated for *mean* Mēsha-saṁkrānti according to the *Present Sūrya-Siddhānta* with the bīja, and is to be used with Table XXVIII A, *Indian Chronography*, just as is Table XXVIII in that work for *apparent* Mēsha-saṁkrānti.

Similarly the present Table XXIX B is for *mean* Mēsha-saṁkrānti by the *First Ārya-Siddhānta* or *Āryabhaṭīya*, and is to be used with Table XXIX A, *Indian Chronography*.

And the present Table XXXI B is for *mean* Mēsha-saṁkrānti by the *Brahma-Siddhānta* and the *Siddhānta-Śirōmaṇi*, and is to be used with Table XXXI A, *Indian Chronography*.

Explanation is fully given in *Indian Chronography* (pp. 52 to 62), and the work is shewn in Examples 55 to 60.

The present Tables XXXI C, D and E are similarly prepared according to the *Second Ārya-Siddhānta*, C for apparent, E for mean Mēsha-saṁkrānti, D being common to both.

212. Table XLII shews at a glance (the numbers in columns 3 to 13 referring to the list at the right side) for every year from A.D. 490-91 to 1915-16 what Jovian name would be given to each solar year according to the Hindu rule of naming the year by the saṁvatsara actually current at Mēsha-saṁkrānti; and this by all the authorities, and both by apparent and mean Mēsha-saṁkrānti. It will be found very useful in testing the accuracy of dates given in inscriptions found in tracts which, as in the north, carried on from year to year the practice of naming the year after the actual astronomical position of Jupiter.

213. Thus, to give an example, suppose we have a date given in a record in the year K. Y. 4606 or Saka 1427 expired (=A.D. 1505-6). Table XLII shews us at a glance that that solar year was called "Aṅgiras" according to the *Sūrya-Siddhānta without the bīja* whether on a basis of apparent or mean Mēsha-saṁkrānti, by the *Sūrya-Siddhānta with the bīja* also on either base, and (if they had been in use) also by the *Original Sūrya* on a mean base, and by the *Second Ārya-Siddhānta* on either base; whereas according to the *First Ārya-Siddhānta* on either base, or according to the *Brahma-Siddhānta* and *Siddhānta-Śirōmaṇi* on either base the name of the year was "Śrīmukha."

CYCLE OF JUPITER. ELEMENTS ON BASIS OF MEAN MĒSHA-SAṀKRĀNTI.

Table XXVII B. By the Sūrya-Siddhānta without the bīja.

214. [*Calculation on the basis of apparent Mēsha-saṁkrānti is fully explained in Indian Chronography, pp. 49-51.*] At the epoch of the Kaliyuga, or in K. Y. 0 expired, B.C. 3102-1, the saṁvatsara 26 Nandana ended and 27 Vijaya began exactly at the moment of mean Mēsha-saṁkrānti, Jupiter being then assumed to be precisely in long. 0°. Since Vijaya ended before the end of the solar year it was suppressed, and did not give its name to any year. From the end of 26 Nandana 34 saṁvatsaras passed before the moment of beginning of 1 Prabhava of the next cycle. Using the letters of the List of elements of this Siddhānta on p. 49, *Indian Chronography*,[1] we calculate the interval between the end of 26 Nandana and the beginning of 1 Prabhava by the formula E—(F × 34). (E) 365·258756481 days—(F × 34) 143·889205368 days=221·369551113 days. This is the time after mean Mēsha-saṁkrānti of K. Y. 33, B.C 3069-8, when 1 Prabhava began. Between this 1 Prabhava and the 1 Prabhava of K. Y. 3117 there were exactly 52 whole saṁvatsara cycles. I × 52=5789·504726772 days. E × 16=5844·140103703 days. (This is a multiple of the length in days of one solar year.) Deduct the latter from the former, and add 221·369551113 days (the beginning time of 1 Prabhava of K. Y. 33), and the result is 166·734174182 days. At this distance of time, therefore, after mean Mēsha-saṁkrānti No. 1 Prabhava began in K. Y. 3117, A.D. 16-17. Calculation for the following cycles follows in order by adding for each the element "I."

---

[1] "D" is the length of one saṁvatsara of Jupiter.
"E" is the length of the sidereal solar year.
"F" = E – D, or the difference between E and D.
"H" = this difference for an entire cycle, or, F × 60.
"I" = E – H, or additive difference for beginnings of successive cycles.

Table XXVIII B. By the Surya-Siddhānta with the bīja.

215. [*Calculation on the basis of apparent Mēsha-saṁkrānti is explained in Indian Chronography, pp. 52-53.*] Although the bīja, or correction, was not introduced till A.D. 1478 still, since it involved the change in some respects of the elements of the Siddhānta (*compare the Lists, pp. 19 and 52, Indian Chronography*), calculation had to be made afresh from the epoch of the Kaliyuga, K. Y. 0 expired. At the moment of mean Mēsha-saṁkrānti in that year 26 Nandana ended and 27 Vijaya began. Vijaya was suppressed (*kshaya*) in that year. Using the elements at the top of p. 53, *Ind. Chron.*, we find E—(F × 34)=221·639172313 days. This is the time measured from mean Mēsha-saṁkrānti, when 1 Prabhava began in K. Y. 33, B.C. 3069-68. From the beginning of this Prabhava to the beginning of the 1 Prabhava in K. Y. 4540, A.D. 1439-40, there were exactly 76 cycles of saṁvatsaras. "I" × 76=8197·744791036 days. E × 23 (a multiple of the solar year length)=8400·951399063 days. Deduct the latter from the former and add 221·639172313 days as above, and the result is 318·432564286 days. In K. Y. 4540, A.D. 1439-40, therefore, 1 Prabhava began 318·4326 days after mean Mēsha-saṁkrānti. For the beginning-moment of each successive cycle we add the element "I," or 111·812131461 days.

Table XXIX B. By the First Ārya-Siddhānta or Āryabhaṭīya.

216. [*For method of calculation on the basis of apparent Mēsha-saṁkrānti see Indian Chronography, pp. 53-55.*] At the epoch of the Kaliyuga 26 Nandana is assumed to have ended, and 27 Vijaya to have begun, precisely at the moment of mean Mēsha-saṁkrānti. The year was K. Y. 0, A.D. 3102-1. Vijaya was suppressed. We use the same formula as before, *viz.* E—(F × 34), to find the number of days by which 1 Prabhava began after mean Mēsha-saṁkrānti in K. Y. 33. E=365·258680555 days; F × 34=144·023981572 days. Result 221·234698983 days. There were exactly 52 cycles between this Prabhava and the Prabhava which began in K. Y. 3117, A.D. 16-17. We therefore add the above result to ("I" × 52) and deduct a multiple of the solar-year length, *i.e.* (E × 16). ("I" × 52)=5777·133079900. Adding for the beginning of Prabhava 221·234698983 we have 5998·367778883. Deduct (E × 16) or 5844·138888880, and the remainder is 154·228890003. This is the number of days by which 1 Prabhava began after mean Mēsha-saṁkrānti in K. Y. 3117, A.D. 16. The calculation begins regularly from that figure, adding the value of "I" for each cycle.

Table XXXI B. By the Brahma-Siddhānta and Siddhānta-Śirōmaṇi.

217. [*For method of calculation on the basis of apparent Mēsha-saṁkrānti see Indian Chronography, pp. 58-62.*] It has already been determined (*see Indian Chronography, p. 59, § 165*) that in K. Y. 0 Jupiter reached long. 0° 0·49836 days after mean Mēsha-saṁkrānti. At that moment 27 Vijaya began and 26 Nandana ended. In the following year, K. Y. 1 expired, 28 Jaya began ("F" =) 4·23840944 days earlier in the year than 27 Vijaya. Hence in that year 28 Jaya began 2·259929956 days after mean Mēsha-saṁkrānti, and as it ended about 361 days later ("D") it ended before the end of the solar year and was suppressed not giving its name to any year. To find the beginning-moment of the No. 1 Prabhava

of the next cycle we add as before E—(F × 34) to the ending-moment of 26 Nandana as found above.

E =	365·258437500	days
(F × 34) =	−144·106621496	do.
	221·151816004	do.
+	6·498360000	do.
	227·650176004	do.

Therefore 1 Prabhava began 227·650176004 days after mean Mēsha-saṁkrānti in the year K. Y. 33, B.C. 3069-68.

Add this to " I " × 52, and deduct a multiple of the solar year length, or E × 16, and we have the datum for K. Y. 3117, A.D. 16-17.

" I " × 52.	5769·537012720
+	227·650176004
	5997·187188724
E × 16.	—5844·135000000
	153·052188724

This last is the number of days by which 1 Prabhava began in that year after mean Mēsha-saṁkrānti.

From that moment we proceed regularly as before, adding the cycle difference " I " for each cycle.

Calculation by the Second Ārya-Siddhānta on basis of (i) apparent, (ii) mean Mēsha-saṁkrānti

218. (*Cancelled.*)

219. The date of the *Second or Mahā Ārya-Siddhānta* is believed to be about A.D. 950 ; and according to the opinion of the late Mr. Sankara Balkrishna Dikshit, it does not seem to have been anywhere in use for a long time. It was, however, known to Bhāskarāchārya in A.D. 1150 and such being the case I have considered it advisable to prepare the Tables for the whole period covered by the other tables referred to. Though this is certainly useless for later years it is dangerous to draw a line and it is best to be on the safe side, as we know as yet neither the tract where this Siddhānta was used nor the date when its use ceased. As regards the saṁvatsaras of Jupiter this Siddhānta could never have been received as an authority in the South of India because there the astronomically calculated succession of saṁvatsaras, in the matter of the application of their names to the solar years, was neglected after the year A.D. 906 ; every year being afterwards serially connected with the name of a saṁvatsara without regard to any suppression. The presumption is that the use of the *Second Ārya-Siddhānta* was confined to the north, or at least to those tracts where suppressions of saṁvatsaras were attended to.

Table XXXIC. Apparent Mēsha-saṁkrānti as basis.

220. The process of calculation for Table **XXXIC** is as follows :—

According to the *Second Ārya Siddhānta* the position of Jupiter at the moment of mean Mēsha-saṁkrānti in K. Y. 0 expired or 1 current, that is to say at the epoch of the Kaliyuga era or the moment of mean sunrise on Friday, 18, B. C. 3102, was 357° 7′ 12″[1] (*Indian Chronography, p. 63*). Jupiter did not reach the point 0° till he had travelled 2° 52′ 48″[1] of arc. Calculating by his mean motion this journey occupied 34d. 15 h. 45 m. or 34·65624537 days (*Table XXXIV*). He reached long. 0° therefore at that length of time after the moment of mean Mēsha-saṁkrānti, and when he reached it the saṁvatsara 27 Vijaya began. The time-interval between mean and apparent Mēsha-saṁkrānti in K. Y. 0, *i.e.* the interval which we call the "sōdhya", was determined by Dr. Schram (*op. cit. p. 16*) as 2·171973 days or 2·171972 days after calculation by two separate methods the results shewing a minute difference of 0·09 of a second. I have halved this difference, and calculated with a sōdhya of 2·1719725 days, or 2d. 4h. 7m. 38·424s. Jupiter therefore reached long. 0°, 26 Nandana ended, and 27 Vijaya began, (34·65624537 + 2·1719725 days=) 36·82821787 days, or (34d. 15h. 45m. + 2d. 4t.7m, 38.424s. =) 36d. 19h. 52m. 38·424s. after apparent Mēsha-saṁkrānti in K. Y. 0 expired.

221. Next has to be ascertained the moment of beginning of the first saṁvatsara "I Prabhava" of the next 60-saṁvatsara cycle. This occurred after the expiration of exactly 34 saṁvatsaras counting from the end of 26 Nandana. The length of the solar year is (E1=) 365·258690278 days. The annual difference between the lengths of the solar year and saṁvatsara is (F =)4·231719473 days. This last multiplied by 34 is 143·878462082 days E—(F × 34)= 221·380228196 days. This, added to the number of days by which 26 Nandana ended after apparent Mēsha-saṁkrānti (*viz.*:36·82821787 days, as found above, *para. 220*) gives us 258·208446066 days. 1 Prabhava therefore began 258·208446066 days after apparent Mēsha-saṁkrānti in the year K. Y. 33 expired or B. C. 3069-68. The reason why the solar year was not K. Y. 34 expired is because in K. Y. 8 expired, B. C. 3094-93, the saṁvatsara 35 Plava was expunged.

222. To arrive at the exact beginning of the "1 Prabhava" which began in A.D. 16-71, between which year and the year K. Y. 33 expired or B.C. 3069-68 there were exactly 52 complete cycles of saṁvatsaras, element "I" must be first calculated. This is the difference in the beginning-time of the saṁvatsara No. 1 Prabhava at the beginning of successive 60-year cycles. The annual difference being (F=) 4·231719473 days, F × 60 is 253·903168380 days. Deduct this from the year-length "E" given above, and the remainder is the value of "I", *viz.* 111·355521898 days. 52 of these cycle-differences ("I" × 52) amount to 5790·487138696 days. To this must be added the time by which the 1 Prabhava began after Mēsha-saṁkrānti in K. Y. 33 expired, or B.C. 3069-68. This was found to be 258·208446066 days. The total is 6048·695584762 days. Deduct from this a multiple of the solar year length E, *viz.* (E × 16=) 5844·139044448, and the remainder is 204·556540314 days.

223. No. 1 Prabhava therefore began in A.D. 16-17 or K. Y. 3117 expired 204·556540314 days after apparent Mēsha-saṁkrānti. From this point the calculation for Table XXXI C is carried regularly forward cycle by cycle the expunged, or *kshaya*, saṁvatsaras being duly noted, with the years in which the expunction took place.

224. It has been mentioned that, in the earliest of the cycles which have been dealt with above, the saṁvatsara 35 Plava was expunged. This occurred in the year K. Y. 8 expired, B.C 3094-3. From 27 Vijaya to 35 Plava is 8 saṁvatsaras. The annual difference "F"

[1] See the list of elements of this Siddhānta on p. 63, *Indian Chronography*, and footnote above p. 4.

multiplied by 8 is 33·853755784 days. Vijaya was found to have begun 36·828217870 days after apparent Mēsha-saṁkrānti in its solar year. Deducting from this 33·853755784 days, *viz.*: the 8 years collective difference, the remainder is 2·974462086 days. 35 Plava, therefore, began at that length of time after apparent Mēsha-saṁkrānti in K. Y. 8 expired or B.C. 3094-3; and since the length of a saṁvatsara is only 361 odd days, it is evident that Plava ended before the expiry of the 365¼ days of the solar year. It has been necessary to work out this point since, if there had been no expunction in the cycle in question, the year connected with 1 Prabhava of the following cycle would not have been, as it is, K. Y. 33 but K. Y. 34 expired.

[For the sake of conformity with the similar Tables for the other Siddhāntas (Tables XXVII to XXXI A, *Indian Chronography*) I have calculated the sōdhya as it has been determined by Dr. Schram for K. Y. 0, *viz.*: 2·1719725 days, leaving it to workers to make the very slight alteration necessary (if a very close case should be discovered) to get perfect accuracy for the century concerned. Dr. Schram's results will be found in *Indian Chronography*, p. 16. The sōdhya in K. Y. 0 was 2·171972 days, in K. Y. 3000 was 2·172707 days, in K. Y. 4000 was 2·172952 days and in K. Y. 5000 was 2·173197 days. Having found by my Tables the beginning-time of a saṁvatsara, if greater accuracy is necessary, deduct from the result after K. Y. 3000, fairly in proportion to the 2000 years' interval, an amount varying from 0·0007 to 0·0012, from 1m. 2s. to 1m. 46s. This last is the greatest possible difference.]

Table XXXI D.

Table XXXI D is to be used, for *Second Ārya-Siddhānta* computation just as Table XXVII A (*Indian Chronography*) is used for computation by the *Sūrya-Siddhānta* without the bīja.

Table XXXI E. Mean Mēsha-saṁkrānti as basis.

225. The method of work for finding the beginning of the saṁvatsara 1 Prabhava in the year A.D. 16-17, K. Y. 3117 expired, on the basis of reference to mean instead of to apparent Mēsha-saṁkrānti, could be explained in exactly the same way as has been already done in the latter case; but it is unnecessary to go into such full details a second time. It suffices to say for a beginning, that with reference to mean Mēsha-saṁkrānti in the year K. Y. 0 expired or at the epoch of the Kaliyuga era it has been shewn that the saṁvatsara 26 Nandana ended, and 27 Vijaya began 34·656245370 days after that moment. We work from this point. 8 saṁvatsaras later 35 Plava began (F × 8) 33·853755784 days earlier than did 27 Vijaya. Deducting the latter from the former figure we find that in the solar year K. Y. 8 expired, B.C. 3069-8 35 Plava began 0·802489586 days after mean Mēsha-saṁkrānti, and therefore ended before the end of the solar year. It was a kshaya, or suppressed, saṁvatsara. Hence, as before so here, the 1 Prabhava of the next cycle began in K. Y. 33 and not in K. Y. 34 expired.

226. No. 27 Vijaya began in K. Y. 0 expired 34·656245370 days after mean Mēsha-saṁkrānti. "E"— ("F" × 34) = 221·380228196 days. (§221 *above*.)

Add these. Then 1 Prabhava in K. Y. 33, B.C. 3069-8, began 256·036473566 days after mean Mēsha-saṁkrānti. Add this to "I" × 52 which = 5790·487138696. Result 6046·523612262 days. Deduct "E" × 16 (a multiple of the solar year length) or 5844·139044448 days and we arrive at 202·384567814 days, which is the number of days by which 1 Prabhava of the cycle began after mean Mesha-saṁkrānti in K. Y. 3117, A.D. 16-17.

This is tabulated as 202·3846 days, and so in succession.

Time-corrections.

227. Calculation by Tables XXXI C and D, or E and D will enable us to ascertain the moment of beginning and ending of any saṁvatsara by the *Second Ārya-Siddhānta* with reference to any Mōsha-saṁkrānti moment, true or mean; but, as in the case of the *Original Sūrya-Siddhānta*, *Brahma-Siddhānta* and *Siddhānta-Śirōmaṇi*, we must, if we use the *Indian Calendar* Table I, for giving us the time of occurrence of Mōsha-saṁkrānti each year (cols. 13 to 17 for the *First Ārya-Siddhānta*) apply a correction in order to get at the exact time of Mōsha-saṁkrānti by the *Second Ārya-Siddhānta* because the length of the year fixed by the *First Ārya* differed slightly from that fixed by the *Second Ārya-Siddhānta*. The two started from the same point, *viz.*: the sunrise epoch of the Kaliyuga, or mean sunrise on Feb. 18 B.C. 3102, but according to the *Second Ārya* the year is 0·84s. longer than the *First Ārya* year (*Ind. Chronography, p. 158, col. 3*). Hence the following Table must be used :—

TABLE A A.

DIFFERENCE BETWEEN THE MOMENTS OF MEAN MĒSHA-SAṀKRĀNTI AS CALCULATED BY (1) THE FIRST ĀRYA-SIDDHĀNTA, (2) THE SECOND ĀRYA-SIDDHĀNTA, THE TWO HAVING BEEN TOGETHER IN K. Y. 0, B.C. 3102.

Having found from Table I, cols. 13 to 17, etc. [*by adding the fixed śōdhya* (*see* §§ *206, 228*) *to the apparent Mēsha-saṁkrānti*] *the moment of mean Mēsha-saṁkrānti by the First Ārya-Siddhānta, add the time difference given in this Table for every expired year of the K. Y. in order to obtain the same by the Second Ārya-Siddhānta.*

Difference in years.	Time difference.			Difference in years.	Time difference.			Difference in years.	Time difference.			Difference in years.	Time difference.		
1	2			1	2			1	2			1	2		
	H.	M.	S.		H.	M.	S.		H.	M.	S.		H.	M.	S.
1	—	—	0·84	10	—	—	8·40	100	—	1	24	1000	—	14	0
2	—	—	1·68	20	—	—	16·80	200	—	2	48	2000	—	28	0
3	—	—	2·52	30	—	—	25·20	300	—	4	12	3000	—	42	0
4	—	—	3·36	40	—	—	33·60	400	—	5	36	4000	—	56	0
5	—	—	4·20	50	—	—	42·0	500	—	7	0	5000	1	10	0
6	—	—	5·04	60	—	—	50·40	600	—	8	24				
7	—	—	5·88	70	—	—	58·80	700	—	9	48				
8	—	—	6·72	80	—	1	7·20	800	—	11	12				
9	—	—	7·56	90	—	1	15·60	900	—	12	36				

N.B.—*To obtain exact time of apparent Mēsha-saṁkrānti by the First Ārya-Siddhānta add 30s. to the time given in Table I, col. 17 of the Indian Calendar in years A. D. whose number is odd ; but not in those whose number is even. See Indian Chronography "Hints for workers," No. 20, p. 79.*

228. Again, to fix the exact moment of apparent Mōsha-saṁkrānti by the *Second Ārya-Siddhānta* we have to note that according to it the śōdhya, or time-difference between mean and apparent Mōsha-saṁkrāntis varies slightly year by year, whereas the śōdhya by the *First Ārya-Siddhānta* is a constant : so that we must for absolute accuracy in *Second Ārya-Siddhānta* time, take note of this varying difference.

Dr. Schram has fixed its value for us (*see Indian Chronography*, 139 D, p. 16) at different millenniums thus—

TABLE B B.

Second Ārya-Siddhānta śōdhya.

K. Y. expired.	Christian year.	Exact value of śōdhya as fixed by Dr. Schram.			
		d.	h.	m.	s.
3000	B.C. 103-02	2	4	8	41·88
4000	A.D. 899-900	2	4	9	3·05
5000	A.D. 1899-1900	2	4	9	24·22

It will be seen that for all ordinary purposes it will suffice to use a constant 2d. 4h. 9m.; but for very close work take the śōdhya-value at K. Y. 3601, A. D. 500, as being 2d. 4h. 8m. 54·582s. and add for every succeeding 100 years 2·117s. and for 1000 years 21·168s.

Rule for work and example.

229. All work formerly necessary for the purpose of ascertaining which Jovian saṁvatsara began in the course of any given year according to any of the principal Siddhāntas, and whether calculated by apparent or mean Mēsha-saṁkrānti, is now obviated by the information given in Table XLII below, which solves the question at a glance. It shews the saṁvatsara current at every Mēsha-saṁkrānti, and we therefore know that the next saṁvatsara of the cycle began during the year. When there is an asterisk shown it means that this latter saṁvatsara both began and ended during the solar year, so that the next again also began during that year and was current at Mēsha-saṁkrānti of next year.

230. But we sometimes desire to know the time of beginning and ending of a saṁvatsara in order to ascertain whether it was current at the time of the event or action chronicled in an inscription.

231. This time is precisely the same whether we calculate from mean or from apparent Mēsha-saṁkrānti; and as the time of these is clearly given in the general working Tables LX, LXI, LXXVI, LXXXII, XC, and as, for the *Second Ārya-Siddhānta* it can be gathered from cols. 13 to 17 or 17a of the *Indian Calendar*, it is easiest to use that information as basis of work. Find this required time, therefore, according to the *Sūrya-Siddhānta* (with or without the bīja), *the First Ārya* or *Āryabhaṭīya*, the *Original Sūrya*, and *Brahma-Siddhāntas*, and the *Siddhānta-Śirōmaṇi* in the manner described in §§ 146, 147, 153, 158, 162 or 167 A and examples 48 to 59 A of *Indian Chronography*, or from the general working Tables below.

232. The work according to the *Second Ārya-Siddhānta* is precisely similar, but we have to use the Tables A A and B B in the text above instead of any of the other Tables in the text of *Indian Chronography*. I proceed with an example.

233. We want to know what saṁvatsara began in K. Y. 4380 expired, A. D. 1279-80 according to the *Second Ārya-Siddhānta*. The answer is given by Table XLII below. 18 Tāraṇa was current both at apparent and mean Mēsha-saṁkrāntis, and therefore in either case gave its name to the solar year; 19 Pārthiva began in the course of the year.

When did Pārthiva begin? and when did it end?

For rough work the following will always suffice, whether we have been calculating by mean or apparent Mēsha-saṁkrānti, the time being the same by both. We will work by apparent Mēsha-saṁkrānti. Table XXXI C below shews that in the cycle concerned 1 Prabhava began 351 days after Mēsha-saṁkrānti, and Table XXXI D shews that in its year 19 Pārthiva began 76 days earlier than did 1 Prabhava; so 19 Pārthiva began (351 − 76) 275 days after apparent Mēsha-saṁkrānti in the given year. We find the time of apparent Mēsha-saṁkrānti in that year from the *Indian Calendar* Table I or Table LXI below, *i.e.* according to the *First Ārya-Siddhānta*, on March 25 on day 84 (*Table IX Ind. Cal. or LXIX below*) at about 21 hours after mean sunrise. Call this day 85.[1] Table AA shews the time-difference between the two Siddhāntas, for the 4380 years since K. Y. 0, as being about one hour. This may be ignored. 19 Pārthiva began 275 days later. 275 + 85 = 360, *i.e.* (Table IX, *Ind. Cal. or LXIX below*) 19 Pārthiva began on December 26, A.D. 1279. This suffices for a rough solution of the problem.

For close work we must calculate more carefully. I give here the closest possible according to our available Tables, following the course prescribed above. For the beginning of 19 Pārthiva (*Table XXXI C and D below*) we have 351·4704 − 76·1710 = 275·2994 = (*Table XXXVI, Ind. Chron.*) 275d. 7h. 11m. 8·16s. after apparent Mēsha-saṁkrānti.

Apparent Mēsha-saṁkrānti by the *First Ārya-Siddhānta* (Table LXI below) was on day 84 at 20h. 57m. 30s. after mean sunrise.

The difference in the śōdhya interval between mean and apparent Mēsha-saṁkrānti has to be taken into account. The *First Ārya-Siddhānta* fixed this interval as always 2d. 3h. 32m. 30s. But according to the *Second Ārya* it varies slightly. (*See above, Table BB, § 228, and accompanying remarks.*) The given K. Y. year is 4380. In K. Y. 4000 it was 2d. 4h. 9m. 3·05s. Add for (say) 400 years 8·47s., at the rate of 2·117s. per 100 years, and we have the śōdhya in the given year by the *Second Ārya-Siddhānta* as 2d. 4h. 9m. 11·52s.

The time-difference between the two authorities (*Table AA above, § 227*) must also be ascertained. This is, for 4000 years, 56m.; for 300 years, 4m. 12s.; for 80 years, 1m. 7·20s.; total 1h. 1m. 19·20s.

Now we make our calculation.

	d.	h.	m.	s.
First Ārya-Siddhānta apparent Mēsha-saṁkrānti ...	84	20	57	30
First Ārya-Siddhānta śōdhya	2	3	32	30
First Ārya mean Mēsha-saṁkrānti	87	0	30	0
Time-difference between *First* and *Second Ārya-Siddhānta* in K. Y. 4380		1	1	19·20
Second Arya-Siddhānta mean Mēsha-saṁkrānti ...	87	1	31	19·20
Second Ārya-Siddhānta śōdhya	−2	4	9	11·52
Apparent Mēsha-saṁkrānti by *Second Ārya-Siddhānta* ..	84	21	22	7·68
19 Pārthiva began after this	275	7	11	8·16
Time of beginning of 19 Pārthiva by the *Second Ārya-Siddhānta*	360	4	33	15·84

360d. = (*Table IX, Indian Calendar, or LXIX below*) December 26.

We have found therefore that 19 Pārthiva according to the *Second Ārya-Siddhānta*, whether based on apparent or mean Mēsha-saṁkrānti (*§ 231 above*) began at 4h. 33m. 15·84s. after mean sunrise on December 26, A.D. 1279.

[1] To suit, that is, the European name of the day, which begins six hours before mean sunrise

TABLE XXVII B.

THE SIXTY-SAṀVATSARA CYCLE OF JUPITER.

Mean-sign system by the SŪRYA-SIDDHĀNTA WITHOUT THE BĪJA, calculated with reference to mean Mēsha-saṁkrānti.

(For all India up to A.D. 906, and for the northern portion alone after and inclusive of that date.)

Year of the Kaliyuga (expired).	Christian year.	Number of days by which 1 Prabhava began after mean Mēsha-saṁkrānti.	Kshaya (expunged) saṁvatsaras.	Year of the Kaliyuga (expired).	Christian year.	Number of days by which 1 Prabhava began after mean Mēsha-saṁkrānti.	Kshaya (expunged) saṁvatsaras.
1	2	3	4	1	2	3	4
	B.C.				A.D.		
(0)	*(3102-01)*	...	27 Vijaya.	*(4009)*	*(908-09)*	...	3 Śukla.
33	3069-8	221·3696		4066	965-66	121·8264	
				(4094)	*(993-94)*	...	29 Manmatha.
	A.D.			4125	1024-25	233·1631	
3117	16-17	166·7342		*(4180)*	*(1079-80)*	...	56 Dundubhi.
(3156)	*(55-56)*	...	40 Parābhava.	4184	1083-84	344·4997	
3176	75-76	278·9708		4244	1143-44	90·5776	
3236	135-36	24·1487		*(4265)*	*(1164-65)*	...	22 Sarvadhārin.
(3241)	*(140-41)*	...	6 Aṅgiras.	4303	1202-03	201·9142	
3295	194-95	135·4853		*(4350)*	*(1249-50)*	...	48 Ānanda.
(3327)	*(226-27)*	...	33 Vikārin.	4362	1261-62	313·2509	
3354	253-54	246·8219		4422	1321-22	59·3287	
(3412)	*(311-12)*	...	59 Krōdhana.	*(4436)*	*(1335-36)*	...	15 Vṛisha.
3413	312-13	358·1586		4481	1380-81	170·6654	
3473	372-73	104·2364		*(4521)*	*(1420-21)*	...	41 Plavaṅga.
(3497)	*(396-97)*	...	25 Khara.	4540	1439-40	282·0020	
3532	431-32	215·5731		*(After this date Tables XXVIII B below and XXVIII A in the Indian Chronography are ordinarily to be used.)*			
(3582)	*(481-82)*	...	51 Piṅgala.				
3591	490-91	326·9097					
3651	550-51	72·9876					
(3668)	*(567-68)*	...	18 Tāraṇa.	4600[2]	1499-1500[2]	28·0799	
3710	609-10	184·3242		*(4606)*	*(1505-06)*	...	7 Śrīmukha.
(3753)	*(652-53)*	...	44 Sādhāraṇa.	4659	1558-59	139·4165	
3769	668-69	295·6608		*(4691)*	*(1590-91)*	...	33 Vikārin.
3829	728-29	41·7387		4718	1617-18	250·7531	
(3838)	*(737-38)*	...	10 Dhātṛi.	*(4777)*	*(1676-77)*	...	60 Kshaya.
3888	787-88	153·0753		4777	1676-77	362·0897	
(3924)	*(823-24)*	...	37 Śōbhana.				
3947	846-47	264·4120					
4007[1]	906-07[1]	16·4898					

[1] In Southern India the expunction of saṁvatsaras was neglected from, and including, the cycle beginning in A.D. 906.

[2] About A.D. 1500 the bīja (correction) was generally introduced, and the beginning moments of the cycles were recalculated from the epoch of the Kaliyuga. For years subsequent to A.D. 1500 Tables XXVIII B below and XXVIII A (*Indian Chronography*) should as a rule be used. But since the bīja was not introduced all over India at the same time calculations for three more cycles have been here given according to the Sūrya-Siddhānta without the bīja.

TABLE XXVIII B.

The Sixty-saṁvatsara Cycle of Jupiter.

Mean-sign system by the Sūrya-Siddhānta with the bīja calculated with reference to mean Mēsha-saṁkrānti.

Year of the Kaliyuga (expired).	Christian year.	Number of days by which 1 Prabhava began after mean Mēsha-saṁkrānti.	Kshaya (expunged) saṁvatsaras.	Year of the Kaliyuga (expired).	Christian year.	Number of days by which 1 Prabhava began after mean Mēsha-saṁkrānti.	Kshaya (expunged) saṁvatsaras.
1	2	3	4	1	2	3	4
	A.D.				A.D.		
4540	1439-40	318·4326		*(4871)*	*(1770-71)*	...	35 Plava.
4600	1499-1500	64·9862		4896	1795-96	258·7896	
(4615)	*(1514-15)*	...	16 Chitrabhānu.	4956	1855-56	5·3433	
4659	1558-59	176·7987		*(4957)*	*(1856-57)*	...	2 Vibhava.
(4700)	*(1599-1600)*	...	42 Kīlaka.	5015	1914-15	117·1557	
4718	1617-18	288·6111		*(5042)*	*(1941-42)*	...	28 Jaya.
4778	1677-78	35·1648		5074	1973-74	228·9682	
(4786)	*(1685-86)*	...	9 Yuvan.	*(5128)*	*(2027-28)*	...	55 Durmati
4837	1736-37	146·9772		5133	2032-33	340·7806	

TABLE XXIX B.

THE SIXTY-SAṀVATSARA CYCLE OF JUPITER.

Mean-sign system by the FIRST ĀRYA-SIDDHĀNTA OR ĀRYABHAṬĪYA.

Calculated with reference to mean Mēsha-saṁkrānti.

Year of the Kaliyuga (expired).	Christian year.	Number of days by which 1 Prabhava began after mean Mēsha-saṁkrānti.	Kshaya (expunged) saṁvatsara.	Year of the Kaliyuga (expired).	Christian year.	Number of days by which 1 Prabhava began after mean Mēsha-saṁkrānti.	Kshaya (expunged) saṁvatsara.
1	2	3	4	1	2	3	4
	B.C.				A.D.		
(0)	*(3102-01)*	...	27 Vijaya.	4066	965-66	105·5149	
33	3069-68	221·2347		*(4090)*	*(989-90)*	...	25 Khara.
				4125	1024-25	216·6136	
				(4176)	*(1075-76)*	...	52 Kālayukta.
	A.D.			4184	1083-84	327·7123	
3117	16-17	154·2289		4244	1143-44	73·5524	
(3153)	*(52-53)*	...	37 Śōbhana.	*(4261)*	*(1160-61)*	...	18 Tāraṇa.
3176	75-76	265·3276		4303	1202-03	184·6511	
3236	135-36	11·1676		*(4346)*	*(1245-46)*	...	44 Sādhāraṇa.
(3238)	*(137-38)*	...	3 Śukla.	4362	1261-62	295·7498	
3295	194-95	122·2663		4422	1321-22	41·5898	
(3323)	*(222-23)*	...	29 Manmatha.	*(4431)*	*(1330-31)*	...	10 Dhātṛi.
3354	253-54	233·3051		4481	1380-81	152·6885	
(3409)	*(308-09)*	...	56 Dundubhi.	*(4517)*	*(1416-17)*	...	37 Śōbhana.
3413	312-13	344·4638		4540	1439-40	263·7872	
3473	372-73	90·3038		4600	1499-1500	9·6273	
(3494)	*(393-94)*	...	22 Sarvadhārin.	*(4602)*	*(1501-02)*	...	3 Śukla.
3532	431-32	201·4025		4659	1558-59	120·7260	
(3579)	*(478-79)*		48 Aranda.	*(4687)*	*(1586-87)*	...	29 Manmatha.
3591	490-91	312·5012		4718	1617-18	231·8247	
3651	550-51	58·3418		*(4772)*	*(1671-72)*	..	55 Durmati.
(3664)	*(563-64)*	...	14 Vikrama.	4777	1676-77	342·9234	
3710	609-10	169·4400		4837	1736-37	88·7634	
(3759)	*(649-50)*	...	41 Plavaṅga.	*(4857)*	*(1756-57)*	...	21 Sarvajit.
3769	668-69	280·5387		4896	1795-96	199·8622	
3829	728-29	26·3787		*(4942)*	*(1841-42)*		47 Pramādin.
(3835)	*(734-35)*	...	7 Śrīmukha.	4955	1854-55	310·9609	
3888	787-88	137·4774		5015	1914-15	56·8609	
(3920)	*(819-20)*	...	33 Vikārin.	*(5028)*	*(1927-28)*	...	14 Vikrama.
3947	846-47	248·5762		5074	1973-74	167·8996	
(4005)	*(904-05)*	...	59 Krōdhana.	*(5113)*	*(2012-13)*	...	40 Parābhava.
4066	905-66	359·6749		5133	2032-33	278·9983	

TABLE XXXI B.

The Sixty-saṁvatsara Cycle of Jupiter.

Mean-sign system by the Brahma-Siddhānta and Siddhānta-Śirōmaṇi.

Calculated with reference to mean Mēsha-saṁkrānti.

Year of the Kaliyuga (expired).	Christian year.	Number of days by which 1 Prabhava began after mean Mēsha-saṁkrānti.	Kshaya (expunged) saṁvatsara.
1	2	3	4
	B.C.		
(1)	*(3101-00)*	...	28 Jaya.
33	3069-68	227·6502	
	A.D.		
3117	16-17	153·0522	
(3153)	*(52-53)*	...	37 Śōbhana.
3176	75-76	264·0048	
3236	135-36	9·6990	
(3238)	*(137-38)*	...	3 Śukla.
3255	194-95	120·6517	
(3323)	*(222-23)*	...	29 Manmatha.
3354	253-54	231·6043	
(3408)	*(307-08)*	...	55 Durmati.
3413	312-13	342·5569	
3473	372-73	88·2511	
(3493)	*(392-93)*	...	21 Sarvajit.
3532	431-32	199·2038	
(3578)	*(477-78)*	...	47 Pramādin.
3591	490-91	310·1564	
3651	550-51	55·8506	
(3664)	*(563-64)*	...	14 Vikrama.
3710	609-10	166·8032	
(3749)	*(648-49)*		40 Parābhava.
3769	668-69	277·7559	
3829	728-29	23·4501	
(3834)	*(733-34)*	...	6 Aṅgiras.
3888	787-88	134·4027	
(3919)	*(818-19)*	...	32 Vilamba.
3947	846-47	245·3553	
(4004)	*(903-04)*	...	58 Raktāksha.
4006	905-06	356·3080	

Year of the Kaliyuga (expired).	Christian year.	Number of days by which 1 Prabhava began after mean Mēsha-saṁkrānti.	Kshaya (expunged) saṁvatsara.
1	2	3	4
	A.D.		
4066	965-66	102·0022	
(4090)	*(989-90)*	...	25 Khara.
4125	1024-25	212·9548	
(4175)	*(1074-75)*	...	51 Piṅgala.
4184	1083-84	323·9074	
4244	1143-44	69·60.6	
(4260)	*(1159-60)*	...	17 Subhānu.
4303	1202-03	180·5543	
(4345)	*(1244-45)*	...	43 Saumya.
4362	1261-62	291·5069	
4422	1321-22	37·2011	
(4430)	*(1329-30)*	...	9 Yuvan.
4481	1380-81	148·1537	
(4515)	*(1414-15)*	...	35 Plava.
4540	1439-40	259·1064	
4600	1499-1500	4·8006	
(4601)	*(1500-01)*	...	2 Vibhava.
4659	1558-59	115·7532	
(4686)	*(1585-86)*	...	28 Jaya.
4718	1617-18	226·7058	
(4771)	*(1670-71)*	...	54 Raudra.
4777	1676-77	337·6585	
4837	1736-37	83·3527	
(4856)	*(1755-56)*	...	20 Vyaya.
4896	1795-96	194·3053	
(4941)	*(1840-41)*	...	46 Paridhāvin.
4955	1854-55	305·2579	
5015	1914-15	50·9521	
(5027)	*(1926-27)*	...	13 Pramāthin.
5074	1973-74	161·9048	
(5112)	*(2011-12)*	...	39 Viśvāvasu.
5133	2032-33	272·8574	

TABLE XXXI C.

The Sixty-saṁvatsara Cycle of Jupiter.

Mean-sign system by the Second Ārya-Siddhānta.

Calculated with reference to apparent Mēsha-saṁkrānti.

Year of the Kaliyuga (expired).	Christian year.	Number of days by which 1 Prabhava began after apparent Mēsha-saṁkrānti.	Kshaya (expunged) saṁvatsara.	Year of the Kaliyuga (expired).	Christian year.	Number of days by which 1 Prabhava began after apparent Mēsha-saṁkrānti.	Kshaya (expunged) saṁvatsara.
1	2	3	4	1	2	3	4
	B.C.				A.D.		
(0)	*(3102-1)*			4007	906-07	48·5959	
(8)	*(3094-3)*	...	35 Plava.	*(4018)*	*(917-18)*	...	12 Bahudhānya.
33	3069-8	258·208446		4066	965-66	159·9514	
	A.D.			*(4103)*	*(1002-03)*	...	38 Krōdhin.
3117	16-17	204·5565		4125	1024-25	271·3070	
(3065)	*(64-65)*	...	49 Rākshasa.	4185	1084-85	17·4038	
3176	75-76	315·9121		*(4189)*	*(1088-89)*	...	5 Prajāpati.
3236	135-36	62·0089		4244	1143-44	128·7523	
(3250)	*(149-50)*	...	15 Vṛisha.	*(4274)*	*(1173-74)*	..	31 Hēmalamba.
3295	194-95	173·3644		4303	1202-03	240·1148	
(3335)	*(234-35)*	...	41 Plavaṅga.	*(4359)*	*(1258-59)*	..	57 Rudhirōdgārin.
3354	253-54	284·7199		4362	1261-62	351·4704	
3414	313-14	30·8168		4422	1321-22	97·5672	
(3421)	*(320-21)*	...	8 Bhāva.	*(4445)*	*(1344-45)*	...	24 Vikṛita.
3473	372-73	142·1723		4481	1380-81	208·9227	
(3506)	*(405-06)*	...	34 Śārvarin.	*(4530)*	*(1429-30)*		50 Anala.
3532	431-32	253·5278		4540	1439-40	320·2782	
(3591)	*(490-91)*	...	60 Kshaya.	4600	1499-1500	66·3751	
3591	490-91	364·8833		*(4615)*	*(1514-15)*	...	16 Chitrabhānu.
3651	550-51	110·9802		4659	1558-59	177·7306	
(3676)	*(575-76)*	...	26 Nandana.	*(4700)*	*(1599-1600)*	...	42 Kīlaka.
3710	609-10	222·3357		4718	167- 18	289·0861	
(3762)	*(661-62)*	...	53 Siddhārthin.	4778	1677-78	35·1829	
3769	668-69	333·6912		*(4786)*	*(1685-86)*	...	9 Yuvan.
3829	728-29	79·7880		4837	1736-37	116·5385	
(3847)	*(746-47)*	...	19 Pārthiya.	*(4871)*	*(1770-71)*	.	35 Plava.
3888	787-88	191·1436		4896	1795-96	257·8940	
(3933)	*(832-33)*	...	46 Paridhāvin.	4956	1855-56	3·9908	
3947	846-47	302·4991		*(4956)*	*(1855-56)*	...	1 Prabhava.
				5015	1914-15	115·3463	
				(5042)	*(1941-42)*	..	28 Jaya.
				5074	1973-74	226·7019	
				(5127)	*(2026-27)*	...	54 Raudra.
				5133	2032-33	338·0574	

N.B.—This table is based on Dr. Schram's valuation of the śōdhya in K. Y. 0, a mean being taken between his two results (see *Indian Chronography*, p. 16) obtained by different modes of calculation, *viz.* 2·171973 days and 2·171972 days. It is taken here as 2·1719725 days. The greatest difference between the śōdhya in K. Y. 0 and that in K. Y. 5000 amounts to no more than 1m. 46½s., or 0·001225 day.

TABLE XXXI D.

THE SIXTY-SAṀVATSARA CYCLE OF JUPITER.

Mean-sign system by the SECOND ĀRYA-SIDDHĀNTA.

The number of days and decimals less than the day given in Table XXXI C by which each saṁvatsara began after apparent Mēsha-saṁkrānti in its solar year.

No.	Saṁvatsara.	Number of days.	No.	Saṁvatsara.	Number of days.
1	2	3	1	2	3
1	Prabhava	0·000	32	Vilamba	131·1833
2	Vibhava	4·2317	33	Vikārin	135·4150
3	Śukla	8·4634	34	Śārvarin	139·6467
4	Pramōda	12·6952	35	Plava	143·8785
5	Prajāpati	16·9269	36	Śubhakṛit	148·1102
6	Aṅgiras	21·1586	37	Śōbhana	152·3419
7	Śrīmukha	25·3903	38	Krōdhin	156·5736
8	Bhāva	29·6220	39	Viśvāvasu	160·8053
9	Yuvan	33·8538	40	Parābhava	165·0371
10	Dhātṛi	38·0855	41	Plavaṅga	169·2688
11	Īśvara	42·3172	42	Kīlaka	173·5005
12	Bahudhānya	46·5489	43	Saumya	177·7322
13	Pramāthin	50·7806	44	Sādhāraṇa	181·9639
14	Vikrama	55·0124	45	Virōdhakṛit	186·1957
15	Vṛisha	59·2441	46	Paridhāvin	190·4274
16	Chitrabhānu	63·4758	47	Pramādin	194·6591
17	Subhānu	67·7075	48	Ānanda	198·8908
18	Tāraṇa	71·9392	49	Rākshasa	203·1225
19	Pārthiva	76·1710	50	Anala	207·3543
20	Vyaya	80·4027	51	Piṅgala	211·5860
21	Sarvajit	84·6344	52	Kālayukta	215·8177
22	Sarvadhārin	88·8661	53	Siddhārthin	220·0494
23	Virōdhin	93·0978	54	Raudra	224·2811
24	Vikṛita	97·3295	55	Durmati	228·5129
25	Khara	101·5613	56	Dundubhi	232·7446
26	Nandana	105·7930	57	Rudhirōdgārin ...	236·9763
27	Vijaya	110·0247	58	Raktāksha	241·2080
28	Jaya	114·2564	59	Krōdhana	245·4397
29	Manmatha	118·4881	60	Kshaya	249·6714
30	Durmukha ...	122·7199	1	Prabhava (*of the following cycle*).	253·9032
31	Hēmalamba	126·9516			

TABLE XXXI E.

The Sixty-saṁvatsara Cycle of Jupiter.

Mean-sign system by the Second Ārya-Siddhānta.

Calculated with reference to mean Mēsha-saṁkrānti.

Year of the Kaliyuga (expired).	Christian year.	Number of days by which 1 Prabhava began after mean Mēsha-saṁkrānti.	Kshaya (expunged) saṁvatsara.
1	2	3	4
	B.C.		
(0)	*(3102-1)*		
(8)	*(3094-3)*	...	35 Plava.
33	3069-68	256·3802	
	A.D.		
3117	1-17	202·3846	
(3064)	*(63-64)*	...	48 Ānanda.
3176	75-76	313·7401	
3236	135-36	59·8369	
(3250)	*(149-50)*	...	15 Vṛisha.
3295	194-95	171·1924	
(3335)	*(234-35)*	...	41 Plavaṅga.
3354	253-54	282·5480	
3414	313-14	28·6448	
(3420)	*(319-20)*	...	7 Śrīmukha.
3473	372-73	140·0003	
(3506)	*(405-06)*	...	34 Śārvarin.
3532	431-32	251·3558	
(3591)	*(490-91)*	...	60 Kshaya.
3591	490-91	362·7114	
3651	550-51	108·8082	
(3676)	*(575-76)*	...	26 Nandana.
3710	609-10	220·1637	
(3762)	*(661-62)*		53 Siddhārthin.
3769	668-69	331·5192	
3829	728-29	77·6161	
(38[illegible])	*(746-47)*	...	19 Pārthiva.
3888	787-88	188·9716	
(3932)	*(831-32)*	...	45 Virōdhakṛit.
3947	846-47	300·3271	
4007	906-07	46·4239	
(4017)	*(916-17)*	...	11 Īśvara.
4066	965-66	157·7795	

Year of the Kaliyuga (expired).	Christian year.	Number of days by which 1 Prabhava began after mean Mēsha-saṁkrānti.	Kshaya (expunged) saṁvatsara.
1	2	3	4
	A.D.		
(4103)	*(1002-03)*	...	38 Krōdhin.
4125	1024-25	269·1350	
4185	1084-85	15·2318	
(4188)	*(1087-88)*	...	4 Pramōda.
4244	1143-44	126·5873	
(4273)	*(1172-73)*	...	30 Durmukha.
4303	1202-03	237·9429	
(4359)	*(1258-59)*	...	57 Rudhirōdgārin.
4362	1261-62	349·2984	
4422	1321-22	95·3952	
(4444)	*(1343-44)*	...	23 Virōdhin.
4481	1380-81	206·7507	
(4529)	*(1428-29)*	...	49 Rākshasa.
4540	1439-40	318·1063	
4600	1499-1500	64·2031	
(4615)	*(1514-15)*	...	6 Chitrabhānu.
4659	1558-59	175·5586	
(4700)	*(1599-1600)*	...	42 Kīlaka.
4718	1617-18	286·9141	
4778	1677-78	33·0110	
(4785)	*(1684-85)*		8 Bhāva.
4837	1736-37	144·3665	
(4871)	*(1770-71)*	...	35 Plava.
4896	1795-96	255·7220	
4956	1855-56	1·8188	
(4956)	*(1855-56)*		1 Prabhava.
5015	1914-15	113·1744	

To determine the beginning and ending times of a saṁvatsara use this Table with Table XXXID.

For śōdhya see foot of Table XXXIC.

TABLE XLII.

The Jovian name of each Hindu Calendar year according to the different Siddhāntas and systems of calculation.

TABLE XLII.

THE JOVIAN NAME OF EACH HINDU CALENDAR YEAR ACCORDING TO THE DIFFERENT SIDDHĀNTAS AND SYSTEMS OF CALCULATION.

An asterisk shews when an expunction of a saṁvatsara occurs, and when, therefore, the following saṁvatsara does not give its name to the next solar year. "S."=Siddhānta; "M. S."=Mēsha-saṁkrānti; numbers in columns 3 to 13 refer to the List of Names of the Jovian saṁvatsaras on the right.

Expired year of Kaliyuga.	Year A.D	NUMBER OF THE SAṀVATSARA CONNECTED WITH EACH SOLAR YEAR ACCORDING TO THE SEVERAL SIDDHĀNTAS, BY REASON OF ITS CURRENCY AT APPARENT, OR AT MEAN, MĒSHA-SAṀKRĀNTI.											Expired year of Kaliyuga.	Year A.D.	NUMBER OF THE SAṀVATSARA CONNECTED WITH EACH SOLAR YEAR ACCORDING TO THE SEVERAL SIDDHĀNTAS, BY REASON OF ITS CURRENCY AT APPARENT, OR AT MEAN, MĒSHA-SAṀKRĀNTI.											Names of the Sixty saṁvatsaras of the cycle of Jupiter.
		SŪRYA-S. NO BĪJA.		SŪRYA-S. WITH BĪJA.		FIRST ĀRYA-S.		ORIG. SŪRYA S.	BRAHMA-S. AND S. ŚIRŌ.		SECOND ĀRYA-S.				SŪRYA-S. NO BĪJA.		SŪRYA-S. WITH BĪJA.		FIRST ĀRYA-S.		ORIG. SŪRYA S.	BRAHMA-S. AND S. ŚIRŌ.		SECOND ĀRYA-S.		
		Apparent M. S.	Mean M. S.	Apparent M. S.	Mean M. S.	Apparent M. S.	Mean M. S.	Mean M. S.	Apparent M. S.	Mean M. S.	Apparent M. S.	Mean M. S.			Apparent M. S.	Mean M. S.	Apparent M. S.	Mean M. S.	Apparent M. S.	Mean M. S.	Mean M. S.	Apparent M. S.	Mean M. S.	Apparent M. S.	Mean M. S.	1. Prabhava.
																										2. Vibhava.
1	2	3	4	5	6	7	8	9	10	11	12	13	1	2	3	4	5	6	7	8	9	10	11	12	13	3. Śukla.
																										4. Pramōda.
																										5. Prajāpati.
3591	490-91	60	60	...	...	60	60	60	60	60	59*	59*	3606	505-06	15	15	...	...	15	15	15	15	15	15	15	6. Aṅgiras.
3592	491-92	1	1	...	...	1	1	1	1	1	1	1	3607	506-07	16	16	...	...	16	16	16	16	16	16	16	7. Śrīmukha.
3593	492-93	2	2	...	...	2	2	2	2	2	2	2	3608	507-08	17	17	...	...	17	17	17	17	17	17	17	8. Bhāva.
3594	493-94	3	3	...	...	3	3	3	3	3	3	3	3609	508-09	18	18	...	...	18	18	18	18	18	18	18	9 Yuvan.
3595	494-95	4	4	...	...	4	4	4	4	4	4	4	3610	509-10	19	19	...	...	19	19	19	19	19	19	19	10. Dhātri.
3596	495-96	5	5	...	...	5	5	5	5	5	5	5	3611	510-11	20	20	...	...	20	20	20	20	20	20	20	11. Īśvara.
3597	496-97	6	6	...	...	6	6	6	6	6	6	6	3612	511-12	21	21	...	...	21	21	21	21	21	21	21	12. Bahudhānya.
3598	497-98	7	7	...	...	7	7	7	7	7	7	7	3613	512-13	22	22	...	...	22	22	22	22	22	22	22	13. Pramāthin.
3599	498-99	8	8	...	...	8	8	8	8	8	8	8	3614	513-14	23	23	...	...	23	23	23	23	23	23	23	14. Vikrama.
3600	499-500	9	9	...	...	9	9	9	9	9	9	9	3615	514-15	24	24	...		24	24	24	24	24	24	24	15. Vṛisha.
3601	500-01	10	10	...	...	10	10	10	10	10	10	10	3616	515-16	25	25	...	...	25	25	25	25	25	25	25	16. Chitrabhānu.
3602	501-02	11	11	...	...	11	11	11	11	11	11	11	3617	516-17	26	26	...	...	26	26	26	26	26	26	26	17. Subhānu.
3603	502-03	12	12	...	...	12	12	12	12	12	12	12	3618	517-18	27	27	...	...	27	27	27	27	27	27	27	18. Tāraṇa.
3604	503-04	13	13	...	...	13	13	13	13	13	13	13	3619	518-19	28	28	...	...	28	28	28	28	28	28	28	19. Pārthiva.
3605	504-05	14	14	...	...	14	14	14	14	14	14	14	3620	519-20	29	29	...	...	29	29	29	29	29	29	29	20. Vyaya.

1	2	3	4	5	6	7	8	9	10	11	12	13	1	2	3	4	5	6	7	8	9	10	11	12	13	
3621	520-21	30	30	...	...	30	30	30	30	30	30	30	3661	560-61	10	10	...	...	10	10	10	10	10	10	10	21. Sarvajit.
3622	521-22	31	31	...	...	31	31	31	31	31	31	31	3662	561-62	11	11	...	...	11	11	11	11	11	11	11	22. Sarvadhārin.
3623	522-23	32	32	...	...	32	32	32	32	32	32	32	3663	562-63	12	12	...	...	12	12	12	12	12	12	12	23. Virōdhin.
3624	523-24	33	33	...	...	33	33	33	33	33	33	33	3664	563-64	13	13	...	...	13	13*	13	13*	13*	13	13	24. Vikrita.
3625	524-25	34	34	...	...	34	34	34	34	34	34	34	3665	564-65	14	14	...	...	14*	15	14	15	15	14	14	25. Khara.
3626	525-26	35	35	...	...	35	35	35	35	35	35	35	3666	565-66	15	15	...	...	16	16	15	16	16	15	15	26. Nandana.
3627	526-27	36	36	...	...	36	36	36	36	36	36	36	3667	566-67	16	16	...	...	17	17	16	17	17	16	16	27. Vijaya.
3628	527-28	37	37	...	...	37	37	37	37	37	37	37	3668	567-68	17*	17*	...	...	18	18	17*	18	18	17	17	28. Jaya.
3629	528-29	38	38	...	...	38	38	38	38	38	38	38	3669	568-69	19	19	...	...	19	19	19	19	19	18	18	29. Manmatha.
3630	529-30	39	39	...	...	39	39	39	39	39	39	39	3670	569-70	20	20	...	...	20	20	20	20	20	19	19	30. Durmukha.
3631	530-31	40	40	...	...	40	40	40	40	40	40	40	3671	570-71	21	21	...	...	21	21	21	21	21	20	20	31. Hēmalamba.
3632	531-32	41	41	...	...	41	41	41	41	41	41	41	3672	571-72	22	22	...	...	22	22	22	22	22	21	21	32. Vilamba.
3633	532-33	42	42	...	...	42	42	42	42	42	42	42	3673	572-73	23	23	...	...	23	23	23	23	23	22	22	33. Vikārin.
3634	533-34	43	43	...	...	43	43	43	43	43	43	43	3674	573-74	24	24	...	...	24	24	24	24	24	23	23	34. Sārvarin.
3635	534-35	44	44	...	...	44	44	44	44	44	44	44	3675	574-75	25	25	...	...	25	25	25	25	25	24	24	35. Plava.
3636	535-36	45	45	...	...	45	45	45	45	45	45	45	3676	575-76	26	26	...	...	26	26	26	26	26	25*	25*	36. Śubhakrit.
3637	536-37	46	46	...	...	46	46	46	46	46	46	46	3677	576-77	27	27	...	...	27	27	27	27	27	27	27	37. Śōbhana.
3638	537-38	47	47	...	...	47	47	47	47	47	47	47	3678	577-78	28	28	...	...	28	28	28	28	28	28	28	38. Krōdhin.
3639	538-39	48	48	...	...	48	48	48	48	48	48	48	3679	578-79	29	29	...	...	29	29	29	29	29	29	29	39 Viśvāvasu.
3640	539-40	49	49	...	...	49	49	49	49	49	49	49	3680	579-80	30	30	...	...	30	30	30	30	30	30	30	40. Parābhava.
3641	540-41	50	50	...	...	50	50	50	50	50	50	50	3681	580-81	31	31	...	...	31	31	31	31	31	31	31	41. Plavaṅga.
3642	541-42	51	51	...	...	51	51	51	51	51	51	51	3682	581-82	32	32	...	...	32	32	32	32	32	32	32	42. Kīlaka.
3643	542-43	52	52	...	...	52	52	52	52	52	52	52	3683	582-83	33	33	...	...	33	33	33	33	33	33	33	43. Saumya.
3644	543-44	53	53	...	...	53	53	53	53	53	53	53	3684	583-84	34	34	...	...	34	34	34	34	34	34	34	44. Sādhārana.
3645	544-45	54	54	...	...	54	54	54	54	54	54	54	3685	584-85	35	35	...	...	35	35	35	35	35	35	35	45. Virōdhakṛit.
3646	545-46	55	55	...	...	55	55	55	55	55	55	55	3686	585-86	36	36	...	...	36	36	36	36	36	36	36	46. Paridhāvin.
3647	546-47	56	56	...	...	56	56	56	56	56	56	56	3687	586-87	37	37	...	...	37	37	37	37	37	37	37	47. Pramādin.
3648	547-48	57	57	...	...	57	57	57	57	57	57	57	3688	587-88	38	38	...	...	38	38	38	38	38	38	38	48. Ānanda.
3649	548-49	58	58	...	...	58	58	58	58	58	58	58	3689	588-89	39	39	...	...	39	39	39	39	39	39	39	49. Rākshasa.
3650	549-50	59	59	...	...	59	59	59	59	59	59	59	3690	589-90	40	40	...	...	40	40	40	40	40	40	40	50. Anala.
3651	550-51	60	60	...	...	60	60	60	60	60	60	60	3691	590-91	41	41	...	...	41	41	41	41	41	41	41	51. Piṅgala.
3652	551-52	1	1	...	...	1	1	1	1	1	1	1	3692	591-92	42	42	...	...	42	42	42	42	42	42	42	52. Kālayukta.
3653	552-53	2	2	...	...	2	2	2	2	2	2	2	3693	592-93	43	43	...	...	43	43	43	43	43	43	43	53. Siddhārthin
3654	553-54	3	3	...	...	3	3	3	3	3	3	3	3694	593-94	44	44	...	...	44	44	44	44	44	44	44	54. Raudra.
3655	554-55	4	4	...	...	4	4	4	4	4	4	4	3695	594-95	45	45	...	...	45	45	45	45	45	45	45	55. Durmati.
3656	555-56	5	5	...	...	5	5	5	5	5	5	5	3696	595-96	46	46	...	...	46	46	46	46	46	46	46	56. Dundubhi.
3657	556-57	6	6	...	...	6	6	6	6	6	6	6	3697	596-97	47	47	...	...	47	47	47	47	47	47	47	57. Rudhirōdgārin.
3658	557-58	7	7	...	...	7	7	7	7	7	7	7	3698	597-98	48	48	...	...	48	48	48	48	48	48	48	58. Raktāksha.
3659	558-59	8	8	...	...	8	8	8	8	8	8	8	3699	598-99	49	49	...	...	49	49	49	49	49	49	49	59. Krōdhana.
3660	559-60	9	9	...	...	9	9	9	9	9	9	9	3700	599-600	50	50	...	...	50	50	50	50	50	50	50	60. Kshaya.

TABLE XLII—*contd.*

Expired year of Kaliyuga.	Year A.D.	Number of the saṁvatsara connected with each solar year according to the several Siddhāntas, by reason of its currency at apparent, or at mean, Mēsha-saṁkrānti. Sūrya-S. no bīja. Apparent M. S.	Sūrya-S. no bīja. Mean M. S.	Sūrya-S. with bīja. Apparent M. S.	Sūrya-S. with bīja. Mean M. S.	First Ārya-S. Apparent M. S.	First Ārya-S. Mean M. S.	Orig. Sūrya S. Mean M. S.	Brahma-S. and S. Śirō. Apparent M. S.	Brahma-S. and S. Śirō. Mean M. S.	Second Ārya-S. Apparent M. S.	Second Ārya-S. Mean M. S.	Expired year of Kaliyuga.	Year A.D.	Number of the saṁvatsara connected with each solar year according to the several Siddhāntas, by reason of its currency at apparent, or at mean, Mēsha-saṁkrānti. Sūrya-S. no bīja. Apparent M. S.	Sūrya-S. no bīja. Mean M. S.	Sūrya-S. with bīja. Apparent M. S.	Sūrya-S. with bīja. Mean M. S.	First Ārya-S. Apparent M. S.	First Ārya-S. Mean M. S.	Orig. Sūrya S. Mean M. S.	Brahma-S. and S. Śirō. Apparent M. S.	Brahma-S. and S. Śirō. Mean M. S.	Second Ārya-S. Apparent M. S.	Second Ārya-S. Mean M. S.	Names of the Sixty saṁvatsaras of the cycle of Jupiter.
1	2	3	4	5	6	7	8	9	10	11	12	13	1	2	3	4	5	6	7	8	9	10	11	12	13	
3701	600-01	51	51	...	...	51	51	51	51	51	51	51	3721	620-21	11	11	...	...	11	11	11	11	11	11	11	1. Prabhava.
3702	601-02	52	52	...	...	52	52	52	52	52	52	52	3722	621-22	12	12	...	...	12	12	12	12	12	12	12	2. Vibhava.
3703	602-03	53	53	...	...	53	53	53	53	53	53	53	3723	622-23	13	13	...	...	13	13	13	13	13	13	13	3. Śukla.
3704	603-04	54	54	...	...	54	54	54	54	54	54	54	3724	623-24	14	14	...	...	14	14	14	14	14	14	14	4. Pramōda.
3705	604-05	55	55	...	...	55	55	55	55	55	55	55	3725	624-25	15	15	...	...	15	15	15	15	15	15	15	5. Prajāpati.
3706	605-06	56	56	...	...	56	56	56	56	56	56	56	3726	625-26	16	16	...	...	16	16	16	16	16	16	16	6. Aṅgiras.
3707	606-07	57	57	...	...	57	57	57	57	57	57	57	3727	626-27	17	17	...	...	17	17	17	17	17	17	17	7. Śrīmukha.
3708	607-08	58	58	...	...	58	58	58	58	58	58	58	3728	627-28	18	18	...	...	18	18	18	18	18	18	18	8. Bhāva.
3709	608-09	59	59	...	...	59	59	59	59	59	59	59	3729	628-29	19	19	...	...	19	19	19	19	19	19	19	9. Yuvan.
3710	609-10	60	60	...	...	60	60	60	60	60	60	60	3730	629-30	20	20	...	...	20	20	20	20	20	20	20	10. Dhātri.
3711	610-11	1	1	...	...	1	1	1	1	1	1	1	3731	630-31	21	21	...	...	21	21	21	21	21	21	21	11. Īśvara.
3712	611-12	2	2	...	...	2	2	2	2	2	2	2	3732	631-32	22	22	...	...	22	22	22	22	22	22	22	12. Bahudhānya.
3713	612-13	3	3	...	...	3	3	3	3	3	3	3	3733	632-33	23	23	...	...	23	23	23	23	23	23	23	13. Pramāthin.
3714	613-14	4	4	...	...	4	4	4	4	4	4	4	3734	633-34	24	24	...	...	24	24	24	24	24	24	24	14. Vikrama.
3715	614-15	5	5	...	...	5	5	5	5	5	5	5	3735	634-35	25	25	...	...	25	25	25	25	25	25	25	15. Vṛisha.
3716	615-16	6	6	...	...	6	6	6	6	6	6	6	3736	635-36	26	26	...	...	26	26	26	26	26	26	26	16. Chitrabhānu.
3717	616-17	7	7	...	...	7	7	7	7	7	7	7	3737	636-37	27	27	...	...	27	27	27	27	27	27	27	17. Subhānu.
3718	617-18	8	8	...	...	8	8	8	8	8	8	8	3738	637-38	28	28	...	...	28	28	28	28	28	28	28	18. Tāraṇa.
3719	618-19	9	9	...	...	9	9	9	9	9	9	9	3739	638-39	29	29	...	...	29	29	29	29	29	29	29	19. Pārthiva.
3720	619-20	10	10	...	...	10	10	10	10	10	10	10	3740	639-40	30	30	...	...	30	30	30	30	30	30	30	20. Vyaya.

1	2	3	4	5	6	7	8	9	10	11	12	13	1	2	3	4	5	6	7	8	9	10	11	12	13	
3741	640-41	31	31	...	...	31	31	31	31	31	31	31	3781	680-81	12	12	...	...	12	12	12	12	12	12	12	21. Sarvajit.
3742	641-42	32	32	...	...	32	32	32	32	32	32	32	3782	681-82	13	13	...	...	13	13	13	13	13	13	13	22. Sarbadhārin.
3743	642-43	33	33	...	...	33	33	33	33	33	33	33	3783	682-83	14	14	...	...	14	14	14	14	14	14	14	23. Virōdhin.
3744	643-44	34	34	...	...	34	34	34	34	34	34	34	3784	683-84	15	15	...	...	15	15	15	15	15	15	15	24. Vikrita.
3745	644-45	35	35	...	...	35	35	35	35	35	35	35	3785	684-85	16	16	...	...	16	16	16	16	16	16	16	25. Khara.
3746	645-46	36	36	...	...	36	36	36	36	36	36	36	3786	685-86	17	17	...	...	17	17	17	17	17	17	17	26. Nandana.
3747	646-47	37	37	...	...	37	37	37	37	37	37	37	3787	686-87	18	18	...	...	18	18	18	18	18	18	18	27. Vijaya.
3748	647-48	38	38	...	...	38	38	38	38	38	38	38	3788	687-88	19	19	...	...	19	19	19	19	19	19	19	28. Jaya.
3749	648-49	39	39	...	...	39	39	39	39*	39*	39	39	3789	688-89	20	20	...	...	20	20	20	20	20	20	20	29. Manmatha.
3750	649-50	40	40	...	...	40*	40*	40	41	41	40	40	3790	689-90	21	21	...	...	21	21	21	21	21	21	21	30. Durmukha.
3751	650-51	41	41	...	...	42	42	41	42	42	41	41	3791	690-91	22	22	...	...	22	22	22	22	22	22	22	31. Hēmalamba.
3752	651-52	42	42	...	...	43	43	42	43	43	42	42	3792	691-92	23	23	...	...	23	23	23	23	23	23	23	32. Vilamba.
3753	652-53	43	43*	...	...	44	44	43*	44	44	43	43	3793	692-93	24	24	...	...	24	24	24	24	24	24	24	33. Vikārin.
3754	653-54	44*	45	...	...	45	45	45	45	45	44	44	3794	693-94	25	25	...	...	25	25	25	25	25	25	25	34. Sārvarin.
3755	654-55	46	46	...	...	46	46	46	46	46	45	45	3795	694-95	26	26	...	...	26	26	26	26	26	26	26	35. Plava.
3756	655-56	47	47	...	...	47	47	47	47	47	46	46	3796	695-96	27	27	...	...	27	27	27	27	27	27	27	36. Śubhakrit.
3757	656-57	48	48	...	...	48	48	48	48	48	47	47	3797	696-97	28	28	...	...	28	28	28	28	28	28	28	37. Sōbhana.
3758	657-58	49	49	...	...	49	49	49	49	49	48	48	3798	697-98	29	29	...	...	29	29	29	29	29	29	29	38. Krōdhin.
3759	658-59	50	50	...	...	50	50	50	50	50	49	49	3799	698-99	30	30	...	...	30	30	30	30	30	30	30	39. Viśvāvasu.
3760	659-60	51	51	...	...	51	51	51	51	51	50	50	3800	699-700	31	31	...	...	31	31	31	31	31	31	31	40. Parābhava.
3761	660-61	52	52	...	...	52	52	52	52	52	51	51	3801	700-01	32	32	...	...	32	32	32	32	32	32	32	41. Plavaṅga.
3762	661-62	53	53	...	...	53	53	53	53	53	52*	52*	3802	701-02	33	33	...	...	33	33	33	33	33	33	33	42. Kīlaka.
3763	662-63	54	54	...	...	54	54	54	54	54	54	54	3803	702-03	34	34	...	...	34	34	34	34	34	34	34	43. Saumya.
3764	663-64	55	55	...	...	55	55	55	55	55	55	55	3804	703-04	35	35	...	...	35	35	35	35	35	35	35	44. Sādhāraṇa.
3765	664-65	56	56	...	...	56	56	56	56	56	56	56	3805	704-05	36	36	...	...	36	36	36	36	36	36	36	45. Virōdhakrit.
3766	665-66	57	57	...	...	57	57	57	57	57	57	57	3806	705-06	37	37	...	...	37	37	37	37	37	37	37	46. Paridhāvin.
3767	666-67	58	58	...	...	58	58	58	58	58	58	58	3807	706-07	38	38	...	...	38	38	38	38	38	38	38	47. Pramādin.
3768	667-68	59	59	...	...	59	59	59	59	59	59	59	3808	707-08	39	39	...	...	39	39	39	39	39	39	39	48. Ānanda.
3769	668-69	60	60	...	...	60	60	60	60	60	60	60	3809	708-09	40	40	...	...	40	40	40	40	40	40	40	49. Rākshasa.
3770	669-70	1	1	...	...	1	1	1	1	1	1	1	3810	709-10	41	41	...	...	41	41	41	41	41	41	41	50. Anala.
3771	670-71	2	2	...	...	2	2	2	2	2	2	2	3811	710-11	42	42	...	...	42	42	42	42	42	42	42	51. Piṅgala.
3772	671-72	3	3	...	...	3	3	3	3	3	3	3	3812	711-12	43	43	...	...	43	43	43	43	43	43	43	52. Kālayukta.
3773	672-73	4	4	...	...	4	4	4	4	4	4	4	3813	712-13	44	44	...	...	44	44	44	44	44	44	44	53. Siddhārthin
3774	673-74	5	5	...	...	5	5	5	5	5	5	5	3814	713-14	45	45	...	...	45	45	45	45	45	45	45	54. Raudra.
3775	674-75	6	6	...	...	6	6	6	6	6	6	6	3815	714-15	46	46	...	...	46	46	46	46	46	46	46	55. Durmati.
3776	675-76	7	7	...	...	7	7	7	7	7	7	7	3816	715-16	47	47	...	...	47	47	47	47	47	47	47	56. Dundubhi.
3777	676-77	8	8	...	...	8	8	8	8	8	8	8	3817	716-17	48	48	...	...	48	48	48	48	48	48	48	57. Rudhirōdgārin.
3778	677-78	9	9	...	...	9	9	9	9	9	9	9	3818	717-18	49	49	...	...	49	49	49	49	49	49	49	58. Raktāksha.
3779	678-79	10	10	...	...	10	10	10	10	10	10	10	3819	718-19	50	50	...	...	50	50	50	50	50	50	50	59. Krōdhana.
3780	679-80	11	11	...	...	11	11	11	11	11	11	11	3820	719-20	51	51	...	...	51	51	51	51	51	51	51	60. Kshaya.

TABLE XLII—*contd.*

Expired year of Kaliyuga.	Year A.D.	NUMBER OF THE SAMVATSARA CONNECTED WITH EACH SOLAR YEAR ACCORDING TO THE SEVERAL SIDDHĀNTAS, BY REASON OF ITS CURRENCY AT APPARENT, OR AT MEAN, MĒSHA-SAMKRĀNTI. Sūrya-S. no bīja. Apparent M. S.	Sūrya-S. no bīja. Mean M. S.	Sūrya-S. with bīja. Apparent M. S.	Sūrya-S. with bīja. Mean M. S.	First Ārya-S. Apparent M. S.	First Ārya-S. Mean M. S.	Orig. Sūrya S. Mean M. S.	Brahma-S. and S. Sirō. Apparent M. S.	Brahma-S. and S. Sirō. Mean M. S.	Second Ārya-S. Apparent M. S.	Second Ārya-S. Mean M. S.
1	2	3	4	5	6	7	8	9	10	11	12	13
3821	720-21	52	52	...	...	52	52	52	52	52	52	52
3822	721-22	53	53	...	...	53	53	53	53	53	53	53
3823	722-23	54	54	...	...	54	54	54	54	54	54	54
3824	723-24	55	55	...	...	55	55	55	55	55	55	55
3825	724-25	56	56	...	...	56	56	56	56	56	56	56
3826	725-26	57	57	...	...	57	57	57	57	57	57	57
3827	726-27	58	58	...	...	58	58	58	58	58	58	58
3828	727-28	59	59	...	...	59	59	59	59	59	59	59
3829	728-29	60	60	...	...	60	60	60	60	60	60	60
3830	729-30	1	1	...	...	1	1	1	1	1	1	1
3831	730-31	2	2	...	...	2	2	2	2	2	2	2
3832	731-32	3	3	...	...	3	3	3	3	3	3	3
3833	732-33	4	4	...	...	4	4	4	4	4	4	4
3834	733-34	5	5	...	...	5	5	5	5	5*	5	5
3835	734-35	6	6	...	...	6*	6"	6	6*	7	6	6
3836	735-36	7	7	...	...	8	8	7	8	8	7	7
3837	736-37	8	8	...	...	9	9	8	9	9	8	8
3838	737-38	9	9*	...	...	10	10	9*	10	10	9	9
3839	738-39	10*	11	...	...	11	11	11	11	11	10	10
3840	739-40	12	12	...	...	12	12	12	12	12	11	11

Expired year of Kaliyuga.	Year A.D.	NUMBER OF THE SAMVATSARA CONNECTED WITH EACH SOLAR YEAR ACCORDING TO THE SEVERAL SIDDHĀNTAS, BY REASON OF ITS CURRENCY AT APPARENT, OR AT MEAN, MĒSHA-SAMKRĀNTI. Sūrya-S. no bīja. Apparent M. S.	Sūrya-S. no bīja. Mean M. S.	Sūrya-S. with bīja. Apparent M. S.	Sūrya-S. with bīja. Mean M. S.	First Ārya-S. Apparent M. S.	First Ārya-S. Mean M. S.	Orig. Sūrya S. Mean M. S.	Brahma-S. and S. Sirō. Apparent M. S.	Brahma-S. and S. Sirō. Mean M. S.	Second Ārya-S. Apparent M. S.	Second Ārya-S. Mean M. S.	Names of the Sixty saṁvatsaras of the cycle of Jupiter.
1	2	3	4	5	6	7	8	9	10	11	12	13	
3841	740-41	13	13	...	...	13	13	13	13	13	12	12	1. Prabhava.
3842	741-42	14	14	...	...	14	14	14	14	14	13	13	2. Vibhava.
3843	742-43	15	15	...	...	15	15	15	15	15	14	14	3. Śukla.
3844	743-44	16	16	...	...	16	16	16	16	16	15	15	4. Pramōda.
3845	744-45	17	17	...	...	17	17	17	17	17	16	16	5. Prajāpati.
3846	745-46	18	18	...	...	18	18	18	18	18	17	17	6. Aṅgiras.
3847	746-47	19	19	...	...	19	19	19	19	19	18*	18*	7. Srīmukha.
3848	747-48	20	20	...	...	20	20	20	20	20	20	20	8. Bhāva.
3849	748-49	21	21	...	...	21	21	21	21	21	21	21	9. Yuvan.
3850	749-50	22	22	...	...	22	22	22	22	22	22	22	10. Dhātri.
3851	750-51	23	23	...	...	23	23	23	23	23	23	23	11. Iśvara.
3852	751-52	24	24	...	...	24	24	24	24	24	24	24	12. Bahudhānya.
3853	752-53	25	25	...	...	25	25	25	25	25	25	25	13. Pramāthin.
3854	753-54	26	26	...	...	26	26	26	26	26	26	26	14. Vikarama.
3855	754-55	27	27	...	...	27	27	27	27	27	27	27	15. Vrisha.
3856	755-56	28	28	...	...	28	28	28	28	28	28	28	16. Chitrabhānu
3857	756-57	29	29	...	...	29	29	29	29	29	29	29	17. Subhānu.
3858	757-58	30	30	...	...	30	30	30	30	30	30	30	18. Tāraṇa.
3859	758-59	31	31	...	...	31	31	31	31	31	31	31	19. Pārthiva.
3860	759-60	32	32	...	...	32	32	32	32	32	32	32	20. Vyaya.

1	2	3	4	5	6	7	8	9	10	11	12	13	1	2	3	4	5	6	7	8	9	10	11	12	13	
3861	760-61	33	33	...	...	33	33	33	33	33	33	33	3901	800-01	13	13	...	...	13	13	13	13	13	13	13	21. Sarvajit.
3862	761-62	34	34	...	...	34	34	34	34	34	34	34	3902	801-02	14	14	...	...	14	14	14	14	14	14	14	22. Sarvadhārin.
3863	762-63	35	35	...	...	35	35	35	35	35	35	35	3903	802-03	15	15	...	...	15	15	15	15	15	15	15	23. Virōdhin.
3864	763-64	36	36	...	...	36	36	36	36	36	36	36	3904	803-04	16	16	...	...	16	16	16	16	16	16	16	24. Vikrita.
3865	764-65	37	37	...	...	37	37	37	37	37	37	37	3905	804-05	17	17	...	...	17	17	17	17	17	17	17	25. Khara.
3866	765-66	38	38	...	...	38	38	38	38	38	38	38	3906	805-06	18	18	...	...	18	18	18	18	18	18	18	26. Nandana.
3867	766-67	39	39	...	...	39	39	39	39	39	39	39	3907	806-07	19	19	...	...	19	19	19	19	19	19	19	27. Vijaya.
3868	767-68	40	40	...	...	40	40	40	40	40	40	40	3908	807-08	20	20	...	...	20	20	20	20	20	20	20	28. Jaya.
3869	768-69	41	41	...	...	41	41	41	41	41	41	41	3909	808-09	21	21	...	...	21	21	21	21	21	21	21	29. Manmatha.
3870	769-70	42	42	...	...	42	42	42	42	42	42	42	3910	809-10	22	22	...	...	22	22	22	22	22	22	22	30. Durmukha.
3871	770-71	43	43	...	...	43	43	43	43	43	43	43	3911	810-11	23	23	...	...	23	23	23	23	23	23	23	31. Hēmalamba.
3872	771-72	44	44	...	...	44	44	44	44	44	44	44	3912	811-12	24	24	...	...	24	24	24	24	24	24	24	32. Vilamba.
3873	772-73	45	45	...	...	45	45	45	45	45	45	45	3913	812-13	25	25	...	...	25	25	25	25	25	25	25	33. Vikārin.
3874	773-74	46	46	...	...	46	46	46	46	46	46	46	3914	813-14	26	26	...	...	26	26	26	26	26	26	26	34. Śārvarin.
3875	774-75	47	47	...	...	47	47	47	47	47	47	47	3915	814-15	27	27	...	...	27	27	27	27	27	27	27	35. Plava.
3876	775-76	48	48	...	...	48	48	48	48	48	48	48	3916	815-16	28	28	...	...	28	28	28	28	28	28	28	36. Śubhakrit.
3877	776-77	49	49	...	...	49	49	49	49	49	49	49	3917	816-17	29	29	...	...	29	29	29	29	29	29	29	37. Śōbhana.
3878	777-78	50	50	...	...	50	50	50	50	50	50	50	3918	817-18	30	30	...	...	30	30	30	30	30	30	30	38. Krōdhin.
3879	778-79	51	51	...	...	51	51	51	51	51	51	51	3919	818-19	31	31	...	...	31	31	31	31	31*	31	31	39. Viśvāvasu.
3880	779-80	52	52	...	...	52	52	52	52	52	52	52	3920	819-20	32	32	...	...	32*	32*	32	32*	33	32	32	40. Parābhava.
3881	780-81	53	53	...	...	53	53	53	53	53	53	53	3921	820-21	33	33	...	...	34	34	33	34	34	33	33	41. Plavanga.
3882	781-82	54	54	...	...	54	54	54	54	54	54	54	3922	821-22	34	34	...	...	35	35	34	35	35	34	34	42. Kīlaka.
3883	782-83	55	55	...	...	55	55	55	55	55	55	55	3923	822-23	35	35	...	...	36	36	35	35	36	35	35	43. Saumya.
3884	783-84	56	56	...	...	56	56	56	56	56	56	56	3924	823-24	36*	36*	...	...	37	37	36*	37	37	36	36	44. Sādhāraṇa.
3885	784-85	57	57	...	...	57	57	57	57	57	57	57	3925	824-25	38	38	...	...	38	38	38	38	38	37	37	45. Virōdhakrit.
3886	785-86	58	58	...	...	58	58	58	58	58	58	58	3926	825-26	39	39	...	...	39	39	39	39	39	38	38	46. Paridhāvin.
3887	786-87	59	59	...	...	59	59	59	59	59	59	59	3927	826-27	40	40	...	...	40	40	40	40	40	39	39	47. Pramādin.
3888	787-88	60	60	...	...	60	60	60	60	60	60	60	3928	827-28	41	41	...	...	41	41	41	41	41	40	40	48. Ânanda.
3889	788-89	1	1	...	...	1	1	1	1	1	1	1	3929	828-29	42	42	...	...	42	42	42	42	42	41	41	49. Rākshasa.
3890	789-90	2	2	...	...	2	2	2	2	2	2	2	3930	829-30	43	43	...	...	43	43	43	43	43	42	42	50. Anala.
3891	790-91	3	3	...	...	3	3	3	3	3	3	3	3931	830-31	44	44	...	...	44	44	44	44	44	43	43	51. Piṅgala.
3892	791-92	4	4	...	...	4	4	4	4	4	4	4	3932	831-32	45	45	...	...	45	45	45	45	45	44	44*	52. Kālayukta.
3893	792-93	5	5	...	...	5	5	5	5	5	5	5	3933	832-33	46	46	...	...	46	46	46	46	46	45*	46	53. Siddhārthin.
3894	793-94	6	6	...	...	6	6	6	6	6	6	6	3934	833-34	47	47	...	...	47	47	47	47	47	47	47	54. Raudra.
3895	794-95	7	7	...	...	7	7	7	7	7	7	7	3935	834-35	48	48	...	...	48	48	48	48	48	48	48	55. Durmati
3896	795-96	8	8	...	...	8	8	8	8	8	8	8	3936	835-36	49	49	...	...	49	49	49	49	49	49	49	56. Dundubhi.
3897	796-97	9	9	...	...	9	9	9	9	9	9	9	3937	836-37	50	50	...	...	50	50	50	50	50	50	50	57. Rudhirōdgārin.
3898	797-98	10	10	...	...	10	10	10	10	10	10	10	3938	837-38	51	51	...	...	51	51	51	51	51	51	51	58. Raktāksha.
3899	798-99	11	11	...	...	11	11	11	11	11	11	11	3939	838-39	52	52	...	...	52	52	52	52	52	52	52	59. Krōdhāna.
3900	799-800	12	12	...	...	12	12	12	12	12	12	12	3940	839-40	53	53	...	...	53	53	53	53	53	53	53	60. Kshaya.

TABLE XLII—contd.

Expired year of Kaliyuga.	Year A.D.	NUMBER OF THE SAṀVATSARA CONNECTED WITH EACH SOLAR YEAR ACCORDING TO THE SEVERAL SIDDHĀNTAS, BY REASON OF ITS CURRENCY AT APPARENT,S OR AT MEAN, MĒSHA AMKRĀNTI.											Expired year of Kaliyuga.	Year A.D.	NUMBER OF THE SAṀVATSARA CONNECTED WITH EACH SOLAR YEAR ACCORDING TO THE SEVERAL SIDDHĀNTAS, BY REASON OF ITS CURRENCY AT APPARENT, OR AT MEAN, MĒSHA-SAṀKRĀNTI.											Names of the Sixty saṁvatsaras of the cycle of Jupiter.
		Sūrya-S. no bīja.		Sūrya-S. with bīja.		First Ārya-S.		Orig. Sūrya S.	Brahma-S. and S. Śirō.		Second Ārya-S.				Sūrya-S. no bīja.		Sūrya-S. with bīja.		First Ārya-S.		Orig. Sūrya S.	Brahma-S. and S. Śirō.		Second Ārya-S.		
		Apparent M. S.	Mean M. S.	Apparent M. S.	Mean M. S.	Apparent M. S.	Mean M. S.	Mean M. S.	Apparent M. S.	Mean M. S.	Apparent M. S.	Mean M. S.			Apparent M. S.	Mean M. S.	Apparent M. S.	Mean M. S.	Apparent M. S.	Mean M. S.	Mean M. S.	Apparent M. S.	Mean M. S.	Apparent M. S.	Mean M. S.	
1	2	3	4	5	6	7	8	9	10	11	12	13	1	2	3	4	5	6	7	8	9	10	11	12	13	
3941	840-41	54	54	...	...	54	54	54	54	54	54	54	3961	860-61	14	14	...	...	14	14	14	14	14	14	14	1. Prabhava.
3942	841-42	55	55	...	...	55	55	55	55	55	55	55	3962	861-62	15	15	...	...	15	15	15	15	15	15	15	2. Vibhava.
3943	842-43	56	56	...	...	56	56	56	56	56	56	56	3963	862-63	16	16	...	...	16	16	16	16	16	16	16	3. Sukla.
3944	843-44	57	57	...	..	57	57	57	57	57	57	57	3964	863-64	17	17	...	...	17	17	17	17	17	17	17	4. Pramōda.
3945	844-45	58	58	...	...	58	58	58	58	58	58	58	3965	864-65	18	18	...	...	18	18	18	18	18	18	18	5. Prajāpati.
3946	845-46	59	59	...	...	59	59	59	59	59	59	59	3966	865-66	19	19	...	...	19	19	19	19	19	19	19	6. Angiras.
3947	846-47	60	60	...	...	60	60	60	60	60	60	60	3967	866-67	20	20	...	...	20	20	20	20	20	20	20	7. Srīmukha.
3948	847-48	1	1	...	...	1	1	1	1	1	1	1	3968	867-68	21	21	...	...	21	21	21	21	21	21	21	8. Bhāva.
3949	848-49	2	2	...	...	2	2	2	2	2	2	2	3969	868-69	22	22	...	...	22	22	22	22	22	22	22	9. Yuvan.
3950	849-50	3	3	...	...	3	3	3	3	3	3	3	3970	869-70	23	23	...	...	23	23	23	23	23	23	23	10. Dhātṛi.
3951	850-51	4	4	...	...	4	4	4	4	4	4	4	3971	870-71	24	24	...	...	24	24	24	24	24	24	24	11. Īśvara.
3952	851-52	5	5	...	...	5	5	5	5	5	5	5	3972	871-72	25	25	...	...	25	25	25	25	25	25	25	12. Bahudhânya.
3953	852-53	6	6	...	...	6	6	6	6	6	5	6	3973	872-73	26	26	...	...	26	26	26	26	26	26	26	13. Pramāthin.
3954	853-54	7	7	...	...	7	7	7	7	7	7	7	3974	873-74	27	27	...	...	27	27	27	27	27	27	27	14. Vikrama.
3955	854-55	8	8	...	...	8	8	8	8	8		8	3975	874-75	28	28	...	...	28	28	28	28	28	28	28	15. Vṛisha.
3956	855-56	9	9	..	...	9	9	9	9	9	9	9	3976	875-76	29	29	...	...	29	29	29	29	29	29	29	16. Chitrabhānu.
3957	856-57	10	10	...	...	10	10	10	10	10	10	10	3977	876-77	30	30	...	...	30	30	30	30	30	30	30	17. Subhānu.
3958	857-58	11	11	...	...	11	11	11	11	11	11	11	3978	877-78	31	31	...	...	31	31	31	31	31	31	31	18. Tāraṇa.
3959	858-59	12	12	...	...	12	12	12	12	12	12	12	3979	878-79	32	32	...	...	32	32	32	32	32	32	32	19. Pārthiva.
3960	859-60	13	13	...	...	13	13	13	13	13	13	13	3980	879-80	33	33	...	...	33	33	33	33	33	33	33	20. Vyaya.

1	2	3	4	5	6	7	8	9	10	11	12	13	1	2	3	4	5	6	7	8	9	10	11	12	13	
3981	880-81	34	34	...	...	34	34	34	34	34	34	34	4021	920-21	15	15	...	...	15	15	15	15	15	15	15	21. Sarvajit.
3982	881-82	35	35	...	...	35	35	35	35	35	35	35	4022	921-22	16	16	...	...	16	16	16	16	16	16	16	22. Sarvadhârin.
3983	882-83	36	36	...	...	36	36	36	36	36	36	36	4023	922-23	17	17	...	...	17	17	17	17	17	17	17	23. Virōdhin.
3984	883-84	37	37	...	...	37	37	37	37	37	37	37	4024	923-24	18	18	...	...	18	18	18	18	18	18	18	24. Vikṛita.
3985	884-85	38	38	...	...	38	38	38	38	38	38	38	4025	924-25	19	19	...	...	19	19	19	19	19	19	19	25. Khara.
3986	885-86	39	39	...	...	39	39	39	39	39	39	39	4026	925-26	20	20	...	...	20	20	20	20	20	20	20	26. Nandana.
3987	886-87	40	40	...	...	40	40	40	40	40	40	40	4027	926-27	21	21	...	...	21	21	21	21	21	21	21	27. Vijaya.
3988	887-88	41	41	...	...	41	41	41	41	41	41	41	4028	927-28	22	22	...	...	22	22	22	22	22	22	22	28. Jaya.
3989	888-89	42	42	...	...	42	42	42	42	42	42	42	4029	928-29	23	23	...	...	23	23	23	23	23	23	23	29. Manmatha.
3990	889-90	43	43	...	...	43	43	43	43	43	43	43	4030	929-30	24	24	...	...	24	24	24	24	24	24	24	30. Durmukha.
3991	890-91	44	44	...	...	44	44	44	44	44	44	44	4031	930-31	25	25	...	...	25	25	25	25	25	25	25	31. Hēmalamba.
3992	891-92	45	45	...	...	45	45	45	45	45	45	45	4032	931-32	26	26	...	...	26	26	26	26	26	26	26	32. Vilamba.
3993	892-93	46	46	...	...	46	46	46	46	46	46	46	4033	932-33	27	27	...	...	27	27	27	27	27	27	27	33. Vikārin.
3994	893-94	47	47	...	...	47	47	47	47	47	47	47	4034	933-34	28	28	...	...	28	28	28	28	28	28	28	34. Śārvarin.
3995	894-95	48	48	...	...	48	48	48	48	48	48	48	4035	934-35	29	29	...	...	29	29	29	29	29	29	29	35. Plava.
3996	895-96	49	49	...	...	49	49	49	49	49	49	49	4036	935-36	30	30	...	...	30	30	30	30	30	30	30	36. Śubhakṛit.
3997	896-97	50	50	...	...	50	50	50	50	50	50	50	4037	936-37	31	31	...	...	31	31	31	31	31	31	31	37. Sōbhana.
3998	897-98	51	51	...	...	51	51	51	51	51	51	51	4038	937-38	32	32	...	...	32	32	32	32	32	32	32	38. Krōdhin.
3999	898-99	52	52	...	...	52	52	52	52	52	52	52	4039	938-39	33	33	...	...	33	33	33	33	33	33	33	39. Viśvâvasu.
4000	899-900	53	53	...	...	53	53	53	53	53	53	53	4040	939-40	34	34	...	...	34	34	34	34	34	34	34	40. Parābhava.
4001	900-01	54	54	...	...	54	54	54	54	54	54	54	4041	940-41	35	35	...	...	35	35	35	35	35	35	35	41. Plavaṅga.
4002	901-02	55	55	...	...	55	55	55	55	55	55	55	4042	941-42	36	36	...	...	36	36	36	36	36	36	36	42. Kīlaka.
4003	902-03	56	56	...	...	56	56	56	56	56	56	56	4043	942-43	37	37	...	...	37	37	37	37	37	37	37	43. Saumya.
4004	903-04	57	57	...	...	57	57	57	57	57*	57	57	4044	943-44	38	38	...	...	38	38	38	38	38	38	38	44. Sādhāraṇa.
4005	904-05	58	58	...	...	58	58*	58	58*	59	58	58	4045	944-45	39	39	...	...	39	39	39	39	39	39	39	45. Virōdhakṛit.
4006	905-06	59	59	...	...	59*	60	59	60	60	59	59	4046	945-46	40	40	...	...	40	40	40	40	40	40	40	46. Paridhāvin.
4007	906-07	60	60	...	...	1	1	60	1	1	60	60	4047	946-47	41	41	...	...	41	41	41	41	41	41	41	47. Pramādin.
4008	907-08	1	1	...	...	2	2	1	2	2	1	1	4048	947-48	42	42	...	...	42	42	42	42	42	42	42	48. Ānanda.
4009	908-09	2*	2*	...	...	3	3	2*	3	3	2	2	4049	948-49	43	43	...	...	43	43	43	43	43	43	43	49. Rākshasa.
4010	909-10	4	4	...	...	4	4	4	4	4	3	3	4050	949-50	44	44	...	...	44	44	44	44	44	44	44	50. Anala.
4011	910-11	5	5	...	...	5	5	5	5	5	4	4	4051	950-51	45	45	...	...	45	45	45	45	45	45	45	51. Piṅgala.
4012	911-12	6	6	...	...	6	6	6	6	6	5	5	4052	951-52	46	46	...	...	46	46	46	46	46	46	46	52. Kālayukta.
4013	912-13	7	7	...	...	7	7	7	7	7	6	6	4053	952-53	47	47	...	...	47	47	47	47	47	47	47	53. Siddhārthin.
4014	913-14	8	8	...	...	8	8	8	8	8	7	7	4054	953-54	48	48	...	...	48	48	48	48	48	48	48	54. Raudra.
4015	914-15	9	9	...	...	9	9	9	9	9	8	8	4055	954-55	49	49	...	...	49	49	49	49	49	49	49	55. Durmati.
4016	915-16	10	10	...	...	10	10	10	10	10	9	9	4056	955-56	50	50	...	...	50	50	50	50	50	50	50	56. Dundubhi.
4017	916-17	11	11	...	...	11	11	11	11	11	10	10*	4057	956-57	51	51	...	...	51	51	51	51	51	51	51	57. Rudhirōdgārin.
4018	917-18	12	12	...	...	12	12	12	12	12	11*	12	4058	957-58	52	52	...	...	52	52	52	52	52	52	52	58. Raktāksha.
4019	918-19	13	13	...	...	13	13	13	13	13	13	13	4059	958-59	53	53	...	...	53	53	53	53	53	53	53	59. Krōdhana.
4020	919-20	14	14	...	...	14	14	14	14	14	14	14	4060	959-60	54	54	...	...	54	54	54	54	54	54	54	60. Kshaya.

TABLE XLII—*contd.*

Expired year of Kaliyuga.	Year A.D.	NUMBER OF THE SAṀVATSARA CONNECTED WITH EACH SOLAR YEAR ACCORDING TO THE SEVERAL SIDDHĀNTAS, BY REASON OF ITS CURRENCY AT APPARENT, OR AT MEAN, MÊSHA-SAṀKRĀNTI. Sūrya-S. no bīja. Apparent M. S.	Sūrya-S. no bīja. Mean M. S.	Sūrya-S. with bīja. Apparent M. S.	Sūrya-S. with bīja. Mean M. S.	First Ārya-S. Apparent M. S.	First Ārya-S. Mean M. S.	Orig. Sūrya S. Mean M. S.	Brahma-S. and S. Sirô. Apparent M. S.	Brahma-S. and S. Sirô. Mean M. S.	Second Ārya-S. Apparent M. S.	Second Ārya-S. Mean M. S.
1	2	3	4	5	6	7	8	9	10	11	12	13
4061	960-61	55	55	...	...	55	55	55	55	55	55	55
4062	961-62	56	56	...	...	56	56	56	56	56	56	56
4063	962-63	57	57	...	...	57	57	57	57	57	57	57
4064	963-64	58	58	...	...	58	58	58	58	58	58	58
4065	964-65	59	59	...	...	59	59	59	59	59	59	59
4066	965-66	60	60	...	...	60	60	60	60	60	60	60
4067	966-67	1	1	...	...	1	1	1	1	1	1	1
4068	967-68	2	2	...	...	2	2	2	2	2	2	2
4069	968-69	3	3	...	...	3	3	3	3	3	3	3
4070	969-70	4	4	...	...	4	4	4	4	4	4	4
4071	970-71	5	5	...	...	5	5	5	5	5	5	5
4072	971-72	6	6	...	...	6	6	6	6	6	6	6
4073	972-73	7	7	...	...	7	7	7	7	7	7	7
4074	973-74	8	8	...	...	8	8	8	8	8	8	8
4075	974-75	9	9	...	...	9	9	9	9	9	9	9
4076	975-76	10	10	...	...	10	10	10	10	10	10	10
4077	976-77	11	11	...	...	11	11	11	11	11	11	11
4078	977-78	12	12	...	...	12	12	12	12	12	12	12
4079	978-79	13	13	...	...	13	13	13	13	13	13	13
4080	979-80	14	14	...	...	14	14	14	14	14	14	14

Expired year of Kaliyuga.	Year A.D.	NUMBER OF THE SAṀVATSARA CONNECTED WITH EACH SOLAR YEAR ACCORDING TO THE SEVERAL SIDDHĀNTAS, BY REASON OF ITS CURRENCY AT APPARENT, OR AT MEAN, MÊSHA-SAṀKRĀNTI. Sūrya-S. no bīja. Apparent M. S.	Sūrya-S. no bīja. Mean M. S.	Sūrya-S. with bīja. Apparent M. S.	Sūrya-S. with bīja. Mean M. S.	First Ārya-S. Apparent M. S.	First Ārya-S. Mean M. S.	Orig. Sūrya S. Mean M. S.	Brahma-S. and S. Sirô. Apparent M. S.	Brahma-S. and S. Sirô. Mean M. S.	Second Ārya-S. Apparent M. S.	Second Ārya-S. Mean M. S.	Names of the Sixty saṁvatsaras of the cycle of Jupiter.
1	2	3	4	5	6	7	8	9	10	11	12	13	
4081	980-81	15	15	...	...	15	15	15	15	15	15	15	1. Prabhava.
4082	981-82	16	16	...	...	16	16	16	16	16	16	16	2. Vibhava.
4083	982-83	17	17	...	...	17	17	17	17	17	17	17	3. Sukla.
4084	983-84	18	18	...	...	18	18	18	18	18	18	18	4. Pramōda.
4085	984-85	19	19	...	...	19	19	19	19	19	19	19	5. Prajāpati.
4086	985-86	20	20	...	...	20	20	20	20	20	20	20	6. Aṅgiras.
4087	986-87	21	21	...	...	21	21	21	21	21	21	21	7. Srīmukha.
4088	987-88	22	22	...	...	22	22	22	22	22	22	22	8. Bhāva.
4089	988-89	23	23	...	...	23	23	23	23	23	23	23	9. Yuvan.
4090	989-90	24	24	...	...	24	24*	24	24*	24*	24	24	10. Dhātṛi.
4091	990-91	25	25	...	...	25*	26	25	26	26	25	25	11. Īsvara.
4092	991-92	26	26	...	...	27	27	26	27	27	26	26	12. Bahudhānya.
4093	992-93	27	27	...	...	28	28	27	28	28	27	27	13. Pramāthin.
4094	993-94	28	28*	...	...	29	29	28*	29	29	28	28	14. Vikrama.
4095	994-95	29*	30	...	...	30	30	30	30	30	29	29	15. Vṛisha.
4096	995-96	31	31	...	...	31	31	31	31	31	30	30	16. Chitrabhānu.
4097	996-97	32	32	...	...	32	32	32	32	32	31	31	17. Subhānu.
4098	997-98	33	33	...	...	33	33	33	33	33	32	32	18. Tāraṇa.
4099	998-99	34	34	...	...	34	34	34	34	34	33	33	19. Pārthiva.
4100	999-1000	35	35	...	...	35	35	35	35	35	34	34	20. Vyaya.

1	2	3	4	5	6	7	8	9	10	11	12	13	1	2	3	4	5	6	7	8	9	10	11	12	13	
4101	1000-01	36	36	...	...	36	6	36	36	36	35	35	4141	1040-41	16	16	...	...	16	16	16	16	16	16	16	21. Sarvajit.
4102	1001-02	37	37	...	...	37	37	37	37	37	36	36	4142	1041-42	17	17	...	...	17	17	17	17	17	17	17	22. Sarvadhārin.
4103	1002-03	38	38	...	...	38	38	38	38	38	37*	37*	4143	1042-43	18	18	...	...	18	18	18	18	18	18	18	23. Virōdhin.
4104	1003-04	39	39	...	...	39	39	39	39	39	39	39	4144	1043-44	19	19	...	...	19	19	19	19	19	19	19	24. Vikrita.
4105	1004-05	40	40	...	...	40	40	40	40	40	40	40	4145	1044-45	20	20	...	...	20	20	20	20	20	20	20	25. Khara.
4106	1005-06	41	41	...	...	41	41	41	41	41	41	41	4146	1045-46	21	21	...	...	21	21	21	21	21	21	21	26. Nandana.
4107	1006-07	42	42	...	...	42	42	42	42	42	42	42	4147	1046-47	22	22	...	...	22	22	22	22	22	22	22	27. Vijaya.
4108	1007-08	43	43	...	...	43	43	43	43	43	43	43	4148	1047-48	23	23	...	...	23	23	23	23	23	23	23	28. Jaya.
4109	1008-09	44	44	...	...	44	44	44	44	44	44	44	4149	1048-49	24	24	...	...	24	24	24	24	24	24	24	29. Manmatha.
4110	1009-10	45	45	...	...	45	45	45	45	45	45	45	4150	1049-50	25	25	...	...	25	25	25	25	25	25	25	30. Durmukha.
4111	1010-11	46	46	...	...	46	46	46	46	46	46	46	4151	1050-51	26	26	...	...	26	26	26	26	26	26	26	31. Hēmalamba.
4112	1011-12	47	47	...	...	47	47	47	47	47	47	47	4152	1051-52	27	27	...	...	27	27	27	27	27	27	27	32. Vilamba.
4113	1012-13	48	48	...	...	48	48	48	48	48	48	48	4153	1052-53	28	28	...	...	28	28	28	28	28	28	28	33. Vikārin.
4114	1013-14	49	49	...	...	49	49	49	49	49	49	49	4154	1053-54	29	29	...	...	29	29	29	29	29	29	29	34. Śārvarin
4115	1014-15	50	50	...	...	50	50	50	50	50	50	50	4155	1054-55	30	30	...	...	30	30	30	30	30	30	30	35. Plava.
4116	1015-16	51	51	...	...	51	51	51	51	51	51	51	4156	1055-56	31	31	...	...	31	31	31	31	31	31	31	36. Subhakrit.
4117	1016-17	52	52	...	...	52	52	52	52	52	52	52	4157	1056-57	32	32	...	...	32	32	32	32	32	32	32	37. Sōbhana.
4118	1017-18	53	53	...	...	53	53	53	53	35	53	53	4158	1057-58	33	33	...	...	33	33	33	33	33	33	33	38. Krōdhin.
4119	1018-19	54	54	...	...	54	54	54	54	54	54	54	4159	1058-59	34	34	...	...	34	34	34	34	34	34	34	39. Viśvāvasu.
4120	1019-20	55	55	...	...	55	55	55	55	55	55	55	4160	1059-60	35	35	...	...	35	35	35	35	35	35	35	40. Parābhava.
4121	1020-21	56	56	...	...	56	56	56	56	56	56	56	4161	1060-61	36	36	...	...	36	36	36	36	36	36	36	41. Playaṅga.
4122	1021-22	57	57	...	...	57	57	57	57	57	57	57	4162	1061-62	37	37	...	...	37	37	37	37	37	37	37	42. Kīlaka.
4123	1022-23	58	58	...	...	58	58	58	58	58	58	58	4163	1062-63	38	38	...	...	38	38	38	38	38	38	38	43. Saumya.
4124	1023-24	59	59	...	...	59	59	59	59	59	59	59	4164	1063-64	39	39	...	...	39	39	39	39	39	39	39	44. Sādhāraṇa.
4125	1024-25	60	60	...	...	60	60	60	60	60	60	60	4165	1064-65	40	40	...	...	40	40	40	40	40	40	40	45. Virōdhakrit
4126	1025-26	1	1	...	...	1	1	1	1	1	1	1	4166	1065-66	41	41	...	...	41	41	41	41	41	41	41	46. Paridhāvin.
4127	1026-27	2	2	...	...	2	2	2	2	2	2	2	4167	1066-67	42	42	...	...	42	42	42	42	42	42	42	47. Pramādin.
4128	1027-28	3	3	...	...	3	3	3	3	3	3	3	4168	1067-68	43	43	...	...	43	43	43	43	43	43	43	48. Ānanda.
4129	1028-29	4	4	...	...	4	4	4	4	4	4	4	4169	1068-69	44	44	...	...	44	44	44	44	44	44	44	49. Rākshasa
4130	1029-30	5	5	...	...	5	5	5	5	5	5	5	4170	1069-70	45	45	...	...	45	45	45	45	45	45	45	50. Anala.
4131	1030-31	6	6	...	...	6	6	6	6	6	6	6	4171	1070-71	46	46	...	...	46	46	46	46	46	46	46	51. Piṅgala.
4132	1031-32	7	7	...	...	7	7	7	7	7	7	7	4172	1071-72	47	47	...	...	47	47	47	47	47	47	47	52. Kālayukta.
4133	1032-33	8	8	...	...	8	8	8	8	8	8	8	4173	1072-73	48	48	...	...	48	48	48	48	48	48	48	53. Siddhārthin.
4134	1033-34	9	9	...	...	9	9	9	9	9	9	9	4174	1073-74	49	49	...	...	49	49	49	49	49	49	49	54. Raudra.
4135	1034-35	10	19	...	...	10	10	10	10	10	10	10	4175	1074-75	50	50	...	...	50	50	50	50*	50*	50	50	55. Durmati.
4136	1035-36	11	11	...	...	11	11	11	11	11	11	11	4176	1075-76	51	51	...	...	51*	51*	51	52	52	51	51	56. Dundubhi.
4137	1036-37	12	12	...	...	12	12	12	12	12	12	12	4177	1076-77	52	52	...	...	53	53	52	53	53	52	52	57. Rudhirōdgārin.
4138	1037-38	13	13	...	...	13	13	13	13	13	13	13	4178	1077-78	53	53	...	...	54	54	53	54	54	53	53	58. Raktāksha.
4139	1038-39	14	14	...	...	14	14	14	14	14	14	14	4179	1078-79	54	54	...	...	55	55	54	55	55	54	54	59. Krōdhana.
4140	1039-40	15	15	...	...	15	15	15	15	15	15	15	4180	1079-80	55*	55*	...	...	56	56	55*	56	56	55	55	60. Kshaya.

TABLE XLII—*contd.*

Expired year of Kaliyuga.	Year A.D.	Number of the Saṁvatsara connected with each solar year according to the several Siddhāntas, by reason of its currency at apparent, or at mean, Mēsha-saṁkrānti. Sūrya-S. no Bīja. Apparent M. S.	Sūrya-S. no Bīja. Mean M. S.	Sūrya-S. with Bīja. Apparent M. S.	Sūrya-S. with Bīja. Mean M. S.	First Ārya-S. Apparent M. S.	First Ārya-S. Mean M. S.	Orig. Sūrya S. Mean M. S.	Brahma-S. and S. Širō. Apparent M. S.	Brahma-S. and S. Širō. Mean M. S.	Second Ārya-S. Apparent M. S.	Second Ārya-S. Mean M. S.	Expired year of Kaliyuga.	Year A.D.	Number of the Saṁvatsara connected with each solar year according to the several Siddhāntas, by reason of its currency at apparent, or at mean, Mēsha-saṁkrānti. Sūrya-S. no Bīja. Apparent M. S.	Sūrya-S. no Bīja. Mean M. S.	Sūrya-S. with Bīja. Apparent M. S.	Sūrya-S. with Bīja. Mean M. S.	First Ārya-S. Apparent M. S.	First Ārya-S. Mean M. S.	Orig. Sūrya S. Mean M. S.	Brahma-S. and S. Širō. Apparent M. S.	Brahma-S. and S. Širō. Mean M. S.	Second Ārya-S. Apparent M. S.	Second Ārya-S. Mean M. S.	Names of the Sixty saṁvatsaras of the cycle of Jupiter.
1	2	3	4	5	6	7	8	9	10	11	12	13	1	2	3	4	5	6	7	8	9	10	11	12	13	
4181	1080-81	57	57	...	...	57	57	57	57	57	56	56	4201	1100-01	17	17	...	...	17	17	17	17	17	17	17	1. Prabhava.
4182	1081-82	58	58	...	...	58	58	58	58	58	57	57	4202	1101-02	18	18	...	...	18	18	18	18	18	18	18	2. Vibhava.
4183	1082-83	59	59	...	...	59	59	59	59	59	58	58	4203	1102-03	19	19	...	...	19	19	19	19	19	19	19	3. Sukla.
4184	1083-84	60	60	...	...	60	60	60	60	60	59	59	4204	1103-04	20	20	...	...	20	20	20	20	20	20	20	4. Pramōda.
4185	1084-85	1	1	...	...	1	1	1	1	1	60	60	4205	1104-05	21	21	...	...	21	21	21	21	21	21	21	5. Prajāpati.
4186	1085-86	2	2	...	...	2	2	2	2	2	1	1	4206	1105-06	22	22	...	...	22	22	22	22	22	22	22	6. Aṅgiras.
4187	1086-87	3	3	...	...	3	3	3	3	3	2	2	4207	1106-07	23	23	...	...	23	23	23	23	23	23	23	7. Srīmukha.
4188	1087-88	4	4	...	...	4	4	4	4	4	3	3*	4208	1107-08	24	24	...	...	24	24	24	24	24	24	24	8. Bhāva.
4189	1088-89	5	5	...	...	5	5	5	5	5	4*	5	4209	1108-09	25	25	...	...	25	25	25	25	25	25	25	9. Yuvan.
4190	1089-90	6	6	...	...	6	6	6	6	6	6	6	4210	1109-10	26	26	...	...	26	26	26	26	26	26	26	10. Dhātṛi.
4191	1090-91	7	7	...	...	7	7	7	7	7	7	7	4211	1110-11	27	27	...	...	27	27	27	27	27	27	27	11. Īśvara.
4192	1091-92	8	8	...	...	8	8	8	8	8	8	8	4212	1111-12	28	28	...	...	28	28	28	28	28	28	28	12. Bahudhānya.
4193	1092-93	9	9	...	...	9	9	9	9	9	9	9	4213	1112-13	29	29	...	...	29	29	29	29	29	29	29	13. Pramāthin.
4194	1093-94	10	10	...	...	10	10	10	10	10	10	10	4214	1113-14	30	30	...	...	30	30	30	30	30	30	30	14. Vikrama.
4195	1094-95	11	11	...	...	11	11	11	11	11	11	11	4215	1114-15	31	31	...	...	31	31	31	31	31	31	31	15. Vṛisha.
4196	1095-96	12	12	...	...	12	12	12	12	12	12	12	4216	1115-16	32	32	...	...	32	32	32	32	32	32	32	16. Chitrabhānu.
4197	1096-97	13	13	...	...	13	13	13	13	13	13	13	4217	1116-17	33	33	...	...	33	33	33	33	33	33	33	17. Subhānu.
4198	1097-98	14	14	...	...	14	14	14	14	14	14	14	4218	1117-18	34	34	...	...	34	34	34	34	34	34	34	18. Tāraṇa.
4199	1098-99	15	15	...	...	15	15	15	15	15	15	15	4219	1118-19	35	35	...	...	35	35	35	35	35	35	35	19. Pārthiva.
4200	1099-1100	16	16	...	...	16	16	16	16	16	16	16	4220	1119-20	36	36	...	...	36	36	36	36	36	36	36	20. Vyaya.

1	2	3	4	5	6	7	8	9	10	11	12	13	1	2	3	4	5	6	7	8	9	10	11	12	13	
4221	1120-21	37	37	...	...	37	37	37	37	37	37	37	4261	1160-61	17	17	...	...	17*	17*	17	18	18	17	17	21. Sarvajit.
4222	1121-22	38	38	...	...	38	38	38	38	38	38	38	4262	1161-62	18	18	...	...	19	19	18	19	19	18	18	22. Sarvadhārin.
4223	1122-23	39	39	...	...	39	39	39	39	39	39	39	4263	1162-63	19	19	...	...	20	20	19	20	20	19	19	23. Virōdhin.
4224	1123-24	40	40	...	...	40	40	40	40	40	40	40	4264	1163-64	20	20	...	...	21	21	20	21	21	20	20	24. Vikṛita.
4225	1124-25	41	41	...	...	41	41	41	41	41	41	41	4265	1164-65	21*	21	...	...	22	22	21*	22	22	21	21	25. Khara.
4226	1125-26	42	42	...	...	42	42	42	42	42	42	42	4266	1165-66	23	23	...	...	23	23	23	23	23	22	22	26. Nandana.
4227	1126-27	43	43	...	...	43	43	43	43	43	43	43	4267	1166-67	24	24	...	...	24	24	24	24	24	23	23	27. Vijaya.
4228	1127-28	44	44	...	...	44	44	44	44	44	44	44	4268	1167-68	25	25	...	...	25	25	25	25	25	24	24	28. Jaya.
4229	1128-29	45	45	...	...	45	45	45	45	45	45	45	4269	1168-69	26	26	...	...	26	26	26	26	26	25	25	29. Manmatha.
4230	1129-30	46	46	...	...	46	46	46	46	46	46	46	4270	1169-70	27	27	...	...	27	27	27	27	27	26	26	30. Durmukha.
4231	1130-31	47	47	...	...	47	47	47	47	47	47	47	4271	1170-71	28	28	...	...	28	28	28	28	28	27	27	31. Hēmalamba.
4232	1131-32	48	48	...	...	48	48	48	48	48	48	48	4272	1171-72	29	29	...	...	29	29	29	29	29	28	28	32. Vilamba.
4233	1132-33	49	49	...	...	49	49	49	49	49	49	49	4273	1172-73	30	30	...	...	30	30	30	30	30	29	29*	33. Vikārin.
4234	1133-34	50	50	...	...	50	50	50	50	50	50	50	4274	1173-74	31	31	...	...	31	31	31	31	31	30*	31	34. Sārvarin.
4235	1134-35	51	51	...	...	51	51	51	51	51	51	51	4275	1174-75	32	32	...	...	32	32	32	32	32	32	32	35. Plava.
4236	1135-36	52	52	...	...	52	52	52	52	52	52	52	4276	1175-76	33	33	...	...	33	33	33	33	33	33	33	36. Śubhakṛit.
4237	1136-37	53	53	...	...	53	53	53	53	53	53	53	4277	1176-77	34	34	...	...	34	34	34	34	34	34	34	37. Śōbhana.
4238	1137-38	54	54	...	...	54	54	54	54	54	54	54	4278	1177-78	35	35	...	...	35	35	35	35	35	35	35	38. Krōdhin.
4239	1138-39	55	55	...	...	55	55	55	55	55	55	55	4279	1178-79	36	36	...	...	36	36	36	36	36	36	36	39. Viśvāvasu.
4240	1139-40	56	56	...	...	56	56	56	56	56	56	56	4280	1179-80	37	37	...	...	37	37	37	37	37	37	37	40. Parābhava.
4241	1140-41	57	57	...	...	57	57	57	57	57	57	57	4281	1180-81	38	38	...	...	38	38	38	38	38	38	38	41. Plavaṅga.
4242	1141-42	58	58	...	...	58	58	58	58	58	58	58	4282	1181-82	39	39	...	...	39	39	39	39	39	39	39	42. Kīlaka.
4243	1142-43	59	59	...	...	59	59	59	59	59	59	59	4283	1182-83	40	40	...	...	40	40	40	40	40	40	40	43. Saumya.
4244	1143-44	60	60	...	...	60	60	60	60	60	60	60	4284	1183-84	41	41	...	...	41	41	41	41	41	41	41	44. Sādhāraṇa.
4245	1144-45	1	1	...	...	1	1	1	1	1	1	1	4285	1184-85	42	42	...	...	42	42	42	42	42	42	42	45. Virōdhakṛit.
4246	1145-46	2	2	...	...	2	2	2	2	2	2	2	4286	1185-86	43	43	...	...	43	43	43	43	43	43	43	46. Paridhāvin.
4247	1146-47	3	3	...	...	3	3	3	3	3	3	3	4287	1186-87	44	44	...	...	44	44	44	44	44	44	44	47. Pramādin.
4248	1147-48	4	4	...	...	4	4	4	4	4	4	4	4288	1187-88	45	45	...	...	45	45	45	45	45	45	45	48. Ānanda.
4249	1148-49	5	5	...	...	5	5	5	5	5	5	5	4289	1188-89	46	46	...	...	46	46	46	46	46	46	46	49. Rākshasa.
4250	1149-50	6	6	...	...	6	6	6	6	6	6	6	4290	1189-90	47	47	...	...	47	47	47	47	47	47	47	50. Anala.
4251	1150-51	7	7	...	...	7	7	7	7	7	7	7	4291	1190-91	48	48	...	...	48	48	48	48	48	48	48	51. Piṅgala.
4252	1151-52	8	8	...	...	8	8	8	8	8	8	8	4292	1191-92	49	49	...	...	49	49	49	49	49	49	49	52. Kālayukta.
4253	1152-53	9	9	...	...	9	9	9	9	9	9	9	4293	1192-93	50	50	...	...	50	50	50	50	50	50	50	53. Siddhārthin.
4254	1153-54	10	10	...	...	10	10	10	10	10	10	10	4294	1193-94	51	51	...	...	51	51	51	51	51	51	51	54. Raudra.
4255	1154-55	11	11	...	...	11	11	11	11	11	11	11	4295	1194-95	52	52	...	...	52	52	52	52	52	52	52	55. Durmati.
4256	1155-56	12	12	...	...	12	12	12	12	12	12	12	4296	1195-96	53	53	...	...	53	53	53	53	53	53	53	56. Dundubhi.
4257	1156-57	13	13	...	...	13	13	13	13	13	13	13	4297	1196-97	54	54	...	...	54	54	54	54	54	54	54	57. Rudhirōdgārin.
4258	1157-58	14	14	...	...	14	14	14	14	14	14	14	4298	1197-98	55	55	...	...	55	55	55	55	55	55	55	58. Raktāksha.
4259	1158-59	15	15	...	...	15	15	15	15	15	15	15	4299	1198-99	56	56	...	...	56	56	56	56	56	56	56	59. Krōdhana.
4260	1159-60	16	16	...	...	16	16	16	16*	16*	16	16	4300	1199-1200	57	57	...	...	58	57	57	57	57	57	57	60. Kshaya.

TABLE XLII--*contd.*

Expired year of Kaliyuga.	Year A.D.	NUMBER OF THE SAṀVATSARA CONNECTED WITH EACH SOLAR YEAR ACCORDING TO THE SEVERAL SIDDHĀNTAS, BY REASON OF ITS CURRENCY AT APPARENT, OR AT MEAN, MĒSHA-SAṀKRĀNTI.											Expired year of Kaliyuga.	Year A.D.	NUMBER OF THE SAṀVATSARA CONNECTED WITH EACH SOLAR YEAR ACCORDING TO THE SEVERAL SIDDHĀNTAS, BY REASON OF ITS CURRENCY AT APPARENT, OR AT MEAN, MĒSHA-SAṀKRĀNTI.											Names of the Sixty saṁvatsaras of the cycle of Jupiter.
		Sūrya-S. no Bīja.		Sūrya-S. with Bīja.		First Ārya-S.		Orig. Sūrya S.	Brahma-S. and S. Śirō.		Second Ārya-S.				Sūrya-S. no Bīja.		Sūrya-S. with Bīja.		First Ārya-S.		Orig. Sūrya S.	Brahma-S. and S. Śirō.		Second Ārya-S.		
		Apparent M. S.	Mean M. S.	Apparent M. S.	Mean M. S.	Apparent M. S.	Mean M. S.	Mean M. S.	Apparent M. S.	Mean M. S.	Apparent M. S.	Mean M. S.			Apparent M. S.	Mean M. S.	Apparent M. S.	Mean M. S.	Apparent M. S.	Mean M. S.	Mean M. S.	Apparent M. S.	Mean M. S.	Apparent M. S.	Mean M. S.	
1	2	3	4	5	6	7	8	9	10	11	12	13	1	2	3	4	5	6	7	8	9	10	11	12	13	
4301	1200-01	58	58	...	...	58	58	58	58	58	58	58	4321	1220-21	18	18	...	...	18	18	18	18	18	18	18	1. Prabhava.
4302	1201-02	59	59	...	...	59	59	59	59	59	59	59	4322	1221-22	19	19	...	...	19	19	19	19	19	19	19	2. Vibhava.
4303	1202-03	60	60	...	...	60	60	60	60	60	60	60	4323	1222-23	20	20	...	...	20	20	20	20	20	20	20	3. Śukla.
4304	1203-04	1	1	...	...	1	1	1	1	1	1	1	4324	1223-24	21	21	...	...	21	21	21	21	21	21	21	4. Pramōda.
4305	1204-05	2	2	...	...	2	2	2	2	2	2	2	4325	1224-25	22	22	...	...	22	22	22	22	22	22	22	5. Prajāpati.
4306	1205-06	3	3	...	...	3	3	3	3	3	3	3	4326	1225-26	23	23	...	...	23	23	23	23	23	23	23	6. Aṅgiras.
4307	1206-07	4	4	...	...	4	4	4	4	4	4	4	4327	1226-27	24	24	...	...	24	24	24	24	24	24	24	7. Srīmukha.
4308	1207-08	5	5	...	...	5	5	5	5	5	5	5	4328	1227-28	25	25	...	...	25	25	25	25	25	25	25	8. Bhāva.
4309	1208-09	6	6	...	...	6	6	6	6	6	6	6	4329	1228-29	26	26	...	...	26	26	26	26	26	26	26	9. Yuvan.
4310	1209-10	7	7	...	...	7	7	7	7	7	7	7	4330	1229-30	27	27	...	...	27	27	27	27	27	27	27	10. Dhātṛi.
4311	1210-11	8	8	...	...	8	8	8	8	8	8	8	4331	1230-31	28	28	...	...	28	28	28	28	28	28	28	11. Iśvara.
4312	1211-12	9	9	...	...	9	9	9	9	9	9	9	4332	1231-32	29	29	...	...	29	29	29	29	29	29	29	12. Bahudhānya.
4313	1212-13	10	10	...	...	10	10	10	10	10	10	10	4333	1232-33	30	30	...	...	30	30	30	30	30	30	30	13. Pramāthin.
4314	1213-14	11	11	...	...	11	11	11	11	11	11	11	4334	1233-34	31	31	...	...	31	31	31	31	31	31	31	14. Vikrama.
4315	1214-15	12	12	...	...	12	12	12	12	12	12	12	4335	1234-35	32	32	...	...	32	32	32	32	32	32	32	15. Vṛisha.
4316	1215-16	13	13	...	...	13	13	13	13	13	13	13	4336	1235-36	33	33	...	...	33	33	33	33	33	33	33	16. Chitrabhānu.
4317	1216-17	14	14	...	...	14	14	14	14	14	14	14	4337	1236-37	34	34	...	...	34	34	34	34	34	34	34	17. Subhānu.
4318	1217-18	15	15	...	...	15	15	15	15	15	15	15	4338	1237-38	35	35	...	...	35	35	35	35	35	35	35	18. Tāraṇa.
4319	1218-19	16	16	...	...	16	16	16	16	16	16	16	4339	1238-39	36	36	...	...	36	36	36	36	36	36	36	19. Pārthiva.
4320	1219-20	17	17	...	...	17	17	17	17	17	17	17	4340	1239-40	37	37	...	...	37	37	37	37	37	37	37	20. Vyaya.

1	2	3	4	5	6	7	8	9	10	11	12	13
4341	1240-41	38	38	...	...	38	38	38	38	38	38	38
4342	1241-42	39	39	...	...	39	39	39	39	39	39	39
4343	1242-43	40	40	...	...	40	40	40	40	40	40	40
4344	1243-44	41	41	...	...	41	41	41	41	41	41	41
4345	1244-45	42	42	...	...	42	42	42	42	42*	42	42
4346	1245-46	43	43	...	...	43	43*	43	43*	44	43	43
4347	1246-47	44	44	...	...	44*	45	44	45	45	44	44
4348	1247-48	45	45	...	...	46	46	45	46	46	45	45
4349	1248-49	46	46	...	...	47	47	46	47	47	46	46
4350	1249-50	47	47*	...	...	48	48	47*	48	48	47	47
4351	1250-51	48*	49	...	...	49	49	49	49	49	48	48
4352	1251-52	50	50	...	...	50	50	50	50	50	49	49
4353	1252-53	51	51	...	...	51	51	51	51	51	50	50
4354	1253-54	52	52	...	...	52	52	52	52	52	51	51
4355	1254-55	53	53	...	...	53	53	53	53	53	52	52
4356	1255-56	54	54	...	...	54	54	54	54	54	53	53
4357	1256-57	55	55	...	...	55	55	55	55	55	54	54
4358	1257-58	56	56	...	...	56	56	56	56	56	55	55
4359	1258-59	57	57	...	...	57	57	57	57	57	56*	56*
4360	1259-60	58	58	...	...	58	58	58	58	58	58	58
4361	1260-61	59	59	...	...	59	59	59	59	59	59	59
4362	1261-62	60	60	...	...	60	60	60	60	60	60	60
4363	1262-63	1	1	...	...	1	1	1	1	1	1	1
4364	1263-64	2	2	...	...	2	2	2	2	2	2	2
4365	1264-65	3	3	...	...	3	3	3	3	3	3	3
4366	1265-66	4	4	...	...	4	4	4	4	4	4	4
4367	1266-67	5	5	...	...	5	5	5	5	5	5	5
4368	1267-68	6	6	...	...	6	6	6	6	6	6	6
4369	1268-69	7	7	...	...	7	7	7	7	7	7	7
4370	1269-70	8	8	...	...	8	8	8	8	8	8	8
4371	1270-71	9	9	...	...	9	9	9	9	9	9	9
4372	1271-72	10	10	...	...	10	10	10	10	10	10	10
4373	1272-73	11	11	...	...	11	11	11	11	11	11	11
4374	1273-74	12	12	...	...	12	12	12	12	12	12	12
4375	1274-75	13	13	...	...	13	13	13	13	13	13	13
4376	1275-76	14	14	...	...	14	14	14	14	14	14	14
4377	1276-77	15	15	...	...	15	15	15	15	15	15	15
4378	1277-78	16	16	...	...	16	16	16	16	16	16	16
4379	1278-79	17	17	...	...	17	17	17	17	17	17	17
4380	1279-80	18	18	...	...	18	18	18	18	18	18	18

1	2	3	4	5	6	7	8	9	10	11	12	13	
4381	1280-81	19	19	...	...	19	19	19	19	19	19	19	21. Sarvajit.
4382	1281-82	20	20	...	...	20	20	20	20	20	20	20	22. Sarvadhārin.
4383	1282-83	21	21	...	...	21	21	21	21	21	21*	21	23. Virōdhin.
4384	1283-84	22	22	...	...	22	22	22	22	22	22	22	24. Vikṛita.
4385	1284-85	23	23	...	...	23	23	23	23	23	23	23	25. Khara.
4386	1285-86	24	24	...	...	24	24	24	24	24	24	24	26. Nandana.
4387	1286-87	25	25	...	...	25	25	25	25	25	25	25	27. Vijaya.
4388	1287-88	26	26	...	...	26	26	26	26	26	26	26	28. Jaya.
4389	1288-89	27	27	...	...	27	27	27	27	27	27	27	29. Manmatha.
4390	1289-90	28	28	...	...	28	28	28	28	28	28	28	30. Durmukha.
4391	1290-91	29	29	...	...	29	29	29	29	29	29	29	31. Hémalamba.
4392	1291-92	30	30	...	...	30	30	30	30	30	30	30	32. Vilamba.
4393	1292-93	31	31	...	...	31	31	31	31	31	31	31	33. Vikārin.
4394	1293-94	32	32	...	...	32	32	32	32	32	32	32	34. Śārvarin.
4395	1294-95	33	33	...	...	33	33	33	33	33	33	33	35. Plava.
4396	1295-96	34	34	...	...	34	34	34	34	34	34	34	36. Śubhakṛit.
4397	1296-97	35	35	...	...	35	35	35	35	35	35	35	37. Sōbhana.
4398	1297-98	36	36	...	...	36	36	36	36	36	36	36	38. Krōdhin.
4399	1298-99	37	37	...	...	37	37	37	37	37	37	37	39. Viśvāvasu.
4400	1299-1300	38	38	...	...	38	38	38	38	38	38	38	40. Parābhava.
4401	1300-01	39	39	...	...	39	39	39	39	39	39	39	41. Plavaṅga.
4402	1301-02	40	40	...	...	40	40	40	40	40	40	40	42. Kīlaka.
4403	1302-03	41	41	...	...	41	41	41	41	41	41	41	43. Saumya.
4404	1303-04	42	42	...	...	42	42	42	42	42	42	42	44. Sādhārana.
4405	1304-05	43	43	...	...	43	43	43	43	43	43	43	45. Virōdhakṛit.
4406	1305-06	44	44	...	...	44	44	44	44	44	44	44	46. Paridhāvin.
4407	1306-07	45	45	...	...	45	45	45	45	45	45	45	47. Pramādin.
4408	1307-08	46	46	...	...	46	46	46	46	46	46	46	48. Ānanda.
4409	1308-09	47	47	...	...	47	47	47	47	47	47	47	49. Rākshasa.
4410	1309-10	48	48	...	...	48	48	48	48	48	48	48	50. Anala.
4411	1310-11	49	49	...	...	49	49	49	49	49	49	49	51. Piṅgala.
4412	1311-12	50	50	...	...	50	50	50	50	50	50	50	52. Kālayukta.
4413	1312-13	51	51	...	...	51	51	51	51	51	51	51	53. Siddhārthin.
4414	1313-14	52	52	...	...	52	52	52	52	52	52	52	54. Raudra.
4415	1314-15	53	53	...	...	53	53	53	53	53	53	53	55. Durmati.
4416	1315-16	54	54	...	...	54	54	54	54	54	54	54	56. Dundubhi.
4417	1316-17	55	55	...	...	55	55	55	55	55	55	55	57. Rudhirōdgārin.
4418	1317-18	56	56	...	...	56	56	56	56	56	56	56	58. Raktāksha.
4419	1318-19	57	57	...	...	57	57	57	57	57	57	57	59. Krōdhana.
4420	1319-20	58	58	...	...	58	58	58	58	58	58	58	60. Kshaya.

TABLE XLII—*contd.*

Expired year of Kaliyuga.	Year A.D.	Number of the Samvatsara connected with each solar year according to the several Siddhāntas, by reason of its currency at apparent, or at mean, Mēsha-saṁkrānti. Sūrya-S. no Bīja. Apparent M. S.	Sūrya-S. no Bīja. Mean M. S.	Sūrya-S. with Bīja. Apparent M. S.	Sūrya-S. with Bīja. Mean M. S.	First Ārya-S. Apparent M. S.	First Ārya-S. Mean M. S.	Orig. Sūrya S. Mean M. S.	Brahma-S. and S. Sirō. Apparent M. S.	Brahma-S. and S. Sirō. Mean M. S.	Second Ārya-S. Apparent M. S.	Second Ārya-S. Mean M. S.	Expired year of Kaliyuga.	Year A.D.	Number of the Samvatsara connected with each solar year according to the several Siddhāntas, by reason of its currency at apparent, or at mean, Mēsha-saṁkrānti. Sūrya-S. no Bīja. Apparent M. S.	Sūrya-S. no Bīja. Mean M. S.	Sūrya-S. with Bīja. Apparent M. S.	Sūrya-S. with Bīja. Mean M. S.	First Ārya-S. Apparent M. S.	First Ārya-S. Mean M. S.	Orig. Sūrya S. Mean M. S.	Brahma-S. and S. Sirō. Apparent M. S.	Brahma-S. and S. Sirō. Mean M. S.	Second Ārya-S. Apparent M. S.	Second Ārya-S. Mean M. S.	Names of the Sixty saṁvatsaras of the cycle of Jupiter.
1	2	3	4	5	6	7	8	9	10	11	12	13	1	2	3	4	5	6	7	8	9	10	11	12	13	
4421	1320-21	59	59	...	...	59	59	59	59	59	59	59	4441	1340-41	20	20	...	...	20	20	20	20	20	19	19	1. Prabhava.
4422	1321-22	60	60	...	...	60	60	60	60	60	60	60	4442	1341-42	21	21	...	...	21	21	21	21	21	20	20	2. Vibhava.
4423	1322-23	1	1	...	...	1	1	1	1	1	1	1	4443	1342-43	22	22	...	...	22	22	22	22	22	21	21	3. Śukla.
4424	1323-24	2	2	...	...	2	2	2	2	2	2	2	4444	1343-44	23	23	...	...	23	23	23	23	23	22	22*	4. Pramōda.
4425	1324-25	3	3	...	...	3	3	3	3	3	3	3	4445	1344-45	24	24	...	...	24	24	24	24	24	23*	24	5. Prajāpati.
4426	1325-26	4	4	...	...	4	4	4	4	4	4	4	4446	1345-46	25	25	...	...	25	25	25	25	25	25	25	6. Angiras.
4427	1326-27	5	5	...	...	5	5	5	5	5	5	5	4447	1346-47	26	26	...	...	26	26	26	26	26	26	26	7. Śrīmukha.
4428	1327-28	6	6	...	...	6	6	6	6	6	6	6	4448	1347-48	27	27	...	...	27	27	27	27	27	27	27	8. Bhāva.
4429	1328-29	7	7	...	...	7	7	7	7	7	7	7	4449	1348-49	28	28	...	...	28	28	28	28	28	28	28	9. Yuvan.
4430	1329-30	8	8	...	...	8	8	8	8	8*	8	8	4450	1349-50	29	29	...	...	29	29	29	29	29	29	29	10. Dhātṛi.
4431	1330-31	9	9	...	...	9	9*	9	9*	10	9	9	4451	1350-51	30	30	...	...	30	30	30	30	30	30	30	11. Iśvara.
4432	1331-32	10	10	...	...	10*	11	10	11	11	10	10	4452	1351-52	31	31	...	...	31	31	31	31	31	31	31	12. Bahudhānya.
4433	1332-33	11	11	...	...	12	12	11	12	12	11	11	4453	1352-53	32	32	...	...	32	32	32	32	32	32	32	13. Pramāthin.
4434	1333-34	12	12	...	...	13	13	12	13	13	12	12	4454	1353-54	33	33	...	...	33	33	33	33	33	33	33	14. Vikrama.
4435	1334-35	13	13	...	...	14	14	13	14	14	13	13	4455	1354-55	34	34	...	...	34	34	34	34	34	34	34	15. Vṛisha.
4436	1335-36	14*	14*	...	...	15	15	14*	15	15	14	14	4456	1355-56	35	35	...	...	35	35	35	35	35	35	35	16. Chitrabhānu.
4437	1336-37	16	16	...	...	16	16	16	16	16	15	15	4457	1356-57	36	36	...	...	36	36	36	36	36	36	36	17. Subhānu.
4438	1337-38	17	17	...	...	17	17	17	17	17	16	16	4458	1357-58	37	37	...	...	37	37	37	37	37	37	37	18. Tārana.
4439	1338-39	18	18	...	...	18	18	18	18	18	17	17	4459	1358-59	38	38	...	...	38	38	38	38	38	38	38	19. Pārthiva.
4440	1339-40	19	19	...	...	19	19	19	19	19	18	18	4460	1359-60	39	39	...	...	39	39	39	39	39	39	39	20. Vyaya.

1	2	3	4	5	6	7	8	9	10	11	12	13	1	2	3	4	5	6	7	8	9	10	11	12	13	
4461	1360-61	40	40	...	...	40	40	40	40	40	40	40	4501	1400-01	20	20	...	...	20	20	20	20	20	20	20	21. Sarvajit.
4462	1361-62	41	41	...	...	41	41	41	41	41	41	41	4502	1401-02	21	21	...	...	21	21	21	21	21	21	21	22. Sarvadhārin.
4463	1362-63	42	42	...	...	42	42	42	42	42	42	42	4503	1402-03	22	22	...	...	22	22	22	22	20	22	22	23. Virōdhin.
4464	1363-64	43	43	...	...	43	43	43	43	43	43	43	4504	1403-04	23	23	...	...	23	23	23	23	23	23	23	24. Vikrita.
4465	1364-65	44	44	...	...	44	44	44	44	44	44	44	4505	1404-05	24	24	...	...	24	24	24	24	24	24	24	25. Khara.
4466	1365-66	45	45	...	...	45	45	45	45	45	45	45	4506	1405-06	25	25	...	...	25	25	25	25	25	25	25	26. Nandana.
4467	1366-67	46	46	...	...	46	46	46	46	46	46	46	4507	1406-07	26	26	...	...	26	26	26	26	26	26	26	27. Vijaya.
4468	1367-68	47	47	...	...	47	47	47	47	47	47	47	4508	1407-08	27	27	...	...	27	27	27	27	27	27	27	28. Jaya.
4469	1368-69	48	48	...	...	48	48	48	48	48	48	48	4509	1408-09	28	28	...	...	28	28	28	28	28	28	28	29. Manmatha.
4470	1369-70	49	49	...	...	49	49	49	49	49	49	49	4510	1409-10	29	29	...	...	29	29	29	29	29	29	29	30. Durmukha.
4471	1370-71	50	50	...	...	50	50	50	50	50	50	50	4511	1410-11	30	30	...	...	30	30	30	30	30	30	30	31. Hēmalamba.
4472	1371-72	51	51	...	...	51	51	51	51	51	51	51	4512	1411-12	31	31	...	...	31	31	31	31	31	31	31	32. Vilamba.
4473	1372-73	52	52	...	...	52	52	52	52	52	52	52	4513	1412-13	32	32	...	...	32	32	32	32	32	32	32	33. Vikārin.
4474	1373-74	53	53	...	...	53	53	53	53	53	53	53	4514	1413-14	33	33	...	...	33	33	33	33	33	33	33	34. Sārvarin.
4475	1374-75	54	54	...	...	54	54	54	54	54	54	54	4515	1414-15	34	34	...	...	34	34	34	34	34*	34	34	35. Plava.
4476	1375-76	55	55	...	...	55	55	55	55	55	55	55	4516	1415-16	35	35	...	...	35	35	35	35*	36	35	35	36. Ṣubhakrit.
4477	1376-77	56	56	...	...	56	56	56	56	56	56	56	4517	1416-17	36	36	...	...	36*	36*	36	37	37	36	36	37. Sōbhana.
4478	1377-78	57	57	...	...	57	57	57	57	57	57	57	4518	1417-18	37	37	...	...	38	38	37	38	38	37	37	38. Krōdhin.
4479	1378-79	58	58	...	...	58	58	58	58	58	58	58	4519	1418-19	38	38	...	...	39	39	38	39	39	38	38	39. Viśvāvasu.
4480	1379-80	59	59	...	...	59	59	59	59	59	59	59	4520	1419-20	39	39	...	...	40	40	39	40	40	39	39	40. Parābhava.
4481	1380-81	60	60	...	...	60	60	60	60	60	60	60	4521	1420-21	40*	40*	...	...	41	41	40*	41	41	40	40	41. Plavaṅga.
4482	1381-82	1	1	...	...	1	1	1	1	1	1	1	4522	1421-22	42	42	...	...	42	42	42	42	42	41	41	42. Kīlaka.
4483	1382-83	2	2	...	...	2	2	2	2	2	2	2	4523	1422-23	43	43	...	...	43	43	43	43	43	42	42	43. Saumya.
4484	1383-84	3	3	...	...	3	3	3	3	3	3	3	4524	1423-24	44	44	...	...	44	44	44	44	44	43	43	44. Sādhāraṇa.
4485	1384-85	4	4	...	...	4	4	4	4	4	4	4	4525	1424-25	45	45	...	...	45	45	45	45	45	44	44	45. Virōḍhakrit.
4486	1385-86	5	5	...	...	5	5	5	5	5	5	5	4526	1425-26	46	46	...	...	46	46	46	46	46	45	45	46. Paridhāvin.
4487	1386-87	6	6	...	...	6	6	6	6	6	6	6	4527	1426-27	47	47	...	...	47	47	47	47	47	46	46	47. Pramādin.
4488	1387-88	7	7	...	...	7	7	7	7	7	7	7	4528	1427-28	48	48	...	...	48	48	48	48	48	47	47	48. Ānanda.
4489	1388-89	8	8	...	...	8	8	8	8	8	8	8	4529	1428-29	49	49	...	...	49	49	49	49	49*	48	48*	49. Rākshasa.
4490	1389-90	9	9	...	...	9	9	9	9	9	9	9	4530	1429-30	50	50	...	...	50	50	50	50	50	49*	50	50. Anala.
4491	1390-91	10	10	...	...	10	10	10	10	10	10	10	4531	1430-31	51	51	...	...	51	51	51	51	51	51	51	51. Piṅgala.
4492	1391-92	11	11	...	...	11	11	11	11	11	11	11	4532	1431-32	52	52	...	...	52	52	52	52	52	52	52	52. Kālayukta.
4493	1392-93	12	12	...	...	12	12	12	12	12	12	12	4533	1432-33	53	53	...	...	53	53	53	53	53	53	53	53. Siddhārthin.
4494	1393-94	13	13	...	...	13	13	13	13	13	13	13	4534	1433-34	54	54	...	...	54	54	54	54	54	[illegible]	54	54. Raudra.
4495	1394-95	14	14	...	...	14	14	14	14	14	14	14	4535	1434-35	55	55	...	...	55	55	55	55	55	55	55	55. Durmati.
4496	1395-96	15	15	...	...	15	15	15	15	15	15	15	4536	1435-36	56	56	...	...	56	56	56	56	56	56	56	56 Dundubhi.
4497	1396-97	16	16	...	...	16	16	16	16	16	16	16	4537	1436-37	57	57	...	...	57	57	57	57	57	57	57	57. Rudhirōdgārin.
4498	1397-98	17	17	...	...	17	17	17	17	17	17	17	4538	1437-38	58	58	...	...	58	58	58	58	58	58	58	58. Raktāksha.
4499	1398-99	18	18	...	...	18	18	18	18	18	18	18	4539	1438-39	59	59	...	...	59	59	59	59	59	59	59	59. Krōdhana.
4500	1399-1400	19	19	...	...	19	19	19	19	19	19	19	4540	1439-40	60	60	...	...	60	60	60	60	60	60	60	60. Kshaya.

TABLE XLII—*contd.*

Expired year of Kaliyuga.	Year A.D.	Sūrya-S. no bīja. Apparent M. S.	Sūrya-S. no bīja. Mean M. S.	Sūrya-S. with bīja. Apparent M. S.	Sūrya-S. with bīja. Mean M. S.	First Ārya-S. Apparent M. S.	First Ārya-S. Mean M. S.	Orig. Sūrya S. Mean M. S.	Brahma-S. and S. Sirō. Apparent M. S.	Brahma-S. and S. Sirō. Mean M. S.	Second Ārya-S. Apparent M. S.	Second Ārya-S. Mean M. S.
1	2	3	4	5	6	7	8	9	10	11	12	13
4541	1440-41	1	1	1	1	1	1	1	1	1	1	1
4542	1441-42	2	2	2	2	2	2	2	2	2	2	2
4543	1442-43	3	3	3	3	3	3	3	3	3	3	3
4544	1443-44	4	4	4	4	4	4	4	4	4	4	4
4545	1444-45	5	5	5	5	5	5	5	5	5	5	5
4546	1445-46	6	6	6	6	6	6	6	6	6	6	6
4547	1446-47	7	7	7	7	7	7	7	7	7	7	7
4548	1447-48	8	8	8	8	8	8	8	8	8	8	8
4549	1448-49	9	9	9	9	9	9	9	9	9	9	9
4550	1449-50	10	10	10	10	10	10	10	10	10	10	10
4551	1450-51	11	11	11	11	11	11	11	11	11	11	11
4552	1451-52	12	12	12	12	12	12	12	12	12	12	12
4553	1452-53	13	13	13	13	13	13	13	13	13	13	13
4554	1453-54	14	14	14	14	14	14	14	14	14	14	14
4555	1454-55	15	15	15	15	15	15	15	15	15	15	15
4556	1455-56	16	16	16	16	16	16	16	16	16	16	16
4557	1456-57	17	17	17	17	17	17	17	17	17	17	17
4558	1457-58	18	18	18	18	18	18	18	18	18	18	18
4559	1458-59	19	19	19	19	19	19	19	19	19	19	19
4560	1459-60	20	20	20	20	20	20	20	20	20	20	20

Expired year of Kaliyuga.	Year A.D.	Sūrya-S. no bīja. Apparent M. S.	Sūrya-S. no bīja. Mean M. S.	Sūrya-S. with bīja. Apparent M. S.	Sūrya-S. with bīja. Mean M. S.	First Ārya-S. Apparent M. S.	First Ārya-S. Mean M. S.	Orig. Sūrya S. Mean M. S.	Brahma-S. and S. Sirō. Apparent M. S.	Brahma-S. and S. Sirō. Mean M. S.	Second Ārya-S. Apparent M. S.	Second Ārya-S. Mean M. S.	Names of the Sixty samvatsaras of the cycle of Jupiter.
1	2	3	4	5	6	7	8	9	10	11	12	13	
4561	1460-61	21	21	21	21	21	21	21	21	21	21	21	1. Prabhava.
4562	1461-62	22	22	22	22	22	22	22	22	22	22	22	2. Vibhava.
4563	1462-63	23	23	23	23	23	23	23	23	23	23	23	3. Sukla.
4564	1463-64	24	24	24	24	24	24	24	24	24	24	24	4. Pramōda.
4565	1464-65	25	25	25	25	25	25	25	25	25	25	25	5. Prajāpati.
4566	1465-66	26	26	26	26	26	26	26	26	26	26	26	6. Angiras.
4567	1466-67	27	27	27	27	27	27	27	27	27	27	27	7. Srīmukha.
4568	1467-68	28	28	28	28	28	28	28	28	28	28	28	8. Bhāva.
4569	1468-69	29	29	29	29	29	29	29	29	29	29	29	9. Yuvan.
4570	1469-70	30	30	30	30	30	30	30	30	30	30	30	10. Dhātri.
4571	1470-71	31	31	31	31	31	31	31	31	31	31	31	11. Īsvara.
4572	1471-72	32	32	32	32	32	32	32	32	32	32	32	12. Bahudhānya.
4573	1472-73	33	33	33	33	33	33	33	33	33	33	33	13. Pramāthin.
4574	1473-74	34	34	34	34	34	34	34	34	34	34	34	14. Vikrama.
4575	1474-75	35	35	35	35	35	35	35	35	35	35	35	15. Vrisha.
4576	1475-76	36	36	36	36	36	36	36	36	36	36	36	16. Chirabhānu.
4577	1476-77	37	37	37	37	37	37	37	37	37	37	37	17. Subhānu.
4578	1477-78	38	38	38	38	38	38	38	38	38	38	38	18. Tāraṇa.
4579	1478-79	39	39	39	39	39	39	39	39	39	39	39	19. Pārthiva.
4580	1479-80	40	40	40	40	40	40	40	40	40	40	40	20. Vyaya.

1	2	3	4	5	6	7	8	9	10	11	12	13	1	2	3	4	5	6	7	8	9	10	11	12	13	
4581	1480-81	41	41	41	41	41	41	41	41	41	41	41	4621	1520-21	22	22	22	22	22	22	22	22	22	22	22	21. Sarvajit.
4582	1481-82	42	42	42	42	42	42	42	42	42	42	42	4622	1521-22	23	23	23	23	23	23	23	23	23	23	23	22. Sarvadhārin.
4583	1482-83	43	43	43	43	43	43	43	43	43	43	43	4623	1522-23	24	24	24	24	24	24	24	24	24	24	24	23. Virōdhin.
4584	1483-84	44	44	44	44	44	44	44	44	44	44	44	4624	1523-24	25	25	25	25	25	25	25	25	25	25	25	24. Vikṛita.
4585	1484-85	45	45	45	45	45	45	45	45	45	45	45	4625	1524-25	26	26	26	26	26	26	26	26	26	26	26	25. Khara.
4586	1485-86	46	46	46	46	46	46	46	46	46	46	46	4626	1525-26	27	27	27	27	27	27	27	27	27	27	27	26. Nandana.
4587	1486-87	47	47	47	47	47	47	47	47	47	47	47	4627	1526-27	28	28	28	28	28	28	28	28	28	28	28	27. Vijaya.
4588	1487-88	48	48	48	48	48	48	48	48	48	48	48	4628	1527-28	29	29	29	29	29	29	29	29	29	29	29	28. Jaya.
4589	1488-89	49	49	49	49	49	49	49	49	49	49	49	4629	1528-29	30	30	30	30	30	30	30	30	30	30	30	29. Manmatha.
4590	1489-90	50	50	50	50	50	50	50	50	50	50	50	4630	1529-30	31	31	31	31	31	31	31	31	31	31	31	30. Durmukha.
4591	1490-91	51	51	51	51	51	51	51	51	51	51	51	4631	1530-31	32	32	32	32	32	32	32	32	32	32	32	31. Hēmalamba.
4592	1491-92	52	52	52	52	52	52	52	52	52	52	52	4632	1531-32	33	33	33	33	33	33	33	33	33	33	33	32. Vilamba.
4593	1492-93	53	53	53	53	53	53	53	53	53	53	53	4633	1532-33	34	34	34	34	34	34	34	34	34	34	34	33. Vikārin.
4594	1493-94	54	54	54	54	54	54	54	54	54	54	54	4634	1533-34	35	35	35	35	35	35	35	35	35	35	35	34. Śārvarin.
4595	1494-95	55	55	55	55	55	55	55	55	55	55	55	4635	1534-35	36	36	36	36	36	36	36	36	36	36	36	35. Plava.
4596	1495-96	56	56	56	56	56	56	56	56	56	56	56	4636	1535-36	37	37	37	37	37	37	37	37	37	37	37	36. Śubhakṛit.
4597	1496-97	57	57	57	57	57	57	57	57	57	57	57	4637	1536-37	38	38	38	38	38	38	38	38	38	38	38	37. Śōbhana.
4598	1497-98	58	58	58	58	58	58	58	58	58	58	58	4638	1537-38	39	39	39	39	39	39	39	39	39	39	39	38. Krōdhin.
4599	1498-99	59	59	59	59	59	59	59	59	59	59	59	4639	1538-39	40	40	40	40	40	40	40	40	40	40	40	39. Viśvāvasu.
4600	1499-1500	60	60	60	60	60	60	60	60	60	60	60	4640	1539-40	41	41	41	41	41	41	41	41	41	41	41	40. Parābhava.
4601	1500-01	1	1	1	1	1	1	1	1*	1*	1	1	4641	1540-41	42	42	42	42	42	42	42	42	42	42	42	41. Plavaṅga.
4602	1501-02	2	2	2	2	2*	2*	2	3	3	2	2	4642	1541-42	43	43	43	43	43	43	43	43	43	43	43	42. Kīlaka.
4603	1502-03	3	3	3	3	4	4	3	4	4	3	3	4643	1542-43	44	44	44	44	44	44	44	44	44	44	44	43. Saumya.
4604	1503-04	4	4	4	4	5	5	4	5	5	4	4	4644	1543-44	45	45	45	45	45	45	45	45	45	45	45	44. Sādhāraṇa.
4605	1504-05	5	5	5	5	6	6	5	6	6	5	5	4645	1544-45	46	46	46	46	46	46	46	46	46	46	46	45. Virōdhakṛit.
4606	1505-06	6	6*	6	6	7	7	6*	7	7	6	6	4646	1545-46	47	47	47	47	47	47	47	47	47	47	47	46. Paridhāvin.
4607	1506-07	7*	8	7	7	8	8	8	8	8	7	7	4647	1546-47	48	48	48	48	48	48	48	48	48	48	48	47. Pramādin.
4608	1507-08	9	9	8	8	9	9	9	9	9	8	8	4648	1547-48	49	49	49	49	49	49	49	49	49	49	49	48. Ānanda.
4609	1508-99	10	10	9	9	10	10	10	10	10	9	9	4649	1548-49	50	50	50	50	50	50	50	50	50	50	50	49. Rākshasa.
4610	1509-10	11	11	10	10	11	11	11	11	11	10	10	4650	1549-50	51	51	51	51	51	51	51	51	51	51	51	50. Anala.
4611	1510-11	12	12	11	11	12	12	12	12	12	11	11	4651	1550-51	52	52	52	52	52	52	52	52	52	52	52	51. Piṅgala.
4612	1511-12	13	13	12	12	13	13	13	13	13	12	12	4652	1551-52	53	53	53	53	53	53	53	53	53	53	53	52. Kālayukta.
4613	1512-13	14	14	13	13	14	14	14	14	14	13	13	4653	1552-53	54	54	54	54	54	54	54	54	54	54	54	53. Siddhārthin.
4614	1513-14	15	15	14	14	15	15	15	15	15	14	14	4654	1553-54	55	55	55	55	55	55	55	55	55	55	55	54. Raudra.
4615	1514-15	16	16	15*	15*	16	16	16	16	16	15*	15*	4655	1554-55	56	56	56	56	56	56	56	56	56	55	56	55. Durmati.
4616	1515-16	17	17	17	17	17	17	17	17	17	17	17	4656	1555-56	57	57	57	57	57	57	57	57	57	57	57	56. Dundubhi.
4617	1516-17	18	18	18	18	18	18	18	18	18	18	18	4657	1556-57	58	58	58	58	58	58	58	58	58	58	58	57. Rudhirōdgārin.
4618	1517-18	19	19	19	19	19	19	19	19	19	19	19	4658	1557-58	59	59	59	59	59	59	59	59	59	59	59	58. Raktāksha.
4619	1518-19	20	20	20	20	20	20	20	20	20	20	20	4659	1558-59	60	60	60	60	60	60	60	60	60	60	60	59. Krōdhana.
4620	1519-20	21	21	21	21	21	21	21	21	21	21	21	4660	1559-60	61	61	61	61	61	61	61	61	61	61	61	60. Kshaya.

TABLE XLII—*contd*

Expired year of Kaliyuga.	Year A.D.	NUMBER OF THE SAṀVATSARA CONNECTED WITH EACH SOLAR YEAR, ACCORDING TO THE SEVERAL SIDDHĀNTAS, BY REASON OF ITS CURRENCY AT APPARENT, OR AT MEAN, MĒSHA-SAṀKRĀNTI.											Expired year of Kali.	Year A.D.	NUMBER OF THE SAṀVATSARA CONNECTED WITH EACH SOLAR YEAR, ACCORDING TO THE SEVERAL SIDDHĀNTAS, BY REASON OF ITS CURRENCY AT APPARENT, OR AT MEAN, MĒSHA-SAṀKRĀNTI.											Names of the Sixty saṁvatsaras of the cycle of Jupiter
		Sūrya-S. no Bīja.		Sūrya-S. with Bīja.		First Ārya-S.		Orig. Sūrya S.	Brahma-S. and S. Śirō.		Second Ārya-S.				Sūrya-S. no Bīja.		Sūrya-S. with Bīja.		First Ārya-S.		Orig. Sūrya S.	Brahma-S. and S. Śirō.		Second Ārya-S.		
		Apparent M.S.	Mean M.S.	Apparent M.S.	Mean M.S.	Apparent M.S.	Mean M.S.	Mean M.S.	Apparent M.S.	Mean M.S.	Apparent M.S.	Mean M.S.			Apparent M.S.	Mean M.S.	Apparent M.S.	Mean M.S.	Apparent M.S.	Mean M.S.	Mean M.S.	Apparent M.S.	Mean M.S.	Apparent M.S.	Mean M.S.	
1	2	3	4	5	6	7	8	9	10	11	12	13	1	2	3	4	5	6	7	8	9	10	11	12	13	
4661	1560-61	2	2	2	2	2	2	2	2	2	2	2	4681	1580-81	22	22	22	22	22	22	22	22	22	22	22	1. Prabhava.
4662	1561-62	3	3	3	3	3	3	3	3	3	3	3	4682	1581-82	23	23	23	23	23	23	23	23	23	23	23	2. Vibhava.
4663	1562-63	4	4	4	4	4	4	4	4	4	4	4	4683	1582-83	24	24	24	24	24	24	24	24	24	24	24	3. Śukla.
4664	1563-64	5	5	5	5	5	5	5	5	5	5	5	4684	1583-84	25	25	25	25	25	25	25	25	25	25	25	4. Pramōda.
4665	1564-65	6	6	6	6	6	6	6	6	6	6	6	4685	1584-85	26	26	26	26	26	26	26	26	26	26	26	5. Prajāpati.
4666	1565-66	7	7	7	7	7	7	7	7	7	7	7	4686	1585-86	27	27	27	27	27	27	27	27*	27*	27	27	6. Aṅgiras.
4667	1566-67	8	8	8	8	8	8	8	8	8	8	8	4687	1586-87	28	28	28	28	28	28*	28	29	29	28	28	7. Śrīmukha.
4668	1567-68	9	9	9	9	9	9	9	9	9	9	9	4688	1587-88	29	29	29	29	29*	30	29	30	30	29	29	8. Bhāva.
4669	1568-69	10	10	10	10	10	10	10	10	10	10	10	4689	1588-89	30	30	30	30	31	31	30	31	31	30	30	9. Yuvan.
4670	1569-70	11	11	11	11	11	11	11	11	11	11	11	4690	1589-90	31	31	31	31	32	32	31	32	32	31	31	10. Dhātri.
4671	1570-71	12	12	12	12	12	12	12	12	12	12	12	4691	1590-91	32	32*	32	32	33	33	32	33	33	32	32	11. Īśvara.
4672	1571-72	13	13	13	13	13	13	13	13	13	13	13	4692	1591-92	33*	34	33	33	34	34	33*	34	34	33	33	12. Bahudhānya.
4673	1572-73	14	14	14	14	14	14	14	14	14	14	14	4693	1592-93	35	35	34	34	35	35	35	35	35	34	34	13. Pramāthin.
4674	1573-74	15	15	15	15	15	15	15	15	15	15	15	4694	1593-94	36	36	35	35	36	36	36	36	36	35	35	14. Vikrama.
4675	1574-75	16	16	16	16	16	16	16	16	16	16	16	4695	1594-95	37	37	36	36	37	37	37	37	37	36	36	15. Vrisha.
4676	1575-76	17	17	17	17	17	17	17	17	17	17	17	4696	1595-96	38	38	37	37	38	38	38	38	38	37	37	16. Chitrabhānu.
4677	1576-77	18	18	18	18	18	18	18	18	18	18	18	4697	1596-97	39	39	38	38	39	39	39	39	39	38	38	17. Subhānu.
4678	1577-78	19	19	19	19	19	19	19	19	19	19	19	4698	1597-98	40	40	39	39	40	40	40	40	40	39	39	18. Tāraṇa.
4679	1578-79	20	20	20	20	20	20	20	20	20	20	20	4699	1598-99	41	41	40	40	41	41	41	41	41	40	40	19. Pārthiva.
4680	1579-80	21	21	21	21	21	21	21	21	21	21	21	4700	1599-1600	42	42	41	41*	42	42	42	42	42	41*	41*	20. Vyaya.

1	2	3	4	5	6	7	8	9	10	11	12	13	1	2	3	4	5	6	7	8	9	10	11	12	13	
4701	1600-01	43	43	42*	43	43	43	43	43	43	43	43	4741	1640-41	23	23	23	23	23	23	23	23	23	23	23	21. Sarvajit.
4702	1601-02	44	44	44	44	44	44	44	44	44	44	44	4742	1641-42	24	24	24	24	24	24	24	24	24	24	24	22. Sarvadhārin.
4703	1602-03	45	45	45	45	45	45	45	45	45	45	45	4743	1642-43	25	25	25	25	25	25	25	25	25	25	25	23. Virōdhin.
4704	1603-04	46	46	46	46	46	46	46	46	46	46	46	4744	1643-44	26	26	26	26	26	26	26	26	26	26	26	24. Vikrita.
4705	1604-05	47	47	47	47	47	47	47	47	47	47	47	4745	1644-45	27	27	27	27	27	27	27	27	27	27	27	25. Khara.
4706	1605-06	48	48	48	48	48	48	48	48	48	48	48	4746	1645-46	28	28	28	28	28	28	28	28	28	28	28	26. Nandana.
4707	1606-07	49	49	49	49	49	49	49	49	49	49	49	4747	1646-47	29	29	29	29	29	29	29	29	29	29	29	27. Vijaya.
4708	1607-08	50	50	50	50	50	50	50	50	50	50	50	4748	1647-48	30	30	30	30	30	30	30	30	30	30	30	28. Jaya.
4709	1608-09	51	51	51	51	51	51	51	51	51	51	51	4749	1648-49	31	31	31	31	31	31	31	31	31	31	31	29. Manmatha.
4710	1609-10	52	52	52	52	52	52	52	52	52	52	52	4750	1649-50	32	32	32	32	32	32	32	32	32	32	32	30. Durmukha.
4711	1610-11	53	53	53	53	53	53	53	53	53	53	53	4751	1650-51	33	33	33	33	33	33	33	33	33	33	33	31. Hēmalamba.
4712	1611-12	54	54	54	54	54	54	54	54	54	54	54	4752	1651-52	34	34	34	34	34	34	34	34	34	34	34	32. Vilamba.
4713	1612-13	55	55	55	55	55	55	55	55	55	55	55	4753	1652-53	35	35	35	35	35	35	35	35	35	35	35	33. Vikārin.
4714	1613-14	56	56	56	56	56	56	56	56	56	56	56	4754	1653-54	36	36	36	36	36	36	36	36	36	36	36	34. Sārvarin.
4715	1614-15	57	57	57	57	57	57	57	57	57	57	57	4755	1654-55	37	37	37	37	37	37	37	37	37	37	37	35. Plava.
4716	1615-16	58	58	58	58	58	58	58	58	58	58	58	4756	1655-56	38	38	38	38	38	38	38	38	38	38	38	36. Subhakrit.
4717	1616-17	59	59	59	59	59	59	59	59	59	59	59	4757	1656-57	39	39	39	39	39	39	39	39	39	39	39	37. Śōbhana.
4718	1617-18	60	60	60	60	60	60	60	60	60	60	60	4758	1657-58	40	40	40	40	40	40	40	40	40	40	40	38. Krōdhin.
4719	1618-19	1	1	1	1	1	1	1	1	1	1	1	4759	1658-59	41	41	41	41	41	41	41	41	41	41	41	39. Viśvāvasu.
4720	1619-20	2	2	2	2	2	2	2	2	2	2	2	4760	1659-60	42	42	42	42	42	42	42	42	42	42	42	40. Parābhava.
4721	1620-21	3	3	3	3	3	3	3	3	3	3	3	4761	1660-61	43	43	43	43	43	43	43	43	43	43	43	41. Plavaṅga.
4722	1621-22	4	4	4	4	4	4	4	4	4	4	4	4762	1661-62	44	44	44	44	44	44	44	44	44	44	44	42. Kīlaka.
4723	1622-23	5	5	5	5	5	5	5	5	5	5	5	4763	1662-63	45	45	45	45	45	45	45	45	45	45	45	43. Saumya.
4724	1623-24	6	6	6	6	6	6	6	6	6	6	6	4764	1663-64	46	46	46	46	46	46	46	46	46	46	46	44. Sādhāraṇa.
4725	1624-25	7	7	7	7	7	7	7	7	7	7	7	4765	1664-65	47	47	47	47	47	47	47	47	47	47	47	45. Virōdhakrit.
4726	1625-26	8	8	8	8	8	8	8	8	8	8	8	4766	1665-66	48	48	48	48	48	48	48	48	48	48	48	46. Paridhāvin.
4727	1626-27	9	9	9	9	9	9	9	9	9	9	9	4767	1666-67	49	49	49	49	49	49	49	49	49	49	49	47. Pramādin.
4728	1627-28	10	10	10	10	10	10	10	10	10	10	10	4768	1667-68	50	50	50	50	50	50	50	50	50	50	50	48. Ānanda.
4729	1628-29	11	11	11	11	11	11	11	11	11	11	11	4769	1668-69	51	51	51	51	51	51	51	51	51	51	51	49. Rākshasa.
4730	1629-30	12	12	12	12	12	12	12	12	12	12	12	4770	1669-70	52	52	52	52	52	52	52	52	52	52	52	50. Anala.
4731	1630-31	13	13	13	13	13	13	13	13	13	13	13	4771	1670-71	53	53	53	53	53	53	53	53	53*	53	53	51. Piṅgala.
4732	1631-32	14	14	14	14	14	14	14	14	14	14	14	4772	1671-72	54	54	54	54	54	54*	54	54*	55	54	54	52. Kālayukta.
4733	1632-33	15	15	15	15	15	15	15	15	15	15	15	4773	1672-73	55	55	55	55	55*	56	55	56	56	55	55	53. Siddhārthin.
4734	1633-34	16	16	16	16	16	16	16	16	16	16	16	4774	1673-74	56	56	56	56	57	57	56	57	57	56	56	54. Raudra.
4735	1634-35	17	17	17	17	17	17	17	17	17	17	17	4775	1674-75	57	57	57	57	58	58	57	58	58	57	57	55. Durmati.
4736	1635-36	18	18	18	18	18	18	18	18	18	18	18	4776	1675-76	58	58	58	58	59	59	58	59	59	58	58	56. Dundubhi.
4737	1636-37	19	19	19	19	19	19	19	19	19	19	19	4777	1676-77	59*	59*	59	59	60	60	59*	60	60	59	59	57. Rudhirōdgārin.
4738	1637-38	20	20	20	20	20	20	20	20	20	20	20	4778	1677-78	1	1	60	60	1	1	1	1	1	60	60	58. Raktāksha.
4739	1638-39	21	21	21	21	21	21	21	21	21	21	21	4779	1678-79	...	...	1	1	2	2	...	2	2	1	1	59. Krōdhana.
4740	1639-40	22	22	22	22	22	22	22	22	22	22	22	4780	1679-80	...	...	2	2	3	3	...	3	3	2	2	60. Kshaya.

TABLE XLII—*contd.*

NUMBER OF THE SAṀVATSARA CONNECTED WITH EACH SOLAR YEAR, ACCORDING TO THE SEVERAL SIDDHĀNTAS, BY REASON OF ITS CURRENCY AT APPARENT, OR AT MEAN, MĒSHA-SAṀKRĀNTI.

Expired year of Kaliyuga.	Year A.D	Sūrya-S. no bīja. Apparent M. S.	Sūrya-S. no bīja. Mean M. S.	Sūrya-S. with Bīja. Apparent M. S.	Sūrya-S. with Bīja. Mean M. S.	First Ārya-S. Apparent M. S.	First Ārya-S. Mean M. S.	Orig. Sūrya S. Mean M. S.	Brahma-S. and S. Śirō. Apparent M. S.	Brahma-S. and S. Śirō. Mean M. S.	Second Ārya-S. Apparent M. S.	Second Ārya-S. Mean M. S.
1	2	3	4	5	6	7	8	9	10	11	12	13
4781	1680-81	...	...	3	3	4	4	...	4	4	3	3
4782	1681-82	...	...	4	4	5	5	...	5	5	4	4
4783	1682-83	...	...	5	5	6	6	...	6	6	5	5
4784	1683-84	...	...	6	6	7	7	...	7	7	6	6
4785	1684-85	...	...	7	7	8	8	...	8	8	7	7*
4786	1685-86	...	...	8*	8*	9	9	...	9	9	8*	9
4787	1686-87	...	...	10	10	10	10	...	10	10	10	10
4788	1687-88	...	...	11	11	11	11	...	11	11	11	11
4789	1688-89	...	...	12	12	12	12	...	12	12	12	12
4790	1689-90	...	...	13	13	13	13	...	13	13	13	13
4791	1690-91	...	...	14	14	14	14	...	14	14	14	14
4792	1691-92	...	...	15	15	15	15	...	15	15	15	15
4793	1692-93	...	...	16	16	16	16	...	16	16	16	16
4794	1693-94	...	...	17	17	17	17	...	17	17	17	17
4795	1694-95	...	...	18	18	18	18	...	18	18	18	18
4796	1695-96	...	...	19	19	19	19	...	19	19	19	19
4797	1696-97	...	...	20	20	20	20	...	20	20	20	20
4798	1679-98	...	...	21	21	21	21	...	21	21	21	21
4799	1698-99	...	...	22	22	22	22	...	22	22	22	22
4800	1699-1700	...	...	23	23	23	23	...	23	23	23	23

NUMBER OF THE SAṀVATSARA CONNECTED WITH EACH SOLAR YEAR, ACCORDING TO THE SEVERAL SIDDHĀNTAS, BY REASON OF ITS CURRENCY AT APPARENT, OR AT MEAN, MĒSHA-SAṀKRĀNTI

Expired year of Kaliyuga	Year A.D.	Sūrya-S. no bīja. Apparent M. S.	Sūrya-S. no bīja. Mean M. S.	Sūrya-S. with bīja. Apparent M. S.	Sūrya-S. with bīja. Mean M. S.	First Ārya-S Apparent M. S.	First Ārya-S Mean M. S.	Orig. Sūrya S. Mean M. S.	Brahma-S. and S. Śirō. Apparent M. S.	Brahma-S. and S. Śirō. Mean M. S.	Second Ārya-S. Apparent M. S.	Second Ārya-S. Mean M. S.	Names of the Sixty saṁvatsaras of the cycle of Jupiter.
1	2	3	4	5	6	7	8	9	10	11	12	13	
4801	1700-01	...	...	24	24	24	24	...	24	24	24	24	1. Prabhava.
4802	1701-02	...	...	25	25	25	25	...	25	25	25	25	2. Vibhava.
4803	1702-03	...	...	26	26	26	26	...	26	26	26	26	3. Śukla.
4804	1703-04	...	...	27	27	27	27	...	27	27	27	27	4. Pramōda.
4805	1704-05	...	...	28	28	28	28	...	28	28	28	28	5. Prajāpati.
4806	1705-06	...	...	29	29	29	29	...	29	29	29	29	6. Aṅgiras.
4807	1706-07	...	...	30	30	30	30	...	30	30	30	30	7. Śrīmukha.
4808	1707-08	...	...	31	31	31	31	...	31	31	31	31	8. Bhāva.
4809	1708-09	...	...	32	32	32	32	...	32	32	32	32	9. Yuvan.
4810	1709-10	...	...	33	33	33	33	...	33	33	33	33	10. Dhātri.
4811	1710-11	...	...	34	34	34	34	...	34	34	34	34	11. Īśvara.
4812	1711-12	...	...	35	35	35	35	...	35	35	35	35	12. Bahudhānya.
4813	1712-13	...	...	36	36	36	36	...	36	36	36	36	13. Pramāthin.
4814	1713-14	...	...	37	37	37	37	...	37	37	37	37	14. Vikrama.
4815	1714-15	...	...	38	38	38	38	...	38	38	38	38	15. Vrisha.
4816	1715-16	...	...	39	39	39	39	...	39	39	39	39	16. Chitrabhānu.
4817	1716-17	...	...	40	40	40	40	...	40	40	40	40	17. Subhānu.
4818	1717-18	...	...	41	41	41	41	...	41	41	41	41	18. Tārana.
4819	1718-19	...	...	42	42	42	42	...	42	42	42	42	19. Pārthiva.
4820	1719-20	...	...	43	43	43	43	...	43	43	43	43	20. Vyaya.

1	2	3	4	5	6	7	8	9	10	11	12	13	1	2	3	4	5	6	7	8	9	10	11	12	13	
4821	1720-21	...	...	44	44	44	44	...	44	44	44	44	4861	1760-61	...	...	24	24	25	25	...	25	25	24	24	21. Sarvajit.
4822	1721-22	...	...	45	45	45	45	...	45	45	45	45	4862	1761-62	...	...	25	25	26	26	...	26	26	25	25	22. Sarvadhârin.
4823	1722-23	...	...	46	46	46	46	...	46	46	46	46	4863	1762-63	...	...	26	26	27	27	...	27	27	26	26	23. Virōdhin.
4824	1723-24	...	...	47	47	47	47	...	47	47	47	47	4864	1763-64	...	...	27	27	28	28	...	28	28	27	27	24. Vikṛita.
4825	1724-25	...	...	48	48	48	48	...	48	48	48	48	4865	1764-65	...	...	28	28	29	29	...	29	29	28	28	25. Khara.
4826	1725-26	...	...	49	49	49	49	...	49	49	49	49	4866	1765-66	...	...	29	29	30	30	...	30	30	29	29	26. Nandana.
4827	1726-27	...	...	50	50	50	50	...	50	50	50	50	4867	1766-67	...	...	30	30	31	31	...	31	31	30	30	27. Vijaya.
4828	1727-28	...	...	51	51	51	51	...	51	51	51	51	4868	1767-68	...	...	31	31	32	32	...	32	32	31	31	28. Jaya.
4829	1728-29	...	...	52	52	52	52	...	52	52	52	52	4869	1768-69	...	...	32	32	33	33	...	33	33	32	32	29. Manmatha.
4830	1729-30	...	...	53	53	53	53	...	53	53	53	53	4870	1769-70	...	...	33	33	34	34	...	34	34	33	33	30. Durmukha.
4831	1730-31	...	...	54	54	54	54	...	54	54	54	54	4871	1770-71	...	...	34	34*	35	35	...	35	35	34*	34*	31. Hēmalamba.
4832	1731-32	...	...	55	55	55	55	...	55	55	55	55	4872	1771-72	...	...	35*	36	36	36	...	36	36	36	36	32. Vilamba.
4833	1732-33	...	...	56	56	56	56	...	56	56	56	56	4873	1772-73	...	...	37	37	37	37	...	37	37	37	37	33. Vikārin.
4834	1733-34	...	...	57	57	57	57	...	57	57	57	57	4874	1773-74	...	...	38	38	38	38	...	38	38	38	38	34. Śārvarin.
4835	1734-35	...	...	58	58	58	58	...	58	58	58	58	4875	1774-75	...	...	39	39	39	39	...	39	39	39	39	35. Plava.
4836	1735-36	...	...	59	59	59	59	...	59	59	59	59	4876	1775-76	...	...	40	40	40	40	...	40	40	40	40	36. Śubhakṛit.
4837	1736-37	...	...	60	60	60	60	...	60	60	60	60	4877	1776-77	...	...	41	41	41	41	...	41	41	41	41	37. Śōbhana.
4838	1737-38	...	...	1	1	1	1	...	1	1	1	1	4878	1777-78	...	...	42	42	42	42	...	42	42	42	42	38. Krōdhin.
4839	1738-39	...	...	2	2	2	2	...	2	2	2	2	4879	1778-79	...	...	43	43	43	43	...	43	43	43	43	39. Viśvāvasu.
4840	1739-40	...	...	3	3	3	3	...	3	3	3	3	4880	1779-80	...	...	44	44	44	44	...	44	44	44	44	40. Parābhava.
4841	1740-41	...	...	4	4	4	4	...	4	4	4	4	4881	1780-81	...	...	45	45	45	45	...	45	45	45	45	41. Plavaṅga.
4842	1741-42	...	...	5	5	5	5	...	5	5	5	5	4882	1781-82	...	...	46	46	46	46	...	46	46	46	46	42. Kīlaka.
4843	1742-43	...	...	6	6	6	6	...	6	6	6	6	4883	1782-83	...	...	47	47	47	47	...	47	47	47	47	43. Saumya.
4844	1743-44	...	...	7	7	7	7	...	7	7	7	7	4884	1783-84	...	...	48	48	48	48	...	48	48	48	48	44. Sādhāraṇa.
4845	1744-45	...	...	8	8	8	8	...	8	8	8	8	4885	1784-85	...	...	49	49	49	49	...	49	49	49	49	45. Virōdhakṛit.
4846	1745-46	...	...	9	9	9	9	...	9	9	9	9	4886	1785-86	...	...	50	50	50	50	...	50	50	50	50	46. Paridhāvin.
4847	1746-47	...	...	10	10	10	10	...	10	10	10	10	4887	1786-87	...	...	51	51	51	51	...	51	51	51	51	47. Pramādin.
4848	1747-48	...	...	11	11	11	11	...	11	11	11	11	4888	1787-88	...	...	52	52	52	52	...	52	52	52	52	48. Ānanda.
4849	1748-49	...	...	12	12	12	12	...	12	12	12	12	4889	1788-89	...	...	53	53	53	53	...	53	53	53	53	49. Rākshasa.
4850	1749-50	...	...	13	13	13	13	...	13	13	13	13	4890	1789-90	...	...	54	54	54	54	...	54	54	54	54	50. Anala.
4851	1750-51	...	...	14	14	14	14	...	14	14	14	14	4891	1790-91	...	...	55	55	55	55	...	55	55	55	55	51. Piṅgala.
4852	1751-52	...	...	15	15	15	15	...	15	15	15	15	4892	1791-92	...	...	56	56	56	56	...	56	56	56	56	52. Kālayukta.
4853	1752-53	...	...	16	16	16	16	...	16	16	16	16	4893	1792-93	...	...	57	57	57	57	...	57	57	57	57	53. Siddhārthin.
4854	1753-54	...	...	17	17	17	17	...	17	17	17	17	4894	1793-94	...	...	58	58	58	58	...	58	58	58	58	54. Raudra.
4855	1754-55	...	...	18	18	18	18	...	18	18	18	18	4895	1794-95	...	...	59	59	59	59	...	59	59	59	59	55. Durmati.
4856	1755-56	...	...	19	19	19	19	...	19	19*	19	19	4896	1795-96	...	...	60	60	60	60	...	60	60	60	60	56. Dundubhi.
4857	1756-57	...	...	20	20	20	20*	...	20*	21	20	20	4897	1796-97	...	...	1	1	1	1	...	1	1	1	1	57. Rudhirōdgārin.
4858	1757-58	...	...	21	21	21*	22	...	22	22	21	21	4898	1797-98	...	...	2	2	2	2	...	2	2	2	2	58. Raktāksha.
4859	1758-59	...	...	22	22	23	23	...	23	23	22	22	4899	1798-99	...	...	3	3	3	3	...	3	3	3	3	59. Krōdhana.
4860	1759-60	...	...	23	23	24	24	...	24	24	23	23	4900	1799-1800	...	...	4	4	4	4	...	4	4	4	4	60. Kshaya.

TABLE XLII—*contd.*

Expired year of Kaliyuga	Year A.D.	NUMBER OF THE SAṀVATSARA CONNECTED WITH EACH SOLAR YEAR, ACCORDING TO THE SEVERAL SIDDHĀNTAS, BY REASON OF ITS CURRENCY AT APPARENT, OR AT MEAN, MĒSHA-SAṀKRĀNTI: Sūrya-S. no Bīja, Apparent M. S.	Sūrya-S. no Bīja, Mean M. S.	Sūrya-S. with Bīja, Apparent M. S.	Sūrya-S. with Bīja, Mean M. S.	First Ārya-S., Apparent M. S.	First Ārya-S., Mean M. S.	Orig. Sūrya S., Mean M. S.	Brahma-S. and S. Śirō., Apparent M. S.	Brahma-S. and S. Śirō., Mean M. S.	Second Ārya-S., Apparent M. S.	Second Ārya-S., Mean M. S.	Expired year of Kaliyuga	Year A.D.	NUMBER OF THE SAṀVATSARA CONNECTED WITH EACH SOLAR YEAR, ACCORDING TO THE SEVERAL SIDDHĀNTAS, BY REASON OF ITS CURRENCY AT APPARENT, OR AT MEAN, MĒSHA-SAṀKRĀNTI: Sūrya-S. no Bīja, Apparent M. S.	Sūrya-S. no Bīja, Mean M. S.	Sūrya-S. with Bīja, Apparent M. S.	Sūrya-S. with Bīja, Mean M. S.	First Ārya-S., Apparent M. S.	First Ārya-S., Mean M. S.	Orig. Sūrya S., Mean M. S.	Brahma-S. and S. Śirō., Apparent M. S.	Brahma-S. and S. Śirō., Mean M. S.	Second Ārya-S., Apparent M. S.	Second Ārya-S., Mean M. S.	Names of the Sixty saṁvatsaras of the cycle of Jupiter.
1	2	3	4	5	6	7	8	9	10	11	12	13	1	2	3	4	5	6	7	8	9	10	11	12	13	
4901	1800-01	...	...	5	5	5	5	...	5	5	5	5	4921	1820-21	...	...	25	25	25	25	...	25	25	25	25	1. Prabhava.
4902	1801-02	...	...	6	6	6	6	...	6	6	6	6	4922	1821-22	...	...	26	26	26	26	...	26	26	26	26	2. Vibhava.
4903	1802-03	...	...	7	7	7	7	...	7	7	7	7	4923	1822-23	...	...	27	27	27	27	...	27	27	27	27	3. Śukla.
4904	1803-04	...	...	8	8	8	8	...	8	8	8	8	4924	1823-24	...	...	28	28	28	28	...	28	28	28	28	4. Pramōda.
4905	1804-05	...	...	9	9	9	9	...	9	9	9	9	4925	1824-25	...	...	29	29	29	29	...	29	29	29	29	5. Prajāpati.
4906	1805-06	...	...	10	10	10	10	...	10	10	10	10	4926	1825-26	...	...	30	30	30	30	...	30	30	30	30	6. Aṅgiras.
4907	1806-07	...	...	11	11	11	11	...	11	11	11	11	4927	1826-27	...	...	31	31	31	31	...	31	31	31	31	7. Śrīmukha.
4908	1807-08	...	...	12	12	12	12	...	12	12	12	12	4928	1827-28	...	...	32	32	32	32	...	32	32	32	32	8. Bhāva.
4909	1808-09	...	...	13	13	13	13	...	13	13	13	13	4929	1828-29	...	...	33	33	33	33	...	33	33	33	33	9. Yuvan.
4910	1809-10	...	...	14	14	14	14	...	14	14	14	14	4930	1829-30	...	...	34	34	34	34	...	34	34	34	34	10. Dhātri.
4911	1810-11	...	...	15	15	15	15	...	15	15	15	15	4931	1830-31	...	...	35	35	35	35	...	35	35	35	35	11. Īśvara.
4912	1811-12	...	...	16	16	16	16	...	16	16	16	16	4932	1831-32	...	...	36	36	36	36	...	36	36	36	36	12. Bahudhānya.
4913	1812-13	...	...	17	17	17	17	...	17	17	17	17	4933	1832-33	...	...	37	37	37	37	...	37	37	37	37	13. Pramāthin.
4914	1813-14	...	...	18	18	18	18	...	18	18	18	18	4934	1833-34	...	...	38	38	38	38	...	38	38	38	38	14. Vikrama.
4915	1814-15	...	...	19	19	19	19	...	19	19	19	19	4935	1834-35	...	...	39	39	39	39	...	39	39	39	39	15. Vrisha.
4916	1815-16	...	...	20	20	20	20	...	20	20	20	20	4936	1835-36	...	...	40	40	40	40	...	40	40	40	40	16. Chitrabhānu.
4917	1816-17	...	...	21	21	21	21	...	21	21	21	21	4937	1836-37	...	...	41	41	41	41	...	41	41	41	41	17. Subhānu.
4918	1817-18	...	...	22	22	22	22	...	22	22	22	22	4938	1837-38	...	...	42	42	42	42	...	42	42	42	42	18. Tāraṇa.
4919	1818-19	...	...	23	23	23	23	...	23	23	23	23	4939	1838-39	...	...	43	43	43	43	...	43	43	43	43	19. Pārthiva.
4920	1819-20	...	...	24	24	24	24	...	24	24	24	24	4940	1839-40	...	...	44	44	44	44	...	44	44	44	44	20. Vyaya.

1	2	3	4	5	6	7	8	9	10	11	12	13	1	2	3	4	5	6	7	8	9	10	11	12	13	
4941	1840-41	...	...	45	45	45	45	...	45	45*	45	45	4981	1880-81	...	...	26	26	26	26	...	26	26	26	26	21. Sarvajit.
4942	1841-42	...	...	46	46	46	46*	...	46*	47	46	46	4982	1881-82	...	...	27	27	27	27	...	27	27	27	27	22. Sarvadhārin.
4943	1842-43	...	...	47	47	47*	48	...	48	48	47	47	4983	1882-83	...	...	28	28	28	28	...	28	28	28	28	23. Virōdhin.
4944	1843-44	...	...	48	48	49	49	...	49	49	48	48	4984	1883-84	...	...	29	29	29	29	...	29	29	29	29	24. Vikrita.
4945	1844-45	...	...	49	49	50	50	...	50	50	49	49	4985	1884-85	...	...	30	30	30	30	...	30	30	30	30	25. Khara.
4946	1845-46	...	...	50	50	51	51	...	51	51	50	50	4986	1885-86	...	...	31	31	31	31	...	31	31	31	31	26. Nandana.
4947	1846-47	...	...	51	51	52	52	...	52	52	51	51	4987	1886-87	...	...	32	32	32	32	...	32	32	32	32	27. Vijaya.
4948	1847-48	...	...	52	52	53	53	...	53	53	52	52	4988	1887-88	...	...	33	33	33	33	...	33	33	33	33	28. Jaya.
4949	1848-49	...	...	53	53	54	54	...	54	54	53	53	4989	1888-89	...	...	34	34	34	34	...	34	34	34	34	29. Manmatha.
4950	1849-50	...	...	54	54	55	55	...	55	55	54	54	4990	1889-90	...	...	35	35	35	35	...	35	35	35	35	30. Durmukha.
4951	1850-51	...	...	55	55	56	56	...	56	56	55	55	4991	1890-91	...	...	36	36	36	36	...	36	36	36	36	31. Hēmalamba.
4952	1851-52	...	...	56	56	57	57	...	57	57	56	56	4992	1891-92	...	...	37	37	37	37	...	37	37	37	37	32. Vilamba.
4953	1852-53	...	...	57	57	58	58	...	58	58	57	57	4993	1892-93	...	...	38	38	38	38	...	38	38	38	38	33. Vikārin.
4954	1853-54	...	...	58	58	59	59	...	59	59	58	58	4994	1893-94	...	...	39	39	39	39	...	39	39	39	39	34. Śārvarin.
4955	1854-55	...	...	59	59	60	60	...	60	60	59	59	4995	1894-95	...	...	40	40	40	40	...	40	40	40	40	35. Plava.
4956	1855-56	...	...	60	60	1	1	...	1	1	60*	60*	4996	1895-96	...	...	41	41	41	41	...	41	41	41	41	36. Śubhakrit.
4957	1856-57	...	...	1*	1*	2	2	...	2	2	2	2	4997	1896-97	...	...	42	42	42	42	...	42	42	42	42	37. Śōbhana.
4958	1857-58	...	...	3	3	3	3	...	3	3	3	3	4998	1897-98	...	...	43	43	43	43	...	43	43	43	43	38. Krōdhin.
4959	1858-59	...	...	4	4	4	4	...	4	4	4	4	4999	1898-99	...	...	44	44	44	44	...	44	44	44	44	39. Viśvāvasu.
4960	1859-60	...	...	5	5	5	5	...	5	5	5	5	5000	1899-1900	...	...	45	45	45	45	...	45	45	45	45	40. Parābhava.
4961	1860-61	...	...	6	6	6	6	...	6	6	6	6	5001	1900-01	...	...	46	46	46	46	...	46	46	46	46	41. Plavanga.
4962	1861-62	...	...	7	7	7	7	...	7	7	7	7	5002	1901-02	...	...	47	47	47	47	...	47	47	47	47	42. Kīlaka.
4963	1862-63	...	...	8	8	8	8	...	8	8	8	8	5003	1902-03	...	...	48	48	48	48	...	48	48	48	48	43. Saumya
4964	1863-64	...	...	9	9	9	9	...	9	9	9	9	5004	1903-04	...	...	49	49	49	49	...	49	49	49	49	44. Sādhāraṇa.
4965	1864-65	...	...	10	10	10	10	...	10	10	10	10	5005	1904-05	...	...	50	50	50	50	...	50	50	50	50	45. Virōdhakrit.
4966	1865-66	...	...	11	11	11	11	...	11	11	11	11	5006	1905-06	...	...	51	51	51	51	...	51	51	51	51	46. Paridhāvin.
4967	1866-67	...	...	12	12	12	12	...	12	12	12	12	5007	1906-07	...	...	52	52	52	52	...	52	52	52	52	47. Pramādin.
4968	1867-68	...	...	13	13	13	13	...	13	13	13	13	5008	1907-08	...	...	53	53	53	53	...	53	53	53	53	48. Ānanda.
4969	1868-69	...	...	14	14	14	14	...	14	14	14	14	5009	1908-09	...	...	54	54	54	54	...	54	54	54	54	49. Rākshasa.
4970	1869-70	...	...	15	15	15	15	...	15	15	15	15	5010	1909-10	...	...	55	55	55	55	...	55	55	55	55	50. Anala.
4971	1870-71	...	...	16	16	16	16	...	16	16	16	16	5011	1910-11	...	...	56	56	56	56	...	56	56	56	56	51. Piṅgala.
4972	1871-72	...	...	17	17	17	17	...	17	17	17	17	5012	1911-12	...	...	57	57	57	57	...	57	57	57	57	52. Kālayukta.
4973	1872-73	...	...	18	18	18	18	...	18	18	18	18	5013	1912-13	...	...	58	58	58	58	...	58	58	58	58	53. Siddārthin.
4974	1873-74	...	...	19	19	19	19	...	19	19	19	19	5014	1913-14	...	...	59	59	59	59	...	59	59	59	59	54. Raudra.
4975	1874-75	...	...	20	20	20	20	...	20	20	20	20	5015	1914-15	...	...	60	60	60	60	...	60	60	60	60	55. Durmati.
4976	1875-76	...	...	21	21	21	21	...	21	21	21	21	5016	1915-16	...	...	1	1	1	1	...	1	1	1	1	56. Dundubhi.
4977	1876-77	...	...	22	22	22	22	...	22	22	22	22														57. Rudhirōdgārin.
4978	1877-78	...	...	23	23	23	23	...	23	23	23	23.														58. Raktāksha.
4979	1878-79	...	...	24	24	24	24	...	24	24	24	24														59. Krōdhana.
4980	1879-80	...	...	25	25	25	25	...	25	25	25	25														60. Kshaya.

THE TRUE LONGITUDE OF THE SUN IN HINDU ASTRONOMY, PART I ĀRYA-AND SŪRYA-SIDDHANTAS.

(Previously published in Epigraphia Indica, Vol. XIV, pp. 1-67.)

234. The exact position of the true or apparent sun at sunrise of each civil day, taken for tabular purposes as mean sunrise, is one of the essential elements of Hindu chronography, and the exact position of the true moon is another. From these positions are calculated the beginning and end of each tithi and nakshatra, with the currency of these at sunrise. All over India for many centuries the civil day has been coupled with the true tithi current at sunrise, the nakshatra in which the true moon stands at sunrise being stated in the local almanacs and constantly mentioned in the dates of historical inscriptions. In Southern India the nakshatra was considered of such importance that from as early as the tenth century it has regularly given its name to the day. For the proper verification of historical inscription-dates, therefore, it is of the highest importance that we should know the precise position of the true sun at any moment and more especially at the moment of mean sunrise [1]

235. Now the process adopted for this purpose in "*The Indian Calendar*" (Sewell and S. B. Dikshit, *1896*), though resulting in a fair approximation, did not, for critical examinations of dates, give a sufficiently close result, as I have already explained in my "*Indian Chronography,*" *§§ 119, 120, pp. 42-43*; something more accurate was required. We want, *for each of the Indian astronomical authorities separately,* extremely accurate determination of the sun's true longitude each day of the year; and there is only one way to obtain this. For each day a calculation must be made of the exact equation of the sun's centre on the basis of the sun's mean anomaly, according to the Hindu method of computation. This was a formidable undertaking; but it has now been accomplished for the *First Ārya-* and *Sūrya-Siddhāntas*, and the Tables are published herewith. It is to be hoped that they are final. They are intended to fix the true longitude of the sun on any day or at any moment of the day, with an accuracy extending to the hundredth part of a second. Similar Tables for the *Siddhānta-Śirōmani* are given in the next section. I give the result in degrees and parts, and in ten-thousandths of the circle. The former, converted as desired, can be adapted to any system of reckoning; the latter are for use by the *Indian Calendar* system.[2]

236. These calculations are, as I have stated, based purely on the Hindu system of reckoning. I have used for the sun's mean anomaly and longitude the mean position and mean motion of the sun as gathered from each Siddhānta separately, and have used the Hindu values of the sines for computing the amount of the equation of the centre, and thence the sun's true position. The Tables are prepared according to the *First Ārya-* and *Present Sūrya-Siddhāntas,* the latter both with and without the bīja. The bija (correction), which came into general use about A.D. 1500, made no change in the length of the solar year or the number of civil days in a mahāyuga, or in the position of the sun's apsis, and therefore none in the sun's longitude whether true or mean.

237. Assuming, since these Tables are not intended for any but the initiated, that the *Indian Calendar* process of calculation, which might be termed Prof. Jacobi's first process and which has the advantage of simplicity, is known to readers of the *Epigraphia*, only one or two remarks need be made before entering on details. Since everything depends on the accuracy of the Table-entries, I must call attention to the great help which I received from M. Louis de Ries of Moscow for many months. He takes the greatest interest in Hindu astronomy, and has prepared certain Tables of his own, the publication of which has been

[1] For calculation affecting all parts of India the basis has to be mean sunrise, and this is always taken as mean sunrise at Lanka, or Ujjain, an imaginary spot on the equator on the meridian of Ujjain, E. long. 75° 46′ 17″.

[2] The *Indian Calendar* system is the system adopted by Prof. Jacobi (of Bonn) in 1888 (*Indian Antiquary*, Vol. XVII), itself founded on Largeteau (*Connaissance des Temps*, 1845)

delayed by the great European war. His processes are characterized by the most painstaking endeavours to obtain extreme accuracy for every result arrived at. Filled with a similar desire, and after my calculations for the sun's exact position (in true longitude for successive 24-hour periods after the true sun's arrival at long. 0°) had been carried out for about one-third of the *Ārya-Siddhānta* year, I asked M. de Ries to calculate some of these positions of the sun by his own method, so that we might compare the results. He most kindly did so; and, when I state that our results, worked in entire independence of one another and by different methods, were found to agree in every respect down to four, and in one case even down to five, decimals of a second, I think that it may be fairly assumed that my Tables may be depended upon.

238. There is more than one reason why the *Indian Calendar* system, though yielding results very fairly approximate, requires some expansion for the purpose of exact calculation. By it we have been in the habit of computing the true moon's place both for the tithi and nakshatra by the *Sūrya-Siddhānta* data, using the same figures for finding the tithi-index, *t*, and nakshatra-index, *n*, for all dates, both for inscriptions known to belong to tracts and times when the *Ārya-Siddhānta* was the authority used by the framers of the record as well as for those which must have been guided by almanacs calculated by the *Sūrya-Siddhānta*. The *c* of the *Indian Calendar* method, i.e. the sun's mean anomaly at any moment, is always the *Sūrya-Siddhānta* "*c*" in thousandths of the circle, and that it differs in various proportions at different times of the year from the "*c*" of the *Ārya-Siddhānta* will be apparent to anyone who compares the entries for the same day given in my new Tables XLVIII A and B, *cols.* 2, 3, in ten thousandths. At the moment of Mēsha-saṁkrānti for instance (the first entry in each Table) the "*c*" by the *Sūrya* is 2794·0642 in ten-thousandths, and is 279 in thousandths in *Indian Calendar* reckoning; but by the *Ārya-Siddhānta* it is 2774·5577, and so for our ordinary reckoning should be stated as 277. In calculation for the tithi-index, "*t*", in ordinary work this difference has no very great effect, though of course it actually has some, and possibly may in some cases alter the value of "*t*" by one unit (4¼ minutes); but it has greater effect when we are calculating the nakshatra, as will presently be explained. As to the difference between the two authorities in the value assigned to the sun's true longitude, "*s*", it will be seen that this varies day by day. About Day 261, i.e. the 261st period of 24 hours each measured from true Mēsha-saṁkrānti, the value of "*s*" is practically the same by the two authorities; about Day 150 the Ārya "*s*" is about 3′ 36″ ahead of the Sūrya "*s*." The difference increases and diminishes regularly throughout the year.

The principal reasons for this difference are that by the *Sūrya-Siddhānta* the position of the sun's perigee-point is different from that assumed by the *Ārya-Siddhānta*, and that there is a difference in the two-year lengths.

239. I have stated above that this difference has only a very slight effect as regards the value of the tithi-index; its effect on the ordinary calculation of the nakshatra and lagnas must now be noticed. In so doing we take first the nakshatra and note the process by which those who have used the *Indian Calendar* have hitherto calculated its index.

Our method of computing the sun's true longitude, "*s*", by the system of the *Indian Calendar* has been to take the "*c*" found for the desired moment, that is to say, the value, in thousandths of the circle, of the sun's mean anomaly according to the *Sūrya-Siddhānta*, making this serve for both Siddhāntas; —to multiply this "*c*" by ten to get its approximate value in ten-thousandths;—to add to it a figure, 7207, representing the longitude of the sun's perigee-point (taken as 7146·3 by the *Sūrya-Siddhānta*) in A.D. 1100 *plus* an addition representing the sun's greatest equation of the centre (roughly 60·4, actually by the *Sūrya-Siddhānta* 60·4214)—an addition which is rendered necessary by the construction of the Tables in order to avoid the necessity for sometimes adding and sometimes subtracting the equation of the centre[1];— to deduct from the result the figure representing this equation;—and so to obtain the sun's true

[1] See *Indian Calendar*, §§ 107, 108; pp. 60, 61

longitude, "*s*". The tithi-index, "*t*"[1] having been already found, we add "*s*" to "*t*" and find the nakshatra-index "*n*" or the longitude of the true moon; this index shews in which nakshatra she stands at the moment. The result is an approximation, but it is not close enough. If we are working for an *Ārya-Siddhānta* date, we have used *Sūrya-Siddhānta* values (which differ slightly), and we have arrived at the value of "*s*" in part by multiplying by 10 a value obtained in thousandths so as to be able to apply it to the other value, that of the moon, which has been obtained in ten-thousandths of the circle. This multiplication by ten creates a possibility of error not inconsiderable. Thus, if we have, in thousandths, the figure "*c*" = 623, this may stand for any value in ten-thousandths between 6225 and 6235, and may lead to a miscalculation amounting to anything under 10 units in our estimate of the nakshatra-index "*n*" and 10 units represent in time-valuation 39 minutes.

240. All these possibilities of error are entirely removed by the present Tables. The exact value of "*s*" by either Siddhānta is easily found—a value which we know to be absolutely correct—, and when we add this "*s*" to the already found "*t*" we know that the result gives the correct nakshatra-index; or at least that the only possibility of error lies in the value "*t*" found for the tithi.

241. These Tables will also be found very useful for calculating the lagna accurately. Hitherto our process for finding, in working for the lagna, the value of the sun's true longitude, "*s*", at mean sunrise of the day concerned has been the same as the not quite perfect process for finding the nakshatra. The present Tables give the exactly accurate "*s*" by either *Siddhānta*, and they give it in degrees, etc., thereby simplifying the calculation.

EXPLANATION OF THE TABLES.

242. Table *XLIII*. The details were worked out with great care by M. L. de Ries from the respective lengths of the sidereal solar year, i.e. the time taken by the true sun to travel from 0° to 0°, according to the several Indian authorities.

Table *XLIV* gives the sun's mean motion per day of 24 hours, and per hour, minute and second, for use in calculation. It is exact for the *Ārya-Siddhānta*, and may be used with care for other authorities, having regard to the footnote.

Table *XLIVA*. See the heading. It explains itself.

Table *XLVA* is for use in calculations. Every valuation given in the main Tables *XLVIIIA* and *B* in ten-thousandths of the circle was made by it.

Table *XLVB* is the reverse of *XLVA*.

Table *XLVI* is a revised nakshatra-Table, shewing the exact ending points of each.

Table *XLVII* is very important, being a revised Table of sines and equations of the sun's centre, given in full after particularly careful calculation. Its preparation is described below, §§ 249-253. The supplementary Table *XLVIIA* gives, for close work, very full details of the exact equations according to authorities other than the *First Ārya-Siddhānta*; and of the differences, in seconds per minute of mean anomaly-arc, between the consecutive base-equations. Table XLVII, cols. 9, 10, may also be used for the *Brahma-Siddhānta*, but not Table XLVIIA.

Tables *XLVIIIA* and *XLVIIIB* are the main working Tables, shewing, by the *First Ārya-* and *Present Sūrya-Siddhāntas* (with or without the bīja), the precise value of the sun's true longitude (*s*) and equation of the centre at each interval of 24 hours measured from true Mēsha-

[1] The tithi-index, "*t*", gives the distance of true moon from true sun, i.e., shews the moon's phase or her true place with reference to the true sun. When this is added to the true sun's longitude, we have the true moon's place in the heavens, "*n*", or the required nakshatra-index.

saṁkrānti, the moment when the true sun arrives each year at celestial longitude 0°; as well as the sun's mean anomaly and mean longitude. There was no possibility of framing a Table which should give these particulars for mean sunrise of each day, the primary requirement for the verification of Indian dates, because the moment of true Mēsha-saṁkrānti varies each year and the starting-point had to be from that moment. These two Tables therefore give the consecutive 24-hour positions of the mean and true sun after that moment.

Tables *XLIX* and *L* enable us to find the sun's true longitude at mean sunrise; the former giving for each group of days the sun's *true* motion per hour, and the latter giving his *mean* motion per minute. It is not necessary for general purposes to give his true motion per minute; if required, this can always be obtained by dividing by 60 the details of Table XLIX for one hour of the day.

243. Tables *XLVIII* to *L* are used in the following way, when we desire to find the "*s*" for mean sunrise. Say that Mēsha-saṁkrānti occurred in the year for which we are working at 12ʰ 15ᵐ after mean sunrise. Then for every day of that year Table XLVIII-A or -B gives us his true longitude, "*s*", *at 12ʰ 15ᵐ after mean sunrise*; and to obtain the "*s*" at *mean sunrise* on the day in question we have to deduct the sun's true motion during 12ʰ and 15ᵐ. We do this by Tables *XLIX* and *L*, and so get the exact "*s*" for mean sunrise on the day in question.

Table *XLIX* for hours is exactly correct for the *Ārya-Siddhānta*. When used for the *Sūrya-Siddhānta*, there may be an error amounting, at the time of year when there is the greatest difference between the two authorities, to about one-third of a second per hour or about seven seconds per day. If anyone desires to be absolutely exact by the *Sūrya-Siddhānta*, he should calculate the true sun's motion during the hours and minutes of the day in question by observing in Table *XLVIIIB* the consecutive 24-hour positions, "*s*" of the sun given in the Table for (i) the day in question and (ii) the previous day, and divide the difference by 24 for each hour's, and this result by 60 for each minute's, true motion. Even this, of course, is not mathematically exact, since the true motion of the sun varies from hour to hour; but it is quite accurate enough.

244. The calculation for the true longitude of the sun each day was made by ascertaining his mean anomaly and then using the sine-Table as finally prepared (Table XLVII) for finding the equation of the centre. The starting point for the year is the value of his mean anomaly at the moment of true Mēsha-saṁkrānti. This had to be computed with great care. The problem is fully discussed below, §§ 254-255.

245. To obtain a correct value of the sun's *mean* longitude at sunrise of any day, take the value given in Table *XLVIIIA* or *B*, as the case may be, cols. 4, 5, and deduct for the intervening hours and minutes (*§ 243, para. 1*) the quantities shewn in Table XLIV for the sun's *mean* motion. Greater accuracy even than this can be obtained by the use of Table XLIII.

246. I do not enter very fully into the difference in the sun's true longitude brought about, according to the *Sūrya-Siddhānta*, by the shift in the apsis of the sun's orbit, because this seems so slight that it may be ignored. It would amount to about 1″ in the last 1500 years (*see below, s. 254, ii*).

Use of the Tables. Rules.

247. That the use of the Tables may be thoroughly understood, I append a few rules of work and examples.

(i) *The nakshatra.* -Work by the usual *Indian Calendar* process for finding "*t*" the tithi-index at mean sunrise of the day in question.[1] Note the serial number of the civil day, ignoring altogether the day of the Hindu solar month. Deduct from this number the serial number of the day on which Mēsha-saṁkrānti occurred (*Table I, or any of the similar general*

[1] Examples are given below, viz. in "the *Siddhānta-Śirōmani*" section, Example 4 (p. 145), and in the section "*First Ārya-Siddhānta*, true system", Examples 4, 5 (pp. 239, 240).

working Tables below, col. 13). The result is the number of the day, or 24-hour period, referred to in col. 1 of the new Tables XLVIIIA and B. Remembering to use the proper Table for the *Siddhānta* concerned, turn to this number in either of those Tables. Against it in col. 9 will be found the correct value of the sun's longitude, "*s*" on that day at a moment as many hours and minutes after mean sunrise as elapsed between mean sunrise and the moment of Mēsha-saṁkrānti at the beginning of the solar year (*Table I or other general Tables, col. 17*). Turn to Table XLIX for hours on the day in question and to Table L for minutes, and deduct from the "*s*" so obtained the values of the sun's motion during those hours and minutes (*above*, § *243*). This gives the sun's exact true longitude at mean sunrise of the day in question. $s + t = n$, the nakshatra-index. For exact ending points of nakshatras, i.e. the points when the true moon passes out of each, consult Table XLVI. (Table VIII of the *Indian Calendar* or Table LXVIII below suffices except in very close cases.) Properly worked, the "*s*" so found yields the correct longitude of the true sun within the hundredth part of a second.

(ii) *The tithi.*—[This may be examined by the new Tables, though probably it will not be liable to change, or at any rate not to any change greater than one unit. Until some new Tables are published, we work for the moon's place by Prof. Jacobi's fixtures, and accept them.] The serial number of the day, or 24-hour period, being found as above, note against it in Table XLVIIIA or B, cols. 2, 3, the value of the sun's mean anomaly; and for the intervening hours and minutes deduct the sun's *mean* motion as given in Table XLIV, observing the remarks in the footnote to that Table. This gives the sun's mean anomaly at mean sunrise of the day in question in ten-thousandths of the circle. Take the value in thousandths of the circle by removing the decimal point one place to the left. Refer to Table VII, *Indian Calendar*, or Table LXVII below, and the corresponding auxiliary Table below each of these for correcting the "equation *c*" of the calculation, if it does not seem necessary to work with greater exactness than by use of units of about $4\frac{1}{4}$ minutes.

We can find the equation more accurately as follows: -It has been noted in § 239 that, in order that "*equation c*" in the *a*, *b*, *c* system may always be additive, the quantity 60·4 was taken from "*a*" (the mean moon's distance from mean sun) and added to the equation of the centre. Hence we shall have the exact "equation *c*", if we deduct from 60·4 the amount of the equation (given in the new Table), when it is *plus* (+), and add to 60·4 the amount of the equation, when it is *minus* (−); the signs are given in the heading of cols. 6, 7, Tables XLVIIIA and B.

The equation can also be obtained with quite sufficient approximation by noting the difference between the equation of the day and the equation of the previous day (*cols. 6, 7*), dividing this difference by 24, and applying to the equation of the day the amount proportionate to the hours intervening from mean sunrise (*see example given below, § 248, D.*).

(iii) *The lagna.*—To find the time of rising of the named sign on the day concerned, calculate the "*s*" for mean sunrise as above explained, but this time use degrees, minutes and seconds (*col. 8 of either Table XLVIIIA or B*). Table XXII, *Indian Chronography*, gives the beginning and ending points of the named sign. Adding to these 360° if necessary, deduct from their value the value of *s* at mean sunrise. The result shews the distance from the sun at mean sunrise of the beginning and ending points of the sign. Multiply the degrees by 4 for minutes, and the minutes by 4 for seconds of time. The result gives the times of rising of the beginning and ending points of the named sign.

Examples of Work.

248. Given an inscription date to be examined, with the details Śaka 1412, Friday, the day of Uttara Bhadrapadā. Chaitra krishṇa 12, Mithuna lagna.

We first examine the date according to the *Indian Calendar* system and Tables ; afterwards verifying, by the new Tables herein given, some of its important elements, such as the sun's mean anomaly, " *c* " the sun's equation of the centre and the value of " equation *c* ", and the sun's true longitude, " *s* " at mean sunrise of the day of the date.

[Let it be remembered that Table I of the *Indian Calendar*, so far as regards calculation for the lunar tithi, uses the postulates of the *Sūrya-Siddhānta* to obtain results for both the *Ārya* and *Sūrya-Siddhāntas* – a course which is sufficiently accurate in most cases but not so in close cases. Its advantage is its simplicity.]

The year in Śaka 1412 expired, or A. D. 1490-91. The day on which the lunnar tithi Chaitra kṛishṇa 12 expires will be about 25 days later than the day on which Chaitra śukla expired. If found not to be so, calculate for a lesser or greater number of days.

	Day.	*Week-day.*	*a.*	*b.*	*c.*
(*Table I (Ind. Cal.), cols. 19-25*) .	81	2	75	430	264
(*Table IV, for 25 days*) . .	25	4	8466	907	68
	106	6	8541	337	332
(*Table VI. Equation b*)			260		
(*Table VII. Equation c*)			7		

(*Table VIII*). Tithi-index (t)=8808=Chait. kṛ. 12.

The day, measured from Jan. 1, was 106, which (*Table IX*) was 16 April 1490. The week-day, 6, was Friday. At mean sunrise that day the current tithi was Chaitra kṛishṇa 12. The nakshatra in which the true moon stood at that moment must now be found, also by the *Indian Calendar* rule.

$c \times 10$ . . .	3320
Constant (*Ind. Cal.*, §§ 135, 156)	+7207
	527
Less equatien c (*above*) . .	−7
Sun's true long., *s* . .	520
Tithi-index, *t* (*above*) . .	+8808
Nakshatra-index, *n* . . .	9328

With this value of *n* Table VIII shews that the true moon stood in the division of the heavens called " Uttara Bhadrapadā " : the date therefore was perfectly sound.

It will now be shewn how the elements of the date may be more closely verified ; and in the end it will be seen that according to the *Ārya-Siddhānta* the nakshatra-index was really 9322, while by the *Sūrya-Siddhānta* it was 9335. Though the differences here are not of great importance, it is manifest that in a close case they would be so, having the effect of placing the moon in a different nakshatra or of altering the number of the tithi current at sunrise, etc. The details of a date require careful examination whenever any final index is found to be close to the border-line between two tithis or two nakshatras or two signs of the zodiac.

A. *Elements of the same date. " c " eqn. c ", and " s " verified by the present Tables.* (i) *The Ārya-Siddhānta.* Before entering on this verification it is advisable to work out the details of the date by the special *Ārya-Siddhānta True System* Tables below (*Tables LXI-LXXV*).

	Day.	Week-day.	a.	b.	c.
(Talbe LXI, cols. 19-25) .	. 81	(2)	63·8714	433·0553	262·5194
(Table LXIV, 25 days) .	. 25	(4)	8465·7968	907·2906	68·4446
	106	(6)	8529·6682	340·3459	330·9640
(Table LXVIA. Eqn. b)			256·6185		
(Table LXVIIA. Eqn. c)			7·5676		
Table LXVIII. Tithi-index (*t*) . .			8793·8543 = Chaitra kṛ. 12.		
For the nakshatra—					
(Above) *c* × 10			3309·6400		
Constant			+ 7226·3542		
			535·9942		
Eqn. *c*			—7·5676		
Sun's true long., (*s*) . .			528·4266		
Tithi-index (*t*), above . .			+ 8793·8543		
Nakshatra-index (*n*) . .			9322·2809		

A close examination of the results thus found, for the sun's mean anom., his true long., and the solar equation of the centre, can be made by the present Tables (**XLVIIIA to L**) thus—

Sun's mean anom., c.

The day of the date was, serially, 106 (i.e. measured from January 1st). Table LXI, cols. 13-17, shews that true Mésha-saṁkrānti took place in the given year on Day 86 at 10^h 55^m after mean sunrise. 106—86 = 20. Turning to the entry for Day 20 (*Table XLVIIIA, col. 1*) it is seen (*col. 3*) that at 10^h 55^m after mean sunrise the sun's mean anom., *c*, was 3322·1148. Deduct from this the sun's mean motion in 10^h 55^m by Table XLIV, viz. for 10^h 11·4074, and for 55^m 1·0457, total 12·4531. Result for mean sunrise on Day 20, *c* = 3309·6617, or, as expressed in thousandths of circle instead of ten-thousandths, *c* = 330·9662.[1]

Sun's equation of centre and "eqn. c".

Table XLVIIIA, col. 7, shews that at 10^h 55^m after mean sunrise on Day 20 the equation of the sun's centre was 51·8996. On the previous day, i.e. exactly 24 hours earlier, it had been 52·3832. The 24-hour difference, therefore, was 0·4836. A 24th part of this is 0·02015. Taking 10^h 55^m as 11^h, which will be sufficiently close, we have the difference for 11^h (0·02015 × 11 =) 0·2216. 51·8996 + 0·2216 = 52·1212. This was the actual equation of the sun's centre at mean sunrise on the day of the date. In our method of calculation by the general Tables "equation *c*" is the amount of the sun's greatest equation of the centre less the actual equation. Here, the sun's greatest equation by the *Ārya-Siddhānta* being 59·6875, this amount less the actual equation, 52·1212, gives us "eqn. *c*" = 7·5663.[2]

Sun's true long., s.

Table XLVIIIA, col. 9, shews that at 10^h 55^m after mean sunrise on Day 20 the sun's true longitude "*s*" was 540·6811 in ten-thousandths of the circle. Deducting from this the sun's *true* motion on Day 20 (*Table XLIX, col. 6,*) for 10^h, viz. 11·2059, and for 55 minutes (*mean motion, Table L*) 1·0457, total 12·2516, we have for the sun's true long. "*s*" at mean sunrise, 528·4295.[3] If, desiring still greater accuracy, we had calculated for the sun's *true* motion in those 55^m instead of utilizing Table L which gives his *mean* motion in minutes, we should have found the result *s* = 528·4483.

[1] As against 330·9640 found by the general verification work carried out before.

[2] As against 7·5676 by the other process.

[3] As against 528·4266.

Another method for finding the value of "*s*" (when the value of the sun's mean anom. "*c*" and of the actual equation of the sun's centre are known) is the following. The sun's true long. "*s*" always = the long. of his perigee-point *plus* his mean anom. "*c*" *plus* or *minus* the actual equation of the centre. The long. of perigee-point according to the *Ārya-Siddhānta* is always 7166·6, in ten-thousandths of the circle. In the present case we have found "*c*" = 3309·6617 and the sun's equation (*plus*) 52·1212. Adding these three together and discarding one whole revolution (10,000) we have as result the sun's true long., "*s*" = 528·4495.

B. *The same elements of the date verified by the present Tables.* (*ii*) *The Sūrya-Siddhānta.*

The general results found by calculation by the ordinary process of the *Indian Calendar* have been given above in whole numbers. The indices found for mean sunrise on the day of the date were sun's mean anom., "*c*" = 332, "equation *c*" = 7, and sun's true long., "*s*" = 520. [Tables for the *Sūrya-Siddhānta* based on circle-measurement and enabling calculation to be made with several places of decimals have not yet been prepared; but the work can be carried out by Prof. Jacobi's Tables in Vol. I of the *Epigraphia Indica*, which are given in degrees, etc., the results being translated into circle-measurement by Table XLVA below.]

Sun's mean anom., c.

For verification of the results by the *Sūrya-Siddhānta* for the elements "*c*", "eqn. *c*", and "*s*" Table XLVIIIB is to be used just as Table XLVIIIA is used for the *Ārya-Siddhānta*. Table I, *Indian Calendar*, shews that the moment of tr[illegible]ēsha-saṁkrānti in the given year was 12ʰ 44ᵐ after mean sunrise on Day 86 (after [illegible] 1st). The day of the date was 106, and was 20 days after the day of true Mēsha-saṁkrānti. Table XLVIIIB gives us (col. 3) for the value of "*c*" at 12ʰ 44ᵐ after mean sunrise on Day 20 the figure 3341·6212 in ten-thousandths of circle. Deduct (Table XLIV) the sun's mean motion during 12 hours, 13·6889, and, for the same during 44ᵐ, 0·8365, total 14·5254. Result, "*c*" at mean sunrise on the given day, = 3327·0958, or in thousandths of circle 332·7096.

Sun's equation of centre and "eqn. c".

Table XLVIIIB, col. 7, shews that on Day 20 at 12ʰ 44ᵐ after mean sunrise the sun's equation of the centre was 52·3475. On the previous day it had been at the same hour, 52·8500. The 24-hour difference was 0·5025, the average diff. per hour being 0·0209. Not to be tediously critical we take 12ʰ 44ᵐ as 13 hours, and obtain the difference for 13 hours as 0·2722. This added to 52·3475 gives us for the sun's equation at mean sunrise 52·6197. This was the actual equation. The greatest equation of the centre by the *Sūrya-Siddhānta* is 60·4244. This less 52·6197 gives us the value of "eqn. *c*" as 7·8047.

Sun's true long., s.

From Table XLVIIIB it is also found (*col. 9*) that at 12ʰ 44ᵐ after mean sunrise on Day 20 (after true Mēsha-saṁkrānti) the sun's true longitude was 540·5000 in ten-thousandths of circle. Deducting from this, by Tables XLIX and L, the sun's true motion on that day for 12ʰ and 44ᵐ, viz. 13·4471 and 0·8365, total 14·2836, it is determined that the sun's true longitude at mean sunrise of the given day was 526·2164. [As shewn above a still more accurate result can be obtained by calculation for true motion in 44ᵐ instead of for mean motion by Table L; but there is not much to be gained by enlarging on this here.]

Worked by the second process, described above in the section relating to the *Ārya-Siddhānta* for finding the sun's true longitude the figures are—

Day 20. ☉'s mean anom. "*c*" (*above*)	3327·0958
Sūrya-Siddh. Long. of ☉'s perigee-point [1]	7146·5313
☉'s equation of the centre (*above*)	52·6197
☉'s true long., "*s*"	526·2468

[1] This was its value in A.D. 1400 (*see* § 254, *ii*, *below*). I have not thought it necessary here to take notice of the change in position of the point of the line of apsides which took place between A.D. 1400 (the base-year of Table XLVIIIB) and 1490, the year of the date under examination. The figure given, 7146·5313, is, in ten-thousandths of the circle, the longitude of the sun's perigee-point in A.D. 1400. In A.D. 1490 it was really 7146·6119.

If now we take these results in thousandths of the circle instead of ten-thousandths and in whole numbers, viz. " *c* " = 333, " equ. *c* " = 8, " *s* " = 526, and substitute them for the equivalent figures in the calculation made by the *Indian Calendar* system at the beginning of this section, it will be seen that by the *Sūrya-Siddhānta* the nakshatra-index, *n*, should be 9335 instead of 9328.

C. The Yōga. By either Siddhānta.—The formula for this is $2s+t$, and, as the value of " *s* " has been correctly found by the above process, no further remark is necessary.

D. The lagna. (i) *By the Ārya-Siddhānta.*—For this we have to find the correct value of " *s* " at mean sunrise in degrees, etc. By Table XLVIIIA, col. 8, the " *s* " for the day in our example above was 19° 27′ 52″·27. Deduct (*Tables XLIX, L*) for, on Day 20, 10 hours 24′ 12″·29, and for 55 minutes 2′ 15″·52, total 26′ 27″·81. Then the " *s* " for mean sunrise was 19° 1′ 24″·46. This was the true sun's longitude at that moment on the meridian of Ujjain. The given lagna was the sign Mithuna. The first point of this is 60°, the last 90°. We take the " *s* " as 19°, which is sufficiently exact. 60° − 19° = 41° and 90° − 19° = 71°. 41 × 4 = 164^m., or 2^h. 44^m. (90° − 19°) × 4 = 284^m, or 4^h 44^m. The first point of Mithuna was 41° distant from the true sun at the moment of mean sunrise, the last point 71°. Mithuna was lagna between 2^h 44^m and 4^h 44^m after mean sunrise on the given day.

(ii) *By the Sūrya-Siddhānta.* " *s* " = (*Table XLVIIIB*) 19° 27′. 28″·80. Deduct for 12 hours (*Tables XLIX, L*) 29′ 2″·74 and for 44 minutes 1′ 48″·42, total 30′ 51″·16. Remainder, or " *s* " for sunrise, 18° 56′ 37″·64. We may call this 19°, and come to the same result as in the former case. The lagna of Mithuna really began twelve seconds later.

(iii) *By the Indian Calendar process, and for both Siddhāntas.*—Here " *s* " was found to be in ten-thousandths, 520. Converted by Table VIIIB, this = 18° 45′. This was the sun's true longitude at mean sunrise. The difference between the actual time of the lagna of Mithuna and that found the *Indian Calendar* is slight.

More accurately worked, the first point of Mithuna was lagna by the *Ārya-Siddhānta* at 2^h 43^m 56^s, by the *Sūrya-Siddhānta* 2^h 44^m 16^s, and by the *Indian Calendar* 2^h 45^m, after mean sunrise on the day in question.

CONSTRUCTION OF THE TABLES.

A detailed explanation is here given of the construction of the principal Tables, in order to satisfy experts as to their accuracy.

249. *The Hindu Sine-Table.*—The *Sūrya-Siddhānta* (ii, 34) gives in minutes the sines of a series of angles, each separated from the other by 3° 45′, twenty-four of these completing the quarter-circle of 90°. These values stand, so far as I can ascertain, for all Indian authorities except the *Brahma-Sidhānta*, which assumes different sine-values. There is no need here to discuss their exact accuracy, as I am concerned solely with chronography as the handmaid of history, and have nothing whatever to do with the casting of horoscopes or any other branch of astrology. The sines, as used in calculations by authorities other than the *Brahma-Siddhānta*, are given in Table XLVII, col. 3, and the differences between them, in minutes, in col. 4. For astronomical purposes the several angles are angles of a planet's mean anomaly, and are so applied to the mean anomaly of both sun and moon.

250. *The equation of the centre.*—For the preparation of the sine and equation Table (XLVII) the equation of the sun's centre for each base-angle of anomaly has been calculated from its sine-value by the proper formula for each *Siddhānta*, the calculation being carried to nine

decimals of a second in order to insure absolute accuracy for the tabulated two decimals. The details for the *First Ārya-Siddhānta* (*Table XLVII, cols. 5, 6*) are complete in themselves; details for the other authorities are given in full in a supplementary Table (*XLVIIA*). Table XLVII differs a little, but only in one or two places, from Jacobi's Table XXIV (*Epig. Ind. I, 459*); I have, however, thought it advisable to record two decimals of seconds in all cases.

251. *Equation of centre.*—In Hindu astronomy the sun is treated as a planet, and in all planetary movement a fundamental principle (Jacobi, *Epig. Ind. I, 441*) is contained in the proportion—sin. equation : sin. mean anomaly : : minutes in the epicycle : minutes in the orbit. The minutes in the sin. anomaly are given in Table XLVII; the minutes in the epicycle are ascertained from statements made in each *Siddhānta*; the minutes in the orbit of 360° are always 21600′. The formula then for all authorities, a being the angle of mean anomaly, is:

$$\text{Equation centre}^{1} = \frac{\text{minutes in epicycle}}{21600'} \times \sin.\ a.$$

252 A. The *First Ārya-Siddhānta* gives for the dimension of the epicycle 13° 30′ or 810′. Hence by that authority:

$$\text{Equation centre} = \frac{810}{21600} \sin.\ a = \frac{3}{80} \sin.\ a.$$

Since there are 3° 45′ between each base-angle, the difference in minutes between each is 225′, and the measure of first or average difference of equation for each intermediate minute of anomaly is the difference between two consecutive equations divided by 225. Taken in seconds, this difference is given in col. 6. Multiply the minutes of difference between the base-angle and the given anomaly-angle by the amount given in col. 6, and, taking the result in seconds, apply it to the base-equation, and you have the correct equation for the given anomaly-angle.

For an example take the 2nd and 3rd sines. The 2nd sine, i.e. of anomaly-angle 7° 30′, is 449′. Multiply by 3 and divide by 80. Result 0° 16′ 50″·25.

The 3rd sine, of anomaly 11° 15′, is 671′. Multiply by 3 and divide by 80. Result 0° 25′ 9″·75.

The difference between the two results is 8′ 19″·50. This is the total difference in 225′ which is the difference between the two anomaly-angles. 8′ 19″·50 divided by 225 gives for each minute of angle the increment 2″·22.

B. *Equation of the centre by the Sūrya-Siddhānta.*—This calculation is made on the same fundamental principle.

The *Sūrya-Siddhānta* (cf. Jacobi, *above, I, 441*) assumes a contraction of the epicycle amounting to 20′ at the end of each of the odd quadrants. If this contraction at any point is called q, we have $q : 20' : : \sin.\ a : \sin.\ 90°$. $\therefore\ q = 20 \frac{\sin.\ a}{\sin.\ 90°}$. Sin. 90° = 3438′ (*Table XLVII*).

Hence $q = \frac{20'}{3438'} \sin.\ a$. The *Sūrya Siddhānta* gives for the dimension of the epicycle 14°

Hence the formula for the equation without the contraction would be $\frac{14'}{360'} \sin.\ a$. With the contraction it is $\frac{14'}{360'} \sin.\ a - \frac{20'}{3438' \times 21600'} \sin.^2 a$; or, finally $\frac{14}{360} \sin.\ a - \frac{\sin.^2 a}{3713040}$.

The best authorities agree that this is the correct formula.

[1] When an angle is very small, as is the case with even the greatest of the equation-angles, which is only about 2° 10′, the sine is taken to be equal to the arc. Hence the presumed equality in the text of "sin. equation" and "equation centre." Table XLVII shews that the sine of 3° 45′ is 225′, the same as the arc. The sine of 1° is 60′, also the same as the arc.

Each equation for the several base angles has been calculated by this formula and fully worked out for nine decimals of a second. The results are given in full in Table XLVIIA, col. 7, and in abbreviated form in Table XLVII, col. 7. The difference in equation per minute of anomaly-arc has been calculated by dividing the difference between consecutive base-equations in minutes by 225, and taking the result in seconds. This is tabulated in full in Table XLVIIA, col. 8, and in abbreviated form in Table XLVII, col. 8.

253 C. *Equation of the centre by the Second Ārya-Siddhānta and Siddhānta-Śirōmaṇi.*—The same fundamental principle holds good. The epicycle is (*Epig. Ind. I, 341*) 13° 40′ or 820′. There is no contraction. Minutes in the orbit, 21600′. Hence the equation is

$\frac{820}{21600}$ sin. a, or $\frac{41}{1080}$ sin. a. The entries are made in abbreviated form in Table XLVII, cols. 9, 10, and in full in Table XLVIIA, cols. 9, 10.

254. *The sun's mean anomaly, and the starting-point for its valuation.*—The sun's daily mean motion, i.e. his mean motion in 24 hours, is given according to the several Hindu authorities in Table XLIII, so that, given his exact mean place at the moment of true Mēsha-saṁkrānti when the true sun was at 0°, his mean position at the end of every 24-hour period is obtained by simple addition. We must, therefore, fix with great care the value of his mean anomaly when the true sun was at 0°.

(i) *By the First Ārya Siddhānta.*—S. B. Dikshit's valuation of the equation by this *Siddhānta*, 2° 6′ 59″·9421, was a trifle too great. Dr. Schram's, 2° 6′ 57″·323495, is exact down to the fifth decimal. M. de Ries with almost painful accuracy has carried it as far as sixteen decimals of a second. Tested by the sine table, his valuation is found exact. The equation (I give nine decimals of a second, the amount which I have generally used in these calculations) is + 2° 6′ 57″·323494885, or, in 10,000ths of the circle, 58·775644170. This is correct for the corresponding mean longitude value 357° 53′ 2″·676505115, or 357° 53′·044608419, or in 10,000ths of circle, 9941·224355830, the two added together amounting to exactly 360°. Thus, the perigee-point of the orbit being by this *Siddhānta* fixed at 258°, or, in 10,000ths of the circle, 7166·6, we have found the sun's mean anomaly at true Mēsha-saṁkrānti to have been 99° 53′ 2″·676505115 or 90° 53′·044608419, or in ten-thousandths of the circle, 2774·557689163 (i.e. 9941·224355830—7166·6). This then is our starting-point for cols. 2, 3, 4, 5, of Table XLVIIIA.

(ii) *By the Present Sūrya Siddhānta.*—In this case we have to deal with an authority which postulates a slight movement in the line of apsides of the sun's obrit, the apogee and perigee-points moving eastwards at the rate of 0″·1161 per ann.; and before working for a correct valuation of the sun's mean anomaly at true Mēsha-saṁkrānti in any year, we have first to decide which year to select as base of operations. I have chosen the year K. Y. 4500 or A.D. 1399-1400, roughly A.D. 1400, for reasons which follow. The period covered by Indian Epigraphy, the historical period, that is, of Indian History, may be taken as the period K.Y. 3500 to 5000, A.D. 400 to 1900, or the last 1500 years, the bulk of the inscriptions belonging to the last millennium K.Y. 4000 to 5000 or A.D. 900 to 1900. I take the central year of this millennium as my base. In K.Y. 4000 the perigee-point was at 257° 15′ 32″·4, and in K.Y. 5000 it was at 257° 17′ 28″·5. Hence in K.Y. 4500, say, A.D. 400, it was 257° 16′ 30″·45, or, in 10,000ths of the circle, 7146·53125.[1] The difference in the sun's equation of the centre and true longitude, caused by this shift of the apsin, is exceedingly small and may well be ignored.

For we are concerned only with the period A.D. 400 to 1900; and calculations by the equation-table on the value of the sun's mean anomaly at the beginning of the Hindu solar

[1] Actually, for nine decimals, 7146·531250000.

year A.D. 400-01 and at the beginning of A.D. 1900-01, allowing for the shift of the perigee-point, proves that the total difference in the equation in the whole period of 1500 years was 1″·0739. This constitutes also the total difference in the sun's true longitude, which is his mean longitude ± the equation, the mean longitude remaining the same whatever may be the shift in the line of apsides.

To assist those interested, however, I append a Table shewing the cumulative change of position of the apsidal points.

The annual shift is a forward one, and, as the longitude of perigee increases, so the mean anomaly decreases. Hence for years earlier than K.Y. 4500, A.D. 1400, the amounts entered in col. 3 must be added to, and for years later deducted from, the sun's mean anomaly as found by calculation.

Change of position of sun's apsidal points according to the Present Sūrya-Siddhānta.

No. of Years.	Change.		No. of Years.	Change.		No. of Years.	Change.		
1	2	3	1	2	3	1	2		3
	″	*10,000ths of circle.*		″	*10,000ths of circle.*		′	″	*10,000ths of circle.*
1	0·1161	0·0009	10	1·161	0·0090	100	0	11·61	0·0896
2	0·2322	0·0018	20	2·322	0·0179	200	0	23·22	0·1792
3	0·3483	0·0027	30	3·483	0·0269	300	0	34·83	0·2687
4	0·4644	0·0036	40	4·644	0·0358	400	0	46·44	0·3583
5	0·5805	0·0045	50	5·805	0·0448	500	0	58·05	0·4479
6	0·6966	0·0054	60	6·966	0·0537	600	1	9·66	0·5375
7	0·8127	0·0063	70	8·127	0·0627	700	1	21·27	0·6271
8	0·9288	0·0072	80	9·288	0·0717	800	1	32·88	0·7167
9	1·0449	0·0081	90	10·449	0·0806	900	1	44·49	0·8062
						1000	1	56·1	0·8958

255 Dr. Schram's valuation of the equation of the centre according to the *Ārya-Siddhānta* was proved to be so accurate that we need not have any hesitation in accepting his similar valuation of the same by the *Sūrya-Siddhānta*. He fixes this for K.Y. 4000 as 2° 8′ 18″·472169, and for K.Y. 5000 as 2° 8′ 19″·1842321. The equation, therefore, in K.Y. 4500, the base-year of my Table, was 2° 8′ 18″·828200553, or in ten-thousands of the circle 59·404538584.

The sun's mean anomaly at the moment of true Mêsha-saṁkrânti is 360° less the combined longitude of perigee and equation of centre, or 360 − (257° 16′ 30″·45 + 2° 8′

18″·828200553). The mean anomaly was therefore 100° 35′ 10″·721799447, or 100° 35′·178696657′ or in ten-thousandths of the circle 2794·064211415. This is the valuation which I have adopted for the starting-point for cols. 2, 3 of Table XLVIIIB.

The sun's mean longitude at the same moment, true Mēsha-saṁkrānti, is his mean anomaly *plus* the longitude of perigee, i.e. 100° 35′ 10″·721799447 + 257° 16′ 30″·45. It was, therefore, 357° 51′ 41″·171799447, or in ten-thousandths of the circle 9940·595461415. Table XLVIIIB, cols. 4, 5, start from this point.

256. In calculating the true sun's correct longitude and equation for each day for the preparation of Tables XLVIIIA and B I have obtained the equation by using the first or average difference in seconds as given in Table XLVII, cols. 6, 8, for each minute of anomaly-angle between the base-angle of the Table and the given angle, in the belief that this represents the practice of the Hindus in bygone centuries. It is possible to calculate with still greater minuteness. We might perhaps be able, by use of some complicated formula, to find out a more exact value of the difference in seconds applicable to the anomaly-angle under consideration; but this system would be so troublesome that it may be reasonably assumed to have never been adopted.

256 *a*. An example will best illustrate how each calculation for the 24-hour periods given in Tables XLVIIIA and XLVIIIB was made. The value of the equation is based on the angle of mean anomaly, "*c*" given in col. 2. The base-equation used is that for the base-angle next lower in the sine-table (*XLVII, col. 5 or 7*), the increment in the equation for the difference in angle between the base-angle and the given angle of anomaly being found by multiplying that difference in minutes and decimals by the amount given (*col. 6 or 8*) in seconds (this being the equation-difference per minute of anomaly-difference). The increment is added to or subtracted from the base-equations according as the consecutive base-equations are increasing or diminishing. The result is the exact equation for the given anomaly-angle, and this is entered in Table XLVIIIA or B, cols. 6, 7. This equation is added to or subtracted from the mean longitude of the sun (*Table XLVIIIA or B, cols. 4, 5*), and the result is the sun's true longitude, "*s*" (*cols. 8, 9*). The heading of the sine-Table (*cols. 2, 11*) shews whether the equation is *plus* or *minus*.

For an example I take Day 27 and work by the *Ārya-Siddhānta*, using only the number of decimals given in my Tables.

Mean anomaly (*Table XLVIIIA, col. 2*) . . .	126° 29′·72124
Next-lower base-anomaly (*Table XLVII, col. 2*) . .	−123 45
Difference	2° 44′·72124

2° 44′ = 164′. The multiplier per minute of difference is (*col. 6*) 1″·31.
164′·72124 × 1″·31 = 215·7848244. 215″ = 3′ 35″. Hence

Base equation for anomaly 123° 45′ (*Table XLVII, col. 5*)	1° 47′ 12″·75
Difference in equation above found, deducted because the values in col. 5 are diminishing	−3 35 ·7848244
Exact equation for given anomaly	1° 43′ 36″·9651756

Sun's mean longitude (*Table XLVIIIA, col. 4*) . .	24° 29′ 43″·27
Equation found (*for sign column-heading*) . .	+1 43 36·97
Exact value of sun's true longitude, "*s*"	26° 13′ 20″·24

This is converted into 10,000ths of the circle by Table XLVA, and both values are entered in cols. 8, 9, of Table XLVIII. Work by the other *Siddhāntas* is precisely the same, the base-equations and multipliers being used, each set for its own authority.

In this way every figure of equation and true longitude has been worked out for every day of the year.

In applying these results to inscription-dates we calculate the "*s*" for mean sunrise as described above, § 238.

If anyone should wish to calculate with a greater number of decimals than the four given in the principal Tables he can work as follows. In § 254 above I have given by both the *Siddhāntas*, with nine decimals of a second, the exact mean anomaly of the sun and mean longitude at true Mēsha-saṁkrānti each year. Add for the intervening days, i.e. from the day on which Mēsha-saṁkrānti occurred down to the day in question (included), the quantity obtained by multiplying the figures given for one day in Table XLIII by the number of intervening days. This gives, with eight decimals of a second, the value of mean anomaly and mean longitude for the day. In calculating for the equation note that the base-equations according to the *Ārya-Siddhānta* are complete as given in Table XLVII. They are given in full for the other authorities in Table XLVIIA.

TABLE XLIII.

MEAN MOTION OF THE SUN IN THE ECLIPTIC
according to the several Hindu astronomical authorities.
(Details worked out by M. Louis de Ries.)

Serial number as in Table XXXVII of "Indian Chronography."	Hindu authority.	Per Day of 24 hours.		Per Hour.	
		Parts of degrees.	10,000ths of circle.	Parts of degrees.	10,000ths of circle.
				′ ″	
5, 6	Original *Sūrya-Siddhānta*. Utpala's *Pauliśa-Siddhānta*.	59 8·16961,948	27·37785,2002	2 27·84040,081	1·14074,3833
7	*First Ārya-Siddhānta* (the *Āryabhaṭīya*).	59 8·17029,407	27·37785,7207	2 27·84042,892	1·14074,4050
8, 13	*Brahma-Siddhānta. Siddhānta-Śirōmaṇi.*	59 8·17265,515	27·37787,5426	2 27·84052,989	1·14074,4829
9	*Parāśara-Siddhānta* . .	59 8·17013,667	27·37785,5993	2 27·84042,236	1·14074,4000
10	*Second Ārya-Siddhānta* . .	59 8·17019,963	27·37785,6479	2 27·84042,498	1·14074,4020
11	*Rājamṛigāṅka* . . .	59 8·17019,064	27·37785,6409	2 27·84042,461	1·14074,4017
12	Present *Sūrya-Siddhānta* (*with or without the bīja*).	59 8·16955,652	27·37785,1516	2 27·84039,819	1·14074,3813

	Hindu authority.	Per minute.		Per second.	
		Parts of degrees.	10,000ths of circle.	Parts of degrees.	10,000ths of circle.
		″		″	
5, 6	Original *Sūrya-Siddhānta*. Utpala's *Pauliśa-Siddhānta*.	2·46400,6680	0·01901,2397	0·04106,6778	
7	*First Ārya-Siddhānta* (the *Āryabhaṭīya*).	2·46400,7149	0·01901,2401	0·04106,6786	
8, 13	*Brahma-Siddhānta. Siddhānta-Śirōmaṇi.*	2·46400,8788	0·01901,2414	0·04106,6813	
9	*Parāśara-Siddhānta*. . .	2·46400,7039	0·01901,24000	0·04106,6784	} 0·00031,6873
10	*Second Ārya-Siddhānta* . .	2·46400,7083	0·01901,24003	0·04106,6785	
11	*Rājamṛigāṅka* . . .	2·46400,7077	0·01901,24003	0·04106,6785	
12	Present *Sūrya-Siddhānta* (*with or without the bīja*).	2·46400,6636	0·01901,2397	0·04106,6777	

TABLE XLIV.

The sun's mean motion

per civil day of 24 hours, hour, minute and second, according to the *First Ārya-Siddhānta*, but generally applicable to all the Indian astronomical *Siddhāntas* (*see foot-note*).

Collective increase per civil day.			Collective increase per hour.			Collective increase per minute.						Collective increase per second.					
No	Degrees, etc.	10,000ths of circle.	No.	Degrees, etc.	10,000ths of circle.	No.	Degrees, etc.	10,000ths of circle.	No.	Degrees, etc.	10,000ths of circle.	No.	Degrees, etc.	10,000ths of circle.	No.	Degrees etc.	10,000ths of circle.
	° ′ ″			′ ″			′ ″			′ ″			″			″	
1	0 59 8·17	27·3779	1	2 27·84	1·1407	1	0 2·46	0·0190	31	1 16·38	0·5894	1	0·04	0·0000	31	1·27	0·0098
2	1 58 16·34	54·7557	2	4 55·68	2·2815	2	0 4·93	0·0380	32	1 18·85	0·6084	2	0·08	0·0006	32	1·31	0·0101
3	2 57 24·51	82·1336	3	7 23·52	3·4222	3	0 7·39	0·0570	33	1 21·31	0·6274	3	0·12	0·0010	33	1·36	0·0105
4	3 56 32·68	109·5114	4	9 51·36	4·5630	4	0 9·86	0·0760	34	1 23·78	0·6464	4	0·16	0·0013	34	1·40	0·0108
5	4 55 40·85	136·8893	5	12 19·20	5·7037	5	0 12·32	0·0951	35	1 26·24	0·6654	5	0·21	0·0016	35	1·44	0·0111
6	5 54 49·02	164·2671	6	14 47·04	6·8445	6	0 14·78	0·1141	36	1 28·70	0·6844	6	0·25	0·0019	36	1·48	0·0114
7	6 53 57·19	191·6450	7	17 14·88	7·9852	7	0 17·29	0·1331	37	1 31·17	0·7035	7	0·29	0·0022	37	1·52	0·0117
8	7 53 5·36	219·0229	8	19 42·72	9·1260	8	0 19·71	0·1521	38	1 33·63	0·7225	8	0·33	0·0025	38	1·56	0·0120
9	8 52 13·53	246·4007	9	22 10·56	10·2667	9	0 22·18	0·1711	39	1 36·10	0·7415	9	0·37	0·0029	39	1·60	0·0124
10	9 51 21·70	273·7786	10	24 38·40	11·4074	10	0 24·64	0·1901	40	1 38·56	0·7605	10	0·41	0·0032	40	1·64	0·0127
11	10 50 29·87	301·1564	11	27 6·24	12·5482	11	0 27·10	0·2091	41	1 41·02	0·7795	11	0·45	0·0035	41	1·68	0·0130
12	11 49 38·04	328·5343	12	29 34·09	13·6889	12	0 29·57	0·2281	42	1 43·49	0·7985	12	0·49	0·0038	42	1·72	0·0133
13	12 48 46·21	355·9121	13	32 1·93	14·8297	13	0 32·03	0·2472	43	1 45·95	0·8175	13	0·53	0·0041	43	1·77	0·0136
14	13 47 54·38	383·2900	14	34 29·77	15·9704	14	0 34·50	0·2662	44	1 48·42	0·8365	14	0·57	0·0044	44	1·81	0·0139
15	14 47 2·55	410·6679	15	36 57·61	17·1112	15	0 36·96	0·2852	45	1 50·88	0·8556	15	0·62	0·0048	45	1·85	0·0143

16	15 46 10·72	438·0457	16	39 25·45	18·2519	16	0 39·42	0·3042	46	1 53·34	0·8746	16	0·66	0·0051	46	1·89	0·0146
17	16 45 18·90	465·4236	17	41 53·29	19·3926	17	0 41·89	0·3232	47	1 55·81	0·8936	17	0·70	0·0054	47	1·93	0·0149
18	17 44 27·07	492·8014	18	44 21·13	20·5334	18	0 44·35	0·3422	48	1 58·27	0·9126	18	0·74	0·0057	48	1·97	0·0152
19	18 43 35·24	520·1793	19	46 48·97	21·6741	19	0 46·82	0·3612	49	2 0·74	0·9316	19	0·78	0·0060	49	2·01	0·0155
20	19 42 43·41	547·5571	20	49 16·81	22·8149	20	0 49·28	0·3802	50	2 3·20	0·9506	20	0·82	0·0063	50	2·05	0·0158
21	20 41 51·58	574·9350	21	51 44·65	23·9556	21	0 51·74	0·3993	51	2 5·66	0·9696	21	0·86	0·0067	51	2·09	0·0162
22	21 40 59·75	602·3129	22	54 12·49	25·0964	22	0 54·21	0·4183	52	2 8·13	0·9886	22	0·90	0·0070	52	2·14	0·0165
23	22 40 7·92	629·6907	23	56 40·33	26·2371	23	0 56·67	0·4373	53	2 10·59	1·0077	23	0·94	0·0073	53	2·18	0·0168
24	23 39 16·09	657·0686				24	0 59·14	0·4563	54	2 13·06	1·0267	24	0·99	0·0076	54	2·22	0·0171
25	24 38 24·26	684·4464				25	1 1·60	0·4753	55	2 15·52	1·0457	25	1·03	0·0079	55	2·26	0·0174
26	25 37 32·43	711·8243				26	1 4·06	0·4943	56	2 17·98	1·0647	26	1·07	0·0082	56	2·30	0·0177
27	26 36 40·60	739·2021				27	1 6·53	0·5133	57	2 20·45	1·0837	27	1·11	0·0086	57	2·34	0·0181
28	27 35 48·77	766·5800				28	1 8·99	0·5323	58	2 22·91	1·1027	28	1·15	0·0089	58	2·38	0·0184
29	28 34 56·94	793·9579				29	1 11·46	0·5514	59	2 25·38	1·1217	29	1·19	0·0092	59	2·42	0·0187
30	29 34 5·11	821·3357				30	1 13·92	0·5704				30	1·23	0·0095			
31	30 33 13·28	848·7136															
100	98 33 37·03	2·37·7857															
200	197 7 14·06	5475·5714															
300	295 40 51·09	8213·3572															
365	359 44 42·16	9992·9179															

Note.—The Table figures are calculated by the *First Ārya-Siddhānta.* The difference between these and the same according to the *Present Sūrya-Siddhānta, Parāśara* and *Second Ārya-Siddhāntas* and the *Rājamṛigāṅka* is negligeable. For the total of 365 days according to the *Brahma-Siddhānta* and *Siddhānta-Śirōmaṇi* the difference amounts to 0″·86 or (in 10,000ths of the circle) 0·0066, by which these are greater than the figures given, their total for 365 days being 359°44′43″·02 or (in 10,000ths of the circle) 9992·9245. It is not necessary for historical purposes to trouble about the *Original Sūrya-* or *Pauliśa-Siddhānta.* Any one desiring to do so can calculate them from Table XLIII.

TABLE XLIVA.

LONGITUDE OF SUN'S APSIS (PERIGEE) AND EQUATION OF CENTRE at different millenniums, according to the Hindu standard authorities.

[*Position of apsis is given according to Jacobi*, Epig. Ind. *I, 440, 450 ; the equation has been calculated by Dr. Schram.*]

First *Ārya-Siddhānta* (*Āryabhaṭīya*).

Kali-yuga.	Christian year (roughly)	Long. of sun's apsis (perigee). °	′	″	Sun's equation of centre at true Mēsha-saṁkrānti. °	′	″
	B.C.						
0	3100						
1000	2100						
2000	1100						
3000	100	258	0	0	2	6	57·323494885
	A.D.						
4000	900						
5000	1900						

Present *Sūrya-Siddhānta*.

Kali-yuga.	Christian year (roughly)	Long. of sun's apsis (perigee). °	′	″	Sun's equation of centre at true Mēsha-saṁkrānti. °	′	″
	B.C.						
0	3100	257	7	48·0	2	8	15·623883311
1000	2100	257	9	44·1	2	8	16·335959734
2000	1100	257	11	40·2	2	8	17·048032824
3000	100	257	13	36·3	2	8	17·760102582
	A.D.						
4000	900	257	15	32·4	2	8	18·472169007
5000	1900	257	17	28·5	2	8	19·184232099

Brahma-Siddhānta.

Kali-yuga.	Christian year (roughly)	°	′	″	°	′	″
	B.C.						
0	3100	257	45	36	2	8	26·527631345
1000	2100	257	48	0	2	8	27·432241607
2000	1100	257	50	24	2	8	28·336851869
3000	100	257	52	48	2	8	29·241462132
	A.D.						
4000	900	257	55	12	2	8	30·146072394
5000	1900	257	57	36	2	8	31·050682657

Siddhānta-Śirōmaṇi.

Kali-yuga.	Christian year (roughly)	°	′	″	°	′	″
	B.C.						
0	3100	257	45	36	2	8	26·527631345
1000	2100	258	3	0	2	8	33·086055747
2000	1100	258	20	24	2	8	39·644480150
3000	100	258	37	48	2	8	46·202904552
	A.D.						
4000	900	258	55	12	2	8	52·761328955
5000	1900	259	12	36	2	8	59·319753357

Second Ārya-Siddhānta.

Kali-yuga.	Christian year (roughly)	°	′	″	°	′	″
	B.C.						
0	3100	257	45	36·0	2	8	26·527631345
1000	2100	257	47	54·3	2	8	27·396434118
2000	1100	257	50	12·6	2	8	28·265236890
3000	100	257	52	30·9	2	8	29·134039663
	A.D.						
4000	900	257	54	49·2	2	8	30·002842436
5000	1900	257	57	7·5	2	8	30·871645200

TABLE XLVA.

FOR CONVERSION OF DEGREES, MINUTES AND SECONDS INTO MEASUREMENT BY 10,000THS OF THE CIRCLE.

Degrees (°)						Minutes (′)				Seconds (″)				Decimals of seconds (″)	
No.	10,000ths of circle.	No.	10,000ths of circle.	No.	10,000ths of circle.	No.	10,000ths of circle.	No.	10,000ths of circle	No.	10,000ths of circle.	No.	10,000ths of circle.	No.	10,000ths of circle.
1	27·7	46	1277·7	91	2527·7	1	0·4629	31	14·3518	1	0·007,716,049	31	0·239,197,531	0·1	0·000,771,605
2	55·5	47	1305·5	92	2555·5	2	0·925	32	14·814	2	0·015,432,099	32	0·246,913,580	0·2	0·001,543,210
3	83·3	48	1333·3	93	2583·3	3	1·38	33	15·27	3	0·023,148	33	0·254,629	0·3	0·002.3148
4	111·1	49	1361·1	94	2611·1	4	1·851	34	15·740	4	0·030,864,197	34	0·262,345,679	0·4	0·003,086,420
5	138·8	50	1388·8	95	2638·8	5	2·3148	35	16·2037	5	0·038,580,247	35	0·270,061,729	0·5	0·003,858,025
6	166·6	51	1416·6	96	2666·6	6	2·7	36	16·6	6	0·046,29	36	0·27	0·6	0·004,629
7	194·4	52	1444·4	97	2694·4	7	3·2407	37	17·1296	7	0·054,012,346	37	0·285,493,827	0·7	0·005,401,235
8	222·2	53	1472·2	98	2722·2	8	3·703	38	17·592	8	0·061,728,395	38	0·293,200,877	0·8	0·006,172,840
9	250	54	1500	99	2750	9	4·16	39	18·05	9	0·0694	39	0·300,925	0·9	0·006,94
10	277·7	55	1527·7	100	2777·7	10	4·629	40	18·518	10	0·077,160,494	40	0·308,641,975		
11	305·5	56	1555·5	110	3055·5	11	5·0925	41	18·9814	11	0·084,876,543	41	0·316,358,025		
12	333·3	57	1583·3	120	3333·3	12	5·5	42	19·4	12	0·0925	42	0·324,07		
13	361·1	58	1611·1	130	3611·1	13	6·0185	43	19·9074	13	0·100,308,642	43	0·331,790,124		
14	388·8	59	1638·8	140	3888·8	14	6·481	44	20·370	14	0·108,024,691	44	0·339,506,173		
15	416·6	60	1666·6	150	4166·6	15	6·94	45	20·83	15	0·115740	45	0·3472		
16	444·4	61	1694·4	160	4444·4	16	7·407	46	21·296	16	0·123,456,790	46	0·354,938,272		
17	472·2	62	1722·2	170	4722·2	17	7·8703.	47	21·7592	17	0·131,172,840	47	0·362,654, 321		
18	500	63	1750	180	5000	18	8·3	48	22·2	18	0·138	48	0·370		
19	527·7	64	1777·7	190	5277·7	19	8·7962	49	22·6851	19	0·146,604,938	49	0·378,086,420		
20	555·5	65	1805·5	200	5555·5	20	9·259	50	23·148	20	0·154,320,988	50	0·385,802,469		
21	583·3	66	1833·3	210	5833·3	21	9·72	51	23·61	21	0·162,037	51	0·393,518		
22	611·1	67	1861·1	220	6111·1	22	10·185	52	24·074	22	0·169,753,086	52	0·401,234,568		
23	638·8	68	1888·8	230	6388·8	23	10·6481	53	24·5370	23	0·177,469,136	53	0·408,950,617		
24	666·6	69	1916·6	240	6666·6	24	11·1	54	25	24	0·185	54	0·416		
25	694·4	70	1944·4	250	6944·4	25	11·5740	55	25·4629	25	0·192,901,235	55	0·424,382,716		
26	722·2	71	1972·2	260	7222·2	26	12·037	56	25·925	26	0·200,617,284	56	0·432,098,765		
27	750	72	2000	270	7500	27	12·5	57	26·38	27	0·2083	57	0·439,814		
28	777·7	73	2027·7	280	7777·7	28	12·962	58	26·851	28	0·216,049,383	58	0·447,530,864		
29	805·5	74	2055·5	290	8055·5	29	13·4259	59	27·3148	29	0·223,765,432	59	0·455,246,914		
30	833·3	75	2083·3	300	8333·3	30	13·8	60	27·7	30	0·231,48	60	0·4629		
31	861·1	76	2111·1	310	8611·1										
32	888·8	77	2138·8	320	8888·8										
33	916·6	78	2166·6	330	9166·6										
34	944·4	79	2194·4	340	9444·4										
35	972·2	80	2222·2	350	9722·2										
36	1000	81	2250	360	10,000										
37	1027·7	82	2277·7												
38	1055·5	83	2305·5												
39	1083·3	84	2333·3												
40	1111·1	85	2361·1												
41	1138·8	86	2388·8												
42	1166·6	87	2416·6												
43	1194·4	88	2444·4												
44	1222·2	89	2472·2												
45	1250	90	1250												

TABLE XLVII.

For conversion of measurement by 10,000ths of the circle into measurement by degrees, minutes and seconds (° ′ ″).

10,000ths of circle.	° ′	10,000ths of circle.	° ′
1000	36 0	100	3 36
2000	72 0	200	7 12
3000	108 0	300	10 48
4000	144 0	400	14 24
5000	180 0	500	18 0
6000	216 0	600	21 36
7000	252 0	700	25 12
8000	288 0	800	28 48
9000	324 0	900	32 24
10,000	360 0	1000	36 0

10,000ths of circle.	°	′	″	10,000ths of circle.	°	′	″	10,000ths of circle.	°	′	″
1	0	2	9·6	41	1	28	33·6	81	2	54	57·6
2	0	4	19·2	42	1	30	43·2	82	2	57	7·2
3	0	6	28·8	43	1	32	52·8	83	2	59	16·8
4	0	8	38·4	44	1	35	2·4	84	3	1	26·4
5	0	10	48·0	45	1	37	12·0	85	3	3	36·0
6	0	12	57·6	46	1	39	21·6	86	3	5	45·6
7	0	15	7·2	47	1	41	31·2	87	3	7	55·2
8	0	17	16·8	48	1	43	40·8	88	3	10	4·8
9	0	19	26·4	49	1	45	50·4	89	3	12	14·4
10	0	21	36·0	50	1	48	0·0	90	3	14	24·0
11	0	23	45·6	51	1	50	9·6	91	3	16	33·6
12	0	25	55·2	52	1	53	19·2	92	3	18	43·2
13	0	28	4·8	53	1	54	28·8	93	3	20	52·8
14	0	30	14·4	54	1	56	38·4	94	3	23	2·4
15	0	32	24·0	55	1	58	48·0	95	3	25	12·0
16	0	34	33·6	56	2	0	57·6	96	3	27	21·6
17	0	36	43·2	57	2	3	7·2	97	3	29	31·2
18	0	38	52·8	58	2	5	16·8	98	3	31	40·8
19	0	41	2·4	59	2	7	26·4	99	3	33	50·4
20	0	43	12·0	60	2	9	36·0	100	3	36	0·0
21	0	44	21·6	61	2	11	45·6				
22	0	46	31·2	62	2	13	55·2				
23	0	48	40·8	63	2	16	4·8				
24	0	50	50·4	64	2	18	14·4				
25	0	54	0·0	65	2	20	24·0				
26	0	56	9·6	66	2	22	33·6				
27	0	58	19·2	67	2	24	43·2				
28	1	0	28·8	68	2	26	52·8				
29	1	2	38·4	69	2	29	2·4				
30	1	4	48·0	70	2	31	12·0				
31	1	6	57·6	71	2	33	21·6				
32	1	9	7·2	72	2	35	31·2				
33	1	11	16·8	73	2	37	40·8				
34	1	13	26·4	74	2	39	50·4				
35	1	15	36·0	75	2	42	0·0				
36	1	17	45·6	76	2	44	9·6				
37	1	19	55·2	77	2	46	19·2				
38	1	22	4·8	78	2	48	28·8				
39	1	24	14·4	79	2	50	38·4				
40	1	26	24·0	80	2	52	48·0				

Decimals of units.

(10,000ths of circle.)

Unit.	′ ″	Unit.	″
0·1	0 12·96	0·01	1·296
0·2	0 25·92	0·02	2·592
0·3	0 38·88	0·03	3·888
0·4	0 51·84	0·04	5·184
0·5	1 4·80	0·05	6·480
0·6	1 17·76	0·06	7·776
0·7	1 30·72	0·07	9·072
0·8	1 43·68	0·08	10·368
0·9	1 56·64	0·09	11·664

For every successive decimal of unit move the decimal point of seconds one place to left.

TABLE XLVI.

INDICES OF NAKSHATRAS AND YOGAS.

To take, for close work, the place of Table VIII, cols. 6 to 13, of the "*Indian Calendar.*"

NAKSHATRA.								YOGA.		
Serial number.	Name.	Ending point by the **Equal-space system.**		Ending point by the system of **Garga.**		Ending point by the **Brahma-Siddhānta.**		Serial number.	Name.	Ending point.
		° ′	10,000ths of circle.	° ′	10,000ths of circle.	° ′ ″	10,000ths of circle.			
1	2	3	4	5	6	7	8	9	10	11
1	Aśvinī*	13 20	370·370	13 20	370·370	13 10 35	366·0108	1	Vishkambha	The ending point is the same as in the case of the equal-space nakshatra (number by number) as given in cols. 3, 4.
2	Bharaṇī	26 40	740·740	20 0	555·5	19 45 52½	549·0051	2	Prīti	
3	Kṛittikā	40 0	1111·1	33 20	925·925	32 56 27½	915·0270	3	Āyushmat	
4	Rōhiṇī	53 20	1481·481	53 20	1481·481	52 42 20	1474·0432	4	Saubhāgya	
5	Mṛigaśiras	66 40	1851·851	66 40	1851·851	65 52 55	1830·0540	5	Sōbhana	
6	Ārdrā	80 0	2222·2	73 20	2037·037	72 28 12½	2013·0594	6	Atigaṇḍa	
7	Punarvasu	93 20	2592·592	93 20	2592·592	92 14 5	2562·0756	7	Sukarman	
8	Pushya	106 40	2962·962	106 40	2962·962	105 24 40	2928·0864	8	Dhṛiti	
9	Aślēshā	120 0	3333·3	113 20	3148·148	111 59 57½	3111·0918	9	Śūla	
10	Maghā	133 20	3703·703	126 40	3518·518	125 10 32½	3477·1026	10	Gaṇḍa	
11	Pūrva Phalgunī	146 40	4074·074	[illegible] 0	3888·8	138 21 7½	3843·1134	11	Vṛiddhi	
12	Uttara Phalgunī	160 0	4444·4	160 0	4444·4	158 7 0	4392·1296	12	Dhruva	
13	Hasta	173 20	4814·814	173 20	4814·814	171 17 35	4758·1404	13	Vyāghāta	
14	Chitrā	186 40	5185·185	186 40	5185·185	184 28 10	5124·1512	14	Harshaṇa	
15	Svāti	200 0	5555·5	193 20	5370·370	191 3 27½	5307·1566	15	Vajra	
16	Viśākhā	213 20	5925·925	213 20	5925·925	210 49 20	5856·1728	16	Siddhi *or* Asṛij.	
17	Anurādhā	226 40	6296·296	226 40	6296·296	223 59 55	6222·1836	17	Vyatīpāta	
18	Jyēshṭhā	240 0	6666·6	233 20	6481·481	230 35 12½	6405·1890	18	Varīyas	
19	Mūla	253 20	7037·037	246 40	6851·851	243 45 47½	6771·1998	19	Parigha	
20	Pūrva Āshāḍhā	266 40	7407·407	260 0	7222·2	256 56 22½	7137·2106	20	Śiva	
21	Uttara Āshāḍhā	280 0	7777·7	280 0	7777·7	276 42 15	7686·2269	21	Siddha	
	Abhijit†	...	...	...	...	280 56 30	7803·9352		...	
22	Śravaṇa	293 20	8148·148	293 20	8148·148	294 7 5	8169·9460	22	Sādhya	
23	Dhanishṭhā or Sravishṭhā.	306 40	8518·518	306 40	8518·518	307 17 40	8535·9568	23	Śubha	
24	Śatabhishaj or Śatatārakā.	320 0	8888·8	313 20	8703·703	313 52 57½	8718·9622	24	Śukla	
25	Pūrva Bhadra-padā.	333 20	9259·259	326 40	9074·074	327 3 32½	9084·9730	25	Brahman	
26	Uttara Bhadra-padā.	346 40	9629·629	346 40	9629·629	346 49 25	9633·9892	26	Indra	
27	Rēvatī	360	10,000	360	10,000	360	10,000	27	Vaidhṛiti	

* Aśvinī begins at 0° by all systems.

† Though properly speaking there is no Abhijit in the equal-space system in ordinary use, sometimes it is referred to as a secondary detail. When this is the case, it has the same limits as fixed by the *Brahma-Siddhānta viz.*, 276° 42′15″ to 285° 56′ 30″, or, in 10,000ths of the circle, 7686·2269 to 7303·9352.

TABLE XLVII.

HINDU SINES, AND EQUATIONS OF SUN'S CENTRE.

N. B. *i.*—The sines, col. 3, stand, it is believed, for all authorities except the *Brahma-Siddhānta* (for this last see Table LXXXIX below).

" *ii.*—" Equation + " or " — " means that the amount of the equation, added to or subtracted from the sun's mean long., gives his true or apparent long.

" *iii.*—This Table is assimilated to that of Prof. Jacobi (*Epig. Ind.*, I. 459).

" *iv.*—*First Ārya* figures are exact. For fuller details see next Table.

Serial number of sine	Sun's mean anomaly. Equation +		Sine of mean anom. angle. Minutes.	Diff.	Sun's equation of the centre according to the First Ārya-Siddhānta. Base-equation.	Diff. per min. of anom.	Present Sūrya-Siddhānta. Base-equation.	Diff. per min. of anom.	Second Ārya- and Siddhānta Sirómaṇi. Base-equation.	Diff. per min. of anom.	Sun's mean anomaly. Equation —		Serial number of sine
1	2		3	4	5	6	7	8	9	10	11		1
	° ′	° ′	′	′	° ′ ″	″	° ′ ″	″	° ′ ″	″	° ′	° ′	
0	0 0	180 0	0		0 0 0		0 0 0		0 0 0		180 0	360 0	0
				225		2·25		2·33		2·28			
1	3 45	176 15	225		0 8 26·25		0 8 44·18		0 8 32·50		183 45	356 15	1
				224		2·24		2·31		2·27			
2	7 30	172 30	449		0 16 50·25		0 17 24·41		0 17 2·72		187 30	352 30	2
				222		2·22		2·28		2·25			
3	11 15	168 45	671		0 25 9·75		0 25 58·39		0 25 28·39		191 15	348 45	3
				219		2·19		2·2[illegible]		2·22			
4	15 0	165 0	890		0 33 22·50		0 34 23·87		0 33 47·22		195 0	345 0	4
				215		2·15		2·20		2·18			
5	18 45	161 15	1105		0 41 26·25		0 42 38·60		0 41 56·94		198 45	341 15	5
				210		2·10		2·14		2·12			
6	22 30	157 30	1315		0 49 18·75		0 50 40·39		0 49 55·28		202 30	337 30	6
				205		2·05		2·08		2·08			
7	26 15	153 45	1520		0 57 0·0		0 58 29·33		0 57 42·22		206 15	333 45	7
				199		1·99		2·02		2·01			
8	30 0	150 0	1719		1 4 27·75		1 6 3·25		1 5 15·50		210 0	330 0	8
				191		1·91		1·93		1·93			
9	33 45	146 15	1910		1 11 37·50		1 13 17·72		1 12 30·56		213 45	326 15	9
				183		1·83		1·85		1·85			
10	37 30	142 30	2093		1 18 29·25		1 20 12·88		1 19 27·39		217 30	322 30	10
				174		1·74		1·75		1·76			
11	41 15	138 45	2267		1 25 0·75		1 26 46·62		1 26 3·72		221 15	318 45	11
				164		1·64		1·65		1·66			
12	45 0	135 0	2431		1 31 9·75		1 32 56·84		1 32 17·28		225 0	315 0	12
				154		1·54		1·54		1·56			
13	48 45	131 15	2585		1 36 56·25		1 38 43·69		1 38 8·06		228 45	311 15	13
				143		1·43		1·43		1·45			
14	52 30	127 30	2728		1 42 18·0		1 44 4·96		1 43 33·78		232 30	307 30	14
				131		1·31		1·31		1·33			
15	56 15	123 45	2859		1 47 12·75		1 48 58·92		1 48 32·17		236 15	303 45	15
				119		1·19		1·18		1·20			
16	60 0	120 0	2978		1 51 40·50		1 53 25·36		1 53 3·22		240 0	300 0	16
				106		1·06		1·05		1·07			
17	63 45	116 15	3084		1 55 39·0		1 57 22·31		1 57 4·67		243 45	296 15	17
				93		0·93		0·92		0·94			
18	67 30	112 30	2177		1 59 8·25		2 0 49·90		2 0 36·50		247 30	292 30	18
				79		0·79		0·78		0·80			
19	71 15	108 45	3256		2 2 6·0		2 3 46·02		2 3 36·44		251 15	288 45	19
				65		0·65		0·64		0·66			
20	75 0	105 0	3321		2 4 32·25		2 6 10·78		2 6 4·50		255 0	285 0	20
				51		0·51		0·50		0·52			
21	78 45	101 15	3372		2 6 27·0		2 8 4·26		2 8 0·67		258 45	281 15	21
				37		0·37		0·37		0·37			
22	82 30	97 30	3409		2 7 50·25		2 9 26·54		2 9 24·94		262 30	277 30	22
				22		0·22		0·22		0·22			
23	86 15	93 45	3431		2 8 39·75		2 10 15·44		2 10 15·06		266 15	273 45	23
				7		0·07		0·07		0·07			
24	90 0	90 6	3438		2 8 55·50		2 10 31·0		2 10 31·0		270 0	270 0	24

TABLE XLVIIA.

(Supplementary to the Sine and Equation Table).

Giving fuller details of the entries in Table XLVII, cols. 7, 8, 9, 10, *viz.*, base-equations and differences per minute of arc, for use in close calculation, according to—

(i) The *Present Sūrya-Siddhānta*,

(ii) The *Second Ārya-Siddhānta* and *Siddhānta-Śirōmaṇi.*

Serial number of sine.	Present Sūrya-Siddhānta.				2nd Ārya-Siddhānta and Siddhānta-Śirōmaṇi.			
	Base-equation.			Diff. per minute of anom. arc.	Base-equation			Diff. per minute of anom. arc.
1	7			8	9			10
	°	′	″	″	°	′	″	″
0	0	0	0·0		0	0	0·0	
				2·3297				2·2777
1	0	8	44·18193720		0	8	32·5	
				2·3121				2·2677
2	0	17	24·40894254		0	17	2·72	
				2·2844				2·24740
3	0	25	58·39110270		0	25	28·38	
				2·2466				2·21703
4	0	34	23·86691232		0	33	47·2	
				2·1988				2·1765
5	0	42	38·60246580		0	41	56·94	
				2·1413				2·1215
6	0	50	40·39032702		0	49	55·27	
				2·0842				2·0753
7	0	58	29·33229918		0	57	42·2	
				2·0174				2·0146
8	1	6	3·25		1	5	15·5	
				1·9310				1·9336
9	1	13	17·71604934		1	12	30·5	
				1·8452				1·8526
10	1	20	12·87859542		1	19	27·38	
				1·7500				1·76148
11	1	26	46·61953014		1	26	3·72	
				1·6454				1·6602
12	1	32	56·83576962		1	32	17·27	
				1·5416				1·5590
13	1	38	43·68681726		1	38	8·05	
				1·4279				1·4477
14	1	44	4·95633636		1	43	33·7	
				1·3065				1·3262
15	1	48	58·91608494		1	48	32·16	
				1·1842				1·2047
16	1	53	25·35847716		1	53	3·2	
				1·0531				1·0731
17	1	57	22·30831878		1	57	4·6	
				0·9226				0·94148
18	2	0	49·89921462		2	0	36·5	
				0·7828				0·7988
19	2	3	46·02029604		2	3	36·4	
				0·6434				0·6580
20	2	6	10·77879576		2	6	4·5	
				0·5044				0·5163
21	2	8	4·26294360		2	8	0·6	
				0·3657				0·3746
22	2	9	26·54196564		2	9	24·94	
				0·2173				0·2727
23	2	10	15·44365260		2	10	15·05	
				0·0691				0·0709
24	2	10	31·0		2	10	31·0	

N. B.—In col. 9 under ″ (seconds) and opposite lines Nos. 1, 8, 18, 20, the last figure, "·5" is not, like the rest, a recurring decimal.

TABLE XLVIII A.

ELEMENTS OF THE SUN'S LONGITUDE FOR THE HINDU SOLAR YEAR
according to the **First Arya-Siddhanta.**
in periods of 24-hours each from the moment of true Mesha-saṁkrānti,
the astronomical beginning of the solar year.
(Exact for all years.)
[True longitude = mean longitude + equation of centre.]

1st Ārya-Siddhānta.

24-hour periods from true Mēsha-saṁkrānti.	Sun's mean anomaly (or mean sun's distance from perigee-point) ("c").			Sun's mean longitude.				Sun's equation of the centre. +				Sun's true longitude ("s").			
1	2		3	4			5	6			7	8			9
	°	′	10,000ths of circle.	°	′	″	10,000ths of circle.	°	′	″	10,000ths of circle.	°	′	″	10,000ths of circle.
(The sun's equation of the centre is + till his mean anomaly reaches 180°).															
At moment of true Mēsha-saṁkrānti.	*99*	*53·04461*	*2774·5577*	*357*	*53*	*2·68*	*9941·2244*	*2*	*6*	*57·32*	*58·7756*	*0*	*0*	*0·0*	*0·0*
1	100	52·18078	2801·[illegible]355	358	52	10·85	9968·6022	2	6	35·44	58·6068	0	58	46·29	27·2090
2	101	51·31695	2829·3134	359	51	19·02	9995·9801	2	6	8·48	58·3987	1	57	27·50	54·3788
3	102	50·45312	2856·6913	0	50	27·19	23·3579	2	5	38·32	58·1660	2	56	5·51	81·5240
4	103	49·58929	2884·0691	1	49	35·36	50·7358	2	5	8·16	57·9333	3	54	43·52	108·6691
5	104	48·72547	2911·4470	2	48	43·53	78·1136	2	4	38·00	57·7006	4	53	21·53	135·8143
6	105	47·86164	2938·8248	3	47	51·70	105·4915	2	4	1·14	57·4162	5	51	52·84	162·9077
7	106	46·99781	2966·2027	4	46	59·87	132·8694	2	3	22·70	57·1196	6	50	22·57	189·9890
8	107	46·13398	2993·5805	5	46	8·04	160·2472	2	2	44·26	56·8230	7	48	52·30	217·0702
9	108	45·27015	3020·9584	6	45	16·21	187·6251	2	2	5·79	56·5261	8	47	22·00	244·1512
10	109	44·40632	3048·3363	7	44	24·38	215·0029	2	1	19·07	56·1657	9	45	43·45	271·1686
11	110	43·54250	3075·7141	8	43	32·55	242·3808	2	0	32·35	55·8052	10	44	4·90	298·1860
12	111	42·67867	3103·0920	9	42	40·72	269·7586	1	59	45·63	55·4447	11	42	26·35	325·2033
13	112	41·81484	3130·4698	10	41	48·89	297·1365	1	58	57·27	55·0715	12	40	46·15	352·2080
14	113	40·95101	3157·8477	11	40	57·06	324·5144	1	58	2·26	54·6471	13	38	59·33	379·1615
15	114	40·08718	3185·2255	12	40	5·23	351·8922	1	57	7·27	54·2228	14	37	12·50	406·1150

1	2		3	4			5	6			7	8			9
16	115	39·22335	3212·6034	13	39	13·40	379·2701	1	56	12·27	53·7984	15	35	25·67	433·0685
17	116	38·35952	3239·9813	14	38	21·57	406·6479	1	55	14·24	53·3506	16	33	35·81	459·9985
18	117	37·49570	3267·3591	15	37	29·74	434·0258	1	54	11.55	52·8669	17	31	41·30	486·8927
19	118	36·63187	3294·7370	16	36	37·91	461·4036	1	53	8·87	52·3832	18	29	46·78	513·7869
20	119	35·76804	3322·1148	17	35	46·08	488·7815	1	52	6·19	51·8996	19	27	52·27	540·6811
21	120	34·90421	3349·4927	18	34	54·25	516·1594	1	50	58·96	51·3808	20	25	53·22	567·5402
22	121	34·04038	3376·8705	19	34	2·42	543·5372	1	49	48·59	50·8379	21	23	51·01	594·3751
23	122	33·17655	3404·2484	20	33	10·59	570·9151	1	48	38·22	50·2949	22	21	48·81	621·2100
24	123	32·31273	3431·6263	21	32	18·76	598·2929	1	47	27·85	49·7519	23	19	46·61	648·0448
25	124	31·44890	3459·0041	22	31	26·93	625·6708	1	46	11·90	49·1659	24	17	38·84	674·8367
26	125	30·58507	3486·3820	23	30	35·10	653·0486	1	44	54·43	48·5682	25	15	29·54	701·6168
27	126	29·72124	3513·7598	24	29	43·27	680·4265	1	43	36·97	47·9704	26	13	20·24	728·3969
28	127	28·85741	3541·1377	25	28	51·44	707·8044	1	42	19·50	47·3727	27	11	10·94	755·1770
29	128	27·99358	3568·5155	26	27	59·62	735·1822	1	40	55·07	46·7212	28	8	54·68	781·9034
30	129	27·12976	3595·8934	27	27	7·79	762·5601	1	39	30·50	46·0687	29	6	38·29	808·6288
At true Vrichabha-saṅk.	*130*	*21·79443*	*3621·2011*	*28*	*21*	*47·67*	*787·8678*	*1*	*38*	*12·33*	*45·4655*	*30*	*0*	*0·0*	*833·5*
31	130	26·26593	3623·2713	28	26	15·96	789·9379	1	38	5·94	45·4162	30	4	21·90	835·3541
32	131	25·40210	3650·6491	29	25	24·13	817·3158	1	36	40·23	44·7548	31	2	4·36	862·0706
33	132	24·53827	3678·0270	30	24	32·30	844·6936	1	35	19·16	44·0522	31	59	41·46	888·7458
34	133	23·67444	3705·4048	31	23	40·47	872·0715	1	33	38·09	43·3495	32	57	18·56	915·4210
35	134	22·81061	3732·7827	32	22	48·64	899·4494	1	32	7·02	42·6468	33	54	55·66	942·0961
36	135	21·94678	3760·1605	33	21	56·81	926·8272	1	30	33·76	41·9271	34	52	30·56	968·7544
37	136	21·08296	3787·5384	34	21	4·98	954·2051	1	28	56·77	41·1788	35	50	1·75	995·3839
38	137	20·21913	3814·9163	25	20	13.15	981·5829	1	27	19·79	40·4305	36	47	32·94	1022·0134
39	138	19·35530	3842·2941	36	19	21·32	1008·9608	1	25	42·81	39·6821	37	45	4·13	1048·6429
40	139	18·49174	3869·6720	37	18	29·49	1036·3386	1	24	2·47	38·9080	38	42	31·96	1075·2466
41	140	17·62764	3897·0498	38	17	37·66	1063·7165	1	22	19·58	38·1140	39	39	57·24	1101·8305
42	141	16·76381	3924·4277	39	16	45·83	1091·0944	1	20	36·68	37·3201	40	37	22·51	1128·4144
43	142	15·89999	3951·8055	40	15	54·00	1118·4722	1	18	53·78	36·5261	41	34	47·78	1154·9983
44	143	15·03616	3979·1834	41	15	2·17	1145·8501	1	17	6·83	35·7009	42	32	9·00	1181·5510
45	144	14·17233	4006·5613	42	14	10·34	1173·2279	1	15	18·61	34·8659	43	29	28·95	1208·0938
46	145	13·30850	4033·9391	43	13	18·51	1200·6058	1	13	30·40	34·0308	44	26	48·91	1234·6366
47	146	12·44467	4061·3170	44	12	26·68	1227·9836	1	11	42·18	33·1958	45	24	8·86	1261·1794
48	147	11·58084	4088·6948	45	11	34·85	1255·3615	1	9	49·43	32·3259	46	21	24·28	1287·6874
49	148	10·71702	4116·0727	46	10	43·02	1282·7394	1	7	56·48	31·4543	47	18	39·50	1314·1937
50	149	9·85319	4143·4505	47	9	51·19	1310·1172	1	6	3·53	30·5828	48	15	54·72	1340·7000

TABLE XLVIII A—*Contd.*

24-hour periods from true Mêsha-saṁkrânti.	Sun's mean anomaly (or mean sun's distance from perigee-point) ("c").		Sun's mean longitude.		Sun's equation of the centre. +		Sun's true longitude ("s").	
1	2	3	4	5	6	7	8	9
	° ′	10,000ths of circle.	° ′ ″	10,000ths of circle.	° ′ ″	10,000ths of circle.	° ′ ″	10,000ths of circle.
51	150 8·98936	4170·8284	48 8 59·36	1337·4951	1 4 9·86	29·7057	49 13 9·22	1367·2008
52	151 8·12553	4198·2063	49 8 7·53	1364·8729	1 2 12·18	28·7977	50 10 19·71	1393·6706
53	152 7·26170	4225·5841	50 7 15·70	1392·2508	1 0 14·50	27·8897	51 7 30·20	1420·1404
54	153 6·39787	4252·9620	51 6 23·87	1419·6286	0 58 16·82	26·9817	52 4 40·69	1446·6103
55	154 5·53404	4280·3398	52 5 32·04	1447·0065	0 56 17·91	26·0641	53 1 49·95	1473·0706
56	155 4·67022	4307·7177	53 4 40·21	1474·3844	0 54 16·68	25·1287	53 58 56·89	1499·5130
57	156 3·80639	4335·0955	54 3 48·38	1501·7622	0 52 15·45	24·1933	54 56 3·83	1525·9555
58	157 2·94256	4362·4734	55 2 56·55	1529·1401	0 50 14·22	23·2579	55 53 10·77	1552·3979
59	158 2·07873	4389·8513	56 2 4·72	1556·5179	0 48 11·38	22·3101	56 50 16·11	1578·8280
60	159 1·21490	4417·2291	57 1 12·89	1583·8958	0 46 7·20	21·3518	57 47 20·09	1605·2476
61	160 0·35107	4444·6070	58 0 21·06	1611·2736	0 44 3·01	20·3936	58 44 24·08	1631·6673
62	160 59·48725	4471·9848	58 59 29·23	1638·6515	0 41 58·83	19·4354	59 41 28·06	1658·0869
At true Mithuna-saṁkrânti.	*161 18·69490*	*4480·8772*	*59 18 41·69*	*1647·5139*	*0 41 18·31*	*19·1227*	*60 0 0·0*	*1666·6*
63	161 58·62342	4499·3627	59 58 37·41	1666·0294	0 39 52·46	18·4603	60 38 29·86	1684·4897
64	162 57·75959	4526·7406	60 57 45·58	1693·4072	0 37 45·32	17·4793	61 35 30·89	1710·8865
65	163 56·89576	4554·1184	61 56 53·75	1720·7851	0 35 38·17	16·4983	62 32 31·92	1737·2833
66	164 56·03193	4581·4963	62 56 1·92	1748·1629	0 33 31·03	15·5172	63 29 32·95	1763·6801
67	165 55·16810	4608·8741	63 55 10·09	1775·5408	0 31 21·68	14·5191	64 26 31·77	1790·0599
68	166 54·30428	4636·2520	64 54 18·26	1802·9186	0 29 12·17	13·5199	65 23 30·43	1816·4385
69	167 53·44045	4663·6298	65 53 26·43	1830·2965	0 27 2·67	12·5206	66 20 29·09	1842·8171
70	168 52·57662	4691·0077	66 52 34·60	1857·6744	0 24 52·93	11·5195	67 17 27·53	1869·1939
71	169 51·71279	4718·3856	67 51 42·77	1885·0522	0 22 41·65	10·5065	68 14 24·41	1895·5588
72	170 50·84896	4745·7634	68 50 50·94	1912·4301	0 20 30·37	9·4936	69 11 21·30	1921·9236

1	2	3	4	5	6	7	8	9
73	171 49·98513	4773·1413	69 49 59·11	1939·8079	0 18 19·08	8·4806	70 8 18·19	1948·2885
74	172 49·12130	4800·5191	70 49 7·28	1967·1858	0 16 7·42	7·4646	71 5 14·70	1974·6504
75	173 48·25748	4827·8970	71 48 15·45	1994·5636	0 13 54·95	6·4425	72 2 10·40	2001·0062
76	174 47·39365	4855·2748	72 47 23·62	2021·9415	0 11 42·49	5·4204	72 59 6·11	2027·3619
77	175 46·52982	4882·6527	73 46 31·79	2049·3194	0 9 30·02	4·3983	73 56 1·81	2053·7177
78	176 45·66599	4910·0306	74 45 39·96	2076·6972	0 7 17·25	3·3739	74 52 57·21	2080·0711
79	177 44·80216	4937·4084	75 44 48·13	2104·0751	0 5 4·20	2·3472	75 49 52·32	2106·4223
80	178 43·93833	4964·7863	76 43 56·30	2131·4529	0 2 51·14	1·3205	76 46 47·44	2132·7734
81	179 43·07451	4992·1641	77 43 4·47	2158·8308	0 0 38·08	0·2938	77 43 42·55	2159·1246
Sun at 78° (apogee)	*180 0·0*	*5000·0*	*78 0 0·0*	*2166·6*	*0 0 0·0*	*0·0*	*78 0 0·0*	*2166·6*

(Sun's equation of centre is minus,—, after his mean anomaly is 180° till it reaches 360° or 0°.)

1	2	3	4	5	6 Sun's equation of the centre	7	8	9
82	180 42·21068	5019·5420	78 42 12·64	2186·2086	0 1 34·97	0·7328	78 40 37·67	2185·4758
83	181 41·34685	5046·9198	79 41 20·81	2213·5865	0 3 48·03	1·7595	79 37 32·78	2211·8270
84	182 40·48302	5074·2977	80 40 28·98	2240·9644	0 6 1·09	2·7862	80 34 27·89	2238·1782
85	183 39·61919	5101·6756	81 39 37·15	2268·3422	0 8 14·14	3·8128	81 31 23·01	2264·5294
86	184 38·75536	5129·0534	82 38 45·32	2295·7201	0 10 26·66	4·8354	82 28 18·66	2290·8847
87	185 37·89153	5156·4313	83 37 53·49	2323·0979	0 12 39·13	5·8575	83 25 14·37	2317·2405
88	186 37·02771	5183·8091	84 37 1·66	2350·4758	0 14 51·59	6·8796	84 22 10·07	2343·5962
89	187 36·16388	5211·1870	85 36 9·83	2377·8536	0 17 3·93	7·9007	85 19 5·90	2369·9529
90	188 35·30005	5238·5648	86 35 18·00	2405·2315	0 19 15·22	8·9137	86 16 2·79	2396·3178
91	189 34·43622	5265·9427	87 34 26·17	2432·6094	0 21 26·50	9·9267	87 12 59·67	2422·6827
92	190 33·57239	5293·3206	88 33 34·34	2459·9872	0 23 37·78	10·9397	88 9 56·56	2449·0476
93	191 32·70856	5320·6984	89 32 42·51	2487·3651	0 25 48·53	11·9485	89 6 53·98	2475·4165
At true Karka-saṁkrânti.	*192 27·82045*	*5 46·2132*	*90 27 49·23*	*2512·8798*	*0 27 49·22*	*12·8798*	*90 0 0·0*	*2500·0*
94	192 31·84474	5384·0763	90 31 50·68	2514·7429	0 27 58·04	12·9478	90 3 52·64	2501·7951
95	193 30·98091	5375·4541	91 30 58·85	2542·1208	0 30 7·55	13·9471	91 0 51·31	2528·1737
96	194 30·11708	5402·8320	92 30 7·02	2569·4986	0 32 17·06	14·9464	91 57 49·97	2554·5522
97	195 29·25325	5430·2098	93 29 15·20	2596·8765	0 34 25·39	15·9367	92 54 49·80	2580·9398
98	196 28·38942	5457·5877	94 28 23·37	2624·2544	0 36 32·54	16·9177	93 51 50·83	2607·3366

TABLE XXVIII A—*Contd.*

1st Ārya-Siddhānta.

24-hour periods from true Mêsha-saṁkrânti.	Sun's mean anomaly or (mean sun's distance from perigee-point) ("c").		Sun's mean longitude.		Sun's equation of the centre. —		Sun's true longitude ("s").	
1	2	3	4	5	6	7	8	9
	° ′	10,000ths of circle.	° ′ ″	10,000ths of circle.	° ′ ″	10,000ths of circle.	° ′ ″	10,000ths of circle.
99	197 27·52559	5484·9656	95 27 31·54	2651·6322	0 38 39·68	17·8988	94 48 51·86	2633·7335
100	198 26·66177	5512·3434	96 26 39·71	2679·0101	0 40 46·82	18·8798	95 45 52·88	2660·1303
101	199 25·79794	5539·7213	97 25 47·88	2706·3879	0 42 51·93	19·8451	96 42 55·95	2686·5428
102	200 24·93411	5567·0991	98 24 56·05	2733·7658	0 44 56·11	20·8033	97 39 59·93	2712·9625
103	201 24·07028	5594·4770	99 24 4·22	2761·1436	0 47 0·30	21·7616	98 37 3·92	2739·3821
104	202 23·20645	5621·8548	100 23 12·39	2788·5215	0 49 4·48	22·7198	99 34 7·90	2765·8017
105	203 22·34262	5649·2327	101 22 20·56	2815·8994	0 51 6·05	23·6578	100 31 14·51	2792·2416
106	204 21·47879	5676·6106	102 21 28·73	2843·2772	0 53 7·28	24·5932	101 28 21·45	2818·6840
107	205 20·61497	5703·9884	103 20 36·90	2870·6551	0 55 8·51	25·5286	102 25 28·30	2845·1264
108	206 19·75114	5731·3663	104 19 45·07	2898·0329	0 57 9·45	26·4618	103 22 35·61	2871·5711
109	207 18·88731	5758·7441	105 18 53·24	2925·4108	0 59 7·14	27·3699	104 19 46·10	2898·0409
110	208 18·02348	5786·1220	106 18 1·41	2952·7886	1 1 4·82	28·2779	105 16 56·59	2924·5107
111	209 17·15965	5813·4998	107 17 9·58	2980·1665	1 3 2·50	29·1859	106 14 7·08	2950·9805
112	210 16·29582	5840·8777	108 16 17·75	3007·5444	1 4 58·88	30·0839	107 11 18·87	2977·4605
113	211 15·43200	5868·2556	109 15 25·92	3034·9222	1 6 51·83	30·9554	108 8 34·09	3003·9668
114	212 14·56817	5895·6334	110 14 34·09	3062·3901	1 8 44·78	31·8270	109 5 49·31	3030·4731
115	213 13·70434	5923·0113	111 13 42·26	3089·6779	1 10 37·73	32·6985	110 3 4·54	3056·9794
116	214 12·84051	5950·3891	112 12 50·43	3117·0558	1 12 28·45	33·5528	111 0 21·98	3083·5030
117	215 11·97668	5977·7670	113 11 58·60	3144·4336	1 14 16·67	34·3879	111 57 41·93	3110·0458
118	216 11·11285	6005·1448	114 11 6·77	3171·8115	1 16 4·89	35·2229	112 55 1·88	3136·5886
119	217 10·24902	6032·5227	115 10 14·94	3199·1894	1 17 53·11	36·0579	113 52 21·83	3163·1314
120	218 9·38520	6059·9006	116 9 23·11	3226·5672	1 19 37·78	36·8656	114 49 45·33	3189·7016
121	219 8·52137	6087·2784	117 8 31·28	3253·9451	1 21 20·68	37·6595	115 47 10·61	3216·2855
122	220 7·65754	6114·6563	118 7 39·45	3281·3229	1 23 3·57	38·4535	116 44 35·88	3242·8694
123	221 6·79371	6142·0341	119 6 47·62	3308·7008	1 24 46·47	39·2475	117 42 1·15	3269·4533
124	222 5·92988	6169·4120	120 5 55·79	3336·0786	1 26 24·27	40·0021	118 39 31·52	3296·0765
125	223 5·06605	6196·7898	121 5 3·96	3363·4565	1 28 1·26	40·7505	119 37 2·71	3322·7061

1	2		3	4			5	6			7	8			9
At true Simha-samkrānti.	*223*	*28·66604*	*6207·7158*	*121*	*28*	*39·96*	*3374·3824*	*1*	*28*	*39·59*	*41·0491*	*120*	*0*	*0·9*	*3333·3*
126	224	4·20223	6224·1677	122	4	12·13	3390·8344	1	29	38·24	41·4988	120	34	33·89	3349·3356
127	225	3·33840	6251·5456	123	3	20·30	3418·2122	1	31	14·89	42·2445	121	32	5·41	3375·9677
128	226	2·47457	6278·9234	124	2	28·47	3445·5901	1	32	45·96	42·9472	122	29	42·51	3402·6429
129	227	1·61074	6306·3013	125	1	[illegible]	3472·9679	1	34	17·03	43·6499	123	27	19·61	3429·3180
130	228	0·74691	6333·6791	126	0	44·31	3500·3458	1	35	48·10	44·3526	124	24	56·71	3455·9932
131	228	59·88308	6361·0570	126	59	52·99	3527·7236	1	37	17·53	45·0427	125	22	35·45	3482·6810
132	229	59·01926	6388·4348	127	59	1·16	3555·1015	1	38	42·10	45·6952	126	20	19·06	3509·4063
133	230	58·15543	6415·8127	128	58	9·33	3582·4794	1	40	6·66	46·3477	127	18	2·66	3536·1317
134	231	57·29160	6443·1906	129	57	17·50	3609·8572	1	41	31·23	47·0002	128	15	46·27	3562·8570
135	232	56·42777	6470·5684	130	56	25·67	3637·2351	1	42	52·62	47·6282	129	13	33·05	3589·6068
136	233	55·56394	6497·9463	131	55	33·84	3664·6129	1	44	10·09	48·2251	130	11	23·75	3616·3869
137	234	54·70011	6525·3241	132	54	42·01	3691·9908	1	45	27·56	48·8238	131	9	14·45	3643·1670
138	235	53·83628	6552·7020	133	53	50·18	3719·3687	1	46	45·03	49·4215	132	7	5·15	3669·9472
139	236	52·97246	6580·0798	134	52	58·35	3746·7465	1	47	57·94	49·9841	133	5	0·41	3696·7624
140	237	52·10863	6607·4577	135	52	6·52	3774·1244	1	49	8·31	50·5271	134	2	58·21	3723·5973
141	238	51·24480	6634·8356	136	51	14·69	3801·5022	1	50	18·68	51·0701	135	0	56·01	3750·4322
142	239	50·38097	6662·2134	137	50	22·86	3828·8801	1	51	29·05	51·6131	135	58	53·80	3777·2670
143	240	49·51714	6689·5913	138	49	31·03	3856·2579	1	52	32·99	52·1064	136	56	58·04	3804·1515
144	241	48·65331	6716·9691	139	48	39·20	3883·6358	1	53	35·67	52·5901	137	55	3·53	3831·0457
145	242	47·78949	6744·3470	140	47	47·37	3911·0137	1	54	38·36	53·0737	138	53	9·01	3857·9399
146	243	46·92566	6771·7248	141	46	55·54	3938·3915	1	55	41·04	53·5574	139	51	14·50	3884·8341
147	244	46·06183	6799·1027	142	46	3·71	3965·7694	1	56	35·79	53·9798	140	49	27·92	3911·7895
148	245	45·19800	6826·4806	143	45	11·88	3993·1472	1	57	30·78	54·4042	141	47	41·10	3938·7430
149	246	44·33417	6853·8584	144	44	20·05	4020·5251	1	58	25·78	54·8286	142	45	54·27	3965·6965
150	247	43·47034	6881·2363	145	43	28·22	4047·9029	1	59	18·89	55·2384	143	44	9·33	3992·6646
151	248	42·60652	6908·6141	146	42	36·39	4075·2808	2	0	5·61	55·5988	144	42	30·78	4019·6820
152	249	41·74269	6935·9920	147	41	44·56	4102·6587	2	0	52·33	55·9593	145	40	52·23	4040·6993
153	250	40·87886	6963·3698	148	40	52·73	4130·0365	2	1	39·04	56·3198	146	39	13·69	4073·7167
154	251	40·01503	6990·7477	149	40	0·90	4157·4144	2	2	22·26	56·6532	147	37	38·64	4100·7611
155	252	39·15120	7018·1256	150	39	9·07	4148·7922	2	3	0·70	56·9498	148	36	8·37	4127·8424
156	253	38·28737	7045·5034	151	38	17·24	4212·1701	2	3	39·14	57·2464	149	34	38·11	4154·9237

TABLE XLVIII A—*Contd.*

24-hour periods from true Mēsha-saṁkrānti.	Sun's mean anomaly (or mean sun's distance from perigee-point) ("c").			Sun's mean longitude.				Sun's equation of the centre. —				Sun's true longitude ("s").			
1	2		3	4			5	6			7	8			9
	°	′	10,000ths of circle.	°	′	″	10,000ths of circle.	°	′	″	10,000ths of circle.	°	′	″	10,000ths of circle.
At true Kanyā-saṁkrānti.	*254*	*3·93008*	*7057·3751*	*152*	*3*	*55·80*	*4224·0417*	*2*	*3*	*55·80*	*57·3751*	*150*	*0*	*0·0*	*4166·6*
157	254	37·42354	7072·8813	152	37	25·41	4239·5479	2	4	17·58	57·5430	150	33	7·84	4182·0049
158	255	36·55972	7100·2591	153	36	33·58	4266·9258	2	4	50·90	57·8001	151	31	42·69	4209·1257
159	256	35·69589	7127·6370	154	35	41·75	4294·3037	2	5	21·05	58·0328	152	30	20·70	4236·2708
160	257	34·83206	7155·0148	155	34	49·92	4321·6815	2	5	51·21	58·2655	153	28	58·71	4263·4160
161	258	33·96823	7182·3927	156	33	58·09	4349·0594	2	6	21·37	58·4983	154	27	36·72	4290·5611
162	259	33·10440	7209·7706	157	33	6·26	4376·4372	2	6	44·80	58·6790	155	26	21·47	4317·7582
163	260	32·24057	7237·1484	158	32	14·43	4403·8151	2	7	6·68	58·8478	156	25	7·76	4344·9672
164	261	31·37675	7264·5263	159	31	22·60	4431·1929	2	7	28·56	59·0167	157	23	54·05	4372·1763
165	262	30·51292	7291·9041	160	30	30·78	4458·5708	2	7	50·36	59·1849	158	22	40·41	4399·3859
166	263	29·64909	7319·2820	161	29	38·95	4485·9487	2	8	3·37	59·2853	159	21	35·57	4426·6634
167	264	28·78526	7346·6598	162	28	47·12	4513·3265	2	8	16·38	59·3857	160	20	30·73	4453·9408
168	265	27·92143	7374·0377	163	27	55·29	4540·7044	2	8	29·39	59·4861	161	19	25·89	4481·2183
169	266	27·05760	7401·4156	164	27	3·46	4568·0822	2	8	40·59	59·5725	162	18	22·86	4508·5097
170	267	26·19377	7428·7934	165	26	11·63	4595·4601	2	8	44·73	59·6044	163	17	26·89	4535·8557
171	268	25·32995	7456·1713	166	25	19·80	4622·8379	2	8	48·87	59·6364	164	16	30·92	4563·2016
172	269	24·46612	7483·5491	167	24	27·97	4650·2158	2	8	53·01	59·6682	165	15	34·95	4590·5475
173	270	23·60229	7510·9270	168	23	36·14	4677·5937	2	8	53·85	59·6748	166	14	42·29	4617·9189
174	271	22·73846	7538·3048	169	22	44·31	4704·9715	2	8	49·71	59·6428	167	13	54·60	4645·3287
175	272	21·87463	7565·6827	170	21	52·48	4732·3494	2	8	45·57	59·6109	168	13	6·91	4672·7385
176	273	21·01080	7593·0606	171	21	0·65	4759·7272	2	8	41·43	59·5789	169	12	19·22	4700·1483
177	274	20·14698	7620·4384	172	20	8·82	4787·1051	2	8	32·02	59·5063	170	11	36·80	4747·5988
178	275	19·28315	7647·8163	173	19	16·99	4814·4829	2	8	19·01	59·4059	171	10	57·93	4755·0770
179	276	18·41932	7675·1941	174	18	25·16	4841·8608	2	8	6·00	59·3055	172	10	19·16	4782·5553
180	277	17·55549	7702·5720	175	17	33·33	4869·2387	2	7	52·99	59·2052	173	9	40·34	4810·0335
181	278	16·69166	7729·9498	176	16	41·50	4896·6165	2	7	32·97	59·0507	174	9	8·53	4837·5658

1	2		3	4			5	6			7	8			9
182	279	15·82783	7757·3277	177	15	49·67	4923·9944	2	7	11·09	58·8819	175	8	38·58	4865·1125
183	280	14·96401	7784·7056	178	14	57·84	4951·3722	2	6	49·21	58·7131	176	8	8·63	4892·6592
184	281	14·10918	7812·0834	179	14	6·01	4978·7501	2	6	27·33	58·5442	177	7	38·68	4920·2058
185	282	13·23635	7839·4613	180	13	14·18	5006·1279	2	5	57·30	58·3125	178	7	16·88	4947·8154
186	283	12·37252	7866·8391	181	12	22·35	5033·5058	2	5	27·14	58·0798	179	6	55·21	4975·4260
At true Tulâ saṁkrânti	*284*	*5·00496*	*7891·2069*	*182*	*5*	*0·30*	*5057·8727*	*2*	*5*	*0·30*	*57·8727*	*180*	*0*	*0·0*	*5000·0*
187	284	11·50869	7894·2170	182	11	30·52	5060·8837	2	4	56·98	57·8471	180	6	33·54	5003·0366
188	285	10·64486	7921·5948	183	10	38·69	5088·2615	2	4	25·33	57·6029	181	6	13·36	5030·6586
189	286	9·78103	7948·9727	184	9	46·86	5115·6394	2	3	46·89	57·3063	182	5	59·97	5058·3331
190	287	8·91721	7976·3506	185	8	55·03	5143·0172	2	3	8·45	57·0097	183	5	46·58	5086·0076
191	288	8·05338	8003·7284	186	8	3·20	5170·3951	2	2	30·02	56·7131	184	5	33·19	5113·6820
192	289	7·18955	8031·1063	187	7	11·37	5197·7729	2	1	48·47	56·3925	185	5	22·90	5141·3804
193	290	6·32572	8058·4841	188	6	19·54	5225·1508	2	1	1·75	56·0320	186	5	17·79	5169·1188
194	291	5·46189	8085·8620	189	5	27·71	5252·5287	2	0	15·04	55·6716	187	5	12·68	5196·8571
195	292	4·59806	8113·2398	190	4	35·88	5279·9065	1	59	28·32	55·3111	188	5	7·57	5224·5954
196	293	3·73423	8140·6177	191	3	44·05	5307·2844	1	48	36·88	54·9142	189	5	7·18	5252·3702
197	294	2·87041	8167·9956	192	2	52·22	5334·6622	1	57	41·88	54·4898	190	5	10·34	5280·1724
198	295	2·00658	8195·3734	193	2	0·39	5362·0401	1	56	46·88	54·0655	191	5	13·51	5307·9746
199	296	1·14275	8222·7513	194	1	8·57	5389·4178	1	55	51·89	53·6411	192	5	16·68	5335·7768
200	297	0·27892	8250·1291	195	0	16·74	5416·7958	1	54	51·00	53·1713	193	5	25·73	5363·6245
201	297	59·41509	8277·5070	195	59	24·91	5444·1737	1	53	48·32	52·6877	194	5	36·59	5391·4860
202	298	58·55127	8304·8848	196	58	33·08	5471·5515	1	52	45·64	52·2040	195	5	47·44	5419·3475
203	299	57·68744	8332·2627	197	57	41·25	5498·9294	1	51	42·95	51·7203	196	5	58·29	5447·2091
204	300	56·82361	8359·6406	198	56	49·42	5526·3072	1	50	32·88	51·1796	197	6	16·54	5475·1276
205	301	55·95978	8387·0184	199	55	57·59	5553·6851	1	49	22·51	50·6366	198	6	35·08	5503·0484
206	302	55·09595	8414·3963	200	55	5·76	5581·0629	1	48	12·14	50·0936	199	6	53·62	5530·9693
207	303	54·23212	8441·7741	201	54	13·93	5608·4408	1	47	0·66	49·5421	200	7	13·27	5558·8987
208	304	53·36829	8469·1520	202	53	22·10	5635·8187	1	45	43·19	48·9443	201	7	38·91	5586·8743
209	305	52·50447	8496·5298	203	52	30·27	5663·1965	1	44	25·72	48·3466	202	8	4·55	5614·8499
210	306	51·64064	8523·9077	204	51	38·44	5690·5744	1	43	8·25	47·7488	203	8	30·19	5642·8255
211	307	50·7768[illegible]	8551·2856	205	50	46·61	5717·9522	1	41	48·29	47·1319	204	8	58·32	5670·8204
212	308	49·91298	8578·6634	206	49	54·78	5745·3301	1	40	23·72	46·4794	205	9	31·05	5698·8507
213	309	49·04915	8606·0413	207	49	2·95	5772·7079	1	38	59·16	45·8268	206	10	3·79	5726·8811
214	310	48·18532	8633·4191	208	48	11·12	5800·0858	1	37	34·59	45·1743	207	10	36·52	5754·9115
215	311	47·32150	8660·7970	209	47	19·29	5827·4637	1	36	6·47	44·4944	208	11	12·81	5782·9693
216	312	46·45767	8688·1748	210	46	27·46	5854·8415	1	34	35·41	43·7917	209	11	52·05	5811·0498

TABLE XLVIII A—*Contd.*

1st Ârya Siddhânta.

24-hour periods from true Mêsha-sâmkranti.	Sun's mean anomaly (or mean sun's distance from perigee-point) ("*g*").		Sun's mean Longitude.		Sun's equation of the centre. —		Sun's true Longitude ("*s*"). ("*c*").	
1	2	3	4	5	6	7	8	9
	° '	10,000ths of circle.	° ' "	10,000ths of circle.	° ' "	10,000ths of circle.	° ' "	10,000ths of circle.
At true Vṛiśchika-saṁk.	*313 33·38561*	*8709·9007*	*211 33 23·14*	*5876·5674*	*1 33 23·14*	*43·2341*	*210 0 0·0*	*5833·3*
217	313 45·59384	8715·5527	211 45 35·63	5882·2194	1 33 4·34	43·0890	210 12 31·29	5839·1304
218	314 44·73001	8742·9306	212 44 43·80	5909·5972	1 31 33·27	42·3863	211 13 10·53	5867·2109
219	315 43·86618	8770·3084	213 43 51·97	5936·9751	1 29 57·81	41·6498	212 13 54·16	5895·3253
220	316 43·00235	8797·6863	214 43 0·14	5964·3529	1 28 20·83	40·9014	213 14 39·32	5923·4515
221	317 42·13853	8825·0641	215 42 8·31	5991·7308	1 26 43·84	40·1531	214 15 24·47	5951·5777
222	318 41·27470	8852·4420	216 41 16·48	6019·1087	1 25 6·86	39·4048	215 16 9·62	5979·7039
223	319 40·41087	8879·8198	217 40 24·65	6046·4865	1 23 24·34	38·6137	216 17 0·32	6007·8728
224	320 39·54704	8907·1977	218 39 32·82	6073·8644	1 21 41·44	37·8197	217 17 51·38	6036·0446
225	321 38·68321	8934·5756	219 38 40·99	6101·2422	1 19 58·54	37·0258	218 18 42·45	6064·2164
226	322 37·81938	8961·9534	220 37 49·16	6128·6201	1 18 14·94	36·2264	219 19 34·22	6092·3937
227	323 36·95555	8989·3313	221 36 57·33	6155·9979	1 16 26·72	35·3914	220 20 30·61	6120·6066
228	324 36·09173	9016·7091	222 36 5·50	6183·3758	1 14 38·50	34·5563	221 21 27·00	6148·8195
229	325 35·22790	9044·0870	223 35 13·67	6210·7537	1 12 50·28	33·7213	222 22 23·39	6177·0323
230	326 34·36407	9071·4648	224 34 21·84	6238·1315	1 11 0·51	32·8743	223 23 21·33	6205·2572
231	327 33·50024	9098·8427	225 33 30·01	6265·5094	1 9 7·56	32·0028	224 24 22·45	6233·5066
232	328 32·63641	9126·2206	226 32 38·18	6292·8872	1 7 14·61	31·1313	225 25 23·57	6261·7559
233	329 31·77258	9153·5984	227 31 46·36	6320·2651	1 5 21·66	30·2598	226 26 24·69	6290·0053
234	330 30·90876	9180·9763	228 30 54·52	6347·6429	1 3 26·24	29·3691	227 27 28·28	6318·2738
235	331 30·04493	9208·3541	229 30 2·70	6375·0208	1 1 28·56	28·4611	228 28 34·13	6346·5597
236	332 29·18110	9235·7320	230 29 10·87	6402·3987	0 59 30·88	27·5531	229 29 39·9[illegible]	6374·8456
237	333 28·31727	9263·1098	231 28 19·04	6429·7765	0 57 33·20	26·6451	230 30 45·84	6403·1315
238	334 27·45344	9290·4877	232 27 27·21	6457·1544	0 55 32·97	25·7174	231 31 54·24	6431·4370
239	335 26·58961	9317·8656	233 26 35·38	6484·5322	0 53 31·74	24·7820	232 33 3·64	6459·7503
240	336 25·72578	9345·2434	234 25 43·55	6511·9101	0 51 30·51	23·8465	233 34 13·03	6488·0635
241	337 24·86196	9372·6213	235 24 51·72	6539·2879	0 49 29·28	22·9111	234 35 22·43	6516·3768

1	2		3	4			5	6			7	8			9
242	338	23·99813	9399·9991	236	23	59·89	6566·6658	0	47	25·35	21·9549	235	36	34·53	6544·7109
243	339	23·13430	9427·3770	237	23	8·06	6594·0437	0	45	21·17	20·9967	236	37	46·89	6573·0470
244	340	22·27047	9454·7548	238	22	16·23	6621·4215	0	43	16·98	20·0384	237	38	59·25	6601·3831
245	341	21·40664	9482·1327	239	21	24·40	6648·7994	0	41	12·48	19·0777	238	40	11·92	6629·7216
246	342	20·54281	9509·5106	240	20	32·57	6676·1772	0	39	5·33	18·0967	239	41	27·24	6658·0805
At true Dhanuh-samkrânti	*342*	*20·41736*	*9517·7997*	*240*	*38*	*26·84*	*6684·4663*	*0*	*38*	*26·84*	*17·7997*	*240*	*0*	*0·0*	*6666·6*
247	343	19·67899	9536·8884	241	19	40·74	6703·5551	0	36	58·19	17·1157	240	42	42·55	6686·4391
248	344	18·81516	9564·2663	242	18	48·91	6730·9329	0	34	51·05	16·1346	241	43	57·86	6714·7983
249	345	17·95133	9591·6441	243	17	57·08	6758·3108	0	32	43·19	15·1480	242	45	13·89	6743·1628
250	346	17·08750	9619·0202	244	17	5·25	6785·6887	0	30	33·68	14·1488	243	46	31·57	6771·5399
251	347	16·22367	9646·3998	245	16	13·42	6813·0665	0	28	24·17	13·1495	244	47	49·25	6799·9171
252	348	15·35984	9673·7777	246	15	21·59	6840·4444	0	26	14·66	12·1502	245	49	6·93	6828·2942
253	349	14·49601	9701·1556	247	14	29·76	6867·8222	0	24	4·27	11·1440	246	50	25·49	6856·6782
254	350	13·63219	9728·5334	248	13	37·93	6895·2001	0	21	52·99	10·1311	247	51	44·94	6885·0690
255	351	12·76836	9755·9113	249	12	46·10	6922·5779	0	19	41·70	9·1181	248	53	4·40	6913·4599
256	352	11·90453	9783·2891	250	11	54·27	6949·9558	0	17	30·42	8·1051	249	54	23·85	6941·8507
257	353	11·04070	9810·6670	251	11	2·44	6977·3337	0	15	18·32	7·0858	250	55	44·12	6970·2479
258	354	10·17687	9838·0448	252	10	10·61	7004·7115	0	13	5·85	6·0637	251	57	4·76	6998·6478
259	355	9·31304	9865·4227	253	9	18·78	7032·0894	0	10	53·39	5·0416	252	58	25·39	7027·0478
260	356	8·44922	9892·8006	254	8	26·95	7059·4672	0	8	40·92	4·0195	253	59	45·03	7055·4478
261	357	7·58539	9920·1784	255	7	35·12	7086·8451	0	6	27·93	2·9933	255	1	7·19	7083·8518
262	358	6·72156	9947·5563	256	6	43·29	7114·2229	0	4	14·88	1·9666	256	2	28·42	7112·2563
263	359	5·85773	9974·9341	257	5	51·46	7141·6008	0	2	1·82	0·9400	257	3	49·64	7140·6608
Sun at 258° (perigee)	*360*	*0·0*	*10000·0*	*258*	*0*	*0·0*	*7166·6*	*0*	*0*	*0·0*	*0·0*	*258*	*0*	*0·0*	*7166·6*

Sun's equation of centre is + (plus) after his mean anom = 360 till it reaches 180°.

1	2		3	4			5	Sun's equation of the centre. +			7	8			9
264	360	4·99390	2·3120	258	4	59·63	7168·9787	0	0	11·24	0·0867	258	5	10·87	7169·0654
265	1	4·13007	29·6898	259	4	7·80	7196·3565	0	2	24·29	1·1134	259	6	32·10	7197·4699
266	2	3·26255	57·0677	260	3	15·97	7223·7344	0	4	37·35	2·1400	260	7	53·32	7225·8744
267	3	2·40242	84·4456	261	2	24·15	7251·1122	0	6	50·41	3·1667	261	9	14·55	7254·2789
268	4	1·53859	111·8238	262	1	32·32	7278·4901	0	9	3·30	4·1921	262	10	35·61	7282·6822

TABLE XLVIII A—*Contd.*

24-hour periods from true Mēsha-saṁkrānti.	Sun's mean anomaly (or mean sun's distance from perigee-point) ("c").			Sun's mean Longitude.				Sun's equation of the centre. +				Sun's true Longitude. ("s").			
1	2		3	4			5	6			7	8			9
	°	′	10,000ths of circle.	°	′	″	10,000ths of circle.	°	′	″	10,000ths of circle.	°	′	″	10,000ths of circle.
269	5	0·67476	139·2013	263	0	40·49	7305·8679	0	11	15·76	5·2142	263	11	56·25	7311·0822
270	5	59·81093	166·5791	263	59	48·66	7333·2458	0	13	28·23	6·2363	264	13	16·88	7339·4821
271	6	58·94710	193·9570	264	58	56·83	7360·6237	0	15	40·69	7·2584	265	14	37·52	7367·8821
272	7	58·08327	221·3348	265	58	5·00	7388·0015	0	17	52·59	8·2762	266	15	57·59	7396·2777
273	8	57·21945	248·7127	266	57	13·17	7415·3794	0	20	3·88	9·2892	267	17	17·04	7424·6684
274	9	56·35162	276·0906	267	56	21·34	7442·7572	0	22	15·16	10·3022	268	18	36·50	7453·0594
275	10	55·49179	303·4684	268	55	29·51	7470·1351	0	24	26·44	11·3151	269	19	55·95	7481·8502
At true Makara-saṁkrānti	*11*	*34·13893*	*321·3607*	*269*	*34*	*8·34*	*7488·0273*	*0*	*25*	*51·66*	*11·9727*	*270*	*0*	*0·0*	*7500·0*
276	11	54·62796	330·8463	269	54	37·68	7497·5129	0	26	36·54	12·3189	270	21	14·21	7509·8319
277	12	53·76413	358·2241	270	53	45·85	7524·8908	0	28	46·04	13·3182	271	22	31·89	7538·2090
278	13	52·90030	385·6020	271	52	54·02	7552·2687	0	30	55·55	14·3175	272	23	49·57	7566·5862
279	14	52·03648	412·9798	272	52	2·19	7579·6465	0	33	5·06	15·3168	273	25	7·25	7594·9633
280	15	51·17265	440·3577	273	51	10·36	7607·0244	0	35	12·52	16·3003	274	26	22·88	7623·3247
281	16	50·30882	467·7356	274	50	18·53	7634·4022	0	37	19·66	17·2814	275	27	38·19	7651·6836
282	17	49·44499	495·1134	275	49	26·70	7661·7801	0	39	26·81	18·2624	276	28	53·51	7680·0425
283	18	48·58116	522·4913	276	48	34·87	7689·1579	0	41	33·77	19·2421	277	30	8·64	7708·4000
284	19	47·71733	549·8691	277	47	43·04	7716·5358	0	43	37·96	20·2003	278	31	21·00	7736·7361
285	20	46·85351	577·2470	278	46	51·21	7743·9137	0	45	42·14	21·1585	279	32	33·35	7765·0722
286	21	45·98968	604·6249	279	45	59·38	7771·2915	0	47	46·33	22·1167	280	33	45·71	7793·4082
287	22	45·12585	632·0027	280	45	7·55	7798·6694	0	49	49·76	23·0691	281	34	57·31	7821·7385
288	23	44·26202	659·3806	281	44	15·72	7826·0472	0	51	50·99	24·0045	282	36	6·71	7850·0518
289	24	43·39819	686·7584	282	43	23·89	7853·4251	0	53	52·22	24·9399	283	37	16·11	7878·3650
290	25	42·53436	714·1363	283	42	32·06	7880·8029	0	55	53·45	25·8754	284	38	25·51	7906·6783

1	2		3	4			5	6			7	8			9
291	26	41·67053	741·5141	284	41	40·23	7908·1808	0	57	53·07	26·7984	285	39	33·31	7934·9792
292	27	40·80671	768·8920	285	40	48·40	7935·5587	0	59	50·76	27·7064	286	40	39·16	7963·2651
293	28	39·94288	796·2699	286	39	56·57	7962·9365	1	1	48·44	28·6145	287	41	45·01	7991·5510
294	29	39·07905	823·6477	287	39	4·74	7990·3144	1	3	46·12	29·5225	288	42	50·86	8019·8369
295	30	38·21522	851·0256	288	38	12·91	8017·6922	1	5	40·74	30·4070	289	43	53·65	8048·0992
296	31	37·35139	878·4034	289	37	21·08	8045·0710	1	7	33·69	31·2785	290	44	54·77	8076·3486
297	32	36·48756	905·7813	290	36	29·25	8072·4479	1	9	26·64	32·1560	291	45	55·90	8104·5980
298	33	35·62374	933·1591	291	35	27·42	8099·8258	1	11	19·59	33·0215	292	46	57·02	8132·8473
299	34	34·75991	960·5370	292	34	45·59	8127·2037	1	13	8·56	33·8623	293	47	54·16	8161·0660
300	35	33·89608	987·9149	293	33	53·76	8154·5815	1	14	56·78	34·6974	294	48	50·54	8189·2789
301	36	33·03225	1015·2927	294	33	1·94	8181·9594	1	16	45·00	35·5324	295	49	46·93	8217·8918
302	37	32·16842	1042·6706	295	32	10·11	8209·3372	1	18	33·02	36·3659	296	50	43·13	8245·7032
303	38	31·30459	1070·0484	296	31	18·28	8236·7151	1	20	15·92	37·1599	297	51	34·29	8273·8750
304	39	30·44077	1097·4268	297	30	26·45	8264·0929	1	21	58·82	37·9538	298	52	25·26	8302·0468
305	40	29·57694	1124·8041	298	29	34·62	8291·4708	1	23	41·71	38·7478	299	53	16·33	8330·2186
At true Kumbha-saṁkrānti	*40*	*36·11516*	*1127·8312*	*298*	*36*	*6·91*	*8294·1978*	*1*	*23*	*53·09*	*38·8355*	*300*	*0*	*0·0*	*8353·3*
306	41	28·71311	1152·1820	299	28	42·79	8318·8487	1	25	23·24	39·5312	300	54	6·03	8358·3798
307	42	27·84928	1179·5599	300	27	50·96	8346·2265	1	27	0·22	40·2795	301	54	51·48	8386·5060
308	43	26·98545	1206·9377	301	26	59·13	8373·6044	1	28	37·21	41·0278	302	55	36·33	8414·6322
309	44	26·12162	1234·3156	302	26	7·30	8400·9822	1	30	14·19	41·7762	303	56	21·49	8442·7584
310	45	25·25779	1261·6934	303	25	15·47	8428·3601	1	31	48·66	42·5051	304	57	4·12	8470·8651
311	46	24·39397	1289·0713	304	24	23·64	8455·7379	1	33	19·72	43·2077	305	57	43·35	8498·9456
312	47	23·53014	1316·4491	305	23	31·81	8483·1158	1	34	50·79	43·9104	306	58	22·59	8527·0262
313	48	22·66631	1343·8270	306	22	39·98	8510·4937	1	36	21·86	44·6131	307	59	1·83	8555·1067
314	49	21·80248	1371·2049	307	21	48·15	8537·8715	1	37	48·88	45·2845	308	59	37·03	8583·1561
315	50	20·93865	1398·5827	308	20	56·32	8565·2494	1	39	13·44	45·9371	310	0	9·76	8611·1864
316	51	20·07482	1425·9666	309	20	4·49	8592·6272	1	40	38·01	46·5896	311	0	42·50	8639·2168
317	52	19·21100	1453·3384	310	19	12·66	8620·0051	1	42	2·57	47·2421	312	1	15·23	8667·2471
318	53	18·34717	1480·7163	311	18	20·83	8647·3829	1	43	21·33	47·8498	313	1	42·16	8695·2328
319	54	17·48334	1508·0941	312	17	29·00	8674·7608	1	44	38·80	48·4476	314	2	7·80	8723·2084
320	55	16·61951	1535·4720	313	16	37·17	8702·1387	1	45	56·27	49·0453	315	2	33·44	8751·1840
321	56	15·75568	1562·8499	314	15	45·34	8729·5165	1	47	13·65	49·6424	316	2	58·99	8779·1589
322	57	14·89185	1590·2277	315	14	53·51	8756·8944	1	48	24·02	50·1853	317	3	17·53	8807·0797
323	58	14·02802	1617·6056	316	14	1·68	8784·2722	1	49	34·39	50·7283	318	3	36·07	8835·0006
324	59	13·16420	1644·9834	317	13	9·85	8811·6501	1	50	44·77	51·2713	319	3	54·62	8862·9214
325	60	12·30037	1672·3613	318	12	18·02	8839·0279	1	51	53·54	51·8020	320	4	11·56	8890·8292

TABLE XLVIII A—*Concld.*

1st Ārya-Siddhānta.

24-hour periods from true Mēsha-saṁkrānti.	Sun's mean anomaly (or mean sun's distance from perigee-point) ("*c*").		Sun's mean Longitude.		Sun's equation of the centre. +		Sun's true Longitude ("*s*").	
1	2	3	4	5	6	7	8	9
	° ′	10,000ths of circle.	° ′ ″	10,000ths of circle.	° ′ ″	10,000ths of circle.	° ′ ″	10,000ths of circle.
326	61 11·43654	1699·7391	319 11 26·19	8866·4958	1 52 56·22	52·2857	321 4 22·42	8918·6915
327	62 10·57271	1727·1170	320 10 34·36	8893·7837	1 53 58·91	52·7693	322 4 33·27	8946·5530
328	63 9·70888	1754·4949	321 9 42·53	8921·1615	1 55 1·59	53·2530	323 4 44·12	8974·4145
229	64 8·84505	1781·8727	322 8 50·70	8948·5394	1 56 1·18	53·7128	324 4 51·88	9002·2529
340	65 7·98123	1809·2506	323 7 58·87	8975·9172	1 56 56·17	54·1371	325 4 55·05	9030·0544
331	66 7·11740	1836·6284	324 7 7·04	9003·2951	1 57 51·17	45·5615	326 4 58·21	9057·8566
332	67 6·25357	1864·0063	325 6 15·21	9030·6729	1 58 46·17	54·9858	327 5 1·38	9085·6588
333	68 5·38974	1891·3841	326 5 23·38	9058·0508	1 55 36·21	55·3720	328 4 59·59	9113·4228
324	69 4·52591	1918·7620	327 4 31·55	9085·4287	2 0 22·93	55·7324	329 4 54·48	9141·1611
At true Mīna-saṁkrānti	*69 58·99196*	*1243·9361*	*327 58 51·12*	*9110·6028*	*2 1 5·88*	*56·0638*	*330 0 0·0*	*9166·6*
335	70 3·66208	1946·1399	328 3 39·73	9112·8065	2 1 9·64	56·0929	330 4 49·37	9168·8994
326	71 2·79826	1973·5177	329 2 47·90	9140·1844	2 1 56·36	56·4534	331 4 44·26	9196·6378
337	72 1·93443	2000·8956	330 1 56·07	9167·5622	2 2 36·51	56·7632	332 4 32·57	9224·3254
338	73 1·07060	2028·2734	331 1 4·24	9194·9401	2 3 14·95	57·0598	333 4 19·18	9251·9999
339	74 0·20677	2055·6513	332 0 12·41	9222·3179	2 3 53·38	57·3564	334 4 5·79	9279·6743
340	74 59·34294	2083·0291	332 59 20·58	9249·6958	2 4 31·82	57·6530	335 3 52·40	9307·3488
541	75 58·47911	2110·4070	333 58 28·75	9277·0737	2 5 2·07	57·8864	336 3 30·82	9334·9600
342	76 57·61528	2137·7849	334 57 36·92	9304·4515	2 5 32·23	58·1191	337 3 9·15	9362·5706
343	77 56·75146	2165·1627	335 56 45·09	9331·8294	2 6 2·39	58·3518	338 2 47·48	9390·1812
544	78 55·88763	2192·5406	336 55 53·26	9359·2072	2 6 31·03	58·5728	339 2 24·29	9417·7800
345	79 55·02380	2219·9184	337 55 1·43	9386·5851	2 6 52·91	58·7416	340 1 54·34	9445·3267
346	80 54·15997	2247·2863	338 54 9·60	9413·9629	2 7 14·79	58·9104	341 1 24·39	9472·8734
347	81 53·29614	2274·6741	339 53 17·77	9441·3408	2 7 36·67	59·0792	342 0 54·44	9500·4200
348	82 52·43231	2302·0520	340 52 25·94	9468·7187	2 7 55·19	59·2221	343 0 21·12	9527·9408
349	83 51·56849	2329·4299	341 51 34·11	9496·0965	2 8 8·20	59·3225	343 59 42·20	9555·4190

1	2		3	4			5	6			7	8			9
350	84	50·70466	2356·8077	342	50	42·28	9523·4744	2	8	21·21	59·4229	344	59	3·48	9582·8973
351	85	49·84083	2384·1856	343	49	50·45	9550·8522	2	8	34·21	59·5233	345	58	24·66	9610·3755
352	86	48·97700	2411·5634	344	48	58·62	9578·2301	2	8	42·13	59·5843	346	57	40·75	9637·8144
353	87	48·11317	2438·9413	345	48	6·79	9605·6080	2	8	46·27	59·6163	347	56	53·06	9665·2242
354	88	47·24934	2466·3191	346	47	14·96	9632·9858	2	8	50·41	59·6482	348	56	5·37	9692·6340
355	89	46·38551	2493·6970	347	46	23·13	9660·3637	2	8	54·55	59·6801	349	55	17·68	9720·0438
356	90	45·52169	2521·0749	348	45	31·30	9687·7415	2	8	52·31	59·6629	350	54	23·61	9747·4044
357	91	44·65786	2548·4527	349	44	39·47	9715·1194	2	8	48·17	59·6310	351	53	27·65	9774·7504
358	92	43·79403	2575·8306	350	43	47·64	9742·[illegible]072	2	8	44·03	59·5990	352	52	31·68	9802·0983
359	93	42·93020	2603·2084	351	42	55·81	9769·8751	2	8	39·89	59·5671	353	51	35·71	9829·4422
360	94	42·06637	2630·5863	352	42	3·98	9797·2530	2	8	27·20	59·4691	354	50	31·18	9856·7221
361	95	41·20254	2657·9641	353	41	12·15	9824·6308	2	8	14·19	59·3687	355	49	26·34	9883·9995
362	96	40·33872	2685·3420	354	40	20·32	9852·0087	2	8	1·18	59·2683	356	48	21·50	9911·2770
363	97	39·47489	2712·7199	355	39	28·49	9879·3865	2	7	46·74	59·1570	357	47	15·24	9938·5435
364	98	38·61106	2740·0977	356	38	36·66	9906·7644	2	7	24·86	58·9881	358	46	1·53	9965·7525
365	99	37·74723	2767·4756	357	37	44·83	9934·1422	2	7	2·98	58·8193	359	44	47·82	9992·9616

TABLE XLVIII-B.

ELEMENTS OF THE SUN'S LONGITUDE FOR THE HINDU SOLAR YEAR
according to the **Present Sūrya-Siddhānta**
in periods of 24-hours each from the moment of true Mēsha-saṁkrānti,
the astronomical beginning of the solar year.

(*Exact for K. Y. 4500, A D 1399—1400. See Text, para. 254, ii.*)

[True longitude=mean longitude ± equation of centre.]

Present Sūrya-Siddhānta.

24-hour periods from true Mēsha-saṁkrānti.	Sun's mean anomaly (or mean sun's distance from perigee-point) ("c").		Sun's mean Longitude.		Sun's equation of the centre. +		Sun's true Longitude ("s").	
1	2	3	4	5	6	7	8	9
	° ′	10,000ths of circle.	° ′ ″	10,000ths of circle.	° ′ ″	10,000ths of circle.	° ′ ″	10,000ths of circle.
(*The sun's equation of the centre is +, plus, till his mean anomaly reaches 180°.*)								
At true Mēsha-saṁkrānti	*100 34·17870*	*2794·0642*	*347 51 41·17*	*9940·5955*	*2 8 18·83*	*59·4045*	*0 0 0·0*	*0·0*
1	101 34·31486	2821·4421	358 50 49·34	9967·9733	2 7 54·61	59·2176	0 58 43·95	27·1909
2	102 33·45102	2848·8199	359 49 57·51	9995·3512	2 7 25·04	58·9895	1 57 22·55	54·3406
3	103 32·58717	2876·1978	0 49 5·68	22·7290	2 6 55·47	58·7613	2 56 1·15	81·4904
4	104 31·72333	2903·5756	1 48 13·85	50·[illegible]069	2 6 25·90	58·5332	3 54 39·75	108·6401
5	105 30·85949	2930·9535	2 47 22·02	77·4847	2 5 51·03	58·2641	4 53 13·05	135·7488
6	106 29·99565	2958·3313	3 46 30·19	104·8626	2 5 13·18	57·9721	5 51 43·37	162·8347
7	107 29·13181	2985·7092	4 45 38·36	132·2404	2 4 35·33	57·6800	6 50 13·69	189·9205
8	108 28·26797	3013·0870	5 44 46·53	159·6183	2 3 57·49	57·3880	7 40 44·02	217·0063
9	109 27·40413	3040·4649	6 43 54·70	186·9961	2 3 12·95	57·0143	8 47 7·64	244·0405
10	110 26·54029	3067·8427	7 43 2·87	214·3740	2 2 26·82	56·6884	9 45 29·69	271·0624
11	111 25·67645	3095·2206	8 42 11·04	241·7518	2 1 40·69	56·3325	10 43 51·73	298·0843
12	112 24·81261	3122·5984	9 41 19·21	269·1297	2 0 54·57	55·9766	11 42 13·77	325·1063
13	113 23·94877	3149·9763	10 40 27·38	296·5075	2 0 0·27	55·5570	12 40 27·64	352·0651
14	114 23·08493	3177·3541	11 39 35·55	323·8854	1 59 5·86	55·1378	13 38 41·41	379·0232
15	115 22·22109	3204·7320	12 38 43·72	351·2632	1 58 11·46	54·7180	14 36 55·17	405·9813

1	2		3	4			5	6			7	8			9
16	116	21·35725	3232·1098	13	37	51·88	378·6411	1	57	15·63	54·2873	15	35	7·52	432,9284
17	117	20·49340	3259·4877	14	37	0·05	406,0189	1	56	13·54	53·8082	16	33	13·59	459·8271
18	118	19·62956	3286·8655	15	36	8·22	433·3968	1	55	11·45	53·3291	17	31	19·67	486·7259
19	119	18·76572	3314·2434	16	35	16·39	460·7746	1	54	9·35	52·8500	18	29	25·75	513·6246
20	120	17·90188	3341·6212	17	34	24·56	488·1525	1	53	4·23	52·3475	19	27	28·80	540·5000
21	121	17·03804	3368·9991	18	33	32·73	515·5303	1	51	54·45	51·8091	20	25	27·19	567·3394
22	122	16·17420	3396·3769	19	32	40·90	542·9082	1	50	44·67	51·2706	21	23	25·57	594·1788
23	123	15·31036	3423·7548	20	31	49·07	570·2860	1	49	34·89	50·7322	22	21	23·96	621·0182
24	124	14·44652	3451·1326	21	30	57·24	597·6639	1	48	20·34	50·1570	23	19	17·58	647·8209
25	125	13·58268	3478·5105	22	30	5·41	625·0417	1	47	2·87	49·5592	24	17	8·28	674·6010
26	126	12·71884	3505·8884	23	29	13·58	652·4196	1	45	45·40	48·9615	25	14	58·98	701·3811
27	127	11·85500	3533·2662	24	28	21·75	679·7975	1	44	27·94	48·3637	26	12	49·69	728·1612
28	128	10·99116	3560·6441	25	27	29·92	707·1753	1	43	6·34	47·7341	27	10	36·26	754·9094
29	129	10·12732	3588·0219	26	26	38·09	734·5532	1	41	41·77	47·0816	28	8	19·86	781·6347
30	130	9·26347	3615·3998	27	25	46·26	761·9310	1	40	17·21	46·4291	29	6	3·47	808·3601
31	131	8·39963	3642·7776	28	24	54·43	789·3089	1	38	52·64	45·7766	30	3	47·07	835·0854
32	132	7·53579	3670·1555	29	24	2·60	816·6867	1	37	22·78	45·0832	31	1	25·39	861·7699
33	133	6·67195	3697·5333	30	23	10·77	844·0646	1	35	51·71	44·3805	31	59	2·48	888·4451
34	134	5·80811	3724·9112	31	22	18·94	871·4424	1	34	20·64	43·6778	32	56	39·58	915·1202
35	135	4·94427	3752·2890	32	21	27·11	898·8203	1	32	48·68	42·9682	33	54	15·78	941·7885
36	136	4·08043	3779·6669	33	20	35·28	926·1981	1	31	11·10	42·2153	34	51	46·38	968·4134
37	137	3·21659	3807·0447	34	19	43·45	953·5760	1	29	33·53	41·4624	35	49	16·97	995·0384
38	138	2·35275	3834·4226	35	18	51·61	980·9538	1	27	55·95	40·7095	36	46	47·57	1021·6633
39	139	1·48891	3861·8004	36	17	59·78	1008·3317	1	26	17·76	39·9519	37	44	17·55	1048·2836
40	140	0·62507	3889·1783	37	17	7·95	1035·7095	1	24	34·28	39·1534	38	41	42·23	1074·8629
41	140	59·76123	3916·5561	38	16	16·12	1063·0874	1	22	50·79	38·3548	39	39	6·91	1101·4422
42	141	58·89739	3943·9340	39	15	24·29	1090·4652	1	21	7·30	37·5563	40	36	31·59	1128·0215
43	142	58·03355	3971·3118	40	14	32·46	1117·8431	1	19	21·02	36·7362	41	33	53·48	1154·5793
44	143	57·16970	3998·6897	41	13	40·63	1145·3209	1	17	33·61	35·8921	42	31	12·25	1181·1130
45	144	56·30586	4026·0675	42	12	48·80	1172·5988	1	15	42·21	35·0479	43	28	31·01	1207·6467
46	145	55·44202	4053·4454	43	11	56·97	1199·9766	1	13	52·81	34·2038	44	25	49·78	1234·1804
47	146	54·57818	4080·8232	44	11	5·14	1227·3545	1	12	1·33	33·3436	45	23	6·47	1260·6981
48	147	53·71434	4108·2011	45	10	13·31	1254·7323	1	10	7·20	32·4629	46	20	20·51	1287·1953
49	148	52·85050	4135·5789	46	9	21·48	1280·1102	1	8	13·66	31·5823	47	17	34·44	1313·6925
50	149	51·98666	4162·9568	47	8	29·65	1309·4880	1	6	18·23	30·7016	48	14	48·58	1340·1897

TABLE XLVIII B—*Contd.*

present Sūrya-Siddhānta.

24-hour periods from true Mēsha-saṁkrānti.	Sun's mean anomaly (or mean sun's distance from perigee-point) ("c").		Sun's mean Longitude		Sun's equation of the centre. +		Sun's true Longitude ("s").	
1	2	3	4	5	6	7	8	9
	° ′	10,000ths of circle.	° ′ ″	10,000ths of circle.	° ′ ″	10,000ths of circle.	° ′ ″	10,000ths of circle.
51	150 51·12282	4190·3346	48 7 37·82	1336·8559	1 4 19·98	29·7838	49 11 57·80	1366·6497
52	151 50·25898	4217·7125	49 6 45·99	1364·2337	1 2 20·53	28·8621	50 9 6·52	1393·1058
53	152 49·39514	4245·0903	50 5 54·16	1391·6116	1 0 21·07	27·9404	51 6 15·23	1419·5620
54	153 48·53130	4272·4682	51 5 2·33	1418·9995	0 58 21·99	27·0215	52 3 24·32	1446·0209
55	154 47·66746	4299·8460	52 4 10·50	1446·3773	0 56 18·98	26·0724	53 0 29·48	1472·4497
56	155 46·80362	4327·2239	53 3 18·67	1473·7551	0 54 15·98	25·1233	53 57 34·65	1498·8785
57	156 45·93978	4354·6017	54 2 26·84	1501·1330	0 52 12·98	24·1742	54 54 39·81	1525·3072
58	157 45·07593	4381·9796	55 1 35·01	1528·5108	0 50 8·13	23·2109	55 51 43·13	1551·7217
59	158 44·21209	4409·3575	56 0 43·18	1555·8887	0 48 1·58	22·2344	56 48 44·75	1578·1231
60	159 43·34825	4436·7353	56 59 51·35	1583·2666	0 45 55·03	21·2579	57 45 46·37	1604·5245
61	160 42·48441	4464·1132	57 58 59·51	1610·6444	0 43 48·47	20·2814	58 42 47·99	1630·9258
62	161 41·62057	4491·4910	58 58 7·68	1638·0223	0 41 40·04	19·2904	59 39 47·72	1657·3127
63	162 40·75673	4518·8689	59 57 16·85	1665·4001	0 39 29·94	18·2866	60 36 45·79	1683·6867
64	163 39·89289	4546·2467	60 56 24·02	1692·7780	0 37 19·84	17·2827	61 33 43·86	1710·0607
65	164 39·02905	4573·6246	61 55 32·19	1720·1558	0 35 9·74	16·2788	62 30 41·93	1736·4347
66	165 38·16521	4601·0024	62 54 40·36	1747·5337	0 32 57·99	15·2623	63 27 38·36	1762·7960
67	166 37·30137	4628·3803	63 53 48·53	1774·9115	0 30 44·94	14·2356	64 24 33·47	1789·1472
68	167 36·43753	4655·7581	64 52 56·70	1802·2894	0 28 31·88	13·2090	65 21 28·58	1815·4983
69	168 35·57369	4683·1360	65 52 4·87	1829·6672	0 26 18·83	12·1823	66 18 23·70	1841·8495
70	169 34·70985	4710·5138	66 51 13·04	1857·0451	0 24 5·05	11·1501	67 15 18·09	1868·1952
71	170 33·84601	4737·8917	67 50 21·21	1884·4229	0 21 50·22	10·1097	68 12 11·43	1894·5327
72	171 32·98216	4765·2695	68 49 29·38	1911·8008	0 19 35·39	9·0694	69 9 4·77	1920·8702
73	172 32·11832	4792·6474	69 48 37·55	1939·1786	0 17 19·52	8·0210	70 5 57·07	1947·1996
74	173 31·25448	4820·0252	70 47 45·72	1966·5565	0 15 2·91	6·9669	71 2 48·63	1973·5234
75	174 30·39064	4847·4031	71 46 53·89	1993·9343	0 12 46·31	5·9129	[illegible]	[illegible]

1	2		3	4			5	6			7	8			9
76	175	29·52680	4874·7809	72	46	2·06	2021·3122	0	10	29·70	4·8588	72	56	31·76	2026·1710
77	176	28·66296	4902·1588	73	45	10·23	2048·6900	0	8	12·35	3·7990	73	53	22·57	2052·4890
78	177	27·79912	4929·5366	74	44	18·40	2076·0679	0	5	54·56	2·7358	74	50	12·96	2078·8037
79	178	26·93528	4956·9145	75	43	26·57	2103·4457	0	3	36·77	1·6726	75	47	3·34	2105 1184
80	179	26·07144	4984·2923	76	42	34·74	2130·8236	0	1	18·99	0·6095	76	43	53·72	2131·4330
Sun in apogee	*180*	*0·0*	*5000·0*	*77*	*16*	*30·45*	*2156·5313*	*0*	*0*	*0·0*	*0·0*	*77*	*16*	*30·45*	*2146·5313*

(The sun's equation of the centre is —, minus, after his mean anom : = 180° till it reaches 360° or 0°)

1	2		3	4			5	6 *Sun's equation of the centre minus, —*			7	8			9
81	180	25·20760	5011·6702	77	41	42·91	2158·2014	0	0	58·73	0·4532	77	40	44·17	2157·7482
82	181	24·34376	5039·0480	78	40	51·08	2185·5793	0	3	16·52	1·5164	78	37	34·55	2184 0629
83	182	23·47992	5066·4259	79	39	59·24	2212·9571	0	5	34·31	2·5795	79	34	24·94	2210·3776
84	183	22·61608	5093·8037	80	39	7·41	2240·3350	0	7	52·10	3·6427	80	31	14·32	2236·6923
85	184	21·75224	5121·1816	81	38	15·58	2267·7128	0	10	9·08	4·6997	81	28	6·50	2263·0132
86	185	20·88839	5148·5594	82	37	23·75	2295·0907	0	12	25·68	5·7537	82	24	58·07	2289·3370
87	186	20·02455	5175·9373	83	36	31·92	2322·4685	0	14	42·29	6·8078	83	21	49·63	2315·6608
88	187	19·16071	5203·3151	84	35	40·09	2349·8464	0	16	58·89	7·8618	84	18	41·20	2341·9846
89	188	18·29687	5230·6930	85	34	48·26	2377·2242	0	19	14·53	8·9084	85	15	33·74	2368·3159
90	189	17·43303	5258·0708	86	33	56·43	2404·6021	0	21	29·36	9·9487	86	12	27·08	2394·6534
91	190	16·56919	5285·4487	87	33	4·60	2431·9799	0	23	44·19	10·9891	87	9	20·41	2420·9909
92	191	15·70535	5312·8266	88	32	12·77	2459·3578	0	25	59·98	12·0369	88	6	12·79	2447·3209
93	192	14·84151	5340·2044	89	31	20·94	2486·7357	0	28	13·03	13·0635	89	3	7·91	2473·6721
94	193	13·97767	5367·5823	90	30	29·11	2514·1135	0	30	26·09	14·0902	90	0	3·02	2500·0233
95	194	13·11383	5394·9601	91	29	37·28	2541·4914	0	32	39·15	15·1169	90	56	58·13	2526·3745
96	195	12·24999	5422·3380	92	28	45·45	2568·8692	0	34	50·82	16·1328	91	53	54·63	2552·7364
97	196	11·38615	5449·7158	93	27	53·62	2596·2471	0	37	0·92	17·1367	92	50	52·70	2579·1104
98	197	10·52231	5477·0937	94	27	1·79	2623·6249	0	39	11·02	18·1406	93	47	50·77	2605·4844
99	198	9·65846	5504·4715	95	26	9·96	2651·0028	0	41	21·12	19·1444	94	44	48·84	2631·8584
100	199	89·79462	5531·8494	96	25	18·13	2678·3806	0	43	29·52	20·1352	95	41	48·60	2658·2454
101	200	7·93078	5559·2272	97	24	26·30	2705·7585	0	45	36·07	21·1117	96	38	50·22	2684·6468
102	201	7·06694	5586·6051	98	23	34·47	2733·1363	0	47	42·63	22·0882	97	35	51·84	2711·0482
103	202	6·20310	5613·9829	99	22	42·64	2760·5142	0	49	49·18	23·0646	98	32	53·46	2737·4495
104	203	5·33926	5641·3608	100	21	50·81	2787·8920	0	51	53·90	24·0270	99	29	56·91	2763·8650
105	204	4·47542	5668·7336	101	20	58·98	2815·2699	0	53	56·90	24·9761	100	27	2·08	2790·2938

TABLE XLVIII B—*Contd.*

Present Sûrya-Siddhânta

24-hour periods from true Mêsha-saṁkrânti.	Sun's mean anomaly (or mean sun's distance from perigee-point) ("c").		Sun's mean Longitude.		Sun's equation of the centre. —		Sun's true Longitude ("s").	
1	2	3	4	5	6	7	8	9
	°	10,000ths of circle.	° ″	10,000ths of circle.	° ′ ″	10,000ths of circle.	° ′ ″	10,000ths of circle.
106	205 3·61158	5696·1165	102 20 7·14	2842·6477	0 55 59·90	25·9252	101 24 7·24	2816·7225
107	206 2·74774	5723·4943	103 19 15·31	2870·0256	0 58 2·91	26·8743	102 21 12·41	2843·1513
108	207 1·88390	5750·8722	104 18 23·48	2897·4034	1 0 4·04	27·8089	103 18 19·45	2869·5945
109	208 1·02006	5778·2500	105 17 31·65	2924·7813	1 2 3·49	28·7307	104 15 28·16	2896·0506
110	209 0·15622	5805·6279	106 16 39·82	2952·1591	1 4 2·95	29·6524	105 12 36·88	2922·5068
111	209 59·29238	5833·0057	107 15 47·99	2979·5370	1 6 2·40	30·5741	106 9 45·59	2948·9629
112	210 58·42854	5860·3836	108 14 56·16	3006·9148	1 7 56·02	31·4507	107 7 0·15	2975·4641
113	211 57·56469	5887·7614	109 14 4·33	3034·2927	1 9 50·15	32·3314	108 4 14·18	3001·9613
114	212 56·70085	5915·1393	110 13 12·50	3061·6705	1 11 44·28	33·2121	109 1 28·22	3028·4585
115	213 55·83701	5942·5171	111 12 20·67	3089·0484	1 13 37·76	34·0877	109 58 42·91	3054·9607
116	214 54·97317	5969·8950	112 11 28·84	3116·4262	1 15 27·17	34·9318	110 56 1·67	3081·4944
117	215 54·10933	5997·2728	113 10 37·01	3143·8041	1 17 16·57	35·7760	111 53 20·44	3108·0281
118	216 53·24549	6024·6507	114 9 45·18	3171·1819	1 19 5·97	36·6201	112 50 39·21	3134·5618
119	217 52·38165	6052·0285	115 8 53·35	3198·5598	1 20 52·05	37·4386	113 48 1·30	3161·1212
120	218 51·51781	6079·4064	116 8 1·52	3225·9376	1 22 35·53	38·2372	114 45 25·98	3187·7005
121	219 50·65397	6106·7842	117 7 9·69	3253·3155	1 24 19·02	39·0357	115 42 50·67	3214·2798
122	220 49·79013	6134·1621	118 6 17·86	3280·6933	1 26 2·51	39·8342	116 40 15·35	3240·8592
123	221 48·92629	6161·5399	119 5 26·03	3308·0712	1 27 42·60	40·6065	117 37 43·43	3267·4647
124	222 48·06245	6188·9178	120 4 34·20	3335·4490	1 29 20·17	41·3594	118 35 14·02	3294·0897
125	223 47·19861	6216·2957	121 3 42·37	3362·8269	1 30 57·75	42·1122	119 42 44·62	3320·7147
126	224 46·33477	6243·6735	122 2 50·54	3390·2048	1 32 35·32	42·8651	120 30 15·21	3347·3396
127	225 45·47092	6271·0514	123 1 58·71	3417·5826	1 34 6·86	43·5715	121 27 52·10	3374·0111
128	226 44·60708	6298·4292	124 1 6·88	3444·9605	1 35 37·93	44·2742	122 25 28·94	3400·6863
129	227 43·74324	6325·8071	125 0 15·04	3472·3383	1 37 9·00	44·9769	123 23 6·04	3427·3615
130	228 42·87940	6353·1849	125 59 23·21	3499·7162	1 38 40·07	45·6796	124 20 43·14	3454·0360

1	2		3	4			5	6			7	8			9
131	229	42·01556	6380·5628	126	58	31·38	3527·0940	1	40	5·22	46·3366	125	18	26·16	3480·7574
132	230	41·15172	6407·9406	127	57	39·55	3554·4719	1	41	29·78	46·9891	126	16	9·77	3507·4828
133	231	40·28788	6435·3185	128	56	47·72	3581·8497	1	42	54·35	47·6416	127	13	53·38	3534·2081
134	232	39·42404	6462·6963	129	55	55·89	3609·2276	1	44	17·30	48·2816	128	11	38·59	3560·9459
135	233	38·56020	6490·0742	130	55	4·06	3636·6054	1	45	34·77	48·8794	129	9	29·29	3587·7266
136	234	37·69636	6517·4520	131	54	12·23	3663·9833	1	46	52·24	49·4771	130	7	19·99	3614·5061
137	235	36·83252	6544·8299	132	53	20·40	3691·3611	1	48	9·71	50·0749	131	5	10·69	3641·2862
138	236	35·96868	6572·2077	133	52	28·57	3718·7390	1	49	23·66	50·6455	132	3	4·91	3668·0935
139	237	25·10484	6599·5856	134	51	36·74	3746·1168	1	50	33·44	51·1839	133	1	3·30	3694·9329
140	238	34·24100	6626·9634	135	50	44·91	3773·4947	1	51	43·22	51·7224	133	59	1·69	3721·7723
141	239	33·37715	6654·3413	136	49	53·08	3800·8725	1	52	53·00	52·2608	134	57	0·08	3748·6117
142	240	32·51331	6681·7191	137	49	1·25	3828·2504	1	53	59·50	52·7739	135	55	1·75	3775·4765
143	241	31·64947	6709·0970	138	48	9·42	3855·6282	1	55	1·59	53·2530	136	53	7·83	3802·3752
144	242	30·78563	6736·4748	139	47	17·59	3883·0061	1	56	3·68	53·7321	137	51	13·90	3829·2740
145	243	29·92179	6763·8527	140	46	25·76	3910·3839	1	57	5·78	54·2112	138	49	19·98	3856·1727
146	244	29·05795	6791·2305	141	45	33·93	3937·7618	1	58	2·84	54·6516	139	47	31·09	3883·1102
147	245	28·19411	6818·6084	142	44	42·10	3965·1396	1	58	57·25	55·0714	140	45	44·85	3910·0683
148	246	27·33027	6845·9862	143	43	50·27	3992·5175	1	59	51·65	55·4912	141	43	58·61	3937·0263
149	247	26·46643	6873·3641	144	42	58·44	4019·8953	2	0	46·06	55·9110	142	42	12·38	3963·9844
150	248	25·60259	6900·7419	145	42	6·61	4047·2732	2	1	33·27	56·2752	143	40	33·34	3990·9980
151	249	24·73875	6928·1198	146	41	14·77	4074·6510	2	2	19·40	56·6311	144	38	55·38	4018·0199
152	250	23·87491	6955·4976	147	40	22·94	4102·0289	2	3	5·52	56·9870	145	37	17·42	4045·0418
153	251	23·01107	6982·8755	148	39	31·11	4129·4067	2	3	51·15	57·3391	146	35	39·97	4072·0676
154	252	22·14723	7010·2533	149	38	39·28	4156·7846	2	4	28·99	57·6311	147	34	10·29	4099·1535
155	253	21·28338	7037·6312	150	37	47·45	4184·1624	2	5	6·84	57·9232	148	32	40·61	4126·2393
156	254	20·41954	7065·0090	151	36	55·62	4211·5403	2	5	44·69	58·2152	149	31	10·93	4153·3251
157	255	19·55570	7092·3869	152	36	3·79	4238·9181	2	6	20·56	58·4919	150	29	43·24	4180·4262
158	256	18·69186	7119·7648	153	35	11·96	4266·2960	2	6	50·12	58·7201	151	28	21·84	4207·5759
159	257	17·82802	7147·1426	154	34	20·13	4293·6739	2	7	19·69	58·9482	152	27	0·44	4234·7256
160	258	16·96418	7174·5205	155	33	28·30	4321·0517	2	7	49·26	59·1764	153	25	39·04	4261·8753
161	259	16·10034	7201·8983	156	32	36·47	4348·4296	2	8	15·77	59·3809	154	24	20·70	4289·0486
162	260	15·23650	7229·2762	157	31	44·64	4375·8074	2	8	37·65	59·5498	155	23	6·99	4316·2576
163	261	14·37266	7256·6540	158	30	52·81	4403·1853	2	8	59·53	59·7186	156	21	53·28	4343·4767
164	262	13·50882	7284·0319	159	30	0·98	4430·5631	2	9	21·41	59·8874	157	20	39·57	4370·6757
165	263	12·64498	7311·4097	160	29	9·15	4457·9410	2	9	35·92	59·9994	158	19	33·22	4397·9415

TABLE XLVIII B—*Contd.*

Present Sûrya-Siddhânta.

24-hour periods from true Mêsha-saṁkrânti.	Sun's mean anomaly (or mean sun's distance from perigee-point) ("c").		Sun's mean Longitude.		Sun's equation of the centre. —		Sun's true Longitude ("s").	
1	2	3	4	5	6	7	8	9
	° ′	10,000th of circle.	° ′ ″	10,000ths of circle.	° ′ ″	10,000ths of circle.	° ′ ″	10,000ths of circle.
166	264 11·78114	7338·7876	161 28 17·32	4485·3188	2 9 48·93	60·0998	159 18 28·38	4425·2190
167	265 10·01730	7366·1654	162 27 25·49	4512·6967	2 10 1·94	60·2002	160 17 23·54	4452·4965
168	266 10·05345	7393·5433	163 26 33·66	4540·0745	2 10 14·95	60·3006	161 16 18·70	4479·7739
169	267 9·18961	7420·9211	164 25 41·83	4567·4524	2 10 19·24	60·3336	162 15 22·59	4507·1187
170	268 8·32577	7448·2990	165 24 50·00	4594·8302	2 10 23·38	60·3656	163 14 26·62	4534·4647
171	269 7·46193	7475·6768	166 23 58·17	4622·2081	2 10 27·52	60·3975	164 13 30·65	4561·8106
172	270 6·59809	7503·0547	167 23 6·34	4649·5859	2 10 20·54	60·4208	165 12 35·80	4589·1651
173	271 5·73425	7530·4325	168 22 14·51	4676·9638	2 10 26·40	60·3889	166 11 48·11	4616·5749
174	272 4·87041	7557·8104	169 21 22·67	4704·3416	2 10 22·26	60·3569	167 11 0·42	4643·9847
175	273 4·00657	7585·1882	170 20 30·84	4731·7195	2 10 18·12	60·3250	168 10 12·73	4671·3945
176	274 3·14273	7612·5661	171 19 39·01	4759·0973	2 10 11·45	60·2736	169 9 27·56	4698·8238
177	275 2·27889	7639·9439	172 18 47·18	4786·4752	2 9 58·44	60·1732	170 8 48·74	4726·3020
178	276 1·41505	7667·3218	173 17 55·35	4813·8530	2 9 45·43	60·0728	171 8 9·92	4753·7803
179	277 0·55121	7694·6996	174 17 3·52	4841·2309	2 9 32·42	59·9724	172 7 31·10	4781·2585
180	277 59·68737	7722·0775	175 16 11·69	4868·6087	2 9 15·56	59·8423	173 6 56·13	4808·7065
181	278 58·82353	7749·4553	176 15 19·86	4895·9866	2 8 53·68	59·6734	174 6 26·18	4836·3132
182	279 57·95968	7776·8332	177 14 28·03	4923·3644	2 8 31·80	59·5046	175 5 56·23	4863·5698
183	280 57·09584	7804·2110	178 13 36·20	4950·7423	2 8 9·92	59·3358	176 5 26·28	4891·4065
184	281 56·23200	7831·5889	179 12 44·37	4978·1201	2 7 43·65	59·1331	177 5 0·72	4918·9871
185	282 55·36816	7858·9667	180 11 52·54	5005·4980	2 7 10·08	58·9049	178 4 38·46	4946·5931
186	283 54·50432	7886·3446	181 11 0·71	5032·8758	2 6 44·51	58·6768	179 4 16·20	4974·1991
187	284 53·64048	7913·7224	182 10 8·88	5060·2537	2 6 14·94	58·4486	180 3 53·94	5001·8051
188	285 52·77664	7941·1003	183 9 17·05	5087·6315	2 5 37·00	58·1559	181 3 40·05	5029·4757
189	286 51·91280	7968·4781	184 8 25·22	5115·0094	2 4 59·15	57·8638	182 3 26·06	5057·1456
190	287 51·04896	7995·8560	185 7 33·39	5142·3872	2 4 21·31	57·5718	183 3 12·08	5084·8154

1	2		3	4			5	6			7	8			9
191	288	50·18512	8023·2339	186	6	41·56	5169·7651	2	3	41·98	57·2683	184	2	59·58	5112·4968
192	289	49·32128	8050·6117	187	5	49·73	5197·1430	2	2	55·85	56·9124	185	2	53·88	5140·2305
193	290	48·45744	8077·9896	188	4	57·90	5224·5208	2	2	9·72	56·5565	186	2	48·17	5167·9643
194	291	47·59360	8105·3674	189	4	6·07	5251·8987	2	1	23·60	56·2006	187	2	42·47	5195·6981
195	292	46·72976	8132·7453	190	3	14·24	5279·2765	2	0	34·51	55·8218	188	2	39·73	5223·4547
196	293	45·86591	8160·1231	191	2	22·40	5306·6544	1	59	40·10	55·4020	189	2	42·30	5251·2523
197	294	45·00207	8187·5010	192	1	30·57	5334·0322	1	58	45·70	54·9822	190	2	44·88	5279·0500
198	295	44·13823	8214·8788	193	0	38·74	5361·4101	1	57	51·29	54·5624	191	2	47·45	5306·8476
199	296	43·27439	8242·2567	193	59	46·91	5388·7879	1	56	52·62	54·1097	192	2	54·29	5334·6782
200	297	42·41055	8269·6345	194	58	55·08	5416·1658	1	55	50·53	53·6306	193	3	4·56	5362·5351
201	298	41·54671	8297·0124	195	58	3·25	5443·5436	1	54	48·43	53·1515	194	3	14·82	5390·3921
202	299	40·68287	8324·3902	196	57	11·42	5470·9215	1	53	46·34	52·6724	195	3	25·08	5418·2491
203	300	39·81903	8351·7681	197	56	19·59	5498·2993	1	52	38·37	52·1479	196	3	41·22	5446·1514
204	301	38·95522	8379·1459	198	55	27·76	5525·6772	1	51	28·59	51·6095	197	3	59·17	5474·0677
205	302	38·09135	8406·5238	199	54	35·93	5553·0550	1	50	18·81	51·0711	198	4	17·12	5501·9840
206	303	37·22751	8433·9016	200	53	44·10	5580·4329	1	49	9·03	50·5326	199	4	35·07	5529·9002
207	304	36·36367	8461·2795	201	52	52·27	5607·8107	1	47	51·63	49·9354	200	5	0·64	5557·8753
208	305	35·49983	8488·6573	202	52	0·44	5635·1886	1	46	34·16	49·3377	201	5	26·28	5585·8509
209	306	34·63599	8516·0352	203	51	8·61	5662·5664	1	45	16·69	48·7399	202	5	51·92	5613·8265
210	307	33·77214	8543·4130	204	50	16·78	5689·9443	1	43	59·56	48·1448	203	6	17·22	5641·7995
211	308	32·90830	8570·7909	205	49	24·95	5717·3221	1	42	35·00	47·4923	204	6	49·95	5669·8299
212	309	32·04446	8598·1687	206	48	32·12	5744·7000	1	41	10·43	46·8398	205	7	22·69	5697·8602
213	310	31·18062	8625·5466	207	47	41·29	5772·0778	1	39	45·87	46·1873	206	7	55·42	5725·8906
214	311	30·31678	8652·9244	208	46	49·46	5799·4557	1	38	20·10	45·5255	207	8	29·36	5753·9302
215	312	29·45294	8680·3023	209	45	57·63	5826·8335	1	36	49·03	44·8228	208	9	8·60	5782·0108
216	313	28·58910	8707·6801	210	45	5·80	5854·2114	1	35	17·96	44·1201	209	9	47·84	5810·0913
217	314	27·72526	8735·0580	211	44	13·97	5881·5892	1	33	46·89	43·4174	210	10	27·08	5838·1719
218	315	26·86142	8762·4358	212	43	22·14	5908·9671	1	32	12·51	42·6892	211	11	9·62	5866·2779
219	316	25·99758	8789·8137	213	42	30·30	5936·3449	1	30	34·94	41·9363	212	11	55·36	5894·4087
220	317	25·13374	8817·1915	214	41	38·47	5963·7228	1	28	57·37	41·1834	213	12	41·11	5922·5394
221	318	24·26990	8844·5694	215	40	46·64	5991·1006	1	27	19·79	40·4305	214	13	26·85	5950·6702
222	319	23·40606	8871·9472	216	39	54·81	6018·4785	1	25	39·41	39·6559	215	14	15·40	5978·8226
223	320	22·54222	8899·3251	217	39	2·98	6045·8563	1	23	55·92	38·8574	216	15	7·06	6006·9989
224	321	21·67837	8926·7030	218	38	11·15	6073·2342	1	22	12·43	38·0589	217	15	58·72	6035·1753
225	322	20·81453	8954·0808	219	37	19·32	6100·6121	1	20	28·94	37·2604	218	16	50·38	6063·3517

TABLE XLVIII B—*Contd.*

Present Sûrya-Siddhânta.

24-hour periods from true Mēsha-saṁkrānti.	Sun's mean anomaly (or mean sun's distance from perigee-point) ("c").			Sun's mean Longitude.				Sun's equation of the centre.				Sun's true Longitude ("s").			
1	2		3	4			5	6			7	8			9
	°	'	10,000ths of circle.	°	'	"	10,000ths of circle.	°	'	"	10,000ths of circle.	°	'	"	10,000ths of circle.
226	323	19·95069	8981·4587	220	36	27·49	6127·9899	1	18	40·47	36·4234	219	17	47·02	6091·5665
227	324	19·08685	9008·8365	221	35	35·66	6155·3678	1	16	51·07	35·5792	220	18	44·59	6119·7885
228	325	18·22301	9036·2144	222	34	43·83	6182·7456	1	15	1·67	34·7351	221	19	42·16	6148·0105
229	326	17·35917	9063·5922	223	33	52·00	6210·1235	1	13	13·16	33·8979	222	20	38·84	6176·2256
230	327	16·49533	9090·9701	224	33	0·17	6237·5013	1	11	19·03	33·0172	223	21	41·14	6204·4841
221	328	15·63148	9118·3479	225	32	8·34	6264·8792	1	9	24·90	32·1366	224	22	43·44	6232·7426
232	329	14·76765	9145·7258	226	31	16·51	6292·2570	1	7	30·76	31·2559	225	23	45·74	6261·0011
233	330	13·90381	9173·1036	227	30	24·68	6319·6349	1	5	35·16	30·3639	226	24	49·51	6289·2709
234	331	13·03997	9200·4815	228	29	32·85	6347·0127	1	3	35·71	29·4422	227	25	57·14	6317·5705
235	332	12·17613	9227·8593	229	28	41·02	6374·3906	1	1	36·25	28·5205	228	27	4·76	6345·8701
236	333	11·31229	9255·2372	230	27	49·19	6401·7684	0	59	36·80	27·5988	229	28	12·39	6374·1697
237	334	10·44844	9282·6150	231	26	57·36	6429·1463	0	57	36·40	26·6697	230	29	20·96	6402·4765
238	335	9·58460	9309·9929	232	26	5·53	6456·5241	0	55	33·40	25·7207	231	30	32·13	6430·8035
239	336	8·72076	9337·3707	233	25	13·70	6483·9020	0	53	30·39	24·7716	232	31	43·30	6459·1304
240	337	7·85692	9364·7486	234	24	21·87	6511·2798	0	51	27·39	23·8225	233	32	54·48	6487·4574
241	338	6·99308	9392·1264	235	23	30·03	6538·6577	0	49	21·23	22·8490	234	34	8·81	6515·8087
242	339	6·12924	9419·5043	236	22	38·20	6566·0355	0	47	14·67	21·8725	235	35	23·53	6544·1630
243	340	5·26540	9446·8821	237	21	46·37	6593·4134	0	45	8·12	20·8960	236	36	38·25	6572·5174
244	341	4·41056	9474·2600	238	20	54·54	6620·7912	0	43	1·57	19·9195	237	37	52·97	6600·8717
245	342	3·53772	9501·6378	239	20	2·71	6648·1691	0	40	51·82	18·9184	238	39	10·89	6629·2507
246	343	2·67388	9529·0157	240	19	10·88	6675·5469	0	38	41·72	17·9145	239	40	29·16	6657·6324
247	344	1·81004	9556·3935	241	18	19·05	6702·9248	0	36	31·62	16·9107	240	41	47·43	6686·0141
248	345	0·94620	9583·7714	242	17	27·22	6730·3026	0	34	21·74	15·9085	241	43	5·48	6714·3942
249	346	0·08236	9611·1492	243	16	35·39	6757·6805	0	32	8·68	14·8818	242	44	26·71	6742·7987
250	346	59·21852	9638·5271	244	15	43·56	6785·0583	0	29	55·63	13·8551	243	45	47·94	6771·2032

1	2		3	4			5	6			7	8			9
251	347	58·35467	9665·9049	245	14	51·73	6812·4362	0	27	42·57	12·8285	244	47	9·16	6799·6077
252	348	57·49083	9693·2828	246	13	59·90	6839·8140	0	25	29·91	11·8049	245	48	29·99	6828·0092
253	349	56·62699	9720·6606	247	13	8·07	6867·1919	0	23	15·08	10·7645	246	49	52·99	6856·4274
254	350	55·76315	9748·0385	248	12	16·24	6894·5697	0	21	0·25	9·7242	247	51	15·99	6884·8456
255	351	54·89931	9775·4163	249	11	24·41	6921·9476	0	18	45·42	8·6838	248	52	38·99	6913·2638
256	352	54·03547	9802·7942	250	10	32·58	6949·3254	0	16	29·89	7·6303	249	54	3·69	6941·6951
257	353	53·17163	9830·1721	251	9	40·75	6976·7033	0	14	12·28	6·5763	250	55	28·47	6970·1270
258	354	52·30779	9857·5499	252	8	48·92	7004·0812	0	11	55·68	5·5222	251	56	53·24	6998·5589
259	355	51·44395	9884·9277	253	7	57·09	7031·4590	0	9	39·07	4·4682	252	58	18·01	7026·9908
260	356	50·58011	9912·3056	254	7	5·26	7058·8369	0	7	21·28	3·4049	253	59	43·98	7055·4319
261	357	49·71627	9939·6835	255	6	13·43	7086·2147	0	5	3·49	2·3418	255	1	9·93	7083·8729
262	358	48·85243	9967·0613	256	5	21·60	7113·5926	0	2	45·71	1·2786	256	2	35·89	7112·3140
263	359	47·98859	9994·4392	257	4	29·77	7140·9704	0	0	27·92	0·2154	257	4	1·85	7140·7550
Sun in perigee	*360*	*0·0*	*10000·0*	*257*	*16*	*30·45*	*7156·5313*	*0*	*0*	*0·0*	*0·0*	*257*	*16*	*30·45*	*7146·5313*

(*The Sun's equation of centre is +, plus, after his mean anomaly=360° till it reaches 180°.*)

1	2		3	4			5	Sun's equation of the centre. +			7	8			9
264	0	47·12475	21·8170	258	3	37·93	7168·3483	0	1	49·80	0·8472	258	5	27·74	7169·1955
265	1	46·26090	49·1949	259	2	46·10	7195·7261	0	4	7·59	1·9104	259	6	53·69	7197·6365
266	2	45·39706	76·5727	260	1	54·27	7223·1040	0	6	25·38	2·9736	260	8	19·65	7226·0775
267	3	44·53222	103·9506	261	1	2·44	7250·4818	0	8	43·16	4·0367	261	9	45·61	7254·5186
268	4	43·66938	131·3284	262	0	10·61	7277·8597	0	10	59·71	5·0903	262	11	10·32	7282·9500
269	5	42·80554	158·7063	262	59	18·78	7305·2375	0	13	16·31	6·1444	263	12	35·10	7311·3819
270	6	41·94170	186·0841	263	58	26·95	7332·6154	0	15	32·92	7·1984	264	13	59·87	7339·8138
271	7	41·07786	213·4620	264	57	35·12	7359·9932	0	17	49·67	8·2536	265	15	24·79	7368·2468
272	8	40·21402	240·8398	265	56	43·29	7387·3711	0	20	4·50	9·2940	266	16	47·79	7396·6650
273	9	39·35018	268·2177	266	55	51·46	7414·7489	0	22	19·33	10·3343	267	18	10·79	7425·0832
274	10	38·48634	295·5955	267	54	59·63	7442·1268	0	24	34·16	11·3747	268	19	33·79	7453·5015
275	11	37·62250	322·9734	268	54	7·80	7469·5046	0	26	49·29	12·4174	269	20	57·09	7481·9220
276	12	36·75866	350·3512	269	53	15·97	7496·8825	0	29	2·35	13·4440	270	22	18·32	7510·3265
277	13	35·89482	377·7291	270	52	24·14	7524·2603	0	31	15·40	14·4707	271	23	39·54	7538·7310
278	14	35·03098	405·1069	271	51	32·31	7551·6382	0	33	28·46	15·4974	272	25	0·77	7567·1356
279	15	34·16713	432·4848	272	50	40·48	7579·0160	0	35	39·03	16·5049	273	26	19·51	7395·5209
280	16	33·30329	459·8626	273	49	48·65	7606·3939	0	37	49·13	17·5088	274	27	37·78	7623·9026
281	17	32·43945	487·2405	274	48	56·82	7633·7717	0	39	59·23	18·5126	275	28	56·05	7652·2843
282	18	31·57561	514·6183	275	48	4·99	7661·1496	0	42	9·33	19·5165	276	30	14·32	7680·6660
283	19	30·71177	541·9962	276	47	13·16	7688·5274	0	44	43·43	20·4971	277	31	29·58	7709·0246

TABLE XLVIII B—*Contd.*

Present Sūrya-Siddhānta.

24 hour periods from true Mēsha-saṁkranti	Sun's mean anomaly (or mean sun's distance from perigee-point) ("c")			Sun's mean Longitude				Sun's equation of the centre +				Sun's true Longitude ("s")			
1	2		3	4			5	6			7	8			9
	°	′	10,000ths of circle.	°	′	″	10,000ths of circle.	°	′	″	10,000ths of circle.	°	′	″	10,000ths of circle.
284	20	29·84793	569·3740	277	46	21·33	7715·9053	0	46	22·98	21·4736	278	32	44·30	7737·3789
285	21	28·98409	596·7519	278	45	29·50	7743·2831	0	48	29·53	22·4501	279	33	59·02	7765·7332
286	22	28·12025	624·1297	279	44	37·66	7770·6610	0	50	36·08	23·4265	280	35	13·74	7794·0875
287	23	27·25641	651·5076	280	43	45·83	7798·0388	0	52	39·48	24·3787	281	36	25·32	7822·4176
288	24	26·39257	678·8854	281	42	54·00	7825·4167	0	54	42·49	25·3278	282	39	36·49	7850·7445
289	25	25·52873	706·2633	282	42	2·17	7852·7945	0	56	45·49	26·2769	283	38	47·66	7879·0715
290	26	24·66489	733·6412	283	41	10·34	7880·1724	0	58	48·86	27·2288	284	39	59·20	7907·4012
291	27	23·80105	761·0190	284	40	18·51	7907·5503	1	0	48·31	28·1505	285	41	6·82	7935·7008
292	28	22·93721	788·3969	285	39	26·68	7934·9281	1	2	47·77	29·0723	286	42	14·45	7964·0004
293	29	22·07336	815·7747	286	38	34·85	7962·3060	1	4	47·22	29·9940	287	43	22·07	7992·2999
294	30	21·20952	843·1526	287	37	43·02	7989·6838	1	6	44·18	30·8965	288	44	27·21	8020·5803
295	31	20·34568	870·5304	288	36	51·19	8017·0617	1	8	38·32	31·7771	289	45	29·51	8048·8388
296	32	19·48184	897·9083	289	35	59·36	8044·4395	1	10	32·45	32·6578	290	46	31·81	8077·0973
297	33	18·61801	925·2861	290	35	7·53	8071·8174	1	12	26·58	33·5384	291	47	34·11	8105·3558
298	34	17·75416	952·6640	291	34	15·70	8099·1952	1	14	18·31	34·4005	292	48	34·01	8133·5958
299	35	16·89032	980·0418	292	33	23·87	8126·5731	1	16	7·71	35·2447	293	49	31·58	8161·8178
300	36	16·02648	1007·4197	293	32	32·04	8153·9509	1	17	57·12	36·0886	294	50	29·15	8190·0398
301	37	15·16264	1034·7975	294	31	40·21	8181·3288	1	19	46·52	36·9330	295	51	26·73	8218·2618
302	38	14·29880	1062·1754	295	30	48·38	8208·7066	1	21	30·40	37·7346	296	52	18·78	8245·4412
303	39	13·43496	1089·5532	296	29	56·55	8236·0845	1	23	13·89	38·5331	297	53	10·44	8274·6176
304	40	12·57112	1116·9311	297	29	4·72	8263·4623	1	24	57·38	39·3316	298	54	2·10	8302·7939
305	41	11·70728	1144·3089	298	28	12·89	8290·8402	1	26	40·87	40·1301	299	54	53·75	8330·9703
306	42	10·84343	1171·6868	299	27	21·06	8318·2180	1	28	18·76	40·8855	300	55	39·82	8359·1035
307	43	9·97959	1199·0646	300	26	29·23	8345·5959	1	29	56·34	41·6384	301	56	25·56	8387·2343
308	44	9·11575	1226·4425	301	25	37·40	8372·9737	1	31	33·91	42·3913	302	57	11·31	8415·3650

1	2		3	4			5	6			7	8			9
309	45	8·25191	1253·8203	302	24	45·56	8400·3516	1	33	9·54	43·1292	303	57	55·11	8443·4808
310	46	7·38807	1281·1982	303	23	53·73	8427·7294	1	34	40·61	43·8319	304	58	34·35	8471·5613
311	47	6·52423	1308·5760	304	23	1·90	8455·1073	1	36	11·68	44·5346	305	59	13·59	8499·6419
312	48	5·66039	1335·9539	305	22	10·07	8482·4851	1	37	42·75	45·2373	306	59	52·83	8527·7224
313	49	4·79655	1363·3317	306	21	18·24	8509·8630	1	39	12·00	45·9259	308	0	30·24	8555·7889
314	50	3·93271	1390·7096	307	20	26·41	8537·2408	1	40	36·56	46·5784	309	1	2·97	8583·8192
315	51	3·06887	1418·0874	308	19	34·58	8564·6187	1	42	1·13	47·2309	310	1	35·71	8611·8496
316	52	2·20503	1445·4653	309	18	42·75	8591·9965	1	43	25·69	47·8834	311	2	8·44	8639·8800
317	53	1·34118	1472·8431	310	17	50·92	8619·3744	1	44	46·01	48·5032	312	2	36·93	8667·8776
318	54	0·47735	1500·2210	311	16	59·09	8646·7522	1	46	3·48	49·1009	313	3	2·57	8695·8532
319	54	59·61351	1527·5988	312	16	7·26	8674·1301	1	47	20·95	49·6987	314	3	28·21	8723·8288
320	55	58·74966	1554·9767	313	15	15·43	8701·5079	1	48	38·42	50·2964	315	3	53·85	8751·8044
321	56	57·88582	1582·3545	314	14	23·60	8728·8858	1	49	49·52	50·8451	316	4	13·12	8779·7309
322	57	57·02198	1609·7324	315	13	13·77	8756·2636	1	50	59·30	51·3835	317	4	31·07	8807·6472
323	58	56·15814	1637·1103	316	12	39·94	8783·6415	1	52	9·08	51·9219	318	4	49·02	8835·5634
324	59	55·29430	1664·4881	317	11	48·11	8811·0194	1	53	18·86	52·4604	319	5	6·97	8863·4797
325	60	54·43046	1691·8660	318	10	56·28	8838·3972	1	54	22·51	52·9515	320	5	18·79	8891·3487
326	61	53·56662	1719·2438	319	10	4·45	8865·7751	1	55	24·60	53·4306	321	5	29·05	8919·2056
327	62	52·70278	1740·6217	320	9	12·62	8893·1529	1	56	26·70	53·9097	322	5	39·31	8947·0626
328	63	51·83894	1773·9995	321	8	20·79	8920·5308	1	57	28·60	54·3873	323	5	49·39	8974·9181
329	64	50·97510	1801·3774	322	7	28·96	8947·9086	1	58	23·01	54·8071	324	5	51·96	9002·7158
330	65	50·11126	1828·7552	323	6	37·13	8975·2865	1	59	17·41	55,2269	325	5	54·54	9030·5134
331	66	49·24742	1856·1331	324	5	45·30	9002·6643	2	0	11·82	55·6467	326	5	57·11	9058·3110
332	67	48·38358	1883·5109	325	-4	53·46	9030·0422	2	1	4·24	56·0512	327	5	57·70	9086·0934
333	68	47·51974	1910·8888	326	4	1·63	9057·4200	2	1	50·36	56·4071	328	5	52·00	9113·8272
334	69	46·65589	1938·2666	327	3	9·80	9084·7979	2	2	36·49	56·7630	329	5	46·29	9141·5609
335	70	45·79205	1965·6445	328	2	17·97	9112·1757	2	3	22·62	57·1190	330	5	40·59	9169·2947
336	71	44·92821	1993·0223	329	1	26·14	9139·5536	2	4	5·17	57·4473	331	5	31·32	9197·0009
337	72	44·06437	2020·4002	330	0	34·31	9166·9314	2	4	43·02	57·7394	332	5	17·73	9224·6708
338	73	43·20053	2047·7780	330	59	42·48	9194·3093	2	5	20·87	58·0314	333	5	3·35	9252·3407
339	74	42·33669	2075·1559	331	58	50·65	9221·6871	2	5	58·72	58·3234	334	4	49·37	9280·0105
340	75	41·47285	2102·5337	332	57	58·82	9249·0650	2	6	31·51	58·5765	335	4	30·34	9307·6415
341	76	40·60901	2129·9116	333	57	6·99	9276·4428	2	7	1·08	58·8047	336	4	8·07	9335·2475
342	77	39·74517	2157·2894	334	56	15·16	9303·8207	2	7	30·65	59·0328	337	3	45·81	9362·8535
343	78	38·88133	2184·6673	335	55	23·33	9331·1985	2	8	0·22	59·2610	338	3	23·55	9390·4595

TABLE XLVIII B—*Contd.*

Present Sûrya-Siddhânta.

24-hour periods from true Mêsha-saṁkrânti.	Sun's mean anomaly (or mean sun's distance from perigee-point) ("c").			Sun's mean Longitude.				Sun's equation of the centre. +				Sun's true Longitude ("s").			
1	2		3	4			5	6			7	8			9
	°	'	10,000ths of circle.	°	'	"	10,000ths of circle.	°	'	"	10,000ths of circle.	°	'	"	10,000ths of circle.
344	79	38·01749	2212·0451	336	54	31·50	9358·5764	2	8	23·88	59·4435	339	2	55·38	9418·0199
345	80	37·15363	2239·4230	337	53	39·67	9385·9542	2	8	45·76	59·6123	340	2	25·43	9445·5666
346	81	36·28981	2266·8008	338	52	47·84	9413·3321	2	9	7·64	59·7812	341	1	55·48	9473·1133
347	82	35·42597	2294·1787	339	51	56·01	9440·7099	2	9	27·74	59·9362	342	1	23·74	9500·6462
348	83	34·56212	2321·5565	340	51	4·18	9468·0878	2	9	40·75	60·0366	343	0	44·92	9528·1244
349	84	33·69828	2348·9344	341	50	12·35	9495·4656	2	9	53·76	60·1370	344	0	6·10	9555·6026
350	85	32·83444	2376·3122	342	49	20·52	9522·8435	2	10	6·77	60·2374	344	59	27·28	9583·0809
351	86	31·97060	2403·6901	343	48	28·69	9550·2213	2	10	16·64	60·3136	345	58	45·33	9610·5349
352	87	31·10676	2431·0679	344	47	36·86	9577·5992	2	10	20·77	60·3455	346	57	57·63	9637·9447
353	88	30·24292	2458·4458	345	46	45·03	9604·9770	2	10	24·91	60·3774	347	57	9·94	9665·3544
354	89	29·37908	2485·8236	346	45	53·19	9632·3549	2	10	29·05	60·4093	348	56	22·24	9692·7642
355	90	28·51524	2513·2015	347	45	1·36	9659·7327	2	10	29·00	60·4090	349	55	30·37	9720·1417
356	91	27·65140	2540·5794	348	44	9·53	9687·1106	2	10	24·86	60·3770	350	54	34·40	9747·4876
357	92	26·78756	2567·9572	349	43	17·70	9714·4885	2	10	20·72	60·3451	351	53	38·43	9774·8336
358	93	25·92372	2595·3351	350	42	25·87	9741·8663	2	10	16·59	60·3132	352	52	42·46	9802·1795
359	94	25·05988	2622·7129	351	41	34·04	9769·2442	2	10	6·63	60·2363	353	51	40·67	9829·4805
360	95	24·19604	2650·0908	352	40	42·21	9796·6220	2	9	53·62	60,1360	354	50	35·83	9856·7580
361	96	23·33216	2677·4686	353	39	50·38	9823·9999	2	9	40·61	60·0356	355	49	30·99	9884·0354
362	97	22·46835	2704·8465	354	38	58·55	9851·3777	2	9	27·60	59·9352	356	48	26·15	9911·3129
363	98	21·60451	2732·2243	355	38	6·72	9878·7556	2	9	7·45	59·7797	357	47	14·17	9938·5353
364	99	20·74067	2759·6022	356	37	14·89	9906·1334	2	8	45·57	59·6109	358	46	0·46	9965·7443
365	100	19·87683	2786·9800	357	36	23·06	9033·5113	2	8	23·69	59·4420	359	44	46·75	9992·9533

TABLE XLIX.

ELEMENTS OF THE SUN'S TRUE LONGITUDE.

HOURS.

N.B.—Column 1 corresponds to the 24-hour periods, measured from true Mēsha-saṁkrānti, entered in Column 1 Tables XLVIII A and B. In the present Table they are grouped in conformity with the Hindu Sine-Table. Figures in Columns 4 to 6 give the actual area travelled on the ecliptic by the true sun in the given number of hours. For minutes see Table L, following. The Table is exact for the FIRST ĀRYA-SIDDHĀNTA, but can be used for all the Hindu authorities.

Grouping of the days.

(*a*) Days 1 to 85 in order, and in reverse order days 86 to 164.
(*b*) Days 165 to 267 in order, and in reverse order days 268 to 363.
(*c*) Days 363 to 365 are grouped with Day 1.

This arrangement had to be adopted to prevent the size of the Table being doubled.

24-hour periods from true Mēsha-saṁkrānti (inclusive).	Arc travelled by true sun in 24 hours.		Arc travelled by true sun per hour.					
	° ′ ″	10,000ths of circle.	No. of Hours.	′ ″	10,000ths of circle.	No. of hours.	′ ″	10,000ths of circle.
1	2	3	4	5	6	4	5	6
363 to 1 } 162 to 164 }	0 58 46·29	27·2090	1	2 26·93	1·1337	13	31 50·07	14·7382
			2	4 53·86	2·2674	14	34 17·00	15·8719
			3	7 20·79	3·4011	15	36 43·93	17·0056
			4	9 47·71	4·5348	16	39 10·86	18·1394
			5	12 14·64	5·6685	17	41 37·79	19·2731
			6	14 41·57	6·8023	18	44 4·72	20·4068
			7	17 8·50	7·9360	19	46 31·65	21·5405
			8	19 35·43	9·0697	20	48 58·57	22·6742
			9	22 2·36	10·2034	21	51 25·50	23·8079
			10	24 29·29	11·3371	22	53 52·43	24·9416
			11	26 56·22	12·4708	23	56 19·36	26·0753
			12	29 23·14	13·6045			
2 to 5 } 158 to 161 }	0 58 38·01	27·1451	1	2 26·58	1·1310	13	31 45·59	14·7036
			2	4 53·17	2·2621	14	34 12·17	15·8347
			3	7 19·75	3·3931	15	36 38·76	16·9657
			4	9 46·34	4·5242	16	39 5·34	18·0968
			5	12 12·92	5·6552	17	41 31·92	19·2278
			6	14 39·50	6·7863	18	43 58·51	20·3589
			7	17 6·09	7·9173	19	46 25·09	21·4899
			8	19 32·67	9·0484	20	48 51·68	22·6210
			9	21 59·25	10·1794	21	51 18·26	23·7520
			10	24 25·84	11·3105	22	53 44·8[illegible]	24·8831
			11	26 52·42	12·4415	23	56 11·43	26·0141
			12	29 19·01	13·5726			

TABLE XLIX—*Contd.*

24-hour periods from true Mēsha-saṁkrānti (inclusive).	Arc travelled by true sun in 24 hours.		Arc travelled by true sun per hour.					
	° ′ ″	10,000ths of circle.	No. of Hours.	′ ″	10,000ths of circle.	No. of hours.	′ ″	10,000ths of circle.
1	2	3	4	5	6	4	5	6
6 to 8 } 154 to 157 }	0 58 29·73	27·0813	1	2 26·24	1·1284	13	31 41·10	14 6690
			2	4 52·48	2·2568	14	34 7·34	15·7974
			3	7 18·72	3·3852	15	36 33·58	16·9258
			4	9 44·96	4 5135	16	38 59 82	18·0542
			5	12 11·19	5·6419	17	41 26 06	19·1826
			6	14 37·43	6·7703	18	43 52·30	20·3109
			7	17 3·67	7·8987	19	46 18·54	21 4393
			8	19 29·91	9·0271	20	48 44·78	22·5677
			9	21 56·15	10·1555	21	51 11·02	23 6961
			10	24 22·39	11·2839	22	53 37·25	24·8245
			11	26 48·63	12·4122	23	56 3·49	25·9529
			12	29 14·87	13·5406			
9 to 12 } 150 to 153 }	0 58 21 45	27·0174	1	2 25·89	1·1257	13	31 36·62	14·6344
			2	4 51·79	2·2514	14	34 2·51	15·7601
			3	7 17·68	3·3772	15	36 28·41	16 8859
			4	9 43 58	4·5029	16	38 54 30	18·0116
			5	12 9·47	5·6286	17	41 20·20	19·1373
			6	14 35·36	6·7543	18	43 46·09	20·2630
			7	17 1·26	7·8801	19	46 11·98	21·3888
			8	19 27·15	9 0058	20	48 37·88	22·5145
			9	21 53·04	10·1315	21	51 3·77	23·6402
			10	24 18·94	11·2572	22	53 29·66	24·7659
			11	26 44·83	12·3830	23	55 55·56	25·8917
			12	29 10 73	13·5087			
13 to 16 } 147 to 149 }	0 58 13·17	26·9535	1	2 25·55	1·1231	13	31 32·14	14 5998
			2	4 51·10	2·2461	14	33 57·68	15·7229
			3	7 16·65	3·3692	15	36 23·23	16·8459
			4	9 42 20	4·4923	16	38 48·78	17·9690
			5	12 7·74	5·6153	17	41 14·33	19·0921
			6	14 33·29	6 7384	18	43 39·88	20·2351
			7	16 58·84	7·8614	19	46 5·43	21·3382
			8	19 24·39	8·9845	20	48 30·98	22·4613
			9	21 49 94	10.1076	21	50 56·53	23·5843
			10	24 15·49	11·2306	22	53 22·08	24·7074
			11	26 41·04	12·3537	23	55 47 62	25·8304
			12	29 6·59	13 4768			
17 to 20 } 143 to 146 }	0 58 5·49	26·8942	1	2 25·23	1·1206	13	31 27·97	14·5677
			2	4 50 46	2·2412	14	33 53·20	15·6883
			3	7 15·69	3·3618	15	36 18·43	16·8089
			4	9 40·91	4·4824	16	38 43·66	17·9295
			5	12 6·14	5.6030	17	41 8·89	19·0500
			6	14 31·37	6·7235	18	43 34·11	20·1706
			7	16 56·60	7·8441	19	45 59·34	21·2912
			8	19 21·83	8·9647	20	48 24·57	22·4118
			9	21 47·06	10·0853	21	50 49·80	23 5324
			10	24 12·29	11·2059	22	53 15·03	24·6530
			11	26 37·51	12·3265	23	55 40·26	26·7736
			12	29 2·74	13 4471			

TABLE XLIX—*Contd.*

24-hour periods from true Mēsha-samkrānti (inclusive).	Arc travelled by true sun in 24 hours.		Arc travelled by true sun per hour.					
	° ′ ″	10,000ths of circle.	No. of hours.	′ ″	10,000ths of circle.	No. of hours.	′ ″	10,000ths of circle.
1	2	3	4	5	6	4	5	6
21 to 24 } 139 to 142 }	0 57 57·80	26·8349	1	2 24.91	1·1181	13	31 23·81	14·5356
			2	4 49·82	2·2362	14	33 48·72	15 6537
			3	7 14·72	3·3544	15	36 13 62	16·7718
			4	9 39·63	4·4725	16	38 38·53	17·8899
			5	12 4·54	5·5906	17	41 3·44	19·0080
			6	14 29 45	6·7087	18	43 28·35	20·1261
			7	16 54·36	7·8268	19	45 53·26	21·2443
			8	19 19·27	8·9450	20	48 18·17	22·3624
			9	21 44·17	10 0631	21	50 43 07	23·4805
			10	24 9·08	11·1812	22	53 7·98	24·5986
			11	26 33 99	12·2993	23	55 32 89	25·7167
			12	28 58 90	13·4174			
25 to 28 } 135 to 138 }	0 57 50·70	26·7801	1	2 24 61	1·1158	13	31 19·96	14 5059
			2	4 49 23	2·2317	14	33 44·58	15·6217
			3	7 13·84	3·3475	15	36 9·19	16·7376
			4	9 38·45	4·4634	16	38 33 80	17·8534
			5	12 3 06	5·5792	17	40 58·41	18 9693
			6	14 27·68	6·6950	18	43 23·03	20·0851
			7	16 52 29	7·8109	19	45 47 64	21 2009
			8	19 16·90	8·9267	20	48 12·25	22·3168
			9	21 41·51	10·0425	21	50 36·86	23·4326
			10	24 6·13	11·1484	22	53 1·48	24·5485
			11	26 30·74	12·2742	23	55 26·09	25·6643
			12	28 55·35	13·3901			
29 to 31 } 131 to 134 }	0 57 43·60	26·7254	1	2 24·32	1·1136	13	31 16·11	14·4762
			2	4 48·63	2·2271	14	33 40·43	15·5898
			3	7 12·95	3·3407	15	36 4·75	16·7033
			4	9 37·27	4·4542	16	38 29·06	17·8169
			5	12 1·58	5·5678	17	40 53·38	18·9305
			6	14 25·90	6·6813	18	43 17·70	20·0440
			7	16 50·22	7·7949	19	45 42·01	21·1576
			8	19 14·53	8·9085	20	48 6·33	22·2711
			9	21 38·85	10·0220	21	50 30·65	23·3847
			10	24 3·17	11·1356	22	52 54·96	24·4982
			11	26 27·48	12·2491	23	55 19·28	25·6118
			12	28 51·80	13·3627			
32 to 35 } 127 to 130 }	0 57 37·10	26·6752	1	2 24·05	1·1115	13	31 12·60	14·4490
			2	4 48 09	2·2229	14	33 36·64	15·5605
			3	7 12·14	3·3344	15	36 0·69	16·6720
			4	9 36·18	4·4459	16	38 24·73	17·7834
			5	12 0·23	5·5573	17	40 48·78	18·8942
			6	14 24·28	6 6688	18	43 12·83	20·0064
			7	16 48·32	7·7803	19	45 36·87	21·1178
			8	19 12·37	8·8917	20	48 0·92	22·2293
			9	21 36·41	10·0032	21	50 24·96	23·3402
			10	24 0·46	11·1147	22	52 49·01	24·4522
			11	26 24·50	12·2261	23	55 13·05	25·5637
			12	28 48·55	13·3376			

TABLE XLIX—*Contd.*

24 hour periods from true Mēsha-saṁkrānti (inclusive).	Arc travelled by true sun in 24 hours.		Arc travelled by true sun per hour.					
	° ′ ″	10,000ths of circle.	No. of hours.	′ ″	10,000ths of circle.	No. of hours.	′ ″	10,000ths of circle.
1	2	3	4	5	6	4	5	6
36 to 39 } 124 to 126 }	0 57 31·19	26·6295	1	2 23·80	1·1096	13	31 9·39	14·4243
			2	4 47·60	2·2191	14	33 33·19	15·5339
			3	7 11·40	3·3287	15	35 56·99	16·6435
			4	9 35·20	4·43[illegible]3	16	38 20·79	17·7530
			5	11 59·00	5·5478	17	40 44·59	18·8626
			6	14 22·80	6·6574	18	43 8·39	19·9721
			7	16 46·60	7·7669	19	45 32·19	21·0817
			8	19 10·40	8·8765	20	47 55·99	22·1913
			9	21 34·20	9·9861	21	50 19·79	23·3008
			10	23 57·99	11·0956	22	52 43·59	24·4104
			11	26 21·79	12·2052	23	55 7·39	25·5200
			12	28 45·59	13·3148			
40 to 43 } 120 to 123 }	0 57 25·27	26·5839	1	2 23·55	1·1077	13	31 6·19	14·3996
			2	4 47·11	2·2153	14	33 29·74	15·5073
			3	7 10·66	3·32[illegible]0	15	35 53·30	16·6149
			4	9 34·21	4·4306	16	38 16·85	17·7226
			5	11 57·77	5·5383	17	40 40·40	18·8303
			6	14 21·32	6·6460	18	43 3·96	19·9379
			7	16 44·87	7·7536	19	45 27·51	21·0456
			8	19 8·42	8·8613	20	47 51·06	22·1532
			9	21 31·98	9·9690	21	50 14·61	23·2609
			10	23 53·53	11·0[illegible]66	22	52 38·17	24·3[illegible]86
			11	26 19·08	12·1[illegible]43	23	55 1·72	25·4762
			12	28 42·64	13·2919			
44 to 47 } 116 to 119 }	0 57 19·95	26·5428	1	2 23·33	1·1060	13	31 3·31	14·3774
			2	4 46·66	2·2119	14	33 26·64	15·4833
			3	7 9·99	3·3179	15	35 49·97	16·5893
			4	9 33·33	4·4238	16	38 13·30	17·6952
			5	11 56·66	5·5298	17	40 36·63	18·8012
			6	14 19·99	6·6457	18	42 59·96	19·9071
			7	16 43·32	7·7417	19	45 23·29	21·0131
			8	19 6·65	8·8476	20	47 46·63	22·1190
			9	21 21·98	9·9536	21	50 9·96	23·2250
			10	23 53·31	11·0595	22	52 33·29	24·3309
			11	26 16·64	12·1655	23	54 56·62	25·4369
			12	28 39·98	13·2714			
48 to 50 } 112 to 1[illegible]5 }	0 57 15·22	26·5063	1	2 23·13	1·1044	13	31 0·74	14·5576
			2	4 46·27	2·2089	14	33 23·88	15·4623
			3	7 9·40	3·3133	15	35 47·01	16·5665
			4	9 32·54	4·4177	16	38 10·15	17·6709
			5	11 55·67	5·5222	17	40 33·28	18·7753
			6	14 18·81	6·6266	18	42 56·42	19·8797
			7	16 41·94	7·7310	19	45 19·55	20·9842
			8	19 5·07	8·8354	20	47 42·68	22·0886
			9	21 28·21	9·9399	21	50 5·82	23·1930
			10	23 51·34	11·0443	22	52 28·95	24·2975
			11	26 14·48	12·1487	23	54 52·09	25·4019
			12	28 37·61	13·2532			

TABLE XLIX—*Contd.*

24-hour periods from true Mēsha-saṁkrānti (inclusive).	Arc travelled by true sun in 24 hours.				Arc travelled by true sun per hour.							
	°	′	″	10,000ths of circle.	No. of hours.	′	″	10,000ths of circle.	No. of hours.	′	″	10,000ths of circle.
1	2			3	4	5		6	4	5		6
51 to 54 } 108 to 111 }	0	57	10·49	26·4698	1	2	22·94	1·1029	13	30	58·18	14·3378
					2	4	45·87	2·2058	14	33	21·12	15·4407
					3	7	8·81	3·3087	15	35	44·06	16·5436
					4	9	31·75	4·4116	16	38	6·99	17·6466
					5	11	54·69	5·5145	17	40	29·93	18·7495
					6	14	17·62	6·6175	18	42	52·87	19·8524
					7	16	40·56	7·7204	19	45	15·80	20·9553
					8	19	3·50	8·8233	20	47	38·74	22·0582
					9	21	26·43	9·9262	21	50	1·68	23·1611
					10	23	49·37	11·0291	22	52	24·62	24·2640
					11	26	12·31	12·1320	23	54	47·55	25·3669
					12	28	35·24	13·2349				
55 to 58 } 105 to 107 }	0	57	6·94	26·4424	1	2	22·79	1·1018	13	30	56·26	14·3230
					2	4	45·58	2·2035	14	33	19·05	15·4248
					3	7	8·37	3·3053	15	35	41·84	16·5265
					4	9	31·16	4·4071	16	38	4·63	17·6283
					5	11	53·95	5·5088	17	40	27·42	18·7301
					6	14	16·74	6·6106	18	42	50·21	19·8318
					7	16	39·52	7·7124	19	45	13·00	20·9336
					8	19	2·31	8·8141	20	47	35·78	22·0354
					9	21	25·10	9·9159	21	49	58·57	23·1371
					10	23	47·89	11·0177	22	52	21·36	24·2389
					11	26	10·68	12·1195	23	54	44·15	25·3407
					12	28	33·47	13·2212				
59 to 62 } 101 to 104 }	0	57	3·98	26·4196	1	2	22·67	1·1008	13	30	54·66	14·3106
					2	4	45·33	2·2016	14	33	17·32	15·4115
					3	7	8·00	3·3025	15	35	39·99	16·5123
					4	9	30·66	4·4033	16	38	2·6[illegible]	17·6131
					5	11	53·33	5·5041	17	40	25·32	18·7139
					6	14	16·00	6·6049	18	42	47·99	19·8147
					7	16	38·66	7·7057	19	45	10·65	20·9155
					8	19	1·33	8·8065	20	47	33·32	22·0164
					9	21	23·99	9·9074	21	49	55·99	23·1172
					10	23	46·66	11·0082	22	52	18·65	24·2180
					11	26	9·33	12·1090	23	54	41·32	25·3188
					12	28	31·99	13·2098				
63 to 66 } 97 to 100 }	0	57	1·03	26·3968	1	2	22·54	1·0999	13	30	53·06	14·2983
					2	4	45·09	2·1997	14	33	15·60	15·3981
					3	7	7·63	3·2996	15	35	38·14	16·4980
					4	9	30·17	4·3995	16	38	0·69	17·5979
					5	11	52·71	5·4993	17	40	23·23	18·6977
					6	14	15·26	6·5992	18	42	45·77	19·7976
					7	16	37·80	7·6991	19	45	8·31	20·8975
					8	19	0·34	8·7989	20	47	30·86	21·9973
					9	21	22·89	9·8988	21	49	53·40	23·0972
					10	23	45·43	10·9987	22	52	15·94	24·1971
					11	26	7·97	12·0985	23	54	38·48	25·2970
					12	28	30·51	13·1984				

TABLE XLIX—*Contd.*

24-hour periods from true Mēsha-saṁkrānti (inclusive).	Arc travelled by true sun in 24 hours.				Arc travelled by true sun per hour.							
	°	'	"	10,000ths of circle.	No. of hours.	'	"	10,000ths of circle.	No. of hours.	'	"	10,000ths of circle.
1	2			3	4	5		6	4	5		6
67 to 69 } 93 to 96 }	0	56	58·66	26·3786	1	2	22·44	1·0991	13	30	51·78	14·2884
					2	4	44·89	2·1982	14	33	14·22	15·3875
					3	7	7·33	3·2973	15	35	6·66	16·4866
					4	9	29·78	4·3964	16	37	59·11	17·5857
					5	11	52·22	5·4955	17	40	21·55	18·6848
					6	14	14·67	6·5946	18	42	44·00	19·7830
					7	16	37·11	7·6937	19	45	6·44	20·8830
					8	18	59·55	8·7929	20	47	28·89	21·9821
					9	21	22·00	9·8920	21	49	51·33	23·0812
					10	23	44·44	10·9911	22	52	13·77	24·1804
					11	26	6·89	12·0902	23	54	36·22	25·2795
					12	28	29·33	13·1893				
70 to 73 } 89 to 92 }	0	56	56·89	26·3649	1	2	22·37	1·0985	13	30	50·81	14·2810
					2	4	44·74	2·1971	14	33	13·18	15·3795
					3	7	7·11	3·2956	15	35	35·55	16·4780
					4	9	29·48	4·3941	16	37	57·93	17·5766
					5	11	51·85	5·4927	17	40	20·30	18·6751
					6	14	14·22	6·5912	18	42	42·67	19·7737
					7	16	36·59	7·6898	19	45	5·04	20·8722
					8	18	58·96	8·7883	20	47	27·41	21·9707
					9	21	21·33	9·8868	21	49	49·78	23·0693
					10	23	43·70	10·9854	22	52	12·15	24·1678
					11	26	6·07	12·0839	23	54	34·52	25·2663
					12	28	28·44	13·1824				
74 to 77 } 86 to 88 }	0	56	55·71	26·3558	1	2	22·32	1·0982	13	30	50·17	14·2760
					2	4	44·64	2·1963	14	33	12·49	15·3742
					3	7	6·96	3·2945	15	35	34·82	16·4723
					4	9	29·28	4·3926	16	37	57·14	17·5705
					5	11	51·61	5·4908	17	40	19·46	18·6687
					6	14	13·93	6·5889	18	42	41·78	19·7668
					7	16	36·25	7·6871	19	45	4·10	20·8650
					8	18	58·57	8·7853	20	47	26·42	21·9631
					9	21	20·89	9·8834	21	49	48·74	23·0613
					10	23	43·21	10·9816	22	52	11·06	24·1594
					11	26	5·53	12·0797	23	54	33·38	25·2576
					12	28	27·85	13·1779				
(78 to 85 [*True sun in apogee on Day 31*). }	0	58	55·11	26·3512	1	2	22·30	1·0980	13	30	49·85	14·2738
					2	4	44·59	2·1959	14	33	12·15	15·3715
					3	7	6·89	3·2939	15	35	34·45	16·4695
					4	9	29·19	4·3919	16	37	56·74	17·5675
					5	11	51·48	5·4898	17	40	19·04	18·6654
					6	14	13·78	6·5878	18	42	41·34	19·7634
					7	16	36·07	7·6858	19	45	3·63	20·8614
					8	18	58·37	8·7837	20	47	29·93	21·9593
					9	21	20·67	9·8817	21	49	48·22	23·0573
					10	23	42·96	10·9797	22	52	10·52	24·1553
					11	26	5·26	12·0776	23	54	32·82	25·2532
					12	28	27·56	13·1756				

TABLE XLIX—*Contd.*

24-hour periods from true Mēsha-saṁkrānti (inclusive).	Arc travelled by true sun in 24 hours.				Arc travelled by true sun per hour.							
	°	′	″	10,000ths of circle	No. of hours.	′	″	10,000ths of circle.	No. of hours.	′	″	10,000ths of circle.
1	2			3	4	5		6	4	5		6
	For all days (Column 1) from 86 to 164 see above, taking the numbers of days backwards.											
165 to 168 360 to 362	0	58	55·16	27·2775	1	2	27·30	1·1366	13	31	54·88	14·7753
					2	4	54·60	2·2731	14	34	22·18	15·9119
					3	7	21·90	3·4097	15	36	49·48	17·0484
					4	9	49·19	4·5462	16	39	16·77	18·1850
					5	12	16·49	5·6828	17	41	44·07	19·3215
					6	14	43·79	6·8914	18	44	11·37	20·4581
					7	17	11·09	7·9559	19	46	38·67	21·5947
					8	19	38·39	9·0925	20	49	5·97	22·7312
					9	22	5·69	10·2291	21	51	33·27	23·8678
					10	24	32·98	11·3656	22	54	0·56	25·0043
					11	27	0·28	12·5022	23	56	27·86	26·1409
					12	29	27·58	13·6387				
169 to 172 356 to 359	0	59	4·03	27·3459	1	2	27·67	1·1394	13	31	59·68	14·8124
					2	4	55·34	2·2788	14	34	27·35	15·9518
					3	7	23·00	3·4182	15	36	55·02	17·0912
					4	9	50·67	4·5577	16	39	22·69	18·2306
					5	12	18·34	5·6971	17	41	50·36	19·3700
					6	14	46·01	6·8365	18	44	18·02	20·5094
					7	17	13·68	7 9759	19	46	45·69	21·6489
					8	19	41·34	9·1153	20	49	13·36	22·7883
					9	22	9·01	10·2547	21	51	41·03	23·9277
					10	24	36·68	11·3941	22	54	8·69	25·0671
					11	27	4·35	12·5335	23	56	36·36	26·2065
					12	29	32·02	13·6730				
173 to 176 352 to 355	0	59	12·31	27·4098	1	2	28·01	1·1421	13	32	4·17	14·8470
					2	4	56·03	2·2841	14	34	32·18	15·9890
					3	7	24·04	3·4262	15	37	0·19	17·1311
					4	9	52·05	4·5683	16	39	28·21	18·2732
					5	12	20·06	5·7104	17	41	56·22	19·4153
					6	14	48·08	6·8524	18	44	24·23	20·5573
					7	17	16·09	7·9945	19	46	52·25	21·6994
					8	19	44·10	9·1366	20	49	20·26	22·8415
					9	22	12·12	10·2787	21	51	48·27	23·9836
					10	24	40·13	11·4207	22	54	16·28	25·1256
					11	27	8·14	12·5628	23	56	44·30	26·2677
					12	29	36·15	13·7049				
177 to 180 348 to 351	0	59	21·18	27·4782	1	2	28·38	1·1449	13	32	8·97	14·8840
					2	4	56·77	2·2899	14	34	37·36	16·0290
					3	7	25·15	3·4348	15	37	5·74	17·1739
					4	9	53·53	4·5797	16	39	34·12	18·3188
					5	12	21·91	5·7246	17	42	2·50	19·4638
					6	14	50·30	6·8696	18	44	30·89	20·6087
					7	17	18·68	8·0145	19	46	59·27	21·7536
					8	19	47·06	9·1594	20	49	27·65	22·8985
					9	22	15·44	10·3043	21	51	56·03	24·0435
					10	24	43·83	11·4493	22	54	24·42	25·1884
					11	27	12·21	12·5942	23	56	52·80	26·3333
					12	29	40·59	13·7391				

TABLE XLIX—*Contd.*

24-hour periods from true Mēsha-saṁkrānti (inclusive).	Arc travelled by true sun in 24 hours.				Arc travelled by true sun per hour.							
	°	′	″	10,000ths of circle.	No. of hours.	′	″	10,000ths of circle.	No. of hours.	′	″	10,000ths of circle.
1	2			3	4	5		6	4	5		6
181 to 184 344 to 347	0	59	30·05	27·5467	1	2	28·75	1·1478	13	32	13·78	14·9211
					2	4	57·50	2·2956	14	34	42·53	16·0689
					3	7	26·26	3·4433	15	37	11·28	17·2167
					4	9	55·01	4·5911	16	39	40·03	18·3645
					5	12	23·76	5·7389	17	42	8·79	19·5122
					6	14	52·51	6·8867	18	44	37·54	20·6600
					7	17	21·26	8·0345	19	47	6·29	21·8078
					8	19	50·02	9·1822	20	49	35·04	22·9556
					9	22	18·77	10·3300	21	52	3·79	24·1034
					10	24	47·52	11·4778	22	54	32·55	25·2511
					11	27	16·27	12·6256	23	57	1·30	26·3989
					12	29	45·03	13·7733				
185 to 187 341 to 343	0	59	38·33	27·6106	1	2	29·10	1·1504	13	32	18·26	14·9557
					2	4	58·19	2·3069	14	34	47·36	16·1062
					3	7	27·29	3·4513	15	37	16·46	17·2567
					4	9	56·39	4·6018	16	39	45·55	18·4070
					5	12	25·49	5·7522	17	42	14·65	19·5575
					6	14	54·58	6·9026	18	44	43·75	20·7079
					7	17	23·68	8·0531	19	47	12·84	21·8584
					8	19	52·78	9·2035	20	49	41·94	23·0088
					9	22	21·87	10·3540	21	52	11·04	24·1592
					10	24	50·97	11·5044	22	54	40·14	25·3097
					11	27	20·07	12·6548	23	57	9·23	26·4601
					12	29	49·16	13·8053				
188 to 191 337 to 340	0	59	46·61	27·6745	1	2	29·44	1·1531	13	32	22·75	14·9903
					2	4	58·88	2·3062	14	34	52·19	16·1434
					3	7	28·33	3·4593	15	37	21·63	17·2965
					4	9	57·77	4·6124	16	39	51·07	18·4496
					5	12	27·21	5·7655	17	42	20·51	19·6027
					6	14	56·65	6·9186	18	44	49·96	20·7558
					7	17	26·09	8·0717	19	47	19·40	21·9089
					8	19	55·54	9·2248	20	49	48·84	23·0620
					9	22	24·98	10·3779	21	52	18·28	24·2151
					10	24	54·42	11·5310	22	54	47·72	25·3682
					11	27	23·86	12·6841	23	57	17·17	26·5213
					12	29	53·30	13·8372				
192 to 195 333 to 336	0	59	54·89	27·7383	1	2	29·79	1·1558	13	32	27·23	15·0249
					2	4	59·57	2·3115	14	34	57·02	16·1807
					3	7	29·36	3·4673	15	37	26·80	17·3365
					4	9	59·15	4·6231	16	39	56·59	18·4922
					5	12	28·93	5·7788	17	42	26·38	19·6480
					6	14	58·72	6·9346	18	44	56·17	20·8037
					7	17	28·51	8·0903	19	47	25·95	21·9595
					8	19	58·30	9·2461	20	49	55·74	23·1153
					9	22	28·08	10·4019	21	52	25·53	24·2710
					10	24	57·87	11·5576	22	54	55·31	25·4268
					11	27	27·66	12·7134	23	57	25·10	26·5826
					12	29	57·44	13·8692				

TABLE XLIX—*Contd.*

24-hour periods from true Mēsha-saṁkrānti (inclusive).	Arc travelled by true sun in 24 hours.				Arc travelled by true sun per hour.							
	°	′	″	10,000ths of circle.	No. of hours.	′	″	10,000ths of circle.	No. of hours.	′	″	10,000ths of circle.
1	2			3	4	5		6	4	5		6
196 to 199 329 to 332	1	0	3·17	27·8022	1	2	30·13	1·1584	13	32	31·72	15·0595
					2	5	0·26	2·3169	14	35	1·85	16·2180
					3	7	30·40	3·4753	15	37	31·98	17·3764
					4	10	0·53	4·6337	16	40	2·11	18·5348
					5	12	30·66	5·7921	17	42	32·24	19·6932
					6	15	0·79	6·9506	18	45	2·38	20·8517
					7	17	30·92	8·1090	19	47	32·51	22·0101
					8	20	1·06	9·2764	20	50	2·64	23·1685
					9	22	31·19	10·4258	21	52	32·77	24·3269
					10	25	1·32	11·5843	22	55	2·90	25·4854
					11	27	31·45	12·7427	23	57	33·03	26·6438
					12	30	1·58	13·9011				
200 to 203 325 to 328	1	0	10·85	27·8615	1	2	30·45	1·1609	13	32	35·88	15·0917
					2	5	0·90	2·3218	14	35	6·33	16·2526
					3	7	31·36	3·4827	15	37	36·78	17·4135
					4	10	1·81	4·6436	16	40	7·24	18·5744
					5	12	32·26	5·8045	17	42	37·69	19·7353
					6	15	2·71	6·9654	18	45	8·14	20·8961
					7	17	33·17	8·1263	19	47	38·59	22·0570
					8	20	3·62	9·2872	20	50	9·05	23.2179
					9	22	34·07	10·4481	21	52	59·50	24·3788
					10	25	4·52	11·6090	22	55	9·95	25·5397
					11	27	34·98	12·7699	23	57	40·40	26·7006
					12	30	5·43	13·9308				
204 to 206 321 to 324	1	0	18·54	27·9209	1	2	30·77	1·1634	13	32	40·04	15·1238
					2	5	1·55	2·3267	14	35	10·82	16·2872
					3	7	32·32	3·4901	15	37	41·59	17·7505
					4	10	3·09	4·6535	16	40	12·36	16·6139
					5	12	33·86	5·8168	17	42	43·13	19·7773
					6	15	4·64	6·9802	18	45	13·91	20·9406
					7	17	35·41	8·1436	19	47	44·68	22·1040
					8	20	6·18	9·3970	20	50	15·45	23·2674
					9	22	36·95	10·4703	21	52	46·22	24·4307
					10	25	7·73	11·6337	22	55	17·00	25·5941
					11	27	38·50	12·7971	23	57	47·77	26·7575
					12	30	9·27	13·9604				
207 to 210 318 to 320	1	0	25·64	27·9756	1	2	31·07	1·1657	13	32	43·89	15·1535
					2	5	2·14	2·3313	14	35	14·76	16·3191
					3	7	33·20	3·4970	15	37	46·02	17·4848
					4	10	4·27	4·6626	16	40	17·09	18·6504
					5	12	35·34	5·8283	17	42	48·16	19·8161
					6	15	6·41	6·9939	18	45	19·23	20·9817
					7	17	37·48	8·1596	19	47	50·30	22·1474
					8	20	8·55	9·3252	20	50	21·37	23·3130
					9	22	39·61	10·4909	21	52	52·43	24·4787
					10	25	10·68	11·6565	22	55	23·50	25·6443
					11	27	41·75	12·8222	23	57	54·57	26·8100
					12	30	12·82	13·9878				

TABLE XLIX—*Contd.*

24-hour periods from true Mēsha-saṁkrānti (inclusive).	Arc travelled by true sun in 24 hours.		Arc travelled by true sun per hour.					
	° ′ ″	10,000ths of circle.	No. of hours.	′ ″	10,000ths of circle.	No. of hours.	′ ″	10,000ths of circle.
1	2	3	4	5	6	4	5	6
211 to 214 314 to 317	1 0 32·74	28·0304	1	2 31·36	1·1679	13	32 47·73	15·1831
			2	5 2·73	2·3359	14	35 19·10	16·3510
			3	7 34·09	3·5038	15	37 50·46	17·5190
			4	10 5·46	4·6717	16	40 21·82	18·6869
			5	12 36·82	5·8397	17	42 53·19	19·8548
			6	15 8·18	7·0076	18	45 24·55	21·0228
			7	17 39·55	8·1755	19	47 55·92	22·1907
			8	20 10·91	9·3435	20	50 27·28	23·3586
			9	22 42·28	10·5114	21	52 58·64	24·5266
			10	25 13·64	11·6793	22	55 30·01	25·6945
			11	27 45·00	12·8472	23	58 1·37	26·8624
			12	30 16·37	14·0152			
215 to 218 310 to 313	1 0 39·24	28·0806	1	2 31·63	1·1700	13	32 51·25	15·2103
			2	5 3·27	2·3400	14	35 22·89	16·3803
			3	7 34·90	3·5101	15	37 54·52	17·5503
			4	10 6·54	4·6801	16	40 26·16	18·7204
			5	12 38·17	5·8501	17	42 57·79	19·8904
			6	15 9·81	7·0201	18	45 29·43	21·0604
			7	17 41·44	8·1902	19	48 1·06	22·2304
			8	20 13·08	9·3602	20	50 32·70	23·4005
			9	22 44·71	10·5302	21	53 4·33	24·5705
			10	25 16·35	11·7002	22	55 35·97	25·7405
			11	27 47·98	12·8703	23	58 7·60	26·9105
			12	30 19·62	14·0403			
219 to 222 306 to 309	1 0 45·15	28·1262	1	2 31·88	1·1719	13	32 54·46	15·2350
			2	5 3·76	2·3438	14	35 26·34	16·4069
			3	7 35·64	3·5158	15	37 58·22	17·5789
			4	10 7·53	4·6877	16	40 30·10	18·7508
			5	12 39·41	5·8596	17	43 1·98	19·9227
			6	15 11·29	7·0315	18	45 33·87	21·0946
			7	17 43·17	8·2035	19	48 5·75	22·2666
			8	20 15·05	9·3754	20	50 37·63	23·4385
			9	22 46·93	10·5473	21	53 9·51	24·6104
			10	25 18·81	11·7192	22	55 41·39	25·7823
			11	27 50·70	12·8912	23	58 13·27	26·9543
			12	30 22·58	14·0631			
223 to 225 302 to 305	1 0 51·07	28·178	1	2 32·13	1·1738	13	32 57·66	15·2597
			2	5 4·26	2·3477	14	35 29·79	16·43[illegible]6
			3	7 36·38	3·5215	15	37 1·92	17·6674
			4	10 8·51	4·6953	16	40 34·04	18·7812
			5	12 40·64	5·8691	17	43 6·17	19·9550
			6	15 12·77	7·0430	18	45 38·30	21·1289
			7	17 44·89	8·2168	19	48 10·43	22·3027
			8	20 17·02	9·3906	20	50 42·56	23·47[illegible]5
			9	22 49·15	10·5644	21	53 14·68	24·65[illegible]3
			10	25 21·28	11·7383	22	55 46·81	25·8242
			11	27 53·41	12·9121	23	58 18·94	26·9980
			12	30 25·53	14·0859			

TABLE XLIX—*Contd.*

24-hour periods from true Mêsha-saṁkrānti (inclusive).	Arc travelled by true sun in 24 hours.				Arc travelled by true sun per hour.								
	°	′	″	10,000ths of circle.	No. of Hours.	′	″	10,000ths of circle.	No. of Hours.	′	″	10,000ths of circle.
1	2			3	4	5		6	4	5		6
226 to 229 } 299 to 301 }	1	0	56·39	28·2129	1	2	32·35	1·1755	13	33	0·54	15·2820
					2	5	4·70	2·3511	14	35	32·89	16·4575
					3	7	37·05	3·5266	15	38	5·24	17·6331
					4	10	9·40	4·7021	16	40	37·59	18·8086
					5	12	41·75	5·8777	17	43	9·94	19·9841
					6	15	14·10	7·0532	18	45	42·29	21·1597
					7	17	46·45	8·2288	19	48	14·64	22·3352
					8	20	18·80	9·4043	20	50	46·99	23·5107
					9	22	51·15	10·5798	21	53	19·34	24·6863
					10	25	23·50	11·7554	22	55	51·69	25·8618
					11	27	55·85	12·9309	23	58	24·04	27·0373
					12	30	28·19	14·1064				
230 to 233 } 295 to 298 }	1	1	1·12	28·2494	1	2	32·55	1·1771	13	33	3·11	15·3108
					2	5	5·09	2·3541	14	35	35·65	16·4788
					3	7	37·64	3·5312	15	38	8·20	17·6559
					4	10	10·19	4·7082	16	40	40·75	18·8329
					5	12	42·73	5·8853	17	43	13·29	20·0100
					6	15	15·28	7·0623	18	45	45·84	21·1870
					7	17	47·83	8·2394	19	48	18·39	22·3641
					8	20	20·37	9·4615	20	50	50·93	23·5412
					9	22	52·92	10·5935	21	53	23·48	24·7182
					10	25	25·47	11·7706	22	55	56·03	25·8953
					11	27	58·01	12·9476	23	58	28·57	27·0723
					12	30	30·56	14·1247				
234 to 237 } 291 to 294 }	1	1	5·85	28·2859	1	2	32·74	1·1786	13	33	5·67	15·3215
					2	5	5·49	2·3572	14	35	38·41	16·5001
					3	7	38·23	3·5357	15	38	11·16	17·6787
					4	10	10·98	4·7143	16	40	43·90	18·8573
					5	12	43·72	5·8929	17	43	16·64	20·0358
					6	15	16·46	7·0715	18	45	49·39	21·2144
					7	17	49·21	8·2501	19	48	22·13	22·3940
					8	20	21·95	9·4286	20	50	54·88	23·5716
					9	22	54·69	10·6072	21	53	27·62	24·7502
					10	25	27·44	11·7858	22	56	0·36	25·9287
					11	28	0·18	12·9644	23	58	3·11	27·1073
					12	30	32·93	14·1429				
238 to 241 } 287 to 290 }	1	1	9·40	28·3133	1	2	32·89	1·1797	13	33	7·59	15·3384
					2	5	5·78	2·3594	14	35	40·48	16·5161
					3	7	38·67	3·5392	15	38	13·37	17·6958
					4	10	11·57	4·7189	16	40	46·27	18·8755
					5	12	44·46	5·8986	17	43	19·16	20·0572
					6	15	17·35	7·0783	18	45	52·05	21·2350
					7	17	50·24	8·2580	19	48	24·94	22·4147
					8	20	23·13	9·4378	20	50	57·83	23·5944
					9	22	56·02	10·6175	21	53	30·72	24·7741
					10	25	28·92	11·7972	22	56	3·62	25·9538
					11	28	1·81	12·9769	23	58	36·51	27·1335
					12	30	34·70	14·1566				

TABLE XLIX—*Contd.*

24-hour periods from true Mêsha-saṁkrânti (inclusive).	Arc travelled by true sun in 24 hours.				Arc travelled by true sun per hour.							
	°	′	″	10,000ths of circle.	No. of Hours.	′	″	10,000ths of circle.	No. of Hours.	′	″	10,000ths of circle.
1	2			3	4	5		6	4	5		6
242 to 244 283 to 286	1	1	12·36	28·3361	1	2	33·01	1·1807	13	33	9·19	15·3487
					2	5	6·03	2·3613	14	35	42·21	16·5294
					3	7	39·04	3 5420	15	38	15·22	17·7101
					4	10	12·06	4·7227	16	40	48·24	18·8907
					5	12	45·07	5·9034	17	43	21·25	20·0714
					6	15	18·09	7·0840	18	45	54·27	21·2521
					7	17	51·10	8·2647	19	48	27·28	22·4327
					8	20	24·12	9·4454	20	51	0·30	23·6134
					9	22	57·13	10·6260	21	53	33·31	24·7941
					10	25	30·15	11·8067	22	56	6·33	25·9747
					11	28	3·16	12·9874	23	58	39·34	27·1554
					12	30	36·18	14·1680				
245 to 248 280 to 282	1	1	15·31	28·3589	1	2	33·14	1·1816	13	33	10·79	15·3611
					2	5	6·28	2·3632	14	35	43·93	16·5427
					3	7	39·41	3·5449	15	38	17·07	17·7243
					4	10	12·55	4·7265	16	40	50·21	18·9059
					5	12	45·69	5·9081	17	43	23·35	20·0876
					6	15	18·83	7·0897	18	45	56·48	21·2692
					7	17	51·97	8·2713	19	48	29·62	22·4508
					8	20	25·10	9·4530	20	51	2·76	23·6324
					9	22	58·24	10·6346	21	53	35·90	24·8140
					10	25	31·38	11 8162	22	56	9·04	25·9957
					11	28	4·52	12·9978	23	58	42·18	27·1773
					12	30	37·66	14·1794				
249 to 252 276 to 279	1	1	16·03	28·3771	1	2	33·17	1·1824	13	33	11·18	15·3710
					2	5	6·34	2·3648	14	35	44·35	16·5533
					3	7	39·50	3·5471	15	38	17·52	17·7357
					4	10	12·67	4·7295	16	40	50·69	18·9181
					5	12	45·84	5·9119	17	43	23·86	20·1005
					6	15	19·01	7·0943	18	45	57·02	21·2829
					7	17	52·18	8·2767	19	48	30·19	22·4652
					8	20	25·34	9·4590	20	51	3·36	23·6476
					9	22	58·51	10·6414	21	53	36·53	24·8300
					10	25	31·68	11·8238	22	56	9·70	26·0124
					11	28	4·85	13·0062	23	58	42·86	27·1948
					12	30	38·02	14·1886				
253 to 256 272 to 275	1	1	19 45	28·3908	1	2	33·31	1·1830	13	33	13·04	15·3784
					2	5	6·62	2·3659	14	35	46·35	16·5613
					3	7	39·93	3·5489	15	38	19·66	17·7443
					4	10	13·24	4·7218	16	40	52·97	18·9272
					5	12	46·55	5·9148	17	43	26·28	20·1102
					6	15	19·86	7·0977	18	45	59·59	21·2931
					7	17	53·17	8·2807	19	48	32·90	22·4761
					8	20	26·48	9·4636	20	51	6·21	23·6590
					9	22	59·79	10·6466	21	53	39·52	24·8420
					10	25	33·11	11·8295	22	56	12·83	26·0248
					11	28	6·42	13·0125	23	58	46·14	27·2078
					12	30	39·73	14·1954				

TABLE XLIX—*Contd.*

24-hour periods from true Mēsha-saṁkrānti (inclusive).	Arc travelled by true sun in 24 hours.		Arc travelled by true sun per hour.					
	° ′ ″	10,000ths of circle.	No. of Hours.	′ ″	10,000ths of circle.	No. of hours.	′ ″	10,000ths of circle.
1	2	3	4	5	6	4	5	6
257 to 260 } 268 to 271 }	1 1 20·64	28·4000	1	2 33·36	1·1833	13	33 13·68	15·3833
			2	5 6·72	2·3667	14	35 47·04	16·5666
			3	7 40·08	3·5500	15	38 20·40	17·7500
			4	10 13·44	4·7333	16	40 53·76	18·9333
			5	12 46·80	5·9167	17	43 27·12	20·1166
			6	15 20·16	7·1000	18	46 0·48	21·3000
			7	17 53·52	8·2833	19	48 33·84	22·4833
			8	20 26·88	9·4667	20	51 7·20	23·6666
			9	23 0·24	10·6500	21	53 40·56	24·8500
			10	25 33·60	11·8333	22	56 13·92	25·0333
			11	28 6·96	13·0166	23	58 47·28	26·2166
			12	30 40·32	14·2000			
261 to 267 (*True sun in perigee, on Day* 263).	1 1 21·23	28·4045	1	2 33·38	1·1835	13	33 14·00	15·3858
			2	5 6·77	2·3670	14	35 47·38	16·5693
			3	7 40·15	3·5506	15	38 20·77	17·7528
			4	10 13·54	4·7341	16	40 54·15	18·9364
			5	12 46·92	5·9176	17	43 27·54	20·1199
			6	15 20·31	7·1011	18	46 0·92	21·3034
			7	17 53·69	8·2847	19	48 34·30	22·4869
			8	20 27·08	9·4682	20	51 7·69	23·6704
			9	23 0·46	10·6517	21	53 41·07	24·8540
			10	25 33·84	11 8352	22	56 14·46	26·0375
			11	28 7·23	13·0187	23	58 47·84	27·2210
			12	30 40·61	14·2023			

TABLE L.

Elements of the Sun's Longtitude.

Minutes.

The figures in Columns 2, 3, show the sun's *mean* movement during the times noted in Column 1.

Time Mins.	′	″	10,000ths of circle.	Time Mins.	′	″	10,000ths of circle.	Time Mins.	′	″	10,000ths of circle.
1	2		3	1	2		3	1	2		3
½	0	1·23	0·0095								
1	0	2·46	0·0190	21	0	51·74	0·3993	41	1	41·02	0·7795
2	0	4·93	0·0380	22	0	54·21	0·4183	42	1	43·49	0·7985
3	0	7·39	0·0570	23	0	56·67	0·4373	43	1	45·95	0·8175
4	0	9·86	0·0760	24	0	59·14	0·4563	44	1	48·42	0·8365
5	0	12·32	0·0951	25	1	1·60	0·4753	45	1	50·88	0·8556
6	0	14·78	0·1141	26	1	4·06	0·4943	46	1	53·34	0·8746
7	0	17·25	0·1331	27	1	6·53	0·5133	47	1	55·81	0·8936
8	0	19·71	0·1521	28	1	8·99	0·5323	48	1	58·27	0·9126
9	0	22·18	0·1711	29	1	11·46	0·5514	49	2	0·74	0·9316
10	0	24·64	0·1901	30	1	13·92	0·5704	50	2	3·20	0·9506
11	0	27·10	0·2091	31	1	16·38	0·5894	51	2	5·66	0·9696
12	0	29·57	0·2281	32	1	18·85	0·6084	52	2	8·13	0·9886
13	0	32·03	0·2472	33	1	21·31	0·6274	53	2	10·59	1·0077
14	0	34·50	0·2662	34	1	23·78	0·6464	54	2	13·06	1·0267
15	0	36·96	0·2852	35	1	26·24	0·6654	55	2	15·52	1·0457
16	0	39·42	0·3042	36	1	28·70	0·6844	56	2	17·98	1·0647
17	0	41·89	0·3232	37	1	31·17	0·7035	57	2	20·45	1·0837
18	0	44·35	0·3422	38	1	33·63	0·7225	58	2	22·91	1·1027
19	0	46·82	0·3612	39	1	36·10	0·7415	59	2	25·38	1·1217
20	0	49·28	0·3802	40	1	38·56	0·7605	60	2	27·84	1·1407

N. B.—Since this Table shows the sun's *mean* motion during the number of minutes indicated, a slight correction must be made in order to ascertain his *true* motion, if very great accuracy is required. The largest possible correction, namely for 59 minutes on the days 81 and 263 (when the sun is in apogee and perigee and is therefore at his slowest and quickest) is, on Day 81, *minus* 5″·4516 or 0·0421, and on Day 263 *plus* the same.

Hence on Day 81 the true sun's journey in 59 m. must be taken as (by the Table, 2′ 25″·38—5″·45=) 2′ 19″·93, or (by the Table, 1·1217—0·0421=) 1·0796; and on Day 263 as (2′ 25″·38+5″·45=) 2′ 30″·83, or (1·1217+0·042=) 1·1638.

It is not necessary to frame a Table to meet corrections less than this. Calculation can always be made by taking from the Hour Table (Table XLIX) the true sun's motion in one hour on the day in question, dividing this by 60, and multiplying the result by the number of minutes concerned.

TABLE L-A.

ELEMENTS OF THE SUN'S LONGITUDE.

Seconds.

Cols. 2, 3, shew the Sun's *mean* movement during times noted in Col. 1.

Time seconds	″	10,000ths of circle.	Time seconds.	″	10,000ths of circle.	Time seconds.	″	10,000ths of circle.
1	2	3	1	2	3	1	2	3
1	0·041	0·0003	21	0·862	0·0067	41	1·684	0·0130
2	0·082	0·0006	22	0·903	0·0070	42	1·725	0·0133
3	0·123	0·0010	23	0·945	0·0073	43	1·766	0·0136
4	0·164	0·0013	24	0·986	0·0076	44	1·807	0·0139
5	0·205	0·0016	25	1·027	0·0079	45	1·848	0·0142
6	0·246	0·0019	26	1·068	0·0082	46	1·889	0·0146
7	0·287	0·0022	27	1·109	0·0086	47	1·930	0·0149
8	0·329	0·0025	28	1·150	0·0089	48	1·971	0·0152
9	0·370	0·0029	29	1·191	0·0092	49	2·012	0·0155
10	0·411	0·0032	30	1·232	0·0095	50	2·053	0·0158
11	0·452	0·0035	31	1·273	0·0098	51	2·094	0·0162
12	0·493	0·0038	32	1·314	0·0101	52	2·135	0·0165
13	0·534	0·0041	33	1·355	0·0105	53	2·177	0·0168
14	0·575	0·0044	34	1·396	0·0108	54	2·218	0·0171
15	0·616	0·0048	35	1·437	0·0111	55	2·259	0·0174
16	0·657	0·0051	36	1·478	0·0114	56	2·300	0·0177
17	0·698	0·0054	37	1·519	0·0117	57	2·341	0·0181
18	0·739	0·0057	38	1·561	0·0120	58	2·382	0·0184
19	0·780	0·0060	39	1 602	0·0124	59	2·423	0·0187
20	0·821	0·0063	40	1·643	0·0127	60	2·464	0·0190

The Table follows M. de Rico's fixture of the sun's *mean* movement in 1 time-minute by the *Siddhānta-Śirōmaṇi* viz. 2″·464,008,788, or 0·019.012,414.

THE TRUE LONGITUDE OF THE SUN IN HINDU ASTRONOMY, PART II. THE SIDDHĀNTA-ŚIRŌMANI.

(Previously published in Epigraphia Indica, Vol. XIV, pp. 241-264.)

257. In my last article I have given Tables for finding the longitude of the sun, both mean and true, at any time of any year according to two of the great Indian astronomical authorities, the *First Ārya-Siddhānta* or *Āryabhaṭīya* of Āryabhaṭa (A.D. 499) and the *Present Sūrya-Siddhānta* (exact date unknown, introduced about A.D. 1100). The present Table affords similar information for the *Siddhānta-Śirōmaṇi* (12th century).

In case my Tables should be considered over-minute in detail, running as the entries do to several decimal points, I would ask readers to remember that they are designed as standard Tables for the settlement of the closest possible cases. Such a case as is mentioned in my former paper (*above, §§ 206, 207, on the Cycle of Jupiter, p. 2*) proves that permanent reference Tables can hardly be too accurate. I have found other cases somewhat similar in calculating the intercalated and suppressed lunar months by the *Siddhānta-Śirōmaṇi*. In ordinary cases it will always suffice to work with merely the whole numbers.

Elements of the Siddhānta-Śirōmaṇi.

258. The *Siddhānta-Śirōmaṇi* by Bhāskarāchārya dates, it is believed, from about A.D. 1150, though Dr. Bhau Dājī (J. R. A. S. *n. s. I. 392*) placed it in about 1105. It was used in some tracts and for some periods—we have yet to learn which—for the preparation of local almanacs.

According to this authority the length of the year from mean Mēsha-saṁkrānti to mean Mēsha-saṁkrānti is $365^d\ 6^h\ 12^m\ 9^s$ or $365^d{\cdot}258437500$.

Its sine-values of angles are the same as in the *Ārya-* and *Sūrya-Siddhāntas*, with radius taken as equal to 3438′.

For the sun's mean motion in days, hours, etc., see Table XLIII above.

The twenty-four base equations are given in col. 9 of Table XLVII above with the differences per minute of anomaly angle (*col. 10*), and in fuller detail in Table XLVII, A, cols. 9-10.

The epicycle of the sun not being considered as contracted at any part of the orbit, as it is in the *Sūrya-Siddhānta*, and the circumference of the epicycle being given as 13° 40′ or 820′, the equation (α being the sun's mean anomaly, or the angular distance of the mean sun from the perigee-point of his orbit) is $\frac{13^\circ\ 40'}{360^\circ}$ sin. α, or $\frac{820'}{21600}$ sin. α, or finally $\frac{41}{1080}$ sin. α.

This Siddhānta postulates a constant forward shift in the line of apsides of the sun's orbit. This shift is more rapid than the *Sūrya-Siddhānta's* shift and amounts to 0′·0174 or 1″·044 per annum, and to 11′ 18″·6 or 11′·31 in the 650 years succeeding A.D. 1100.[1]

According to the *Siddhānta-Śirōmaṇi* the Kaliyuga began, or in other words K. Y. 0 began, with a conjunction at celestial longitude 0° or 360° of mean sun, mean moon and other planets at the moment of mean sunrise or 6 A.M. on Friday 18th February B.C. 3102 or 18th

[1] The shift according to the *Ārya-Siddhānta* is *nil*.
" " *Brahma-Siddhānta* 0″·144 per ann.
" " *Sūrya-Siddhānta* 0″·1161 "
" " *Siddhānta-Śirōmaṇi* 1″·044 "
" " *2nd Ārya-Siddhānta* 0″·1383 "
(Jacobi, *Epig. Ind. I. 441.*)

February $0^h\ 0^m\ 0^s$ **Laṅka** time. This was the moment of mean Mēsha-saṁkrānti in that year. True **Mēsha-saṁkrānti, the** moment when the true or apparent sun touched long. 0°, occurred by the same authority on Tuesday 15th Feb. in that year at $19^h\ 52^m\ 21\frac{1}{2}^s$ after mean sunrise.

The interval between these two occurrences, which we call the *śōdhya*, and which is the time occupied by the sun in travelling over the arc of the equation-angle, was 2^d171971 or $2^d\ 4^h\ 7^m\ 38\frac{1}{2}^s$ in K.Y. 0 according to Dr. Schram's calculation (*see* "*Indian Chronography*," Table, p. 16).[1]

259. In the matter of the sun's equation and true longitude it should be noted that every entry in cols. 6 to 9 of Table XLVIII C has been separately calculated from the value of his mean anomaly at each twenty-four-hour period measured from the moment of true Mēsha-saṁkrānti, by use of the *Siddhānta-Śirōmaṇi* equation Table.

260. The forward shift of the sun's apsis, while leaving the sun's mean longitude unaffected, causes a slight change every year in the sun's mean anomaly (his mean distance from the perigee-point), this becoming each year proportionally less as the perigee-point moves forward. And since the shift induces a corresponding, though very minute, change in the velocity of the sun (considered as a planet) at all times of the year, the sun's equation and true longitude are each year a little different from what they were in the year previous.

The change in mean anomaly is stated in Table LI below.

The change caused by the shift of the apsis in the equation and true longitude of the sun at true Mēsha-saṁkrānti amounts to only 2″ (actually 1″·9675) in the 300 years on either side of K.Y. 4500, which is the base-year of the main Table XLVIII C which follows,—the annual change being at the rate of about 0″·0066 per annum.[2]

The corresponding time-difference, or change in the *śōdhya*-value, is about 0^s·16 per annum (actually 0^s·15975) by which amount the *śōdhya*-value at true Mēsha-saṁkrānti increases every year. In 300 years this amounts to 47^s·925 or about 48^s. (*For particulars see Table LII.*)

261. The length of the solar year from mean Mēsha-saṁkrānti to mean Mēsha-saṁkrānti according to this Siddhānta being $365^d\ 6^h\ 12^m\ 9^s$, it differs from that of the *Ārya-Siddhānta* year of $365^d\ 6^h\ 12^m\ 30^s$ by 21^s every year since K.Y. 0. The difference-Table given in *Indian Chronography*, p. 61, is here reprinted for ready reference (Table LIII). The difference is cumulative from K.Y. 0. In A.D. 1120, which is the very earliest date possible for the *Siddhānta-Śirōmaṇi* to have come into use (it was probably 30 years later), the moment of mean Mēsha-saṁkrānti by that authority was already $1^d\ 0^h\ 37^m\ 21^s$ earlier than the same according to the *Ārya-Siddhānta*, and the difference between them increased with every subsequent year. Consequently both mean and true Mēsha-saṁkrānti by the *Siddhānta-Śirōmaṇi* always fell respectively on the day previous to their occurrence by *Ārya-Siddhānta* reckoning, the time of which is given in the "*Indian Calendar*," Table I, cols. 13 to 17.

When therefore we are examining a date and have worked in the ordinary way for settlement of details by the *Ārya-Siddhānta*, using the *Indian Calendar* process for finding the values *a*, *b*, *c*, *s* and *n*, if we desire to find roughly the value of *s* according to the *Siddhānta-Śirōmaṇi* by use of the new Table XLVIII C below for determination of the nakshatra by that authority we must take the Table value of *s* (*cols. 8-9*) not for the day-number given in the Table, but for the day next following. *E.g.*, if we suppose that preliminary examination of a date by the *Indian Calendar* process proves the record-date to be Day 120 (as measured from 1st Jan.) and that Table I, cols. 13-17, shews that by the *Ārya-Siddhānta* true Mēsha-saṁkrānti took place on Day 85, then in order to ascertain the equation and longitude of the sun by the

[1] For explanation of technical matters see above, §§ 249-255, pp. 52-55.

[2] *Minus* for years earlier, *plus* for years later, than the base-year.

Siddhānta-Śirōmaṇi we must take the details given in Table XVIII C not as given for (120—85) Day 35, but for Day 36, that number of days having elapsed since true Mēsha-saṁkrānti by the latter authority. For accuracy the difference between the times of true Mēsha-saṁkrānti by the two authorities must be allowed for.

262. Since the Table-entries are for each twenty-four-hour period from true Mēsha-saṁkrānti in any year it is necessary to know the number of hours and minutes since sunrise of the occurrence of true Mēsha-saṁkrānti in the year in question, and deduct the sun's movement during those hours and minutes, in order to arrive at his true longitude at mean sunrise of the given day. The hours and minutes are given in Table LX below, cols. 13-17. For the sun's movement it will almost always suffice to use Tables XLIX, L, above. See § 243 above, p. 47, where the remarks regarding the *Sūrya-Siddhānta* apply, *mutatis mutandis*, to the *Siddhānta-Śirōmaṇi* also. The entries in Table LX, cols. 13-17, may be verified in the following manner.

To find time of true Mēsha-saṁkrānti by the *Siddhānta-Śirōmaṇi*; (i) *The longer rule*. Take the moment of true Mēsha-saṁkrānti by the *Ārya-Siddhānta* from Table I of the *Indian Calendar*, cols. 13 to 17, adding 30^s in odd A.D. years, none in even (*Hint 20, p. 79, Indian Chronography*). Add the *śōdhya* by that authority—always 2^d 3^h 32^m 30^s. This gives the time of mean Mēsha-saṁkrānti. Deduct for every year of the Kaliyuga expired at the given date the amount obtained from Table LIII below. This gives the time of mean Mēsha-saṁkrānti by the *Siddhānta-Śirōmaṇi*. Deduct the amount of *śōdhya* noted in Table LII below for the given year; for great exactness it may be found from col. 3, difference for the given year in minutes and seconds being calculated from the entry for the beginning of the century: for close approximation take, without further calculation, the century entry in col. 4. The result is the required time of true Mēsha-saṁkrānti by the *Siddhānta-Śirōmaṇi*.

(ii) *The shorter rule*. Take the *Ārya-Siddhānta* time of true Mēsha-saṁkrānti—*the first process in* (*i*). Add together the amounts gathered from Table LIII—*the third process in* (*i*)—and the number of minutes for the century in col. 5 of Table LII. Deduct the total from the *Ārya-Siddhānta* time of true Mēsha-saṁkrānti. The result gives the required time of true Mēsha-saṁkrānti by the *Siddhānta-Śirōmaṇi* with sufficient exactness for ordinary purposes.

263. Calculation for the correct tithi-index by the *Siddhānta-Śirōmaṇi* may for the present be considered as sufficiently carried out by work according to the *Ārya-Siddhānta*; there will often be a difference between the two. Correction of the equation (*see above*, § *247*, *ii*, *the tithi*) may cause a difference of one unit in the tithi-index, and there may be a slight difference in consequence of a different mean anomaly value requiring the equation to be calculated from a different base-angle.

Construction of the Main-Table XLVIIIC.

264. In order to conform to my similar Tables for the *Ārya-* and *Sūrya-Siddhāntas* (*above, Tables XLVIIIA and B*), I have worked for the year K.Y. 4500 expired, A.D. 1399-1400. The first thing was to fix the exact value of the sun's mean anomaly in that year at the moment of true Mēsha-saṁkrānti.

From Dr. Schram's fixture of the sun's equation of the centre by the *Siddhānta-Śirōmaṇi* at that moment in K.Y. 4000 as 2° 8′ 52″·761328955 and in K.Y. 5000 as 2° 8′ 59″·319753357 we find the equation in K.Y. 4500 to be 2° 8′ 56″·040541156, or, in 10,000ths of the circle, 59·691670842.

From Prof. Jacobi's determination of the position of the sun's apsis (I take perigee, not apogee) at that moment as 258° 55′ 12″ in K.Y. 4000 and 259° 12′ 36″ in K.Y. 5000 we find the perigee-point in K.Y. 4500 to be 259° 3′ 54″, or in 10,000ths of the circle, 7196·250 (exact).

The sun's **mean anomaly at any moment** is 360° *minus* the longitude of perigee and the equation of the **centre.** This, using the above figures, gives us his mean anomaly at that moment in K.Y. 4500 as 98° 47′ 9″·959458844 or, in decimals of a minute for purposes of calculation, 98° 47′·165990981; or, in 10,000ths of the circle, 2744·058329158.

Tested by the sine-and-equation-Table (*above, Tables XLVII and XLVIIA*) with use of the most accurate possible details (*for method see text* § *256, above*) I find that the result of calculation from that amount of mean anomaly gives the sun's true longitude as exactly 360° down to four decimals of a second. The figures, then, are accurate for the moment of true Mēsha-saṁkrānti in K.Y. 4500.

The sun's mean longitude at any moment is his true longitude less the equation of the centre, here 360° − 2° 8′ 56″·040541156 or 357° 51′ 3″·959458844, or, in 10,000ths, 9940·308329158.

These figures are given for the moment of true Mēsha-saṁkrānti at the head of the main Table.

EXAMPLE.

265. An inscription is found the date of which is stated as "Śaka 1571, Virōdhin, Margaśira kṛishṇa 30, Sunday, (*nakshatra*) Uttara Ashāḍhā, 25 Dhanus."

Worked out by the Tables below for calculation by the *Siddhānta-Śirōmaṇi* (*Tables XLIVA-LX*) the date is found to be perfectly sound. The resulting tithi-index ($t = 9868·4370$ by calculation) proves that the tithi Mārgaś. kṛ. 30 was properly connected with Sunday, 23 December, A.D. 1649, which corresponded with the year Virōdhin, Śaka 1571 expired. That Sunday was the 357th day after January 1st. Work for the solar month and day shews that this Sunday, the 357th day after January 1st, was the 25th day of Dhanus.

But the value of the nakshatra-index, n, found in the course of calculation points to the true moon's place in the heavens at mean sunrise of that Sunday having been so close to the point of junction of two nakshatras that it is advisable to test the essential details as closely as possible.

The true sun's longitude, "s," at mean sunrise of the 357th day after January 1st, is found by the present Table XLVIIIC. The solar year began (*Table LX, cols. 13-17*) on the 86th day after January 1st at 9^h 32^m after mean sunrise. That was the moment of true Mēsha-saṁkrānti. 357 − 86 = 271. For the purpose of the Table the Sunday in question was "Day 271" after true Mēsha-saṁkrānti.

Table XLVIIIC shews that at 9^h 32^m after mean sunrise on Day 271 the sun's true long., in ten-thousandths of the circle, was 7365·9104. From this must be deducted the sun's true motion during 9 hours on Day 271 (*Table XLIX above, p. 107*) and 32 minutes (taken for convenience in mean motion by Table L, p. 108), respectively, 10·6500 and 0·6084, total 11·2584, 7365·9104 − 11·2584 = 7354·6520. This was the value of "s" at mean sunrise of the given day.

The tithi-index, t, was found to be at the same moment 9868·4370; and since $s + t = n$, the index of the nakshatra, the value of "n" is found to be 7223·0890. Turning to Table XLVI above it is seen that by the equal-space division of the heavens the true moon was in the nakshatra Pūrva Ashāḍhā, but that by the systems of Garga and the Brahma-Siddhānta she was in Uttara Ashāḍhā, the former beginning at 7222·2 and the latter at 7137·2106.

If the framers of an almanac computed it on the principles of the *Siddhānta-Śirōmaṇi*, an authority of the Brāhma school of astronomy, they would naturally be supposed to follow the *Brahma-Siddhānta* system of nakshatras. Hence the date is proved to be correct in every particular.

NOTE.

The figures in the following Table are correct for K. Y. 4500, A.D. 1399-1400. In rdinary work for computation of the sun's true longitude ("*s*" in the *Indian Calendar* system) they may be taken as applicable to all years during which the *Siddhānta-Śirōmaṇi* was in use.

But for very great accuracy in other calculations the figures are subject to the following alterations:—

(*Cols.* 2, 3, 4, 5).—Sun's mean anomaly and mean longitude. For every 100 years earlier than A.D. 1400 add (*cols.* 2, 4) 1′ 45″·0558, or (*cols.* 3, 5) 0·8106. For every 100 years later deduct the same.

(*Cols* 6, 7).—Sun's equation of the centre. For every 100 years earlier than A.D. 1400 deduct (*col.* 6) 0″·6558, or (*col.* 7) 0·0051. For every 100 years later add the same.

TABLE XLVIII-C.

ELEMENTS OF THE SUN'S LONGITUDE FOR THE HINDU SOLAR YEAR,

according to the **Siddhânta-Śirōmaṇi.**

in periods of 24 hours each from the moment of the true Mēsha-saṁkrānti,

the astronomical beginning of the solar year.

(Exact for K. Y. 4500, A.D. 1399-1400. See Text §§ 260, 264.)

Siddhânta-Śirōmaṇi

24-hour periods from true Mēsha-saṁkrānti.	Sun's mean anomaly (or mean sun's distance from perigee-point) ("c").		Sun's mean longitude.		Sun's equation of the centre. +		Sun's true longitude ("s").	
1	2	3	4	5	6	7	8	9
	° ′	10,000ths of circle.	° ′ ″	10,000ths of circle.	° ′ ″	10,000ths of circle.	° ′ ″	10,000ths of circle
	(The sun's equation is +, plus, till his mean anomaly reaches 180°.)							
At true Mēsha-saṁkrānti .	*98 47·16599*	*2744·0583*	*357 51 3·96*	*9940·3083*	*2 8 56·04*	*59·6917*	*360 0 0·0*	*0·0*
1	99 46·30220	2771·4362	358 50 12·13	9967·6862	2 8 34·01	59·5217	0 58 46·14	27·2079
2	100 45·42841	2798·8141	359 49 20·30	9995·0641	2 8 12·13	59·3529	1 57 32·44	54·4170
3	101 44·57462	2826·1920	0 48 28·48	22·4420	2 7 45·29	59·1457	2 56 13·77	81·5877
4	102 43·71083	2853·5698	1 47 36·65	49·8198	2 7 14·54	58·9085	3 54 51·19	108·7283
5	103 42·84705	2880·9477	2 46 44·82	77·1977	2 6 43·79	58·6712	4 53 28·61	135·8689
6	104 41·98326	2908·3256	3 45 53·00	104·5756	2 6 13·04	58·4339	5 52 6·03	163·0095
7	105 41·11947	2935·7035	4 45 1·17	131·9535	2 5 37·36	58·1587	6 50 38·53	190·1121
8	106 40·25568	2963·0813	5 44 9·34	159·3313	2 4 58·33	57·8575	7 49 7·67	217·1888
9	107 39·39189	2990·4592	6 43 17·51	186·7092	2 4 19·30	57·5563	8 47 36·81	244·2655
10	108 38·52810	3017·8371	7 42 25·69	214·0871	2 3 40·27	57·2552	9 46 5·96	271·3423
11	109 37·66431	3045·2150	8 41 33·86	241·4650	2 2 54·31	56·9006	10 44 28·17	298·3655
12	110 36·80052	3072·5928	9 40 42·03	268·8428	2 2 7·00	56·5355	11 42 49·04	325·3784
13	111 35·93673	3099·9707	10 39 50·20	296·2207	2 1 19·70	56·1705	12 41 9·90	352·3912
14	112 35·07294	3127·3486	11 38 58·38	323·5986	2 0 31·73	55·8004	13 39 30·11	379·3990
15	113 34·20915	3154·7265	12 38 6·55	350·9765	1 59 36·14	55·3715	14 37 42·69	406·3479

TABLE XLVIII-C—*Contd.*

Siddhānta-Śirōmaṇi.

24-hour periods from true Mēsha-saṁkrānti.	Sun's mean anomaly (or mean sun's distance from perigee-point) ("c").			Sun's mean longitude.				Sun's equation of the centre +				Sun's true longitude ("s").			
1	2		3	4			5	6			7	8			9
	°	′	10,000ths of circle.	°	′	″	10,000ths of circle.	°	′	″	10,000ths of circle.	°	′	″	10,000ths of circle.
16	114	33·34537	3182·1043	13	37	14·72	378·3543	1	58	40·56	54·9426	15	35	55·28	433·2969
17	115	32·48158	3209·4822	14	36	22·89	405·7322	1	57	44·97	54·5136	16	34	7·86	460·2458
18	116	31·61779	3236·8601	15	35	31·07	433·1101	1	56	46·89	54·0655	17	32	17·95	487·1756
19	117	30·75400	3264·2380	16	34	39·24	460·4880	1	55	43·61	53·5772	18	30	22·85	514·0652
20	118	29·89021	3291·6158	17	33	47·41	487·8658	1	54	40·33	53·0890	19	28	27·75	540·9548
21	119	29·02642	3318·9937	18	32	55·59	515·2437	1	53	37·06	52·6008	20	26	32·64	567·8445
22	120	28·16263	3346·3716	19	32	3·76	542·6216	1	52	29·43	52·0789	21	24	33·18	594·7005
23	121	27·29884	3373·7495	20	31	11·93	569·9995	1	51	18·46	51·5314	22	22	30·39	621·5308
24	122	26·43505	3401·1273	21	30	20·10	597·3773	1	50	7·50	50·9838	23	20	27·60	648·3611
25	123	25·57126	3428·5052	22	29	28·28	624·7552	1	48	56·54	50·4362	24	18	24·81	675·1915
26	124	24·70747	3455·8831	23	28	36·45	652·1331	1	47	39·36	49·8407	25	16	15·80	701·9738
27	125	23·84369	3483·2610	24	27	44·62	679·5510	1	46	20·70	49·2338	26	14	5·33	728·7448
28	126	22·97990	3510·6388	25	26	52·79	706·8888	1	45	2·05	48·7260	27	11	54·85	755·5158
29	127	22·11611	3538·0167	26	26	0·97	734·2667	1	43	43·40	48·0201	28	9	44·37	782·2688
30	128	21·25232	3565·3946	27	25	9·14	761·6446	1	42	19·46	47·3724	29	7	28·60	809·0170
31	129	20·38853	3592·7725	28	24	17·31	789·0225	1	40	46·71	46·7108	30	5	11·03	835·7332
32	130	19·52474	3620·1503	29	23	25·48	816·4003	1	39	27·97	46·0491	31	2	53·45	862·4495
33	131	18·66095	3647·5282	30	22	33·66	843·7782	1	38	2·34	45·3885	32	0	36·00	889·1667
34	132	17·79716	3674·9061	31	21	41·83	871·1561	1	36	30·09	44·6766	32	58	11·92	915·8327
35	133	16·93337	3702·2840	32	20	50·00	898·5340	1	34	57·84	43·9648	33	55	47·84	942·4988
36	134	16·06958	3729·6618	33	19	58·18	925·9118	1	33	25·59	43·2530	34	53	23·76	969·1648
37	135	15·20579	3757·0397	34	19	6·35	953·2897	1	31	52·04	42·5311	35	50	58·38	995·8209
38	136	14·34201	3784·4176	35	18	14·52	980·6676	1	30	13·87	41·7737	36	48	28·39	1022·4413
39	137	13·47822	3811·7955	36	17	22·69	1008·0455	1	28	35·70	41·0162	37	45	58·40	1049·0617
40	138	12·61443	3839·1733	37	16	30·87	1035·4233	1	26	57·54	40·2588	38	43	28·40	1075·6821

1	2		3	4			5	6			7	8			9
41	139	11·75064	3866·5512	38	15	39·04	1062·8012	1	25	16·64	39·4803	39	40	55·68	1102·2815
42	140	10·88685	3893·9291	39	14	47·21	1090·1791	1	23	32·56	38·6772	40	38	19·77	1128·8563
43	141	10·02306	3921·3070	40	13	55·38	1117·5570	1	21	48·48	37·8741	41	35	43·87	1155·4311
44	142	9·15927	3948·6848	41	13	3·56	1144·9348	1	20	4·40	37·0710	42	33	7·96	1182·0059
45	143	8·29548	3976·0627	42	12	11·73	1172·3127	1	18	16·54	36·2388	43	30	28·27	1208·5515
46	144	7·43169	4003·4406	43	11	19·90	1199·6906	1	16	27·14	35·3946	44	27	47·04	1235·0852
47	145	6·56790	4030·8185	44	10	28·07	1227·0685	1	14	37·74	34·5504	45	25	5·81	1261·6189
48	146	5·70412	4058·1963	45	9	36·25	1254·4463	1	12	48·34	33·7063	46	22	24·58	1288·1526
49	147	4·84033	4085·5742	46	8	44·42	1281·8242	1	10	54·36	32·8269	47	19	38·78	1314·6511
50	148	3·97654	4112·9521	47	7	52·59	1309·2021	1	9	0·23	31·9462	48	16	52·82	1341·1483
51	149	3·11275	4140·3300	48	7	0·76	1336·5800	1	7	6·10	31·0656	49	14	6·86	1367·6455
52	150	2·24896	4167·7079	49	6	8·94	1363·9579	1	5	10·98	30·1773	50	11	19·92	1394·1352
53	151	1·38517	4195·0857	50	5	17·11	1391·3357	1	3	12·12	29·2602	51	8	29·23	1420·5959
54	152	0·52138	4222·4636	51	4	25·28	1418·7136	1	1	13·25	28·3430	52	5	38·53	1447·0566
55	152	59·65759	4249·8415	52	3	33·46	1446·0915	0	59	14·39	27·4258	53	2	47·84	1473·5173
56	153	58·79380	4277·2194	53	2	41·63	1473·4694	0	57	13·53	26·4933	53	59	55·16	1499·9626
57	154	57·93001	4304·5972	54	1	49·80	1500·8472	0	55	10·53	25·5442	54	57	0·33	1526·3914
58	155	57·06622	4331·9751	55	0	57·97	1528·2251	0	53	7·52	24·5951	55	54	5·50	1552·8202
59	156	56·20244	4359·3530	56	0	6·15	1555·6030	0	51	4·52	23·6460	56	51	10·67	1579·2490
60	157	55·33865	4386·7309	56	59	14·32	1582·9809	0	49	1·56	22·6972	57	48	15·88	1605·6781
61	158	54·47486	4414·1087	57	58	22·49	1610·3587	0	46	56·19	21·7299	58	45	18·68	1632·0886
62	159	53·61107	4441·4866	58	57	30·66	1637·7366	0	44	50·82	20·7625	59	42	21·49	1638·4991
63	160	52·74728	4468·8645	59	56	38·84	1665·1145	0	42	45·45	19·7952	60	39	24·29	1684·9096
64	161	51·88349	4496·2424	60	55	47·01	1692·4924	0	40	36·54	18·8004	61	36	23·55	1711·2928
65	162	51·01970	4523·6202	61	54	55·18	1719·8702	0	38	27·62	17·8057	62	33	22·80	1737·6760
66	163	50·15591	4550·9981	62	54	3·35	1747·2481	0	36	18·70	16·8110	63	30	22·06	1764·0591
67	164	49·29212	4578·3760	63	53	11·53	1774·6260	0	34	9·79	15·8163	64	27	21·31	1790·4422
68	165	48·42833	4605·7539	64	52	19·70	1802·0039	0	31	59·71	14·8126	65	24	19·41	1816·8164
69	166	47·56454	4633·1317	65	51	27·87	1829·3817	0	29	48·43	13·7996	66	21	16·30	1843·1813
70	167	46·70076	4660·5096	66	50	36·05	1856·7596	0	27	37·15	12·7866	67	18	13·19	1869·5462
71	168	45·83697	4687·8875	67	49	44·22	1884·1375	0	25	26·51	11·7786	68	15	10·72	1895·9161
72	169	44·97318	4715·2654	68	48	52·39	1911·5154	0	23	13·45	10·7519	69	12	5·84	1922·2673
73	170	44·10939	4742·6432	69	48	0·56	1938·8932	0	21	0·39	9·7253	70	9	0·96	1948·6185
74	171	43·24560	4770·0211	70	47	8·74	1966·2711	0	18	47·34	8·6986	71	5	56·07	1974·9697
75	172	42·38181	4797·3990	71	46	16·91	1993·6490	0	16	34·62	7·6745	72	2	51·52	2001·3235

TABLE XLVIII-C—*Contd.*

Siddhānta-Śirōmaṇi.

24-hour periods from true Mēsha-saṁkrānti.	Sun's mean anomaly (or mean sun's distance from perigee-point) ("*c*").			Sun's mean longitude.				Sun's equation of the centre. +				Sun's true longitude ("*s*").			
1	2		3	4			5	6			7	8			9
	°	′	10,000ths of circle.	°	′	″	10,000ths of circle.	°	′	″	10,000ths of circle.	°	′	″	10,000ths of circle.
76	173	41·51802	4824·7769	72	45	25·08	2021·0269	0	14	20·38	6·6387	72	59	45·46	2027·6656
77	174	40·65423	4852·1547	73	44	33·25	2048·4047	0	12	6·14	5·6029	73	56	39·39	2054·0076
78	175	39·79044	4879·5326	74	43	41·43	2075·7826	0	9	51·90	4·5671	74	53	33·32	2080·3497
79	176	38·92665	4906·9105	75	42	49·60	2103·1605	0	7	37·95	3·5335	75	50	27·55	2106·6940
80	177	38·06286	4934·2884	76	41	57·77	2130·5384	0	5	23·12	2·4932	76	47	20·89	2133·0315
81	178	37·19908	4961·6662	77	41	5·94	2157·9162	0	3	8·29	1·4528	77	44	14·23	2159·3691
82	179	36·33529	4989·0441	78	40	14·12	2185·2941	0	0	53·46	0·4125	78	41	7·57	2185·7066
Sun in apogee	*180*	*0·0*	*5000·0*	*79*	*3*	*54·00*	*2195·8333*	*0*	*0*	*0·0*	*0·0*	*79*	*3*	*54·00*	*2195·3333*
The sun's equation of the centre is —, minus, after his mean anomaly = 180° till it reaches 360°.															
83	180	35·47150	5016·4220	79	39	22·29	2212·6720	0	1	20·88	0·6240	79	38	1·41	2212·0480
84	181	34·60771	5043·7999	80	38	30·46	2240·0499	0	3	35·71	1·6644	80	34	54·76	2238·3855
85	182	33·74392	5071·1777	81	37	38·64	2267·4277	0	5	50·54	2·7048	81	31	48·10	2264·7230
86	183	32·88013	5098·5556	82	36	46·81	2294·8056	0	8	5·37	3·7451	82	28	41·44	2291·0605
87	184	32·01634	5125·9335	83	35	54·98	2322·1840	0	10	19·23	4·7780	83	25	35·65	2317·4055
88	185	31·15255	5153·3114	84	35	3·15	2349·5614	0	12	33·47	5·8138	84	22	29·69	2343·7476
89	186	30·28876	5180·6892	85	34	11·33	2376·9392	0	14	47·71	6·8496	85	19	23·62	2370·0897
90	187	29·42497	5208·0671	86	33	19·50	2404·3171	0	17	1·94	7·8854	86	16	17·55	2396·4317
91	188	28·56118	5235·4450	87	32	27·67	2431·6950	0	19	14·48	8·9081	87	13	13·19	2422·7869
92	189	27·69740	5262·8229	88	31	35·84	2459·0729	0	21	27·54	9·9347	88	10	8·30	2449·1381
93	190	26·83361	5290·2007	89	30	44·02	2486·4507	0	23	40·60	10·9614	89	7	3·42	2475·4893
94	191	25·96982	5317·5786	90	29	52·19	2513·8286	0	25	52·74	11·9810	90	3	59·45	2491·8476
95	192	25·10603	5344·9565	91	29	0·36	2541·2065	0	28	4·02	12·9940	91	0	56·34	2528·2125
96	193	24·24224	5372·3344	92	28	8·53	2568·5844	0	30	15·31	14·0070	91	57	53·23	2554·5774
97	194	23·37845	5399·7122	93	27	16·71	2595·9622	0	32	26·59	15·0200	92	54	50·12	2580·9423

1	2		3	4			5	6			7	8			9
98	195	22·51466	5427·0901	94	26	24·88	2623·3401	0	34	36·30	16·0209	93	51	48·58	2607·3193
99	196	21·65087	5454·4680	95	25	33·05	2650·7180	0	36	45·22	17·0156	94	48	47·83	2633·7024
100	197	20·78708	5481·8459	96	24	41·22	2678·0959	0	38	54·14	18·0103	95	45	47·09	2660·0855
101	198	19·92329	5509·2237	97	23	49·40	2705·4737	0	41	3·06	19·0051	96	42	46·34	2686·4687
102	199	19·05950	5536·6016	98	22	57·57	2732·8516	0	43	9·15	19·9780	97	39	48·42	2712·8736
103	200	18·19572	5563·9795	99	22	5·74	2760·2295	0	45	14·52	20·9454	98	36	51·22	2739·2841
104	201	17·33193	5591·3574	100	21	13·92	2787·6074	0	47	19·89	21·9127	99	33	54·03	2765·6947
105	202	16·46814	5618·7352	101	20	22·09	2814·9852	0	49	25·26	22·8801	100	30	56·83	2792·1052
106	203	15·60435	5646·1131	102	19	30·26	2842·3631	0	51	30·13	23·8436	101	28	0·13	2818·5195
107	204	14·74056	5673·4910	103	18	38·43	2869·7410	0	53	33·14	24·7927	102	25	5·30	2844·9483
108	205	13·87677	5700·8689	104	17	46·61	2897·1189	0	55	36·14	25·7418	103	22	10·47	2871·3770
109	206	13·01298	5728·2468	105	16	54·78	2924·4968	0	57	39·14	26·6909	104	19	15·63	2897·8058
110	207	12·14919	5755·6246	106	16	2·95	2951·874[illegible]	0	59	37·09	27·6010	105	16	25·86	2924·2736
111	208	11·28540	5783·0025	107	15	11·12	2979·2525	1	1	35·96	28·5182	106	13	35·17	2950·7343
112	209	10·42161	5810·3804	108	14	19·30	3006·6304	1	3	34·82	29·4353	107	10	44·48	2977·1960
113	210	9·55782	5837·7583	109	13	27·47	3034·0083	1	5	33·95	30·3545	108	7	53·52	3003·6537
114	211	8·69404	5865·1361	110	12	35·64	3061·3861	1	7	28·08	31·2352	109	5	7·56	3030·1509
115	212	7·83025	5892·5140	111	11	43·81	3088·7640	1	9	22·21	32·1158	110	2	21·60	3056·6482
116	213	6·96646	5919·8919	112	10	51·99	3116·1419	1	11	16·35	32·9965	110	59	35·64	3083·1454
117	214	6·10267	5947·2698	113	10	0·16	3143·5198	1	13	9·60	33·8703	111	56	50·56	3109·6494
118	215	5·23888	5974·6476	114	9	8·33	3170·8976	1	14	59·00	34·7145	112	54	9·34	3136·1831
119	216	4·37509	6002·0255	115	8	16·51	3198·2755	1	16	48·40	35·5586	113	51	28·11	3162·7169
120	217	3·51130	6029·4034	116	7	24·68	3225·6534	1	18	37·80	36·4028	114	48	46·88	3189·2506
121	218	2·64751	6056·7813	117	6	32·85	3253·0313	1	20	24·85	37·2288	115	46	8·00	3215·8025
122	219	1·78372	6084·1591	118	5	51·02	3280·4091	1	22	8·93	38·0319	116	43	32·10	3242·3773
123	220	0·91993	6111·5370	119	4	49·20	3307·7870	1	23	53·01	38·8349	117	40	56·19	3268·9521
124	221	0·05614	6138·9149	120	3	57·37	3335·1649	1	25	37·09	39·6380	118	38	20·28	3295·5269
125	221	59·19236	6166·2928	121	3	5·54	3362·5428	1	27	17·08	40·4096	119	35	48·46	3322·1332
126	222	58·32857	6193·6706	122	2	13·71	3389·9206	1	28	55·25	41·1670	120	33	18·47	3348·7536
127	223	57·46478	6221·0485	123	1	21·89	3417·2985	1	30	33·41	41·9245	121	30	48·47	3375·3740
128	224	56·60099	6248·4264	124	0	30·06	3444·6764	1	32	11·58	42·6819	122	28	18·48	3401·9944
129	225	55·73720	6275·8043	124	59	38·23	3472·0543	1	35	44·23	43·3968	123	25	54·00	3428·6574
130	226	54·87341	6303·1821	125	58	46·40	3499·4321	1	35	16·48	44·1086	124	23	29·92	3455·3235
131	227	54·00962	6330·5600	126	57	54·58	3526·8100	1	36	48·73	44·8205	125	21	5·84	3481·9895
132	228	53·14583	635 ·9379	127	57	2·75	3554·1879	1	38	19·87	45·5237	126	18	42·88	3508·6642

TABLE XLVIII-C—*Contd.*

Siddhânta-Śirômaṇi.

24-hour periods from true Mêsha-saṁkrânti.	Sun's mean anomaly (or mean sun's distance from perigee-point) ("c").		Sun's mean longitude.		Sun's equation of the centre. —		Sun's true longitude ("s").	
1	2	3	4	5	6	7	8	9
	° ′	10,000ths of circle.	° ′ ″	10,000ths of circle.	° ′ ″	10,000ths of circle.	° ′ ″	10,000ths of circle.
133	229 52·28204	6385·3158	128 56 10·92	3581·5658	1 39 45·61	46·1853	127 16 25·31	3535·3805
134	230 51·41825	6412·6936	129 55 19·10	3608·9436	1 41 11·36	46·8469	128 14 7·73	3562·0967
135	231 50·55447	6440·0715	130 54 27·27	3636·3215	1 42 37·11	47·5086	129 11 50·16	3588·8130
136	232 49·69068	6467·4494	131 53 35·44	3663·6994	1 43 59·97	48·1479	130 9 35·47	3615·5515
137	233 48·82689	6494·8273	132 52 43·61	3691·0773	1 45 18·62	48·7548	131 7 25·00	3642·3225
138	234 47·96310	6522·2051	133 51 51·79	3718·4551	1 46 37·27	49·3616	132 5 14·52	3669·0935
139	235 47·09931	6549·5830	134 50 59·96	3745·8330	1 47 55·92	49·9685	133 3 4·04	3695·8645
140	236 46·23552	6576·9609	135 50 8·13	3773·2109	1 49 9·65	50·5374	134 0 58·48	3722·6735
141	237 45·37173	6604·3388	136 49 16·30	3800·5888	1 50 20·61	51·0850	134 58 55·69	3749·5038
142	238 44·50794	6631·7166	137 48 24·48	3827·9666	1 51 31·58	51·6325	135 56 52·90	3776·3341
143	239 43·64415	6659·0945	138 47 32·65	3855·3445	1 52 42·54	52·1801	136 54 50·11	3803·1644
144	240 42·78036	6686·4724	139 46 40·82	3882·7224	1 53 49·00	52·6929	137 52 51·82	3830·0295
145	241 41·91657	6713·8503	140 45 48·99	3910·1003	1 54 52·27	53·1811	138 50 56·72	3856·9191
146	242 41·05279	6741·2281	141 44 57·17	3937·4781	1 55 55·55	53·6694	139 49 1·62	3883·8088
147	243 40·19000	6768·6060	142 44 5·34	3964·8560	1 56 58·83	54·1576	140 47 6·51	3910·6984
148	244 39·32521	6795·9839	143 43 13·51	3992·2339	1 57 55·73	54·5967	141 45 17·78	3937·6372
149	245 38·46142	6823·3618	144 42 21·69	4019·6118	1 58 51·32	55·0256	142 43 30·36	3964·5862
150	246 37·59763	6850·7396	145 41 29·86	4046·9896	1 59 46·91	55·4545	143 41 42·95	3991·5351
151	247 36·73384	6878·1175	146 40 38·03	4074·3675	2 0 41·89	55·8788	144 39 56·14	4018·4888
152	248 35·87005	6905·4954	147 39 46·20	4101·7454	2 1 29·20	56·2438	145 38 17·01	4045·5016
153	249 35·00626	6932·8733	148 38 54·38	4129·1233	2 2 16·51	56·6088	146 36 37·87	4072·5144
154	250 34·14247	6960·2511	149 38 2·55	4156·5011	2 3 3·81	56·9739	147 34 58·73	4099·5273
155	251 33·27868	6987·6290	150 37 10·72	4183·8790	2 3 48·51	57·3187	148 33 22·21	4126·5603
156	252 32·41489	7015·0069	151 36 18·89	4211·2569	2 4 27·54	57·6199	149 31 51·36	4153·6370
157	253 31·55111	7042·3848	152 35 27·07	4238·6348	2 5 6·57	57·9211	150 30 20·50	4180·7137

TABLE XLVIII-C—*Contd.*

24-hour periods from true Mēsha-saṁkrānti.	Sun's mean anomaly (or mean sun's distance from perigee-point) ("c").			Sun's mean longitude,				Sun's equation of the centre —				Sun's true longitude, ("s").			
1	2		3	4			5	6			7	8			9
	°	'	10,000ths of circle.	°	'	"	10,000ths of circle.	°	'	"	10,000ths of circle.	°	'	"	10,000ths of circle.
193	289	0·45470	8027·9883	188	4	21·28	5224·2383	2	3	24·08	57·1303	186	0	57·20	5167·1080
194	289	59·59090	8055·3662	189	3	29·45	5251·6162	2	2	36·77	56·7652	187	0	52·68	5194·8509
195	290	58·72712	8082·7440	190	2	37·63	5278·9940	2	1	49·64	56·4002	188	0	48·16	5222·5939
196	291	57·86333	8110·1219	191	1	45·80	5306·3719	2	1	2·15	56·0351	189	0	43·65	5250·3368
197	292	56·99954	8137·4998	192	0	53·97	5333·7498	2	0	11·12	55·6414	190	0	42·85	5278·1084
198	293	56·13575	8164·8777	193	0	2·15	5361·1277	1	59	15·53	55·2124	191	0	46·61	5305·9152
299	294	55·27196	8192·2555	193	59	10·32	5388·5055	1	58	19·94	54·7835	192	0	50·37	5333·7220
200	295	54·40817	8219·6334	194	58	18·49	5415·8834	1	57	24·36	54·3546	193	0	54·13	5371·5288
201	296	53·54439	8247·0113	195	57	26·66	5443·2613	1	56	23·42	53·8844	194	1	3·24	5389·3768
202	297	52·68060	8274·3892	196	56	34·84	5470·6392	1	55	20·15	53·3962	195	1	14·69	5417·2430
203	298	51·81681	8301·7670	197	55	43·01	5498·0170	1	54	16·87	52·9080	196	1	26·14	5445·1091
204	299	50·95302	8329·1449	198	54	51·18	5525·3949	1	53	13·60	52·4197	197	1	37·58	5472·9752
205	300	50·08923	8356·5228	199	53	59·35	5552·7728	1	52	3·12	51·8759	198	1	56·24	5500·8969
206	301	49·22544	8383·9007	200	53	7·53	5580·1507	1	50	52·15	51·3283	199	2	15·37	5528·8223
207	302	48·36165	8411·2785	201	52	15·70	5607·5285	1	49	41·19	50·7808	200	2	34·51	5556·7478
208	303	47·49786	8438·6564	202	51	23·87	5634·9064	1	48	28·84	50·2226	201	2	55·03	5584·6839
209	304	46·63407	8466·0343	203	50	32·04	5662·2843	1	47	10·19	49·6157	202	3	21·85	5612·6686
210	305	45·77028	8493·4122	204	49	40·22	5689·6622	1	45	51·54	49·0088	203	3	48·67	5640·6534
211	306	44·90649	8520·7900	205	48	48·39	5717·0400	1	44	32·89	48·4019	204	4	15·50	5668·6381
212	307	44·04271	8548·1679	206	47	56·56	5744·4179	1	43	13·42	47·7887	205	4	43·15	5696·6292
213	308	43·17892	8575·5458	207	47	4·74	5771·7958	1	41	47·67	47·1271	206	5	17·17	5724·6687
214	309	42·31513	8602·9237	208	46	12·91	5799·1737	1	40	21·92	46·4654	207	5	50·99	5752·7082
215	310	41·45134	8630·3015	209	45	21·08	5826·5515	1	38	56·17	45·8038	208	6	24·91	5780·7477
216	311	40·58755	8657·6794	210	44	29·25	5853·9294	1	37	28·14	45·1245	209	7	1·11	5808·8049
217	312	39·72376	8685·0573	211	43	37·43	5881·3073	1	35	55·89	44·4127	210	7	41·54	5836·8946

1	2		3	4			5	6			7	8			9
218	313	38·85997	8712·4352	212	42	45·60	5908·6852	1	34	23·63	43·7009	211	8	21·96	5864·9843
219	314	37·99618	8739·8130	213	41	53·77	5936·0630	1	32	51·38	42·9891	212	9	2·39	5893·0740
220	315	37·13239	8767·1909	214	41	1·94	5963·4409	1	31	15·64	42·2503	213	9	46·31	5921·1906
221	316	36·26860	8794·5688	215	40	10·12	5990·8188	1	29	37·47	41·4928	214	10	32·64	5949·3260
222	317	35·40482	8821·9467	216	39	18·29	6018·1967	1	27	59·31	40·7354	215	11	18·98	5977·4613
223	318	34·54103	8849·3245	217	38	26·46	6045·5745	1	26	21·14	39·9779	216	12	5·32	6005·5966
224	319	33·67724	8876·7024	218	37	34·63	6072·9524	1	24	38·05	39·1825	217	12	56·58	6033·7399
225	320	32·81345	8904·0803	219	36	42·81	6100·3303	1	22	53·97	38·3794	218	13	48·84	6061·9509
226	321	31·94966	8931·4582	220	35	50·98	6127·7082	1	21	9·89	37·5763	219	14	41·09	6090·1319
227	322	31·08587	8958·8361	221	34	59·15	6155·0861	1	19	25·38	36·7699	220	15	33·77	6118·3161
228	323	30·22208	8986·2139	222	34	7·32	6182·4639	1	17	35·98	35·9258	221	16	31·35	6146·5382
229	324	29·35829	9013·5918	223	33	15·50	6209·8418	1	15	46·58	35·0816	222	17	28·92	6174·7602
230	325	28·49450	9040·9697	224	32	23·67	6237·2197	1	13	57·17	34·2375	223	18	26·50	6202·9822
231	326	27·63071	9068·3476	225	31	31·84	6264·5976	1	12	6·18	33·3810	224	19	25·66	6231·2165
232	327	26·76692	9095·7254	226	30	40·02	6291·9754	1	10	12·05	32·5004	225	20	27·97	6259·4751
233	328	25·90314	9123·1033	227	29	48·19	6319·3533	1	8	17·91	31·6197	226	21	30·27	6287·7336
234	329	25·03935	9150·4812	228	28	56·36	6346·7312	1	6	23·78	30·7390	227	22	32·58	6315·9921
235	330	24·17556	9177·8591	229	28	4·53	6374·1091	1	4	26·91	29·8372	228	23	37·63	6344·2718
236	331	23·31177	9205·2369	230	27	12·71	6401·4869	1	2	28·04	28·9201	229	24	44·66	6372·5668
237	332	22·44798	9232·6148	231	26	20·88	6428·8648	1	0	29·18	28·0029	230	25	51·70	6400·8619
238	333	21·58419	9259·9927	232	25	29·05	6456·2427	0	58	30·32	27·0858	231	26	58·74	6429·1569
239	334	20·72040	9287·3606	233	24	37·22	6483·6206	0	56	27·92	26·1414	232	28	9·30	6457·4752
240	335	19·85661	9314·7484	234	23	45·40	6510·9984	0	54	24·92	25·1923	233	29	20·48	6485·8061
241	336	18·99282	9342·1263	235	22	53·57	6538·3763	0	52	21·92	24·2432	234	30	31·65	6514·1331
242	337	18·12903	9369·5042	236	22	1·74	6565·7542	0	50	18·91	23·2933	235	31	42·83	6542·4608
243	338	17·26524	9396·8821	237	21	9·91	6593·1321	0	48	15·08	22·3385	236	32	54·84	6570·7935
244	339	16·40146	9424·2599	238	20	18·09	6620·5099	0	46	9·71	21·3712	237	34	8·38	6599·1387
245	340	15·53767	9451·6378	239	19	26·26	6647·8878	0	44	4·34	20·4038	238	35	21·92	6627·4840
246	341	14·67388	9479·0157	240	18	34·43	6675·2567	0	41	58·97	19·4365	239	36	35·46	6655·8292
247	342	13·81009	9506·3936	241	17	42·61	6702·6436	0	39	48·74	18·4316	240	37	53·87	6684·2119
248	343	12·94630	9533·7714	242	16	50·78	6730·0214	0	37	39·82	17·4369	241	39	10·96	6721·5845
249	344	12·08251	9561·1493	243	15	58·95	6757·3993	0	35	30·90	16·4422	242	40	28·05	6740·9571
250	345	11·21872	9588·5272	244	15	7·12	6784·7772	0	33	22·32	15·4500	243	41	44·81	6769·3272
251	346	10·35493	9615·9051	245	14	15·30	6812·1551	0	31	11·03	14·4370	244	43	4·26	6797·7181
252	347	9·49114	9643·2829	246	13	23·47	6839·5329	0	28	59·75	13·4240	245	44	23·72	6826·1089

TABLE XLVIII-C—*Contd.*

Siddhânta-Śirômaṇi.

24-hour periods from true Mêsha-saṁkrânti.	Sun's mean anomaly (or mean sun's distance from perigee-point) ("*c*").			Sun's mean longitude.				Sun's equation of the centre. —				Sun's true longitude ("*s*").			
1	2		3	4			5	6			7	8			9
	°	′	10,000ths of circle.	°	′	″	10,000ths of circle.	°	′	″	10,000ths of circle.	°	′	″	10,000ths of circle.
253	348	8·62735	9670·6608	247	12	31·64	6866·9108	0	26	48·47	12·4110	246	45	43·17	6854·4998
254	349	7·76356	9698·0387	248	11	39·81	6894·2887	0	24	37·17	11·3979	247	47	2·64	6882·8908
255	350	6·89978	9725·4166	249	10	47·99	6921·6666	0	22	24·11	10·3713	248	48	23·87	6911·2953
256	351	6·03599	9752·7944	250	9	56·16	6949·0444	0	20	11·06	9·3446	249	49	45·10	6939·6999
257	352	5·17220	9780·1723	251	9	4·33	6976·4223	0	17	58·00	8·3179	250	51	6·33	6968·1044
258	353	4·30841	9807·5502	252	8	12·50	7003·8002	0	15	44·84	7·2904	251	52	26·66	6996·5097
259	354	3·44462	9834·9281	253	7	20·68	7031·1781	0	13	30·60	6·2547	252	53	50·07	7024·9234
260	355	2·58083	9862·3059	254	6	28·85	7058·5559	0	11	16·36	5·2189	253	55	12·49	7053·3371
261	356	1·71704	9889·6838	255	5	37·02	7085·9338	0	9	2·12	4·1831	254	56	34·90	7081·7508
262	357	0·85325	9917·0617	256	4	45·20	7113·3117	0	6	47·95	3·1478	255	57	57·24	7110·1639
263	357	59·98946	9944·4396	257	3	53·37	7140·6896	0	4	33·12	2·1074	256	59	20·24	7138·5821
264	358	59·12567	9971·8174	258	3	1·54	7168·0674	0	2	18·29	1·0671	258	0	43·25	7167·0004
265	359	58·26188	9999·1953	259	2	9·71	7195·4453	0	0	2·46	0·0267	259	2	6·25	7195·4186
Sun in perigee	*360*	*0·0*	*10000·0*	*259*	*3*	*54·00*	*7195·8333*	*0*	*0*	*0·0*	*0·0*	*259*	*3*	*54·00*	*7195·8333*
(The sun's equation of the centre is +, plus, after his mean anomaly=360°, till it reaches 180°).								Sun's equation of the centre. +							
266	0	57·39810	26·5732	260	1	17·89	7222·8232	0	2	10·87	1·0098	260	3	38·75	7223·8330
267	1	56·53431	53·9511	261	0	26·06	7250·2011	0	4	25·70	2·0501	261	4	51·76	7252·2512
268	2	55·67052	81·3289	261	59	34·23	7277·5789	0	6	40·53	3·0905	262	6	14·76	7280·6694
269	3	54·80673	108·7068	262	58	42·40	7304·9568	0	8	54·76	4·1262	263	7	37·16	7309·0831
270	4	53·94294	136·0847	263	57	50·58	7332·2347	0	11	9·00	5·1620	264	8	59·58	7337·4967
271	5	53·07915	163·4626	264	56	58·75	7359·7126	0	13	23·24	6·1978	265	10	21·99	7365·9104
272	6	52·21536	190·8404	265	56	6·92	7387·0904	0	15	37·48	7·2336	266	11	44·40	7394·3241
273	7	51·35157	218·2183	266	55	15·09	7414·4083	0	17	50·76	8·2621	267	13	5·86	7422·7304
274	8	50·48778	245·5962	267	54	23·27	7441·8462	0	20	3·82	9·2887	268	14	27·09	7451·1349
275	9	49·62399	272·9741	268	53	31·44	7469·2241	0	22	16·83	10·3154	269	15	48·32	7479·5395

1	2		3	4			5	6			7	8			9
276	10	48·76020	300·3519	269	52	39·61	7496·6019	0	24	29·93	11·3421	270	17	9·54	7507·9440
277	11	47·89642	327·7298	270	51	47·78	7523·9798	0	26	41·42	12·3566	271	18	29·20	7536·3364
278	12	47·03263	355·1077	271	50	55·96	7551·3577	0	28	52·70	13·3696	272	19	48·66	7564·7273
279	13	46·16884	382·4856	272	50	4·13	7578·7356	0	31	3·98	14·3826	273	21	8·11	7593·1182
280	14	45·30505	409·8634	273	49	12·30	7606·1134	0	33	15·27	15·3956	274	22	27·57	7621·5090
281	15	44·44126	437·2413	274	48	20·48	7633·4913	0	35	24·10	16·3897	275	23	44·58	7649·8810
282	16	43·57747	464·6192	275	47	28·65	7660·8692	0	37	33·02	17·3844	276	25	1·67	7678·2536
283	17	42·71368	491·9971	276	46	36·82	7688·2471	0	39	41·94	18·3792	277	26	18·76	7706·6262
284	18	41·84989	519·3750	277	45	44·99	7715·6250	0	41	50·85	19·3739	278	27	35·85	7734·9988
285	19	40·98610	546·7528	278	44	53·17	7743·0028	0	43	55·63	20·3367	279	28	48·80	7763·3395
286	20	40·12231	574·1307	279	44	1·34	7770·3807	0	46	1·00	21·3040	280	30	2·34	7791·6847
287	21	39·25852	601·5086	280	43	9·51	7797·7586	0	48	6·37	22·2714	281	31	15·88	7820·0300
288	22	38·39474	628·8865	281	42	17·68	7825·1365	0	50	12·74	23·2464	282	32	30·42	7848·3829
289	23	37·53095	656·2643	282	41	25·86	7852·5143	0	52	15·74	24·1955	283	33	41·60	7876·7099
290	24	36·66716	683·6422	283	40	34·03	7879·8922	0	54	18·75	25·1439	284	34	52·77	7905·0368
291	25	35·80337	711·0201	284	39	42·20	7907·2701	0	56	21·75	26·0937	285	36	3·95	7933·3638
292	26	34·93958	738·3980	285	38	50·37	7934·6480	0	58	22·30	27·0239	286	37	12·68	7961·6719
293	27	34·07579	765·7758	286	37	58·55	7962·0258	1	0	21·16	27·9411	287	38	19·71	7989·9696
294	28	33·21200	793·1537	287	37	6·72	7989·4037	1	2	20·03	28·8582	288	39	26·75	8018·2619
295	29	32·34821	820·5316	288	36	14·89	8016·7816	1	4	18·89	29·7754	289	40	33·78	8046·5570
296	30	31·48442	847·9095	289	35	23·07	8044·1595	1	6	16·26	30·6811	290	41	39·33	8074·8405
297	31	30·62063	875·2873	290	34	31·24	8071·5373	1	8	10·40	31·5617	291	42	41·64	8013·0990
298	32	29·75684	902·6652	291	33	39·41	8098·9152	1	10	4·53	32·4424	292	43	43·94	8131·3576
299	33	28·89306	930·0431	292	32	47·58	8126·2931	1	11	58·66	33·3230	293	44	46·25	8159·6161
300	34	28·02927	957·4210	293	31	55·76	8153·6710	1	13	50·16	34·1833	294	45	45·92	8187·8543
301	35	27·16548	984·7988	294	31	3·93	8181·0488	1	15	39·56	35·0275	295	46	43·49	8216·0763
302	36	26·30169	1012·1767	295	30	12·10	8208·4267	1	17	28·96	35·8716	296	47	41·06	8244·2983
303	37	25·43790	1039·5546	296	29	20·27	8235·8046	1	19	18·37	36·7158	297	48	38·64	8272·5204
304	38	24·57411	1066·9325	297	28	28·45	8263·1825	1	21	3·44	37·5265	298	49	31·89	8300·7090
305	39	23·71032	1094·3103	298	27	36·62	8290·5603	1	22	47·52	38·3296	299	50	24·14	8328·8900
306	40	22·84653	1121·6882	299	26	44·79	8117·9382	1	24	31·60	39·1327	300	51	16·39	8357·0709
307	41	21·98274	1149·0661	300	25	52·96	8345·3161	1	26	15·31	39·9330	301	52	8·28	8385·2491
308	42	21·11895	1176·4440	301	25	1·14	8372·6940	1	27	53·48	40·6904	302	52	54·62	8413·3844
309	43	20·25516	1203·8218	302	24	9·31	8400·0718	1	29	31·65	41·4479	303	53	40·96	8441·5197
310	44	19·39138	1231·1997	303	23	17·48	8427·4497	1	31	9·81	42·2053	304	54	27·29	8469·6550

TABLE XLVIII-C—*Concld.*

Siddhânta-Śirōmaṇi.

24-hour periods from true Mēsha-saṁkrânti.	Sun's mean anomaly (or mean sun's distance from perigee-point) ("*c*").			Sun's mean longitude.				Sun's equation of the centre. +				Sun's true longitude ("*s*").			
1	2		3	4			5				7	8			9
	°	′	10,000ths of circle.	°	′	″	10,000ths of circle.	°	′	″	10,000ths of circle.	°	′	″	10,000ths of circle.
311	45	18·52759	1258·5776	304	22	25·66	8454·8276	1	32	46·18	42·9489	305	55	11·84	8497·7765
312	46	17·66380	1285·9555	305	21	33·83	8482·2055	1	34	18·43	43·6008	306	55	52·26	8525·8662
313	47	16·80001	1313·3333	306	20	42·00	8509·5833	1	35	50·69	44·3726	307	56	32·69	8553·9559
314	48	15·93622	1340·7112	307	19	50·17	8536·9612	1	37	22·94	45·0844	308	57	13·11	8582·0456
315	49	15·07243	1368·0891	308	18	58·35	8564·3391	1	38	51·66	45·7690	309	57	50·01	8610·1081
316	50	14·20864	1395·4670	309	18	6·52	8591·7170	1	40	17·41	46·4306	310	58	23·93	8638·1476
317	51	13·34485	1422·8448	310	17	14·69	8619·0948	1	41	43·16	47·0922	311	58	57·85	8666·1871
318	52	12·48106	1450·2227	311	16	22·86	8646·4727	1	43	8·90	47·7539	312	59	31·77	8694·2266
319	53	11·61727	1477·6006	312	15	31·04	8673·8506	1	44	29·13	48·3729	314	0	0·17	8722·2235
320	54	10·75349	1504·9785	313	14	39·21	8701·2285	1	45	47·78	48·9798	315	0	26·99	8750·2082
221	55	9·89970	1532·3563	314	13	47·38	8728·6063	1	47	6·43	49·5867	316	0	53·81	8778·1930
222	56	9·02591	1559·7342	315	12	55·55	8755·9842	1	48	25·08	50·1935	317	1	20·64	8806·1778
323	57	8·16212	1587·1121	316	12	3·73	8783·3621	1	49	35·96	50·7404	318	1	39·69	8834·1025
324	58	7·29823	1614·4900	317	11	11·90	8810·7400	1	50	46·92	51·2880	319	1	58·82	8862·0280
325	59	6·43454	1641·8678	318	10	20·07	8838·1178	1	51	57·89	51·8356	320	2	17·96	8889·9534
326	60	5·57075	1669·2457	319	9	28·25	8865·4957	1	53	9·18	52·3857	321	2	37·43	8917·8814
327	61	4·70696	1696·6236	320	8	36·42	8892·8736	1	54	12·46	52·8739	322	2	48·88	8945·7475
328	62	3·84317	1724·0015	321	7	44·59	8920·2515	1	55	15·73	53·3621	323	3	0·32	8973·6136
329	63	2·97938	1751·3793	322	6	52·76	8947·6293	1	56	19·01	53·8504	324	3	11·77	9001·4797
330	64	2·11559	1778·7572	323	6	0·94	8975·0072	1	57	20·76	54·3268	325	3	21·69	9029·3340
331	65	1·25181	1806·1351	324	5	9·11	9002·3851	1	58	16·34	54·7557	326	3	25·45	9057·1408
332	66	0·38802	1833·5130	325	4	17·28	9029·7630	1	59	11·93	55·1847	327	3	29·21	9084·9476
333	66	59·52423	1860·8908	326	3	25·45	9057·1408	2	0	7·52	55·6136	328	3	32·97	9112·7544
335	67	58·66044	1888·2687	327	2	33·63	9084·5187	2	0	59·43	56·0141	329	3	33·05	9140·5328
335	68	57·79665	1915·6466	328	1	41·80	9111·8966	2	1	46·74	56·3791	330	3	28·54	9168·2757

1	2		3	4			5	6			7	8			9
336	69	56·93286	1943·0245	329	0	49·97	9139·2745	2	2	34·05	56·7442	331	3	24·02	9196·0187
337	70	56·06907	1970·4023	329	59	58·14	9166·6523	2	3	21·36	57·1002	332	3	19·50	9223·7616
338	71	55·20528	1997·7802	330	59	6·32	9194·0302	2	4	2·98	57·4304	333	3	9·30	9251·4606
339	72	54·34149	2025·1581	331	58	14·49	9221·4081	2	4	42·01	57·7316	334	2	56·50	9279·1397
340	73	53·47770	2052·5360	332	57	22·66	9248·7860	2	5	21·04	58·0327	335	2	43·70	9306·8187
341	74	52·61391	2079·9138	333	56	30·83	9276·1638	2	6	0·07	53·3339	336	2	30·90	9334·4977
342	75	51·75013	2107·2917	334	55	39·01	9303·5417	2	6	31·41	58·5757	337	2	10·42	9362·1174
343	76	50·88634	2134·6696	335	54	47·18	9330·9196	2	7	2·16	58·8130	338	1	49·34	9389·7326
344	77	50·02255	2162·0475	336	53	55·35	9358·2975	1	7	32·91	59·0502	339	1	28·26	9417·3477
345	78	49·15876	2189·4254	337	53	3·53	9385·6754	2	8	2·21	59·2763	340	1	5·73	9444·9516
346	79	48·29497	2216·8032	338	52	11·70	9413·0532	2	8	24·09	59·4451	341	0	35·78	9472·4983
347	80	47·43118	2244·1811	339	51	19·87	9440·4311	2	8	45·97	59·6139	342	0	5·84	9500·0450
348	81	46·56739	2271·5590	340	50	28·04	9467·8090	2	9	7·85	59·7828	342	59	35·89	9527·5917
349	82	45·70360	2298·9369	341	49	36·22	9495·1869	2	9	28·40	59·9414	343	59	4·62	9555·1282
350	83	44·83981	2326·3147	342	48	44·39	9522·5647	2	9	41·41	60·0417	344	58	25·80	9582·6065
351	84	43·97602	2353·6926	343	47	52·56	9549·9426	2	9	54·42	60·1421	345	57	46·98	9610·0847
352	85	43·11223	2381·0705	344	47	0·73	9577·3205	2	10	7·43	60·2425	346	57	8·16	9637·5630
353	86	42·24845	2408·4484	345	46	8·91	9604·6984	2	10	16·96	60·3161	347	56	25·87	9665·0144
354	87	41·38466	2435·8262	346	45	17·08	9632·0762	2	10	21·10	60·3480	348	55	38·18	9692·4242
355	88	40·52087	2463·2041	347	44	25·25	9658·4541	2	10	25·24	60·3800	349	54	50·49	9719·8341
356	89	39·65708	2490·5820	348	43	33·42	9686·8320	2	10	29·38	60·4119	350	54	2·81	9747·2439
357	90	38·79329	2517·9599	349	42	41·60	9714·2099	2	10	28·28	60·4034	351	53	9·88	9774·6133
358	91	37·92950	2545·3377	350	41	49·77	9741·5787	2	10	24·14	60·3715	352	52	13·91	9801·9592
359	92	37·06571	2572·7156	351	40	57·94	9768·9656	2	10	20·01	60·3395	353	51	17·95	9829·3051
360	93	36·20192	2600·0935	352	40	6·12	9796·3435	2	10	15·87	60·3076	354	50	21·98	9856·6511
361	94	35·33813	2627·4714	353	39	14·29	9823·7214	2	10	3·98	60·2159	355	49	18·27	9883·9373
362	95	34·47434	2654·8492	354	38	22·46	9851·0992	2	9	50·97	60·1155	356	48	13·43	9911·2148
363	96	33·61055	2682·2271	355	37	30·63	9878·4771	2	9	37·96	60·0151	357	47	8·59	9938·4922
364	97	32·74677	2709·6050	356	36	38·81	9905·8550	2	9	23·93	59·9069	358	46	2·73	9965·7618
365	98	31·88298	2736·9829	357	35	46·98	9933·2329	2	9	2·05	59·7380	359	44	47·03	9992·9709

TABLE LI.

THE CHANGE IN THE VALUE OF THE SUN'S MEAN ANOMALLY FROM THE VALUE GIVEN IN TABLE XLVIII-C, COLS. 2, 3, FOR THE BASE-YEAR OF THAT TABLE.

caused by the annual shift of the apsis of the sun's orbit postulated by the SIDDHĀNTA-ŚIRŌMAṆI.

[Add for years earlier, deduct for years later, than K. Y. 4500, A. D. 1399-1400.]

Years.	CHANGE.		Years.	CHANGE.		Years.	CHANGE.	
	Minutes and seconds.	10,000ths of circle.		Minutes and seconds.	10,000ths of circle.		Minutes and seconds.	10,000ths of circle.
1	0′·0174, or 1″·044	0·00805̇	10	0′·174, or 10″·44	0·0805̇	100	1′·74, or 1′ 44″·4	0·805̇
2	0′·0348, or 2″·088	0·0161̇	20	0′·348, or 20″·88	0·161	200	3′·48, or 3′ 28″·8	1·61
3	0′·0522, or 3″·132	0·02416̇	30	0′·522, or 31″·32	0·2416̇	300	5′·22, or 5′ 13″·2	2·416
4	0′·0696, or 4″·176	0·032̇	40	0′·696, or 41″·76	0·32̇			
5	0′·0870, or 5″·220	0·04027̇	50	0′·870, or 52″·20	0·4027̇			
6	0′·1044, or 6″·264	0·0483̇	60	1′·044, or 1′ 2″·64	0·483̇			
7	0′·1218, or 7″·308	0·05638	70	1′·218, or 1′ 13″·08	0·5638̇			
8	0′·1392, or 8″·352	0·064̇	80	1′·392, or 1′ 23″·52	0·64̇			
9	0′·1566, or 9″·396	0·07249̇	90	1′·566, or 1′ 33″·96	0·7249̇			

TABLE LII.

VALUE OF SŌDHYA, OR TIME-DIFFERENCE BETWEEN THE MOMENTS OF "TRUE MĒSHA-SAMKRĀNTI" (TRUE SUN AT O°) AND "MEAN MĒSHA-SAMKRĀNTI" (MEAN SUN AT O°) BY THE SIDDHĀNTA-ŚIRŌMAṆI,

as fixed by Dr. Schram for seven centuries. And Table of difference between that authority and the *First Ārya-Siddhānta.*

In the year K. Y. expired.	In A.D.	Exact value of *śōdhya.*				Value to be used in calculation by the longer rule.			Diff. between *Ārya Siddh.* and *Siddh. Śirōmaṇi* values of *śōdhya*; for use by the shorter rule.
1	2	3				4			5
		d.	*h.*	*m.*	*s.*	*d.*	*h.*	*m.*	Minutes
4200	1099-1100	2	4	18	49·0	2	4	19	46
4300	1199-1200	2	4	19	4·975	2	4	19	47
4400	1299-1300	2	4	19	20·95	2	4	19	47
4500	1399-1400	2	4	19	36·925	2	4	20	47
4600	1499-1500	2	4	19	52·9	2	4	20	47
4700	1599-1600	2	4	20	8·875	2	4	20	48
4800	1699-1700	2	4	20	24·85	2	4	20	48
4900	1799-1800	2	4	20	40·825	2	4	20	48
5000	1899-1900	2	4	20	56·800	2	4	21	48

The śōdhya *increases annually in amount by about 0s. 16. actually 0s. 15975.*

TABLE LIII.

DIFFERENCE BETWEEN THE MOMENTS OF MEAN MĒSHA-SAMKRĀNTI AS CALCULATED BY (i) THE FIRST ĀRYA-SIDDHĀNTA, (ii) THE SIDDHĀNTA-ŚIRŌMAṆI,

the two having been together at K. Y. 0 or B.C. 3102.

[*The moment of mean Mēsha-saṁkranti by the Ārya-Siddhānta having been found, deduct from this the time-difference for the Kaliyuga year of the given date. Result is moment of mean Mēsha-saṁkrānti by the Siddhānta Śirōmaṇi.*]

Diff. in years.	Time Difference.			Diff. in years.	Time Difference.			Diff. in years.	Time Difference.			Diff. in years.	Time Difference.		
1	2			1	2			1	2			1	2		
	h.	m.	s.		h.	m.	s.		h.	m.	s.		h.	m.	s.
1	0	0	21	10	0	3	30	100	0	35	0	1000	5	50	0
2	0	0	42	20	0	7	0	200	1	10	0	2000	11	40	0
3	0	1	3	30	0	10	30	300	1	45	0	3000	17	30	0
4	0	1	24	40	0	14	0	400	2	20	0	4000	23	20	0
5	0	1	45	50	0	17	30	500	2	55	0				
6	0	2	6	60	0	21	0	600	3	30	0				
7	0	2	27	70	0	24	30	700	4	5	0				
8	0	2	48	80	0	28	0	800	4	40	0				
9	0	3	9	90	0	31	30	900	5	15	0				

THE SIDDHANTA-SIROMANI.

GENERAL TABLES FOR CALCULATION BY THE TRUE OR APPARENT MOTION OF SUN AND MOON.

(Previously published in Epigraphia Indica, Vol. XV, pp. 159 to 245.)

267. The present article deals with the exact calculation of dates by the *Siddhānta-Śirōmaṇi,* which is believed to have been largely followed in some parts of India from the 12th century A.D. It provides complete Tables for the settlement of all the elements of the date, the solar month and day, the luni-solar month and *tithi,* the intercalated or suppressed month, and so on. They are framed on the *a, b, c* system of Largeteau and Jacobi, and follow the general lines of the *Indian Calendar.*

268. Since the name given to the whole year sometimes differs from that assigned by the *Ārya-* and *Sūrya-Siddhāntas* (*see above, Table XLII, cols. 10, 11*) and since the day of the solar month always differs, while the tithi, the intercalated lunar month,[1] and nakshatra very often differ, the necessity for these Tables will be apparent.

To give an example. Professor L. D. Barnett has called attention to a record found in the village of Hulgūr, Bankāpur Taluq, Dhārwār District, Bombay, which is dated in the year Ananda (A.D. 1254-55), Monday, Phālguna full-moon day, the day of a saṁkrānti. Worked by the *Sūrya-Siddhānta,* the date is found to be irregular, inasmuch as the saṁkrānti occurred not on the Monday in question (22 Feb. 1255), but at 8^h 52^m after mean sunrise on Tuesday, 23 Feb. But it is perfectly correct by the *Siddhānta-Śirōmaṇi,* according to which the moment of the saṁkrānti was 6^h 10^m after mean sunrise on the Monday. The document, therefore, if otherwise acceptable, should be given full historical weight.

269. Before we proceed a word of caution is necessary. While I hope that use of the Tables will yield exactly correct scientific results according to the requirements of the *Siddhānta-Śirōmaṇi,* we have at present no knowledge of how closely or how loosely the mediæval framers of local pañchāṅgs, (almanacs) followed the rules. If they only used whole numbers and disregarded fractions, which is probably the case, epigraphists must be prepared for occasional differences in close cases. If, again, they calculated time only in *ghaṭikās* and *palas,* it must not be forgotten that the *pala* is a division of 24 seconds, while my Tables give results down to a fraction of a second. This affords rather a wide margin for possible differences. The moral is that dates with slight differences should not be hastily rejected. Each should be treated on its merits and reasonable allowance made. Notes of close cases in the matter of intercalated and suppressed lunar months will be found inserted before Table LX. Differences in *tithis* must be examined, each on its merits.

270. It may at first sight seem absurd to work so closely as to state the value of "*a*," "*b*," "*c*" in nine decimals of a second, as I have done in the heading of Table LIVB; but let it be remembered that this value may be worked up into years and centuries for purposes of Tables LVIIA, B. In the body of the Tables four decimal places are given for all values.

For general verification of dates the whole numbers should first be used, as in the *Indian Calendar,* decimals being resorted to only in close cases.

I give full explanation of all my processes and calculations, so that these may be clear to experts, and that they may be corrected if in error.

---

[1] In the whole period of 800 years comprised in Table LX it will be found that out of a total of 304 years in which intercalations and suppressions of lunar months occurred there are differences between the *Sūrya-Siddhānta* and the *Siddhānta-Śirōmaṇi* in 234 years. The difference also of a whole day in every solar year implies a corresponding difference in the sun's longitude and leads to constant differences in the *nakshatra.*

Elements of the Siddhānta-Śirōmaṇi.

271. The *Siddhānta-Śirōmaṇi* is believed to have been composed by Bhāskarāchārya in A.D. 1150, when he was 36 years of age. The late Dr. James Burgess[1] states that the date is "supported by the evidence of an inscription near Chalisgām." Dr. Bhau Daji placed its appearance in A.D. 1105.[2]

The late Sankara Balkrishna Dikshit pronounced[3] that the *Rājamṛigāṅka*, a work composed in A.D. 1042, was the same as the *Siddhānta-Śirōmaṇi* in the matter of the calculation of an almanac, and if so, all the Tables which follow would apply to the former as well as to the latter. But up to the present I have no certainty about this. If my information is correct, the length of the solar year according to the two authorities differs; though in some other respects they may well be similar. For the *Rājamṛigāṅka*, while following the *Brahma-Siddhānta* of Brahmagupta (A.D. 628), introduced changes in it, which changes were adopted in the *Siddhānta-Śirōmaṇi*. Only one complete copy of the *Rājamṛigāṅka* has come to light. This is in the Deccan College Library at Poona, which also possesses a fragment consisting of two chapters. Professor N. K. Majumdar of the Calcutta University, who has kindly made enquiries for me, writes that, although there seems to be frequent reference to a table of sines, such a Table is not to be found in either of the copies. It seems therefore somewhat premature to assert that Tables adapted for computation by the *Siddhānta-Śirōmaṇi* will apply in all respects to work by the *Rājamṛigāṅka*.

272. According to the *Siddhānta-Śirōmaṇi* the length of the mean solar sidereal year, on the basis of 1,577,916,450 civil days to a *yuga* of 4,320,000 years, is 365·2584375 days or $365^d\ 6^h\ 12^m\ 9^s$, a quantity less than that of the *Ārya-Siddhānta* by 21^s.

The sines of the 24 base angles of anomaly have the same value as in the *Ārya-* and *Sūrya-Siddhāntas*, with sin. 90°, or radius, = 3438′. [See Table XLVII above for these sines and equations of sun's centre. For the moon see Table LIX below.]

For the sun's mean motions per day, hour, etc., see Table XLIV above, and footnote to it.

The epicycles of sun and moon are not contracted at any point. That of the sun has a circumference of 13° 40′; that of the moon 31° 36′ (*Jacobi, above, Vol. I, p. 441*). The sun and moon are always treated as planets.

The line of apsides of the sun's orbit has a constant slight forward shift, the movement amounting to 0′·0174 or 1″·044 per annum. In the total period of 800 years embraced by my Table LX this shift amounts to 13′ 55″·2.

The greatest equation of the sun (i.e. eqn. 90°) is 2° 10′ 31″, or in ten-thousandths of circle 60·424382715. That of the moon is 5° 2′ 7″·3661 or 139·871652005. The two together = 200 296034720.

The epoch of the Kaliyuga was the moment of mean sunrise, or 6 A.M. Laṅkā time, on Friday 18 Feb. B.C. 3102, a moment which for purposes of computation is treated as K.Y. 0 expired, $0^h\ 0^m\ 0^s$. This was the moment of occurrence of mean Mēsha-saṁkrānti in that year, when mean moon, mean sun and mean Jupiter were all considered to be in exact conjunction at the 0° point of celestial longitude. True Mēsha-saṁkrānti in that year, i.e. the moment when the true sun touched that point, occurred on Tuesday 15 Feb. at $19^h\ 52^m\ 21\frac{1}{2}^s$ after mean sunrise.

We have given the term "*śōdhya*" to the interval in time between true and mean Mēsha-saṁkrānti. In K.Y. 0 expired this was $2^d\ 4^h\ 7^m\ 38^s{\cdot}50$, or $2^d{\cdot}171971$ (*Indian Chronography, Table, p. 16; Dr. Schram's valuation*). For later centuries see Table B in § 273 below, p. 133.

[1] J. R. A. S., Oct. 1893, p. 751, § 31.

[2] J. R. A. S., n. s., I. 392.

[3] *Indian Calendar*, p. 8.

The position of the moon's apsis at K.Y. 0 was 305° 29′ 46″. Mean moon being at 0°, her mean anom. at that moment was (360° − 305° 29′ 46″ =) 54° 30′ 14″ (*Jacobi, above, I, 442*).

The position of the sun's apsis, perigee-point, at that moment was 257° 45′ 36″, and his mean anom. was (360° − 257° 45′ 36″ =) 102° 14′ 24″ (*Jacobi, above, I, 442*). For later centuries see Table XLIVA above.

EFFECTS OF THESE ELEMENTS.

273. (i) *Length of the mean solar year.* Since, as above stated, the *Siddhānta-Śirōmaṇi* year is less by 21ˢ than the *Ārya-Siddhānta* year, and since this divergence is annual and began in B.C. 3102 at the epoch of the Kaliyuga, when the two were together, it had, by the year A.D. 1100 when my Table LX begins, increased to more than 24 hours. Hence the moments of both mean and true Mēsha-saṁkrānti according to the *Siddhānta-Śirōmaṇi* are always a day earlier than they are by the *Ārya-Siddhānta*, the times of the occurrence of which are given in Table I of the *Indian Calendar* and Table LXI below. (*See also Table LIII above.*) The moment of true Mēsha-saṁkrānti each year can be calculated from Table LIII and Table B below, but it is not necesary to do so, unless to check my fixtures as all details are given in Table LX below.

(ii) The Table given in *Indian Chronography*, p. 27, for calculating the *śōdhya* at different dates, during the period covered by Table LX below according to the *Siddhānta-Śirōmaṇi*, is here reproduced to save reference.

TABLE B.

VALUE OF ŚŌDHYA BY THE SIDDHĀNTA-ŚIRŌMAṆI.

Dr. Schram's fixtures.

In K.Y. year expired.	In A.D.	Exact value of *śōdhya*.				Difference between *Siddh.-Śirō*: and *Ārya-Siddh*: value of *śōdhya*. For work by shorter rule.	
1	2	3				4	
		d.	h.	m.	s.	m.	s.
4200	1099-1100	2	4	18	49·000	46	19·000
4300	1199-1200	2	4	19	4·975	46	34·975
4400	1299-1300	2	4	19	20·950	46	50·950
4500	1399-1400	2	4	19	36·925	47	6·925
4600	1499-1500	2	4	19	52·900	47	22·900
4700	1599-1600	2	4	20	8·875	47	38·875
4800	1699-1700	2	4	20	24·850	47	54·850
4900	1799-1800	2	4	20	40·825	48	10·825
5000	1899-1900	2	4	20	56·800	48	26·800

Longer rule. Take time of true Mēsha-saṁkrānti by the *Ārya-Siddhānta* from Table I, *Indian Calendar*, or Table LXI below, adding 30ˢ in odd A.D. years (*Indian Chronography*, Hint 20, p. 79). Add *Ārya-Siddhānta śōdhya* (constant) 2ᵈ 3ʰ 32ᵐ 30ˢ. This gives mean Mēsha saṁkrānti by *Ārya-Siddhānta*. Deduct time-difference (*Table A, above*) for interval of years from K.Y. 0, and so find mean Mēsha-saṁkrānti by *Siddhānta-Śirōmaṇi*. Deduct *Siddhānta-Śirōmaṇi śōdhya* (*Table B, col. 3*). This gives the required true Mēsha-saṁkrānti time by *Siddhānta-Śirōmaṇi*.

Work approximately, if this is considered sufficient, by whole minutes, ignoring seconds and decimals.

Shorter rule. Take time of true Mēsha-saṁkrānti by the *Ārya-Siddhānta* as above. From this deduct the sum of the amounts for time-difference for interval of years (*Table LIII above*) and the difference given in col. 4, *Table B*.

For examples of work see *Indian Chronography*, p. 27, § 62 H, where the system is shewn approximately in whole minutes. It can be extended into seconds and decimals, if necessary.

(iii) *The shift of the sun's apsis.* The constant forward shift of the sun's apsis slightly affects the moment in each year when the true sun reaches 0°, the moment of true Mēsha-saṁkrānti; and creates a small change in the lengths of the true solar months owing to the change in the times of his reaching the points of the signs, and in their collective duration as measured from 0°. The sun is always regarded as a planet in Hindu astronomy, and his orbit is geocentric. His velocity is, in each year, in consequence of the shift of apsis a little greater than in the year previous in some parts of his orbit and a little less in others. For the purpose of correct calculation in very close cases these differences are detailed in Table LVIII-D below; but as they are very slight, they may in ordinary cases be ignored. And let it be always borne in mind that, as yet, we do not know how far the local almanac-makers of mediæval times paid any attention to them.

As regards the time of the true sun reaching long. 0°, since his velocity is greatest at the perigee-point and since this point is annually moving forward, he reaches long. 0° every year a trifle earlier than in the year before. The change is 0s·15975 per annum. And for the same reason every year his mean anomaly at that point grows slightly less and the equation of the centre slightly greater. The change in the equation amounts to 0″·65584 or, in 1,000ths of circle, 0·0005058 per century.

The shift of the apsis being 1″·044 per annum, it amounts to 1′ 44″·4 in a century, or, calculated in 1,000ths of the circle, to 0·0805. The sun's mean anom. at true Mēsha-saṁkrānti therefore decreases every century by this amount, and every year in proportion.

The *Siddhānta-Śirōmaṇi* length of year is 365d 6h 12m 9s, and therefore the length of the year as measured between two consecutive true Mēsha-saṁkrāntis is this amount less 0s·15975, or is 365d 6h 12m 8s·84025. On this basis, which agrees exactly with Dr. Schram's determination of the value of the *śōdhya* in different millenniums (*Indian Chronography*, p. 16), the moments of true Mēsha-saṁkrānti given in Table LX below have been computed.

(iv) *Note on work for the nakshatra.* The constant given in the *Indian Calendar* (*pp.* 65, 97), in the formala for verifying the *nakshatra*-index, is 7207. It is made up of the long. of the sun's perigee *plus* the amount of the sun's greatest equation. The amount 7207 represents the *Sūrya-Siddhānta* value, which varies from 7206·5077 in A.D. 900 to 7207·4035 in A.D. 1900.

The *Ārya-Siddhānta* value is a constant, and is always 7226·3542, roughly 7226.

By the *Siddhānta-Śirōmaṇi*, owing to its greater increase in the shift of the sun's apsis year by year, the variation in this factor is more pronounced. The long. of the apsis in A.D. 900 was 258° 55′ 12″, or, in 10,000ths of circle, 7192·2, and in A.D. 1900 it was 259° 12′ 36″ or 7200·27. The difference in 100 years is 0·805, in circle measurement, or in 1,000 years 8·05.

The greatest equation of the sun's centre (§ *272 above*) is, in circle notation, 60·4244,—the same as by the *Sūrya-Siddhānta*. The factor therefore in the formula referred to for finding the *nakshatra*-index is, for the beginning of the K.Y. year 4000, (7192·2 + 60·4244 =) 7252·6466. And for later centries is as shown in the following Table:—

K.Y. cent.	A.D. cent.	Exact factor.	Roughly
4000	900	7252·6466	7253
4100	1000	7253·4522	7253
4200	1100	7254·2577	7254
4300	1200	7255·0633	7255
4100	1300	7255·8688	7256
4500	1400	7256·6744	7257
4600	1500	7257·4799	7257
4700	1600	7258·2855	7258
4800	1700	7259·0910	7259
4900	1800	7259·8965	7260
5000	1900	7260·7023	7262

In very close work intermediate quantities must be taken for intermediate years. See Table LI above, which gives the quantities for the change in the sun's mean anom. The same figures can be applied to this factor.

Construction of the Tables.

Tables LIV A and B. Advance of "a," "b," "c" for days, hours, minutes and seconds.

274. These Tables are to be used in calculation by the *Siddhānta-Śirōmaṇi* in the same way as Tables IV, V of the *Indian Calendar* are used for the *Sūrya-Siddhānta*; working first with whole numbers and resorting to the decimals only in close cases. The values of "a," "b," "c" at mean sunrise on Chaitra śukla 1 of any year being taken down from Table LX below, addition of the figures given in Tables LIV A, LIV B for the intervening days, hours, etc., up to the given date furnishes the "a," "b," "c" at any moment of any subsequent day, i.e. gives us for that moment (a) the distance between mean moon and mean sun, (b) the moon's mean anom., (c) the sun's mean anom. The figures are parts of the circle,—a ten thousandths, b and c thousandths.

To arrive at an exact estimate of the value of these quantities an examination was made of Prof. Jacobi's fixtures for their value at mean sunrise of the first day of the 42nd century K.Y., a moment, that is, separated from the epoch of the Kaliyuga or mean Mēsha-saṁkrānti K.Y. 0, by exactly 4200 years K.Y. Mean Mēsha-saṁkrānti K.Y. 4200 (expired) took place on Friday 25 March A.D. 1099 at 10^{h} 30^{m} after mean sunrise and therefore 13^{h} 30^{m} before the mean sunrise of Saturday. 13^{h} 30^{m}=33^{gh} 44^{p}, the amount of Jacobi's "Cor.," or correction (*Epig. Ind., Vol. I, Table XIII, p. 450*). In that Table he gives the figures for the beginning of century 42 K.Y. as "a" (Dist. ☾ − ☉)=14° 18′ 0″, "b" (☾'s anom.)=51° 24′ 13″, "c" (☉'s anom.)=281° 1′ 19″. Owing to his arrangement of Tables, by which he gives only one Table for calculation of solar days (*Table XXI*) applicable to all Siddhāntas, whereas the date of occurrence of mean Mēsha-saṁkrānti by the *Siddhānta-Śirōmaṇi* is always a day earlier than by the other authorities, we have, for comparison with his tabular figures, to add a day's increase to the above valuation. This gives us "a"=26° 29′ 27″, "b"=64° 28′ 7″, "c"=282° 0′ 27″. Adding the increase in 13^{h} 30^{m} or 33^{gh} 44^{p} (Jacobi's "Cor.," Table XXII), we have finally for the values at mean sunrise of Sunday "a"=33° 20′ 40″·4, "b"=71° 48′ 50″·86, "c"=282° 33′ 41″·36. In 10,000ths of the circle (a), and 1,000ths (b and c), these values shew the increase in 4,200 years to have been a=926·237654, b=199·483677, c=784·893163. From "a" has to be deducted in accordance with our *Indian Calendar* working-system the sum of the greatest equations of moon and sun, viz. 200·296035, and hence "a"=725·9416.

Prof. Jacobi, however, has, since his valuation published in Vol. I, slightly modified his estimate of this value of "a." In Vol. XI above (*Table IX, B*) he states the three values as a=7263, b=1995, c=7849. In my notation these figures are a=726·3, b=699·5, c=284·9.[1] The difference being very small (0·4), I conclude to accept his later estimate of the increase of "a."

On this basis then, viz. the exact amount of increase of "a," "b," "c," in 4200 K.Y. years, has been calculated the increase per civil day (*Table LIV A*), per hour, minute and second (*Table LIV B*), per year and per century (*heading of Table LIV A*), according to the *Siddhānta-Śirōmaṇi*. The valuation of increase of "a" differs from that of the *Sūrya-Siddhānta* by about 2 units in a century. Note that a common century consists of 36,526 days, a defective century of 36,525 days. The whole period consisted of 1,534,087 civil days.

To assist in the calculation the yearly increases of "a," "b," "c" given, from year to year, in Prof. Jacobi's *Special Tables (above, Vol. I, Tables XVI, XIX)* were also referred to. It would have been easier had these contained decimals of seconds.

[1] I measure the ☾'s and ☉'s anom. from perigee, Jacobi from apogee.

Tables LV, LVI. Equations of the centre—moon and sun.

275. The values of "*a*," "*b*," "*c*" at any moment, which fix the positions of mean moon and mean sun, having been found by use of Tables LIV A and B, the *tithi*, or the position of the true moon with reference to the true sun, is ascertained by applying the equations of moon (*eqn.* "*b*") and sun (*eqn.* "*c*") to the value of "*a*." Tables LV, LVI give these equations in closer detail than heretofore (*compare Tables VI, VII, "Indian Calendar"*), enabling great accuracy to be obtained.

Each equation (*col. 3*) is the exact value (the value, that is, to be used in our system of work), in 10,000ths of the circle, of the equation of the mean anomaly angle ("Arg.") stated on either side in cols. 2*a*, 2*b*. Col. 1 gives the number of the base-equation, that is to say, the serial number of the equation of each of the 24 base-angles of anom.; each such angle separated from the next by 3° 45′, the whole forming the quadrant of 90°. Each section of 3° 45′ is divided into five equal parts, the whole forming a group within the limits of which, following universal Hindu practice, the equation is computed by the fixed value of the sine of the base-angle. In 10,000ths of the circle 3° 45′=10·416, and one-fifth of this is 2·083. The difference, col. 4, is the difference between the equations of each of the five parts of the group.

When examining a date Tables VI, VII of the *Indian Calendar* or Tables LXXXIV, LXXXV below may be used for obtaining approximate results, or the new Tables may be used with whole numbers only. The latter form a sort of eye-Table. Absolute accuracy, or very close approximation, can be obtained by using the decimals as a whole or in part. Thus—

(*Rule*) Take the difference between the value of anom., ('*b*' or "*c*"), found in work for a date, and the nearest to it, greater or less, in col. 2*a* or 2*b* of Table LV or LVI respectively. Multiply this difference by the group-difference (*col. 4*), and divide the quotient by 2·083. Add, or subtract, the result to, or from, the next equation. This gives the exact value of equation "*b*" or equation "*c*." For an approximation use only one or two decimals, and instead of dividing by 2·083 divide by 2 or by 2·1.

The amount of "equation *b*" or "equation *c*" is a compound of the actual equation for the given anom. and the greatest equation (which is the actual equation for anom. 90°). The first half of each of the equation-Tables LV and LVI concerns the quantity of anom. 0° to 180°, or, in 1,000ths of circle, 0 to 500. Here the tabulated "equation *b*" (Table LV) is the moon's greatest equation *plus* the actual equation of the given anom. The tabulated "equation *b*" in the second half of Table LV deals with the moon's anom. 180° to 360° or, in 1,000ths of circle, 500 to 1000; and the tabulated equation is the greatest equation *minus* the actual equation of the given anom. In the first half of Table LVI (for sun's anom. 0° to 180°, or, in 1,000ths of circle, 0 to 500) the tabulated "equation *c*" is the sun's greatest equation *minus* the actual equation. In the second half (for sun's anom. 180° to 360°, or, in 1,000ths of circle, 500 to 1000) the tabulated equation is the sun's greatest equation *plus* the actual equation of the given anom.

The actual equation-Table for the moon is given below—Table LIX. That of the sun in Table XLVII above. All details have been fully worked out by the proper formulæ.

For method of work see Example 3 below.

TABLES LVIIA, B, C.

Value of "a," "b," "c" for centuries, years and days.

276. These Tables enable us to ascertain the value of "*a*," "*b*," "*c*," and so to determine the exact position of mean moon and mean sun at the beginning of any year with which the general Table LX is concerned. Table LVIIA gives the "*a*," "*b*," "*c*" of mean sunrise, i.e. mean sunrise of the day on which mean Mêsha-saṁkrānti occurred at the beginning of the century; Table

LVIIB the same for the beginning of the given year; Table LVII C the same for the days on which true Mêsha-saṁkrânti occurred and on which began the luni-solar year. The respective week-days for the beginning of the solar and luni-solar year are given in Table LX, but can be found also by these Tables.

In the case of a date in the solar year the values of "a", "b", "c" in Table LVIIA are added to those of Table LVIIB, and the sum of these is added to the values of the day of true Mêsha-saṁkrânti in Table LVIIC. The values for the interval of days between true Mêsha-saṁkrânti and the day given in the date in question are obtained from Table LIVA, and thus are found the positions of moon and sun at mean sunrise of the latter day. For any subsequent moment of that day the values in Table LIVB are added to the result.

In the case of a date given in the *luni-solar year* (the most usual method) Table LX provides the "a", "b", "c" for mean sunrise on the initial day of the luni-solar year, while Tables LIVA and B enable the calculation to be completed. The values given in Table LX can be checked by Tables LVIIA, B, C.

From the "a", "b", "c" of true Mêsha-saṁkrânti in any year, found by Tables LVIIA, B, C, the "a", "b", "c" of each true *saṁkrânti* in the year are found by addition of the values given in Table LVIIIA; and by the result it is ascertained whether there was any intercalation or suppression of a lunar month in the given year.

277. *Table LVIIA.* The most important point here is the settlement of the values of "a", "b", "c" at the moment of mean sunrise of the day on which the 42nd K.Y. century began. This was the day on which occurred mean Mêsha-saṁkrânti of K.Y. 4200, or A.D. 1099-1100. In § 274 above details are given explaining Prof. Jacobi's values for the moment in question. Enough has been said about the value of "a". The following notes about the respective values of "b" and "c" may be found helpful.

The value of "b", the moon's mean anom. for K.Y. 4200. In my notation this was stated as in 1,000ths of the circle, 699·4837. Working the calculation by the values given in the heading of Table LIV for the mean moon's movement in 4,200 years, consisting of 37 common and 5 defective centuries, the total is found to be, excluding whole revolutions, 548·145255. To this has to be added the moon's mean anom. at the epoch of the Kaliyuga. At that moment the moon's apsis (perigee) stood at 305° 29′ 46″,—apogee being at 125° 29′ 46″ [1]—and the mean moon was at 0°. Therefore her mean anom. was (360°—305° 29′ 46″) 54° 30′ 14″. This in 1,000ths of the circle is 151·3997. Adding this to the above, her mean anom., "b", at mean sunrise of the day on which mean Mêsha-saṁkrânti occurred in K.Y. 4200 expired is found to be 699·5449. The difference between the two calculations is 0·0612. Both agree with Jacobi's valuation 699·5.

The value of "c" the sun's mean anom. At the epoch of the Kaliyuga the sun's apsis (perigee) was at long. 257° 45′ 36″. Mean sun being at long. 0°, the sun's mean anom. was (360°—257° 45′ 36″) 102° 14′ 24″. This, in 1,000ths of circle, is 284·0. The increase of "c" (*Table LIVA, heading*) in 37 common and 5 defective centuries, total 42, is, excluding whole revolutions, 4·278478. This, added to the value of "c" in K.Y. 0, viz. 284·0, gives the value of "c" at beginning of K.Y. 4200 expired as 288·278478. From this has to be deducted the amount of the decrease in the sun's mean anom. due to the forward shift of the apsis. This was shown above (§ 273, ii) to be, in 1,000ths of the circle, 0·0805 per century. In 42 centuries the decrease amounts to 3·383. 288·278478—3·383=284·8951. In § 274 the valuation was given as 284·893163. The difference between the two is less than 0·002, and both agree with Jacobi's valuation 284·9.

[1] Jacobi, *Epig. Ind.* Vol. I, pp. 440, 441. See also E. Burgess's "*Sûrya-Siddhânta.*"

Comparing the two sets of results I have decided to adhere to Prof. Jacobi's own fixtures, as given in § 274; and, fully worked out, the figures for mean sunrise on Sunday 27 March A.D. 1099 are a=726·307704844, b=699·483676555, c=284·893163057. For two days earlier, namely for mean sunrise on Friday 25 March A.D. 1099, on which day mean Mēsha-saṁkrānti of K.Y. 4200 expired took place at 10ʰ 30ᵐ after mean sunrise, the correct details, obtained by deduction of 2 days' value (*Table LIVA*) from the above, are—

(6) Friday
a=49·043734020
b=626·900376983
c=279·417587971.

This explains the first entry in Table LVIIA. The rest follow by addition of the century values given in the heading of Table LIVA. Century 42 was a defective one of 36,525 days. The rest were common ones, each of 36,526 days.

36,525 divided by 7 leaves remainder 6. Mean Mēsha-saṁkrānti in K.Y. 4200 took place on 6 Friday. 6+6=(week-day) 5. Hence the day of the week of mean Mēsha-saṁkrānti in K.Y. 4300 was 5 Thursday; and since 36,526 divided by 7 leaves no remainder, mean Mēsha-saṁkrānti at the beginning of each of the following centuries took place on a Thursday.

Coupling the arrangement made in Table LVIIA for centuries with the arrangement for days made in Table LVIIC, the result of calculations made by these Tables coincides precisely with those obtained by use of Jacobi's Tables. Such arrangement is the one best suited to the requirements of the *Siddhānta-Śirōmaṇi.* An example will best illustrate this.

Given that it is desired to find the "a", "b", "c" at mean sunrise of the day on which true Mēsha-saṁkrānti took place in K.Y. 4806 expired, A.D. 1705·6. This day was (*see Table LX*) Tuesday 27 March A.D. 1705.

Worked by Jacobi's Tables IX, X, XIII of Epig. Ind., Vol. XI, we have—

	w-d.	*a.*	*b.*	*c.*
For cent. 48	0	3619·0	696·0	784·1
,, year 6	0	1942·7	515·0	998·5
True Mēsha-saṁk. day[1] .	3	8645·5	854·8	989·0
	3 (Tues.)	4207·2	65·8	771·6

In my reckoning, "b" and "c" being calculated from perigee instead of from apogee, these are a=4207·2, b=565·8, c=271·6.

Worked, with only one decimal, by Tables LVII A, B, C below, the result is the same; thus—

	w-d.	*a.*	*b.*	*c.*
For cent. 48	5	2941·8	123·5	278·7
,, year 6	0	1942·7	515·0	998·5
True Mēsha-saṁk. day .	5	9322·7	927·4	994·5
	3 (Tues.)	4207·2	565·9	271·7

278. *Table LVIIB.* This Table shews the increase of a, b, c for each year of a century corresponding with Prof. Jacobi's (Epig. Ind., Vol. XI) Table X, but in greater detail, derived from use of the figures given in the heading of Table LIVA.

[1] Jacobi's Table XIII is framed to suit all Siddhāntas. By the *Ārya-* and *Sūrya-Siddhāntas* the day on which true Mēsha saṁkrānti occurred is shewn as "0 Vaiśākha," 4 Wednesday. By the *Siddhānta-Śirōmaṇi* that day was a day earlier (*above*, § 273, i), namely the day tabulated by Jacobi as "29 Mīna," 3 Tuesday.

279. *Table LVIIC.* Col. 1 shews the number of day's interval between mean sunrise of true Mēsha-saṁkrānti day, " Mēsha 0," and mean sunrise of the day which in each year was coupled with the first *tithi* of the luni-solar year and was called the day of " Chaitra śukla 1." Col. 2 gives the number of the day of the solar month Mīna (Panguni in the Tamil country); col. 3, the week-day; cols. 4, 5, 6, the value of " *a* ", " *b* ", " *c* " at mean sunrise of that day. The " *a*," " *b*," " *c* " of mean sunrise on the first day of the luni-solar year called " Chaitra śukla 1 " are found by adding to the " *a* ", " *b* ", " *c* " of the K.Y. century (*Table LVIIA*) and of the year (*Table LVIIB*) the values of " *a* ", " *b* ", " *c* " given in Table LVIIC for the number of days intervening between the day of Chaitra śukla 1 in the given year and the day of true Mēsha-saṁkrānti (*Table LX, cols. 13, 19,—figures in brackets*). This work, however, need not be carried out by epigraphists, since the required values of " *a* ", " *b* ", " *c* " for Chaitra śukla 1 in each year are stated in Table LX, cols. 23, 24, 25.

These values being known, the *tithi*-index at mean sunrise on any day in the given year is easily found, as in work by the *Indian Calendar*, by addition to them of the " *a* ", " *b* ", " *c* " for intervening days given in Table LIVA; and for any moment of any day by use of Table LIVP.

Tables LVIII-A,-B,-C,-D. Duration of true solar months.

280. Table LVIII-A is, for the *Siddhānta-Śirōmaṇi*, what Tables XVIII-A,-B in my *Indian Chronography* are for the *Ārya-* and *Sūrya-Siddhāntas.* It states the duration of each true solar month from *saṁkrānti* to *saṁkrānti*, and the collective duration from true Mēsha-saṁkrānti to each true *saṁkrānti*, with the corresponding increases of " *a* ", " *b* ", " *c* ". By the aid of this Table are calculated the solar elements of the date and the intercalations and suppressions of lunar months. The Table is designed to suit the year K.Y. 4500 expired, A.D. 1399-1400,—the year of my Table XLVIII-C above. The differences in the duration of months in other years, caused by the shift of the sun's apsis, are dealt with in Table LVIII-D.

Tables LVIII-B and -C are supplementary and explain themselves. They will be found very useful in calculation for the sun's mean anom., " *c* ", and the corresponding " equation *c* " at the several *saṁkrāntis* and at true Mēsha-saṁkrānti in different years.

Table LVIIID shews how the shift of the sun's apsis affects the duration of the several solar months in different years, and the " *a* ", " *b* ", " *c* " at the several solar saṁkrāntis. The change given in the Table is that for an interval of three centuries on either side of K.Y. 4500, and in very close cases should be applied to the figures arrived at by use of the other Tables—cases that is, where after use of those figures it seems doubtful whether a certain lunar month was intercalated or suppressed.

For an example of its use. Compare the positions of sun and moon at the moment when the true sun reached the Dhanus-saṁkrānti in K.Y. 4200 (A.D. 1100) and in K.Y. 4800 (A.D. 1700). Table LVIII-A shews that in K.Y. 4500 the sun took $246^{d}\ 9^{h}\ 9^{m}\ 34^{s}$ to travel from Mēsha-saṁkrānti, long. 0°, to the Dhanus-saṁkrānti, long. 240°, while the increase of " *a*," " *b* ", " *c* " during this interval was—" *a* "=3432·7047, " *b* "=941·5957, " *c* "=674·5407. To ascertain what this respective increase was in K.Y. 4200 we use the correction given in Table LVIII-D—thus

	a.	*b.*	*c.*
$246^{d}\ 9^{h}\ 9^{m}\ 34^{s}$.	3432·7047	941·5957	674·5407
$-4\ \ 55$.	− 1·1563	− 0·1239	− 0·0092
$246^{d}\ 9^{h}\ 4^{m}\ 39^{s}$.	3431·5484	941·4718	674·5315

These last are the correct figures for the year K.Y. 4200, A.D. 1100. For the year K.Y. 4800, using the figures of Table LVIII-D with reversed sign, the correct figures are found to be $249^{d}\ 9^{h}\ 14^{m}\ 29^{s}$, " *a* "=3433·8610, " *b* "=941·7196, " *c* "=674·5499. In a close case this difference in value of " *a* ", " *b* ", " *c* " may prove the intercalation or suppression of a different lunar month.

Changes for years less than 300 may be taken proportionally. The Table need seldom be used, as it is only very occasionally required.

281. The determination of the exact lengths of the several solar months and their collective duration (*Table LVIIIA*) has been a matter of considerable difficulty, and in publishing the quantities given in the Table I must not be held to assert that the mediæval Hindu used those lengths and no others. He may have calculated roughly, or, if scientifically, then by several different processes.

Take as an example the time of the true sun's arrival, say in K.Y. 4500, at the Vṛishabha saṁkrānti, 30°, in order to determine the length of the solar month Mēsha.

(i) One method of reckoning is that which was used in the preparation of Table XLVII-C (*above*), viz. by applying to the mean long. of the sun (*col. 4*) the equation of the centre (*col. 6*) as found by computation from the Hindu equation-Table (*Table XLVII*), which is based on a series of groups of angles; and so obtaining the sun's true long. According to this system it is found that in the first 30 whole days from true Mēsha-saṁkrānti the sun travelled 29° 7′ 28″·60 (*Table XLVIII-C, col. 8*). Before he reached 30°, therefore, he had to travel 52′ 31″·40.

(ii) Another method is, discarding the group system of the equation-Table, to ascertain directly the value of the sine of the mean anom. angle at the beginning of the 30th day after the moment of true Mēsha-saṁkrānti, and to work the equation of that sine-value; afterwards calculating for the remaining hours and minutes taken by the sun to reach 30°. The value of the sine is obtained by the method described in § 282.

Thus we find from Table XLVIII-C that the sun's mean anom. at the beginning of the 30th day was 128° 21′·25232, or 7701′·25232. This divided by 225 is 34 with remainder 51′·25232. The 34th sine is, counting down and then up on the left side of the equation-Table, the base sine No. 14, or the sine of 127° 30′. This is 2728′ (*col. 3*). The difference between this and the next base sine is 143′ (*col. 4*). 51′·25232 × 143 = 7329′·08176, and this divided by 225 is 32′·57369. 2728′ − 32′·57369 = 2695′·4263; and this, therefore, is the sine of the given anom. angle 128° 21′·25232.

The equation-formula is sin. eqn. = $\frac{41}{1080}$ sin. anom. (§ *258 above*) and the result is (the angle being a small one) that the equation = 1° 42′ 21″·578. The sun's mean long. (*Table XLVIII-C, col. 4*) at the beginning of this 30th day was 27° 25′ 9″·14; and, adding the equation we find that his true long. at that moment was 29° 7′ 28″·72. The true sun, before he reached long. 30°, therefore, had to travel (30° − 29° 7′ 28″·72) 52′ 31″·28.

In either of the above cases how long did he take to accomplish the journey?

To ascertain this we may either use the sun's mean motion (*Table XLIV*); or we may use the true motion in hours for the particular 30th day (*Table XLIX*), as fixed by the group system of the equation-Table, with his mean motion in minutes and seconds (*Table L, LI*); or we may carefully work out his true motion for that 30th day by dividing his motion during that day by 24 for hours, and again by 60 for minutes, and each minute by 60 for seconds; or, yet again, even still more accurately, by calculating his real motion during the particular hours of the day actually concerned, and so the rest.

Thus it is clear that we can calculate the length of Mēsha in a number of ways, with slight differences in the result of each; and so with all the solar months and their collective lengths. These differences in the lengths of months may amount to two or three seconds in each, and at some parts of the orbit the cumulative difference may amount to perhaps a quarter of a minute, but probably not more than that.

I have tried all the methods noted above, except the last, which it seemed unnecessary to attempt, in order to arrive at the exact lengths of the months, and believe that my Table LVIII-A is sufficiently accurate. Since it is not known how the mediæval Hindu astronomers carried out their computation, no better course presented itself.

Let it be noted that any little difference that may exist will have no effect whatever on the value of the *tithi*; and as regards the intercalated and suppressed months care has been taken to avoid any possibility of error by a special note of every close case in the page preceding the body of Table LX.

Table LIX. The moon's equation of the centre.

282. The Table itself requires no explanation. The equations have been calculated by the proper formula, viz. sin. eqn. $= \frac{\text{sin. } \alpha \times \text{mins. in epicycle}}{\text{mins. in orbit}}$, here $\frac{\text{sin. } \alpha \times 1896'}{21600'}$, or $\frac{79'}{900'}$ sin. α. (*§ 251 above; and especially § 272, para. 3. Moon's epicycle 31° 96′=1896′.*)

It has to be noted, however, that—whereas, when (as in the case of the equation of the sun) the sine of the equation-angle is less than 3° 45′, the equation is the same as the sine and therefore the formula may be read as "eqn. $= \frac{79}{900}$ sin. α"—here, in the lower half of the Table of the moon's equations, the sine of the equation-angle is greater than 3° 45′. Thus *sin. eqn. 90°* is 5° 1′ 46″·8, but *eqn. 90°* is 5° 2′ 7″·366.

The rule for finding the equation, when sin. eqn. is greater than 3° 45′ and less than 7° 30′ (it is always less in the present case), is as follows. First ascertain the value of sin. eqn. by the above formula. Deduct 225′ from this value; either multiply the remainder by 225 and divide the product by 224 or add to the same remainder a 224th part of itself (*see cols. 2, 3, 4, Table LIX*). Add to the result 225′ (*col. 3*).

Thus for the given moon's mean anom. 90°. Sin. 90°=3438′ (*col. 3*), and $\frac{79' \times 3438'}{900'} =$ 301′·78, or 5° 1′ 46″·8, as stated above. This is the value of the *sin. eqn.* For the equation we work with 301′·78 as the given angle. This *minus* 225′=76′·78. 76′·78 × 225 = 17275′·50, and this divided by 224 is 77′·122768. 77′·122768 + 225′=302′·122768, and this = 5° 2′ 7″·366068, which is the correct *equation* of the moon's centre when his mean anom. is 90°. Worked in the other way, a 224th part of 76·78 is 0·342768, and this added to 76·78 gives the same result, viz. 77·122768.

283. It is advisable here to state the *Hindu rule for finding the sine of any angle*,[1] viz.:—Ascertain the number of minutes contained in the given arc. Divide these by 225′ (= 3° 45′). The quotient is the serial number of the preceding base-sine as given in Table LIX, col. 1. Multiply the *remainder* by the difference between the preceding and succeeding base-sines (*col. 4*) and divide by 225. Add the result to the preceding base-sine. Thus with arc 24° or 1440′. 1440°÷225 yields quotient 6, remainder 90. 6 is the serial number of the sine of 22° 30′ (*col. 1*). The difference between the base-sine No. 6 and base-sine No. 7 is (*col. 4*) 205. 90 × 205 = 18450, and this divided by 225 = 82, with no remainder. The preceding base-sine, No. 6, is 1315′, and this *plus* 82 = 1397′. 1397′ is the sine of 24°.

283 A. The equation-Table for the moon's centre given below (*Table LIX*) is practically the same as that of Prof. Jacobi's Table XXIV-A (*Vol. I, p. 458, above*); but in the former decimal points are given which are omitted in the latter. We agree also in our equation-Tables for the sun (mine in Table XLVII, above, his in Table XXIV-B, *Epigraphia Indica* Vol. I, p. 459). But there seems to be some mistake in the figures entered by him, stated in parts of the circle, in his equivalent Table of the equations of the sun's centre given in *Epig. Ind.* Vol. XI

[1] The Hindu rule laid down in the *Sûrya-Siddhânta* (vv. 31-33) is, as interpreted by Spottiswoode (*Journal of the Royal Asiatic Society* for 1863, Vol. XX), sin. n (225 + θ) = $\theta \left\{ \frac{\text{sin. } (n+1)\ 225' \} - \text{sin. } n\ 225'}{225'} \right.$ The sine is a line, not a ratio.

(*Table XII, p. 169, col.* "△ 10") for differences in consecutive equations. For instance, the equation for anom. 221° 15′ is 1° 26′ 3″·72 (base-equation No. 11) and for anom. 225′ is 1° 32′ 17″·28 (base-equation No. 12). Difference 6′ 13″·56. There is a difference of 225′ in the anomalies, and 6′ 13″·56 ÷ 225 gives the difference per minute of anom. as 1″·66. In this we both agree.

Now 6′ 13″·56, in 10,000ths of the circle, is 2·8824, or, with two decimals only, 2·88, but Prof. Jacobi in Vol. XI quotes "2·78" as the figure. It stands between his "arg. c" 1146 and 1250, which are the equivalents in his notation of the anom. angles corresponding to 221° 15′ and 225°—serial numbers 11 and 12 in the equation-Table.

One-fifth of 2·8824 = 0·5765, and this is the entry given in col. 4 of Table LVI below as the group-difference for all anom. angles between those of the serial numbers 11 and 12.

I venture to suggest the following amendments to all the entries in Prof. Jacobi's col. "△ 10," reading from top to bottom of his Table XII (Vol. XI):—

For	3·75			For	3·26	read	3·36	For	1·83	read	1·86
„	3·85	„	3·94	„	3·07	„	3·22	„	1·53	„	1·63
„	3·75	„	3·90	„	2·88	„	3·06	„	1·34	„	1·39
„	3·65	„	3·85	„	2·78	„	2·88	„	1·15	„	1·14
„	3·65	„	3·78	„	2·59	„	2·71	„	0·86	„	0·90
„	3·56	„	3·69	„	2 40	„	2·51	„	0·58	„	0·65
„	3·46	„	3·61	„	2·21	„	2·30	„	0·38	„	0·39
„	3·36	„	3·50	„	2·02	„	2·09	„	0·10	„	0·12

These differences stand in regular progression. It is possible that the Professor's first entry "3·75" is a printer's error for 3·95; but even so our agreement is only in that one out of 24 entries.

Table LX. Working Table for computation of dates.

284. Table LX is the principal working Table by which the *tithi*, lunar and solar month and day, *nakshatra* and *yoga* given in the date of a document or inscription and based on the *Siddhānta-Śirōmaṇi* can be verified and converted into European reckoning; the *nakshatra*, *yoga* and *lagna* being still more accurately computed by use of Table XLVIII-C above. Table LX is to be used exactly as Table I of the *Indian Calendar* is used for *Ārya-* and *Sūrya-Siddhānta* reckoning. In the latter whole numbers only are given. Here four places of decimals are added (they need not of course be used, unless necessary), and seconds of time are given as well as minutes. For further explanation see the page of note preceding the Table.

To be entirely on the safe side, and for convenience of working from the beginning of a century of the Kaliyuga, as well as for guidance in studying the working of the Metonic cycle according to this authority, the Table begins with K.Y. 4200 expired (A.D. 1099-1100); though in all probability the *Siddhānta-Śirōmaṇi* was not used in India for the preparation of almanacs till A.D. 1150 at earliest.

A date should be first computed approximately by use of whole numbers only, and the equation-Tables LV and LVI used merely as eye-Tables. Very great accuracy can be obtained by close work in greater detail.

Each intercalation and suppression of a lunar month has been carefully calculated. For the process reference may be made to my *Indian Chronography*, §§ 95-103, and Examples 27-32. The months are true months, as it is almost certain that calculation by mean months was never resorted to at so late a date as that when our authority came into use.

(*Cols. 13, 14, 17.*) See the last para. of § 273 above. The true sun arrives at 0° every year after a journey lasting $365^d\ 6^h\ 12^m\ 8^s{\cdot}84025$. The moment of this arrival, *i.e.* the moment of true Mēsha-saṁkrānti in the first year of the Table, was fixed by calculation from Dr. Schram's determination of the *śōdhya* and the sun's equation at that instant (*above*, § 273). For all later years the time-interval was added to this. The result accords exactly with Dr. Schram's fixtures.

(*Cols. 19-20.*) The luni-solar date, week-day and "*a*", "*b*", "*c*" have each been separately calculated. For process see Example 2 below. The date and week-day are generally the same as those found by *Sūrya-Siddhānta* computation, but differ from these in occasional close cases, and where the intercalations and suppressions of lunar months differ.

The 19-year Metonic sequence.

285. [For a note as to this see *Indian Calendar*, § 50, p. 29.] This sequence, in work by the *Siddhānta-Śirōmaṇi*, proceeds with the same general regularity as when computed by the *Ārya-* and *Sūrya-Siddhāntas*. In the period of 650 years dealt with in Table LX the intercalated lunar months are, in seven cases, the month next to that expected by the sequence, not that month itself (*see note preceding the Table*). The rest are regular. Suppressions follow the sequence in all cases. In the same period there are six such irregularities by *Sūrya-Siddhānta* and two by *Ārya-Siddhānta* work.

Future research will no doubt settle the question whether the irregularity of seven out of 260 cases of intercalations and suppressions in the period embraced is attributable to the postulates of the *Siddhānta-Śirōmaṇi* or to any defect in my calculations. All possibility of error, however, in computation of dates of records by these Tables is removed by the footnotes entered in each case and the Remarks embodied in the page preceding Table LX. Whenever a record-date belonging to either of these seven years is examined, it should be tested both ways.

EXAMPLES.

Example 1. To find the value of "a", "b", "c" for the moment of true Mēsha-saṁkrānti in any year, the beginning of the solar year.

Rule. Note in Table LX the number of the **expired** year of the Kaliyuga (*col. 2.*) [In this column the K.Y. year is that *current* in the corresponding A.D. year. The *expired* K.Y. year is the next earlier]. Note (*cols. 13-17*) the day, week-day, and time of occurrence of true Mēsha-saṁkrānti in that year. Take from Table LVII-A the week-day and "*a*", "*b*", "*c*" for the beginning of the K.Y. century; from Table LVII-B the same for the expired K.Y. year of the century; from Table LVII-C the same for the day marked "Mēsha 0" (*col. 2*), or the day next to it, being guided by the given week-day (*Table LX, col. 14*); and add together the three sets of values so obtained. The sum of these shews the positions of the moon and sun (*a*, *b*, *c*) at mean sunrise of the day on which true Mēsha-saṁkrānti occurred. For the moment of the saṁkrānti add to these values of "*a*", "*b*", "*c*" those for the hours, minutes and seconds elapsed since mean sunrise (*col. 17*), obtaining them from Table LIV-B.

Work. Given that the values of "*a*", "*b*", "*c*" are wanted for the moment of mean sunrise of the day on which true Mēsha-saṁkrānti occurred in K.Y. 4492 expired, A.D. 1391-2; and at the moment of that saṁkrānti.

Table LX shews that the day was (0) Saturday 25 March A.D. 1391, and that the saṁkrānti occurred on that day at $17^h\ 18^m\ 12^s$.

(i) *Approximate calculation, by whole numbers.*

	w-d.	*a.*	*b.*	*c.*
Table LVII-A. K.Y. cent. 44	5	7454	768	277
,, LVII-B. Year 92	4	9389	545	1
,, LVII-C. Mēsha 0	5	9323	927	995
At mean sunrise of Sat. 25 March . .	0 (Sat.)	6166	240	273
Table LIV-B. 17 hours		240	26	2
18 minutes		4	0	0
At moment of saṁkrānti		6410	266	27[illegible]

(ii) *Full calculation.* Worked to the full extent, with use of decimals and including the value of "*a*", "*b*", "*c*" for seconds the result is—

For mean sunrise, *a*=6166·1839, *b*=240·2250, *c*=272·5113.

For moment of Mēsha-saṁkrānti, *a*=6410·3281, *b*=266·3902, *c*=274·4852.

Note. The value found for "*c*" will always be a guide as to whether the calculation has been made for the right day (*see Table LVIII-C below*); for at true Mēsha-saṁkrānti "*c*" is always 274 or 275. In this case let it be observed that 8 years later than the given year, viz. in K.Y. 4500, the value of "*c*" at true Mēsha-saṁkrānti was 274·4058. The change in "*c*" at that moment, owing to shift of sun's apsis (§ *273, ii*), being 0·0805 per century, and our calculation having been based on the value for K.Y. 4400, we should, for extreme accuracy, deduct from 274·4852 the proportional change for 92 years, which amounts to 0·0741, leaving our *c* for A.D. 1391=274·4111.

Example 2. Required to find the value of *a*, *b*, *c* at mean sunrise of the civil day called Chaitra śukla 1, the civil beginning of the luni-solar year K.Y. 4492 expired, A.D. 1931-2.

Rule. (i) If the *a*, *b*, *c* of mean sunrise on the day on which true Mēsha-saṁkrānti occurred in the year in question has already been found, as above, note the interval of days between mean sunrise on the day of Chaitra śukla 1 (*Table LX, col. 19*) and on the day of true Mēsha-saṁkrānti in the given year (*col. 13*), both in brackets. With that interval of days turn to Table LIV-A and find it in *col. 1*. Take the week-day and "*a*", "*b*", "*c*" values stated against it, and deduct the amount from the ascertained value of "*a*", "*b*", "*c*" for the Mēsha-saṁkrānti day (mean sunrise). Thus—

In Example 1 we have determined the "*a*", "*b*", "*c*" values for mean sunrise on 25 March A.D. 1391, Day 84 (*Table LX, col. 13*). The day of Chaitra śukla 1 was 7 March, Day 66 (*col. 19*). Interval 18 days. We deduct 18 days' "*a*", "*b*", "*c*" from the former by Table LIV-A.

	w-d.	*a.*	*b.*	*c.*
Mēsha 0 mean sunrise	0	6166·1839	240·2250	272·5113
For 18 days' interval (Table LIV-A)	−4	−6095·3757	−653·2496	−49·2802
	3 (Tues.)	70·8082	586·9754	223·2311

These were the values of "*a*," "*b*." "*c*" on Tuesday 7 March A.D. 1391. (*Compare entry in Table LX.*)

(ii) If the "*a*," "*b*", "*c*" of mean sunrise on Mēsha-saṁkrānti day has not already been found, add together as in Example 1 the week-day and "*a*", "*b*", "*c*" of the K.Y. century and the year (*Tables LVII A, B*), and to the sum of these add the week-day and the "*a*", "*b*", "*c*" stated in Table LVII-C against the interval of days (*as above*). Here the K.Y. century is 44, the year is 92, the interval of days is 18.

	w-d.	*a.*	*b.*	*c.*
Table LVII-A. Cent[illegible] .	5	7454·2101	768·2089	277·3743
,, LVII-B. Year 92 .	4	9389·2378	544·5994	0·6126
,, LVII-C. 18 days .	1	3227·3603	274·1671	945·2442
	3 (Tues.)	70·8082	586·9754	223·2311

The result is the same as by process (i).

Owing to the formation of the Tables the week-day will sometimes be found to be different by one from the week-day noted in Table LX, col. 19. In such case the week-day and "*a*", "*b*", "*c*" in Table LVIIC to be applied must be that of the altered interval, the week-day always being that stated in Table LX.

Thus in A.D. 1390-91, K.Y. 4491, the interval (*Table LX, cols. 13, 19*) is (84—77) 7 days. When we come to work, we find (*Table LVII-A*) given the week-day 5, and (*Table LVII-B*) week-day 2, Total 7, or 0. Now in Table LVII-C against 7 days' interval (*col. 3*) we find week-day 5; but, as we have to arrive at the entry in Table LX (*col. 20*), i.e. at the "*a*", "*b*," "*c*" for 6 Friday, we add the week-day (6) and the "*a*", "*b*", "*c*" for it (standing for 6 days' interval instead of 7) in Table LVII-C. Such change is never more than one day.

Example 3. Given the moon's mean anom. "b", or the sun's mean anom. "c", as found in work for verifying a date, required to find "equ. b," or "equ. c."

The work is similar in either case. We will take an instance of a case where "*c*", the sun's mean anom., has been found to be 146·3264.

By Table LVI we see that the equation for anom. values between 145·83 and 147·916 lies between 12·4786 and 12·0181, the difference between them being 0·4605. For rule of work see § 275 above.

Approximation. A glance at Table LVI shews that equ. *c* must be 12 and a small fraction.

Closer work. The difference between 145·3 and the next figure of Arg. in the Table (col. 2 *a*), viz. 147·9, is 1·6. The group-difference (col. 4) is 0·4605. Call this 0·5. The invariable difference between successive entries of arc ("Arg.") is 2·08$\dot{3}$. Call this 2. $1·6 \times 0·5 = 0·8$. This divided by 2 is 0·4. Add this to the equation stated for Arg. 147·9, viz. 12·0. Result 12·4.

Still closer work. The actual anom. difference (147·91$\dot{6}$—146·3264) is 1·5902. This multiplied by the group-difference, 0·4605=0·7323. This divided by 2·08$\dot{3}$ is 0·3515. And this, added to 12·0181 (the equation of anom. 147·91$\dot{6}$), gives us the exact equation of anom. 146·3264 as 12·3696.

Example 4. To find the tithi current at mean sunrise of any civil day, or at any moment of that day.

Rule. Take the European date, serial number of the day (in brackets measured from Jan. 1st of the A.D. year) and "*a*", "*b*", "*c*" of Chaitra śukla 1 of the luni-solar year, from cols. 19 to 25 of Table LX. Find the interval of days to the given day and add to the "*a*", "*b*", "*c*" of Chaitra śukla 1 the "*a*", "*b*", "*c*" for that number of days given in Table LIV-A. This gives the "*a*", "*b*", "*c*" of sunrise on the given day.

For subsequent hours, minutes and seconds add the "*a*", "*b*", "*c*" given in Tablo LIV-B.

Find *equ. b* and *equ. c* from Tables LV and LVI, and add them to the "*a*" already found. The result is the *tithi*-index; with which find the current *tithi* in Table VIII, *Indian Calendar* or Table LXVIII below.

Compare Example 4 in the section on the *First Ārya-Siddhānta—True System.* Work in similar manner, but with the use of *Siddhānta-Śirōmaṇi* Tables.

Example 5. Calculation for intercalated (adhika) and suppressed (kshaya) lunar months.

This is the same as in work by the *Indian Calendar* or *Indian Chronography*, but the lengths of the solar months, their collective duration, week-days and "*a*", "*b*", "*c*" must be taken from Table LVIII below when working by the *Siddh.-Śirōmaṇi*. In a very close case use may be made of Table LVIII-D. But even so, in work for the *tithi*, or for intercalations and suppressions of months, the correction in the value of "*a*" need alone be taken into account, since the change in the *tithi*-index, "*t*", is governed by the value of *eqn. b* and *eqn. c*, not of "*b*" and "*c*"; and the difference in these equations is infinitesimal.

An example is here given of work by the Tables in a very close case, viz. the intercalation of a lunar month in K.Y. 4850 expired, A.D. 1749-50.

In that year, according to the *Sūrya-Siddhānta* Bhādrapada was the added month. Was it so according to the *Siddhānta-Śirōmaṇi*?

In that year (*Table LX, cols. 13-17*) true Mēsha-saṁkrānti occurred on Tuesday, 28 March, A.D. 1749, at 5^h 46^m 57^s after mean sunrise. First must be ascertained the position of mean moon and mean sun at that moment, individually and relatively, i.e. the values of "*a*", "*b*", "*c*". For this process see Example 1

Approximate calculation with whole numbers.

	w-d.	*a.*	*b.*	*c.*
(*Table LVII-A*) For K. Y. cent. 48 . .	5	2942	123	279
(,, *LVII-B*) ,, ,, year 50 . .	0	4436	794	0
(,, *LVII-C*) ,, 0 Mēsha mean sunrise .	5	9323	927	995
(,, *LIV-B*) { ,, 5 hours . . .	...	71	8	1
(,, *LIV-B*) { ,, 47 minutes . . .	...	11	1	0
At true Mēsha-saṁkrānti	3 (Tues.)	6783	853	275
(*Table LVIII-A, cols. 6, 7, 8*) Interval to Siṁha-saṁkrānti .	...	2471	552	343
At true Siṁha-saṁkrānti . . .		9254	405	618
(*Table LV*) *Eqn. b*	...	218		
(,, *LVI*) *Eqn. c*	...	101		
	"*t*" =	9573		

Hence the moon was waning at the Siṁha-saṁkrānti. At the next (Kanyā) saṁkrānti was she waning or waxing?

	a.	*b.*	*c.*
(*Above*) At Siṁha-saṁkrānti	9254	405	618
(*Table LVIII-A, cols. 13, 14, 15*) Interval to Kanyā-saṁkrānti .	518	127	85
At Kanyā-saṁkrānti	9772	532	703
(*Table LV*) *Eqn. b*	111		
(,, *LVI*) *Eqn. c*	118		
t =	10001		

This is so close to 10000, or 0, that it seems doubtful whether new moon took place before or after the Kanyā-saṁkrānti, whether, that is, at that moment the moon was still waning or had begun to wax. It is certain that she was waning at the previous Siṁha-saṁkrānti, and therefore we can calculate direct from the Mēsha to the Kanyā-saṁkrānti. For greater accuracy we use one decimal place and guess a little more carefully the values of "*equ. b*" and "*equ. c*" at the latter saṁkrānti.

	a.	b.	c.
K. Y. cent. 48	2941·8	123·4	278·8
" year 50	4435·9	794·4	0·2
Mēsha-saṁkrānti day (mean sunrise) . . .	9322·7	927·4	994·5
5 hours	70·5	7·6	0·6
47 minutes	11·1	1·2	0·1
At Mēsha-saṁkrānti	6782·0	854·0	274·2
Interval to Kanyā-saṁk. (Table LVIII-A, cols. 6, 7, 8)	2989·5	679·0	428·4
At Kanyā-saṁkrānti	9771·5	533·0	702·6[1]
Equ. b	110·9		
Equ. c	118·2		
t =	10000·6	or 0·6	

On a still closer examination, using the full number of given decimals and calculating the *equations b* and *c* thoroughly, it is found that at the Kanyā-saṁkrānti the *tithi*-index was 10000·9421. It is not necessary to give the full working figures. It is certain that at that saṁkrānti the moon was waxing, so far as we have gone, and therefore the intercalated lunar month was (*Table LVIIIA, cols. 1, 2*) 6 Bhādrapada.

But since the date K.Y. 4850 is 350 years subsequent to the base-year K.Y. 4500, and the lengths of the solar months have in the interval slightly changed in consequence of the shift of the sun's apsis, it is necessary to find out whether this change would make any difference in the result. We therefore correct the "*a*" of the Kanyā-saṁkrānti by Table LVIII-D. At the Kanyā-saṁkrānti 300 years after K.Y. 4500 the change in "*a*" (*col. 3*) was −0·0901. Increase this by one-sixth for another 50 years' change. Total change −0·1051. Hence the rea *tithi*-index, "*t*", at Kanyā-saṁkrānti was (0·9421−0·1051=) 0·8370. Bhādrapada was certainly intercalated.

In § 274 above (*Para. 3, p. 130*) I stated that I accepted Prof. Jacobi's figures for the value of *a* in K.Y. 4200, although by my own estimate his was too large by 0·4. If, in this very close case, we reduce the value of "*a*" (found to be 9771·5 at Kanyā-saṁkrānti) by 0·4, making a=9771·1, we find that the state of the true moon at the Kanyā-saṁkrānti was (t=) 10000·2; or with the correction applied as in the last para. 10000·4370. Thus the moon was really waxing at that moment (new moon occurring at the point 10,000 or 0), but had only begun to do so about two minutes before the sun entered Kanyā.

[1] In all cases the value of "*c*" at *saṁkrāntis* should be compared with the values given in Table LVIII-B below, and the equation taken therefrom should be used.

TABLE LIV-A.

INCREASE OF "a", "b", "c" IN DAYS.

(a in 10,000ths; b and c in 1,000ths of circle.)

Increase in 1 day $a = 338·631985412$; $b = 36·291649786$; $c = 2·737787543$.
Do. in 1 year of 365 days $a = 3600·674675380$; $b = 246·452171890$; $c = 999·292153195$.
Do. in " 366 " $a = 3939·306660782$; $b = 282·743821676$; $c = 2·030240738$.
Do. in 1 cent. of 36525 " $a = 8533·267173300$; $b = 552·508433650$; $c = 997·609452520$.
Do. in " 36526 " $a = 8871·899158712$; $b = 588·800083436$; $c = 0·347240063$.

N.B.—By first calculation, "c" for a cent. of 36525 days is 997·690008075, and for a cent. of 36526 days is 0·427795618. Each of these quantities is reduced by 0·0805 on account of shift of ⊙*'s apsis. (See Text, § 273, 57.)*

This Table answers to Table IV, *Indian Calendar.*

DAYS OF 24 HOURS EACH.

No.	Week-day.	a.	b.	c.	No.	Week-day.	a.	b.	c.
1	2	3	4	5	1	2	3	4	5
1	1	338·6320	36·2916	2·7378	41	6	3883·9114	487·9576	112·2493
2	2	677·2640	72·5833	5·4756	42	0	4222·5434	524·2493	114·9871
3	3	1015·8960	108·8749	8·2134	43	1	4561·1754	560·5409	117·7249
4	4	1354·5279	145·1666	10·9512	44	2	4899·8073	596·8326	120·4627
5	5	1693·1599	181·4582	13·6889	45	3	5238·4393	633·1242	123·2004
6	6	2031·7919	217·7499	16·4267	46	4	5577·0713	669·4159	125·9382
7	0	2370·4239	254·0415	19·1645	47	5	5915·7033	705·7075	128·6760
8	1	2709·0559	290·3332	21·9023	48	6	6254·3353	741·9992	131·4138
9	2	3047·6879	326·6248	24·6401	49	0	6592·9673	778·2908	134·1516
10	3	3386·3199	362·9165	27·3779	50	1	6931·5993	814·5825	136·8894
11	4	3724·9518	399·2081	30·1157	51	2	7270·2312	850·8741	139·6272
12	5	4063·5838	435·4998	32·8535	52	3	7608·8632	887·1658	142·3650
13	6	4402·2158	471·7914	35·5912	53	4	7947·4952	923·4574	145·1027
14	0	4740·8478	508·0831	38·3290	54	5	8286·1272	959·7491	147·8405
15	1	5079·4798	544·3747	41·0668	55	6	8624·7592	996·0407	150·5783
16	2	5418·1118	580·6664	43·8046	56	0	8963·3912	32·3324	153·3161
17	3	5756·7437	616·9580	46·5424	57	1	9302·0232	68·6240	156·0539
18	4	6095·3757	653·2496	49·2802	58	2	9640·6551	104·9157	158·7917
19	5	6434·0077	689·5413	52·0180	59	3	9979·2871	141·2073	161·5295
20	6	6772·6397	725·8329	54·7558	60	4	317·9191	177·4990	164·2673
21	0	7111·2717	762·1246	57·4935	61	5	656·5511	213·7906	167·0050
22	1	7449·9037	798·4162	60·2313	62	6	995·1831	250·0823	169·7428
23	2	7788·5357	834·7079	62·9691	63	0	1333·8151	286·3739	172·4806
24	3	8127·1676	870·9995	65·7069	64	1	1672·4471	322·6656	175·2184
25	4	8465·7996	907·2912	68·4447	65	2	2011·0790	358·9572	177·9562
26	5	8804·4316	943·5828	71·1825	66	3	2349·7110	395·2489	180·6940
27	6	9143·0636	979·8745	73·9203	67	4	2688·3430	431·5405	183·4318
28	0	9481·6956	16·1661	76·6581	68	5	3026·9750	467·8322	186·1696
29	1	9820·3276	52·4578	79·3958	69	6	3365·6070	504·1238	188·9073
30	2	158·9596	88·7495	82·1336	70	0	3704·2390	540·4155	191·6451
31	3	497·5915	125·0411	84·8714	71	1	4042·8709	576·7071	194·3829
32	4	836·2235	161·3328	87·6092	72	2	4381·5029	612·9988	197·1207
33	5	1174·8555	197·6244	90·3470	73	3	4720·1349	649·2904	199·8585
34	6	1513·4875	233·9161	93·0848	74	4	5058·7669	685·5821	202·5963
35	0	1852·1195	270·2077	95·8226	75	5	5397·3989	721·8737	205·3341
36	1	2190·7515	306·4994	98·5604	76	6	5736·0309	758·1654	208·0719
37	2	2529·3834	342·7910	101·2981	77	0	6074·6629	794·4570	210·8096
38	3	2868·0154	379·0827	104·0359	78	1	6413·2948	830·7487	213·5474
39	4	3206·6474	415·3743	106·7737	79	2	6751·9268	867·0403	216·2852
40	5	3545·2794	451·6660	109·5115	80	3	7090·5588	903·3320	219·0230

TABLE LIV-A—*contd.*

DAYS OF 24 HOURS EACH.

No.	Week-day.	*a.*	*b.*	*c.*	No.	Week-day.	*a.*	*b.*	*c.*
1	2	3	4	5	1	2	3	4	5
81	4	7429·1908	939·6236	221·7608	136	3	6053·9500	935·9644	372·3391
82	5	7767·8228	975·9153	224·4986	137	4	6392·5820	971·9560	375·0769
83	6	8106·4548	12·2069	227·2364	138	5	6731·2140	8·2477	377·8147
84	0	8445·0867	48·4986	229·9742	139	6	7069·8460	44·5393	380·5525
85	1	8784·7187	84·7902	232·7119	140	0	7408·4780	80·8310	383·2903
86	2	9122·3507	121·0819	235·4497	141	1	7747·1099	117·1226	386·0281
87	3	9460·9827	157·3735	238·1875	142	2	8085·7419	153·4143	388·7658
88	4	9799·6147	193·6652	240·9253	143	3	8424·3739	189·7059	391·5036
89	5	138·2467	229·9568	243·6631	144	4	8763·0059	225·9976	394·2414
90	6	476·8787	266·2485	246·4009	145	5	9101·6379	262·2892	396·9792
91	0	815·5106	302·5401	249·1387	146	6	9440·2699	298·5809	399·7170
92	1	1154·1426	338·8318	251·8765	147	0	9778·9019	334·8725	402·4548
93	2	1492·7746	375·1234	254·6142	148	1	117·5338	371·1642	405·1926
94	3	1831·4066	411·4151	257·3520	149	2	456·1658	407·4558	407·9304
95	4	2170·0386	447·7067	260·0898	150	3	794·7978	443·7475	410·6681
96	5	2508·6706	483·9984	262·8276	151	4	1133·4298	480·0391	413·4059
97	6	2847·3026	520·2900	265·5654	152	5	1472·0618	516·3308	416·1437
98	0	3185·9346	556·5817	268·3032	153	6	1810·6938	552·6224	418·8815
99	1	3524·5666	592·8733	271·0410	154	0	2149·3258	588·9141	421·6193
100	2	3863·1985	629·1650	273·7788	155	1	2487·9577	625·2057	424·3571
101	3	4201·8305	665·4566	276·5165	156	2	2826·5897	661·4974	427·0949
102	4	4540·4625	701·7483	279·2543	157	3	3165·2217	697·7890	429·8327
103	5	4879·0945	738·0399	281·9921	158	4	3503·8537	734·0807	432·5705
104	6	5217·7265	774·3316	284·7299	159	5	3842·4857	770·3723	435·3082
105	0	5556·3585	810·6232	287·4677	160	6	4181·1177	806·6640	438·0460
106	1	5894·9905	846·9149	290·2055	161	0	4519·7497	842·9556	440·7838
107	2	6233·6224	883·2065	292·9433	162	1	4858·3816	879·2473	443·5216
108	3	6572·2544	919·4982	295·6811	163	2	5197·0136	915·5389	446·2594
109	4	6910·8864	955·7898	298·4189	164	3	5535·6456	951·8306	448·9972
110	5	7249·5184	992·0815	301·1566	165	4	5874·2776	988·1222	451·7350
111	6	7588·1504	28·3731	303·8944	166	5	6212·9096	24·4139	454·4728
112	0	7926·7824	64·6648	306·6322	167	6	6551·5416	60·7055	457·2105
113	1	8265·4144	100·9564	309·3700	168	0	6890·1735	96·9972	459·9483
114	2	8604·0463	137·2481	312·1078	169	1	7228·8055	133·2888	462·6861
115	3	8942·6783	173·5397	314·8456	170	2	7567·4375	169·5805	465·4239
116	4	9281·3103	209·8314	317·5834	171	3	7906·0695	205·8721	468·1617
117	5	9619·9423	246·1230	320·3212	172	4	8244·7015	242·1638	470·8995
118	6	9958·5743	282·4147	323·0590	173	5	8583·3335	278·4554	473·6373
119	0	297·2063	318·7063	325·7967	174	6	8921·9655	314·7471	476·3750
120	1	635·8382	354·9980	328·5345	175	0	9260·5974	351·0387	479·1128
121	2	974·4702	391·2896	331·2723	176	1	9599·2294	387·3304	481·8506
122	3	1313·1022	427·5813	334·0101	177	2	9937·8614	423·6220	484·5884
123	4	1651·7342	463·8729	336·7479	178	3	276·4934	459·9137	487·3262
124	5	1990·3662	500·1646	339·4857	179	4	615·1254	496·2053	490·0640
125	6	2328·9982	536·4562	342·2235	180	5	953·7574	532·4970	492·8018
126	0	2667·6302	572·7479	344·9613	181	6	1292·3894	568·7886	495·5396
127	1	3006·2621	609·0395	347·6990	182	0	1631·0213	605·0803	498·2773
128	2	3344·8941	645·3312	350·4368	183	1	1969·6533	641·3719	501·0151
129	3	3683·5261	681·6228	353·1746	184	2	2308·2853	677·6636	503·7529
130	4	4022·1581	717·9145	355·9124	185	3	2646·9173	713·9552	506·4907
131	5	4360·7901	754·2061	358·6502	186	4	2985·5493	750·2469	509·2285
132	6	4699·4221	790·4978	361·3880	187	5	3324·1813	786·5385	511·9663
133	0	5038·0541	826·7894	364·1258	188	6	3662·8133	822·8302	514·7041
134	1	5376·6860	863·0811	366·8635	189	0	4001·4452	859·1218	517·4419
135	2	5715·3180	899·3727	369·60[illegible]	190	1	4340·0772	895·4135	520·1796

TABLE LIV-A—*contd.*

Days of 24 hours each.

No.	Week-day.	*a.*	*b.*	*c.*	No.	Week-day.	*a.*	*b.*	*c.*
1	2	3	4	5	1	2	3	4	5
191	2	4678·7092	931·7051	522·9174	241	3	1610·3085	746·2876	659·8068
192	3	5017·3412	967·9968	525·6552	242	4	1948·9405	782·5793	662·5446
193	4	5355·9732	4·2884	528·3930	243	5	2287·5725	818·8709	665·2824
194	5	5694·6052	40·5801	531·1308	244	6	2626·2044	855·1626	668·0202
195	6	6033·2372	76·8717	533·8686	245	0	2964·8364	891·4542	670·7580
196	0	6371·8691	113·1634	536·6064	246	1	3303·4684	927·7459	673·4958
197	1	6710·5011	149·4550	539·3442	247	2	3642·1004	964·0375	676·2335
198	2	7049·1331	185·7467	542·0820	248	3	3980·7324	0·3292	678·9713
199	3	7387·7651	222·0383	544·8197	249	4	4319·3644	36·6208	681·7091
200	4	7726·3971	258·3300	547·5575	250	5	4657·9964	72·9125	684·4469
201	5	8065·0291	294·6216	550·2953	251	6	4996·6283	109·2041	687·1847
202	6	8403·6611	330·9132	553·0331	252	0	5335·2603	145·4958	689·9225
203	0	8742·2930	367·2049	555·7709	253	1	5673·8923	181·7874	692·6603
204	1	9080·9250	403·4966	558·5087	254	2	6012·5243	218·0791	695·3980
205	2	9419·5570	439·7882	561·2465	255	3	6351·1563	254·3707	698·1358
206	3	9758·1890	476·0799	563·9843	256	4	6689·7883	290·6624	700·8726
207	4	96·8210	512·3715	566·7220	257	5	7028·4203	326·9540	703·6114
208	5	435·4530	548·6632	569·4598	258	6	7367·0522	363·2457	706·3492
209	6	774·0850	584·9548	572·1976	259	0	7705·6842	399·5373	709·0870
210	0	1112·7169	621·2465	574·9354	260	1	8044·3162	435·8289	711·8248
211	1	1451·3489	657·5381	577·6732	261	2	8382·9482	472·1206	714·5626
212	2	1789·9809	693·8298	580·4110	262	3	8721·5802	508·4122	717·3003
213	3	2128·6129	730·1214	583·1488	263	4	9060·2122	544·7039	720·0381
214	4	2467·2449	766·4131	585·8865	264	5	9398·8441	580·9955	722·7759
215	5	2805·8769	802·7047	588·6243	265	6	9737·4761	617·2872	725·5137
216	6	3144·5088	838·9964	591·3621	266	0	76·1081	653·5788	728·2515
217	0	3483·1408	875·2880	594·0999	267	1	414·7401	689·8705	730·9893
218	1	3821·7728	911·5797	596·8377	268	2	753·3721	726·1621	733·7271
219	2	4160·4048	947·8713	599·5755	269	3	1092·0041	762·4538	736·4649
220	3	4499·0368	984·1630	602·3133	270	4	1430·6361	798·7454	739·2026
221	4	4837·6688	20·4546	605·0510	271	5	1769·2680	835·0371	741·9404
222	5	5176·3008	56·7463	607·7888	272	6	2107·9000	871·3287	744·6782
223	6	5514·9327	93·0379	610·5266	273	0	2446·5320	907·6204	747·4160
224	0	5853·5647	129·3296	613·2644	274	1	2785·1640	943·9120	750·1538
225	1	6192·1967	165·6212	616·0022	275	2	3123·7960	980·2037	752·8916
226	2	6530·8287	201·9129	618·7400	276	3	3462·4280	16·4953	755·6294
227	3	6869·4607	238·2045	621·4778	277	4	3801·0600	52·7870	758·3672
228	4	7208·0927	274·4962	624·2156	278	5	4139·6919	89·0786	761·1050
229	5	7546·7247	310·7878	626·9534	279	6	4478·3239	125·3703	763·8428
230	6	7885·3566	347·0795	629·6911	280	0	4816·9559	161·6619	766·5805
231	0	8223·9886	383·3711	632·4289	281	1	5155·5879	197·9536	769·3183
232	1	8562·6206	419·6628	635·1667	282	2	5494·2199	234·2452	772·0561
233	2	8901·2526	455·9544	637·9045	283	3	5832·8519	270·5369	774·7939
234	3	9239·8846	492·2461	640·6423	284	4	6171·4839	306·8285	777·[illegible]
235	4	9578·5166	528·5377	643·3801	285	5	6510·1158	343·1202	780·2695
236	5	9917·1486	564·8294	646·1179	286	6	6848·7478	379·4118	783·0073
237	6	255·7805	601·1210	648·8557	287	0	7187·3798	415·7035	785·7451
238	0	594·4125	637·4127	651·5935	288	1	7526·0118	451·9951	788·4829
239	1	933·0445	673·7043	654·3312	289	2	7864·6438	488·2868	791·2207
240	2	1271·6765	709·9960	657·0690	290	3	8203·2758	524·5784	793·9584

TABLE LIV-A—*concld.*

DAYS OF 24 HOURS EACH.

No.	Week-day.	*a.*	*b.*	*c.*	No.	Week-day.	*a.*	*b.*	*c.*
1	2	3	4	5	1	2	3	4	5
291	4	8541·9078	560·8701	796·6962	341	5	5473·5070	375·4526	933·5856
292	5	8880·5397	597·1617	799·4340	342	6	5812·1390	411·7442	936·3233
293	6	9219·1717	633·4534	802·1718	343	0	6150·7710	448·0359	939·0611
294	0	9557·8037	669·7450	804·9096	344	1	6489·4030	484·3275	941·7989
295	1	9896·4357	706·0367	807·6473	345	2	6828·0350	520·6192	944·5367
296	2	235·0677	742·3283	810·3851	346	3	7166·6670	556·9108	947·2745
297	3	573·6997	778·6200	813·1229	347	4	7505·2989	593·2025	950·0123
298	4	912·3317	814·9116	815·8607	348	5	7843·9309	629·4941	952·7501
299	5	1250·9636	851·2033	818·5985	349	6	8182·5629	665·7858	955·4879
300	6	1589·5956	887·4949	821·3363	350	0	8521·1949	702·0774	958·2256
301	0	1928·2276	923·7866	824·0741	351	1	8859·8269	738·3691	960·9634
302	1	2266·8596	960·0782	826·8118	352	2	9189·4589	774·6607	963·7012
303	2	2605·4916	996·3699	829·5496	353	3	9537·0909	810·9524	966·4390
304	3	2944·1236	32·6615	832·2874	354	4	9875·7228	847·2440	969·1768
305	4	3282·7556	68·9532	835·0252	355	5	214·3548	883·5357	971·9146
306	5	3621·3875	105·2448	837·7630	356	6	552·9868	919·8273	974·6524
307	6	3960·0195	141·5365	840·5008	357	0	891·6188	956·1190	977·3902
308	0	4298·6515	177·8281	843·2386	358	1	1230·2508	992·4106	980·1280
309	1	4637·2835	214·1198	845·9764	359	2	1568·8828	28·7023	982·8658
310	2	4975·9155	250·4114	848·7141	360	3	1907·5147	64·9939	985·6035
311	3	5314·5475	286·7031	851·4519	361	4	2246·1467	101·2856	988·3413
312	4	5653·1794	322·9947	854·1897	362	5	2584·7787	137·5772	991·0791
313	5	5991·8114	359·2864	856·9275	363	6	2923·4107	173·8689	993·8169
314	6	6330·4434	395·5780	859·6653	364	0	3262·0427	210·1605	996·5547
315	0	6669·0754	431·8697	862·4031	365	1	3600·6747	246·4522	999·2925
316	1	7007·7074	468·1613	865·1409	366	2	3939·3067	282·7438	2·0302
317	2	7346·3394	504·4530	867·8787	367	3	4277·9386	319·0355	4·7680
318	3	7684·9714	540·7446	870·6165	368	4	4616·5706	355·3271	7·5058
319	4	8023·6033	577·0363	873·3543	369	5	4955·2026	391·6188	10·2436
320	5	8362·2353	613·3279	876·0920	370	6	5293·8346	427·9104	12·9814
321	6	8700·8673	649·6196	878·8298	371	0	5632·4666	464·2021	15·7192
322	0	9039·4993	685·9112	881·5676	372	1	5971·0986	500·4937	18·4570
323	1	9378·1313	722·2029	884·3054	373	2	6309·7306	536·7854	21·1948
324	2	9716·7633	758·4945	887·0432	374	3	6648·3625	573·0770	23·9326
325	3	55·3953	794·7862	889·7810	375	4	6986·9945	609·3687	26·6703
326	4	394·0272	831·0778	892·5188	376	5	7325·6265	645·6603	29·4081
327	5	732·6592	867·3695	895·2565	377	6	7364·2585	681·9520	32·1459
328	6	1071·2912	903·6611	897·9943	378	0	8002·8905	718·2436	34·8837
329	0	1409·9332	939·9528	900·7321	379	1	8341·5225	754·5353	37·6215
330	1	1748·5552	976·2444	903·4699	380	2	8680·1545	790·8269	40·3593
331	2	2087·1872	12·5361	906·2077	381	3	9018·7864	827·1186	43·0971
332	3	2425·8192	48·8277	908·9455	382	4	9357·4184	863·4102	45·8349
333	4	2764·4511	85·1194	911·6833	383	5	9696·0504	899·7019	48·5726
334	5	3103·0831	121·4110	914·4211	384	6	34·6824	935·9935	51·3104
335	6	3441·7151	157·7027	917·1588	385	0	373·3144	972·2852	54·0482
336	0	3780·3471	193·9943	919·8966					
337	1	4118·9791	230·2860	922·6344					
338	2	4457·6111	266·5776	925·3722					
339	3	4796·2431	302·8693	928·1100					
340	4	5134·8750	339·1609	930·8477					

TABLE LIV-B.

INCREASE OF *a*, *b*, *c* IN HOURS, MINUTES AND SECONDS.

(*a* in 10,000ths of circle, *b* and *c* in 1,000ths.)

These Tables correspond to Table V, *Indian Calendar*, for hours and minutes.

Increase in 1 hour—*a*, 14·109666059 ; *b*, 1·512150744 ; *c*, 0·114074481.

Increase in 1 minute—*a*, 0·235161101 ; *b*, 0·025202533 ; *c*, 0·001901220.

Increase in 1 second—*a*, 0·003919352 ; *b*, 0·000420042 ; *c*, 0·000031687.

HOURS.

No.	*a*.	*b*.	*c*.	No.	*a*.	*b*.	*c*.	No.	*a*.	*b*.	*c*.
1	14·1097	1·5122	0·1141	9	126·9870	13·6049	1·0267	17	239·8643	25·7056	1·9393
2	28·2193	3·0243	0·2281	10	141·0967	15·1215	1·1407	18	253·9740	27·2187	2·0533
3	42·3290	4·5365	0·3422	11	155·2063	16·6337	1·2548	19	268·0837	28·7309	2·1674
4	56·4387	6·0486	0·4563	12	169·3160	18·1458	1·3689	20	282·1933	30·2430	2·2815
5	70·5483	7·5608	0·5704	13	183·4257	19·6580	1·4830	21	296·3030	31·7552	2·3956
6	84·6580	9·0729	0·6844	14	197·5353	21·1701	1·5970	22	310·4127	33·2673	2·5096
7	98·7677	10·5851	0·7985	15	211·6450	22·6823	1·7111	23	324·5223	34·7795	2·6237
8	112·8773	12·0972	0·9126	16	225·7547	24·1944	1·8252	24	338·6320	36·2916	2·7378

MINUTES.

No.	*a*.	*b*.	*c*.	No.	*a*.	*b*.	*c*.	No.	*a*.	*b*.	*c*.
1	0·2352	0·0252	0·0019	21	4·9384	0·5293	0·0399	41	9·9416	1·0333	0·0780
2	0·4703	0·0504	0·0038	22	5·1735	0·5545	0·0418	42	9·8768	1·0585	0·0799
3	0·7055	0·0756	0·0057	23	5·4087	0·5797	0·0437	43	10·1119	1·0837	0·0818
4	0·9406	0·1008	0·0076	24	5·6439	0·6049	0·0456	44	10·3471	1·1089	0·0837
5	1·1758	0·1260	0·0095	25	5·8790	0·6301	0·0475	45	10·5822	1·1341	0·0856
6	1·4110	0·1512	0·0114	26	6·1142	0·6553	0·0494	46	10·8174	1·1593	0·0875
7	1·6461	0·1764	0·0133	27	6·3493	0·6805	0·0513	47	11·0526	1·1845	0·0894
8	1·8813	0·2016	0·0152	28	6·5845	0·7057	0·0532	48	11·2877	1·2097	0·0913
9	2·1164	0·2268	0·0171	29	6·8197	0·7309	0·0551	49	14·5229	1·2349	0·0932
10	2·3516	0·2520	0·0190	30	7·0548	0·7561	0·0570	50	11·7581	1·2601	0·0951
11	2·5868	0·2772	0·0209	31	7·2900	0·7813	0·0589	51	11·9932	1·2853	0·0970
12	2·8219	0·3024	0·0228	32	7·5252	0·8065	0·0608	52	12·2284	1·3105	0·0989
13	3·0571	0·3276	0·0247	33	7·7603	0·8317	0·0627	53	12·4635	1·3357	0·1008
14	3·2923	0·3528	0·0266	34	7·9955	0·8569	0·0646	54	12·6987	1·3609	0·1027
15	3·5274	0·3780	0·0285	35	8·2306	0·8821	0·0665	55	12·9339	1·3861	0·1046
16	3·7626	0·4032	0·0304	36	8·4658	0·9073	0·0684	56	13·1690	1·4113	0·1065
17	3·9977	0·4284	0·0323	37	8·7010	0·9325	0·0703	57	13·4042	1·4365	0·1084
18	4·2329	0·4536	0·0342	38	8·9361	0·9577	0·0722	58	13·6393	1·4617	0·1103
19	4·4681	0·4788	0·0361	39	9·1713	0·9829	0·0741	59	13·8745	1·4869	0·1122
20	4·7032	0·5041	0·0380	40	9·4064	1·0081	0·0760	60	14·1097	1·5122	0·1141

TABLE LIV B—*contd.*

SECONDS.

No.	*a.*	*b.*	*c.*	No.	*a.*	*b.*	*c.*	No.	*a.*	*b.*	*c.*
1	0·0039	0·0004	0·0000	21	0·0823	0·0088	0·0007	41	0·1607	0·0172	0·0013
2	0·0078	0·0008	0·0001	22	0·0862	0·0092	0·0007	42	0·1646	0·0176	0·0013
3	0·0118	0·0013	0·0001	23	0·0901	0·0097	0·0007	43	0·1685	0·0181	0·0014
4	0·0157	0·0017	0·0001	24	0·0941	0·0101	0·0008	44	0·1725	0·0185	0·0014
5	0·0196	0·0021	0·0002	25	0·0980	0·0105	0·0008	45	0·1764	0·0189	0·0014
6	0·0235	0·0025	0·0002	26	0·1019	0·0109	0·0008	46	0·1803	0·0193	0·0015
7	0·0274	0·0029	0·0002	27	0·1058	0·0113	0·0009	47	0·1842	0·0197	0·0015
8	0·0314	0·0034	0·0003	28	0·1097	0·0118	0·0009	48	0·1881	0·0202	0·0015
9	0·0353	0·0038	0·0003	29	0·1137	0·0122	0·0009	49	0·1920	0·0206	0·0016
10	0·0392	0·0042	0·0003	30	0·1176	0·0126	0·0010	50	0·1960	0·0210	0·0016
11	0·0431	0·0046	0·0003	31	0·1215	0·0130	0·0010	51	0·1999	0·0214	0·0016
12	0·0470	0·0050	0·0004	32	0·1254	0·0134	0·0010	52	0·2038	0·0218	0·0016
13	0·0510	0·0055	0·0004	33	0·1293	0·0139	0·0010	53	0·2077	0·0223	0·0017
14	0·0549	0·0059	0·0004	34	0·1333	0·0143	0·0011	54	0·2116	0·0227	0·0017
15	0·0588	0·0063	0·0005	35	0·1372	0·0147	0·0011	55	0·2156	0·0231	0·0017
16	0·0627	0·0067	0·0005	36	0·1411	0·0151	0·0011	56	0·2195	0·0235	0·0018
17	0·0666	0·0071	0·0005	37	0·1450	0·0155	0·0012	57	0·2234	0·0239	0·0018
18	0·0705	0·0076	0·0006	38	0·1489	0·0160	0·0012	58	0·2273	0·0244	0·0018
19	0·0745	0·0080	0·0006	39	0·1529	0·0164	0·0012	59	0·2312	0·0248	0·0019
20	0·0784	0·0084	0·0006	40	0·1568	0·0168	0·0013	60	0·2352	0·0252	0·0019

TABLE

THE MOON'S

Corresponding to "*Equation b*"

For either of the mean anomaly values given in cols. 2*a*, 2*b*, the equation and difference are as stated in cols. 3, 4. The equation col. 3, from "Arg. *b*" 0 to 500 or 0° to 180°, is the moon's greatest equation of the centre *plus* the actual equation, in 10,000ths of circle. (*For the 24 base equations in degrees, etc., see Table LIX.*)

Base Eqn. No.	Arg. *b*.	"Equation *b*."	Diff.	Arg. *b*.	Base Eqn. No.	Arg. *b*.	"Equation *b*."	Diff.	Arg. *b*.
1	2*a*	3	4	2*b*	1	2*a*	3	4	2*b*
0	0·0	139·8717		500·0	12	125·0	238·6631		375·0
	2·083̇	141·7004		497·916̇		127·083̇	239·9153		372·916̇
	4·16̇	143·5291	1·8287	495·83̇		129·16̇	241·1676	1·2523	370·83̇
	6·25	145·3578		493·75		131·25	242·4199		268·75
	8·3̇	147·1865		491·6̇		133·3̇	243·6722		366·6̇
1	10·416̇	149·0152		489·583̇	13	135·416̇	244·9244		364·583̇
	12·5	150·8357		487·5		137·5	246·0919		362·5
	14·583̇	152·6563	1·8206	485·416̇		139·583̇	247·2593	1·1674	360·416̇
	16·6̇	154·4769		482·3̇		141·6̇	248·4268		358·3̇
	18·75	156·2975		481·25		143·75	249·5942		356·25
2	20·83̇	158·1180		479·16̇	14	145·83̇	250·7616		354·16̇
	22·916̇	159·9224		477·083̇		147·916̇	251·8311		352·083̇
	25·0	161·7267	1·8043	475·0		150·0	252·9006	1·0695	350·0
	27·083̇	163·5310		472·916̇		152·083̇	253·9701		347·916̇
	29·16̇	165·3553		470·83̇		154·16̇	255·0396		345·83̇
3	31·25	167·1397		468·75	15	156·25	256·1090		343·75
	33·3̇	168·9196		466·6̇		158·3̇	257·0805		341·6̇
	35·416̇	170·6995	1·7799	464·583̇		160·416̇	258·0520	0·9715	339·583̇
	37·5	172·4795		462·5		162·5	259·0235		337·5
	39·583̇	174·2594		460·416̇		164·583̇	259·9950		335·416̇
4	41·6̇	176·0393		458·3̇	16	166·6̇	260·9664		333·3̇
	43·75	177·7868		456·25		168·75	261·8322		331·25
	45·83̇	179·5342	1·7474	454·16̇		170·83̇	262·6980	0·8658	329·16̇
	47·916̇	181·2816		452·083̇		172·916̇	263·5638		327·083̇
	50·0	183·0291		450·0		175·0	264·4296		325·0
5	52·083̇	184·7765		447·916̇	17	177·083̇	265·2953		322·916̇
	54·16̇	186·4833		445·83̇		179·16̇	266·0541		320·83̇
	56·25	188·1901	1·7068	443·75		181·25	266·8129	0·7588	318·75
	58·3̇	189·8969		441·6̇		183·3̇	267·5717		316·6̇
	60·416̇	191·6036		439·583̇		185·416̇	268·3305		314·583̇
6	62·5	193·3104			18	187·5	269·0893		312·5
	64·583̇	194·9766		435·416̇		189·583̇	269·7332		310·416̇
	66·6̇	196·6427	1·6662	433·3̇		191·6̇	270·3772	0·6440	308·3̇
	68·75	198·3089		431·25		193·75	271·0211		306·25
	70·83̇	199·9750		429·16̇		195·83̇	271·6651		304·16̇
7	72·916̇	201·6412		427·083̇	19	197·916̇	272·3090		302·083̇
	75·0	203·2586		425·0		200·0	272·8417		300·0
	77·083̇	204·8761	1·6175	422·916̇		202·083̇	273·3745	0·5327	297·916̇
	79·16̇	206·4936		420·83̇		204·6̇	273·9072		295·83̇
	81·25	208·1110		418·75		206·25	274·4399		293·75
8	83·3̇	209·7285		416·6̇	20	208·3̇	274·9726		291·6̇
	85·416̇	211·2808		414·583̇		210·416̇	275·3879		289·583̇
	87·5	212·8331	1·5523	412·5		212·5	275·8033	0·4153	287·5
	89·583̇	214·3854		410·416̇		214·583̇	276·2186		285·416̇
	91·6̇	215·9377		408·3̇		216·6̇	276·6339		283·3̇
9	93·75	217·4900		406·25	21	218·75	277·0492		281·25
	95·83̇	218·9773		404·16̇		220·83̇	277·3513		279·16̇
	97·916̇	220·4646	1·4873	402·083̇		222·916̇	277·6534	0·3021	277·083̇
	100·0	221·9519		400·0		225·0	277·9554		275·0
	102·083̇	223·4393		397·916̇		227·083̇	278·2575		272·916̇
10	104·16̇	224·9266		395·83̇	22	229·16̇	278·5595		270·83̇
	106·25	226·3408		393·75		241·25	278·7391		268·75
	108·3̇	227·7550	1·4142	391·6̇		233·3̇	278·9188	0·1796	266·6̇
	110·416̇	229·1693		389·583̇		235·416̇	279·0984		264·583̇
	112·5	230·5835		387·5		237·5	279·2780		262·5
11	114·583̇	231·9977		385·416̇	23	239·583̇	279·4576		260·416̇
	116·6̇	233·3308		383·3̇		241·6̇	279·5147		258·3̇
	118·75	234·6638	1·3331	381·25		243·75	279·5719	0·0571	256·25
	120·83̇	235·9969		379·16̇		245·83̇	279·6290		254·16̇
	122·916̇	237·3300		377·083̇		247·916̇	279·6862		252·083̇
					24	250·0	279·7433		250·0

LV.

"EQUATION *b*"

in Table VI, "*Indian Calendar.*"

The equation, col. 3, from "Arg. *b*" 500 to 1000, or 180° to 360°, is the moon's greatest equation of the centre *minus* the actual equation, stated in 10,000ths of the circle.

Base Eqn. No.	Arg. *b*.	"Equation *b*."	Diff.	Arg. *b*.	Base Eqn. No.	Arg. *b*.	"Equation *b*."	Diff.	Arg. *b*.
1	2*a*	3	4	2*b*	1	2*a*	3	4	2*b*
0	500·0	139·8717		1000·0	12	625·0	41·0802		875·0
	502·083	138·0429		997·916̇		627·083̇	39·8280		872·916̇
	504·16̇	136·2142	1·8287	995·83̇		629·16̇	38·5757	1·2523	870·83̇
	506·25	134·3855		993·75		631·25	37·3234		868·75
	508·3̇	132·5568		991·6̇		633·3̇	36·0711		866·6̇
1	510·416̇	130·7281		989·583̇	13	635·416̇	34·8188		864·583̇
	512·5	128·9076		987·5		637·5	33·6514		862·5
	514·583̇	127·0870	1·8206	985·416̇		639·583̇	32·4840	1·1674	860·416̇
	516·6̇	125·2664		983·3̇		641·6̇	31·3165		858·3̇
	518·75	123·4458		981·25		643·75	30·1491		856·25
2	520·83̇	121·6253		979·16̇	14	645·83̇	28·9817		854·16̇
	522·916̇	119·8209		977·083̇		647·916̇	27·9122		852·083̇
	525·0	118·0166	1·8043	975·0		650·0	26·8427	1·0695	850·0
	527·083̇	116·2123		972·916̇		652·083̇	25·7732		847·916̇
	529·16̇	114·4080		970·83̇		654·16̇	24·7037		845·83̇
3	531·25	112·6036		968·75	15	656·25	23·6343		843·75
	533·3̇	110·8237		966·6̇		658·3̇	22·6628		841·6̇
	535·416̇	109·0438	1·7800	964·583̇		660·416̇	21·6913	0·9715	839·583̇
	537·5	107·2638		962·5		662·5	20·7198		837·5
	539·583̇	105·4839		960·416̇		664·583̇	19·7483		835·416̇
4	541·6̇	103·7040		958·3̇	16	666·6̇	18·7769		833·3̇
	543·75	101·9565		956·25		668·75	17·9111		831·25
	545·83̇	100·2091	1·7474	954·16̇		670·83̇	17·0453	0·8658	829·16̇
	547·916̇	98·4617		952·083̇		672·916̇	16·1795		827·083̇
	550·0	96·7142		950·0		675·0	15·3137		825·0
5	552·083̇	94·9668		947·916̇	17	677·083̇	14·4480		822·916̇
	554·16̇	93·2600		945·83̇		679·16̇	13·6892		820·83̇
	556·25	91·5532	1·7068	943·75		681·25	12·9304	0·7588	818·75
	558·3̇	89·8464		941·6̇		683·3̇	12·1716		816·6̇
	560·416̇	88·1397		939·583		685·416̇	11·4128		814·583̇
6	562·5	86·4329		937·5	18	687·5	10·6540		812·5
	564·583̇	84·7667		935·416̇		689·583̇	10·0101		810·416̇
	566·6̇	83·1006	1·6662	933·3̇		691·6̇	9·3661	0·6440	808·3̇
	568·75	81·4344		931·25		693·75	8·7222		806·25
	570·83̇	79·7683		929·16̇		695·83̇	8·0782		804·16̇
7	572·916̇	78·1021		927·083̇	19	697·916	7·4343		802·083̇
	575·0	76·4847		925·0		700·0	6·9016		800·0
	577·083̇	74·8672	1·6175	922·916̇		702·083̇	6·3688	0·5327	797·916̇
	579·16	73·2497		920·83̇		704·6̇	5·8361		795·83̇
	581·25	71·6323		918·75		706·25	5·3034		793·75
8	583·3̇	70·0148		916·6̇	20	708·3̇	4·7707		791·6̇
	585·416̇	68·4625		914·583̇		710·416̇	4·3554		789·583̇
	587·5	66·9102	1·5523	912·5		712·5	3·9430	0·4153	787·5
	589·583̇	65·3579		910·416̇		714·583̇	3·5247		785·416̇
	591·6̇	63·8057		908·3̇		716·6̇	3·1094		783·3̇
9	593·75	62·2533		906·25	21	718·75	2·6941		781·25
	595·83̇	60·7660		904·16̇		720·83̇	2·3920		779·16̇
	597·916̇	59·2787	1·4873	902·083̇		722·916	2·0859	0·3021	777·083
	600·0	57·7914		900·0		725·0	1·7879		775·0
	602·083̇	56·3040		897·916̇		727·083̇	1·4858		772·916̇
10	604·16̇	54·8167		895·83̇	22	729·16̇	1·1838		770·83
	606·25	53·4025		893·75		731·25	1·0042		768·75
	608·3̇	51·9883	1·4142	891·6̇		733·3̇	0·8245	0·1796	766·6̇
	610·416	50·5741		889·583̇		735·416̇	0·6449		764·583̇
	612·5	49·1598		887·5		737·5	0·4653		762·5
11	614·583	47·7456		885·416̇	23	739·583̇	0·2857		760·416̇
	616·6̇	46·4125		883·3̇		741·6̇	0·2286		758·3̇
	618·75	45·0795	1·3331	881·25		743·75	0·1714	0·0571	756·25
	620·83̇	43·7464		879·16̇		745·83̇	0·1143		754·16̇
	622·916	42·4133		877·083̇		747·916	0·0571		752·083
					24	750·0	0·0		750·0

TABLI

THE SUN'

Corresponding to " Equation c"

For either of the mean anom. values given in cols. 2a or 2b the equation and difference are as stated in cols. 3, 4. The equation, col. 3, from "Arg. b" 0 to 500 or 0 to 180, is the Sun's greatest equation of the centre *minus* the actual equation, in 10,000ths of circle. (*For the 24 base-equations see Table XLVII above, Vol. XIV (also Prof. Jacobi's Table XXIV, Epig. Ind. I, p. 512).*)

Base Eqn. No.	Arg. c.	Equation c.	Diff.	Arg. c.
1	a	3	4	2b
0	0·0	60·4244		500·0
	2·083̇	59·6335		497·916̇
	4·16̇	58·8426	0·7909	495·83̇
	6·25	58·0517		493·75
	8·3̇	57·2608		491·6̇
1	10·416̇	56·4699		489·583̇
	12·5	55·6825		487·5
	14·583̇	54·8951	0·7874	485·416̇
	16·6̇	54·1078		483·3̇
	18·75	53·3204		481·25
2	20·83̇	52·5330		479·16̇
	22·916̇	51·7527		477·083̇
	25·0	50·9723	0·7804	475·0
	27·083̇	50·1920		472·916̇
	29·16̇	49·4116		470·83̇
3	31·25	48·6313		468·75
	33·3̇	47·8615		466·6̇
	35·416̇	47·0916	0·7698	464·583̇
	37·5	46·3218		462·5
	39·583̇	45·5520		460·416̇
4	41·6̇	44·7822		458·3̇
	43·75	44·0265		456·25
	45·83̇	43·2707	0·7557	454·16̇
	47·916̇	42·5150		452·083̇
	50·0	41·7593		450·0
5	52·083̇	41·0035		447·916̇
	54·16̇	40·2653		445·83̇
	56·25	39·5272	0·7382	443·75
	58·3̇	38·7890		441·6̇
	60·416̇	38·0508		439·583̇
6	62·5	37·3127		437·5
	64·583̇	36·5921		435·416̇
	66·6̇	35·8715	0·7206	433·3̇
	68·75	35·1509		431·25
	70·83̇	34·4303		429·16̇
7	72·916̇	33·7097		427·083̇
	75·0	33·1012		425·0
	77·083̇	32·3107	0·6995	422·916̇
	79·16̇	31·6112		420·83̇
	81·25	30·9117		418·75
8	83·3̇	30·2122		416·6̇
	85·416̇	29·5408		414·583̇
	87·5	28·8694	0·6714	412·5
	89·583̇	28·1980		410·416̇
	91·6̇	27·5267		408·3̇
9	93·75	26·8553		406·25
	95·83̇	26·2120		404·16̇
	97·916̇	25·5688	0·6433	402·083̇
	100·0	24·9255		400·0
	102·083̇	24·2822		397·916̇
10	104·16̇	23·6390		395·83̇
	106·25	23·0274		393·75
	108·3̇	22·4157	0·6116	391·6̇
	110·416̇	21·8041		389·583̇
	112·5	21·1925		387·5
11	114·583̇	20·5808		385·416̇
	116·6̇	20·0044		383·3̇
	118·75	19·4279	0·5765	381·25
	120·83̇	18·8514		379·16̇
	122·916̇	18·2750		377·083̇

Base Eqn. No.	Arg. c.	Equation c.	Diff.	Arg. c.
1	2a	3	4	2b
12	125·0	17·6985		375·0
	127·083̇	17·1564		372·916̇
	129·16̇	16·6143	0·5421	370·83̇
	131·25	16·0722		368·75
	133·3̇	15·5301		366·6̇
13	135·416̇	14·9880		364·583̇
	137·5	14·4861		362·5
	139·583̇	13·9842	0·5019	360·416̇
	141·6̇	13·4823		358·3̇
	143·75	12·9805		356·25
14	145·83̇	12·4786		354·16̇
	147·916̇	12·0181		352·083̇
	150·0	11·5576	0·4605	350·0
	152·083̇	11·0971		347·916̇
	154·16̇	10·6367		345·83̇
15	156·25	10·1762		343·75
	158·3̇	9·7579		341·6̇
	160·416̇	9·3396	0·4183	339·583̇
	162·5	8·9213		337·5
	164·583̇	8·5030		335·416̇
16	166·6̇	8·0847		333·3̇
	168·75	7·7121		331·25
	170·83̇	7·3395	0·3726	329·16̇
	172·916̇	6·9669		327·083̇
	175·0	6·5943		325·0
17	177·083̇	6·2217		322·916̇
	179·16̇	5·8948		320·83̇
	181·25	5·5679	0·3269	318·75
	183·3̇	5·2410		316·6̇
	185·416̇	4·9141		314·583̇
18	187·5	4·5872		312·5
	189·583̇	4·3095		310·416̇
	191·6̇	4·0318	0·2777	308·3̇
	193·75	3·7541		306·25
	195·83̇	3·4764		304·16̇
19	197·916̇	3·1987		302·083̇
	200·0	2·9703		300·0
	202·083̇	2·7418	0·2285	297·916̇
	204·16̇	2·5133		295·83̇
	206·25	2·2848		293·75
20	208·3̇	2·0563		291·6̇
	210·416̇	1·8771		289·583̇
	212·5	1·6978	0·1793	287·5
	214·583̇	1·5185		285·416̇
	216·6̇	1·3393		283·3̇
21	218·75	1·1600		281·25
	220·83̇	1·0299		279·16̇
	222·916̇	0·8999	0·1301	277·083̇
	225·0	0·7698		275·0
	227·083̇	0·6397		272·916̇
22	229·16̇	0·5097		270·83̇
	231·25	0·4324		268·75
	233·3̇	0·3550	0·0773	266·6̇
	235·416̇	0·2777		264·583̇
	237·5	0·2004		262·5
23	239·583̇	0·1230		260·416̇
	241·6̇	0·0984		258·3̇
	243·75	0·0738	0·0246	256·25
	245·83̇	0·0492		254·16̇
	247·916̇	0·0246		252·083̇
24	250·0	0·0		250·0

LVI.

"EQUATION *c*"

in Table VII, "*Indian Calendar.*"

From "Arg. *c*" 500 to 1000 or 180° to 360° the equation (col. 3) is the Sun's greatest equation of the centre *plus* the actual equation, stated in 10,000ths of the circle.

Base Eqn. No.	Arg. c.	Equation c.	Diff.	Arg. c.	Base Eqn. No.	Arg. c.	Equation c.	Diff.	Arg. c.
1	2a	3	4	2b	1	2a	3	4	2b
0	500·0	60·4244		1000·0	12	625·0	103·1503		875·0
	502·083̇	61·2153		997·916̇		627·083̇	103·6924		872·916̇
	504·16	62·0062	0·7909	995·83̇		629·16	104·2345	0·5421	870·83̇
	506·25	62·7971		993·75		631·25	104·7766		868·75
	508·3	63·5880		991·6̇		633·3	105·3187		866·6
1	510·416̇	64·3789		989·583̇	13	635·416̇	105·8608		864·583̇
	512·5	65·1662		987·5		637·5	106·3627		862·5
	514·583̇	65·9536	0·7874	985·416̇		639·583̇	106·8645	0·5019	860·416̇
	516·6̇	66·7410		983·3		641·6	107·3664		858·3
	518·75	67·5284		981·25		643·75	107·8683		856·25
2	520·83̇	68·3158		979·16̇	14	645·83̇	108·3702		854·16
	522·916̇	69·0961		977·083̇		647·916̇	108·8307		852·083̇
	525·0	69·8765	0·7804	975·0		650·0	109·2912	0·4605	850·0
	527·083̇	70·6568		972·916̇		652·083̇	109·7516		847·916̇
	529·16̇	71·4372		970·83̇		654·16̇	110·2121		845·83
3	531·25	72·2175		968·75	15	656·25	110·6726		843·75
	533·3̇	72·9873		966·6̇		658·3	111·0909		841·6̇
	535·416̇	73·7571	0·7698	964·583̇		660·416̇	111·5092	0·4183	839·583̇
	537·5	74·5269		962·5		662·5	111·9275		837·5
	539·583	75·2967		960·416		664·583̇	112·3458		835·416
4	541·6̇	76·0665		958·3	16	666·6	112·7641		833·3
	543·75	76·8223		956·25		668·75	113·1367		831·25
	545·83̇	77·5780	0·7557	954·16̇		670·83	113·5093	0·3726	829·16̇
	547·916̇	78·3338		952·083̇		672·916̇	113·8819		827·083̇
	550·0	79·0895		950·0		675·0	114·2545		825·0
5	552·083̇	79·8452		947·916̇	17	677·083̇	114·6271		822·916̇
	554·16̇	80·5834		945·83̇		679·16̇	114·9540		820·83̇
	556·25	81·3216	0·7382	943·75		681·25	115·2809	0·3269	818·75
	558·3̇	82·0598		941·6̇		683·3	115·6078		816·6
	560·416̇	82·7979		939·583̇		685·416̇	115·9347		814·583
6	562·5	83·5361		937·5	18	687·5	116·2616		812·5
	564·583̇	84·2567		935·416̇		689·583̇	116·5393		810·416
	566·6̇	84·9773	0·7206	933·3̇		691·6̇	116·8170	0·27[illegible]	808·3
	568·75	85·6979		931·25		693·75	117·0946		806·25
	570·83̇	86·4185		929·16̇		695·83̇	117·3723		804·16
7	572·916̇	87·1391		927·08?	19	697·916̇	117·6500		802·083̇
	575·0	87·8386		925·0		700·0	117·8785		800·0
	577·083̇	88·5381	0·6995	922·916̇		702·083̇	118·4070	0·2285	797·916̇
	579·16	89·2376		920·83̇		704·16̇	118·3355		795·83
	581·25	89·9371		918·75		706·25	118·5640		793·75
8	583·3̇	90·6366		916·6̇	20	708·3̇	118·7924		791·6̇
	585·416	91·3080		914·583̇		710·416̇	118·9717		789·583̇
	587·5	91·9793	0·6174	912·5		712·5	119·1510	0·1793	787·5
	589·583̇	92·6507		910·416̇		714·583̇	119·3502		785·416̇
	591·6̇	93·3221		908·3̇		716·6	119·5095		783·3
9	593·75	93·9935		906·25	21	718·75	119·6888		781·25
	595·83̇	94·6367		904·16̇		720·83̇	119·8188		779·16̇
	597·916̇	95·2800	0·6433	902·083̇		722·916̇	119·9489	0·1301	777·083̇
	600·0	95·9233		900·0		725·0	120·0790		775·0
	602·682	96·5665		897·916		727·083̇	120·2091		772·916
10	604·16̇	97·2098		895·83̇	22	729·16̇	120·3391		770·83̇
	606·25	97·8214		893·75		731·25	120·4164		768·75
	608·3̇	98·4330	0·6116	891·6̇		733·3̇	120·4937	0·0773	766·3̇
	610·416̇	99·0447		889·583̇		735·416̇	120·5711		764·583̇
	612·5	99·6563		887·5		737·5	120·6134		762·5
11	614·583̇	100·2679		885·416	23	739·583̇	120·7257		760·416
	616·6̇	100·8444		883·3̇		741·6̇	120·7503		758·3
	618·75	101·4209	0·5765	881·25		743·75	120·7749	0·0246	756·25
	620·83̇	101·9973		879·16̇		745·83̇	120·7996		754·16̇
	622·916̇	102·5738		877·083̇		747·916	120·8242		752·083̇
					24	750·0	120·8488		750

TABLE LVII A.

VALUE OF *a*, *b*, *c* AT BEGINNING OF KALIYUGA CENTURIES.

Corresponding to Prof. Jacobi's Table IX B (Vol. XI above.) but framed for two days earlier in each century.

Century K. Y.	Week day.	*a*.	*b*.	*c*.
42	6	49·0437	626·9004	276·4176
43	5	8582·3109	179·4088	277·0270
44	5	7454·2101	768·2089	277·3743
45	5	6326·1092	357·0090	277·7215
46	5	5198·0084	945·8091	278·0688
47	5	4069·9075	534·6091	278·4160
48	5	2941·8067	123·4092	278·7632
49	5	1831·7059	712·2093	279·1104
50	4	346·9731	264·7177	279·7199

TABLE LVII B.

INCREASE OF *a*, *b*, *c* FOR YEARS OF THE KALIYUGA.

Corresponding to Prof. Jacobi's Table X *Epig. Ind.*, Vol. XI, p. 168

Years of 366 days.

Year.	Week day.	*a*.	*b*.	*c*.	Year.	Week day.	*a*.	*b*.	*c*.
0	0	0	0	0	31	4	4329·9708	930·3505	999·9683
1	1	3600·6747	246·4522	999·2925	32	5	7930·6455	176·8027	999·2608
*2	2	7201·3494	492·9043	998·5849	*33	6	1531·3202	423·2549	998·5533
3	4	1140·6560	775·6482	0·6151	34	1	5470·6268	705·9987	0·5835
4	5	4741·3307	22·1003	999·9076	35	2	9071·3015	952·4509	999·8759
5	6	8242·0054	268·5525	999·2001	36	3	2671·9762	198·9030	999·1684
*6	0	1942·6800	515·0047	998·4925	*37	4	6272·6509	445·3552	998·4609
7	2	5881·9867	797·7485	0·5227	38	6	211·9575	728·0990	0·4911
8	3	9482·6614	44·2007	999·8152	39	0	3812·6322	974·5512	999·7836
9	4	3083·3360	290·6528	999·1077	40	1	7413·3069	221·0034	999·0760
*10	5	6684·0107	537·1050	998·4001	*41	2	1013·9815	467·4555	998·3685
11	0	623·3174	819·8488	0·4303	42	4	4953·2882	750·1994	0·3987
12	1	4223·9921	66·3010	999·7228	43	5	8553·9629	996·6515	999·6912
*13	2	7824·6667	312·7532	999·0153	*44	6	2154·6376	243·1037	998·9836
14	4	1763·9734	595·4970	1·0455	45	1	6093·9442	525·8475	1·0128
15	5	5364·6481	841·9492	0·3379	46	2	9694·6189	772·2997	0·3063
16	6	8965·3227	88·4013	999·6304	47	3	3295·2936	18·7519	999·5988
*17	0	2565·9974	334·8535	998·9229	*48	4	6895·9682	265·2040	998·8912
18	2	6505·3041	617·5973	0·9531	49	6	835·2749	547·9479	0·9214
19	3	105·9788	864·0495	0·2455	50	0	4435·9496	794·4000	0·2139
20	4	3706·6534	110·5017	999·5380	51	1	8036·6243	40·8522	999·5064
*21	5	7307·3281	356·9539	998·8305	*52	2	1637·2989	287·3044	998·7988
22	0	1246·6348	639·6977	0·8607	53	4	5576·6056	570·0482	0·8290
23	1	4847·3094	886·1499	0·1531	54	5	9177·2803	816·5004	0·1215
24	2	8447·9841	132·6020	999·4456	55	6	2777·9549	62·9526	999·4140
*25	3	2048·6588	379·0542	998·7381	*56	0	6378·6296	309·4047	998·7064
26	5	5987·9655	661·7980	0·7683	57	2	317·9363	592·1485	0·7366
27	6	9588·6401	908·2502	0·0607	58	3	3918·6110	838·6007	0·0291
28	0	3189·3148	154·7024	999·3532	59	4	7519·2856	85·0529	999·3216
*29	1	6789·9895	401·1545	998·6457	*60	5	1119·9603	331·5051	998·6140
30	3	729·2961	683·8984	0·6759	61	0	5059·2670	614·2489	0·6442

TABLE LVII B—*contd.*

Year.	Week day.	*a.*	*b.*	*c.*
62	1	8659·9416	860·7011	999·9367
63	2	2260·6163	107·1532	999·2292
*64	3	5861·2910	353·6054	998·5216
65	5	9800·5977	636·3492	0·5518
66	6	3401·2723	882·8014	999·8443
67	0	7001·9470	129·2536	999·1368
*68	1	602·6217	375·7057	998·4292
69	3	4541·9283	658·4496	0·4594
70	4	8142·6030	904·9017	999·7519
*71	5	1743·2777	151·3539	999·0444
72	0	5682·5844	434·0977	1·0746
73	1	9283·2590	680·5499	0·3670
74	2	2883·9337	927 0021	999·6595
*75	3	6484·6084	173·4542	998·9520
76	5	423·9150	456·1981	0·9822
77	6	4054·5897	702·6502	0·2746
78	0	7625·2644	949·1024	999·5671
*79	1	1225·9391	195·5546	998·8596
80	3	5165·2457	478·2984	0·8898
81	4	8765·9204	724·7506	0·1822
82	5	2366·5951	971·2027	999·4747
*83	6	5967·2698	217·6549	998·7672
84	1	9906·5764	500·3987	0·7974
85	2	3507·2511	746·8509	0·0898
86	3	7107·9258	993·3031	999·3823
*87	4	708·6004	239·7552	998·6748
88	6	4647·9071	522·4991	0·7050
89	0	8248·5818	768·9512	999·9974
90	1	1849·2565	15·4034	999·2899
*91	2	5449·9311	261.8556	998·5824
92	4	9389·2378	544·5994	0.6126
93	5	2989·9125	791·0516	999·9050
94	6	6590·5871	37·5038	999·1975
*95	0	191·2618	283·9559	998·4000
96	2	4130·5685	566·6997	0.5202
97	3	7731·2434	813·1519	999·8126
98	4	1331·9178	59·6041	999·1051
*99	5	4932·5925	306·0563	998·3976
100	0	8871·8992	588·8001	0·4278

TABLE LVII-C.

VALUES OF "*a*," "*b*," "*c*" ON DAYS FROM MĪNA 1 TO MĒSHA 2, THE DAY OF MEAN MĒSHA-SAMKRĀNTI.

Corresponding to the first part of Prof. Jacobi's Table XIII (*of Epig. Ind., Vol. XI, 170*) but arranged for the *Siddhānta-Śirōmaṇi*.

No. of days interval from Mēsha 0	Month and day.	Week day.	*a.*	*b.*	*c.*
1	2	3	4	5	6
29	Mīna 1	4	9502·4085	874·9589	915·1286
28	,, 2	5	9841·0404	911·2506	917·8664
27	,, 3	6	179·6724	947·5422	920·6042
26	,, 4	0	518·3044	983·8339	923·3419
25	,, 5	1	856·9364	20·1255	926·0797
24	,, 6	2	1195·5684	46·4172	928·8175
23	,, 7	3	1534·2004	92·7088	931·5553
22	,, 8	4	1872·8324	129·0005	934·2931
21	,, 9	5	2211·4643	165·2921	937·0309
20	,, 10	6	2550·0963	201·5838	939·7687
19	,, 11	0	2888·7283	237·8754	942·5065
18	,, 12	1	3227·3603	274·1671	945·2442
17	,, 13	2	3565·9923	310·4587	947·9820
16	,, 14	3	3904·6243	346·7504	950·7198
15	,, 15	4	4243·2563	383·0420	953 4576
14	,, 16	5	4581·8882	419·3336	956·1954
13	,, 17	6	4920·5202	455·6253	958·9332
12	,, 18	0	5259·1522	491·9169	961·6710
11	,, 19	1	5597·7842	528·2086	964·4088
10	,, 20	2	5936·4162	564·5002	967·1465
9	,, 21	3	6275·0482	600·7919	969·8843
8	,, 22	4	6613·6801	637·0835	972·6221
7	,, 23	5	6952·3121	637·3752	975·3599
6	,, 24	6	7290·9441	709·6668	978·0977
5	,, 25	0	7629·5761	745·9585	980·8355
4	,, 26	1	7968·2081	782·2501	983·5733
3	,, 27	2	8306·8401	818·5418	986·3111
2	,, 28	3	8645·4721	854·8334	989·0488
1	,, 29	4	8984·1040	891·1251	991·7866
	Mēsha 0	5	9322·7360	927·4167	994·5244
	,, 1	6	9661·3680	963·7084	997·2622
	,, 2	0	0·0	0·0	0·0

By this Table, the *a, b, c* of the civil day coupled with Chaitra Sukla, 1 is easily found

TABLE LVIII-A.

DURATION AND COLLECTIVE DURATION OF TRUE SOLAR MONTHS; WITH INCREASE OF "*a*," "*b*," "*c*" AT EACH SAṀKRĀNTI

Calculated for the year K. Y. 4500, expired, A.D. 1399-1400.

"a" in 10,000ths of circle; "b" and "c" in 1,000ths.

Luni-solar month (ending after the second of the two solar saṁkrāntis connected with it).	True solar saṁkrānti.	Collective duration in days, hours, etc.; and collective increase of *a, b, c* from true Mêsha-saṁkrānti to each true saṁkrānti.						True solar saṁkrānti.	Length of month preceding each true saṁkrānti and increase of *a, b, c* between each such saṁkrānti.					
		Day	Week day.	H. M. S.	*a.*	*b.*	*c.*		Day.	Week day.	H. M. S.	*a.*	*b.*	*c.*
1	2	3	4	5	6	7	8	9	10	11	12	13	14	15
1. Chaitra	Mîna-saṁ. (*of previous year*)													
2. Vaiśākha	Mêsha-saṁ.	0	...	0 0 0	0·0	0·0	0·0	Mêsha-saṁ.	0	...	0 0 0	0·0	0·0	0·0
3. Jyêshṭha	Vṛishabha-saṁ.	30	(2)	21 50 45	467·1970	121·7837	84·6258	Vṛishabha-saṁ.	30	(2)	21 50 45	467·1970	121·7837	84·6258
4. Āshāḍha	Mithuna-saṁ.	62	(6)	7 25 16	1099·8924	261·3940	170·5896	Mithuna-saṁ.	31	(3)	9 34 31	632·6954	139·5203	85·9638
5. Śrāvaṇa	Karka-saṁ.	93	(2)	22 18 58	1807·6473	408·8685	257·1601	Karka-saṁ.	31	(3)	14 53 42	707·7549	147·5645	86·5705
6. Bhādrapada	Siṁha-saṁ.	125	(6)	10 5 44	2471·4428	551·7219	343·3753	Siṁha-saṁ.	31	(3)	11 46 46	663·7955	142·8534	86·2152
7. Āśvina	Kanyā-saṁ.	156	(2)	11 32 47	2989·5051	678·9569	428·4122	Kanyā-saṁ.	31	(3)	1 27 3	518·0623	127·2350	85·0369
8. Kārttika	Tulā-saṁ.	186	(4)	22 59 48	3310·0242	785·0209	511·8519	T lā-saṁ.	30	(2)	11 27 1	320·5191	106·0640	83·4397
9. Mārgaśira	Vṛiśchika-saṁ.	216	(6)	20 57 12	3440·1530	870·6805	593·7525	Vṛiśchika-saṁ.	29	(1)	21 57 24	130·1288	85·6596	81·9006
10. Pausha	Dhanus-saṁ.	246	(1)	9 9 34	3432·7047	941·5957	674·5407	Dhanus-saṁ.	29	(1)	12 12 22	9992·5517	70·9152	80·7882
11. Māgha	Makara-saṁ.	275	(2)	17 16 58	3367·6498	6·3372	754·8633	Makara-saṁ.	29	(1)	8 7 24	9034·9451	64·7415	80·3226
12. Phālguna	Kumbha saṁ.	305	(4)	3 46 43	3336·0701	74·6663	835·4563	Kumbha-saṁ.	29	(1)	10 29 45	9968·4203	68·3291	80·5930
1. Chaitra (*of following year*)	Mîna-saṁ.	334	(5)	22 36 7	3421·9886	155·5878	916·9994	Mîna-saṁ.	29	(1)	18 49 24	85·9185	80·9215	81·5431
	Mêsha-saṁ. (*of following year*)	365	(1)	6 12 9	3688·1894	255·8304	1000·0	Mêsha-saṁ. (*of following year*).	30	(2)	7 36 2	266·2008	100·2426	83·0006

TABLE LVIII-B.

VALUE OF " *c* " AND OF " EQUATION *c* " AT THE SEVERAL TRUE SAṀKRĀNTIS.

Correct for K. Y. 4500, A.D. 1399-1400.

"c" in 1,000ths of circle ; " equation c " in 10,000ths.

Saṁkranti.	*c.*	Equation *c.*
Mésha-saṁk. . .	274·4058	0·7327
Vṛishabha-saṁk. .	359·0316	13·6505
Mithuna-saṁk. .	444·9954	39·9684
Karka-saṁk. . .	531·5659	72·3342
Siṁha-saṁk. .	617·7811	101·1528
Kanyā-saṁk. .	703·8180	118·1876
Tulā-saṁk. . .	786·2577	119·2579
Vriśchika-saṁk. .	868·1583	104·9306
Dhanus-saṁk. .	948·9465	79·4803
Makara-saṁk. .	29·2691	49·3732
Kumbha-saṁk. .	109·8621	21·9669
Mina-saṁk. . .	191·4052	4·0666

TABLE LVIII-C.

EXACT VALUE OF "*c*" AND OF "EQUATION *c*" AT THE MOMENT OF TRUE MĒSHA-SAṀKRĀNTI AT BEGINNING OF EACH CENTURY K. Y.

"c" in 1,000ths of circle ; "equation c" in 10,000ths.

K. Y.	A.D.	*c.*	Equation *c.*
4200	1099—1100	274·6475	0·7312
4300	1199—1200	274·5669	0·7317
4400	1299—1300	274·4864	0·7332
4500	1399—1400	274·4058	0·7327
4600	1499—1500	274·3253	0·7322
4700	1599—1600	274·2447	0·7337
4800	1699—1700	274·1642	0·7342

TABLE LVIII-D.

CHANGES IN LENGTHS OF TRUE SOLAR MONTHS, AND IN VALUE OF *a, b, c* DUE TO THE FORWARD SHIFT OF THE SUN'S APSIS POSTULATED BY THE *Siddhānta-Śirōmaṇi.*

The entries shew differences from standard (Table LVIII-A, for K. Y. 4500, A.D. 1400) for a year 300 years earlier or later ; *i.e.*, for K. Y. 4200 (A.D. 1100) or 4800 (A.D. 1700). Change for intermediate years to be taken proportionately.

(For years earlier than A.D. 1400 use + or — signs as given. For later years reverse the signs.)

At true solar *saṁkranti.*	Change in collective duration and in collective increase of a. b, c from Mēsha saṁkrānti to each *saṁkranti.*					Change in length of each month between true solar saṁkrantis, and increase of *a, b, c* between each.				
	M.	S.	*a.*	*b.*	*c.*	M.	S.	*a.*	*b.*	*c.*
1	2		3			4		5		
Mēsha-saṁ. .	0	0	0·0	0·0	0·0	0	0	0·0	0·0	0·0
Vrishabha-saṁ	+0	34	+0·1333	+0·0143	+0·0011	+0	34	+0·1333	+0·0143	+0·0011
Mithuna-saṁ.	+2	46	+0·6506	+0·0697	+0·0053	+2	12	+0·5173	+0·0554	+0·0042
Karka-saṁ. .	+2	27	+0·5761	+0·0617	+0·0047	—0	19	—0·0745	—0·0080	—0·0006
Siṁha-saṁ	+2	34	+0·6035	+0·0646	+0·0049	+0	7	+0·0274	+0·0029	+0·0002
Kanyā-saṁ.	+0	23	+0·0901	+0·0096	+0·0008	—2	11	—0·5134	—0·0550	—0·0041
Tulā-saṁ. .	—1	2	—0·2431	—0·0261	—0·0019	—1	25	—0·3332	—0·0357	—0·0027
Vriśchika-saṁ	—2	3	—0·4822	—0·0517	—0·0038	—1	1	—0·2391	—0·0256	—0·0019
Dhanus-saṁ.	—4	55	—1·1563	—0·1239	—0·0092	—2	52	—0·6741	—0·0722	—0·0054
Makara-saṁ.	—4	9	—0·9760	—0·1046	—0·0077	+0	46	+0·1803	+0·0193	+0·0015
Kumbha-saṁ.	—2	47	—0·6546	—0·0702	—0·0051	+1	22	+0·3214	+0·0344	+0·0026
Mina-saṁ. .	—1	31	—0·3567	—0·0383	—0·0027	+1	16	+0·2979	+0·0319	+0·0024
Mēsha-saṁ. (*of following year.*)	—0	12	—0·0470	—0·0050	—0·0004	+1	19	+0·3097	+0·0332	+0·0025

TABLE LIX.

THE MOON'S EQUATION OF THE CENTRE BY THE *Siddhânta-Śiromaṇi*.

(For equation of the Sun's centre see Table XLVII, above, p. 23.)

Serial No. of Sine.	MOON'S MEAN ANOM. Moon's equation —				SINE OF MEAN ANOM. ANGLE. Value in minutes.	Diff.	EQUATION. Equation in degrees.	Diff. per min. of anom.	Equation in 10,000ths of circle.	MOON'S MEAN ANOM. Moon's equation +				Serial No. of Sine.
1	2				3	4	5	6	7	8				1
	°	′	°	′		′	° ′ ″	″		°		°	′	
0	0	0	180	0	0		0 0 0		0·0	180	0	360	0	0
						225		5·26̇						
1	3	45	176	15	225		0 19 45·00		9·1435	183	45	356	15	1
						224		5·2433						
2	7	30	172	30	449		0 39 24·73		18·2564	187	30	352	30	2
						222		5·1967						
3	11	15	168	15	671		0 58 53·93		27·2680	191	15	318	45	3
						219		5·1262						
4	15	0	165	0	890		1 18 7·3		36·1677	195	0	345	0	4
						215		5·0326						
5	18	45	161	15	1105		1 36 59·6̇		44·9048	198	45	341	15	5
						210		4·915						
6	22	30	157	30	1315		1 55 25·6̇		53·4388	202	30	337	30	6
						205		4·7985						
7	26	15	153	45	1520		2 13 25·3		61·7695	206	15	333	45	7
						199		4·6581						
8	30	0	150	0	1719		2 30 53·40		69·8568	210	0	330	0	8
						191		4·4708						
9	33	45	146	15	1910		2 47 39·3		77·6183	213	45	326	15	9
						183		4·2835						
10	37	30	142	30	2093		3 3 43·12̇		85·0550	217	30	322	30	10
						174		4·0728						
11	41	15	138	45	2267		3 18 59·53̇		92·1260	221	15	318	45	11
						164		3·8383						
12	45	0	135	0	24[illegible]		3 33 23·36̇		98·7914	225	0	315	0	12
						154		3·6070						
13	48	45	131	15	2585		3 46 54·8428		105·0528	228	45	311	15	13
						143		3·3622						
14	52	30	127	30	2728		3 59 31·3393		110·8900	232	30	307	30	14
						131		3·0801						
15	56	15	123	45	2859		4 11 4·3661		116·2374	236	15	303	45	15
						119		2·7979						
16	60	0	120	0	2978		4 21 33·8839		121·0948	240	0	300	0	16
						106		2·4890						
17	63	45	116	15	3084		4 30 54·9107		125·4237	243	45	296	15	17
						93		2·1853						
18	67	30	112	30	3177		4 39 6·6027		129·2176	247	30	292	30	18
						79		1·8546						
19	71	15	108	45	3256		4 46 3·8839		132·4374	251	15	288	45	19
						65		1·5342						
20	75	0	105	0	3321		4 51 49·0848		135·1010	255	0	285	0	20
						51		1·1961						
21	78	45	101	15	3372		4 56 18·2143		137·1776	258	45	281	15	21
						37		0·8699						
22	82	30	97	30	3409		4 59 33·9509		138·6879	262	30	277	30	22
						22		0·5173						
23	86	15	93	45	3431		5 1 30·3348		139·5859	266	15	273	45	23
						7		0·1646						
24	90	0	90	0	3438		5 2 7·3661		139·8717	270	0	270	0	24

TABLE LX.

CONSTRUCTION OF TABLE.

The Table is constructed on the lines of Table I of the *Indian Calendar*, and columns are similarly numbered, so as to facilitate comparison of details by the *Ārya*-and *Sūrya-Siddhāntas* with those of the *Siddhānta-Śiromaṇi*, to which the present Table applies.

Cols. 1, 2.—In conformity with this the Kaliyuga and Śaka years stated are current years, not expired years. For years of other eras refer to Tables I and II, Part III, *Indian Calendar*.

Col. 5.—Years A.D. marked with an asterisk are leap-years.

Col. 7.—The *samvatsara*-name—*i.e.*, the name of the Jovian cycle—of the year is given as determined by my previous calculations. See Table XLII above. Entries in italics shew cases where the *samvatsara*-name of the year differs from that fixed by *Sūrya-Siddhānta* calculation.

Col. 8.—Months entered in roman characters are intercalated (*adhika*) lunar months. Those in italics are suppressed (*kshaya*) months.

Cols. 13, 19.—Figures in brackets give the serial number of the day measured from January 1.

Col. 23. "*a*" = distance mean moon from mean sun, stated in 10,000ths of circle.

Col. 24. "*b*" = mean anomaly of moon, or moon's mean distance from perigee-point of apsis, stated in 1,000ths of circle.

Col. 25. "*c*" = sun's mean anomaly, or sun's mean distance from perigee-point of apsis, stated in 1,000ths of circle.

REMARKS.

A.D. 1128-29.—Close case. Possibly 9 Mārgaśira, intercalated (*adhika*), 10 Pausha suppressed (*kshaya*), 12 Phālguna *adhika*.

" 1183-84.—According to the 19-year sequence the *adhika* month should have been 3 Jyeshtha.

" 1242-43.—The *adhika* month should have been 6 Bhādrapada by sequence.

" 1316-17.—Close case. By sequence 2 Vaiśākha expected as *adhika*.

" 1410-11.—By sequence 7 Āśvina expected as *adhika*.

" 1429-30.—By sequence 7 Āśvina expected as *adhika*.

" 1679-80, 1698-99, 1717-18, 1736-37.—By the 19-year sequence in the two former years 4 Āshāḍha expected as *adhika*; or else in the two latter years 3 Jyeshtha expected as *adhika*. But the result in each case by work from the Tables is as tabulated.

" 1749.—Close case. See Text example 6 at end.

TABLE

CONCURRENT YEAR.								Intercalated and suppressed (*ksh.*) lunar months.
Kali.	Śaka.	Chaitrādi Vikrama.	Mēshādi (solar) year in Bengal.	Kollam.	A.D.	Jovian Samvatsara. Southern system.	Jovian Samvatsara. Northern system.	
1	2	3	3*a*	4	5	6	7	8
4201	1022	1157	506	274-75	1099-1100	13 Pramāthin .	16 Chitrabhānu .	3 Jyēshṭha .
4202	1023	1158	507	275-76	*1100-01	14 Vikrama .	17 Subhānu .	...
4203	1024	1159	508	276-77	1101-02	15 Vrisha . .	18 Tāraṇa . .	7 Āśvina .
4204	1025	1160	509	277-78	1102-03	16 Chitrabhānu .	19 Pārthiva .	...
4205	1026	1161	510	278-79	1103-04	17 Subhānu .	20 Vyaya .	...
4206	1027	1162	511	279-80	*1104-05	18 Tāraṇa . .	21 Sarvajit .	4 Āshāḍha .
4207	1028	1163	512	280-81	1105-06	19 Pārthiva .	22 Sarvadhārin .	...
4208	1029	1164	513	281-82	1106-07	20 Vyaya . .	23 Virōdhin .	...
4209	1030	1165	514	282-83	1107-08	21 Sarvajit .	24 Vikrita . .	3 Jyēshṭha
4210	1031	1166	515	283-84	*1108-09	22 Sarvadhārin .	25 Khara . .	...
4211	1032	1167	516	284-85	1109-10	23 Virōdhin .	26 Nandana .	8 Kārttika 10 *Pausha* (*ksh.*) 12 Phālguna
4212	1033	1168	517	285-86	1110-11	24 Vikrita . .	27 Vijaya . .	...
4213	1034	1169	518	286-87	1111-12	25 Khara . .	28 Jaya . .	...
4214	1035	1170	519	287-88	*1112-13	26 Nandana .	29 Manmatha .	5 Śrāvaṇa .
4215	1036	1171	520	288-89	1113-14	27 Vijaya . .	30 Durmukha .	...
4216	1037	1172	521	289-90	1114-15	28 Jaya . .	31 Hēmalamba .	...
4217	1038	1173	522	290-91	1115-16	29 Manmatha .	32 Vilamba .	4 Āshāḍha .
4218	1039	1174	523	291-92	*1116-17	30 Durmukha .	33 Vikārin .	...
4219	1040	1175	524	292-93	1117-18	31 Hēmalamba .	34 Śārvarin .	...
4220	1041	1176	525	293-94	1118-19	32 Vilamba .	35 Plava . .	2 Vaiśākha .
4221	1042	1177	526	294-95	1119-20	33 Vikārin .	36 Śubhakrit .	...
4222	1043	1178	527	295-96	*1120-21	34 Śārvarin .	37 Śōbhana .	6 Bhādrapada
4223	1044	1179	528	296-97	1121-22	35 Plava . .	38 Krōdhin .	...
4224	1045	1180	529	297-98	1122-23	36 Śubhakrit .	39 Viśvāvasu .	...
4225	1046	1181	530	298-99	1123-24	37 Śōbhana .	40 Parābhava .	4 Āshāḍha .

LX.

COMMENCEMENT OF THE											
Solar year.					Luni-solar year (mean sunrise of day on which Chaitra Śukla 1 ends).						Kali year.
Day and month, A. D.	Week-day.	Time of true Mēsha-samkrānti.			Day and month, A. D.	Week-day.	a.	b	c.		
13	14	17			19	20	23	24	25		1
		H.	M.	S.							
23 Mar. (82)	4 Wed.	6	11	11	24 Feb. (55)	5 Thur.	228·7161	574·4426	200·0218		4201
22 Mar. (82)	5 Thur.	12	23	20	13 Mar. (73)	3 Tues.	9924·7666	474·1445	248·5944		4202
22 Mar. (81)	6 Fri.	18	35	29	2 Mar. (61)	0 Sat.	9800·4894	321·3885	217·7712		4203
23 Mar. (82)	1 Sun.	0	47	38	21 Mar. (80)	6 Fri.	9835·1718	256·3820	269·0815		4204
23 Mar. (82)	2 Mon.	6	59	46	11 Mar. (70)	4 Wed.	49·5266	140·9176	240·9962		4205
22 Mar. (82)	3 Tues.	13	11	55	28 Feb. (59)	1 Sun.	9925·2495	988·1617	210·1700		4206
22 Mar. (81)	4 Wed.	19	24	4	18 Mar. (77)	0 Sat.	9959·9318	924·1552	261·4834		4207
23 Mar. (82)	6 Fri.	1	36	13	8 Mar. (67)	5 Thur.	174·2867	807·6909	233·3979		4208
23 Mar. (82)	0 Sat.	7	48	22	25 Feb. (56)	2 Mon.	50·0095	654·9350	202·5747		4209
22 Mar. (82)	1 Sun.	14	0	31	15 Mar. (75)	1 Sun.	84·6918	590·9284	253·8852		4210
22 Mar. (81)	2 Mon.	20	12	39	4 Mar. (63)	5 Thur.	9960·4147	438·1725	223·0619		4211
23 Mar. (82)	4 Wed.	2	24	48	23 Mar. (82)	4 Wed.	9995·0971	374·1659	274·3723		4212
23 Mar. (82)	5 Thur.	8	36	57	12 Mar. (71)	1 Sun.	9870·8200	221·4100	243·5492		4213
22 Mar. (82)	6 Fri.	14	49	6	1 Mar. (61)	6 Fri.	85·1747	104·9457	215·4638		4214
22 Mar. (81)	0 Sat.	21	1	15	20 Mar. (79)	5 Thur.	119·8572	40·9392	266·7742		4215
23 Mar. (82)	2 Mon.	3	13	24	9 Mar. (68)	2 Mon.	9995·5800	888·1832	235·9509		4216
23 Mar. (82)	3 Tues.	9	25	32	27 Feb. (58)	0 Sat.	209·9348	771·7279	207·8655		4217
22 Mar. (82)	4 Wed.	15	37	41	17 Mar. (77)	6 Fri.	244·6172	707·7124	259·1760		4218
22 Mar. (81)	5 Thur.	21	49	50	6 Mar. (65)	3 Tues.	120·3401	554·9564	228·3527		4219
23 Mar. (82)	0 Sat.	4	1	59	23 Feb. (54)	0 Sat.	9996·0629	402·2005	197·5295		4220
23 Mar. (82)	1 Sun.	10	14	8	14 Mar. (73)	6 Fri.	30·7453	338·1940	248·8399		4221
22 Mar. (82)	2 Mon.	16	26	17	2 Mar. (62)	3 Tues.	9906·4681	185·4382	218·0168		4222
22 Mar. (81)	3 Tues.	22	38	25	21 Mar. (80)	2 Mon.	9941·1506	121·4315	269·3271		4222
23 Mar. (82)	5 Thur.	4	5	34	11 Mar. (70)	0 Sat.	155·5053	4·9672	241·2417		4224
23 Mar. (82)	6 Fri.	11	2	43	28 Feb. (59)	4 Wed.	31·2282	851·6634	209·7110		4225

TABLE

CONCURRENT YEAR.								Intercalated and suppressed (*l.sh.*) lunar months.
						Jovian Samvatsara.		
Kali.	Śaka.	Chaitrâdi Vikrama.	Mêshâdi (solar) year in Bengal.	Kollam.	A.D.	Southern system.	Northern system.	
1	2	3	3a	4	5	6	7	8
4226	1047	1182	531	299-00	*1124-25	38 Krōdhin	41 Plavaṅga	...
4227	1048	1183	532	300-01	1125-26	39 Viśvāvasu	42 Kīlaka	...
4228	1049	1184	533	301-02	1126-27	40 Parābhava	43 Saumya	3 Jyēshṭha
4229	1050	1185	534	302-03	1127-28	41 Plavaṅga	44 Sādhāraṇa	...
4230	1051	1186	535	303-04	*1128-29	42 Kīlaka	45 Virōdhakrit	12 Phālguna†
4231	1052	1187	536	304-05	1129-30	43 Saumya	46 Paridhāvin	...
4232	1053	1188	537	305-06	1130-31	44 Sādhāraṇa	47 Pramādin	...
4233	1054	1189	538	306-07	1131-32	45 Virōdhakrit	48 Ānanda	5 Śrāvaṇa
4234	1055	1190	539	307-08	*1132-33	46 Paridhāvin	49 Rākshasa	...
4235	1056	1191	540	308-09	1133-34	47 Pramādin	50 Anala	...
4236	1057	1192	541	309-10	1134-35	48 Ānanda	51 Piṅgala	4 Āshāḍha
4237	1058	1193	542	310-11	1135-36	49 Rākshasa	52 Kālayukta	...
4238	1059	1194	543	311-12	*1136-37	50 Anala	53 Siddhārthin	...
4239	1060	1195	544	312-13	1137-38	51 Piṅgala	54 Raudra	2 Vaiśākha
4240	1061	1196	545	313-14	1138-39	52 Kālayukta	55 Durmati	...
4241	1062	1197	546	314-15	1139-40	53 Siddhārthin	56 Dundubhi	6 Bhādrapada
4242	1063	1198	547	315-16	*1140-41	54 Raudra	57 Rudhirōdgārin	...
4243	1064	1199	548	316-17	1141-42	55 Durmati	58 Raktāksha	...
4244	1065	1200	549	317-18	1142-43	56 Dundubhi	59 Krōdhana	4 Āshāḍha
4245	1066	1201	550	318-19	1143-44	57 Rudhirōdgārin	60 Kshaya	...
4246	1067	1202	551	319-20	*1144-45	58 Raktāksha	1 Prabhava	...
4247	1068	1203	552	320-21	1145-46	59 Krōdhana	2 Vibhava	3 Jyēshṭha
4248	1069	1204	553	321-22	1146-47	60 Kshaya	3 Śukla	...
4249	1070	1205	554	322-23	1147-48	1 Prabhava	4 Pramōda	8 Kārttika 10 *Pausha* (*l.sh.*) 12 Phālguna
4250	1071	1206	555	323-24	*1148-49	2 Vibhava	5 Prajāpati	...

* See Remarks, p. 163 above.

LX—*Contd.*

COMMENCEMENT OF THE									Kali year.
SOLAR YEAR.			LUNI-SOLAR YEAR (MEAN SUNRISE OF DAY ON WHICH CHAITRA ŚUKLA 1 ENDS).						
Day and month, A.D.	Week-day.	Time of true Mēsha-saṁkrānti.	Day and month, A.D.	Week-day.	*a.*	*b.*	*c.*		
13	14	17	19	20	23	24	25		1
		H. M. S.							
22 Mar. (82)	0 Sat. .	17 14 52	18 Mar. (78)	3 Tues.	65·9106	788·2047	261·7290		4226
22 Mar. (81)	1 Sun. .	23 27 1	8 Mar. (67)	1 Sun. .	280·2655	671·7404	233·6435		4227
23 Mar. (82)	3 Tues.	5 39 10	25 Feb. (56)	5 Thur.	155·9882	518·9845	202·8202		4228
23 Mar. (82)	4 Wed.	11 51 19	15 Mar. (74)	3 Tues.	9852·0386	418·6[illegible]63	251·3929		4229
22 Mar. (82)	5 Thur.	18 3 27	3 Mar. (63)	0 Sat. .	9727·7615	265·9303	220·5698		4230
23 Mar. (82)	0 Sat. .	0 15 36	22 Mar. (81)	6 Fri.	9762·4438	201·9239	271·8801		4231
23 Mar. (82)	1 Sun. .	6 27 45	12 Mar. (71)	4 Wed.	9976·7987	85·4595	243·7947		4232
23 Mar. (82)	2 Mon. .	12 39 54	2 Mar. (61)	2 Mon. .	191·1545	968·9952	215·7093		4233
22 Mar. (82)	3 Tues.	18 52 3	20 Mar. (80)	1 Sun. .	225·8360	904·9887	267·0197		4234
23 Mar. (82)	5 Thur.	1 4 12	9 Mar. (68)	5 Thur.	101·5587	751·2327	236·1965		4235
23 Mar. (82)	6 Fri. .	7 16 20	26 Feb. (57)	2 Mon. .	9977·2816	599·4768	205·3732		4236
23 Mar. (82)	0 Sat. .	13 28 29	17 Mar. (76)	1 Sun. .	11·9640	535·4702	256·6837		4237
22 Mar. (82)	1 Sun. .	19 40 38	5 Mar. (65)	5 Thur.	9887·6769	382·7143	225·8[illegible]05		4238
23 Mar. (82)	3 Tues.	1 52 47	22 Feb. (53)	2 Mon. .	9763·4097	229·9583	195·0373		4239
23 Mar. (82)	4 Wed.	8 4 56	13 Mar. (72)	1 Sun. .	9798·0921	165·9518	246·3477		4240
23 Mar. (82)	5 Thur.	14 17 5	3 Mar. (62)	5 Fri. .	12·4469	49·4876	218·2623		4241
22 Mar. (82)	6 Fri. .	20 29 13	21 Mar. (81)	5 Thur.	47·1292	985·4840	269·5727		4242
23 Mar. (82)	1 Sun. .	2 41 22	11 Mar. (70)	3 Tues.	261·4841	869·0167	241·4873		4243
23 Mar. (82)	2 Mon. .	8 53 31	28 Feb. (59)	0 Sat.	137·2070	716·2597	210·6641		4244
23 Mar. (82)	3 Tues.	15 5 40	19 Mar. (78)	6 Fri.	171·8894	652·1542	261·9745		4245
22 Mar. (82)	4 Wed.	21 17 49	7 Mar. (67)	3 Tues.	47·6122	499·4983	231·1512		4246
23 Mar. (82)	6 Fri. .	3 29 58	24 Feb. (55)	0 Sat.	9923·3350	346·7423	200·3281		4247
23 Mar. (82)	0 Sat. .	9 42 7	15 Mar. (74)	6 Fri. .	9958·0174	282·7358	251·6385		4248
23 Mar. (82)	1 Sun. .	15 54 15	4 Mar. (63)	3 Tues.	9833·7402	129·9798	220·8153		4249
22 Mar. (82)	2 Mon. .	22 6 24	22 Mar. (82)	2 Mon. .	9868·4226	65·9734	272·1256		4250

TABLE

CONCURRENT YEAR.								
						Jovian Samvatsara.		Intercalated and suppressed (*ksh.*) lunar months.
Kali.	Śaka.	Chaitrādi Vikrama.	Mēshādi (solar) year in Bengal.	Kollam.	A.D.	Southern system.	Northern system.	
1	2	3	3*a*	4	5	6	7	8
4251	1072	1207	556	324-25	1149-50	3 Śukla	6 Aṅgiras	...
4252	1073	1208	557	325-26	1150-51	4 Pramōda	7 Śrīmukha	5 Śrāvaṇa
4253	1074	1209	558	326-27	1151-52	5 Prajāpati	8 Bhāva	...
4254	1075	1210	559	327-28	*1152-53	6 Aṅgiras	9 Yuvan	...
4255	1076	1211	560	328-29	1153-54	7 Śrīmukha	10 Dhātṛi	4 Āshāḍha
4256	1077	1212	561	329-30	1154-55	8 Bhāva	11 Īśvara	...
4257	1078	1213	562	330-31	1155-56	9 Yuvan	12 Bahudhānya	...
4258	1079	1214	563	331-32	*1156-57	10 Dhātṛi	13 Pramāthin	2 Vaiśākha
4259	1080	1215	564	332-33	1157-58	11 Īsvara	14 Vikrama	...
4260	1081	1216	565	333-34	1158-59	12 Bahudhānya	15 Vṛisha	6 Bhādrapada
4261	1082	1217	566	334-35	1159-60	13 Pramāthin	16 Chitrabhānu	...
4262	1083	1218	567	335-36	*1160-61	14 Vikrama	18 *Tāraṇa*†	...
4263	1084	1219	568	336-37	1161-62	15 Vṛisha	19 *Pārthiva*	4 Āshāḍha
4264	1085	1220	569	337-38	1162-63	16 Chitrabhānu	20 *Vyaya*	...
4265	1086	1221	570	338-39	1163-64	17 Subhānu	21 *Śarvajit*	...
4266	1087	1222	571	339-40	*1164-65	18 Tāraṇa	22 *Sarvadhārin*	3 Jyēshṭha
4267	1088	1223	572	340-41	1165-66	19 Pārthiva	23 Virōdhin	...
4268	1089	1224	573	341-42	1166-67	20 Vyaya	24 Vikṛita	7 Āśvina 10 *Pausha* (*ksh.*) 12 Phālguna
4269	1090	1225	574	342-43	1167-68	21 Sarvajit	25 Khara	...
4270	1091	1226	575	343-44	*1168-69	22 Sarvadhārin	26 Nandana	...
4271	1092	1227	576	344-45	1169-70	23 Virōdhin	27 Vijaya	5 Śrāvaṇa
4272	1093	1228	577	345-46	1170-71	24 Vikṛita	28 Jaya	...
4272	1094	1229	578	246-47	1171-72	25 Khara	29 Manmatha	...
4274	1095	1239	579	347-48	*1172-73	26 Nandana	30 Durmukha	4 Āshāḍha
4275	1096	1231	580	348-49	1173-74	27 Vijaya	31 Hēmalamba	...

† 17 Subbānu was suppressed in the north.

LX—*Contd.*

COMMENCEMENT OF THE											
Solar year.					Luni-solar year (mean sunrise of day on which Chaitra śukla 1 ends).						Kali year.
Day and month, A.D.	Week-day.	Time of true Mēsha-saṁkrānti.			Day and month, A.D.	Week-day.	*a.*	*b.*	*c.*		
13	14	17			19	20	23	24	25		1
		H.	M.	S.							
23 Mar. (82)	4 Wed.	4	18	33	12 Mar. (71)	0 Sat. .	82·7775	949·5090	244·0403		4251
23 Mar. (82)	5 Thur.	10	30	42	2 Mar. (61)	5 Thur.	297·1322	833·0447	215·9549		4252
23 Mar. (82)	6 Fri. .	16	42	51	21 Mar. (80)	4 Wed.	331·8147	769·0742	267·2662		4253
22 Mar. (82)	0 Sat. .	22	55	0	9 Mar. (69)	1 Sun. .	207·5375	616·2822	236·4420		4254
23 Mar. (82)	2 Mon. .	5	7	8	26 Feb. (57)	5 Thur.	83·2604	463·5263	205·6188		4255
23 Mar. (82)	3 Tues.	11	19	17	16 Mar. (75)	3 Tues.	9779·3107	363·2282	254·1915		4256
23 Mar. (82)	4 Wed.	17	31	26	6 Mar. (65)	1 Sun. .	9993·6656	246·7638	226·1060		4257
22 Mar. (82)	5 Thur.	23	43	35	23 Feb. (54)	5 Thur.	9869·3885	94·0078	195·2928		4258
23 Mar. (82)	0 Sat. .	5	55	44	13 Mar. (72)	4 Wed.	9904·0709	30·0013	246·5932		4259
23 Mar. (82)	1 Sun. .	12	7	53	3 Mar. (62)	2 Mon. .	118·4256	913·5371	218·5079		4260
23 Mar. (82)	2 Mon. .	18	20	1	22 Mar. (81)	1 Sun. .	153·1080	849·5306	269·7796		4261
23 Mar. (83)	4 Wed.	0	32	10	10 Mar. (70)	5 Thur.	28·8309	696·7746	238·9950		4262
23 Mar. (82)	5 Thur.	6	44	19	27 Feb. (58)	2 Mon. .	9904·5537	544·0187	208·1718		4263
23 Mar. (82)	6 Fri. .	12	56	28	18 Mar. (77)	1 Sun. .	9939·2361	480·0121	259·4823		4264
23 Mar. (82)	0 Sat. .	19	8	37	7 Mar. (66)	5 Thur.	9814·9590	327·2562	228·6590		4265
23 Mar. (83)	2 Mon. .	1	20	46	25 Feb. (56)	3 Tues.	29·3138	210·7918	200·5736		4266
23 Mar. (82)	3 Tues.	7	32	54	15 Mar. (74)	2 Mon. .	63·9961	146·7853	251·8740		4267
23 Mar. (82)	4 Wed.	13	45	3	4 Mar. (63)	6 Fri. .	9939·7190	994·0294	221·0609		4268
23 Mar. (82)	5 Thur.	19	57	12	23 Mar. (82)	5 Thur.	9974·4014	930·0228	272·3713		4269
23 Mar. (83)	0 Sat. .	2	9	21	12 Mar. (72)	3 Tues.	188·7562	813·5586	244·2858		4270
23 Mar. (82)	1 Sun. .	8	21	30	1 Mar. (60)	0 Sat. .	64·4791	660·8026	213·4626		4271
23 Mar. (82)	2 Mon.	14	33	39	20 Mar. (79)	6 Fri. .	99·1615	596·7961	264·7731		4272
23 Mar. (82)	3 Tues.	20	45	47	9 Mar. (68)	3 Tues.	9974·8844	444·0401	233·9498		4273
23 Mar. (83)	5 Thur.	2	57	56	26 Feb. (57)	0 Sat. .	9850·6071	291·2842	203·1265		4274
23 Mar. (82)	6 Fri. .	9	10	5	16 Mar. (75)	6 Fri. .	9885·2895	227·2777	254·4370		4275

TABLE

CONCURRENT YEAR.								Intercalated and suppressed (*ksh.*) lunar months.
Kali.	Śaka.	Chaitrādi Vikrama.	Mēshādi (solar) year in Bengal.	Kollam.	A.D.	Jovian Samvatsara. Southern system.	Jovian Samvatsara. Northern system.	
1	2	3	3a	4	5	6	7	8
4276	1097	1232	581	349-50	1174-75	28 Jaya . .	32 Vilamba .	...
4277	1098	1233	582	350-51	1175-76	29 Manmatha .	33 Vikārin	2 Vaiśākha .
4278	1099	1234	583	351-52	*1176-77	30 Durmukha .	34 Śārvarin .	...
4279	1100	1235	584	352-53	1177-78	31 Hāmalamba	35 Plava .	6 Bhādrapada
4280	1101	1236	585	353-54	1178-79	32 Vilamba .	36 Śubhakṛit .	...
4281	1102	1237	586	354-55	1179-80	33 Vikārin .	37 Śōbhana .	...
4282	1103	1238	587	355-56	*1180-81	34 Śārvarin .	38 Krōdhin .	4 Āshāḍha .
4283	1104	1239	588	356-57	1181-82	35 Plava . .	39 Viśvāvasu .	...
4284	1105	1240	589	357-58	1182-83	36 Śubhakṛit .	40 Parābhava .	...
4285	1106	1241	590	358-59	1183-84	37 Śōbhana .	41 Plavaṅga .	2 Vaiśākha†
4286	1107	1242	591	359-60	*1184-85	38 Krōdhin .	42 Kīlaka . .	...
4287	1108	1243	592	360-61	1185-86	39 Viśvāvasu .	43 Saumya .	6 Bhādrapada
4288	1109	1244	593	361-62	1186-87	40 Parābhava .	44 Sādhāraṇa	...
4289	1110	1245	594	362-63	1187-88	41 Plavaṅga .	45 Virōdhakṛit .	...
4290	1111	1246	595	363-64	*1188-89	42 Kīlaka . .	46 Paridhāvin .	5 Śrāvaṇa
4291	1112	1247	596	364-65	1189-90	43 Saumya .	47 Pramādin .	...
4292	1113	1248	597	365-66	1190-91	44 Sādhāraṇa .	48 Ananda .	...
4293	1114	1249	598	366-67	1191-92	45 Virōdhakṛit .	49 Rākshasa .	3 Jyēshṭha .
4294	1115	1250	599	367-68	*1192-93	46 Paridhāvin .	50 Anala . .	...
4295	1116	1251	600	368-69	1193-94	47 Pramādin .	51 Piṅgala .	...
4296	1117	1252	601	369-70	1194-95	48 Ānanda .	52 Kālayukta .	2 Vaiśākha
4297	1118	1253	602	570-71	1195-96	49 Rākshasa .	53 Siddhārthin .	...
4298	1119	1254	603	371-72	*1196-97	50 Anala . .	54 Raudra .	6 Bhādrapada
4299	1120	1255	304	372-73	1197-98	51 Piṅgala .	55 Durmati	...
4300	1121	1256	605	373-74	1198-99	52 Kālayukta .	56 Dundubhi .	...

† See Remarks, p. 163 above.

LX—*Contd.*

COMMENCEMENT OF THE									
Solar year.			Luni-solar year (mean sunrise of day on which Chaitra śukla 1 ends).					Kali year.
Day and month, A.D.	Week-day.	Time of true Mêsha-samkrānti.			Day and month, A.D.	Week-day.	*a.*	*b.*	*c.*	
13	14	17			19	20	23	24	25	1
		H.	M.	S.						
23 Mar. (82)	0 Sat. .	15	22	14	6 Mar. (65)	4 Wed.	99·6444	110·8133	226·3516	4276
23 Mar. (82)	1 Sun. .	21	34	23	23 Feb. (54)	1 Sun. .	9975·3672	958·0573	195·5284	4277
23 Mar. (83)	3 Tues. .	3	46	32	13 Mar. (73)	0 Sat. .	10·0496	894·0508	246·8387	4278
23 Mar. (82)	4 Wed. .	9	58	41	3 Mar. (62)	5 Thur.	224·4044	777·5866	218·7534	4279
23 Mar. (82)	5 Thur.	16	10	49	22 Mar. (81)	4 Wed. .	259·0868	713·5801	270·0638	4280
23 Mar. (82)	6 Fri. .	22	22	58	11 Mar. (70)	1 Sun. .	134·8096	560·8241	239·2406	4281
23 Mar. (83)	1 Sun. .	4	35	7	28 Feb. (59)	5 Thur.	10·5325	408·0682	208·4173	4282
23 Mar. (82)	2 Mon. .	10	47	16	18 Mar. (77)	4 Wed. .	45·2149	344·0616	259·7278	4283
23 Mar. (82)	3 Tues. .	16	59	25	7 Mar. (66)	1 Sun. .	9920·9377	191·3017	228·9046	4284
23 Mar. (82)	4 Wed.	23	11	34	24 Feb. (55)	5 Thur.	9796·6605	38·5497	198·0814	4285
23 Mar. (83)	6 Fri. .	5	23	42	15 Mar. (75)	5 Thur.	169·9748	10·8348	252·1295	4286
23 Mar. (82)	0 Sat. .	11	35	51	4 Mar. (63)	2 Mon.	45·6978	858·0789	221·3064	4287
23 Mar. (82)	1 Sun. .	17	48	0	23 Mar. (82)	1 Sun. .	80·3801	794·0717	272·6168	4288
24 Mar. (83)	3 Tues.	0	0	9	13 Mar. (72)	6 Fri. .	294·7350	677·5180	244·5314	4289
23 Mar. (83)	4 Wed.	6	12	8	1 Mar. (61)	3 Tues.	170·4579	524·8521	213·7081	4290
23 Mar. (82)	5 Thur.	12	24	27	19 Mar. (78)	1 Sun. .	9866·5083	421·5529	262·2808	4291
23 Mar. (82)	6 Fri. .	18	36	35	8 Mar. (67)	5 Thur.	9742·2311	271·7980	231·4576	4292
24 Mar. (83)	1 Sun.	0	48	44	26 Feb. (57)	3 Tues.	9956·5859	155·3337	203·3721	4293
23 Mar. (83)	2 Mon.	7	0	53	16 Mar. (76)	2 Mon.	9991·2683	91·3272	254·6825	4294
23 Mar. (82)	3 Tues.	13	13	2	6 Mar. (65)	0 Sat. .	205·6231	974·8629	226·5971	4295
23 Mar. (82)	4 Wed.	19	25	11	23 Feb. (54)	4 Wed.	81·3459	822·1069	195·7740	4296
24 Mar. (83)	6 Fri. .	1	37	20	14 Mar. (73)	3 Tues.	116·0284	758·1003	247·0843	4297
23 Mar. (83)	0 Sat. .	7	49	28	2 Mar. (62)	0 Sat. .	9991·7511	605·2444	216·2611	4298
23 Mar. (82)	1 Sun. .	14	1	37	21 Mar. (80)	6 Fri. .	26·4336	541·3379	267·5715	4299
23 Mar. (82)	2 Mon. .	20	13	46	10 Mar. (69)	3 Tues.	9902·1564	388·5820	236·7484	4300

TABLE

CONCURRENT YEAR.								Intercalated and suppressed (*ksh.*) lunar months.
Kali.	Śaka.	Chaitrādi Vikrama.	Mēshādi (solar) year in Bengal.	Kollam.	A.D.	Jovian Samvatsara. Southern system.	Jovian Samvatsara. Northern system.	
1	2	3	3a	4	5	6	7	8
4301	1122	1257	606	374-75	1199-00	53 Siddhārthin .	57 Rudhirōdgārin	4 Āshāḍha.
4302	1123	1258	607	375-76	*1200-01	54 Raudra . .	58 Raktāksha .	...
4303	1124	1259	608	376-77	1201-02	55 Durmati .	59 Krōdhana .	...
4304	1125	1260	609	377-78	1202-03	56 Dundubhi .	60 Kshaya . .	3 Jyēshṭha .
4305	1126	1261	610	378-79	1203-04	57 Rudhirōdgārin	1 Prabhava .	...
4306	1127	1262	611	379-80	*1204-05	58 Raktāksha .	2 Vibhava .	6 Bhādrapada .
4307	1128	1263	612	380-81	1205-06	59 Krōdhana .	3 Śukla .	...
4308	1129	1264	613	381-82	1206-07	60 Kshaya . .	4 Pramōda .	...
4309	1130	1265	614	382-83	1207-08	1 Prabhava .	5 Prajāpati .	5 Śrāvaṇa .
4310	1131	1266	615	383-84	*1208-09	2 Vibhava . .	6 Aṅgiras . .	...
4311	1132	1267	616	384-85	1209-10	3 Śukla . .	7 Śrīmukha .	...
4312	1133	1268	617	385-86	1210-11	4 Pramōda .	8 Bhāva . .	3 Jyēshṭha .
4313	1134	1269	618	386-87	1211-12	5 Prajāpati .	9 Yuvan . .	...
4314	1135	1270	619	387-88	*1212-13	6 Aṅgiras . .	10 Dhātṛi .	8 Kārttika 9 *Mārgaś.(ksh)*
4315	1136	1271	620	388-89	1213-14	7 Śrīmukha .	11 Īśvara .	2 Vaiśākha .
4316	1137	1272	621	389-90	1214-15	8 Bhāva . .	12 Bahudhānya	...
4317	1138	1273	622	390-91	1215-16	9 Yuvan . .	13 Pramāthin .	6 Bhādrapada .
4318	1139	1274	623	391-92	*1216-17	10 Dhātṛi . .	14 Vikrama .	...
4319	1140	1275	624	392-93	1217-18	11 Iśvara . .	15 Vṛisha .	..
4320	1141	1276	625	393-94	1218-19	12 Bahudhānya .	16 Chitrabhānu .	4 Āshāḍha .
4321	1142	1277	626	394-95	1219-20	13 Pramāthin .	17 Subhānu .	...
4322	1143	1278	627	395-96	*1220-21	14 Vikrama .	18 Tāraṇa . .	...
4323	1144	1279	628	396-97	1221-22	15 Vṛisha . .	19 Pārthiva .	3 Jyēshṭha
4324	1145	1280	629	397-98	1222-23	16 Chitrabhānu .	20 Vyaya . .	...
4325	1146	1281	630	398-99	1223-24	17 Subhānu .	21 Sarvajit .	6 Bhādrapada .

LX—*Contd.*

COMMENCEMENT OF THE									
SOLAR YEAR.			LUNI-SOLAR YEAR (MEAN SUNRISE OF DAY ON WHICH CHAITRA ŚUKLA 1 ENDS).						Kali year.
Day and month, A.D.	Week-day	Time of true Mēsha-saṁkrānti.	Day and month, A.D.	Week-day.	*a.*	*b.*	*c.*		
13	14	17	19	20	23	24	25		1
		H. M. S.							
24 Mar. (83)	4 Wed.	2 25 55	27 Feb. (58)	0 Sat.	9777·8793	236·2261	205·8446		4301
23 Mar. (83)	5 Thur.	8 38 4	17 Mar. (77)	6 Fri.	9812·5617	171·8196	257·1551		4302
23 Mar. (82)	6 Fri.	14 50 13	7 Mar. (66)	4 Wed.	26·9166	55·3552	229·0696		4303
23 Mar. (82)	0 Sat.	21 2 22	25 Feb. (56)	2 Mon.	241·2713	938·8910	200·9741		4304
24 Mar. (83)	2 Mon.	3 14 30	16 Mar. (75)	1 Sun.	275·9537	874·8844	252·2946		4305
23 Mar. (83)	3 Tues	9 26 39	4 Mar. (64)	5 Thur.	151·6766	722·1285	221·4714		4306
23 Mar. (82)	4 Wed.	15 38 48	23 Mar. (82)	4 Wed.	186·3589	658·1220	272·7818		4307
23 Mar. (82)	5 Thur.	21 50 57	12 Mar. (71)	1 Sun.	62·0918	505·3660	241·9586		4308
24 Mar. (83)	0 Sat.	4 3 6	1 Mar. (60)	5 Thur.	9937·8047	352·6101	211·1354		4309
23 Mar. (83)	1 Sun.	10 15 15	19 Mar. (79)	4 Wed.	9972·4870	288·6035	262·4459		4310
23 Mar. (82)	2 Mon.	16 27 23	8 Mar. (67)	1 Sun.	9848·2098	135·8475	231·6226		4311
23 Mar. (82)	3 Tues.	22 39 32	26 Feb. (57)	6 Fri.	62·5647	19·3832	203·5371		4312
24 Mar. (83)	5 Thur.	4 51 41	17 Mar. (76)	5 Thur.	97·2471	955·3767	254·8476		4313
23 Mar. (83)	6 Fri.	11 3 50	5 Mar. (65)	2 Mon.	9972·9699	802·6209	224·0244		4314
23 Mar. (82)	0 Sat.	17 15 59	23 Feb. (54)	0 Sat.	187·3447	686·1565	195·9390		4315
23 Mar. (82)	1 Sun.	23 28 8	14 Mar. (73)	6 Fri.	222·0072	622·1500	247·2493		4316
24 Mar. (83)	3 Tues.	5 40 16	3 Mar. (62)	3 Tues.	97·7299	468·4030	216·4262		4317
23 Mar. (83)	4 Wed.	11 52 25	20 Mar. (80)	1 Sun.	9793·7804	369·0958	264·9988		4318
23 Mar. (82)	5 Thur.	18 4 34	10 Mar. (69)	6 Fri.	8·1352	252·6315	236·9134		4319
24 Mar. (83)	0 Sat.	0 16 43	27 Feb. (58)	3 Tues.	9883·8581	99·8756	205·3826		4320
24 Mar. (83)	1 Sun.	6 28 52	18 Mar. (77)	2 Mon.	9918·5404	35·8691	257·4906		4321
23 Mar. (83)	2 Mon	12 41 1	7 Mar. (67)	0 Sat.	132·8953	919·4048	229·3152		4322
23 Mar (82)	3 Tues	18 53 10	24 Feb. (55)	4 Wed.	8·6181	766·6488	198·4920		4323
24 Mar. (83)	5 Thur.	1 5 18	15 Mar. (74)	3 Tues.	43·3004	702·6423	249·8023		4324
24 Mar. (83)	6 Fri.	7 17 27	4 Mar. (63)	0 Sat.	9919·0233	549·88[illegible]3	218·9792		4325

TABLE

CONCURRENT YEAR.								Intercalated and suppressed (*ksh.*) lunar months.
Kali.	Śaka.	Chaitrādi Vikrama.	Mēshādi (solar) year in Bengal.	Kollam.	A.D.	Jovian Samvatsara. Southern system.	Jovian Samvatsara. Northern system.	
1	2	3	3*a*	4	5	6	7	8
4326	1147	1282	631	399-00	*1224-25	18 Tāraṇa . .	22 Sarvadhārin .	...
4327	1148	1283	632	400-01	1225-26	19 Pārthiva .	23 Virōdhin .	...
4328	1149	1284	633	401-02	1226-27	20 Vyaya . .	24 Vikṛita . .	5 Śrāvaṇa .
4329	1150	1285	634	402-03	1227-28	21 Sarvajit .	25 Khara . .	...
4330	1151	1286	635	403-04	*1228-29	22 Sarvadhārin .	26 Nandana .	...
4331	1152	1287	636	404-05	1229-30	23 Virōdhin .	27 Vijaya . .	3 Jyēshṭha .
4332	1153	1288	637	405-06	1230-31	24 Vikṛita . .	28 Jaya . .	...
4333	1154	1289	638	406-07	1231-32	25 Khara . .	29 Manmatha	{ 8 Kārttika } { 10 *Pausha*(*ksh.*) }
4334	1155	1290	639	407-08	*1232-33	26 Nandana .	30 Durmukha .	1 Chaitra .
4335	1156	1291	640	408-09	1233-34	27 Vijaya . .	31 Hēmalamba .	...
4336	1157	1292	641	409-10	1234-35	28 Jaya . .	32 Vilamba .	5 Śrāvaṇa .
4337	1158	1293	642	410-11	1235-36	29 Manmatha .	33 Vikārin .	...
4338	1159	1294	643	411-12	*1236-37	30 Durmukha .	34 Śārvarin .	...
4339	1160	1295	644	412-13	1237-38	31 Hēmalamba .	35 Plava . .	4 Āshāḍha .
4340	1161	1296	645	413-14	1238-39	32 Vilamba .	36 Śubhakṛit .	...
4341	1162	1297	646	414-15	1239-40	33 Vikārin .	37 Śōbhana .	...
4342	1163	1298	647	415-16	*1240-41	34 Śārvarin .	38 Krōdhin .	3 Jyēshṭha .
4343	1164	1299	648	416-17	1241-42	35 Plava . .	39 Viśvāvasu .	...
4344	1165	1300	649	417-18	1242-43	36 Śubhakṛit .	40 Parābhava .	7 Āśvina‡ .
4345	1166	1301	650	418-19	1243-44	37 Śōbhana .	41 Plavaṅga .	...
4346	1167	1302	651	419-20	*1244-45	38 Krōdhin .	42 Kīlaka . .	...
4347	1168	1303	652	420-21	1245-46	39 Viśvāvasu .	43 Saumya .	4 Āshāḍha .
4348	1169	1304	653	421-22	1246-47	40 Parābhava .	45 *Virōdhakrit*† .	...
4349	1170	1305	654	422-23	1247-48	41 Plavaṅga .	46 *Paridhāvin* .	...
4350	1171	1306	655	423-24	*1248-49	42 Kīlaka . .	47 *Pramādin* .	3 Jyēshṭha .

† 44 Sādhāraṇa was suppressed in the south.
‡ See Remarks, p. 163 above.

LX—*Contd.*

COMMENCEMENT OF THE											
SOLAR YEAR.					LUNI-SOLAR YEAR (MEAN SUNRISE OF DAY ON WHICH CHAITRA ŚUKLA 1 ENDS).					Kali year.
Day and month, A.D.	Week-day.	Time of true Mēsha-saṁkrānti.			Day and month, A.D.	Week-day.	*a.*	*b.*	*c.*	
13	14	17			19	20	23	24	25	1
		H.	M.	S.						
23 Mar. (83)	0 Sat.	13	29	36	22 Mar. (82)	6 Fri.	9953·7057	485·8798	270·2896	4326
23 Mar. (82)	1 Sun.	19	41	45	11 Mar. (70)	3 Tues.	9829·4286	333·1238	239·4664	4327
24 Mar. (83)	3 Tues.	1	53	54	1 Mar. (60)	1 Sun.	43·7834	216·6596	211·3809	4328
24 Mar. (83)	4 Wed.	8	6	3	20 Mar. (79)	0 Sat.	78·4658	152·6531	262·6914	4329
23 Mar. (83)	5 Thur.	14	18	11	8 Mar. (68)	4 Wed.	9954·1886	999·8970	231·8682	4330
23 Mar. (82)	6 Fri.	20	30	20	26 Feb. (57)	2 Mon.	168·5434	883·4328	203·7827	4331
24 Mar. (83)	1 Sun.	2	42	29	17 Mar. (76)	1 Sun.	203·2258	819·4262	255·0931	4332
24 Mar. (83)	2 Mon.	8	54	38	6 Mar. (65)	5 Thur.	78·9487	666·6703	224·2699	4333
23 Mar. (83)	3 Tues.	15	6	47	23 Feb. (54)	2 Mon.	9954·6715	513·9144	193·4468	4334
23 Mar. (82)	4 Wed.	21	18	56	13 Mar. (72)	1 Sun.	9989·3539	449·9078	244·7571	4335
24 Mar. (83)	6 Fri.	3	31	4	2 Mar. (61)	5 Thur.	9865·0767	297·1519	213·9339	4336
24 Mar. (83)	0 Sat.	9	43	13	21 Mar. (80)	4 Wed.	9899·7592	233·1453	265·2439	4337
23 Mar. (83)	1 Sun.	15	55	22	9 Mar. (69)	1 Sun.	9775·4720	80·3894	234·4212	4338
23 Mar. (82)	2 Mon.	22	7	31	27 Feb. (58)	6 Fri.	9989·8369	963·9251	206·3357	4339
24 Mar. (83)	4 Wed.	4	19	40	18 Mar. (77)	5 Thur.	24·5192	899·9186	257·6462	4340
24 Mar. (83)	5 Thur.	10	31	49	8 Mar. (67)	3 Tues.	238·8741	783·4543	229·5607	4341
23 Mar. (83)	6 Fri.	16	43	57	25 Feb. (56)	0 Sat.	114·5968	630·6983	198·7375	4342
23 Mar. (82)	0 Sat.	22	56	6	15 Mar. (74)	6 Fri.	149·2792	566·6918	250·0479	4343
24 Mar. (83)	2 Mon.	5	18	5	4 Mar. (63)	3 Tues.	25·0021	413·9358	219·2248	4344
24 Mar. (83)	3 Tues.	11	20	24	23 Mar. (82)	2 Mon.	59·9845	349·9293	270·5351	4345
23 Mar. (83)	4 Wed.	17	32	33	11 Mar. (71)	6 Fri.	9935·4073	197·1733	239·7119	4346
23 Mar. (82)	5 Thur.	23	41	42	28 Feb. (59)	3 Tues.	9811·1302	44·4174	208·8887	4347
24 Mar. (83)	0 Sat.	5	56	51	19 Mar. (78)	2 Mon.	9845·8126	980·4109	260·1992	4348
24 Mar. (83)	1 Sun.	12	8	59	9 Mar. (68)	0 Sat.	60·1673	863·9465	232·1137	4349
23 Mar (83)	2 Mon.	18	21	8	27 Feb. (58)	5 Thur.	274·5222	747·4823	204·0282	4350

TABLE

CONCURRENT YEAR.								Intercalated and suppressed (*ksh.*) lunar months.
						Jovian Samvatsara.		
Kali.	Saka.	Chaitrādi Vikrama.	Mēshādi (solar) year in Bengal.	Kollam.	A.D.	Southern system.	Northern system.	
1	2	3	3a	4	5	6	7	8
4351	1172	1307	656	424-25	1249-50	43 Saumya .	48 *Ānanda* .	...
4352	1173	1308	657	425-26	1250-51	44 Sādhāraṇa .	49 *Rākshasa* .	8 Kārttika .
4353	1174	1309	658	426-27	1251-52	45 Virōdhakṛit .	50 Anala . .	...
4354	1175	1310	659	427-28	*1252-53	46 Paridhāvin .	51 Piṅgala . .	...
4355	1176	1311	660	428-29	1253-54	47 Pramādin .	52 Kālayukta .	5 Srāvaṇa .
4356	1177	1312	661	429-30	1254-55	48 Ānanda .	53 Siddhārthin .	...
4357	1178	1313	662	430-31	1255-56	49 Rākshasa .	54 Raudra .	...
4358	1179	1314	663	431-32	*1256-57	50 Anala .	55 Durmati .	4 Āshāḍha .
4359	1180	1315	664	432-33	1257-58	51 Piṅgala .	56 Dundubhi .	...
4360	1181	1316	665	433-34	1258-59	52 Kālayukta .	57 Rudhirōdgārin	...
4361	1182	1317	666	434-35	1259-60	53 Siddhārthin .	58 Raktāksha .	2 Vaiśākha .
4362	1183	1318	667	435-36	*1260-61	54 Raudra .	59 Krōdhana .	...
4363	1184	1319	668	436-37	1261-62	55 Durmati .	60 Kshaya .	6 Bhādrapada
4364	1185	1320	669	437-38	1262-63	56 Dundubhi .	1 Prabhava .	...
4365	1186	1321	670	438-39	1263-64	57 Rudhirōdgārin	2 Vibhava .	...
4366	1187	1322	671	439-40	*1264-65	58 Raktāksha .	3 Śukla .	4 Āshāḍha .
4367	1188	1323	672	440-41	1265-66	59 Krōdhana .	4 Pramōda .	...
4368	1189	1324	673	441-42	1266-67	60 Kshaya .	5 Prajāpati .	...
4369	1190	1325	674	442-43	1267-68	1 Prabhava .	6 Aṅgiras .	3 Jyēshtha
4370	1191	1326	675	443-44	*1268-69	2 Vibhava .	7 Śrīmukha .	...
4371	1192	1327	676	444-45	1269-70	3 Śukla .	8 Bhāva .	8 Kārttika .
4372	1193	1328	677	445-46	1270-71	4 Pramōda .	9 Yuvan . .	...
4373	1194	1329	678	446-47	1271-72	5 Prajāpati .	10 Dhātṛi . .	...
4374	1195	1330	679	447-48	*1272-73	6 Aṅgiras .	11 Īśvara . .	5 Srāvaṇa .
4375	1196	1331	680	448-49	1273-74	7 Śrīmukha .	12 Bahudhānya .	...

LX—*Contd.*

COMMENCEMENT OF THE									
SOLAR YEAR.			LUNI-SOLAR YEAR (MEAN SUNRISE OF DAY ON WHICH CHAITRA ŚUKLA 1 ENDS).						Kali year.
Day and month, A.D.	Week-day.	Time of true Mēsha-saṁkrānti.	Day and month, A.D.	Week-day.	*a.*	*b.*	*c.*		
13	14	17	19	20	23	24	25		1
		H. M. S.							
24 Mar. (83)	4 Wed.	0 33 17	17 Mar. (76)	4 Wed.	309·2046	683·4757	255·3387		4351
24 Mar. (83)	5 Thur.	6 45 26	6 Mar. (65)	1 Sun.	184·9274	530·7198	224·4769		4352
24 Mar. (83)	6 Fri.	12 57 35	24 Mar. (83)	6 Fri.	9880·9778	430·4577	273·0881		4353
23 Mar. (83)	0 Sat.	19 9 44	12 Mar. (72)	3 Tues.	9756·7007	277·6657	242·2263		4354
24 Mar. (83)	2 Mon.	1 21 52	2 Mar. (61)	1 Sun.	9971·0555	161·2014	214·1795		4355
24 Mar. (83)	3 Tues.	7 34 1	21 Mar. (80)	0 Sat.	5·7379	97·1948	265·4799		4356
24 Mar. (83)	4 Wed.	13 46 10	19 Mar. (69)	4 Wed.	9881·4607	944·4389	234·6667		4357
23 Mar. (83)	5 Thur.	19 58 19	28 Feb. (59)	2 Mon.	95·8156	827·9746	206·5812		4358
24 Mar. (83)	0 Sat.	2 10 28	18 Mar. (77)	1 Sun.	130·4880	763·9681	257·8917		4359
24 Mar. (83)	1 Sun.	8 22 37	7 Mar. (66)	5 Thur.	6·2208	611·2122	227·0685		4360
24 Mar. (83)	2 Mon.	14 34 45	24 Feb. (55)	2 Mon.	9881·9436	458·4562	196·2453		4361
23 Mar. (83)	3 Tues.	20 46 54	14 Mar. (74)	1 Sun.	9916·6261	394·4497	247·5556		4362
24 Mar. (83)	5 Thur.	2 59 3	3 Mar. (62)	5 Thur.	9792·3488	241·6938	216·7225		4363
24 Mar. (83)	6 Fri.	9 11 12	22 Mar. (81)	4 Wed.	9827·0312	177·6872	268·0439		4364
24 Mar. (83)	0 Sat.	15 23 21	12 Mar. (71)	2 Mon.	41·3861	61·2229	239·9575		4365
23 Mar. (83)	1 Sun.	21 35 30	29 Feb. (60)	6 Fri.	9917·1090	908·4669	209·1342		4366
24 Mar. (83)	3 Tues.	3 47 38	19 Mar. (78)	5 Thur.	9951·7913	844·4605	260·4447		4367
24 Mar. (83)	4 Wed.	9 59 47	9 Mar. (68)	3 Tues.	166·1461	727·9961	232·3593		4368
24 Mar. (83)	5 Thur.	16 11 56	26 Feb. (57)	0 Sat.	41·8690	575·2401	201·5360		4369
23 Mar. (83)	6 Fri.	22 24 5	16 Mar. (76)	6 Fri.	76·5513	511·2337	252·8464		4370
24 Mar. (83)	1 Sun.	4 36 14	5 Mar. (64)	3 Tues.	9952·2742	358·4777	222·0232		4371
24 Mar. (83)	2 Mon.	10 28 23	24 Mar. (83)	2 Mon.	9986·9566	294·4712	273·3337		4372
24 Mar. (83)	3 Tues.	17 0 32	13 Mar. (72)	6 Fri.	9862·6795	141·7152	242·5105		4373
23 Mar. (83)	4 Wed.	23 17 40	2 Mar. (62)	4 Wed.	77·0342	25·2509	214·4256		43[illegible]4
24 Mar. (83)	6 Fri.	5 24 49	21 Mar. (80)	3 Tues.	111·7167	931·2444	265·7354		[illegible]

TABLE

CONCURRENT YEAR.								Intercalated and suppressed (*ksh.*) lunar months.
						Jovian Samvatsara.		
Kali.	Saka.	Chaitrādi Vikrama.	Mēshādi (solar) year in Bengal.	Kollam.	A. D.	Southern system.	Northern system.	
1	2	3	3a	4	5	6	7	8
4376	1197	1332	681	449-50	1274-75	8 Bhāva . .	13 Pramāthin .	...
4377	1198	1333	682	450-51	1275-76	9 Yuvan . .	14 Vikrama .	4 Āshāḍha
4378	1199	1334	683	451-52	*1276-77	10 Dhātṛi . .	15 Vṛisha . .	...
4379	1200	1335	684	452-53	1277-78	11 Īśvara . .	16 Chitrabhānu .	...
4380	1201	1336	685	453-54	1278-79	12 Bahudhānya .	17 Subhānu .	2 Vaiśākha .
4381	1202	1337	686	454-55	1279-80	13 Pramāthin .	18 Tāraṇa . .	...
4382	1203	1338	687	455-56	*1280-81	14 Vikrama .	19 Pārthiva .	6 Bhādrapada
4383	1204	1339	688	456-57	1281-82	15 Vṛisha . .	20 Vyaya . .	...
4384	1205	1340	689	457-58	1282-83	16 Chitrabhānu	21 Sarvajit .	...
4385	1206	1341	690	458-59	1283-84	17 Subhānu .	22 Sarvadhārin .	4 Āshāḍha
4386	1207	1342	691	459-60	*1284-85	18 Tāraṇa . .	23 Virōdhin .	...
4387	1208	1343	692	460-61	1285-86	19 Pārthiva .	24 Vikṛita . .	...
4388	1209	1344	693	461-62	1286-87	20 Vyaya .	25 Khara . .	3 Jyēshṭha
4389	1210	1345	694	462-63	1287-88	21 Sarvajit .	26 Nandana .	...
4390	1211	1346	695	463-64	*1288-89	22 Sarvadhārin .	27 Vijaya . .	8 Kārttika .
4391	1212	1347	696	464-65	1289-90	23 Virōdhin .	28 Jaya . .	...
4392	1213	1348	697	465-66	1290-91	24 Vikṛita . .	29 Manmatha .	...
4393	1214	1349	698	466-67	1291-92	25 Khara . .	30 Durmukha .	5 Śrāvaṇa .
4394	1215	1350	699	467-68	*1292-93	26 Nandana .	31 Hēmalamba .	...
4395	1216	1351	700	468-69	1293-94	27 Vijaya . .	32 Vilamba .	...
4396	1217	1352	701	469-70	1294-95	28 Jaya . .	33 Vikārin . .	4 Āshāḍha .
4397	1218	1353	702	470-71	1295-96	29 Manmatha .	34 Śārvarin .	...
4398	1219	1354	703	471-72	*1296-97	30 Durmukha .	35 Plava . .	...
4399	1220	1355	704	472-73	1297-98	31 Hēmalamba .	36 Subhakṛit .	2 Vaiśākha
4400	1221	1356	705	473-74	1298-99	32 Vilamba .	37 Sōbhana .	...

LX—*Contd.*

COMMENCEMENT OF THE											
SOLAR YEAR.					LUNI-SOLAR YEAR (MEAN SUNRISE OF DAY ON WHICH CHAITRA ŚUKLA 1 ENDS).					Kali year.
Day and month, A.D.	Week-day.	Time of true Mésha-samkrânti.			Day and month, A.D.	Week-day.	*a.*	*b.*	*c.*	
13	14	17			19	20	23	24	25	1
		H.	M.	S.						
24 Mar. (83)	0 Sat.	11	36	58	10 Mar. (69)	0 Sat.	9987·4395	808·4884	234·9123	4376
24 Mar. (83)	1 Sun.	17	49	7	28 Feb. (59)	5 Thur.	201·7943	622·0241	206·8266	4377
24 Mar. (84)	3 Tues.	0	1	16	18 Mar. (78)	4 Wed.	236·4767	628·0176	258·1372	4378
24 Mar. (83)	4 Wed.	6	13	25	7 Mar. (66)	1 Sun.	112·1996	475·2617	227·3140	4379
24 Mar. (83)	5 Thur.	12	25	33	24 Feb. (55)	5 Thur.	9987·9224	322·5057	196·4909	4380
24 Mar. (83)	6 Fri.	18	37	42	15 Mar. (74)	4 Wed.	22·6048	258·4092	247·8012	4381
24 Mar. (84)	1 Sun.	0	49	51	3 Mar. (63)	1 Sun.	9898·3276	105·7433	216·9780	4382
24 Mar. (83)	2 Mon.	7	2	0	22 Mar. (81)	0 Sat.	9933·0100	41·7367	268·2884	4383
24 Mar. (83)	3 Tues.	13	14	9	12 Mar. (71)	5 Thur.	147·3648	925·2684	240·2031	4384
24 Mar. (83)	4 Wed.	19	26	18	1 Mar. (60)	2 Mon.	23·0877	772·5164	209·3798	4385
24 Mar. (84)	6 Fri.	1	38	26	19 Mar. (79)	1 Sun.	57·7700	707·5099	260·6902	4386
24 Mar. (83)	0 Sat.	7	50	35	8 Mar. (67)	5 Thur.	9933·4930	555·7540	229·8670	4387
24 Mar. (83)	1 Sun.	14	2	44	25 Feb. (56)	2 Mon.	9809·2157	402·9980	199·0438	4388
24 Mar. (83)	2 Mon.	20	14	53	16 Mar. (75)	1 Sun.	9843·8981	338·9914	250·4042	4389
24 Mar. (84)	4 Wed.	2	27	2	4 Mar. (64)	5 Thur.	9719·6210	186·2355	219·5310	4390
24 Mar. (83)	5 Thur.	8	39	11	23 Mar. (82)	4 Wed.	9754·3934	122·2308	270·8414	4391
24 Mar. (83)	6 Fri.	14	51	19	13 Mar. (72)	2 Mon.	9968·6582	5·7647	242·7560	4392
24 Mar. (83)	0 Sat.	21	3	28	2 Mar. (62)	0 Sat.	183·0130	889·3004	214·6706	4393
24 Mar. (84)	2 Mon.	3	15	37	21 Mar. (81)	6 Fri.	217·6855	825·2939	265·9809	4394
24 Mar. (83)	3 Tues.	9	27	46	10 Mar. (69)	3 Tues.	93·4182	672·5380	235·1578	4395
24 Mar. (83)	4 Wed.	15	39	55	27 Feb. (58)	0 Sat.	9969·1412	519·7820	204·3346	4396
24 Mar. (83)	5 Thur.	21	52	4	18 Mar. (77)	6 Fri.	3·8235	455·7754	255·6450	4397
24 Mar. (84)	0 Sat.	4	4	12	6 Mar. (66)	3 Tues.	9879·5463	303·0195	224·8217	4398
24 Mar. (83)	1 Sun.	10	16	21	23 Feb. (54)	0 Sat.	9755·2691	150·2635	193·9986	4399
24 Mar. (83)	2 Mon.	16	28	30	14 Mar. (73)	6 Fri.	9789·9516	86·2571	245·7090	4400

TABLE

CONCURRENT YEAR.								Intercalated and suppressed (*ksh.*) lunar months.
						Jovian Samvatsara.		
Kali.	Śaka.	Chaitrādi Vikrama.	Mēshādi (solar) year in Bengal.	Kollam.	A.D.	Southern system.	Northern system.	
1	2	3	3a	4	5	6	7	8
4401	1222	1357	706	474-75	1299-1300	33 Vikārin	38 Krōdhin	6 Bhādrapada
4402	1223	1358	707	475-76	*1300-01	34 Śārvarin	39 Viśvāvasu	...
4403	1224	1359	708	476-77	1301-02	35 Plava	40 Parābhava	...
4404	1225	1360	709	477-78	1302-03	36 Śubhakṛit	41 Plavaṅga	4 Āshāḍha
4405	1226	1361	710	478-79	1303-04	37 Śōbhana	42 Kīlaka	...
4406	1227	1362	711	479-80	*1304-05	38 Krōdhin	43 Saumya	...
4407	1228	1363	712	480-81	1305-06	39 Viśvāvasu	44 Sādhārana	3 Jyēshṭha
4408	1229	1364	713	481-82	1306-07	40 Parābhava	45 Virōdhakṛit	...
4409	1230	1365	714	482-83	1307-08	41 Plavaṅga	46 Paridhāvin	{ 7 Āśvina 11 *Māgha(ksh).* 12 Phālguna
4410	1231	1366	715	483-84	*1308-09	42 Kīlaka	47 Pramādin	...
4411	1232	1367	716	484-85	1309-10	43 Saumya	48 Ānanda	...
4412	1233	1368	717	485-86	1310-11	44 Sādhāraṇa	49 Rākshasa	5 Śrāvaṇa
4413	1234	1369	718	486-87	1311-12	45 Virōdhakṛit	50 Anala	...
4414	1235	1370	719	487-88	*1312-13	46 Paridhāvin	51 Piṅgala	...
4415	1236	1371	720	488-89	1313-14	47 Pramādin	52 Kālayukta	4 Āshāḍha
4416	1237	1372	721	489-90	1314-15	48 Ānanda	53 Siddhārthin	...
4417	1238	1373	722	490-91	1315-16	49 Rākshasa	54 Raudra	...
4418	1239	1274	723	491-92	*1316-17	50 Anala	55 Durmati	1 Chaitra†
4419	1240	1375	724	492-93	1317-18	51 Piṅgala	56 Dundubhi	...
4420	1241	1376	725	493-94	1318-19	52 Kālayukta	57 Rudhirōdgārin	6 Bhādrapada
4421	1242	1377	726	494-95	1319-20	53 Siddhārthin	58 Raktāksha	...
4422	1243	1378	727	495-96	*1320-21	54 Raudra	59 Krōdhana	...
4423	1244	1379	728	496-97	1321-22	55 Durmati	60 Kshaya	4 Āshāḍha
4424	1245	1380	729	497-98	1322-23	56 Dundubhi	1 Prabhava	...
4425	1246	1381	730	498-99	1323-24	57 Rudhirōdgārin	2 Vibhava	...

† See Remarks, p. 163 above.

LX—*Contd.*

COMMENCEMENT OF THE											
SOLAR YEAR.					LUNI-SOLAR YEAR (MEAN SUNRISE OF DAY ON WHICH CHAITRA ŚUKLA 1 ENDS).						Kali year.
Day and month, A.D.	Week-day.	Time of true Mēsha-saṁkrānti.			Day and month, A.D.	Week-day.	*a*.	*b*.	*c*.		
13	14	17			19	20	23	24	25		1
		H.	M.	S.							
24 Mar. (83)	3 Tues.	22	40	39	4 Mar. (63)	4 Wed.	4·3064	969·7928	217·1430		4401
24 Mar. (84)	5 Thur.	4	52	48	22 Mar. (82)	3 Tues.	38·9888	905·7863	263·4534		4402
24 Mar. (83)	6 Fri.	11	4	57	12 Mar. (71)	1 Sun.	253·3437	789·3219	240·3680		4403
24 Mar. (83)	0 Sat.	17	17	6	1 Mar. (60)	5 Thur.	129·0665	636·5660	209·5447		4404
24 Mar. (83)	1 Sun.	23	19	14	20 Mar. (79)	4 Wed.	163·7489	572·5594	260·8552		4405
24 Mar. (84)	3 Tues.	5	41	23	8 Mar. (68)	1 Sun.	39·4718	419·8035	230·0320		4406
24 Mar. (83)	4 Wed.	11	53	32	25 Feb. (56)	5 Thur.	9915·1945	267·0476	199·2089		4407
24 Mar. (83)	5 Thur.	18	5	41	16 Mar. (75)	4 Wed.	9949·8769	203·0410	250·5181		4408
25 Mar. (84)	0 Sat.	0	17	50	5 Mar. (64)	1 Sun.	9825·5998	50·2851	219·6960		4409
24 Mar. (84)	1 Sun.	6	26	59	23 Mar. (83)	0 Sat.	9860·2821	986·2785	271·0064		4410
24 Mar. (83)	2 Mon.	12	42	7	13 Mar. (72)	5 Thur.	74·6370	869·8142	242·9209		4411
24 Mar. (83)	3 Tues.	18	54	16	3 Mar. (62)	3 Tues.	288·9918	753·3499	215·8355		4412
25 Mar. (84)	5 Thur.	1	6	25	21 Mar. (80)	1 Sun.	9985·0423	653·0518	263·4082		4413
24 Mar. (84)	6 Fri.	7	18	34	10 Mar. (70)	6 Fri.	199·3970	536·5875	235·3128		4414
24 Mar. (83)	0 Sat.	13	30	43	27 Feb. (58)	3 Tues.	75·1199	383·8315	204·4995		4415
24 Mar. (83)	1 Sun.	19	42	52	17 Mar. (76)	1 Sun.	9771·1703	283·5334	253·0721		4416
25 Mar. (84)	3 Tues.	1	55	0	7 Mar. (66)	6 Fri.	9985·5251	167·0780	224·9867		4417
24 Mar. (84)	4 Wed.	8	7	9	24 Feb. (55)	3 Tues.	9861·2476	14·3131	194·1636		4418
24 Mar. (83)	5 Thur.	14	19	18	14 Mar. (73)	2 Mon.	9895·9304	950·3066	245·4739		4419
24 Mar. (83)	6 Fri.	20	31	27	4 Mar. (63)	0 Sat.	110·2852	833·8423	217·2885		4420
25 Mar. (84)	1 Sun.	2	43	36	23 Mar. (82)	6 Fri.	144·9675	769·8358	268·6989		4421
24 Mar. (84)	2 Mon.	8	55	45	11 Mar. (71)	3 Tues.	20·7024	617·7098	237·8758		4422
24 Mar. (83)	3 Tues.	15	7	54	28 Feb. (59)	0 Sat.	9896·3133	464·3239	207·0525		4423
24 Mar. (83)	4 Wed.	21	20	2	19 Mar. (78)	6 Fri.	9931·0956	500·3174	258·3619		4424
25 Mar. (84)	6 Fri.	3	12	11	8 Mar. (67)	3 Tues.	9806·8185	247·5614	227·5397		4425

TABLE

CONCURRENT YEAR.								Intercalated and suppressed (*ksh.*) lunar months.
						Jovian Samvatsara.		
Kali.	Saka.	Chaitrādi Vikrama.	Mēshādi (solar) year in Bengal.	Kollam.	A.D.	Southern system.	Northern system.	
1	2	3	3*a*	4	5	6	7	8
4426	1247	1382	731	499-500	*1324-25	58 Raktāksha .	3 Śukla . .	2 Vaiśākha .
4427	1248	1383	732	500-01	1325-26	59 Krōdhana .	4 Pramōda .	...
4428	1249	1384	733	501-02	1326-27	60 Kshaya .	5 Prajāpati .	6 Bhādrapada
4429	1250	1385	734	502-03	1327-28	1 Prabhava .	6 Aṅgiras .	...
4430	1251	1386	735	503-04	*1328-29	2 Vibhava .	7 Śrīmukha .	...
4431	1252	1387	736	504-05	1329-30	3 Śukla . .	8 Bhāva . .	5 Śrāvaṇa .
4432	1253	1388	737	505-06	1330-31	4 Pramōda .	9 Yuvan† .	...
4433	1254	1389	738	506-07	1331-32	5 Prajāpati .	11 *Īśvara* . .	...
4434	1255	1390	739	507-08	*1332-33	6 Aṅgiras .	12 *Bahudhānya* .	3 Jyēshṭha .
4435	1256	1391	740	508-09	1333-34	7 Śrimukha .	13 *Pramāthin* .	...
4436	1257	1392	741	509-10	1334-35	8 Bhāva . .	14 *Vikrama* .	...
4437	1258	1393	742	510-11	1335-36	9 Yuvan . .	15 *Vṛisha* . .	2 Vaiśākha :
4438	1259	1394	743	511-12	*1336-37	10 Dhātri . .	16 Chitrabhānu .	...
4439	1260	1395	744	512-13	1337-38	11 Īśvara . .	17 Subhānu .	6 Bhādrapada
4440	1261	1396	745	513-14	1338-39	12 Bahudhānya .	18 Tāraṇa . .	...
4441	1262	1397	746	514-15	1339-40	13 Pramāthin .	19 Pārthiva .	...
4442	1263	1398	747	515-16	*1340-41	14 Vikrama .	20 Vyaya . .	4 Āshāḍha .
4443	1264	1399	748	516-17	1341-42	15 Vṛisha . .	21 Sarvajit . .	...
4444	1265	1400	749	517-18	1342-43	16 Chitrabhānu .	22 Sarvadhārin .	...
4445	1266	1401	750	518-19	1343-44	17 Subhānu .	23 Virōdhin .	2 Vaiśākha .
4446	1267	1402	751	519-20	*1344-45	18 Tāraṇa . .	24 Vikṛita . .	...
4447	1268	1403	752	520-21	1345-46	19 Pārthiva .	25 Khara . .	6 Bhādrapada
4448	1269	1404	753	521-22	1346-47	20 Vyaya . .	26 Nandana .	...
4449	1270	1405	754	522-23	1347-48	21 Sarvajit .	27 Vijaya . .	...
4450	1271	1406	755	523-24	*1348-49	22 Sarvadhārin .	28 Jaya . .	5 Śrāvaṇa .

† 10 Dhātṛi was suppressed in the north.

LX—*Contd.*

COMMENCEMENT OF THE										
Solar year.					Luni-solar year (mean sunrise of day on which Chaitra śukla 1 ends).					Kali year.
Day and month, A.D.	Week-day.	Time of true Mēsha-saṁkrānti.			Day and month, A.D.	Week-day.	*a.*	*b.*	*c.*	
13	14	17			19	20	23	24	25	1
		H.	M.	S.						
24 Mar. (84)	0 Sat.	9	44	20	26 Feb. (57)	1 Sun.	21·1733	131·0971	199·4543	4426
24 Mar. (83)	1 Sun.	15	56	29	16 Mar. (75)	0 Sat.	58·8557	67·0905	250·7647	4427
24 Mar. (83)	2 Mon.	22	8	38	5 Mar. (64)	4 Wed.	9931·5785	914·3346	219·9415	4428
25 Mar. (84)	4 Wed.	4	20	47	24 Mar. (83)	3 Tues.	9966·2609	850·3281	271·2519	4429
24 Mar. (84)	5 Thur.	10	32	55	13 Mar. (73)	1 Sun.	180·6158	733·8637	243·1665	4430
24 Mar. (83)	6 Fri.	16	45	4	2 Mar. (61)	5 Thur.	56·3286	581·1079	212·3433	4431
24 Mar. (83)	0 Sat.	22	57	13	21 Mar. (80)	4 Wed.	91·0210	517·1013	263·7537	4432
25 Mar. (84)	2 Mon.	5	9	22	10 Mar (69)	1 Sun.	9966·7438	364·3453	232·8305	4433
24 Mar. (84)	3 Tues.	11	21	31	27 Feb. (58)	5 Thur.	9842·4667	211·5894	202·0073	4434
24 Mar. (83)	4 Wed.	17	33	40	17 Mar. (76)	4 Wed.	9877·1490	147·5829	253·3177	4435
24 Mar. (83)	5 Thur.	23	45	48	7 Mar. (66)	2 Mon.	91·5129	31·1186	225·2422	4436
25 Mar. (84)	0 Sat.	5	57	57	24 Feb. (55)	6 Fri.	9967·2267	878·3626	194·4091	4437
24 Mar. (84)	1 Sun.	12	10	6	14 Mar. (74)	5 Thur	·8992	814·3561	245·7195	4438
24 Mar. (83)	2 Mon.	18	22	15	4 Mar. (63)	3 Tues	216·2639	697·8918	217·5941	4439
25 Mar. (84)	4 Wed.	0	34	24	23 Mar. (82)	2 Mon.	250·9463	634·8853	268·9445	4440
25 Mar. (84)	5 Thur.	6	46	33	12 Mar. (71)	6 Fri.	126·6692	481·1293	238·1213	4441
24 Mar. (84)	6 Fri.	12	58	42	29 Feb. (60)	3 Tues.	2·3920	328·3733	207·2981	4442
24 Mar. (83)	0 Sat.	19	10	50	19 Mar. (78)	2 Mon.	37·0744	264·3669	258·6085	4443
25 Mar. (84)	2 Mon.	1	22	59	8 Mar. (67)	6 Fri.	9912·7973	111·6109	227·7853	4444
25 Mar. (84)	3 Tues.	7	35	8	26 Feb. (57)	4 Wed.	127·1521	995·1466	199·6995	4445
24 Mar. (84)	4 Wed.	13	47	17	16 Mar. (76)	3 Tues.	161·8344	931·1400	251·0102	4446
24 Mar. (83)	5 Thur.	19	59	26	5 Mar. (64)	0 Sat.	37·5573	778·3841	220·1871	4447
25 Mar. (84)	0 Sat.	2	11	35	24 Mar. (83)	6 Fri.	72·2397	714·3776	271·4975	4448
25 Mar. (84)	1 Sun.	8	23	43	13 Mar. (72)	3 Tues.	9947·9625	561·6216	240·6743	4449
24 Mar. (84)	2 Mon.	14	35	52	1 Mar. (61)	0 Sat.	9823·6854	408·8657	209·8510	4450

TABLE

CONCURRENT YEAR.								Intercalated and suppressed (*ksh.*) lunar months.
						Jovian Samvatsara.		
Kali.	Śaka.	Chaitrādi Vikrama.	Mēshādi (solar) year in Bengal	Kollam.	A.D.	Southern system.	Northern system.	
1	2	3	3a	4	5	6	7	8
4451	1272	1407	756	524-25	1349-50	23 Virōdhin .	29 Manmatha .	...
4452	1273	1408	757	525-26	1350-51	24 Vikṛita . .	30 Durmukha	...
4453	1274	1409	758	526-27	1351-52	25 Khara . .	31 Hēmalamba .	3 Jyēshṭha
4454	1275	1410	759	527-28	*1352-53	26 Nandana .	32 Vilamba .	...
4455	1276	1411	760	528-29	1353-54	27 Vijaya . .	33 Vikārin	{ 8 Kārttika } { 9 *Mārgaś (ksh.)* }
4456	1277	1412	761	529-30	1354-55	28 Jaya . .	34 Śārvarin .	2 Vaiśākha .
4457	1278	1413	762	530-31	1355-56	29 Manmatha .	35 Plava . .	...
4458	1279	1414	763	531-32	*1356-57	30 Durmukha .	36 Śubhakṛit .	6 Bhādrapada .
4459	1280	1415	764	532-33	1357-58	31 Hēmalamba .	37 Śōbhana	...
4460	1281	1416	765	533-34	1358-59	32 Vilamba .	38 Krōdhin	...
4461	1282	1417	766	534-35	1359-60	33 Vikārin .	39 Viśvāvasu .	4 Āshāḍha .
4462	1283	1418	767	535-36	*1360-61	34 Śārvarin .	40 Parābhava .	...
4463	1284	1419	768	536-37	1361-62	35 Plava . .	41 Plavaṅga .	...
4464	1285	1420	769	537-38	1362-63	36 Śubhakṛit .	42 Kīlaka .	2 Vaiśākha .
4465	1286	1421	770	538-39	1363-64	37 Śōbhana .	43 Saumya .	...
4466	1287	1422	771	539-40	*1364-65	38 Krōdhin .	44 Sādhāraṇa .	6 Bhādrapada .
4467	1288	1423	772	540-41	1365-66	39 Viśvāvasu .	45 Virōdhakṛit .	...
4468	1289	1424	773	541-42	1366-67	40 Parābhava .	46 Paridhāvin .	...
4469	1290	1425	774	542-43	1367-68	41 Plavaṅga .	47 Pramādin .	5 Śrāvaṇa .
4470	1291	1426	775	543-44	*1368-69	42 Kīlaka . .	48 Ānanda .	...
4471	1292	1427	776	544-45	1369-70	43 Saumya .	49 Rākshasa .	...
4472	1293	1428	777	545-46	1370-71	44 Sādhāraṇa .	50 Anala .	3 Jyēshṭha .
4473	1294	1429	778	546-47	1371-72	45 Virōdhakṛit .	51 Piṅgala .	...
4474	1295	1430	779	547-48	*1372-73	46 Paridhāvin .	52 Kālayukta* .	{ 7 Āśvina } { 10 *Pausha (ksh.)* }
4475	1296	1431	780	548-49	1373-74	47 Pramādin .	53 Siddhārthin .	1 Chaitra .

LX—*Contd.*

COMMENCEMENT OF THE											
SOLAR YEAR.					LUNI-SOLAR YEAR (MEAN SUNRISE OF DAY ON WHICH CHAITRA ŚUKLA 1 ENDS).					Kali year.
Day and month, A.D.	Week-day.	Time of true Mēsha-samkrānti			Day and month, A.D.	Week-day.	*a.*	*b.*	*c.*	
13	14	17			19	20	21	24	25	1
		H.	M.	S.						
24 Mar. (83)	3 Tues. .	20	48	1	20 Mar. (79)	6 Fri. .	9858·3678	344·8591	261·1615	4451
25 Mar. (84)	5 Thur.	3	0	10	9 Mar. (68)	3 Tues. .	9734·0906	192·0932	230·3383	4452
25 Mar. (84)	6 Fri. .	9	12	19	27 Feb. (58)	1 Sun. .	9948·4454	75·6749	202·2528	4453
24 Mar. (84)	0 Sat. .	15	24	28	17 Mar. (77)	0 Sat. .	9983·1278	11·6324	253·5632	4454
24 Mar. (83)	1 Sun. .	21	36	36	7 Mar. (66)	5 Thur.	197·4827	895·1681	225·4778	4455
25 Mar. (84)	3 Tues. .	3	48	45	24 Feb. (55)	2 Mon. .	73·2054	742·4122	194·6547	4456
25 Mar. (84)	4 Wed. .	10	0	54	15 Mar. (74)	1 Sun .	107·8879	678·4056	245·9650	4457
24 Mar. (84)	5 Thur.	16	13	3	3 Mar. (63)	5 Thur	9983·6107	526·6596	215·1418	4458
24 Mar. (83)	6 Fri. .	22	25	12	22 Mar. (81)	4 Wed. .	18·2932	461·6431	266·4522	4459
25 Mar. (84)	1 Sun. .	4	37	21	11 Mar. (70)	1 Sun. .	9894·0159	309·8872	235·6291	4460
25 Mar. (84)	2 Mon. .	10	49	29	28 Feb. (59)	5 Thur.	9769·7388	156·1313	204·8058	4461
24 Mar. (84)	3 Tues. .	17	1	38	18 Mar. (78)	4 Wed. .	9804·4212	92·1247	256·1162	4462
24 Mar. (83)	4 Wed. .	23	13	47	8 Mar. (67)	2 Mon. .	18·7760	975·6605	228·0308	4463
25 Mar. (84)	6 Fri. .	5	25	56	26 Feb. (57)	0 Sat. .	233·1308	859·1961	199·9454	4464
25 Mar. (84)	0 Sat. .	11	38	5	17 Mar. (76)	6 Fri. .	267·8132	795·1896	251·2558	4465
24 Mar. (84)	1 Sun. .	17	50	14	5 Mar. (65)	3 Tues. .	143·5361	642·4536	220·4326	4466
25 Mar. (84)	3 Tues. .	0	2	23	24 Mar. (83)	2 Mon. .	178·2184	578·4271	271·7430	4467
25 Mar. (84)	4 Wed. .	6	14	31	13 Mar. (72)	6 Fri. .	53·9413	425·6712	249·9199	4468
25 Mar. (84)	5 Thur.	12	26	40	2 Mar. (61)	3 Tues. .	9929·6642	272·9152	210·0966	4469
24 Mar. (84)	6 Fri. .	18	38	49	20 Mar. (80)	2 Mon. .	9964·3465	208·9087	261·4070	4470
25 Mar. (84)	1 Sun. .	0	50	58	9 Mar. (68)	6 Fri. .	9840·0694	56·1527	230·5838	4471
25 Mar. (84)	2 Mon. .	7	3	7	27 Feb. (58)	4 Wed.	54·4242	939·6884	202·4984	4472
25 Mar. (84)	3 Tues. .	13	15	16	18 Mar. (77)	3 Tues. .	89·1066	875·6819	253·8088	4473
24 Mar. (84)	4 Wed. .	19	27	24	7 Mar. (67)	1 Sun. .	303·4614	759·2176	225·7233	4474
25 Mar. (84)	6 Fri. .	1	39	33	24 Feb. (55)	5 Thur.	179·1842	606·4617	194·9062	4475

TABLE

CONCURRENT YEAR.								
Kali.	Śaka.	Chaitrādi Vikrama.	Mēshādi (solar) year in Bengal.	Kollam.	A.D.	Jovian Samvatsara. Southern system.	Jovian Samvatsara. Northern system.	Intercalated and suppressed (*ksh.*) lunar months.
1	2	3	3*a*	4	5	6	7	8
4476	1297	1432	781	549-50	1374-75	48 Ānanda .	54 Raudra .	...
4477	1298	1433	782	550-51	1375-76	49 Rākshasa .	55 Durmati .	6 Bhādrapada
4478	1299	1434	783	551-52	*1376-77	50 Anala † . .	56 Dundubhi .	...
4479	1300	1435	784	552-53	1377-78	51 Piṅgala .	57 Rudhirōdgārin	...
4480	1301	1436	785	553-54	1378-79	52 Kālayukta .	58 Raktāksha .	4 Āshāḍha .
4481	1302	1437	786	554-55	1379-80	53 Siddhārthin .	59 Krōdhana .	...
4482	1303	1438	787	555-56	*1380-81	54 Raudra .	60 Kshaya .	...
4483	1304	1439	788	556-57	1381-82	55 Durmati .	1 Prabhava .	2 Vaiśākha .
4484	1305	1440	789	557-58	1382-83	56 Dundubhi .	2 Vibhava	...
4485	1306	1441	790	558-59	1383-84	57 Rudhirōdgārin	3 Śukla . .	6 Bhādrapada
4486	1307	1442	791	559-60	*1384-85	58 Raktāksha .	4 Pramōda .	...
4487	1308	1443	792	560-61	1385-86	69 Krōdhana .	5 Prajāpati .	...
4488	1309	1444	793	561-62	1386-87	60 Kshaya .	6 Aṅgiras .	4 Āshāḍha .
4489	1310	1445	794	562-63	1387-88	1 Prabhava .	7 Śrīmukha .	...
4490	1311	1446	795	563-64	*1388-89	2 Vibhava .	8 Bhāva .	...
4491	1312	1447	796	564-65	1389-90	3 Śukla . .	9 Yuvan .	3 Jyēshṭha .
4492	1313	1448	797	565-66	1390-91	4 Pramōda .	10 Dhātṛi .	...
4493	1314	1449	798	566-67	1391-92	5 Prajāpati .	11 Īśvara . .	7 Āśvina. .
4494	1315	1450	799	567-68	*1392-93	6 Aṅgiras .	12 Bahudhānya .	...
4495	1316	1451	800	568-69	1393-94	7 Śrīmukha .	13 Pramāthin .	...
4496	1317	1452	801	569-70	1394-95	8 Bhāva . .	14 Vikrama .	5 Śrāvaṇa .
4497	1318	1453	802	570-71	1395-96	9 Yuvan . .	15 Vṛisha .	...
4498	1319	1454	803	571-72	*1396-97	10 Dhātṛi . .	16 Chitrabhānu .	...
4499	1320	1455	804	572-73	1397-98	11 Īśvara . .	17 Subhānu .	4 Āshāḍha .
4500	1321	1456	805	573-74	1398-99	12 Bahudhānya .	18 Tāraṇa . .	...

† The moment of new moon was 15 hours 26 minutes before mean sunrise on 25th March, which was sunrise. The case is peculiar, since in general all days

LX—*Contd.*

COMMENCEMENT OF THE											
SOLAR YEAR					LUNI-SOLAR YEAR (MEAN SUNRISE OF DAY ON WHICH CHAITRA ŚUKLA 1 ENDS).						Kali year.
Day and month, A.D.	Week-day.	Time of true Mēsha-saṁkrānti.			Day and month, A.D.	Week-day.	*a.*	*b.*	*c.*		
13	14	17			19	20	23	24	25		1
		H.	M.	S.							
25 Mar. (84)	0 Sat.	7	51	42	15 Mar. (74)	4 Wed.	215·8667	542·4551	246·2106		4476
25 Mar. (84)	1 Sun.	14	3	51	4 Mar. (63)	1 Sun.	89·6894	[illegible]39·6991	215·3874		4477
24 Mar. (84)	2 Mon.	20	16	0	21 Mar. (81)	6 Fri.	9785·6399	288·4010	263·9600		4478
25 Mar. (84)	4 Wed.	2	28	9	11 Mar. (70)	4 Wed.	9999·9947	172·9367	235·8746		4479
25 Mar. (84)	5 Thur.	8	40	17	28 Feb. (59)	1 Sun.	9875·7176	20·1808	205·0514		4480
25 Mar. (84)	6 Fri.	14	52	26	19 Mar. (78)	0 Sat.	9910·3999	956·1742	256·3618		4481
24 Mar. (84)	0 Sat.	21	4	35	8 Mar. (68)	5 Thur.	124·7548	839·7100	[illegible]28·2763		4482
25 Mar. (84)	2 Mon.	3	16	44	25 Feb. (56)	2 Mon.	0·4776	686·9539	197·4532		4483
25 Mar. (84)	3 Tues.	9	28	53	16 Mar. (75)	1 Sun.	35·1599	622·9434	24[illegible]·7636		4484
25 Mar. (84)	4 Wed.	15	41	2	5 Mar. (64)	5 Thur.	9910·8828	470·1915	217·9404		4485
24 Mar. (84)	5 Thur.	21	53	10	23 Mar. (83)	4 Wed.	9945·5651	406·1850	269·2507		4486
25 Mar. (84)	0 Sat.	4	5	19	12 Mar. (71)	1 Sun.	9821·2881	253·4290	238·4276		4487
25 Mar. (84)	1 Sun.	10	17	28	2 Mar. (61)	6 Fri.	35·6429	136·9647	210·3422		4488
25 Mar. (84)	2 Mon.	16	29	37	21 Mar. (80)	5 Thur.	70·3253	72·9581	261·6526		4489
24 Mar. (84)	3 Tues.	22	41	46	9 Mar. (69)	2 Mon.	9946·0482	920·2004	230·8293		4490
25 Mar. (84)	5 Thur.	4	53	55	27 Feb. (58)	0 Sat.	160·4030	803·7379	202·7439		4491
25 Mar. (84)	6 Fri.	11	6	4	18 Mar. (77)	6 Fri.	195·0853	739·7314	[illegible]54·0544		4492
25 Mar. (84)	0 Sat.	17	18	12	7 Mar. (66)	3 Tues.	70·8082	586·9755	223·2311		4493
24 Mar. (84)	1 Sun.	23	30	21	25 Mar. (85) †	2 Mon.	105·4906	522·9690	274·5415		4494
25 Mar. (84)	3 Tues.	5	42	30	14 Mar. (73)	6 Fri.	9981·2134	370·2130	243·7183		4495
25 Mar. (84)	4 Wed.	11	54	39	3 Mar. (62)	3 Tues.	9856·9362	217·4570	212·8952		4496
25 Mar. (84)	5 Thur.	18	6	48	22 Mar. (81)	2 Mon.	9891·6187	153·4505	264·2056		4497
25 Mar. (85)	0 Sat.	0	18	57	11 Mar. (71)	0 Sat.	105·9734	36·9862	236·1201		4498
25 Mar. (84)	1 Sun.	6	31	5	28 Feb. (59)	4 Wed.	9981·6963	884·2303	205·2969		4499
25 Mar. (84)	2 Mon.	12	43	14	19 Mar. (78)	3 Tues.	16·3787	820·2228	256·6074		4500

therefore, the day "Chaitra śukla 1." The moment of true Mēsha-saṁkrānti was 30 minutes before that
in column 19 are earlier than those in column 12.

TABLE

CONCURRENT YEAR.								Intercalated and suppressed (*ksh.*) lunar months.
						Jovian Samvatsara.		
Kali.	Saka.	Chaitrādi Vikrama.	Mēshādi (solar) year in Bengal.	Kollam.	A.D.	Southern system.	Northern system.	
1	2	3	3a	4	5	6	7	8
4501	1322	1457	806	574-75	1399-00	13 Pramāthin .	19 Pārthiva .	...
4502	1323	1458	807	575-76	*1400-01	14 Vikrama .	20 Vyaya .	2 Vaiśākha .
4503	1324	1459	808	576-77	1401-02	15 Vṛisha .	21 Sarvajit .	...
4504	1325	1460	809	577-78	1402-03	16 Chitrabhānu .	22 Sarvadhārin .	6 Bhādrapada .
4505	1326	1461	810	578-79	1403-04	17 Subhānu .	23 Virōdhin .	...
4506	1327	1462	811	579-80	*1404-05	18 Tāraṇa .	24 Vikṛita .	...
4507	1328	1463	812	580-81	1405-06	19 Pārthiva .	25 Khara .	4 Āshāḍha .
4508	1329	1464	813	581-82	1406-07	20 Vyaya .	26 Nandana .	...
4509	1330	1465	814	582-83	1407-08	21 Sarvajit .	27 Vijaya .	...
4510	1331	1466	815	583-84	*1408-09	22 Sarvadhārin .	28 Jaya .	3 Jyēshṭha .
4511	1332	1467	816	584-85	1409-10	23 Virōdhin .	29 Manmatha .	...
4512	1333	1468	817	585-86	1410-11	24 Vikṛita .	30 Durmukha .	8 Kārttika‡ .
4513	1334	1469	818	586-87	1411-12	25 Khara .	31 Hēmalamba .	...
4514	1335	1470	819	587-88	*1412-13	26 Nandana .	32 Vilamba .	...
4515	1336	1471	820	588-89	1413-14	27 Vijaya .	33 Vikārin .	5 Śrāvaṇa .
4516	1337	1472	821	589-90	1414-15	28 Jaya .	34 Sārvarin .	...
4517	1338	1473	822	590-91	1415-16	29 Manmatha .	35 Plava† .	...
4518	1339	1474	823	591-92	*1416-17	30 Durmukha .	37 *Śōbhana* .	4 Āshāḍha .
4519	1340	1475	824	592-93	1417-18	31 Hēmalamba .	38 *Krōdhin* .	...
4520	1341	1476	825	593-94	1418-19	32 Vilamba .	39 *Viśvāvasu* .	...
4521	1342	1477	826	594-95	1419-20	33 Vikārin .	40 *Parābhava* .	2 Vaiśākha .
4522	1343	1478	827	595-96	*1420-21	34 Śārvarin .	41 *Plavaṅga* .	...
4523	1344	1479	828	596-97	1421-22	35 Plava .	42 Kīlaka .	6 Bhādrapada .
4524	1345	1480	829	597-98	1422-23	36 Śubhakṛit .	43 Saumya .	...
4525	1346	1481	830	598-99	1423-24	37 Śōbhana .	44 Sādhāraṇa .	...

LX—*Contd.*

COMMENCEMENT OF THE											
SOLAR YEAR.					LUNI-SOLAR YEAR (MEAN SUNRISE OF DAY ON WHICH CHAITRA ŚUKLA 1 ENDS).					Kali year.
Day and month, A.D.	Week-day.	Time of true Mēsha-saṁkrānti.			Day and month, A.D.	Week-day.	*a.*	*b.*	*c.*	
13	14	17			19	20	23	24	25	1
		H	M.	S.						
25 Mar. (84)	3 Tues.	18	55	23	9 Mar. (68)	1 Sun.	230·7335	703·7594	228·4414	4501
25 Mar. (85)	5 Thur.	1	7	32	26 Feb. (57)	5 Thur.	106·4563	551·1034	197·6283	4502
25 Mar. (84)	6 Fri.	7	19	41	16 Mar. (75)	4 Wed.	141·1387	186·9968	248·9286	4503
25 Mar. (84)	0 Sat.	13	31	50	5 Mar. (64)	1 Sun.	16·8615	334·2410	218·1054	4504
25 Mar. (84)	1 Sun.	19	43	58	24 Mar. (83)	0 Sat.	51·5439	270·2344	269·4158	4505
25 Mar. (85)	3 Tues.	1	56	7	12 Mar. (72)	4 Wed.	9927·2668	117·4784	238·5927	4506
25 Mar. (84)	4 Wed.	8	8	16	2 Mar. (61)	2 Mon.	141·6216	1·0142	210·5072	4507
25 Mar. (84)	5 Thur.	14	20	25	21 Mar. (80)	1 Sun.	176·3040	937·0076	261·8176	4508
25 Mar. (84)	6 Fri.	20	32	34	10 Mar. (69)	5 Thur.	52·0269	784·2517	230·9944	4509
25 Mar. (85)	1 Sun.	2	44	43	28 Feb. (59)	3 Tues.	266·3816	667·7673	202·9090	4510
25 Mar. (84)	2 Mon.	8	56	51	17 Mar. (76)	1 Sun.	9962·4320	567·4892	251·4816	4511
25 Mar. (84)	3 Tues.	15	9	0	6 Mar. (65)	5 Thur.	9838·1549	414·7332	220·6584	4512
25 Mar. (84)	4 Wed.	21	21	9	25 Mar. (84)	4 Wed.	9872·8373	350·7267	271·9668	4513
25 Mar. (85)	6 Fri.	3	33	18	13 Mar. (73)	1 Sun.	9748·5601	197·9690	241·1457	4514
25 Mar. (84)	0 Sat.	9	45	27	3 Mar. (62)	6 Fri.	9962·9150	81·5065	213·0602	4515
25 Mar. (84)	1 Sun.	15	57	36	22 Mar. (81)	5 Thur.	9997·5980	17·5000	264·3706	4516
25 Mar. (84)	2 Mon	22	9	45	12 Mar. (71)	3 Tues.	211·9521	901·0446	236·2862	4517
25 Mar. (85)	4 Wed.	4	21	53	29 Feb. (60)	0 Sat.	87·6750	748·2797	205·4630	4518
25 Mar. (84)	5 Thur.	10	34	2	19 Mar. (78)	6 Fri.	122·3574	684·2731	256·7734	4519
25 Mar. (84)	6 Fri.	16	46	11	8 Mar. (67)	3 Tues.	9998·0803	531·5172	255·9491	4520
25 Mar. (84)	0 Sat.	22	58	20	25 Feb. (56)	0 Sat.	9873·8030	378·7613	195·1260	4521
25 Mar. (85)	2 Mon.	5	10	29	15 Mar. (75)	6 Fri.	9908·4855	314·7548	246·4364	4522
25 Mar. (84)	3 Tues.	11	22	38	4 Mar. (63)	3 Tues.	9784·2083	161·9988	215·6132	4523
25 Mar. (84)	4 Wed.	17	34	46	23 Mar. (82)	2 Mon.	9818·8907	97·9923	266·9235	4524
25 Mar. (84)	5 Thur.	23	46	55	13 Mar. (72)	0 Sat.	33·2455	981·5279	238·8382	4525

TABLE

CONCURRENT YEAR.								Intercalated and suppressed (*ksh.*) lunar months.
						Jovian Samvatsara.		
Kali.	Śaka.	Chaitrādi Vikrama.	Mēshādi (solar) year in Bengal.	Kollam.	A.D.	Southern system.	Northern system.	
1	2	3	3a	4	5	6	7	8
4526	1347	1482	831	599-600	*1424-25	38 Krōdhin .	45 Virōdhakṛit .	4 Āshāḍha
4527	1348	1483	832	600-01	1425-26	39 Viśvāvasu .	46 Paridhāvin .	...
4528	1349	1484	833	601-02	1426-27	40 Parābhava .	47 Pramādin .	...
4529	1350	1485	834	602-03	1427-28	41 Plavaṅga .	48 Ānanda .	3 Jyēshṭha .
4530	1351	1486	835	603-04	*1428-29	42 Kīlaka .	49 Rākshasa .	...
4531	1352	1487	836	604-05	1429-30	43 Saumya .	50 Anala .	8 Kārttika ‡ .
4532	1353	1488	837	605-06	1430-31	44 Sādhāraṇa .	51 Piṅgala .	...
4533	1354	1489	838	606-07	1431-32	45 Virōdhakṛit .	52 Kālayukta .	...
4534	1355	1490	839	607-08	*1432-33	46 Paridhāvin .	53 Siddhārthin .	5 Śrāvana .
4535	1356	1491	840	608-09	1433-34	47 Pramādin .	54 Raudra .	...
4536	1357	1492	841	609-10	1434-35	48 Ānanda .	55 Durmati .	...
4537	1358	1493	842	610-11	1435-36	49 Rākshasa .	56 Dundubhi .	4 Āshāḍha .
4538	1359	1494	843	611-12	*1436-37	50 Anala .	57 Rudhirōdgārin	...
4539	1360	1495	844	612-13	1437-38	51 Piṅgala .	58 Raktāksha .	...
4540	1361	1496	845	613-14	1438-39	52 Kālayukta .	59 Krōdhana .	1 Chaitra
4541	1362	1497	846	614-15	1439-40	53 Siddhārthin .	60 Kshaya .	...
4542	1363	1498	847	615-16	*1440-41	54 Raudra .	1 Prabhava .	6 Bhādrapada .
4543	1364	1499	848	616-17	1441-42	55 Durmati .	2 Vibhava .	...
4544	1365	1500	849	617-18	1442-43	56 Dundubhi .	3 Śukla .	...
4545	1366	1501	850	618-19	1443-44	57 Rudhirōdgārin	4 Pramōda .	4 Āshāḍha
4546	1367	1502	851	619-20	*1444-45	58 Raktāksha .	5 Prajāpati .	...
4547	1368	1503	852	620-21	1445-46	59 Krōdhana .	6 Aṅgiras .	...
4548	1369	1504	853	621-22	1446-47	60 Kshaya .	7 Srīmukha .	3 Jyēshṭha
4549	1370	1505	854	622-23	1447-48	1 Prabhava .	8 Bhāva .	...
4550	1371	1506	855	623-24	*1448-49	2 Vibhava .	9 Yuvan .	7 Āśvina .

Remarks, p. 163 above.

LX—*Contd.*

COMMENCEMENT OF THE								
SOLAR YEAR.			LUNI-SOLAR YEAR (MEAN SUNRISE OF DAY ON WHICH CHAITRA ŚUKLA 1 ENDS).					Kali year.
Day and month, A.D.	Week-day.	Time of true Mesha-sankrānti.	Day and month, A.D.	Week-day.	a	b	c	
13	14	17	19	20	23	24	25	1
		H. M. S.						
25 Mar. (85)	0 Sat.	5 59 4	2 Mar. (62)	5 Thur.	247·6004	965·0627	210·7228	4526
25 Mar. (84)	1 Sun.	12 11 13	21 Mar. (80)	4 Wed.	282·2828	801·0671	262·0632	4527
25 Mar. (84)	2 Mon.	18 23 22	10 Mar. (69)	1 Sun.	158·0056	648·3012	231·2300	4528
26 Mar. (85)	4 Wed.	0 35 31	27 Feb. (58)	5 Thur.	33·7284	495·5452	200·4167	4529
25 Mar. (85)	5 Thur.	6 47 39	17 Mar. (77)	4 Wed.	68·4108	431·5387	251·7272	4530
25 Mar. (84)	6 Fri.	12 59 41	6 Mar. (65)	1 Sun.	9944·1336	278·7828	220·9040	4531
25 Mar. (84)	0 Sat.	19 11 57	25 Mar. (84)	0 Sat.	9978·8160	214·7762	272·2143	4532
26 Mar. (85)	2 Mon.	1 24 6	14 Mar. (73)	4 Wed.	9854·5389	62·0203	241·3912	4533
25 Mar. (85)	3 Tues.	7 36 15	3 Mar. (62)	2 Mon.	89·8937	945·4560	213·3058	4534
25 Mar. (84)	4 Wed.	13 48 24	22 Mar. (81)	1 Sun.	103·3761	881·5495	264·6182	4535
25 Mar. (84)	5 Thur.	20 0 32	12 Mar. (71)	6 Fri.	317·9309	765·0832	236·5307	4536
26 Mar. (85)	0 Sat.	2 12 41	1 Mar. (60)	3 Tues.	193·6538	612·3262	205·7075	4537
26 Mar. (85)	1 Sun.	8 24 50	19 Mar. (79)	2 Mon.	227·3362	548·3227	257·0180	4538
25 Mar. (84)	2 Mon.	14 36 59	7 Mar. (66)	5 Thur.	9785·4279	359·2751	223·4309	4539
25 Mar. (84)	3 Tues.	20 49 8	25 Feb. (56)	2 Tues.	9979·7818	242·8108	193·3716	4540
26 Mar. (85)	5 Thur.	3 1 17	16 Mar. (75)	2 Mon.	14·4642	178·8042	246·6819	4541
25 Mar. (85)	6 Fri.	9 13 26	4 Mar. (64)	6 Fri.	9890·1870	26·0483	215·8588	4542
25 Mar. (84)	0 Sat.	15 25 34	23 Mar. (82)	5 Thur.	9924·8695	962·0414	267·1601	4543
25 Mar. (84)	1 Sun.	21 37 43	13 Mar. (72)	3 Tues.	139·2242	845·5774	239·0636	4544
26 Mar. (85)	3 Tues.	3 49 52	2 Mar. (61)	0 Sat.	14·9472	692·8215	208·2605	4545
25 Mar. (85)	4 Wed.	10 2 1	20 Mar. (80)	6 Fri.	49·6296	628·8050	259·5700	4546
25 Mar. (84)	5 Thur.	16 14 10	9 Mar. (68)	3 Tues.	9925·3524	475·0661	228·7001	4547
25 Mar. (84)	6 Fri.	22 26 19	26 Feb. (57)	0 Sat.	9801·0752	323·3021	197·9244	4548
26 Mar. (85)	1 Sun.	4 38 27	17 Mar. (76)	6 Fri.	9835·7575	259·2361	249·2339	4549
25 Mar. (85)	2 Mon.	10 50 36	6 Mar. (66)	4 Wed.	50·1124	142·8233	221·1436	4550

TABLE

CONCURRENT YEAR.								Intercalated and suppressed (*ksh.*) lunar months.
						Jovian Samvatsara.		
Kali.	Śaka.	Chaitrādi Vikrama.	Mēshādi (solar) year in Bengal.	Kollam.	A.D.	Southern system.	Northern system.	
1	2	3	3a	4	5	6	7	8
4551	1372	1507	856	624-25	1449-50	3 Śukla . .	10 Dhātri . .	...
4552	1373	1508	857	625-26	1450-51	4 Pramōda .	11 Īśvara . .	...
4553	1374	1509	858	626-27	1451-52	5 Prajāpati .	12 Bahudhānya .	5 Śrāvaṇa
4554	1375	1510	859	627-28	*1452-53	6 Aṅgiras . .	13 Pramāthin .	...
4555	1376	1511	860	628-29	1453-54	7 Śrīmukha .	14 Vikrama .	...
4556	1377	1512	861	629-30	1454-55	8 Bhāva . .	15 Vṛisha .	4 Āshāḍha .
4557	1378	1513	862	630-31	1455-56	9 Yuvan . .	16 Chitrabhānu .	...
4558	1379	1514	863	631-32	*1456-57	10 Dhātṛi . .	17 Subhānu .	...
4559	1380	1515	864	632-33	1457-58	11 Īśvara . .	18 Tāraṇa . .	1 Chaitra .
4560	1381	1516	865	633-34	1458-59	12 Bahudhānya .	19 Pārthiva .	...
4561	1382	1517	866	634-35	1459-60	13 Pramāthin .	20 Vyaya . .	5 Śrāvaṇa .
4562	1383	1518	867	635-36	*1460-61	14 Vikrama .	21 Sarvajit .	...
4563	1384	1519	868	636-37	1461-62	15 Vṛisha . .	22 Sarvadhārin .	...
4564	1385	1520	869	637-38	1462-63	16 Chitrabhānu .	23 Virōdhin .	4 Āshāḍha .
4565	1386	1521	870	638-39	1463-64	17 Subhānu .	24 Vikṛita .	...
4566	1387	1522	871	639-40	*1464-65	18 Tāraṇa . .	25 Khara .	...
4567	1388	1523	872	640-41	1465-66	19 Pārthiva .	26 Nandana .	2 Vaiśākha .
4568	1389	1524	873	641-42	1466-67	20 Vyaya . .	27 Vijaya . .	...
4569	1390	1525	874	642-43	1467-68	21 Sarvajit .	28 Jaya . .	6 Bhādrapada .
4570	1391	1526	875	643-44	*1468-69	22 Sarvadhārin .	29 Manmatha .	...
4571	1392	1527	876	644-45	1469-70	23 Virōdhin .	30 Durmukha .	...
4572	1393	1528	877	645-46	1470-71	24 Vikṛita . .	31 Hēmalamba .	5 Śrāvaṇa .
4573	1394	1529	878	646-47	1471-72	25 Khara . .	32 Vilamba . .	...
4574	1395	1530	879	647-48	*1472-73	26 Nandana .	33 Vikārin . .	...
4575	1393	1531	880	648-49	1473-74	27 Vijaya . .	34 Śārvarin .	3 Jyēshṭha .

LX—*Contd.*

COMMENCEMENT OF THE											
SOLAR YEAR.					LUNI-SOLAR YEAR (MEAN SUNRISE OF DAY ON WHICH CHAITRA ŚUKLA 1 ENDS).					Kali year.
Day and month, A.D.	Week-day.	Time of true Mēsha-saṁkrānti.			Day and month, A.D.	Week-day.	*a.*	*b.*	*c.*	
13	14	17			19	20	23	24	25	1
		H.	M.	S.						
25 Mar. (84)	3 Tues. .	17	2	45	25 Mar. (84)	3 Tues. .	84·7948	78·8257	272·4599	4551
25 Mar. (84)	4 Wed. .	23	14	54	14 Mar. (73)	0 Sat. .	9960·5176	926·0698	241·6368	4552
26 Mar. (85)	6 Fri. .	5	27	3	4 Mar. (63)	5 Thur	174·8724	809·5415	213·5513	4553
25 Mar. (85)	0 Sat, .	11	39	12	22 Mar. (82)	4 Wed. .	209·5549	745·5990	264·8617	4554
25 Mar. (84)	1 Sun. .	17	51	20	11 Mar. (70)	1 Sun. .	85·2777	592·84[illegible]0	234·0385	4555
26 Mar. (85)	3 Tues. .	0	3	29	28 Feb. (59)	5 Thur.	9961·9005	440·0871	203·2153	4556
26 Mar. (85)	4 Wed. .	6	15	38	19 Mar. (78)	4 Wed. .	9995·6829	376·0805	254·5257	4557
25 Mar. (85)	5 Thur.	12	27	47	7 Mar. (67)	1 Sun. .	9871·4058	223·3246	223·7024	4558
25 Mar. (84)	6 Fri. .	18	39	56	25 Feb. (56)	6 Fri. .	85·7606	106·8603	195·6171	5559
26 Mar. (85)	1 Sun. .	0	52	5	16 Mar. (75)	5 Thur.	120·4430	42·8538	246·9275	4560
26 Mar. (85)	2 Mon. .	7	4	13	5 Mar. (64)	2 Mon. .	9996·1658	890·0978	216·1053	4561
25 Mar. (85)	3 Tues. .	13	16	22	23 Mar. (83)	1 Sun. .	30·8483	826·0913	267·4146	4562
25 Mar. (84)	4 Wed. .	19	28	31	13 Mar. (72)	6 Fri. .	245·2030	709·6270	239·3293	4563
26 Mar. (85)	6 Fri. .	1	40	40	2 Mar. (61)	3 Tues. .	120·9259	556·8710	208·5061	4564
26 Mar. (85)	0 Sat. .	7	52	49	21 Mar. (80)	2 Mon. .	155·6083	492·8645	259·8165	4565
25 Mar. (85)	1 Sun. .	14	4	58	9 Mar. (69)	6 Fri. .	31·3312	340·1086	228·9942	4566
25 Mar. (84)	2 Mon. .	20	17	7	26 Feb. (57)	3 Tues. .	9907·0539	187·3526	158·1701	4567
26 Mar. (85)	4 Wed. .	2	29	15	17 Mar. (76)	2 Mon. .	9941·7363	123·3461	249·4805	4568
26 Mar. (85)	5 Thur.	8	41	24	7 Mar. (66)	0 Sat. .	156·0912	6·8818	221·3950	4569
25 Mar. (85)	6 Fri. .	14	53	33	25 Mar. (85)	6 Fri. .	190·7735	942·8753	272·7054	4570
25 Mar. (84)	0 Sat. .	21	5	42	14 Mar. (73)	3 Tes. .	66·4964	790·1193	241·8823	4571
26 Mar. (85)	2 Mon. .	3	17	51	4 Mar. (63)	1 Sun. .	280·8512	673·6550	213·7969	4572
26 Mar. (85)	3 Tues. .	9	30	0	22 Mar. (81)	6 Fri. .	9976·9017	573·3568	262·3695	4573
25 Mar. (85)	4 Wed. .	15	42	8	10 Mar. (70)	3 Tues. .	9852·6245	420·6009	231·5662	4574
25 Mar. (84)	5 Thur.	21	54	17	27 Feb. (58)	0 Sat	9723·3473	267·8450	200·7230	4575

TABLE

Concurrent year.						Jovian Samvatsara.		Intercalated and suppressed (*ksh.*) lunar months.
Kali.	Saka.	Chaitrâdi Vikrama.	Mêshâdi (solar) year in Bengal.	Kollam.	A.D.	Southern system.	Northern system.	
1	2	3	3a	4	5	6	7	8
4576	1397	1532	881	649-50	1474-75	28 Jaya . .	35 Plava .	...
4577	1398	1533	882	650-51	1475-76	29 Manmatha .	36 Śubhakṛit .	...
4578	1399	1534	883	651-52	*1476-77	30 Durmukha .	37 Śōbhana .	1 Chaitra .
4579	1400	1535	884	652-53	1477-78	31 Hēmalamba .	38 Krōdhin .	...
4580	1401	1536	885	653-54	1478-79	32 Vilamba . .	39 Visvāvasu .	5 Śrāvaṇa .
4581	1402	1537	886	654-55	1479-80	33 Vikārin .	40 Parābhava .	...
4582	1403	1538	887	655-56	*1480-81	34 Śārvarin .	41 Plavaṅga .	...
4583	1404	1539	888	656-57	1481-82	35 Plava . .	42 Kīlaka . .	4 Āshāḍha .
4584	1405	1540	889	657-58	1482-83	36 Śubhakṛit .	43 Saumya .	...
4585	1406	1541	890	658-59	1483-84	37 Śōbhana .	44 Sādhāraṇa .	...
4586	1407	1542	891	659-60	*1484-85	38 Krōdhin . .	45 Virōdhakṛit .	2 Vaiśākha .
4587	1408	1543	892	660-61	1485-86	39 Visvāvasu .	46 Paridhāvin .	...
4588	1409	1544	893	661-62	1486-87	40 Parābhava .	47 Pramādin .	6 Bhādrapada .
4589	1410	1545	894	662-63	1487-88	41 Plavaṅga .	48 Ānanda . .	...
4590	1411	1546	895	663-64	*1488-89	42 Kīlaka . .	49 Rākshasa .	...
4591	1412	1547	896	664-65	1489-90	43 Saumya .	50 Anala .	5 Śrāvaṇa .
4592	1413	1548	897	665-66	1490-91	44 Sādhāraṇa .	51 Piṅgala . .	...
4593	1414	1549	898	666-67	1491-92	45 Virōdhakṛit .	52 Kālayukta .	...
4594	1415	1550	899	667-68	*1492-93	46 Paridhāvin .	53 Siddhārthin .	3 Jyēshṭha .
4595	1416	1551	900	668-69	1493-94	47 Pramādin .	54 Raudra . .	...
4596	1417	1552	901	669-70	1494-95	48 Ānanda . .	55 Durmati .	...
4597	1418	1553	902	670-71	1495-96	49 Rākshasa .	56 Dundubhi .	1 Chaitra .
4598	1419	1554	903	671-72	*1496-97	50 Anala . .	57 Rudhirōdgārin	...
4599	1420	1555	904	672-73	1497-98	51 Piṅgala . .	58 Raktāksha .	5 Śrāvaṇa .
4600	1421	1556	905	673-74	1498-99	52 Kālayukta .	59 Krōdhana .	...

LX—*Contd.*

COMMENCEMENT OF THE									Kali. year.
SOLAR YEAR.			LUNI-SOLAR YEAR (MEAN SUNRISE OF DAY ON WHICH CHAITRA ŚUKLA 1 ENDS).						
Day and month, A.D.	Week-day.	Time of true Mēsha saṁkrānti.	Day and month, A.D.	Week-day.	*a.*	*b.*	*c.*		
13	14	17	19	20	23	24	25		I
		H. M. S.							
26 Mar. (85)	0 Sat.	4 6 26	18 Mar. (77)	6 Fri.	9763·0297	203·8384	252·0335		4576
26 Mar. (85)	1 Sun.	10 18 35	8 Mar. (67)	4 Wed.	9977·3845	87·3741	223·9480		4577
25 Mar. (85)	2 Mon.	16 30 44	26 Feb. (57)	2 Mon.	191·7393	970·9068	195·8626		4578
25 Mar. (84)	3 Tues.	22 42 53	16 Mar. (75)	1 Sun.	226·4218	906·9033	247·1730		4579
26 Mar. (85)	5 Thur.	4 55 1	5 Mar. (64)	5 Thur.	102·1446	754·0474	216·3499		4580
26 Mar. (85)	6 Fri.	11 7 10	24 Mar. (83)	4 Wed.	136·8270	690·1408	267·6602		4581
25 Mar. (85)	0 Sat.	17 19 19	12 Mar. (72)	1 Sun.	12·5498	537·3849	236·8370		4582
25 Mar. (84)	1 Sun.	23 31 28	1 Mar. (60)	5 Thur.	9888·2727	384·6289	206·0138		4583
26 Mar. (85)	3 Tues.	5 43 37	20 Mar. (79)	4 Wed.	9922·9550	320·6184	257·3[illegible]43		4584
26 Mar. (85)	4 Wed.	11 55 46	9 Mar. (68)	1 Sun.	9798·6779	167·8664	226·[illegible]010		[illegible]585
25 Mar. (85)	5 Thur.	18 7 54	27 Feb. (58)	6 Fri.	13·0326	51·4021	198·4156		4586
26 Mar (85)	0 Sat.	0 20 3	17 Mar. (76)	5 Thur.	47·7151	987·3956	249·7260		4587
26 Mar. (85)	1 Sun.	6 32 12	7 Mar. (66)	3 Tues.	262·0699	870·9313	221·6416		4588
26 Mar. (85)	2 Mon.	12 44 21	26 Mar. (85)	2 Mon.	296·7523	806·9247	272·9510		4589
25 Mar. (85)	3 Tues.	18 56 30	14 Mar. (74)	6 Fri.	172·4752	654·1688	242·1278		4590
26 Mar. (85)	5 Thur	1 8 39	3 Mar. (62)	3 Tues.	48·1981	501·4129	211·3046		4591
26 Mar. (85)	6 Fri.	7 20 48	22 Mar. (81)	2 Mon.	82·8804	437·4064	262·6151		4592
26 Mar. (85)	0 Sat.	13 32 56	11 Mar. (70)	6 Fri	9958·5833	284·6504	231·7918		4593
25 Mar. (85)	1 Sun.	19 45 5	28 Feb. (59)	3 Tues.	9834·3261	131·8945	200·9685		4594
26 Mar. (85)	3 Tues.	1 57 14	18 Mar. (77)	2 Mon.	9869·0084	67·8880	252·2790		4595
26 Mar. (85)	4 Wed.	8 9 23	8 Mar. (67)	0 Sat.	83·3633	951·4236	2241936		4596
26 Mar. (85)	5 Thur	14 21 32	26 Feb. (57)	5 Thur.	297·7181	834·9593	196·1082		4597
25 Mar. (85)	6 Fri.	20 33 41	15 Mar. (75)	3 Tues.	9993·7685	734·6612	244·6807		4598
26 Mar. (85)	1 Sun.	2 45 49	5 Mar. (64)	1 Sun.	208·1233	618·1969	216·5954		4599
26 Mar. (85)	2 Mon.	8 57 58	23 Mar. (82)	6 Fri.	9904·1738	[illegible]17·8977	265·1680		4600

TABL

CONCURRENT YEAR.								
						Jovian Samvatsara.		Intercalated and suppressed (*ksh.*) lunar months.
Kali.	Śaka.	Chaitrādi Vikrama.	Mēshādi (solar) year in Bengal	Kollam	A.D.	Southern system.	Northern system.	
1	2	3	3a	4	5	6	7	8
4601	1422	1557	906	674-75	1499-1500	53 Siddhārthin .	60 Kshaya .	...
4602	1423	1558	907	675-76	*1500-01	54 Raudra .	1 Prabhava† .	4 Āshāḍha .
4603	1424	1559	908	676-77	1501-02	55 Durmati .	3 *Śukla* .	...
4604	1425	1560	909	677-78	1502-03	56 Dundubhi .	4 *Pramōda* .	...
4605	1426	1561	910	678-79	1503-04	57 Rudhirōdgārin	5 *Prajāpati* .	2 Vaiśākha .
4606	1427	1562	911	679-80	*1504-05	58 Raktāksha .	6 *Aṅgiras* .	...
4607	1428	1563	912	680-81	1505-06	59 Krōdhana .	7 *Śrīmukha* .	6 Bhādrapada
4608	1429	1564	913	681-82	1506-07	60 Kshaya .	8 *Bhāva* .	...
4609	1430	1565	914	682-83	1507-08	1 Prabhava .	9 *Yuvan* .	...
4610	1431	1566	915	683-84	*1508-09	2 Vibhava .	10 *Dhātṛi* .	5 Śrāvaṇa .
4611	1432	1567	916	684-85	1509-10	3 Śukla .	11 *Īśvara* .	...
4612	1433	1568	917	685-86	1510-11	4 Pramōda .	12 *Bahudhānya* .	...
4613	1434	1569	918	686-87	1511-12	5 Prajāpati .	13 *Pramāthin* .	3 Jyēshṭha .
4614	1435	1570	919	687-88	*1512-13	6 Aṅgiras .	14 *Vikrama* .	...
4615	1436	1571	920	688-89	1513-14	7 Śrīmukha .	15 *Vṛisha* .	7 Āsvina 10 *Pausha* (*ksh*)
4616	1437	1572	921	689-90	1514-15	8 Bhāva .	16 *Chitrabhānu* .	1 Chaitra .
4617	1438	1573	922	690-91	1515-16	9 Yuvan .	17 Subhānu	...
4618	1439	1574	923	691-92	*1516-17	10 Dhātṛi .	18 Tāraṇa .	5 Śrāvaṇa .
4619	1440	1575	924	692-93	1517-18	11 Īśvara .	19 Pārthiva .	...
4620	1441	1576	925	693-94	1518-19	12 Bahudhānya .	20 Vyaya .	...
4621	1442	1577	926	694-95	1519-20	13 Pramāthin	21 Sarvajit .	4 Āshāḍha .
4622	1443	1578	927	695-96	*1520-21	14 Vikrama	22 Sarvadhārin .	...
4623	1444	1579	928	696-97	1521-22	15 Vṛisha .	23 Virōdhin	
4624	1445	1580	929	697-98	1522-23	16 Chitrabhānu .	24 Vikrita .	2 Vaiśākha .
4625	1446	1581	930	698-99	1523-24	17 Subhānu .	25 Khara .	...

† Vibhava was suppressed in the north

LX—*Contd.*

COMMENCEMENT OF THE											
SOLAR YEAR.					LUNI-SOLAR YEAR (MEAN SUNRISE OF DAY ON WHICH CHAITRA ŚUKLA 1 ENDS).						Kali year.
Day and month, A.D.	Week-day	Time of true Mēsha-saṁkrānti.			Day and month, A.D.	Week-day.	*a.*	*b.*	*c.*		
13	14	17			19	20	23	24	25		1
		H.	M.	S.							
26 Mar. (85)	3 Tues.	15	10	7	12 Mar. (71)	3 Tues	9779·8966	365·1427	234·2642		4601
25 Mar. (85)	4 Wed.	21	22	16	1 Mar. (61)	1 Sun. .	9994·2515	248·6785	206·1788		4602
26 Mar. (85)	6 Fri. .	3	34	25	20 Mar. (79)	0 Sat. .	29·0339	184·6719	257·4892		4603
26 Mar. (85)	0 Sat. .	9	46	34	9 Mar. (68)	4 Wed.	9904·6567	31·9160	226·6659		4604
26 Mar. (85)	1 Sun.	15	58	42	27 Feb. (58)	2 Mon. .	119·0115	915·4516	198·5806		4605
25 Mar. (85)	2 Mon. .	22	10	51	17 Mar. (77)	1 Sun. .	153·6939	851·4451	249·8910		4606
26 Mar. (85)	4 Wed.	4	23	0	6 Mar. (65)	5 Thur.	29·4167	698·6892	219·0678		4607
26 Mar. (85)	5 Thur.	10	35	9	25 Mar. (84)	4 Wed.	64·0991	634·6827	270·3781		4608
26 Mar. (85)	6 Fri. .	16	47	18	14 Mar. (73)	1 Sun. .	9939·8220	481·9267	239·5550		4609
25 Mar. (85)	0 Sat. .	22	59	27	2 Mar. (62)	5 Thur.	9816·5448	329·1707	208·7318		4610
26 Mar. (85)	2 Mon .	5	11	36	21 Mar. (80)	4 Wed.	9850·2272	265·1642	260·0422		4611
26 Mar. (85)	3 Tues.	11	23	44	11 Mar. (70)	2 Mon. .	64·5821	148·6999	231·9567		4612
26 Mar. (85)	4 Wed.	17	35	53	28 Feb. (59)	6 Fri. .	9940·3049	995·9440	201·1335		4613
25 Mar. (85)	5 Thur.	23	48	2	18 Mar. (78)	5 Thur. •	9974·9872	931·9375	252·4440		4614
26 Mar. (85)	0 Sat. .	6	0	11	8 Mar. (67)	3 Tues.	189·3421	815·4732	224·3585		4615
26 Mar. (85)	1 Sun. .	12	12	20	25 Feb. (56)	0 Sat. .	65·0650	662·7172	193·5353		4616
26 Mar. (85)	2 Mon. .	18	24	29	16 Mar. (75)	6 Fri. .	99·7473	598·7196	244·8457		4617
26 Mar (86)	4 Wed.	0	36	37	4 Mar. (64)	3 Tues.	9975·4701	445·9547	214·0226		4618
26 Mar. (85)	5 Thur.	6	48	46	23 Mar. (82)	2 Mon. .	10·1526	381·9482	265·3350		4619
26 Mar. (85)	6 Fri. .	13	0	55	12 Mar. (71)	6 Fri. .	9885·8754	229·1922	234·5097		4620
26 Mar. (85)	0 Sat. .	19	13	4	2 Mar. (61)	4 Wed.	100·2302	112·7280	206·4243		4621
26 Mar. (86)	2 Mon. .	1	25	13	20 Mar. (80)	3 Tues.	134·9126	48·7215	257·7349		4622
26 Mar. (85)	3 Tues.	7	37	22	9 Mar. (68)	0 Sat. .	10·6355	895·9655	226·9115		4623
26 Mar. (85)	4 Wed.	13	49	30	27 Feb. (58)	5 Thur.	224·9902	779·5012	198·8261		4624
26 Mar. (85)	5 Thur.	20	1	39	18 Mar. (77)	4 Wed.	259·6726	715·4946	250·1365		4625

TABLE

CONCURRENT YEAR.								Intercalated and suppressed (*ksh.*) lunar months.
						Jovian Samvatsara.		
Kali.	Śaka.	Chaitrādi Vikrama.	Mēshādi (solar) year in Bengal.	Kollam.	A.D.	Southern system.	Northern system.	
1	2	3	3a	4	5	6	7	8
4626	1447	1582	931	699-700	*1524-25	18 Tāraṇa	26 Nandana	6 Bhādrapada
4627	1448	1583	932	700-01	1525-26	19 Pārthiva	27 Vijaya	...
4628	1449	1584	933	701-02	1526-27	20 Vyaya	28 Jaya	...
4629	1450	1585	934	702-03	1527-28	21 Sarvajit	29 Manmatha	4 Āshāḍha
4630	1451	1586	935	703-04	*1528-29	22 Sarvadhārin	30 Durmukha	...
4631	1452	1587	936	704-05	1529-30	23 Virōdhin	31 Hēmalamba	...
4632	1453	1588	937	705-06	1530-31	24 Vikrita	32 Vilamba	3 Jyōshtha
4633	1454	1589	938	706-07	1531-32	25 Khara	33 Vikārin	...
4634	1455	1590	939	707-08	*1532-33	26 Nandana	34 Śārvarin	7 Āśvina
4635	1456	1591	940	708-09	1533-34	27 Vijaya	35 Plava	...
4636	1457	1592	941	709-10	1534-35	28 Jaya	36 Śubhakrit	...
4637	1458	1593	942	710-11	1535-36	29 Manmatha	37 Śōbhana	5 Śrāvaṇa
4638	1459	1594	943	711-12	*1536-37	30 Durmukha	38 Krōdhin	...
4639	1460	1595	944	712-13	1537-38	31 Hēmalamba	39 Viśvāvasu	...
4640	1461	1596	945	713-14	1538-39	32 Vilamba	40 Parābhava	4 Āshāḍha
4641	1462	1597	946	714-15	1539-40	33 Vikārin	41 Plavaṅga	...
4642	1463	1598	947	715-16	*1540-41	34 Śārvarin	42 Kīlaka	...
4643	1464	1599	948	716-17	1541-42	35 Plava	43 Saumya	2 Vaiśākha
4644	1465	1600	949	717-18	1542-43	36 Śubhakrit	44 Sādhāraṇa	...
4645	1466	1601	950	718-19	1543-44	37 Śōbhana	45 Virōdhakrit	6 Bhādrapada
4646	1467	1602	951	719-20	*1544-45	38 Krōdhin	46 Paridhāvin	...
4647	1468	1603	952	720-21	1545-46	39 Viśvāvasu	47 Pramādin	...
4648	1469	1604	953	721-22	1546-47	40 Parābhava	48 Ānanda	4 Āshāḍha
4649	1470	1605	954	722-23	1547-48	41 Plavaṅga	49 Rākshasa	...
4650	1471	1606	955	723-24	*1548-49	42 Kīlaka	50 Anala	...

LX—*Contd.*

COMMENCEMENT OF THE											
SOLAR YEAR.					LUNI-SOLAR YEAR (MEAN SUNRISE OF DAY ON WHICH CHAITRA ŚUKLA 1 ENDS).						Kali year.
Day and month, A. D.	Week-day.	Time of true Mēsha-saṁkrānti.			Day and month, A.D.	Week-day.	*a.*	*b.*	*c.*		
13	14	17			19	20	23	24	25		1
		H.	M.	S.							
26 Mar. (86)	0 Sat. .	2	13	48	6 Mar. (66)	1 Sun. .	135·3955	562·7387	219·3134		4626
26 Mar. (85)	1 Sun. .	8	25	57	25 Mar. (84)	0 Sat. .	170·0779	498·7322	270·6237		4627
26 Mar. (85)	2 Mon. .	14	38	6	14 Mar. (73)	4 Wed. .	45·8007	345·9762	239·8005		4628
26 Mar. (85)	3 Tues. .	20	50	15	3 Mar. (62)	1 Sun. .	9921·5236	193·2203	208·9773		4629
26 Mar. (86)	5 Thur.	3	2	23	21 Mar. (81)	0 Sat. .	9956·3060	129·2137	260·2878		4630
26 Mar. (85)	6 Fri. .	9	14	32	11 Mar. (70)	5 Thur. .	170·5608	12·7495	232·2023		4631
26 Mar. (85)	0 Sat. .	15	26	41	28 Feb. (59)	2 Mon. .	46·2836	860·0035	201·3790		4632
26 Mar. (85)	1 Sun. .	21	38	50	19 Mar. (78)	1 Sun. .	80·9660	795·9870	252·6895		4633
26 Mar. (86)	3 Tues. .	3	50	59	8 Mar. (68)	6 Fri. .	295·3209	679·5227	224·6041		4634
26 Mar. (85)	4 Wed. .	10	3	8	26 Mar. (85)	4 Wed. .	9991·3712	579·1945	273·1767		4635
26 Mar. (85)	5 Thur. .	16	15	17	15 Mar. (74)	1 Sun. .	9867·0941	426·4686	242·3535		4636
26 Mar. (85)	6 Fri. .	22	27	25	4 Mar. (63) .	5 Thur.	9742·8170	273·7126	211·5303		4637
26 Mar. (86)	1 Sun. .	4	39	34	22 Mar. (82)	4 Wed. .	9777·4894	209·7061	262·8408		4638
26 Mar. (85)	2 Mon. .	10	51	43	12 Mar. (71)	2 Mon. .	9991·8551	93·2417	234·7553		4639
26 Mar. (85)	3 Tues. .	17	3	52	2 Mar. (61)	0 Sat. .	206·2090	976·7775	206·6699		4640
26 Mar. (85)	4 Wed. .	23	16	1	21 Mar. (80)	6 Fri. .	240·8914	912·7710	258·6803		4641
26 Mar. (86)	6 Fri. .	5	28	10	9 Mar. (69)	3 Tues. .	116·6132	760·0151	227·1571		4642
26 Mar. (85)	0 Sat. .	11	40	18	26 Feb. (57)	0 Sat. .	9992·3370	607·2591	196·3339		4643
26 Mar. (85)	1 Sun. .	17	52	27	17 Mar. (76)	6 Fri. .	27·0195	543·2525	247·6443		4644
27 Mar. (86)	3 Tues. .	0	4	36	6 Mar. (65)	3 Tues. .	9902·7423	390·4966	216·8211		4645
26 Mar. (86)	4 Wed. .	6	12	45	24 Mar. (84)	2 Mon. .	9937·4247	326·4900	268·1214		4646
26 Mar. (85)	5 Thur.	12	28	54	13 Mar. (72)	6 Fri. .	9813·1475	173·7341	237·3083		4647
26 Mar. (85)	6 Fri. .	18	41	3	3 Mar. (62)	4 Wed. .	27·5024	57·2698	209·2229		4648
27 Mar. (86)	1 Sun. .	0	53	11	22 Mar. (81)	3 Tues. .	62·1847	993·2632	260·5333		4649
26 Mar. (86)	2 Mon. .	7	5	20	11 Mar. (71)	1 Sun. .	276·5396	876·7990	232·4478		4650

TABLE

CONCURRENT YEAR.								Intercalated and suppressed (*ksh.*) lunar months.
						Jovian Samvatsara.		
Kali.	Śaka.	Chaitrādi Vikrama	Mēshādi (solar) year in Bengal.	Kollam.	A.D.	Southern system.	Northern system.	
1	2	3	3*a*	4	5	6	7	8
4651	1472	1607	956	724-25	1549-50	43 Saumya .	51 Piṅgala . .	3 Jyēshṭha .
4652	1473	1608	957	725-26	1550-51	44 Sādhāraṇa .	52 Kālayukta .	...
4653	1474	1609	958	726-27	1551-52	45 Virōdhakṛit .	53 Siddhārthin .	7 Āśvina .
4654	1475	1610	959	727-28	*1552-53	46 Paridhāvin .	54 Raudra . .	...
4655	1476	1611	960	728-29	1553-54	47 Pramādin .	55 Durmati .	...
4656	1477	1612	961	729-30	1554-55	48 Ānanda . .	56 Dundubhi .	5 Śrāvaṇa
4657	1478	1613	962	730-31	1555-56	49 Rākshasa .	57 Rudhirōdgārin	...
4658	1479	1614	963	731-32	*1556-57	50 Anala . .	58 Raktāksha .	...
4659	1480	1615	964	732-33	1557-58	51 Piṅgala . .	59 Krōdhana .	4 Āshāḍha .
4660	1481	1616	965	733-34	1558-59	52 Kālayukta .	60 Kshaya . .	...
4661	1482	1617	966	734-35	1559-60	53 Siddhārthin .	1 Prabhava .	...
4662	1483	1618	967	735-36	*1560-61	54 Raudra . .	2 Vibhava . .	2 Vaiśākha .
4663	1484	1619	968	736-37	1561-62	55 Durmati .	3 Śukla . .	...
4664	1485	1620	969	737-38	1562-63	56 Dundubhi .	4 Pramōda .	6 Bhādrapada .
4665	1486	1621	970	738-39	1563-64	57 Rudhirōdgārin	5 Prajāpati .	...
4666	1487	1622	971	739-40	*1564-65	58 Raktāksha .	6 Aṅgiras . .	...
4667	1488	1623	972	740-41	1565-66	59 Krōdhana .	7 Śrīmukha .	4 Āshāḍha .
4668	1489	1624	973	741-42	1566-67	60 Kshaya . .	8 Bhāva . .	...
4669	1490	1625	974	742-43	1567-68	1 Prabhava .	9 Yuvan . .	...
4670	1491	1626	975	743-44	*1568-69	2 Vibhava . .	10 Dhātṛi . .	3 Jyēshṭha .
4671	1492	1627	976	744-45	1569-70	3 Śukla . .	11 Īśvara . .	...
4672	1493	1628	977	745-46	1570-71	4 Pramōda .	12 Bahudhānya .	7 Āśvina .
4673	1494	1629	978	746-47	1571-72	5 Prajāpati .	13 Pramāthin .	...
4674	1495	1630	979	747-48	*1572-73	6 Aṅgiras . .	14 Vikrama .	...
4375	1496	1631	980	748-49	1573-74	7 Śrīmukha .	15 Vṛisha . .	5 Śrāvaṇa .

LX—*Contd.*

COMMENCEMENT OF THE

Solar year.			Luni-solar year (mean sunrise of day on which Chaitra Śukla 1 ends).					Kali year.
Day and month, A.D.	Week-day.	Time of true Mēsha-saṁkrānti.	Day and month, A.D.	Week-day.	*a.*	*b.*	*c.*	
13	14	17	19	20	23	24	25	1
		H. M. S						
26 Mar. (85)	3 Tues. .	13 17 29	28th Feb. (59)	5 Thur.	152·2624	724·0430	201·6246	4651
26 Mar. (85)	4 Wed. .	19 29 38	19 Mar. (78)	4 Wed. .	186·9447	660·0365	252·9851	4652
27 Mar. (86)	6 Fri. .	1 41 47	8 Mar. (67)	1 Sun. .	62·6676	507·3166	222·1018	4653
26 Mar. (86)	0 Sat. .	7 52 56	26 Mar. (86)	0 Sat. .	97·3500	443·2740	273·4222	4654
26 Mar. (85)	1 Sun. .	14 6 4	15 Mar. (74)	4 Wed. .	9973·0729	290·5181	242·5991	4655
26 Mar. (85)	2 Mon. .	20 18 13	4 Mar. (63)	1 Sun. .	9848·7957	137·7622	212·2759	4656
27 Mar. (86)	4 Wed. .	2 30 22	23 Mar. (82)	0 Sat. .	9883·4781	73·7556	263·0863	4657
26 Mar. (86)	5 Thur.	8 42 31	12 Mar. (72)	5 Thur.	97·8329	957·2912	235·0008	4658
26 Mar. (85)	6 Fri. .	14 54 40	2 Mar. (61)	3 Tues. .	312·1878	840·8270	206·9154	46[illegible]9
26 Mar. (85)	0 Sat. .	21 6 49	20 Mar. (79)	1 Sun. .	8·2381	740·5288	255·4881	4660
27 Mar. (86)	2 Mon. .	3 18 58	10 Mar. (69)	6 Fri. .	222·5930	624·0646	227·4026	4661
26 Mar. (86)	3 Tues. .	9 31 6	27 Feb. (58)	3 Tues. .	98·3158	471·3086	196·5794	4662
26 Mar. (85)	4 Wed. .	15 43 15	16 Mar. (75)	1 Sun. .	9794·3672	371·0104	245·1420	4663
26 Mar. (85)	5 Thur.	21 55 24	6 Mar. (65)	6 Fri. .	8·7210	254·5461	217·0667	4664
27 Mar. (86)	0 Sat. .	4 7 33	25 Mar. (84)	5 Thur.	43·4034	190·5396	268·3770	4665
26 Mar. (86)	1 Sun. .	10 19 42	13 Mar. (73)	2 Mon. .	9919·1263	37·7836	237·5538	4666
26 Mar. (85)	2 Mon. .	16 31 51	3 Mar. (62)	0 Sat. .	133·4811	921·3193	209·4684	4667
26 Mar. (85)	3 Tues. .	22 43 59	22 Mar. (81)	6 Fri. .	168·1635	857·3128	260·7789	4668
27 Mar. (86)	5 Thur.	4 56 8	11 Mar. (70)	3 Tues. .	43·8864	704·5568	229·9556	4669
26 Mar. (86)	6 Fri. .	11 8 17	28 Feb. (59)	0 Sat. .	9919·6901	551·8009	199·1324	4670
26 Mar. (85)	0 Sat. .	17 20 26	18 Mar. (77)	6 Fri. .	9954·2915	487·7943	250·4428	4671
26 Mar. (85)	1 Sun. .	23 32 35	7 Mar. (66)	3 Tues. .	9831·6144	375·5[illegible]84	219·6197	4672
27 Mar. (86)	3 Tues. .	5 44 44	26 Mar. (85)	2 Mon. .	9864·6968	271·0319	270·9300	4673
26 Mar. (86)	4 Wed. .	11 56 52	15 Mar. (75)	0 Sat. .	79·0516	154·5676	242·8446	4674
26 Mar. (85)	5 Thur.	18 9 1	4 Mar. (63)	4 Wed. .	9954·7745	1·8117	212·0214	4675

TABLE

CONCURRENT YEAR.								
Kali.	Śaka.	Chaitrādi Vikrama.	Mēshādi (solar) year in Bengal.	Kollam.	A.D.	Jovian Samvatsara. Southern system.	Jovian Samvatsara. Northern system.	Intercalated and suppressed (*ksh.*) lunar months.
1	2	3	3a	4	5	6	7	8
4676	1497	1632	981	749-50	1574-75	8 Bhāva	16 Chitrabhānu	...
4677	1498	1633	982	750-51	1575-76	9 Yuvan	17 Subhānu	...
4678	1499	1634	983	751-52	*1576-77	10 Dhātri	18 Tāraṇa	4 Āshāḍha
4679	1500	1635	984	752-53	1577-78	11 Īśvara	19 Pārthiva	...
4680	1501	1636	985	753-54	1578-79	12 Bahudhānya	20 Vyaya	...
4681	1502	1637	986	754-55	1579-80	13 Pramāthin	21 Sarvajit	1 Chaitra
4682	1503	1638	987	755-56	*1580-81	14 Vikrama	22 Sarvadhārin	...
4683	1504	1639	988	756-57	1581-82	15 Vṛisha	23 Virōdhin	6 Bhādrapada
4684	1505	1640	989	757-58	1582-83	16 Chitrabhānu	24 Vikṛita	...
4685	1506	1641	990	758-59	1583-84	17 Subhānu	25 Khara	...
4686	1507	1642	991	759-60	*1584-85	18 Tāraṇa	26 Nandana	4 Āshāḍha
4687	1508	1643	992	760-61	1585-86	19 Pārthiva	27 Vijaya†	...
4688	1509	1644	993	761-62	1586-87	20 Vyaya	29 *Manmatha*	...
4689	1510	1645	994	762-63	1587-88	21 Sarvajit	30 *Durmukha*	3 Jyēshṭha
4690	1511	1646	995	763-64	*1588-89	22 Sarvadhārin	31 *Hēmalamba*	...
4691	1512	1647	996	764-65	1589-90	23 Virōdhin	32 *Vilamba*	7 Āśvina
4692	1513	1648	997	765-66	1590-91	24 Vikṛita	33 *Vikārin*	...
4693	1514	1649	998	766-67	1591-92	25 Khara	34 *Śārvarin*	...
4694	1515	1650	999	767-68	*1592-93	26 Nandana	35 *Plava*	5 Śrāvaṇa
4695	1516	1651	1000	768-69	1593-94	27 Vijaya	36 *Śubhakṛit*	...
4696	1517	1652	1001	769-70	1594-95	28 Jaya	37 *Śōbhana*	...
4697	1518	1653	1002	770-71	1595-96	29 Manmatha	38 *Krōdhin*	4 Āshāḍha
4698	1519	1654	1003	771-72	*1596-97	30 Durmukha	39 *Viśvāvasu*	...
4699	1520	1655	1004	772-73	1597-98	31 Hēmalamba	40 *Parābhava*	...
4700	1521	1656	1005	773-74	1598-99	32 Vilamba	41 *Plavaṅga*	1 Chaitra

† 28 Jaya was suppressed in the north.

LX—*Contd.*

COMMENCEMENT OF THE									Kali year.
SOLAR YEAR.			LUNI-SOLAR YEAR (MEAN SUNRISE OF DAY ON WHICH CHAITRA SUKLA 1 ENDS).						
Day and month A.D.	Week-day.	Time of true Mēsha-samkrānti.	Day and month, A.D.	Week-day.	*a.*	*b.*	*c.*		
13	14	17	19	20	23	24	25		1
		H. M. S.							
27 Mar. (86)	0 Sat.	0 21 10	23 Mar. (82)	3 Tues.	9989·4569	937·8051	263·3319		4676
2 Mar. (86)	1 Sun.	6 13 19	13 Mar. (72)	1 Sun.	203·8116	821·3407	235·2464		4677
26 Mar. (86)	2 Mon.	12 45 28	1 Mar. (61)	5 Thur.	79·5345	668·5848	204·4231		4678
26 Mar. (85)	3 Tues.	18 57 37	20 Mar. (79)	4 Wed.	114·2169	604·5783	255·7336		4679
27 Mar. (86)	5 Thur.	1 9 45	9 Mar. (68)	1 Sun.	9989·9398	451·8224	224·9104		4680
27 Mar. (86)	6 Fri.	7 21 54	26 Feb. (57)	5 Thur.	9865·6626	299·0664	194·0872		4681
2 Mar. (86)	0 Sat.	13 34 3	16 Mar. (76)	4 Wed.	9900·3450	235·0599	245·3975		4682
26 Mar. (85)	1 Sun.	19 46 12	5 Mar. (64)	1 Sun.	9776·0678	82·3039	214·5744		4683
27 Mar. (86)	3 Tues.	1 58 21	24 Mar. (83)	0 Sat.	9810·7501	18·2935	265·8848		4684
27 Mar. (86)	4 Wed.	8 10 30	14 Mar. (73)	5 Thur.	25·1950	901·8331	237·7994		4685
26 Mar. (86)	5 Thur.	14 22 39	3 Mar. (63)	3 Tues.	239·4598	785·3688	209·7139		4686
26 Mar. (85)	6 Fri.	20 34 47	22 Mar. (81)	2 Mon.	274·1423	721·3623	261·0244		4687
27 Mar. (86)	1 Sun.	2 46 56	11 Mar. (70)	6 Fri.	149·8651	568·6063	230·2012		4688
27 Mar. (86)	2 Mon.	8 59 5	28 Feb. (59)	3 Tues.	25·5879	415·8503	199·3780		4689
26 Mar. (86)	3 Tues.	15 11 14	[illegible] Mar. (78)	2 Mon.	60·2703	351·8438	250·6883		4690
26 Mar. (85)	4 Wed.	21 23 23	7 Mar. (66)	6 Fri.	9935·9932	199·0879	219·8652		4691
27 Mar. (86)	6 Fri.	3 35 32	26 Mar. (85)	5 Thur.	9970·6755	135·0814	271·1756		4692
27 Mar. (86)	0 Sat.	9 47 40	15 Mar. (74)	2 Mon.	9846·3985	982·3255	240·3524		4693
26 Mar. (86)	1 Sun.	15 59 49	4 Mar. (64)	0 Sat.	60·7533	865·8612	212·2669		4694
26 Mar. (85)	2 Mon.	22 11 58	23 Mar. (82)	6 Fri.	95·4356	802·8547	263·5774		4695
27 Mar. (86)	4 Wed.	4 24 8	13 Mar. (72)	4 Wed.	309·7904	685·3903	235·4917		4696
27 Mar. (86)	5 Thur.	10 36 16	2 Mar. (61)	1 Sun.	185·5133	532·6343	204·6687		4697
26 Mar. (86)	6 Fri.	16 48 25	19 Mar. (79)	6 Fri.	9881·5636	432·3362	253·2413		4698
26 Mar. (85)	0 Sat.	23 0 33	8 Mar. (67)	3 Tues.	9757·2865	279·5803	222·4181		4699
27 Mar. (86)	2 Mon.	5 12 42	26 Feb. (57)	1 Sun.	9971·6413	163·1160	194·3328		4700

TABLE

CONCURRENT YEAR								
Kali.	Saka.	Caitrādi Vikrama.	Mēshādi (solar) year in Bengal.	Kollam.	A.D.	Jovian Samvatsara. Southern system.	Jovian Samvatsara. Northern system.	Intercalated and suppressed (*ksh.*) lunar months.
1	2	3	3a	4	5	6	7	8
4701	1522	1657	1006	774-75	1599-00	33 Vikārin .	42 *Kīlaka* .	...
4702	1523	1658	1007	775-76	*1600-01	34 Śārvarin .	43 *Saumya* .	5 Śrāvaṇa .
4703	1524	1659	1008	776-77	1601-02	35 Plava .	44 Sādhāraṇa .	...
4704	1525	1660	1009	777-78	1602-03	36 Subhakṛit .	45 Virōdhakṛit .	...
4705	1526	1661	1010	778-79	1603-04	37 Śōbhana .	46 Paridhāvin .	4 Āshāḍha .
4706	1527	1662	1011	779-80	*1604-05	38 Krōdhin .	47 Pramādin .	...
4707	1528	1663	1012	780-81	1605-06	39 Viśvāvasu .	48 Ānanda .	...
4708	1529	1664	1013	781-82	1606-07	40 Parābhava .	49 Rākshasa .	2 Vaiśākha .
4709	1530	1665	1014	782-83	1607-08	41 Plavaṅga .	50 Ānala .	...
4710	1531	1666	1015	783-84	*1608-09	42 Kīlaka .	51 Piṅgala .	6 Bhādrapada .
4711	1532	1667	1016	784-85	1609-10	43 Saumya .	52 Kālayukta .	...
4712	1533	1668	1017	785-86	1610-11	44 Sādhāraṇa .	53 Siddhārthin .	...
4713	1534	1669	1018	786-87	1611-12	45 Virōdhakṛit .	54 Raudra .	5 Śrāvaṇa .
4714	1535	1670	1019	787-88	*1612-13	46 Paridhāvin .	55 Durmati .	...
4715	1536	1671	1020	788-89	1613-14	47 Pramādin .	56 Dundubhi .	...
4716	1537	1672	1021	789-90	1614-15	48 Ananda .	57 Rudhirōdgārin	3 Jyēshṭha .
4717	1538	1673	1022	790-91	1615-16	49 Rākshasa .	58 Raktāksha .	...
4718	1539	1674	1023	791-92	*1616-17	50 Anala .	59 Krōdhana .	...
4719	1540	1675	1024	792-93	1617-18	51 Piṅgala .	60 Kshaya .	1 Chaitra .
4720	1541	1676	1025	793-94	1618-19	52 Kālayukta .	1 Prabhava .	... .
4721	1542	1677	1026	794-95	1619-20	53 Siddhārthin .	2 Vibhava .	5 Sravana
4722	1543	1678	1027	795-96	*1620-21	54 Raudra .	3 Sukla .	...
4723	1544	1679	1028	796-97	1621-22	55 Durmati .	4 Pramōda .	... .
4724	1545	1680	1029	797-98	1622-23	56 Dundubhi .	5 Prajāpati .	4 Ashāḍha
4725	1546	1681	1030	798-99	1623-24	57 Rudhirōdgārin	6 Aṅgiras .	...

LX—*Contd.*

COMMENCEMENT OF THE									Kali year.
Solar year			Luni-solar year (mean sunrise of day on which Chaitra śukla 1 ends).						
Day and month, A.D.	Week-day.	Time of true Mēsha-saṁkrānti.	Day and month, A.D.	Week-day.	*a.*	*b.*	*c.*		
13	14	17	19	20	23	24	25		1
		H. M. S.							
27 Mar. (86)	3 Tues.	11 24 51	17 Mar. (76)	0 Sat.	6·3237	99·1094	245·5239		4701
26 Mar. (86)	4 Wed.	17 37 0	6 Mar. (66)	5 Thur.	220·6785	982·6452	217·4772		4702
26 Mar. (85)	5 Thur.	23 49 9	25 Mar. (84)	4 Wed.	255·3609	918·6386	268·7875		4703
27 Mar. (86)	0 Sat.	6 1 18	14 Mar. (73)	1 Sun.	131·0837	765·8827	237·9643		4704
27 Mar. (86)	1 Sun.	12 13 26	3 Mar. (62)	5 Thur.	6·8066	613·1267	207·1411		4705
26 Mar. (86)	2 Mon.	18 25 35	21 Mar. (81)	4 Wed.	41·4890	549·1202	258·4516		4706
27 Mar. (86)	4 Wed.	0 37 44	10 Mar. (69)	1 Sun.	9917·2118	396·3643	227·6283		4707
27 Mar. (86)	5 Thur.	6 49 53	27 Feb. (58)	5 Thur.	9792·9346	243·6083	196·8051		4708
27 Mar. (86)	6 Fri.	13 2 2	18 Mar. (77)	4 Wed.	9827·8171	179·6018	248·1155		4709
26 Mar. (86)	0 Sat.	19 14 11	7 Mar. (67)	2 Mon.	41·9718	63·1374	220·0302		4710
27 Mar. (86)	2 Mon.	1 26 20	26 Mar. (85)	1 Sun.	76·7452	999·1309	271·3405		4711
27 Mar. (86)	3 Tues.	7 38 28	16 Mar. (75)	6 Fri.	291·0091	882·6666	243·2551		4712
27 Mar. (86)	4 Wed.	13 50 37	5 Mar. (64)	3 Tues.	166·7320	729·9107	212·4319		4713
26 Mar. (86)	5 Thur.	20 2 46	23 Mar. (83)	2 Mon.	201·4143	665·9042	263·7424		4714
27 Mar. (86)	0 Sat.	2 14 55	12 Mar. (71)	6 Fri.	77·1372	513·1482	232·9181		4715
27 Mar. (86)	1 Sun.	8 27 4	1 Mar. (60)	3 Tues.	9952·8600	360·3923	202·0958		4716
27 Mar. (86)	2 Mon.	14 39 13	20 Mar. (79)	2 Mon.	9987·5423	296·4047	253·4063		4717
26 Mar. (86)	3 Tues.	20 51 21	8 Mar. (68)	6 Fri.	9863·2652	143·6298	222·5831		4718
27 Mar. (86)	5 Thur.	3 3 30	26 Feb. (57)	4 Wed.	77·6201	27·1654	194·4977		4719
27 Mar. (86)	6 Fri.	9 15 39	17 Mar. (76)	3 Tues.	112·3025	963·1589	245·8080		4720
27 Mar. (86)	0 Sat.	15 27 48	6 Mar. (65)	0 Sat.	9988·0252	810·4030	214·9849		4721
26 Mar. (86)	1 Sun.	21 39 57	24 Mar. (84)	6 Fri.	22·7077	746·3965	266·2953		4722
27 Mar. (86)	3 Tues.	3 52 6	14 Mar. (73)	4 Wed.	237·0625	629·9332	238·2099		4723
27 Mar. (86)	4 Wed.	10 4 14	3 Mar. (62)	1 Sun	112·7853	477·1763	207·3865		4724
27 Mar. (86)	5 Thur.	16 16 23	21 Mar. (80)	6 Fri.	9808·8357	376·8780	255·9593		4725

TABLE

CONCURRENT YEAR.								Intercalated and suppressed (*ksh.*) lunar months.
						Jovian Samvatsara.		
Kali.	Śaka.	Chaitrādi Vikrama.	Mēshādi (solar) year in Bengal.	Kollam.	A.D.	Southern system.	Northern system	
1	2	3	3*a*	4	5	6	7	8
4726	1547	1682	1031	799-00	*1624-25	58 Raktāksha	7 Śrīmukha	...
4727	1548	1683	1032	800-01	1625-26	59 Krōdhana	8 Bhāva	2 Vaiśākha
4728	1549	1684	1033	801-02	1626-27	60 Kshaya	9 Yuvan	...
4729	1550	1685	1034	802-03	1627-28	1 Prabhava	10 Dhātṛi	6 Bhādrapada
4730	1551	1686	1035	803-04	*1628-29	2 Vibhava	11 Īśvara	...
4731	1552	1687	1036	804-05	1629-30	3 Śukla	12 Bahudhānya	...
4732	1553	1688	1037	805-06	1630-31	4 Pramōda	13 Pramāthin	5 Śrāvaṇa
4733	1554	1689	1038	806-07	1631-32	5 Prajāpati	14 Vikrama	...
4734	1555	1690	1039	807-08	*1632-33	6 Aṅgiras	15 Vṛisha	...
4735	1556	1691	1040	808-09	1633-34	7 Śrīmukha	16 Chitrabhānu	3 Jyēshṭha
4736	1557	1692	1041	809-10	1634-35	8 Bhāva	17 Subhānu	...
4737	1558	1693	1042	810-11	1635-36	9 Yuvan	18 Tāraṇa	...
4738	1559	1694	1043	811-12	*1636-37	10 Dhātṛi	19 Pārthiva	1 Chaitra
4739	1560	1695	1044	812-13	1637-38	11 Īśvara	20 Vyaya	...
4740	1561	1696	1045	813-14	1638-39	12 Bahudhānya	21 Sarvajit	5 Śrāvaṇa
4741	1562	1697	1046	814-15	1639-40	13 Pramāthin	22 Sarvadhārin	...
4742	1563	1698	1047	815-16	*1640-41	14 Vikrama	23 Virōdhin	...
4743	1564	1699	1048	816-17	1641-42	15 Vṛisha	24 Vikṛita	4 Āshāḍha
4744	1565	1700	1049	817-18	1642-43	16 Chitrabhānu	25 Khara	...
4745	1566	1701	1050	818-19	1643-44	17 Subhānu	26 Nandana	...
4746	1567	1702	1051	819-20	*1644-45	18 Tāraṇa	27 Vijaya	2 Vaiśākha
4747	1568	1703	1052	820-21	1645-46	19 Pārthiva	28 Jaya	...
4748	1569	1704	1053	821-22	1646-47	20 Vyaya	29 Manmatha	6 Bhādrapada
4749	1570	1705	1054	822-23	1647-48	21 Sarvajit	30 Durmukha	...
4750	1571	1706	1055	823-24	*1648-49	22 Sarvadhārin	31 Hēmalamba	...

LX—*Contd.*

COMMENCEMENT OF THE											Kali year.
SOLAR YEAR.					LUNI-SOLAR YEAR (MEAN SUNRISE OF DAY ON WHICH CHAITRA SUKLA 1 ENDS).						
Day and month, A.D.	Week-day.	Time of true Mēsha-saṁkrānti.			Day and month, A.D.	Week-day.	*a.*	*b.*	*c.*		
13	14	17			19	20	23	24	25		1
		H.	M.	S							
26 Mar. (86)	6 Fri. .	22	28	32	10 Mar. (70)	4 Wed	23·1906	260·4138	227·8739		4726
27 Mar. (86)	1 Sun.	4	40	41	27 Feb. (58)	1 Sun. .	9898·9134	107·6578	197·0507		4727
27 Mar. (86)	2 Mon. .	10	52	50	18 Mar. (77)	0 Sat. .	9933·5958	43·6413	248·3610		4728
27 Mar. (86)	3 Tues.	17	4	59	8 Mar. (67)	5 Thur	147·9506	927·1870	220·2757		4729
26 Mar. (86)	4 Wed.	23	17	7	26 Mar. (86)	4 Wed.	183·6330	862·1804	271·5861		4730
27 Mar. (86)	6 Fri. .	5	29	16	15 Mar. (74)	1 Sun. .	58·3558	710·4245	240·7629		4731
27 Mar. (86)	0 Sat. .	11	41	25	5 Mar. (64)	6 Fri. .	272·7107	593·9602	212·6774		4732
27 Mar. (86)	1 Sun. .	17	53	34	23 Mar. (82)	4 Wed.	9968·7611	493·6620	261·2501		4733
27 Mar. (87)	3 Tues.	0	5	43	11 Mar (71)	1 Sun. .	9844·4840	340·9061	230·4269		4734
27 Mar. (86)	4 Wed.	6	17	52	28 Feb. (59)	5 Thur.	9720·2067	188·1500	199·6037		4735
27 Mar. (86)	5 Thur.	12	30	1	19 Mar. (78	4 Wed.	9754·8891	124·14[illegible]6	250·9140		4736
27 Mar. (86)	6 Fri .	18	42	9	9 Mar. (68,	2 Mon. .	9969·2440	7·6793	222·8286		4737
27 Mar. (87)	1 Sun. .	0	54	18	27 Feb. (58)	0 Sat. .	183·5888	891·2150	194·7433		4738
27 Mar. (86)	2 Mon. .	7	6	27	17 Mar. (76)	6 Fri. .	218·2812	827·2084	246·0536		4739
27 Mar. (86)	3 Tues.	13	18	36	6 Mar. (65)	3 Tues.	94·0040	674·4525	215·2305		4740
27 Mar. (86)	4 Wed.	19	30	45	25 Mar. (84)	2 Mon .	128·6865	610·4460	266·5408		4741
27 Mar. (87)	6 Fri. .	1	42	54	13 Mar. (73)	6 Fri. .	4·3092	457·6800	235·7177		4742
27 Mar. (86)	0 Sat. .	7	55	2	2 Mar. (61)	3 Tues.	9880·1321	304·9341	204·8934		4743
27 Mar. (86)	1 Sun. .	14	7	11	21 Mar. (80)	2 Mon. .	9914·8145	240·9275	256·2049		4744
27 Mar. (86)	2 Mon. .	20	19	20	10 Mar. (69)	6 Fri. .	9790·5374	88·1716	225·3816		4745
27 Mar. (87)	4 Wed.	2	31	29	28 Feb. (59)	4 Wed	4·8921	971·7073	197·2962		4746
27 Mar. (86)	5 Thur.	8	43	38	18 Mar. (77)	3 Tues.	39·5746	907·7008	248·6066		4747
27 Mar. (86)	6 Fri. .	14	55	47	8 Mar. (67)	1 Sun. .	253·9294	791·2365	220·4233		4748
27 Mar. (86)	0 Sat. .	21	7	55	27 Mar. (86)	0 Sat. .	288·6117	727·2299	271·8316		4749
27 Mar. (87)	2 Mon. .	3	20	4	15 Mar. (75)	4 Wed.	164·4346	574·4740	241·0081		4750

TABLE

CONCURRENT YEAR.								Intercalated and suppressed (*ksh.*) lunar months.
						Jovian Samvatsara.		
Kali.	Śaka.	Chaitrādi Vikrama.	Mēshādi (solar) year in Bengal.	Kollam.	A.D.	Southern system.	Northern system.	
1	2	3	3a	4	5	6	7	8
4751	1572	1707	1056	824-25	1649-50	23 Virōdhin .	32 Vilamba .	5 Śrāvaṇa .
4752	1573	1708	1057	825-26	1650-51	24 Vikṛita .	33 Vikārin .	...
4753	1574	1709	1058	826-27	1651-52	25 Khara .	34 Śārvarin .	...
4754	1575	1710	1059	827-28	*1652-53	26 Nandana .	35 Plava .	3 Jyēshṭha
4755	1576	1711	1060	828-29	1653-54	27 Vijaya .	36 Śubhakṛit .	...
4756	1577	1712	1061	829-30	1654-55	28 Jaya .	37 Śōbhana	{ 7 Āsvina 10 *Pausha* (*ksh.*) }
4757	1578	1713	1062	830-31	1655-56	29 Manmatha .	38 Krōdhin	1 Chaitra
4758	1579	1714	1063	831-32	*1656-57	30 Durmukha .	39 Visvāvasu .	...
4759	1580	1715	1064	832-33	1657-58	31 Hēmalamba .	40 Parābhava .	5 Śrāvaṇa .
4760	1581	1716	1065	833-34	1658-59	32 Vilamba .	41 Plavaṅga .	...
4761	1582	1717	1066	834-35	1659-60	33 Vikārin .	42 Kīlaka .	...
4762	1583	1718	1067	835-36	*1660-61	34 Śārvarin .	43 Saumya .	4 Āshāḍha .
4763	1584	1719	1068	836-37	1661-62	35 Plava .	44 Sādhārana .	...
4764	1585	1720	1069	837-38	1662-63	36 Śubhakṛit .	45 Virōdhakṛit .	...
4765	1586	1721	1070	838-39	1663-64	37 Śōbhana .	46 Paridhāvin	2 Vaisākha .
4766	1587	1722	1071	839-40	*1664-65	38 Krōdhin .	47 Pramādin .	...
4767	1588	1723	1072	840-41	1665-66	39 Visvāvasu .	48 Ānanda .	6 Bhādrapada .
4768	1589	1724	1073	841-42	1666-67	40 Parābhava .	49 Rākshasa .	...
4769	1590	1725	1074	842-43	1667-68	41 Plavaṅga .	50 Anala .	...
4770	1591	1726	1075	843-44	*1668-69	42 Kīlaka .	51 Piṅgala .	4 Āshāḍha .
4771	1592	1727	1076	844-45	1669-70	43 Saumya .	52 Kālayukta .	...
4772	1593	1728	1077	845-46	1670-71	44 Sādhārana .	53 Siddhārthin .	...
4773	1594	1729	1078	846-47	1671-72	45 Virōdhakṛit .	54 Raudra† .	3 Jyēshtha
4774	1595	1730	1079	847-48	*1672-73	46 Paridhāvin .	56 *Dundubhi* .	...
4775	1596	1731	1080	848-49	1673-74	47 Pramādin .	57 *Rudhirōdgārin*	{ 7 Āsvina 11 *Māgha* (*ksh.*) }

† 55 Durmati was suppressed in the north.

LX—*Contd.*

COMMENCEMENT OF THE										
SOLAR YEAR.					LUNI-SOLAR YEAR (MEAN SUNRISE OF DAY ON WHICH CHAITA SUKLA 1 ENDS).					Kali year.
Day and month, A.D.	Week-day.	Time of true Mēsha-saṁkrānti.			Day and month, A.D.	Week-day.	*a.*	*b.*	*c.*	
13	14	17			19	20	23	24	25	1
		H.	M.	S.						
27 Mar. (86)	3 Tues.	9	32	13	4 Mar. (63)	1 Sun. .	40·0575	421·6930	210·1852	475
27 Mar. (86)	4 Wed.	15	44	22	23 Mar. (82)	0 Sat. .	74·7398	357·6915	261·4957	4752
27 Mar. (86)	5 Thur.	21	56	31	19 Mar. (71)	4 Wed.	9950·4627	204·9916	230·6724	4753
27 Mar. (87)	0 Sat. .	4	8	41	29 Feb. (60)	1 Sun. .	9826·1855	52·1996	199·8492	4754
27 Mar. (86)	1 Sun. .	10	20	49	19 Mar. (78)	0 Sat. .	9860·8679	988·1931	251·1596	4755
27 Mar. (86)	2 Mon. .	16	32	58	9 Mar. (68)	5 Thur.	75·2227	871·7289	223·0742	4756
27 Mar. (86)	3 Tues.	22	45	7	26 Feb. (57)	2 Mon. .	9950·9456	718·9728	192·2510	4757
27 Mar. (87)	5 Thur.	4	57	16	16 Mar. (76)	1 Sun. .	9985·6280	654·9663	243·5614	4758
27 Mar. (86)	6 Fri.	11	9	25	6 Mar. (65)	6 Fri. .	199·9828	538·5020	215·4762	4759
27 Mar. (86)	0 Sat.	17	21	34	24 Mar. (83)	4 Wed.	9896·0332	438·2039	264·0487	4760
27 Mar. (86)	1 Sun.	23	33	43	13 Mar. (72)	1 Sun. .	9771·7560	285·5479	233·2254	4761
27 Mar. (87)	3 Tues.	5	45	50	2 Mar. (62)	6 Fri. .	9986·1109	168·9836	205·1399	4762
27 Mar. (86)	4 Wed.	11	57	59	21 Mar. (80)	5 Thur.	20·7932	104·9771	256·4504	4763
27 Mar. (86)	5 Thur.	18	10	8	10 Mar. (69)	2 Mon. .	9896·5161	952·2211	225·6272	4764
28 Mar. (87)	0 Sat. .	0	22	17	28 Feb. (59)	0 Sat. .	110·8709	835·7568	197·5418	4765
27 Mar. (87)	1 Sun. .	6	34	26	18 Mar. (78)	6 Fri. .	145·5534	771·7503	248·8521	4766
27 Mar. (83)	2 Mon. .	12	46	35	7 Mar. (66)	3 Tues.	21·2761	618·9944	218·0290	4767
27 Mar. (86)	3 Tues.	18	58	43	26 Mar. (85)	2 Mon. .	55·9585	554·9879	269·3394	4768
28 Mar. (87)	5 Thur.	1	10	52	15 Mar. (74)	6 Fri. .	9931·6814	402·2319	238·5162	4769
27 Mar. (87)	6 Fri. .	7	23	1	3 Mar. (65)	3 Tues.	9807·4042	249·4760	207·6929	4770
27 Mar. (86)	0 Sat. .	13	35	10	22 Mar. (81)	2 Mon. .	9842·0866	185·4694	259·0034	4771
27 Mar. (86)	1 Sun. .	19	47	19	12 Mar. (71)	0 Sat. .	56·4415	69·0051	230·9180	4772
28 Mar. (87)	3 Tues.	1	59	28	1 Mar. (60)	4 Wed.	9932·1643	916·2491	200·0948	4773
27 Mar. (87)	4 Wed.	8	11	36	19 Mar. (79)	3 Tues.	9966·[illegible]	852·2426	251·4051	4774
27 Mar. (86)	5 Thur.	14	23	45	9 Mar. (68)	1 Sun. .	181·2015	735·7788	223·3197	4775

TABLE

CONCURRENT YEAR.								Intercalated and suppressed (*ksh.*) lunar months.
						Jovian Samvatsara.		
Kali.	Śaka.	Chaitrādi Vikrama.	Mēshādi (solar) year in Bengal.	Kollam	A. D.	Southern system.	Northern system.	
1	2	3	3a	4	5	6	7	8
4776	1597	1732	1081	849-50	1674-75	48 Ānanda .	58 *Raktāksha* .	1 Chaitra
4777	1598	1733	1082	850-51	1675-76	49 Rākshasa .	59 *Krōdhana* .	...
4778	1599	1734	1083	851-52	*1676-77	50 Anala .	60 *Kshaya* .	5 Śrāvana .
4779	1600	1735	1084	852-53	1677-78	51 Piṅgala .	1 *Prabhava* .	...
4780	1601	1736	1085	853-54	1678-79	52 Kālayukta .	2 *Vibhava* .	...
4781	1602	1737	1086	854-55	1679-80	53 Siddhārthin .	3 *Śukla* .	3 Jyēshṭha ‡ .
4782	1603	1738	1087	855-56	*1680-81	54 Raudra .	4 *Pramōda* .	...
4783	1604	1739	1088	856-57	1681-82	55 Durmati .	5 *Prajāpati* .	...
4784	1605	1740	1089	857-58	1682-83	56 Dundubhi .	6 *Aṅgiras* .	2 Vaiśākha .
4785	1606	1741	1090	858-59	1683-84	57 Rudhirōdgārin	7 *Śrīmukha* .	...
4786	1607	1742	1091	859-60	*1684-85	58 Raktāksha .	8 *Bhāva* .	6 Bhādrapada .
4787	1608	1743	1092	860-61	1685-86	59 Krōdhana .	9 *Yuvan* .	...
4788	1609	1744	1093	861-62	1686-87	60 Kshaya .	10 Dhātri .	...
4789	1610	1745	1094	862-63	1687-88	1 Prabhava .	11 Īśvara .	4 Āshāḍha .
4790	1611	1746	1095	863-64	*1688-89	2 Vibhava .	12 Bahudhānya .	...
4791	1612	1747	1096	864-65	1689-90	3 Śukla .	13 Pramathin .	...
4792	1613	1748	1097	865-66	1690-91	4 Pramōda .	14 Vikrama .	3 Jyēshṭha .
4793	1614	1749	1098	866-67	1691-92	5 Prajāpati .	15 Vṛisha .	...
4794	1615	1750	1099	867-68	*1692-93	6 Aṅgiras .	16 Chitrabhānu .	7 Āśvina .
4795	1616	1751	1100	868-69	1693-94	7 Śrīmukha .	17 Subhānu .	...
4796	1617	1752	1101	869-70	1694-95	8 Bhāva .	18 Tārana .	...
4797	1618	1753	1102	870-71	1695-96	9 Yuvan .	19 Pārthiva .	5 Śrāvaṇa .
4798	1619	1754	1103	871-72	*1696-97	10 Dhātri .	20 Vyaya .	...
4799	1620	1755	1104	872-73	1697-98	11 Īśvara .	21 Sarvajit .	...
4800	1621	1756	1105	873-74	1698-99	12 Bahudhānya .	22 Sarvadhārin .	3 Jyēshṭha .

‡ See Remarks, p. 163 above.

LX—*Contd.*

COMMENCEMENT OF THE											
Solar year.					Luni-solar year (mean sunrise of day on which Chaitra Śukla 1 ends).					Kali year.
Day and month, A.D.	Week-day.	Time of true Mēsha-saiṅkrānti.			Day and month, A.D.	Week-day.	*a.*	*b.*	*c.*	
13	14	17			19	20	23	24	25	1
		H.	M.	S.						
27 Mar. (86)	6 Fri.	20	35	54	26 Feb. (57)	5 Thur.	56·9244	583·0221	192·4966	4776
28 Mar. (87)	1 Sun.	2	48	3	17 Mar. (76)	4 Wed.	91·6067	519·0158	243·8070	4777
27 Mar. (87)	2 Mon.	9	0	12	5 Mar. (65)	1 Sun.	9967·3296	366·2599	212·9837	4778
27 Mar. (86)	3 Tues.	15	12	21	24 Mar. (83)	0 Sat.	2·0120	302·2534	264·2942	4779
27 Mar. (86)	4 Wed.	21	24	30	13 Mar. (72)	4 Wed.	9877·7348	149·4947	233·4710	4780
28 Mar. (87)	6 Fri.	3	36	38	3 Mar. (62)	2 Mon.	92·0896	33·0331	205·3855	4781
27 Mar (87)	0 Sat.	9	48	47	21 Mar. (81)	1 Sun.	126·7720	969·0266	256·6959	4782
27 Mar (86)	1 Sun.	16	0	56	10 Mar. (69)	5 Thur.	2·4949	816·2706	225·8727	4783
27 Mar (86)	2 Mon.	22	13	5	28 Feb. (59)	3 Tues.	216·8496	699·8023	197·7874	4784
28 Mar. (87)	4 Wed.	4	25	14	19 Mar. (78)	2 Mon.	251·5321	635·7998	249·0977	4785
27 Mar. (87)	5 Thur.	10	37	23	7 Mar. (67)	6 Fri.	127·2548	483·0439	218·2745	4786
27 Mar. (86)	6 Fri.	16	49	31	25 Mar. (84)	4 Wed.	9823·3054	382·7457	266·8471	4787
27 Mar. (86)	0 Sat.	23	1	40	15 Mar. (74)	2 Mon.	37·6601	266·2813	238·7618	4788
28 Mar. (87)	2 Mon.	5	13	49	4 Mar. (63)	6 Fri.	9913·3830	113·5254	207·9385	4789
27 Mar. (87)	3 Tues.	11	25	58	22 Mar. (82)	5 Thur.	9948·0654	49·5189	259·2489	4790
27 Mar. (86)	4 Wed.	17	38	7	12 Mar. (71)	3 Tues.	162·4203	933·0536	231·1635	4791
27 Mar. (86)	5 Thur.	23	50	16	1 Mar. (60)	0 Sat.	38·1430	780·2987	200·3403	4792
28 Mar. (87)	0 Sat.	6	2	24	20 Mar. (79)	6 Fri.	72·8254	716·2821	251·6507	4793
27 Mar. (87)	1 Sun.	12	14	33	8 Mar. (68)	3 Tues.	9948·5483	563·5362	220·8275	4794
27 Mar. (86)	2 Mon.	18	26	42	27 Mar. (86)	2 Mon.	9983·2306	499·5297	272·1379	4795
28 Mar. (87)	4 Wed.	0	38	51	16 Mar. (75)	6 Fri.	9858·9535	346·7737	241·3148	4796
28 Mar. (87)	5 Thur.	6	51	0	5 Mar. (64)	3 Tues.	9734·6764	194·0177	210·4915	4797
27 Mar (87)	6 Fri	13	3	9	23 Mar. (83)	2 Mon.	9769·3587	130·0112	261·8019	4798
27 Mar. (86)	0 Sat.	19	15	17	13 Mar. (72)	0 Sat.	9983·7135	13·5469	233·7165	4799
28 Mar. (87)	2 Mon.	1	27	26	3 Mar. (62)	5 Thur.	198·0684	897·0827	205·6311	4800

TABLE

CONCURRENT YEAR.						Jovian Samvatsara.		Intercalated and suppressed (*ksh.*) lunar months.
Kali.	Śaka.	Chaitrādi Vikrama.	Mēshādi (solar) year in Bengal.	Kollam.	A.D.	Southern system.	Northern system.	
1	2	3	3a	4	5	6	7	8
4801	1622	1757	1106	874-75	1699-00	13 Pramāthin	23 Virōdhin	...
4802	1623	1758	1107	875-76	*1700-01	14 Vikrama	24 Vikṛita	...
4803	1624	1759	1108	876-77	1701-02	15 Vṛisha	25 Khara	2 Vaiśākha
4804	1625	1760	1109	877-78	1702-03	16 Chitrabhānu	26 Nandana	...
4805	1626	1761	1110	878-79	1703-04	17 Subhānu	27 Vijaya	6 Bhādrapada
4806	1627	1762	1111	879-80	*1704-05	18 Tāraṇa	28 Jaya	...
4807	1628	1763	1112	880-81	1705-06	19 Pārthiva	29 Manmatha	...
4808	1629	1764	1113	881-82	1706-07	20 Vyaya	30 Durmukha	4 Āshāḍha
4809	1630	1765	1114	882-83	1707-08	21 Sarvajit	31 Hēmalamba	...
4810	1631	1766	1115	883-84	*1708-09	22 Sarvadhārin	32 Vilamba	...
4811	1632	1767	1116	884-85	1709-10	23 Virōdhin	33 Vikārin	3 Jyēshṭha
4812	1633	1768	1117	885-86	1710-11	24 Vikṛita	34 Śārvarin	...
4813	1634	1769	1118	886-87	1711-12	25 Khara	35 Plava	7 Āśvina
4814	1635	1770	1119	887-88	*1712-13	26 Nandana	36 Śubhakṛit	...
4815	1636	1771	1120	888-89	1713-14	27 Vijaya	37 Śōbhana	...
4816	1637	1772	1121	889-90	1714-15	28 Jaya	38 Krōdhin	5 Śrāvaṇa
4817	1638	1773	1122	890-91	1715-16	29 Manmatha	39 Viśvāvasu	...
4818	1639	1774	1123	891-92	*1716-17	30 Durmukha	40 Parābhava	...
4819	1640	1775	1124	892-93	1717-18	31 Hēmalamba	41 Plavaṅga	4 Āshāḍha†
4820	1641	1776	1125	893-94	1718-19	32 Vilamba	42 Kīlaka	...
4821	1642	1777	1126	894-95	1719-20	33 Vikārin	43 Saumya	...
4822	1643	1778	1127	895-96	*1720-21	34 Śārvarin	44 Sādhāraṇa	1 Chaitra
4823	1644	1779	1128	896-97	1721-22	35 Plava	45 Virōdhakṛit	...
4824	1645	1780	1129	897-98	1722-23	36 Śubhakṛit	46 Paridhāvin	6 Bhādrapada
4825	1646	1781	1130	898-99	1723-24	37 Śōbhana	47 Pramādin	...

† See Remarks, p. 163 above.

LX—*Contd.*

COMMENCEMENT OF THE									Kali year.
Solar year.			Luni-solar year (mean sunrise of day on which Chaitra śukla 1 ends).						
Day and month, A.D.	Week-day.	Time of true Mēsha saṁkrānti.	Day and month, A.D.	Week-day.	*a.*	*b.*	*c.*		
13	14	17	19	20	23	24	25		1
		H. M. S.							
28 Mar. (87)	3 Tues.	7 39 35	22 Mar. (81)	4 Wed.	232·7508	833·0761	256·8610		4801
27 Mar. (87)	4 Wed.	13 51 44	10 Mar (70)	1 Sun.	108·4737	680·3202	226·0378		4802
27 Mar. (86)	5 Thur.	20 3 53	27 Feb. (58)	5 Thur.	9984·1965	527·5642	195·2146		4803
28 Mar. (87)	0 Sat.	2 16 2	18 Mar. (77)	4 Wed.	18·8789	463·5577	246·5249		4804
28 Mar. (87)	1 Sun.	8 28 11	7 Mar. (66)	1 Sun.	9894·6017	310·8017	215·7018		4805
27 Mar. (87)	2 Mon.	14 40 19	25 Mar. (85)	0 Sat.	9929·2842	246·7952	267·0122		4806
27 Mar. (86)	3 Tues.	20 52 28	14 Mar. (73)	4 Wed.	9805·0069	94·9493	236·1890		4807
28 Mar. (87)	5 Thur.	3 4 37	4 Mar. (63)	2 Mon.	19·3618	977·5750	208·1035		4808
28 Mar. (87)	6 Fri.	9 16 46	23 Mar. (82)	1 Sun.	54·0442	913·5685	259·4140		4809
27 Mar (87)	0 Sat.	15 28 55	12 Mar. (72)	6 Fri.	268·3990	797·1041	231·3286		4810
27 Mar. (86)	1 Sun.	21 41 4	1 Mar. (60)	3 Tues.	144·1218	644·3482	200·5053		4811
28 Mar. (87)	3 Tues.	3 53 12	20 Mar. (79)	2 Mon.	178·8042	580·3416	251·8157		4812
28 Mar. (87)	4 Wed.	10 5 21	9 Mar. (68)	6 Fri.	54·5271	427·5857	220·9926		4813
27 Mar. (87)	5 Thur.	16 17 30	26 Mar. (86)	4 Wed.	9750·5774	327·2876	269·5652		4814
27 Mar. (86)	6 Fri.	22 29 39	16 Mar. (75))	2 Mon.	9964·0323	210·8232	241·4798		4815
28 Mar. (87)	1 Sun.	4 41 48	5 Mar. (64)	6 Fri.	9840·6552	58·0673	210·6565		4816
28 Mar. (87)	2 Mon.	10 53 57	24 Mar. (83)	5 Thur.	9875·3375	994·0697	261·9670		4817
27 Mar. (87)	3 Tues.	17 6 5	13 Mar. (73)	3 Tues.	89·6923	877·5964	233·8816		4818
27 Mar (86)	4 Wed.	23 18 14	3 Mar. (62)	1 Sun.	304·0472	761·1321	205·7961		4819
28 Mar. (87)	6 Fri.	5 30 23	21 Mar. (80)	6 Fri.	0·0976	660·8340	254·3677		4820
28 Mar (87)	0 Sat.	11 42 32	11 Mar. (70))	4 Wed.	214·4524	544·3697	226·2833		4821
27 Mar. (87)	1 Sun.	17 54 41	28 Feb. (59))	1 Sun.	90·1752	391·6138	196·4602		4822
28 Mar (87)	3 Tues.	0 6 50	17 Mar. (76)	6 Fri.	9756·2257	291·3156	244·0328		4823
28 Mar. (87)	4 Wed.	6 18 58	7 Mar. (66)	4 Wed.	0·5804	174·8513	215·9473		4824
28 Mar (87)	5 Thur.	12 31 7	26 Mar (85)	3 Tues.	35·2629	110·8447	267·2577		4825

TABL

CONCURRENT YEAR.								Intercalated and suppressed (*ksh.*) lunar months.
Kali.	Śaka.	Chaitrādi Vikrama.	Mēshādi (solar) year in Bengal.	Kollam.	A.D.	Jovian Samvatsara. Southern system.	Jovian Samvatsara. Northern system.	
1	2	3	3*a*	4	5	6	7	8
4826	1647	1782	1131	899-00	*1724-25	38 Krōdhin	48 Ānanda	...
4827	1648	1783	1132	900-01	1725-26	39 Visvāvasu	49 Rākshasa	4 Āshāḍha
4828	1649	1784	1133	901-02	1726-27	40 Parābhava	50 Anala	...
4829	1650	1785	1134	902-03	1727-28	41 Plavaṅga	51 Piṅgala	...
4830	1651	1786	1135	903-04	*1728-29	42 Kīlaka	52 Kālayukta	3 Jyēshṭha
4831	1652	1787	1136	904-05	1729-30	43 Saumya	53 Siddhārthin	...
4832	1653	1788	1137	905-06	1730-31	44 Sādhāraṇa	54 Raudra	7 Āśvina
4833	1654	1789	1138	906-07	1731-32	45 Virōdhakṛit	55 Durmati	...
4834	1655	1790	1139	907-08	*1732-33	46 Paridhāvin	56 Dundubhi	...
4835	1656	1791	1140	908-09	1733-34	47 Pramādin	57 Rudhirōdgārin	5 Śrāvaṇa
4836	1657	1792	1141	909-10	1734-35	48 Ānanda	58 Raktāksha	...
4837	1658	1793	1142	910-11	1735-36	49 Rākshasa	59 Krōdhana	...
4838	1659	1794	1143	911-12	*1736-37	50 Anala	60 Kshaya	4 Āshāḍha†
4839	1660	1795	1144	912-13	1737-38	51 Piṅgala	1 Prabhava	...
4840	1661	1796	1145	913-14	1738-39	52 Kālayukta	2 Vibhava	...
4841	1662	1797	1146	914-15	1739-40	53 Siddhārthin	3 Śukla	1 Chaitra
4842	1663	1798	1147	915-16	*1740-41	54 Raudra	4 Pramōda	...
4843	1664	1799	1148	916-17	1741-42	55 Durmati	5 Prajāpati	5 Śrāvaṇa
4844	1665	1800	1149	917-18	1742-43	56 Dundubhi	6 Aṅgiras	...
4845	1666	1801	1150	918-19	1743-44	57 Rudhirōdgārin	7 Śrīmukha	...
4846	1667	1802	1151	919-20	*1744-45	58 Raktāksha	8 Bhāva	4 Āshāḍha
4847	1668	1803	1152	920-21	1745-46	59 Krōdhana	9 Yuvan	...
4848	1669	1804	1153	921-22	1746-47	60 Kshaya	10 Dhātṛi	...
4849	1670	1805	1154	922-23	1747-48	1 Prabhava	11 Īśvara	2 Vaiśākha
4850	1671	1806	1155	923-24	*1748-49	2 Vibhava	12 Bahudhānya	...
4851	1672	1807	1156	924-25	1749-50	3 Śukla	13 Pramāthin	6 Bhādrapada†
4852	1673	1808	1157	925-26	1750-51	4 Pramōda	14 Vikrama	...

† See Remarks, p. 163 above.

LX—*Contd.*

COMMENCEMENT OF THE										
Solar year.					Luni-solar year (mean sunrise of day on which Chaita śukla 1 ends).					Kali year.
Day and month, A.D.	Week-day.	Time of true Mēsha-saṁkrānti.			Day and month, A.D.	Week-day.	*a.*	*b.*	*c.*	
13	14	17			19	20	23	24	25	1
		H.	M.	S.						
27 Mar (87)	6 Fri. .	18	43	16	14 Mar. (74)	0 Sat. .	9910·9857	958·0888	236·4346	4826
28 Mar (87)	1 Sun. .	0	55	25	4 Mar. (63)	5 Thur.	125·3406	841·6245	208·3491	4827
28 Mar. (87)	2 Mon.	7	7	34	23 Mar. (82)	4 Wed.	160·0229	777·6180	259·6595	4828
28 Mar. (87)	3 Tues.	13	19	43	12 Mar (71)	1 Sun.	35·7458	624·8621	228·8363	4829
27 Mar (87)	4 Wed.	19	31	52	29 Feb. (60)	5 Thur.	9911·4686	472·1060	198·0132	4830
28 Mar. (87)	6 Fri. .	1	44	0	19 Mar. (78))	4 Wed.	9946·1510	408·0996	249·3235	4831
28 Mar. (87)	0 Sat. .	7	56	9	8 Mar. (67)	1 Sun.	9821·8738	255·3436	218·5003	4832
28 Mar. (87)	1 Sun. .	14	8	18	27 Mar (86)	0 Sat. .	9856·5562	191·3371	269·8107	4833
27 Mar. (87)	2 Mon. .	20	20	27	16 Mar. (76)	5 Thur.	70·9111	74·8718	241·7254	4834
28 Mar. (87)	4 Wed.	2	32	36	5 Mar. (64)	2 Mon.	9946·6339	922·0868	210·9021	4835
28 Mar. (87)	5 Thur.	8	44	45	24 Mar. (83)	1 Sun. .	9981·3163	858·1103	262·2125	4836
28 Mar. (87)	6 Fri. .	14	56	53	14 Mar. (73)	6 Fri. .	195·6711	741·6459	234·1271	4837
27 Mar. (87)	0 Sat. .	21	9	2	2 Mar. (62))	3 Tues.	71·3840	588·8900	203·3039	4838
28 Mar. (87)	2 Mon.	3	21	11	21 Mar. (80)	2 Mon.	106·0763	524·8835	254·6143	4839
28 Mar. (87)	3 Tues.	9	33	20	10 Mar. (69)	6 Fri. .	9981·7992	372·1276	22[illegible]·7911	4840
28 Mar (87)	4 Wed.	15	45	29	27 Feb (58))	3 Tues	9857·5221	219·3716	192·9679	4841
27 Mar. (87)	5 Thur.	21	57	38	17 Mar. (77)	2 Mon. .	9892·2044	155·3650	244·2783	4842
28 Mar. (87)	0 Sat. .	4	9	46	7 Mar. (66)	0 Sat. .	106·5592	38·9008	216·1929	4843
28 Mar. (87)	1 Sun. .	10	21	55	26 Mar. (85)	6 Fri. .	141·2417	974·8942	267·5033	4844
28 Mar. (87)	2 Mon. .	16	34	4	15 Mar. (74)	3 Tues.	16·9645	822·1383	236·6801	4845
27 Mar. (87)	3 Tues.	22	46	13	4 Mar. (64)	1 Sun. .	231·3193	705·6740	208·5946	4846
28 Mar. (87)	5 Thur.	4	58	22	23 Mar. (82)	0 Sat. .	266·0017	641·6675	259·9051	4847
28 Mar. (87)	6 Fri. .	11	10	31	12 Mar. (71)	4 Wed.	141·7246	488·9116	229·0819	4848
28 Mar. (87)	0 Sat .	17	22	39	1 Mar. (60)	1 Sun.	17·4473	336·1555	198·2587	4849
27 Mar. (87)	1 Sun. .	23	34	48	19 Mar. (79)	0 Sat. .	52·1298	272·1491	249·5690	4850
28 Mar (87)	3 Tues.	5	46	57	8 Mar (67)	4 Wed.	9928·8526	119·3931	218·7459	4851
28 Mar (87)	4 Wed.	11	59	6	27 Mar. (86)	3 Tues.	9962·5349	55·3866	270·0563	4852

TABLE

CONCURRENT YEAR.									Intercalated and suppressed (*ksh.*) lunar months.
Kali	Śaka.	Chaitrādi Vikrama.	Meshādi (solar) year in Bengal.	Kollam.	A.D.	Jovian Samvatsara. Southern system.	Jovian Samvatsara. Northern system.		
1	2	3	3*a*	4	5	6	7		8
4853	1674	1809	1158	926-27	1751-52	5 Prajāpati	15 Vṛisha		...
4854	1675	1810	1159	927-28	*1752-53	6 Aṅgiras	16 Chitrabhānu		5 Śrāvaṇa
4855	1676	1811	1160	928-29	1753-54	7 Śrīmukha	17 Subhānu		...
4856	1677	1812	1161	929-30	1754-55	8 Bhāva	18 Tāraṇa		...
4857	1678	1813	1162	930-31	1755-56	9 Yuvan	19 Pārthiva		3 Jyēshṭha
4858	1679	1814	1163	931-32	*1756-57	10 Dhātṛi	20 Vyaya*		...
4859	1680	1815	1164	932-33	1757-58	11 Īśvara	22 *Sarvadhārin*		...
4860	1681	1816	1165	933-34	1758-59	12 Bahudhānya	23 *Virōdhin*		1 Chaitra
4861	1682	1817	1166	934-35	1759-60	13 Pramāthin	24 *Vikṛita*		...
4862	1683	1818	1167	935-36	*1760-61	14 Vikrama	25 *Khara*		5 Śrāvaṇa
4863	1684	1819	1168	936-37	1761-62	15 Vṛisha	26 *Nandana*		...
4864	1685	1820	1169	937-38	1762-63	16 Chitrabhānu	27 *Vijaya*		...
4865	1686	1821	1170	938-39	1763-64	17 Subhānu	28 *Jaya*		4 Āshāḍha
4866	1687	1822	1171	939-40	*1764-65	18 Tāraṇa	29 *Manmatha*		...
4867	1688	1823	1172	940-41	1765-66	19 Pārthiva	30 *Durmukha*		...
4868	1689	1824	1173	941-42	1766-67	20 Vyaya	31 *Hēmalamba*		2 Vaiśākha
4869	1690	1825	1174	942-43	1767-68	21 Sarvajit	32 *Vilamba*		...
4870	1691	1826	1175	943-44	*1768-69	22 Sarvadhārin	33 *Vikārin*		6 Bhādrapada
4871	1692	1827	1176	944-45	1769-70	23 Virōdhin	34 *Śārvarin*		...
4872	1693	1828	1177	945-46	1770-71	24 Vikṛita	35 *Plava*		...
4873	1694	1829	1178	946-47	1771-72	25 Khara	36 *Śubhakṛit*		5 Śrāvaṇa
4874	1695	1830	1179	947-48	*1772-73	26 Nandana	37 Śōbhana		...
4875	1696	1831	1180	948-49	1773-74	27 Vijaya	38 Krōdhin		...
4876	1697	1832	1181	949-50	1774-75	28 Jaya	39 Viśvāvasu		3 Jyēshṭha
4877	1698	1833	1182	950-51	1775-76	29 Manmatha	40 Parābhava		...

* 21 Sarvajit was suppressed in the north.

LX—*Contd.*

COMMENCEMENT OF THE									
SOLAR YEAR.			LUNI-SOLAR YEAR (MEAN SUNRISE OF DAY ON WHICH CHAITRA ŚUKLA 1 ENDS).						Kali year.
Day and month, A.D.	Week-day.	Time of true Mēsha-saṁkrānti.	Day and month, A.D.	Week-day.	*a.*	*b.*	*c.*		
13	14	17	19	20	23	24	25		1
28 Mar. (87)	5 Thur. .	18 11 15	17 Mar. (76)	1 Sun. .	176·8898	938·9222	241·9708		4853
28 Mar. (88)	0 Sat. .	0 23 24	5 Mar. (65)	5 Thur.	52·6127	876·1662	211·1475		4854
8 Apr. (98)*	1 Sun. .	6 35 33	4 Apr. (94)*	4 Wed. .	87·2951	722·1597	262·4580		4855
8 Apr. (98)	2 Mon. .	12 47 42	24 Mar. (83)	1 Sun. .	9963·0179	569·4038	231·6348		4856
8 Apr. (98)	3 Tues. .	18 59 50	13 Mar. (72)	5 Thur.	9838·7407	416·6478	200·8115		4857
8 Apr. (99)	5 Thur.	1 11 59	31 Mar. (91)	4 Wed. .	9873·4231	352·6412	252·1219		4858
8 Apr. (98)	6 Fri. .	7 24 8	20 Mar. (79)	1 Sun. .	9749·1460	199·8853	221·2988		4859
8 Apr. (98)	0 Sat. .	13 36 17	10 Mar. (69)	6 Fri. .	9963·5007	83·4211	193·2123		4860
8 Apr. (98)	1 Sun. .	19 48 26	29 Mar. (88)	5 Thur.	9998·1832	19·4145	244·5237		4861
8 Apr. (99)	3 Tues. .	2 0 35	18 Mar. (78)	3 Tues. .	212·5380	902·9502	216·4383		4862
8 Apr. (98)	4 Wed. .	8 12 43	6 Apr. (96)	2 Mon. .	247·2204	838·9437	277·7387		4863
8 Apr. (98)	5 Thur.	14 24 52	26 Mar. (85)	6 Fri. .	121·9432	686·1877	236·9256		4864
8 Apr. (98)	6 Fri. .	20 37 1	15 Mar. (74)	3 Tues. .	9998·6661	533·4318	206·1023		4865
8 Apr. (99)	1 Sun. .	2 49 10	2 Apr. (93)	2 Mon. .	33·3485	469·5252	257·4127		4866
8 Apr. (98)	2 Mon. .	9 1 19	22 Mar. (81)	6 Fri. .	9909·0713	316·6693	226·5895		4867
8 Apr. (98)	3 Tues. .	15 13 28	11 Mar. (70)	3 Tues. .	9784·7941	163·9134	195·7664		4868
8 Apr. (98)	4 Wed. .	21 25 36	30 Mar. (89)	2 Mon. .	9819·4766	99·9068	247·0767		4869
8 Apr. (99)	6 Fri. .	3 37 45	19 Mar. (79)	0 Sat. .	33·8313	983·4426	218·9913		4870
8 Apr. (98)	0 Sat. .	9 49 54	7 Apr. (97)	6 Fri. .	68·5137	919·4360	270·3017		4871
8 Apr. (98)	1 Sun. .	16 2 3	28 Mar. (87)	4 Wed. .	282·8685	802·9717	242·2164		4872
8 Apr. (98)	2 Mon. .	22 14 12	17 Mar. (76)	1 Sun. .	158·5915	650·2158	211·3931		4873
8 Apr. (99)	4 Wed. .	4 26 21	4 Apr. (95)	0 Sat. .	193·2738	586·2692	262·7035		4874
8 Apr. (98)	5 Thur.	10 38 29	25 Mar. (83)	4 Wed. .	67·9967	433·4533	231·8803		4875
8 Apr. (98)	6 Fri. .	16 50 38	13 Mar. (72)	1 Sun. .	9944·7195	290·6973	201·0571		4876
8 Apr. (98)	0 Sat. .	23 2 47	1 Apr. (91)	0 Sat. .	9979·5018	216·6908	252·3675		4877

* From here (inclusive) forward the dates A. D. are New Style.

TABLE

CONCURRENT YEAR.								Intercalated and suppressed (*ksh.*) lunar months.
Kali.	Śaka.	Chaitrādi Vikrama.	Mēshādi (solar) year in Bengal.	Kollam.	A.D.	Jovian Samvatsara. Southern system.	Jovian Samvatsara. Northern system.	
1	2	3	3*a*	4	5	6	7	8
4878	1699	1834	1183	951-52	*1776-77	30 Durmukha	41 Plavaṅga	...
4879	1700	1835	1184	952-53	1777-78	31 Hēmalamba	42 Kīlaka	1 Chaitra
4880	1701	1836	1185	953-54	1778-79	32 Vilamba	43 Saumya	...
4881	1702	1837	1186	954-55	1779-80	33 Vikārin	44 Sādhāraṇa	5 Śrāvaṇa
4882	1703	1838	1187	955-56	*1780-81	34 Śārvarin	45 Virōdhakṛit	...
4883	1704	1839	1188	956-57	1781-82	35 Plava	46 Paridhāvin	...
4884	1705	1840	1189	957-58	1782-83	36 Śubhakṛit	47 Pramādin	4 Āshāḍha
4885	1706	1841	1190	958-59	1783-84	37 Śōbhana	48 Ānanda	...
4886	1707	1842	1191	959-60	*1784-85	38 Krōdhin	49 Rākshasa	...
4887	1708	1843	1192	960-61	1785-86	39 Viśvāvasu	50 Anala	2 Vaiśākha
4888	1709	1844	1193	961-62	1786-87	40 Parābhava	51 Piṅgala	...
4889	1710	1845	1194	962-63	1787-88	41 Plavaṅga	52 Kālayukta	6 Bhādrapada
4890	1711	1846	1195	963-64	*1788-89	42 Kīlaka	53 Siddhārthin	...
4891	1712	1847	1196	964-65	1789-90	43 Saumya	54 Raudra	...
4892	1713	1848	1197	965-66	1790-91	44 Sādhāraṇa	55 Durmati	5 Śrāvaṇa
4893	1714	1849	1198	966-67	1791-92	45 Virōdhakṛit	56 Dundubhi	...
4894	1715	1850	1199	967-68	*1792-93	46 Paridhāvin	57 Rudhirōdgārin	...
4895	1716	1851	1200	968-69	1793-94	47 Pramādin	58 Raktāksha	3 Jyēshṭha
4896	1717	1852	1201	969-70	1794-95	48 Ānanda	59 Krōdhana	...
4897	1718	1853	1202	970-71	1795-96	49 Rākshasa	60 Kshaya	7 Āśvina 10 *Pausha*(*Ksh.*)
4898	1719	1854	1203	971-72	*1796-97	50 Anala	1 Prabhava	1 Chaitra
4899	1720	1855	1204	972-73	1797-98	51 Piṅgala	2 Vibhava	...
4900	1721	1856	1205	973-74	1798-99	52 Kālayukta	3 Śukla	5 Śrāvaṇa
4901	1722	1857	1206	974-75	1799-00	53 Siddhārthin	4 Pramōda	...
4902	1723	1858	1207	975-76	1800-01†	54 Raudra	5 Prajāpati	...

† The year 1800 A. D. was not a Leap-year.

LX—*Contd.*

COMMENCEMENT OF THE									
SOLAR YEAR.			LUNI-SOLAR YEAR (MEAN SUNRISE OF DAY ON WHICH CHAITRA ŚUKLA 1 ENDS).						Kali year.
Day and month, A.D.	Week-day.	Time of true Mēsha-saṁkrānti.	Day and month, A.D.	Week-day.	*a.*	*b.*	*c.*		
13	14	17	19	20	23	24	25	1	
8 Apr. (99)	2 Mon. .	5 14 56	20 Mar. (80)	4 Wed. .	9855·1247	63·9348	221·5443	4878	
8 Apr. (98)	3 Tues. .	1[illegible] 27 5	10 Mar. (69)	2 Mon. .	69·4795	947·4706	193·4578	4879	
8 Apr. (98)	4 Wed. .	17 39 14	29 Mar. (88)	1 Sun. .	104·1620	883·4640	244·7693	4880	
8 Apr. (98)	5 Thur.	2[illegible] 51 23	19 Mar. (78)	6 Fri. .	318·5167	766·9997	216·6839	4881	
8 Apr. (99)	0 Sat. .	6 3 31	5 Apr. (96)	4 Wed. .	14·5672	666·7016	265·2565	4882	
8 Apr. (98)	1 Sun. .	12 15 40	25 Mar. (84)	1 Sun. .	9890·2900	513·9455	234·4333	4883	
8 Apr. (98)	2 Mon. .	18 27 49	14 Mar. (73)	5 Thur.	9766·0129	361·1896	203·6101	4884	
9 Apr. (99)	4 Wed. .	0 39 58	2 Apr. (92)	4 Wed. .	9800·7952	297·1831	254·9205	4885	
8 Apr. (99)	5 Thur.	6 52 6	22 Mar. (82)	2 Mon. .	15·0501	180·7188	226·8350	4886	
8 Apr. (98)	6 Fri. .	13 4 16	11 Mar. (70)	6 Fri. .	9890·7729	27·9629	196·0119	4887	
8 Apr. (98)	0 Sat. .	19 16 24	30 Mar. (89)	5 Thur.	9925·4553	963·9563	247·3223	4888	
9 Apr. (99)	2 Mon. .	1 28 33	20 Mar. (79)	3 Tues. .	139·8101	847·4921	219·2369	4889	
8 Apr. (99)	3 Tues. .	7 40 42	7 Apr. (98)	2 Mon. .	174·4925	783·4855	270·5472	4890	
8 Apr. (98)	4 Wed. .	13 52 51	27 Mar. (86)	6 Fri. .	50·2154	630·7295	239·7241	4891	
8 Apr. (98)	5 Thur.	20 5 0	16 Mar. (75)	3 Tues. .	9925·9382	477·9736	208·9009	4892	
9 Apr. (99)	0 Sat. .	2 17 9	4 Apr. (94)	2 Mon. .	9960·6206	413·9671	260·2113	4893	
8 Apr. (99)	1 Sun. .	8 29 17	23 Mar. (83)	6 Fri. .	9836·3435	261·2112	229·3880	4894	
8 Apr. (98)	2 Mon. .	14 41 26	13 Mar. (72)	4 Wed. .	50·6982	144·7469	201·3026	4895	
8 Apr. (98)	3 Tues. .	20 53 35	1 Apr. (91)	3 Tues. .	85·3806	80·7303	252·6131	4896	
9 Apr. (99)	5 Thur.	3 5 44	21 Mar. (80)	0 Sat. .	9961·1035	927·9843	221·7899	4897	
8 Apr. (99)	6 Fri. .	9 17 53	10 Mar. (69)	4 Wed. .	175·4582	811·5201	193·8033	4898	
8 Apr. (98)	0 Sat. .	15 30 2	29 Mar. (88)	4 Wed. .	210·1407	747·5135	245·0148	4899	
8 Apr. (98)	1 Sun. .	21 42 10	18 Mar. (77)	1 Sun. .	85·8635	594·7576	214·1917	4900	
9 Apr. (99)	3 Tues. .	3 54 19	6 Apr. (96)	0 Sat. .	120·5460	530·7511	265·5021	4901	
9 Apr. (99)	4 Wed. .	10 6 28	26 Mar. (85)	4 Wed. .	9996·2688	377·9950	234·5983	4902	

TABLE

CONCURRENT YEAR.									
						Jovian Samvatsara.		Intercalated and suppressed (*ksh.*) lunar months.
Kali.	Śaka.	Chaitrādi Vikrama.	Mēshādi (solar) year in Bengal.	Kollam.	A.D.	Southern system.	Northern system.	
1	2	3	3*a*	4	5	6	7	8
4903	1724	1859	1208	976-77	1801-02	55 Durmati .	6 Angiras . .	4 Āshāḍha .
4904	1725	1860	1209	977-78	1802-03	56 Dundubhi .	7 Śṛīmukha .	...
4905	1726	1861	1210	978-79	1803-04	57 Rudhirōdgārin	8 Bhāva . .	...
4906	1727	1862	1211	979-80	*1804-05	58 Raktāksha .	9 Yuvan . .	2 Vaiśākha .
4907	1728	1863	1212	980-81	1805-06	59 Krōdhana .	10 Dhātṛi . .	...
4908	1729	1864	1213	981-82	1806-07	60 Kshaya . .	11 Īśvara . .	6 Bhādrapada .
4909	1730	1865	1214	982-83	1807-08	1 Prabhava .	12 Bahudhānya .	...
4910	1731	1866	1215	983-84	*1808-09	2 Vibhava . .	13 Pramāthin .	...
4911	1732	1867	1216	984-85	1809-10	3 Śukla . .	14 Vikrama .	4 Ashāḍha .
4912	1733	1868	1217	985-86	1810-11	4 Pramōda .	15 Vṛisha . .	...
4913	1734	1869	1218	986-87	1811-12	5 Prajāpati .	16 Chitrabhānu .	...
4914	1735	1870	1219	987-88	*1812-13	6 Aṅgiras . .	17 Subhānu .	3 Jyēshtha .
4915	1736	1871	1220	988-89	1813-14	7 Srīmukha .	18 Tāraṇa . .	...
4916	1737	1872	1221	989-90	1814-15	8 Bhāva . .	19 Pārthiva .	{ 7 Āśvina . { 11 *Māgha* (*ksh*)
4917	1738	1873	1222	990-91	1815-16	9 Yuvan . .	20 Vyaya . .	1 Chaitra .
4918	1739	1874	1223	991-92	*1816-17	10 Dhātṛi . .	21 Sarvajit .	...
4919	1740	1875	1224	992-93	1817-18	11 Īśvara . .	22 Sarvadhārin .	5 Śrāvaṇa .
4920	1741	1876	1225	993-94	1818-19	12 Bahudhānya .	23 Virōdhin .	...
4921	1742	1877	1226	994-95	1819-20	13 Pramāthin .	24 Vikṛita . .	...
4922	1743	1878	1227	995-96	*1820-21	14 Vikrama .	25 Khara . .	3 Jyēshṭha .
4923	1744	1879	1228	996-97	1821-22	15 Vṛisha . .	26 Nandana .	...
4924	1745	1880	1229	997-98	1822-23	15 Chitrabhānu .	27 Vijaya . .	...
4925	1746	1881	1230	998-99	1823-24	17 Subhānu .	28 Jaya . .	2 Vaiśākha .
4926	1747	1882	1231	999-1000	*1824-25	18 Tāraṇa . .	29 Manmatha .	...
4927	1748	1883	1232	1000-01	1825-26	19 Pārthiva .	30 Durmukha .	6 Bhādrapada .

LX—*Contd.*

COMMENCEMENT OF THE								
SOLAR YEAR.			LUNI-SOLAR YEAR (MEAN SUNRISE OF DAY ON WHICH CHAITRA ŚUKLA 1 ENDS).					Kali year.
Day and month, A.D.	Week-day.	Time of true Mēsha-saṁkrānti.	Day and month, A.D.	Week-day.	*a.*	*b.*	*c.*	
13	14	17	19	20	23	24	25	1
9 Apr. (99)	5 Thur.	16 18 37	15 Mar. (74)	1 Sun.	9871·9917	225·2391	203·7750	4903
9 Apr. (99)	6 Fri.	22 30 46	3 Apr. (93)	0 Sat.	9906·6740	161·2327	255·0754	4904
10 Apr. (100)	1 Sun.	4 42 55	24 Mar. (83)	5 Thur.	121·0289	44·7683	227·0000	4905
9 Apr. (100)	2 Mon.	10 55 4	12 Mar. (72)	2 Mon.	9996·7517	892·0124	196·1769	4906
9 Apr. (99)	3 Tues.	17 7 12	31 Mar. (90)	1 Sun.	31·4341	828·0059	247·4872	4907
9 Apr. (99)	4 Wed.	23 19 21	21 Mar. (80)	6 Fri.	245·7889	711·5416	219·4018	4908
10 Apr. (100)	6 Fri.	5 31 30	9 Apr. (99)	5 Thur.	280·4713	647·5351	270·7122	4909
9 Apr. (100)	0 Sat.	11 43 39	28 Mar. (88)	2 Mon.	156·1941	494·7790	239·8891	4910
9 Apr. (99)	1 Sun.	17 55 48	17 Mar. (76)	6 Fri.	31·9170	342·0231	209·0658	4911
10 Apr. (100)	3 Tues.	0 7 57	4 Apr. (94)	4 Wed.	9727·9674	241·7150	257·6384	4912
10 Apr. (100)	4 Wed.	6 20 5	25 Mar. (84)	2 Mon.	9942·3223	125·2607	229·5530	4913
9 Apr. (100)	5 Thur.	12 32 14	14 Mar. (74)	0 Sat.	156·6770	8·7964	201·4676	4914
9 Apr. (99)	6 Fri.	18 44 23	2 Apr. (92)	6 Fri.	191·3594	944·7898	252·7780	4915
10 Apr. (100)	1 Sun.	0 56 32	22 Mar. (81)	3 Tues.	67·0823	792·0339	221·9548	4916
10 Apr. (100)	2 Mon.	7 8 41	12 Mar. (71)	1 Sun.	281·4370	675·5705	193·8694	4917
9 Apr. (100)	3 Tues.	13 20 50	29 Mar. (89)	6 Fri.	9977·4875	575·2714	242·4421	4918
9 Apr. (99)	4 Wed.	19 32 58	18 Mar. (77)	3 Tues.	9853·2104	422·5154	211·6188	4919
10 Apr. (100)	6 Fri.	1 45 7	6 Apr. (96)	2 Mon.	9887·8928	358·5089	262·9292	4920
10 Apr. (100)	0 Sat.	7 57 16	26 Mar. (85)	6 Fri.	9763·6156	205·7530	232·1060	4921
9 Apr. (100)	1 Sun.	14 9 25	15 Mar. (75)	4 Wed.	9977·9704	89·2887	203·9206	4922
9 Apr. (99)	2 Mon.	20 21 34	3 Apr. (93)	3 Tues.	12·6528	25·2822	255·3309	4923
10 Apr. (100)	4 Wed.	2 33 43	24 Mar. (83)	1 Sun.	227·0076	908·8179	227·2456	4924
10 Apr. (100)	5 Thur.	8 45 52	13 Mar. (72)	5 Thur.	102·7304	756·0619	196·4224	4925
9 Apr. (100)	6 Fri.	14 58 0	31 Mar. (91)	4 Wed.	137·4129	692·0554	247·7328	4926
9 Apr. (99)	0 Sat.	21 10 9	20 Mar. (79)	1 Sun.	13·1357	539·2994	216·9096	4927

TABLE

CONCURRENT YEAR.								Intercalated and suppressed (*ksh.*) lunar months.
Kali.	Śaka.	Chaitrādi Vikrama.	Mēshādi (solar) year in Bengal.	Kollam.	A.D.	Jovian Samvatsara. Southern system.	Jovian Samvatsara. Northern system.	
1	2	3	3a	4	5	6	7	8
4928	1749	1884	1233	1001-02	1826-27	20 Vyaya . .	31 Hēmalamba .	...
4929	1750	1885	1234	1002-03	1827-28	21 Sarvajit .	32 Vilamba .	...
4930	1751	1886	1235	1003-04	*1828-29	22 Sarvadhārin .	33 Vikārin . .	4 Āshāḍha .
4931	1752	1887	1236	1004-05	1829-30	23 Virōdhin .	34 Śārvarin .	...
4932	1753	1888	1237	1005-06	1830-31	24 Vikṛita . .	35 Plava . .	...
4933	1754	1889	1238	1006-07	1831-32	25 Khara . .	36 Śubhakṛit .	3 Jyēshṭha .
4934	1755	1890	1239	1007-08	*1832-33	26 Nandana .	37 Śōbhana .	...
4935	1756	1891	1240	1008-09	1833-34	27 Vijaya . .	38 Krōdhin .	7 Āśvina .
4936	1757	1892	1241	1009-10	1834-35	28 Jaya . .	39 Viśvāvasu .	...
4937	1758	1893	1242	1010-11	1835-36	29 Manmatha .	40 Parābhava .	...
4938	1759	1894	1243	1011-12	*1836-37	30 Durmukha .	41 Plavaṅga .	5 Śrāvana .
4939	1760	1895	1244	1012-13	1837-38	31 Hēmalamba .	42 Kīlaka . .	...
4940	1761	1896	1245	1013-14	1838-39	32 Vilamba .	43 Saumya .	...
4941	1762	1897	1246	1014-15	1839-40	33 Vikārin . .	44 Sādhāraṇa .	3 Jyēshṭha .
4942	1763	1898	1247	1015-16	*1840-41	34 Śārvarin .	45 Virōdhakṛit .	...
4943	1764	1899	1248	1016-17	1841-42	35 Plava . .	46 Paridhāvin† .	...
4944	1765	1900	1249	1017-18	1842-43	36 Śubhakṛit .	48 *Ānanda* . .	2 Vaiśākha .
4945	1766	1901	1250	1018-19	1843-44	37 Śōbhana .	49 *Rākshasa* .	...
4946	1767	1902	1251	1019-20	*1844-45	38 Krōdhin .	50 *Anala* . .	6 Bhādrapada .
4947	1768	1903	1252	1020-21	1845-46	39 Viśvāvasu .	51 *Piṅgala* . .	...
4948	1769	1904	1253	1021-22	1846-47	40 Parābhava .	52 *Kālayukta* .	...
4949	1770	1905	1254	1022-23	1847-48	41 Plavaṅga .	53 *Siddhārthin* .	4 Āshāḍha .
4950	1771	1906	1255	1023-24	*1848-49	42 Kīlaka . .	54 *Raudra* . .	...
4951	1772	1907	1256	1024-25	1849-50	43 Saumya .	55 *Durmati* .	...
4952	1773	1908	1257	1025-26	1850-51	44 Sādhāraṇa .	56 *Dundubhi* .	3 Jyēshṭha .

† 47 Pramādin was suppressed in the north.

LX—*Contd.*

COMMENCEMENT OF THE									
Solar year.			Luni-solar year (mean sunrise of day on which Chaitra Śukla 1 ends).						Kali year.
Day and month, A.D.	Week-day.	Time of true Mēsha-saṁkrānti.	Day and month, A.D.	Week-day.	*a.*	*b*	*c.*		
13	14	17	19	20	23	24	25		1
10 Apr. (100)	2 Mon. .	3 22 18	8 Apr. (98)	0 Sat. .	47·8131	475·2929	268·2199		4928
10 Apr. (100)	3 Tues. .	9 34 27	28 Mar. (87)	4 Wed. .	9923·5409	322·5370	237·3968		4929
9 Apr. (100)	4 Wed. .	15 46 36	16 Mar. (76)	1 Sun. .	9799·2638	169·7810	206·5736		4930
9 Apr. (99)	5 Thur.	21 58 45	4 Apr. (94)	0 Sat. .	9833·9461	105·7745	257·8840		4931
10 Apr. (100)	0 Sat. .	4 10 53	25 Mar. (84)	5 Thur. .	48·3010	989·3102	229·7985		4932
10 Apr. (100)	1 Sun. .	10 23 2	15 Mar. (74)	3 Tues. .	262·6558	872·8459	201·7131		4983
9 Apr. (100)	2 Mon. .	16 35 11	2 Apr. (93)	2 Mon. .	297·3382	808·8394	253·0236		4934
9 Apr. (99)	3 Tues. .	22 47 20	22 Mar. (81)	6 Fri. .	173·0610	656·0834	222·2004		4935
10 Apr. (100)	5 Thur.	4 59 29	10 Apr. (100)	5 Thur.	207·7434	592·0769	273·6107		4936
10 Apr. (100)	6 Fri. .	11 11 38	30 Mar. (89)	2 Mon. .	83·4663	439·3209	242·6876		4937
9 Apr. (100)	0 Sat. .	17 23 46	18 Mar. (78)	6 Fri. .	9959·1892	286·5650	211·8644		4938
9 Apr. (99)	1 Sun. .	23 35 55	9 Apr. (96)	5 Thur.	9993·8715	222·5584	263·1748		4939
10 Apr. (100)	3 Tues. .	5 48 4	26 Mar. (85)	2 Mon. .	9869·5944	69·8025	232·3516		4940
10 Apr. (100)	4 Wed. .	12 0 13	16 Mar. (75)	0 Sat. .	83·9492	953·3382	204·2661		4941
9 Apr. (100)	5 Thur.	18 12 22	3 Apr. (94)	6 Fri. .	118·6315	889·3316	255·5766		4942
10 Apr. (100)	0 Sat. .	0 24 31	23 Mar. (82)	3 Tues. .	9994·3544	736·5758	224·7533		4943
10 Apr. (100)	1 Sun. .	6 36 39	13 Mar. (72)	1 Sun. .	208·7092	620·1114	196·6680		4944
10 Apr. (100)	2 Mon. .	12 48 48	31 Mar. (90)	6 Fri. .	9904·7597	519·8132	245·2405		4945
9 Apr. (100)	3 Tues. .	19 0 57	19 Mar. (79)	3 Tues. .	9770·4824	366·0573	214·4173		4946
10 Apr. (100)	5 Thur.	1 13 6	7 Apr. (97)	2 Mon. .	9815·1649	303·0508	265·7278		4947
10 Apr. (100)	6 Fri. .	7 25 15	28 Mar. (87)	0 Sat. .	29·5197	186·5865	237·6424		4948
10 Apr. (100)	0 Sat. .	13 37 24	17 Mar. (76)	4 Wed. .	9905·2425	33·8305	206·8191		4949
9 Apr. (100)	1 Sun. .	19 49 33	4 Apr. (95)	3 Tues. .	9939·9249	969·8[illegible]40	257·1295		4950
10 Apr. (100)	3 Tues. .	2 1 41	25 Mar. (84)	1 Sun. .	154·2798	853·3597	230·0441		4951
10 Apr. (100)	4 Wed. .	8 13 50	14 Mar. (73)	5 Thur.	30·0026	700·6037	199·2210		4952

TABLE

CONCURRENT YEAR.								
						Jovian Samvatsara.		Intercalated and suppressed (*ksh.*) lunar months.
Kali.	Saka.	Chaitrādi Vikrama.	Mēshādi (solar) year in Bengal.	Kollam.	A.D.	Southern system.	Northern system.	
1	2	3	3*a*	4	5	6	7	8
4953	1774	1909	1258	1026-27	1851-52	45 Virōdhakrit	57 *Rudhirōdgārin*	...
4954	1775	1910	1259	1027-28	*1852-53	46 Paridhāvin	58 *Raktāksha*	7 Āśvina
4955	1776	1911	1260	1028-29	1853-54	47 Pramādin	59 *Krōdhana*	...
4956	1777	1912	1261	1029-30	1854-55	48 Ānanda	60 Kshaya	...
4957	1778	1913	1262	1030-31	1855-56	49 Rākshasa	1 Prabhava	5 Śrāvaṇa
4958	1779	1914	1263	1031-32	*1856-57	50 Anala	2 Vibhava	...
4959	1780	1915	1264	1032-33	1857-58	51 Piṅgala	3 Śukla	...
4960	1781	1916	1265	1033-34	1858-59	52 Kālayukta	4 Pramōda	3 Jyēshṭha
4961	1782	1917	1266	1034-35	1859-60	53 Siddhārthin	5 Prajāpati	...
4962	1783	1918	1267	1035-36	*1860-61	54 Raudra	6 Aṅgiras	...
4963	1784	1919	1268	1036-37	1861-62	55 Durmati	7 Śrīmukha	2 Vaiśākha
4964	1785	1920	1269	1037-38	1862-63	56 Dundubhi	8 Bhāva	...
4965	1786	1921	1270	1038-39	1863-64	57 Rudhirōdgārin	9 Yuvan	6 Bhādrapada
4966	1787	1922	1271	1039-40	*1864-65	58 Raktāksha	10 Dhātṛi	...
4967	1788	1923	1272	1040-41	1865-66	59 Krōdhana	11 Īśvara	...
4968	1789	1924	1273	1041-42	1866-67	60 Kshaya	12 Bahudhānya	4 Āshāḍha
4969	1790	1925	1274	1042-43	1867-68	1 Prabhava	13 Pramāthin	...
4970	1791	1926	1275	1043-44	*1868-69	2 Vibhava	14 Vikrama	...
4971	1792	1927	1276	1044-45	1869-70	3 Śukla	15 Vṛisha	3 Jyēshṭha
4972	1793	1928	1277	1045-46	1870-71	4 Pramōda	16 Chitrabhānu	...
4973	1794	1929	1278	1046-47	1871-72	5 Prajāpati	17 Subhānu	7 Āśvina
4974	1795	1930	1279	1047-48	*1872-73	6 Aṅgiras	18 Tāraṇa	...
4975	1796	1931	1280	1048-49	1873-74	7 Śrīmukha	19 Pārthiva	...
4976	1797	1932	1281	1049-50	1874-75	8 Bhāva	20 Vyaya	5 Śrāvaṇa
4977	1798	1933	1282	1050-51	1875-76	9 Yuvan	21 Sarvajit	...

LX—*Contd.*

COMMENCEMENT OF THE											
SOLAR YEAR.					LUNI-SOLAR YEAR (MEAN SUNRISE OF DAY ON WHICH CHAITRA ŚUKLA 1 ENDS).						Kali year.
Day and month, A.D.	Week-day.	Time of true Mēsha-saṁkrānti.			Day and month, A.D.	Week-day.	*a.*	*b.*	*c.*		
13	14	17			19	20	23	24	25		1
		H.	M.	S.							
10 Apr. (100)	5 Thur.	14	25	59	2 Apr. (92)	4 Wed. .	64·6849	636·5972	250·5313		4953
9 Apr. (100)	6 Fri. .	20	38	8	21 Mar. (81)	1 Sun. .	9940·4078	483·8413	219·7081		4954
10 Apr. (100)	1 Sun. .	2	50	17	9 Apr. (99)	0 Sat. .	9975·0902	419·8348	271·0185		4955
10 Apr. (100)	2 Mon. .	9	2	26	29 Mar. (88)	4 Wed.	9850·8130	267·0788	240·1954		4956
10 Apr. (100)	3 Tues.	15	13	34	19 Mar. (78)	2 Mon. .	65·1679	150·6145	212·1099		4957
9 Apr. (100)	4 Wed.	21	26	43	6 Apr. (97)	1 Sun. .	99·8503	86·6079	263·4203		4958
10 Apr. (100)	6 Fri. .	3	38	52	26 Mar. (85)	5 Thur.	9975·5732	933·8520	232·5971		4959
10 Apr. (100)	0 Sat. .	9	51	1	16 Mar. (75)	3 Tues.	139·9279	817·3877	204·5117		4960
10 Apr. (100)	1 Sun. .	16	3	10	4 Apr. (94)	2 Mon. .	224·6103	753·3812	255·8221		4961
9 Apr. (100)	2 Mon. .	22	15	19	23 Mar. (83)	6 Fri. .	100·3332	600·6253	224·9988		4962
10 Apr. (100)	4 Wed.	4	27	27	12 Mar. (71)	3 Tues.	9976·0559	447·8693	194·1757		4963
10 Apr. (100)	5 Thur.	10	39	36	31 Mar. (90)	2 Mon. .	10·7384	383·8627	245·4861		4964
10 Apr. (100)	6 Fri. .	16	51	45	20 Mar. (79)	6 Fri. .	9886·4612	231·1068	214·6629		4965
9 Apr. (100)	0 Sat. .	23	3	54	7 Apr. (98)	5 Thur.	9921·1437	167·1003	265·9733		4966
10 Apr. (100)	2 Mon. .	5	16	3	28 Mar. (87)	3 Tues.	135·4984	50·6360	237·8879		4967
10 Apr. (100)	3 Tues.	11	28	12	17 Mar. (76)	0 Sat. .	11·2213	898·8801	207·0647		4968
10 Apr. (100)	4 Wed.	17	40	20	5 Apr. (95)	6 Fri. .	45·9037	833·8735	258·3751		4969
9 Apr. (100)	5 Thur.	23	52	29	25 Mar. (85)	4 Wed.	260·2585	717·4093	230·2896		4970
10 Apr. (100)	0 Sat.	6	4	38	14 Mar. (73)	1 Sun.	135·9813	564·6532	199·4665		4971
10 Apr. (100)	1 Sun.	12	16	47	2 Apr. (92)	0 Sat.	170·6639	500·6467	250·7769		4972
10 Apr. (100)	2 Mon. .	18	28	56	22 Mar. (81)	4 Wed.	46·3866	347·8908	219·9537		4973
10 Apr. (101)	4 Wed.	0	41	5	8 Apr. (99)	2 Mon.	9742·4370	247·5926	268·5262		4974
10 Apr (100)	5 Thur.	6	53	14	29 Mar. (88)	0 Sat. .	9956·7918	131·1283	240·4409		4975
10 Apr. (100)	6 Fri. .	13	5	22	19 Mar. (78)	5 Thur.	171·1467	14·6640	212·3555		4976
10 Apr. (100)	0 Sat. .	19	17	31	7 Apr. (97)	4 Wed.	205·8290	950·6575	263·6659		4977

TABLE

CONCURRENT YEAR.								Intercalated and suppressed (*ksh.*) lunar months.
						Jovian Samvatsara.		
Kali.	Śaka.	Chaitrādi Vikrama.	Mēshādi (solar) year in Bengal	Kollam.	A.D.	Southern system.	Northern system.	
1	2	3	3*a*	4	5	6	7	8
4978	1799	1934	1283	1051-52	*1876-77	10 Dhātṛi . .	22 Sarvadhārin .	...
4979	1800	1935	1284	1052-53	1877-78	11 Īśvara . .	23 Virōdhin .	3 Jyēshṭha .
4980	1801	1936	1285	1053-54	1878-79	12 Bahudhānya .	24 Vikṛita . .	...
4981	1802	1937	1286	1054-55	1879-80	13 Pramāthin .	25 Khara . .	...
4982	1803	1938	1287	1055-56	*1880-81	14 Vikrama .	26 Nandana .	1 Chaitra .
4983	1804	1939	1288	1056-57	1881-82	15 Vṛisha . .	27 Vijaya . .	...
4984	1805	1940	1289	1057-58	1882-83	16 Chitrabhānu .	28 Jāya . .	5 Śrāvaṇa .
4985	1806	1941	1290	1058-59	1883-84	17 Subhānu .	29 Manmatha .	...
4986	1807	1942	1291	1059-60	*1884-85	18 Tāraṇa . .	30 Durmukha .	...
4977	1808	1943	1292	1060-61	1885-86	19 Pārthiva .	31 Hēmalamba .	4 Āshāḍha .
4988	1809	1944	1293	1061-62	1886-87	20 Vyaya . .	32 Vilamba .	...
4989	1810	1945	1294	1062-63	1887-88	21 Sarvajit .	33 Vikārin .	...
4990	1811	1946	1295	1063-64	*1888-89	22 Sarvadhārin .	34 Śārvarin .	2 Vaiśākha .
4991	1812	1947	1296	1064-65	1889-90	23 Virōdhin .	35 Plava . .	...
4992	1813	1948	1297	1065-66	1890-91	24 Vikṛita . .	36 Śubhakṛit .	7 Āśvina .
4993	1814	1949	1298	1066-67	1891-92	25 Khara . .	37 Śōbhana .	...
4994	1815	1950	1299	1067-68	*1892-93	26 Nandana .	38 Krōdhin .	...
4995	1816	1951	1300	1068-69	1893-94	27 Vijaya . .	39 Viśvāvasu .	5 Śravāṇa .
4996	1817	1952	1301	1069-70	1894-95	28 Jaya . .	40 Parābhava .	...
4997	1818	1953	1302	1070-71	1895-96	29 Manmatha .	41 Plavaṅga .	...
4998	1819	1954	1303	1071-72	*1896-97	30 Durmukha .	42 Kīlaka . .	3 Jyeshṭha .
4999	1820	1955	1304	1072-73	1897-98	31 Hēmalamba .	43 Saumya .	
5000	1821	1956	1305	1073-74	1898-99	32 Vilamba .	44 Sādhāraṇa .	...
5001	1822	1957	1306	1074-75	1899-1900	33 Vikārin .	45 Virōdhakrit .	1 Chaitra
5002	1823	1958	1307	1075-76	1900-01†	34 Śārvarin .	46 Paridhāvin .	...

* The year A. D. 1900 was not a Leap-year.

LX—*Contd.*

COMMENCEMENT OF THE								
SOLAR YEAR.			LUNI-SOLAR YEAR (MEAN SUNRISE OF DAY ON WHICH CHAITRA ŚUKLA 1 ENDS).					Kali year.
Day and month, A.D.	Week-day.	Time of true Mēsha-saṁkrānti.	Day and month, A.D.	Week-day.	*a*.	*b*.	*c*.	
13	14	17	19	20	23	24	25	1
10 Apr. (101)	2 Mon. .	1 29 40	26 Mar. (86)	1 Sun. .	81·5519	797·9015	232·8426	4978
10 Apr. (100)	3 Tues.	7 41 49	16 Mar. (75)	6 Fri. .	295·9067	681·4372	205·7472	4979
10 Apr. (100)	4 Wed.	13 53 58	3 Apr. (93)	4 Wed. .	9991·9571	581·1391	253·3299	4980
10 Apr. (100)	5 Thur.	20 6 7	23 Mar. (82)	1 Sun. .	9867·6799	428·3831	222·5067	4981
10 Apr. (101)	0 Sat. .	2 18 15	11 Mar. (71)	5 Thur. .	9743·4027	285·6272	191·6834	4982
10 Apr. (100)	1 Sun. .	8 30 24	30 Mar. (89)	4 Wed.	9978·0852	211·6206	242·9939	4983
10 Apr. (100)	2 Mon. .	14 42 33	20 Mar. (79)	2 Mon. .	9992·4400	95·1563	214·9085	4984
10 Apr. (100)	3 Tues.	20 54 42	8 Apr. (98)	1 Sun. .	27·1224	31·1498	266·2189	4985
10 Apr. (101)	5 Thur.	3 6 51	28 Mar. (88)	6 Fri. .	241·4772	914·6855	238·1334	4986
10 Apr. (100)	6 Fri. .	9 19 0	17 Mar. (76)	3 Tues. .	115·2001	761·9296	207·3102	4987
10 Apr. (100)	0 Sat. .	15 30 8	5 Apr. (95)	2 Mon. .	151·8824	697·9230	258·6207	4988
10 Apr. (100)	1 Sun. .	21 42 17	25 Mar. (84)	6 Fri. .	27·6053	545·1671	227·7974	4989
10 Apr. (101)	3 Tues.	3 54 26	13 Mar. (73)	3 Tues. .	9903·3281	392·4111	196·9742	4990
10 Apr. (100)	4 Wed.	10 6 35	1 Apr. (91)	2 Mon. .	9938·0106	328·4046	248·2846	4991
10 Apr. (100)	5 Thur.	16 19 44	21 Mar. (80)	6 Fri. .	9813·7333	175·6487	218·4615	4992
10 Apr. (100)	6 Fri. .	22 31 53	9 Apr. (99)	5 Thur . .	9848·4158	111·6421	268·7718	4993
10 Apr. (101)	1 Sun. .	4 44 1	29 Mar. (89)	3 Tues. .	62·7706	995·1778	240·6864	4994
10 Apr. (100)	2 Mon. .	10 56 10	19 Mar. (78)	1 Sun. .	277·1254	878·7136	212·6010	4995
10 Apr. (100)	3 Tues.	17 18 19	7 Apr. (97)	0 Sat. .	311·8078	814·7070	263·9115	4993
10 Apr. (100)	4 Wed.	23 20 28	27 Mar. (86)	4 Wed. .	187·5307	661·9510	233·0882	4997
10 Apr. (101)	6 Fri. .	5 32 37	15 Mar. (75)	1 Sun. .	63·2537	509·1951	202 2649	4998
10 Apr. (100)	0 Sat. .	11 44 46	3 Apr. (93)	0 Sat. .	97·9358	445·1886	253·5754	4999
10 Apr. (100)	1 Sun. .	17 56 55	23 Mar. (82)	4 Wed. .	9973·6587	292·4327	[illegible]·7522	5000
11 Apr. (101)	3 Tues.	0 9 3	12 Mar. (71)	1 Sun. .	9849·3815	139·6767	[illegible]·9290	5001
11 Apr. (101)	4 Wed.	6 21 12	31 Mar. (80)	0 Sat. .	9884·0640	75·6701	[illegible]·1589	5002

THE FIRST ARYA-SIDDHANTA

THE "ARYABHATIYA," OR "LAGHU ARYA-SIDDHANTA", OF ARYABHATA, A.D. 499.

WORKING TABLES FOR CALCULATION BY THE TRUE, OR APPARENT, MOTIONS OF SUN AND MOON.

(*Previously published in Epigraphia Indica, Vol. XVI, pp. 100 to 221.*)

286. My last article provided working Tables for verifying dates according to the requirements of the *Siddhānta-Śirōmani* on the basis of the "true" or apparent motions of the sun and moon; the present one provides similar Tables for the *First Ārya-Siddhānta*. These Tables are framed so as to correspond to those published in the *Indian Calendar*, which, for luni-solar computation, generally followed the *Sūrya-Siddhānta*.

No pains have been spared to render the information that follows scientifically correct. But we do not yet know how far, or in what tracts or in what periods, the by-gone framers of local almanacs adhered strictly to rule; or used other sets of Tables for their guidance; or worked by whole numbers alone, discarding fractions; or made their calculations in true or apparent time instead of, as in these Tables, in mean or clock time. We have, moreover, as yet no definite information as to how late a date calculations were made by the sun's and moon's mean movements as opposed to their true or apparent movements; nor do we know with any certainty the boundaries of the tracts within which the different rules governing the civil beginnings of solar months were adhered to (*Ind. Calendar*, § 28). Such matters are problems of the future, only to be solved after protracted enquiry and investigation. Dewan Bahadur L. D. Swamikannu Pillai gives it as his opinion (*Indian Chronology, p. 70, § 169*) that, while the *Ārya-Siddhānta* was used for solar computation, the authors of South-Indian pañchāngs carried out their lunar calculations for the tithi, nakshatra, etc., by *Sūrya-Siddhānta* rule.

287. It is easy to understand how dates of documents, the details of which dates depend on the position of sun and moon, must often differ when calculated by different authorities. Taking only the *Ārya-* and *Sūrya-Siddhāntas* into consideration, it will be seen by Table A at the end of the text (p. 248 below) that in 142 years out of the 1900 with which the main Table LXI is concerned there were radical differences. In 95 of these years the saṁvatsara cycle-name of the whole year was different; in 30 years the intercalation and suppression of lunar months were different; and the day on which the luni-solar year began was different in 21 years.

Consider the year A.D. 1418-19, for instance, or Saka 1340 expired. This year was, according to the northern system of nomenclature, called "Viśvāvasu" by the followers of the *Ārya-* but "Krōdhin" by those of the *Sūrya-Siddhānta*. In the same year there was, by the *Ārya-Siddhānta*, a suppression of the lunar month Māgha and an intercalation of Phālguna, while by the *Sūrya-Siddhānta* there was none such; so that a date correctly expressed in *Ārya-Siddhānta* reckoning in that year would seem entirely inaccurate when tested by *Sūrya-Siddhānta* Tables.

ARRANGEMENTS OF THE TABLES.

288. The principal working Tables for computation of dates expressed in *First Ārya-Siddhānta* reckoning are Tables LXI to LXXI below. Tables LXI to LXX are disposed so as to correspond in rotation with Tables I to X of the "*Indian Calendar*," and have been framed i..

similar manner. This arrangement is adopted for the convenience of those who, during the last twenty-five years, have become accustomed to the processes of that publication.

Table LXI corresponds to Table I, "*Indian Calendar.*"[1]
,, LXII ,, ,, ,, II, Part II, "*Indian Calendar.*"
,, LXIII-A ,, ,, ,, III, Part I, ,, ,,
,, LXIII-B ,, ,, ,, ,, Part II, ,, ,,

[This Table is framed in a similar manner to Table XVIII-A, "*Indian Chronography,*" which it is intended to supersede.]

Table LXIV corresponds to Table IV, "*Indian Calendar.*"
,, LXV ,, ,, ,, V, ,, ,,
,, LXVI ,, ,, ,, VI, ,, ,,
,, LXVII ,, ,, ,, VII, ,, ,,

[Tables LXVI-A, LXVII-A give closer details than do Tables LXVI, LXVII, and are to be used for very accurate calculation in doubtful cases.]

Table LXVIII corresponds to Table VIII, "*Indian Calendar.*"
,, LXIX ,, ,, ,, IX, ,, ,,
,, LXX ,, ,, ,, X, ,, ,,

Table LXXI is taken from Tables XLI-A and B, "*Indian Chronography*" (*pp.* 176, 177). It enables the week-day corresponding to the Hindu date under examination to be determined according to European computation.

Then follow three Tables by which the details given in the main Table LXI have been calculated. These are Table LXXII, which fixes the values of "*a*", "*b*", "*c*" (mean distance of moon from sun, moon's mean anom., sun's mean anom.) at the beginning of the centuries concerned; Table LXXIII which gives the same information for the beginnings of odd years of centuries; and Table LXXIV, which provides, in combination with Tables LXXII and LXXIII, an easy method of arriving at the values of "*a*", "*b*", "*c*", or the mean positions of sun and moon at mean sunrise on the first civil day of each luni-solar year. The system of work is the same as that of Prof. Jacobi.

Full particulars of the moon's equation of the centre will be found in the last Table LXXV.

ELEMENTS OF THE FIRST ĀRYA-SIDDHĀNTA.

289. This work was composed by Āryabhaṭa at Kusumapura in A.D. 499, or the year 3600 (expired) of the Kaliyuga. About A.D. 638 a treatise called the *Dhi-vriddhida* was written by Lalla, who introduced a *bīja*, or correction, affecting three of the principal elements of the Siddhānta. He seems to have reduced by about 10′ in a century the moon's increase in her mean distance from mean sun (our "*a*"); and he added about 36′ in a century to the moon's mean anomaly (our "*b*"); his third correction had reference to the planet Jupiter, with which at present we are not concerned. He did not make any change in the sun's mean anomaly (our "*c*"). The *Karaṇa-prakāśa*, of date A.D. 1092, an authority largely used in Southern India, is based on Āryabhaṭa's *Siddhānta* as amended by Lalla.

[1]Because of this intentional correspondence the years of Indian eras quoted in cols. 1 to 4 are [illegible] years, as in the "*Indian Calendar.*"

The Tables given below, which deal with the period A.D. 899-900 (K.Y. 4000 expired) A.D. 1900-01 (K.Y. 5001 expired), include Lalla's corrections.

290. (i) The length of the sidereal solar year, according to the *Ārya-Siddhānta*, is 365·2586805 days, or 365ᵈ 6ʰ 12ᵐ 30ˢ

(ii) Sines of angles are the same as those of the *Sūrya-Siddhānta*, based on a radius of (sin. 90° =) 3438′. The 24 base sines and equations of the sun's centre are given in my Table XLVII above. Those of the moon's centre in Table LXXV below.

(iii) For the sun's mean motion per day, hour, minute and second, see Table XLIV above.

(iv) The circumference of the sun's epicycle is 13° 30′; that of the moon 31° 30′. There is no contraction of the epicycle in either case. (*Jacobi, Epig. Ind., Vol. I, p. 441.*)

(v) There is no shift of the sun's apsis. The longitude of his perigee-point is always 258°; apogee 78°. In ten-thousandths of the circle the perigee is 7166·6̇.

(vi) The sun's equation of the centre at the moment of true Mēsha-saṁkrānti in every year, i.e. the moment when the true sun reaches celestial longitude 0°, is, according to Dr. Schram's calculation, 2° 6′ 57″·323494885, or, in ten-thousandths of circle, 58·775644170[1]; the sun's mean longitude at the same moment being 357° 53′ 2″·676505115, or, in ten-thousandths of circle, 9941·224355830; and his mean anomaly 99° 53′ 2″·676505115, or, in ten-thousandths of circle, 2774·557689163.

(vii) For the sun's mean and true long. for every consecutive 24-hour period measured from the same moment (true Mēsha-saṁkrānti) readers are referred to Table XLVIII-A above.

(viii) The sun's equation of the centre (*see above, Table XLVII*) is obtained by the formula $\frac{3}{80}$ sin. a. For[2] sin. eqn. $= \frac{\text{minutes in epicycle}}{\text{minutes in orbit}} \times$ sin. a, where a is the sun's mean anom.; and here the minutes in the epicycle are 810′, the circumference being 13° 30′ and those of the orbit are 21600′ (360°). Hence sin. eqn. $= \frac{810}{21600}$ sin. a, or $\frac{3}{80}$ sin. a. In all equations of the sun's centre, the angle being less than 3° 45′, the eqn. is the same as the sin. eqn. (*below*, § *294 ii*).

(ix) The moon's equation of the centre (*below, Table LXXV*) is obtained by a similar proportion. The circumference of the epicycle being 31° 30′ or 1890′, the working formula is sin eqn. $= \frac{1890'}{21600}$ sin. a, or $\frac{7}{80}$ sin. a. In this case, however, for all angles in the quadrant lying between 3° 45′ and 7° 30′, the equation does not equal the sin. eqn. The process for obtaining the former from the latter is fully set forth in § 294 below.

(x) The *śōdhya*, or time-equivalent of the equation of the centre—in other words the interval of time between the moments of the true sun reaching long. 0° (true Mēsha-saṁkrānti) and mean sun reaching the same point (mean Mēsha-saṁkrānti)—is calculated by Dr. Schram as 2·146831 days, or 2ᵈ 3ʰ 31ᵐ 26ˢ·1984. This differs a little from the accepted Hindu valuation 2ᵈ 3ʰ 32ᵐ 30ˢ. As the latter is believed to have been always taken in India as the śōdhya value according to the *First Ārya-Siddhānta*, it is the value adopted in the present work.

[1] M. de Ries has worked this out quite independently, and his calculation agrees with that of Dr. Schram as far as the 6th decimal.

[2] Above, p. 54, §§ 251 252; Jacobi, *Epig. Ind.*, Vol. I, p. 441.

(xi) According to this *Siddhānta* the Kaliyuga era began, or in other words K.Y. 0 expired or K. Y. 1 current began, with a conjunction at celestial longitude 0° of mean moon, mean sun and the principal planets at the moment of mean sunrise at Laṅka on Friday, 18 February B.C. 3102. That was the moment of mean Mēsha-saṁkrānti in that year. It was 0^h 0^m Laṅka time on that morning.

(xii) At that moment, and the same in every succeeding year, the sun's apsis (perigee) being at long. 258°, his mean anom. (our "*c*") is (360° − 258°) 102°, or, in thousandths of circle (our notation), $283\dot{3}$.

(xiii) The moon's mean anom. (our "*b*") was 90°, or, in thousandths of circle, 250.

(xiv) Since mean moon and mean sun were at that moment in conjunction, the distance between them was *nil*. This is represented in ten-thousandths of circle by the completed circle 10,000. From this, in order to arrive at the exact value of our "*a*," must be deducted the sum of the greatest equations of ☾ and ☉. These are deducted for convenience of calculation, the respective quantities being added to "eqn. *b*" and "eqn. *c*," so that the working values may always be additive. The sum of these greatest equations I estimate at 199·115048361, in ten-thousandths of circle (*below*, § 296). 10,000 less this quantity = 9800·884951639. Hence at the beginning of the Kaliyuga—

$$a = 9800{\cdot}884951639$$
$$b = 250$$
$$c = 283{\cdot}\dot{3}$$

Construction of the Tables.

291. No special remarks are necessary except with reference to Tables LXIII-B (lengths of solar months), LXVI-A and LXVII-A (Detailed "Equation *b*" and "Equation *c*"), LXVIII (Indices of tithis, etc.), and the three Tables LXXII, LXXIII, LXIV. The remainder are only duplicates of the similar Tables in the "*Indian Calendar*." (*See "Arrangement of Tables," above*, § 288.)

Table LXIII-B.—Lengths of the true solar months.

292. M. Louis de Ries has been repeatedly quoted in these pages as a most careful calculator. Several years ago he kindly worked out for me an estimate of the lengths of the true solar months according to the *First Ārya-Siddhānta*, but did not inform me of the process by which he obtained his results. An entirely independent calculation has now been carried out, based on my own Table of the sun's true longitude for each 24-hour period of the solar year (*above, Table XLVIII-A*)—a Table, let it be understood, prepared some years subsequent to M. de Ries' communication and to which he has never had access.[1] Comparison of results proves the accuracy of M. de Ries' figures, and these have been adopted without alteration in my Table. The complete agreement of our respective fixtures is really remarkable.

For example, M. de Ries found that the true sun, according to Āryabhaṭa as corrected by Lalla, reaches 180° of celestial long., the moment of the Tulā-saṁkranti, 186^d 21^h 21^m $37^s{\cdot}82$ after the moment of true Mēsha-saṁkrānti, the astronomical beginning of the true solar year.

My own work for solution of this problem is as follows:—It will be seen from Table XLVIII-A above that on that 186th day, i.e. after 186 periods of 24 hours each from the moment of true Mēsha-saṁkrānti, the true sun has to travel (180° − 179° 6′ 55″·21 =) 53′ 4″·79 before reaching the Tulā-saṁkrānti point, 180°. Calculating by his actual velocity on Day 186

[1] It was published during the war.

(*Table XLIX*), the time required for him to accomplish this journey (using his true, not mean, velocity in minutes and seconds as well as in hours[1]) is found to be $21^h\ 21^m\ 37^s{\cdot}82$,—precisely M. de Ries' fixture. All the details given by M. de Ries have been similarly examined, and found correct.

Dewan Bahadur L. D. Swamikannu Pillai's estimate of the lengths of these months (*Indian Chronology, Table II*) differs somewhat from ours, the sun according to him arriving at each saṁkrānti always a little later than it does by our determination. The greatest difference between us is at the Tulā-saṁkrānti, which his Table shews to occur $3^m\ 34^s{\cdot}18$ later than the time yielded by our Table. Adding together the lengths of the twelve solar months as given by him, the length of the *Ārya-Siddhānta* year appears to be $365^d\ 6^h\ 12^m\ 37^s$, or 7 seconds longer than its accepted length.

Tables LXVIA, LXVIIA.—" Equation b" and " Equation c."

293. In order to obtain the correct working equations of ☾ and ☉ from their respective mean anomalies it is only necessary in ordinary cases to use Tables LXVI, LXVII, which give the values of "eqn. *b*" and "eqn. *c*" roughly in whole numbers. For very close calculation, however, Tables LXVI-A and LXVII-A are provided, which give the exact equations with four decimal places for a large number of anomaly angles. For an explanation as to the construction of these Tables see § 275 above.

294. It is advisable to explain clearly my reason for differing from Prof. Jacobi as to the amount of the greatest equation of the moon, which he values, in ten-thousandths of the circle, at 139·0 as against my 139·4.

"*Eqn. b.*" The general formula (§ 290, ix) for the equation of the moon's centre is, a being the angle of mean anom., sin. eqn. $=\frac{7}{80}$ sin. a. To obtain the equation from the sine of the equation-angle the proportion eqn. : sin. eqn. :: diff. in angle : diff. in sine is used. The Hindu astronomers always worked by sections of anomaly-arc, each measuring 3° 45′, or 225′. Reference to the Equation-Table LXXV will shew that in the case of the first group, anom. 0° to 3° 45′, the diff. in anom. is 225′ and the diff. in sine is also 225′. Hence, in the case of all anom. angles between 0° and 3° 45′, eqn.=sin. eqn. But in the case of all anom. angles between 3° 45′ and 7° 30′—and no equation angle of the moon's anom. exceeds the latter quantity—the diff. in angle is 225′ and the diff. in sine is 224′; so that the formula to be used for all angles coming into this second group is eqn. $=\frac{225'}{224'}$ sin. eqn. This applies only to the excess in the angle over 3° 45′. The working rule, therefore, for finding the equation of angles lying between 3° 45′ and 7° 30′ is as follows:—

With the formula $\frac{7}{80}$ sin. a, find the sin. eqn. From the sin. eqn deduct 225′. Multiply the remainder by 225′ and divide the product by 224′. Add 225′ to the result.

Or, a little more simply,—From the sin. eqn. deduct 225′. Divide the remainder by 224′. Add the result + 225′ to the sin. eqn.

For an example let us suppose that it is required to find the moon's eqn. for anom. 67° 30′. Sin. 67° 30′ = (*Table LXXV*) 3177′. $\frac{7 \times 3177'}{80}$ = 277′·9875, or 4° 37′ 59″·25, an angle

[1] That is to say, dividing up the velocity per hour (Table XLIX) on that day into minutes and seconds, and not using Table L—which only states the sun's *mean* velocity.

between 3° 45′ and 7° 30′. 277′·9875 − 225′=52′·9875, and this divided by 224′=0′·236551. 52′·9875+0′·236551+225′=278′·224051, or 4° 38′ 13″·44306. This is the correct "equation *b*" for the given anom. It is stated by Prof. Jacobi (*Epig. Ind. Vol. I, Table XXIVA*) shortly as 4° 38′ 13″.

Turning now to the equation of 90°, the greatest equation ☾, and working in the same way, sin. 90°=3438′. $\frac{7 \times 3438'}{80}$=300′·825. This less 225′=75′·825, and this divided by 224′= 0′·338504464. 75′·825+0′·338504464+225′=301′·163504464, or 5° 1′ 9″·810268, which is the exact equation required. In ten-thousandths of circle this=139·427548361.

295. "*Eqn. c.*" [Working similarly for the greatest equation ☉ or the equation of sun's anom. 90°.] The formula for finding sin. eqn. in this case is (see § *290, viii*) $\frac{3}{80}$ sin. *a* Sin. 90° =3438′. Sin. eqn.=$\frac{3 \times 3438'}{80}$=128′·925, or 2° 8′ 55″·5, or, in ten-thousandths of circle, 59·6875; and, because this angle is one in the first group, being less than 3° 45′, the eqn.=sin. eqn. Hence ☉'s eqn. 90°= 59·6875 This is the same as Prof. Jacobi's valuation, which he gives in degrees as 2° 8′ 56″ and in circle measurement (my notation) as 59·7

296. *Greatest equations* ☾ *and* ☉. My estimate, therefore, of the sum of the greatest equations ☾ and ☉ is—

☾	.	139·427548361
☉	.	59·687500000
TOTAL	.	199·115048361

The difference between us causes a slight difference in our respective Tables of equation.[1]

Table LXVIII.—Indices of tithis, etc.

296-A. In this Table the indices are given with decimal points for guidance in close cases. Otherwise they correspond exactly to those in Table VIII, "*Indian Calendar.*" The indices of yogas (col. 6) are the same as those of nakshatras (col. 8).

Tables LXXII, LXXIII, LXIV.

297. Prof. Jacobi (*Epig. Ind. Vol. I, p. 450*) has provided a Table, XIII, shewing for four of the Indian astronomical authorities the places of the sun and moon at the beginning of centuries,[2] and another, XV, shewing their increases through the years of a century according to the *Ārya-Siddhānta* with Lalla's corrections. These corrections were to be applied to the *First Ārya-Siddhānta* from the year Śaka 420 expired, or from A.D. 498-99, i.e. from the date of its compilation by Āryabhaṭa. (See his *Śishyadhīvṛiddhida*, Benares Edit. of S. Dvivedi, p. 10 v.v. 59, 60; p. 50, v.v. 18, 19.)

If, therefore, we establish by Āryabhaṭa alone the values of "*a*", "*b*", "*c*" for 36 centuries of the Kaliyuga and add to these their values at the beginning of that era as given above

[1] For the information of those who wish to compare the two it is desirable to point out that in Prof. Jacobi's Table VII (*Epig. Ind. Vol. XI*), under head "Equation" on left side, the tenth entry from the top "626" is probably a misprint for "616"; and in the same column, the eighth entry from the bottom, "152" should preferably be read "112."

[2] There appears to be one misprint in Jacobi's Table XIII. Under head "Dist. ☾ - ☉ uncorrected." in the section dealing with the *Ārya-Siddhānta*, against K.Y. century 4300, the number of minutes should be "14" not "24."

we shall arrive at their values (positions of sun and moon) at the beginning of K.Y. 3600—values, that is, recognized by Lalla; and Tables giving Lalla's estimate of the periodic changes in position of the sun and moon for centuries, years, and days will enable us to ascertain their position at any later date when computed by the *Ārya-Siddhānta* with the *bīja*.

298. (i) First to find the century increase of "*a*", "*b*", "*c*" respectively according to Āryabhaṭa uncorrected. We work for mean sunrise values only, not for values at moments of Mēsha-saṁkrānti. We require, that is, the several increases in a common century of 36526 civil days and in a defective century of 36525 such days. In the 36 Kaliyuga centuries concerned there were 31 of the former and 5 of the latter.

(ii) As regards the time-interval between the moments of mean Mēsha-saṁkrānti and the nearest mean sunrises at the beginning of each century, Prof. Jacobi's column headed "Cor." in Table XIII states these clearly in ghaṭikas and palas. Mean Mēsha-saṁkrānti always occurs $2^d\ 3^h\ 32^m\ 30^s$ after true Mēsha-saṁkrānti, and the moment of the latter's occurrence every year is given in hours and minutes in col 17, Table I, "*Indian Calendar.*" There is no difference between us in this respect.

(iii) The advances in the values of "*a*", "*b*", "*c*" respectively during a *common century of 36526 civil days* according to Āryabhaṭa uncorrected, excluding whole revolutions, are—"*a*" (mean moon's distance from mean sun) 319° 24′ 30″·645, "*b*" (☾'s mean anom.) 211° 1′ 55″·775, "*c*" (☉'s mean anom.) 0° 7′ 48″·139. These in circle measurement (our notation) are—

$$a = 8872{\cdot}458680555$$
$$b = 586{\cdot}100443673$$
$$c = 0{\cdot}361215706$$

(iv) Taking only the circle measurement, the respective increases for *one day of 24-hours* are—

$$a = 338{\cdot}632000730$$
$$b = 36{\cdot}291575876$$
$$c = 2{\cdot}737785720$$

(v) Deducting one day's increase from the former fixtures we have for a *defective century of 36525 civil days*—

$$a = 8533{\cdot}826679825$$
$$b = 549{\cdot}808867797$$
$$c = 997{\cdot}623429986$$

We now have to work out the correct details for the first 36 centuries of the Kaliyuga, 31 common and 5 defective.

299. (i) "*a*". Using the above figures it is found that the advance of "*a*" in that period (omitting quantities of 10,000 or whole revolutions[1]) was 7715·352496330; and since at the epoch of the Kaliyuga the distance between mean moon and mean sun was *nil* (*above, § 290, xiv*), the same represents their relation at the beginning of K.Y. 3600. But for tabulation purposes we have to deduct from this the sum of the greatest equations ☾ and ☉ (*§ 290 xiv; and 295*). This sum, as already stated, I estimate at 199·115048361. Therefore the tabular "*a*" for the beginning of K.Y. 3600 is 7516·237447969. Prof. Jacobi gives this figure, as I interpret him,[2] in our notation as 7516·6. The difference between us is due to his estimation of the greatest equations ☾ and ☉ as 198·7 (*margin of Table quoted in footnote below*) instead of 199·1. But I adhere to my figure, the reason for which has been fully explained.

[1] There are 1236 synodical revolutions of the moon in a century.

[2] In both sections of his Table V (*Epig. Ind.*, Vol. XI, A and B) Prof. Jacobi's entry "76166" is manifestly a misprint for "75166." In the same Table, Section A, opposite "cent. 41" the entry "19789" should be "18789."

(ii) "*b*". The advance of "*b*" in the first 36 centuries, omitting whole revolutions,[1] was, in thousandths of the circle, 918·158092848. Adding to this the value of "*b*" at K.Y. 0 (§ 290, xiii), namely 250, we have for the moon's mean anom. at the beginning of K.Y. 3600, "*b*"=168·158092848.

(iii) Now in this matter Prof. Jacobi and myself are not quite in accord. He states the value (*Epig. Ind.*, *Vol. XI, Table V-B*) as in his notation 6718. This in my notation, measuring from perigee instead of apogee, is 171·8. This figure corresponds to his valuation of "*b*" at that moment, in degrees, etc., as given in *Epig. Ind.*, Vol. I, Special Table XIII, where it is fixed, for the moment of mean Mēsha-saṁkrānti, as 245° 6′, 0″. The correction for mean sunrise value is the moon's change in 15 ghaṭikas, or 3° 15′ 58″·5; making the position of ☾ at mean sunrise 241° 50′ 1″·5, which, in thousandths of circle, is 171·760416667. Not being absolutely certain in this case that my valuation is more accurate than his, I defer to him, and accept his figure as correct.

(iv) *In any very close case arising from the use of the Tables which follow, the difference between us in the value of "b", namely 3·6, may be deducted from the resulting "b", and the date tested by my own estimate.*

(v) "*c*". The change in the sun's mean anom. (our "*c*"), similarly calculated for the 36 centuries,[2] was 999·314836816. Adding 283·3, the value of "*c*" at K.Y. 0 (§ *290, xii*), we have for K.Y. 3600 "*c*"=282·648170149. But here again there is a minute difference between my estimate and that of Prof. Jacobi. He gives, for the sun's mean anom. (measured from apogee) at the beginning of K.Y. 3600 (mean Mēsha-saṁkrānti), 282°—a value certainly correct. To obtain mean sunrise value 14′ 47″ has to be deducted,[3] with the result 281° 45′ 13″, which in thousandths of circle=782·648919753, and in my notation (measurement from perigee) =282·648919753. I let this stand.

(vi) The values, then, adopted in this work for the positions of ☉ and ☾ at *mean sunrise at the beginning of K.Y. 3600 are*—

$$a=7516{\cdot}237447969$$
$$b=\ 171{\cdot}760416667$$
$$c=\ 282{\cdot}648919753$$

300. (i) Table LXI below, however, the main working Table, starts from the year K.Y. 4000, and we have to add to the above figures the respective increases of "*a*", "*b*", "*c*" for four centuries, these increases being assessed by Lalla's values and not by the original values of Āryabhaṭa (§ *289*).

(ii) The increases of "*a*", "*b*", "*c*" in one day, one year and one century according to Lalla are given in the heading of Table LXIV below. The four centuries are all common ones, and, adding the necessary quantities, we have *for the beginning of K.Y. 4000,—mean sunrise value*—

$$a=2987{\cdot}553682533$$
$$b=\ 523{\cdot}155092591$$
$$c=\ 284{\cdot}093782577$$[4]

[1] There are 1325 anomalistic revolutions of the moon in a century.

[2] Omitting 100 whole sidereal revolutions.

[3] 14′ 47″, or actually 14′ 47″·04, is the ☉'s mean motion in 6 hours, the difference in time between mean sunrise and the moment of mean Mēsha-saṁkrānti on the day when, astronomically, K.Y. 3600 began.

[4] We may estimate the value of "*c*" on the Sunday at the beginning of K.Y. 4000 in another way. The sun's mean anom. at the moment of mean Mēsha-saṁkrānti is always 283·3̇, or 102° (§ *290, xii*). In the year in question, A.D. 899, true Mēsha-saṁkrānti took place (*Indian Calendar, Table I*) at $13^h\ 47^m\ 30^s$ after mean sunrise on Thurs., 22 March, while the moment of mean Mēsha-saṁkrānti was (§ *290, x*) $2^d\ 3^h\ 32^m\ 30^s$ later, or $6^h\ 40^m$ before mean sunrise on Sunday, 25 March. Adding the sun's motion for $6^h\ 40^m$ from Table LXV below, viz. 0·760495686 to 283·3̇, the "*c*" for mean sunrise on that Sunday is found to be 284·093829019

These agree, *mutatis mutandis*, with Prof. Jacobi's figures (*Epig. Ind., Vol. XI, Table V*), which, in my notation, are $a=2988{\cdot}0$, $b=523{\cdot}2$, $c=284{\cdot}1$.

(iii) Now these values are, as will be seen from the entry "1" in Jacobi's column for the week-day (w.), the figures for mean sunrise on Sunday, that is to say, on Sunday 25 March, A.D. 899, mean Mēsha-saṁkrānti having taken place on the previous day, Saturday, at $17^h\ 20^m$ after mean sunrise. Following general practice I work for mean sunrise on the day on which the mean saṁkrānti occurred, i.e. for the Saturday, and deduct one day's values from the above.

Finally then the working, Tabular, values for the beginning of K.Y. 4000 (Sat. 24 Mar. A.D. 899, mean sunrise) are—

$$a = 2648{\cdot}921808551$$
$$b = 486{\cdot}863468853$$
$$c = 281{\cdot}355996857$$

301. The century Table LXXII below is prepared from these details by addition of century increases. All the centuries concerned except century 42, which was defective, are common ones, each of 36526 days.

Table LXXIII gives the increases of "a", "b", "c" for each year of the century,[1] following Lalla's *bīja* (correction).

Table LXXIV gives the values to be added for the days intervening between that on which true Mēsha-saṁkrānti occurred in each year and the day of the corresponding beginning of the luni-solar year, i.e. the civil day called "Chaitra śukla 1." This Table is prepared for the purpose of assisting workers to check the main Table entries giving the values of "a", "b", "c" (*Table LXI, cols. 23-25*). The week-day stated in the main Table will always serve as a guide. Compare the similar Table in my article on the *Siddhānta-Śirōmaṇi* above, where instructions for its use are given (§ 279).

The Nakshatra.

302. A special note must be made regarding the working of the "*Indian Calendar*" rule (§ *156, p. 97*) for obtaining approximately the index of the nakshatra.

It will be observed there that part of the process (see § *133, Ind. Cal.*) consists of the addition to the value of "c", the sun's mean long., of a constant, viz. 7207, as stated in 10,000ths of circle. This is the *Sūrya-Siddhānta* quantity. For work by the *Ārya-Siddhānta* we require the *Ārya-Siddhānta* quantity

The *Sūrya-Siddhānta* figure is made up of (i) long. of sun's perigee-point[2] (257° 15′ 55″·7=) 7146·3 and (ii) 60·4, the greatest equation of the sun's centre.

Now (i) the long. of the sun's perigee-point according to the *Ārya-Siddhānta* is always 258°, or, in 10,000ths of circle, 7166·6 (§ *290, v, above*); and (ii) the greatest equation of the sun's centre (§§ *295, 296*) is 59·6875 Hence the *Ārya-Siddhānta* constant for calculating the nakshatra is (7166·6 + 59·6875 =) 7226·3542 ; and for approximate calculation is 7226, not 7207.

[1] There appear to be two misprints in Prof. Jacobi's Table VI (*Epig. Ind., Vol. XI, p. 165*) in which he gives similar annual increases. Against year 3, under "c," "61" should be "6"; and against year 52, under "a," "16312" should be "18352."

[2] This is its position in A.D. 1100, a date about the middle of the period, A.D. 300—1900, dealt with in Table I of the *Indian Calendar*. In ten-thousands of circle the long. of perigee by the *Sūrya-Siddhānta* varies from 7145·54583 in A.D. 300 to 7146.97916 in A.D. 1900.

Thus the rules for finding the nakshatra by the *Ārya-Siddhānta* are as follows :—

A. Roughly. Find "*a*", "*b*", "*c*" and "*t*" in whole numbers; multiply "*c*" by 10; add 7226 to the result; from this subtract "equation *c*." The result is "*s*", the sun's true longitude.

B. More closely. Find "*a*", "*b*", "*c*" and "*t*" with the fractions in decimals; to the value of "*c*" multiplied by 10, or with the decimal point one place to the right, add the constant 7226·3542; from the result deduct (including decimals) the amount of "equation *c*." The result is "*s*" in full detail. $s+t=n$, the index of the nakshatra, with which turn to Table LXVIII, which gives the name of the nakshatra and fixes the true moon's place in the ecliptic circle.

The work is shewn in Example 7 below.

EXAMPLES.

Example 1. To find the "a", "b", "c" values for mean sunrise on the first civil day of the luni-solar year.

Rule. Add together the entries in Tables LXXII and LXXIII for the corresponding expired year of the Kaliyuga, and those in Table LXXIV for the number of days' interval from true Mēsha-saṁkrānti (*Table LXI, col. 13, bracket-number*) to the first civil day of the luni-solar year, called "Chaitra śukla 1" (*col. 19, bracket-number*). Note specially the week-day of Chaitra śukla 1, and work for that day.[1] *Decimals need not be used except in close cases.*

For an example I take the year A.D. 1110-11. It corresponds (*Table LXI*) to K.Y. 4211 expired. The entries shew that true Mēsha-saṁkrānti occurred on Day 83 (Thursday 24 March A.D. 1110), and Chaitra śukla 1 on Day 82, the day previous. Interval between them 1 day.

Full work with the decimals :—

	w.-d.	*a.*	*b.*	*c.*
(*Table LXXII*) Beginning of K.Y. cent. 42 . .	(0)	384·5799	662·5608	282·0784
(*Table LXXIII*) Beginning of year 11	(0)	622·8697	819·7442	0·4230
(*Table LXXIV*) Interval of days, 1	(4)	8984·1044	891·1251	991·7866
At mean sunrise on Day 82, or on (4) Wednesday 23 March, A.D. 1110	(4)	9991·5540	373·4301	274·2880

These are the entries for that day in Table LXI.

The same result can be obtained by first finding the "*a*", "*b*", "*c*" for mean sunrise on the day on which true Mēsha-saṁkrānti took place, and then deducting the values for the intervening days as given in Table LXIV. [The day on which true Mēsha-saṁkrānti took place is, in Table LXXIV, the day "Mēsha 0" (*col. 2*).]

[1] Owing to the formation of the several Tables the interval of days measured by their bracket-numbers in Table LXI, cols. 13, 19, sometimes differs by 1, but never by more than 1. But this leads to no difficulty when the desired week-day is duly noted. The point to remember is that the resulting week-day in our addition must be the correct one as given in Table LXI, and that we must use the entries in Table LXXIV for such number of days as will make the final week-day the one we work for.

Thus :—

	w.-d.	a.	b.	c.
(*Table LXXII*) As before . .	(0)	384·5799	662·5608	282·0784
(*Table LXXIII*) Do. . .	(0)	622·8697	819·7442	0·4230
(*Table LXXIV*) "Mēsha 0" .	(5)	9322·7363	927·4168	994·5244
At mean sunrise on day of true Mēsha-saṁkrānti, (5) Thursday, 24 March (Day 83) . .	(5)	330·1859	409·7218	277·0258
(*Table LXIV*) Less 1 day interval	−1	−338·6319	−36·2916	−2·7378
At mean sunrise on Day 82, (4) Wed. 23 March . . .	(4)	9991·5540	373·4302	274·2880

The result is the same as above.

Example 2. The same for a year with a greater interval of days between Mēsha-saṁkrānti and Chaitra śukla 1.

Take the year A.D. 1603, K.Y. 4704 expired. The interval of days from true Mēsha-saṁkrānti (*Table LXI, col. 13*) back to Chaitra śukla 1 (*col. 19*) (mean sunrise in both cases) is (87 − 62) 25.

First process—with full decimals :—

	w.-d.	a.	b.	c.
(*Table LXXII*) Cent. 47 . .	(6)	4385·0933	565·5125	281·1467
(*Table LXXIII*) Year 4 . .	(5)	4741·1679	22·0623	999·9049
(*Table LXXIV*) Interval 25 days	(1)	856·9394	20·1262	926·0798
At mean sunrise on Day 62, or Chaitra śukla 1, (5) Thursday 3 March, A.D. 1603 . .	(5)	9983·2006	607·7010	207·1314

These are the entries in Table LXI.

Second process :—

	w.-d.	a.	b.	c.
(*Table LXXII*) Cent. 47 . .	(6)	4385·0933	565·5125	281·1467
(*Table LXXIII*) Year 4 . .	(5)	4741·1679	22·0623	999·9049
(*Table LXXIV*) "Mēsha 0" .	(5)	9322·7363	927·4168	994·5244
At mean sunrise of (*Day 87*) Mēsha-saṁkrānti day, (2) Mon. 28 March, A.D. 1603 . .	(2)	8448·9975	514·9916	275·5760
(*Table LXIV*) Less for 25 days' interval . .	−(4)	−8465·7968	−907·2906	−68·4446
At mean sunrise on Day 62 . .	(5)	9983·2007	607·7010	207·1314

Result, the same.

Computation of a Date.

Example 3. We will now take a suppositious Record-date, and in the following examples explain the complete method of work for proving the accuracy of all its details; and for settling some other matters.

The date is "Śaka 1148 expired, K.Y. 4327, Vyaya, Saturday, Bhādrapada śukla 5, Kanyā 1, Bāva karaṇa, nakshatra Viśākhā, yoga Vaidhṛiti, Kanyā lagna."

Table LXI shews that the year corresponded to A.D. 1226-27; that in that year true Mēsha-saṁkrānti took place 3^h 55^m after mean sunrise on Wed. 25 March (Day 84 from 1 Jan.); that the civil day "Chaitra śukla 1" was Sunday 1 March (Day 60 from 1 Jan.); and that (*col. 8*) the lunar month Āshāḍha was intercalated in that year. The year was called "Vyaya" in South India, "Vikṛita" in the North.

The interval of days between the initial days of the solar and luni-solar year was (84 − 60) 24.

In this example we work for the values of "*a*", "*b*", "*c*" and "*t*" at mean sunrise of the day "Chaitra śukla 1", which is stated in Table LXI to have been (*col. 20*) a Sunday. We work by the first process shewn above, and with full decimals. In using Table LXXIV for the interval of days— 24 as already stated—it is observed that the week-day number (*col. 3*) for that number of days' interval (*col. 1*) is 2, and that, since the week-days obtained for the year from Tables LXXII, LXXIII are respectively 6 and 6, total 12, the addition of 2 will make total 14, or 0, or a Saturday, whereas the day we are working for was Sunday. Hence we use the figures for 23 days' interval, week-day 3, which gives us the correct "*a*", "*b*", "*c*" for 1 Sunday. (*See note to Example 1.*)

	w.-d.	*a.*	*b.*	*c.*
(*Table LXXII*) K.Y. Cent. 43 .	(6)	8913·7771	214·1179	279·7019
(*Table LXXIII*) Year 27 . .	(6)	9587·5412	907·9933	0·0428
(*Table LXXIV*) 23 days' interval	(3)	1534·2032	92·7094	931·5554
At mean sunrise on (1) Sunday 1 March, A.D. 1226, i.e. the day "Chaitra śukla 1" . . .	(1)	35·5215	214·8206	211·3001

The above work has been thus fully carried out in order to prove the correctness of the entries in Table LXI, cols. 23, 24, 25, which are the same. This work is not required to be done in practice as the Table provides the information.

Now, knowing the Table entry to be accurate, we proceed

The tithi. Ordinary work

Example 4. The true tithi.[1] The given date is Bhādrapada śukla 5. Table LXIII-A shews that, Āshāḍha having been intercalated in the year in question and Bhādrapada being therefore the seventh and not the sixth lunar month of the year, it began about 177 days after the day "Chaitra śukla 1"; consequently "Bhādr. śuk. 5" was about 181 days after. Having

[1] The mean tithi (and probably the mean nakshatra and yoga also) was used in earlier years—to how late a date is not yet known. The mean tithi is the mean moon's distance from mean sun, our *a*. To find it, add to the ascertained value of *a* (as in Example 3) for the day the sum of the greatest equations of moon and sun, i.e. 199·1150. The total gives the *a* of the mean tithi (= *t* of the true tithi). Thus for the day in question the mean tithi-index is (36 + 199) 235, or (35·5215 + 199·1150) 234·6365. This was its value at mean sunrise of the given day.

added the values of "a", "b", "c" for 181 days to those already found for Chaitra śukla 1. the equations of "b" and "c" are added from Tables LXVI, LXVII approximately, or from Tables LXVI-A, LXVII-A in very close and doubtful cases, to the resulting value of "a" for the day; thus "t", the true tithi-index, is found.

In this example we work approximately.

The serial number of the day Chaitra śukla 1 (in March A.D. 1226) is 60 and the week day 1 Sunday (*Example 3*). The *a*, *b*, *c* for mean sunrise have been settled in Example 3.

	d.	w.-d.	a.	b	c.
Table LXI, cols. 19-23 . .	(60)	(1)	36	215	211
(*Table LXIV*) . . .	(181)	(6)	1292	569	496
	(241)	(0)	1328	784	707
(*Table LXVI*) "Eqn. *b*" . .			3		
(*Table LXVII*) "Eqn. *c*" .			117		

At mean sunrise on day 241, $t=1448=$(*Table LXVIII*) śukla 5.

Day 241 was (Table LXIX) August 29. Week-day 0=Saturday. Reference to Table LXXI confirms this as the right week-day.

The given Hindu date then is so far correct. The 5th śukla tithi of Bhādrapada ended on and gave its name to, Sat. 29 Aug. A.D. 1226. For historical purposes it is seldom necessary, unless the karaṇa is mentioned, to find the time of beginning and ending of the tithi, but if required this is obtained approximately from Tables LXVIII, col. 3, and LXIX. At mean sunrise the tithi-index was 1448. It began (1448−1333 =) 115, or (Table LXX) $8^h\ 9^m$ before, and ended (1667−1448 =) 219, or $15^h\ 31^m$ after mean sunrise on that Saturday.

The tithi. Exact work.

Example 5. Working the same date with the full decimals, we have—

	d.	w.-d.	a.	b.	c.
As in Example 3 . .	(60)	(1)	35·5215	214·8206	211·3001
Table LXIV . . .	(181)	(6)	1292·3692	568·7839	495·5392
	(241)	(0)	1327·8907	783·6045	706·8393

For either "equation *b*" or "equation *c*" note the difference between the values of "*b*" or "*c*" thus found and the nearest value respectively in Table LXVI-A or LXVII-A, *cols. 2a, 2b*. Multiply this difference by the group-difference (*col. 4*). Divide the result roughly by 2 or exactly by 2·083; and add or subtract the result to or from the standard equation-value given in the Table (*col. 3*) as necessity demands.

[This is the complete process, but it almost always suffices to arrive very near to the truth merely by the exercise of common sense, using Tables LXVI-A, LXVII-A as Eye-Tables.]

Here the moon's anom. "*b*" is 783·6045, and the nearest amount of "Argument *b*" in Table LXVI-A is 783·3, whose exact equation is 3·1006 (*col. 3*). As the difference in anom. is only about 0·3, viz. 0·2712, and the group-difference only 0·4150 we may take 3·1006 as the required equation of the given anom. Or we may work roughly by a multiplication of the first two decimals of the anom. diff. (0·27) by those of the group-diff. (0·42) and a division of the result by 2—yielding 0·0567, which, added to 3·1006, makes "equation *b*"=3·1573; or we may work completely with all four decimals, arriving at the absolutely correct result ·1546.

The sun's anom. "*c*" is 706·8393. The equation is similarly found by use of Tables LXVII or LXVII-A. The nearest amount of "Argument" in Table LXVII-A is 706·2500. Full work is as follows.—Diff. in anom. is 0·5893. This, multiplied by the group-difference (*col. 4*) 0·2257, is 0·133005. This, divided by 2·083, is 0·0638. The equation of anom. 706·2500 is (*col. 3*) 117·1181. This plus 0·0638 = 117·1819, the exact equation required.

Applying, as before, these exact equations of the values of anom. "*b*" and "*c*" to the value of "*a*", we have—

	a
As already found . .	1327·8907
Eqn. *b*	3·1546
Eqn. *c*	117·1819
The tithi-index, t =	1448·2272

By the work as in Example 4 the tithi-index (t) at mean sunrise was 1448.

The karaṇa.

Example 6. The karaṇa is half a tithi. See Table LXVIII, cols. 4, 5. For the date we are examining (*Examples 3, 4, 5*), viz. śukla 5 (*Table, col. 2*), the two karaṇas are Bāva and Bālava. The tithi began (*end of Example 4*) $8^h\ 9^m$ before and ended $15^h\ 31^m$ after mean sunrise on 29 Aug. A.D. 1226. Its length was $23^h\ 40^m$. Half of this is $11^h\ 50^m$. Thus Bāva was the karaṇa from $9^h\ 9^m$ before to $3^h\ 41^m$ after mean sunrise on 29 Aug., and Bālava was the karaṇa from $3^h\ 41^m$ to $15^h\ 31^m$ on that day. Since the karaṇa mentioned in the given date was Bāva the action referred to in the record must have taken place between mean sunrise and $3^h\ 41^m$ later, on 29 Aug. 1226, i.e. roughly between 6·0 and 9·41 A.M. on that day.

The nakshatra.

Example 7. Required the nakshatra of the same day, month and year as in Examples 3, 4, 5, 6.

A nakshatra, or lunar mansion, is, in the equal-space system, a 27th part of the complete journey of the moon in a lunar month through the circle of the stars. Our nakshatra-index shews in which of these parts the moon was at any given moment. In these examples we are working for the true, not mean, moon's place. Each of these 27 parts has its own nakshatra-name and yoga-name (*see Table LXVIII*). In the systems of Garga and the *Brahma-Siddhānta* the divisions of the constellation-circle are unequal, being designed more nearly to suit the positions of the principal stars,[1] but the names of the divisions are the same as in the equal-space system.

The indices of the beginning and ending points of the nakshatras are stated, in 10,000ths of the circle, in Table LXVIII. The same in degrees are given, together with those of the zodiacal solar signs, in "*Indian Chronography,*" Table XXII.

(A) The rule for finding the nakshatra roughly when working with only whole numbers is as follows:—Take the "*c*" of the date; multiply it by 10; add the constant 7226 (*see* § *302 above*); and deduct the amount of "equation *c*." This gives "*s*", the sun's true longitude at mean sunrise of the given day. Add "*s*" to "*t*" and the result is "*n*", the nakshatra-index. Reference with this index to Table LXVIII (*col. 8, or 9, or 10*) shews the nakshatra required

[1] Mr. G. R. Kaye, in his "*Astronomical Observatories of Jai Singh*" (p. 117), gives the actual lat. and long. of the stars after which the nakshatras were named.

i.e. the true moon's place amongst the constellations at mean sunrise, stated in 10,000ths of the circle. The moon's place in degrees, minutes, and seconds can be found by Table XLV-B, above.

Thus, by the figures in Example 4 :—

$c \times 10$	$=7070$
Constant	$+7226$
	4296
Less eqn. c	$-$ 117
Sun's true long., s	$=4179$
Tithi-index, t,	$+1448$

Nakshatra-index n $=5627=$ (*Table LXVIII, cols. 8, 9, 10*) Viśākhā by all systems.

This is approximately correct.

(B) Greater exactness can be obtained by using the decimals as in example 5, thus—

$c \times 10$	$=7068{\cdot}3930$
Constant	$+7226{\cdot}3542$
	4294·7472
Less eqn. c	$-$ 117·1819
s	$=4177{\cdot}5653$
t	$+1448{\cdot}2272$
n	$=5625{\cdot}7925$

There is here a little difference in the resulting nakshatra-index, which may in some cases be as great as nearly 10 units owing to the roughness of the earlier method.

(C) The value of "s" at mean sunrise of the day in question can also be obtained easily by my Tables for the sun's true longitude for each day of the solar year given above (*pp. 45—130*). The following shews the method of work :—

In the present case the serial number of the day in question was 241 (*Example 4*). True Mēsha-saṁkrānti took place (*see Example 3*) on Day 84 at 3^h 55^m after mean sunrise. The day of our date was (241—84) the 157th period (each of 24 hours) after the moment of true Mēsha-saṁkrānti. On this 157th day at 3^h 55^m after mean sunrise the sun's true longitude, "s" was, in 10,000ths of circle, 4182·0049 (*Table XLVIII-A, above, p. 74, col. 9*). Deduct the values for 3 hours (*Table XLIX, p. 96, sun's true motion on that 157th day*) and 55^m (*Table L, mean motion in minutes*), viz., respectively, 3·3852 and 1·0457, total 4·4309.

	4182·0049
	$-$ 4·4309
At mean sunrise "s"	$=4177{\cdot}5740$

This is the value of "s" at mean sunrise of the 29 August of our date, and, added to "t" (1448·2272), it gives us the correct nakṣhatra-index 5625·8012, shewing a slight difference of ·0087 in results

If, for even greater accuracy, instead of using the value of the sun's *mean* motion in 55^m we had worked by his *true* motion on that 157th day, viz. by dividing by 60 his true motion in 1 hour (*Table XLIX, p. 96*) and multiplying the result by 55, we should have found "*n*" =5625·8092.

This method C, for finding the sun's longitude "*s*", is believed to be absolutely accurate and should be relied on in case of doubt.

The yoga.

Example 8. The nakshatra (*Example 7*), as quoted in the given date shews in which of the 27 sidereal divisions the moon stood at the moment in question, or the extent of the moon's journey from celestial long. 0°. The yoga deals with the combined journeys of both sun and moon.

To find, therefore, the index of the yoga at mean sunrise of the given day we have to add the long. of the true sun to the long. of the true moon at that moment. But the long. of the true moon is the index "*n*", i.e. the nakshatra-index already found. And the long. of the sun is the index "*s*", also already found (Example 7).

Hence the yoga-index "*y*" $= s + n$; or, since $n = s + t$ (*Example 7*), $y = 2s + t$. The latter formula makes it easy to find the yoga when it is unnecessary to find the nakshatra.

At mean sunrise of 29 Aug. A.D. 1226 we have found that "*s*" = 4177·5653 and that "*n*" = 5625·7925; hence the yoga-index "*y*" = 9803·3578, and (Table LXVIII) the yoga of the day was 27 Vaidhṛiti. If we had not already ascertained the amount of the nakshatra-index "*n*", but knew that "*s*" = 4177·5653, we could have multiplied this value of "*s*" by 2 and added the quantity to the amount of the tithi-index "*t*". The result is the same.

The several saṁkrāntis.

Example 9. To find the values of "*a*", "*b*", "*c*" *and* "*t*" *at the moments of the several solar saṁkrāntis in the given year*, and thereby to find whether a lunar month was common, intercalary (*adhika*), or suppressed (*kshaya*)

A saṁkrānti takes place when the sun touches the point of a zodiacal sign, i.e. when he reaches long. 30°, 60°, etc. When, at the first of two such successive occurrences, the true moon is waning and at the second is also waning, or at the first is waxing and at the second is also waxing, the lunar month is common. If the moon is waning at the first and waxing at the second, the lunar month is repeated. It is intercalary (*adhika*). When the moon is waxing at the first and waning at the second the lunar month is altogether suppressed (*kshaya*).

Thus it is necessary to find the "*a*", "*b*", "*c*" for the moment of the astronomical beginning of the solar year, the actual moment, that is, of the true Mēsha-saṁkrānti, and add to their values their respective increases during the several true solar months, thus obtaining the "*a*", "*b*", "*c*" for the moments of the true saṁkrāntis concerned. Adding to the value of "*a*" at the moment of a saṁkrānti the values of "equation *b*" and "equation *c*" (as in the former examples), we find the index of the tithi "*t*", which shews whether the true moon was waxing or waning at the moment.

The date and time of the true Mēsha-saṁkrānti is given in Table LXI, cols. 13, 14, 17. The intervals in time to each subsequent saṁkrānti, and the collective intervals to each, are given in Table LXIII-B, cols. 8 and 3; and the corresponding increases in the values of "*a*", "*b*", "*c*" are given in the same Table, cols. 9, 10, 11 and 4, 5, 6.

We will consider the conditions for the first few saṁkrāntis of the same year as in Examples 3-8, viz. A.D. 1226 27, K. Y. 4327, Śaka 1148.

First we have to ascertain the values of "*a*", "*b*", "*c*" at the moment of true Mēsha-saṁkrānti, which took place (*Table LXI, cols. 13, 14, 17*) at 3^h 55^m after mean sunrise on Day 84, namely Wednesday 25 March A.D 1226. The "*a*", "*b*", "*c*" for mean sunrise of Day 60, Sunday, 1 March, the day of Chaitra śukla 1, are given in cols. 23, 24, 25 of the same Table. Interval between the two, whole days, (84—60=) 24. Taking down the "*a*", "*b*", "*c*" for 25 March and adding their increase for 24^d 3^h 55^m from Tables LXIV, LXV, we find the values of "*a*", "*b*", "*c*" at the moment of true Mēsha-saṁkrānti, as required.

Table LXIII-B gives us the exact interval in time and the amount of increase of "*a*", "*b*", "*c*", during that interval, up to the moment of every subsequent saṁkrānti in the year. In close cases, of course, full decimals can be used and the equation-values very carefully examined, but in general it is only necessary to use whole numbers, as in this example. Only in a doubtful case need we do more.

We desire, let us suppose, to ascertain, from the values of "*t*" at the respective Mithuna and Karka-saṁkrāntis, whether the moon was waxing or waning at the moments of their occurrence. The work is as follows :—

	d.	*w.-d.*	*a.*	*b.*	*c.*
Mean sunrise Chait. śuk. 1 (*Table LXI*) .	60	1	36	215	211
24 days' increase (*Table LXIV*) . . .	24	3	8127	871	66
3 hours' do. (*Table LXV*) . . .			42	5	0
55 minutes' do. (*do.*) . . .			13	1	0
At moment of true Mēsha-saṁkrānti . .	84	4	8218	92	277
Interval to Mithuna saṁk. (*T. LXIII-B, left side*)			+1105	262	171
At moment of Mithuna-saṁkrānti . .			9323	354	448 [1]
Eqn. b (*Table LXVI*) . . .			250		
Eqn. c (*Table LXVII*) . . .			41 [1]		
Index, at moment of Mithuna-saṁk., of true moon .			*t* = 9614		

This value of "*t*" shews that at the Mithuna-saṁkrānti the moon had not reached the point of new moon when "*t*" = 10,000. She was still waning.

	a.	*b.*	*c.*
At moment of Mithuna-saṁkrānti, as above . . .	9323	354	448
Interval to Karka saṁk. (*T. LXIII-B, cols. 9, 10, 11*) .	703	147	47
At moment of Karka-saṁkranti	26	501	535 [1]
Eqn. b (*Table LXVI*)	138		
Eqn. c (*Table LXVII*)	73 [1]		
Tithi-index . .	*t* = 237		

[It is not really necessary, when it is seen that "*a*" (here 26) is greater than 0, to add the equations, because the value of "*a*" proves that the moon had begun a new synodical revolution and was waxing.]

The value of "*t*" (and "*a*") shews that the moon was waxing at the Karka-saṁkrānti. Thus the lunar month Āshāḍha (*see cols. 1, 2, Table LXIII-B*) was intercalated in the given year.

The place of the moon at the moments of the later saṁkrāntis is obtained, if required, by a continuation of similar work and the use of Table LXIII-B

1 See note to Table LXIII B. These values are given in the auxiliary Table. At the Mithuna-saṁkrānti "*c*" is always 448·0877 and "eqn. *c*" always 40·5649. At the Karka-saṁkrānti "*c*" is always 534·6213 and "eqn. *c*" always 72·5193.

Days of the solar year.

Example 10. To find the day and week-day of the solar year corresponding to any given day in the luni-solar year.

The moment of true Mēsha-saṁkrānti, as given in Table LXI, cols. 13, 14, 17, marks the astronomical beginning of the solar year. In different parts of India (see *Indian Calendar*, § 28, p. 12, and *Indian Chronography*, § 43, pp. 18, 19) there are different rules for fixing the first day of the solar month, which is sometimes the same day, sometimes the next day, sometimes (in Bengal) the third day. In the present case we imagine the record to have come from the Tamil country and we work by the Tamil rule.

In the given year (*Example 3*), A.D. 1226, true Mēsha-saṁkrānti took place on Day 84 (measured from Jan. 1), Wednesday 25 March, at 3^h 55^m after mean sunrise, and the Wednesday was the day "1 Mēsha" since the saṁkrānti occurred before sunset.

The days in Mēsha follow regularly. But to find the first civil day of each successive month in the year we must establish the moment when each saṁkrānti took place. This information is obtained from Table LXIII-B.

We have determined the given date to be (*see Examples 4, 5*) the serial day 241 measured from Jan. 1, and the 157th day after the day on which Mēsha-saṁkrānti occurred, which was Day 84. Turn to Table LXIII-B. Kanyā began 156 days after true Mēsha-saṁkrānti so our date will be in the solar month Kanyā. Calculate the moment of occurrence of the Kanyā-saṁkrānti in the given year from the same Table.

	d.	w.-d.	h.	m.	s.
(*Table LXI*) True Mēsha-saṁkrānti . .	(84)	(4)	3	55	0
(*Table LXIII-B*) Interval to Kanyā-saṁk. .	(156)	(2)	10	24	25
Moment of Kanyā-saṁkrānti . . .	(240)	(6)	14	19	25

By Tamil rule, since the saṁkrānti took place after sunset, or 12^h Laṅka time, viz. at 14^h 19^m 25^s after mean sunrise, the civil day "1 Kanyā" was not (6) Friday (Day 240) 28 August, the day of the saṁkrānti, but was Saturday (Day 241), 29 August.

And this Saturday happens to have been the very day of our record, which day was in solar-year reckoning "1 Kanyā."

[Observe that if the record had come from Bengal its solar date would have been the same, since the saṁkranti occurred before midnight on Friday, and the Saturday was therefore "1 Kanyā." Had it come from Orissa, the Saturday would have been "2 Kanyā," since the first day of the solar month is, in that country, always the day of the saṁkrānti, and so "1 Kanyā" was the Friday. By the Malabar Rule "1 Kanyā" was Saturday.]

The lagna.

Example 11. On the day in question (*Example 7*) it has been established that at mean sunrise the sun's true long. "*s*", in 10,000ths of the circle, was 4177·5653. To calculate the lagna we must have "*s*" in degrees, etc., which can be calculated by Table XLV-B, above, or by Tables XLVIII-A, XLIX, L. We work by the latter.

The day of the record was the 157th after true Mēsha-saṁkrānti, which took place 3^h 55^m after mean sunrise on the day of its occurrence. Table XLVIII-A (p. 74, *col. 9*) shews that at 3^h 55^m after mean sunrise 157 days later the sun's true long. "*s*", was 150° 32′ 7″·84. Deduct his motion (true) for 3^h by Table XLIX (p. 96), viz. 7′ 18″·72, and (mean) for 55^m by Table L, viz. 2′ 15″·52, total 9′ 34″·24. Then "*s*" at mean sunrise was 150° 23′ 33″·60.

The long. of the point of rising of Kanyā is (*Indian Chronography, Table XXII*) 150°, and that sign ends at 180°. Take the ending-point and calculate the distance between it and the sun at mean sunrise. 180° − 150° 23′ 33″·60 = 29° 36′ 26″·40. There is no need here for great accuracy, and we take this as 29° 36′. Turn this into time by multiplying the degrees by 4^m, and the minutes by 4^s. Resu ^{h} 58^m 24^s.

Thus on the given day Kanyā was lagna from very shortly before till about 1^h 58^m after mean sunrise.

In examining the given date in the matter of the karaṇa (*Example 6*) we found that the action referred to in the record must have taken place between mean sunrise and 3^h 41^m later, or between 6.0 and 9.41 A.M., on Sat., 29 Aug., A.D. 1226. The mention of the lagna still further reduces the time and shews that the action referred to must have taken place between mean sunrise and a time 1^h 58 later; or between 6.0 and 7.58 A.M. on that day.

NOTE.

The above examples may perhaps, strike the uninitiated as involving an immense amount of complicated work in order to obtain the desired result. But such is by no means the case. Every date can be calculated in whole numbers at first, and it is very seldom that the decimals need be resorted to. They are provided for the purpose of deciding doubtful cases where very great accuracy is required.

For all the details of the given date,—and it is very seldom that so many are stated in an inscription or grant,—the following exemplifies all the work necessary to be done to put us in full possession of the facts. In about a quarter of an hour we learn everything that has to be learned; and when less details are given their accuracy can be proved or disproved in a few minutes. What follows shews the ordinary work to be done for the date given in Examples 3-10.

Given year = Śaka 1148, K.Y. 4327, Vyaya, A.D. 1226-27.

	d	*w.-d.*	*h.*	*m.*	*s.*	*d.*	*w.-d.*	*a.*	*b.*	*c.*
Mēsha samk: =	(84)	(4)	3	55	0	(60)	(1)	36	215	211
	(156)	(2)	10	24	25	(181)	(6)	1292	569	496
	(240)	(6)	14	19	25	(241)	(0)	1328	784	707
1 Kanyā =	(241)	0 Sat. 29 Aug.						3		
								117		

1448 − 1333 ; 1667 − 1448 . . . . t = 1448 Bhādr. śuk. 5

tithi began 115 = 8^h 9^m. 219 = 15^h 31^m (end of tithi.)

c = 7070
7226
4296
− 117
s = 4179
t = 1448
∴ n = 5627

4179
2
8358
+ 1448
y = 9806

8^h 9^m
15 31
2)23 40

Total Bāva . 11 50
− 8 9

Bava on Sat. 29 Aug. = 3 41

n = Viśākhā y = Vaidhṛiti.

(The lagna requires a short calculation by itself.)

The above decides the solar month, day and week-day.

" " " " luni-solar month, day and week-day.

" " " " tithi.

" " " " karaṇa.

" " " " nakshatra.

" " " " yōga.

" " " " the positions of sun and moon, their longitudes, and distance from one another.

" " " " the time of day referred to, within 2 hours.

TABLE A.

DIFFERENCES IN THE CALENDAR BETWEEN ĀRYA AND SŪRYA SIDDHĀNTA FIXTURES.

Cols. 1, 2.—The number of the year here given is the one generally used in records of the year A.D. noted in column 3, and is stated here so as to catch the eye readily. In referring to the main Table LXI the number of the year in columns 1, 2 therein is the present number advanced by 1, being the corresponding concurrent year.

Col. 4, Class A.—Saṁvatsara-names given to solar and luni-solar years by northern system.

Col. 4, Class B.—Intercalations and suppressions of different lunar months. "adh."=an intercalated (adhika) month; "ksh," a suppressed (kshaya) month.

Col. 4, Class C.—Differences in the civil day called "Chaitra Śukla 1," the civil beginning of the luni-solar year. The figure in brackets in columns 5, 6 is the number of the civil day measured from January 1st.

K. Y. expired.	Śaka expired.	A. D.	Class.	Fixtures according to the First Ārya-Siddhānta.	Fixtures according to the Sūrya-Siddhānta.
1	2	3	4	5	6
4007	828	906-7	A	1 "Prabhava"	60 "Kshaya."
4008	829	907-8	A	2 "Vibhava"	1 "Prabhava."
4009	830	908-9	A	3 "Sukla"	2 "Vibhava."
4075	896	974-75	B	4 Āshāḍha (adh.)	3 Jyēshṭha (adh.)
4080	901	979-80	B	6 Bhādrapada (adh.)	3 Jyēshṭha (adh.).
4092	913	991-92	A	27 "Vijaya"	26 "Nandana."
4093	914	992-93	A	28 "Jaya"	27 "Vijaya."
4094	915	993-94	A	29 "Manmatha"	28 "Jaya"
4095	916	994-95	A	30 "Durmukha"	29 "Manmatha."
4159	980	1058-59	B	4 Āshāḍha (adh.)	3 Jyēshṭha (adh.).
4177	998	1076-77	A	53 "Siddhārthin"	52 "Kālayukta."
4178	999	1077-78	A	54 "Raudra"	53 "Siddhārthin."
4179	1000	1078-79	A	55 "Durmati"	54 "Raudra."
4180	1001	1079-80	A	56 "Dundubhi"	55 "Durmati."
4193	1014	1092-93	C	11 Mar. (71), 5 Thur.	12 Mar. (72), 6 Fri.
4232	1053	1131-32	B	5 Śrāvaṇa (adh.)	4 Āshāḍha (adh.).
4251	1072	1150-51	B	5 Śrāvaṇa (adh.)	4 Āshāḍha (adh.).
4256	1077	1155-56	B	*Nil*	12 Phālguna (adh.).
4257	1078	1156-57	B	1 Chaitra (adh.)	*Nil.*
id.	id.	id.	C	23 Feb. (54), 5 Thur.	24 Mar. (84), 0 Sat.
4262	1083	1161-62	A	19 "Pārthiva"	18 "Tāraṇa."
4263	1084	1162-63	A	20 "Vyaya"	19 "Pārthiva."
4264	1085	1163-64	A	21 "Sarvajit"	20 "Vyaya."
4265	1086	1164-65	A	22 "Sarvadhārin"	21 "Sarvajit."
4313	1134	1212-13	B	7 Āśvina (adh.) 11 *Māgha (ksh.)* 12 Phālguna (adh.)	7 Āśvina (adh.)
4348	1169	1247-48	A	46 "Paridhāvin"	45 "Virōdhakṛit."
4349	1170	1248-49	A	47 "Pramādin."	46 "Paridhāvin."
4350	1171	1249-50	A	48 "Ānanda"	47 "Pramādin."
4351	1172	1250-51	A	49 "Rākshasa"	48 "Ānanda.
4356	1177	1255-56	C	11 Mar. (70) 5 Thur.	10 Mar. (69), 4 Wed.
4378	1199	1277-78	B	9 Mārgaśira (adh.) 10 *Pausha (ksh.)* 12 Phālguna (adh.)	8 Kārttika (adh.). 10 *Pausha (ksh.).* 12 Phālguna (adh.).
4397	1218	1296-97	B	12 Phālguna (adh.)	9 Mārgaśira (adh.). 10 *Pausha (ksh.).* 12 Phālguna (adh.)

TABLE A—*Contd.*

K. Y. expired.	Saka expired.	A. D.	Class.	Fixtures according to the First Ārya-Siddhānta.	Fixtures according to the Sūrya-Siddhānta.
1	2	3	4	5	6
4416	1237	1315-16	B	12 Phālguna (adh.)	8 Kārttika (adah.) 9 *Mārgaśira (ksh.)*. 12 Phālguna (adh.).
4433	1254	1332-33	A	12 "Bahudhānya"	11 "Īsvara."
4434	1255	1333-34	A	13 "Pramāthin"	12 "Bahudhānya."
4435	1256	1334-35	A	14 "Vikrama"	13 "Pramāthin."
4436	1257	1335-36	A	15 "Vṛisha"	14 "Vikrama."
4454	1275	1353-54	B	7 Āśvina (adh.) 11 *Māgha (ksh.)* 12 Phālguna (adh.)	6 Bhādrapada (adh).
4471	1292	1370-71	B	3 Jyēshṭha (adh.)	2 Vaiśākha (adh.)
4481	1302	1380-81	B	*Nil.*	8 Kārttika (adh.). 9 *Mārgaśira (ksh.)*.
4492	1313	1391-92	B	7 Āśvina (adh.)	6 Bhādrapada (adh.).
4509	1330	1408-9	B	3 Jyēshṭha (adh.)	2 Vaisākha (adh.).
4511	1332	1410-11	B	7 Āśvina (adh.)	6 Bhādrapada (adh).
4518	1339	1417-18	A	38 "Krōdhin"	37 "Sōbhana."
4519	1340	1418-19	A	39 "Viśvāvasu"	38 "Krōdhin."
id.	id.	id.	B	8 Kārttika (adh.) 11 *Māgha (ksh.)* 12 Phālguna (adh.)	8 Kārttika (adh.)
4520	1341	1419-20	A	40 "Parābhava"	39 "Viśvāvasu."
4521	1342	1420-21	A	41 "Plavaṅga"	40 "Parābhava."
4537	1358	1436-37	C	18 Mar. (78), 1 Sun.	19 Mar. (79), 2 Mon.
4557	1378	1456-57	B	8 Kārttika (adh.) 10 *Pausha (ksh.)* 12 Phālguna (adh.)	8 Kārttika (adh.).
4566	1387	1465-66	B	2 Vaiśākha (adh.)	1 Chaitra (adh.).
4574	1395	1473-74	C	28 Feb. (59), 1 Sun.	27 Feb. (58), 0 Sat.
4576	1397	1475-76	B	7 Āśvina (adh.) 10 *Pausha (ksh.)* 12 Phālguna (adh.)	7 Āśvina (adh.). 11 *Māgha (ksh.)*. 12 Phālguna (adh.).
4587	1408	1486-87	B	6 Bhādrapada (adh.)	5 Śrāvaṇa (ad .).
4603	1424	1502-3	A	4 "Pramōda"	3 "Śukla."
4604	1425	1503-4	A	5 "Prajāpati"	4 "Pramōda."
id.	id.	id.	B	2 Vaisākha (adh.)	1 Chaitra (adh.).
4605	1426	1504-5	A	6 "Aṅgiras"	5 "Prajāpati."
4606	1427	1505-6	A	7 "Śrīmukha"	6 "Aṅgiras."
id.	id.	id.	B	6 Bhādrapada (adh.)	5 Śrāvaṇa (adh.).
4607	1428	1506-7	A	8 "Bhāva"	7 "Śrīmukha."
4608	1429	1507-8	A	9 "Yuvan"	8 "Bhāva."
4609	1430	1508-9	A	10 "Dhātri"	9 "Yuvan."
4610	1431	1509-10	A	11 "Īśvara"	10 "Dhātri."
4611	1432	1510-11	A	12 "Bahudhānya"	11 "Īśvara."
4612	1433	1511-12	A	13 "Pramāthin"	12 "Bahudhānya."
4613	1434	1512-13	A	14 "Vikrama"	13 "Pramāthin."
4614	1435	1513-14	A	15 "Vrisha"	14 "Vikrama."
4615	1436	1514-15	A	16 "Chitrabhānu"	15 "Vṛisha."
4622	1443	1521-22	B	*Nil.*	8 Kārttika (adh.). 9 *Mārgaśira (ksh.)*.

TABLE A—*Contd.*

. Y. expired.	Śaka expired.	A. D.	Class.	Fixtures according to the First Ārya-Siddhānta.	Fixtures according to the Sūrya-Siddhānta.
1	2	3	4	5	6
4644	1465	1543-44	B	6 Bhādrapada (adh.) . .	5 Śrāvaṇa (adh.).
4659	1480	1558-59	C	21 Mar. (80), 2 Mon. . .	20 Mar. (79), 1 Sun.
4660	1481	1559-60	B	8 Kārttika (adh.) . . . 11 *Māgha (ksh.)* . . 12 Phālguna (adh.) . .	7 Āśvina (adh.).
4679	1500	1578-79	B	8 Kārttika (adh.) . . .	7 Āśvina (adh.).
4682	1503	1581-82	C	6 Mar. (65), 2 Mon. . .	5 Mar. (64), 1 Sun.
4689	1510	1588-89	A	31 " Hēmalamba" . .	30 " Durmukha."
4690	1511	1589-90	A	32 " Vilamba" . . .	31 " Hēmalamba."
4691	1512	1590-91	A	33 " Vikārin" . . .	32 " Vilamba."
4692	1513	1591-92	A	34 " Śārvarin" . . .	33 " Vikārin."
4693	1514	1592-93	A	35 " Plava"	34 " Śārvarin."
4694	1515	1593-94	A	36 " Śubhakṛit" . . .	35 " Plava."
4695	1516	1594-95	A	37 " Śōbhana" . . .	36 " Śubhakṛit."
4696	1517	1595-96	A	38 " Krōdhin" . . .	37 " Śōbhana."
4697	1518	1596-97	A	39 " Viśvāvasu" . . .	38 " Krōdhin."
4698	1519	1597-98	A	40 " Parābhava" . . .	39 " Viśvāvasu."
id.	id.	id.	B	8 Kārttika (adh.) . . .	7 Āśvina (adh.).
4699	1520	1598-99	A	41 " Plavaṅga" . . .	40 " Parābhava."
4700	1521	1599-1600	A	42 " Kīlaka" . . .	41 " Plavaṅga."
4701	1522	1600-1	A	43 " Saumya" . . .	42 " Kīlaka."
4720	1541	1619-20	C	7 Mar. (66), 1 Sun. . .	6 Mar. (65), 0 Sat.
4731	1552	1630-31	C	4 Mar. (63), 5 Thur. . .	5 Mar. (64), 6 Fri.
4754	1575	1653-54	C	20 Mar. (79), 1 Sun. . .	19 Mar. (78), 0 Sat.
4757	1578	1656-57	C	17 Mar. (77), 2 Mon. . .	16 Mar. (76), 1 Sun.
4773	1594	1672-73	C	20 Mar. (80), 4 Wed. . .	19 Mar. (79), 3 Tues.
4774	1595	1673-74	A	57 " Rudhirōdgārin" . .	56 " Dundubhi."
4775	1596	1674-75	A	58 " Raktāksha" . . .	57 " Rudhirōdgārin."
4776	1597	1675-76	A	59 " Krōdhana" . . .	58 " Raktāksha."
4777	1598	1676-77	A	60 " Kshaya" . . .	59 " Krōdhana."
4778	1599	1677-78	A	1 " Prabhava" . . .	60 " Kshaya."
4779	1600	1678-79	A	2 " Vibhava" . . .	1 " Prabhava."
4780	1601	1679-80	A	3 " Śukla"	2 " Vibhava."
4781	1602	1680-81	A	4 " Pramōda" . . .	3 " Śukla."
4782	1603	1681-82	A	5 " Prajāpati" . . .	4 " Pramōda."
4783	1604	1682-83	A	6 " Aṅgiras" . . .	5 " Prajāpati."
4784	1605	1683-84	A	7 " Śrīmukha" . . .	6 " Aṅgiras."
4785	1606	1684-85	A	8 " Bhāva" . . .	7 " Śrīmukha."
4786	1607	1685-86	A	9 " Yuvan" . . .	8 " Bhāva."
4801	1622	1700-1	B	7 Āśvina (adh.) . . . 11 *Māgha (ksh.)* . . .	7 Āśvina (adh.).
4802	1623	1701-2	B	1 Chaitra (adh.) . . .	*Nil.*
id.	id.	id.	C	27 Feb. (58), 5 Thur. . .	29 Mar. (88), 0 Sat.
4807	1628	1706-7	B	4 Āshāḍha (adh.) . .	3 Jyēshṭha (adh.).
4819	1640	1718-19	C	22 Mar. (81), 0 Sat. .	21 Mar. (80), 6 Fri.
4826	1647	1725-26	B	4 Āshāḍha (adh.) . .	3 Jyēshṭha (adh.)
4858	1679	1757-58	A	22 " Sarvadhārin" .	21 " Sarvajit."
4859	1680	1758-59	A	23 " Virōdhin" . . .	22 " Sarvadhārin."
4860	1681	1759-60	A	24 " Vikṛita" . . .	23 " Virōdhin."

TABLE A—*Contd.*

K. Y. expired.	Saka expired.	A. D.	Class.	Fixtures according to the First Ārya-Siddhānta.	Fixtures according to the Sūrya-Siddhānta.
1	2	3	4	5	6
4861	1682	1660-61	A	25 "Khara"	24 "Vikṛita"
4862	1683	1761-62	A	26 "Nandana"	25 "Khara"
4863	1684	1762-63	A	27 "Vijaya"	26 "Nandana"
4864	1685	1763-64	A	28 "Jaya"	27 "Vijaya"
Do.	Do.	Do.	B	4 Āshādha (adh.)	3 Jyēshṭha (adh.)
4865	1686	1764-65	A	29 "Manmatha"	28 "Jaya"
4866	1687	1765-66	A	30 "Durmukha"	29 "Manmatha"
4867	1688	1766-67	A	31 "Hēmalamba"	30 "Durmukha"
4868	1689	1767-68	A	32 "Vilamba"	31 "Hēmalamba"
4869	1690	1768-69	A	33 "Vikārin"	32 "Vilamba"
4870	1691	1769-70	A	34 "Śārvarin"	33 "Vikārin"
4871	1692	1770-71	A	35 "Plava"	34 "Śārvarin"
4872	1693	1771-72	A	36 "Śubhakṛit"	35 "Plava"
4877	1698	1776-77	B	7 Āsvina (adh.)	6 Bhādrapada (adh.)
4882	1703	1781-82	C	26 Mar. (85), 2 Mon.	25 Mar. (84), 1 Sun.
4883	1704	1782-83	C	15 Mar. (74, 6 Fri.)	Mar. (73), 5 Thur.
4942	1763	1841-42	B	7 Āśvina (adh.) 11 Magha (ksh.)	7 Āśvina (adh.)
4943	1764	1842-43	B	1 Chaitra (adh.	*Nil.*
Do.	Do.	Do.	C	13 Mar. (72), 1 Sun.	11 Apr. (101), 2 Mon.
4944	1765	1843-44	A	49 "Rākshasa"	48 "Ānanda"
Do.	Do.	Do.	C	1 Apr. (91), 0 Sat.	31 Mar. (90) 6 Fri
4945	1766	1844-45	A	50 "Anala"	49 "Rākshasa"
Do.	Do.	Do.	C	20 Mar. (80), 4 Wed	19 Mar. (79), 3 Tues.
4946	1767	1845-46	A	51 "Pingala."	50 "Anala"
4947	1768	1846-47	A	52 "Kālayukta"	51 "Pingala"
4948	1769	1847-48	A	53 "Siddhārthin"	52 "Kālayukta"
4949	1770	1848-49	A	54 "Raudra"	53 "Siddhārthin"
4950	1771	1849-50	A	55 "Durmati"	54 "Raudra"
4951	1772	1850-51	A	56 "Dundubhi"	55 "Durmati"
4952	1773	1851-52	A	57 "Rudhirōdgārin	56 "Dundubhi"
4953	1774	1852-53	A	58 "Raktāksha"	57 "Rudhirōdgārin"
4954	1775	1853-54	A	59 "Krōdhana"	58 "Raktāksha"
4955	1776	1854-55	A	60 "Kshaya"	59 "Krōdhana"
4956	1777	1855-56	A	1 "Prabhava"	60 "Kshaya"
4957	1778	1856-57	A	2 "Vibhava"	1 "Prabhava"
4973	1794	1872-73	C	9 Apr. (100), 3 Tues	8 Apr. (99), 2 Mon.

TABLE LXI.

NOTES.

Cols. 1 to 4.—The present Table states the *concurrent* years so as exactly to correspond with Table I of the "Indian Calendar" and in that respect to save trouble for those who have become accustomed to use that publication. The year usually quoted in inscriptions is the *expired* year, though sometimes the concurrent year is given; *e.g.*, the year A.D. 899-900 corresponds to the concurrent years K. Y. 4001, Śaka 822, but to the expired years K. Y. 4000, Śaka 821.

Col. 8.—All the entries are of intercalated (adhika) months, except those in italics, which are suppressed (*kshaya*) months.

A List of instances wherein important details the Ārya and Sūrya Siddhāntas differ is given in Table A, pages 248-251.

It has not been thought necessary to include in this Table the years between A.D. 499 and 899. This paper concerns computation by the true motions of sun and moon, and it is practically certain that prior, at least, to the latter date all calculations for almanacs in India were made by mean planetary motions.

TABLE

GENERAL TABLE FOR CALCULATION

Conforming to Table I "Indian Calendar,"

Entries in italics in Column 7 shew where, in the Northern system, saṁvatsara-

* = *Leap-years of 366 days.*

CONCURRENT YEAR.								
Kali.	Śaka.	Chaitrādi Vikrama.	Mēshādi solar year in Bengal.	Kollam.	A.D.	Jovian Saṁvatsara. Southern system.	Jovian Saṁvatsara. Northern system.	Intercalated (*adhika*) and suppressed (*kshaya*) Lunar months (true).
1	2	3	3*a*	4	5	6	7	8
4001	822	957	306	74-75	899-900	53 Siddhārthin	53 Siddhārthin	...
4002	823	958	307	75-76	*900-01	54 Raudra	54 Raudra	...
4003	824	959	308	76-77	901-02	55 Durmati	55 Durmati	2 Vaiśākha
4004	825	960	309	77-78	902-03	56 Dundubhi	56 Dundubhi	...
4005	826	961	310	78-79	903-04	57 Rudhirōdgārin	57 Rudhirōdgārin	6 Bhādrapada
4006	827	962	311	79-80	*904-05	58 Raktāksha	58 Raktāksha	...
4007	828	963	312	80-81	905-06	59 Krōdhana	59 Krōdhana†	...
4008	829	964	313	81-82	906-07	60 Kshaya	1 *Prabhava*	5 Śrāvaṇa
4009	830	965	314	82-83	907-08	1 Prabhava	2 *Vibhava*	...
4010	831	966	315	83-84	*908-09	2 Vibhava	3 *Śukla*	...
4011	832	967	316	84-85	909-10	3 Śukla	4 Pramōda	3 Jyēshṭha
4012	833	968	317	85-86	910-11	4 Pramōda	5 Prajāpati	...
4013	834	969	318	86-87	911-12	5 Prajāpati	6 Aṅgiras	7 Āśvina 10 *Pausha* (*ksh.*)
4014	835	970	319	87-88	*912-13	6 Aṅgiras	7 Śrīmukha	1 Chaitra
4015	836	971	320	88-89	913-14	7 Śrīmukha	8 Bhāva	...
4016	837	972	321	89-90	914-15	8 Bhāva	9 Yuvan	5 Śrāvaṇa
4017	838	973	322	90-91	915-16	9 Yuvan	10 Dhātri	...
4018	839	974	323	91-92	*916-17	10 Dhātri	11 Īśvara	...
4019	840	975	324	92-93	917-18	11 Īśvara	12 Bahudhānya	4 Āshāḍha
4020	841	976	325	93-94	918-19	12 Bahudhānya	13 Pramāthin	...
4021	842	977	326	94-95	919-20	13 Pramāthin	14 Vikrama	...

† 60 Kshaya was suppressed in the north.

LXI.

BY THE FIRST ARYA-SIDDHĀNTA.

the columns being similarly numbered.

names of solar years differ from those given by followers of the Sūrya-Siddhānta.

Cols. 13, 19.—Figures in brackets=number of civil days measured from January 1st.

COMMENCEMENT OF THE									Kali.
SOLAR YEAR.			LUNI-SOLAR YEAR (MEAN SUNRISE OF CIVIL DAY ON WHICH CHAITRA ŚUKLA 1 ENDS).						
Day and month, A.D.	Week-day.	Time of true Mēsha-saṁkrānti.	Day and month, A.D.	Week-day.	*a.*	*b.*	*c.*		
13	14	17	19	20	23	24	25		1
		H. M. S.							
22 Mar. (81)	5 Thur.	13 47 30	16 Mar. (75)	6 Fri.	9939·8668	196·5305	259·4537		4001
21 Mar. (81)	6 Fri.	20 0 0	4 Mar. (64)	3 Tues.	9815·5502	43·7653	228·6299		4002
22 Mar. (81)	1 Sun.	2 12 30	22 Feb. (53)	1 Sun.	29·8654	927·2917	200·5438		4003
22 Mar. (81)	2 Mon.	8 25 0	13 Mar. (72)	0 Sat.	64·5051	863·2752	251·8535		4004
22 Mar. (81)	3 Tues.	14 37 30	3 Mar. (62)	5 Thur.	278·8203	746·8017	223·7674		4005
21 Mar. (81)	4 Wed.	20 50 0	20 Mar. (80)	3 Tues.	9974·8281	646·4936	272·3393		4006
22 Mar. (81)	6 Fri.	3 2 30	10 Mar. (69)	1 Sun.	189·1433	530·0200	244·2533		4007
22 Mar. (81)	0 Sat.	9 15 0	27 Feb. (58)	5 Thur.	64·8268	377·2548	213·4295		4008
22 Mar. (81)	1 Sun.	15 27 30	17 Mar. (76)	3 Tues.	9760·8345	276·9467	262·0014		4009
21 Mar. (81)	2 Mon.	21 40 0	6 Mar. (66)	1 Sun.	9975·1497	160·4731	233·9153		4010
22 Mar. (81)	4 Wed.	3 52 30	23 Feb. (54)	5 Thur.	9850·8331	7·7079	203·0914		4011
22 Mar. (81)	5 Thur.	10 5 0	14 Mar. (73)	4 Wed.	9885·4728	943·6915	254·4011		4012
22 Mar. (81)	6 Fri.	16 17 30	4 Mar. (63)	2 Mon.	99·7880	827·2178	226·3151		4013
21 Mar. (81)	0 Sat.	22 30 0	22 Feb. (53)	0 Sat.	314·1033	710·7442	198·2290		4014
22 Mar. (81)	2 Mon.	4 42 30	11 Mar. (70)	5 Thur.	10·1109	610·4362	246·8010		4015
22 Mar. (81)	3 Tues.	10 55 0	28 Feb. (59)	2 Mon.	9885·7943	457·6710	215·9771		4016
22 Mar. (81)	4 Wed.	17 7 30	19 Mar. (78)	1 Sun.	9920·4340	393·6545	267·2868		4017
21 Mar. (81)	5 Thur.	23 20 0	7 Mar. (67)	5 Thur.	9796·1174	240·8893	236·4209		4018
22 Mar. (81)	0 Sat.	5 32 30	25 Feb. (56)	3 Tues.	10·4326	124·4158	208·3769		4019
22 Mar. (81)	1 Sun.	11 45 0	16 Mar. (75)	2 Mon.	45·0722	60·3992	259·5866		4020
22 Mar. (81)	2 Mon.	17 57 30	5 Mar. (64)	6 Fri.	9920·7556	907·6340	228·5628		4021

TABLE

CONCURRENT YEAR.								
Kali.	Śaka.	Chaitrādi Vikrama.	Mēshādi solar year in Bengal.	Kollam.	A.D.	Jovian Saṁvatsara. Southern system.	Jovian Saṁvatsara. Northern system.	Intercalated (*adhika*) and suppressed (*kshaya*) Lunar months (true).
1	2	3	3*a*	4	5	6	7	8
4022	843	978	327	95-96	*920-21	14 Vikrama	15 Vṛisha	2 Vaiśākha
4023	844	979	328	96-97	921-22	15 Vṛisha	16 Chitrabhānu	...
4024	845	980	329	97-98	922-23	16 Chitrabhānu	17 Subhānu	6 Bhādrapada
4025	846	981	330	98-99	923-24	17 Subhānu	18 Tāraṇa	...
4026	847	982	331	99-100	*924-25	18 Tāraṇa	19 Pārthiva	...
4027	848	983	332	100-01	925-26	19 Pārthiva	20 Vyaya	4 Āshāḍha
4028	849	984	333	101-02	926-27	20 Vyaya	21 Sarvajit	...
4029	850	985	334	102-03	927-28	21 Sarvajit	22 Sarvadhārin	...
4030	851	986	335	103-04	*928-29	22 Sarvadhārin	23 Virōdhin	3 Jyēshṭha
4031	852	987	336	104-05	929-30	23 Virōdhin	24 Vikṛita	...
4032	853	988	337	105-06	930-31	24 Vikṛita	25 Khara	7 Āśvina
4033	854	989	338	106-07	931-32	25 Khara	26 Nandana	...
4034	855	990	339	107-08	*932-33	26 Nandana	27 Vijaya	...
4035	856	991	340	108-09	933-34	27 Vijaya	28 Jaya	5 Śrāvaṇa
4036	857	992	341	109-10	934-35	28 Jaya	29 Manmatha	...
4037	858	993	342	110-11	935-36	29 Manmatha	30 Durmukha	...
4038	859	994	343	111-12	*936-37	30 Durmukha	31 Hēmalamba	3 Jyēshṭha
4039	860	995	344	112-13	937-38	31 Hēmalamba	32 Vilamba	...
4040	861	996	345	113-14	938-39	32 Vilamba	33 Vikārin	...
4041	862	997	346	114-15	939-40	33 Vikārin	34 Śārvarin	2 Vaiśākha
4042	863	998	347	115-16	*940-41	34 Śārvarin	35 Plava	...
4043	864	999	348	116-17	941-42	35 Plava	36 Śubhakrit	6 Bhādrapada
4044	865	1000	349	117-18	942-43	36 Śubhakṛit	37 Śōbhana	...
4045	866	1001	350	118-19	943-44	37 Śōbhana	38 Krōdhin	...
4046	867	1002	351	119-20	*944-45	38 Krōdhin	39 Viśvāvasu	4 Āshāḍha

LXI—*Cont'd.*

COMMENCEMENT OF THE									Kali.
SOLAR YEAR.			LUNI-SOLAR YEAR (MEAN SUNRISE OF CIVIL DAY ON WHICH CHAITRA ŚUKLA 1 ENDS).						
Day and month, A.D.	Week-day.	Time of true Mēsha-saṁkrānti.	Day and month, A.D.	Week-day.	*a.*	*b.*	*c.*		
13	14	17	19	20	23	24	25		1
		H. M. S.							
22 Mar. (82)	4 Wed.	0 10 0	23 Feb. (54)	4 Wed.	135·0709	791·1625	200·7767		4022
22 Mar. (81)	5 Thur.	6 22 30	13 Mar. (72)	3 Tues.	169·7105	727·1460	252·0864		4023
22 Mar. (81)	6 Fri.	12 35 0	2 Mar. (61)	0 Sat.	45·3939	574·3808	221·2635		2024
22 Mar. (81)	0 Sat.	18 47 30	21 Mar. (80)	6 Fri.	80·0335	510·3623	272·5722		4025
22 Mar. (82)	2 Mon.	1 0 0	9 Mar. (69)	3 Tues.	9955·7169	357·5972	241·7524		4026
22 Mar. (81)	3 Tues.	7 12 30	26 Feb. (57)	0 Sat.	9831·4003	204·8339	210·9246		4027
22 Mar. (81)	4 Wed.	13 25 0	17 Mar. (76)	6 Fri.	9866·0399	140·8154	262·2323		4028
22 Mar. (81)	5 Thur.	19 37 30	7 Mar. (66)	4 Wed.	80·3551	24·3419	234·1482		4029
22 Mar. (82)	0 Sat.	1 50 0	24 Feb. (55)	1 Sun.	9956·0385	871·5766	203·3243		4030
22 Mar. (81)	1 Sun.	8 2 30	14 Mar. (73)	0 Sat.	9990·6782	807·5702	254·6340		4031
22 Mar. (81)	2 Mon.	14 15 0	4 Mar. (63)	5 Thur.	204·9934	691·0866	226·5480		4032
22 Mar. (81)	3 Tues.	20 27 30	23 Mar. (82)	4 Wed.	239·6331	627·0701	277·8577		4033
22 Mar. (82)	5 Thur.	2 40 0	11 Mar. (71)	1 Sun.	115·3164	474·3049	247·0339		4034
22 Mar. (81)	6 Fri.	8 52 30	28 Feb. (59)	5 Thur.	9990·9998	321·5397	216·2100		4035
22 Mar. (81)	0 Sat.	15 5 0	19 Mar. (78)	4 Wed.	25·6394	257·8149	270·2575		4036
22 Mar. (81)	1 Sun.	21 17 30	8 Mar. (67)	1 Sun.	9901·3228	104·7580	236·6958		4037
22 Mar. (82)	3 Tues.	3 30 0	26 Feb. (57)	6 Fri.	115·6381	988·2845	208·6098		4038
22 Mar. (81)	4 Wed.	9 42 30	16 Mar. (75)	5 Thur.	150·2777	924·2680	259·9195		4039
22 Mar. (81)	5 Thur.	15 55 0	5 Mar. (64)	2 Mon.	25·9611	771·5027	229·0957		4040
22 Mar. (81)	6 Fri.	22 7 30	23 Feb. (54)	0 Sat.	240·2763	655·0292	201·0096		4041
22 Mar. (82)	1 Sun.	4 20 0	12 Mar. (72)	5 Thur.	9936·2841	554·7211	249·5816		4042
22 Mar. (81)	2 Mon.	10 32 30	1 Mar. (60)	2 Mon.	9811·9675	401·9560	218·7576		4043
22 Mar. (81)	3 Tues.	16 45 0	20 Mar. (79)	1 Sun.	9846·6072	337·9394	270·0674		4044
22 Mar. (81)	4 Wed.	22 57 30	9 Mar. (68)	5 Thur.	9722·3005	185·1742	239·9517		4045
22 Mar. (82)	6 Fri.	5 10 0	27 Feb. (58)	3 Tues.	9936·6057	68·7007	211·1575		4046

TABLE

CONCURRENT YEAR.								Intercalated (*adhika*) and suppressed (*kshaya*) Lunar months (true).
Kali.	Śaka.	Chaitrādi Vikrama.	Mêshādi solar year in Bengal.	Kollam.	A.D.	Jovian Samvatsara. Southern system.	Jovian Samvatsara. Northern system.	
1	2	3	3a	4	5	6	7	8
4047	868	1003	352	120-21	945-46	39 Viśvāvasu .	40 Parābhava .	...
4048	869	1004	353	121-22	946-47	40 Parābhava .	41 Plavaṅga .	...
4049	870	1005	354	122-23	947-48	41 Plavaṅga .	42 Kīlaka . .	3 Jyēshṭha .
4050	871	1006	355	123-24	*948-49	42 Kīlaka . .	43 Saumya .	...
4051	872	1007	356	124-25	949-50	43 Saumya .	44 Sādhāraṇa .	7 Āśvina .
4052	873	1008	357	125-26	950-51	44 Sādhāraṇa .	45 Virōdhakṛit .	...
4053	874	1009	358	126-27	951-52	45 Virōdhakṛit .	46 Paridhāvin .	...
4054	875	1010	359	127-28	*952-53	46 Paridhāvin .	47 Pramādin .	5 Śrāvaṇa .
4055	876	1011	360	128-29	953-54	47 Pramādin .	48 Ānanda .	...
4056	877	1012	361	129-30	954-55	48 Ānanda .	49 Rākshasa .	...
4057	878	1013	362	130-31	955-56	49 Rākshasa .	50 Anala . .	3 Jyēshṭha .
4058	879	1014	363	131-32	*956-57	50 Anala . .	51 Piṅgala .	...
4059	880	1015	364	132-33	957-58	51 Piṅgala .	52 Kālayukta .	...
4060	881	1016	365	133-34	958-59	52 Kālayukta .	53 Siddhārthin .	2 Vaiśākha .
4061	882	1017	366	134-35	959-60	53 Siddhārthin .	54 Raudra .	...
4062	883	1018	367	135-36	*960-61	54 Raudra .	55 Durmati .	6 Bhādrapada
4063	884	1019	368	136-37	961-62	55 Durmati .	56 Dundubhi .	...
4064	885	1020	369	137-38	962-63	56 Dundubhi .	57 Rudhirōdgārin	...
4065	886	1021	370	138-39	963-64	57 Rudhirōdgārin	58 Raktāksha .	4 Āshādha
4066	887	1022	371	139-40	*964-65	58 Raktāksha .	59 Krōdhana .	...
4067	888	1023	372	140-41	965-66	59 Krōdhana .	60 Kshaya .	...
4068	889	1024	373	141-42	966-67	60 Kshaya .	1 Prabhava .	3 Jyēshṭha .
4069	890	1025	374	142-43	967-68	1 Prabhava .	2 Vibhava .	...
4070	891	1026	375	143-44	*968-69	2 Vibhava .	3 Śukla . .	7 Āśvina .
4071	892	1027	376	144-45	969-70	3 Śukla . .	4 Pramōda .	...

LXI—*Contd.*

COMMENCEMENT OF THE									
SOLAR YEAR.			LUNI-SOLAR YEAR (MEAN SUNRISE OF CIVIL DAY ON WHICH CHAITRA ŚUKLA 1 ENDS).					Kali.
Day and month, A.D.	Week-day.	Time of true Mēsha-saṁkrānti.	Day and month, A.D.	Week-day.	*a.*	*b.*	*c.*	
13	14	17	19	20	23	24	25	1
		H. M. S.						
22 Mar. (81)	0 Sat. .	11 22 30	17 Mar. (76)	2 Mon. .	9972·2453	4·6841	262·4672	4047
22 Mar. (81)	1 Sun. .	17 35 0	7 Mar. (66)	0 Sat. .	185·5605	888·2106	234·3811	4048
22 Mar. (81)	2 Mon. .	23 47 30	24 Feb. (55)	4 Wed.	61·2440	735·4454	203·5584	4049
22 Mar. (82)	4 Wed.	6 0 0	14 Mar. (74)	3 Tues.	95·8836	671·4290	254·8669	4050
22 Mar. (81)	5 Thur.	12 12 30	3 Mar. (62)	0 Sat. .	9971·5669	518·6637	224·0431	4051
22 Mar. (81)	6 Fri. .	18 25 0	22 Mar. (81)	6 Fri. .	6·2066	454·6473	275·3528	4052
23 Mar. (82)	1 Sun. .	0 37 30	11 Mar. (70)	3 Tues. .	9881·8899	301·8921	244·5290	4053
22 Mar. (82)	2 Mon. .	6 50 0	28 Feb. (59)	0 Sat. .	9757·5734	149·1168	213·7052	4054
22 Mar. (81)	3 Tues.	13 2 30	18 Mar. (77)	6 Fri. .	9792·2130	85·1004	265·0148	4055
22 Mar. (81)	4 Wed.	19 15 0	8 Mar. (67)	4 Wed. .	6·5282	968·6268	236·9287	4056
23 Mar. (82)	6 Fri. .	1 27 30	26 Feb. (57)	2 Mon. .	220·8435	852·1532	208·8427	4057
22 Mar. (82)	0 Sat. .	7 40 0	16 Mar. (76)	1 Sun. .	255·4831	788·1367	260·1524	4058
22 Mar. (81)	1 Sun. .	13 52 30	5 Mar. (64)	5 Thur.	131·1665	635·3715	229·3286	4059
22 Mar. (81)	2 Mon. .	20 5 0	22 Feb. (53)	2 Mon. .	6·8499	482·6064	198·5047	4060
23 Mar. (82)	4 Wed.	2 17 30	13 Mar. (72)	1 Sun. .	41·4895	418·5898	249·8145	4061
22 Mar. (82)	5 Thur.	8 30 0	1 Mar. (61)	5 Thur.	9917·1729	265·8247	218·9905	4062
22 Mar. (81)	6 Fri. .	14 42 30	20 Mar. (79)	4 Wed.	9951·8125	201·8082	270·3003	4063
22 Mar. (81)	0 Sat. .	20 55 0	9 Mar. (68)	1 Sun. .	9827·4959	49·0429	239·4764	4064
23 Mar. (82)	2 Mon. .	3 7 30	27 Feb. (58)	6 Fri. .	41·8112	932·5694	211·3904	4065
22 Mar. (82)	3 Tues.	9 20 0	17 Mar. (77)	5 Thur.	75·4508	868·5529	262·7001	4066
22 Mar. (81)	4 Wed.	15 32 30	7 Mar. (66)	3 Tues.	290·7660	752·0794	234·6440	4067
22 Mar. (81)	5 Thur.	21 45 0	24 Feb. (55)	0 Sat. .	166·4494	599·3141	203·7901	4068
23 Mar. (82)	0 Sat. .	3 57 30	15 Mar. (74)	6 Fri. .	201·0890	535·2977	255·0998	4069
22 Mar. (82)	1 Sun.	10 10 0	3 Mar (63)	3 Tues.	76·7724	382·5385	224·2760	4070
22 Mar. (81)	2 Mon. .	16 22 30	21 Mar. (80)	1 Sun. .	9772·7802	282·2243	272·8479	4071

TABLE

CONCURRENT YEAR.								Intercalated (*adhika*) and suppressed (*kshaya*) Lunar months (true).
						Jovian Saṁvatsara.		
Kali.	Śaka.	Chaitrādi Vikrama.	Mēshādi solar year in Bengal.	Kollam.	A.D.	Southern system.	Northern system.	
1	2	3	3*a*	4	5	6	7	8
4072	893	1028	377	145-46	970-71	4 Pramōda	5 Prajāpati	...
4073	894	1029	378	146-47	971-72	5 Prajāpati	6 Aṅgiras	5 Śrāvaṇa
4074	895	1030	379	147-48	*972-73	6 Aṅgiras	7 Śrīmukha	...
4075	896	1031	380	148-49	973-74	7 Śrīmukha	8 Bhāva	...
4076	897	1032	381	149-50	974-75	8 Bhāva	9 Yuvan	4 Āshāḍha
4077	898	1033	382	150-51	975-76	9 Yuvan	10 Dhātṛi	...
4078	899	1034	383	151-52	*976-77	10 Dhātṛi	11 Īśvara	...
4079	900	1035	384	152-53	977-78	11 Īśvara	12 Bahudhānya	1 Chaitra
4080	901	1036	385	153-54	978-79	12 Bahudhānya	13 Pramāthin	...
4081	902	1037	386	154-55	979-80	13 Pramāthin	14 Vikrama	6 Bhādrapada
4082	903	1038	387	155-56	*980-81	14 Vikrama	15 Vṛisha	...
4083	904	1039	388	156-57	981-82	15 Vṛisha	16 Chitrabhānu	...
4084	905	1040	389	157-58	982-83	16 Chitrabhānu	17 Subhānu	4 Āshāḍha
4085	906	1041	390	158-59	983-84	17 Subhānu	18 Tāraṇa	...
4086	907	1042	391	159-60	*984-85	18 Tāraṇa	19 Pārthiva	...
4087	908	1043	392	160-61	985-86	19 Pārthiva	20 Vyaya	3 Jyēshṭha
4088	909	1044	393	161-62	986-87	20 Vyaya	21 Sarvajit	...
4089	910	1045	394	162-63	987-88	21 Sarvajit	22 Sarvadhārin	7 Āśvina
4090	911	1046	395	163-64	*988-89	22 Sarvadhārin	23 Virōdhin	...
4091	912	1047	396	164-65	989-90	23 Virōdhin	24 Vikṛita	...
4092	913	1048	397	165-66	990-91	24 Vikṛita	25 Khara†	5 Śrāvaṇa
4093	914	1049	398	166-67	991-92	25 Khara	27 *Vijaya*	...
4094	915	1050	399	167-68	*992-93	26 Nandana	28 *Jaya*	...
4095	916	1051	400	168-69	993-94	27 Vijaya	29 *Manmatha*	3 Jyēshṭha
4096	917	1052	401	169-70	994-95	28 Jaya	30 *Durmukha*	...

† 26 Nandana was suppressed in the north

LXI—*Contd.*

COMMENCEMENT OF THE											
Solar year.					Luni-solar year (mean sunrise of civil day on which Chaitra śukla 1 ends).						Kali.
Day and month, A.D.	Week-day.	Time of true Mēsha-saṁkrānti.			Day and month, A.D.	Week-day.	*a.*	*b.*	*c.*		
13	14	17			19	20	23	24	25		1
		H.	M.	S.							
22 Mar. (81)	3 Tues.	22	35	0	11 Mar. (70)	6 Fri.	9987·0954	165·7508	244·7619		4072
23 Mar. (82)	5 Thur.	4	47	30	28 Feb. (59)	3 Tues.	9862·7789	12·9856	213·9381		4073
22 Mar. (82)	6 Fri.	11	0	0	18 Mar. (78)	2 Mon.	9897·4185	948·9692	265·2477		4074
22 Mar. (81)	0 Sat.	17	12	30	8 Mar. (67)	0 Sat.	111·7337	832·4955	237·1616		4075
22 Mar. (81)	1 Sun.	23	25	0	25 Feb. (56)	4 Wed.	9987·4171	679·7304	206·3378		4076
23 Mar. (82)	3 Tues.	5	37	30	16 Mar. (75)	3 Tues.	22·0566	615·7139	257·6475		4077
22 Mar. (82)	4 Wed.	11	50	0	4 Mar. (64)	0 Sat.	9897·7400	462·9486	226·8237		4078
22 Mar. (81)	5 Thur.	18	2	30	21 Feb. (52)	4 Wed.	9773·4234	310·1835	195·9998		4079
23 Mar. (82)	0 Sat.	0	15	0	12 Mar. (71)	3 Tues.	9808·0631	246·1670	247·3096		4080
23 Mar. (82)	1 Sun.	6	27	30	2 Mar. (61)	1 Sun.	22·3783	129·6934	219·2234		4081
22 Mar. (82)	2 Mon.	12	40	0	20 Mar. (80)	0 Sat.	57·0179	65·6869	270·5332		4082
22 Mar. (81)	3 Tues.	18	52	30	9 Mar. (68)	4 Wed.	9932·7013	912·9117	239·7093		4083
23 Mar. (82)	5 Thur.	1	5	0	27 Feb. (58)	2 Mon.	147·0166	796·4381	211·6233		4084
23 Mar. (82)	6 Fri.	7	17	30	18 Mar. (77)	1 Sun.	181·6562	732·4216	262·9330		4085
22 Mar. (82)	0 Sat.	13	30	0	6 Mar. (66)	5 Thur.	57·3396	579·6565	232·1091		4086
22 Mar. (81)	1 Sun.	19	42	30	23 Feb. (54)	2 Mon.	9933·0229	426·8913	201·2852		4087
23 Mar. (82)	3 Tues.	1	55	0	14 Mar. (73)	1 Sun.	9967·6626	362·8648	252·5949		4088
23 Mar. (82)	4 Wed.	8	7	30	3 Mar. (62)	5 Thur.	9843·3460	210·1096	221·7711		4089
22 Mar. (82)	5 Thur.	14	20	0	21 Mar. (81)	4 Wed.	9877·9856	146·0931	273·0808		4090
22 Mar. (81)	6 Fri.	20	32	30	11 Mar. (70)	2 Mon.	92·3008	29·6195	244·9948		4091
23 Mar. (82)	1 Sun.	2	45	0	28 Feb. (59)	6 Fri.	9967·9842	876·8543	214·1709		4092
23 Mar. (82)	2 Mon.	8	57	30	19 Mar. (78)	5 Thur.	3·6239	812·8379	265·4806		4093
22 Mar. (82)	3 Tues.	15	10	0	8 Mar. (68)	3 Tues.	216·9391	696·3643	237·3945		4094
22 Mar. (81)	4 Wed.	21	22	30	25 Feb. (56)	0 Sat.	9[illegible]·6225	543·5991	206·5707		4095
23 Mar. (82)	6 Fri.	3	35	0	16 Mar. (76)	6 Fri.	127·2621	479·5826	257·8804		4096

TABLE

CONCURRENT YEAR.								Intercalated (*adhika*) and suppressed (*kshaya*) Lunar months (true)
						Jovian Samvatsara.		
Kali.	Saka.	Chaitrādi Vikrama.	Mēshādi solar year in Bengal.	Kollam.	A.D.	Southern system.	Northern system.	
1	2	3	3a	4	5	6	7	8
4097	918	1053	402	170-71	995-96	29 Manmatha .	31 Hēmalamba .	...
4098	919	1054	403	171-72	*996-97	30 Durmukha .	32 Vilamba .	1 Chaitra .
4099	920	1055	404	172-73	997-98	31 Hēmalamba .	33 Vikārin . .	...
4100	921	1056	405	173-74	998-99	32 Vilamba .	34 Śārvarin .	5 Śrāvaṇa .
4101	922	1057	406	174-75	999-1000	33 Vikārin . .	35 Plava . .	...
4102	923	1058	407	175-76	*1000-01	34 Śārvarin .	36 Śubhakṛit .	...
4103	924	1059	408	176-77	1001-02	35 Plava . .	37 Śōbhana .	4 Āshāḍha
4104	925	1060	409	177-78	1002-03	36 Śubhakṛit .	38 Krōdhin .	...
4105	926	1061	410	178-79	1003-04	37 Śōbhana .	39 Viśvāvasu .	...
4106	927	1062	411	179-80	*1004-05	38 Krōdhin .	40 Parābhava .	2 Vaiśākha .
4107	928	1063	412	180-81	1005-06	39 Viśvāvasu .	41 Plavaṅga .	...
4108	929	1064	413	181-82	1006-07	40 Parābhava .	42 Kīlaka . .	6 Bhādrapada
4109	930	1065	414	182-83	1007-08	41 Plavaṅga .	43 Saumya .	...
4110	931	1066	415	183-84	*1008-09	42 Kīlaka . .	44 Sādhāraṇa .	...
4111	932	1067	416	184-85	1009-10	43 Saumya .	45 Virōdhakṛit .	5 Śrāvaṇa .
4112	933	1068	417	185-86	1010-11	44 Sādhāraṇa .	46 Paridhāvin .	...
4113	934	1069	418	186-87	1011-12	45 Virōdhakṛit .	47 Pramādin .	...
4114	935	1070	419	187-88	*1012-13	46 Paridhāvin .	48 Ānanda .	3 Jyēshṭha .
4115	936	1071	420	188-89	1013-14	47 Pramādin .	49 Rākshasa .	...
4116	937	1072	421	189-90	1014-15	48 Ānanda .	50 Anala . .	...
4117	938	1073	422	190-91	1015-16	49 Rākshasa .	51 Piṅgala .	1 Chaitra .
4118	939	1074	423	191-92	*1016-17	50 Anala . .	52 Kālayukta .	...
4119	940	1075	424	192-93	1017-18	51 Piṅgala .	53 Siddhārthin .	5 Śrāvaṇa .
4120	941	1076	425	193-94	1018-19	52 Kālayukta .	54 Raudra .	...
4121	942	1077	426	194-95	1019-20	53 Siddhārthin .	55 Durmati .	...

LXI—*Contd.*

COMMENCEMENT OF THE											Kali.
SOLAR YEAR.					LUNI-SOLAR YEAR (MEAN SUNRISE OF CIVIL DAY ON WHICH CHAITRA ŚUKLA 1 ENDS).						
Day and month, A.D.	Week-day.	Time of true Mēsha-saṁkrānti.			Day and month, A.D.	Week-day.	*a.*	*b.*	*c.*		
13	14	17			19	20	23	24	25		1
		H.	M.	S.							
23 Mar. (82)	0 Sat.	9	47	30	5 Mar. (64)	3 Tues.	2·9455	326·8174	227·0566		4097
22 Mar. (82)	1 Sun.	16	0	0	22 Feb. (53)	0 Sat.	9878·6289	174·0522	196·2327		4098
22 Mar. (81)	2 Mon.	22	12	30	12 Mar. (71)	6 Fri.	9913·2685	110·0357	247·5424		4099
23 Mar. (82)	4 Wed.	4	25	0	2 Mar. (61)	4 Wed.	127·5838	993·5622	219·4563		4100
23 Mar. (82)	5 Thur.	10	37	30	21 Mar. (80)	3 Tues.	162·2234	929·5456	270·7661		4101
22 Mar. (82)	6 Fri.	16	50	0	9 Mar. (69)	0 Sat.	37·9068	776·7804	239·9422		4102
22 Mar. (81)	0 Sat.	23	2	30	27 Feb. (58)	5 Thur.	252·2221	660·3068	211·8562		4103
23 Mar. (82)	2 Mon.	5	15	0	17 Mar. (76)	3 Tues.	9948·2298	559·9987	260·4280		4104
23 Mar. (82)	3 Tues.	11	27	30	6 Mar. (65)	0 Sat.	9823·9122	407·2335	229·6042		4105
22 Mar. (82)	4 Wed.	17	40	0	24 Feb. (55)	5 Thur.	38·2274	290·7599	201·5181		4106
22 Mar. (81)	5 Thur.	23	52	30	13 Mar. (72)	3 Tues.	9734·2362	190·4518	250·0901		4107
23 Mar. (82)	0 Sat.	6	5	0	3 Mar. (62)	1 Sun.	9948·5515	73·9783	222·0040		4108
23 Mar. (82)	1 Sun.	12	17	30	22 Mar. (81)	0 Sat.	9983·1911	9·9618	274·3137		4109
22 Mar. (82)	2 Mon.	18	30	0	11 Mar. (71)	5 Thur.	197·5063	893·4882	245·2277		4110
23 Mar. (82)	4 Wed.	0	42	30	28 Feb. (59)	2 Mon.	73·1897	740·7230	214·4037		4111
23 Mar. (82)	5 Thur	6	55	0	19 Mar. (78)	1 Sun.	107·8294	676·7066	265·7135		4112
23 Mar. (82)	6 Fri.	13	7	30	8 Mar. (67)	5 Thur.	9983·5127	523·9413	234·8896		4113
22 Mar. (82)	0 Sat.	19	20	0	25 Feb. (56)	2 Mon.	9859·1961	371·1761	204·0658		4114
23 Mar. (82)	2 Mon.	1	32	30	15 Mar. (74)	1 Sun.	9893·8357	307·4513	258·1133		4115
23 Mar. (82)	3 Tues.	7	45	0	4 Mar. (63)	5 Thur.	9769·5190	154·3945	224·5517		4116
23 Mar. (82)	4 Wed.	13	57	30	22 Feb. (53)	3 Tues.	9983·8344	37·9209	196·5655		4117
22 Mar. (82)	5 Thur.	20	10	0	12 Mar. (72)	2 Mon.	18·4740	973·9044	247·7753		4118
23 Mar. (82)	0 Sat.	2	22	30	2 Mar. (61)	0 Sat.	232·7892	857·4309	219·6892		4119
23 Mar. (82)	1 Sun.	8	35	0	21 Mar. (80)	6 Fri.	267·4288	793·4143	270·9990		4120
23 Mar. (82)	2 Mon.	14	47	30	10 Mar. (69)	3 Tues.	143·1122	640·6491	240·1751		4121

TABLE

CONCURRENT YEAR.								INTERCALATED (*adhika*) and SUPPRESSED (*kshaya*) LUNAR MONTHS (true).
Kali.	Śaka.	Chaitrādi Vikrama.	Mēshādi solar year in Bengal.	Kollam.	A.D.	Jovian Saṁvatsara. Southern system.	Jovian Saṁvatsara. Northern system.	
1	2	3	3*a*	4	5	6	7	8
4122	943	1078	427	195-96	*1020-21	54 Raudra .	56 Dundubhi .	4 Āshāḍha .
4123	944	1079	428	196-97	1021-22	55 Durmati .	57 Rudhirōdgārin	...
4124	945	1080	429	197-98	1022-23	56 Dundubhi .	58 Raktāksha .	...
4125	946	1081	430	198-99	1023-24	57 Rudhirōdgārin	59 Krōdhana .	2 Vaiśākha .
4126	947	1082	431	199-200	*1024-25	58 Raktāksha .	60 Kshaya .	...
4127	948	1083	432	200-01	1025-26	59 Krōdhana .	1 Prabhava .	6 Bhādrapada
4128	949	1084	433	201-02	1026-27	60 Kshaya .	2 Vibhava .	...
4129	950	1085	434	202-03	1027-28	1 Prabhava .	3 Śukla . .	...
4130	951	1086	435	203-04	*1028-29	2 Vibhava .	4 Pramōda .	5 Śrāvaṇa .
4131	952	1087	436	204-05	1029-30	3 Śukla . .	5 Prajāpati .	...
4132	953	1088	437	205-06	1030-31	4 Pramōda .	6 Aṅgiras .	...
4133	954	1089	438	206-07	1031-32	5 Prajāpati .	7 Śrīmukha .	3 Jyēshṭha .
4134	955	1090	439	207-08	*1032-33	6 Aṅgiras .	8 Bhāva . .	...
4135	956	1091	440	208-09	1033-34	7 Śrīmukha .	9 Yuvan . .	...
4136	957	1092	441	209-10	1034-35	8 Bhāva . .	10 Dhātṛi . .	1 Chaitra .
4137	958	1093	442	210-11	1035-36	9 Yuvan . .	11 Īśvara . .	...
4138	959	1094	443	211-12	*1036-37	10 Dhātṛi . .	12 Bahudhānya .	5 Śrāvaṇa .
4139	960	1095	444	212-13	1037-38	11 Īśvara . .	13 Pramāthin .	...
4140	961	1096	445	213-14	1038-39	12 Bahudhānya .	14 Vikrama .	...
4141	962	1097	446	214-15	1039-40	13 Pramāthin .	15 Vṛisha . .	4 Āshāḍha .
4142	963	1098	447	215-16	*1040-41	14 Vikrama .	16 Chitrabhānu .	...
4143	964	1099	448	216-17	1041-42	15 Vṛisha . .	17 Subhānu .	...
4144	965	1100	449	217-18	1042-43	16 Chitrabhānu .	18 Tāraṇa .	2 Vaiśākha .
4145	966	1101	450	218-19	1043-44	17 Subhānu .	19 Pārthiva .	...
4146	967	1102	451	219-20	*1044-45	18 Tāraṇa . .	20 Vyaya . .	6 Bhādrapada

LXI—*Contd.*

COMMENCEMENT OF THE											Kali.
Solar year.					Luni-solar year (mean sunrise of civil day on which Chaitra śukla 1 ends).						
Day and month, A.D.	Week-day.	Time of true Mēsha-saṁkrānti.			Day and month, A.D.	Week-day.	*a.*	*b.*	*c.*		
13	14	17			19	20	23	24	25		1
		H.	M.	S.							
22 Mar. (82)	3 Tues.	21	0	0	27 Feb. (58)	0 Sat.	18·6956	487·8840	209·3513		4122
23 Mar. (82)	5 Thur.	3	12	30	17 Mar. (76)	6 Fri.	53·4352	423·8675	260·6609		4123
23 Mar. (82)	6 Fri.	9	25	0	6 Mar. (65)	3 Tues.	9929·1186	271·1022	229·8371		4124
23 Mar. (82)	0 Sat.	15	37	30	23 Feb. (54)	0 Sat.	9804·8020	118·3371	199·0132		4125
22 Mar. (82)	1 Sun.	21	50	0	13 Mar. (73)	6 Fri.	9839·4416	54·3206	250·3230		4126
23 Mar. (82)	3 Tues.	4	2	30	3 Mar. (62)	4 Wed.	53·7569	937·8470	222·2369		4127
23 Mar. (82)	4 Wed.	10	15	0	22 Mar. (81)	3 Tues.	88·3965	873·8305	273·5466		4128
23 Mar. (82)	5 Thur.	16	27	30	12 Mar. (71)	1 Sun.	302·7117	757·3570	245·4606		4129
22 Mar. (82)	6 Fri.	22	40	0	29 Feb. (60)	5 Thur.	178·3951	604·5917	214·6366		4130
23 Mar. (82)	1 Sun.	4	52	30	18 Mar. (77)	3 Tues.	9874·4029	504·2837	263·2086		4131
23 Mar. (82)	2 Mon.	11	5	0	7 Mar. (66)	0 Sat.	9750·0862	351·5185	232·3847		4132
23 Mar. (82)	3 Tues.	17	17	30	25 Feb. (56)	5 Thur.	9964·4015	235·0448	204·2987		4133
22 Mar. (82)	4 Wed.	23	30	0	15 Mar. (75)	4 Wed.	9999·0411	171·0284	255·6084		4134
23 Mar. (82)	6 Fri.	5	42	30	4 Mar. (63)	1 Sun.	9874·7245	18·2632	224·7846		4135
23 Mar. (82)	0 Sat.	11	55	0	22 Feb. (53)	6 Fri.	89·0398	901·7897	196·6984		4136
23 Mar. (82)	1 Sun.	18	7	30	13 Mar. (72)	5 Thur.	123·6794	837·7731	248·0082		4137
23 Mar. (83)	3 Tues.	0	20	0	1 Mar. (61)	2 Mon.	9999·3628	685·0080	217·1843		4138
23 Mar. (82)	4 Wed.	6	32	30	20 Mar. (79)	1 Sun.	34·0024	620·9915	268·4941		4139
23 Mar. (82)	5 Thur.	12	45	0	9 Mar. (68)	5 Thur.	9909·6858	468·2262	237·6702		4140
23 Mar. (82)	6 Fri.	18	57	30	26 Feb. (67)	2 Mon.	9785·3692	315·4611	206·8464		4141
23 Mar. (83)	1 Sun.	1	10	0	16 Mar. (76)	1 Sun.	9820·0088	251·4446	258·1561		4142
23 Mar. (82)	2 Mon.	7	22	30	6 Mar. (65)	6 Fri.	34·3241	134·9710	230·0700		4143
23 Mar. (82)	3 Tues.	13	35	0	23 Feb. (54)	3 Tues.	9910·0075	982·2058	199·2461		4144
23 Mar. (82)	4 Wed.	19	47	30	1 Mar. (73)	2 Mon.	9944·6471	918·1893	250·5559		4145
23 Mar. (83)	6 Fri.	2	0	0	3 Mar. (63)	0 Sat.	158·9623	801·7158	222·4698		4146

TABLE

CONCURRENT YEAR.								INTERCALATED (*adhika*) and SUPPRESSED (*kshaya*) LUNAR MONTHS (true).
						JOVIAN SAMVATSARA.		
Kali.	Śaka.	Chaitrādi Vikrama.	Mēshādi solar year in Bengal.	Kollam.	A.D.	Southern system.	Northern system.	
1	2	3	3*a*	4	5	6	7	8
4147	968	1103	452	220-21	1045-46	19 Pārthiva .	21 Sarvajit .	...
4148	969	1104	453	221-22	1046-47	20 Vyaya . .	22 Sarvadhārin .	...
4149	970	1105	454	222-23	1047-48	21 Sarvajit .	23 Virōdhin .	5 Śrāvaṇa .
4150	971	1106	455	223-24	*1048-49	22 Sarvadhārin .	24 Vikṛita . .	...
4151	972	1107	456	224-25	1049-50	23 Virōdhin .	25 Khara . .	...
4152	973	1108	457	225-26	1050-51	24 Vikṛita . .	26 Nandana .	3 Jyēshṭha .
4153	974	1109	458	226-27	1051-52	25 Khara . .	27 Vijaya . .	...
4154	975	1110	459	227-28	*1052-53	26 Nandana .	28 Jaya . .	7 Āśvina 10 *Pausha* (*ksh*)
4155	976	1111	460	228-29	1053-54	27 Vijaya . .	29 Manmatha .	1 Chaitra .
4156	977	1112	461	229-30	1054-55	28 Jaya . .	30 Durmukha .	...
4157	978	1113	462	230-31	1055-56	29 Manmatha .	31 Hēmalamba .	5 Śrāvaṇa .
4158	979	1114	463	231-32	*1056-57	30 Durmukha .	32 Vilamba .	...
4159	980	1115	464	232-33	1057-58	31 Hēmalamba .	33 Vikārin . .	...
4160	981	1116	465	233-34	1058-59	32 Vilamba .	34 Śārvarin .	4 Āshāḍha .
4161	982	1117	466	234-35	1059-60	33 Vikārin .	35 Plava . .	...
4162	983	1118	467	235-36	*1060-61	34 Śārvarin .	36 Śubhakṛit .	...
4163	984	1119	468	236-37	1061-62	35 Plava . .	37 Śōbhana .	2 Vaiśākha .
4164	985	1120	469	237-38	1062-63	36 Śubhakṛit .	38 Krōdhin .	...
4165	986	1121	470	238-39	1063-64	37 Śōbhana .	39 Viśvāvasu .	6 Bhādrapada
4166	987	1122	471	239-40	*1064-65	38 Krōdhin .	40 Parābhava .	...
4167	988	1123	472	240-41	1065-66	39 Visvāvasu .	41 Plavaṅga .	...
4168	989	1124	473	241-42	1066-67	40 Parābhava .	42 Kīlaka . .	4 Āshāḍha .
4169	990	1125	474	242-43	1067-68	41 Plavaṅga .	43 Saumya .	...
4170	991	1126	475	243-44	*1068-69	42 Kīlaka . .	44 Sādhāraṇa .	...
4171	992	1127	476	244-45	1069-70	43 Saumya .	45 Virōdhakṛit .	3 Jyēshṭha .

LXI—*Contd.*

COMMENCEMENT OF THE								Kali.
SOLAR YEAR.			LUNI-SOLAR YEAR (MEAN SUNRISE OF CIVIL DAY ON WHICH CHAITRA ŚUKLA 1 ENDS).					
Day and month, A.D.	Week-day.	Time of true Mēsha-saṁkrānti.	Day and month, A.D.	Week-day.	*a.*	*b.*	*c.*	
13	14	17	19	20	23	24	25	1
		H. M. S.						
23 Mar. (82)	0 Sat.	8 12 30	22 Mar. (81)	6 Fri.	193·6019	737·6992	273·7795	4147
23 Mar. (82)	1 Sun.	14 25 0	11 Mar. (70)	3 Tues.	69·2853	584·9341	242·9557	4148
23 Mar. (82)	2 Mon.	20 37 30	28 Feb. (59)	0 Sat.	9944·9688	432·1689	212·1318	4149
23 Mar. (83)	4 Wed.	2 50 0	18 Mar. (78)	6 Fri.	9979·6083	368·1524	263·4415	4150
23 Mar. (82)	5 Thur.	9 2 30	7 Mar. (66)	3 Tues.	9855·2917	215·3872	232·6177	4151
23 Mar. (82)	6 Fri.	15 15 0	25 Feb. (56)	1 Sun.	69·6069	98·9136	204·5316	4152
23 Mar. (82)	0 Sat.	21 27 30	16 Mar. (75)	0 Sat.	104·2465	34·8972	255·8413	4153
23 Mar. (83)	2 Mon.	3 40 0	4 Mar. (64)	4 Wed.	9979·9299	882·1319	225·0175	4154
23 Mar. (82)	3 Tues.	9 52 30	22 Feb. (53)	2 Mon.	194·2452	765·6584	196·9313	4155
23 Mar. (82)	4 Wed.	16 5 0	13 Mar. (72)	1 Sun.	228·8848	701·6419	248·2411	4156
23 Mar. (82)	5 Thur.	22 17 30	2 Mar. (61)	5 Thur.	104·5682	548·8767	217·4172	4157
23 Mar. (83)	0 Sat.	4 30 0	20 Mar. (80)	4 Wed.	139·2078	484·8602	268·7270	4158
23 Mar. (82)	1 Sun.	10 42 30	9 Mar. (68)	1 Sun.	14·8912	332·0950	237·9031	4159
23 Mar. (82)	2 Mon.	16 55 0	26 Feb. (57)	5 Thur.	9890·5746	179·3299	207·0793	4160
23 Mar. (82)	3 Tues.	23 7 30	17 Mar. (76)	4 Wed.	9925·2142	115·3133	258·3890	4161
23 Mar. (83)	5 Thur.	5 20 0	6 Mar. (66)	2 Mon.	139·5295	998·8397	230·3029	4162
23 Mar. (82)	6 Fri.	11 32 30	23 Feb. (54)	6 Fri.	15·2129	846·0746	199·4790	4163
23 Mar. (82)	0 Sat.	17 45 0	14 Mar. (73)	5 Thur.	49·8525	782·0580	250·7888	4164
23 Mar. (82)	1 Sun.	23 57 30	4 Mar. (63)	3 Tues.	264·1677	665·5845	222·7027	4165
23 Mar. (83)	3 Tues.	6 10 0	21 Mar. (81)	1 Sun.	9960·1755	565·2764	271·2747	4166
23 Mar. (82)	4 Wed.	12 22 30	10 Mar. (69)	5 Thur.	9835·8589	412·5112	240·5508	4167
23 Mar. (82)	5 Thur.	18 35 0	28 Feb. (59)	3 Tues.	50·1742	296·0396	212·3647	4168
24 Mar. (83)	0 Sat.	0 47 30	18 Mar. (77)	1 Sun.	9746·1819	195·7275	260·9366	4169
23 Mar. (83)	1 Sun.	7 0 0	7 Mar. (67)	6 Fri.	9960·4972	79·2560	232·8506	4170
23 Mar. (82)	2 Mon.	13 12 30	25 Feb. (56)	4 Wed.	174·3124	962·7823	204·7645	4171

TABLE

CONCURRENT YEAR.								Intercalated (*adhika*) and suppressed (*kshaya*) Lunar months (true).
Kali.	Śaka.	Chaitrādi Vikrama.	Mēshādi solar year in Bengal.	Kollam.	A.D.	Jovian Saṁvatsara. Southern system.	Jovian Saṁvatsara. Northern system.	
1	2	3	3a	4	5	6	7	8
4172	993	1128	477	245-46	1070-71	44 Sādhāraṇa .	46 Paridhāvin .	...
4173	994	1129	478	246-47	1071-72	45 Virōdhakṛit .	47 Pramādin .	7 Āśvina .
4174	995	1130	479	247-48	*1072-73	46 Paridhāvin .	48 Ānanda .	...
4175	996	1131	480	248-49	1073-74	47 Pramādin .	49 Rākshasa .	...
4176	997	1132	481	249-50	1074-75	48 Ānanda .	50 Anala . .	5 Śrāvaṇa .
4177	998	1133	482	250-51	1075-76	49 Rākshasa .	51 Piṅgala† .	...
4178	999	1134	483	251-52	*1076-77	50 Anala . .	53 *Siddhārthin* .	...
4179	1000	1135	484	252-53	1077-78	51 Piṅgala .	54 *Raudra* . .	3 Jyēshṭha .
4180	1001	1136	485	253-54	1078-79	52 Kālayukta .	55 *Durmati* .	...
4181	1002	1137	486	254-55	1079-80	53 Siddhārthin .	56 *Dundubhi* .	...
4182	1003	1138	487	255-56	*1080-81	54 Raudra .	57 Rudhirōdgārin	2 Vaiśākha .
4183	1004	1139	488	256-57	1081-82	55 Durmati .	58 Raktāksha .	...
4184	1005	1140	489	257-58	1082-83	56 Dundubhi .	59 Krōdhana .	6 Bhādrapada
4185	1006	1141	490	258-59	1083-84	57 Rudhirōdgārin	60 Kshaya .	...
4186	1007	1142	491	259-60	*1084-85	58 Raktāksha .	1 Prabhava .	...
4187	1008	1143	492	260-61	1085-86	59 Krōdhana .	2 Vibhava .	4 Āshāḍha .
4188	1009	1144	493	261-62	1086-87	60 Kshaya .	3 Śukla . .	...
4189	1010	1145	494	262-63	1087-88	1 Prabhava .	4 Pramōda .	...
4190	1011	1146	495	263-64	*1088-89	2 Vibhava .	5 Praājpati .	3 Jyēshṭha .
4191	1012	1147	496	264-65	1089-90	3 Śukla . .	6 Aṅgiras .	...
4192	1013	1148	497	265-66	1090-91	4 Pramōda .	7 Srīmukha .	7 Āśvina .
4193	1014	1149	498	266-67	1091-92	5 Prajāpati .	8 Bhāva . .	...
4194	1015	1150	499	267-68	*1092-93	6 Aṅgiras .	9 Yuvan . .	...
4195	1016	1151	500	268-69	1093-94	7 Śrīmukha .	10 Dhātri . .	5 Śrāvaṇa .
4196	1017	1152	501	269-70	1094-95	8 Bhāva . .	11 Īśvara . .	...

† 52 Kālayukta was suppressed in the north.

LXI—*Contd.*

COMMENCEMENT OF THE										
SOLAR YEAR.					LUNI-SOLAR YEAR (MEAN SUNRISE OF CIVIL DAY ON WHICH CHAITRA ŚUKLA 1 ENDS).					Kali.
Day and month, A.D.	Week-day.	Time of true Mēsha-saṁkrānti.			Day and month, A.D.	Week-day.	*a.*	*b.*	*c.*	
13	14	17			19	20	23	24	25	1
		H.	M.	S.						
23 Mar. (82)	3 Tues.	19	25	0	16 Mar. (75)	3 Tues.	209·4520	898·7659	256·0742	4172
24 Mar. (83)	5 Thur.	1	37	30	5 Mar. (64)	0 Sat.	85·1354	746·0007	225·2504	4173
23 Mar. (83)	6 Fri.	7	50	0	23 Mar. (83)	6 Fri.	119·7751	681·9843	276·5600	4174
23 Mar. (82)	0 Sat.	14	2	30	12 Mar. (71)	3 Tues.	9995·4584	529·2190	245·7362	4175
23 Mar. (82)	1 Sun.	20	15	0	1 Mar. (60)	0 Sat.	9871·1418	376·4538	214·9123	4176
24 Mar. (83)	3 Tues.	2	27	30	20 Mar. (79)	6 Fri.	9905·7814	312·4374	266·2221	4177
23 Mar. (83)	4 Wed.	8	40	0	8 Mar. (68)	3 Tues.	9781·4647	159·6721	235·3982	4178
23 Mar. (82)	5 Thur.	14	52	30	26 Feb. (57)	1 Sun	9995·7800	43·1986	207·3122	4179
23 Mar. (82)	6 Fri.	21	5	0	17 Mar. (76)	0 Sat.	30·4197	979·1821	258·6219	4180
24 Mar. (83)	1 Sun.	3	17	30	7 Mar. (66)	5 Thur.	244·7349	862·7084	230·5358	4181
23 Mar. (83)	2 Mon.	9	30	0	24 Feb. (55)	2 Mon.	120·4183	709·9433	199·7119	4182
23 Mar. (82)	3 Tues.	15	42	30	14 Mar. (73)	1 Sun.	155·0579	645·9268	251·0217	4183
23 Mar. (82)	4 Wed.	21	55	0	3 Mar. (62)	5 Thur.	30·7413	493·1616	220·1978	4184
24 Mar. (83)	6 Fri.	4	7	30	22 Mar. (81)	4 Wed.	65·3809	429·1451	271·5066	4185
23 Mar. (83)	0 Sat.	10	20	0	10 Mar. (70)	1 Sun.	9941·0643	276·3799	240·6836	4186
23 Mar. (82)	1 Sun.	16	32	30	27 Feb. (58)	5 Thur.	9816·7477	123·6148	209·8598	4187
23 Mar. (82)	2 Mon.	22	45	0	18 Mar. (77)	4 Wed.	9851·3873	59·5982	261·1695	4188
24 Mar. (83)	4 Wed.	4	57	30	8 Mar. (67)	2 Mon.	65·7026	943·1247	233·0835	4189
23 Mar. (83)	5 Thur.	11	10	0	26 Feb. (57)	0 Sat.	280·0178	826·6511	204·9974	4190
23 Mar. (82)	6 Fri.	17	22	30	16 Mar. (75)	6 Fri.	314·6574	762·6346	256·3071	4191
23 Mar. (82)	0 Sat.	23	35	0	5 Mar. (64)	3 Tues.	190·3408	608·8694	225·4833	4192
24 Mar. (83)	2 Mon.	5	47	30	23 Mar. (82)	1 Sun.	9886·3486	509·5613	274·0551	4193
23 Mar. (83)	3 Tues.	12	0	0	11 Mar. (71)	5 Thur.	9762·0319	356·7962	243·2313	4194
23 Mar. (82)	4 Wed.	18	12	30	1 Mar. (60)	3 Tues.	9976·3472	240·3225	215·1452	4195
24 Mar. (83)	6 Fri.	0	25	0	20 Mar. (79)	2 Mon.	10·9868	176·3061	266·4550	4196

TABLE

CONCURRENT YEAR.								
Kali.	Śaka.	Chaitrādi Vikrama.	Mēshādi solar year in Bengal.	Kollam.	A.D.	JOVIAN SAMVATSARA. Southern system.	JOVIAN SAMVATSARA. Northern system.	INTERCALATED (*adhika*) and SUPPRESSED (*kshaya*) LUNAR MONTHS (true).
1	2	3	3*a*	4	5	6	7	8
4197	1018	1153	502	270-71	1095-96	9 Yuvan	12 Bahudhānya	...
4198	1019	1154	503	271-72	*1096-97	10 Dhātri	13 Pramāthin	3 Jyēshṭha
4199	1020	1155	504	272-73	1097-98	11 Īśvara	14 Vikrama	...
4200	1021	1156	505	273-74	1098-99	12 Bahudhānya	15 Vṛisha	...
4201	1022	1157	506	274-75	1099-1100	13 Pramāthin	16 Chitrabhānu	2 Vaiśākha
4202	1023	1158	507	275-76	*1100-01	14 Vikrama	17 Subhānu	...
4203	1024	1159	508	276-77	1101-02	15 Vṛisha	18 Tāraṇa	6 Bhādrapada
4204	1025	1160	509	277-78	1102-03	16 Chitrabhānu	19 Pārthiva	...
4205	1026	1161	510	278-79	1103-04	17 Subhānu	20 Vyaya	...
4206	1027	1162	511	279-80	*1104-05	18 Tāraṇa	21 Sarvajit	4 Āshāḍha
4207	1028	1163	512	280-81	1105-06	19 Pārthiva	22 Sarvadhārin	...
4208	1029	1164	513	281-82	1106-07	20 Vyaya	23 Virōdhin	...
4209	1030	1165	514	282-83	1107-08	21 Sarvajit	24 Vikṛita	3 Jyēshṭha
4210	1031	1166	515	283-84	*1108-09	22 Sarvadhārin	25 Khara	...
4211	1032	1167	516	284-85	1109-10	23 Virōdhin	26 Nandana	7 Āśvina
4212	1033	1168	517	285-86	1110-11	24 Vikṛita	27 Vijaya	...
4213	1034	1169	518	286-87	1111-12	25 Khara	28 Jaya	...
4214	1035	1170	519	287-88	*1112-13	26 Nandana	29 Manmatha	5 Śrāvaṇa
4215	1036	1171	520	288-89	1113-14	27 Vijaya	30 Durmukha	...
4216	1037	1172	521	289-90	1114-15	28 Jaya	31 Hēmalamba	...
4217	1038	1173	522	290-91	1115-16	29 Manmatha	32 Vilamba	3 Jyēshṭha
4218	1039	1174	523	291-92	*1116-17	30 Durmukha	33 Vikārin	...
4219	1040	1175	524	292-93	1117-18	31 Hēmalamba	34 Śārvarin	...
4220	1041	1176	525	293-94	1118-19	32 Vilamba	35 Plava	1 Chaitra
4221	1042	1177	526	294-95	1119-20	33 Vikārin	36 Śubhakṛit	...

LXI—*Contd.*

COMMENCEMENT OF THE											Kali.
SOLAR YEAR.					LUNI-SOLAR YEAR (MEAN SUNRISE OF CIVIL DAY ON WHICH CHAITRA ŚUKLA 1 ENDS).						
Day and month, A.D.	Week-day.	Time of true Mēsha-saṁkrānti.			Day and month, A.D.	Week-day.	*a.*	*b.*	*c.*		
13	14	17			19	20	23	24	25		1
		H.	M.	S.							
24 Mar. (83)	0 Sat.	6	37	30	9 Mar. (68)	6 Fri.	9886·6702	23·5409	235·6311		4197
23 Mar. (83)	1 Sun.	12	50	0	27 Feb. (58)	4 Wed.	100·9855	907·0673	207·5451		4198
23 Mar. (82)	2 Mon.	19	2	30	17 Mar. (76)	3 Tues.	135·6251	843·0508	258·8547		4199
24 Mar. (83)	4 Wed.	1	15	0	6 Mar. (65)	0 Sat.	11·3085	690·2856	227·9309		4200
24 Mar. (83)	5 Thur.	7	27	30	24 Feb. (55)	5 Thur.	225·6237	573·8121	199·9448		4201
23 Mar. (83)	6 Fri.	13	40	0	13 Mar. (73)	3 Tues.	9921·6314	473·5040	248·5168		4202
23 Mar. (82)	0 Sat.	19	52	30	2 Mar. (61)	0 Sat.	9767·3148	320·7388	217·6929		4203
24 Mar. (83)	2 Mon.	2	5	0	21 Mar. (80)	6 Fri.	9831·9544	256·7233	269·0026		4204
24 Mar. (83)	3 Tues.	8	17	30	11 Mar. (70)	4 Wed.	46·2697	140·2487	240·9165		4205
23 Mar. (83)	4 Wed.	14	30	0	28 Feb. (59)	1 Sun.	9921·9531	987·4835	216·0927		4206
23 Mar. (82)	5 Thur	20	42	30	18 Mar. (77)	0 Sat.	9956·5927	923·4670	261·4024		4207
24 Mar. (83)	0 Sat.	2	55	0	8 Mar. (67)	5 Thur.	170·9080	806·9935	233·3163		4208
24 Mar. (83)	1 Sun.	9	7	30	25 Feb. (56)	2 Mon.	46·5913	654·2283	202·4925		4209
23 Mar. (83)	2 Mon.	15	20	0	15 Mar. (75)	1 Sun.	81·2310	590·2118	253·8022		4210
23 Mar. (82)	3 Tues.	21	32	30	4 Mar. (63)	5 Thur.	9956·9143	437·4466	222·9783		4211
24 Mar. (83)	5 Thur.	3	45	0	23 Mar. (82)	4 Wed.	9991·5540	373·4301	274·2880		4212
24 Mar. (83)	6 Fri.	9	57	30	12 Mar. (71)	1 Sun.	9867·2374	220·6649	243·4642		4213
23 Mar. (83)	0 Sat.	16	10	0	1 Mar. (61)	6 Fri.	81·5526	104·1913	215·3781		4214
23 Mar. (82)	1 Sun.	22	22	30	20 Mar. (79)	5 Thur.	116·1922	40·1749	266·6879		4215
24 Mar. (83)	3 Tues.	4	35	0	9 Mar. (68)	2 Mon.	9991·8755	887·4097	235·8740		4216
24 Mar. (83)	4 Wed.	10	47	30	27 Feb. (58)	0 Sat.	206·1909	770·9361	207·7779		4217
23 Mar. (83)	5 Thur.	17	0	0	17 Mar. (77)	6 Fri.	240·8305	706·9196	259·0866		4218
23 Mar. (82)	6 Fri.	23	12	30	6 Mar. (65)	3 Tues.	116·5133	554·1544	228·2638		4219
24 Mar. (83)	1 Sun.	5	25	0	23 Feb. (54)	0 Sat.	9992·1972	401·3892	197·4398		4220
24 Mar. (83)	2 Mon.	11	37	30	14 Mar. (73)	6 Fri.	26·8368	337·3727	248·7497		4221

TABLE

CONCURRENT YEAR.								INTERCALATED (*adhika*) and SUPPRESSED (*kshaya*) LUNAR MONTHS (true).
Kali.	Śaka.	Chaitrādi Vikrama.	Mēshādi solar year in Bengal.	Kollam.	A.D.	JOVIAN SAMVATSARA. Southern system.	JOVIAN SAMVATSARA. Northern system.	
1	2	3	3*a*	4	5	6	7	8
4222	1043	1178	527	295-96	*1120-21	34 Śārvarin .	37 Śōbhana .	6 Bhādrapada .
4223	1044	1179	528	296-97	1121-22	35 Plava . .	38 Krōdhin .	...
4224	1045	1180	529	297-98	1122-23	36 Śubhakṛit	39 Viśvāvasu .	...
4225	1046	1181	530	298-99	1123-24	37 Śōbhana .	40 Parābhava .	4 Āshāḍha .
4226	1047	1182	531	299-300	*1124-25	38 Krōdhin .	41 Plavaṅga .	...
4227	1048	1183	532	300-01	1125-26	39 Viśvāvasu .	42 Kīlaka . .	...
4228	1049	1184	533	301-02	1126-27	40 Parābhava .	43 Saumya .	3 Jyēshṭha .
4229	1050	1185	534	302-03	1127-28	41 Plavaṅga .	44 Sādhāraṇa .	...
4230	1051	1186	535	303-04	*1128-29	42 Kīlaka . .	45 Virōdhakṛit .	7 Āśvina .
4231	1052	1187	536	304-05	1129-30	43 Saumya .	46 Paridhāvin .	...
4232	1053	1188	537	305-06	1130-31	44 Sādhāraṇa .	47 Pramādin .	...
4233	1054	1189	538	306-07	1131-32	45 Virōdhakṛit .	48 Ānanda .	5 Śrāvaṇa .
4234	1055	1190	539	307-08	*1132-33	46 Paridhāvin .	49 Rākshasa .	...
4235	1056	1191	540	308-09	1133-34	47 Pramādin .	50 Anala . .	...
4236	1057	1192	541	309-10	1134-35	48 Ānanda .	51 Piṅgala .	3 Jyēshṭha .
4237	1058	1193	542	310-11	1135-36	49 Rākshasa .	52 Kālayukta .	...
4238	1059	1194	543	311-12	*1136-37	50 Anala . .	53 Siddhārthin .	...
4239	1060	1195	544	312-13	1137-38	51 Piṅgala .	54 Raudra .	1 Chaitra .
4240	1061	1196	545	313-14	1138-39	52 Kālayukta .	55 Durmati .	...
4241	1062	1197	546	314-15	1139-40	53 Siddhārthin .	56 Dundubhi .	5 Śrāvaṇa
4242	1063	1198	547	315-16	*1140-41	54 Raudra .	57 Rudhirōdgārin	...
4243	1064	1199	548	316-17	1141-42	55 Durmati .	58 Raktāksha .	...
4244	1065	1200	549	317-18	1142-43	56 Dundubhi .	59 Krōdhana .	4 Āshāḍha .
4245	1066	1201	550	318-19	1143-44	57 Rudhirōdgārin	60 Kshaya .	...
4246	1067	1202	551	319-20	*1144-45	58 Raktāksha .	1 Prabhava .	...

LXI—*Contd.*

COMMENCEMENT OF THE									Kali.
SOLAR YEAR.			LUNI-SOLAR YEAR (MEAN SUNRISE OF CIVIL DAY ON WHICH CHAITRA ŚUKLA 1 ENDS).						
Day and month, A.D.	Week-day.	Time of true Mēsha-saṁkrānti.	Day and month, A.D.	Week-day.	*a.*	*b.*	*c.*		
13	14	17	19	20	23	24	25		1
		H. M. S.							
23 Mar. (83)	3 Tues.	17 50 0	2 Mar. (62)	3 Tues.	9902·5202	184·6076	217·9258		4222
24 Mar. (83)	5 Thur.	0 2 30	21 Mar. (80)	2 Mon.	9937·1598	120·5911	269·2355		4223
24 Mar. (83)	6 Fri.	6 15 0	11 Mar. (70)	0 Sat.	151·4751	4·1174	241·1494		4224
24 Mar. (83)	0 Sat.	12 27 30	28 Feb. (59)	4 Wed.	27·1585	851·3523	210·3256		4225
23 Mar. (83)	1 Sun.	18 40 0	18 Mar. (78)	3 Tues.	61·7981	787·3358	261·6353		4226
24 Mar. (83)	3 Tues.	0 52 30	8 Mar. (67)	1 Sun.	276·1134	670·8622	233·5493		4227
24 Mar. (83)	4 Wed.	7 5 0	25 Feb. (56)	5 Thur.	151·7967	518·0970	202·7254		4228
24 Mar. (83)	5 Thur.	13 17 30	15 Mar. (74)	3 Tues.	9847·8045	416·7889	251·2974		4229
23 Mar. (83)	6 Fri.	19 30 0	3 Mar. (63)	0 Sat.	9723·4879	265·0237	220·4734		4230
24 Mar. (83)	1 Sun.	1 42 30	22 Mar. (81)	6 Fri.	9758·1275	201·0072	271·7832		4231
24 Mar. (83)	2 Mon.	7 55 0	12 Mar. (71)	4 Wed.	9972·4428	84·5337	243·7071		4232
24 Mar. (83)	3 Tues.	14 7 30	2 Mar. (61)	2 Mon.	186·7580	968·0600	215·6120		4233
23 Mar. (83)	4 Wed.	20 20 0	20 Mar. (80)	1 Sun.	221·3976	904·0436	266·9208		4234
24 Mar. (83)	6 Fri.	2 32 30	9 Mar. (68)	5 Thur.	97·0810	751·2784	236·0969		4235
24 Mar. (83)	0 Sat.	8 45 0	26 Feb. (57)	2 Mon.	9972·7644	598·5132	205·2730		4236
24 Mar. (83)	1 Sun.	14 57 30	17 Mar. (76)	1 Sun.	7·4040	534·4967	256·5727		4237
23 Mar. (83)	2 Mon.	21 10 0	5 Mar. (65)	5 Thur.	9883·0874	381·7315	225·7589		4238
24 Mar. (83)	4 Wed.	3 22 30	22 Feb. (53)	2 Mon.	9758·7708	228·9664	194·9350		4239
24 Mar. (83)	5 Thur.	9 35 0	13 Mar. (72)	1 Sun.	9793·4104	164·9498	246·2448		4240
24 Mar. (83)	6 Fri.	15 47 30	3 Mar. (62)	6 Fri.	7·7257	48·4763	218·1587		4241
23 Mar. (83)	0 Sat.	22 0 0	21 Mar. (81)	5 Thur.	42·3653	934·4598	269·4685		4242
24 Mar. (83)	2 Mon.	4 12 30	11 Mar. (70)	3 Tues.	256·6806	867·9862	241·3823		4243
24 Mar. (83)	3 Tues.	10 25 0	28 Feb. (59)	0 Sat.	132·3640	715·2210	210·5585		4244
24 Mar. (83)	4 Wed.	16 37 30	19 Mar. (78)	6 Fri.	167·0036	651·2045	261·9682		4245
23 Mar. (83)	5 Thur.	22 50 0	8 Mar. (67)	3 Tues.	42·6869	498·4393	231·0444		4246

TABLE

CONCURRENT YEAR.								Intercalated (*adhika*) and suppressed (*kshaya*) Lunar months (true).
						Jovian Saṁvatsara.		
Kali.	Śaka.	Chaitrādi Vikrama.	Mēshādi solar year in Bengal.	Kollam.	A.D.	Southern system.	Northern system.	
1	2	3	3*a*	4	5	6	7	8
4247	1068	1203	552	320-21	1145-46	59 Krōdhana .	2 Vibhava .	2 Vaiśākha .
4248	1069	1204	553	321-22	1146-47	60 Kshaya .	3 Śukla . .	...
4249	1070	1205	554	322-23	1147-48	1 Prabhava .	4 Pramōda .	6 Bhādrapada
4250	107	1206	555	323-24	*1148-49	2 Vibhava .	5 Prajāpati .	...
4251	1072	1207	556	324-25	1149-50	3 Śukla . .	6 Aṅgiras .	...
4252	1073	1208	557	325-26	1150-51	4 Pramōda .	7 Śrīmukha .	5 Śrāvaṇa .
4253	1074	1209	558	326-27	1151-52	5 Prajāpati .	8 Bhāva . .	...
4254	1075	1210	559	327-28	*1152-53	6 Aṅgiras .	9 Yuvan . .	...
4255	1076	1211	560	328-29	1153-54	7 Śrīmukha .	10 Dhātṛi . .	3 Jyēshṭha .
4256	1077	1212	561	329-30	1154-55	8 Bhāva . .	11 Īśvara . .	...
4257	1078	1213	562	330-31	1155-56	9 Yuvan . .	12 Bahudhānya .	...
4258	1079	1214	563	331-32	*1156-57	10 Dhātṛi . .	13 Pramāthin .	1 Chaitra .
4259	1080	1215	564	332-33	1157-58	11 Īśvara . .	14 Vikrama .	...
4260	1081	1216	565	333-34	1158-59	12 Bahudhānya .	15 Vṛisha . .	5 Śrāvaṇa .
4261	1082	1217	566	334-35	1159-60	13 Pramāthin .	16 Chitrabhānu .	...
4262	1083	1218	567	335-36	*1160-61	14 Vikrama .	17 Subhānu† .	...
4263	1084	1219	568	336-37	1161-62	15 Vrisha . .	19 *Pārthiva* .	4 Āshāḍha .
4264	1085	1220	569	337-38	1162-63	16 Chitrabhānu .	20 *Vyaya* . .	...
4265	1086	1221	570	338-39	1163-64	17 Subhānu .	21 *Sarvajit* .	...
4266	1087	1222	571	339-40	*1164-65	18 Tāraṇa . .	22 *Sarvadhārin* .	2 Vaiśākha .
4267	1088	1223	572	340-41	1165-66	19 Pārthiva .	23 Virōdhin .	...
4268	1089	1224	573	341-42	1166-67	20 Vyaya . .	24 Vikrita . .	6 Bhādrapada
4269	1090	1225	574	342-43	1167-68	21 Sarvajit .	25 Khara . .	...
4270	1091	1226	575	343-44	*1168-69	22 Sarvadhārin .	26 Nandana .	...
4271	1092	1227	576	344-45	1169-70	23 Virōdhin .	27 Vijaya . .	5 Śrāvaṇa .

† 18 Tāraṇa was suppressed in the north

LXI—*Contd.*

COMMENCEMENT OF THE											
SOLAR YEAR.					LUNI-SOLAR YEAR (MEAN SUNRISE OF CIVIL DAY ON WHICH CHAITRA ŚUKLA 1 ENDS).					Kali.
Day and month, A.D.	Week-day.	Time of true Mēsha-saṁkrānti.			Day and month, A.D.	Week-day.	*a.*	*b.*	*c.*	
13	14	17			19	20	23	24	25	1
		H.	M.	S.						
24 Mar. (83)	0 Sat. .	5	2	30	24 Feb. (55)	0 Sat. .	9918·3703	345·6741	200·2205	4247
24 Mar. (83)	1 Sun. .	11	15	0	15 Mar. (74)	6 Fri. .	9953·0099	281·6576	251·4803	4248
[illegible] Mar. (83)	2 Mon. .	17	27	30	4 Mar. (63)	3 Tues.	9828·6934	128·8925	220·7063	4249
23 Ma (83)	3 Tues.	23	40	0	22 Mar. (82)	2 Mon. .	9863·3329	64·8760	271·2161	4250
24 Mar. (83)	5 Thur.	5	52	30	12 Ma.. (71)	0 Sat. .	77·6481	948·4024	243·9300	4251
24 Mar. (83)	6 Fri. .	12	5	0	2 Mar. (61)	5 Thur.	291·9634	831·9288	215·8439	4252
24 Mar. (83)	0 Sat. .	18	17	30	21 Mar. (80)	4 Wed.	326·6030	767·9126	267·1537	4253
24 Mar. (84)	2 Mon. .	0	30	0	9 Mar. (69) .	1 Sun. .	202·2864	615·1471	236·3298	4254
24 Mar. (83)	3 Tues.	6	42	30	26 Feb. (57)	5 Thur.	77·9698	462·3819	205·5071	4255
24 Mar. (83)	4 Wed.	12	55	0	16 Mar. (75)	3 Tues.	9773·9776	362·0739	254·0778	4256
24 Mar. (83)	5 Thur.	19	7	30	6 Mar. (65)	1 Sun. .	9988·2928	245·6002	225·9918	4257
24 Mar. (84)	0 Sat. .	1	20	0	23 Feb. (54)	5 Thur.	9863·9762	92·8351	195·1679	4258
24 Mar. (83)	1 Sun. .	7	32	30	13 Mar. (72)	4 Wed.	9899·0158	29·8186	246·4777	4259
24 Mar. (83)	2 Mon. .	13	45	0	3 Mar. (62)	2 Mon. .	112·9311	912·3451	218·3916	4260
24 Mar. (83)	3 Tues.	19	57	30	22 Mar. (81)	1 Sun. .	147·5707	848·3285	269·7014	4261
24 Mar. (84)	5 Thur.	2	10	0	10 Mar. (70)	5 Thur.	23·2541	695·5633	238·8774	4262
24 Mar. (83)	6 Fri. .	8	22	30	27 Feb. (58)	2 Mon. .	9899·3375	542·7982	208·0536	4263
24 Mar. (83)	0 Sat. .	14	35	0	18 Mar. (77)	1 Sun. .	9933·5672	478·7816	259·3633	4264
24 Mar. (83)	1 Sun. .	20	47	30	7 Mar. (66)	5 Thur.	9809·2605	326·0164	228·5395	4265
24 Mar. (84)	3 Tues	3	0	0	25 Feb. (56)	3 Tues.	23·5758	209·5429	200·4534	4266
24 Mar. (83)	4 Wed.	9	12	30	15 Mar. (74)	2 Mon.	58·2354	145·5264	251·7632	4267
24 Mar. (83)	5 Thur.	15	25	0	4 Mar. (63)	6 Fri. .	9933·8988	992·7612	220·9392	4268
24 Mar. (83)	6 Fri. .	21	37	30	23 Mar. (82)	5 Thur.	9968·5284	928·7447	272·2489	4269
24 Mar. (84)	1 Sun. .	3	50	0	12 Mar. (72)	3 Tues	182·8537	812·2712	244·1626	427[illegible]
24 Mar. (83)	2 Mon. .	10	2	30	1 Mar. (60)	0 Sat. .	58·5371	629·5059	213·3391	4271

TABLE

CONCURRENT YEAR.								Intercalated (*adhika*) and suppressed (*kshaya*) Lunar months (true).
						Jovian Saṁvatsara.		
Kali.	Śaka.	Chaitrādi Vikrama.	Mēshādi solar year in Bengal.	Kollam.	A.D.	Southern system.	Northern system.	
1	2	3	3a	4	5	6	7	8
4272	1093	1228	577	345-46	1170-71	24 Vikṛita . .	28 Jaya . .	...
4273	1094	1229	578	346-47	1171-72	25 Khara . .	29 Manmatha .	...
4274	1095	1230	579	347-48	*1172-73	26 Nandana .	30 Durmukha .	3 Jyēshṭha .
4275	1096	1231	580	348-49	1173-74	27 Vijaya . .	31 Hēmalamba .	...
4276	1097	1232	581	349-50	1174-75	28 Jaya . .	32 Vilamba .	...
4277	1098	1233	582	350-51	1175-76	29 Manmatha .	33 Vikārin . .	1 Chaitra .
4278	1099	1234	583	351-52	*1176-77	30 Durmukha .	34 Śārvarin .	...
4279	1100	1235	584	352-53	1177-78	31 Hēmalamba .	35 Plava . .	5 Śrāvaṇa .
4280	1101	1236	585	353-54	1178-79	32 Vilamba .	36 Śubhakṛit .	...
4281	1102	1237	586	354-55	1179-80	33 Vikārin .	37 Śōbhana .	...
4282	1103	1238	587	355-56	*1180-81	34 Śārvarin .	38 Krōdhin .	4 Āshādha .
4283	1104	1239	588	356-57	1181-82	35 Plava . .	39 Viśvāvasu .	...
4284	1105	1240	589	357-58	1182-83	36 Subhakṛit .	40 Parābhava .	...
4285	1106	1241	590	358-59	1183-84	37 Śōbhana .	41 Plavaṅga .	2 Vaiśākha .
4286	1107	1242	591	359-60	*1184-85	38 Krōdhin .	42 Kīlaka . .	...
4287	1108	1243	592	360-61	1185-86	39 Viśvāvasu .	43 Saumya .	6 Bhādrapada
4288	1109	1244	593	361-62	1186-87	40 Parābhava .	44 Sādhāraṇa .	...
4289	1110	1245	594	362-63	1187-88	41 Plavaṅga .	45 Virōdhakṛit .	...
4290	1111	1246	595	363-64	*1188-89	42 Kīlaka . .	46 Paridhāvin .	5 Śrāvaṇa .
4291	1112	1247	596	364-65	1189-90	43 Saumya. .	47 Pramādin .	...
4292	1113	1248	597	365-66	1190-91	44 Sādhāraṇa .	48 Ānanda .	...
4293	1114	1249	598	366-67	1191-92	45 Virōdhakṛit .	49 Rākshasa .	3 Jyēshṭha .
4294	1115	1250	599	367-68	*1192-93	46 Paridhāvin .	50 Anala . .	...
4295	1116	1251	600	368-69	1193-94	47 Pramādin .	51 Pingala	{ 7 Āśvina 10 *Paushа* (*ksh.*) }
4296	1117	1252	601	369-70	1194-95	48 Ānanda .	52 Kālayukta .	1 Chaitra .

* Tāraṇa was suppressed in the north.

LXI—*Contd.*

COMMENCEMENT OF THE											
SOLAR YEAR.					LUNI-SOLAR YEAR (MEAN SUNRISE OF CIVIL DAY ON WHICH CHAITRA ŚUKLA 1. ENDS).					Kali.
Day of month, A.D.	Week-day.	Time of true Mēsha-saṁkrānti.			Day and month, A.D.	Week-day.	*a.*	*b.*	*c.*	
13	14	17			19	20	23	24	25	1
		H.	M.	S.						
24 Mar. (83)	3 Tues.	16	15	0	20 Mar. (79)	6 Fri. .	93·1767	595·4895	264·6488	4272
24 Mar. (83)	4 Wed.	22	27	30	9 Mar. (68)	3 Tues.	9968·8601	442·7243	233·8250	4273
24 Mar. (84)	6 Fri. .	4	40	0	26 Feb. (57)	0 Sat. .	9844·5534	289·9591	203·0010	4274
24 Mar. (83)	0 Sat. .	10	52	30	16 Mar. (75)	6 Fri. .	9879·1831	225·9426	254·3107	4275
24 Mar. (83)	1 Sun. .	17	5	0	6 Mar. (65)	4 Wed.	93·4983	109·4690	226·2247	4276
24 Mar. (83)	2 Mon. .	23	17	30	23 Feb. (54)	1 Sun. .	9969·1816	956·7039	195·4008	4277
24 Mar. (84)	4 Wed.	5	30	0	13 Mar. (73)	0 Sat. .	3·8212	892·6873	246·7106	4278
24 Mar. (83)	5 Thur.	11	42	30	3 Mar. (62)	5 Thur.	218·1365	776·2138	218·6245	4279
24 Mar. (83)	6 Fri. .	17	55	0	22 Mar. (81)	4 Wed.	252·7762	712·1973	269·9343	4280
25 Mar. (84)	1 Sun. .	0	7	30	11 Mar. (70)	1 Sun. .	128·4595	559·4320	239·1103	4281
24 Mar. (84)	2 Mon. .	6	20	0	28 Feb. (59)	5 Thur.	4·1429	406·6669	208·2851	4282
24 Mar. (83)	3 Tues.	12	32	30	18 Mar. (77)	4 Wed.	38·7825	342·6504	259·5962	4283
24 Mar. (83)	4 Wed.	18	45	0	7 Mar. (66)	1 Sun. .	9914·4659	189·8851	228·7724	4284
25 Mar. (84)	6 Fri. .	0	57	30	24 Feb. (55)	5 Thur.	9790·1493	37·1200	197·9485	4285
24 Mar. (84)	0 Sat. .	7	10	0	15 Mar. (75)	5 Thur.	163·4208	9·3951	251·9960	4286
24 Mar. (83)	1 Sun. .	13	22	30	4 Mar. (63)	2 Mon. .	39·1042	856·6300	221·1721	4287
24 Mar. (83)	2 Mon. .	19	35	0	23 Mar. (82)	1 Sun. .	73·7438	792·6134	272·4618	4288
25 Mar. (84)	4 Wed.	1	47	30	13 Mar. (72)	6 Fri. .	288·0591	676·1399	244·3958	4289
24 Mar. (84)	5 Thur.	8	0	0	1 Mar. (61)	3 Tues.	163·7425	523·2747	213·5720	4290
24 Mar. (83)	6 Fri. .	14	12	30	19 Mar. (78)	1 Sun.	9859·7302	423·0665	262·1439	4291
24 Mar. (83)	0 Sat. .	20	25	0	8 Mar. (67)	5 Thur.	9735·4336	270·3014	231·3201	4292
25 Mar. (84)	2 Mon. .	2	37	30	26 Feb. (57)	3 Tues.	9949·7488	153·8278	203·2339	4293
24 Mar. (84)	3 Tues.	8	50	0	16 Mar. (76)	2 Mon. .	9984·3885	89·8114	254·5436	4294
24 Mar. (83)	4 Wed.	15	2	30	6 Mar. (65)	0 Sat. .	198·7037	973·3377	226·4576	4295
24 Mar. (83)	5 Thur.	21	15	0	23 Feb. (54)	4 Wed.	74·3871	820·5726	195·6337	4296

TABLE

CONCURRENT YEAR.								Intercalated (*adhika*) and suppressed (*kshaya*) Lunar months (true).
						Jovian Samvatsara.		
Kali.	Śaka.	Chaitrādi Vikrama.	Mēshādi solar year in Bengal.	Kollam.	A.D.	Southern system.	Northern system.	
1	2	3	3*a*	4	5	6	7	8
4297	1118	1253	602	370-71	1195-96	49 Rākshasa .	53 Siddhārthin	...
4298	1119	1254	603	371-72	*1196-97	50 Anala . .	54 Raudra	5 Śrāvaṇa .
4299	1120	1255	604	372-73	1197-98	51 Piṅgala . .	55 Durmati .	...
4300	1121	1256	605	373-74	1198-99	52 Kālayukta .	56 Dundubhi .	...
4301	1122	1257	606	374-75	1199-1200	53 Siddhārthin .	57 Rudhirōdgārin	4 Āshāḍha .
4302	1123	1258	607	375-76	*1200-01	54 Raudra . .	58 Raktāksha .	...
4303	1124	1259	608	376-77	1201-02	55 Durmati .	59 Krōdhana .	...
4304	1125	1260	609	377-78	1202-03	56 Dundubhi .	60 Kshaya .	2 Vaiśākha .
4305	1126	1261	610	378-79	1203-04	57 Rudhirōdgārin	1 Prabhava .	...
4306	1127	1262	611	379-80	*1204-05	58 Raktāksha .	2 Vibhava .	6 Bhādrapada
4307	1128	1263	612	380-81	1205-06	59 Krōdhana .	3 Śukla . .	...
4308	1129	1264	613	381-82	1206-07	60 Kshaya .	4 Pramōda .	...
4309	1130	1265	614	382-83	1207-08	1 Prabhava .	5 Prajāpati .	4 Āshāḍha .
4310	1131	1266	615	383-84	*1208-09	2 Vibhava .	6 Aṅgiras .	...
4311	1132	1267	616	384-85	1209-10	3 Śukla . .	7 Śrīmukha .	...
4312	1133	1268	617	385-86	1210-11	4 Pramōda .	8 Bhāva . .	3 Jyēshṭha .
4313	1134	1269	618	386-87	1211-12	5 Prajāpati .	9 Yuvan . .	...
4314	1135	1270	619	387-88	*1212-13	6 Aṅgiras .	10 Dhātṛi . .	7 Āśvina 11 *Māgha* (*ksh.*) 12 Phālguna
4315	1136	1271	620	388-89	1213-14	7 Śrīmukha .	11 Īśvara . .	...
4316	1137	1272	621	389-90	1214-15	8 Bhāva .	12 Bahudhānya .	...
4317	1138	1273	622	390-91	1215-16	9 Yuvan . .	13 Pramāthin .	5 Śrāvaṇa .
4318	1139	1274	623	391-92	*1216-17	10 Dhātṛi . .	14 Vikrama .	...
4319	1140	1275	624	392-93	1217-18	11 Īśvara . .	15 Vṛisha . .	...
4320	1141	1276	625	393-94	1218-19	12 Bahudhānya .	16 Chitrabhānu	3 Jyēshṭha .
4321	1142	1277	626	394-95	1219-20	13 Pramāthin .	17 Subhānu	...

LXI—*Contd.*

COMMENCEMENT OF THE									
SOLAR YEAR.			LUNI-SOLAR YEAR (MEAN SUNRISE OF CIVIL DAY ON WHICH CHAITRA ŚUKLA 1 ENDS).						Kali.
Day and month, A.D.	Week-day.	Time of true Mēsha-saṁkrānti.	Day and month, A.D.	Week-day.	a.	b.	c.		
13	14	17	19	20	23	24	25		1
		H. M. S.							
25 Mar. (84)	0 Sat. .	3 27 30	14 Mar. (73)	3 Tues.	109·0267	756·5561	246·9435		4297
24 Mar. (84)	1 Sun. .	9 40 0	2 Mar. (62)	0 Sat. .	9984·7101	603·7908	216·1196		4298
24 Mar. (83)	2 Mon. .	15 52 30	21 Mar. (80)	6 Fri. .	19·3497	539·7744	267·4293		4299
24 Mar. (83)	3 Tues.	22 5 0	10 Mar. (69)	3 Tues.	9895·0331	387·0092	236·6054		4300
25 Mar. (84)	5 Thur.	4 17 30	27 Feb. (58)	0 Sat. .	9770·7165	234·2441	205·7817		4301
24 Mar. (84)	6 Fri. .	10 30 0	17 Mar. (77)	6 Fri. .	9805·3561	170·2276	257·0914		4302
24 Mar. (83)	0 Sat. .	16 42 30	7 Mar. (66)	Wed.	19·6714	53·7540	229·0054		4303
24 Mar. (83)	1 Sun. .	22 55 0	25 Feb. (56)	2 Mon. .	233·9866	937·2894	200·9192		4304
25 Mar. (84)	3 Tues.	5 7 30	16 Mar. (75)	1 Sun. .	268·6263	873·2640	252·2289		4305
24 Mar. (84)	4 Wed.	11 20 0	4 Mar. (64)	5 Thur.	144·3096	720·4987	221·4051		4306
24 Mar. (83)	5 Thur.	17 32 30	23 Mar. (82)	4 Wed.	178·9493	656·4823	272·7148		4307
24 Mar. (83)	6 Fri. .	23 45 0	12 Mar. (71)	1 Sun.	54·6327	503·7171	241·8910		4308
25 Mar. (84)	1 Sun. .	5 57 30	1 Mar. (60)	5 Thur.	9930·3161	350·9519	211·0672		4309
24 Mar. (84)	2 Mon.	12 10 0	19 Mar. (79)	4 Wed.	9964·9557	236·9354	262·3769		4310
24 Mar. (83)	3 Tues.	18 22 30	8 Mar. (67)	1 Sun. .	9840·6399	134·1702	231·5529		4311
25 Mar. (84)	5 Thur.	0 35 0	26 Feb. (57)	6 Fri. .	54·9544	13·6966	203·4669		4312
25 Mar. (84)	6 Fri. .	6 47 30	17 Mar. (76)	5 Thur.	89·5939	953·6801	254·7766		4313
24 Mar. (84)	0 Sat. .	13 0 0	6 Mar. (66)	3 Tues.	303·9092	837·2065	226·6906		4314
24 Mar. (83)	1 Sun.	19 12 30	24 Mar. (83)	1 Sun. .	9999·9169	736·8985	275·2625		4315
25 Mar. (84)	3 Tues.	1 25 0	14 Mar. (73)	6 Fri. .	214·2321	620·4249	247·1765		4316
25 Mar. (84)	4 Wed.	7 37 30	3 Mar. (62)	3 Tues.	89·9156	467·6597	215·3526		4317
24 Mar. (84)	5 Thur.	13 50 0	20 Mar. (80)	1 Sun. .	9785·9233	367·3616	264·9245		4318
24 Mar. (83)	6 Fri. .	20 2 30	10 Mar. (69)	6 Fri. .	0·2385	250·8780	236·8384		4319
25 Mar. (84)	1 Sun. .	2 15 0	27 Feb. (58)	3 Tues.	9875·9219	98·1128	206·0146		4320
25 Mar. (84)	2 Mon. .	8 27 30	18 Mar. (77)	2 Mon. .	9910·5615	34·0963	257·3243		4321

TABLE

CONCURRENT YEAR.								INTERCALATED (*adhika*) and SUPPRESSED (*kshaya*) LUNAR MONTHS (true).
Kali.	Saka.	Chaitrādi Vikrama.	Mēshādi solar year in Bengal.	Kollam.	A.D.	Jovian Saṁvatsara. Southern system.	Jovian Saṁvatsara. Northern system.	
1	2	3	3a	4	5	6	7	8
4322	1143	1278	627	395-96	*1220-21	14 Vikrama	18 Tāraṇa . .	...
4023	1144	1279	628	396-97	1221-22	15 Vṛisha . .	19 Pārthiva .	2 Vaiśākha .
4324	1145	1280	629	397-98	1222-23	16 Chitrabhānu .	20 Vyaya . .	...
4325	1146	1281	630	398-99	1223-24	17 Subhānu .	21 Sarvajit .	6 Bhādrapada
4326	1147	1282	631	399-400	*1224-25	18 Tāraṇa . .	22 Sarvadhārin .	...
4327	1148	1283	632	400-01	1225-26	19 Pārthiva .	23 Virōdhin .	...
4328	1149	1284	633	401-02	1226-27	20 Vyaya . .	24 Vikṛita . .	4 Āshāḍha
4329	1150	1285	634	402-03	1227-28	21 Sarvajit .	25 Khara . .	...
4330	1151	1286	635	403-04	*1228-29	22 Sarvadhārin .	26 Nandana .	...
4331	1152	1287	636	404-05	1229-30	23 Virōdhin .	27 Vijaya . .	3 Jyēshṭha .
4332	1153	1288	637	405-06	1230-31	24 Vikṛita . .	28 Jaya . .	...
4333	1154	1289	638	406-07	1231-32	25 Khara . .	29 Manmatha .	7 Āśvina
4334	1155	1290	639	407-08	*1232-33	26 Nandana .	30 Durmukha .	...
4335	1156	1291	640	408-09	1233-34	27 Vijaya . .	31 Hēmalamba .	...
4336	1157	1292	641	409-10	1234-35	28 Jaya . .	32 Vilamba .	5 Śrāvaṇa .
4337	1158	1293	642	410-11	1235-36	29 Manmatha .	33 Vikārin . .	...
4338	1159	1294	643	411-12	*1236-37	30 Durmukha .	34 Śārvarin .	...
4339	1160	1295	644	412-13	1237-38	31 Hēmalamba .	35 Plava . .	3 Jyēshṭha .
4340	1161	1296	645	413-14	1238-39	32 Vilamba .	36 Śubhakṛit .	...
4341	1162	1297	646	414-15	1239-40	33 Vikārin . .	37 Śōbhana .	...
4342	1163	1298	647	415-16	*1240-41	34 Śārvarin .	38 Krōdhin .	2 Vaiśākha
4343	1164	1299	648	416-17	1241-42	35 Plava . .	39 Viśvāvasu .	...
4344	1165	1300	649	417-18	1242-43	36 Śubhakrit .	40 Parābhava .	6 Bhādrapada
4345	1166	1301	650	418-19	1243-44	37 Śōbhana .	41 Plavanga .	...
4346	1167	1302	651	419-20	*1244-45	38 Krōdhin .	42 Kīlaka . .	...

LXI—*Contd.*

COMMENCEMENT OF THE										
SOLAR YEAR.					LUNI-SOLAR YEAR (MEAN SUNRISE OF CIVIL DAY ON WHICH CHAITRA ŚUKLA 1 ENDS).					Kali.
Day and month, A.D.	Week-day.	Time of true Mêsha-saṁkrânti.			Day and month, A.D.	Week-day.	*a.*	*b.*	*c.*	
13	14	17			19	20	23	24	25	1
		H.	M.	S.						
24 Mar. (84)	3 Tues.	14	40	0	7 Mar. (67)	0 Sat. .	124·8768	917·6228	229·2383	4322
24 Mar. (83)	4 Wed.	20	52	30	24 Feb. (55)	4 Wed.	0·5602	754·8576	198·4143	4323
25 Mar. (84)	6 Fri. .	3	5	0	15 Mar. (74)	3 Tues.	35·1998	700·8410	249·7241	4324
25 Mar. (84)	0 Sat. .	9	17	30	4 Mar. (63)	0 Sat. .	9910·8832	548·0759	218·9002	4325
24 Mar. (84)	1 Sun. .	15	30	0	22 Mar. (82)	6 Fri. .	9945·5228	484·0594	270·2099	4326
24 Mar. (83)	2 Mon. .	21	42	30	11 Mar. (70)	3 Tues.	9821·2062	331·2941	239·3861	4327
25 Mar. (84)	4 Wed.	3	55	0	1 Mar. (60)	1 Sun. .	35·5215	214·8206	211·3001	4328
25 Mar. (84)	5 Thur.	10	7	30	20 Mar. (79)	0 Sat. .	70·1611	150·8142	262·6098	4329
24 Mar. (84)	6 Fri. .	16	20	0	8 Mar. (68)	4 Wed.	9945·8444	998·0389	231·7858	4330
24 Mar. (83)	0 Sat. .	22	32	30	26 Feb. (57)	2 Mon. .	160·1597	881·5653	203·6998	4331
25 Mar. (84)	2 Mon. .	4	45	0	17 Mar. (76)	1 Sun. .	194·7993	817·5489	255·0095	4332
25 Mar. (84)	3 Tues.	10	57	30	6 Mar. (65)	5 Thur.	70·4827	664·7836	224·1857	4333
24 Mar. (84)	4 Wed.	17	10	0	24 Mar. (84)	4 Wed.	105·1223	600·7672	275·4954	4334
24 Mar. (83)	5 Thur.	23	22	30	13 Mar. (72)	1 Sun. .	9980·8057	448·0020	244·6716	4335
25 Mar. (84)	0 Sat. .	5	35	0	2 Mar. (61)	5 Thur.	9856·4891	295·2368	213·8476	4336
25 Mar. (84)	1 Sun. .	11	47	30	21 Mar. (80)	4 Wed.	9891·1287	231·2203	265·1574	4337
24 Mar. (84)	2 Mon. .	18	0	0	9 Mar. (69)	1 Sun. .	9766·8121	78·4551	234·3335	4338
25 Mar. (84)	4 Wed.	0	12	30	27 Feb. (58)	6 Fri. .	9981·1274	961·9816	206·2475	4339
25 Mar. (84)	5 Thur.	6	25	0	18 Mar. (77)	5 Thur.	15·7670	897·9640	257·5572	4340
25 Mar. (84)	6 Fri. .	12	37	30	8 Mar. (67)	3 Tues.	230·0823	781·4915	229·4612	4341
24 Mar. (84)	0 Sat. .	18	50	0	25 Feb. (56)	0 Sat. .	105·7656	628·7263	198·6473	4342
25 Mar. (84)	2 Mon. .	1	2	30	15 Mar. (74)	6 Fri. .	140·4053	564·7098	249·9570	4343
25 Mar. (84)	3 Tues.	7	15	0	4 Mar. (63)	3 Tues.	16·0887	411·9446	219·1331	4344
25 Mar. (84)	4 Wed.	13	27	30	23 Mar. (82)†	2 Mon. .	50·7283	347·9281	270·4428	4345
25 Mar. (84)	5 Thur.	19	40	0	11 Mar. (71)	3 Fri. .	9926·4116	195·1629	239·6190	4346

TABLE

CONCURRENT YEAR.								INTERCALATED (*adhika*) and SUPPRESSED (*kshaya*) LUNAR MONTHS (true).
						JOVIAN SAMVATSARA.		
Kali	Śaka.	Chaitrādi Vikrama.	Mēshādi solar year in Bengal.	Kollam.	A.D.	Southern system.	Northern system.	
1	2	3	3*a*	4	5	6	7	8
4347	1168	1303	652	420-21	1245-46	39 Viśvāvasu	43 Saumya.	4 Āshāḍha
4348	1169	1304	653	421-22	1246-47	40 Parābhava	44 Sādhāraṇa†	...
4349	1170	1305	654	422-23	1247-48	41 Plavaṅga	46 *Paridhāvin*	...
4350	1171	1306	655	423-24	*1248-49	42 Kīlaka	47 *Pramādin*	3 Jyēshtha
4351	1172	1307	656	424-25	1249-50	43 Saumya	48 *Ānanda*	...
4352	1173	1308	657	425-26	1250-51	44 Sādhāraṇa	49 *Rākshasa*	7 Āśvina
4353	1174	1309	658	426-27	1251-52	45 Virōdhakṛit	50 Anala	...
4354	1175	1310	659	427-28	*1252-53	46 Paridhāvin	51 Piṅgala	...
4355	1176	1311	660	428-29	1253-54	47 Pramādin	52 Kālayukta	5 Śrāvaṇa
4356	1177	1312	661	429-30	1254-55	48 Ānanda	53 Siddhārthin	...
4357	1178	1313	662	430-31	1255-56	49 Rākshasa	54 Raudra	...
4358	1179	1314	663	431-32	*1256-57	50 Anala	55 Durmati	3 Jyēshtha
4359	1180	1315	664	432-33	1257-58	51 Piṅgala	56 Dundubhi	...
4360	1181	1316	665	433-34	1258-59	52 Kālayukta	57 Rudhirōd-gārin	8 Kārttika 10 *Pausha* (*ksh.*)
4361	1182	1317	666	434-35	1259-60	53 Siddhārthin	58 Raktāksha	1 Chaitra
4362	1183	1318	667	435-36	*1260-61	54 Raudra	59 Krōdhaṇa	...
4363	1184	1319	668	436-37	1261-62	55 Durmati	60 Kshaya	6 Bhādrapada
4364	1185	1320	669	437-38	1262-63	56 Dundubhi	1 Prabhava	...
4365	1186	1321	670	438-39	1263-64	57 Rudhirōdgārin	2 Vibhava	...
4366	1187	1322	671	439-40	*1264-65	58 Raktāksha	3 Śukla	4 Āshāḍha
4367	1188	1323	672	440-41	1265-66	59 Krōdhana	4 Pramōda	...
4368	1189	1324	673	441-42	1266-67	60 Kshaya	5 Prājāpati	...
4369	1190	1325	674	442-43	1267-68	1 Prabhava	6 Aṅgiras	3 Jyēshṭha
4370	1191	1326	675	443-44	*1268-69	2 Vibhava	7 Śrīmukha	...
4371	1192	1327	676	444-45	1269-70	3 Śukla	8 Bhāva	7 Āśvina

† 45 Virōdhakṛit was suppressed in the north

LXI—*Contd.*

COMMENCEMENT OF THE								Kali.
SOLAR YEAR.			LUNI-SOLAR YEAR (MEAN SUNRISE OF CIVIL DAY ON WHICH CHAITRA ŚUKLA 1 ENDS).					
Day and month, A.D.	Week-day.	Time of true Mēsha-saṃkrānti.	Day and month, A.D.	Week-day.	*a.*	*b.*	*c*	
13	14	17	19	20	23	24	25	1
		H. M. S.						
25 Mar. (84)	0 Sat.	1 52 30	28 Feb. (59)	3 Tues.	9802·0950	42·3977	208·7952	4347
25 Mar. (84)	1 Sun.	8 5 0	20 Mar. (79)	3 Tues.	175·3365	14·6728	262·8427	4348
25 Mar. (84)	2 Mon.	14 17 30	9 Mar. (68)	0 Sat.	51·0499	861·9077	232·0187	4349
24 Mar. (84)	3 Tues.	20 30 0	27 Feb. (58)	5 Thur.	265·3651	745·4341	203·9327	4350
25 Mar. (84)	5 Thur.	2 42 30	17 Mar. (76)	4 Wed.	300·0047	681·4176	255·2424	4351
25 Mar. (84)	6 Fri.	8 55 0	6 Mar. (65)	1 Sun.	175·6881	528·6524	224·4186	4352
25 Mar. (84)	0 Sat.	15 7 30	24 Mar. (83)	6 Fri.	9871·6959	428·3444	274·9905	4353
24 Mar. (84)	1 Sun.	21 20 0	12 Mar. (72)	3 Tues.	9747·3793	275·5791	242·1667	4354
25 Mar. (84)	3 Tues.	3 32 30	2 Mar. (61)	1 Sun.	9961·6945	159·1655	214·6805	4355
25 Mar. (84)	4 Wed.	9 45 0	21 Mar. (80)	0 Sat.	9996·3341	95·0891	265·3903	4356
25 Mar. (84)	5 Thur.	15 57 30	11 Mar. (70)	5 Thur.	210·6494	978·6154	237·3042	4357
24 Mar. (84)	6 Fri.	22 10 0	28 Feb. (59)	2 Mon.	86·3328	825·8503	206·4804	4358
25 Mar. (84)	1 Sun.	4 22 30	18 Mar. (77)	1 Sun.	120·9724	761·8338	257·7901	4359
25 Mar. (84)	2 Mon.	10 35 0	7 Mar. (66)	5 Thur.	9996·6558	609·0686	226·9663	4360
25 Mar. (84)	3 Tues.	16 47 30	24 Feb. (55)	2 Mon.	9872·3392	456·3034	196·1424	4361
24 Mar. (84)	4 Wed.	23 0 0	4 Mar. (74)	1 Sun.	9906·9788	392·2869	247·4521	4362
25 Mar. (84)	6 Fri.	5 12 30	3 Mar. (62)	5 Thur.	9782·6622	239·5218	216·6282	4363
25 Mar. (84)	0 Sat.	11 25 0	22 Mar. (81)	4 Wed.	9817·3018	175·5052	267·9380	4364
25 Mar. (84)	1 Sun.	17 37 30	12 Mar. (71)	2 Mon.	31·6171	59·0317	239·8519	4365
24 Mar. (84)	2 Mon.	23 50 0	29 Feb. (60)	6 Fri.	9907·3005	906·2665	209·0281	4366
25 Mar. (84)	4 Wed.	6 2 30	20 Mar. (79)	6 Fri.	280·5720	878·5417	263·0756	4367
25 Mar. (84)	5 Thur.	12 15 0	9 Mar. (68)	3 Tues.	156·2553	725·7764	232·2516	4368
25 Mar. (84)	6 Fri.	18 27 30	26 Feb. (57)	0 Sat.	31·9387	573·0112	201·4278	4369
25 Mar. (85)	1 Sun.	0 40 0	16 Mar. (76)	6 Fri.	66·5784	509·2864	255·4753	4370
25 Mar. (84)	2 Mon.	6 52 30	5 Mar. (64)	3 Tues.	9942 2617	56·2295	221·9137	4371

TABLE

CONCURRENT YEAR.								INTERCALATED (*adhika*) and SUPPRESSED (*kshaya*) LUNAR MONTHS (true).
						JOVIAN SAMVATSARA.		
Kali.	Śaka.	Chaitrādi Vikrama.	Mēshādi solar year in Bengal.	Kollam.	A.D.	Southern system.	Northern system.	
1	2	3	3*a*	4	5	6	7	8
4372	1193	1328	677	445-46	1270-71	4 Pramōda	9 Yuvan	...
4373	1194	1329	678	446-47	1271-72	5 Prajāpati	10 Dhātṛi	...
4374	1195	1330	679	447-48	*1272-73	6 Aṅgiras	11 Īśvara	4 Āshāḍha
4375	1196	1331	680	448-49	1273-74	7 Śrīmukha	12 Bahudhānya	...
4376	1197	1332	681	449-50	1274-75	8 Bhāva	13 Pramāthin	...
4377	1198	1333	682	450-51	1275-76	9 Yuvan	14 Vikrama	3 Jyēshṭha
4378	1199	1334	683	451-52	*1276-77	10 Dhātṛi	15 Vṛisha	...
4379	1200	1335	684	452-53	1277-78	11 Īśvara	16 Chitrabhānu	9 Mārgasira 10 *Pausha* (*ksh.*) 12 Phālguna
4380	1201	1336	685	453-54	1278-79	12 Bahudhānya	17 Subhānu	...
4381	1202	1337	686	454-55	1279-80	13 Pramāthin	18 Tāraṇa	...
4382	1203	1338	687	455-56	*1280-81	14 Vikrama	19 Pārthiva	5 Śrāvaṇa
4383	1204	1339	688	456-57	1281-82	15 Vṛisha	20 Vyaya	...
4384	1205	1340	689	457-58	1282-83	16 Chitrabhānu	21 Sarvajit	...
4385	1206	1341	690	458-59	1283-84	17 Subhānu	22 Sarvadhārin	4 Āshāḍha
4386	1207	1342	691	459-60	*1284-85	18 Tāraṇa	23 Virōdhin	...
4387	1208	1343	692	460-61	1285-86	19 Pārthiva	24 Vikṛita	...
4388	1209	1344	693	461-62	1286-87	20 Vyaya	25 Khara	2 Vaiśākha
4389	1210	1345	694	462-63	1287-88	21 Sarvajit	26 Nandana	...
4390	1211	1346	695	463-64	*1288-89	22 Sarvadhārin	27 Vijaya	6 Bhādrapada
4391	1212	1347	696	464-65	1289-90	23 Virōdhin	28 Jaya	...
4392	1213	1348	697	465-66	1290-91	24 Vikṛita	29 Manmatha	...
4393	1214	1349	698	466-67	1291-92	25 Khara	30 Durmukha	4 Āshāḍha
4394	1215	1350	699	467-68	*1292-93	26 Nandana	31 Hēmalamba	...
4395	1216	1351	700	468-69	1293-94	27 Vijaya	32 Vilamba	...
4396	1217	1352	701	469-70	1294-95	28 Jaya	33 Vikārin	3 Jyēshṭha

LXI—*Contd.*

COMMENCEMENT OF THE												
SOLAR YEAR.					LUNI-SOLAR YEAR (MEAN SUNRISE OF CIVIL DAY ON WHICH CHAITRA ŚUKLA 1 ENDS).					Kali.
Day and month, A.D.	Week-day.	Time of true Mēsha-saṁkrānti.			Day and month, A.D.	Week-day.	*a.*	*b.*	*c.*	
13	14	17			19	20	23	24	25	1
		H.	M.	S.						
25 Mar. (84)	3 Tues.	13	5	0	24 Mar. (83)	2 Mon. .	9976·9014	292·2121	273·2234	4372
25 Mar. (84)	4 Wed.	19	17	30	13 Mar. (72)	6 Fri. .	9852·5848	139·4479	242·3996	4373
25 Mar. (85)	6 Fri. .	1	30	0	2 Mar. (62)	4 Wed.	66·9000	22·9743	214·3134	4374
25 Mar. (84)	0 Sat. .	7	42	30	21 Mar. (80)	3 Tues.	101·5396	958·9578	265·6232	4375
25 Mar. (84)	1 Sun. .	13	55	0	10 Mar. (69)	0 Sat. .	9977·2230	806·1926	234·7993	4376
25 Mar. (84)	2 Mon. .	20	7	30	28 Feb. (59)	5 Thur.	191·5382	689·7191	206·7133	4377
25 Mar. (85)	4 Wed.	2	20	0	18 Mar. (78)	4 Wed.	226·1778	624·7025	258·0230	4378
25 Mar. (84)	5 Thur.	8	32	30	7 Mar. (66)	1 Sun. .	101·8612	472·9373	227·1992	4379
25 Mar. (84)	6 Fri. .	14	45	0	25 Mar. (84)	6 Fri. .	9797·8690	372·6293	275·7711	4380
25 Mar. (84)	0 Sat. .	20	57	30	15 Mar. (74)	4 Wed.	12·1842	256·1556	247·6750	4381
25 Mar. (85)	2 Mon. .	3	10	0	3 Mar. (63)	1 Sun. .	9887·8676	103·3905	216·8611	4382
25 Mar. (84)	3 Tues.	9	22	30	22 Mar. (81)	0 Sat. .	9922·5072	39·3740	268·1709	4383
25 Mar. (84)	4 Wed.	15	35	0	12 Mar. (71)	5 Thur.	136·8225	922·9004	240·0848	4384
25 Mar. (84)	5 Thur.	21	47	30	1 Mar. (60)	2 Mon. .	12·5059	770·1352	209·2610	4385
25 Mar. (85)	0 Sat. .	4	0	0	19 Mar. (79)	1 Sun. .	47·1455	706·1187	260·5706	4386
25 Mar. (84)	1 Sun. .	10	12	30	8 Mar. (67)	5 Thur.	9922·8289	553·3536	229·7458	4387
25 Mar. (84)	2 Mon. .	16	25	0	25 Feb. (56)	2 Mon. .	9798·5122	400·5883	198·9229	4388
25 Mar. (84)	3 Tues.	22	37	30	16 Mar. (75)	1 Sun. .	9833·1519	336·5718	250·1827	4389
25 Mar. (85)	5 Thur.	4	50	0	5 Mar. (65)	6 Fri. .	47·4671	220·0983	222·1466	4390
25 Mar. (84)	6 Fri. .	11	2	30	23 Mar. (82)	4 Wed.	9743·4749	119·7901	270·7185	4391
25 Mar. (84)	0 Sat. .	17	15	0	13 Mar. (72)	2 Mon. .	9957·7901	3·3166	242·6325	4392
25 Mar. (84)	1 Sun. .	23	27	30	3 Mar. (62)	0 Sat. .	172·1054	886·8430	214·5463	4393
25 Mar. (85)	3 Tues.	5	40	0	21 Mar. (81)	6 Fri. .	206·7450	822·8266	265·8561	4394
25 Mar. (84)	4 Wed.	11	52	30	10 Mar. (69)	3 Tues.	82·4284	670·0613	235·0522	[illegible]
25 Mar. (84)	5 Thur.	18	5	0	27 Feb. (58)	0 Sat. .	9958·1118	517·2962	204·2084	4396

TABLE

CONCURRENT YEAR.								INTERCALATED (*adhika*) and SUPPRESSED (*kshaya*) LUNAR MONTHS (true).
Kali.	Saka.	Chaitrādi Vikrama.	Mēshādi solar year in Bengal.	Kollam.	A.D.	Jovian Saṁvatsara. Southern system.	Jovian Saṁvatsara. Northern system.	
1	2	3	3a	4	5	6	7	8
4397	1218	1353	702	470-71	1295-96	29 Manmatha .	34 Sārvarin .	...
4398	1219	1354	703	471-72	*1296-97	30 Durmukha .	35 Plava . .	12 Phālguna .
4399	1220	1355	704	472-73	1297-98	31 Hēmalamba .	36 Subhakṛit .	...
4400	1221	1356	705	473-74	1298-99	32 Vilamba .	37 Sōbhana .	...
4401	1222	1357	706	474-75	1299-1300	33 Vikārin . .	38 Krōdhin .	5 Srāvaṇa .
4402	1223	1358	707	475-76	*1300-01	34 Sārvarin .	39 Viśvāvasu .	...
4403	1224	1359	708	476-77	1301-02	35 Plava . .	40 Parābhava .	..
4404	1225	1360	709	477-78	1302-03	36 Subhakṛit .	41 Plavaṅga .	4 Āshāḍha .
4405	1226	1361	710	478-79	1303-04	37 Sōbhana .	42 Kīlaka . .	...
4406	1227	1362	711	479-80	*1304-05	38 Krōdhin .	43 Saumya .	...
4407	1228	1363	712	480-81	1305-06	39 Viśvāvasu .	44 Sādhāraṇa .	2 Vaiśākha .
4408	1229	1364	713	481-82	1306-07	40 Parābhava .	45 Virōdhakṛit .	...
4409	1230	1365	714	482-83	1307-08	41 Plavaṅga .	46 Paridhāvin .	6 Bhadrapada.
4410	1231	1366	715	483-84	*1308-09	42 Kīlaka . .	47 Pramādin .	...
4411	1232	1367	716	484-85	1309-10	43 Saumya .	48 Ānanda .	...
4412	1233	1368	717	485-86	1310-11	44 Sādhāraṇa .	49 Rākshasa .	4 Āshāḍha .
4413	1234	1369	718	486-87	1311-12	45 Virōdhakṛit .	50 Anala . .	...
4414	1235	1370	719	487-88	*1312-13	46 Paridhāvin .	51 Piṅgala .	...
4415	1236	1371	720	488-89	1313-14	47 Pramādin .	52 Kālayukta .	3 Jyēshṭha .
4416	1237	1372	721	489-90	1314-15	48 Ānanda .	53 Siddhārthin .	...
4417	1238	1373	722	490-91	1315-16	49 Rākshasa .	54 Raudra .	12 Phālguna .
4418	1239	1374	723	491-92	*1316-17	50 Anala . .	55 Durmati .	...
4419	1240	1375	724	492-93	1317-18	51 Piṅgala .	56 Dundubhi .	...
4420	1241	1376	725	493-94	1318-19	52 Kālayukta .	57 Rudhirōdgārin	5 Srāvaṇa
4421	1242	1377	726	494-95	1319-20	53 Siddhārthin .	58 Raktāksha .	...

[TABLE] LXI.—*Contd.*

COMMENCEMENT OF THE											
SOLAR YEAR.					LUNI-SOLAR YEAR (MEAN SUNRISE OF CIVIL DAY ON WHICH CHAITRA ŚUKLA 1 ENDS).					Kali.
Day and month, A.D.	Week-day.	Time of true Mēsha-saṁkrānti.			Day and month, A.D.	Week-day.	*a.*	*b.*	*c.*	
13	14	17			19	20	23	24	25	1
		H.	M.	S.						
26 Mar. (85)	0 Sat. .	0	17	30	18 Mar. (77)	6 Fri. .	9992·7514	453·2797	255·5181	4397
25 Mar. (85)	1 Sun. .	6	30	0	6 Mar. (66)	3 Tues.	9868·4348	300·5144	224·6943	4398
25 Mar. (84)	2 Mon. .	12	42	30	25 Mar. (84)	2 Mon. .	9903·0744	236·4980	276·0039	4399
25 Mar. (84)	3 Tues.	18	55	0	14 Mar. (73)	6 Fri. .	9778·7578	83·7328	245·1801	4400
26 Mar. (85)	5 Thur.	1	7	30	4 Mar. (63)	4 Wed.	9993·0731	967·2592	217·0940	4401
25 Mar. (85)	6 Fri. .	7	20	0	22 Mar. (82)	3 Tues.	27·7127	903·2427	258·4038	4402
25 Mar. (84)	0 Sat. .	13	32	30	12 Mar. (71)	1 Sun. .	242·0280	786·7691	240·3177	4403
25 Mar. (84)	1 Sun. .	19	45	0	1 Mar. (60)	5 Thur.	117·7114	634·0039	209·4938	4404
26 Mar. (85)	3 Tues.	1	57	30	20 Mar. (79)	4 Wed.	152·3510	569·9874	260·8035	4405
25 Mar. (85)	4 Wed.	8	10	0	8 Mar. (68)	1 Sun. .	28·0344	417·2222	229·9797	4406
25 Mar. (84)	5 Thur.	14	22	30	25 Feb. (56)	5 Thur.	9903·7177	264·4570	199·1558	4407
25 Mar. (84)	6 Fri. .	20	35	0	16 Mar. (75)	4 Wed.	9938·3574	200·4405	250·4656	4408
26 Mar. (85)	1 Sun. .	2	47	30	5 Mar. (64)	1 Sun. .	9814·0408	47·6754	219·6417	4409
25 Mar. (85)	2 Mon. .	9	0	0	23 Mar. (83)	0 Sat. .	9848·6804	983·7588	270·9514	4410
25 Mar. (84)	3 Tues.	15	12	30	13 Mar. (72)	5 Thur.	62·9956	867·1853	242·8653	4411
25 Mar. (84)	4 Wed.	21	25	0	3 Mar. (62)	3 Tues.	277·3109	750·7117	214·7792	4412
26 Mar. (85)	6 Fri. .	3	37	30	21 Mar. (80)	1 Sun. .	9973·3187	650·4036	263·3512	4413
25 Mar. (85)	0 Sat. .	9	50	0	10 Mar. (70)	6 Fri. .	187·6339	533·9300	235·2651	4414
25 Mar. (84)	1 Sun. .	16	2	30	27 Feb. (58)	3 Tues.	63·3172	381·1648	204·4413	4415
25 Mar. (84)	2 Mon .	22	15	0	17 Mar. (76)	1 Sun. .	9759·3250	280·8568	253·0132	4416
26 Mar. (85)	4 Wed.	4	27	30	7 Mar. (66)	6 Fri. .	9973·6403	164·3831	224·9271	4417
25 Mar. (85)	5 Thur.	10	40	0	25 Mar. (85)	5 Thur.	8·2799	100·3667	276·2368	4418
25 Mar. (84)	6 Fri. .	16	52	30	14 Mar. (73)	2 Mon. .	9883·9632	947·6015	245·4130	4419
25 Mar. (84)	0 Sat. .	23	5	0	4 Mar. (63)	0 Sat. .	98·2785	831·1279	217·3269	4420
26 Mar. (85)	2 Mon .	5	17	30	23 Mar. (82)	6 Fri. .	132·9181	767·1114	268·6367	4421

TABLE

CONCURRENT YEAR.								Intercalated (*adhika*) and suppressed (*kshaya*) Lunar months (true).
						Jovian Samvatsara.		
Kali.	Śaka.	Chaitrādi Vikrama.	Mēshādi solar year in Bengal.	Kollam.	A.D.	Southern system.	Northern system.	
1	2	3	3a	4	5	6	7	8
4422	1243	1378	727	495-96	*1320-21	54 Raudra .	59 Krōdhana .	...
4423	1244	1379	728	496-97	1321-22	55 Durmati .	60 Kshaya .	4 Āshāḍha .
4424	1245	1380	729	497-98	1322-23	56 Dundubhi .	1 Prabhava .	...
4425	1246	1381	730	498-99	1323-24	57 Rudhirōdgārin	2 Vibhava .	...
4426	1247	1382	731	499-500	*1324-25	58 Raktāksha .	3 Śukla . .	2 Vaiśākha .
4427	1248	1383	732	500-01	1325-26	59 Krōdhana .	4 Pramōda .	...
4428	1249	1384	733	501-02	1326-27	60 Kshaya .	5 Prajāpati .	6 Bhādrapada
4429	1250	1385	734	502-03	1327-28	1 Prabhava .	6 Aṅgiras .	...
4430	1251	1386	735	503-04	*1328-29	2 Vibhava .	7 Śrīmukha .	...
4431	1252	1387	736	504-05	1329-30	3 Śukla . .	8 Bhāva . .	4 Āshāḍha .
4432	1253	1388	737	505-06	1330-31	4 Pramōda .	9 Yuvan . .	...
4433	1254	1389	738	506-07	1331-32	5 Prajāpati .	10 Dhātṛi † .	...
4434	1255	1390	739	507-08	*1332-33	6 Aṅgiras .	12 *Bahudhānya* .	3 Jyēshṭha .
4435	1256	1391	740	508-09	1333-34	7 Śrīmukha .	13 *Pramāthin* .	...
4436	1257	1392	741	509-10	1334-35	8 Bhāva . .	14 *Vikrama*	{ 7 Āsvina 10 *Pausha* (*ksh.*) 12 Phālguna }
4437	1258	1393	742	510-11	1335-36	9 Yuvan . .	15 *Vṛisha* . .	...
4438	1259	1394	743	511-12	*1336-37	10 Dhātṛi . .	16 Chitrabhānu .	...
4439	1260	1395	744	512-13	1337-38	11 Īśvara . .	17 Subhānu .	5 Śrāvaṇa .
4440	1261	1396	745	513-14	1338-39	12 Bahudhānya .	18 Tāraṇa . .	...
4441	1262	1397	746	514-15	1339-40	13 Pramāthin .	19 Pārthiva .	...
4442	1263	1398	747	515-16	*1340-41	14 Vikrama .	20 Vyaya . .	4 Āshāḍha .
4443	1264	1399	748	516-17	1341-42	15 Vṛisha . .	21 Sarvajit .	...
4444	1265	1400	749	517-18	1342-43	16 Chitrabhānu .	22 Sarvadhārin .	...
4445	1266	1401	750	518-19	1343-44	17 Subhānu .	23 Virōdhin .	2 Vaiśākha .
4446	1267	1402	751	519-20	*1344-45	18 Tāraṇa . .	24 Vikṛita . .	...

† 11 Īśvara was suppressed in the north.

LXI—*Contd.*

COMMENCEMENT OF THE									Kali.
SOLAR YEAR.			LUNI-SOLAR YEAR (MEAN SUNRISE OF CIVIL DAY ON WHICH CHAITRA ŚUKLA 1 ENDS).						
Day and month, A.D.	Week-day.	Time of true Mēsha-saṁkrānti.	Day and month, A.D.	Week-day.	*a.*	*b.*	*c.*		
13	14	17	19	20	23	24	25		1
		H. M. S.							
25 Mar. (85)	3 Tues.	11 30 0	11 Mar. (71)	3 Tues.	8·6015	614·3462	237·8628		4422
25 Mar. (84)	4 Wed.	17 42 30	28 Feb. (59)	0 Sat.	9884·2849	461·5811	206·9889		4423
25 Mar. (84)	5 Thur.	23 55 0	19 Mar. (78)	6 Fri.	9918·9245	397·5645	258·2986		4424
26 Mar. (85)	0 Sat.	6 7 30	8 Mar. (67)	3 Tues.	9794·6078	244·7993	227·4748		4425
25 Mar. (85)	1 Sun.	12 20 0	26 Feb. (57)	1 Sun.	8·9231	128·3258	199·3887		4426
25 Mar. (84)	2 Mon.	18 32 30	16 Mar. (75)	0 Sat.	43·5628	64·3092	250·6985		4427
26 Mar. (85)	4 Wed.	0 45 0	5 Mar. (64)	4 Wed.	9919·2462	911·5441	219·8746		4428
26 Mar. (85)	5 Thur.	6 57 30	24 Mar. (83)	3 Tues.	9953·8858	847·5276	271·1843		4429
25 Mar. (85)	6 Fri.	13 10 0	13 Mar. (73)	1 Sun.	168·3010	731·0530	243·0982		4430
25 Mar. (84)	0 Sat.	19 22 30	2 Mar. (61)	5 Thur.	45·8845	578·2878	212·2744		4431
26 Mar. (85)	2 Mon.	1 35 0	21 Mar. (80)	4 Wed.	78·5241	514·2714	263·5841		4432
26 Mar. (85)	3 Tues.	7 47 30	10 Mar. (69)	1 Sun.	9954·2074	361·5061	232·7602		4433
25 Mar. (85)	4 Wed.	14 0 0	27 Feb. (58)	5 Thur.	9829·8908	208·7409	202·1364		4434
25 Mar. (84)	5 Thur.	20 12 30	17 Mar. (76)	4 Wed.	9864·5305	144·7245	253·2461		4435
26 Mar. (85)	0 Sat.	2 25 0	7 Mar. (66)	2 Mon.	78·8457	28·2509	225·1600		4436
26 Mar. (85)	1 Sun.	8 37 30	26 Mar. (85)	1 Sun.	113·4853	964·2344	276·4697		4437
25 Mar. (85)	2 Mon.	14 50 0	14 Mar. (74)	5 Thur.	9989·1687	811·4702	245·6459		4438
25 Mar. (84)	3 Tues.	21 2 30	4 Mar. (63)	3 Tues.	203·4840	694·9967	217·5598		4439
26 Mar. (85)	5 Thur.	3 15 0	23 Mar. (82)	2 Mon.	238·1236	629·9801	268·8696		4440
26 Mar. (85)	6 Fri.	9 27 30	12 Mar. (71)	6 Fri.	113·8081	478·2149	238·0457		4441
25 Mar. (85)	0 Sat.	15 40 0	29 Feb. (60)	3 Tues.	9989·4904	325·4498	207·2219		4442
25 Mar. (84)	1 Sun.	21 52 30	19 Mar. (78)	2 Mon.	24·1200	261·4333	259·5315		4443
26 Mar. (85)	3 Tues.	4 5 0	8 Mar. (67)	6 Fri.	9899·8134	108·6680	227·7077		4444
26 Mar. (85)	4 Wed.	10 17 30	26 Feb. (57)	4 Wed.	114·1286	992·1945	199·6316		4445
25 Mar. (85)	5 Thur.	16 30 0	16 Mar. (76)	3 Tues.	148·7682	928·1780	250·9314		4446

TABLE

CONCURRENT YEAR								INTERCALATED (*adhika*) and SUPPRESSED (*kshaya*) LUNAR MONTHS (true).
						JOVIAN SAMVATSARA.		
Kali.	Śaka.	Chaitrādi Vikrama.	Mēshādi solar year in Bengal.	Kollam.	A.D.	Southern system.	Northern system.	
1	2	3	3*a*	4	5	6	7	8
4447	1268	1403	752	520-21	1345-46	19 Pārthiva	25 Khara	6 Bhādrapada
4448	1269	1404	753	521-22	1346-47	20 Vyaya	26 Nandana	...
4449	1270	1405	754	522-23	1347-48	21 Sarvajit	27 Vijaya	...
4450	1271	1406	755	523-24	*1348-49	22 Sarvadhārin	28 Jaya	4 Āshāḍha
4451	1272	1407	756	524-25	1349-50	23 Virōdhin	29 Manmatha	...
4452	1273	1408	757	525-26	1350-51	24 Vikṛita	30 Durmukha	...
4453	1274	1409	758	526-27	1351-52	25 Khara	31 Hēmalamba	2 Vaiśākha
4454	1275	1410	759	527-28	*1352-53	26 Nandana	32 Vilamba	...
4455	1276	1411	760	528-29	1353-54	27 Vijaya	33 Vikārin	7 Āśvina 11 *Māgha* (*ksh.*) 12 Phālguna
4456	1277	1412	761	529-30	1354-55	28 Jaya	34 Śārvarin	...
4457	1278	1413	762	530-31	1355-56	29 Manmatha	35 Plava	...
4458	1279	1414	763	531-32	*1356-57	30 Durmukha	36 Śubhakṛit	5 Śrāvaṇa
4459	1280	1415	764	532-33	1357-58	31 Hēmalamba	37 Śōbhana	...
4460	1281	1416	765	533-34	1358-59	32 Vilamba	38 Krōdhin	...
4461	1282	1417	766	534-35	1359-60	33 Vikārin	39 Viśvāvasu	3 Jyēshṭha
4462	1283	1418	767	535-36	*1360-61	34 Śārvarin	40 Parābhava	...
4463	1284	1419	768	536-37	1361-62	35 Plava	41 Plavaṅga	...
4464	1285	1420	769	537-38	1362-63	36 Śubhakṛit	42 Kīlaka	2 Vaiśākha
4465	1286	1421	770	538-39	1363-64	37 Śōbhana	43 Saumya	...
4466	1287	1422	771	539-40	*1364-65	38 Krōdhin	44 Sādhāraṇa	6 Bhādrapada
4467	1288	1423	772	540-41	1365-66	39 Viśvāvasu	45 Virōdhakṛit	...
4468	1289	1424	773	541-42	1366-67	40 Parābhava	46 Paridhāvin	...
4469	1290	1425	774	542-43	1367-68	41 Plavaṅga	47 Pramādin	4 Āshāḍha
4470	1291	1426	775	543-44	*1368-69	42 Kīlaka	48 Ānanda	...
4471	1292	1427	776	544-45	1369-70	43 Saumya	49 Rākshasa	...

LXI—*Contd.*

COMMENCEMENT OF THE									
SOLAR YEAR.			LUNI-SOLAR YEAR (MEAN SUNRISE OF CIVIL DAY ON WHICH CHAITRA ŚUKLA 1 ENDS).						Kali.
Day and month, A.D.	Week-day.	Time of true Mēsha-saṁkrānti.	Day and month, A.D.	Week-day.	*a.*	*b.*	*c.*		
13	14	17	19	20	23	24	25		1
		H. M. S.							
25 Mar. (84)	6 Fri.	22 42 30	5 Mar. (64)	0 Sat.	24·4516	775·4128	220·1075		4447
26 Mar. (85)	1 Sun.	4 55 0	24 Mar. (83)	6 Fri.	59·0912	711·3963	271·4172		4448
26 Mar. (85)	2 Mon.	11 7 30	13 Mar. (72)	3 Tues	9934·7747	558·6312	240·5933		4449
25 Mar. (85)	3 Tues.	17 20 0	1 Mar. (61)	0 Sat.	9810·4580	405·8660	209·7695		4450
25 Mar. (84)	4 Wed.	23 32 30	20 Mar. (79)	6 Fri.	9845·0976	341·8494	261·0792		4451
26 Mar. (85)	6 Fri.	5 45 0	9 Mar. (68)	3 Tues.	9720·7810	189·0843	230·2554		4452
26 Mar. (85)	0 Sat.	11 57 30	27 Feb. (58)	1 Sun.	9935·0962	72·6107	202·1693		4453
25 Mar. (85)	1 Sun.	18 10 0	17 Mar. (77)	0 Sat.	9969·7359	8·5942	253·4790		4454
26 Mar. (85)	3 Tues.	0 22 30	7 Mar. (66)	5 Thur.	184·0511	892·1206	225·3929		4455
26 Mar. (85)	4 Wed.	6 35 0	26 Mar. (85)	4 Wed.	218·6907	828·1042	276·7026		4456
26 Mar. (85)	5 Thur.	12 47 30	15 Mar. (74)	1 Sun.	94·3741	675·3389	245·8788		4457
25 Mar. (85)	6 Fri.	19 0 0	3 Mar. (63)	5 Thur.	9970·0575	522·5737	215·4549		4458
26 Mar. (85)	1 Sun.	1 12 30	22 Mar. (81)	4 Wed.	4·6971	458·5573	266·3647		4459
26 Mar. (85)	2 Mon.	7 25 0	11 Mar. (70)	1 Sun.	9880·3805	305·7921	235·5408		4460
26 Mar. (85)	3 Tues.	13 37 30	28 Feb. (59)	5 Thur.	9756·0639	153·0269	204·7170		4461
25 Mar. (85)	4 Wed.	19 50 0	18 Mar. (78)	4 Wed.	9790·7035	89·0104	256·0266		4462
26 Mar. (85)	6 Fri.	2 2 30	8 Mar. (67)	2 Mon.	5·0188	972·5368	227·9406		4463
26 Mar. (85)	0 Sat.	8 15 0	26 Feb. (57)	0 Sat.	219·3338	856·0632	199·8545		4464
26 Mar. (85)	1 Sun	14 27 30	17 Mar. (76)	6 Fri.	253·9737	792·0468	251·1642		4465
25 Mar. (85)	2 Mon.	20 40 0	5 Mar. (65)	3 Tues.	129·6571	639·2816	220·3404		4466
26 Mar. (85)	4 Wed.	2 52 30	24 Mar. (83)	2 Mon.	164·2967	575·2651	271·6501		4467
26 Mar. (85)	5 Thur.	9 5 0	13 Mar. (72)	6 Fri.	39·9801	422·4999	241·1180		4468
26 Mar. (85)	6 Fri.	15 17 30	2 Mar. (61)	3 Tues	9915·6635	269·7347	210·0024		4469
25 Mar. (85)	0 Sat.	21 30 0	20 Mar. (80)	2 Mon.	9950·3031	205·7182	261·3121		4470
26 Mar. (85)	2 Mon.	3 42 30	9 Mar. (68)	6 Fri	9825·9865	52·9530	230·4883		4471

TABLE

CONCURRENT YEAR.								Intercalated (*adhika*) and suppressed (*kshaya*) Lunar months (true).
Kali.	Śaka.	Chaitrādi Vikrama.	Mēshādi solar year in Bengal.	Kollam.	A.D.	Jovian Samvatsara. Southern system.	Jovian Samvatsara. Northern system.	
1	2	3	3*a*	4	5	6	7	8
4472	1293	1428	777	545-46	1370-71	44 Sādhāraṇa	50 Anala	3 Jyēshṭha
4473	1294	1429	778	546-47	1371-72	45 Virōdhakṛit	51 Piṅgala	...
4474	1295	1430	779	547-48	*1372-73	46 Paridhāvin	52 Kālayukta	6 Bhādrapada
4475	1296	1431	780	548-49	1373-74	47 Pramādin	53 Siddhārthin	...
4476	1297	1432	781	549-50	1374-75	48 Ānanda	54 Raudra	...
4477	1298	1433	782	550-51	1375-76	49 Rākshasa	55 Durmati	5 Śrāvaṇa
4478	1299	1434	783	551-52	*1376-77	50 Anala	56 Dundubhi	...
4479	1300	1435	784	552-53	1377-78	51 Piṅgala	57 Rudhirōdgārin	...
4480	1301	1436	785	553-54	1378-79	52 Kālayukta	58 Raktāksha	3 Jyēshṭha
4481	1302	1437	786	554-55	1379-80	53 Siddhārthin	59 Krōdhana	...
4482	1303	1438	787	555-56	*1380-81	54 Raudra	60 Kshaya	...
4483	1304	1439	788	556-57	1381-82	55 Durmati	1 Prabhava	2 Vaiśākha
4484	1305	1440	789	557-58	1382-83	56 Dundubhi	2 Vibhava	...
4485	1306	1441	790	558-59	1383-84	57 Rudhirōdgārin	3 Śukla	6 Bhādrapada
4486	1307	1442	791	559-60	*1384-85	58 Raktāksha	4 Pramōda	...
4487	1308	1443	792	560-61	1385-86	59 Krōdhana	5 Prajāpati	...
4488	1309	1444	793	561-62	1386-87	60 Kshaya	6 Aṅgiras	4 Āshāḍha
4489	1310	1445	794	562-63	1387-88	1 Prabhava	7 Śrīmukha	...
4490	1311	1446	795	563-64	*1388-89	2 Vibhava	8 Bhāva	...
4491	1312	1447	796	564-65	1389-90	3 Śukla	9 Yuvan	3 Jyēshṭha
4492	1313	1448	797	565-66	1390-91	4 Pramōda	10 Dhātṛi	...
4493	1314	1449	798	566-67	1391-92	5 Prajāpati	11 Īśvara	7 Āśvin
4494	1315	1450	799	567-68	*1392-93	6 Aṅgiras	12 Bahudhānya	...
4495	1316	1451	800	568-69	1393-94	7 Śrīmukha	13 Pramāthin	...
4496	1317	1452	801	569-70	1394-95	8 Bhāva	14 Vikrama	5 Śrāvaṇa

LXI—*Contd.*

COMMENCEMENT OF THE									
SOLAR YEAR.			LUNI-SOLAR YEAR (MEAN SUNRISE OF CIVIL DAY ON WHICH CHAITRA ŚUKLA 1 ENDS).						Kali.
Day and month, A.D.	Week-day.	Time of true Mēsha-samkrānti.	Day and month, A.D.	Week-day.	*a.*	*b.*	*c.*		
13	14	17	19	20	23	24	25		1
		H. M. S.							
26 Mar. (85)	3 Tues.	9 55 0	27 Feb. (58)	4 Wed.	40·3017	936·4794	202·4022		4472
26 Mar. (85)	4 Wed.	16 7 30	18 Mar. (77)	3 Tues.	74·9414	872·4630	253·7119		4473
25 Mar. (85)	5 Thur.	22 20 0	7 Mar. (67)	1 Sun.	289·2566	755·9894	225·6258		4474
26 Mar. (85)	0 Sat.	4 32 30	25 Mar. (84)	6 Fri.	9985·2644	655·6813	274·1977		4475
26 Mar. (85)	1 Sun.	10 45 0	15 Mar. (74)	4 Wed.	199·5796	539·2077	246·1117		4476
26 Mar. (85)	2 Mon.	16 57 30	4 Mar. (63)	1 Sun.	75·2629	386·4425	215·2878		4477
25 Mar. (85)	3 Tues.	23 10 0	21 Mar. (81)	6 Fri.	9771·2707	286·1344	263·8598		4478
26 Mar. (85)	5 Thur.	5 22 30	11 Mar. (70)	4 Wed.	9985·5859	169·6608	235·7737		4479
26 Mar. (85)	6 Fri.	11 35 0	28 Feb. (59)	1 Sun.	9861·2694	16·8957	204·9499		4480
26 Mar. (85)	0 Sat.	17 47 30	19 Mar. (78)	0 Sat.	9895·9080	952·8791	256·2595		4481
26 Mar. (86)	2 Mon.	0 0 0	8 Mar. (68)	5 Thur.	110·2242	836·4055	228·1735		4482
26 Mar. (85)	3 Tues.	6 12 30	25 Feb. (56)	2 Mon.	9985·9076	683·6404	197·6414		4483
26 Mar. (85)	4 Wed.	12 25 0	16 Mar. (75)	1 Sun.	20·5472	619·6238	248·6594		4484
26 Mar. (85)	5 Thur.	18 37 30	5 Mar. (64)	5 Thur.	9896·2306	466·8587	217·8355		4485
26 Mar. (86)	0 Sat.	0 50 0	23 Mar. (83)	4 Wed.	9930·8702	402·8422	269·1452		4486
26 Mar. (85)	1 Sun.	7 2 30	12 Mar. (71)	1 Sun.	9806·5536	250·0770	238·3213		4487
26 Mar. (85)	2 Mon.	13 15 0	2 Mar. (61)	6 Fri.	20·8689	133·6034	210·2353		4488
26 Mar. (85)	3 Tues.	19 27 30	21 Mar. (80)	5 Thur.	55·5085	69·5869	261·5430		4489
26 Mar. (86)	5 Thur.	1 40 0	9 Mar. (69)	2 Mon.	9931·1919	916·8218	230·7212		4490
26 Mar. (85)	6 Fri.	7 52 30	27 Feb. (58)	0 Sat.	145·5071	800·3481	202·6351		4491
26 Mar. (85)	0 Sat.	14 5 0	18 Mar. (77)	6 Fri.	180·1467	736·0401	251·2070		4492
26 Mar. (85)	1 Sun.	20 17 30	7 Mar. (66)	3. Tues.	55·8301	583·5665	223·1209		4493
26 Mar. (86)	3 Tues.	2 30 0	25 Mar. (85)	2 Mon.	90·4698	519·5501	274·4306		4494
26 Mar. (85)	4 Wed.	8 42 30	14 Mar. (73)	6 Fri.	9966·1531	366·7848	243·6068		4495
26 Mar. (85)	5 Thur.	14 55 0	3 Mar. (62)	3 Tues.	9841·8365	214·0196	212·7829		4496

TABLE

CONCURRENT YEAR.								Intercalated (*adhika*) and suppressed (*kshaya*) Lunar months (true).
Kali.	Śaka.	Chaitrādi Vikrama.	Mêshâdi solar year in Bengal.	Kollam.	A.D.	Jovian Saṁvatsara. Southern system.	Jovian Saṁvatsara. Northern system.	
1	2	3	3a	4	5	6	7	8
4497	1318	1453	802	570-71	1395-96	9 Yuvan . .	15 Vṛisha . .	...
4498	1319	1454	803	571-72	*1396-97	10 Dhātṛi . .	16 Chitrabhānu .	...
4499	1320	1455	804	572-73	1397-98	11 Īsvara . .	17 Subhānu .	3 Jyēshṭha .
4500	1321	1456	805	573-74	1398-99	12 Bahudhānya .	18 Tāraṇa . .	...
4501	1322	1457	806	574-75	1399-1400	13 Pramāthin .	19 Pārthiva	{ 8 Kārttika 10 *Pausha* (*ksh.*) }
4502	1323	1458	807	575-76	*1400-01	14 Vikrama .	20 Vyaya . .	1 Chaitra .
4503	1324	1459	808	576-77	1401-02	15 Vṛisha . .	21 Sarvajit	...
4504	1325	1460	809	577-78	1402-03	16 Chitrabhānu .	22 Sarvadhārin .	6 Bhādrapada
4505	1326	1461	810	578-79	1403-04	17 Subhānu .	23 Virōdhin .	...
4506	1327	1462	811	579-80	*1404-05	18 Tāraṇa . .	24 Vikṛita . .	...
4507	1328	1463	812	580-81	1405-06	19 Pārthiva .	25 Khara . .	4 Āshāḍha .
4508	1329	1464	813	581-82	1406-07	20 Vyaya . .	26 Nandana .	...
4509	1330	1465	814	582-83	1407-08	21 Sarvajit .	27 Vijaya . .	...
4510	1331	1466	815	583-84	*1408-09	22 Sarvadhārin .	28 Jaya . .	3 Jyēshṭha .
4511	1332	1467	816	584-85	1409-10	23 Virōdhin .	29 Manmatha .	...
4512	1333	1468	817	585-86	1410-11	24 Vikṛita . .	30 Durmukha .	7 Āśvina .
4513	1334	1469	818	586-87	1411-12	25 Khara . .	31 Hēmalamba .	...
4514	1335	1470	819	587-88	*1412-13	26 Nandana .	32 Vilamba .	...
4515	1336	1471	820	588-89	1413-14	27 Vijaya . .	33 Vikārin . .	4 Āshāḍha .
4516	1337	1472	821	589-90	1414-15	28 Jaya . .	34 Śārvarin .	...
4517	1338	1473	822	590-91	1415-16	29 Manmatha .	35 Plava . .	...
4518	1339	1474	823	591-92	*1416-17	30 Durmukha .	36 Śubhakṛit† .	3 Jyēshṭha .
4519	1340	1475	824	592-93	1417-18	31 Hēmalamba .	38 *Krōdhin* .	...
4520	1341	1476	825	593-94	1418-19	32 Vilamba .	39 *Viśvāvasu*	{ 8 Kārttika 11 *Māgha* (*ksh.*) 12 Phālguna }
4521	1342	1477	826	594-95	1419-20	33 Vikārin .	40 *Parābhava* .	...

† 37 Śōbhana was suppressed in the north.

LXI—*Contd.*

COMMENCEMENT OF THE											
Solar year.					Luni-solar year (mean sunrise of civil day on which Chaitra śukla 1 ends).					Kali.
Day and month, A.D.	Week-day.	Time of true Mēsha-saṁkrānti.			Day and month, A.D.	Week-day.	*a.*	*b.*	*c.*	
13	14	17			19	20	23	24	25	1
		H.	M.	S.						
26 Mar. (85)	6 Fri. .	21	7	30	22 Mar. (81)	2 Mon. .	9876·4762	150·0032	264·0927	4497
26 Mar. (86)	1 Sun. .	3	20	0	11 Mar. (71)	0 Sat. .	90·7914	33·5295	236·0066	4498
26 Mar. (85)	2 Mon. .	9	32	30	28 Feb. (59)	4 Wed.	9966·4748	880·7644	205·1827	4499
26 Mar. (85)	3 Tues.	15	45	0	19 Mar. (78)	3 Tues.	1·1144	816·7479	256·4924	4500
26 Mar. (85)	4 Wed.	21	57	30	9 Mar. (68)	1 Sun. .	215·4296	700·2743	228·4064	4501
26 Mar. (86)	6 Fri. .	4	10	0	26 Feb. (57)	5 Thur.	91·1130	547·5092	197·5825	4502
26 Mar. (85)	0 Sat. .	10	22	30	16 Mar. (75)	4 Wed.	125·7526	483·4926	248·8923	4503
26 Mar. (85)	1 Sun. .	16	35	0	5 Mar. (64)	1 Sun. .	1·4360	330·7275	218·0683	4504
26 Mar. (85)	2 Mon.	22	47	30	24 Mar. (83)	0 Sat. .	36·0756	266·7110	269·3781	4505
26 Mar. (86)	4 Wed.	5	0	0	12 Mar. (72)	4 Wed.	9911·7590	113·9457	238·5542	4506
26 Mar. (85)	5 Thur.	11	12	30	2 Mar. (61)	2 Mon. .	126·0743	997·4722	210·4682	4507
26 Mar. (85)	6 Fri. .	17	25	0	21 Mar. (80)	1 Sun. .	160·7139	933·4557	261·7779	4508
26 Mar. (85)	0 Sat. .	23	37	30	10 Mar. (69)	5 Thur.	36·3973	780·6906	230·9541	4509
26 Mar. (86)	2 Mon .	5	50	0	28 Feb. (59)	3 Tues.	250·7125	664·2169	202·8680	4510
26 Mar. (85)	3 Tues.	12	2	30	17 Mar. (76)	1 Sun. .	9946·7203	563·9089	251·4308	4511
26 Mar. (85)	4 Wed.	18	15	0	6 Mar. (65)	5 Thur.	9822·4037	411·1437	220·6160	4512
27 Mar. (86)	6 Fri. .	0	27	30	25 Mar. (84)	4 Wed.	9857·0433	347·1271	271·9257	4513
26 Mar. (86)	0 Sat. .	6	40	0	13 Mar. (73)	1 Sun. .	9732·7267	194·3620	241·1019	4514
26 Mar. (85)	1 Sun. .	12	52	30	3 Mar. (62)	6 Fri. .	9947·0419	77·8884	213·7161	4515
26 Mar. (85)	2 Mon. .	19	5	0	22 Mar. (81)	5 Thur.	9981·6815	13·8720	264·3256	4516
27 Mar. (86)	4 Wed.	1	17	30	12 Mar. (71)	3 Tues.	195·9968	897·3983	236·2394	4517
26 Mar. (86)	5 Thur.	7	30	0	29 Feb. (66)	0 Sat. .	71·6802	744·6332	205·4156	4518
26 Mar. (85)	5 Fri. .	13	42	30	19 Mar. (78)	6 Fri. .	106·3197	680·6167	256·7253	4519
26 Mar. (85)	0 Sat. .	19	55	0	8 Mar. (67)	3 Tues.	9982·0031	527·8514	225·9015	4520
27 Mar. (86)	2 Mon. .	2	7	30	27 Mar. (86)	2 Mon.	16·6427	363·8350	277·2112	4521

TABLE

CONCURRENT YEAR.								INTERCALATED (*adhika*) and SUPPRESSED (*kshaya*) LUNAR MONTHS (true).
Kali.	Śaka.	Chaitrādi Vikrama.	Mēshādi solar year in Bengal.	Kollam.	A.D.	JOVIAN SAMVATSARA. Southern system.	JOVIAN SAMVATSARA. Northern system.	
1	2	3	3*a*	4	5	6	7	8
4522	1343	1478	827	595-96	*1420-21	34 Śārvarin .	41 *Plavaṅga* .	...
4523	1344	1479	828	596-97	1421-22	35 Plava . .	42 Kīlaka . .	5 Śrāvaṇa .
4524	1345	1480	829	597-98	1422-23	36 Śubhakṛit .	43 Saumya. .	...
4525	1346	1481	830	598-99	1423-24	37 Śōbhana .	44 Sādhāraṇa .	...
4526	1347	1482	831	599-600	*1424-25	38 Krōdhin .	45 Virōdhakṛit .	4 Āshāḍha .
4527	1348	1483	832	600-01	1425-26	39 Viśvāvasu .	46 Paridhāvin .	...
4528	1349	1484	833	601-02	1426-27	40 Parābhava .	47 Pramādin .	...
4529	1350	1485	834	602-03	1427-28	41 Plavaṅga .	48 Ānanda .	2 Vaiśākha .
4530	1351	1486	835	603-04	*1428-29	42 Kīlaka .	49 Rākshasa .	...
4531	1352	1487	836	604-05	1429-30	43 Saumya .	50 Anala . .	6 Bhādrapada
4532	1353	1488	837	605-06	1430-31	44 Sādhāraṇa .	51 Piṅgala .	...
4533	1354	1489	838	606-07	1431-32	45 Virōdhakṛit .	52 Kālayukta .	...
4534	1355	1490	839	607-08	*1432-33	46 Paridhāvin .	53 Siddhārthin .	4 Āshāḍha
4525	1356	1491	840	608-09	1433-34	47 Pramādin .	54 Raudra .	...
4536	1357	1492	841	609-10	1434-35	48 Ānanda .	55 Durmati .	...
4537	1358	1493	842	610-11	1435-36	49 Rākshasa .	56 Dundubhi .	3 Jyēshṭha .
4538	1359	1494	843	611-12	*1436-37	50 Anala . .	57 Rudhirōdgārin	
4539	1360	1495	844	612-13	1437-38	51 Piṅgala .	58 Raktāksha .	3 Kārttika .
4540	1361	1496	845	613-14	1438-39	52 Kālayukta .	59 Krōdhana .	...
4541	1362	1497	846	614-15	1439-40	53 Siddhārthin .	60 Kshaya .	...
4542	1363	1498	847	615-16	*1440-41	54 Raudra .	1 Prabhava .	5 Śrāvaṇa .
4543	1364	1499	848	616-17	1441-42	55 Durmati .	2 Vibhava .	...
4544	1365	1500	849	617-18	1442-43	56 Dundubhi .	3 Śukla . .	...
4545	1366	1501	850	618-19	1443-44	57 Rudhirōdgārin	4 Pramōda .	4 Āshāḍha .
4546	1367	1502	851	619-20	*1444-45	58 Raktāksha .	5 Prajāpati .	...

LXI—*Contd.*

COMMENCEMENT OF THE									
SOLAR YEAR.			LUNI-SOLAR YEAR (MEAN SUNRISE OF CIVIL DAY ON WHICH CHAITRA ŚUKLA 1 ENDS).						*Kali.*
Day and month, A.D.	Week-day.	Time of true Mēsha-samkrānti.	Day and month, A.D.	Week-day.	*a.*	*b.*	*c.*		
13	14	17	19	20	23	24	25		1
		H. M. S.							
26 Mar. (86)	3 Tues.	8 20 0	15 Mar. (75)	6 Fri.	9892·3261	311·0698	246·3894		4522
26 Mar. (85)	4 Wed.	14 32 30	4 Mar. (63)	3 Tues.	9768·0095	157·3046	215·5634		4523
26 Mar. (85)	5 Thur.	20 45 0	23 Mar. (82)	2 Mon.	9802·6491	94·2881	266·8732		4524
27 Mar. (86)	0 Sat.	2 57 30	13 Mar. (72)	0 Sat.	16·9644	977·8145	238·7871		4525
26 Mar. (86)	1 Sun.	9 10 0	2 Mar. (62)	5 Thur.	231·2797	861·3410	210·7011		4526
26 Mar. (85)	2 Mon.	15 22 30	21 Mar. (80)	4 Wed.	265·9193	796·3241	262·0208		4527
26 Mar. (85)	3 Tues.	21 35 0	10 Mar. (69)	1 Sun.	141·6027	644·5593	231·1870		4528
27 Mar. (86)	5 Thur.	3 47 30	27 Feb. (58)	5 Thur.	17·2860	491·7941	200·3631		4529
26 Mar. (86)	6 Fri.	10 0 0	17 Mar. (77)	4 Wed.	51·9257	427·7776	251·6727		4530
26 Mar. (85)	0 Sat.	16 12 30	6 Mar. (65)	1 Sun.	9927·6091	275·0124	220·8489		4531
26 Mar. (85)	1 Sun.	22 25 0	25 Mar. (84)	0 Sat.	9962·2487	210·9959	272·1586		4532
27 Mar. (86)	3 Tues.	4 37 30	14 Mar. (73)	4 Wed.	9837·1321	58·2307	241·3348		4533
26 Mar. (86)	4 Wed.	10 50 0	3 Mar. (63)	2 Mon.	52·2473	941·7571	213·2487		4534
26 Mar. (85)	5 Thur.	17 2 30	22 Mar. (81)	1 Sun.	86·8870	877·7407	264·5585		4535
26 Mar. (85)	6 Fri.	23 15 0	12 Mar. (71)	6 Fri.	301·2022	761·2671	236·4723		4536
27 Mar. (86)	1 Sun.	5 27 30	1 Mar. (60)	3 Tues.	176·8856	608·5019	205·6485		4537
26 Mar. (86)	2 Mon.	11 40 0	18 Mar. (78)	1 Sun.	9872·8933	508·1938	254·2204		4538
26 Mar. (85)	3 Tues.	17 52 30	8 Mar. (67)	6 Fri.	87·2086	391·7202	226·1344		4539
27 Mar. (86)	5 Thur.	0 5 0	26 Mar. (85)	4 Wed.	9783·2164	291·4121	274·7063		4540
27 Mar. (86)	6 Fri.	6 17 30	16 Mar. (75)	2 Mon.	9997·5316	174·9385	246·6203		4541
26 Mar. (86)	0 Sat.	12 30 0	4 Mar. (64)	6 Fri.	9873·2150	22·1734	216·7964		4542
26 Mar. (85)	1 Sun.	18 42 30	23 Mar. (82)	5 Thur.	9907·8546	958·1569	267·[illegible]061		4543
27 Mar. (86)	3 Tues.	0 55 0	13 Mar. (72)	3 Tues.	122·4699	841·6932	239·0200		4544
27 Mar. (86)	4 Wed.	7 7 30	2 Mar. (61)	0 Sat.	9997·8533	688·9181	208·1962		4545
26 Mar. (86)	5 Thur.	13 20 0	20 Mar. (80)	6 Fri.	32·4928	624·9016	259·5059		4546

TABLE

CONCURRENT YEAR.								
Kali.	Śaka.	Chaitrādi Vikrama.	Mēshādi solar year in Bengal.	Kollam.	A.D.	Jovian Samvatsara. Southern system.	Jovian Samvatsara. Northern system.	Intercalated (*adhika*) and suppressed (*kshaya*) Lunar months (true).
1	2	3	3a	4	5	6	7	8
4547	1368	1503	852	620-21	1445-46	59 Krodhana	6 Aṅgiras	...
4548	1369	1504	853	621-22	1446-47	60 Kshaya	7 Śrīmukha	2 Vaiśākha
4549	1370	1505	854	622-23	1447-48	1 Prabhava	8 Bhāva	...
4550	1371	1506	855	623-24	*1448-49	2 Vibhava	9 Yuvan	6 Bhādrapada
4551	1372	1507	856	624-25	1449-50	3 Śukla	10 Dhātṛi	...
4552	1373	1508	857	625-26	1450-51	4 Pramōda	11 Īśvara	...
4553	1374	1509	858	626-27	1451-52	5 Prajāpati	12 Bahudhānya	4 Āshāḍha
4554	1375	1510	859	627-28	*1452-53	6 Aṅgiras	13 Pramāthin	...
4555	1376	1511	860	628-29	1453-54	7 Śrīmukha	14 Vikrama	...
4556	1377	1512	861	629-30	1454-55	8 Bhāva	15 Vṛisha	3 Jyēshṭha
4557	1378	1513	862	630-31	1455-56	9 Yuvan	16 Chitrabhānu	...
4558	1379	1514	863	631-32	*1456-57	10 Dhātṛi	17 Subhānu	8 Kārttika 10 *Pausha* (*ksh.*) 12 Phālguna
4559	1380	1515	864	632-33	1457-58	11 Īśvara	18 Tāraṇa	...
4560	1381	1516	865	633-34	1458-59	12 Bahudhānya	19 Pārthiva	...
4561	1382	1517	866	634-35	1459-60	13 Pramāthin	20 Vyaya	5 Śrāvaṇa
4562	1383	1518	867	635-36	*1460-61	14 Vikrama	21 Sarvajit	...
4563	1384	1519	868	636-37	1461-62	15 Vṛisha	22 Sarvadhārin	...
4564	1385	1520	869	637-38	1462-63	16 Chitrabhānu	23 Virōdhin	4 Āshāḍha
4565	1386	1521	870	638-39	1463-64	17 Subhānu	24 Vikṛita	...
4566	1387	1522	871	639-40	*1464-65	18 Tāraṇa	25 Khara	...
4567	1388	1523	872	640-41	1465-66	19 Pārthiva	26 Nandana	2 Vaiśākha
4568	1389	1524	873	641-42	1466-67	20 Vyaya	27 Vijaya	...
4569	1390	1525	874	642-43	1467-68	21 Sarvajit	28 Jaya	6 Bhādrapada
4570	1391	1526	875	643-44	*1468-69	22 Sarvadhārin	29 Manmatha	...
4571	1392	1527	876	644-45	1469-70	23 Virōdhin	30 Durmukha	...

LXI—*Contd.*

COMMENCEMENT OF THE											Kali.
SOLAR YEAR.					LUNI-SOLAR YEAR (MEAN SUNRISE OF CIVIL DAY ON WHICH CHAITRA ŚUKLA 1 ENDS).						
Day and month, A.D.	Week-day.	Time of true Mēsha-saṁkrānti.			Day and month, A.D.	Week-day.	*a.*	*b.*	*c.*		
13	14	17			19	20	23	24	25		1
		H.	M.	S.							
26 Mar. (85)	6 Fri.	19	32	30	9 Mar. (68)	3 Tues.	9908·1762	472·1363	228·6821		4547
27 Mar. (86)	1 Sun.	1	45	0	26 Feb. (57)	0 Sat.	9784·8596	319·3712	197·8582		4548
27 Mar. (86)	2 Mon.	7	57	30	17 Mar. (76)	6 Fri.	9818·4993	255·3547	249·1679		4549
26 Mar. (86)	3 Tues.	14	10	0	6 Mar. (66)	4 Wed.	32·8145	138·8812	22[illegible]·0818		4550
26 Mar. (85)	4 Wed.	20	22	30	25 Mar. (84)	3 Tues.	67·4541	74·8646	272·3915		4551
27 Mar. (86)	6 Fri.	2	35	0	14 Mar. (73)	0 Sat.	9943·1375	922·0995	241·5677		4552
27 Mar. (86)	0 Sat.	8	47	30	4 Mar. (63)	5 Thur.	157·4527	805·6259	213·4816		4553
26 Mar. (86)	1 Sun	15	0	0	22 Mar. (82)	4 Wed.	192·0924	741·6094	264·7914		4554
26 Mar. (85)	2 Mon	21	12	30	11 Mar. (70)	1 Sun.	67·7757	588·8442	233·9674		4555
27 Mar. (86)	4 Wed.	3	25	0	28 Feb (59)	5 Thur.	9943·4591	436·0790	203·1436		4556
27 Mar. (86)	5 Thur.	9	37	30	19 Mar. (78)	4 Wed.	9978·0987	372·0625	254·4533		4557
26 Mar. (86)	6 Fri.	15	50	0	7 Mar. ([illegible]7)	1 Sun.	9853·7821	219·2973	223·6295		4558
26 Mar. (85)	0 Sat.	22	2	30	25 Mar. (85)	0 Sat.	9888·4218	155·2809	274·9392		4559
27 Mar. (86)	2 Mon.	4	15	0	16 Mar. (75)	5 Thur.	102·7370	38·8073	246·8532		4560
27 Mar. (86)	3 Tues.	10	27	30	5 Mar. (64)	2 Mon.	9978·4204	885·0421	216·0293		4561
26 Mar. (86)	4 Wed.	16	40	0	23 Mar. (83)	1 Sun.	13·0600	822·0256	267·3390		4562
26 Mar. (85)	5 Thur.	22	52	30	13 Mar. (72)	6 Fri.	227·3753	705·5520	239·2529		4563
27 Mar. (86)	0 Sat.	5	5	0	2 Mar. (61)	3 Tues.	103·0587	552·7868	208·4291		4564
27 Mar. (86)	1 Sun.	11	17	30	21 Mar. (80)	2 Mon.	137·6983	488·7703	259·7388		4565
26 Mar. (86)	2 Mon.	17	30	0	9 Mar. (69)	6 Fri.	13·3817	336·0051	228·9150		4566
26 Mar. (85)	3 Tues.	23	42	30	26 Feb. (57)	3 Tues.	9889·0651	183·2400	198·6911		4567
27 Mar. (86)	5 Thur.	5	55	0	17 Mar. (76)	2 Mon.	9923·7047	119·2214	249·4608		4568
27 Mar. (86)	6 Fri.	12	7	30	7 Mar. (66)	0 Sat.	138·0199	2·7499	221·3147		4569
26 Mar. (86)	0 Sat.	18	20	0	25 Mar. (85)	6 Fri.	172·6596	938·7334	272·6244		4570
27 Mar. (86)	2 Mon.	0	32	30	14 Mar. (73)	3 Tues.	48·3430	785·9682	241·8066		4571

TABLE

CONCURRENT YEAR.								INTERCALATED (*adhika*) and SUPPRESSED (*kshaya*) LUNAR MONTHS (true)
						JOVIAN SAMVATSARA.		
Kali.	Śaka.	Chaitrādi Vikrama.	Mēshādi solar year in Bengal.	Kollam.	A.D.	Southern system.	Northern system.	
1	2	3	3*a*	4	5	6	7	8
4572	1393	1528	877	645-46	1470-71	24 Vikṛita . .	31 Hēmalamba .	4 Āshāḍha .
4573	1394	1529	878	646-47	1471-72	25 Khara . .	32 Vilamba .	...
4574	1395	1530	879	647-48	*1472-73	26 Nandana .	33 Vikārin . .	...
4575	1396	1531	880	648-49	1473-74	27 Vijaya . .	34 Śārvarin .	3 Jyēshṭha .
4576	1397	1532	881	649-50	1474-75	28 Jaya . .	35 Plava . .	...
4577	1398	1533	882	650-51	1475-76	29 Manmatha .	36 Śubhakṛit .	7 Āśvina 10 *Pausha* (*ksh.*) 12 Phālguna
4578	1399	1534	883	651-52	*1476-77	30 Durmukha .	37 Śōbhana .	...
4579	1400	1535	884	652-53	1477-78	31 Hēmalamba .	38 Krōdhin .	...
4580	1401	1536	885	653-54	1478-79	32 Vilamba .	39 Viśvāvasu .	5 Śrāvaṇa .
4581	1402	1537	886	654-55	1479-80	33 Vikārin . .	40 Parābhava .	...
4582	1403	1538	887	655-56	*1480-81	34 Śārvarin .	41 Plavaṅga .	...
4583	1404	1539	888	656-57	1481-82	35 Plava . .	42 Kīlaka . .	4 Āshāḍha .
4584	1405	1540	889	657-58	1482-83	36 Śubhakṛit .	43 Saumya .	...
4585	1406	1541	890	658-59	1483-84	37 Śōbhana .	44 Sādhāraṇa .	...
4586	1407	1542	891	659-60	*1484-85	38 Krōdhin .	45 Virōdhakṛit .	1 Chaitra .
4587	1408	1543	892	660-61	1485-86	39 Viśvāvasu .	46 Paridhāvin .	...
4588	1409	1544	893	661-62	1486-87	40 Parābhava .	47 Pramādin .	6 Bhādrapada
4589	1410	1545	894	662-63	1487-88	41 Plavaṅga .	48 Ānanda .	...
4590	1411	1546	895	663-64	*1488-89	42 Kīlaka . .	49 Rākshasa .	...
4591	1412	1547	896	664-65	1489-90	43 Saumya .	50 Anala . .	4 Āshāḍha .
4592	1413	1548	897	665-66	1490-91	44 Sādhāraṇa .	51 Piṅgala .	...
4593	1414	1549	898	666-67	1491-92	45 Virōdhakṛit .	52 Kālayukta	...
4594	1415	1550	899	667-68	*1492-93	46 Paridhāvin .	53 Siddhārthin .	2 Vaiśākha .
4595	1416	1551	900	668-69	1493-94	47 Pramādin .	54 Raudra .	...
96	1417	1552	901	669-70	1494-95	48 Ānanda .	55 Durmati .	6 Bhādrapada

LXI—*Contd.*

COMMENCEMENT OF THE									
SOLAR YEAR.			LUNI-SOLAR YEAR (MEAN SUNRISE OF CIVIL DAY ON WHICH CHAITRA ŚUKLA 1 ENDS).						Kali.
Day and month, A.D.	Week-day.	Time of true Mēsha-saṁkrānti.	Day and month, A.D.	Week-day.	*a.*	*b.*	*c*		
13	14	17	19	20	23	24	25		1
		H. M. S.							
27 Mar. (86)	3 Tues.	6 45 0	4 Mar. (63)	1 Sun. .	262·6582	669·4946	213·7145		4572
27 Mar. (86)	4 Wed.	12 57 30	22 Mar. (81)	6 Fri. .	9958·6660	569·1865	262·2865		4573
26 Mar. (86)	5 Thur.	19 10 0	10 Mar. (70)	3 Tues.	9838·3494	416·4214	231·4626		4574
27 Mar. (86)	0 Sat. .	1 22 30	28 Feb. (59)	1 Sun. .	48·6646	299·9477	203·3765		4575
27 Mar. (86)	1 Sun. .	7 35 0	18 Mar. (77)	6 Fri. .	9744·6724	199·6397	251·9484		4576
27 Mar. (86)	2 Mon. .	13 47 30	8 Mar. (67)	4 Wed.	9958·9875	83·1661	223·8624		4577
26 Mar. (86)	3 Tues.	20 0 0	26 Mar. (86)	3 Tues.	9993·6272	19·1496	275·1721		4578
27 Mar. (86)	5 Thur.	2 12 30	16 Mar. (75)	1 Sun. .	207·9424	902·6760	247·0861		4579
27 Mar. (86)	6 Fri. .	8 25 0	5 Mar. (64)	5 Thur.	83·6259	749·9109	216·2622		4580
27 Mar. (86)	0 Sat. .	14 37 30	24 Mar. (83)	4 Wed.	118·2654	685·8943	267·5720		4581
26 Mar. (86)	1 Sun. .	20 50 0	12 Mar. (72)	1 Sun. .	9993·9488	533·1291	236·7480		4582
27 Mar. (86)	3 Tues.	3 2 30	1 Mar. (60)	5 Thur.	9869·6322	380·3640	205·9242		4583
27 Mar. (86)	4 Wed.	9 15 0	20 Mar. (79)	4 Wed.	9904·2718	316·3474	257·2339		4584
27 Mar. (86)	5 Thur.	15 27 30	9 Mar. (68)	1 Sun. .	9779·9552	163·5822	226·4101		4585
26 Mar. (86)	6 Fri. .	21 40 0	27 Feb. (58)	6 Fri. .	9994·2705	17·1087	198·3239		4586
27 Mar. (86)	1 Sun. .	3 52 30	17 Mar. (76)	5 Thur.	28·9101	983·0922	249·6337		4587
27 Mar. (86)	2 Mon. .	10 5 0	7 Mar. (66)	3 Tues.	243·2253	866·6186	221·5476		4588
27 Mar. (86)	3 Tues.	16 17 30	26 Mar. (85)	2 Mon. .	277·8650	802·6021	272·8573		4589
26 Mar. (86)	4 Wed.	22 30 0	14 Mar. (74)	6 Fri. .	153·5484	649·8370	242·0335		4590
27 Mar. (86)	6 Fri. .	4 42 30	3 Mar. (62)	3 Tues.	29·2318	497·0717	211·2097		4591
27 Mar. (86)	0 Sat. .	10 55 0	22 Mar. (81)	2 Mon. .	63·8714	433·0553	262·5194		4592
27 Mar. (86)	1 Sun. .	17 7 30	11 Mar. (70)	6 Fri. .	9939·5548	280·2901	231·6955		4593
26 Mar. (86)	2 Mon. .	23 20 0	28 Feb. (59)	3 Tues.	9815·2381	127·5249	200·8716		4594
27 Mar. (86)	4 Wed.	5 32 30	18 Mar. (77)	2 Mon. .	9849·8778	63·5684	252·1813		4595
27 Mar. (86)	5 Thur.	1 45 0	8 Mar. (67)	0 Sat. .	64·1930	247·0348	224·095		4596

TABLE

CONCURRENT YEAR.								INTERCALATED (*adhika*) and SUPPRESSED (*kshaya*) LUNAR MONTHS (true).
Kali.	Śaka.	Chaitrādi Vikrama.	Mēshādi solar year in Bengal.	Kollam.	A.D.	Jovian Saṁvatsara. Southern system.	Jovian Saṁvatsara. Northern system.	
1	2	3	3*a*	4	5	6	7	8
4597	1418	1553	902	670-71	1495-96	49 Rākshas .	56 Dundubhi .	...
4598	1419	1554	903	671-72	*1496-97	50 Anala . .	57 Rudhirōdgārin	...
4599	1420	1555	904	672-73	1497-98	51 Piṅgala .	58 Raktāksha .	5 Śrāvaṇa
4600	1421	1556	905	673-74	1498-99	52 Kālayukta .	59 Krōdhana .	...
4601	1422	1557	906	674-75	1499-1500	53 Siddhārthin .	60 Kshaya .	...
4602	1423	1558	907	675-76	*1500-01	54 Raudra .	1 Prabhava .	3 Jyēshṭha
4603	1424	1559	908	676-77	1501-02	55 Durmati .	2 Vibhava† .	...
4604	1425	1560	909	677-78	1502-03	56 Dundubhi .	4 *Pramōda* .	...
4605	1426	1561	910	678-79	1503-04	57 Rudhirōdgārin	5 *Prajāpati* .	2 Vaiśākha .
4606	1427	1562	911	679-80	*1504-05	58 Raktāksha .	6 *Aṅgiras* .	...
4607	1428	1563	912	680-81	1505-06	59 Krōdhana .	7 *Śrīmukha* .	6 Bhādrapada
4608	1429	1564	913	681-82	1506-07	60 Kshaya .	8 *Bhāva* . .	...
4609	1430	1565	914	682-83	1507-08	1 Prabhava .	9 *Yuvan* . .	...
4610	1431	1566	915	683-84	*1508-09	2 Vibhava .	10 *Dhātṛi* . .	4 Āshāḍha .
4611	1432	1567	916	684-85	1509-10	3 Śukla . .	11 *Īśvara* . .	...
4612	1433	1568	917	685-86	1510-11	4 Pramōda .	12 *Bahudhānya* .	...
4613	1434	1569	918	686-87	1511-12	5 Prajāpati .	13 *Pramāthin* .	2 Vaiśākha .
4614	1435	1570	919	687-88	*1512-13	6 Aṅgiras .	14 *Vikrama* .	...
4615	1436	1571	920	688-89	1513-14	7 Śrīmukha .	15 *Vṛisha* . .	6 Bhādrapada
4616	1437	1572	921	689-90	1514-15	8 Bhāva . .	16 *Chitrabhānu* .	...
4617	1438	1573	922	690-91	1515-16	9 Yuvan . .	17 Subhānu .	...
4618	1439	1574	923	691-92	*1516-17	10 Dhātri . .	18 Tāraṇa . .	5 Śrāvaṇa .
4619	1440	1575	924	692-93	1517-18	11 Īśvara . .	19 Pārthiva .	...
4620	1441	1576	925	693-94	1518-19	12 Bahudhānya .	20 Vyaya . .	...
4621	1442	1577	926	694-95	1519-20	13 Pramāthin .	21 Sarvajit .	3 Jyēshṭha .

† No. 3 Śukla was suppressed in the north.

LXI—*Contd.*

COMMENCEMENT OF THE									Kali.
SOLAR YEAR.			LUNI-SOLAR YEAR (MEAN SUNRISE OF CIVIL DAY ON WHICH CHAITRA ŚUKLA 1 ENDS).						
Day and month, A.D.	Week-day.	Time of true Mêsha-saṁkrânti.	Day and month, A.D.	Week-day.	*a.*	*b.*	*c.*		
13	14	17	19	20	23	24	25		1
		H. M. S.							
27 Mar. (86)	6 Fri. .	17 57 30	27 Mar. (86)	6 Fri. .	98·8327	883·0184	275·4050		4597
27 Mar. (87)	1 Sun. .	0 10 0	16 Mar. (76)	4 Wed. .	313·1479	766·5447	247·319[illegible]		4598
27 Mar. (86)	2 Mon. .	6 22 30	5 Mar. (64)	1 Sun. .	188·8313	613·7796	216·4950		4599
27 Mar. (86)	3 Tues. .	12 35 0	23 Mar. (82)	6 Fri. .	9884·8390	513·4715	265·0670		4600
27 Mar. (86)	4 Wed. .	18 47 30	12 Mar. (71)	3 Tues. .	9760·5224	360·7063	234·2431		4601
27 Mar. (87)	6 Fri. .	1 0 0	1 Mar. (61)	1 Sun. .	9974·8377	244·2328	206·1571		4602
27 Mar. (86)	0 Sat. .	7 12 30	20 Mar. (79)	0 Sat. .	9·4773	180·2162	257·4668		4603
27 Mar. (86)	1 Sun. .	13 25 0	9 Mar. (68)	4 Wed. .	9885·1607	27·4510	226·6429		4604
27 Mar. (86)	2 Mon. .	19 37 30	27 Feb. (58)	2 Mon. .	99·4760	910·9775	198·5568		4605
27 Mar. (87)	4 Wed. .	1 50 0	17 Mar. (77)	1 Sun. .	134·1156	846·9609	249·8666		4606
27 Mar. (86)	5 Thur. .	8 2 30	6 Mar. (65)	5 Thur. .	9·7990	694·1958	219·0427		4607
27 Mar. (86)	6 Fri. .	14 15 0	25 Mar. (84)	4 Wed. .	44·4386	630·1793	270·3525		4608
27 Mar. (86)	0 Sat. .	20 27 30	14 Mar. (73)	1 Sun. .	9920·1220	477·4141	239·5286		4609
27 Mar. (87)	2 Mon. .	2 40 0	2 Mar. (62)	5 Thur. .	9795·8054	324·6489	208·7048		4610
27 Mar. (86)	3 Tues. .	8 52 30	21 Mar. (80)	4 Wed. .	9830·4450	260·6324	260·0144		4611
27 Mar. (86)	4 Wed. .	15 5 0	11 Mar. (70)	2 Mon. .	44·7603	144·1589	231·9284		4612
27 Mar. (86)	5 Thur. .	21 17 30	28 Feb. (59)	6 Fri. .	9920·4426	991·3[illegible]36	201·1045		4613
27 Mar. (87)	0 Sat. .	3 30 0	18 Mar. (78)	5 Thur. .	9955·0933	927·3772	252·4142		4614
27 Mar. (86)	1 Sun. .	9 42 30	8 Mar. (67)	3 Tues. .	169·3984	810·9036	224·3282		4615
27 Mar (86)	2 Mon. .	15 55 0	27 Mar. (86)	2 Mon. .	202·0381	746·8872	275 6379		4616
27 Mar. (86)	3 Tues. .	22 7 30	16 Mar. (75)	6 Fri. .	79·7215	594·1219	244·8140		4617
27 Mar. (87)	5 Thur. .	4 20 0	4 Mar. (64)	3 Tues .	9955·4049	441·3567	213·9901		4618
27 Mar. (86)	6 Fri. .	10 32 30	23 Mar. (82)	2 Mon. .	9990·0445	377·3403	265·2999		4619
27 Mar. (86)	0 Sat. .	16 45 0	12 Mar. (71)	6 Fri.	9865·7278	224·5750	234·4760		4620
27 Mar. (86)	1 Sun.	22 57 30	2 Mar. (61)	4 Wed. .	80·0431	108·1015	206·3800		4621

TABLE

CONCURRENT YEAR.								INTERCALATED (*adhika*) and SUPPRESSED (*kshaya*) LUNAR MONTHS (true).
Kali.	Saka.	Chaitrādi Vikrama.	Mēshādi solar year in Bengal.	Kollam.	A.D.	JOVIAN SAMVATSARA. Southern system.	JOVIAN SAMVATSARA. Northern system.	
1	2	3	3*a*	4	5	6	7	8
4622	1443	1578	927	695-96	*1520-21	14 Vikrama	22 Sarvadhārin	...
4623	1444	1579	928	696-97	1521-22	15 Vṛisha	23 Virōdhin	...
4624	1445	1580	929	697-98	1522-23	16 Chitrabhānu	24 Vikṛita	2 Vaiśākha
4625	1446	1581	930	698-99	1523-24	17 Subhānu	25 Khara	...
4626	1447	1582	931	699-700	*1524-25	18 Tāraṇa	26 Nandana	6 Bhādrapada
4627	1448	1583	932	700-01	1525-26	19 Pārthiva	27 Vijaya	...
4628	1449	1584	933	701-02	1526-27	20 Vyaya	28 Jaya	...
4629	1450	1585	934	702-03	1527-28	21 Sarvajit	29 Manmatha	4 Āshāḍha
4630	1451	1586	935	703-04	*1528-29	22 Sarvadhārin	30 Durmukha	...
4631	1452	1587	936	704-05	1529-30	23 Virōdhin	31 Hēmalamba	...
4632	1453	1588	937	705-06	1530-31	24 Vikrita	32 Vilamba	2 Vaiśākha
4633	1454	1589	938	706-07	1531-32	25 Khara	33 Vikārin	...
4634	1455	1590	939	707-08	*1532-33	26 Nandana	34 Śārvarin	6 Bhādrapada
4635	1456	1591	940	708-09	1533-34	27 Vijaya	35 Plava	...
4636	1457	1592	941	709-10	1534-35	28 Jaya	36 Śubhakṛit	...
4637	1458	1593	942	710-11	1535-36	29 Manmatha	37 Śōbhana	5 Śrāvaṇa
4638	1459	1594	943	711-12	*1536-37	30 Durmukha	38 Krōdhin	...
4639	1460	1595	944	712-13	1537-38	31 Hēmalamba	39 Viśvāvasu	...
4640	1461	1596	945	713-14	1538-39	32 Vilamba	40 Parābhava	3 Jyēshṭha
4641	1462	1597	946	714-15	1539-40	33 Vikārin	41 Plavaṅga	...
4642	1463	1598	947	715-16	*1540-41	34 Śārvarin	42 Kīlaka	7 Āsvina† / 10 *Pausha* (*ksh.*)
4643	1464	1599	948	716-17	1541-42	35 Plava	43 Saumya	1 Chaitra
4644	1465	1600	949	717-18	1542-43	36 Śubhakṛit	44 Sādhāraṇa	...
4645	1466	1601	950	718-19	1543-44	37 Śōbhana	45 Virōdhakṛit	6 Bhādrapada
4646	1467	1602	951	719-20	*1544-45	38 Krōdhin	46 Paridhāvin	

† A close case. At the Tulā-saṁkrānti the moon had been waxing for less than 2 minutes.

LXI—*Contd.*

COMMENCEMENT OF THE									
SOLAR YEAR.			LUNI-SOLAR YEAR (MEAN SUNRISE OF CIVIL DAY ON WHICH CHAITRA ŚUKLA 1 ENDS).					Kali.
Day and month. A.D.	Week-day.	Time of true Mēsha-saṁkrānti.	Day and month, A.D.	Week-day.	*a.*	*b.*	*c.*	
13	14	17	19	20	23	24	25	1
		H. M. S.						
27 Mar. (87)	3 Tues.	5 10 0	20 Mar. (80)	3 Tues.	114·6827	44·0850	257·6997	4622
27 Mar. (86)	4 Wed.	11 22 30	9 Mar. (68)	0 Sat.	9990·3661	891·3198	226·8758	4623
27 Mar. (86)	5 Thur.	17 35 0	27 Feb. (58)	5 Thur.	204·6814	774·8462	198·7897	4624
27 Mar. (86)	6 Fri.	23 47 30	18 Mar. (77)	4 Wed.	239·3210	710·8297	250·0995	4625
27 Mar. (87)	1 Sun.	6 0 0	6 Mar. (66)	1 Sun.	115·0044	558·0646	219·2756	4626
27 Mar. (86)	2 Mon.	12 12 30	25 Mar. (84)	0 Sat.	149·6440	494·0480	270·5854	4627
27 Mar. (86)	3 Tues.	18 25 0	14 Mar. (73)	4 Wed.	25·3274	341·2828	239·7615	4628
28 Mar. (87)	5 Thur.	0 37 30	3 Mar. (62)	1 Sun.	9901·0108	188·5177	208·9577	4629
27 Mar. (87)	6 Fri.	6 50 0	21 Mar. (81)	0 Sat.	9935·6504	124·5011	160·2473	4630
27 Mar. (86)	0 Sat.	13 2 30	11 Mar. (70)	5 Thur.	149·9657	8·0276	232·1613	4631
27 Mar. (86)	1 Sun.	19 15 0	28 Feb. (59)	2 Mon.	25·6490	855·2624	201·3374	4632
28 Mar (87)	3 Tues.	1 27 30	19 Mar. (78)	1 Sun.	60·2887	791·2459	252·6471	4633
27 Mar. (87)	4 Wed.	7 40 0	8 Mar. (68)	6 Fri.	274·6009	674·7723	224·5641	4634
27 Mar. (86)	5 Thur.	13 52 30	26 Mar. (85)	4 Wed.	9970·6117	574·4642	273·1330	4635
27 Mar. (86)	6 Fri.	20 5 0	15 Mar. (74)	1 Sun.	9846·2851	421·6991	242·3091	4636
28 Mar. (87)	1 Sun.	2 17 30	4 Mar. (63)	5 Thur.	9721·9785	268·9338	211·4853	4637
27 Mar. (87)	2 Mon.	8 30 0	22 Mar. (82)	4 Wed.	9756·6181	204·9174	262·7950	4638
27 Mar. (86)	3 Tues.	14 42 30	12 Mar. (71)	2 Mon.	9970·9333	88·4438	234·7089	4639
27 Mar. (86)	4 Wed.	20 55 0	2 Mar. (61)	0 Sat.	185·2486	971·8702	206·6229	4640
28 Mar. (87)	6 Fri.	3 7 30	21 Mar. (80)	6 Fri.	219·8882	907·9537	257·9326	4641
27 Mar. (87)	0 Sat.	9 20 0	9 Mar. (69)	3 Tues.	95·5716	755·1885	227·1088	4642
27 Mar. (86)	1 Sun.	15 32 30	26 Feb. (57)	0 Sat.	9971·2550	602·4234	196·2848	4643
27 Mar. (86)	2 Mon.	21 45 0	17 Mar. (76)	6 Fri.	5·8946	538·406[illegible]	247·5946	4644
28 Mar. (87)	4 Wed.	3 57 30	6 Mar. (65)	3 Tues.	9881·5780	385·6417	216·7707	4645
27 Mar. (87)	5 Thur.	10 10 0	24 Mar. (84)	2 Mon.	9916·2175	521·62[illegible]	268·0805	4646

TABLE

CONCURRENT YEAR.								
						Jovian Saṁvatsara.		Intercalated (*adhika*) and suppressed (*kshaya*) Lunar months (true).
Kali.	Śaka.	Chaitrādi Vikrama.	Mēshādi solar year in Bengal.	Kollam.	A.D.	Southern system.	Northern system.	
1	2	3	3*a*	4	5	6	7	8
4647	1468	1603	952	720-21	1545-46	39 Viśvāvasu .	47 Pramādin .	...
4648	1469	1604	953	721-22	1546-47	40 Parābhava .	48 Ānanda .	4 Āshādha
4649	1470	1605	954	722-23	1547-48	41 Plavaṅga .	49 Rākshasa .	...
4650	1471	1606	955	723-24	*1548-49	42 Kīlaka . .	50 Anala .	...
4651	1472	1607	956	724-25	1549-50	43 Saumya .	51 Piṅgala .	2 Vaiśākha
4652	1473	1608	957	725-26	1550-51	44 Sādhāraṇa .	52 Kālayukta .	...
4653	1474	1609	958	726-27	1551-52	45 Virōdhakṛit .	53 Siddhārthin .	6 Bhādrapada
4654	1475	1610	959	727-28	*1552-53	46 Paridhāvin .	54 Raudra .	...
4655	1476	1611	960	728-29	1553-54	47 Pramādin .	55 Durmati .	...
4656	1477	1612	961	729-30	1554-55	48 Ānanda .	56 Dundubhi .	4 Āshāḍha
4657	1478	1613	962	730-31	1555-56	49 Rākshasa .	57 Rudhirōdgārin	...
4658	1479	1614	963	731-32	*1556-57	50 Anala . .	58 Raktāksha .	...
4659	1480	1615	964	732-33	1557-58	51 Piṅgala .	59 Krōdhana .	3 Jyēshṭha
4660	1481	1616	965	733-34	1558-59	52 Kālayukta .	60 Kshaya .	...
4661	1482	1617	966	734-35	1559-60	53 Siddhārthin .	1 Prabhava	8 Kārttika 11 *Māgha* (*ksh.*) 12 Phālguna
4662	1483	1618	967	735-36	*1560-61	54 Raudra .	2 Vibhava .	...
4663	1484	1619	968	736-37	1561-62	55 Durmati .	3 Śukla . .	...
4664	1485	1620	969	737-38	1562-63	56 Dundubhi .	4 Pramōda .	5 Śrāvaṇa .
4665	1486	1621	970	738-39	1563-64	57 Rudhirōdgārin	5 Prajāpati .	...
4666	1487	1622	971	739-40	*1564-65	58 Raktāksha .	6 Angiras .	...
4667	1488	1623	972	740-41	1565-66	59 Krōdhana .	7 Śrīmukha .	4 Āshāḍha .
4668	1489	1624	973	741-42	1566-67	60 Kshaya .	8 Bhāva .	...
4669	1490	1625	974	742-43	1567-68	1 Prabhava .	9 Yuvan . .	...
4670	1491	1626	975	743-44	*1568-69	2 Vibhava .	10 Dhātri . .	2 Vaiśākha .
4671	1492	1627	976	744-45	1569-70	3 Sukla . .	11 Īśvara . .	...

LXI—*Contd.*

COMMENCEMENT OF THE									Kali.
SOLAR YEAR.			LUNI-SOLAR YEAR (MEAN SUNRISE OF CIVIL DAY ON WHICH CHAITRA ŚUKLA 1 ENDS).						
Day and month, A.D.	Week-day.	Time of true Mēsha-saṁkrānti.	Day and month, A.D.	Week-day.	*a.*	*b.*	*c.*		
13	14	17	19	20	23	24	25		1
		H. M. S.							
27 Mar. (86)	6 Fri.	16 22 30	13 Mar. (72)	6 Fri.	9791·9009	168·8599	237·2566		4647
27 Mar. (86)	0 Sat.	22 35 0	3 Mar. (62)	4 Wed.	6·2162	52·3864	209·1706		4648
28 Mar. (87)	2 Mon.	4 47 30	22 Mar. (81)	3 Tues.	40·9559	988·3699	260·4802		4649
27 Mar. (87)	3 Tues.	11 0 0	11 Mar. (71)	1 Sun.	255·1711	871·8964	232·3942		4650
27 Mar. (86)	4 Wed.	17 12 30	28 Feb. (59)	5 Thur.	130·8544	719·1311	201·5703		4651
27 Mar. (86)	5 Thur.	23 25 0	19 Mar. (78)	4 Wed.	165·4941	655·1147	252·8800		4652
28 Mar. (87)	0 Sat.	5 37 30	8 Mar. (67)	1 Sun.	41·1774	502·3495	222·0562		4653
27 Mar. (87)	1 Sun.	11 50 0	26 Mar. (86)	0 Sat.	75·8171	438·3329	273·3659		4654
27 Mar. (86)	2 Mon.	18 2 30	15 Mar. (74)	4 Wed.	9952·5005	285·5678	242·5420		4655
28 Mar. (87)	4 Wed.	0 15 0	4 Mar. (63)	1 Sun.	9827·1839	132·8021	211·7182		4656
28 Mar. (87)	5 Thur.	6 27 30	23 Mar. (82)	0 Sat.	9861·8235	68·7856	263·0279		4657
27 Mar. (87)	6 Fri.	12 40 0	12 Mar. (72)	5 Thur.	76·1387	952·3120	234·9418		4658
27 Mar. (86)	0 Sat.	18 52 30	2 Mar. (61)	3 Tues.	290·4540	835·8385	206·8558		4659
28 Mar. (87)	2 Mon.	1 5 0	21 Mar. (80)	2 Mon.	325·0936	760·8220	258·1655		4660
28 Mar. (87)	3 Tues.	7 17 30	10 Mar. (69)	6 Fri.	200·7771	619·0567	227·3417		4661
27 Mar. (87)	4 Wed.	13 30 0	27 Mar. (87)	4 Wed.	9896·7848	518·7487	275·9135		4662
27 Mar. (86)	5 Thur.	19 42 30	16 Mar. (75)	1 Sun.	9772·4681	365·9835	245·0897		4663
28 Mar. (87)	0 Sat.	1 55 0	6 Mar. (65)	6 Fri.	9986·7834	249·5104	217·0033		4664
28 Mar. (87)	1 Sun.	8 7 30	25 Mar. (84)	5 Thur.	21·4230	185·4939	268·3134		4665
27 Mar. (87)	2 Mon.	14 20 0	13 Mar. (73)	2 Mon.	9897·1064	32·7287	237·4895		4666
27 Mar. (86)	3 Tues.	20 32 30	3 Mar. (62)	0 Sat.	111·4197	916·2552	209·4035		4667
28 Mar. (87)	5 Thur.	2 45 0	22 Mar. (81)	6 Fri.	146·0613	852·2386	260·7131		4668
28 Mar. (87)	6 Fri.	8 57 30	11 Mar. (70)	3 Tues.	21·7447	69[illegible]·4735	229·8883		4669
27 Mar. (87)	0 Sat.	15 10 0	28 Feb. (59)	0 Sat.	9897·4281	546·7083	199·1654		4670
27 Mar. (86)	1 Sun.	21 22 30	18 Mar. (77)	3 Fri.	9912·0677	482·6917	250·3782		4671

TABLE

CONCURRENT YEAR.								INTERCALATED (*adhika*) and SUPPRESSED (*kshaya*) LUNAR MONTHS (true).
Kali	Śaka.	Chaitrādi Vikrama.	Mēshādi solar year in Bengal.	Kollam.	A.D.	JOVIAN SAMVATSARA. Southern system.	JOVIAN SAMVATSARA. Northern system.	
1	2	3	3*a*	4	5	6	7	8
4672	1493	1628	977	745-46	1570-71	4 Pramōda .	12 Bahudhānya .	6 Bhādrapada
4673	1494	1629	978	746-47	1571-72	5 Prajāpati .	13 Pramāthin .	...
4674	1495	1630	979	747-48	*1572-73	6 Aṅgiras .	14 Vikrama .	...
4675	1496	1631	980	748-49	1573-74	7 Śrīmukha .	15 Vṛisha . .	4 Āshāḍha .
4676	1497	1632	981	749-50	1574-75	8 Bhāva . .	16 Chitrabhānu .	...
4677	1498	1633	982	750-51	1575-76	9 Yuvan . .	17 Subhānu .	...
4678	1499	1634	983	751-52	*1576-77	10 Dhātṛi . .	18 Tāraṇa . .	3 Jyēshṭha .
4679	1500	1635	984	752-53	1577-78	11 Īśvara . .	19 Pārthiva .	...
4680	1501	1636	985	753-54	1578-79	12 Bahudhānya .	20 Vyaya . .	8 Kārttika .
4681	1502	1637	986	754-55	1579-80	13 Pramāthin .	21 Sarvajit .	...
4682	1503	1638	987	755-56	*1580-81	14 Vikrama .	22 Sarvadhārin .	...
4683	1504	1639	988	756-57	1581-82	15 Vṛisha . .	23 Virōdhin .	5 Śrāvaṇa .
4684	1505	1640	989	757-58	1582-83	16 Chitrabhānu .	24 Vikṛita . .	...
4685	1506	1641	990	758-59	1583-84	17 Subhānu .	25 Khara . .	...
4686	1507	1642	991	759-60	*1584-85	18 Tāraṇa . .	26 Nandana .	4 Āshāḍha .
4687	1508	1643	992	760-61	1585-86	19 Pārthiva .	27 Vijaya . .	...
4688	1509	1644	993	761-62	1586-87	20 Vyaya . .	28 Jaya . .	...
4689	1510	1645	994	762-63	1587-88	21 Sarvajit .	29 Manmatha† .	2 Vaiśākha .
4690	1511	1646	995	763-64	*1588-89	22 Sarvadhārin .	31 *Hēmalamba* .	...
4691	1512	1647	996	764-65	1589-90	23 Virōdhin .	32 *Vilamba* .	6 Bhādrapada
4692	1513	1648	997	765-66	1590-91	24 Vikṛita . .	33 *Vikārin* . .	...
4693	1514	1649	998	766-67	1591-92	25 Khara . .	34 *Śārvarin* .	...
4694	1515	1650	999	767-68	*1592-93	26 Nandana .	35 *Plava* . .	4 Āshāḍha .
4695	1516	1651	1000	768-69	1593-94	27 Vijaya . .	36 *Śubhakrit* .	...
4696	1517	1652	1001	769-70	1594-95	28 Jaya . .	37 *Śōbhana* .	...

† No. 30 Durmati was suppressed in the north

LXI—*Contd.*

COMMENCEMENT OF THE											
SOLAR YEAR.					LUNI-SOLAR YEAR (MEAN SUNRISE OF CIVIL DAY ON WHICH CHAITRA ŚUKLA 1 ENDS).						Kali.
Day and month, A.D.	Week-day.	Time of true Mēsha-saṁkrānti.			Day and month, A.D.	Week-day.	*a.*	*b.*	*c.*		
13	14	17			19	20	23	24	25		1
		H.	M.	S.							
28 Mar. (87)	3 Tues. .	3	35	0	7 Mar. (66)	3 Tues. .	9807·7511	330·2366	219·5513		4672
28 Mar. (87)	4 Wed. .	9	47	30	26 Mar. (85)	2 Mon. .	9842·3907	265·9101	270·8611		4673
27 Mar. (87)	5 Thur. .	16	0	0	15 Mar. (75)	0 Sat. .	56·7060	149·4366	242·7749		4674
27 Mar. (86)	6 Fri. .	22	12	30	4 Mar. (63)	4 Wed. .	9932·3894	996·6713	211·9511		4675
28 Mar. (87)	1 Sun. .	4	25	0	23 Mar. (82)	3 Tues. .	9967·0290	932·6549	263·2608		4676
28 Mar. (87)	2 Mon. .	10	37	30	13 Mar. (72)	1 Sun. .	181·3441	816·1813	235·1747		4677
27 Mar. (87)	3 Tues. .	16	50	0	1 Mar. (61)	5 Thur. .	57·0275	663·4160	204·3509		4678
27 Mar. (86)	4 Wed. .	23	2	30	20 Mar. (79)	4 Wed. .	91·6671	599·3996	255·9524		4679
28 Mar. (87)	6 Fri. .	5	15	0	9 Mar. (68)	1 Sun. .	9967·3506	446·6344	224·8368		4680
28 Mar. (87)	0 Sat. .	11	27	30	28 Mar. (87)	0 Sat. .	1·9902	382·6179	276·1464		4681
27 Mar. (87)	1 Sun. .	17	40	0	16 Mar. (76)	4 Wed. .	9877·6755	229·8527	245·3226		4682
27 Mar. (86)	2 Mon. .	23	52	30	6 Mar. (65)	2 Mon. .	91·9888	113·3791	217·2365		4683
28 Mar. (87)	4 Wed. .	6	5	0	25 Mar. (84)	1 Sun. .	126·6284	49·3626	268·5463		4684
28 Mar. (87)	5 Thur. .	12	17	30	14 Mar. (73)	5 Thur. .	2·3118	896·5974	237·7224		4685
27 Mar. (87)	6 Fri. .	18	30	0	3 Mar. (63)	3 Tues. .	216·6271	780·1239	209·6363		4686
28 Mar. (87)	1 Sun. .	0	42	30	22 Mar. (81)	2 Mon. .	251·2667	716·1074	260·9460		4687
28 Mar. (87)	2 Mon. .	6	55	0	11 Mar. (70)	6 Fri. .	126·9501	563·3422	230·1222		4688
28 Mar. (87)	3 Tues. .	13	7	30	28 Feb. (59)	3 Tues. .	2·6335	410·5770	199·2983		4689
27 Mar. (87)	4 Wed. .	19	20	0	18 Mar. (78)	2 Mon. .	37·2731	346·5605	250·6081		4690
28 Mar. (87)	6 Fri. .	1	32	30	7 Mar. (66)	6 Fri. .	9912·9565	193·7953	219·7842		4691
28 Mar. (87)	0 Sat. .	7	45	0	26 Mar. (85)	5 Thur. .	9947·5961	129·7788	271·0939		4692
28 Mar. (87)	1 Sun. .	13	57	30	16 Mar. (75)	3 Tues. .	161·9114	13·2053	243·0078		4693
27 Mar. (87)	2 Mon. .	20	10	0	4 Mar. (64)	0 Sat. .	37·5948	860·5401	21[illegible]·1840		4694
28 Mar. (87)	4 Wed. .	2	22	30	23 Mar. (82)	6 Fri. .	72·2344	796·5236	263·4937		4695
28 Mar. (87)	5 Thur. .	8	35	0	13 Mar. (72)	4 Wed. .	286·5496	680·[illegible]	23[illegible]·4076		4696

TABLE

CONCURRENT YEAR.								Intercalated (*adhika*) and suppressed (*kshaya*) Lunar Months (true).
Kali.	Śaka.	Chaitrādi Vikrama.	Mēshādi solar year in Bengal.	Kollam.	A.D.	Jovian Saṁvatsara. Southern system.	Jovian Saṁvatsara. Northern system.	
1	2	3	3*a*	4	5	6	7	8
4697	1518	1653	1002	770-71	1595-96	29 Manmatha	38 *Krōdhin*	3 Jyēshṭha
4698	1519	1654	1003	771-72	*1596-97	30 Durmukha	39 *Viśvāvasu*	...
4699	1520	1655	1004	772-73	1597-98	31 Hēmalamba	40 *Parābhava*	8 Kārttika
4700	1521	1656	1005	773-74	1598-99	32 Vilamba	41 *Plavaṅga*	...
4701	1522	1657	1006	774-75	1599-1600	33 Vikārin	42 *Kīlaka*	...
4702	1523	1658	1007	775-76	*1600-01	34 Śārvarin	43 *Saumya*	5 Śrāvaṇa
4703	1524	1659	1008	776-77	1601-02	35 Plava	44 Sādhāraṇa	...
4704	1525	1660	1009	777-78	1602-03	36 Śubhakṛit	45 Virōdhakṛit	...
4705	1526	1661	1010	778-79	1603-04	37 Śōbhana	46 Paridhāvin	4 Āshāḍha
4706	1527	1662	1011	779-80	*1604-05	38 Krōdhin	47 Pramādin	...
4707	1528	1663	1012	780-81	1605-06	39 Viśvāvasu	48 Ānanda	...
4708	1529	1664	1013	781-82	1606-07	40 Parābhava	49 Rākshasa	1 Chaitra
4709	1530	1665	1014	782-83	1607-08	41 Plavaṅga	50 Anala	...
4710	1531	1666	1015	783-84	*1608-09	42 Kīlaka	51 Piṅgala	6 Bhādrapada
4711	1532	1667	1016	784-85	1609-10	43 Saumya	52 Kālayukta	...
4712	1533	1668	1017	785-86	1610-11	44 Sādhāraṇa	53 Siddhārthin	...
4713	1534	1669	1018	786-87	1611-12	45 Virōdhakṛit	54 Raudra	4 Āshāḍha
4714	1535	1670	1019	787-88	*1612-13	46 Paridhāvin	55 Durmati	...
4715	1536	1671	1020	788-89	1613-14	47 Pramādin	56 Dundubhi	...
4716	1537	1672	1021	789-90	1614-15	48 Ānanda	57 Rudhirōdgārin	3 Jyēshṭha
4717	1538	1673	1022	790-91	1615-16	49 Rākshasa	58 Raktāksha	...
4718	1539	1674	1023	791-92	*1616-17	50 Anala	59 Krōdhana	7 Āśvina
4719	1540	1675	1024	792-93	1617-18	1 Pingala	60 Kshaya	...
4720	1541	1676	1025	793-94	1618-19	52 Kālayukta	1 Prabhava	...
4721	1542	1677	1026	794-95	1619-20	53 Siddhārthin	2 Vibhava	5 Śrāvaṇa

LXI—*Contd.*

COMMENCEMENT OF THE											
SOLAR YEAR.					LUNI-SOLAR YEAR (MEAN SUNRISE OF CIVIL DAY ON WHICH CHAITRA ŚUKLA 1 ENDS).						Kali.
Day and month, A.D.	Week-day.	Time of true Mēsha-saṁkrānti.			Day and month, A.D.	Week-day.	*a.*	*b.*	*c.*		
13	14	17			19	20	23	24	25		1
		H.	M.	S.							
28 Mar. (87)	6 Fri.	14	47	30	2 Mar. (61)	1 Sun.	162·2330	527·2848	204·5838		4697
27 Mar. (87)	0 Sat.	21	0	0	19 Mar. (79)	6 Fri.	9858·2408	426·9767	253·1557		4698
28 Mar. (87)	2 Mon.	3	12	30	8 Mar. (67)	3 Tues.	9733·9241	274·2115	222·3318		4699
28 Mar. (87)	3 Tues.	9	25	0	27 Mar. (86)	2 Mon.	9768·5638	210·1951	273·6415		4700
28 Mar. (87)	4 Wed.	15	37	30	17 Mar. (76)	0 Sat.	9982·8789	93·7214	245·5555		4701
27 Mar. (87)	5 Thur.	21	50	0	6 Mar. (66)	5 Thur.	197·1942	977·2479	218·4694		4702
28 Mar. (87)	0 Sat.	4	2	30	25 Mar. (84)	4 Wed.	231·8338	913·2313	268·7792		4703
28 Mar. (87)	1 Sun.	10	15	0	14 Mar. (73)	1 Sun.	107·5172	760·4661	237·9552		4704
28 Mar. (87)	2 Mon.	16	27	30	3 Mar. (62)	5 Thur.	9983·2006	607·7010	207·1314		4705
27 Mar. (87)	3 Tues.	22	40	0	21 Mar. (81)	4 Wed.	17·8402	543·6844	258·4411		4706
28 Mar. (87)	5 Thur.	4	52	30	10 Mar. (69)	1 Sun.	9893·5236	390·9192	227·6173		4707
28 Mar. (87)	6 Fri.	11	5	0	27 Feb. (58)	5 Thur.	9769·2070	238·1541	196·7934		4708
28 Mar. (87)	0 Sat.	17	17	30	18 Mar. (77)	4 Wed.	9803·8466	174·1376	248·1032		4709
27 Mar. (87)	1 Sun.	23	30	0	7 Mar. (67)	2 Mon.	18·1619	57·6640	220·0171		4710
28 Mar. (87)	3 Tues.	5	42	30	26 Mar. (85)	1 Sun.	52·8015	993·6475	271·3267		4711
28 Mar. (87)	4 Wed.	11	55	0	16 Mar. (75)	6 Fri.	267·1178	877·1740	243·2407		4712
28 Mar. (87)	5 Thur.	18	7	30	5 Mar. (64)	3 Tues.	142·8002	724·4087	212·4169		4713
28 Mar. (88)	0 Sat.	0	20	0	23 Mar. (83)	2 Mon.	177·4398	660·3923	263·7266		4714
28 Mar. (87)	1 Sun.	6	32	30	12 Mar. (71)	6 Fri.	53·1233	507·6271	232·9028		4715
28 Mar. (87)	2 Mon.	12	45	0	1 Mar. (60)	3 Tues.	9928·8064	354·8619	202·0789		4716
28 Mar. (87)	3 Tues.	18	57	30	20 Mar. (79)	2 Mon.	9962·4462	290·8454	253·3885		4717
28 Mar. (88)	5 Thur.	1	10	0	8 Mar. (68)	6 Fri.	9839·1305	138·0802	222·5647		4718
28 Mar. (87)	6 Fri.	7	22	30	27 Mar. (86)	5 Thur.	9874·7691	74·0637	273·8744		4719
28 Mar. (87)	0 Sat.	13	35	0	17 Mar. (76)	3 Tues.	88·0843	957·5901	245·7884		4720
28 Mar. (87)	1 Sun.	19	47	30	7 Mar. (66)	1 Sun.	302·3996	841·1165	217·7023		4721

TABLE

CONCURRENT YEAR.								INTERCALATED (*adhika*) and SUPPRESSED (*kshaya*) LUNAR MONTHS (true).
						JOVIAN SAMVATSARA.		
Kali.	Śaka.	Chaitrādi Vikrama.	Mēshādi solar year in Bengal.	Kollam.	A.D.	Southern system.	Northern system.	
1	2	3	3*a*	4	5	6	7	8
4722	1543	1678	1027	795-96	*1620-21	54 Raudra .	3 Śukla . .	...
4723	1544	1679	1028	796-97	1621-22	55 Durmati .	4 Pramōda .	...
4724	1545	1680	1029	797-98	1622-23	56 Dundubhi .	5 Prajāpati .	4 Āshādha .
4725	1546	1681	1030	798-99	1623-24	57 Rudhirōdgārin	6 Aṅgiras .	...
4726	1547	1682	1031	799-800	*1624-25	58 Raktāksha .	7 Śrīmukha .	...
4727	1548	1683	1032	800-01	1625-26	59 Krōdhana .	8 Bhāva . .	1 Chaitra .
4728	1549	1684	1033	801-02	1626-27	60 Kshaya .	9 Yuvan . .	...
4729	1550	1685	1034	802-03	1627-28	1 Prabhava .	10 Dhātṛi . .	5 Śrāvaṇa .
4730	1551	1686	1035	803-04	*1628-29	2 Vibhava .	11 Īśvara . .	...
4731	1552	1687	1036	804-05	1629-30	3 Śukla . .	12 Bahudhānya .	...
4732	1553	1688	1037	805-06	1630-31	4 Pramōda .	13 Pramāthin .	4 Āshāḍha .
4733	1554	1689	1038	806-07	1631-32	5 Prajāpati .	14 Vikrama .	...
4734	1555	1690	1039	807-08	*1632-33	6 Angiras .	15 Vṛisha . .	
4735	1556	1691	1040	808-09	1633-34	7 Śrīmukha .	16 Chitrabhānu .	2 Vaiśākha .
4736	1557	1692	1041	809-10	1634-35	8 Bhāva . .	17 Subhānu .	...
4737	1558	1693	1042	810-11	1635-36	9 Yuvan . .	18 Tāraṇa . .	6 Bhādrapada
4738	1559	1694	1043	811-12	*1636-37	10 Dhātṛi . .	19 Pārthiva .	...
4739	1560	1695	1044	812-13	1637-38	11 Īśvara . .	20 Vyaya . .	...
4740	1561	1696	1045	813-14	1638-39	12 Bahudhānya .	21 Sarvajit . .	5 Śrāvaṇa .
4741	1562	1697	1046	814-15	1639-40	13 Pramāthin .	22 Sarvadhārin .	...
4742	1563	1698	1047	815-16	*1640-41	14 Vikrama .	23 Virōdhin .	...
4743	1564	1699	1048	816-17	1641-42	15 Vṛisha .	24 Vikṛita . .	3 Jyēshṭha .
4744	1565	1700	1049	817-18	1642-43	16 Chitrabhānu .	25 Khara . .	...
4745	1566	1701	1050	818-19	1643-44	17 Subhānu .	26 Nandana .	...
4746	1567	1702	1051	819-20	*1644-45	18 Tāraṇa . .	27 Vijaya . .	1 Chaitra .

LXXI—*Contd.*

COMMENCEMENT OF THE											
SOLAR YEAR.					LUNI-SOLAR YEAR (MEAN SUNRISE OF CIVIL DAY ON WHICH CHAITRA ŚUKLA 1 ENDS).					Kali.
Day and month, A.D.	Week-day.	Time of true Mēsha-saṁkrānti.			Day and month, A.D.	Week-day.	*a.*	*b.*	*c.*	
13	14	17			19	20	23	24	25	1
		H.	M.	S.						
28 Mar. (88)	3 Tues.	2	0	0	24 Mar. (84)	6 Fri. .	9998·4073	740·8085	266·2743	4722
28 Mar. (87)	4 Wed.	8	12	30	14 Mar. (73)	4 Wed.	212·7226	624·3349	238·1881	4723
28 Mar. (87)	5 Thur.	14	25	0	3 Mar. (62)	1 Sun. .	88·4060	471·5697	207·3643	4724
28 Mar. (87)	6 Fri. .	20	37	30	21 Mar. (80)	6 Fri. .	9784·4137	371·2616	255·0362	4725
28 Mar. (88)	1 Sun. .	2	50	0	10 Mar. (70)	4 Wed.	9998·7290	254·7880	227·8502	4726
28 Mar. (87)	2 Mon. .	9	2	30	27 Feb. (58)	1 Sun. .	9874·4124	102·0228	197·0263	4727
28 Mar. (87)	3 Tues.	15	15	0	18 Mar. (77)	0 Sat. .	9909·0520	38·0063	248·3361	4728
28 Mar. (87)	4 Wed.	21	27	30	8 Mar. (67)	5 Thur.	123·3673	921·5328	220·2500	4729
28 Mar. (88)	6 Fri. .	3	40	0	26 Mar. (86)	4 Wed.	158·0079	857·5162	271·4596	4730
28 Mar. (87)	0 Sat. .	9	52	30	15 Mar. (74)	1 Sun. .	33·6902	704·7511	240·7358	4731
28 Mar. (87)	1 Sun. .	16	5	0	4 Mar. (63)	5 Thur.	9909·3737	551·9859	209·9120	4732
28 Mar. (87)	2 Mon. .	22	17	30	23 Mar. (82)	4 Wed.	9944·0133	487·9693	261·2217	4733
28 Mar. (88)	4 Wed.	4	30	0	11 Mar. (71)	1 Sun. .	9819·6967	335·2042	230·3979	4734
28 Mar. (87)	5 Thur.	10	42	30	1 Mar. (60)	6 Fri. .	34·0119	218·7306	202·3118	4735
28 Mar. (87)	6 Fri. .	16	55	0	20 Mar. (79)	5 Thur.	68·6516	154·7141	253·1575	4736
28 Mar. (87)	0 Sat. .	23	7	30	9 Mar. (68)	2 Mon. .	9944·3349	1·9489	222·7976	4737
28 Mar. (88)	2 Mon. .	5	20	0	27 Mar. (87)	1 Sun. .	9978·9746	937·9325	274·1073	4738
28 Mar. (87)	3 Tues.	11	32	30	17 Mar. (76)	6 Fri. .	193·2898	821·4589	246·0213	4739
28 Mar. (87)	4 Wed.	17	45	0	6 Mar. (65)	3 Tues.	68·9732	668·6936	215·1974	4740
28 Mar. (87)	5 Thur.	23	57	30	25 Mar. (84)	2 Mon. .	103·6128	604·6772	266·5072	4741
28 Mar. (88)	0 Sat. .	6	10	0	13 Mar. (73)	6 Fri. .	9979·2962	451·9120	235·6833	4742
28 Mar. (87)	1 Sun. .	12	22	30	2 Mar. (61)	3 Tues.	9854·9796	299·1468	204·8594	4743
28 Mar. (87)	2 Mon. .	18	35	0	21 Mar. (80)	2 Mon. .	9890·6192	235·1303	256·1691	4744
29 Mar. (88)	4 Wed.	0	47	30	10 Mar. (69)	6 Fri. .	9765·3026	82·3651	225·3453	4745
28 Mar. (88)	5 Thur.	7	0	0	28 Feb. (59)	4 Wed.	9979·6178	965·8916	197·2592	4746

TABLE

CONCURRENT YEAR.								INTERCALATED (*adhika*) and SUPPRESSED (*kshaya*) LUNAR MONTHS (true).
						JOVIAN SAMVATSARA.		
Kali.	Śaka.	Chaitrādi Vikrama.	Mēshādi solar year in Bengal.	Kollam.	A.D.	Southern system.	Northern system.	
1	2	3	3a	4	5	6	7	8
4747	1568	1703	1052	820-21	1645-46	19 Pārthiva .	28 Jaya .	...
4748	1569	1704	1053	821-22	1646-47	20 Vyaya . .	29 Manmatha .	5 Śrāvaṇa .
4749	1570	1705	1054	822-23	1647-48	21 Sarvajit .	30 Durmukha .	...
4750	1571	1706	1055	823-24	*1648-49	22 Sarvadhārin .	31 Hēmalamba .	...
4751	1572	1707	1056	824-25	1649-50	23 Virōdhin .	32 Vilamba .	4 Āshāḍha .
4752	1573	1708	1057	825-26	1650-51	24 Vikṛita . .	33 Vikārin . .	...
4753	1574	1709	1058	826-27	1651-52	25 Khara . .	34 Śārvarin .	...
4754	1575	1710	1059	827-28	*1652-53	26 Nandana .	35 Plava . .	2 Vaiśākha .
4755	1576	1711	1060	828-29	1653-54	27 Vijaya . .	36 Śubhakṛit .	...
4756	1577	1712	1061	829-30	1654-55	28 Jaya . .	37 Śōbhana .	6 Bhādrapada
4757	1578	1713	1062	830-31	1655-56	29 Manmatha .	38 Krōdhin .	...
4758	1579	1714	1063	831-32	*1656-57	30 Durmukha .	39 Viśvāvasu .	...
4759	1580	1715	1064	832-33	1657-58	31 Hēmalamba .	40 Parābhava .	5 Śrāvaṇa .
4760	1581	1716	1065	833-34	1658-59	32 Vilamba .	41 Plavaṅga .	...
4761	1582	1717	1066	834-35	1659-60	33 Vikārin . .	42 Kīlaka .	...
4762	1583	1718	1067	835-36	*1660-61	34 Śārvarin .	43 Saumya .	3 Jyēshṭha .
4763	1584	1719	1068	836-37	1661-62	35 Plava . .	44 Sādhāraṇa .	...
4764	1585	1720	1069	837-38	1662-63	36 Śubhakṛit .	45 Virōdhakṛit .	...
4765	1586	1721	1070	838-39	1663-64	37 Śōbhana .	46 Paridāvin .	1 Chaitra .
4766	1587	1722	1071	839-40	*1664-65	38 Krōdhin .	47 Pramādin .	...
4767	1588	1723	1072	840-41	1665-66	39 Viśvāvasu .	48 Ānanda .	5 Śrāvaṇa .
4768	1589	1724	1073	841-42	1666-67	40 Parābhava .	49 Rākshasa .	...
4769	1590	1725	1074	842-43	1667-68	41 Plavaṅga .	50 Anala .	...
4770	1591	1726	1075	843-44	*1668-69	42 Kīlaka . .	51 Piṅgala .	4 Āshāḍha .
4771	1592	1727	1076	844-45	1669-70	43 Saumya .	52 Kālayukta .	...

LXI—*Contd.*

COMMENCEMENT OF THE									
SOLAR YEAR.			LUNI-SOLAR YEAR (MEAN SUNRISE OF DAY ON WHICH CHAITRA ŚUKLA 1 ENDS).						Kali.
Day and month, A.D.	Week-day.	Time of true Mēsha-saṁkrānti.			Day and month, A.D.	Week-day.	*a.*	*b.*	*c.*	
13	14	17			19	20	23	24	25	1
		H.	M.	S.						
28 Mar. (87)	6 Fri. .	13	12	30	18 Mar. (77)	3 Tues.	14·26[illegible]4	901·8750	248·4690	4747
28 Mar. (87)	0 Sat. .	19	25	0	8 Mar. (67)	1 Sun. .	229·5727	785·4015	220·4829	4748
29 Mar. (88)	2 Mon. .	1	37	30	27 Mar. (86)	0 Sat. .	263·2124	720·3850	271·7925	4749
28 Mar. (88)	3 Tues.	7	50	0	15 Mar. (75)	4 Wed.	138·8957	568·6198	240·9687	4750
28 Mar. (87)	4 Wed.	14	2	30	4 Mar. (63)	1 Sun. .	14·5791	415·8546	210·1449	4751
28 Mar. (87)	5 Thur.	20	15	0	23 Mar. (82)	0 Sat. .	49·2187	351·8381	261·4546	4752
29 Mar. (88)	0 Sat. .	2	27	30	12 Mar. (71)	4 Wed.	9924·9021	199·0730	230·6308	4753
28 Mar. (88)	1 Sun. .	8	40	0	29 Feb. (60)	1 Sun. .	9800·5855	46·3077	199·8269	4754
28 Mar. (87)	2 Mon. .	14	52	30	20 Mar. (79)	1 Sun. .	173·8570	18·5828	254·8044	4755
28 Mar. (87)	3 Tues.	21	5	0	9 Mar. (68)	5 Thur.	49·5403	865·8177	223·0305	4756
29 Mar. (88)	5 Thur.	3	17	30	28 Mar. (87)	4 Wed.	84·1800	801·8012	274·3402	4757
28 Mar. (88)	6 Fri. .	9	30	0	17 Mar. (77)	2 Mon. .	298·4953	685·3276	246·2542	4758
28 Mar. (87)	0 Sat. .	15	42	30	6 Mar. (65)	6 Fri. .	174·7786	532·5624	215·4303	4759
28 Mar. (87)	1 Sun. .	21	55	0	24 Mar. (83)	4 Wed	9870·7864	432·2544	264·0023	4760
29 Mar. (88)	3 Tues.	4	7	30	13 Mar. (72)	1 Sun. .	9746·4697	279·4893	233·1784	4761
28 Mar. (88)	4 Wed.	10	20	0	2 Mar. (62)	6 Fri. .	9960·7850	163·0155	205·0743	4762
28 Mar. (87)	5 Thur.	16	32	30	21 Mar. (80)	5 Thur.	9995·4246	98·9991	256·4020	4763
28 Mar. (87)	6 Fri. .	22	45	0	10 Mar. (69)	2 Mon. .	9871·1080	946·2338	225·5782	4764
29 Mar. (88)	1 Sun. .	4	57	30	28 Feb. (59)	0 Sat. .	84·8233	829·7603	197·4921	4765
28 Mar. (88)	2 Mon. .	11	10	0	18 Mar. (78)	6 Fri .	119·4629	765·8038	248·8019	4766
28 Mar. (87)	3 Tues.	17	22	30	7 Mar. (66)	3 Tues.	9996·1463	612·9787	217·9780	4767
28 Mar. (87)	4 Wed.	23	35	0	26 Mar. (85)	2 Mon. .	29·7859	548·9621	259·2877	4768
29 Mar. (88)	6 Fri .	5	47	30	15 Mar. (74)	6 Fri. .	9905·4693	396·1969	238·4628	4769
28 Mar. (88)	0 Sat. .	12	0	0	3 Mar. (63)	3 Tues.	9781·1527	243·4318	207·6460	4770
28 Mar. (87)	1 Sun. .	18	12	30	22 Mar. (81)	2 Mon. .	9815·7923	179·4152	258·9497	4771

TABLE

CONCURRENT YEAR.								INTERCALATED (*adhika*) and SUPPRESSED (*kshaya*) LUNAR MONTHS (true).
						JOVIAN SAMVATSARA.		
Kali.	Śaka.	Chaitrādi Vikrama.	Mēshādi solar year in Bengal.	Kollam.	A.D.	Southern system.	Northern system.	
1	2	3	3*a*	4	5	6	7	8
4772	1593	1728	1077	845-46	1670-71	44 Sādhāraṇa .	53 Siddhārthin .	...
4773	1594	1729	1078	846-47	1671-72	45 Virōdhakṛit .	54 Raudra .	2 Vaiśākha .
4774	1595	1730	1079	847-48	*1672-73	46 Paridhāvin .	55 Durmati† .	...
4775	1596	1731	1080	848-49	1673-74	47 Pramādin .	57 *Rudhirōdgārin*	6 Bhādrapada
4776	1597	1732	1081	849-50	1674-75	48 Ānanda .	58 *Raktāksha* .	...
4777	1598	1733	1082	850-51	1675-76	49 Rākshasa .	59 *Krōdhana* .	...
4778	1599	1734	1083	851-52	*1676-77	50 Anala .	60 *Kshaya* . .	5 Śrāvaṇa .
4779	1600	1735	1084	852-53	1677-78	51 Piṅgala .	1 *Prabhava* .	...
4780	1601	1736	1085	853-54	1678-79	52 Kālayukta .	2 *Vibhava* .	...
4781	1602	1737	1086	854-55	1679-80	53 Siddhārthin .	3 *Śukla* . .	3 Jyēshṭha .
4782	1603	1738	1087	855-56	*1680-81	54 Raudra .	4 *Pramōda* .	...
4783	1604	1739	1088	856-57	1681-82	55 Durmati .	5 *Prajāpati*	7 Āśvina 10 *Pausha* (*ksh.*)
4784	1605	1740	1089	857-58	1682-83	56 Dundubhi .	6 *Aṅgiras* .	1 Chaitra
4785	1606	1741	1090	858-59	1683-84	57 Rudhirōdgārin	7 *Śrīmukha* .	...
4786	1607	1742	1091	859-60	*1684-85	58 Raktāksha .	8 *Bhāva* . .	5 Śrāvaṇa .
4787	1608	1743	1092	860-61	1685-86	59 Krōdhana .	9 *Yuvan* . .	...
4788	1609	1744	1093	861-62	1686-87	60 Kshaya .	10 Dhātṛi . .	...
4789	1610	1745	1094	862-63	1687-88	1 Prabhava .	11 Īśvara . .	4 Āshāḍha .
4790	1611	1746	1095	863-64	*1688-89	2 Vibhava .	12 Bahudhānya .	...
4791	1612	1747	1096	864-65	1689-90	3 Śukla .	13 Pramāthin	...
4792	1613	1748	1097	865-66	1690-91	4 Pramōda .	14 Vikrama .	2 Vaiśākha .
4793	1614	1749	1098	866-67	1691-92	5 Prajāpati .	15 Vṛisha . .	...
4794	1615	1750	1099	867-68	*1692-93	6 Aṅgiras .	16 Chitrabhānu .	6 Bhādrapada
4795	1616	1751	1100	868-69	1693-94	7 Śrīmukha .	17 Subhānu .	...
4796	1617	1752	1101	869-70	1694-95	8 Bhāva .	18 Tāraṇa . .	...

† No. 56 Dundubhi was suppressed in the north.

LXI—*Contd.*

COMMENCEMENT OF THE											Kali.
SOLAR YEAR.					LUNI-SOLAR YEAR (MEAN SUNRISE OF CIVIL DAY ON WHICH CHAITRA ŚUKLA 1 ENDS).						
Day and month, A.D.	Week-day.	Time of true Mēsha-saṁkrānti.			Day and month, A.D.	Week-day.	*a.*	*b.*	*c.*		
13	14	17			19	20	23	24	25		1
		H.	M.	S.							
29 Mar. (88)	3 Tues.	0	25	0	12 Mar. (71)	0 Sat.	30·1076	62·9417	230·8637		4772
29 Mar. (88)	4 Wed.	6	37	30	1 Mar. (60)	4 Wed.	9905·7910	910·1765	200·0398		4773
28 Mar. (88)	5 Thur.	12	50	0	20 Mar. (80)	4 Wed.	279·0625	882·4516	254·0873		4774
28 Mar. (87)	6 Fri.	19	2	30	9 Mar. (68)	1 Sun.	154·7458	729·6864	223·2634		4775
29 Mar. (88)	1 Sun.	1	15	0	28 Mar. (87)	0 Sat.	189·3855	665·6700	274·5731		4776
29 Mar. (88)	2 Mon.	7	27	30	17 Mar. (76)	4 Wed.	65·0688	512·9048	253·7493		4777
28 Mar. (88)	3 Tues.	13	40	0	5 Mar. (65)	1 Sun.	9940·7522	360·1395	212·9255		4778
28 Mar. (87)	4 Wed.	19	52	30	24 Mar. (83)	0 Sat.	9975·3918	296·1231	264·2352		4779
29 Mar. (88)	6 Fri.	2	5	0	13 Mar. (72)	4 Wed.	9851·0752	143·3579	233·4113		4780
29 Mar. (88)	0 Sat.	8	17	30	3 Mar. (62)	2 Mon.	65·3904	26·8842	205·3252		4781
28 Mar. (88)	1 Sun.	14	30	0	21 Mar. (81)	1 Sun.	100·0300	962·8678	256·6349		4782
28 Mar. (87)	2 Mon.	20	42	30	10 Mar. (69)	5 Thur.	9975·7134	810·1026	225·8111		4783
29 Mar. (88)	4 Wed.	2	55	0	28 Feb. (59)	3 Tues.	190·0287	693·6290	197·7250		4784
29 Mar. (88)	5 Thur.	9	7	30	19 Mar. (78)	2 Mon.	224·6683	629·6125	249·0348		4785
28 Mar. (88)	6 Fri.	15	20	0	7 Mar. (67)	6 Fri.	100·3517	476·8474	218·2108		4786
28 Mar. (87)	0 Sat.	21	32	30	25 Mar. (84)	4 Wed.	9796·3594	376·5391	266·7828		4787
29 Mar. (88)	2 Mon.	3	45	0	15 Mar. (74)	2 Mon.	10·6747	260·0656	238·6967		4788
29 Mar. (88)	3 Tues.	9	57	30	4 Mar. (63)	6 Fri.	9886·3581	107·3005	207·8729		4789
28 Mar. (88)	4 Wed.	16	10	0	22 Mar. (82)	5 Thur.	9920·9977	43·2840	259·1826		4790
28 Mar. (87)	5 Thur.	22	22	30	12 Mar. (71)	3 Tues.	135·3130	926·8104	231·0966		4791
29 Mar. (88)	0 Sat.	4	35	0	1 Mar. (60)	0 Sat.	10·9963	774·0452	200·2727		4792
29 Mar. (88)	1 Sun.	10	47	30	20 Mar. (79)	6 Fri.	45·6360	710·0287	251·5824		4793
28 Mar. (88)	2 Mon.	17	0	0	8 Mar. (68)	3 Tues.	9921·3194	557·2636	220·7585		4794
28 Mar. (87)	3 Tues.	23	12	30	27 Mar. (86)	2 Mon.	9955·9590	493·2471	272·0682		4795
29 Mar. (88)	5 Thur.	5	25	0	16 Mar. (75)	6 Fri.	9831·6424	340·4819	241·2444		4796

TABLE

CONCURRENT YEAR.								
						Jovian Samvatsara.		Intercalated (*adhika*) and suppressed (*kshaya*) Lunar months (true).
Kali.	Śaka.	Chaitrādi Vikrama.	Mēshādi solar year in Bengal.	Kollam.	A.D.	Southern system.	Northern system.	
1	2	3	3*a*	4	5	6	7	8
4797	1618	1753	1102	870-71	1695-96	9 Yuvan	19 Pārthiva	4 Āshādha
4798	1619	1754	1103	871-72	*1696-97	10 Dhātṛi	20 Vyaya	...
4799	1620	1755	1104	872-73	1697-98	11 Īśvara	21 Sarvajit	...
4800	1621	1756	1105	873-74	1698-99	12 Bahudhānya	22 Sarvadhārin	3 Jyēshṭha
4801	1622	1757	1106	874-75	1699-1700	13 Pramāthin	23 Virōdhin	...
4802	1623	1758	1107	875-76	*1700-01	14 Vikrama	24 Vikṛita	7 Āśvina 11 *Magha* (*ksh.*)
4803	1624	1759	1108	876-77	1701-02	15 Vṛisha	25 Khara	1 Chaitra
4804	1625	1760	1109	877-78	1702-03	16 Chitrabhānu	26 Nandana	...
4805	1626	1761	1110	878-79	1703-04	17 Subhānu	27 Vijaya	5 Śrāvaṇa
4806	1627	1762	1111	879-80	*1704-05	18 Tāraṇa	28 Jaya	...
4807	1628	1763	1112	880-81	1705-06	19 Pārthiva	29 Manmatha	...
4808	1629	1764	1113	881-82	1706-07	20 Vyaya	30 Durmukha	4 Āshādha
4809	1630	1765	1114	882-83	1707-08	21 Sarvajit	31 Hēmalamba	...
4810	1631	1766	1115	883-84	*1708-09	22 Sarvadhārin	32 Vilamba	...
4811	1632	1767	1116	884-85	1709-10	23 Virōdhin	33 Vikārin	2 Vaiśākha
4812	1633	1768	1117	885-86	1710-11	24 Vikṛita	34 Śārvarin	...
4813	1634	1769	1118	886-87	1711-12	25 Khara	35 Plava	6 Bhādrapada
4814	1635	1770	1119	887-88	*1712-13	26 Nandana	36 Subhakṛit	...
4815	1636	1771	1120	888-89	1713-14	27 Vijaya	37 Śōbhana	...
4816	1637	1772	1121	889-90	1714-15	28 Jaya	38 Krōdhin	4 Āshādha
4817	1638	1773	1122	890-91	1715-16	29 Manmatha	39 Viśvāvasu	...
4818	1639	1774	1123	891-92	*1716-17	30 Durmukha	40 Parābhava	...
4819	1640	1775	1124	892-93	1717-18	31 Hēmalamba	41 Plavaṅga	3 Jyēshṭha
4820	1641	1776	1125	893-94	1718-19	32 Vilamba	42 Kīlaka	...
4821	1642	1777	1126	894-95	1719-20	33 Vikārin	43 Saumya	7 Āśvina

LXI—*Contd.*

COMMENCEMENT OF THE											
SOLAR YEAR.					LUNI-SOLAR YEAR (MEAN SUNRISE OF CIVIL DAY ON WHICH CHAITRA ŚUKLA 1 ENDS).					Kali.
Day and month, A.D.	Week-day.	Time of true Mēsha-saṁkrānti.			Day and month, A.D.	Week-day.	*a.*	*b.*	*c.*	
13	14	17			19	20	23	24	25	1
		H.	M.	S.						
29 Mar. (88)	6 Fri.	11	37	30	6 Mar. (65)	4 Wed.	45·9577	224·0083	213·1584	4797
28 Mar. (88)	0 Sat.	17	50	0	23 Mar. (83)	2 Mon.	9741·9654	123·7001	261·7303	4798
29 Mar. (88)	2 Mon.	0	2	30	13 Mar. (72)	0 Sat.	9956·2806	7·2266	233·6441	4799
29 Mar. (88)	3 Tues.	6	15	0	3 Mar. (62)	5 Thur.	170·5959	890·7531	205·5581	4800
29 Mar. (88)	4 Wed.	12	27	30	22 Mar. (81)	4 Wed.	205·2355	826·7366	256·8678	4801
28 Mar. (88)	5 Thur.	18	40	0	10 Mar. (70)	1 Sun.	80·9189	673·9714	226·0440	4802
29 Mar. (88)	0 Sat.	0	52	30	27 Feb. (58)	5 Thur.	9956·6022	521·2062	195·2191	4803
29 Mar. (88)	1 Sun.	7	5	0	18 Mar. (77)	4 Wed.	9991·2419	357·1897	246·5298	4804
29 Mar. (88)	2 Mon.	13	17	30	7 Mar. (66)	1 Sun.	9866·9253	304·4245	215·7059	4805
28 Mar. (88)	3 Tues.	19	30	0	25 Mar. (85)	0 Sat.	9901·5649	240·4080	267·0157	4806
29 Mar. (88)	5 Thur.	1	42	30	14 Mar. (73)	4 Wed.	9777·2483	87·6428	236·1918	4807
29 Mar. (88)	6 Fri.	7	55	0	4 Mar. (63)	2 Mon.	9991·5636	971·1693	208·1058	4808
29 Mar. (88)	0 Sat.	14	7	30	23 Mar. (82)	1 Sun.	26·2032	907·1528	259·4155	4809
28 Mar. (88)	1 Sun.	20	20	0	12 Mar. (72)	6 Fri.	240·5185	790·6792	231·2295	4810
29 Mar. (88)	3 Tues.	2	32	30	1 Mar. (60)	3 Tues.	116·2018	637·9140	200·5055	4811
29 Mar. (88)	4 Wed.	8	45	0	20 Mar. (79)	2 Mon.	150·8415	573·8975	251·8153	4812
29 Mar. (88)	5 Thur.	14	57	30	9 Mar. (68)	6 Fri.	26·5249	421·1323	220·9914	4813
28 Mar. (88)	6 Fri.	21	10	0	27 Mar. (87)	5 Thur.	61·1645	357·1158	272·3011	4814
29 Mar. (88)	1 Sun.	3	22	30	16 Mar. (75)	2 Mon.	9936·8478	204·3506	241·4773	4815
29 Mar. (88)	2 Mon.	9	35	0	5 Mar. (64)	6 Fri.	9812·5312	51·5855	210·6535	4816
29 Mar. (88)	3 Tues.	15	47	30	24 Mar. (83)	5 Thur.	9847·1709	987·5689	261·9631	4817
28 Mar. (88)	4 Wed.	22	0	0	13 Mar. (73)	3 Tues.	61·4864	871·0954	233·8770	4818
29 Mar. (88)	6 Fri.	4	12	30	3 Mar. (62)	1 Sun.	275·8013	754·6218	205·7910	4819
29 Mar. (88)	0 Sat.	10	[illegible]	0	22 Mar. (81)	0 Sat.	310·4410	691·6053	257·1007	4820
29 Mar. (88)	1 Sun.	16	37	30	11 Mar. (70)	4 Wed.	186·1243	537·8401	226·2769	4821

TABLE

CONCURRENT YEAR.								INTERCALATED (*adhika*) and SUPPRESSED (*kshaya*) LUNAR MONTHS (true).
Kali.	Śaka.	Chaitrādi Vikrama.	Mēshādi solar year in Bengal.	Kollam.	A.D.	JOVIAN SAṀVATSARA. Southern system.	JOVIAN SAṀVATSARA. Northern system.	
1	2	3	3*a*	4	5	6	7	8
4822	1643	1778	1127	895-96	*1720-21	34 Śārvarin .	44 Sādhāraṇa .	...
4823	1644	1779	1128	896-97	1721-22	35 Plava .	45 Virōdhakṛit .	...
4824	1645	1780	1129	897-98	1722-23	36 Śubhakṛit .	46 Paridhāvin .	5 Śrāvaṇa .
4825	1646	1781	1130	898-99	1723-24	37 Śōbhana .	47 Pramādin .	...
4826	1647	1782	1131	899-900	*1724-25	38 Krōdhin .	48 Ānanda .	...
4827	1648	1783	1132	900-01	1725-26	39 Viśvāvasu .	49 Rākshasa .	4 Āshāḍha .
4828	1649	1784	1133	901-02	1726-27	40 Parābhava .	50 Anala . .	...
4829	1650	1785	1134	902-03	1727-28	41 Plavaṅga .	51 Piṅgala .	...
4830	1651	1786	1135	903-04	*1728-29	42 Kīlaka .	52 Kālayukta .	2 Vaiśākha .
4831	1652	1787	1136	904-05	1729-30	43 Saumya .	53 Siddhārthin .	...
4832	1653	1788	1137	905-06	1730-31	44 Sādhāraṇa .	54 Raudra .	6 Bhādrapada
4833	1654	1789	1138	906-07	1731-32	45 Virōdhakṛit .	55 Durmati .	...
4834	1655	1790	1139	907-08	*1732-33	46 Paridhāvin .	56 Dundubhi .	...
4835	1656	1791	1140	908-09	1733-34	47 Pramādin .	57 Rudhirōdgārin	4 Āshāḍha .
4836	1657	1792	1141	909-10	1734-35	48 Ānanda .	58 Raktāksha .	...
4837	1658	1793	1142	910-11	1735-36	49 Rākshasa .	59 Krōdhana .	...
4838	1659	1794	1143	911-12	*1736-37	50 Anala .	60 Kshaya .	3 Jyēshṭha .
4839	1660	1795	1144	912-13	1737-38	51 Piṅgala .	1 Prabhava .	...
4840	1661	1796	1145	913-14	1738-39	52 Kālayukta .	2 Vibhava .	7 Āśvina .
4841	1662	1797	1146	914-15	1739-40	53 Siddhārthin .	3 Śukla .	...
4842	1663	1798	1147	915-16	*1740-41	54 Raudra .	4 Pramōda .	...
4843	1664	1799	1148	916-17	1741-42	55 Durmati .	5 Prajāpati .	5 Śrāvaṇa .
4844	1665	1800	1149	917-18	1742-43	56 Dundubhi .	6 Aṅgiras .	...
4845	1666	1801	1150	918-19	1743-44	57 Rudhirōdgārin	7 Śrīmukha .	...
4846	1667	1802	1151	919-20	*1744-45	58 Raktāksha .	8 Bhāva .	4 Āshāḍha .

LXI—*Contd.*

COMMENCEMENT OF THE									Kali.
SOLAR YEAR			LUNI-SOLAR YEAR (MEAN SUNRISE OF DAY ON WHICH CHAITRA ŚUKLA 1 ENDED).						
Day and month, A.D.	Week-day.	Time of true Mēsha-samkrānti.	Day and month, A.D.	Week-day.	*a.*	*b.*	*c.*		
13	14	17	19	20	23	24	25		1
		H. M. S.							
28 Mar. (88)	2 Mon. .	22 50 0	28 Mar. (88)	2 Mon. .	9882·1321	437·5321	274·8488		4822
29 Mar. (88)	4 Wed. .	5 2 30	17 Mar. (76)	6 Fri. .	9757·8155	284·7669	244·0249		4823
29 Mar. (88)	5 Thur.	11 15 0	7 Mar. (66)	4 Wed. .	9972·1307	168·2932	215·9388		4824
29 Mar. (88)	6 Fri. .	17 27 30	26 Mar. (85)	3 Tues. .	6·7703	104·2768	267·2486		4825
28 Mar. (88)	0 Sat. .	23 40 0	14 Mar. (74)	0 Sat. .	9882·4537	951·5116	236·4247		4826
29 Mar. (88)	2 Mon. .	5 52 30	4 Mar. (63)	5 Thur.	96·7690	835·0380	208·3387		4827
29 Mar. (88)	3 Tues. .	12 5 0	23 Mar. (82)	4 Wed. .	131·4086	771·0215	259·6484		4828
29 Mar. (88)	4 Wed. .	18 17 30	12 Mar. (71)	1 Sun. .	7·0920	618·2563	228·8246		4829
29 Mar. (89)	6 Fri. .	0 30 0	29 Feb. (60)	5 Thur.	9882·7754	465·4911	198·0006		4830
29 Mar. (88)	0 Sat. .	6 42 30	19 Mar. (78)	4 Wed. .	9917·4150	401·4746	248·3104		4831
29 Mar. (88)	1 Sun. .	12 55 0	8 Mar. (67)	1 Sun. .	9793·0984	248·7095	218·4865		4832
29 Mar. (88)	2 Mon. .	19 7 30	27 Mar. (86)	0 Sat. .	9827·7380	184·6929	208·7963		4833
29 Mar. (89)	4 Wed. .	1 20 0	16 Mar. (76)	5 Thur.	42·0533	68·2194	241·7102		4834
29 Mar. (88)	5 Thur.	7 32 30	5 Mar. (64)	2 Mon. .	9917·7367	915·4542	210·8864		4835
29 Mar. (88)	6 Fri. .	13 45 0	24 Mar. (83)	1 Sun. .	9952·3763	851·4377	260·1960		4836
29 Mar. (88)	0 Sat. .	19 57 30	14 Mar. (73)	6 Fri. .	166·6915	734·9641	234·1099		4837
29 Mar. (89)	2 Mon. .	2 10 0	2 Mar. (62)	3 Tues. .	42·3749	582·1989	203·2861		4838
29 Mar. (88)	3 Tues. .	8 22 30	21 Mar. (80)	2 Mon. .	77·0146	518·1725	254·5958		4839
29 Mar. (88)	4 Wed. .	14 35 0	10 Mar. (69)	6 Fri. .	9952·6979	365·4172	223·7720		4840
29 Mar. (88)	5 Thur.	20 47 30	29 Mar. (88)	5 Thur.	9987·3376	301·4008	275·1017		4841
29 Mar. (89)	0 Sat. .	3 0 0	17 Mar. (77)	2 Mon. .	9863·0209	148·6356	244·2579		4842
29 Mar. (88)	1 Sun. .	9 12 30	7 Mar. (66)	0 Sat. .	77·3362	31·1320	216·17[illegible]7		4843
29 Mar. (88)	2 Mon. .	15 25 0	26 Mar. (85)	6 Fri. .	111·9758	968·1455	267·4815		4844
29 Mar. (88)	3 Tues. .	21 37 30	15 Mar. (74)	3 Tues. .	9987·6592	815·3803	236·6576		4845
29 Mar. (89)	5 Thur.	3 50 0	4 Mar. (64)	1 Sun. .	201·9744	698·9068	208·5707		4846

TABLE

CONCURRENT YEAR.								INTERCALATED (*adhika*) and SUPPRESSED (*kshaya*) LUNAR MONTHS (true).
Kali.	Saka.	Chaitrādi Vikrama.	Mēshādi solar year in Bengal.	Kollam.	A.D.	JOVIAN SAMVATSARA. Southern system.	JOVIAN SAMVATSARA. Northern system.	
1	2	3	3*a*	4	5	6	7	8
4847	1668	1803	1152	920-21	1745-46	59 Krōdhana	9 Yuvan	...
4848	1669	1804	1153	921-22	1746-47	60 Kshaya	10 Dhātṛi	...
4849	1670	1805	1154	922-23	1747-48	1 Prabhava	11 Īśvara	1 Chaitra
4850	1671	1806	1155	923-24	1748-49	2 Vibhava	12 Bahudhānya	...
4851	1672	1807	1156	924-25	1749-50	3 Śukla	13 Pramāthin	6 Bahudhānya
4852	1373	1808	1157	925-26	1750-51	4 Pramōda	14 Vikrama	...
4853	1674	1809	1158	926-27	1751-52	5 Prajāpati	15 Vṛisha	...
4854	1675	1810	1159	927-28	*1752-53	6 Aṅgiras	16 Chitrabhānu	4 Āshāḍha
4855	1676	1811	1160	928-29	1753-54	7 Śrīmukha	17 Subhānu	...
4856	1677	1812	1161	929-30	1754-55	8 Bhāva	18 Tāraṇa	...
4857	1678	1813	1162	930-31	1755-56	9 Yuvan	19 Pārthiva	3 Jyēshṭha
4858	1679	1814	1163	931-32	*1756-57	10 Dhātṛi	20 Vyaya	...
4859	1680	1815	1164	932-33	1757-58	11 Īśvara	21 Sarvajit†	7 Āsvina
4860	1681	1816	1165	933-34	1758-59	12 Bahudhānya	23 *Virōdhin*	...
4861	1682	1817	1166	934-35	1759-60	13 Pramāthin	24 *Vikṛita*	...
4862	1683	1818	1167	935-36	*1760-61	14 Vikrama	25 *Khara*	5 Śrāvaṇa
4863	1684	1819	1168	936-37	1761-62	15 Vṛisha	26 *Nandana*	...
4864	1685	1820	1169	937-38	1762-63	16 Chitrabhānu	27 *Vijaya*	...
4865	1686	1821	1170	938-39	1763-64	17 Subhānu	28 *Jaya*	4 Āshāḍha
4866	1687	1822	1171	939-40	*1764-65	18 Tāraṇa	29 *Manmatha*	...
4867	1688	1823	1172	940-41	1765-66	19 Pārthiva	30 *Durmukha*	...
4868	1689	1824	1173	941-42	1766-67	20 Vyaya	31 *Hēmalamba*	1 Chaitra
4869	1690	1825	1174	942-43	1767-68	21 Sarvajit	32 *Vilamba*	...
4870	1691	1826	1175	943-44	*1768-69	22 Sarvadhārin	33 *Vikārin*	5 Śrāvaṇa
4971	1692	1827	1176	944-45	1769-70	23 Virōdhin	34 *Sarvarin*	...

† 22 Sarvadhārin was suppressed in the North.

LXI—*Contd.*

COMMENCEMENT OF THE								Kali.
SOLAR YEAR.			LUNI-SOLAR YEAR (MEAN SUNRISE OF DAY ON WHICH CHAITRA ŚUKLA 1 ENDED).					
Day and month, A.D.	Week-day.	Time of true Mēsha-saṃkrānti.	Day and month, A.D.	Week-day.	*a.*	*b.*	*c.*	
13	14	17	19	20	23	24	25	1
		H. M. S.						
29 Mar. (88)	6 Fri.	10 2 30	23 Mar. (82)	0 Sat.	236·6140	634·8902	259·8813	4847
29 Mar. (88)	0 Sat.	16 15 0	12 Mar. (71)	4 Wed.	112·2974	482·1250	229·0575	4848
29 Mar. (88)	1 Sun.	22 27 30	1 Mar. (60)	1 Sun.	9987·9809	329·3599	198·2335	4849
29 Mar. (89)	3 Tues.	4 40 0	19 Mar. (79)	0 Sat.	22·6204	265·3434	249·5433	4850
29 Mar. (88)	4 Wed.	10 52 30	8 Mar. (67)	4 Wed.	9898·3038	112·5782	219·7194	4851
29 Mar. (88)	5 Thur.	17 5 0	27 Mar. (86)	3 Tues.	9932·9434	48·5617	270·0292	4852
29 Mar. (88)	6 Fri.	23 17 30	17 Mar. (76)	1 Sun.	147·2587	932·0882	241·9431	4853
29 Mar. (89)	1 Sun.	5 30 0	5 Mar. (65)	5 Thur.	22·9421	779·3229	211·1193	4854
9 Apr. (99)‡	2 Mon.	11 42 30	4 Apr. (94)‡	4 Wed.	57·5817	715·3058	262·4289	4855
9 Apr. (99)	3 Tues.	17 55 0	24 Mar. (83)	1 Sun.	9933·2651	562·5413	231·6051	4856
10 Apr. (100)	5 Thur.	0 7 30	13 Mar. (72)	5 Thur.	9808·9484	409·7760	200·7812	4857
9 Apr. (100)	6 Fri.	6 20 0	31 Mar. (91)	4 Wed.	9843·5881	345·7595	252·0910	4858
9 Apr. (99)	0 Sat.	12 32 30	20 Mar. (79)	1 Sun.	9719·2715	192·9944	221·2671	4859
9 Apr. (99)	1 Sun.	18 45 0	8 Apr. (98)	0 Sat.	9753·9111	128·9779	272·5768	4860
10 Apr. (100)	3 Tues.	0 57 30	29 Mar. (88)	5 Thur.	9968·2263	12·5043	244·4908	4861
9 Apr. (100)	4 Wed.	7 10 0	18 Mar. (78)	3 Tues.	182·5416	896·0307	216·4046	4862
9 Apr. (99)	5 Thur.	13 22 30	6 Apr. (96)	2 Mon.	217·1812	832·0143	267·7144	4863
9 Apr. (99)	6 Fri.	19 35 0	26 Mar. (85)	6 Fri.	92·8646	679·2490	236·8905	4864
10 Apr. (100)	1 Sun.	1 47 30	15 Mar. (74)	3 Tues.	9968·5480	526·4839	206·0667	4865
9 Apr. (100)	2 Mon.	8 0 0	2 Apr. (93)	2 Mon.	3·1876	462·4674	257·3764	4866
9 Apr. (99)	3 Tues.	14 12 30	22 Mar. (81)	6 Fri.	9878·8710	309·7022	226·5526	4867
9 Apr. (99)	4 Wed.	20 25 0	11 Mar. (70)	3 Tues.	9754·5544	156·9370	195·7286	4868
10 Apr. (100)	6 Fri.	2 37 30	30 Mar. (89)	2 Mon.	9789·1940	92·9205	247·0384	4869
9 Apr. (100)	0 Sat.	8 50 0	19 Mar. (79)	0 Sat.	3·5093	976·4470	218·9523	4870
9 Apr. (99)	1 Sun.	15 2 30	7 Apr. (97)	6 Fri.	38·1489	912·4304	270·2621	4871

‡ From here, inclusive, the dates A. D. are in New Style.

TABLE

CONCURRENT YEAR.									Intercalated (*adhika*) and suppressed (*kashaya*) lunar months (true).
						Jovian Saṁvatsara.			
Kali.	Śaka.	Chaitrādi Vikrama.	Mēshādi solar year in Bengal.	Kollam.	A.D.	Southern system.		Northern system.	
1	2	3	3a	4	5	6		7	8
4872	1693	1828	1177	945-46	1770-71	24 Vikṛita . .		35 *Plava* . .	..
4873	1694	1829	1178	946-47	1771-72	25 Khara . .		36 Śubhakṛit .	4 Āshāḍha .
4874	1695	1830	1179	947-48	*1772-73	26 Nandana .		37 Śōbhana .	..
4875	1696	1831	1180	948-49	1773-74	27 Vijaya . .		38 Krōdhin .	..
4876	1697	1832	1181	949-50	1774-75	28 Jaya . .		39 Viśvāvasu .	2 Vaiśākha .
4877	1698	1833	1182	950-51	1775-76	29 Manmatha .		40 Parābhava .	..
4878	1699	1834	1183	951-52	*1776-77	30 Durmukha .		41 Plavanga .	7 Āśvina .
4879	1700	1835	1184	952-53	1777-78	31 Hēmalamba .		42 Kīlaka . .	..
4880	1701	1836	1185	953-54	1778-79	32 Vilamba .		43 Saumya . .	..
4881	1702	1837	1186	954-55	1779-80	33 Vikārin .		44 Sādhāraṇa .	5 Śrāvaṇa .
4882	1703	1838	1187	955-56	*1780-81	34 Śārvarin .		45 Virōdhakṛit .	..
4883	1704	1839	1188	956-57	1781-82	35 Plava . .		46 Paridhāvin .	..
4884	1705	1840	1189	957-58	1782-83	36 Śubhakṛit .		47 Pramādin .	3 Jyēshṭha .
4885	1706	1841	1190	958-59	1783-84	37 Śōbhana .		48 Ānanda .	..
4886	1707	1842	1191	959-60	*1784-85	38 Krōdhin .		49 Rākshasa .	..
4887	1708	1843	1192	960-61	1785-86	39 Viśvāvasu .		50 Anala . .	1 Chaitra .
4888	1709	1844	1193	961-62	1786-87	40 Parābhava .		51 Pingala .	..
4889	1710	1845	1194	962-63	1787-88	41 Plavanga .		52 Kalayukta .	5 Śrāvaṇa .
4890	1711	1846	1195	963-64	*1788-89	42 Kīlaka . .		53 Siddhārthin .	..
4891	1712	1847	1196	964-65	1789-90	43 Saumya . .		54 Raudra .	..
4892	1713	1848	1197	965-66	1790-91	44 Sādhāraṇa .		55 Durmati .	4 Āshāḍha .
4893	1714	1849	1198	966-67	1791-92	45 Virōdhakṛit .		56 Dundubhi .	..
4894	1715	1850	1199	967-68	*1792-93	46 Paridhāvin .		57 Rudhirōdgārin	..
4895	1716	1851	1200	968-69	1793-94	47 Pramādin .		58 Raktāksha .	2 Vaiśākha .
4896	1717	1852	1201	969-70	1794-95	48 Ānanda .		59 Krōdhana .	..

LXI—*Contd.*

COMMENCEMENT OF THE									
SOLAR YEAR.			LUNI-SOLAR YEAR (MEAN SUNRISE OF DAY ON WHICH CHAITRA ŚUKLA 1 ENDED).						Kali.
Day and month, A.D.	Week-day.	Time of true Mēsha-saṁkrānti.	Day and month, A.D.	Week-day.	*a.*	*b.*	*c.*		
13	14	17	19	20	23	24	25		1
		H. M. S.							
9 Apr. (99)	2 Mon. .	21 15 0	28 Mar. (87)	4 Wed.	252·4642	795·9569	242·1760		4872
10 Apr. (100)	4 Wed. .	3 27 30	17 Mar. (76)	1 Sun. .	128·1476	643·1917	211·3522		4873
9 Apr. (100)	5 Thur.	9 40 0	4 Apr. (95)	0 Sat. .	162·7872	579·1752	262·6618		4874
9 Apr. (99)	6 Fri. .	15 52 30	24 Mar. (83)	4 Wed. .	38·4706	426·4100	231·8380		4875
9 Apr. (99)	0 Sat. .	22 5 0	13 Mar. (72)	1 Sun. .	9914·1539	273·6448	201·0141		4876
10 Apr. (100)	2 Mon. .	4 17 30	1 Apr. (91)	0 Sat. .	9948·7935	209·6283	252·3239		4877
9 Apr. (100)	3 Tues. .	10 30 0	20 Mar. (80)	4 Wed. .	9824·4769	56·8631	221·5000		4878
9 Apr. (99)	4 Wed. .	16 42 30	8 Apr. (98)	3 Tues. .	9859·1165	992·8466	272·8097		4879
9 Apr. (99)	5 Thur.	22 55 0	29 Mar. (88)	1 Sun. .	73·4318	876·3731	244·7237		4880
10 Apr. (100)	0 Sat. .	5 7 30	19 Mar. (78)	6 Fri. .	287·7470	759·8994	216·6375		4881
9 Apr. (100)	1 Sun. .	11 20 0	5 Apr. (96)	4 Wed. .	9983·7548	659·5914	265·2095		4882
9 Apr. (99)	2 Mon. .	17 32 30	26 Mar. (85)	2 Mon. .	198·0700	544·1178	237·1234		4883
9 Apr. (99)	3 Tues. .	23 45 0	15 Mar. (74)	6 Fri. .	73·7534	390·3525	206·2996		4884
10 Apr. (100)	5 Thur.	5 57 30	2 Apr. (92)	4 Wed. .	9769·7612	290·0445	254·8715		4885
9 Apr. (100)	6 Fri. .	12 10 0	22 Mar. (82)	2 Mon. .	9984·0764	173·5709	226·7854		4886
9 Apr. (99)	0 Sat. .	18 22 30	11 Mar. (70)	6 Fri. .	9859·7598	20·8058	195·9615		4887
10 Apr. (100)	2 Mon. .	0 35 0	30 Mar. (89)	5 Thur.	9894·3994	956·7892	247·2713		4888
10 Apr. (100)	3 Tues. .	6 47 30	20 Mar. (79)	3 Tues. .	108·7147	840·3157	219·1852		4889
9 Apr. (100)	4 Wed. .	13 0 0	7 Apr. (98)	2 Mon. .	143·3443	776·2992	270·4950		4890
9 Apr. (99)	5 Thur.	19 12 30	27 Mar. (86)	6 Fri. .	19·0377	623·5339	239·6711		4891
10 Apr. (100)	0 Sat. .	1 25 0	16 Mar. (75)	3 Tues. .	9894·7211	470·7688	208·8473		4892
10 Apr. (100)	1 Sun. .	7 37 30	4 Apr. (94)	2 Mon. .	9929·3607	406·7523	260·1569		4893
9 Apr. (100)	2 Mon. .	13 50 0	23 Mar. (83)	6 Fri .	9805·0441	253·9871	229·3332		4894
9 Apr. (99)	3 Tues. .	20 2 30	13 Mar. (72)	4 Wed. .	19·3593	137·5135	201·2470		4895
10 Apr. (100)	5 Thur.	2 15 0	1 Apr. (91)	3 Tues. .	53·9990	73·4971	252·5567		4896

TABLE

CONCURRENT YEAR.								INTERCALATED (*adhika*) and SUPPRESSED (*kshaya*) LUNAR MONTHS (true).
Kali.	Śaka.	Chaitrādi Vikrama.	Mēshādi solar year in Bengal.	Kollam.	A.D.	Jovian Saṁvatsara. Southern system.	Jovian Saṁvatsara. Northern system.	
1	2	3	3*a*	4	5	6	7	8
4897	1718	1853	1202	970-71	1795-96	49 Rākshasa	60 Kshaya	6 Bhādrapada
4898	1719	1854	1203	971-72	*1796-97	50 Anala	1 Prabhava	...
4899	1720	1855	1204	972-73	1797-98	51 Piṅgala	2 Vibhava	...
4900	1721	1856	1205	973-74	1798-99	52 Kālayukta	3 Śukla	5 Śrāvaṇa
4901	1722	1857	1206	974-75	1799-1800	53 Siddhārthin	4 Pramōda	...
4902	1723	1858	1207	975-76	1800-01§	54 Raudra	5 Prajāpati	...
4903	1724	1859	1208	976-77	1801-02	55 Durmati	6 Aṅgiras	3 Jyēshṭha
4904	1725	1860	1209	977-78	1802-03	56 Dundubhi	7 Śrīmukha	...
4905	1726	1861	1210	978-79	1803-04	57 Rudhirōdgārin	8 Bhāva	...
4906	1727	1862	1211	979-80	*1804-05	58 Raktāksha	9 Yuvan	1 Chaitra
4907	1728	1863	1212	980-81	1805-06	59 Krōdhana	10 Dhātṛi	...
4908	1729	1864	1213	981-82	1806-07	60 Kshaya	11 Īśvara	5 Śrāvaṇa
4909	1730	1865	1214	982-83	1807-08	1 Prabhava	12 Bahudhānya	...
4910	1731	1866	1215	983-84	*1808-09	2 Vibhava	13 Pramāthin	...
4911	1732	1867	1216	984-85	1809-10	3 Śukla	14 Vikrama	4 Āshāḍha
4912	1733	1868	1217	985-86	1810-11	4 Pramōda	15 Vṛisha	...
4913	1734	1869	1218	986-87	1811-12	5 Prajāpati	16 Chitrabhānu	...
4914	1735	1870	1219	987-88	*1812-13	6 Aṅgiras	17 Subhānu	2 Vaiśākha
4915	1736	1871	1220	988-89	1813-14	7 Śrīmukha	18 Tāraṇa	...
4916	1737	1872	1221	989-90	1814-15	8 Bhāva	19 Pārthiva	6 Bhādrapada
4917	1738	1873	1222	990-91	1815-16	9 Yuvan	20 Vyaya	...
4918	1739	1874	1223	991-92	*1816-17	10 Dhātṛi	21 Sarvajit	...
4919	1740	1875	1224	992-93	1817-18	11 Īśvara	22 Sarvadhārin	5 Śrāvaṇa
4920	1741	1876	1225	993-94	1818-19	12 Bahudhānya	23 Virōdhin	...
4921	1742	1877	1226	994-95	1819-20	13 Pramāthin	24 Vikṛita	...

§ The year A. D. 1800 was not a Leap-year.

LXI—*Contd.*

COMMENCEMENT OF THE									
SOLAR YEAR.			LUNI-SOLAR YEAR (MEAN SUNRISE OF DAY ON WHICH CHAITRA ŚUKLA 1 ENDED).						Kali.
Day and month, A.D.	Week-day.	Time of true Mēsha-saṁkrānti.	Day and month. A.D.	Week-day.	*a.*	*b.*	*c.*		
13	14	17	19	20	23	24	25		1
		H. M. S.							
10 Apr. (100)	6 Fri.	8 27 30	21 Mar. (80)	0 Sat.	9929·6824	920·7319	221·7329		4897
9 Apr (100)	0 Sat.	14 40 0	8 Apr. (99)	6 Fri.	9964·3220	856·7153	273·0426		4898
9 Apr. (99)	1 Sun.	20 52 30	29 Mar. (88)	4 Wed.	178·6372	740·2418	244·9565		4899
10 Apr. (100)	3 Tues.	3 5 0	18 Mar. (77)	1 Sun.	54·3206	587·4766	214·1326		4900
10 Apr. (100)	4 Wed.	9 17 30	6 Apr. (96)	0 Sat.	88·9603	522·4602	265·4424		4901
10 Apr. (100)	5 Thur.	15 30 0	26 Mar. (85)	4 Wed.	9964·6436	370·6950	234·6186		4902
10 Apr. (100)	6 Fri.	21 42 30	15 Mar. (74)	1 Sun.	9840·3270	217·9297	203·7948		4903
11 Apr. (101)	1 Sun.	3 55 0	3 Apr. (93)	0 Sat.	9874·9667	153·9133	255·1044		4904
11 Apr. (101)	2 Mon.	10 7 30	24 Mar. (83)	5 Thur.	89·2819	37·4397	227·0184		4905
10 Apr. (101)	3 Tues.	16 20 0	12 Mar. (72)	2 Mon.	9964·9653	884·6745	196·1945		4906
10 Apr. (100)	4 Wed.	22 32 30	31 Mar. (90)	1 Sun.	9999·7049	820·6580	247·5043		4907
11 Apr. (101)	6 Fri.	4 45 0	21 Mar. (80)	6 Fri.	213·9202	704·1845	219·4182		4908
11 Apr. (101)	0 Sat.	10 57 30	9 Apr. (99)	5 Thur.	248·5598	640·1680	270·7280		4909
10 Apr. (101)	1 Sun.	17 10 0	28 Mar. (88)	2 Mon.	124·2432	487·4027	239·9041		4910
10 Apr. (100)	2 Mon.	23 22 30	17 Mar. (76)	6 Fri.	9999·9266	334·6376	209·0802		4911
11 Apr. (101)	4 Wed.	5 35 0	5 Apr. (95)	5 Thur.	34·5662	270·6211	260·3899		4912
11 Apr. (101)	5 Thur.	11 47 30	25 Mar. (84)	2 Mon.	9910·2496	117·8558	229·5661		4913
10 Apr. (101)	6 Fri.	18 0 0	14 Mar. (74)	0 Sat.	124·5648	1·3823	201·4800		4914
11 Apr. (101)	1 Sun.	0 12 30	2 Apr. (92)	6 Fri.	159·2044	937·3658	252·7898		4915
11 Apr. (101)	2 Mon.	6 25 0	22 Mar. (81)	3 Tues.	34·8878	784·7007	221·9659		4916
11 Apr. (101)	3 Tues.	12 37 30	10 Apr. (100)	2 Mon.	69·5275	720·5841	273·2755		4917
10 Apr. (101)	4 Wed.	18 50 0	29 Mar. (89)	6 Fri.	9945·2109	566·8190	242·4517		4918
11 Apr. (101)	6 Fri.	1 2 30	18 Mar. (77)	3 Tues.	9820·8942	415·0538	211·3279		4919
11 Apr. (101)	0 Sat.	7 15 0	6 Apr. (96)	2 Mon.	9855·5338	351·0372	262·0376		4920
11 Apr. (101)	1 Sun.	13 27 30	26 Mar. (85)	6 Fri.	9731·2172	199·2721	232·1158		4921

TABLE

CONCURRENT YEAR.								Intercalated (*adhika*) and suppressed (*kshaya*) lunar months (true)
						Jovian Saṁvatsara.		
Kali.	Śaka.	Chaitrādi Vikrama.	Mēshādi solar year in Bengal.	Kollam.	A.D.	Southern system.	Northern system.	
1	2	3	3*a*	4	5	6	7	8
4922	1743	1878	1227	995-96	*1820-21	14 Vikrama	25 Khara	3 Jyēshṭha
4923	1744	1879	1228	996-97	1821-22	15 Vṛisha	26 Nandana	...
4924	1745	1880	1229	997-98	1822-23	16 Chitrabhānu	27 Vijaya	7 Āśvina 10 Pausha (Ksh)
4925	1746	1881	1230	998-99	1823-24	17 Subhānu	28 Jaya	1 Chaitra
4926	1747	1882	1231	999-1000	*1824-25	18 Tāraṇa	29 Manmatha	...
4927	1748	1883	1232	1000-01	1825-26	19 Pārthiva	30 Durmukha	5 Śrāvaṇa
4928	1749	1884	1233	1001-02	1826-27	20 Vyaya	31 Hēmalamba	...
4929	1750	1885	1234	1002-03	1827-28	21 Sarvajit	32 Vilamba	...
4930	1751	1886	1235	1003-04	*1828-29	22 Sarvadhārin	33 Vikārin	4 Āshāḍha
4931	1752	1887	1236	1004-05	1829-30	23 Virōdhin	34 Śārvarin	...
4932	1753	1888	1237	1005-06	1830-31	24 Vikṛita	35 Plava	...
4933	1754	1889	1238	1006-07	1831-32	25 Khara	36 Śubhakṛit	2 Vaiśākha
4934	1755	1890	1239	1007-08	*1832-33	26 Nandana	37 Śōbhana	...
4935	1756	1891	1240	1008-09	1833-34	27 Vijaya	38 Krōdhin	6 Bhādrapada
4936	1757	1892	1241	1009-10	1834-35	28 Jaya	39 Viśvāvasu	...
4937	1758	1893	1242	1010-11	1835-36	29 Manmatha	40 Parābhava	...
4938	1759	1894	1243	1011-12	*1836-37	30 Durmukha	41 Plavaṅga	4 Āshāḍha
4939	1760	1895	1244	1012-13	1837-38	31 Hēmalamba	42 Kīlaka	...
4940	1761	1896	1245	1013-14	1838-39	32 Vilamba	43 Saumya	...
4941	1762	1897	1246	1014-15	1839-40	33 Vikārin	44 Sādhāraṇa	3 Jyēshṭha
4942	1763	1898	1247	1015-16	*1840-41	34 Śārvarin	45 Virōdhakṛit	...
4943	1764	1899	1248	1016-17	1841-42	35 Plava	46 Paridhāvin	7 Āśvina 11 Māgha (Ksh)
4944	1765	1900	1249	1017-18	1842-43	36 Śubhakṛit	47 Pramādin†	1 Chaitra
4945	1766	1901	1250	1018-19	1843-44	37 Śōbhana	49 *Rakshasa*	...
4946	1767	1902	1251	1019-20	*1844-45	38 Krōdhin	50 *Anala*	5 Śrāvaṇa

† 48 Ananda was suppressed in the North.

LXI—*Contd.*

COMMENCEMENT OF THE									
SOLAR YEAR.			LUNI-SOLAR YEAR (MEAN SUNRISE OF DAY ON WHICH CHAITRA ŚUKLA 1 ENDED).						Kali.
Day and month. A.D.	Week-day.	Time of true Mēsha saṁkrānti.	Day and month, A.D.	Week-day.	*a.*	*b.*	*c.*		
13	14	17	19	20	23	24	25	1
		H. M. S.						
10 Apr. (101)	2 Mon.	19 40 0	15 Mar. (75)	4 Wed.	9945·5324	81·7985	204·0277	4922
11 Apr. (101)	4 Wed.	1 52 30	3 Apr. (93)	3 Tues.	9980·1723	17·7821	255·3373	4923
11 Apr. (101)	5 Thur.	8 5 0	24 Mar. (83)	1 Sun.	194·4873	901·3084	227·2513	4924
11 Apr. (101)	6 Fri.	14 17 30	13 Mar. (72)	5 Thur.	70·1767	748·5433	196·4274	4925
10 Apr. (101)	0 Sat.	20 30 0	31 Mar. (91)	4 Wed.	104·8103	684·5268	247·7372	4926
11 Apr. (101)	2 Mon.	2 42 30	20 Mar. (79)	1 Sun.	9980·4937	531·7615	216·9133	4927
11 Apr. (101)	3 Tues.	8 55 0	8 Apr. (98)	0 Sat.	15·1333	467·7451	268·2231	4928
11 Apr. (101)	4 Wed.	15 7 30	28 Mar. (87)	4 Wed.	9890·8167	314·9799	237·3992	4929
10 Apr. (101)	5 Thur.	21 20 0	16 Mar. (76)	1 Sun.	9766·5001	162·2147	206·5753	4930
11 Apr. (101)	0 Sat.	3 32 30	4 Apr. (94)	0 Sat.	9801·1397	98·1982	257·8848	4931
11 Apr. (101)	1 Sun.	9 45 0	25 Mar. (84)	5 Thur.	15·4550	981·7246	229·7990	4932
11 Apr. (101)	2 Mon.	15 57 30	15 Mar. (74)	3 Tues.	229·7702	865·2510	201·7129	4933
10 Apr. (101)	3 Tues.	22 10 0	2 Apr. (93)	2 Mon.	264·4099	801·2346	253·0226	4934
11 Apr. (101)	5 Thur.	4 22 30	22 Mar. (81)	6 Fri.	140·0933	648·4694	222·1988	4935
11 Apr. (101)	6 Fri.	10 35 0	10 Apr. (100)	5 Thur.	174·7329	584·4529	273·5084	4936
11 Apr. (101)	0 Sat.	16 47 30	30 Mar. (89)	2 Mon.	50·4163	431·6877	242·6846	4937
10 Apr. (101)	1 Sun.	23 0 0	18 Mar. (78)	6 Fri.	9926·0997	279·9225	211·8608	4938
11 Apr. (101)	3 Tues.	5 12 30	6 Apr. (96)	5 Thur.	9960·7393	214·9060	263·1705	4939
11 Apr. (101)	4 Wed.	11 25 0	26 Mar. (85)	2 Mon.	9836·4227	62·1408	232·3467	4940
11 Apr. (101)	5 Thur.	17 37 30	16 Mar. (75)	0 Sat.	50·7379	945·6672	204·2606	4941
10 Apr. (101)	6 Fri.	23 50 0	3 Apr. (94)	6 Fri.	85·3775	881·6508	255·5703	4942
11 Apr. (101)	1 Sun.	6 2 30	24 Mar. (83)	4 Wed.	299·6928	765·1772	327·4342	4943
11 Apr. (101)	2 Mon.	12 15 0	13 Mar. (72)	1 Sun.	175·3762	612·4120	196 6603	4944
11 Apr. (101)	3 Tues.	18 27 30	1 Apr. (91)	0 Sat.	210·0338	548·3955	247·9701	4945
11 Apr. (102)	5 Thur.	0 40 0	20 Mar. (80)	4 Wed.	85·6992	395·6303	217·1463	4946

TABLE

CONCURRENT YEAR.								INTERCALATED (*adhika*) and SUPPRESSED (*kshaya*) LUNAR MONTHS (true).
						JOVIAN SAMVATSARA.		
Kali.	Śaka.	Chaitrādi Vikrama.	Mēshādi solar year in Bengal.	Kollam.	A.D.	Southern system.	Northern system.	
1	2	3	3*a*	4	5	6	7	8
4947	1768	1903	1252	1020-21	1845-46	39 Viśvāvasu	51 *Piṅgala*	...
4948	1769	1904	1253	1021-22	1846-47	40 Parābhava	52 *Kālayukta*	...
4949	1770	1905	1254	1022-23	1847-48	41 Plavaṅga	53 *Siddhārthin*	3 Jyēshṭha
4950	1771	1906	1255	1023-24	*1848-49	42 Kīlaka	54 *Raudra*	...
4951	1772	1907	1256	1024-25	1849-50	43 Saumya	55 *Durmati*	...
4952	1773	1908	1257	1025-26	1850-51	44 Sādhāraṇa	56 *Dundubhi*	2 Vaiśākha
4953	1774	1909	1258	1026-27	1851-52	45 Virōdhakṛit	57 *Rudhirōdgārin*	...
4954	1775	1910	1259	1027-28	*1852-53	46 Paridhāvin	58 *Raktāksha*	6 Bhādrapada
4955	1776	1911	1260	1028-29	1853-54	47 Pramādin	59 *Krōdhana*	...
4956	1777	1912	1261	1029-30	1854-55	48 Ānanda	60 *Kshaya*	...
4957	1778	1913	1262	1030-31	1855-56	49 Rākshasa	1 *Prabhava*	4 Āshāḍha
4958	1779	1914	1263	1031-32	*1856-57	50 Anala	2 *Vibhava*	...
4959	1780	1915	1264	1032-33	1857-58	51 Piṅgala	3 Śukla	...
4960	1781	1916	1265	1033-34	1858-59	52 Kālayukta	4 Pramōda	3 Jyēshṭha
4961	1782	1917	1266	1034-35	1859-60	53 Siddhārthin	5 Prajāpati	...
4962	1783	1918	1267	1035-36	*1860-61	54 Raudra	6 Aṅgiras	7 Āśvina
4963	1784	1919	1268	1036-37	1861-62	55 Durmati	7 Śrīmukha	...
4964	1785	1920	1269	1037-38	1862-63	56 Dundubhi	8 Bhāva	...
4965	1786	1921	1270	1038-39	1863-64	57 Rudhirōdgārin	9 Yuvan	5 Srāvana
4966	1787	1922	1271	1039-40	*1864-65	58 Raktāksha	10 Dhātṛi	...
4967	1788	1923	1272	1040-41	1865-66	59 Krōdhana	11 Īśvara	...
4968	1789	1924	1273	1041-42	1866-67	60 Kshaya	12 Bahudhānya	3 Jyēshṭha
4969	1790	1925	1274	1042-43	1867-68	1 Prabhava	13 Pramāthin	...
4970	1791	1926	1275	1043-44	*1868-69	2 Vibhava	14 Vikrama	...
4971	1792	1927	1276	1044-45	1869-70	3 Śukla	15 Vṛisha	2 Vaiśākha

LXI—*Contd.*

COMMENCEMENT OF THE											
SOLAR YEAR					LUNI-SOLAR YEAR (MEAN SUNRISE OF DAY ON WHICH CHAITRA ŚUKLA 1 ENDED).					Kali.
Day and month, A.D.	Week-day.	Time of true Mēsha saṁkrānti.			Day and month, A.D.	Week-day.	*a.*	*b.*	*c*	
13	14	17			19	20	23	24	25	1
		H.	M.	S.						
11 Apr. (101)	6 Fri.	6	52	30	7 Apr. (97)	2 Mon.	9781·7069	295·3222	265·7182	4947
11 Apr. (101)	0 Sat.	13	5	0	28 Mar. (87)	0 Sat.	9996·0221	178·8486	237·6321	4948
11 Apr. (101)	1 Sun.	19	17	30	17 Mar. (76)	4 Wed.	9871·7056	26·0835	206·8082	4949
11 Apr. (102)	3 Tues.	1	30	0	4 Apr. (95)	3 Tues.	9906·3451	962·0670	258·1179	4950
11 Apr. (101)	4 Wed.	7	42	30	25 Mar. (84)	1 Sun.	120·6604	845·5933	230·0319	4951
11 Apr. (101)	5 Thur.	13	55	0	14 Mar. (73)	5 Thur.	9996·3438	692·8282	199·2080	4952
11 Apr. (101)	6 Fri.	20	7	30	2 Apr. (92)	4 Wed.	30·9834	628·8117	249·5178	4953
11 Apr. (102)	1 Sun.	2	20	0	21 Mar. (81)	1 Sun.	9906·6668	476·0465	219·6939	4954
11 Apr. (101)	2 Mon.	8	32	30	9 Apr. (99)	0 Sat.	9941·3064	412·0390	271·0036	4955
11 Apr. (101)	3 Tues.	14	45	0	29 Mar. (88)	4 Wed.	9816·9898	259·2645	240·1797	4956
11 Apr. (101)	4 Wed.	20	57	30	19 Mar. (78)	2 Mon.	31·3051	142·7912	212·0937	4957
11 Apr. (102)	6 Fri.	3	10	0	6 Apr. (97)	1 Sun.	65·9447	78·7747	263·4034	4958
11 Apr. (101)	0 Sat.	9	22	30	26 Mar. (85)	5 Thur.	9941·5281	926·0096	232·5796	4959
11 Apr. (101)	1 Sun.	15	35	0	16 Mar. (75)	3 Tues.	155·9433	809·5360	204·4935	4960
11 Apr. (101)	2 Mon.	21	47	30	4 Apr. (94)	2 Mon.	190·5929	745·5195	255·8032	4961
11 Apr. (102)	4 Wed.	4	0	0	23 Mar. (83)	6 Fri.	66·2663	592·7543	224·9793	4962
11 Apr. (101)	5 Thur.	10	12	30	11 Apr. (101)	5 Thur.	100·9060	528·7379	276·2890	4963
11 Apr. (101)	6 Fri.	16	25	0	31 Mar. (90)	2 Mon.	9976·5893	375·9726	245·4652	4964
11 Apr. (101)	0 Sat.	22	37	30	20 Mar. (79)	6 Fri.	9852·2927	223·2074	214·6413	4965
11 Apr. (102)	2 Mon.	4	50	0	7 Apr. (98)	5 Thur.	9886·9124	159·1910	265·9511	4966
11 Apr. (101)	3 Tues.	11	2	30	28 Mar. (87)	3 Tues.	101·2276	42·7174	237·8650	4967
11 Apr. (101)	4 Wed.	17	15	0	17 Mar. (76)	0 Sat.	9976·9110	889·9522	207·0411	4968
11 Apr. (101)	5 Thur.	23	27	30	5 Apr. (95)	6 Fri.	11·5506	825·9357	258·3508	4969
11 Apr. (102)	0 Sat.	5	40	0	25 Mar. (85)	4 Wed.	225·8659	709·4621	230·2648	4970
11 Apr. (101)	1 Sun.	11	52	30	14 Mar. (73)	1 Sun.	101·5493	556·6969	199·4409	4971

TABLE

CONCURRENT YEAR.								INTERCALATED (*adhika*) and SUPPRESSED (*kshaya*) LUNAR MONTHS (true).
Kali.	Śaka.	Chaitrādi Vikrama.	Mēshādi solar year in Bengal.	Kollam.	A.D.	Jovian Saṁvatsara. Southern system.	Jovian Saṁvatsara. Northern system	
1	2	3	3a	4	5	6	7	8
4972	1793	1928	1277	1045-46	1870-71	4 Pramōda	16 Chitrabhānu	...
4973	1794	1929	1278	1046-47	1871-72	5 Prajāpati	17 Subhānu	6 Bhādrapada
4974	1795	1930	1279	1047-48	*1872-73	6 Aṅgiras	18 Tāraṇa	...
4975	1796	1931	1280	1048-49	1873-74	7 Srīmukha	19 Pārthiva	...
4976	1797	1932	1281	1049-50	1874-75	8 Bhāva	20 Vyaya	4 Āshāḍha
4977	1798	1933	1282	1050-51	1875-76	9 Yuvan	21 Sarvajit	...
4978	1799	1934	1283	1051-52	*1876-77	10 Dhātṛi	22 Sarvadhārin	...
4979	1800	1935	1284	1052-53	1877-78	11 Iśvara	23 Virōdhin	3 Jyēshṭha
4980	1801	1936	1285	1053-54	1878-79	12 Bahudhānya	24 Vikṛita	...
4981	1802	1937	1286	1054-55	1879-80	13 Pramāthin	25 Khara	7 Āśvina
4982	1803	1938	1287	1055-56	*1880-81	14 Vikrama	26 Nandana	...
4983	1804	1939	1288	1056-57	1881-82	15 Vṛisha	27 Vijaya	...
4984	1805	1940	1289	1057-58	1882-83	16 Chitrabhānu	28 Jaya	5 Śrāvaṇa
4985	1806	1941	1290	1058-59	1883-84	17 Subhānu	29 Manmatha	...
4986	1807	1942	1291	1059-60	*1884-85	18 Tāraṇa	30 Durmukha	...
4987	1808	1943	1292	1060-61	1885-86	19 Pārthiva	31 Hēmalamba	3 Jyēshtha
4988	1809	1944	1293	1061-62	1886-87	20 Vyaya	32 Vilamba	...
4989	1810	1945	1294	1062-63	1887-88	21 Sarvajit	33 Vikārin	...
4990	1811	1946	1295	1063-64	*1888-89	22 Sarvadhārin	34 Śārvarin	1 Chaitra
4991	1812	1947	1296	1064-65	1889-90	23 Virōdhin	35 Plava	...
4992	1813	1948	1297	1065-66	1890-91	24 Vikṛita	36 Śubhakṛit	6 Bhādrapada
4993	1814	1949	1298	1066-67	1891-92	25 Khara	37 Śōbhana	...
4994	1815	1950	1299	1067-68	*1892-93	26 Nandana	38 Krōdhin	...
4995	1816	1951	1300	1068-69	1893-94	27 Vijaya	39 Viśvāvasu	4 Āshāḍha
4996	1817	1952	1301	1069-70	1894-95	28 Jaya	40 Parābhava	...
4997	1818	1953	1302	1070-71	1895-96	29 Manmatha	41 Plavaṅga	...
4998	1819	1954	1303	1071-72	*1896-97	30 Durmukha	42 Kīlaka	3 Jyēshṭha
4999	1820	1955	1304	1072-73	1897-98	31 Hēmalamba	43 Saumya	...
5000	1821	1956	1305	1073-74	1898-99	32 Vilamba	44 Śādhāraṇa	7 Āśvina
5001	1822	1957	1306	1074-75	1899-1900	33 Vikārin	45 Virōdhakṛit	...
5002	1823	1958	1307	1075-76	§1900-01	34 Śārvarin	46 Paridhāvin	...

§ The year 1900 A.D. was not a Leap-year.

LXI—*Concld.*

COMMENCEMENT OF THE											
SOLAR YEAR.					LUNI-SOLAR YEAR (MEAN SUNRISE OF DAY ON WHICH CHAITRA ŚUKLA 1 ENDED).						Kali.
Day and month, A.D.	Week-day.	Time of true Mēsha saṁkrānti.			Day and month, A.D.	Week-day.	*a.*	*b.*	*c.*		
13	14	17			19	20	23	24	25		1
		H.	M.	S.							
11 Apr. (101)	2 Mon. .	18	5	0	2 Apr. (92)	0 Sat. .	136·1889	492·6804	250·7517		4972
12 Apr. (102)	4 Wed. .	0	17	30	22 Mar. (81)	4 Wed. .	11·8733	339·9153	219·9268		4973
11 Apr. (102)	5 Thur.	6	30	0	9 Apr. (100)	3 Tues.	46·5119	275·8988	271·2365		4974
11 Apr. (101)	6 Fri. .	12	42	30	29 Mar. (88)	0 Sat. .	9922·1953	123·1335	240·4126		4975
11 Apr. (101)	0 Sat. .	18	55	0	19 Mar. (78)	5 Thur.	136·5106	6·6600	212·3266		4976
12 Apr. (102)	2 Mon. .	1	7	30	7 Apr. (97)	4 Wed. .	171·1501	942·6435	263·6363		4977
11 Apr. (102)	3 Tues.	7	20	0	26 Mar. (86)	1 Sun. .	46·8335	789·8783	232·8125		4978
11 Apr. (101)	4 Wed. .	13	32	30	16 Mar. (75)	6 Fri. .	261·1487	673·4047	204·8264		4979
11 Apr. (101)	5 Thur.	19	45	0	3 Apr. (93)	4 Wed. .	9957·1566	573·0967	253·0983		4980
12 Apr. (102)	0 Sat. .	1	57	30	23 Mar. (82)	1 Sun. .	9832·8399	420·3314	222·4744		4981
11 Apr. (102)	1 Sun. .	8	10	0	10 Apr. (101)	0 Sat. .	9867·4795	356·3149	273·7841		4982
11 Apr. (101)	2 Mon. .	14	22	30	30 Mar. (89)	4 Wed. .	9743·1629	203·5498	242·9603		4983
11 Apr. (101)	3 Tues.	20	35	0	20 Mar. (79)	2 Mon. .	9957·4781	87·0761	214·8742		4984
12 Apr. (102)	5 Thur.	2	47	30	8 Apr. (98)	1 Sun. .	9992·1178	23·0597	266·1840		4985
11 Apr. (102)	6 Fri. .	9	0	0	28 Mar. (88)	6 Fri. .	206·4330	906·5861	238·0978		4986
11 Apr. (101)	0 Sat. .	15	12	30	17 Mar. (76)	3 Tues.	82·1164	753·8210	207·2730		4987
11 Apr. (101)	1 Sun. .	21	25	0	5 Apr. (95)	2 Mon. .	116·7560	689·8044	258·5837		4988
12 Apr. (102)	3 Tues.	3	37	30	25 Mar. (84)	6 Fri. .	9992·4394	537·0392	227·7599		4989
11 Apr. (102)	4 Wed. .	9	50	0	13 Mar. (73)	3 Tues.	9868·1228	384·2741	196·9360		4990
11 Apr. (101)	5 Thur.	16	2	30	1 Apr. (91)	2 Mon. .	9902·7624	320·2575	248·2457		4991
11 Apr. (101)	6 Fri. .	22	15	0	21 Mar. (80)	6 Fri. .	9778·4458	167·4924	217·4219		4992
12 Apr. (102)	1 Sun. .	4	27	30	9 Apr. (99)	5 Thur.	9813·0854	103·4759	268·7316		4993
11 Apr. (102)	2 Mon. .	10	40	0	29 Mar. (89)	3 Tues.	27·4007	987·0023	240·6455		4994
11 Apr. (101)	3 Tues.	16	52	30	19 Mar. (78)	1 Sun. .	241·7160	870·5287	212·5595		4995
11 Apr. (101)	4 Wed. .	23	5	0	7 Apr. (97)	0 Sat. .	276·3556	806·5123	263·8692		4996
12 Apr. (102)	6 Fri. .	5	17	30	27 Mar. (86)	4 Wed. .	152·0390	653·7471	233·0454		4997
11 Apr. (102)	0 Sat. .	11	30	0	15 Mar. (75)	1 Sun. .	27·7223	500·9718	202·2215		4998
11 Apr. (101)	1 Sun. .	17	42	30	3 Apr. (93)	0 Sat. .	62·3620	436·9653	253·5311		4999
11 Apr. (101)	2 Mon. .	23	55	0	23 Mar. (82)	4 Wed. .	9938·0453	284·2062	222·7073		5000
12 Apr. (102)	4 Wed. .	6	7	30	11 Apr. (101)	3 Tues.	9971·6850	220·1837	274·0170		5001
12 Apr. (102)	5 Thur.	12	20	0	31 Mar. (90)	0 Sat. .	9848·3683	67·4185	243·1932		5002

TABLE LXII.

Names of months and nakshatras.

(*Corresponding to Table II, Part II, "Indian Calendar.)"*

Lunar months.			Solar months.					
No.	Usual name.	Tamil name.	No.	Sign name.	Bengal name.	Tamil name.	Malayālam name.	Orissa name.
1	2	3	4	5	6	7	8	9
1	Chaitra	Paggu	1	Mēsha	Vaiśākha	Chittirai	Mēḍam	Baiśāk.
2	Vaiśākha	Bēsa	2	Vrishabha	Jyēshṭha	Vaikāsi[1]	Edavam	Joisṭhō.
3	Jyēshṭha	Kārtelu	3	Mithuna	Āshāḍha	Āni	Midunam	Āssar.
4	Āshāḍha	Āti	4	Karka	Śrāvaṇa	Āḍi	Karkaḍagam	Sāwun.
5	Śrāvaṇa	Sōna	5	Simha	Bhādrapada	Āvani	Chiṅgam	Bhādro.
6	Bhādrapada	Nirmāla	6	Kanyā	Āśvina	Purattādi[2]	Kanni	Āssin.
7	Āśvina	Bontelu	7	Tulā	Kārttika	Aippaśi[3]	Tuḷām	Kārttik.
8	Kārttika	Jārde	8	Vriśchika	Mārgaśira	Kārttigai	Vriśchikam	Āghrān.
9	Mārgaśira	Perārde	9	Dhanus	Pausha	Mārgaḷi	Dhanu	Paus.
10	Pausha	Puntelu	10	Makara	Māgha	Tai	Makaram	Māgha.
11	Māgha	Māyi	11	Kumbha	Phālguna	Māsi	Kumbham	Falgun.
12	Phālguna	Suggi	12	Mīna	Chaitra	Paṅguni	Mīnam	Choitro.

[1] or Vaiyāsi. [2] or Purattāsi. [3] or Ārppisi, or Appisi.

Nakshatras.[1]

No.	Name.	Tamil name.	Deity.	No.	Name.	Tamil name.	Deity.
1	Aśvinī	Asuvati	Aśvin.	15	Svāti	Sōdi	Vāyu.
2	Bharaṇī	Bharanī	Yama.	16	Viśākhā	Viśākam	Indrāgnī.
3	Kṛittikā	Kiruttigai	Agni.	17	Anurādhā	Anusham, or Anilum.	Mitra.
4	Rohiṇī	Rohinī	Prajāpati.	18	Jyēshṭhā	Kēṭṭai	Indra.
5	Mṛigaśira	Mirugusīram	Soma.	19	Mūla	Mūlam	Nirriti.
6	Ārdrā	Ārudra, or Tiruvādirai.	Rudra.	20	Purva Ashādhā.	Pūrāḍam	Āpaḥ.
7	Punarvasu	Punarpūsam	Aditi.	21	Uttara Ashādhā.	Uttirāḍam	Viśvadevaḥ.
8	Pushya	Pūsam	Bṛihaspati.		Abhijit.	...	Brahma.
				22	Śravaṇa	Tiruvōnam	Vishṇu.
9	Āślēshā	Āyilyam	Sarpāḥ.	23	Dhanishṭhā or Śravishṭhā.	Aviṭṭam	Vasavaḥ.
10	Maghā	Magham	Pitaraḥ.				
11	Pūrva-Phalguni.	Pūram	Bhaga.	24	Śatabhishaj or Śatatārakā.	Sadayam	Varuṇa.
12	Uttara Phalguni.	Uttiram	Āryaman.	25	Pūrva Bhadrapadā.	Pūraṭṭādi	Aja Ekapād.
13	Hasta	Hastam or Āttam.	Savitṛi.	26	Uttara Bhadrapadā.	Uttiraṭṭādi	Ahi Budhnya.
14	Chitrā	Chittirai	Tvashṭri.	27	Rēvatī	Rēvatī	Pūshan.

[1] Tamil names and those of Deities are borrowed from Dewan Bahadur L. D. Swamikannu Pillai's "*Indian Chronology*."

TABLE LXIII A.

(Corresponding to Table III, Part I, "Indian Calendar.")

COLLECTIVE DURATION OF MEAN LUNAR MONTHS.

LUNI-SOLAR YEAR (CHAITRĀDI).				
Serial number.	Name of month.	COLLECTIVE DURATION FROM BEGINNING OF YEAR TO END OF EACH *mean* LUNAR MONTH.		
		Exactly in Tithis.	In civil days.	
			Approximate.	Exact.
1	2	3	3*a*	3*b*
1	Chaitra . . .	30	30	29·53
2	Vaiśākha . .	60	59	59·06
3	Jyēshṭha . .	90	89	88·59
4	Āshāḍha . .	120	118	118·12
5	Śrāvaṇa . .	150	148	147·65
6	Bhādrapada . .	180	177	177·18
7	Āśvina . . .	210	207	206·71
8	Kārttika . .	240	236	236·24
9	Mārgaśira . .	270	266	265·77
10	Pausha . . .	300	295	295·30
11	Māgha . . .	330	325	324·83
12	Phālguna . .	360	354	354·36
	In intercalary years.	390	384	383·89

TABLE

DURATION AND COLLECTIVE DURATION OF TRUE SOLAR MONTHS, WITH INCREASE

The values are tho e

"W. D."—Week-day. "a" in 10,000ths

(This Table supersedes Table XVIII A "Indian

Luni-solar month ending at the second of the two solar saṁkrāntis with which it is connected.	At the true solar saṁkrānti.	Collective duration in days, hours, etc., and collective increase of *a, b, c* from true Mēsha saṁkrānti to each true saṁkrānti.							
		D.	W.-D.	H.	M.	S.	*a*	*b*	*c*
1	2	3					4	5	6
	Mīna S. (*of previous year*).								
1. Chaitra									
	Mēsha saṁkr.	0	(0)	0	0	0	0	0	0
2. Vaiśākha									
	Vṛishabha saṁkr.	30	(2)	22	11	6·99	471·9831	122·2961	84·6643
3. Jyēshṭha									
	Mithuna saṁkr.	62	(6)	7	47	43·05	1105·1653	261·8682	170·6319
4. Āshāḍha									
	Karka saṁkr.	93	(2)	22	22	0·37	1808·3520	408·9426	257·1654
5. Śrāvaṇa									
	Siṁha saṁkr.	125	(6)	9	34	40·40	2464·1251	550·9358	343·3157
6. Bhādrapada									
	Kanyā saṁkr.	156	(2)	10	24	24·88	2973·4105	677·2297	428·2817
7. Āśvina									
	Tulā saṁkr.	186	(4)	21	21	37·82	3286·9182	782·5419	511·6648
8. Kārttika									
	Vṛiśchika saṁkr.	216	(6)	19	2	43·34	3413·2087	867·7898	593·5344
9. Margaśīrsha									
	Dhanus saṁkr.	246	(1)	7	15	59·08	3405·9677	938·7268	674·3243
10. Pausha									
	Makara saṁkr.	275	(2)	15	41	4·81	3345·0707	3·9135	754·6804
11. Māgha									
	Kumbha saṁkr.	305	(4)	2	39	12·57	3320·1612	72·9570	835·3275
12. Phālguna									
	Mīna saṁkr.	334	(5)	22	4	5·29	3414·4196	154·7719	916·9379
1. Chaitra (*of following year*)									
	Mēsha saṁkr. (*of following year*)	365	(1)	6	12	30·0	3688·2315	255·8299	1000·0

NOTE.

EXACT VALUE OF "*c*" AND OF "EQUATION *c*" AT THE SEVERAL TRUE SAṀKRĀNTIS IN EACH YEAR.

Saṁkrānti.	*c*.	Eqn. *c*.
1. Mēsha saṁkr.	277·4558	0·9119
2. Vṛishabha saṁkr.	362·1201	14·2168
3. Mithuna saṁkr.	448·0877	40·5649
4. Karka saṁkr.	534·6212	72·5193
5. Siṁha saṁkr.	620·7715	100·7366
6. Kanyā saṁkr.	705·7375	117·0626
7. Tulā saṁkr.	789·1206	117·5601
8. Vṛiśchika saṁkr.	870·9902	102·9215
9. Dhanus saṁkr.	951·7801	77·4872
10. Makara saṁkr.	32·1362	47·7147
11. Kumbha saṁkr.	112·7833	20·8518
12. Mīna saṁkr.	194·3937	3·6236

LXIII B.

OF *a*, *b*, *c*, AT EACH SAṀKRĀNTI BY THE FIRST ARYA-SIDDHĀNTA.

fixed by M. de Ries.

of circle; "b" and "c" in 1,000ths.

Chronography," p. 132) and "Indian Calendar," Table III, Part II.

At true solar saṁkrānti.	Length of month preceding each true saṁkrānti and increase of *a*, *b*, *c*, between each true saṁkrānti.							
	D.	W-D.	H.	M.	S.	*a*.	*b*.	*c*.
7	8					9	10	11
Mēsha saṁkr.	0	0	0	0	0	0	0	0
Vṛishabha saṁkr.	30	(2)	22	11	6·99	471·9831	122·296	84·6643
Mithuna saṁkr.	31	(3)	9	36	36·06	633·1822	139·5721	85·9676
Karka saṁkr.	31	(3)	14	34	17·32	703·1867	147·0744	86·5335
Siṁha saṁkr.	31	(3)	11	12	40·02	655·7731	141·9932	86·150?
Kanyā saṁkr.	31	(3)	0	49	44·48	509·2854	126·2939	84·9660
Tulā saṁkr.	30	(2)	10	57	12·94	313·5077	105·3122	83·3831
Vṛischika saṁkr.	29	(1)	21	41	5·52	126·2905	85·2479	81·8696
Dhanus saṁkr.	29	(1)	12	13	15·74	9992·7590	70·9370	30·7899
Makara saṁkr.	29	(1)	8	25	5·73	9939·1030	65·1867	80·3561
Kumbha saṁkr.	29	(1)	10	58	7·76	9975·0905	69·0435	80·6471
Mīna saṁkr.	29	(1)	19	24	52·72	94·2584	81·8149	81·6104
Mēsha saṁkr. (*of following year*).	30	(2)	8	8	24·71	273·8119	101·0580	83·0822

TABLE LXIV.

INCREASE OF a, b, c IN DAYS OF 24 HOURS EACH BY THE FIRST ĀRYA SIDDHĀNTA WITH LALLA'S BIJA.

"a" in 10,000ths; "b" and "c" in 1,000ths of circle.

This Table corresponds to Table IV, "Indian Calendar."

Increase in	a.	b.	c.
One day	338·631873982	36·291623738	2·737785720
One year of 365 days	3600·634003430	246·442664370	999·291787800
One year of 366 days	3939·265877412	282·734288108	2·029573520
One century of 36,525 days . . .	8529·197184659	551·557045243	997·623429983
One century of 36,526 days . . .	8867·829058641	587·848668981	0·361215706

DAYS OF 24 HOURS EACH.

No.	Week-day.	a	b.	c.	No.	Week-day	a.	b.	c.
1	2	3	4	5	1	2	3	4	5
1	1	338·6319	36·2916	2·7378	31	[illegible]	497·5881	125·0103	84·871
2	2	677·2637	72·5832	5·4756	32	[illegible]	836·2200	161·3320	87·609
3	3	1015·8956	108·8749	8·2134	33	[illegible]	1174·8518	197·6236	90·346
4	4	1354·5275	145·1665	10·9511	34	6	1513·4837	233·9152	93·084
5	5	1693·1594	181·4581	13·6889	35	0	1852·1156	270·2068	95·822
6	6	2031·7912	217·7497	16·4267	36	1	2190·7475	306·4985	98·560
7	0	2370·4231	254·0414	19·1645	37	2	2529·3793	342·7901	101·298
8	1	2709·0550	290·3330	21·9023	38	3	2868·0112	379·0817	104·035
9	2	3047·6869	326·6246	24·6401	39	4	3206·6431	415·3733	106·773
10	3	3386·3187	362·9162	27·3779	40	5	3545·2750	451·6649	109·511
11	4	3724·9506	399·2079	30·1156	41	6	3883·9068	487·9566	112·2492
12	5	4063·5825	435·4995	32·8534	42	0	4222·5387	524·2482	114·9870
13	6	4402·2144	471·7911	35·5912	43	1	4561·1706	560·5398	117·7248
14	0	4740·8462	508·0827	38·3290	44	2	4899·8025	596·8314	120·4626
15	1	5079·4781	544·3744	41·0668	45	3	5238·4343	633·1231	123·2004
16	2	5418·1100	580·6660	43·8046	46	4	5577·0662	669·4147	125·9381
17	3	5756·7419	616·9576	46·5424	47	5	5915·6981	705·7063	128·6759
18	4	6095·3737	653·2492	49·2801	48	6	6254·3300	741·9979	131·4137
19	5	6434·0056	689·5409	52·0179	49	0	6592·9618	778·2896	134·1515
20	6	6772·6375	725·8325	54·7557	50	1	6931·5937	814·5812	136·8893
21	0	7111·2694	762·1241	57·4935	51	2	7270·2256	850·8728	139·6271
22	1	7449·9012	798·4157	60·2313	52	3	7608·8574	887·1644	142·3649
23	2	7788·5331	834·7073	62·9691	53	4	7947·4893	923·4561	145·1026
24	3	8127·1650	870·9990	65·7069	54	5	8286·1212	959·7477	147·8404
25	4	8465·7968	907·2906	68·4446	55	6	8624·7531	996·0393	150·5782
26	5	8804·4287	943·5822	71·1824	56	0	8963·3849	32·3309	153·3160
27	6	9143·0606	979·8738	73·9202	57	1	9302·0168	68·6226	156·0538
28	0	9481·6925	16·1655	76·6580	58	2	9640·6487	104·9142	158·7916
29	1	9820·3243	52·4571	79·3958	59	3	9979·2806	141·2058	161·5294
30	2	158·9562	88·7487	82·1236	60	4	317·9124	177·4974	164·2671

TABLE LXIV—*Contd.*

DAYS—*Contd.*

No.	Week-day.	a.	b.	c.	No.	Week-day.	a.	b.	c.
1	2	3	4	5	1	2	3	4	5
61	5	656·5443	213·7890	167·0049	111	6	7588·1380	28·3702	303·8942
62	6	995·1762	250·0807	169·7427	112	0	7926·7699	64·6619	306·6320
63	0	1333·8081	286·3723	172·4805	113	1	8265·4018	100·9535	309·3698
64	1	1672·4399	322·6639	175·2183	114	2	8604·0336	137·2451	312·1076
65	2	2011·0718	358·9555	177·9561	115	3	8942·6655	173·5367	314·8454
66	3	2349·7037	395·2472	180·6939	116	4	9281·2974	209·8284	317·5831
67	4	2688·3356	431·5388	183·4316	117	5	9619·9293	246·1200	320·3209
68	5	3026·9674	467·8304	186·1694	118	6	9958·5611	282·4116	323·0587
69	6	3365·5993	504·1220	188·9072	119	0	297·1930	318·7032	325·7965
70	0	3704·2312	540·4137	191·6450	120	1	635·8249	354·9948	328·5343
71	1	4042·8631	576·7053	194·3828	121	2	974·4568	391·2865	331·2721
72	2	4381·4949	612·9969	197·1206	122	3	1313·0886	427·5781	334·0099
73	3	4720·1268	649·2885	199·8584	123	4	1651·7205	463·8697	336·7476
74	4	5058·7587	685·5801	202·5961	124	5	1990·3524	500·1613	339·4854
75	5	5397·3905	721·8718	205·3339	125	6	2328·9842	536·4530	342·2232
76	6	5736·0224	758·1634	208·0717	126	0	2667·6161	572·7446	344·9610
77	0	6074·6543	794·4550	210·8095	127	1	3006·2480	609·0362	347·6988
78	1	6413·2862	830·7467	213·5473	128	2	3344·8799	645·3278	350·4366
79	2	6751·9180	867·0383	216·2851	129	3	3683·5117	681·6195	353·1744
80	3	7090·5499	903·3299	219·0229	130	4	4022·1436	717·9111	355·9121
81	4	7429·1818	939·6215	221·7606	131	5	4360·7755	754·2027	358·6499
82	5	7767·8137	975·9131	224·4984	132	6	4699·4074	790·4943	361·3877
83	6	8106·4455	12·2048	227·2362	133	0	5038·0392	826·7860	364·1255
84	0	8445·0774	48·4964	229·9740	134	1	5376·6711	863·0776	366·8633
85	1	8783·7093	84·7880	232·7118	135	2	5715·3030	899·3692	369·6011
86	2	9122·3412	121·0796	235·4496	136	3	6053·9349	935·66[illegible]3	372·3389
87	3	9460·9730	157·3713	238·1874	137	4	6392·5667	971·9525	375·0766
88	4	9799·6049	193·6629	240·9251	138	5	6731·1986	8·2441	377·8144
89	5	138·2367	229·9545	243·6629	139	6	7069·8305	44·5357	380·5522
90	6	476·8687	266·2461	246·4007	140	0	7408·4624	80·8273	383·2900
91	0	815·5005	302·5378	249·1385	141	1	7747·0942	117·1189	386·0278
92	1	1154·1324	338·8294	251·8763	142	2	8085·7261	153·4106	388·7656
93	2	1492·7643	375·1210	254·6141	143	3	8424·3580	189·7022	391·5034
94	3	1831·3962	411·4126	257·3519	144	4	8762·9899	225·9938	394·2411
95	4	2170·0280	447·7043	260·0896	145	5	9101·6217	262·2854	396·9789
96	5	2508·6599	483·9959	262·8274	146	6	9440·2536	298·5771	399·7167
97	6	2847·2918	520·2875	265·5652	147	0	9778·8855	334·8686	402·4545
98	0	3185·9237	556·5791	268·3030	148	1	117·5173	371·1603	405·1923
99	1	3524·5555	592·8708	271·0408	149	2	456·1492	407·4519	407·9301
100	2	3863·1874	629·1624	273·7786	150	3	794·7811	443·7436	410·6679
101	3	4201·8193	665·4540	276·5164	151	4	1133·4130	480·0352	413·4056
102	4	4540·4511	701·7456	279·2541	152	5	1472·0448	516·3268	416·1434
103	5	4879·0830	738·0372	281·9919	153	6	1810·6767	552·6184	418·8812
104	6	5217·7149	774·3289	284·7297	154	0	2149·3086	588·9101	421·6190
105	0	5556·3468	810·6205	287·4675	155	1	2487·9405	625·2017	424·3568
106	1	5894·9786	846·9121	290·2053	156	2	2826·5723	661·4933	427·0946
107	2	6233·6105	883·2037	292·9431	157	3	3165·2042	697·7849	429·8324
108	3	6572·2424	919·4954	295·6809	158	4	3503·8361	734·0766	432·5701
109	4	6910·8743	955·7870	298·4186	159	5	3842·4680	770·3682	435·3079
110	5	7249·5061	992·0786	301·1564	160	6	4181·0998	806·6598	438·0475

TABLE LXIV—*Contd.*

DAYS—*Contd.*

No.	Week-day.	*a.*	*b.*	*c.*	No.	Week-day.	*a.*	*b.*	*c.*
1	2	3	4	5	1	2	3	4	5
161	0	4519·7317	842·9514	440·7835	211	1	1451·3254	657·5326	577·6728
162	1	4858·3636	879·2430	443·5213	212	2	1789·9572	693·8242	580·4106
163	2	5196·9955	915·5347	446·2591	213	3	2128·5892	729·1159	583·1484
164	3	5535·6273	951·8263	448·9969	214	4	2467·2210	766·4075	585·8861
165	4	5874·2592	988·1179	451·7346	215	5	2805·8529	802·6991	588·6239
166	5	6212·8911	24·4095	454·4724	216	6	3144·4848	838·9907	591·3617
167	6	6551·5230	60·7012	457·2102	217	0	3483·1167	875·2824	594·0995
168	0	6890·1548	96·9928	459·9480	218	1	3821·7485	911·5740	596·8373
169	1	7228·7867	133·2844	462·6858	219	2	4160·3804	947·8656	599·5751
170	2	7567·4186	169·5760	465·4236	220	3	4499·0123	984·1572	602·3129
171	3	7906·0505	205·8677	468·1613	221	4	4837·6442	20·4488	605·0506
172	4	8244·6823	242·1593	470·8991	222	5	5176·2760	56·7405	607·7884
173	5	8583·3142	278·4509	473·6369	223	6	5514·9079	93·0321	610·5262
174	6	8921·9461	314·7425	476·3747	224	0	5853·5398	129·3237	613·2640
175	0	9260·5779	351·0342	479·1125	225	1	6192·1716	165·6153	616·0018
176	1	9599·2098	387·3258	481·8503	226	2	6530·8035	201·9070	618·7396
177	2	9937·8417	423·6174	484·5881	227	3	6869·4354	238·1986	621·4774
178	3	276·4736	459·9090	487·3259	228	4	7208·0673	274·4902	624·2151
179	4	615·1054	496·2006	490·0636	229	5	7546·6991	310·7818	626·9529
180	5	953·7373	532·4923	492·8014	230	6	7885·3310	347·0735	629·6907
181	6	1292·3692	568·7839	495·5392	231	0	8223·9629	383·3651	632·4285
182	0	1631·0011	605·0755	498·2770	232	1	8562·5948	419·6567	635·1663
183	1	1969·6329	641·3671	501·0148	233	2	8901·2266	455·9483	637·9041
184	2	2308·2648	677·6588	503·7526	234	3	9239·8585	492·2400	640·6419
185	3	2646·8967	713·9504	506·4904	235	4	9578·4904	528·5316	643·3796
186	4	2985·5286	750·2420	509·2281	236	5	9917·1223	564·8232	646·1174
187	5	3324·1604	786·5336	511·9659	237	6	255·7541	601·1148	648·8552
188	6	3662·7923	822·8253	514·7037	238	0	594·3860	637·4064	651·5930
189	0	4001·4242	858·1169	517·4415	239	1	933·0179	673·6981	654·3308
190	1	4340·0561	895·4085	520·1793	240	2	1271·6498	709·9897	657·0686
191	2	4678·6879	931·7001	522·9171	241	3	1610·2816	746·2813	659·8064
192	3	5017·3198	967·9918	525·6549	242	4	1948·9135	782·5729	662·5441
193	4	5355·9517	4·2834	528·3926	243	5	2287·5454	818·8646	665·2819
194	5	5694·5836	40·5750	531·1304	244	6	2626·1773	855·1562	668·0197
195	6	6033·2154	76·8666	533·8682	245	0	2964·8091	891·4478	670·7575
196	0	6371·8473	113·1583	536·6060	246	1	3303·4410	927·7394	673·4953
197	1	6710·4792	149·4499	539·3438	247	2	3642·0729	964·0311	676·2331
198	2	7049·1110	185·7415	542·0816	248	3	3980·7047	0·3227	678·9709
199	3	7387·7429	222·0331	544·8194	249	4	4319·3366	36·6143	681·7086
200	4	7726·3748	258·3247	547·5571	250	5	4657·9685	72·9059	684·4464
201	5	8065·0067	294·6164	550·2949	251	6	4996·6004	109·1976	687·1842
202	6	8403·6385	330·9080	553·0327	252	0	5335·2322	145·4892	689·9220
203	0	8742·2704	367·1996	555·7705	253	1	5673·8641	181·7808	692·6598
204	1	9080·9023	403·4912	558·5083	254	2	6012·4960	218·0724	695·3975
205	2	9419·5342	439·7829	561·2461	255	3	6351·1279	254·3641	698·1354
206	3	9758·1660	476·0745	563·9839	256	4	6689·7597	290·6557	700·8731
207	4	96·7979	512·3661	566·7216	257	5	7028·3916	326·9473	703·6109
208	5	335·4298	548·6577	569·4594	258	6	7367·0235	363·2389	706·3487
209	6	774·0617	584·9494	572·1972	259	0	7705·6554	399·5305	709·0865
210	9	1112·6935	621·2410	574·9350	260	1	8044·2872	435·8222	711·8243

TABLE LXIV—*Contd.*

Days—*Contd.*

No.	Week-day.	a.	b.	c.	No.	Week-day.	a	b.	c.
1	2	3	4	5	1	2	3	4	5
261	2	8382·9191	472·1138	714·5621	311	3	5314·5128	286·6950	851·4514
262	3	8721·5510	508·4054	717·2999	312	4	5653·1446	322·9866	854·1891
263	4	9060·1829	544·6970	720·0376	313	5	5991·7766	359·2782	856·9269
264	5	9398·8147	580·9887	722·7754	314	6	6330·4084	395·5699	859·6647
265	6	9737·4466	617·2803	725·5132	315	0	6669·0403	431·8615	862·4025
266	0	76·0785	653·5719	728·2510	316	1	7007·6722	468·1531	865·1403
267	1	414·7104	689·8635	730·9888	317	2	7346·3041	504·4447	867·8781
268	2	753·3422	726·1552	733·7266	318	3	7684·9359	540·7363	870·6159
269	3	1091·9741	762·4468	736·4644	319	4	8023·5678	577·0280	873·3536
270	4	1430·6060	798·7384	739·2021	320	5	8362·1997	613·3196	876·0914
271	5	1769·2378	835·0300	741·9399	321	6	8700·8315	649·6112	878·8292
272	6	2107·8697	871·3217	744·6777	322	0	9039·4634	685·9028	881·5670
273	0	2446·5016	907·6133	747·4155	323	1	9378·0953	722·1945	884·3048
274	1	2785·1335	943·9049	750·1533	324	2	9716·7272	758·4861	887·0426
275	2	3123·7653	979·1965	752·8911	325	3	55·3590	794·7777	889·7804
276	3	3462·3972	16·4882	755·6289	326	4	394·9909	831·0693	892·5181
277	4	3801·0291	52·7798	758·3666	327	5	732·6228	867·3610	895·2559
278	5	4139·6610	89·0714	761·1044	328	6	1071·2547	903·6526	897·9937
279	6	4478·2928	125·3630	763·8422	329	0	1409·8865	939·9442	900·7315
280	0	4816·9247	161·6546	766·5800	330	1	1748·5184	976·2358	903·4693
281	1	5155·5566	197·9463	769·3178	331	2	2087·1503	12·5275	906·2071
282	2	5494·1885	234·2379	772·0556	332	3	2425·7822	48·8191	908·9449
283	3	5832·8203	270·5295	774·7934	333	4	2764·4140	85·1107	911·6826
284	4	6171·4522	306·8211	777·5311	334	5	3103·0459	121·4023	914·4204
285	5	6510·0841	343·1128	780·2689	335	6	3441·6778	157·6940	917·1582
286	6	6848·7160	379·4044	783·0067	336	0	3780·3097	193·9856	919·8960
287	0	7187·3478	415·6960	785·7445	337	1	4118·9415	230·2772	922·6338
288	1	7525·9797	451·9876	788·4823	338	2	4457·5734	266·5688	925·3716
289	2	7864·6116	488·2793	791·2201	339	3	4796·2053	302·8604	928·1094
290	3	8203·2435	524·5709	793·9579	340	4	5134·8372	339·1521	930·8471
291	4	8541·8753	560·8625	796·6956	341	5	5473·4690	375·4437	933·5849
292	5	8880·5072	597·1541	799·4334	342	6	5812·1009	411·7353	936·3227
293	6	9219·1391	633·4458	802·1712	343	0	6150·7328	448·0269	939·0605
294	0	9557·7710	669·7374	804·9090	344	1	6489·3646	484·3186	941·7983
295	1	9896·4028	706·0290	807·6468	345	2	6827·9965	520·6102	944·5361
296	2	235·0347	742·3206	810·3846	346	3	7166·6284	556·9018	947·2739
297	3	573·6666	778·6123	813·1224	347	4	7505·2603	593·1934	950·0116
298	4	912·2984	814·9039	815·8601	348	5	7843·8921	629·4851	952·7494
299	5	1250·9303	851·1955	818·5979	349	6	8182·5240	665·7767	955·4872
300	6	1589·5622	887·4871	821·3357	350	0	8521·1559	702·0683	958·2250
301	0	1928·1941	923·7787	824·0735	351	1	8859·7878	738·3599	960·9828
302	1	2266·8259	960·0704	826·8113	352	2	9198·4196	774·6516	963·7006
303	2	2605·4578	996·3620	829·5491	353	3	9537·0515	810·9432	966·4384
304	3	2944·0897	32·6536	832·2869	354	4	9875·6834	847·2348	969·1761
305	4	3282·7216	68·9452	835·0246	355	5	214·3153	883·5264	971·9136
306	5	3621·3534	105·2369	837·7624	356	6	552·9471	919·8181	974·6517
307	6	3959·9853	141·5285	840·5002	357	0	891·5790	956·1097	977·3895
308	0	4298·6172	177·8201	843·2380	358	1	1230·2109	992·4013	980·1273
309	1	4637·2491	214·1117	845·9758	359	2	1568·8428	28·6929	982·8651
310	2	4975·8809	250·4034	848·7136	360	3	1907·4746	64·9845	985·6029

TABLE LXIV—*Concld.*

DAYS—*Concld.*

No.	Week-day.	*a.*	*b.*	*c.*
1	2	3	4	5
361	4	2246·1065	101·2762	988·3406
362	5	2584·7384	137·5678	991·0784
363	6	2923·3703	173·8594	993·8162
364	0	3262·0021	210·1510	996·5540
365	1	3600·6340	246·4427	999·2918
366	2	3939·2659	282·7343	2·0296
367	3	4277·8978	319·0259	4·7674
368	4	4616·5296	355·3175	7·5051
369	5	4955·1615	391·6092	10·2429
370	6	5293·7934	427·9008	12·9807
371	0	5632·4252	464·1924	15·7185
372	1	5971·0571	500·4840	18·4563
373	2	6309·6890	536·7757	21·1941
374	3	6648·3209	573·0673	23·9319
375	4	6986·9527	609·3589	26·6696
376	5	7325·5846	645·6505	29·4074
377	6	7664·2165	681·9421	32·1452
378	0	8002·8484	718·2338	34·8830
379	1	8341·4802	754·5254	37·6208
380	2	8680·1121	790·8170	40·3586
381	3	9018·7440	827·1086	43·0964
382	4	9357·3759	863·4003	45·8341
383	5	9696·0077	899·6919	48·5719
384	6	34·6396	935·9835	51·3097
385	0	373·2715	972·2751	54·0475

TABLE LXV.

INCREASE OF *a*, *b*, *c* BY THE FIRST ĀRYA-SIDDHĀNTA WITH LALLA'S BIJA.

HOURS, MINUTES AND SECONDS.

("*a*" in 10,000ths of circle; "*b*" and "*c*" in 1,000ths.)

This Table corresponds to Table V, "Indian Calendar."

Increase in	*a.*	*b.*	*c.*
One hour	14·109661416	1·512150989	0·114074405
One minute	0·235161024	0·025202517	0·001901210
One second	0·003919350	0·000420042	0·000031687

HOURS.

No.	*a.*	*b.*	*c.*	No.	*a.*	*b.*	*c.*
1	14·1097	1·5122	0·1141	13	183·4256	19·6580	1·4830
2	28·2193	3·0243	0·2281	14	197·5353	21·1701	1·5970
3	42·3290	4·5365	0·3422	15	211·6449	22·6823	1·7111
4	56·4386	6·0486	0·4563	16	225·7546	24·1944	1·8252
5	70·5483	7·5608	0·5704	17	239·8642	25·7066	1·9393
6	84·6580	9·0729	0·6844	18	253·9739	27·2187	2·0533
7	98·7676	10·5851	0·7985	19	268·0836	28·7309	2·1674
8	112·8773	12·0972	0·9126	20	282·1932	30·2430	2·2815
9	126·9870	13·6094	1·0267	21	296·3029	31·7552	2·3956
10	141·0966	15·1215	1·1407	22	310·4126	33·2673	2·5096
11	155·2063	16·6337	1·2548	23	324·5222	34·7795	2·6237
12	169·3159	18·1458	1·3689	24	338·6319	36·2916	2·7378

MINUTES.

No.	*a.*	*b.*	*c.*	No.	*a.*	*b.*	*c.*	No.	*a.*	*b.*	*c.*
1	0·2352	0·0252	0·0019	21	4·9384	0·5293	0·0399	41	9·6416	1·0333	0·0780
2	0·4703	0·0504	0·0038	22	5·1735	0·5545	0·0418	42	9·8768	1·0585	0·0799
3	0·7055	0·0756	0·0057	23	5·4087	0·5797	0·0437	43	10·1119	1·0837	0·0818
4	0·9406	0·1008	0·0076	24	5·6439	0·6049	0·0456	44	10·3471	1·1089	0·0837
5	1·1758	0·1260	0·0095	25	5·8790	0·6301	0·0475	45	10·5822	1·1341	0·0856
6	1·4110	0·1512	0·0114	26	6·1142	0·6553	0·0494	46	10·8174	1·1593	0·0875
7	1·6461	0·1764	0·0133	27	6·3493	0·6805	0·0513	47	11·0526	1·1845	0·0894
8	1·8813	0·2016	0·0152	28	6·5845	0·7057	0·0532	48	11·2877	1·2097	0·0913
9	2·1164	0·2268	0·0171	29	6·8197	0·7309	0·0551	49	11·5229	1·2349	0·0932
10	2·3516	0·2520	0·0190	30	7·0548	0·7561	0·0570	50	11·7581	1·2601	0·0951
11	2·5868	0·2772	0·0209	31	7·2900	0·7813	0·0589	51	11·9932	1·2853	0·0970
12	2·8219	0·3024	0·0228	32	7·5252	0·8065	0·0608	52	12·2284	1·3105	0·0989
13	3·0571	0·3276	0·0247	33	7·7603	0·8317	0·0627	53	12·4635	1·3357	0·1008
14	3·2923	0·3528	0·0266	34	7·9955	0·8569	0·0646	54	12·6987	1·3609	0·1027
15	3·5274	0·3780	0·0285	35	8·2306	0·8821	0·0665	55	12·9339	1·3861	0·1046
16	3·7626	0·4032	0·0304	36	8·4658	0·9073	0·0684	56	13·1690	1·4113	0·1065
17	3·9977	0·4284	0·0323	37	8·7010	0·9325	0·0703	57	13·4042	1·4365	0·1084
18	4·2329	0·4536	0·0342	38	8·9361	0·9577	0·0722	58	13·6393	1·4617	0·1103
19	4·4681	0·4788	0·0361	39	9·1713	0·9829	0·0741	59	13·8745	1·4869	0·1122
20	4·7032	0·5041	0·0380	40	9·4064	1·0081	0·0760	60	14·1097	1·5122	0·1141

TABLE LXV—*Contd.*

SECONDS.

No.	a.	b.	c.	No.	a.	b.	c.	No.	a.	b.	c.
1	0·0039	0·0004	0·0000	21	0·0823	0·0088	0·0007	41	0·1607	0·0172	0·0013
2	0·0078	0·0008	0·0001	22	0·0862	0·0092	0·0007	42	0·1646	0·0176	0·0013
3	0·0118	0·0013	0·0001	23	0·0901	0·0097	0·0007	43	0·1685	0·0181	0·0014
4	0·0157	0·0017	0·0001	24	0·0941	0·0101	0·0008	44	0·1725	0·0185	0·0014
5	0·0196	0·0021	0·0002	25	0·0980	0·0105	0·0008	45	0·1764	0·0189	0·0014
6	0·0235	0·0025	0·0002	26	0·1019	0·0109	0·0008	46	0·1803	0·0193	0·0015
7	0·0274	0·0029	0·0002	27	0·1058	0·0113	0·0009	47	0·1842	0·0197	0·0015
8	0·0314	0·0034	0·0003	28	0·1097	0·0118	0·0009	48	0·1881	0·0202	0·0015
9	0·0353	0·0038	0·0003	29	0·1137	0·0122	0·0009	49	0·1920	0·0206	0·0016
10	0·0392	0·0042	0·0003	30	0·1176	0·0126	0·0010	50	0·1960	0·0210	0·0016
11	0·0431	0·0046	0·0003	31	0·1215	0·0130	0·0010	51	0·1999	0·0214	0·0016
12	0·0470	0·0050	0·0004	32	0·1254	0·0134	0·0010	52	0·2038	0·0218	0·0016
13	0·0510	0·0055	0·0004	33	0·1293	0·0139	0·0010	53	0·2077	0·0223	0·0017
14	0·0549	0·0059	0·0004	34	0·1333	0·0143	0·0011	54	0·2116	0·0227	0·0017
15	0·0588	0·0063	0·0005	35	0·1372	0·0147	0·0011	55	0·2156	0·0231	0·0017
16	0·0627	0·0067	0·0005	36	0·1411	0·0151	0·0011	56	0·2195	0·0235	0·0018
17	0·0666	0·0071	0·0005	37	0·1450	0·0155	0·0012	57	0·2234	0·0239	0·0018
18	0·0705	0·0076	0·0006	38	0·1489	0·0160	0·0012	58	0·2273	0·0244	0·0018
19	0·0745	0·0080	0·0006	39	0·1525	0·0164	0·0012	59	0·2312	0·0248	0·0019
20	0·0784	0·0084	0·0006	40	0·1568	0·0168	0·0013	60	0·2352	0·0252	0·0019

TABLES LXVI, LXVII.

"EQUATION *b*" AND "EQUATION *c*" IN WHOLE NUMBERS BY THE FIRST ĀRYA-SIDDHĀNTA

(corresponding to Tables VI, VII, "Indian Calendar").

Tables LXVI-A and LXVII-A state the values of "equation *b*" and "equation *c*" in detail.

TABLE LXVI.

LUNAR "EQUATION *b*."

Arg.	Eqn.	Arg.	Arg.	Eqn.	Arg.
0	139	500	500	139	1000
10	148	490	510	130	990
20	157	480	520	121	980
30	165	470	530	114	970
40	174	460	540	105	960
50	182	450	550	96	950
60	191	440	560	88	940
70	199	430	570	80	930
80	206	420	580	72	920
90	214	410	590	65	910
100	221	400	600	58	900
110	228	390	610	51	890
120	235	380	620	44	880
130	241	370	630	38	870
140	247	360	640	32	860
150	252	350	650	27	850
160	257	340	660	22	840
170	262	330	670	17	830
180	265	320	680	13	820
190	269	310	690	10	810
200	272	300	700	7	800
210	274	290	710	4	790
220	276	280	720	2	780
230	277	270	730	1	770
240	278	260	740	0	760
250	279	250	750	0	750

TABLE LXVII.

SOLAR "EQUATION *c*."

Arg.	Eqn.	Arg.	Arg.	Eqn.	Arg.
0	60	500	500	60	1000
10	56	490	510	63	990
20	52	480	520	67	980
30	49	470	530	71	970
40	45	460	540	75	960
50	41	450	550	78	950
60	38	440	560	81	940
70	34	430	570	85	930
80	31	420	580	88	920
90	28	410	590	92	910
100	25	400	600	95	900
110	21	390	610	98	890
120	18	380	620	101	880
130	16	370	630	103	870
140	14	360	640	106	860
150	11	350	650	108	850
160	9	340	660	110	840
170	7	330	670	112	830
180	6	320	680	113	820
190	4	310	690	115	810
200	3	300	700	116	800
210	2	290	710	117	790
220	1	280	720	118	780
230	1	270	730	119	770
240	0	260	740	119	760
250	0	250	750	119	750

Diff. in equation.	Last figure of argument.								
	9	8	7	6	5	4	3	2	1
	Add or subtract.								
9	8	7	6	5	4 or 5	4	3	2	1
8	7	6	6	5	4	3	2	2	1
7	6	6	5	4	3 or 4	3	2	1	1
6	5	5	4	4	3	2	2	1	1
5	4 or 5	4	3 or 4	3	2 or 3	2	1 or 2	1	0 or 1
4	4	3	3	2	2	2	1	1	0
3	3	2	2	2	1 or 2	1	1	1	1
2	2	2	1	1	1	1	1	0	1
1	1	1	1	1	0 or 1	0	0	0	0

TABLE LXVI A.

(A) Moon's equation of the centre ("Equation *b*.") by the First Ārya-Siddhānta from ☾'s mean anomaly ("Arg. *b*.") 0—500 (0°—180°).

Cols. 3, 4.—Equation and difference stand for either of the mean anom. values in cols. 2a, 2b. For the 24 base-equations see Table LXX.

"Arg. b" is ☾'s mean anom. in 1,000ths of circle.

Col. 3.—The equation is ☾'s greatest equation *plus* the actual equation, in 10,000ths of circle.

Serial No. of sine.	Arg. *b*.	Equation *b*.	Diff.	Arg. *b*.	Serial No of sine.	Arg. *b*.	Equation *b*.	Diff.	Arg. *b*.
•	2*a*	3	4	2*b*	1	2*a*	3	4	2*b*
0	0·0	139·4275		500·0	12	125·0	237·9056		375·0
	2·083̇	141·2505		497·916̇		127·083̇	239·1537		372·916̇
	4·16̇	143·0734	1·8229	495·83̇		129·16̇	240·4019	1·2482	370·83̇
	6·25	144·8963		493·75		131·25	241·6501		368·75
	8·3̇	146·7192		491·6̇		133·3̇	242·8983		366·6̇
1	10·416̇	148·5421		489·583̇	13	135·416̇	244·1464		364·583̇
	12·5	150·3569		487·5		137·5	245·3102		362·5
	14·583̇	152·1718	1·8148	485·416̇		139·583̇	246·4739	1·1637	360·416̇
	16·6̇	153·9866		483·3̇		141·6̇	247·6376		358·3̇
	18·75	155·8014		481·25		143·75	248·8014		356·25
2	20·83̇	157·6162		479·16̇	14	145·83̇	249·9651		354·16̇
	22·916̇	159·4148		477·083̇		147·916̇	251·0312		352·083̇
	25·0	161·2134	1·7986	475·0		150·0	252·0973	1·0661	350·0
	27·083̇	163·0120		472·916̇		152·083̇	253·1634		347·916̇
	29·16̇	164·8106		470·83̇		154·16̇	254·2294		345·83̇
3	31·25	166·6093		468·75	15	156·25	255·2955		343·75
	33·3̇	168·3836		466·6̇		158·3̇	256·2640		341·6̇
	35·416̇	170·1579	1·7743	464·583̇		160·416̇	257·2324	0·9684	339·583̇
	37·5	171·9322		462·5		162·5	258·2008		337·5
	39·583̇	173·7065		460·416̇		164·583̇	259·1692		335·416̇
4	41·6̇	175·4808		458·3̇	16	166·6̇	260·1376		333·3̇
	43·75	177·2227		456·25		168·75	261·0003		331·25
	45·83̇	178·9649	1·7419	454·16̇		170·83̇	261·8629	0·8626	329·16̇
	47·916̇	180·7065		452·083̇		172·916̇	262·7255		327·083̇
	50·0	182·4484		450·0		175·0	263·5882		325·0
5	52·083̇	184·1903		447·916̇	17	177·083̇	264·4508		322·916̇
	54·16̇	185·8917		445·83̇		179·16̇	265·2076		320·83̇
	56·25	187·5931	1·7014	443·75		181·25	265·9645	0·7568	318·75
	58·3̇	189·2944		441·6̇		183·3̇	266·7213		316·6̇
	69·416̇	190·9958		439·583̇		185·416̇	267·4781		314·583̇
6	62·5	192·6972		437·5	18	187·5	268·2350		312·5
	64·583̇	194·3581		435·416̇		189·583̇	268·8779		310·416̇
	66·6̇	196·0190	1·6609	433·3̇		191·6̇	269·5208	0·6429	308·3̇
	68·75	197·6799		431·25		193·75	270·1637		306·25
	70·83̇	199·3407		429·16̇		195·83̇	270·8066		304·16̇
7	72·916̇	201·0016		427·083̇	19	197·916̇	271·4495		302·083̇
	75·0	202·6139		425·0		200·0	271·9785		300·0
	77·083̇	204·2262	1·6123	422·916̇		202·083̇	272·5074	0·5290	297·916̇
	79·16̇	205·8384		420·83̇		204·6̇	273·0364		295·83̇
	81·25	207·4507		418·75		206·25	273·5654		293·75
8	83·3̇	209·0630		416·6̇	20	208·3̇	274·0944		291·6̇
	85·416̇	210·6104		414·583̇		210·416̇	274·5094		289·583̇
	87·5	212·1579	1·5475	412·5		212·5	274·9244	0·4150	287·5
	89·583̇	213·7053		410·416̇		214·583̇	275·3395		285·416̇
	91·6̇	215·2528		408·3̇		216·6̇	275·7545		283·3̇
9	93·75	216·8002		406·25	21	218·75	276·1695		281·25
	95·83̇	218·2829		404·16̇		220·83̇	276·4707		279·16̇
	97·916̇	219·7655	1·4826	402·083̇		222·916̇	276·7718	0·3011	277·083̇
	100·0	221·2481		400·0		225·0	277·0729		275·0
	102·083̇	222·7308		397·916̇		227·083̇	277·3740		272·916̇
10	104·16̇	224·2134		395·83̇	22	229·16̇	277·6751		270·83̇
	106·25	225·6231		393·75		231·25	277·8741		268·75
	108·3̇	227·0329	1·4097	391·6̇		233·3̇	278·0332	0·1790	266·6̇
	110·416̇	228·4426		389·583̇		235·416̇	278·2122		264·583̇
	112·5	229·8523		387·5		237·5	278·3912		262·5
11	114·583̇	231·2620		385·416̇	23	239·583̇	278·5703		260·416̇
	116·6̇	232·5907		383·3̇		241·6̇	278·6272		258·3̇
	118·75	233·9194	1·3287	381·25		243·75	278·6842	0·0570	256·25
	120·83̇	235·2482		379·16̇		245·83̇	278·7412		254·16̇
	122·916̇	236·5769		377·083̇		247·916̇	278·7981		252·083̇
					24	250·0	278·8551		250·0

TABLE LXVI A—*Contd.*

(B) Moon's equation of the centre ("Equation *b.*") by the First Ārya-Siddhānta from ☾'s mean anomaly ("Arg. *b.*") 500–1000 (180°—360°).

Col. 3.—The equation is ☾'s greatest equation *minus* the actual equation, in 10,000ths of circle.

Serial No. of sine.	Arg. *b.*	Equation *b.*	Diff.	Arg. *b.*	Serial No. of sine.	Arg. *b.*	Equation *b.*	Diff.	Arg. *b*
1	2*a*	3	4	2*b*	1	2*a*	3	4	2*b*
0	500·0	139·4275		1000·0	12	625·0	40·9495		875·0
	502·083̇	137·6046		997·916̇		627·083̇	39·7014		872·918
	504·16̇	135·7817	1·8229	995·83̇		629·16̇	38·4532	1·2482	870·83̇
	506·25	133·9588		993·75		631·25	37·2050		868·75
	508·3̇	132·1359		991·6̇		633·3̇	35·9568		866·6̇
1	510·416̇	130·3130		989·583̇	13	635·416̇	34·7087		864·583̇
	512·5	128·4982		987·5		637·5	33·5449		862·5
	514·583̇	126·6833	1·8148	985·416̇		639·583̇	32·3812	1·1637	860·416̇
	516·6̇	124·8685		983·3̇		641·6̇	31·2175		858·3̇
	518·75	123·0537		981·25		643·75	30·0537		856·25
2	520·83̇	121·2389		979·16̇	14	645·83̇	28·8900		854·16̇
	522·916̇	119·4403		977·083̇		647·916̇	27·8239		852·083̇
	525·0	117·6417	1·7986	975·0		650·0	26·7578	1·0661	850·0
	527·083̇	115·8431		972·916̇		652·083̇	25·6917		847·916̇
	529·16̇	114·0444		970·83̇		654·16̇	24·6257		845·83̇
3	531·25	112·2458		968·75	15	656·25	23·5596		843·75
	533·3̇	110·4715		966·6̇		658·3̇	22·5911		841·6̇
	535·416̇	108·6972	1·7743	964·583̇		660·416̇	21·6227	0·9684	839·583̇
	537·5	106·9229		962·5		662·5	20·6543		837·5
	539·583̇	105·1486		960·416̇		664·583̇	19·6859		835·416̇
4	541·6̇	103·3743		958·3̇	16	666·6̇	18·7175		833·3̇
	543·75	101·6324		956·25		668·75	17·8548		831·25
	545·83̇	99·8905	1·7419	954·16̇		670·83̇	16·9922	0·8626	829·16̇
	547·916̇	98·1486		952·083̇		672·916̇	16·1296		827·083̇
	550·0	96·4067		950·0		675·0	15·2669		825·0
5	552·083̇	94·6648		947·916̇	17	677·083̇	14·4043		822·916̇
	554·16̇	92·9634		945·83̇		679·16̇	13·6475		820·83̇
	556·25	91·2620	1·7014	943·75		681·25	12·8906	0·7568	818·75
	558·3̇	89·5697		941·6̇		683·3̇	12·1338		816·6̇
	560·416̇	87·8593		939·583̇		685·416̇	11·3770		814·583̇
6	562·5	86·1579		937·5	18	687·5	10·6201		812·5
	564·583̇	84·4970		935·416̇		689·583̇	9·9772		810·416̇
	566·6̇	82·8361	1·6609	933·3̇		691·6̇	9·3343	0·6429	808·3̇
	568·75	81·1752		931·25		693·75	8·6914		806·25
	570·83̇	79·5144		929·16̇		695·83̇	8·0485		804·16̇
7	572·916̇	77·8535		927·083̇	19	697·916̇	7·4056		802·083̇
	575·0	76·2412		925·0		700·0	6·8766		800·0
	577·083̇	74·6289	1·6123	922·916̇		702·083̇	6·3477	0·5290	797·916̇
	579·16̇	73·0167		920·83̇		704·6̇	5·8187		795·83̇
	581·25	71·4044		918·75		706·25	5·2897		793·75
8	583·3̇	69·7921		916·6̇	20	708·3̇	4·7607		791·6̇
	585·416̇	68·2447		914·583̇		710·416̇	4·3457		789·583̇
	587·5	66·6972	1·5475	912·5		712·5	3·9307	0·4150	787·5
	589·583̇	65·1498		910·416̇		714·583̇	3·5156		785·416̇
	591·6̇	63·6023		908·3̇		716·6̇	3·1006		783·3̇
9	593·75	62·0549		906·25	21	718·75	2·6855		781·25
	595·83̇	60·5722		904·16̇		720·83̇	2·3844		779·16̇
	597·916̇	59·0896	1·4826	902·083̇		722·916̇	2·0833	0·3011	777·083̇
	600·0	57·6069		900·0		725·0	1·7822		775·0
	602·083̇	56·1243		897·916̇		727·083̇	1·4811		772·916̇
10	604·16̇	54·6417		895·83̇	22	729·16̇	1·1800		770·83̇
	606·25	53·2319		893·75		731·25	1·0010		768·75
	608·3̇	51·8222	1·4097	891·6̇		733·3̇	0·8219	0·1790	766·6̇
	610·416̇	50·4125		889·583̇		735·416̇	0·6429		764·583̇
	612·5	49·0028		887·5		737·5	0·4639		762·5
11	614·583̇	47·5931		885·416̇	23	739·583̇	0·2848		760·416̇
	616·6̇	46·2644		883·3̇		741·6̇	0·2279		758·3̇
	618·75	44·9357	1·3287	881·25		743·75	0·1709	0·0570	756·25
	620·83̇	43·6069		879·16̇		745·83̇	0·1139		754·16̇
	622·916̇	42·2782		877·083̇		747·916̇	0·0570		752·083̇
					24	750·0	0·0000		750·0

TABLE LXVII A.

(A) SUN'S EQUATION OF THE CENTRE ("EQUATION *c.*") BY THE FIRST ĀRYA-SIDDHĀNTA FROM ☉'S MEAN ANOMALY ("Arg. *c.*") 0—500 (0°—180°).

Cols. 3, 4.—Equation and Difference stand for either of the mean anom. values in cols. 2a, 2b. For the 24 base-equations see Table LXVII, above.

"Arg. c" is ☉'s mean anomaly in 1,000ths of circle.

Col. 3.—The equation is ☉'s greatest equation *minus* the actual equation, in 10,000ths of circle.

Serial No. of sine.	Arg. *c.*	Equation *c.*	Diff.	Arg. *c.*	Serial No. of sine.	Arg. *c.*	Equation *c.*	Diff.	Arg *c.*
1	2*a*	3	4	2*b*	1	2*a*	3	4	2*b*
0	0·0	59·6875		500·0	12	125·0	17·4826		375·0
	2·083̇	58·9078		497·916̇		127·083̇	16·9479		372·916̇
	4·16̇	58·1281	0·7797	495·83̇		129·16̇	16·4132	0·5347	370·83
	6·25	57·3484		493·75		131·25	15·8785		368·75
	8·3̇	56·5687		491·6̇		133·3̇	15·3438		366·6̇
1	10·416̇	55·7890		489·583̇	13	135·416̇	14·8090		364·583̇
	12·5	55·0096		487·5		137·5	14·3125		362·5
	14·583̇	54·2303	0·7793	485·416̇		139·583̇	13·8160	0·4965	360·416̇
	16·6̇	53·4510		483·3̇		141·6̇	13·3194		358·3̇
	18·75	52·6717		481·25		143·75	12·8229		356·25
2	20·83̇	51·8924		479·16̇	14	145·83̇	12·3264		354·16̇
	22·916̇	51·1215		477·083̇		147·916̇	11·8715		352·083̇
	25·0	50·3507	0·7708	475·0		150·0	11·4167	0·4549	350.0
	27·083̇	49·5799		472·916̇		152·083̇	10·9618		347·916
	29·16̇	48·8090		470·83̇		154·16̇	10·5069		345·83̇
3	31·25	48·0382		468·75	15	156·25	10·0521		343·75
	33·3̇	47·2778		466·6̇		158·3̇	9·6389		341·6̇
	35·416̇	46·5174	0·7604	464·583̇		160·416̇	9·2257	0·4132	339·583̇
	37·5	45·7569		462·5		162·5	8·8125		337·5
	39·583̇	44·9965		460·416̇		164·583̇	8·3993		335·416̇
4	41·6̇	44·2361		458·3̇	16	166·6̇	7·9861		333·3̇
	43·75	43·4896		456·25		168·75	7·6181		331·25
	45·83̇	42·7431	0·7465	454·16		170·83̇	7·2500	0·3681	329·16̇
	47·916̇	41·9965		452·083̇		172·916̇	6·8819		327·083̇
	50·0	41·2500		450·0		175·0	6·5139		325·0
5	52·083̇	40·5035		447·916̇	17	177·083̇	6·1458		322·916̇
	54·16̇	39·7743		445·83̇		179·16	5·8229		320·83̇
	56·25	39·0451	0·7292	443·75		181·25	5·5000	0·3229	318·75
	58·3̇	38·3160		441·6̇		183·3̇	5·1771		316·6̇
	60·416̇	37·5868		439·583̇		185·416̇	4·8542		314·58
6	62·5	36·8576		437·5	18	187·5	4·5313		312·5
	64·583	36·1458		435·416̇		189·583̇	4·2569		310·416̇
	66·6̇	35·4340	0·7118	433·3̇		191·6̇	3·9826	0·2743	308·3̇
	68·75	34·7222		431·25		193·75	3·7083		306·25
	70·83̇	34·0104		429·16̇		195·83̇	3·4340		304·16̇
7	72·916̇	33·2986		427·083̇	19	197·916̇	3·1597		302·083̇
	75·0	32·6076		425·0		200·0	2·9340		300·0
	77·083̇	31·9167	0·6910	422·916̇		202·083̇	2·7083	0·2257	297·916̇
	79·16̇	31·2257		420·83̇		204·16̇	2·4826		295·83̇
	81·25	30·5347		418·75		206·25	2·2569		293·75
8	83·3̇	29·8438		416·6̇	20	208·3̇	2·0312		291·6̇
	85·416̇	29·1806		414·583̇		210·416̇	1·8542		289·583̇
	87·5	28·5174	0·6632	412·5		212·5	1·6771	0·1771	287·5
	89·583̇	27·8542		410·416̇		214·583̇	1·5000		285·416
	91·6̇	27·1910		408·3̇		216·6̇	1·3229		283·3
9	93·75	26·5278		406·25	21	218·75	1·1458		281·25
	95·83̇	25·8924		404·16̇		220·83̇	1·0174		279·16̇
	97·916̇	25·2569	0·6354	402·083̇		222·916̇	0·8889	0·1285	277·083̇
	100·0	24·6215		400·0		225·0	0·7604		275·0
	102·083	23·9861		397·916̇		227·083̇	0·6319		272·916̇
10	104·16̇	23·3507		395·83̇	22	229·16̇	0·5035		270·83̇
	106·25	22·7465		393·75		231·25	0·4279		268·75
	108·3̇	22·1424	0·6042	391·6̇		233·3	0·3522	0·0756	266·6̇
	110·416̇	21·5382		389·583̇		235·416̇	0·2766		264·583̇
	112·5	20·9341		387·5		237·5	0·2010		262·5
11	114·583	20·3299		385·416̇	23	239·583̇	0·1254		260·416
	116·6̇	19·7604		383·3̇		241·6	0·1003		258·3̇
	118·75	19·1910	0·5694	381·25		243·75	0·0752	0·0251	256·25
	120·83̇	18·6215		379·16̇		245·83	0·0502		254·16̇
	122·916̇	18·0521		377·083̇		247·916̇	0·0251		252·083̇
					24	250·0	0·0		250·0

TABLE LXVII A—*Contd.*

(B) SUN'S EQUATION OF THE CENTRE ("EQUATION c.") BY THE FIRST ĀRYA-SIDDHĀNTA. FROM ⊙'S MEAN ANOMALY ("Arg. c.") 500—1000 (180°—360°).

Col. 3.—The equation is ⊙'s greatest equation *plus* the actual equation, in 10,000ths of circle.

Serial No. of sine.	Arg. c.	Equation c.	Diff.	Arg. c.	Serial No. of sine.	Arg. c.	Equation c.	Diff.	Arg. c.
1	2a	3	4	2b	1	2a	3	4	2b
0	500·0	59·6875		1000·0	12	625·0	101·8924		875·0
	502·083̇	60·4672		997·916̇		627·083̇	102·4271		872·916̇
	504·16̇	61·2469	0·7797	995·83̇		629·16̇	102·9618	0·5347	870·83̇
	506·25	62·0266		993·75		631·25	103·4965		868·75
	508·3̇	62·8063		991·6̇		633·3̇	104·0312		866·6̇
1	510·416̇	63·5860		989·583̇	13	635·416̇	104·5660		864·583̇
	512·5	64·3654		987·5		637·5	105·0625		862·5
	514·583̇	65·1447	0·7793	985·416̇		639·583̇	105·5590	0·4965	860·416̇
	516·6̇	65·9240		983·3̇		641·6̇	106·0556		858·3̇
	518·75	66·7033		981·25		643·75	106·5521		856·25
2	520·83̇	67·4826		979·16̇	14	645·83̇	107·0486		854·16̇
	522·916̇	68·2535		977·083̇		647·916̇	107·5035		852·083̇
	525·0	69·0243	0·7708	975·0		650·0	107·9583	0·4549	850·0
	527·083̇	69·7951		972·916̇		652·083̇	108·4132		847·916̇
	529·16̇	70·5660		970·83̇		654·16̇	108·8681		845·83̇
3	531·25	71·3368		968·75	15	656·25	109·3229		843·75
	533·3̇	72·0972		966·6̇		658·3̇	109·7361		841·6̇
	535·416̇	72·8576	0·7604	964·583̇		660·416̇	110·1493	0·4132	839·583̇
	537·5	73·6181		962·5		662·5	110·5265		837·5
	539·583̇	74·3785		960·416̇		664·583̇	110·9767		835·416̇
4	541·6̇	75·1389		958·3̇	16	666·6̇	111·3889		833·3̇
	543·75	75·8854		956·25		668·75	111·7569		831·25
	545·83̇	76·6319	0·7465	954·16̇		670·83̇	112·1250	0·3681	829·16̇
	547·916̇	77·3785		952·083̇		672·916̇	112·4931		827·083̇
	550·0	78·1250		950·0		675·0	112·8611		825·0
5	552·083̇	78·8715		947·916̇	17	677·083̇	113·2292		822·916̇
	554·16̇	79·6007		945·83̇		679·16̇	113·5521		820·83̇
	556·25	80·3299	0·7292	943·75		681·25	113·8750	0·3229	818·75
	558·3̇	81·0590		941·6̇		683·3̇	114·1979		816·6̇
	560·416̇	81·7882		939·583̇		685·416̇	114·5208		814·583̇
6	562·5	82·5174		937·5	18	687·5	114·8438		812·5
	564·583̇	83·2292		935·416̇		689·583̇	115·1181		810·416̇
	566·6̇	83·9410	0·7118	933·3̇		691·6̇	115·3924	0·2743	808·3̇
	568·75	84·6528		931·25		693·75	115·6667		806·25
	570·83̇	85·3646		929·16̇		695·83̇	115·9410		804·16̇
7	572·916̇	86·0764		927·083̇	19	697·916̇	116·2153		802·083̇
	575·0	86·7674		925·0		700·0	116·4410		800·0
	577·083̇	87·4583	0·6910	922·916̇		702·083̇	116·6667	0·2257	797·916̇
	579·16̇	88·1493		920·83̇		704·16̇	116·8924		795·83̇
	581·25	88·8403		918·75		706·25	117·1181		793·75
8	583·3̇	89·5312		916·6̇	20	708·3̇	117·3438		791·6̇
	585·416̇	90·1944		914·583̇		710·416̇	117·5208		789·583̇
	587·5	90·8576	0·6632	912·5		712·5	117·6979	0·1771	787·5
	589·583̇	91·5208		910·416̇		714·583̇	117·8750		785·416̇
	591·6̇	92·1840		908·3̇		716·6̇	118·0521		783·3̇
9	593·75	92·8472		906·25	21	718·75	118·2292		781·25
	595·83̇	93·4826		904·16̇		720·83̇	118·3576		779·16̇
	597·916̇	94·1181	0·6354	902·083̇		722·916̇	118·4861	0·1285	777·083̇
	600·0	94·7535		900·0		725·0	118·6146		775·0
	602·083̇	95·3889		897·916̇		727·083̇	118·7431		772·916̇
10	604·16̇	96·0243		895·83̇	22	729·16̇	118·8715		770·83̇
	606·25	96·6285		893·75		731·25	118·9471		768·75
	608·3̇	97·2326	0·6042	891·6̇		733·3̇	119·0228	0·0756	766·6̇
	610·416̇	97·8368		889·583̇		735·416̇	119·0984		764·583̇
	612·5	98·4410		887·5		737·5	119·1740		762·5
11	614·583̇	99·0451		885·416̇	22	739·583̇	119·2495		760·416̇
	616·6̇	99·6146		883·3̇		741·6̇	119·2747		758·3̇
	618·75	100·1840	0·56 4	881·25		743·75	119·2998	0·0251	756·25
	620·83̇	100·7535		879·16̇		745·83̇	119·3248		754·16̇
	622·916̇	101·3229		877·083̇		747·916̇	119·3499		752·083̇
					24	750·0	119·3750		750·0

TABLE LXVIII.

INDICES OF TITHIS, KARAṆAS, YŌGAS AND NAKSHATRAS, IN 10,000THS OF CIRCLE MEASUREMENT.

Indices of yōgas "(y)" are numerically the same as those of nakshatras "(n)."

This Table corresponds to Table VIII, "Indian Calendar."

TITHI AND KARANA.					YŌGA.		NAKSHATRA.			
Serial number	No. in *paksha* or (lunar fortnight).	Tithi-index (*t*).	KARAṆA. First half of Tithi.	KARAṆA. Second half of Tithi.	Name.	No. of Yōga or Nakshatra.	Name.	Index of Nakshatra ("n") and Yōga ("y"). Ordinary (equal-space) system.	Index of ending point of Nakshatra and Yōga by the unequal space systems of Garga.	Brahma Siddhānta.
1	2	3	4	5	6		7	8	9	10
1	Śukla. 1	0 — 333·3̇	Kiṁstughna*	1 Bava	Vishkambha	1	Aśvinī	0 — 370·37̇0̇	370·37̇0̇	366·0108
2	2	333·3̇— 666·6̇	2 Bālava	3 Kaulava	Prīti	2	Bharaṇī	370·37̇0̇— 740·74̇0̇	555·5̇	549·0051
3	3	666·6̇—1000	4 Taitila	5 Gara	Āyushmat	3	Kṛittikā	740·74̇0̇—1111·1̇	925·92̇5̇	915·0270
4	4	1000 —1333·3̇	6 Vaṇij	7 Vishṭi†	Saubhāgya	4	Rōhiṇī	1111·1̇ —1481·48̇1̇	1481·48̇1̇	1464·0432
5	5	1333·3̇—1666·6̇	1 Bava	2 Bālava	Śōbhana	5	Mṛigaśiras	1481·48̇1̇—1851·85̇1̇	1851·85̇1̇	1830·0540
6	6	1666·6̇—2000	3 Kaulava	4 Taitila	Atigaṇḍa	6	Ārdrā	1851·85̇1̇—2222·2̇	2037·03̇7̇	2013·0594
7	7	2000 —2333·3̇	5 Gara	6 Vaṇij	Sukarman	7	Punarvasu	2222·2̇ —2592·59̇2̇	2592·59̇2̇	2562·0756
8	8	2333·3̇—2666·6̇	7 Vishṭi†	1 Bava	Dhṛiti	8	Pushya	2592·59̇2̇—2962·96̇2̇	2962·96̇2̇	2928·0864
9	9	2666·6̇—3000	2 Bālava	3 Kaulava	Śūla	9	Aślēshā	2962·96̇2̇—3333·3̇	3148·14̇8̇	3111·0918
10	10	3000 —3333·3̇	4 Taitila	5 Gara	Gaṇḍa	10	Maghā	3333·3̇ —3703·70̇3̇	3518·51̇8̇	3477·1026
11	11	3333·3̇—3666·6̇	5 Vaṇij	7 Vishṭi	Vṛiddhi	11	Pūrva-Phalgunī	3703·70̇3̇—4074·07̇4̇	3888·8̇	3843·1134

12	12	3666·6—4000	1 Bava .	2 Bālava .	Dhruva . .	12	Uttara-Phalgunī .	4074·074—4444·4	4444·4	4392·1296
13	13	4000 —4333·3	3 Kaulava .	4 Taitila .	Vyāghāta . .	13	Hasta . .	4444·4 —4814·814	4814·814	4758·1404
14	14	4333·3—4666·6	5 Gara .	6 Vaṇij .	Harshaṇa .	14	Chitrā . .	4814·814—5185·185	5185·185	5124·1512
15	15	4666·6—5000	7 Vishṭi .	1 Bava .	Vajra .	15	Svāti . .	5185·185—5555·5	5570·370	5307·1566
	Kṛishṇa.									
16	1	5000 —5333·3	2 Bālava .	3 Kaulava .	Siddhi‡ . .	16	Viśākhā . .	5555·5 —5925·925	5925·925	5856·1728
17	2	5333·3—5666·6	4 Taitila .	5 Gara .	Vyatīpāta .	17	Anurādhā . .	5925·925—6296·296	6296·296	6222·1836
18	3	5666·6—6000	6 Vaṇij .	7 Vishṭi .	Variyas . .	18	Jyēshṭhā . .	6292·296—6666·6	6481·481	6405·1890
19	4	6000 —6333·3	1 Bava .	2 Bālava .	Parigha . .	19	Mūla . . .	6666·6 —7037·037	6852·852	6771·1998
20	5	6333·3—6666·6	3 Kaulava .	4 Taitila .	Śiva . .	20	Pūrva-Ashāḍhā .	7037·037—7407·407	7222·2	7137·2106
21	6	6666·6—7000	5 Gara .	6 Vaṇij .	Siddha . .	21	Uttara-Ashāḍhā .	7407·407—7777·7	7777·7	7686·2269§
22	7	7000 —7333·3	7 Vishṭi .	1 Bava .			Abhijit§ . .	...	...	7803·935·§
23	8	7333·3—7666·6	2 Bālava .	3 Kaulava .	Sādhya . .	22	Śravaṇa . .	7777·7 —8148·148	8148·148	8169·9460
24	9	7666·6—8000	4 Taitila .	5 Gara .	Śubha . .	23	Dhanishṭhā§§ .	8148·148—8518·518	8518·518	8535·9568
25	10	8000 —8333·3	6 Vaṇij .	7 Vishṭi .	Śukla . .	24	Śatabhishaj¶ .	8518·518—8888·8	8703·703	8718·9322
26	11	8333·3—8666·6	1 Bava .	2 Bālava .	Brahman .	25	Pūrva-Bhadrapadā	8888·8 —9259·259	9074·074	9084·9730
27	12	8666·6—9000	3 Kaulava .	4 Taitila .	Indra . .	26	Uttara-Bhadrapadā	9259·259—9629·629	9629·629	9633·9892
28	13	9000 —9333·3	5 Gara .	6 Vaṇij .	Vaidhṛiti .	27	Rēvatī . .	9629·629—10,000	10,000	10,000
29	14	9333·3—9666·6	7 Vishṭi .	Śakuni . .	...					
30	15	9666·6—10,000	Chatushpada .	Nāga . .	...					

* or Kiṁtughna. † Vishṭi is also called Bhadrā, or Kalyāṇī. ‡ or Asrij.

§ The figures given in Col. 10 follow the limits of Abhijit as given in the "Indian Calandar," p. 22, *viz.*, from 276° 42′ 15″ to 280° 56′ 30″. Professor Jacobi and Dr. Burgess, however, give these limits as from 270° 40′ to 281° 40′ (*Epig. Ind. I.*, *p.* 449; *Journal R. A. S.*, 1893, *p.* 755). If they are correct, Abhijit (Col. 10) should be read as beginning at 7685·1852 and ending at 7824·074.

§§ or Śravishṭhā. ¶ or Śatatārakā.

TABLE LXIX.

Serial number of days in a year A.D. for two consecutive years.

N.B.—The numbers given are those in a common year. In Leap-years, after February 29, the day of the month must be reduced by 1. Thus Day 153, in a Leap-year, is not June 2, but June 1.

The Table is the same as Table IX, "Indian Calendar."

PART I.

Day of month.	Number of days reckoned from 1st January of the same year.												Day of month.
	Jan.	Feb.	Mar.	April.	May.	June.	July.	Aug.	Sept.	Oct.	Nov.	Dec.	
1	1	32	60	91	121	152	182	213	244	274	305	335	1
2	2	33	61	92	122	153	183	214	245	275	306	336	2
3	3	34	62	93	123	154	184	215	246	276	307	337	3
4	4	35	63	94	124	155	185	216	247	277	308	338	4
5	5	36	64	95	125	156	186	217	248	278	309	339	5
6	6	37	65	96	126	157	187	218	249	279	310	340	6
7	7	38	66	97	127	158	188	219	250	280	311	341	7
8	8	39	67	98	128	159	189	220	251	281	312	342	8
9	9	40	68	99	129	160	190	221	252	282	313	343	9
10	10	41	69	100	130	161	191	222	253	283	314	344	10
11	11	42	70	101	131	162	192	223	254	284	315	345	11
12	12	43	71	102	132	163	193	224	255	285	316	346	12
13	13	44	72	103	133	164	194	225	256	286	317	347	13
14	14	45	73	104	134	165	195	226	257	287	318	348	14
15	15	46	74	105	135	166	196	227	258	288	319	349	15
16	16	47	75	106	136	167	197	228	259	289	320	350	16
17	17	48	76	107	137	168	198	229	260	290	321	351	17
18	18	49	77	108	138	169	199	230	261	291	322	352	18
19	19	50	78	109	139	170	200	231	262	292	323	353	19
20	20	51	79	110	140	171	201	232	263	293	324	354	20
21	21	52	80	111	141	172	202	233	264	294	325	355	21
22	22	53	81	112	142	173	203	234	265	295	326	356	22
23	23	54	82	113	143	174	204	235	266	296	327	357	23
24	24	55	83	114	144	175	205	236	267	297	328	358	24
25	25	56	84	115	145	176	206	237	268	298	329	359	25
26	26	57	85	116	146	177	207	238	269	299	330	360	26
27	27	58	86	117	147	178	208	239	270	300	331	361	27
28	28	59	87	118	148	179	209	240	271	301	332	362	28
29	29	60	88	119	149	180	210	241	272	302	333	363	29
30	30	...	89	120	150	181	211	242	273	303	334	364	30
31	31	...	90	...	151	...	212	243	...	304	...	365	31
	Jan.	Feb.	Mar.	April.	May.	June.	July.	Aug.	Sept.	Oct.	Nov.	Dec.	

TABLE LXIX—*Contd.*

SERIAL NUMBER OF DAYS IN A YEAR A.D. FOR TWO CONSECUTIVE YEARS.

N. B.—When the previous year was a Leap-year, the days of the month must all be reduced by 1; and so all those after February 29, when the given year is a Leap-year.

PART II.

Day of month.	NUMBER OF DAYS RECKONED FROM 1 JANUARY OF THE PRECEDING YEAR.												Day of month.
	Jan.	Feb.	Mar.	April.	May.	June.	July.	Aug.	Sept.	Oct.	Nov.	Dec	
1	366	397	425	456	486	517	547	578	609	639	670	700	1
2	367	398	426	457	487	518	548	579	610	640	671	701	2
3	368	399	427	458	488	519	549	580	611	641	672	702	3
4	369	400	428	459	489	520	550	581	612	642	673	703	4
5	370	401	429	460	490	521	551	582	613	643	674	704	5
6	371	402	430	461	491	522	552	583	614	644	675	705	6
7	372	403	431	462	492	523	553	584	615	645	676	706	7
8	373	404	432	463	493	524	554	585	616	646	677	707	8
9	374	405	433	464	494	525	555	586	617	647	678	708	9
10	375	406	434	465	495	526	556	587	618	648	679	709	10
11	376	407	435	466	496	527	557	588	619	649	680	710	11
12	377	408	436	467	497	528	558	589	620	650	681	711	12
13	378	409	437	468	498	529	559	590	621	651	682	712	13
14	379	410	438	469	499	530	560	591	622	652	683	713	14
15	380	411	439	470	500	531	561	592	623	653	684	714	15
16	381	412	440	471	501	532	562	593	624	654	685	715	16
17	382	413	441	472	502	533	563	594	625	655	686	716	17
18	383	414	442	473	503	534	564	595	626	656	687	717	18
19	384	415	443	474	504	535	565	596	627	657	688	718	19
20	385	416	444	475	505	536	566	597	628	658	689	719	20
21	386	417	445	476	506	537	567	598	629	659	690	720	21
22	387	418	446	477	507	538	568	599	630	660	691	721	22
23	388	419	447	478	508	539	569	600	631	661	692	722	23
24	389	420	448	479	509	540	570	601	632	662	693	723	24
25	390	421	449	480	510	541	571	602	633	663	694	724	25
26	391	422	450	481	511	542	572	603	634	664	695	725	26
27	392	423	451	482	512	543	573	604	635	665	696	726	27
28	393	424	452	483	513	544	574	605	636	666	697	727	28
29	394	425	453	484	514	545	575	606	637	667	698	728	29
30	395	...	454	485	515	546	576	607	638	668	699	729	30
31	396	...	455	...	516	...	577	608	...	669	...	730	31
	Jan.	Feb.	Mar.	April.	May.	June.	July.	Aug.	Sept.	Oct.	Nov.	Dec.	

TABLE LXX.

Conversion of Tithi-parts and indices of Tithis, Nakshatras and Yōgas into time.

(Corresponding to Table X, "Indian Calendar.")

1 unit of the "Argument" = 4m·2524 (a trifle over $4\frac{1}{4}$ minutes of time), in the case of the tithi-index (t), and 4m·3831 in the case of the nakshatra-index (n).

Argument.	Time equivalent of: Tithi-parts. H. M.	Tithi-index (t). H. M.	Nakshatra index (n). H. M.	Yōga-index (y). H. M.
1	0 1	0 4	0 4	0 4
2	0 3	0 9	0 8	0 7
3	0 4	0 13	0 12	0 11
4	0 6	0 17	0 16	0 15
5	0 7	0 21	0 20	0 18
6	0 9	0 26	0 24	0 22
7	0 10	0 30	0 28	0 26
8	0 11	0 34	0 31	6 29
9	0 13	0 38	0 35	0 33
10	0 14	0 43	0 39	0 37
11	0 16	0 47	0 43	0 40
12	0 17	0 51	0 47	0 44
13	0 18	0 55	0 51	0 48
14	0 20	1 0	0 55	0 51
15	0 21	1 4	0 59	0 55
16	0 23	1 8	1 3	0 59
17	0 24	1 12	1 7	1 2
18	0 26	1 17	1 11	1 6
19	0 27	1 21	1 15	1 10
20	0 28	1 25	1 19	1 13
21	0 30	1 29	1 23	1 17
22	0 31	1 34	1 27	1 21
23	0 33	1 38	1 30	1 24
24	0 34	1 42	1 34	1 28
25	0 35	1 46	1 38	1 32
26	0 37	1 51	1 42	1 35
27	0 38	1 55	1 46	1 39
28	0 40	1 59	1 50	1 42
29	0 41	2 3	1 54	1 46
30	0 43	2 8	1 58	1 50
31	0 44	2 12	2 2	1 53
32	0 45	2 16	2 6	1 57
33	0 47	2 20	2 10	2 1
34	0 48	2 25	2 14	2 4
35	0 50	2 29	2 18	2 8
36	0 51	2 33	2 22	2 12
37	0 52	2 37	2 26	2 15
38	0 54	2 42	2 30	2 19
39	0 55	2 46	2 33	2 23
40	0 57	2 50	2 37	2 26

Argument.	Time equivalent of: Tithi-parts. H. M.	Tithi-index (t). H. M.	Nakshatra index (n). H. M.	Yōga-index (y). H. M.
41	0 58	2 54	2 41	2 30
42	1 0	2 59	2 45	2 34
43	1 1	3 3	2 49	2 37
44	1 2	3 7	2 53	2 41
45	1 4	3 11	2 57	2 45
46	1 5	3 16	3 1	2 48
47	1 7	3 20	3 5	2 52
48	1 8	3 24	3 9	2 56
49	1 9	3 28	3 13	2 59
50	1 11	3 33	3 17	3 3
51	1 12	3 37	3 21	3 7
52	1 14	3 41	3 25	3 10
53	1 15	3 45	3 29	3 14
54	1 17	3 50	3 32	3 18
55	1 18	3 54	3 36	3 21
56	1 19	3 58	3 40	3 25
57	1 21	4 2	3 44	3 29
58	1 22	4 7	3 48	3 32
59	1 24	4 11	3 52	3 36
60	1 25	4 15	3 56	3 40
61	1 26	4 19	4 0	3 43
62	1 28	4 24	4 4	3 47
63	1 29	4 28	4 8	3 51
64	1 31	4 32	4 12	3 54
65	1 32	4 36	4 16	3 58
66	1 34	4 41	4 20	4 2
67	1 35	4 45	4 24	4 5
68	1 36	4 49	4 28	4 9
69	1 38	4 53	4 31	4 13
70	1 39	4 58	4 35	4 16
71	1 41	5 2	4 39	4 20
72	1 42	5 6	4 43	4 24
73	1 43	5 10	4 47	4 27
74	1 45	5 15	4 51	4 31
75	1 46	5 13	4 55	4 35

Argument.	Time equivalent of: Tithi-parts. H. M.	Tithi-index (t). H. M.	Nakshatra index (n). H. M.	Yōga-index (y). H. M.
76	1 48	5 23	4 59	4 38
77	1 49	4 27	5 3	4 42
78	1 51	5 32	5 7	4 46
79	1 52	5 36	5 11	4 49
80	1 53	5 40	5 15	4 53
81	1 55	5 44	5 19	4 57
82	1 56	5 49	5 23	5 0
83	1 58	5 53	5 27	5 4
84	1 59	5 57	5 30	5 7
85	2 0	6 1	5 34	5 11
86	2 2	6 6	5 38	5 15
87	2 3	6 19	5 42	5 18
88	2 5	6 14	5 46	5 22
89	2 6	6 18	5 50	5 26
90	2 8	6 23	5 54	5 29
91	2 9	6 27	5 58	5 33
92	2 10	6 31	6 2	5 37
93	2 12	6 35	6 6	5 40
94	2 13	6 40	6 10	5 44
95	2 15	6 44	6 14	5 48
96	2 16	6 48	6 18	5 51
97	2 17	6 52	6 22	5 55
98	2 19	6 57	6 26	5 59
99	2 20	7 1	6 29	6 2
100	2 22	7 5	6 33	6 6
200	4 43	14 10	13 7	12 18
300	7 5	21 16	19 40	18 12
400	9 27	28 21		
500	11 49	35 26		
600	14 10	42 31		
700	16 32	49 37		
800	18 54	56 42		
900	21 16	63 47		
1000	23 37	70 52		

TABLE LXXI.

The European Calendar.

A. Initial days of centuries, Julian and Gregorian Calendars.

Odd years of centuries.	Centuries A.D. Old Style. 0 / 700 / 1400	100 / 800 / 1500	200 / 900 / 1600	300 / 1000 / 1700	400 / 1100 / 1800	500 / 1200 / 1900	600 / 1300 / 2000	New Style. Sa 1600 / 2000	Fr 1700 / 2100	W 1800 / 2200	Mo 1500 / 1900 / 2300	Leap-years.
	Initial days.											
0 28 56 84	Th	W	Tu	Mo	S	Sa	Fr	Sa	Th	Tu	S	L.Y.
1 29 57 85	Sa	Fr	Th	W	Tu	Mo	S	Mo	Sa	Th	Tu	
2 30 58 86	S	Sa	Fr	Th	W	Tu	Mo	Tu	S	Fr	W	
3 31 59 87	Mo	S	Sa	Fr	Th	W	Tu	W	Mo	Sa	Th	
4 32 60 88	Tu	Mo	S	Sa	Fr	Th	W	Th	Tu	S	Fr	L.Y.
5 33 61 89	Th	W	Tu	Mo	S	Sa	Fr	Sa	Th	Tu	S	
6 34 62 90	Fr	Th	W	Tu	Mo	S	Sa	S	Fr	W	Mo	
7 35 63 91	Sa	Fr	Th	W	Tu	Mo	S	Mo	Sa	Th	Tu	
8 36 64 92	S	Sa	Fr	Th	W	Tu	Mo	Tu	S	Fr	W	L.Y.
9 37 65 93	Tu	Mo	S	Sa	Fr	Th	W	Th	Tu	S	Fr	
10 38 66 94	W	Tu	Mo	S	Sa	Fr	Th	Fr	W	Mo	Sa	
11 39 67 95	Th	W	Tu	Mo	S	Sa	Fr	Sa	Th	Tu	S	
12 40 68 96	Fr	Th	W	Tu	Mo	S	Sa	S	Fr	W	Mo	L.Y.
13 41 69 97	S	Sa	Fr	Th	W	Tu	Mo	Tu	S	Fr	W	
14 42 70 98	Mo	S	Sa	Fr	Th	W	Tu	W	Mo	Sa	Th	
15 43 71 99	Tu	Mo	S	Sa	Fr	Th	W	Th	Tu	S	Fr	
16 44 72 100	W	Tu	Mo	S	Sa	Fr	Th	Fr	W	Mo	Sa	L.Y.
17 45 73	Fr	Th	W	Tu	Mo	S	Sa	S	Fr	W	Mo	
18 46 74	Sa	Fr	Th	W	Tu	Mo	S	Mo	Sa	Th	Tu	
19 47 75	S	Sa	Fr	Th	W	Tu	Mo	Tu	S	Fr	W	
20 48 76	Mo	S	Sa	Fr	Th	W	Tu	W	Mo	Sa	Th	L.Y.
21 49 77	W	Tu	Mo	S	Sa	Fr	Th	Fr	W	Mo	Sa	
22 50 78	Th	W	Tu	Mo	S	Sa	Fr	Sa	Th	Tu	S	
23 51 79	Fr	Th	W	Tu	Mo	S	Sa	S	Fr	W	Mo	
24 52 80	Sa	Fr	Th	W	Tu	Mo	S	Mo	Sa	Th	Tu	L.Y.
25 53 81	Mo	S	Sa	Fr	Th	W	Tu	W	Mo	Sa	Th	
26 54 82	Tu	Mo	S	Sa	Fr	Th	W	Th	Tu	S	Fr	
27 55 83	W	Tu	Mo	S	Sa	Fr	Th	Fr	W	Mo	Sa	

B. Week-days for one year. Applicable to both Old and New Style dates.

Months in Common years.								Months in Leap-years.
	S	Mo	Tu	W	Th	Fr	Sa	
	Mo	Tu	W	Th	Fr	Sa	S	
	Tu	W	Th	Fr	Sa	S	Mo	
	W	Th	Fr	Sa	S	Mo	Tu	
	Th	Fr	Sa	S	Mo	Tu	W	
	Fr	Sa	S	Mo	Tu	W	Th	
	Sa	S	Mo	Tu	W	Th	Fr	
	1	2	3	4	5	6	7	
	8	9	10	11	12	13	14	Jan.
Jan.	15	16	17	18	19	20	21	April.
Oct.	22	23	24	25	26	27	28	July.
	29	30	31					
				1	2	3	4	
Feb.	5	6	7	8	9	10	11	Feb.
Mar.	12	13	14	15	16	17	18	Aug.
Nov.	19	20	21	22	23	24	25	
	26	27	28	29	30	31		
							1	
	2	3	4	5	6	7	8	
April	9	10	11	12	13	14	15	Sept.
July	16	17	18	19	20	21	22	Dec.
	23	24	25	26	27	28	29	
	30	31						
			1	2	3	4	5	
	6	7	8	9	10	11	12	
Aug.	13	14	15	16	17	18	19	May
	20	21	22	23	24	25	26	
	27	28	29	30	31			
						1	2	
	3	4	5	6	7	8	9	
Sept.	10	11	12	13	14	15	16	June
Dec.	17	18	19	20	21	22	23	
	24	25	26	27	28	29	30	
	31							
		1	2	3	4	5	6	
	7	8	9	10	11	12	13	
May.	14	15	16	17	18	19	20	Oct.
	21	22	23	24	25	26	27	
	28	29	30	31				
					1	2	3	
	4	5	6	7	8	9	10	Mar.
June	11	12	13	14	15	16	17	Nov.
	18	19	20	21	22	23	24	
	25	26	27	28	29	30	31	

To find the initial day of a given year A.D. take the day marked in Section A, perpendicular under the given century and horizontal opposite the given year. Note this initial day in column 2 of the heading of Section B. Find the given day of month in the body of Section B. Run up to the week-day in horizontal line with the initial day in the heading. The day so found is the week-day of the given day of month and year.

E.g. Wanted week-day of 23rd March, A.D. 645. At junction of century 600 (perpendicular) and 45 (horizontal) in Section A is Saturday. This was the initial day of A.D. 645. The year was common. The week-day noted in the heading of Section B at the junction of 23rd March (perpendicular) and of "Sa." in column 2 of heading (horizontal) is "W," Wednesday. Therefore 23rd March, A.D. 645, was a Wednesday.

In common years work with the month on left, in leap-years with that on right.

N. B.—In the New Style the years 1600 and 2000 are leap-years, but 1700, 1800, 1900 are common years. The initial week-day of the first year of each New Style century is given *above* it in heading of Section A. For the initial week-day of other years of the century look for the day in the junction of columns as mentioned above; *e.g.*, A.D. 1900 began (top) on Monday. 1901 began (junction of columns) on Tuesday. 1928 begins on Sunday, 1919 began on Wednesday

TABLE LXXII.

VALUE OF a, b, c AT BEGINNING OF CENTURIES OF THE KALIYUGA BY THE FIRST ARYA-SIDDHĀNTA AT MEAN SUNRISE ON DAY OF OCCURRENCE OF MEAN MĒSHA-SAṀKRĀNTI, WHICH IS THE MOMENT WHEN MEAN SUN REACHES LONGITUDE 0°.

Century.	Week-day.	a.	b.	c.
36	0	7177·6056	135·4688	279·9111
37	0	6045·4346	723·3175	280·2723
38	6	4913·2637	311·1661	280·6336
39	0	3781·0927	899·0148	280·9948
40	0	2648·9218	486·8635	281·3560
41	0	1516·7509	74·7121	281·7172
42	0	384·5799	662·5608	282·0784
43	6	8913·7771	214·1179	279·7019
44	6	7781·6062	801·9665	280·0631
45	6	6649·4352	389·8152	280·4243
46	6	5517·2643	977·6639	280·7855
47	6	4385·0933	565·5125	281·1467
48	6	3252·9224	153·3612	281·5079
49	6	2120·7515	741·2099	281·8692
50	5	649·9486	292·7669	279·492

N. B.—The value of "b", the ☾'s mean anomaly, is given as estimated by Professor Jacobi. The present author estimates its value as less than the given amount by 3·6. In a very close case both valuations may be tried.

TABLE LXXIII.

INCREASE OF *a*, *b*, *c* FOR YEARS OF THE K. Y. CENTURY BY THE ĀRYA-SIDDHĀNTA.

* *Years thus marked are years of 366 days, the rest of 365 each.*

Year.	W-d.	*a.*	*b.*	*c.*	Year.	W-d.	*a.*	*b.*	*c.*
0	0	0	0	0	50	0	4433·9145	793·9243	0·1806
1	1	3600·6340	246·4427	999·2918	51	1	8034·5485	40·3670	999·4724
*2	2	7201·2680	492·8853	998·5836	*52	2	1635·1825	286·8097	998·7642
3	4	1140·5339	775·6196	0·6131	53	4	5574·4484	569·5439	0·7938
4	5	4741·1679	22·0623	999·9049	54	5	9175·0824	815·9866	0·0855
5	6	8341·8019	268·5049	999·1967	55	6	2775·7164	62·4293	999·3773
*6	0	1942·4359	514·9476	998·4885	*56	0	6376·3504	308·8719	998·6691
7	2	5881·7018	797·6819	0·5181	57	2	315·6163	591·6062	0·6987
8	3	9482·3358	44·1246	999·8099	58	3	3916·2503	838·0489	999·9905
9	4	3082·9698	290·5672	999·1017	59	4	7516·8843	84·4916	999·2823
*10	5	6683·6038	537·0099	998·3934	*60	5	1117·5183	330·9342	998·5741
11	0	622·8697	819·7442	0·4230	61	0	5056·7842	613·6685	0·6036
12	1	4223·5037	66·1868	999·7148	62	1	8657·4182	860·1112	999·8954
*13	2	7824·1377	312·6295	999·0066	63	2	2258·0522	106·5538	999·1872
14	4	1763·4035	595·3638	1·0362	*64	3	5858·6862	352·9965	998·4790
15	5	5364·0375	841·8065	0·3280	65	5	9797·9521	635·7308	0·5086
16	6	8964·6715	88·2491	999·6197	66	6	3398·5861	882·1735	999·8004
*17	0	2565·3056	334·6918	998·9115	67	0	6999·2201	128·6161	999·0921
18	2	6504·5714	617·4261	0·9411	*68	1	599·8541	375·0588	998·3839
19	3	105·2054	863·8687	0·2329	69	3	4539·1200	657·7931	0·4135
20	4	3705·8394	110·3114	999·5247	70	4	8139·7540	904·2357	999·7053
*21	5	7306·4734	356·7541	998·8165	*71	5	1740·3880	150·6784	998·9971
22	0	1245·7393	639·4884	0·8460	72	0	5679·6539	433·4127	1·0267
23	1	4846·3733	885·9310	0·1378	73	1	9280·2879	679·8554	0·3184
24	2	8447·0073	132·3737	999·4296	74	2	2880·9219	926·2980	999·6102
*25	3	2047·6413	378·8164	998·7214	*75	3	6481·5559	172·7407	998·9020
26	5	5986·9072	661·5506	0·7510	76	5	420·8217	455·4750	0·9316
27	6	9587·5412	907·9933	0·0428	77	6	4021·4557	701·9176	0·2234
28	0	3188·1752	154·4360	999·3346	78	0	7622·0897	948·3603	999·5152
*29	1	6788·8092	400·8786	998·6263	*79	1	1222·7238	194·8030	998·8070
30	3	728·0751	683·6129	0·6559	80	3	5161·9896	477·5372	0·8365
31	4	4328·7091	930·0556	999·9477	81	4	8762·6236	723·9799	0·1283
32	5	7929·3431	176·4982	999·2395	82	5	2363·2576	970·4226	999·4201
*33	6	1529·9771	422·9409	998·5313	*83	6	5963·8916	216·8652	998·7119
34	1	5469·2430	705·6752	0·5609	84	1	9903·1575	499·5995	0·7415
35	2	9069·8770	952·1179	999·8526	85	2	3503·7915	746·0422	0·0332
36	3	2670·5110	198·5605	999·1444	86	3	7104·4255	992·4849	999·3250
*37	4	6271·1450	445·0032	998·4362	*87	4	705·0595	238·9275	998·6168
38	6	210·4109	727·7375	0·4658	88	6	4644·3254	521·6618	0·6464
39	0	3811·0449	974·1801	999·7576	89	0	8244·9594	768·1045	999·9382
40	1	7411·6789	220·6228	999·0494	90	1	1845·5934	14·5471	999·2300
*41	2	1012·3129	467·0655	998·3412	*91	2	5446·2274	260·9898	998·5218
42	4	4951·5788	749·7998	0·3707	92	4	9385·4933	543·7241	0·5513
43	5	8552·2128	996·2424	999·6625	93	5	2986·1273	790·1668	999·8431
*44	6	2152·8468	242·6851	998·9543	94	6	6586·7613	36·6094	999·1349
45	1	6092·1126	525·4194	0·9839	*95	0	187·3953	283·0521	998·4267
46	2	9692·7466	771·8620	0·2757	96	2	4126·6612	565·7864	0·4563
47	3	3293·3806	18·3047	999·5675	97	3	7727·2952	812·2290	999·7481
*48	4	6894·0147	264·7474	998·8592	98	4	1327·9292	58·6717	999·0398
49	6	833·2805	547·4817	0·8888	*99	5	4928·5632	305·1144	998·3316
					100	0	8867·8291	587·8487	0·3412

TABLE LXXIV.

Daily values of *a*, *b*, *c* from 0 Mīna to 2 Mēsha.

For calculation of their value at mean sunrise on the day Chaitra Sukla 1.

Interval of days from true Mēsha-saṁkranti.	Day of Solar month.		Week day.	*a*.	*b*.	*c*.
1	2		3	4	5	6
30	Mīna	0	3	9163·7800	838·6681	912·3908
29	,,	1	4	9502·4119	874·9597	915·1286
28	,,	2	5	9841·0438	911·2513	917·8664
27	,,	3	6	179·6756	947·5429	920·6042
26	,,	4	0	518·3075	983·8345	923·3420
25	,,	5	1	856·9394	20·1262	926·0798
24	,,	6	2	1195·5713	56·4178	928·8176
23	,,	7	3	1534·2032	92·7094	931·5554
22	,,	8	4	1872·8350	129·0010	934·2931
21	,,	9	5	2211·4669	165·2927	937·0309
20	,,	10	6	2550·0988	201·5843	939·7687
19	,,	11	0	2888·7306	237·8759	942·5065
18	,,	12	1	3227·3625	274·1675	945·2443
17	,,	13	2	3565·9944	310·4591	947·9821
16	,,	14	3	3904·6263	346·7508	950·7199
15	,,	15	4	4243·2581	383·0424	953·4576
14	,,	16	5	4581·8900	419·3340	956·1954
13	,,	17	6	4920·5219	455·6256	958·9332
12	,,	18	0	5259·1538	491·9173	961·6710
11	,,	19	1	5597·7856	528·2089	964·4088
10	,,	20	2	5936·4175	564·5005	967·1466
9	,,	21	3	6275·0494	600·7921	969·8844
8	,,	22	4	6613·6813	637·0838	972·6221
7	,,	23	5	6952·3131	673·3754	975·3599
6	,,	24	6	7290·9450	709·6670	978·0977
5	,,	25	0	7629·5769	745·9586	980·8555
4	,,	26	1	7968·2088	782·2503	983·5733
3	,,	27	2	8306·8406	818·5419	986·3111
2	,,	28	3	8645·4725	854·8335	989·0489
1	,,	29	4	8984·1044	891·1251	991·7866
	Mēsha	0	5	9322·7363	927·4168	994·5244
	,,	1	6	9661·3681	963·7084	997·2622
	,,	2	0	0	0	0

The figures for Mēsha 0 are those for mean sunrise on the day when true Mēsha-saṁkrānti occurred, *i.e.*, on the day when true sun reached long. 0°.

The table serves equally for calculation from the day of mean Mēsha-saṁkrānti by noting the interval of days.

TABLE LXXV.

MOON'S EQUATION OF CENTRE BY THE FIRST ARYA-SIDDHANTA.

(For equation of sun's centre see Table XLVII, above.)

Serial No. of sine.	Moon's mean anomaly.		Sine of mean anom. angle. Value in minutes.	Sine of mean anom. angle. Difference.	Equation. Equation in degrees.	Equation. Diff. per minute of anom.	Equation. Equation in 10,000th of circle.	Moon's mean anomaly.		Serial No. of sine.
1	2		3	4	5	6	7	8		1
0	0° 0′	180° 0′	0′	″	0° 6′ 0″	″	0	180° 0′	360° 0′	0
				225		5·250				
1	3 45	176 15	225		0 19 41·25		9·114583	183 45	356 15	1
				224		5·22$\dot{6}$				
2	7 30	172 30	449		0 39 17·25		18·188657	187 30	352 30	2
				222		5·180				
3	11 15	168 45	671		0 58 42·75		27·181713	191 15	348 45	3
				219		5·110				
4	15 0	165 0	890		1 17 52·5		36·053240	195 0	345 0	4
				215		5·01$\dot{6}$				
5	18 45	161 15	1105		1 36 41·25		44·762730	198 45	341 15	5
				210		4·909				
6	22 30	157 30	1315		1 55 3·75		53·269675	202 30	337 30	6
				205		4·78$\dot{3}$				
7	26 15	153 45	1520		2 13 0·0		61·574074	206 15	333 45	7
				199		4·643				
8	30 0	150 0	1719		2 30 24·75		69·635415	210 0	330 0	8
				191		4·45$\dot{6}$				
9	33 45	146 15	1910		2 47 7·5		77·372684	213 45	326 15	9
				183		4·270				
10	37 30	142 30	2093		3 3 8·25		84·785878	217 30	322 30	10
				174		4·060				
11	41 15	138 45	2267		3 18 21·75		91·834490	221 15	318 45	11
				164		3·92$\dot{6}$				
12	45 0	135 0	2431		3 32 42·75		98·478009	225 0	315 0	12
				154		3·5947				
13	48 45	131 15	2585		3 46 11·5681		104·718890	228 45	311 15	13
				143		3·3516				
14	52 30	127 30	2728		3 58 45·6696		110·537572	232 30	307 30	14
				131		3·0603				
15	56 15	123 45	2859		4 10 16·4900		115·867978	236 15	303 45	15
				119		2·7979				
16	60 0	120 0	2978		4 20 44·0290		120·710099	240 0	300 0	16
				106		2·4844				
17	63 45	116 15	3084		4 30 3·0134		125·023250	243 45	296 15	17
				93		2·1797				
18	67 30	112 30	3177		4 38 13·4431		128·807432	247 30	292 30	18
				79		1·8416				
19	71 15	108 45	3256		4 45 10·0446		132·021949	251 15	288 45	19
				65		1·5234				
20	75 0	105 0	3321		4 50 52·8179		134·666805	255 0	285 0	20
				51		1·1953				
21	78 45	101 15	3372		4 55 21·7634		136·742001	258 45	281 15	21
				37		0·8672				
22	82 30	97 30	3409		4 58 36·8804		138·247533	262 30	277 30	22
				22		0·5156				
23	86 15	93 45	3431		5 0 32·8962		139·142717	266 15	273 45	23
				7		0·1641				
24	90 0	90 0	3438		5 1 9·8103		139·427548	270 0	270 0	24

THE FIRST ĀRYA-SIDDHĀNTA. MEAN SYSTEM.

303. It has long been known that in earlier years the Pañchāṅg Brahmans in India framed their local almanacs on calculations made by the use of the mean, as opposed to the true or apparent, motions of the sun and moon. The change from the mean to the true systems of calculation was advocated by Śrīpathi (A.D. 1040), and the latter system may have been adopted in some places about that time; becoming more general from about A.D. 1100 onwards. India, however, is a very conservative country, and the late Dr. Fleet was of opinion that the mean system may have been adhered to, in some tracts at least, till a far later date.

304. With this opinion in mind I have prepared the Tables which follow, so as to cover the period of nine centuries from Āryabhaṭa's date, K.Y. 3600 (A.D. 499-500), to 4500 (A.D. 1399-1400). It would be well if all dates of inscriptions that have hitherto been set aside as irregular by Epigraphists could be re-examined, seeing that the difference between the two systems of the *Ārya Siddhānta* constantly leads to differences in the computed positions of the sun and moon on the same civil day, and consequently to differences in the almanac; let alone the differences caused by the use of different Siddhāntas.

Thus, to give an example. The civil day, Monday, 21 October A.D. 1090, was by the Ārya Siddhānta true system described as "Monday, 25 Tulā, nija Āśvina kr. 10," while by the mean system it was "Monday, 27 Tulā, Kārttika kr. 10." Thursday, 31 Oct., in the same year was by the true system "Thursday, 5 Vṛiśchika, Kārttika śukla 6," while by the mean system it was "Thursday, 7 Vṛiśchika, Mārgaśira śukla 5."

305. The present Tables are based on the First Ārya Siddhānta as amended by Lalla. The principal Table LXXVI is framed on the lines of the *Indian Calendar*, Table I, so as to meet the convenience of Epigraphists who have become accustomed to the use of that work. The numbers of the columns are made to correspond in both Tables.

Results of calculation carried out by the present Tables will be found to correspond with hose worked by use of Prof essor H. Jacobi's skeleton Tables published in *Epig. Ind.* Vol. XI. There is no need for me to dwell on the great services he has rendered to the cause of Indian history and epigraphy. These are well known. All I have done is to follow in his footsteps, verify his figures to the best of my ability and apply the results to practical use. Some little differences that exist between us have been fully set forth and their cause explained.

Elements. Ārya Siddhānta, mean system.

306. (i) The length of the mean sidereal solar year is $365^d\ 6^h\ 12^m\ 30^s$, or $365^d.2586805$.

(ii) For the sun's mean motion per day, hour, etc., see Tables XLIII, XLIV, above.

(iii) The distance of mean moon from mean sun (our "*a*"), measured in 10,000ths of the circle, i.e. 10,000ths of the mean synodical revolution of the moon and excluding 12 whole revolutions, increases, during one sidereal solar year, from 0 to 3688·231484714. That is the advance of "*a*" in the year. Table LXIV-A above col. 3, shews this advance per day, and Table LXV the advance per hour etc.

(iv) The value of "*a*" in mean reckoning corresponds to that of "*t*", the tithi-index, in true reckoning. It shews what mean tithi was current at the moment in question.[1] In general calculation by the Tables this moment is the moment of mean sunrise at Laṅkā, taken as 6 A.M.

(v) In reckoning by 10,000ths of the circle the advance of "*a*" in one mean solar month is 307·352623726.

(vi) Each mean solar month consists of 30ᵈ 10ʰ 31ᵐ 2½ˢ. The collective duration from the moment of mean Mēsha-saṁkrānti (the beginning of the mean solar year when the mean sun is at celestial long. 0°) to each separate saṁkrānti, or the moment when the mean sun enters each of the signs, is given in Table LXXVII.

(vii) The length of each mean lunar month is 29ᵈ 12ʰ 44ᵐ 2ˢ·79 or 29ᵈ·530587946, during which the mean moon's distance from mean sun, "*a*" increases, in our circle reckoning, from 0 to 10,000. The length of one mean tithi, or one-thirtieth of the mean lunar synodic month, is 23ʰ 37ᵐ 28ˢ·09, or 0ᵈ·984352931; during which, in circle reckoning the increase of "*a*" is 333·3̇.

(viii) The *sōdhya*, or time-difference between the moments of arrival at celestial long. 0° of the true and mean suns, which moments are known respectively as the true and mean Mēsha saṁkrāntis, is 2ᵈ 3ʰ 32ᵐ 30ˢ, true Mēsha-saṁkrānti being the earlier. This is invariable.

The time of occurrence of mean Mēsha-saṁkrānti in every year is given in Table LXXVI cols. 13 to 17.

(ix) The *saṁvatsara* name of the solar year is the same by both true and mean reckonings, except in the years A.D. 564-5, 905-6, 990-1, 1246-7 and 1331-2. A special footnote is appended to the main Table LXXVI in each case.

(x) There can be no suppression of a lunar month when calculation is made by the mean system; for the length of a mean solar month is greater than that of a mean lunar month, so that two mean solar saṁkrāntis cannot take place within the limits of one mean lunar month.

(xi) Let it be noted that no intercalation of a lunar month can take place unless, at mean sunrise of the day on which mean Mēsha-saṁkrānti took place, the value of "*a*" is more than 6280·4892, or unless at the moment of mean Mēsha-saṁkrānti the value of "*a*" is more than 6619·1211; the latter value being 10,000 − 3380·8789, the total increase of "*a*" from Mēsha- to Mīna-saṁkrānti, and the former being 6619·1211 − 338·6319, this last being the increase of "*a*" in 24-hours.

The 19-year intercalation cycle

307. (*See Indian Calendar*, § 56, *p.* 29.) By the mean system the cycle-sequence is found to work with almost perfect regularity. After four successive intercalations at intervals of 19 years each the intercalated lunar month gives way to the month preceding it. But there are two exceptions in the nine centuries embraced in Table LXXVI. Between A.D. 751 and 827 there is a run of five intercalary mean Pausha months, and between A.D. 1242 and 1318 there is a run of five intercalary mean Āśvina months.

In eleven instances the names of the mean intercalary months given in Table LXXVI differ from those stated in the *Indian Calendar*. These differences are due to the former calculations having been based on Professor Jacobi's earliest Tables published 35 years ago, while the present ones agree with the results of calculation made by his more recent elementary fixtures. Each difference is specially noted at foot of Table LXXVI.

[1] The equations of sun and moon are not taken into account in mean reckoning.

The nakshatra.

308. In the mean system the position at any moment of the mean moon in the ecliptic circle, *i.e.*, the mean moon's nakshatra, is found by adding her mean distance from the mean sun to the latter's longitude; that is to say, by adding to the value of "*s*" (the mean sun's longitude) the value of "*a*" at the same moment as found by calculation for the mean tithi. All work by the Tables being in the first instance for the mean positions of sun and moon at mean sunrise of any day, Table LXXX provides the sun's mean long. (*s*) in 10,000ths of the circle, for each period of 24-hours measured from the moment of mean Mēsha-saṁkrānti, while Table LXXXI states the same increase for fractions of the day. To obtain the value of "*s*" for mean sunrise of any day it is necessary to note first its value after the interval of days between the day of Mēsha-saṁkrānti and the given day (Table LXXX), and, since that value is measured from the moment of Mēsha-saṁkrānti and not from mean sunrise, afterwards to deduct from the value so obtained the increase during that fraction of the day (Table LXXXI). The result is the required "*s*", or the mean sun's long. at mean sunrise of the given day. Then $s + a = n$, the nakshatra index required, or the mean moon's place in the ecliptic circle at mean sunrise of that day.

The Rule for work, then, is as follows. Find the value of a $(= t)$, the mean tithi-index at mean sunrise of the given day (*Example 2 below*). Note the serial number of the day as measured from Jan. 1. Deduct from this the serial number of the day of mean Mēsha-saṁkrānti (*Table LXXVI, col. 13, in brackets*). This gives the number of intervening days. Turn to Table LXXX and note the value of "*s*" against that interval of days. Deduct from this the mean sun's movement given in Table LXXXI during the hours and minutes stated in Table LXXVI, col. 17. The result is the required value of "*s*" at mean sunrise of the given day. Add s to a. This $= n$, the required nakshatra-index. Table LXVIII above, or Table VIII, *Indian Calendar*, gives the name of the nakshatra.

The Tables.

309. Table LXXVI corresponds to Table I *Indian Calendar* in formation and is to be used in the same way. Here the value of "*a*" is the value of "*t*". It gives the tithi-index direct without further calculation.[1]

Table LXXVII shews the duration and collective duration of mean solar months, and the increase in the moon's phase, "*a*", during each such month.

Table LXXVIII gives the value of "*a*" at the beginning of each Kaliyuga century.

Table LXXIX corresponds, with a necessary shift of position, to Table LXXIV above, the use of which is fully explained in my former paper, 301.

Tables LXXVIII and LXXIX, with Table LXXIII above (under heading "*a*"), which gives the value of "*a*" at the beginning of each year of the Kaliyuga century, enable us to find the value of "*a*" at mean sunrise of the civil day Chaitra śukla 1 at the beginning of each luni-solar year. Tables LXXVIII and LXXIII yield the value of "*a*" at mean sunrise of the day on

[1] To find the value of "*a*", or "*t*", *i.e.*, the exact moon's phase, in 10,000ths of the circle, at any moment of any day, note its value at mean sunrise of the first civil day of the luni-solar year, as given in Table LXXVI (col. 23), and add its value for intervening days, hours, etc. (*Tables LXIV, LXV under heading* "*a*").

which mean Mêsha-saṁkrânti occurred; and Table LXXIX enables, by addition, the "*a*" for the interval of days between that day and the day Chaitra śukla 1 to be ascertained. [The same can be found by subtracting from the sum of the values obtained from Tables LXXVIII and LXXIII (col. *a*) the value for those intervening days given in Table LXIV above (*see Example 1*).]

The use of Tables LXXX and LXXXI is explained above (§ 308). They correspond *mutatis mutandis*, with Tables XLVIII A, XLIX above used in calculation for the sun's true longitude.

310. The century-Table LXXVIII requires some further explanation. Its object is to determine the mean moon's phase, "*a*", at mean sunrise of the opening civil day of each Kaliyuga century, *i.e.*, the day on which mean Mêsha-saṁkrânti occurred at some time later on that day. Reference to Table LXXVI shews that this opening day occurred at the beginnings of centuries 36 and 37 K.Y. on a Sunday, and in centuries 38 to 45 on a Saturday. From Table I, *Indian Calendar*, by adding the *sôdhya* interval (*above*, § *306, viii*) to the date and time there given for the moment of true Mêsha-saṁkrânti, we find that in centuries 46 to 48 it fell on a Friday. In the mean system, therefore, centuries 37 and 45 were defective centuries, while the rest were common.

Table LXXVIII corresponds to Table LXXII above, which concerns true solar years, and by the true system, *i.e.*, calculation by the movements of true sun, the only defective century was century 42. This accounts for the difference between the two Tables.

It has been shewn above (§ 299, i) that the actual value of "*a*" at mean sunrise of Sunday, 21 March A.D. 499, on which day, 6 hours later, occurred the moment of mean Mêsha-saṁkrânti (mean sun at 0°) at the beginning of Kaliyuga century 36, was, in notation in 10,000ths of the circle, 7715·352496330. The values of *a* for later century-beginnings are found by addition to this of the century increases of *a*, common and defective as required.

EXAMPLES.

Example I. To find the European day, week-day, and phase of mean moon, i.e., the mean tithi-index "a" (*which* = "t", *the true moon's index*) *at mean sunrise of the first civil day of the luni-solar year; that is to say, of the day called "Chaitra śukla 1" of the year in question.*

[This example is given in order to enable any student to verify the entries in Table LXXVI, cols. 19-23. For ordinary date work the entries themselves afford all information.]

The mean new moon which marks the astronomical beginning of any mean lunar year is the new moon at the end of the lunar month Phâlguna of the previous year. The moment of its occurrence is always earlier than the moment in the current year of mean Mêsha-saṁkrânti, the beginning of the mean solar year. The civil day next following the moment of the initial mean new moon of the year is called "Chaitra śukla 1," that tithi being current at mean sunrise of that civil day. Our tabular calculations being for mean sunrise, the value of "*a*" in Table LXXVI, col. 23, must always be between 0 and 333·3, the last being the limit of the tithi.

To find its value for any year we must first calculate the value of "*a*" at mean sunrise on the day of occurrence of mean Mêsha-saṁkrânti from Tables LXXVIII and LXXIII (above) under heading "*a*"

This done there are two processes by which the mean sunrise value of "*a*" on the day Chaitra śukla 1 can be obtained. One is to use Table LXIV, which, by *deducting* from the "*a*" of mean Mêsha-saṁkrânti-day mean sunrise (already found) the next lower value of "*a*" in the Table as given for the first 30 days, yields at once the interval of days between Chaitra śukla 1 and

Mēsha-saṁkrānti, the value of "*a*" at mean sunrise of the former, and the required week-day. The second process is, using Table LXXIX, to find such earlier day as by *adding* its "*a*" to the "*a*" of Mēsha-saṁkrānti, already found, will yield a result between 0 and 333·3. The Table then shews the interval of days between the two sunrises, and the week-day corresponding to Chaitra śukla 1.

A. Take for instance the year K.Y. 3725 expired, A.D. 624-25. Mean Mēsha-saṁkrānti occurred in that year (*Table LXXVI, cols. 13-17*) on Wed. 21 Mar.,—serial day 81, from Jan. 1. We take the value of "*a*" at mean sunrise at the beginning of the Kaliyuga century and at the beginning of the expired year from Tables LXXVIII and LXXIII, respectively. The result gives the value of "*a*" at mean sunrise of Mēsha-saṁkrānti day in the given year.

	w-d.	*a.*
(*Table LXXVIII*). K.Y. cent. 37	(1)	6583·1816
(*Table LXXIII above*). K.Y. year 25 . . .	(3)	2047·6413
At mean sunrise on Wed. 21 Mar., the day of occurrence of mean Mēsha-saṁkrānti	(4)	8630·8229

Process 1.

(*Table LXIV above*). Next lower value of "*a*" in the first 30 days of the Table, *i.e.*, that for 25 days .	—(4)	—8465·7968
At mean sunrise of the day Chaitra śukla 1 . .	(0)	165·0261

This Chaitra śukla 1 civil day was (81—25=) Day 56, or (*Table IX, Indian Calendar, or LXIX above*) Sat. 25 Feb. A.D. 624.

Process 2.

	w-d.	*a.*
At mean sunrise on Wed. 21 Mar., the day of mean Mēsha-saṁkrānti (*as above*)	(4)	8630·8229
(*Table LXXIX*). The only value of "*a*" which yields result between 0 and 333·3	+(3)	+1534·2032
At mean sunrise of the day Chaitra śukla 1 . .	(0)	165·0261

Table LXXIX shews that the interval of days was 25, and the result is in all respects the same as the former.

B. Calculation for the mean sunrise value of "*a*" on the day of mean Mēsha-saṁkrānti, the first step shewn in the above, by use of Tables LXXVIII and LXXIII often results in the day found being not the actual day on which Mēsha-saṁkrānti took place but the day next to it. This is inevitable, seeing that only one Table has to stand for the odd years of all centuries. In such case the necessary adjustment must be made for one day's difference. The entries in Table LXXVI, cols. 13 to 17, are conclusive as to the actual day.

Take the year A.D. 625-26, K.Y. 3726 expired. In that year mean Mēsha-saṁkrānti occurred on *Thurs.* 21 Mar., serial day 80.

	w-d.	*a.*
(*Table LXXVIII*). K.Y. century 37	(1)	6583·1816
(Table LXXIII). K.Y. year 26	(5)	5986·9072
At mean sunrise of *Friday*, 22 Mar.	(6)	2570·0888
Deduct value for one day (*Table LXIV*)	—(1)	—338·6319
At m. sunrise of *Thurs.* 21 Mar, the day of mean Mēsha-saṁkrānti	(5)	2231·4569

For the "*a*" of Chaitra śukla 1 and its day and week-day we use either of the two processes

Process 1	*w-d.*	*a.*
At m. sunrise of m. M. S.-day, Thurs. 21 Mar. (*Table LXIV above*).	(5)	2231·4569
Next lower value of "*a*" in the first 30 days of the Table, *viz.*, for 6 days' interval .	−(6)	−2031·7912
At mean sunrise of Fri. 15 Mar., being the day Chaitra śukla 1	(6)	199·6657

Or, *Process 2.*	*w-d.*	*a*
At m. sunrise of m. Mēsha-saṁk. day (*as above*)	(5)	2231·4569
Add (*Table LXXIX* for 6 days earlier) . . .	+(1)	+7968·2088
Result (*same as above*)	(6)	199·6657

Example 2. To find the mean tithi-index "a" *for any day in the year, or any moment of any day.*

Table LXXVI, cols. 19-23, states the civil day, Chaitra śukla 1, for each year, its serial number from Jan. 1, its week-day, and its tithi-index "*a*" at mean sunrise. Calculate, from Table III *Indian Calendar* or Table LXIII above, the interval of whole days to mean sunrise on the given day, and, if necessary, the fraction of day subsequent to that sunrise. Add the increment of "*a*" for whole days from Table LXIV, and for fractions of the day from Table LXV, to the "*a*" given in Table LXXVI.

Whole numbers may always be used for whole days, the decimals being only resorted to for close cases and when the calculation includes a fraction of a day.

E.g. Required the tithi-index at mean sunrise on Āshāḍha śukla 4 in the year corresponding to A.D. 625-26; and at $8^h\ 20^m\ 15^s$ after m. sunrise on that day.

	d.	*w-d.*	*a.*
Table LXXVI. Chait. śuk. 1, mean sunrise (*Example 1*)	(74)	(6)	199·6657
Tables LXIII A, LXIV. Interval to Āsh. śuk. 4, and increase of "*a*"	(91)	(0)	815·5005
At mean sunrise on the day Āsh. śuk. 4 . .	(165)	(6)	1015·1662

Day 165 was (Table IX, *Indian Calendar, or Table LXIX above*) 14 June A.D. 625. (6)=Friday. *a*=1015 shews (*Table VIII or LXVIII*) that śukla 4 was current at mean sunrise of that day.

For the specific hour mentioned—

		a.
At mean sunrise on that day		1015·1662
(*Table LXV*)	8^h	112·8773
	20^m	4·7032
	15^s	0·0588
At $8^h\ 20^m\ 15^s$ after mean sunrise	*a* =	1132·8055

Example 3. To find "a" *(the tithi-index, or phase of mean moon) at each of the solar saṁkrāntis in the year (the moments of the mean sun's entrance into the several signs), and to determine whether an intercalation of a lunar month took place during the year.*

Table LXXVI, cols. 13, 14, 17, shews the day and time of occurrence of mean Mēsha-saṁkrānti (mean sun at long. 0°) in each year, and Example 1 shews how to find the value of "a" at mean sunrise of that day. To that value must be added from Table LXV the increment of "a" during the interval from mean sunrise to moment of saṁkrānti. The advance of "a" during each mean solar month, *i.e.*, from each mean saṁkrānti to the next (*Table LXXVII*) is 307·3526. The work may be carried out by use of whole numbers, except when a case is very close. This occurs when a waning moon is very near 10,000, or when a waxing moon is very near 0.

Required the above details for the years noted in Examples 1, 2, viz. A.D. 624-5 and 625-6.

In A.D. 624-25 mean Mēsha-saṁkrānti took place 14^h 2^m 30^s after mean sunrise. In A.D. 625-26 it took place 20^h 15^m 0^s after mean sunrise (*Table LXXVI, cols. 13-17*).

	a.
A.D. 624-25. Value of "*a*" at m. sunrise on mean Mēsha-saṁkrānti-day, as already found (Example 1) . . .	8630·8229
(*Table LXV*). Increase of "*a*" in 14^h	197·5353
Ditto 2^m	0·4703
Ditto 30^s	0·1176
Exact value of "*a*" at moment of mean Mēsha-saṁkrānti .	8828·9461
A.D. 625-26. Value of "*a*" at m. sunrise of mean Mēsha-saṁkrānti-day as found	2231·4569
(*Table LXV*). Increase of "*a*" in 20^h	282·1932
Ditto 15^m	3·5274
Exact value of "*a*" at moment of mean Mēsha-saṁkrānti .	2517·1775

For the several saṁkrāntis in each year we work here roughly with whole numbers only, adding successively the increase of *a* in 1 solar month.

	A.D. 624-25	A.D. 625-26
At Mēsha-saṁkr. . .	*a*=8829	2517
	307	307
At Vṛishabha-saṁkr. . .	9136	2824
	307	307
At Mithuna-saṁkr. . .	9443	3131
	307	307
At Karka-saṁkr. . .	9750	3438
	307	307
At Siṁha-saṁkr. . .	10,057	3745
	etc	etc.

In A.D. 624-25 it is seen that the mean moon was waning at the Karka-saṁkrānti and waxing at the Siṁha-saṁkrānti, proving an intercalation of a lunar month, which month (*see Table LXXVII, col. 1*) was Śrāvaṇa. Actually "*a*" at Siṁha-saṁkrānti was 58·36.

In A.D. 625-26 the small value of *a* at the moment of Mēsha-saṁkrānti shews that there could have been no intercalation in that year (*see above*, § *306*, *xi*).

Example 4. To find the mean moon's nakshatra, or her place in the ecliptic circle at any moment.

(*See* § *308 above.*) We have to find the value of "*s*", the sun's mean long., at the given moment and the value at the same moment of "*a*", the index of the mean tithi. $s + a = n$, the index of the nakshatra. I assume that, as usual, the values wanted are those at mean sunrise on the given day; for later moments they can easily be found, from Table LXV for "*a*", and from Table LXXXI for "*s*". The example here given will shew the process of work.

Required the nakshatra at mean sunrise on the day referred to in Example 2, viz. Ashāḍha śukla 4 in K.Y. 3726, which was proved to be 14 June A.D. 625, and on which day at mean sunrise the value of "*a*" was found to be 1015·1662. The day, measured from Jan. 1, was serial number 165. In that year mean Mēsha-saṁkrānti took place (*Table LXXVI*) on Day 80 at 20^h 15^m after mean sunrise. The interval of whole days between 20^h 15^m after mean sunrise on the day of Mēsha-saṁkrānti and 20^h 15^m after mean sunrise on the given day is (165 – 80 =) 85.

		s.
(*Table LXXX*). Interval of 85 days		2327·1179
Less (*Table LXXXI*) for 20^h . .	22·8149	
for 15^m . .	0·2852	
	23·1001 . .	– 23·1001
At mean sunrise on the day Āshāḍha śuk. 4, "*s*" = . .		2304·0178
Add "*a*", as found for that mean sunrise		1015·1662
At mean sunrise on that day (=14 June) "*n*" = . . .		3319·1840

Table VIII *Indian Calendar*, or Table LXVIII above, shews that the moon was then in the nakshatra Aślēshā by the equal-space system and by Garga, but in Maghā by the Brahma Siddhānta.[1]

The value of "*n*", 3319·1840, in 10,000ths of the circle, can be converted into degrees, if required, by Table XLV B, above. It = 119° 29′ 26″. That was the mean moon's place.

Example 5. The lagna. (*See Indian Chronography*, § *193*, *p. 74*, *and Example 63*, *p. 127*.) Required to ascertain at what hour on the day Āshāḍha śuk 4 K.Y. 3726, or 14 June A.D. 625, the sign Tulā became lagna.

At mean sunrise the sun's mean long. "*s*" was (*Example 4*) 2304·0178, roughly (*Table XLV above*) 82° 57′. The first point of Tulā (Libra) (*Indian Chronography, Table XXII*) is 180° 180° – 82° 57′ = 97° 3′. 97° × 4 = 388^m, or 6^h 28^m, 3′ × 4 = 12^s. The first point of Tulā, therefore, was lagna at 6^h 28^m 12^s after mean sunrise on the day in question. It lasted for 2 hours, when Vṛiśchika (Scorpio) became lagna.

---

[1] As to these systems see *Indian Calendar* § 38 p. 21; *Indian Chronography* § 112, etc.

TABLE

MEAN SYSTEM TABLE,

Numbers of columns conform

(Cols. 1 to 4.)—The years herein stated are the *current* years corresponding

(Cols. 6 and 7.)—*Saṁvatsara-names of mean solar years in italics shew where*

CONCURRENT YEAR.								
						Jovian samvatsara.		Mean Intercalated (adhika) lunar month.
Kali.	Śaka.	Chaitrâdi Vikrama.	Mêshâdi solar year in Bengal.	Kollam.	A.D.	Southern system.	Northern system.	
1	2	3	3a	4	5	6	7	8a
3601	422	557			499-500	9 Yuvan . . .		9 Mârgaśira .
3602	423	558			*500-01	10 Dhâtṛi . . .		...
3603	424	559			501-02	11 Îśvara . . .		...
3604	425	560			502-03	12 Bahudhânya . .		5 Śrâvaṇa .
3605	426	561			503-04	13 Pramâthin . .		...
3606	427	562			*504-05	14 Vikrama . . .		...
3607	428	563			505-06	15 Vṛisha . . .		2 Vaiśâkha .
3608	429	564			506-07	16 Chitrabhânu . .		...
3609	430	565			507-08	17 Subhânu . . .		10 Pausha .
3610	431	566			*508-09	18 Târaṇa . . .		...
3611	432	567			509-10	19 Pârthiva . . .		...
3612	433	568			510-11	20 Vyaya . . .		7 Âśvina .
3613	434	569			511-12	21 Sarvajit . . .		...
3614	435	570			*512-13	22 Sarvadhârin . .		...
3615	436	571			513-14	23 Virôdhin . . .		3 Jyêshṭha .
3616	437	572			514-15	24 Vikṛita . . .		...
3617	438	573			515-16	25 Khara . . .		12 Phâlguna .
3618	439	574			*516-17	26 Nandana . .		...
3619	440	575			517-18	27 Vijaya . . .		...
3620	441	576			518-19	28 Jaya . . .		8 Kârttika .

LXXVI.

FIRST ARYA SIDDHĀNTA.

to Table I, "Indian Calendar."

to the A.D. years in col. 5 ; as in Table I, "Indian Calendar."

differences exist from Sūrya Siddhānta nomenclature in true solar years.

1 Ārya Siddhānta, mean system.

COMMENCEMENT OF THE									
MEAN SOLAR YEAR.					MEAN LUNI-SOLAR YEAR (MEAN SUNRISE OF CIVIL DAY ON WHICH CHAITRA ŚUKLA 1 ENDS).			Kali year.
Day and month, A.D.	Week-day.	Time of mean Mēsha-saṁkrānti.			Day and month, A.D.	Week-day.	*a* (here=*t*, the index of the tithi).	
13	14	17			19	20	23	1
		H.	M.	S.				
21 Mar. (80) . .	1 Sun. .	6	0	0	27 Feb. (58) .	0 Sat. .	265·4513	3601
20 Mar. (80) . .	2 Mon. .	12	12	30	17 Mar. (77) .	6 Fri. .	300·0909	3602
20 Mar. (79) . .	3 Tues. .	18	25	0	6 Mar. (65) .	3 Tues. .	175·7743	3603
21 Mar. (80) . .	5 Thur. .	0	37	30	23 Feb. (54) .	0 Sat. .	51·4577	3604
21 Mar. (80) . .	6 Fri. .	6	50	0	14 Mar. (73) .	6 Fri. .	86·0973	3605
20 Mar. (80) . .	0 Sat. .	13	2	30	3 Mar. (63) .	4 Wed. .	300·4125	3606
20 Mar. (79) . .	1 Sun. .	19	15	0	20 Feb. (51) .	1 Sun. .	176·0959	3607
21 Mar. (80) . .	3 Tues. .	1	27	30	11 Mar. (70) .	0 Sat. .	210·7356	3608
21 Mar. (80) . .	4 Wed. .	7	40	0	28 Feb. (59) .	4 Wed. .	86·4189	3609
20 Mar. (80) . .	5 Thur. .	13	52	30	18 Mar. (78) .	3 Tues. .	121·0586	3610
20 Mar. (79) . .	6 Fri. .	20	5	0	7 Mar. (66) .	0 Sat. .	9996·7419†	3611
21 Mar. (80) . .	1 Sun. .	2	17	30	25 Feb. (56) .	5 Thur. .	211·0572	3612
21 Mar. (80) . .	2 Mon. .	8	30	0	16 Mar. (75) .	4 Wed. .	245·6968	3613
20 Mar. (80) . .	3 Tues. .	14	42	30	4 Mar. (64) .	1 Sun. .	121·3802	3614
20 Mar. (79) . .	4 Wed. .	20	55	0	21 Feb. (52) .	5 Thur. .	9997·0635†	3615
21 Mar. (80) . .	6 Fri. .	3	7	30	12 Mar. (71) .	4 Wed. .	31·7031	3616
21 Mar. (80) . .	0 Sat. .	9	20	0	2 Mar. (61) .	2 Mon. .	246·0185	3617
20 Mar. (80) . .	1 Sun. .	15	32	30	20 Mar. (80) .	1 Sun. .	280·6581	3618
20 Mar. (79) . .	2 Mon. .	21	45	0	9 Mar. (68) .	5 Thur. .	156·3414	3619
21 Mar. (80) . .	4 Wed. .	3	57	30	26 Feb. (57) .	2 Mon. .	32·0248	3620

† As a mean tithi Chaitra Śukla 1 was suppressed. The civil day corresponding to it, *i.e.*, the first day of the mean luni-solar year, was as given in cols. 19, 20.

TABLE

CONCURRENT YEAR.								
Kali.	Śaka.	Chaitrādi Vikrama.	Mēshādi solar year in Bengal.	Kollam.	A.D.	JOVIAN SAṀVATSARA. Southern system.	Northern system.	Mean Intercalated (adhika) lunar month.
1	2	3	3a	4	5	6	7	8a
3621	442	577			519-20	29 Manmatha . .		...
3622	443	578			*520-21	30 Durmukha . .		...
3623	444	579			521-22	31 Hēmalamba . .		5 Śrāvaṇa .
3624	445	580			522-23	32 Vilamba . . .		...
3625	446	581			523-24	33 Vikārin . .		...
3626	447	582			*524-25	34 Śārvarin . . .		1 Chaitra .
3627	448	583			525-26	35 Plava . . .		...
3628	449	584			526-27	36 Śubhakṛit . .		10 Pausha .
3629	450	585			527-28	37 Sōbhana . . .		...
3630	451	586			*528-29	38 Krōdhin . . .		...
3631	452	587			529-30	39 Viśvāvasu . . .		7 Āśvina .
3632	453	588			530-31	40 Parābhava . .		...
3633	454	589			531-32	41 Plavaṅga . . .		...
3634	455	590			*532-33	42 Kīlaka . . .		3 Jyēshṭha .
3635	456	591			533-34	43 Saumya . . .		...
3636	457	592			534-35	44 Sādhāraṇa . .		12 Phālguna .
3637	458	593			535-36	45 Virōdhakṛit . .		...
3638	459	594			*536-37	46 Paridhāvin . .		...
3639	460	595			537-38	47 Pramādin . .		8 Kārttika
3640	461	596			538-39	48 Ānanda . . .		...
3641	462	597			539-40	49 Rākshasa . . .		...
3642	463	598			*540-41	50 Anala . . .		5 Śrāvaṇa
3643	464	599			541-42	51 Piṅgala . . .		...
3644	465	600			542-43	52 Kālayukta . .		...
3645	466	601			543-44	53 Siddhārthin . .		1 Chaitra .

LXXVI—*Contd.*

I Ārya Siddhānta, mean system.

COMMENCEMENT OF THE						
MEAN SOLAR YEAR			MEAN LUNI-SOLAR YEAR (MEAN SUNRISE OF CIVIL DAY ON WHICH CHAITRA ŚUKLA 1 ENDS).			Kali year.
Day and month, A.D.	Week-day.	Time of mean Mēsha-saṁkranti.	Day and month, A.D.	Week-day.	*a* (here = *t*, the index of the tithi).	
13	14	17	19	20	23	1
		H. M. S.				
21 Mar. (80) . .	5 Thur. .	10 10 0	17 Mar. (76) .	1 Sun. .	66·6644	3621
20 Mar. (80) . .	6 Fri. .	16 22 30	6 Mar. (66) .	6 Fri. .	280·9797	3622
20 Mar. (79) . .	0 Sat. .	22 35 0	23 Feb. (54) .	3 Tues. .	156·6631	3623
21 Mar. (80) . .	2 Mon. .	4 47 30	14 Mar. (73) .	2 Mon. .	191·3027	3624
21 Mar. (80) . .	3 Tues. .	11 0 0	3 Mar. (62) .	6 Fri. .	66·9860	3625
20 Mar. (80) . .	4 Wed. .	17 12 30	21 Feb. (52) .	4 Wed. .	281·3013	3626
20 Mar. (79) . .	5 Thur. .	23 25 0	11 Mar. (70) .	3 Tues. .	315·9409	3627
21 Mar. (80) . .	0 Sat. .	5 37 30	28 Feb. (59) .	0 Sat. .	191·6243	3628
21 Mar. (80) . .	1 Sun. .	11 50 0	19 Mar. (78) .	6 Fri. .	226·2640	3629
20 Mar. (80) . .	2 Mon. .	18 2 30	7 Mar. (67) .	3 Tues. .	101·9473	3630
21 Mar. (80) . .	4 Wed. .	0 15 0	25 Feb. (56) .	1 Sun. .	316·2626	3631
21 Mar. (80) . .	5 Thur. .	6 27 30	15 Mar. (74) .	6 Fri. .	12·2703	3632
21 Mar. (80) . .	6 Fri. .	12 40 0	5 Mar. (64) .	4 Wed	226·5856	3633
20 Mar. (80) . .	0 Sat. .	18 52 30	22 Feb. (53) .	1 Sun. .	102·2690	3634
21 Mar. (80) . .	2 Mon. .	1 5 0	12 Mar. (71) .	0 Sat. .	136·9086	3635
21 Mar. (80) . .	3 Tues. .	7 17 30	1 Mar. (60) .	4 Wed. .	12·5920	3636
21 Mar. (80) . .	4 Wed. .	13 30 0	20 Mar. (79) .	3 Tues. .	47·2316	3637
20 Mar. (80) . .	5 Thur. .	19 42 30	9 Mar. (69) .	1 Sun. .	261·5469	3638
21 Mar. (80) . .	0 Sat. .	1 55 0	26 Feb. (57) .	5 Thur. .	137·2303	3639
21 Mar. (80) . .	1 Sun. .	8 7 30	17 Mar. (76) .	4 Wed. .	171·8699	3640
21 Mar. (80) . .	2 Mon. .	14 20 0	6 Mar. (65) .	1 Sun. .	47·5533	3641
20 Mar. (80) . .	3 Tues. .	20 32 30	24 Feb. (55) .	6 Fri .	261·8686	3642
21 Mar. (80) . .	5 Thur. .	2 45 0	14 Mar. (73) .	5 Thur. .	296·5082	3643
21 Mar. (80) . .	6 Fri. .	8 57 30	3 Mar. (62) .	2 Mon. .	172·1916	3644
21 Mar. (80) . .	0 Sat. .	15 10 0	20 Feb. (51) .	6 Fri. .	47·8749	3645

TABLE

CONCURRENT YEAR.								Mean Intercalated (adhika) lunar month.
Kali	Śaka.	Chaitrādi Vikrama.	Mēshādi solar year in Bengal.	Kollam.	A.D.	Jovian saṁvatsara. Southern system.	Jovian saṁvatsara. Northern system.	
1	2	3	3a	4	5	6	7	8a
3646	467	602			*544-45	54 Raudra . . .		...
3647	468	603			545-46	55 Durmati . . .		10 Pausha .
3648	469	604			546-47	56 Dundubhi . .		...
3649	470	605			547-48	57 Rudhirōdgārin . .		...
3650	471	606			*548-49	58 Raktāksha . .		6 Bhādrapada
3651	472	607			549-50	59 Krōdhana . .		...
3652	473	608			550-51	60 Kshaya . . .		...
3653	474	609			551-52	1 Prabhava . . .		3 Jyēshṭha .
3654	475	610			*552-53	2 Vibhava . . .		...
3655	476	611			553-54	3 Śukla . . .		11 Māgha .
3656	477	612			554-55	4 Pramōda . . .		...
3657	478	613			555-56	5 Prajāpati . .		...
3658	479	614			*556-57	6 Aṅgiras . . .		8 Kārttika .
3659	480	615			557-58	7 Śrīmukha . . .		...
3660	481	616			558-59	8 Bhāva . . .		...
3661	482	617			559-60	9 Yuvan . . .		4 Āshāḍha .
3662	483	618			*560-61	10 Dhātṛi . . .		...
3663	484	619			561-62	11 Īśvara . . .		...
3664	485	620			562-63	12 Bahudhānya . .		1 Chaitra .
3665	486	621			563-64	13 Pramādin † . .		...
3666	487	622			*564-65	15 *Vṛisha* . . .		10 Pausha .
3667	488	623			565-66	16 *Chitrabhānu* . .		...
3668	489	624			566-67	17 *Subhānu* . . .		...
3669	490	625			567-68	18 *Tāraṇa* . . .		6 Bhādrapada.
3670	491	626			*568-69	19 Pārthiva . . .		...

† By the First Ārya Siddhānta *mean* system 14 Vikrama was expunged, and A.D. 564-65 corresponded to 15 Vrisha. By the same authority *true* system A.D. 564-65 corresponded to 14 Vikrama, and 15 Vrisha was expunged. A.D. 565-66 was 16 Chitrabhānu by both systems.

LXXVI—*Contd.*

1 Ārya Siddhānta, mean system.

COMMENCEMENT OF THE							
MEAN SOLAR YEAR.			MEAN LUNI-SOLAR YEAR (MEAN SUNRISE OF CIVIL DAY ON WHICH CHAITRA ŚUKLA 1 ENDS).			Kali year.
Day and month, A.D.	Week-day.	Time of mean Mēsha saṁkrānti.	Day and month, A.D.	Week-day.	*a* (here = *t*, the index of the tithi).	
13	14	17	19	20	23	1
		H. M. S.				
20 Mar. (80) . .	1 Sun. .	21 22 30	10 Mar. (70) .	5 Thur. .	82·5145	3646
21 Mar. (80) . .	3 Tues. .	3 35 0	28 Feb. (59) .	3 Tues. .	296·8298	3647
21 Mar. (80) . .	4 Wed. .	9 47 30	19 Mar. (78) .	2 Mon. .	331·4694	3648
21 Mar. (80) . .	5 Thur. .	16 0 0	8 Mar. (67) .	6 Fri. .	207·1528	3649
20 Mar. (80) . .	6 Fri. .	22 12 30	25 Feb. (56) .	3 Tues. .	82·8361	3650
21 Mar. (80) . .	1 Sun. .	4 25 0	15 Mar. (74) .	2 Mon. .	117·4757	3651
21 Mar. (80) . .	2 Mon. .	10 37 30	5 Mar. (64) .	0 Sat. .	331·7910	3652
21 Mar. (80) . .	3 Tues. .	16 50 0	22 Feb. (53) .	4 Wed. .	207·4744	3653
20 Mar. (80) . .	4 Wed. .	23 2 30	12 Mar. (72) .	3 Tues. .	242·1140	3654
21 Mar. (80) . .	6 Fri. .	5 15 0	1 Mar. (60) .	0 Sat. .	117·7974	3655
21 Mar. (80) . .	0 Sat. .	11 27 30	20 Mar. (79) .	6 Fri. .	152·4370	3656
21 Mar. (80) . .	1 Sun. .	17 40 0	9 Mar. (68) .	3 Tues. .	28·1204	3657
20 Mar. (80) . .	2 Mon. .	23 52 30	27 Feb. (58) .	1 Sun. .	242·4357	3658
21 Mar. (80) . .	4 Wed. .	6 5 0	17 Mar. (76) .	0 Sat. .	277·0753	3659
21 Mar. (80) . .	5 Thur. .	12 17 30	6 Mar. (65) .	4 Wed. .	152·7587	3660
21 Mar. (80) . .	6 Fri. .	18 30 0	23 Feb. (54) .	1 Sun. .	28·4421	3661
21 Mar. (81) . .	1 Sun. .	0 42 30	13 Mar. (73) .	0 Sat. .	63·0817	3662
21 Mar. (80) . .	2 Mon. .	6 55 0	3 Mar. (62) .	5 Thur. .	277·3970	3663
21 Mar. (80) . .	3 Tues. .	13 7 30	20 Feb. (51) .	2 Mon. .	153·0803	3664
21 Mar. (80) . .	4 Wed. .	19 20 0	11 Mar. (70) .	1 Sun. .	187·7200	3665
21 Mar. (81) . .	6 Fri. .	1 32 30	28 Feb. (59) .	5 Thur. .	63·4034	3666
21 Mar. (80) . .	0 Sat. .	7 45 0	18 Mar. (77) .	4 Wed. .	98·0430	3667
21 Mar. (80) . .	1 Sun. .	13 57 30	8 Mar. (67) .	2 Mon. .	312·3582	3668
21 Mar. (80) . .	2 Mon. .	20 10 0	25 Feb. (56) .	6 Fri. .	188·0416	3669
21 Mar. (81) . .	4 Wed. .	2 22 30	15 Mar. (75) .	5 Thur. .	222·6813	3670

TABLE

CONCURRENT YEAR.								Mean Intercalated (adhika) lunar month.
						Jovian samvatsara.		
Kali.	Śaka.	Chaitrādi Vikrama.	Mēshādi solar year in Bengal.	Kollam.	A.D.	Southern system.	Northern system.	
1	2	3	3a	4	5	6	7	8a
3671	492	627			569-70	20 Vyaya . . .		...
3672	493	628			570-71	21 Sarvajit . . .		3 Jyēshṭha .
3673	494	629			571-72	22 Sarvadhārin . .		...
3674	495	630			*572-73	23 Virōdhin . . .		11 Māgha .
3675	496	631			573-74	24 Vikṛita . . .		...
3676	497	632			574-75	25 Khara . . .		...
3677	498	633			575-76	26 Nandana . . .		8 Kārttika .
3678	499	634			*576-77	27 Vijaya . . .		...
3679	500	635			577-78	28 Jaya . . .		...
3680	501	636			578-79	29 Manmatha . .		4 Āshāḍha .
3681	502	637			579-80	30 Durmukha .		...
3682	503	638			*580-81	31 Hēmalamba . .		...
3683	504	639			581-82	32 Vilamba . . .		1 Chaitra .
3684	505	640			582-83	33 Vikārin . . .		...
3685	506	641			583-84	34 Śārvarin . . .		9 Mārgaśira .
3686	507	642			*584-85	35 Plava . . .		...
3687	508	643			585-86	36 Śubhakṛit . .		...
3688	509	644			586-87	37 Śōbhana . . .		6 Bhādrapada.
3689	510	645			587-88	38 Krōdhin . . .		...
3690	511	646			*588-89	39 Viśvāvasu . .		...
3691	512	647			589-90	40 Parābhava . .		2 Vaiśākha .
3692	513	648			590-91	41 Plavaṅga . .		...
3693	514	649			591-92	42 Kīlaka . . .		11 Māgha .
3694	515	650			*592-93	43 Saumya . . .		...
3695	516	651			593-94	44 Sādhāraṇa . .		...

LXXVI—*Contd.*

1 Arya Siddhānta, mean system.

COMMENCEMENT OF THE							
MEAN SOLAR YEAR.			MEAN LUNI-SOLAR YEAR (MEAN SUNRISE OF CIVIL DAY ON WHICH CHAITRA ŚUKLA 1 ENDS).			Kali year.
Day and month A.D.	Week-day.	Time of mean Mēsha-saṁkrānti.	Day and month, A.D.	Week-day.	*a* (here = *t*, the index of the tithi).	
13	14	17	19	20	23	1
		H. M. S.				
21 Mar. (80)	5 Thur.	8 35 0	4 Mar. (63)	2 Mon.	98·3646	3671
21 Mar. (80)	6 Fri.	14 47 30	22 Feb. (53)	0 Sat.	312·6799	3672
21 Mar. (80)	0 Sat.	21 0 0	12 Mar. (71)	5 Thur.	8·6876	3673
21 Mar. (81)	2 Mon.	3 12 30	1 Mar. (61)	3 Tues.	223·0029	3674
21 Mar. (80)	3 Tues.	9 25 0	20 Mar. (79)	2 Mon.	257·6425	3675
21 Mar. (80)	4 Wed.	15 37 30	9 Mar. (68)	6 Fri.	133·3259	3676
21 Mar. (80)	5 Thur.	21 50 0	26 Feb. (57)	3 Tues.	9·0092	3677
21 Mar. (81)	0 Sat.	4 2 30	16 Mar. (76)	2 Mon.	43·6488	3678
21 Mar. (80)	1 Sun.	10 15 0	6 Mar. (65)	0 Sat.	257·9641	3679
21 Mar. (80)	2 Mon.	16 27 30	23 Feb. (54)	4 Wed.	133·6476	3680
21 Mar. (80)	3 Tues.	22 40 0	14 Mar. (73)	3 Tues.	168·2871	3681
21 Mar. (81)	5 Thur.	4 52 30	2 Mar. (62)	0 Sat.	43·9705	3682
21 Mar. (80)	6 Fri.	11 5 0	20 Feb. (51)	5 Thur.	258·2857	3683
21 Mar. (80)	0 Sat.	17 17 30	11 Mar. (70)	4 Wed.	292·9254	3684
21 Mar. (80)	1 Sun.	23 30 0	28 Feb. (59)	1 Sun.	168·6087	3685
21 Mar. (81)	3 Tues.	5 42 30	18 Mar. (78)	0 Sat.	203·2484	3686
21 Mar. (80)	4 Wed.	11 55 0	7 Mar. (66)	4 Wed.	78·9317	3687
21 Mar. (80)	5 Thur.	18 7 30	25 Feb. (56)	2 Mon.	293·2470	3688
22 Mar. (81)	0 Sat.	0 20 0	16 Mar (75)	1 Sun.	327·8867	3689
21 Mar. (81)	1 Sun.	6 32 30	4 Mar. (64)	5 Thur.	293·5700	3690
21 Mar. (80)	2 Mon.	12 45 0	21 Feb. (52)	2 Mon.	79·2534	3691
21 Mar. (80)	3 Tues.	15 57 30	12 Mar. (71)	1 Sun.	113·8930	3692
22 Mar. (81)	5 Thur.	1 10 0	2 Mar. (61)	6 Fri.	328·2083	3693
21 Mar. (81)	6 Fri.	7 22 30	19 Mar. (79)	4 Wed.	24·2160	3694
21 Mar. (80)	0 Sat.	13 35 0	9 Mar. (68)	2 Mon.	238·5313	3695

TABLE

CONCURRENT YEAR.								
						Jovian samvatsara.		Mean Intercalated (adhika) lunar month.
Kali.	Śaka.	Chaitrādi Vikrama.	Mēshādi solar year in Bengal.	Kollam.	A.D.	Southern system.	Northern system.	
1	2	3	3a	4	5	6	7	8a
3696	517	652	1		594-95	45 Virōdhakṛit		7 Āśvina
3697	518	653	2		595-96	46 Paridhāvin		...
3698	519	654	3		*596-97	47 Pramādin		...
3699	520	655	4		597-98	48 Ānanda		4 Āshādha
3700	521	656	5		598-99	49 Rākshasa		...
3701	522	657	6		599-600	50 Anala		12 Phālguna
3702	523	658	7		*600-01	51 Piṅgala		...
3703	524	659	8		601-02	52 Kālayukta		...
3704	525	660	9		602-03	53 Siddhārthin		9 Mārgaśira
3705	526	661	10		603-04	54 Raudra		...
3706	527	662	11		*604-05	55 Durmati		...
3707	528	663	12		605-06	56 Dundubhi		6 Bhādrapada.
3708	529	664	13		606-07	57 Rudhirōdgārin		...
3709	530	665	14		607-08	58 Raktāksha		...
3710	531	666	15		*608-09	59 Krōdhana		2 Vaiśākha
3711	532	667	16		609-10	60 Kshaya		...
3712	533	668	17		610-11	1 Prabhava		11 Māgha
3713	534	669	18		611-12	2 Vibhava		...
3714	535	670	19		*612-13	3 Śukla		...
3715	536	671	20		613-14	4 Pramōda		7 Āśvina
3716	537	672	21		614-15	5 Prajāpati		...
3717	538	673	22		615-16	6 Aṅgiras		...
3718	539	674	23		*616-17	7 Śrīmukha		4 Āshādha
3719	540	675	24		617-18	8 Bhāva		...
3720	541	676	25		618-19	9 Yuvan		12 Phālguna

LXXVI—*Contd.*

I Ārya Siddhānta, mean system.

COMMENCEMENT OF THE									
MEAN SOLAR YEAR.					MEAN LUNI-SOLAR YEAR (MEAN SUNRISE OF CIVIL DAY ON WHICH CHAITRA ŚUKLA 1 ENDS).			Kali year.
Day and month, A.D.	Week-day.	Time of mean Mēsha saṁkrānti.			Day and month, A.D.	Week-day.	*a* (here = *t*, the index of the tithi).	
13	14	17			19	20	23	1
		H.	M.	S.				
21 Mar. (80) . .	1 Sun. .	19	47	30	26 Feb. (57) .	6 Fri. .	114·2147	3696
22 Mar. (81) . .	3 Tues. .	2	0	0	17 Mar. (76) .	5 Thur. .	148·8543	3697
21 Mar. (81) . .	4 Wed. .	8	12	30	5 Mar. (65) .	2 Mon. .	24·5377	3698
21 Mar. (80) . .	5 Thur. .	14	25	0	23 Feb. (54) .	0 Sat. .	238·8530	3699
21 Mar. (80) . .	6 Fri. .	20	37	30	14 Mar. (73) .	6 Fri. .	273·4926	3700
22 Mar. (81) . .	1 Sun. .	2	50	0	3 Mar. (62) .	3 Tues .	149·1760	3701
21 Mar. (81) . .	2 Mon. .	9	2	30	21 Mar. (81) .	2 Mon. .	183·8156	3702
21 Mar. (80) . .	3 Tues. .	15	15	0	10 Mar. (69) .	6 Fri. .	59·4990	3703
21 Mar. (80) . .	4 Wed. .	21	27	30	28 Feb. (59) .	4 Wed. .	273·8142	3704
22 Mar. (81) . .	6 Fri. .	3	40	0	19 Mar. (78) .	3 Tues. .	308·4539	3705
21 Mar. (81) . .	0 Sat. .	9	52	30	7 Mar. (67) .	0 Sat. .	184·1373	3706
21 Mar. (80) . .	1 Sun. .	16	5	0	24 Feb. (55) .	4 Wed. .	59·8207	3707
21 Mar. (80) . .	2 Mon. .	22	17	30	15 Mar. (74) .	3 Tues. .	94·4603	3708
22 Mar. (81) . .	4 Wed. .	4	30	0	5 Mar. (64) .	1 Sun. .	308·7756	3709
21 Mar. (81) . .	5 Thur. .	10	42	30	22 Feb. (53) .	5 Thur. .	184·4589	3710
21 Mar. (80) . .	6 Fri. .	16	55	0	12 Mar. (71) .	4 Wed. .	219·0985	3711
21 Mar. (80) . .	0 Sat. .	23	7	30	1 Mar. (60) .	1 Sun. .	94·7819	3712
22 Mar. (81) . .	2 Mon. .	5	20	0	20 Mar. (79) .	0 Sat. .	129·4215	3713
21 Mar. (81) . .	3 Tues. .	11	32	30	8 Mar. (68) .	4 Wed. .	5·1049	3714
21 Mar. (80) . .	4 Wed. .	17	45	0	26 Feb. (57) .	2 Mon. .	219·4201	3715
21 Mar. (80) . .	5 Thur. .	23	57	30	17 Mar. (76) .	1 Sun. .	254·0597	3716
22 Mar. (81) . .	0 Sat. .	6	10	0	6 Mar. (65) .	5 Thur. .	129·7432	3717
21 Mar. (81) . .	1 Sun. .	12	22	30	23 Feb. (54) .	2 Mon. .	5·4266	3718
21 Mar. (80) . .	2 Mon. .	18	35	0	13 Mar. (72) .	1 Sun. .	40·0681	3719
22 Mar. (81) . .	4 Wed.	0	47	30	3 Mar. (62) .	6 Fri. .	2[illegible]4·3814	3720

TABLE

CONCURRENT YEAR.								Mean Intercalated (adhika) lunar month.
						Jovian Saṁvatsara.		
Kali.	Śaka.	Chaitrādi Vikrama.	Mēshādi solar year in Bengal.	Kollam.	A.D.	Southern system.	Northern system.	
1	2	3	3a	4	5	6	7	8a
3721	542	677	26		619-20	10 Dhātṛi . . .		...
3722	543	678	27		*620-21	11 Īśvara . . .		...
3723	544	679	28		621-22	12 Bahudhānya . .		9 Mārgaśira .
3724	545	680	29		622-23	13 Pramādin . .		...
3725	546	681	30		623-24	14 Vikrama . . .		...
3726	547	682	31		*624-25	15 Vṛisha . . .		5 Śrāvaṇa .
3727	548	683	32		625-26	16 Chitrabhānu . .		...
3728	549	684	33		626-27	17 Subhānu . . .		...
3729	550	685	34		627-28	18 Tāraṇa . . .		2 Vaiśākha .
3730	551	686	35		*628-29	19 Pārthiva . . .		...
3731	552	687	36		629-30	20 Vyaya . . .		10 Pausha .
3732	553	688	37		630-31	21 Sarvajit . . .		...
3733	554	689	38		631-32	22 Sarvadhārin . .		...
3734	555	690	39		*632-33	23 Virōdhin . . .		7 Āśvina .
3735	556	691	40		633-34	24 Vikṛita . . .		...
3736	557	692	41		634-35	25 Khara . . .		...
3737	558	693	42		635-36	26 Nandana . . .		3 Jyēshṭha .
3738	559	694	43		*636-37	27 Vijaya . . .		...
3739	560	695	44		637-38	28 Jaya . .		12 Phālguna .
3740	561	696	45		638-39	29 Manmatha . .		...
3741	562	697	46		639-40	30 Durmukha . .		...
3742	563	698	47		*640-41	31 Hēmalamba . .		9 Mārgaśira
3743	564	699	48		641-42	32 Vilamba . . .		...
3744	565	700	49		642-43	33 Vikārin . . .		...
3745	566	701	50		643-44	34 Śārvarin . .		5 Śrāvaṇa

LXXVI—*Contd.*

I Ārya Siddhānta, mean system.

COMMENCEMENT OF THE							
MEAN SOLAR YEAR.			MEAN LUNI-SOLAR YEAR (MEAN SUNRISE OF CIVIL DAY ON WHICH CHAITRA ŚUKLA 1 ENDS).			Kali year.
Day and month, A.D.	Week-day.	Time of mean Mēsha-saṁkrānti.	Day and month, A.D.	Week-day.	*a* (here = *t*, the index of the tithi).	
13	14	17	19	20	23	1
		H. M. S.				
22 Mar. (81) . .	5 Thur. .	7 0 0	22 Mar. (81) .	5 Thur. .	289·0209	3721
21 Mar. (81) . .	6 Fri. .	13 12 30	10 Mar. (70) .	2 Mon. .	164·7044	3722
21 Mar. (80) . .	0 Sat. .	19 25 0	27 Feb. (58) .	6 Fri. .	40·3877	3723
22 Mar. (81) . .	2 Mon. .	1 37 30	18 Mar. (77) .	5 Thur. .	75·0274	3724
22 Mar. (81) . .	3 Tues. .	7 50 0	8 Mar. (67) .	3 Tues. .	289·3427	3725
21 Mar. (81) . .	4 Wed. .	14 2 30	25 Feb. (56) .	0 Sat. .	165·0261	3726
21 Mar. (80) . .	5 Thur. .	20 15 0	15 Mar. (74) .	6 Fri. .	199·6657	3727
22 Mar. (81) . .	0 Sat. .	2 27 30	4 Mar. (63) .	3 Tues. .	75·3491	3728
22 Mar. (81) . .	1 Sun. .	8 40 0	22 Feb. (53) .	1 Sun. .	289·6643	3729
21 Mar. (81) . .	2 Mon. .	14 52 30	12 Mar. (72) .	0 Sat. .	324·3039	3730
21 Mar. (80) . .	3 Tues. .	21 5 0	1 Mar. (60) .	4 Wed. .	199·9873	3731
22 Mar. (81) . .	5 Thur. .	3 17 30	20 Mar. (79) .	3 Tues. .	234·6269	3732
22 Mar. (81) . .	6 Fri. .	9 30 0	9 Mar. (68) .	0 Sat. .	110·3103	3733
21 Mar. (81) . .	0 Sat. .	15 42 30	27 Feb. (58) .	5 Thur. .	324·6256	3734
21 Mar. (80) . .	1 Sun. .	21 55 0	16 Mar. (75) .	3 Tues. .	20·6333	3735
22 Mar. (81) . .	3 Tues. .	4 7 30	6 Mar. (65) .	1 Sun. .	234·9486	3736
22 Mar. (81) . .	4 Wed. .	10 20 0	23 Feb. (54) .	5 Thur. .	110·6320	3737
21 Mar. (81) . .	5 Thur. .	16 32 30	13 Mar. (73) .	4 Wed. .	145·2716	3738
21 Mar. (80) . .	6 Fri. .	22 45 0	2 Mar. (61) .	1 Sun. .	20·9550	3739
22 Mar. (81) . .	1 Sun. .	4 57 30	21 Mar. (80) .	0 Sat. .	55·5946	3740
22 Mar. (81) . .	2 Mon. .	11 10 0	11 Mar. (70) .	5 Thur. .	269·9099	3741
21 Mar. (81) . .	3 Tues. .	17 22 30	28 Feb. (59) .	2 Mon. .	145·5933	3742
21 Mar. (80) . .	4 Wed. .	23 35 0	18 Mar. (77) .	1 Sun. .	180·2329	3743
22 Mar. (81) . .	6 Fri. .	5 47 30	7 Mar. (66) .	5 Thur. .	56·9163	3744
22 Mar. (81) . .	0 Sat. .	12 0 0	25 Feb. (56) .	3 Tues. .	270·2316	3745

TABLE

CONCURRENT YEAR.								Mean Intercalated (adhika) lunar month.
Kali.	Śaka.	Chaitrādi Vikrama.	Mēshādi solar year in Bengal.	Kollam.	A.D.	JOVIAN SAMVATSARA. Southern system.	JOVIAN SAMVATSARA. Northern system.	
1	2	3	3a	4	5	6	7	8a
3746	567	702	51		*644-45	35 Plava . . .		...
3747	568	703	52		645-46	36 Śubhakṛit . .		...
3748	569	704	53		646-47	37 Śōbhana . . .		2 Vaiśākha .
3749	570	705	54		647-48	38 Krōdhin . .		...
3750	571	706	55		*648-49	39 Viśvāvasu . .		10 Pausha .
3751	572	707	56		649-50	40 Parābhava† . .		...
3752	573	708	57		650-51	42 *Kīlaka* . . .		...
3753	574	709	58		651-52	43 *Saumya* . . .		7 Āśvina .
3754	575	710	59		*652-53	44 *Sādhāraṇa* . .		...
3755	576	711	60		653-54	45 *Virōdhakṛit* . .		...
3756	577	712	61		654-55	46 Paridhāvin . .		3 Jyēshṭha .
3757	578	713	62		655-56	47 Pramādin . .		...
3758	579	714	63		*656-57	48 Ānanda . . .		12 Phālguna .
3759	580	715	64		657-58	49 Rākshasa . . .		...
3760	581	716	65		658-59	50 Anala . . .		...
3761	582	717	66		659-60	51 Piṅgala . .		8 Kārttika .
3762	583	718	67		*660-61	52 Kālayukta . .		...
3763	584	719	68		661-62	53 Siddhārthin . .		...
3764	585	720	69		662-63	54 Raudra . . .		5 Śrāvana .
3765	586	721	70		663-64	55 Durmati . . .		...
3766	587	722	71		*664-65	56 Dundubhi . .		...
3767	588	723	72		665-66	57 Rudhirōdgārin . .		1 Chaitra .
3768	589	724	73		666-67	58 Raktāksha . .		...
3769	590	725	74		667-68	59 Krōdhana . .		10 Pausha .
3770	591	726	75		*668-69	60 Kshaya . . .		...

† By the mean system 41 Plavaṅga was expunged. as also by the true system.

LXXVI—*Contd.*

1 Ārya Siddhānta, mean system.

COMMENCEMENT OF THE									
MEAN SOLAR YEAR.					MEAN LUNI-SOLAR YEAR (MEAN SUNRISE OF CIVIL DAY ON WHICH CHAITRA ŚUKLA 1 ENDS).			Kali year.
Day and month, A.D.	Week-day.	Time of mean Mēsha-saṁkrānti.			Day and month, A.D.	Week-day.	*a* (here = *t*, the index of the tithi).	
13	14	17			19	20	23	1
		H.	M.	S.				
21 Mar. (81)	1 Sun.	18	12	30	15 Mar. (75)	2 Mon.	304·8711	3746
22 Mar. (81)	3 Tues.	0	25	0	4 Mar. (63)	6 Fri.	180·5545	3747
22 Mar. (81)	4 Wed.	6	37	30	21 Feb. (52)	3 Tues.	56·2378	3748
22 Mar. (81)	5 Thur.	12	50	0	12 Mar. (71)	2 Mon.	90·8775	3749
21 Mar. (81)	6 Fri.	19	2	30	1 Mar. (61)	0 Sat.	305·1927	3750
22 Mar. (81)	1 Sun.	1	15	0	19 Mar. (78)	5 Thur.	1·2005	3751
22 Mar. (81)	2 Mon.	7	27	30	9 Mar. (68)	3 Tues.	215·5157	3752
22 Mar. (81)	3 Tues.	13	40	0	26 Feb. (57)	0 Sat.	91·1991	3753
21 Mar. (81)	4 Wed.	19	52	30	16 Mar. (76)	6 Fri.	125·8387	3754
22 Mar. (81)	6 Fri.	2	5	0	5 Mar. (64)	3 Tues.	1·5221	3755
22 Mar. (81)	0 Sat.	8	17	30	23 Feb. (54)	1 Sun.	215·8374	3756
22 Mar. (81)	1 Sun.	14	30	0	14 Mar. (73)	0 Sat.	250·4770	3757
21 Mar. (81)	2 Mon.	20	42	30	2 Mar. (62)	4 Wed.	126·1604	3758
22 Mar. (81)	4 Wed.	2	55	0	21 Mar. (80)	3 Tues.	160·8000	3759
22 Mar. (81)	5 Thur.	9	7	30	10 Mar. (69)	0 Sat.	36·4834	3760
22 Mar. (81)	6 Fri.	15	20	0	28 Feb. (59)	5 Thur.	250·7987	3761
21 Mar. (81)	0 Sat.	21	32	30	18 Mar. (78)	4 Wed.	285·4383	3762
22 Mar. (81)	2 Mon.	3	45	0	7 Mar. (66)	1 Sun.	161·1217	3763
22 Mar. (81)	3 Tues.	9	57	30	24 Feb. (55)	5 Thur.	36·8051	3764
22 Mar. (81)	4 Wed.	16	10	0	15 Mar. (74)	4 Wed.	71·4447	3765
21 Mar. (81)	5 Thur.	22	22	30	4 Mar. (64)	2 Mon.	285·7599	3766
22 Mar. (81)	0 Sat.	4	35	0	21 Feb. (52)	6 Fri.	131·4433	3767
22 Mar. (81)	1 Sun.	10	47	30	12 Mar. (71)	5 Thur.	196·0830	3768
22 Mar. (81)	2 Mon.	17	0	0	1 Mar. (60)	2 Mon.	71·7663	3769
21 Mar. (81)	3 Tues.	23	12	30	18 Mar. (78)	1 Sun.	106·4030	3770

TABLE

CONCURRENT YEAR.						JOVIAN SAMVATSARA.		Mean Intercalated (adhika) lunar month.
Kali.	Śaka.	Chaitrādi Vikrama.	Mēshādi solar year in Bengal.	Kollam.	A.D.	Southern system.	Northern system.	
1	2	3	3a	4	5	6	7	8a
3771	592	727	76		669-70	1 Prabhava . .		...
3772	593	728	77		670-71	2 Vibhava . . .		6 Bhādrapada
3773	594	729	78		671-72	3 Śukla . . .		...
3774	595	730	79		*672-73	4 Pramōda . . .		...
3775	596	731	80		673-74	5 Prajāpati . . .		3 Jyēshṭha .
3776	597	732	81		674-75	6 Aṅgiras . . .		...
3777	598	733	82		675-76	7 Śrīmukha . . .		11 Māgha .
3778	599	734	83		*676-77	8 Bhāva . . .		...
3779	600	735	84		677-78	9 Yuvan . . .		...
3780	601	736	85		678-79	10 Dhātṛi . . .		8 Kārttika .
3781	602	737	86		679-80	11 Īśvara . . .		...
3782	603	738	87		*680-81	12 Bahudhānya . .		...
3783	604	739	88		681-82	13 Pramādin . .		5 Śrāvaṇa .
3784	605	740	89		682-83	14 Vikrama . . .		...
3785	606	741	90		683-84	15 Vṛisha . . .		...
3786	607	742	91		*684-85	16 Chitrabhānu . .		1 Chaitra .
3787	608	743	92		685-86	17 Subhānu . . .		...
3788	609	744	93		686-87	18 Tāraṇa . . .		10 Pausha .
3789	610	745	94		687-88	19 Pārthiva . . .		...
3790	611	746	95		*688-89	20 Vyaya . . .		...
3791	612	747	96		689-90	21 Sarvajit . . .		6 Bhādrapada
3792	613	748	97		690-91	22 Sarvadhārin . .		...
3793	614	749	98		691-92	23 Virōdhin . . .		...
3794	615	750	99		*692-93	24 Vikṛita . . .		3 Jyēshṭha .
3795	616	751	100		693-94	25 Khara . . .		...

LXXVI—*Contd.*

I Ārya Siddhānta, mean system

COMMENCEMETN OF THE							
MEAN SOLAR YEAR.			MEAN LUNI-SOLAR YEAR (MEAN SUNRISE OF CIVIL DAY ON WHICH CHAITRA ŚUKLA 1 ENDS).				Kali year
Day and month. A.D.	Week-day.	Time of mean Mēsha-samkrānti.	Day and month. A.D.	Week-day.	*a* (here = *t*, the index of the tithi).		
13	14	17	19	20	23		1
		H. M. S.					
22 Mar. (81) . .	5 Thur. .	5 25 0	9 Mar. (68) .	6 Fri. .	320·7213		3771
22 Mar. (81) . .	6 Fri. .	11 37 30	26 Feb. (57) .	3 Tues. .	196·4046		3772
22 Mar. (81) . .	0 Sat. .	17 50 0	17 Mar. (76)	2 Mon. .	231·0442		3773
22 Mar. (82) . .	2 Mon. .	0 2 30	5 Mar. (65) .	6 Fri. .	106·7276		3774
22 Mar. (81) . .	3 Tues. .	6 15 0	23 Feb. (54) .	4 Wed. .	321·0429		3775
22 Mar. (81) . .	4 Wed. .	12 27 30	13 Mar. (72) .	2 Mon. .	17·0506		3776
22 Mar. (81) . .	5 Thur. .	18 40 0	3 Mar. (62) .	0 Sat .	231·3658		3777
22 Mar. (82) . .	0 Sat. .	0 52 30	21 Mar. (81) .	6 Fri. .	266·0054		3778
22 Mar. (81) . .	1 Sun. .	7 5 0	10 Mar. (69) .	3 Tues. .	141·6888		3779
22 Mar. (81) . .	2 Mon. .	13 17 30	27 Feb. (58) .	0 Sat. .	17·3723		3780
22 Mar. (81) . .	3 Tues. .	19 30 0	18 Mar. (77) .	6 Fri. .	52·0118		3781
22 Mar. (82) . .	5 Thur. .	1 42 30	7 Mar. (67) .	4 Wed. .	266·3271		3782
22 Mar. (81) . .	6 Fri. .	7 55 0	24 Feb. (55) .	1 Sun. .	142·0105		3783
22 Mar. (81) . .	0 Sat. .	14 7 30	15 Mar. (74) .	0 Sat. .	176·6501		3784
22 Mar. (81) . .	1 Sun. .	20 20 0	4 Mar. (63) .	4 Wed. .	52·3334		3785
22 Mar. (82) . .	3 Tues. .	2 32 30	22 Feb. (53) .	2 Mon. .	266·6487		3786
22 Mar. (81) . .	4 Wed. .	8 45 0	12 Mar. (71) .	1 Sun. .	301·2884		3787
22 Mar. (81) . .	5 Thur. .	14 57 30	1 Mar. (60) .	5 Thur. .	176·9717		3788
22 Mar. (81) . .	6 Fri. .	21 10 0	20 Mar. (79) .	4 Wed. .	211·6114		3789
22 Mar. (82) . .	1 Sun. .	3 22 30	8 Mar. (68) .	1 Sun. .	87·2948		3790
22 Mar. (81) . .	2 Mon. .	9 35 0	26 Feb. (57) .	6 Fri. .	301·6100		3791
22 Mar. (81) . .	3 Tues. .	15 47 30	16 Mar. (75) .	4 Wed. .	9997·6177†		3792
22 Mar. (81) . .	4 Wed. .	22 0 0	6 Mar. (65) .	2 Mon. .	211·9330		3793
22 Mar. (82) . .	6 Fri. .	4 12 30	23 Feb. (54) .	6 Fri. .	87·6164		3794
22 Mar. (81) . .	0 Sat. .	10 25 0	13 Mar. (72) .	5 Thur. .	122·2560		3795

† As a mean tithi Chaitra Śukla 1 was expunged. The civil day corresponding to it, i.e., the first day of the mean luni-solar year was as given in cols. 19, 20.

TABLE

CONCURRENT YEAR.								
						Jovian samvatsara.		Mean Intercalated (adhika) lunar month.
Kali.	Śaka.	Chaitrādi Vikrama.	Mēshādi solar year in Bengal.	Kollam.	A.D.	Southern system.	Northern system.	
1	2	3	3a	4	5	6	7	8a
3796	617	752	101		694-95	26 Nandana . . .		11 Māgha .
3797	618	753	102		695-96	27 Vijaya . . .		...
3798	619	754	103		*696-97	28 Jaya . . .		...
3799	620	755	104		697-98	29 Manmatha . .		8 Kārttika .
3800	621	756	105		698-99	30 Durmukha . .		...
3801	622	757	106		699-700	31 Hēmalamba . .		...
3802	623	758	107		*700-01	32 Vilamba . . .		4 Āshāḍha .
3803	624	759	108		701-02	33 Vikārin . . .		...
3804	625	760	109		702-03	34 Śārvarin . . .		...
3805	626	761	110		703-04	35 Plava . . .		1 Chaitra .
3806	627	762	111		*704-05	36 Śubhakṛit . .		...
3807	628	763	112		705-06	37 Śōbhana . . .		9 Mārgaśira .
3808	629	764	113		706-07	38 Krōdhin . . .		...
3809	630	765	114		707-08	39 Viśvāvasu . .		...
3810	631	766	115		*708-09	40 Parābhava . .		6 Bhādrapada
3811	632	767	116		709-10	41 Plavaṅga . . .		...
3812	633	768	117		710-11	42 Kīlaka . . .		...
3813	634	769	118		711-12	43 Saumya . . .		2 Vaiśākha .
3814	635	770	119		*712-13	44 Sādhāraṇa . .		...
3815	636	771	120		713-14	45 Virōdhakṛit . .		11 Māgha .
3816	637	772	121		714-15	46 Paridhāvin . .		...
3817	638	773	122		715-16	47 Pramādin . .		...
3818	639	774	123		*716-17	48 Ānanda . . .		8 Kārttika† .
3819	640	775	124		717-18	49 Rākshasa . . .		...
3820	641	776	125		718-19	50 Anala . . .		...

† By the "Indian Calendar" 7 Āśvina was intercalated but the case was a close one.

LXXVI—*Contd.*

1 Ārya Siddhānta, mean system.

COMMENCEMENT OF THE						
MEAN SOLAR YEAR.			MEAN LUNI-SOLAR YEAR (MEAN SUNRISE OF CIVIL DAY ON WHICH CHAITRA ŚUKLA 1 ENDS).			Kali year
Day and month, A.D.	Week-day.	Time of mean Mēsha-samkrānti.	Day and month, A.D.	Week-day.	*a* (here=*t*, the index of the tithi).	
13	14	17	19	20	23	1
		H. M. S.				
22 Mar. (81) . .	1 Sun. .	16 37 30	2 Mar. (61) .	2 Mon. .	9997·9394†	3796
22 Mar. (81) . .	2 Mon. .	22 50 0	21 Mar. (80) .	1 Sun. .	32·5790	3797
22 Mar. (82) . .	4 Wed. .	5 2 30	10 Mar. (70) .	6 Fri. .	246·8943	3798
22 Mar. (81) . .	5 Thur. .	11 15 0	27 Feb. (58) .	3 Tues. .	122·5777	3799
22 Mar. (81) . .	6 Fri. .	17 27 30	18 Mar. (77) .	2 Mon. .	157·2173	3800
22 Mar. (81) . .	0 Sat. .	23 40 0	7 Mar. (66) .	6 Fri. .	32·9006	3801
22 Mar. (82) . .	2 Mon. .	5 52 30	25 Feb. (56) .	4 Wed. .	247·2159	3802
22 Mar. (81) . .	3 Tues. .	12 5 0	15 Mar. (74) .	3 Tues. .	281·8555	3803
22 Mar. (81) . .	4 Wed. .	18 37 30	4 Mar. (63) .	0 Sat. .	157·5389	3804
23 Mar. (82) . .	6 Fri. .	0 30 0	21 Feb. (52) .	4 Wed. .	33·2223	3805
22 Mar. (82) . .	0 Sat. .	6 42 30	11 Mar. (71) .	3 Tues. .	67·8619	3806
22 Mar. (81) . .	1 Sun. .	12 55 0	1 Mar. (60) .	1 Sun. .	282·1771	3807
22 Mar. (81) . .	2 Mon. .	19 7 30	20 Mar. (79) .	0 Sat. .	316·8168	3808
23 Mar. (82) . .	4 Wed. .	1 20 0	9 Mar. (68) .	4 Wed. .	192·5002	3809
22 Mar. (82) . .	5 Thur. .	7 32 30	26 Feb. (57) .	1 Sun. .	68·1835	3810
22 Mar. (81) . .	6 Fri. .	13 45 0	16 Mar. (75) .	0 Sat. .	102·8231	3811
22 Mar. (81) . .	0 Sat. .	19 57 30	6 Mar. (65) .	5 Thur. .	317·1384	3812
23 Mar. (82) . .	2 Mon. .	2 10 0	23 Feb. (54) .	2 Mon. .	192·8218	3813
22 Mar. (82) . .	3 Tues. .	8 22 30	13 Mar. (73) .	1 Sun. .	227·4614	3814
22 Mar. (81) . .	4 Wed. .	14 35 0	2 Mar. (61) .	5 Thur. .	103·1447	3815
22 Mar. (81) . .	5 Thur. .	20 47 30	21 Mar. (80) .	4 Wed. .	137·7843	3816
23 Mar. (82) . .	0 Sat. .	3 0 0	10 Mar. (69) .	1 Sun. .	13·4678	3817
22 Mar. (82) . .	1 Sun. .	9 12 30	28 Feb. (59) .	6 Fri. .	227·7831	3818
22 Mar. (81) . .	2 Mon. .	15 25 0	18 Mar. (77) .	5 Thur. .	262·4226	3819
22 Mar. (81) . .	3 Tues. .	21 37 30	7 Mar. (66) .	2 Mon. .	138·1060	3820

† As a mean tithi Chaitra śukla 1 was suppressed. The civil day corresponding to it, i.e., the first day of the mean luni-solar year, was as given in cols. 19, 20.

TABLE

CONCURRENT YEAR.								Mean Intercalated (adhika) lunar month.
Kali.	Saka.	Chaitrādi Vikrama.	Mēshādi solar year in Bengal.	Kollam.	A.D.	Jovian samvatsara. Southern system.	Jovian samvatsara. Northern system.	
1	2	3	3a	4	5	6	7	8a
3821	642	777	126		719-20	51 Piṅgala . . .		4 Āshādha .
3822	643	778	127		*720-21	52 Kālayukta . .		...
3823	644	779	128		721-22	53 Siddhārthin . .		...
3824	645	780	129		722-23	54 Raudra . . .		1 Chaitra .
3825	646	781	130		723-24	55 Durmati . . .		...
3826	647	782	131		*724-25	56 Dundubhi . .		9 Mārgaśira .
3827	648	783	132		725-26	57 Rudhirōdgārin . .		...
3828	649	784	133		726-27	58 Raktāksha . .		...
3829	650	785	134		727-28	59 Krōdhana . .		6 Bhādrapada
3830	651	786	135		*728-29	60 Kshaya . . .		...
3831	652	787	136		729-30	1 Prabhava . . .		...
3832	653	788	137		730-31	2 Vibhava . . .		2 Vaiśākha .
3833	654	789	138		731-32	3 Śukla . . .		...
3834	655	790	139		*732-33	4 Pramōda . . .		11 Māgha .
3835	656	791	140		733-34	5 Prajāpati . . .		...
3836	657	792	141		734-35	6 Aṅgiras† . . .		...
3837	658	793	142		735-36	8 *Bhāva* . . .		7 Āśvina .
3838	659	794	143		*736-37	9 *Yuvan* . . .		...
3839	660	795	144		737-38	10 *Dhātṛi* . .		...
3840	661	796	145		738-39	11 *Īśvara* . . .		4 Āshādha .
3841	652	797	146		739-40	12 Bahudhānya . .		...
3842	663	798	147		*740-41	13 Pramāthin . .		12 Phālguna .
3843	664	799	148		741-42	14 Vikrama . .		...
3844	665	800	149		742-43	15 Vrisha . . .		...
3845	666	801	150		743-44	16 Chitrabhānu . .		9 Mārgaśira .

† By the mean system, as well as by the true system, 7 Śrīmukha was expunged.

LXXVI—*Contd.*

1 Arya Siddhânta, mean system.

COMMENCEMENT OF THE							
MEAN SOLAR YEAR.			MEAN LUNI-SOLAR YEAR (MEAN SUNRISE OF CIVIL DAY ON WHICH CHAITRA ŚUKLA 1 ENDS).			Kali year.	
Day and month, A.D.	Week-day.	Time of mean Mēsha-saṁkrānti.		Day and month, A.D.	Week-day.	*a* (here *t*, the index of the tithi).	
13	14	17		19	20	23	1
		H. M.	S.				
23 Mar. (82) . .	5 Thur. .	3 50	0	24 Feb. (55) .	6 Fri. .	13·7894	3821
22 Mar. (82) . .	6 Fri. .	10 2	30	14 Mar. (74) .	5 Thur. .	48·4290	3822
22 Mar. (81) . .	0 Sat. .	16 15	0	4 Mar. (63) .	3 Tues. .	262·7443	3823
22 Mar. (81) . .	1 Sun. .	22 27	30	21 Feb. (52) .	0 Sat. .	138·4276	3824
23 Mar. (82) . .	3 Tues. .	4 40	0	12 Mar. (71) .	6 Fri. .	173·0673	3825
22 Mar. (82) . .	4 Wed. .	10 52	30	29 Feb. (60) .	3 Tues. .	48·7506	3826
22 Mar. (81) . .	5 Thur. .	17 5	0	19 Mar. (78) .	2 Mon. .	83·3903	3827
22 Mar. (81) . .	6 Fri. .	23 17	30	9 Mar. (68) .	0 Sat. .	2[illegible]5	3828
23 Mar. (82) . .	1 Sun. .	5 30	0	26 Feb. (57) .	4 Wed. .	173·3890	3829
22 Mar. (82) . .	2 Mon. .	11 42	30	16 Mar. (76) .	3 Tues. .	208·0286	3830
22 Mar. (81) . .	3 Tues. .	17 55	0	5 Mar. (64) .	0 Sat. .	83·7119	3831
23 Mar. (82) . .	5 Thur. .	0 7	30	23 Feb. (54) .	5 Thur. .	298·0272	3832
23 Mar. (82) . .	6 Fri. .	6 20	0	14 Mar. (73) .	4 Wed. .	332·6669	3833
22 Mar. (82) . .	0 Sat. .	12 32	30	2 Mar. (62) .	1 Sun. .	208·3502	3834
22 Mar. (81) . .	1 Sun. .	18 45	0	21 Mar. (80) .	0 Sat. .	242·9898	3835
23 Mar. (82) . .	3 Tues. .	0 57	30	10 Mar. (69) .	4 Wed. .	118·6732	3836
23 Mar. (82) . .	4 Wed. .	7 10	0	28 Feb. (59) .	2 Mon. .	332·9885	3837
22 Mar. (82) . .	5 Thur. .	13 22	30	17 Mar. (77) .	0 Sat. .	28·9962	3838
22 Mar. (81) . .	6 Fri. .	19 35	0	7 Mar. (66) .	5 Thur. .	243·3115	3839
23 Mar. (82) . .	1 Sun. .	1 47	30	24 Feb. (55) .	2 Mon. .	118·9949	3840
23 Mar. (82) . .	2 Mon. .	8 0	0	15 Mar. (74) .	1 Sun. .	153·6345	3841
22 Mar. (82) . .	3 Tues. .	14 12	30	3 Mar. (63) .	5 Thur. .	29·3179	3842
22 Mar. (81) . .	4 Wed. .	20 25	0	22 Mar. (81) .	4 Wed. .	63·9575	3843
23 Mar. (82) . .	6 Fri. .	2 37	30	12 Mar. (71) .	2 Mon. .	278·2728	3844
23 Mar. (82) . .	0 Sat. .	8 50	0	1 Mar. (60) .	6 Fri. .	153·9561	3845

TABLE

CONCURRENT YEAR.								
						Jovian saṁvatsara.		Mean Intercalated (adhika) lunar month.
Kali.	Śaka.	Chaitrādi Vikrama.	Mēshādi solar year in Bengal.	Kollam.	A.D.	Southern system.	Northern system.	
1	2	3	3a	4	5	6	7	8a
3846	667	802	151		*744-45	17 Subhānu . . .		...
3847	668	803	152		745-46	18 Tāraṇa . . .		...
3848	669	804	153		746-47	19 Pārthiva . . .		5 Śrāvaṇa .
3849	670	805	154		747-48	20 Vyaya . . .		...
3850	671	806	155		*748-49	21 Sarvajit . . .		...
3851	672	807	156		749-50	22 Sarvadhārin . .		2 Vaiśākha .
3852	673	808	157		750-51	23 Virōdhin . . .		...
3853	674	809	158		751-52	24 Vikṛita . . .		10 Pausha .
3854	675	810	159		*752-53	25 Khara . . .		...
3855	676	811	160		753-54	26 Nandana . . .		...
3856	677	812	161		754-55	27 Vijaya . . .		7 Āśvina .
3857	678	813	162		755-56	28 Jaya . . .		...
3858	679	814	163		*756-57	29 Manmatha . .		...
3859	680	815	164		757-58	30 Durmukha . .		4 Āshāḍha .
3860	681	816	165		758-59	31 Hēmalamba . .		...
3861	682	817	166		759-60	32 Vilamba . . .		12 Phālguna .
3862	683	818	167		*760-61	33 Vikārin . . .		...
3863	684	819	168		761-62	34 Śārvarin . . .		...
3864	685	820	169		762-63	35 Plava . . .		9 Mārgaśira .
3865	686	821	170		763-64	36 Śubhakṛit . . .		...
3866	687	822	171		*764-65	37 Śōbhana . . .		...
3867	688	823	172		765-66	38 Krōdhin . . .		5 Śrāvaṇa .
3868	689	824	173		766-67	39 Viśvāvasu . .		...
3869	690	825	174		767-68	40 Parābhava . .		...
3870	691	826	175		*768-69	41 Plavanga . . .		2 Vaiśākha .

LXXVI—*Contd.*

1 Ārya Siddhānta, mean system

COMMENCEMENT OF THE									Kali year.
MEAN SOLAR YEAR.					MEAN LUNI-SOLAR YEAR (MEAN SUNRISE OF CIVIL DAY ON WHICH CHAITRA ŚUKLA 1 ENDS).				
Day and month. A.D.	Week-day.	Time of mean Mêsha-saṁkrānti.			Day and month. A.D.	Week-day.	*a* (here = *t*, the index of the tithi).		
13	14	17			19	20	23		1
		H.	M.	S.					
22 Mar. (82) .	1 Sun.	15	2	30	19 Mar. (79)	5 Thur.	188·5957		3846
22 Mar. (81) .	2 Mon.	21	15	0	8 Mar. (67)	2 Mon.	64·2790		3847
23 Mar. (82) .	4 Wed.	3	27	30	26 Feb. (57)	0 Sat.	278·5944		3848
23 Mar. (82) .	5 Thur.	9	40	0	17 Mar. (76)	6 Fri.	313·2341		3849
22 Mar. (82) .	6 Fri.	15	52	30	5 Mar. (65)	3 Tues.	188·9173		3850
22 Mar. (81) .	0 Sat.	22	5	0	22 Feb. (53)	0 Sat.	64·6007		3851
23 Mar. (82) .	2 Mon.	4	17	30	13 Mar. (72)	6 Fri.	99·2404		3852
23 Mar. (82) .	3 Tues.	10	30	0	3 Mar. (62)	4 Wed.	313·5556		3853
22 Mar. (82) .	4 Wed.	16	42	30	20 Mar. (80)	2 Mon.	9·5633		3854
22 Mar. (81) .	5 Thur.	22	55	0	10 Mar. (69)	0 Sat.	223·8786		3855
23 Mar. (82) .	0 Sat.	5	7	30	27 Feb. (58)	4 Wed.	99·5620		3856
23 Mar. (82) .	1 Sun.	11	20	0	18 Mar. (77)	3 Tues.	134·2016		3857
22 Mar. (82) .	2 Mon.	17	32	30	6 Mar. (66)	0 Sat.	9·8850		3858
22 Mar. (81) .	3 Tues.	23	45	0	24 Feb. (55)	5 Thur.	224·2003		3859
23 Mar. (82) .	5 Thur.	5	57	30	15 Mar. (74)	4 Wed.	258·8399		3860
23 Mar. (82) .	6 Fri.	12	10	0	4 Mar. (63)	1 Sun.	134·5233		3861
22 Mar. (82) .	0 Sat.	18	22	30	22 Mar. (82)	0 Sat.	169·1628		3862
23 Mar. (82) .	2 Mon.	0	35	0	11 Mar. (70)	4 Wed.	44·8463		3863
23 Mar. (82) .	3 Tues.	6	47	30	1 Mar. (60)	2 Mon.	259·1616		3864
23 Mar. (82) .	4 Wed.	13	0	0	20 Mar. (79)	1 Sun.	293·8012		3865
22 Mar. (82) .	5 Thur.	19	12	30	8 Mar. (68)	5 Thur.	169·4846		3866
23 Mar. (82) .	0 Sat.	1	25	0	25 Feb. (56)	2 Mon.	45·1680		3867
23 Mar. (82) .	1 Sun.	7	37	30	16 Mar. (75)	1 Sun.	79·8076		3868
23 Mar. (82) .	2 Mon.	13	50	0	6 Mar. (65)	6 Fri.	294·1228		3869
22 Mar. (82) .	3 Tues.	20	2	30	22 Feb. (51)	3 Tues.	169·8062		3870

TABLE

CONCURRENT YEAR.						Jovian saṁvatsara.		Mean Intercalated (adhika) lunar month.
Kali.	Śaka	Chaitrâdi Vikrama.	Mêshâdi solar year in Bengal.	Kollam.	A.D.	Southern system.	Northern system.	
1	2	3	3a	4	5	6	7	8a
3871	692	827	176		769-70	42 Kîlaka . . .		...
3872	693	828	177		770-71	43 Saumya . . .		10 Pausha .
3873	694	829	178		771-72	44 Sâdhâraṇa . .		...
3874	695	830	179		*772-73	45 Virôdhakṛit . .		...
3875	696	831	180		773-74	46 Paridhâvin . .		7 Âśvina .
3876	697	832	181		774-75	47 Pramâdin . .		...
3877	698	833	182		775-76	48 Ânanda . . .		...
3878	699	834	183		*776-77	49 Râkshasa . . .		3 Jyêshṭha .
3879	700	835	184		777-78	50 Anala . . .		...
3880	701	836	185		778-79	51 Piṅgala . . .		12 Phâlguna .
3881	702	837	186		779-80	52 Kâlayukta . .		...
3882	703	838	187		*780-81	53 Siddhârthin . .		...
3883	704	839	188		781-82	54 Raudra . . .		8 Kârttika .
3884	705	840	189		782-83	55 Durmati . .		...
3885	706	841	190		783-84	56 Dundubhi . .		...
3886	707	842	191		*784-85	57 Rudhirôdgârin . .		5 Srâvaṇa .
3887	708	843	192		785-86	58 Raktâksha . .		...
3888	709	844	193		786-87	59 Krôdhana . .		...
3889	710	845	194		787-88	60 Kshaya . . .		1 Chaitra .
3890	711	846	195		*788-89	1 Prabhava . . .		...
3891	712	847	196		789-90	2 Vibhava . . .		10 Pausha .
3892	713	848	197		790-91	3 Sukla . . .		...
3893	714	849	198		791-92	4 Pramôda . . .		...
3894	715	850	199		*792-93	5 Prajâpati . . .		7 Âśvina† .
3895	716	851	200		793-94	6 Angiras . . .		...

† By the "Indian Calendar" 6 Bhâdrapada was intercalated.

LXXVI—*Contd.*

1 Ārya Siddhānta, mean system.

COMMENCEMENT OF THE									
MEAN SOLAR YEAR.					MEAN LUNI-SOLAR YEAR (MEAN SUNRISE OF CIVIL DAY ON WHICH CHAITRA ŚUKLA 1 ENDS).			Kali year.
Day and month, A.D.	Week-day	Time of mean Mēsha-saṁkrānti.			Day and month, A.D.	Week-day.	*a* (here = *t*, the index of the tithi).	
13	14	17			19	20	23	1
		H.	M.	S.				
23 Mar. (82) . .	5 Thur. .	2	15	0	13 Mar. (72) .	2 Mon. .	204·4459	3871
23 Mar. (82) . .	6 Fri. .	8	27	30	2 Mar. (61) .	6 Fri. .	80·1292	3872
23 Mar. (82) . .	0 Sat. .	14	40	0	21 Mar. (80) .	5 Thur. .	114·7688	3873
22 Mar. (82) . .	1 Sun. .	20	52	30	10 Mar. (70) .	3 Tues. .	329·0841	3874
23 Mar. (82) . .	3 Tues. .	3	5	0	27 Feb. (58) .	0 Sat. .	204·7675	3875
23 Mar. (82) . .	4 Wed. .	9	17	30	18 Mar. (77) .	6 Fri. .	239·4071	3876
23 Mar. (82) . .	5 Thur. .	15	30	0	7 Mar. (66) .	3 Tues. .	115·0904	3877
22 Mar. (82) . .	6 Fri. .	21	42	30	25 Feb. (56) .	1 Sun. .	329·4057	3878
23 Mar. (82) . .	1 Sun. .	3	55	0	14 Mar. (73) .	6 Fri. .	25·4134	3879
23 Mar. (82) . .	2 Mon. .	10	7	30	4 Mar. (63) .	4 Wed. .	239·7288	3880
23 Mar. (82) . .	3 Tues. .	16	20	0	23 Mar. (82) .	3 Tues. .	274·3682	3881
22 Mar. (82) . .	4 Wed. .	22	32	30	11 Mar. (71) .	0 Sat. .	150·0517	3882
23 Mar. (82) . .	6 Fri. .	4	45	0	28 Feb. (59) .	4 Wed. .	25·7351	3883
23 Mar. (82) . .	0 Sat. .	10	57	30	19 Mar. (78) .	3 Tues. .	60·3747	3884
23 Mar. (82) . .	1 Sun. .	17	10	0	9 Mar. (68) .	1 Sun. .	274·6900	3885
22 Mar. (82) . .	2 Mon. .	23	22	30	26 Feb. (57) .	5 Thur. .	150·3734	3886
23 Mar. (82) . .	4 Wed. .	5	35	0	16 Mar. (75) .	4 Wed. .	185·0130	3887
23 Mar. (82) . .	5 Thur. .	11	47	30	5 Mar. (64) .	1 Sun. .	60·6963	3888
23 Mar. (82) . .	6 Fri. .	18	0	0	23 Feb. (54) .	6 Fri. .	275·0116	3889
23 Mar. (83) . .	1 Sun. .	0	12	30	13 Mar. (73) .	5 Thur. .	309·6513	3890
23 Mar. (82) . .	2 Mon. .	6	25	0	2 Mar. (61) .	2 Mon. .	185·3346	3891
23 Mar. (82) . .	3 Tues. .	12	37	30	21 Mar. (80) .	1 Sun. .	219·9743	3892
23 Mar. (82) . .	4 Wed. .	18	50	0	10 Mar. (69) .	5 Thur. .	95·6576	3893
23 Mar. (83) . .	6 Fri. .	1	2	30	23 Feb. (59) .	3 Tues. .	309·9730	3894
23 Mar. (82) . .	0 Sat. .	7	15	0	17 Mar. (76) .	1 Sun. .	5·9807	3895

TABLE

CONCURRENT YEAR.								Mean Intercalated (adhika) lunar month.
						Jovian Saṁvatsara.		
Kali.	Śaka.	Chaitrādi Vikrama.	Mēshādi solar year in Bengal.	Kollam.	A.D.	Southern system.	Northern system.	
1	2	3	3a	4	5	6	7	8a
3896	717	852	201		794-95	7 Śrīmukha . . .		
3897	718	853	202		795-96	8 Bhāva . . .		3 Jyēshṭha .
3898	719	854	203		*796-97	9 Yuvan . . .		...
3899	720	855	204		797-98	10 Dhātṛi . . .		12 Phālguna .
3900	721	856	205		798-99	11 Īśvara . . .		...
3901	722	857	206		799-800	12 Bahudhānya . .		...
3902	723	858	207		*800-01	13 Pramāthin . .		8 Kārttika .
3903	724	859	208		801-02	14 Vikrama . . .		...
3904	725	860	209		802-03	15 Vṛisha . . .		...
3905	726	861	210		803-04	16 Chitrabhānu . .		5 Śrāvaṇa .
3906	727	862	211		*804-05	17 Subhānu . . .		...
3907	728	863	212		805-06	18 Tāraṇa . . .		...
3908	729	864	213		806-07	19 Pārthiva . . .		1 Chaitra .
3909	730	865	214		807-08	20 Vyaya . . .		...
3910	731	866	215		*808-09	21 Sarvajit . . .		10 Pausha .
3911	732	867	216		809-10	22 Sarvadhārin . .		...
3912	733	868	217		810-11	23 Virōdhin . . .		...
3913	734	869	218		811-12	24 Vikṛita . . .		6 Bhādrapada.
3914	735	870	219		*812-13	25 Khara . . .		...
3915	736	871	220		813-14	26 Nandana . . .		...
3916	737	872	221		814-15	27 Vijaya . . .		3 Jyēshṭha .
3917	738	873	222		815-16	28 Jaya . . .		...
3918	739	874	223		*816-17	29 Manmatha . .		11 Māgha .
3919	740	875	224		817-18	30 Durmukha . .		...
3920	741	876	225		818-19	31 Hēmalamba . .		...

LXXVI—*Contd.*

I Ārya Siddhānta, mean system.

COMMENCEMENT OF THE							Kali year.
MEAN SOLAR YEAR.			MEAN LUNI-SOLAR YEAR (MEAN SUNRISE OF CIVIL DAY ON WHICH CHAITRA ŚUKLA 1 ENDS).				
Day and month. A.D.	Week-day.	Time of mean Mesha-saṁkrānti.	Day and month. A.D.	Week-day.	*a* (here=*t*, the index of the tithi).		
13	14	17	19	20	23		1
		H. M. S.					
23 Mar. (82) . .	1 Sun. .	13 27 30	7 Mar. (66) .	6 Fri. .	220·2959		3896
23 Mar. (82) . .	2 Mon. .	19 40 0	24 Feb. (55) .	3 Tues. .	95·9793		3897
23 Mar. (83) . .	4 Wed. .	1 52 30	14 Mar. (74) .	2 Mon. .	130·6189		3898
23 Mar. (82) . .	5 Thur. .	8 5 0	3 Mar. (62) .	6 Fri. .	6·3023		3899
23 Mar. (82) . .	6 Fri. .	14 17 30	22 Mar. (81) .	5 Thur. .	40·9419		3900
23 Mar. (82) . .	0 Sat. .	20 30 0	12 Mar. (71) .	3 Tues. .	255·2572		3901
23 Mar. (83) . .	2 Mon. .	2 42 30	29 Feb. (60) .	0 Sat. .	130·9406		3902
23 Mar. (82) . .	3 Tues. .	8 55 0	19 Mar. (78) .	6 Fri. .	165·5802		3903
23 Mar. (82) . .	4 Wed. .	15 7 30	8 Mar. (67) .	3 Tues. .	41·2636		3904
23 Mar. (82) . .	5 Thur. .	21 20 0	26 Feb. (57) .	1 Sun. .	255·5789		3905
23 Mar. (83) . .	0 Sat. .	3 32 30	16 Mar. (76) .	0 Sat. .	290·2185		3906
23 Mar. (82) . .	1 Sun. .	9 45 0	5 Mar. (64) .	4 Wed. .	165·5018		3907
23 Mar. (82) . .	2 Mon. .	15 57 30	22 Feb. (53) .	1 Sun. .	41·5852		3908
23 Mar. (82) . .	3 Tues. .	22 10 0	13 Mar. (72) .	0 Sat. .	76·2248		3909
23 Mar. (83) . .	5 Thur. .	4 22 30	2 Mar. (62) .	5 Thur. .	290·5401		3910
23 Mar. (82) .	6 Fri. .	10 35 0	21 Mar. (80) .	4 Wed. .	325·1798		3911
23 Mar. (82) . .	0 Sat. .	16 47 30	10 Mar. (69) .	1 Sun. .	200·8631		3912
23 Mar. (82) . .	1 Sun. .	23 0 0	27 Feb. (58) .	5 Thur. .	76·5465		3913
23 Mar. (83) . .	3 Tues. .	5 12 30	17 Mar. (77) .	4 Wed. .	111·1862		3914
23 Mar. (82) . .	4 Wed. .	11 25 0	7 Mar. (66) .	2 Mon. .	325·5013		3915
23 Mar. (82) . .	5 Thur. .	17 37 30	24 Feb. (55) .	6 Fri. .	201·1847		3916
23 Mar. (82) . .	6 Fri. .	23 50 0	15 Mar. (74) .	5 Thur. .	235·8244		3917
23 Mar. (83) . .	1 Sun. .	6 2 30	3 Mar. (63) .	2 Mon. .	111·5978		3918
23 Mar. (82) . .	2 Mon.	12 15 0	22 Mar. (81) .	1 Sun. .	146·1473		3919
23 Mar. (82) . .	3 Tues. .	18 27 30	11 Mar. (70) .	5 Thur. .	21·8397		3920

TABLE

CONCURRENT YEAR.								Mean Intercalated (adhika) lunar month.
						Jovian Saṁvatsara.		
Kali.	Saka.	Chaitrādi Vikrama.	Mēshādi solar year in Bengal.	Kollam.	A. D.	Southern system.	Northern system.	
1	2	3	3a	4	5	6	7	8a
3921	742	877	226		819-20	32 Vilamba† . . .		8 Kārttika .
3922	743	878	227		*820-21	34 *Sārvarin* . . .		...
3923	744	879	228		821-22	35 *Plava* . . .		...
3924	745	880	229		822-23	36 *Śubhakṛit* . . .		4 Āshāḍha .
3925	746	881	230		823-24	37 *Sōbhana* . . .		...
3926	747	882	231		*824-25	38 Krōdhin . . .		...
3927	748	883	232	0-1	825-26	39 Viśvāvasu . .		1 Chaitra .
3928	749	884	233	1-2	826-27	40 Parābhava .		...
3929	750	885	234	2-3	827-28	41 Plavaṅga . . .		10 Pausha
3930	751	886	235	3-4	*828-29	42 Kīlaka . . .		...
3931	752	887	236	4-5	829-30	43 Saumya . . .		...
3932	753	888	237	5-6	830-31	44 Sādhāraṇa .		6 Bhādrapada.
3933	754	889	238	6-7	831-32	45 Virōdhakṛit . .		...
3934	755	890	239	7-8	*832-33	46 Paridhāvin . .		...
3935	756	891	240	8-9	833-34	47 Pramādin . .		3 Jyēshṭha .
3936	757	892	241	9-10	834-35	48 Ānanda . . .		...
3937	758	893	242	10-11	835-36	49 Rākshasa . .		11 Māgha .
3938	759	894	243	11-12	*836-37	50 Anala . . .		...
3939	760	895	244	12-13	837-38	51 Piṅgala . . .		...
3940	761	896	245	13-14	838-39	52 Kālayukta . .		8 Kārttika .
3941	762	897	246	14-15	839-40	53 Siddhārthin . .		...
3942	763	898	247	15-16	*840-41	54 Raudra . . .		...
3943	764	899	248	16-17	841-42	55 Durmati . . .		4 Āshāḍha .
3944	765	900	249	17-18	842-43	56 Dundubhi . .		...
3945	766	901	250	18-19	843-44	57 Rudhirōdgārin . .		...

† By both mean and true systems 33 Vikārin was expunged.

LXXVI—*Contd.*

COMMENCEMENT OF THE							
MEAN SOLAR YEAR.			MEAN LUNI-SOLAR YEAR (MEAN SUNRISE OF CIVIL DAY ON WHICH CHAITRA ŚUKLA 1 ENDS).				Kali year
Day and month, A.D.	Week-day.	Time of mean Mēsha-saṁkrānti.		Day and month, A.D.	Week-day.	*a* (here=*t*, the index of the tithi).	
13	14	17		19	20	23	1
		H. M. S.					
24 Mar. (83) .	5 Thur. .	0 40 0		1 Mar. (60) .	3 Tues. .	236·1460	3921
23 Mar. (83) . .	6 Fri. .	6 52 30		19 Mar. (79) .	2 Mon. .	270·7856	3922
23 Mar. (82) . .	0 Sat. .	13 5 0		8 Mar. (67) .	6 Fri. .	146·4690	3923
23 Mar. (82) .	1 Sun. .	19 17 30		25 Feb. (56) .	3 Tues. .	22·1524	3924
24 Mar. (83) .	3 Tues. .	1 30 0		16 Mar. (75) .	2 Mon. .	56·7920	3925
23 Mar. (83) . .	4 Wed. .	7 42 30		5 Mar. (65) .	0 Sat. .	271·1073	3926
23 Mar. (82) . .	5 Thur. .	13 55 0		22 Feb. (53) .	4 Wed. .	146·7906	3927
23 Mar. (82) . .	6 Fri. .	20 7 30		13 Mar. (72) .	3 Tues. .	181·4303	3928
24 Mar. (83) . .	1 Sun. .	2 20 0		2 Mar. (61) .	0 Sat. .	57·1137	3929
23 Mar. (83) . .	2 Mon. .	8 32 30		20 Mar. (80) .	6 Fri. .	91·7533	3930
23 Mar. (82) . .	3 Tues. .	14 45 0		10 Mar. (69) .	4 Wed. .	306·0686	3931
23 Mar. (82) . .	4 Wed. .	20 57 30		27 Feb. (58) .	1 Sun. .	181·7519	3932
24 Mar. (83) . .	6 Fri. .	3 10 0		18 Mar. (77) .	0 Sat. .	216·3916	3933
23 Mar. (83) . .	0 Sat. .	9 22 30		6 Mar. (66) .	4 Wed. .	92·0749	3934
23 Mar. (82) . .	1 Sun. .	15 35 0		24 Feb. (55) .	2 Mon. .	306·3902	3935
23 Mar. (82) . .	2 Mon. .	21 47 30		14 Mar. (73) .	0 Sat. .	2·3979	3936
24 Mar. (83) . .	4 Wed. .	4 0 0		4 Mar. (63) .	5 Thur. .	216·7132	3937
23 Mar. (83) . .	5 Thur. .	10 12 30		22 Mar. (82) .	4 Wed. .	251·3528	3938
23 Mar. (82) . .	6 Fri. .	16 25 0		11 Mar. (70) .	1 Sun. .	127·0362	3939
23 Mar. (82) . .	0 Sat. .	22 37 30		28 Feb. (59) .	5 Thur. .	2·7176	3940
24 Mar. (83) . .	2 Mon .	4 50 0		19 Mar. (78) .	4 Wed. .	37·3592	3941
23 Mar. (83) . .	3 Tues. .	11 2 30		8 Mar. (68) .	2 Mon. .	251·6745	3942
23 Mar. (82) . .	4 Wed. .	17 15 0		25 Feb. (56) .	6 Fri. .	127·3579	3943
23 Mar. (82) . .	5 Thur. .	23 17 30		16 Mar. (75) .	5 Thur. .	161·9975	3944
24 Mar. (83) . .	0 Sat. .	5 40 0		5 Mar. (64) .	2 Mon. .	37·6809	3945

TABLE

CONCURRENT YEAR.						Jovian Saṁvatsara.		Mean Intercalated (adhika) lunar month.
Kali.	Śaka.	Chaitrādi Vikrama.	Mēshādi solar year in Bengal.	Kollam.	A.D.	Southern system.	Northern system.	
1	2	3	3a	4	5	6	7	8a
3946	767	902	251	19-20	*844-45	58 Raktāksha		1 Chaitra
3947	768	903	252	20-21	845-46	59 Krōdhana		...
3948	769	904	253	21-22	846-47	60 Kshaya		9 Mārgaśira
3949	770	905	254	22-23	847-48	1 Prabhava		...
3950	771	906	255	23-24	*848-49	2 Vibhava		...
3951	772	907	256	24-25	849-50	3 Śukla		6 Bhādrapada.
3952	773	908	257	25-26	850-51	4 Pramōda		...
3953	774	909	258	26-27	851-52	5 Prajāpati		...
3954	775	910	259	27-28	*852-53	6 Aṅgiras		2 Vaiśākha
3955	776	911	260	28-29	853-54	7 Śrīmukha		...
3956	777	912	261	29-30	854-55	8 Bhāva		11 Māgha
3957	778	913	262	30-31	855-56	9 Yuvan		...
3958	779	914	263	31-32	*856-57	10 Dhātri		...
3959	780	915	264	32-33	857-58	11 Īśvara		7 Āśvina
3960	781	916	265	33-34	858-59	12 Bahudhānya		...
3961	782	917	266	34-35	859-60	13 Pramādin		...
3962	783	918	267	35-36	*860-61	14 Vikrama		4 Āshāḍha
3963	784	919	268	36-37	861-62	15 Vrisha		...
3964	785	920	269	37-38	862-63	16 Chitrabhānu		12 Phālguna
3965	786	921	270	38-39	863-64	17 Subhānu		...
3966	787	922	271	39-40	*864-65	18 Tāraṇa		...
3967	788	923	272	40-41	865-66	19 Pārthiva		9 Mārgaśira
3968	789	924	273	41-42	866-67	20 Vyaya		...
3969	790	925	274	42-43	867-68	21 Sarvajit		...
3970	791	92[illegible]	275	43-44	*868-69	22 Sarvadhārin		6 Bhādrapada.†

† By the "Indian Calendar" 5 Śrāvaṇa was intercalated.

LXXVI.—*Contd.*

Arya Siddhānta, mean system.

COMMENCEMENT OF THE									Kali year.
MEAN SOLAR YEAR.					MEAN LUNI-SOLAR YEAR (MEAN SUNRISE OF CIVIL DAY ON WHICH CHAITRA ŚUKLA 1 ENDS).				
Day and month, A.D.	Week-day.	Time of mean Mēsha-samkrānti.			Day and month, A.D.	Week-day.	*a* (here = *t*, the index of the tithi).		
13	14	17			19	20	23		1
		H.	M.	S.					
23 Mar. (83)	1 Sun.	11	52	30	23 Feb. (54)	0 Sat.	251·9960		3946
23 Mar. (82)	2 Mon.	18	5	0	13 Mar. (72)	6 Fri.	286·6357		3947
24 Mar. (83)	4 Wed.	0	17	30	2 Mar. (61)	3 Tues.	162·3191		3948
24 Mar. (83)	5 Thur	6	30	0	21 Mar. (80)	2 Mon.	196·9588		3949
23 Mar. (83)	6 Fri.	12	42	30	9 Mar. (69)	6 Fri.	72·6421		3950
23 Mar. (82)	0 Sat.	18	55	0	27 Feb. (58)	4 Wed.	286·9573		3951
24 Mar. (83)	2 Mon.	1	7	30	18 Mar. (77)	3 Tues.	321·5970		3952
24 Mar. (83)	3 Tues.	7	20	0	7 Mar. (66)	0 Sat.	197·2803		3953
23 Mar. (83)	4 Wed.	13	32	30	24 Feb. (55)	4 Wed.	72·9637		3954
23 Mar. (82)	5 Thur.	19	45	0	14 Mar. (73)	3 Tues.	107·6033		3955
24 Mar. (83)	0 Sat.	1	57	30	4 Mar. (63)	1 Sun.	321·9186		3956
24 Mar. (83)	1 Sun.	8	10	0	22 Mar. (81)	6 Fri.	17·9263		3957
23 Mar. (83)	2 Mon.	14	22	30	11 Mar. (71)	4 Wed.	232·2416		3958
23 Mar. (82)	3 Tues.	20	25	0	28 Feb. (59)	1 Sun.	107·9250		3959
24 Mar. (83)	5 Thur.	2	47	30	19 Mar. (78)	0 Sat.	142·5646		3960
24 Mar. (83)	6 Fri.	9	0	0	8 Mar. (67)	4 Wed.	18·2480		3961
23 Mar. (83)	0 Sat.	15	12	30	26 Feb. (57)	2 Mon.	232·5633		3962
23 Mar. (82)	1 Sun.	21	25	0	16 Mar. (75)	1 Sun.	267·2029		3963
24 Mar. (83)	3 Tues.	3	37	30	5 Mar. (64)	5 Thur.	142·8863		3964
24 Mar. (83)	4 Wed.	9	50	0	24 Mar. (83)	4 Wed.	177·5259		3965
23 Mar. (83)	5 Thur.	16	2	30	12 Mar. (72)	1 Sun.	53·2093		3966
23 Mar. (82)	6 Fri.	22	15	0	2 Mar. (61)	6 Fri.	267·5245		3967
24 Mar. (83)	1 Sun.	4	27	30	21 Mar. (80)	5 Thur.	302·1642		3968
24 Mar. (83)	2 Mon.	10	40	0	10 Mar. (69)	2 Mon	177·8476		3969
23 Mar. (83)	3 Tues.	16	52	30	27 Feb. (58)	6 Fri.	53·5303		3970

TABLE

CONCURRENT YEAR.								Mean Intercalated (adhika) lunar month.
Kali.	Saka.	Chaitrâdi Vikrama.	Mêshâdi solar year in Bengal.	Kollam.	A.D.	Jovian Samvatsara. Southern system.	Jovian Samvatsara. Northern system.	
1	2	3	3a	4	5	6	7	8a
3971	792	927	276	44-45	869-70	23 Virôdhin . . .		...
3972	793	928	277	45-46	870-71	24 Vikṛita . . .		...
3973	794	929	278	46-47	871-72	25 Khara . . .		2 Vaiśâkha .
3974	795	930	279	47-48	*872-73	26 Nandana . . .		...
3975	796	931	280	48-49	873-74	27 Vijaya . . .		11 Mâgha .
3976	797	932	281	49-50	874-75	28 Jaya . . .		...
3977	798	933	282	50-51	875-76	29 Manmatha . .		...
3978	799	934	283	51-52	*876-77	30 Durmukha . .		7 Âśvina .
3979	800	935	284	52-53	877-78	31 Hēmalamba . .		...
3980	801	936	285	53-54	878-79	32 Vilamba . . .		...
3981	802	937	286	54-55	879-80	33 Vikârin . . .		4 Âshâḍha .
3982	803	938	287	55-56	*880-81	34 Śârvarin . . .		...
3983	804	939	288	56-57	881-82	35 Plava . . .		12 Phâlguna .
3984	805	940	289	57-58	882-83	36 Śubhakrit . .		...
3985	806	941	290	58-59	883-84	37 Śôbhana . . .		...
3986	807	942	291	59-60	*884-85	38 Krôdhin . . .		9 Mârgaśira .
3987	808	943	292	60-61	885-86	39 Viśvâvasu. . .		...
3988	809	944	293	61-62	886-87	40 Parâbhava		...
3989	810	945	294	62-63	887-88	41 Plavaṅga . . .		5 Śrâvaṇa .
3990	811	946	295	63-64	*888-89	42 Kîlaka . . .		...
3991	812	947	296	64-65	889-90	43 Saumya . . .		...
3992	813	948	297	65-66	890-91	44 Sâdhâraṇa		2 Vaiśâkha
3993	814	949	298	66-67	891-92	45 Virôdhakṛit		...
3994	815	950	299	67-68	*892-93	46 Paridhâvin		10 Pausha
3995	816	951	300	68-69	893-94	47 Pramâdin		...

LXXVI—*Contd.*

I Ārya Siddhānta, mean system.

COMMENCEMENT OF THE							
MEAN SOLAR YEAR.			MEAN LUNI-SOLAR YEAR (MEAN SUNRISE OF CIVIL DAY ON WHICH CHAITRA ŚUKLA 1 ENDS).			Kali year.
Day and month, A.D.	Week-day.	Time of mean Mêsha-saṁkrânti.	Day and month, A.D.	Week-day.	*a* (hereat, the index of the tithi).	
13	14	17	19	20	23	1
		H. M. S.				
23 Mar. (82) . .	4 Wed.	23 5 0	17 Mar. (76) .	5 Thur. .	88·1705	3971
24 Mar. (83) . .	6 Fri. .	5 17 30	7 Mar. (66) .	3 Tues. .	302·4858	3972
24 Mar. (83) . .	0 Sat. .	11 30 0	24 Feb. (55) .	0 Sat. .	178·1692	3973
23 Mar. (83) . .	1 Sun. .	17 42 30	14 Mar. (74) .	6 Fri. .	212·8088	3974
23 Mar. (82) . .	2 Mon. .	23 55 0	3 Mar. (62) .	3 Tues. .	88·4922	3975
24 Mar. (83) . .	4 Wed. .	6 7 30	22 Mar. (81) .	2 Mon. .	123·1218	3976
24 Mar. (83) . .	5 Thur. .	12 20 0	11 Mar. (70) .	6 Fri. .	9998·8151†	3977
23 Mar. (83) . .	6 Fri. .	18 32 30	29 Feb. (60) .	4 Wed. .	213·1304	3978
24 Mar. (83) . .	1 Sun. .	0 45 0	19 Mar. (78) .	3 Tues. .	247·7700	3979
24 Mar. (83) . .	2 Mon. .	6 57 30	8 Mar. (67) .	0 Sat. .	123·4535	3980
24 Mar. (83) . .	3 Tues. .	13 10 0	25 Feb. (56) .	4 Wed. .	9999·1368†	3981
23 Mar. (83) . .	4 Wed. .	19 22 30	15 Mar. (75) .	3 Tues. .	53·7764	3982
24 Mar. (83) . .	6 Fri. .	1 35 0	5 Mar. (64) .	1 Sun. .	248·6917	3983
24 Mar. (83) . .	0 Sat. .	7 47 30	24 Mar. (83) .	0 Sat. .	282·7313	3984
24 Mar. (83) . .	1 Sun. .	14 0 0	13 Mar. (72) .	4 Wed. .	158·4147	3985
23 Mar. (83) . .	2 Mon. .	20 12 30	1 Mar. (61) .	1 Sun. .	34·0980	3986
24 Mar. (83) . .	4 Wed. .	2 25 0	20 Mar. (79) .	0 Sat. .	68·7377	3987
24 Mar. (83) . .	5 Thur. .	8 37 30	10 Mar. (69) .	5 Thur. .	232·0530	3988
24 Mar. (83) . .	6 Fri. .	14 50 0	27 Feb. (58) .	2 Mon. .	158·7364	3989
23 Mar. (83) . .	0 Sat. .	21 2 30	17 Mar. (77) .	1 Sun. .	193·3769	3990
24 Mar. (83) . .	2 Mon. .	3 15 0	6 Mar. (65) .	5 Thur. .	69·0594	3991
24 Mar. (83) . .	3 Tues. .	9 27 30	24 Feb. (55) .	3 Tues. .	283·3746	3992
24 Mar. (83) . .	4 Wed. .	15 40 0	15 Mar. (74) .	2 Mon. .	318·0143	3993
23 Mar. (83) . .	5 Thur. .	21 52 30	3 Mar. (63) .	6 Fri. .	193·6976	3994
24 Mar. (83) . .	0 Sat. .	4 5 0	22 Mar. (81) .	5 Thur. .	228·3372	3995

† As a mean tithi Chaitra Śukla 1 was suppressed. The civil day corresponding to it, *i.e.*, the first day of the mean luni-solar year, was as given in cols. 19, 20.

TABLE

CONCURRENT YEAR.								
Kali.	Saka.	Chaitrādi Vikrama.	Mēshādi solar year in Bengal.	Kollam.	A.D.	Jovian Saṁvatsara. Southern system.	Jovian Saṁvatsara. Northern system.	Mean Intercalated (adhika) lunar month.
1	2	3	3a	4	5	6	7	8a
3996	817	952	301	69-70	894-95	48 Ānanda . . .		...
3997	818	953	302	70-71	895-96	49 Rākshasa . . .		7 Āśvina .
3998	819	954	303	71-72	*896-97	50 Anala . . .		...
3999	820	955	304	72-73	897-98	51 Piṅgala . . .		...
4000	821	956	305	73-74	898-99	52 Kālayukta . .		3 Jyēshṭha .
4001	822	957	306	74-75	899-900	53 Siddhārthin .		...
4002	823	958	307	75-76	*900-01	54 Raudra . . .		12 Phālguna .
4003	824	959	308	76-77	901-02	55 Durmati . . .		...
4004	825	960	309	77-78	902-03	56 Dundubhi . .		...
4005	826	961	310	78-79	903-04	57 Rudhirōdgārin .		9 Mārgaśira .
4006	827	962	311	79-80	*904-05	58 Raktāksha† . .		...
4007	828	963	312	80-81	905-06	59 Krōdhana .	60 *Kshaya* .	...
4008	829	964	313	81-82	906-07	60 Kshaya‡ .	1 *Prabhava* .	5 Śrāvaṇa .
4009	830	965	314	82-83	907-08	1 Prabhava .	2 *Vibhava* .	...
4010	831	966	315	83-84	*908-09	2 Vibhava .	3 *Śukla* .	...
4011	832	967	316	84-85	909-10	3 Śukla .	4 Pramōda .	2 Vaiśākha .
4012	833	968	317	85-86	910-11	4 Pramōda .	5 Prajāpati .	...
4013	834	969	318	86-87	911-12	5 Prajāpati .	6 Aṅgiras .	10 Pausha .
4014	835	970	319	87-88	*912-13	6 Aṅgiras .	7 Śrīmukha .	...
4015	836	971	320	88-89	913-14	7 Śrīmukha .	8 Bhāva .	...
4016	837	972	321	89-90	914-15	8 Bhāva .	9 Yuvan .	7 Āśvina .
4017	838	973	322	90-91	915-16	9 Yuvan . .	10 Dhātṛi . .	...
4018	839	974	323	91-92	*916-17	10 Dhātṛi . .	11 Īśvara . .	...
4019	840	975	324	92-93	917-18	11 Īśvara . .	12 Bahudhānya .	3 Jyēshṭha .
4020	841	976	325	93-94	918-19	12 Bahudhānya .	13 Pramādin .	...

† By the mean system 59 Krōdhana was expunged; by the true system 60 Kshaya was the expunged saṁvatsara and the year A.D. 905-6 was called "Krōdhana."

‡ By southern reckoning there was no suppression after this year.

§ By the "*Indian Calendar*" 8 Kārttika was intercalated.

LXXVI—*Contd.*

1 Ārya Siddhānta, mean system.

COMMENCEMENT OF THE							
MEAN SOLAR YEAR.			MEAN LUNI-SOLAR YEAR (MEAN SUNRISE OF CIVIL DAY ON WHICH CHAITRA ŚUKLA 1 ENDS).			Kali year.
Day and month, A.D.	Week-day.	Time of mean Mēsha-saṁkrānti.	Day and month, A.D.	Week-day.	*a* (here = *t*, the index of the tithi).	
13	14	17	19	20	23	1
		H. M. S.				
24 Mar. (83) . .	1 Sun. .	10 17 30	11 Mar. (70) .	2 Mon. .	104·0206	3996
24 Mar. (83) . .	2 Mon. .	16 30 0	1 Mar. (60) .	0 Sat. .	318·3359	3997
23 Mar. (83) . .	3 Tues. .	22 42 30	18 Mar. (78) .	5 Thur. .	14·3436	3998
24 Mar. (83) . .	5 Thur. .	4 55 0	8 Mar. (67) .	3 Tues. .	228·6589	3999
24 Mar. (83) . .	6 Fri. .	11 7 30	25 Feb. (56) .	0 Sat. .	104·3423	4000
24 Mar. (83) . .	0 Sat. .	17 20 0	16 Mar. (75) .	6 Fri. .	138·9819	4001
23 Mar. (83) . .	1 Sun. .	23 32 30	4 Mar. (64) .	3 Tues. .	14·6653	4002
24 Mar. (83) . .	3 Tues. .	5 45 0	23 Mar. (82) .	2 Mon. .	49·3049	4003
24 Mar. (83) . .	4 Wed. .	11 57 30	13 Mar. (72) .	0 Sat. .	263·6202	4004
24 Mar. (83) . .	5 Thur. .	18 10 0	2 Mar. (61) .	4 Wed. .	139·3034	4005
24 Mar. (84) . .	0 Sat. .	0 22 30	20 Mar. (80) .	3 Tues. .	173·9431	4006
24 Mar. (83) . .	1 Sun. .	6 35 0	9 Mar. (68) .	0 Sat. .	49·6264	4007
24 Mar. (83) . .	2 Mon. .	12 47 30	27 Feb. (58) .	5 Thur. .	263·9418	4008
24 Mar. (83) . .	3 Tues. .	19 0 0	18 Mar. (77) .	4 Wed. .	298·5814	4009
24 Mar. (84) . .	5 Thur. .	1 12 30	6 Mar. (66) .	1 Sun. .	174·2647	4010
24 Mar. (83) . .	6 Fri. .	7 25 0	23 Feb. (54) .	5 Thur. .	49·9481	4011
24 Mar. (83) . .	0 Sat. .	13 37 30	14 Mar. (73) .	4 Wed. .	84·5878	4012
24 Mar. (83) . .	1 Sun. .	19 50 0	4 Mar. (63) .	2 Mon. .	298·9030	4013
24 Mar. (84) . .	3 Tues. .	2 2 30	21 Mar. (81) .	0 Sat. .	9994·9109†	4014
24 Mar. (83) . .	4 Wed. .	8 15 0	11 Mar. (70) .	5 Thur. .	209·2259	4015
24 Mar. (83) . .	5 Thur. .	14 27 30	28 Feb. (59) .	2 Mon. .	84·9093	4016
24 Mar. (83) . .	6 Fri. .	20 40 0	19 Mar. (78) .	1 Sun. .	119·5490	4017
24 Mar. (84) . .	1 Sun. .	2 52 30	7 Mar. (67) .	5 Thur. .	9995·2324†	4018
24 Mar. (83) . .	2 Mon. .	9 5 0	25 Feb. (56) .	3 Tues. .	209·5476	4019
24 Mar. (83) . .	3 Tues. .	15 17 30	16 Mar. (75) .	2 Mon. .	244·1872	4020

† As a mean tithi Chaitra Śukla 1 was suppressed. The civil day corresponding to it, *i.e.*, the first day of he luni-solar year was as given in cols. 19, 20.

TABLE

Kali.	Śaka.	Chaitrādi Vikrama.	Mēshādi solar year in Bengal.	Kollam.	A.D	Jovian Saṁvatsara. Southern system.	Jovian Saṁvatsara. Northern system.	Mean Intercalated (adhika) lunar month.
						CONCURRENT YEAR.		
1	2	3	3a	4	5	6	7	8a
4021	842	977	326	94-95	919-20	13 Pramādin .	14 Vikrama .	12 Phālguna .
4022	843	978	327	95-96	*920-21	14 Vikrama .	15 Vṛisha . .	...
4023	844	979	328	96-97	921-22	15 Vṛisha . .	16 Chitrabhānu .	...
4024	845	980	329	97-98	922-23	16 Chitrabhānu .	17 Subhānu .	8 Kārttika .
4025	846	981	330	98-99	923-24	17 Subhānu .	18 Tāraṇa . .	...
4026	847	982	331	99-00	*924-25	18 Tāraṇa . .	19 Pārthiva .	...
4027	848	983	332	100-01	925-26	19 Pārthiva .	20 Vyaya . .	5 Śrāvaṇa .
4028	849	984	333	101-02	926-27	20 Vyaya . .	21 Sarvajit .	...
4029	850	985	334	102-03	927-28	21 Sarvajit .	22 Sarvadhārin .	...
4030	851	986	335	103-04	*928-29	22 Sarvadhārin .	23 Virōdhin .	1 Chaitra .
4031	852	987	336	104-05	929-30	23 Virōdhin .	24 Vikṛita . .	...
4032	853	988	337	105-06	930-31	24 Vikṛita . .	25 Khara . .	10 Pausha .
4033	854	989	338	106-07	931-32	25 Khara . .	26 Nandana .	...
4034	855	990	339	107-08	*932-33	26 Nandana .	27 Vijaya . .	...
4035	856	991	340	108-09	933-34	27 Vijaya . .	28 Jaya . .	6 Bhādrapada
4036	857	992	341	109-10	934-35	28 Jaya . .	29 Manmatha .	...
4037	858	993	342	110-11	935-36	29 Manmatha .	30 Durmukha .	...
4038	859	994	343	111-12	*936-37	30 Durmukha .	31 Hēmalamba .	3 Jyēshṭha .
4039	860	995	344	112-13	937-38	31 Hēmalamba .	32 Vilamba .	...
4040	861	996	345	113-14	938-39	32 Vilamba .	33 Vikārin .	11 Māgha .
4041	862	997	346	114-15	939-40	33 Vikārin .	34 Śārvarin .	...
4042	863	998	347	115-16	*940-41	34 Śārvarin .	35 Plava . .	...
4043	864	999	348	116-17	941-42	35 Plava . .	36 Śubhakṛit .	8 Kārttika .
4044	865	1000	349	117-18	942-43	36 Śubhakṛit .	37 Śōbhana	...
4045	866	1001	350	118-19	943-44	37 Śōbhana .	38 Krōdhin .	...

LXXVI—*Contd.*

1 Ārya Siddhānta, mean system.

COMMENCEMENT OF THE									Kali year.
MEAN SOLAR YEAR.					MEAN LUNI-SOLAR YEAR (MEAN SUNRISE OF CIVIL DAY ON WHICH CHAITRA ŚUKLA 1 ENDS).				
Day and month, A.D.	Week-day.	Time of mean Mēsha-saṁkrānti.			Day and month, A.D.	Week-day.	*a* (here = *t*, the index of the tithi).		
13	14	17			19	20	23		1
		H.	M.	S.					
24 Mar. (83) . .	4 Wed. .	21	30	0	5 Mar. (64) .	6 Fri. .	119·8706		4021
24 Mar. (84) . .	6 Fri. .	3	42	30	23 Mar. (83) .	5 Thur. .	154·5102		4022
24 Mar. (83) . .	0 Sat. .	9	55	0	12 Mar. (71) .	2 Mon. .	30·1936		4023
24 Mar. (83) . .	1 Sun. .	16	7	30	2 Mar. (61) .	0 Sat. .	244·5089		4024
24 Mar. (83) . .	2 Mon. .	22	20	0	21 Mar. (80) .	6 Fri. .	279·1485		4025
24 Mar. (84) . .	4 Wed. .	4	32	30	9 Mar. (69) .	3 Tues. .	154·8319		4026
24 Mar. (83) . .	5 Thur. .	10	45	0	26 Feb. (57) .	0 Sat. .	30·5153		4027
24 Mar. (83) . .	6 Fri. .	16	57	30	17 Mar. (76) .	6 Fri. .	65·1549		4028
24 Mar. (83) . .	0 Sat. .	23	10	0	7 Mar. (66) .	4 Wed. .	279·4701		4029
24 Mar. (84) . .	2 Mon. .	5	22	30	24 Feb. (55) .	1 Sun. .	155·1535		4030
24 Mar. (83) . .	3 Tues. .	11	35	0	14 Mar. (73) .	0 Sat. .	189·7932		4031
24 Mar. (83) . .	4 Wed. .	17	47	30	3 Mar. (62) .	4 Wed. .	65·4765		4032
25 Mar. (84) . .	6 Fri. .	0	0	0	22 Mar. (81) .	3 Tues. .	100·1162		4033
24 Mar. (84) . .	0 Sat. .	6	12	30	11 Mar. (71) .	1 Sun. .	314·4314		4034
24 Mar. (83) . .	1 Sun. .	12	25	0	28 Feb. (59) .	5 Thur. .	190·1148		4035
24 Mar. (83) . .	2 Mon. .	18	37	30	19 Mar. (78) .	4 Wed. .	224·7544		4036
25 Mar. (84) . .	4 Wed. .	0	50	0	8 Mar. (67) .	1 Sun. .	100·4378		4037
24 Mar. (84) . .	5 Thur. .	7	2	30	26 Feb. (57) .	6 Fri. .	314·7531		4038
24 Mar. (83) . .	6 Fri. .	13	15	0	15 Mar. (74) .	4 Wed. .	10·7698		4039
24 Mar. (83) . .	0 Sat. .	19	27	30	5 Mar. (64) .	2 Mon. .	225·0661		4040
25 Mar. (84) . .	2 Mon. .	1	49	0	24 Mar. (83) .	1 Sun. .	259·7156		4041
24 Mar. (84) . .	3 Tues. .	7	52	30	12 Mar. (72) .	5 Thur. .	13[illegible]·3991		4042
24 Mar. (83) . .	4 Wed	14	5	0	1 Mar. (60) .	2 Mon. .	11·0825		4043
24 Mar. (83) . .	5 Thur .	20	17	30	20 Mar. (79) .	1 Sun. .	45·7222		4044
25 Mar. (84) . .	0 Sat.	2	30	0	19 Mar. (69) .	6 Fri. .	2[illegible]0·0474		4045

TABLE

CONCURRENT YEAR.								Mean Intercalated (adhika) lunar month.
Kali.	Śaka.	Chaitrādi Vikrama.	Mēshādi solar year in Bengal.	Kollam.	A.D.	Jovian Saṁvatsara. Southern system.	Jovian Saṁvatsara. Northern system.	
1	2	3	3a	4	5	6	7	8a
4046	867	1002	351	119-20	*944-45	38 Krōdhin .	39 Viśvāvasu .	5 Śrāvaṇa† .
4047	868	1003	352	120-21	945-46	39 Viśvāvasu .	40 Parābhava .	...
4048	869	1004	353	121-22	946-47	40 Parābhava .	41 Plavaṅga .	...
4049	870	1005	354	122-23	947-48	41 Plavaṅga .	42 Kīlaka . .	1 Chaitra .
4050	871	1006	355	123-24	*948-49	42 Kīlaka . .	43 Saumya . .	...
4051	872	1007	356	124-25	949-50	43 Saumya . .	44 Sādhāraṇa .	10 Pausha .
4052	873	1008	357	125-26	950-51	44 Sādhāraṇa .	45 Virōdhakṛit .	...
4053	874	1009	358	126-27	951-52	45 Virōdhakṛit .	46 Paridhāvin .	...
4054	875	1010	359	127-28	*952-53	46 Paridhāvin .	47 Pramādin .	6 Bhādrapada
4055	876	1011	360	128-29	953-54	47 Pramādin .	48 Ānanda . .	...
4056	877	1012	361	129-30	954-55	48 Ānanda .	49 Rākshasa .	...
4057	878	1013	362	130-31	955-56	49 Rākshasa .	50 Anala . .	3 Jyēshṭha .
4058	879	1014	363	131-32	*956-57	50 Anala . .	51 Piṅgala .	...
4059	880	1015	364	132-33	957-58	51 Piṅgala .	52 Kālayukta .	11 Māgha .
4060	881	1016	365	133-34	958-59	52 Kālayukta .	53 Siddhārthin .	...
4061	882	1017	366	134-35	959-60	53 Siddhārthin .	54 Raudra .	...
4062	883	1018	367	135-36	*960-61	54 Raudra .	55 Durmati .	8 Kārttika .
4063	884	1019	368	136-37	961-62	55 Durmati .	56 Dundubhi .	...
4064	885	1020	369	137-38	962-63	56 Dundubhi .	57 Rudhirōdgārin	...
4065	886	1021	370	138-39	963-64	57 Rudhirōdgārin	58 Raktāksha .	4 Āshāḍha
4066	887	1022	371	139-40	*964-65	58 Raktāksha .	59 Krōdhana .	...
4067	888	1023	372	140-41	965-66	59 Krōdhana .	60 Kshaya .	...
4068	889	1024	373	141-42	966-67	60 Kshaya .	1 Prabhava .	1 Chaitra
4069	890	1025	374	142-43	967-68	1 Prabhava .	2 Vibhava .	...
4070	891	1026	375	143-44	*968-69	2 Vibhava .	3 Śukla . .	9 Mārgaśira .

† By the "*Indian Calendar*" the intercalated month was 4 Āshāḍha.

LXXVI—*Contd.* **1 Ārya Siddhānta, mean system.**

COMMENCEMENT OF THE							
Mean solar year.			Mean luni-solar year (mean sunrise of civil day on which Chaitra śukla 1 ends).			Kali year.
Day and month, A.D.	Week-day.	Time of mean Mêsha-saṁkrānti.	Day and month, A.D.	Week-day.	*a* (here = *t*, the index of the tithi).	
13	14	17	19	20	23	1
		H. M. S.				
24 Mar. (84) . .	1 Sun. .	8 42 30	27 Feb. (58) .	3 Tues. .	135·7207	4046
24 Mar. (83) . .	2 Mon. .	14 55 0	17 Mar. (76) .	2 Mon. .	170·3603	4047
24 Mar. (83) . .	3 Tues. .	21 7 30	6 Mar. (65) .	6 Fri. .	46·0436	4048
25 Mar. (84) . .	5 Thur. .	3 20 0	24 Feb. (55) .	4 Wed. .	260·3590	4049
24 Mar. (84) . .	6 Fri. .	9 32 30	14 Mar. (74) .	3 Tues. .	294·9986	4050
24 Mar. (83) . .	0 Sat. .	15 45 0	3 Mar. (62) .	0 Sat. .	170·6819	4051
24 Mar. (83) . .	1 Sun. .	21 57 30	22 Mar. (81) .	6 Fri. .	205·3216	4052
25 Mar. (84) . .	3 Tues. .	4 10 0	11 Mar. (70) .	3 Tues. .	81·0049	4053
24 Mar. (84) . .	4 Wed. .	10 22 30	29 Feb. (60) .	1 Sun. .	295·3203	4054
24 Mar. (83) . .	5 Thur. .	16 35 0	19 Mar. (78) .	0 Sat. .	329·9599	4055
24 Mar. (83) . .	6 Fri. .	22 47 30	8 Mar. (67) .	4 Wed. .	205·6432	4056
25 Mar. (84) . .	1 Sun. .	5 0 0	25 Feb. (56) .	1 Sun. .	81·3266	4057
24 Mar. (84) . .	2 Mon. .	11 12 30	15 Mar. (75) .	0 Sat. .	115·9662	4058
24 Mar. (83) . .	3 Tues. .	17 25 0	5 Mar. (64) .	5 Thur. .	330·2815	4059
24 Mar. (83) . .	4 Wed. .	23 37 30	23 Mar. (82) .	3 Tues. .	26·2892	4060
25 Mar. (84) . .	6 Fri. .	5 50 0	13 Mar. (72) .	1 Sun. .	240·6045	4061
24 Mar. (84) . .	0 Sat. .	12 2 30	1 Mar. (61) .	5 Thur. .	116·2879	4062
24 Mar. (83) . .	1 Sun. .	18 15 0	20 Mar. (79) .	4 Wed. .	150·9275	4063
25 Mar. (84) . .	3 Tues. .	0 27 30	9 Mar. (68) .	1 Sun. .	26·6109	4064
25 Mar. (84) . .	4 Wed. .	6 40 0	27 Feb. (58) .	6 Fri. .	240·9262	4065
24 Mar. (84) . .	5 Thur. .	12 52 30	17 Mar. (77) .	5 Thur. .	275·5658	4066
24 Mar. (83) . .	6 Fri. .	19 5 0	6 Mar. (65) .	2 Mon. .	151·2491	4067
25 Mar. (84) . .	1 Sun. .	1 17 30	23 Feb. (51) .	6 Fri. .	26·9325	4068
25 Mar. (84) . .	2 Mon. .	7 30 0	14 Mar. (73) .	5 Thur. .	61·5721	4069
24 Mar. (84) . .	3 Tues. .	13 42 30	3 Mar. (63) .	3 Tues. .	275·8874	4070

TABLE

CONCURRENT YEAR.								Mean Intercalated (adhika) lunar month.
Kali	Śaka.	Chaitrādi Vikrama.	Mēshādi solar year in Bengal.	Kollam.	A.D.	Jovian Saṁvatsara. Southern system.	Jovian Saṁvatsara. Northern system.	
1	2	3	3a	4	5	6	7	8a
4071	892	1027	376	144-45	969-70	3 Śukla . .	4 Pramōda .	
4072	893	1028	377	145-46	970-71	4 Pramōda .	5 Prajāpati .	...
4073	894	1029	378	146-47	971-72	5 Prajāpati .	6 Aṅgiras .	6 Bhādrapada
4074	895	1030	379	147-48	*972-73	6 Aṅgiras .	7 Śrīmukha .	...
4075	896	1031	380	148-49	973-74	7 Śrīmukha .	8 Bhāva . .	...
4076	897	1032	381	149-50	974-75	8 Bhāva . .	9 Yuvan .	2 Vaiśākha .
4077	898	1033	382	150-51	975-76	9 Yuvan . .	10 Dhātṛi . .	...
4078	899	1034	383	151-52	*976-77	10 Dhātṛi . .	11 Īśvara . .	11 Māgha .
4079	900	1035	384	152-53	977-78	11 Īsvara . .	12 Bahudhānya .	...
4080	901	1036	385	153-54	978-79	12 Bahudhānya .	13 Pramādin .	...
4081	902	1037	386	154-55	979-80	13 Pramādin	14 Vikrama .	8 Kārttika † .
4082	903	1038	387	155-56	*980-81	14 Vikrama .	15 Vṛisha . .	...
4083	904	1039	388	156-57	981-82	15 Vṛisha . .	16 Chitrabhānu .	...
4084	905	1040	389	157-58	982-83	16 Chitrabhānu .	17 Subhānu .	4 Āshādha .
4085	906	1041	390	158-59	983-84	17 Subhānu .	18 Tāraṇa . .	...
4086	907	1042	391	159-60	*984-85	18 Tāraṇa . .	19 Pārthiva .	...
4087	908	1043	392	160-61	985-86	19 Pārthiva .	20 Vyaya . .	1 Chaitra .
4088	909	1044	393	161-62	986-87	20 Vyaya . .	21 Sarvajit .	...
4089	910	1045	394	162-63	987-88	21 Sarvajit .	22 Sarvadhārin .	9 Mārgaśira .
4090	911	1046	395	163-64	*988-89	22 Sarvadhārin .	23 Virōdhin .	...
4091	912	1047	396	164-65	989-90	23 Virōdhin .	24 Vikṛita ‡ .	...
4092	913	1048	397	165-66	990-91	24 Vikrita . .	26 *Nandana* .	6 Bhādrapada
4093	914	1049	398	166-67	991-92	25 Khara . .	27 *Vijaya* . .	...
4094	915	1050	399	167-68	*992-93	26 Nandana .	28 *Jaya* . .	...
4095	916	1051	400	168-69	993-94	27 Vijaya . .	29 *Manmatha* .	2 Vaiśākha .

† By the "*Indian Calendar*" 7 Āśvina was intercalated.

‡ 25 Khara was expunged in the north by the mean system, but 26 Nandana by the true system. By the true system the year A.D. 990-91 was, in the north, called "Khara."

LXXVI—*Contd.* **1 Ārya Siddhānta, mean system.**

COMMENCEMENT OF THE									Kali year.
MEAN SOLAR YEAR.					MEAN LUNI-SOLAR YEAR (MEAN SUNRISE OF CIVIL DAY ON WHICH CHAITRA ŚUKLA 1 ENDS).				
Day and month, A.D.	Week-day.	Time of mean Mēsha-saṁkrānti.			Day and month, A.D.	Week-day.	*a* (here=*t*, the index of the tithi).		
13	14	17			19	20	23		1
		H.	M.	S.					
24 Mar. (83)	4 Wed.	19	55	0	22 Mar. (81)	2 Mon.	310·5271		4071
25 Mar. (84)	6 Fri.	2	7	30	11 Mar. (70)	6 Fri.	186·2104		4072
25 Mar. (84)	0 Sat.	8	20	0	28 Feb. (59)	3 Tues.	61·8939		4073
24 Mar. (84)	1 Sun.	14	32	30	18 Mar. (78)	2 Mon.	96·5335		4074
24 Mar. (83)	2 Mon.	20	45	0	8 Mar. (67)	0 Sat.	310·8487		4075
25 Mar. (84)	4 Wed.	2	57	30	25 Feb. (56)	4 Wed.	186·5321		4076
25 Mar. (84)	5 Thur.	9	10	0	16 Mar. (75)	3 Tues.	221·1716		4077
24 Mar. (84)	6 Fri.	15	22	30	4 Mar. (64)	0 Sat.	96·8550		4078
24 Mar. (83)	0 Sat.	21	35	0	23 Mar. (82)	6 Fri.	131·4946		4079
25 Mar. (84)	2 Mon.	3	47	30	12 Mar. (71)	3 Tues.	7·1781		4080
25 Mar. (84)	3 Tues.	10	0	0	2 Mar. (61)	1 Sun.	221·4933		4081
24 Mar. (84)	4 Wed.	16	12	30	20 Mar. (80)	0 Sat.	256·1329		4082
24 Mar. (83)	5 Thur.	22	25	0	9 Mar. (68)	4 Wed.	131·8163		4083
25 Mar. (84)	0 Sat.	4	37	30	26 Feb. (57)	1 Sun.	7·4998		4084
25 Mar. (84)	1 Sun.	10	50	0	17 Mar. (76)	0 Sat.	41·1393		4085
24 Mar. (84)	2 Mon.	17	2	30	6 Mar. (66)	5 Thur.	256·4546		4086
24 Mar. (83)	3 Tues.	23	15	0	23 Feb. (54)	2 Mon.	132·1379		4087
25 Mar. (84)	5 Thur.	5	27	30	14 Mar. (73)	1 Sun.	166·7776		4088
25 Mar. (84)	6 Fri.	11	40	0	3 Mar. (62)	5 Thur.	42·4610		4089
24 Mar. (84)	0 Sat.	17	52	30	21 Mar. (81)	4 Wed.	77·1006		4090
25 Mar. (84)	2 Mon.	0	5	0	11 Mar. (70)	2 Mon.	291·4158		4091
25 Mar. (84)	3 Tues.	6	17	30	28 Feb. (59)	6 Fri.	167·0992		4092
25 Mar. (84)	4 Wed.	12	30	0	19 Mar. (78)	5 Thur.	201·7389		4093
24 Mar. (84)	5 Thur.	18	42	30	7 Mar. (67)	2 Mon.	77·4222		4094
25 Mar. (84)	0 Sat.	0	55	0	25 Feb. (56)	0 Sat.	291·7375		4095

TABLE

CONCURRENT YEAR								Mean Intercalated (adhika) lunar month.
						Jovian Saṁvatsara.		
Kali.	Śaka.	Chaitrādi Vikrama.	Mēshādi solar year in Bengal.	Kollam.	A.D.	Southern system.	Northern system.	
1	2	3	3a	4	5	6	7	8a
4096	917	1052	401	169-70	994-95	28 Jaya .	30 *Durmukha* .	...
4097	918	1053	402	170-71	995-96	29 Manmatha .	31 Hēmalamba .	11 Māgha .
4098	919	1054	403	171-72	*996-97	30 Durmukha .	32 Vilamba .	...
4099	920	1055	404	172-73	997-98	31 Hēmalamba .	33 Vikārin .	...
4100	921	1056	405	173-74	998-99	32 Vilamba .	34 Śārvarin .	7 Āsvina .
4101	922	1057	406	174-75	999-000	33 Vikārin .	35 Plava . .	...
4102	923	1058	407	175-76	*1000-01	34 Śārvarin .	36 Śubhakṛit .	...
4103	924	1059	408	176-77	1001-02	35 Plava . .	37 Śōbhana .	4 Āshāḍha .
4104	925	1060	409	177-78	1002-03	36 Śubhakṛit .	38 Krōdhin .	...
4105	926	1061	410	178-79	1003-04	37 Śōbhana .	39 Viśvāvasu .	12 Phālguna .
4106	927	1062	411	179-80	*1004-05	38 Krōdhin .	40 Parābhava .	...
4107	928	1063	412	180-81	1005-06	39 Viśvāvasu .	41 Plavaṅga .	...
4108	929	1064	413	181-82	1006-07	40 Parābhava .	42 Kīlaka . .	9 Mārgasira .
4109	930	1065	414	182-83	1007-08	41 Plavaṅga .	43 Saumya .	...
4110	931	1066	415	183-84	*1008-09	42 Kīlaka .	44 Sādhāraṇa .	...
4111	932	1067	416	184-85	1009-10	43 Saumya .	45 Virōdhakṛit .	5 Śrāvaṇa .
4112	933	1068	417	185-86	1010-11	44 Sādhāraṇa .	46 Paridhāvin .	...
4113	934	1069	418	186-87	1011-12	45 Virōdhakṛit .	47 Pramādin .	...
4114	935	1070	419	187-88	*1012-13	46 Paridhāvin .	48 Ānanda .	2 Vaiśākha .
4115	936	1071	420	188-89	1013-14	47 Pramādin .	49 Rākshasa .	...
4116	937	1072	421	189-90	1014-15	48 Ānanda .	50 Anala . .	10 Pausha .
4117	938	1073	422	190-91	1015-16	49 Rākshasa .	51 Piṅgala .	...
4118	939	1074	423	191-92	*1016-17	50 Anala . .	52 Kālayukta .	...
4119	940	1075	424	192-93	1017-18	51 Piṅgala .	53 Siddhārthin .	7 Āśvina .
4120	941	1076	425	193-94	1018-19	52 Kālayukta .	54 Raudra .	...

LXXVI—*Contd.*

1 Ārya Siddhānta, mean system.

COMMENCEMENT OF THE						
MEAN SOLAR YEAR.			MEAN LUNI-SOLAR YEAR (MEAN SUNRISE OF CIVIL DAY ON WHICH CHAITRA SUKLA 1 ENDS).			Kali year.
Day and month, A.D.	Week-day.	Time of mean Mésha-saṁkrānti.	Day and month, A.D.	Week-day.	*a* (here = *t*, the index of the tithi).	
13	14	17	19	20	23	1
		H. M. S.				
25 Mar. (84)	1 Sun.	7 7 30	16 Mar. (75)	6 Fri.	326·3771	4096
25 Mar. (84)	2 Mon.	13 20 0	5 Mar. (64)	3 Tues.	202·0605	4097
24 Mar. (84)	3 Tues.	19 32 30	23 Mar. (83)	2 Mon.	236·7001	4098
25 Mar. (84)	5 Thur.	1 45 0	12 Mar. (71)	6 Fri.	112·3825	4099
25 Mar. (84)	6 Fri.	7 57 30	2 Mar. (61)	4 Wed.	326·6988	4100
25 Mar. (84)	0 Sat.	14 10 0	20 Mar. (79)	2 Mon.	22·7065	4101
24 Mar. (84)	1 Sun.	20 22 30	9 Mar. (69)	0 Sat.	237·0218	4102
25 Mar. (84)	3 Tues.	2 35 0	26 Feb. (57)	4 Wed.	112·7052	4103
25 Mar. (84)	4 Wed.	8 47 30	17 Mar. (76)	3 Tues.	147·3448	4104
25 Mar. (84)	5 Thur.	15 0 0	6 Mar. (65)	0 Sat.	23·0272	4105
24 Mar. (84)	6 Fri.	21 12 30	24 Mar. (84)	6 Fri.	57·6667	4106
25 Mar. (84)	1 Sun.	3 25 0	14 Mar. (73)	4 Wed.	271·9831	4107
25 Mar. (84)	2 Mon.	9 37 30	3 Mar. (62)	1 Sun.	147·6665	4108
25 Mar. (84)	3 Tues.	15 50 0	22 Mar. (81)	0 Sat.	182·3061	4109
24 Mar. (84)	4 Wed.	22 2 30	10 Mar. (70)	4 Wed.	57·[illegible]894	4110
25 Mar. (84)	6 Fri.	4 15 0	28 Feb. (59)	2 Mon.	272·3047	4111
25 Mar. (84)	0 Sat.	10 27 30	19 Mar. (78)	1 Sun.	306·9444	4112
25 Mar. (84)	1 Sun.	16 40 0	8 Mar. (67)	5 Thur.	182·6277	4113
24 Mar. (84)	2 Mon.	22 52 30	25 Feb. (56)	2 Mon.	58·3111	4114
25 Mar. (84)	4 Wed.	5 5 0	15 Mar. (74)	1 Sun.	92·9507	4115
25 Mar. (84)	5 Thur.	11 17 30	5 Mar. (64)	6 Fri.	307·2659	4116
25 Mar. (84)	6 Fri.	17 30 0	23 Mar. (82)	4 Wed.	3·2737	4117
24 Mar. (84)	0 Sat.	23 42 30	12 Mar. (72)	2 Mon.	217·5890	4118
25 Mar. (84)	2 Mon.	5 55 0	1 Mar. (60)	6 Fri.	93·2723	4119
25 Mar. (84)	3 Tues.	12 7 30	20 Mar. (79)	5 Thur.	127·9119	4120

TABLE

CONCURRENT YEAR.								Mean Intercalated (adhika) lunar month.
Kali.	Śaka.	Chaitrādi Vikrama.	Mēshādi solar year in Bengal.	Kollam.	A.D.	Jovian Saṁvatsara. Southern system.	Jovian Saṁvatsara. Northern system.	
1	2	3	3a	4	5	6	7	8a
4121	942	1077	426	194-95	1019-20	53 Siddhārthin .	55 Durmati .	...
4122	943	1078	427	195-96	*1020-21	54 Raudra .	56 Dundubhi .	4 Āshāḍha ‡ .
4123	944	1079	428	196-97	1021-22	55 Durmati .	57 Rudhirōdgārin	...
4124	945	1080	429	197-98	1022-23	56 Dundubhi .	58 Raktāksha .	12 Phālguna .
4125	946	1081	430	198-99	1023-24	57 Rudhirōdgārin	59 Krōdhana .	...
4126	947	1082	431	199-200	*1024-25	58 Raktāksha .	60 Kshaya .	...
4127	948	1083	432	200-01	1025-26	59 Krōdhana .	1 Prabhava .	9 Mārgaśira .
4128	949	1084	433	201-02	1026-27	60 Kshaya .	2 Vibhava .	...
4129	950	1085	434	202-03	1027-28	1 Prabhava .	3 Śukla . .	...
4130	951	1086	435	203-04	*1028-29	2 Vibhava .	4 Pramōda .	5 Śrāvaṇa .
4131	952	1087	436	204-05	1029-30	3 Śukla . .	5 Prajāpati .	...
4132	953	1088	437	205-06	1030-31	4 Pramōda .	6 Aṅgiras .	...
4133	954	1089	438	206-07	1031-32	5 Prajāpati .	7 Śrīmukha .	2 Vaiśākha .
4134	955	1090	439	207-08	*1032-33	6 Aṅgiras .	8 Bhāva . .	...
4135	956	1091	440	208-09	1033-34	7 Śrīmukha .	9 Yuvan . .	10 Pausha .
4136	957	1092	441	209-10	1034-35	8 Bhāva . .	10 Dhātṛi . .	...
4137	958	1093	442	210-11	1035-36	9 Yuvan . .	11 Īśvara . .	...
4138	959	1094	443	211-12	*1036-37	10 Dhātṛi . .	12 Bahudhānya .	7 Āśvina .
4139	960	1095	444	212-13	1037-38	11 Īśvara . .	13 Pramādin .	...
4140	961	1096	445	213-14	1038-39	12 Bahudhānya .	14 Vikrama .	...
4141	962	1097	446	214-15	1039-40	13 Pramādin .	15 Vṛisha . .	3 Jyēshṭha .
4142	963	1098	447	215-16	*1040-41	14 Vikrama .	16 Chitrabhānu .	...
4143	964	1099	448	216-17	1041-42	15 Vṛisha . .	17 Subhānu .	12 Phālguna
4144	965	1100	449	217-18	1042-43	16 Chitrabhānu .	18 Tāraṇa . .	...
4145	966	1101	450	218-19	1043-44	17 Subhānu .	19 Pārthiva .	

‡ By the "*Indian Calendar*" 3 Jyēshṭha was intercalated.

LXXVI—*Contd.*

1 Ārya Siddhānta, mean system.

COMMENCEMENT OF THE								
MEAN SOLAR YEAR.					MEAN LUNI-SOLAR YEAR (MEAN SUNRISE OF CIVIL DAY ON WHICH CHAITRA ŚUKLA 1 ENDS).			Kali year.
Day and month, A.D.	Week-day.	Time of mean Mésha-saṁkrānti.			Day and month, A.D.	Week-day.	*a* (here = *t*, the index of the tithi).	
13	14	17			19	20	23	1
		H.	M.	S.				
25 Mar. (84) . .	4 Wed. .	18	20	0	9 Mar. (68) .	2 Mon. .	3·5953	4121
25 Mar. (85) . .	6 Fri. .	0	32	30	27 Feb. (58) .	0 Sat. .	217·8106	4122
25 Mar. (84) . .	0 Sat. .	6	45	0	17 Mar. (76) .	6 Fri. .	252·5502	4123
25 Mar. (84) . .	1 Sun. .	12	57	30	6 Mar. (65) .	3 Tues. .	128·2336	4124
25 Mar. (84) . .	2 Mon. .	19	10	0	25 Mar. (84) .	2 Mon. .	162·8732	4125
25 Mar. (85) . .	4 Wed. .	1	22	30	13 Mar. (73) .	6 Fri. .	38·5566	4126
25 Mar. (84) . .	5 Thur. .	7	35	0	3 Mar. (62) .	4 Wed. .	252·8719	4127
25 Mar. (84) . .	6 Fri. .	13	47	30	22 Mar. (81) .	3 Tues. .	287·5115	4128
25 Mar. (84) . .	0 Sat. .	20	0	0	11 Mar. (70) .	0 Sat. .	163·1948	4129
25 Mar. (85) . .	2 Mon. .	2	12	30	28 Feb. (59) .	4 Wed. .	38·8782	4130
25 Mar. (84) . .	3 Tues. .	8	25	0	18 Mar. (77) .	3 Tues. .	73·5179	4131
25 Mar. (84) . .	4 Wed. .	14	37	30	8 Mar. (67) .	1 Sun. .	287·8331	4132
25 Mar. (84) . .	5 Thur. .	20	50	0	25 Feb. (56) .	5 Thur. .	163·5165	4133
25 Mar. (85) . .	0 Sat. .	3	2	30	15 Mar. (75) .	4 Wed. .	198·1561	4134
25 Mar. (84) . .	1 Sun. .	9	15	0	4 Mar. (63) .	1 Sun. .	73·8395	4135
25 Mar. (84) . .	2 Mon. .	15	27	30	23 Mar. (82) .	0 Sat. .	108·4791	4136
25 Mar. (84) . .	3 Tues. .	21	40	0	13 Mar. (72) .	5 Thur. .	322·7944	4137
25 Mar. (85) . .	5 Thur. .	3	52	30	1 Mar. (64) .	2 Mon. .	198·4778	4138
25 Mar. (84) . .	6 Fri. .	10	5	0	20 Mar. (79) .	1 Sun. .	233·1174	4139
25 Mar. (84) . .	0 Sat. .	16	17	30	9 Mar. (68) .	5 Thur. .	108·8008	4140
25 Mar. (84) . .	1 Sun. .	22	30	0	27 Feb. (58) .	3 Tues. .	323·1161	4141
25 Mar. (85) . .	3 Tues. .	4	42	30	16 Mar. (76) .	1 Sun. .	19·1238	4142
25 Mar. (84) . .	4 Wed. .	10	55	0	6 Mar. (65) .	6 Fri. .	233·4391	4143
25 Mar. (84) .	5 Thur. .	17	7	30	25 Mar. (84) .	5 Thur. .	268·0787	4144
25 Mar. (84) .	6 Fri. .	23	20	0	14 Mar. (73) .	2 Mon. .	143·7621	4145

TABLE

CONCURRENT YEAR.								Mean Intercalated (adhika) lunar month.
Kali.	Śaka.	Chaitrādi Vikrama.	Mēshādi solar year in Bengal.	Kollam.	A.D.	Jovian Saṁvatsara. Southern system.	Jovian Saṁvatsara. Northern system.	
1	2	3	3a	4	5	6	7	8a
4146	967	1102	451	219-20	*1044-45	18 Tāraṇa . .	20 Vyaya . .	8 Kārttika .
4147	968	1103	452	220-21	1045-46	19 Pārthiva .	21 Sarvajit .	...
4148	969	1104	453	221-22	1046-47	20 Vyaya . .	22 Sarvadhārin .	...
4149	970	1105	454	222-23	1047-48	21 Sarvajit .	23 Virōdhin .	5 Śrāvaṇa .
4150	971	1106	455	223-24	*1048-49	22 Sarvadhārin .	24 Vikṛita . .	...
4151	972	1107	456	224-25	1049-50	23 Virōdhin .	25 Khara . .	...
4152	973	1108	457	225-26	1050-51	24 Vikṛita . .	26 Nandana .	1 Chaitra .
4153	974	1109	458	226-27	1051-52	25 Khara . .	27 Vijaya . .	...
4154	975	1110	459	227-28	*1052-53	26 Nandana .	28 Jaya . .	10 Pausha .
4155	976	1111	460	228-29	1053-54	27 Vijaya . .	29 Manmatha .	...
4156	977	1112	461	229-30	1054-55	28 Jaya . .	30 Durmukha .	...
4157	978	1113	462	230-31	1055-56	29 Manmatha .	31 Hēmalamba .	7 Āśvina† .
4158	979	1114	463	231-32	*1056-57	30 Durmukha .	32 Vilamba .	...
4159	980	1115	464	232-33	1057-58	31 Hēmalamba .	33 Vikārin .	...
4160	981	1116	465	233-34	1058-59	32 Vilamba .	34 Śārvarin .	3 Jyēshṭha .
4161	982	1117	466	234-35	1059-60	33 Vikārin .	35 Plava . .	...
4162	983	1118	467	235-36	*1060-61	34 Śārvarin .	36 Śubhakṛit .	12 Phālguna .
4163	984	1119	468	236-37	1061-62	35 Plava . .	37 Śōbhana .	...
4164	985	1120	469	237-38	1062-63	36 Śubhakṛit .	38 Krōdhin .	...
4165	986	1121	470	238-39	1063-64	37 Śōbhana .	39 Viśvāvasu .	8 Kārttika .
4166	987	1122	471	239-40	*1064-65	38 Krōdhin .	40 Parābhava .	...
4167	988	1123	472	240-41	1065-66	39 Viśvāvasu .	41 Plavaṅga .	...
4168	989	1124	473	241-42	1066-67	40 Parābhava .	42 Kīlaka . .	5 Śrāvaṇa .
4169	990	1125	474	242-43	1067-68	41 Plavaṅga .	43 Saumya .	...
4170	991	1126	475	243-44	*1068-69	42 Kīlaka . .	44 Sādhāraṇa	...

† By the "*Indian Calendar*" 6 Bhādrapada was the intercalated month

LXXVI—*Contd.*

1 Ārya Siddhānta, mean system.

COMMENCEMENT OF THE									Kali year.
Mean solar year.					Mean luni-solar year (mean sunrise of civil day on which Chaitra Śukla 1 ends).				
Day and month, A.D.	Week-day.	Time of mean Mēsha-saṁkrānti.			Day and month, A.D.	Week-day.	a (here $=t$, the index of the tithi).		
13	14	17			19	20	23		1
		H.	M.	S.					
25 Mar. (85) . .	1 Sun. .	5	32	30	2 Mar. (62) .	6 Fri. .	19·4454		4146
25 Mar. (84) . .	2 Mon. .	11	45	0	21 Mar. (80) .	5 Thur. .	54·0850		4147
25 Mar. (84) . .	3 Tues. .	17	57	30	11 Mar. (70) .	3 Tues. .	268·4003		4148
26 Mar. (85) . .	5 Thur. .	0	10	0	28 Feb. (59) .	0 Sat. .	144·0838		4149
25 Mar. (85) . .	6 Fri. .	6	22	30	18 Mar. (78) .	6 Fri. .	178·7233		4150
25 Mar. (84) . .	0 Sat. .	12	35	0	7 Mar. (66) .	3 Tues. .	54·4067		4151
25 Mar. (84) . .	1 Sun. .	18	47	30	25 Feb. (56) .	1 Sun. .	268·7219		4152
26 Mar. (85) . .	3 Tues. .	1	0	0	16 Mar. (75) .	0 Sat. .	303·3615		4153
25 Mar. (85) . .	4 Wed. .	7	12	30	4 Mar. (64) .	4 Wed. .	179·0449		4154
25 Mar. (84) . .	5 Thur. .	13	25	0	23 Mar. (82) .	3 Tues. .	213·6845		4155
25 Mar. (84) . .	6 Fri. .	19	37	30	12 Mar. (71) .	0 Sat. .	89·3679		4156
26 Mar. (85) . .	1 Sun. .	1	50	0	2 Mar. (61) .	5 Thur. .	303·6832		4157
25 Mar. (85) . .	2 Mon. .	8	2	30	19 Mar. (79) .	3 Tues. .	9999·6909	§	4158
25 Mar. (84) . .	3 Tues. .	14	15	6	9 Mar. (68) .	1 Sun. .	214·0062		4159
25 Mar. (84) . .	4 Wed. .	20	27	30	26 Feb. (57) .	5 Thur. .	89·6896		4160
26 Mar. (85) . .	6 Fri. .	2	40	0	17 Mar. (76) .	4 Wed. .	124·3292		4161
25 Mar. (85) . .	0 Sat. .	8	52	30	5 Mar. (65) .	1 Sun. .	0·0126		4162
25 Mar. (84) . .	1 Sun. .	15	5	0	24 Mar. (83) .	0 Sat. .	34·6522		4163
25 Mar. (84) . .	2 Mon. .	21	17	30	14 Mar. (73) .	5 Thur. .	218·9675		4164
26 Mar. (85) . .	4 Wed. .	3	30	0	3 Mar. (62) .	2 Mon. .	124·6508		4165
25 Mar. (85) . .	5 Thur. .	9	42	30	21 Mar. (81) .	1 Sun. .	159·2905		4166
25 Mar. (84) . .	6 Fri. .	15	55	0	10 Mar. (69) .	5 Thur. .	34·9739		4167
25 Mar. (84) . .	0 Sat. .	22	7	30	28 Feb. (59) .	3 Tues. .	249·2892		4168
26 Mar. (85) . .	2 Mon. .	4	20	0	19 Mar. (78) .	2 Mon. .	283·9288		4169
25 Mar. (85) . .	3 Tues. .	10	32	30	7 Mar. (67) .	6 Fri. .	159·6122		4170

§ As a mean tithi Chaitra Śukla 1 was expunged. The civil day corresponding to it, *i.e.*, the first day of the luni-solar year was as given in cols. 19, 20.

TABLE

CONCURRENT YEAR								
						Jovian Saṁvatsara.		Mean Intercalated (adhika) lunar month.
Kali.	Śaka.	Chaitrādi Vikrama.	Mēshādi solar year in Bengal.	Kollam.	A.D.	Southern system.	Northern system.	
1	2	3	3a	4	5	6	7	8a
4171	992	1127	476	244-45	1069-70	43 Saumya	45 Virōdhakṛit	1 Chaitra
4172	993	1128	477	245-46	1070-71	44 Sādhāraṇa	46 Paridhāvin	...
4173	994	1129	478	246-47	1071-72	45 Virōdhakṛit	47 Pramādin	10 Pausha
4174	995	1130	479	247-48	*1072-73	46 Paridhāvin	48 Ānanda	...
4175	996	1131	480	248-49	1073-74	47 Pramādin	49 Rakshasa	...
4176	997	1132	481	249-50	1074-75	48 Ānanda	50 Anala	6 Bhādrapada
4177	998	1133	482	250-51	1075-76	49 Rākshasa	51 Piṅgala †	...
4178	999	1134	483	251-52	*1076-77	50 Anala	53 *Siddhārthin*	...
4179	1000	1135	484	252-53	1077-78	51 Piṅgala	54 *Raudra*	3 Jyēshṭha
4180	1001	1136	485	253-54	1078-79	52 Kālayukta	55 *Durmati*	...
4181	1002	1137	486	254-55	1079-80	53 Siddhārthin	56 *Dundubhi*	11 Māgha
4182	1003	1138	487	255-56	*1080-81	54 Raudra	57 Rudhirōdgārin	...
4183	1004	1139	488	256-57	1081-82	55 Durmati	58 Raktāksha	...
4184	1005	1140	489	257-58	1082-83	56 Dundubhi	59 Krōdhana	8 Kārttika
4185	1006	1141	490	258-59	1083-84	57 Rudhirōdgārin	60 Kshaya	...
4186	1007	1142	491	259-60	*1084-85	58 Raktāksha	1 Prabhava	...
4187	1008	1143	492	260-61	1085-86	59 Krōdhana	2 Vibhava	4 Āshāḍha
4188	1009	1144	493	261-62	1086-87	60 Kshaya	3 Śukla	...
4189	1010	1145	494	262-63	1087-88	1 Prabhava	4 Pramōda	...
4190	1011	1146	495	263-64	*1088-89	2 Vibhava	5 Prajāpati	1 Chaitra
4191	1012	1147	496	264-65	1089-90	3 Śukla	6 Aṅgiras	...
4192	1013	1148	497	265-66	1090-91	4 Pramōda	7 Śrīmukha	9 Mārgaśira
4193	1014	1149	498	266-67	1091-92	5 Prajāpati	8 Bhava	...
4194	1015	1150	499	267-68	*1092-93	6 Aṅgiras	9 Yuvan	...
4195	1016	1151	500	268-69	1093-94	7 Śrīmukha	10 Dhātri	6 Bhādrapada

† 52 Kālayukta was suppressed in the north.

LXXVI—*Contd.*

1 Ārya Siddhānta, mean system.

COMMENCEMENT OF THE							
MEAN SOLAR YEAR.			MEAN LUNI-SOLAR YEAR (MEAN SUNRISE OF CIVIL DAY ON WHICH CHAITRA ŚUKLA 1 ENDS).				Kali year.
Day and month, A.D.	Week-day.	Time of mean Mēsha-saṁkrānti.		Day and month, A.D.	Week-day.	a (here $=t$, the index of the tithi).	
13	14	17		19	20	23	1
		H. M. S.					
25 Mar. (84) . .	4 Wed. .	16 45 0		24 Feb. (55) .	3 Tues. .	35·2955	4171
25 Mar. (84) . .	5 Thur. .	22 57 30		15 Mar. (74) .	2 Mon. .	69·9351	4172
26 Mar. (85) .	0 Sat. .	5 10 0		5 Mar. (64) .	0 Sat. .	284·2504	4173
25 Mar. (85) . .	1 Sun. .	11 22 30		23 Mar. (83) .	6 Fri. .	318·8901	4174
25 Mar. (84) . .	2 Mon. .	17 35 0		12 Mar. (71) .	3 Tues. .	194·5734	4175
25 Mar. (84) . .	3 Tues. .	23 47 30		1 Mar. (60) .	0 Sat. .	70·2568	4176
26 Mar. (85) . .	5 Thur. .	6 0 0		20 Mar. (79) .	6 Fri. .	104·8964	4177
25 Mar. (85) . .	6 Fri. .	12 12 30		9 Mar. (69) .	4 Wed. .	319·2116	4178
25 Mar. (84) . .	0 Sat. .	18 25 0		26 Feb. (57) .	1 Sun. .	194·8950	4179
26 Mar. (85) . .	2 Mon. .	0 37 30		17 Mar. (76) .	0 Sat. .	229·5347	4180
26 Mar. (85) . .	3 Tues. .	6 50 0		6 Mar. (65) .	4 Wed. .	105·2180	4181
25 Mar. (85) . .	4 Wed. .	13 2 30		24 Mar. (84) .	3 Tues. .	139·8576	4182
25 Mar. (84) . .	5 Thur. .	19 15 0		13 Mar. (72) .	0 Sat. .	15·5410	4183
26 Mar. (85) . .	0 Sat. .	1 27 30		3 Mar. (62) .	5 Thur. .	229·8563	4184
26 Mar. (85) . .	1 Sun. .	7 40 0		22 Mar. (81) .	4 Wed. .	264·4959	4185
25 Mar. (85) . .	2 Mon. .	13 52 30		10 Mar. (70) .	1 Sun. .	140·1793	4186
25 Mar. (84) . .	3 Tues. .	20 5 0		27 Feb. (58) .	5 Thur. .	15·8627	4187
26 Mar. (85) . .	5 Thur. .	2 17 30		18 Mar. (77) .	4 Wed. .	50·5023	4188
26 Mar. (85) . .	6 Fri. .	8 30 0		8 Mar. (67) .	2 Mon. .	264·8176	4189
25 Mar. (85) . .	0 Sat. .	14 42 30		25 Feb. (56) .	6 Fri. .	140·5009	4190
25 Mar. (84) . .	1 Sun. .	20 55 0		15 Mar. (74) .	5 Thur.	175·1465	4191
26 Mar. (85) . .	3 Tues. .	3 7 30		4 Mar. (63) .	2 Mon.	50·8239	4192
26 Mar. (85) . .	4 Wed. .	9 20 0		23 Mar. (82) .	1 Sun.	85·4636	4193
25 Mar. (85) . .	5 Thur. .	15 32 30		12 Mar. (72) .	6 Fri.	229·7788	4194
25 Mar. (84) . .	6 Fri. .	21 45 0		1 Mar. (60) .	3 Tues.	175·4622	4195

TABLE

CONCURRENT YEAR.								Mean Intercalated (adhika) lunar month.
Kali.	Śaka.	Chaitrādi Vikrama.	Mēshādi solar year in Bengal.	Kollam.	A.D.	Jovian Saṁvatsara. Southern system.	Jovian Saṁvatsara. Northern system.	
1	2	3	3a	4	5	6	7	8a
4196	1017	1152	501	269-70	1094-95	8 Bhāva . .	11 Īśvara . .	...
4197	1018	1153	502	270-71	1095-96	9 Yuvan . .	12 Bahudhānya .	...
4198	1019	1154	503	271-72	*1096-97	10 Dhātṛi . .	13 Pramādin .	3 Jyēshṭha † .
4199	1020	1155	504	272-73	1097-98	11 Īśvara . .	14 Vikrama .	...
4200	1021	1156	505	273-74	1098-99	12 Bahudhānya .	15 Vṛisha . .	11 Māgha
4201	1022	1157	506	274-75	1099-00	13 Pramādin .	16 Chitrabhānu .	...
4202	1023	1158	507	275-76	*1100-01	14 Vikrama .	17 Subhānu .	...
4203	1024	1159	508	276-77	1101-02	15 Vṛisha . .	18 Tāraṇa .	8 Kārttika .
4204	1025	1160	509	277-78	1102-03	16 Chitrabhānu .	19 Pārthiva .	...
4205	1026	1161	510	278-79	1103-04	17 Subhānu .	20 Vyaya . .	...
4206	1027	1162	511	279-80	*1104-05	18 Tāraṇa . .	21 Sarvajit .	4 Āshāḍha .
4207	1028	1163	512	280-81	1105-06	19 Pārthiva .	22 Sarvadhārin .	...
4208	1029	1164	513	281-82	1106-07	20 Vyaya . .	23 Virōdhin .	...
4209	1030	1165	514	282-83	1107-08	21 Sarvajit .	24 Vikṛita . .	1 Chaitra
4210	1031	1166	515	283-84	*1108-09	22 Sarvadhārin .	25 Khara . .	...
4211	1032	1167	516	284-85	1109-10	23 Virōdhin .	26 Nandana .	9 Mārgaśira
4212	1033	1168	517	285-86	1110-11	24 Vikṛita . .	27 Vijaya . .	...
4213	1034	1169	518	286-87	1111-12	25 Khara . .	28 Jaya . .	...
4214	1035	1170	519	287-88	*1112-13	26 Nandana .	29 Manmatha .	6 Bhādrapada
4215	1036	1171	520	288-89	1113-14	27 Vijaya . .	30 Durmukha .	...
4216	1037	1172	521	289-90	1114-15	28 Jaya . .	31 Hēmalamba	...
4217	1038	1173	522	290-91	1115-16	29 Manmatha .	32 Vilamba .	2 Vaiśākha
4218	1039	1174	523	291-92	*1116-17	30 Durmukha .	33 Vikārin .	...
4219	1040	1175	524	292-93	1117-18	31 Hēmalamba .	34 Śārvarin .	11 Māgha
4220	1041	1176	525	293-94	1118-19	32 Vilamba .	35 Plava . .	...

† By the "*Indian Calendar*" 2 Vaiśākha was intercalated.

LXXVI—*Contd.*

1 Ārya Siddhānta, mean system.

COMMENCEMENT OF THE								
MEAN SOLAR YEAR.					MEAN LUNI-SOLAR YEAR (MEAN SUNRISE OF CIVIL DAY ON WHICH CHAITRA ŚUKLA 1 ENDS).			Kali year.
Day and month, A.D.	Week-day.	Time of mean Mēsha-saṁkrānti.			Day and month, A.D.	Week-day.	*a* (here=*t*, the index of the tithi).	
13	14	17			19	20	23	1
		H.	M.	S.				
26 Mar. (85) . .	1 Sun. .	3	57	30	20 Mar. (79) .	2 Mon. .	210·1018	4196
26 Mar. (85) . .	2 Mon. .	10	10	0	9 Mar. (68) .	6 Fri. .	85·7852	4197
25 Mar. (85) . .	3 Tues. .	16	22	30	27 Feb. (58) .	4 Wed. .	300·1005	4198
25 Mar. (84) . .	4 Wed. .	22	35	0	16 Mar. (75) .	2 Mon. .	9996·1082†	4199
26 Mar. (85) . .	6 Fri. .	4	47	30	6 Mar. (65) .	0 Sat. .	210·4235	4200
26 Mar. (85) . .	0 Sat. .	11	0	0	25 Mar. (84) .	6 Fri. .	245·0630	4201
25 Mar. (85) . .	1 Sun. .	17	12	30	13 Mar. (73) .	3 Tues. .	120·7464	4202
25 Mar. (84) . .	2 Mon. .	23	25	0	2 Mar. (61) .	0 Sat. .	9996·4298†	4203
26 Mar. (85) . .	4 Wed. .	5	37	30	21 Mar. (80) .	6 Fri. .	31·0694	4204
26 Mar. (85) . .	5 Thur. .	11	50	0	11 Mar. (70) .	4 Wed. .	245·3847	4205
25 Mar. (85) . .	6 Fri. .	18	2	30	28 Feb. (59) .	1 Sun. .	121·0681	4206
26 Mar. (85) . .	1 Sun. .	0	15	0	18 Mar. (77) .	0 Sat. .	155·7077	4207
26 Mar. (85) . .	2 Mon. .	6	27	30	7 Mar. (66) .	4 Wed. .	31·3911	4208
26 Mar. (85) . .	3 Tues. .	12	40	0	25 Feb. (56) .	2 Mon. .	245·7063	4209
25 Mar. (85) . .	4 Wed. .	18	52	30	15 Mar. (75) .	1 Sun. .	280·3460	4210
26 Mar. (85) . .	6 Fri. .	1	5	0	4 Mar. (63) .	5 Thur. .	156·0293	4211
26 Mar. (85) . .	0 Sat. .	7	17	30	23 Mar. (82) .	4 Wed. .	190·6690	4212
26 Mar. (85) . .	1 Sun. .	13	30	0	12 Mar. (71) .	1 Sun. .	66·3524	4213
25 Mar. (85) . .	2 Mon. .	19	42	30	1 Mar. (61) .	6 Fri. .	280·6676	4214
26 Mar. (85) . .	4 Wed. .	1	55	0	20 Mar. (79) .	5 Thur. .	315·3072	4215
26 Mar. (85) . .	5 Thur. .	8	7	30	9 Mar. (68) .	2 Mon. .	190·9905	4216
26 Mar. (85) . .	6 Fri. .	14	20	0	26 Feb. (57) .	6 Fri. .	66·6740	4217
25 Mar. (85) . .	0 Sat. .	20	32	30	16 Mar. (76) .	5 Thur. .	101·3136	4218
26 Mar. (85) . .	2 Mon. .	2	45	0	6 Mar. (65) .	3 Tues. .	315·6288	4219
26 Mar. (85) . .	3 Tues. .	8	57	30	24 Mar. (83) .	1 Sun. .	11·6365	4220

† As a mean tithi Chaitra śukla 1 was expunged. The civil day corresponding to it, *i.e.*, the first day of the luni-solar year, was as given in cols. 19, 20.

TABLE

CONCURRENT YEAR.								Mean Intercalated (adhika) lunar month.
						Jovian Saṁvatsara.		
Kali.	Śaka.	Chaitrādi Vikrama.	Mēshādi solar year in Bengal.	Kollam.	A.D.	Southern system.	Northern system.	
1	2	3	3a	4	5	6	7	8a
4221	1042	1177	526	294-95	1119-20	33 Vikārin .	36 Śubhakṛit .	...
4222	1043	1178	527	295-96	*1120-21	34 Śārvarin .	37 Śōbhana .	7 Āśvina
4223	1044	1179	528	296-97	1121-22	35 Plava . .	38 Krōdhin .	...
4224	1045	1180	529	297-98	1122-23	36 Śubhakṛit .	39 Viśvāvasu .	...
4225	1046	1181	530	298-99	1123-24	37 Śōbhana .	40 Parābhava .	4 Āshāḍha .
4226	1047	1182	531	299-00	*1124-25	38 Krōdhin .	41 Plavaṅga .	...
4227	1048	1183	532	300-01	1125-26	39 Viśvāvasu .	42 Kīlaka . .	12 Phālguna .
4228	1049	1184	533	301-02	1126-27	40 Parābhava .	43 Saumya .	...
4229	1050	1185	534	302-03	1127-28	41 Plavaṅga .	44 Sādhāraṇa .	...
4230	1051	1186	535	303-04	*1128-29	42 Kīlaka . .	45 Virōdhakṛit .	9 Mārgaśira ...
4231	1052	1187	536	304-05	1129-30	43 Saumya .	46 Paridhāvin .	...
4232	1053	1188	537	305-06	1130-31	44 Sādhāraṇa .	47 Pramādin .	...
4233	1054	1189	538	306-07	1131-32	45 Virōdhakṛit .	48 Ānanda .	6 Bhādrapada
4234	1055	1190	539	307-08	*1132-33	46 Paridhāvin .	49 Rākshasa .	...
4235	1056	1191	540	308-09	1133-34	47 Pramādin .	50 Anala . .	...
4236	1057	1192	541	309-10	1134-35	48 Ānanda .	51 Piṅgala .	2 Vaiśākha .
4237	1058	1193	542	310-11	1135-36	49 Rākshasa .	52 Kālayukta .	...
4238	1059	1194	543	311-12	*1136-37	50 Anala .	53 Siddhārthin .	11 Māgha .
4239	1060	1195	544	312-13	1137-38	51 Piṅgala .	54 Raudra .	...
4240	1061	1196	545	313-14	1138-39	52 Kālayukta .	55 Durmati .	...
4241	1062	1197	546	314-15	1139-40	53 Siddhārthin .	56 Dundubhi .	7 Āśvina
4242	1063	1198	547	315-16	*1140-41	54 Raudra .	57 Rudhirōdgārin	...
4243	1064	1199	548	316-17	1141-42	55 Durmati .	58 Raktākshi .	...
4244	1065	1200	549	317-18	1142-43	56 Dundubhi .	59 Krōdhana .	4 Āshāḍha .
4245	1066	1201	550	318-19	1143-44	57 Rudhirōdgārin	60 Kshaya	...

LXXVI—*Contd.* **1 Ārya Siddhānta, mean system.**

COMMENCEMENT OF THE									
MEAN SOLAR YEAR.					MEAN LUNI-SOLAR YEAR (MEAN SUNRISE OF CIVIL DAY ON WHICH CHAITRA ŚUKLA 1 ENDS).			Kali year.
Day and month, A.D.	Week-day.	Time of mean Mēsha-saṁkrānti.			Day and month, A.D.	Week-day.	*a* (here = *t*, the index of the tithi).	
13	14	17			19	20	23	1
		H.	M.	S.				
26 Mar. (85) . .	4 Wed. .	15	10	0	14 Mar. (73) .	6 Fri. .	225·9518	4221
25 Mar. (85) . .	5 Thur. .	21	22	30	2 Mar. (62) .	3 Tues. .	101·6352	4222
26 Mar. (85) . .	0 Sat. .	3	35	0	21 Mar. (80) .	2 Mon. .	136·2748	4223
26 Mar. (85) . .	1 Sun. .	9	47	30	10 Mar. (69) .	6 Fri. .	11·9582	4224
26 Mar. (85) . .	2 Mon. .	16	0	0	28 Feb. (59) .	4 Wed. .	226·2735	4225
25 Mar. (85) . .	3 Tues. .	22	12	30	18 Mar. (78) .	3 Tues. .	260·9131	4226
26 Mar. (85) . .	5 Thur. .	4	25	0	7 Mar. (66) .	0 Sat. .	136·5965	4227
26 Mar. (85) . .	6 Fri. .	10	37	30	26 Mar. (85) .	6 Fri. .	171·2360	4228
26 Mar. (85) . .	0 Sat. .	16	50	0	15 Mar. (74) .	3 Tues. .	46·9195	4229
25 Mar. (85) . .	1 Sun. .	23	2	30	4 Mar. (64) .	1 Sun. .	261·2348	4230
26 Mar. (85) . .	3 Tues. .	5	15	0	23 Mar. (82) .	0 Sat. .	295·8744	4231
26 Mar. (85) . .	4 Wed. .	11	27	30	12 Mar. (71) .	4 Wed. .	171·5578	4232
26 Mar. (85) . .	5 Thur. .	17	40	0	1 Mar. (60) .	1 Sun. .	47·2411	4233
25 Mar. (85) . .	6 Fri. .	23	52	30	19 Mar. (79) .	0 Sat. .	81·8807	4234
26 Mar. (85) . .	1 Sun. .	6	5	0	9 Mar. (68) .	5 Thur. .	296·1960	4235
26 Mar. (85) . .	2 Mon. .	12	17	30	26 Feb. (57) .	2 Mon. .	171·8794	4236
26 Mar. (85) . .	3 Tues. .	18	30	0	17 Mar. (76) .	1 Sun. .	206·5190	4237
26 Mar. (86) . .	5 Thur. .	0	42	30	5 Mar. (65) .	5 Thur. .	82·2024	4238
26 Mar. (85) . .	6 Fri. .	6	55	0	24 Mar. (83) .	4 Wed. .	116·8420	4239
26 Mar. (85) . .	0 Sat. .	13	7	30	14 Mar. (73) .	2 Mon. .	331·1573	4240
26 Mar. (85) . .	1 Sun. .	19	20	0	3 Mar. (62) .	6 Fri. .	206·8407	4241
26 Mar. (86) . .	3 Tues. .	1	32	30	21 Mar. (81) .	5 Thur. .	241·4803	4242
26 Mar. (85) . .	4 Wed. .	7	45	0	19 Mar. (69) .	2 Mon. .	117·1637	4243
26 Mar. (85) . .	5 Thur. .	13	57	30	28 Feb. (59) .	0 Sat. .	331·4790	4244
26 Mar. (85) . .	6 Fri. .	20	10	0	18 Mar. (77) .	5 Thur. .	27·4867	4245

TABLE

CONCURRENT YEAR.								Mean Intercalated (adhika) lunar month.
Kali.	Saka.	Chaitrâdi Vikrama.	Mêshâdi solar year in Bengal.	Kollam.	A.D.	Jovian Saṁvatsara. Southern system.	Jovian Saṁvatsara. Northern system.	
1	2	3	3a	4	5	6	7	8a
4246	1067	1202	551	319-20	*1144-45	58 Raktāksha .	1 Prabhava .	12 Phālguna .
4247	1068	1203	552	320-21	1145-46	59 Krōdhana .	2 Vibhava .	...
4248	1069	1204	553	321-22	1146-47	60 Kshaya .	3 Śukla . .	...
4249	1070	1205	554	322-23	1147-48	1 Prabhava .	4 Pramōda .	9 Mārgaśira .
4250	1071	1206	555	323-24	*1148-49	2 Vibhava .	5 Prajāpati .	...
4251	1072	1207	556	324-25	1149-50	3 Śukla . .	6 Aṅgiras .	...
4252	1073	1208	557	325-26	1150-51	4 Pramōda .	7 Śrīmukha .	5 Srāvaṇa .
4253	1074	1209	558	326-27	1151-52	5 Prajāpati .	8 Bhāva . .	...
4254	1075	1210	559	327-28	*1152-53	6 Aṅgiras .	9 Yuvan . .	...
4255	1076	1211	560	328-29	1153-54	7 Śrīmukha .	10 Dhātṛi . .	2 Vaiśākha .
4256	1077	1212	561	329-30	1154-55	8 Bhāva . .	11 Īśvara . .	...
4257	1078	1213	562	330-31	1155-56	9 Yuvan . .	12 Bahudhānya .	10 Pausha .
4258	1079	1214	563	331-32	*1156-57	10 Dhātṛi . .	13 Pramādin .	...
4259	1080	1215	564	332-33	1157-58	11 Iśvara . .	14 Vikrama .	...
4260	1081	1216	565	333-34	1158-59	12 Bahudhānya .	15 Vṛisha . .	7 Āśvina .
4261	1082	1217	566	334-35	1159-60	13 Pramādin .	16 Chitrabhānu .	...
4262	1083	1218	567	335-36	*1160-61	14 Vikrama .	17 Subhānu† .	...
4263	1084	1219	568	336-37	1161-62	15 Vṛisha . .	19 *Pārthiva* .	3 Jyēshtha .
4264	1085	1220	569	337-38	1162-63	16 Chitrabhānu .	20 *Vyaya* . .	...
4265	1086	1221	570	338-39	1163-64	17 Subhānu .	21 *Sarvajit* .	12 Phālguna .
4266	1087	1222	571	339-40	*1164-65	18 Tāraṇa . .	22 *Sarvadhārin* .	...
4267	1088	1223	572	340-41	1165-66	19 Pārthiva .	23 Virōdhin .	...
4268	1089	1224	573	341-42	1166-67	20 Vyaya . .	24 Vikṛita . .	8 Kārttika .
4269	1090	1225	574	342-43	1167-68	21 Sarvajit .	25 Khara . .	...
4270	1091	1226	575	343-44	*1168-69	22 Sarvadhārin .	26 Nandana .	...

† 18 Tāraṇa was suppressed in the north.

LXXVI—*Contd.*

1 Ārya Siddhānta, mean system.

COMMENCEMENT OF THE									
MEAN SOLAR YEAR.					MEAN LUNI-SOLAR YEAR (MEAN SUNRISE OF CIVIL DAY ON WHICH CHAITRA ŚUKLA 1 ENDS).			Kali year.
Day and month, A.D.	Week-day.	Time of mean Mesha-samkrānti.			Day and month, A.D.	Week-day.	*a* (here = t, the index of the tithi).	
13	14	17			19	20	23	1
		H.	M.	S.				
26 Mar. (86) . .	1 Sun. .	2	22	30	7 Mar. (67) .	3 Tues. .	241·8019	4246
26 Mar. (85) . .	2 Mon. .	8	35	0	26 Mar. (85) .	2 Mon. .	276·4415	4247
26 Mar. (85) . .	3 Tues. .	14	47	30	15 Mar. (74) .	6 Fri. .	152·1249	4248
26 Mar. (85) . .	4 Wed. .	21	0	0	4 Mar. (63) .	3 Tues. .	27·8084	4249
26 Mar. (86) . .	6 Fri. .	3	12	30	22 Mar. (82) .	2 Mon. .	62·4479	4250
26 Mar. (85) . .	0 Sat. .	9	25	0	12 Mar. (71) .	0 Sat. .	276·7631	4251
26 Mar. (85) . .	1 Sun. .	15	37	30	1 Mar. (60) .	4 Wed. .	152·4465	4252
26 Mar. (85) . .	2 Mon. .	21	50	0	20 Mar. (79) .	3 Tues. .	187·0861	4253
26 Mar. (86) . .	4 Wed. .	4	2	30	8 Mar. (68) .	0 Sat. .	62·7695	4254
26 Mar. (85) . .	5 Thur. .	10	15	0	26 Feb. (57) .	5 Thur. .	277·0848	4255
26 Mar. (85) . .	6 Fri. .	16	27	30	17 Mar. (76) .	4 Wed. .	311·7245	4256
26 Mar. (85) . .	0 Sat. .	22	40	0	6 Mar. (65) .	1 Sun. .	187·4078	4257
26 Mar. (86) . .	2 Mon. .	4	52	30	24 Mar. (84) .	0 Sat. .	222·0474	4258
26 Mar. (85) . .	3 Tues. .	11	5	0	13 Mar. (72) .	4 Wed. .	98·1308	4259
26 Mar. (85) . .	4 Wed. .	17	17	30	3 Mar. (62) .	2 Mon. .	312·0461	4260
26 Mar. (85) . .	5 Thur. .	23	30	0	21 Mar. (80) .	0 Sat. .	8·0538	4261
26 Mar. (86) . .	0 Sat. .	5	42	30	10 Mar. (70) .	5 Thur. .	222·3691	4262
26 Mar. (85) . .	1 Sun. .	11	55	0	27 Feb. (58) .	2 Mon. .	98·4525	4263
26 Mar. (85) . .	2 Mon. .	18	7	30	18 Mar. (77) .	1 Sun. .	132·6822	4264
27 Mar. (86) . .	4 Wed. .	0	20	0	7 Mar. (66) .	5 Thur. .	8·3755	4265
26 Mar. (86) . .	5 Thur. .	6	32	30	25 Mar. (85) .	4 Wed. .	43·0151	4266
26 Mar. (85) . .	6 Fri. .	12	45	0	15 Mar. (74) .	2 Mon. .	257·3504	4267
26 Mar. (85) . .	0 Sat. .	18	57	30	4 Mar. (63) .	6 Fri. .	133·0138	4268
27 Mar. (86) . .	2 Mon. .	1	10	0	23 Mar. (82) .	5 Thur. .	167·6454	4269
26 Mar. (86) . .	3 Tues. .	7	22	30	11 Mar. (71) .	2 Mon. .	43·3368	4270

TABLE

CONCURRENT YEAR.								Mean Intercalated (adhika) lunar month.
						Jovian Saṁvatsara.		
Kali.	Saka.	Chaitrādi Vikrama.	Mēshādi solar year in Bengal	Kollam.	A.D.	Southern system.	Northern system.	
1	2	3	3a	4	5	6	7	8a
4271	1092	1227	576	344-45	1169-70	23 Virōdhin .	27 Vijaya . .	5 Śrāvaṇa .
4272	1093	1228	577	345-46	1170-71	24 Vikṛita . .	28 Jaya . .	...
4273	1094	1229	578	346-47	1171-72	25 Khara . .	29 Manmatha .	...
4274	1095	1230	579	347-48	*1172-73	26 Nandana .	30 Durmukha .	2 Vaiśākha .
4275	1096	1231	580	348-49	1173-74	27 Vijaya . .	31 Hēmalamba .	...
4276	1097	1232	581	349-50	1174-75	28 Jaya . .	32 Vilamba .	10 Pausha .
4277	1098	1233	582	350-51	1175-76	29 Manmatha .	33 Vikārin .	...
4278	1099	1234	583	351-52	*1176-77	30 Durmukha .	34 Śārvarin .	...
4279	1100	1235	584	352-53	1177-78	31 Hēmalamba .	35 Plava . .	7 Āsvina .
4280	1101	1236	585	353-54	1178-79	32 Vilamba .	36 Śubhakṛit .	...
4281	1102	1237	586	354-55	1179-80	33 Vikārin .	37 Śōbhana .	...
4282	1103	1238	587	355-56	*1180-81	34 Sārvarin .	38 Krōdhin .	3 Jyēshtha .
4283	1104	1239	588	356-57	1181-82	35 Plava . .	39 Viśvāvasu .	...
4284	1105	1240	589	357-58	1182-83	36 Subhakṛit .	40 Parābhava .	12 Phālguna .
4285	1106	1241	590	358-59	1183-84	37 Sōbhana .	41 Plavaṅga .	...
4286	1107	1242	591	359-60	*1184-85	38 Krōdhin .	42 Kīlaka . .	...
4287	1108	1243	592	360-61	1185-86	39 Viśvāvasu .	43 Saumya .	8 Kārttika .
4288	1109	1244	593	361-62	1186-87	40 Parābhava .	44 Sādhāraṇa .	...
4289	1110	1245	594	362-63	1187-88	41 Plavaṅga .	45 Virōdhakṛit .	...
4290	1111	1246	595	363-64	*1188-89	42 Kīlaka . .	46 Paridhāvin .	5 Srāvaṇa .
4291	1112	1247	596	364-65	1189-90	43 Saumya .	47 Pramādin .	...
4292	1113	1248	597	365-66	1190-91	44 Sādhāraṇa .	48 Ānanda .	...
4293	1114	1249	598	366-67	1191-92	45 Virōdhakṛit .	49 Rākshasa .	1 Chaitra .
4294	1115	1250	599	367-68	*1192-93	46 Paridhāvin .	50 Anala . .	...
4295	1116	1251	600	368-69	1193-94	47 Pramādin .	51 Piṅgala .	10 Pausha .

LXXVI—*Contd.*

1 Ārya Siddhānta, mean system.

COMMENCEMENT OF THE						Kali year.
Mean solar year.			Mean luni-solar year (mean sunrise of civil day on which Chaitra śukla 1 ends).			
Day and month, A.D.	Week-day.	Time of mean Mēsha-saṁkrānti.	Day and month, A.D.	Week-day.	a (here $=t$, the index of the tithi).	
13	14	17	19	20	23	1
		H. M. S.				
26 Mar. (85) . .	4 Wed. .	13 35 0	1 Mar. (60) .	0 Sat. .	257·6521	4271
26 Mar. (85) . .	5 Thur. .	19 47 30	20 Mar. (79) .	6 Fri. .	292·2917	4272
27 Mar. (86) . .	0 Sat. .	2 0 0	9 Mar. (68) .	3 Tues. .	167·9751	4273
26 Mar. (86) . .	1 Sun. .	8 12 30	26 Feb. (57) .	0 Sat. .	43·6684	4274
26 Mar. (85) . .	2 Mon. .	14 25 0	16 Mar. (75) .	6 Fri. .	78·2981	4275
26 Mar. (85) . .	3 Tues. .	20 37 30	6 Mar. (65) .	4 Wed. .	292·6133	4276
27 Mar. (86) . .	5 Thur. .	2 50 0	25 Mar. (84) .	3 Tues. .	327·2528	4277
26 Mar. (86) . .	6 Fri. .	9 2 30	13 Mar. (73) .	0 Sat. .	202·9372	4278
26 Mar. (85) . .	0 Sat. .	15 15 0	2 Mar. (61) .	4 Wed. .	78·6196	4279
26 Mar. (85) . .	1 Sun. .	21 27 30	21 Mar. (80) .	3 Tues. .	113·2593	4280
27 Mar. (86) . .	3 Tues. .	3 40 0	11 Mar. (70) .	1 Sun .	327·5745	4281
26 Mar. (86) . .	4 Wed. .	9 52 30	28 Feb. (59) .	5 Thur. .	203·2579	4282
26 Mar. (85) . .	5 Thur. .	16 5 0	18 Mar. (77) .	4 Wed. .	237·8975	4283
26 Mar. (85) . .	6 Fri. .	22 17 30	7 Mar. (66) .	1 Sun. .	113·5809	4284
27 Mar. (86) . .	1 Sun. .	4 30 0	26 Mar. (85) .	0 Sat. .	148·2205	4285
26 Mar. (86) . .	2 Mon. .	10 42 30	14 Mar. (74) .	4 Wed. .	23·9039	4286
26 Mar. (85) . .	3 Tues. .	16 55 0	4 Mar. (63) .	2 Mon .	238·2192	4287
26 Mar. (85) . .	4 Wed. .	23 7 30	23 Mar. (82) .	1 Sun. .	272·8588	4288
27 Mar. (86) . .	6 Fri. .	5 20 0	12 Mar. (71) .	5 Thur. .	148·5422	4289
26 Mar. (86) . .	0 Sat. .	11 32 30	29 Feb. (60) .	2 Mon. .	14·2256	4290
26 Mar. (85) . .	1 Sun. .	17 45 0	19 Mar. (78) .	1 Sun .	58·8452	4291
26 Mar. (85) . .	2 Mon. .	23 57 30	9 Mar. (68) .	6 Fri .	273·1805	4292
27 Mar. (86) . .	4 Wed. .	6 10 0	26 Feb. (57) .	3 Tues. .	148·8638	4293
26 Mar. (86) . .	5 Thur. .	12 22 30	16 Mar. (76) .	2 Mon. .	183·5035	4294
26 Mar. (85) . .	6 Fri. .	18 35 0	5 Mar. (64) .	6 Fri. .	59·1868	4295

TABLE

CONCURRENT YEAR.								Mean Intercalated (adhika) lunar month.
Kali.	Śaka.	Chaitrādi Vikrama.	Mēshādi solar year in Bengal.	Kollam.	A.D.	Jovian Saṁvatsara.		
						Southern system.	Northern system.	
1	2	3	3a	4	5	6	7	8a
4296	1117	1252	601	369-70	1194-95	48 Ānanda	52 Kālayukta	...
4297	1118	1253	602	370-71	1195-96	49 Rākshasa	53 Siddhārthin	...
4298	1119	1254	603	371-72	*1196-97	50 Anala	54 Raudra	6 Bhādrapada
4299	1120	1255	604	372-73	1197-98	51 Piṅgala	55 Durmati	...
4300	1121	1256	605	373-74	1198-99	52 Kālayukta	56 Dundubhi	...
4301	1122	1257	606	374-75	1199-00	53 Siddhārthin	57 Rudhirōdgārin	3 Jyēshṭha
4302	1123	1258	607	375-76	*1200-01	54 Raudra	58 Raktāksha	...
4303	1124	1259	608	376-77	1201-02	55 Durmati	59 Krōdhana	11 Māgha
4304	1125	1260	609	377-78	1202-03	56 Dundubhi	60 Kshaya	...
4305	1126	1261	610	378-79	1203-04	57 Rudhirōdgārin	1 Prabhava	...
4306	1127	1262	611	379-80	*1204-05	58 Raktāksha	2 Vibhava	8 Kārttika
4307	1128	1263	612	380-81	1205-06	59 Krōdhana	3 Śukla	...
4308	1129	1264	613	381-82	1206-07	60 Kshaya	4 Pramōda	...
4309	1130	1265	614	382-83	1207-08	1 Prabhava	5 Prajāpati	5 Śrāvaṇa
4310	1131	1266	615	383-84	*1208-09	2 Vibhava	6 Aṅgiras	...
4311	1132	1267	616	384-85	1209-10	3 Śukla	7 Śrīmukha	...
4312	1133	1268	617	385-86	1210-11	4 Pramōda	8 Bhāva	1 Chaitra
4313	1134	1269	618	386-87	1211-12	5 Prajāpati	9 Yuvan	...
4314	1135	1270	619	387-88	*1212-13	6 Aṅgiras	10 Dhātṛi	10 Pausha
4315	1136	1271	620	388-89	1213-14	7 Śrīmukha	11 Īśvara	...
4316	1137	1272	621	389-90	1214-15	8 Bhāva	12 Bahudhānya	...
4317	1138	1273	622	390-91	1215-16	9 Yuvan	13 Pramādin	6 Bhādrapada
4318	1139	1274	623	391-92	*1216-17	10 Dhātṛi	14 Vikrama	...
4319	1140	1275	624	392-93	217-18	11 Īśvara	15 Vṛisha	...
4320	1141	1276	625	393-94	1218-19	12 Bahudhānya	16 Chitrabhānu	3 Jyēshṭha

LXXVI—*Contd.*

1 Ārya Siddhānta, mean system.

COMMENCEMENT OF THE									
MEAN SOLAR YEAR.					MEAN LUNI-SOLAR YEAR (MEAN SUNRISE OF CIVIL DAY ON WHICH CHAITRA ŚUKLA 1 ENDS).			Kali year.
Day and month, A.D.	Week-day.	Time of mean Mēsha-saṁkrānti.			Day and month, A.D.	Week-day.	*a* (here=*t*, the index of the tithi).	
13	14	17			19	20	23	1
		H.	M.	S.				
27 Mar. (86)	1 Sun.	0	47	30	21 Mar. (83)	5 Thur.	93·8264	4296
27 Mar. (86)	2 Mon.	7	0	0	14 Mar. (73)	3 Tues.	308·1417	4297
26 Mar. (86)	3 Tues.	13	12	30	2 Mar. (62)	0 Sat	183·8251	4298
26 Mar. (85)	4 Wed.	19	25	0	21 Mar. (80)	6 Fri.	218·4647	4299
27 Mar. (86)	6 Fri.	1	37	30	10 Mar. (69)	3 Tues.	94·1481	4300
27 Mar. (86)	0 Sat.	7	50	0	28 Feb. (59)	1 Sun.	308·4634	4301
26 Mar. (86)	1 Sun.	14	2	30	17 Mar. (77)	6 Fri.	4·4711	4302
26 Mar. (85)	2 Mon.	20	15	0	7 Mar. (66)	4 Wed.	218·7864	4303
27 Mar. (86)	4 Wed.	2	27	30	26 Mar. (85)	3 Tues.	253·4359	4304
27 Mar. (86)	5 Thur.	8	40	0	15 Mar. (74)	0 Sat.	129·1094	4305
26 Mar. (86)	6 Fri.	14	52	30	3 Mar. (63)	4 Wed.	4·7927	4306
26 Mar. (85)	0 Sat.	21	5	0	22 Mar. (81)	3 Tues.	39·4324	4307
27 Mar. (86)	2 Mon.	3	17	30	12 Mar. (71)	1 Sun.	253·7477	4308
27 Mar. (86)	3 Tues.	9	30	0	1 Mar. (60)	5 Thur.	129·4311	4309
26 Mar. (86)	4 Wed.	15	42	30	19 Mar. (79)	4 Wed.	164·0707	4310
26 Mar. (85)	5 Thur.	21	55	0	8 Mar. (67)	1 Sun.	39·7540	4311
27 Mar. (86)	0 Sat.	4	7	30	26 Feb. (57)	6 Fri.	254·0693	4312
27 Mar. (86)	1 Sun.	10	20	0	17 Mar. (76)	5 Thur.	288·7089	4313
26 Mar. (86)	2 Mon.	16	32	30	5 Mar. (65)	2 Mon.	164·3923	4314
26 Mar. (85)	3 Tues.	22	45	0	24 Mar. (83)	1 Sun.	199·0319	4315
27 Mar. (86)	5 Thur.	4	57	30	13 Mar. (72)	5 Thur.	74·7152	4316
27 Mar. (86)	6 Fri.	11	10	0	3 Mar. (62)	3 Tues.	289·0306	4317
26 Mar. (86)	0 Sat.	17	22	30	21 Mar. (81)	2 Mon.	323·6702	4318
26 Mar. (85)	1 Sun.	23	35	0	10 Mar. (69)	6 Fri.	199·3535	4319
27 Mar. (86)	3 Tues.	5	47	30	27 Feb. (58)	3 Tues.	75·0369	4320

TABLE

CONCURRENT YEAR.								Mean Intercalated (adhika) lunar month.
Kali.	Śaka.	Chaitrādi Vikrama.	Mēshādi solar year in Bengal.	Kollam.	A.D.	Jovian Saṁvatsara. Southern system.	Jovian Saṁvatsara. Northern system.	
1	2	3	3a	4	5	6	7	8a
4321	1142	1277	626	394-95	1219-20	13 Pramāthin .	17 Subhānu .	...
4322	1143	1278	627	395-96	*1220-21	14 Vikrama .	18 Tāraṇa . .	11 Māgha .
4323	1144	1279	628	396-97	1221-22	15 Vṛisha . .	19 Pārthiva .	...
4324	1145	1280	629	397-98	1222-23	16 Chitrabhānu .	20 Vyaya . .	...
4325	1146	1281	630	398-99	1223-24	17 Subhānu .	21 Sarvajit .	8 Kārttika .
4326	1147	1282	631	399-400	*1224-25	18 Tāraṇa . .	22 Sarvadhārin .	...
4327	1148	1283	632	400-01	1225-26	19 Pārthiva .	23 Virōdhin .	...
4328	1149	1284	633	401-02	1226-27	20 Vyaya . .	24 Vikṛita . .	4 Āshāḍha .
4329	1150	1285	634	402-03	1227-28	21 Sarvajit .	25 Khara . .	...
4330	1151	1286	635	403-04	*1228-29	22 Sarvadhārin .	26 Nandana .	...
4331	1152	1287	636	404-05	1229-30	23 Virōdhin .	27 Vijaya . .	1 Chaitra .
4332	1153	1288	637	405-06	1230-31	24 Vikṛita . .	28 Jaya .	...
4333	1154	1289	638	406-07	1231-32	25 Khara . .	29 Manmatha .	9 Mārgaśira .
4334	1155	1290	639	407-08	*1232-33	26 Nandana .	30 Durmukha .	...
4335	1156	1291	640	408-09	1233-34	27 Vijaya . .	31 Hēmalamba .	...
4336	1157	1292	641	409-10	1234-35	28 Jaya . .	32 Vilamba .	6 Bhādrapada
4337	1158	1293	642	410-11	1235-36	29 Manmatha .	33 Vikārin .	...
4338	1159	1294	643	411-12	*1236-37	30 Durmukha .	34 Śārvarin .	...
4339	1160	1295	644	412-13	1237-38	31 Hēmalamba .	35 Plava . .	2 Vaiśākha .
4340	1161	1296	645	413-14	1238-39	32 Vilamba .	36 Śubhakṛit .	...
4341	1162	1297	646	414-15	1239-40	33 Vikārin .	37 Śōbhana .	11 Māgha .
4342	1163	1298	647	415-16	*1240-41	34 Śārvarin .	38 Krōdhin .	...
4343	1164	1299	648	416-17	1241-42	35 Plava .	39 Viśvāvasu .	...
4344	1165	1300	649	417-18	1242-43	36 Śubhakrit	40 Parābhava .	7 Āśvina .
4345	1166	1301	650	418-19	1243-44	37 Śōbhana	41 Plavaṅga .	...

LXXVI—*Contd.*

1 Ārya Siddhānta, mean system.

COMMENCEMENT OF THE							
MEAN SOLAR YEAR.			MEAN LUNI-SOLAR YEAR (MEAN SUNRISE OF CIVIL DAY ON WHICH CHAITRA ŚUKLA 1 ENDS).			Kali year.
Day and month, A.D.	Week-day.	Time of mean Mēsha-saṁkrānti.	Day and month, A.D.	Week-day.	a (here $= t$, the index of the tithi).	
13	14	17	19	20	23	1
		H. M. S.				
27 Mar. (86) . .	4 Wed. .	12 0 0	18 Mar. (77) .	2 Mon. .	109·6765	4321
26 Mar. (86) . .	5 Thur. .	18 12 30	7 Mar. (67) .	0 Sat. .	323·9918	4322
27 Mar. (86) . .	0 Sat. .	0 25 0	25 Mar. (84) .	5 Thur. .	19·9995	4323
27 Mar. (86) . .	1 Sun. .	6 37 30	15 Mar. (74) .	3 Tues. .	234·3148	4324
27 Mar. (86) . .	2 Mon. .	12 50 0	4 Mar. (63) .	0 Sat. .	109·9982	4325
26 Mar. (86) . .	3 Tues. .	19 2 30	22 Mar. (82) .	6 Fri. .	144·6378	4326
27 Mar. (86) . .	5 Thur. .	1 15 0	11 Mar. (70) .	3 Tues. .	20·3212	4327
27 Mar. (86) . .	6 Fri. .	7 27 30	1 Mar. (60) .	1 Sun. .	234·6365	4328
27 Mar. (86) . .	0 Sat. .	13 40 0	20 Mar. (79) .	0 Sat. .	269·2761	4329
26 Mar. (86) . .	1 Sun. .	19 52 30	8 Mar. (68) .	4 Wed. .	144·9594	4330
27 Mar. (86) . .	3 Tues. .	2 5 0	25 Feb. (56) .	1 Sun. .	20·6428	4331
27 Mar. (86) . .	4 Wed. .	8 17 30	16 Mar. (75) .	0 Sat. .	55·2824	4332
27 Mar. (86) . .	5 Thur. .	14 30 0	6 Mar. (65) .	5 Thur. .	269·5977	4333
26 Mar. (86) . .	6 Fri. .	20 42 30	24 Mar. (84) .	4 Wed. .	304·2373	4334
27 Mar. (86) . .	1 Sun. .	2 55 0	13 Mar. (72) .	1 Sun. .	179·9207	4335
27 Mar. (86) . .	2 Mon. .	9 7 30	2 Mar. (61) .	5 Thur. .	55·6041	4336
27 Mar. (86) . .	3 Tues. .	15 20 0	21 Mar. (80) .	4 Wed. .	90·2437	4337
26 Mar. (86) . .	4 Wed. .	21 32 30	10 Mar. (70) .	2 Mon. .	304·5590	4338
27 Mar. (86) . .	6 Fri. .	3 45 0	27 Feb. (58) .	6 Fri. .	180·2424	4339
27 Mar. (86) . .	0 Sat. .	9 57 30	18 Mar. (77) .	5 Thur. .	214·8820	4340
27 Mar. (86) . .	1 Sun. .	16 10 0	7 Mar. (66) .	2 Mon. .	90·5654	4341
26 Mar. (86) . .	2 Mon. .	22 22 30	25 Mar. (85) .	1 Sun. .	125·2049	4342
27 Mar. (86) . .	4 Wed. .	4 35 0	14 Mar. (73) .	5 Thur. .	0·8884	4343
27 Mar. (86) . .	5 Thur. .	10 47 30	4 Mar. (63) .	3 Tues. .	215·2037	4344
27 Mar. (86) .	6 Fri. .	17 0 0	23 Mar. (82) .	2 Mon. .	249·3433	4345

TABLE

CONCURRENT YEAR.								Mean Intercalated (adhika) lunar month.
						Jovian Saṁvatsara.		
Kali.	Śaka.	Chaitrādi Vikrama.	Mēshādi solar year in Bengal.	Kolam.	A.D.	Southern system.	Northern system.	
1	2	3	3a	4	5	6	7	8a
4346	1167	1302	651	419-20	*1244-45	38 Krōdhin .	42 Kīlaka . .	...
4347	1168	1303	652	420-21	1245-46	39 Viśvāvasu .	43 Saumya† .	4 Āshādha .
4348	1169	1304	653	421-22	1246-47	40 Parābhava .	45 *Virōdhakṛit* .	...
4349	1170	1305	654	422-23	1247-48	41 Plavaṅga .	46 *Paridhāvin* .	...
4350	1171	1306	655	423-24	*1248-49	42 Kīlaka . .	47 *Pramādin* .	1 Chaitra .
4351	1172	1307	656	424-25	1249-50	43 Saumya .	48 *Ānanda* .	...
4352	1173	1308	657	425-26	1250-51	44 Sādhāraṇa .	49 *Rākshasa* .	9 Mārgaśira .
4353	1174	1309	658	426-27	1251-52	45 Virōdhakṛit .	50 Anala . .	...
4354	1175	1310	659	427-28	*1252-53	46 Paridhāvin .	51 Piṅgala .	...
4355	1176	1311	660	428-29	1253-54	47 Pramādin .	52 Kālayukta .	6 Bhādrapada
4356	1177	1312	661	429-30	1254-55	48 Ānanda .	53 Siddhārthin .	...
4357	1178	1313	662	430-31	1255-56	49 Rākshasa .	54 Raudra .	...
4358	1179	1314	663	431-32	*1256-57	50 Anala . .	55 Durmati .	2 Vaiśākha .
4359	1180	1315	664	432-33	1257-58	51 Piṅgala .	56 Dundubhi .	...
4360	1181	1316	665	433-34	1258-59	52 Kālayukta .	57 Rudhirōdgārin	11 Māgha .
4361	1182	1317	666	434-35	1259-60	53 Siddhārthin .	58 Raktāksha .	...
4362	1183	1318	667	435-36	*1260-61	54 Raudra .	59 Krōdhana .	...
4363	1184	1319	668	436-37	1261-62	55 Durmati .	60 Kshaya .	7 Āśvina .
4364	1185	1320	669	437-38	1262-63	56 Dundubhi .	1 Prabhava .	...
4365	1186	1321	670	438-39	1263-64	57 Rudhirōdgārin	2 Vibhava .	...
4366	1187	1322	671	439-40	*1264-65	58 Raktāksha .	3 Śukla . .	4 Āshādha .
4367	1188	1323	672	440-41	1265-66	59 Krōdhana .	4 Pramōda .	...
4368	1189	1324	673	441-42	1266-67	60 Kshaya .	5 Prajāpati .	12 Phālguna
4369	1190	1325	674	442-43	1267-68	1 Prabhava .	6 Aṅgiras .	...
4370	1191	1326	675	443-44	1268-69	2 Vibhava .	7 Śrīmukha .	...

† 44 Sādhāraṇa was suppressed in the north by the mean system, but 45 Virōdhakṛit by the true system. By the latter system the year A.D. 1246-47 was called in the north "Sādhāraṇa."

LXXVI—*Contd.*

1 Ārya Siddhānta, mean system.

COMMENCEMENT OF THE									
MEAN SOLAR YEAR.					MEAN LUNI-SOLAR YEAR (MEAN SUNRISE OF CIVIL DAY ON WHICH CHAITRA ŚUKLA 1 ENDS).			Kali year.
Day and month, A.D.	Week-day.	Time of mean Mēsha-saṁkrānti.			Day and month, A.D.	Week-day.	*a* (here=*t*, the index of the tithi).	
13	14	17			19	20	23	1
		H.	M.	S.				
26 Mar. (86)	0 Sat.	23	12	30	11 Mar. (71)	6 Fri.	125·5266	4346
27 Mar. (86)	2 Mon.	5	25	0	28 Feb. (59)	3 Tues.	1·2100	4347
27 Mar. (86)	3 Tues.	11	37	30	19 Mar. (78)	2 Mon.	35·8196	4348
27 Mar. (86)	4 Wed.	17	50	0	9 Mar. (68)	0 Sat.	250·1649	4349
27 Mar. (87)	6 Fri.	0	2	30	26 Feb. (57)	4 Wed.	125·8482	4350
27 Mar. (86)	0 Sat.	6	15	0	16 Mar. (75)	3 Tues.	160·4878	4351
27 Mar. (86)	1 Sun.	12	27	30	5 Mar. (64)	0 Sat.	36·1712	4352
27 Mar. (86)	2 Mon.	18	40	0	24 Mar. (83)	6 Fri.	70·8109	4353
27 Mar. (87)	4 Wed.	0	52	30	13 Mar. (73)	4 Wed.	285·1262	4354
27 Mar. (86)	5 Thur.	7	5	0	2 Mar. (61)	1 Sun.	160·8095	4355
27 Mar. (86)	6 Fri.	13	17	30	21 Mar. (80)	0 Sat.	195·4491	4356
27 Mar. (86)	0 Sat.	[illegible]	30	0	10 Mar. (69)	4 Wed.	71·1325	4357
27 Mar. (87)	2 Mon	1	42	30	28 Feb. (59)	2 Mon.	285·4478	4358
27 Mar. (86)	3 Tues.	7	55	0	18 Mar. (77)	1 Sun.	320·0874	4359
27 Mar. (86)	4 Wed.	14	7	30	7 Mar. (66)	5 Thur.	195·7708	4360
27 Mar. (86)	5 Thur.	20	20	0	26 Mar. (85)	4 Wed.	230·4104	4361
27 Mar. (87)	0 Sat.	2	32	30	14 Mar. (74)	1 Sun.	106·0938	4362
27 Mar. (86)	1 Sun.	8	45	0	4 Mar. (63)	6 Fri.	320·4091	4363
27 Mar. (86)	2 Mon.	14	57	30	22 Mar. (81)	4 Wed.	16·4168	4364
27 Mar. (86)	3 Tues.	21	10	0	12 Mar. (71)	2 Mon.	230·7321	4365
27 Mar. (87)	5 Thur.	3	22	30	29 Feb. (60)	6 Fri.	106·4155	4366
27 Mar. (86)	6 Fri.	9	35	0	19 Mar. (78)	5 Thur.	141·0551	4367
27 Mar. (86)	0 Sat.	15	47	30	8 Mar. (67)	2 Mon.	16·7384	4368
27 Mar. (86)	1 Sun.	22	0	0	27 Mar. (86)	1 Sun.	51·3780	4369
27 Mar. (87)	3 Tues.	4	12	30	16 Mar. (76)	6 Fri.	265·5934	4370

TABLE

CONCURRENT YEAR.								Mean Intercalated (adhika) lunar month.
						Jovian Saṁvatsara.		
Kali.	Śaka.	Chaitrādi Vikrama.	Mēshādi solar year in Bengal.	Kollam.	A.D.	Southern system.	Northern system.	
1	2	3	3a	4	5	6	7	8a
4371	1192	1327	676	444-45	1269-70	3 Śukla . .	8 Bhāva . .	9 Mārgaśira .
4372	1193	1328	677	445-46	1270-71	4 Pramōda .	9 Yuvan . .	...
4373	1194	1329	678	446-47	1271-72	5 Prajāpati .	10 Dhātṛi . .	...
4374	1195	1330	679	447-48	*1272-73	6 Aṅgiras .	11 Īśvara . .	5 Śrāvaṇa .
4375	1196	1331	680	448-49	1273-74	7 Śrīmukha .	12 Bahudhānya .	...
4376	1197	1332	681	449-50	1274-75	8 Bhāva . .	13 Pramādin .	...
4377	1198	1333	682	450-51	1275-76	9 Yuvan . .	14 Vikrama .	2 Vaiśākha .
4378	1199	1334	683	451-52	*1276-77	10 Dhātṛi . .	15 Vṛisha . .	...
4379	1200	1335	684	452-53	1277-78	11 Īśvara . .	16 Chitrabhānu .	10 Pausha
4380	1201	1336	685	453-54	1278-79	12 Bahudhānya .	17 Subhānu .	...
4381	1202	1337	686	454-55	1279-80	13 Pramāthin .	18 Tāraṇa . .	...
4382	1203	1338	687	455-56	*1280-81	14 Vikrama .	19 Pārthiva .	7 Āśvina .
4383	1204	1339	688	456-57	1281-82	15 Vṛisha . .	20 Vyaya . .	...
4384	1205	1340	689	457-58	1282-83	16 Chitrabhānu .	21 Sarvajit .	...
4385	1206	1341	690	458-59	1283-84	17 Subhānu .	22 Sarvadhārin .	4 Āshāḍha .
4386	1207	1342	691	459-60	*1284-85	18 Tāraṇa . .	23 Virōdhin .	...
4387	1208	1343	692	460-61	1285-86	19 Pārthiva .	24 Vikṛita . .	12 Phālguna .
4388	1209	1344	693	461-62	1286-87	20 Vyaya . .	25 Khara . .	...
4389	1210	1345	694	462-63	1287-88	21 Sarvajit .	26 Nandana .	...
4390	1211	1346	695	463-64	*1288-89	22 Sarvadhārin .	27 Vijaya . .	9 Mārgaśira .
4391	1212	1347	696	464-65	1289-90	23 Virōdhin .	28 Jaya . .	...
4392	1213	1348	697	465-66	1290-91	24 Vikṛita . .	29 Manmatha .	...
4393	1214	1349	698	466-67	1291-92	25 Khara . .	30 Durmukha .	5 Śrāvaṇa .
4394	1215	1350	699	467-68	*1292-93	26 Nandana .	31 Hēmalamba .	...
4395	1216	1351	700	468-69	1293-94	27 Vijaya . .	32 Vilamba .	...

LXXVI—*Contd.*

1 Ārya Siddhānta, mean system.

COMMENCEMENT OF THE									Kali year.
MEAN SOLAR YEAR.					MEAN LUNI-SOLAR YEAR (MEAN SUNRISE OF CIVIL DAY ON WHICH CHAITRA ŚUKLA 1 ENDS).				
Day and month, A.D.	Week-day.	Time of mean Mēsha-saṁkrānti.			Day and month, A.D.	Week-day.	a (here $=t$, the index of the tithi).		
13	14	17			19	20	23		1
		H.	M.	S.					
27 Mar. (86) . .	4 Wed. .	10	25	0	5 Mar. (64) .	3 Tues. .	141·3767		4371
27 Mar. (86) . .	5 Thur. .	16	37	30	24 Mar. (83) .	2 Mon. .	176·0164		4372
27 Mar. (86) . .	6 Fri. .	22	50	0	13 Mar. (72) .	6 Fri. .	51·6998		4373
27 Mar. (87) . .	1 Sun. .	5	2	30	2 Mar. (62) .	4 Wed. .	266·0150		4374
27 Mar. (86) . .	2 Mon. .	11	15	0	21 Mar. (80) .	3 Tues. .	300·6546		4375
27 Mar. (86) . .	3 Tues. .	17	27	30	10 Mar. (69) .	0 Sat. .	176·3380		4376
27 Mar. (86) . .	4 Wed. .	23	40	0	27 Feb. (58) .	4 Wed. .	52·0213		4377
27 Mar. (87) . .	6 Fri. .	5	52	30	17 Mar. (77) .	3 Tues. .	86·6609		4378
27 Mar. (86) . .	0 Sat. .	12	5	0	7 Mar. (66) .	1 Sun. .	300·9762		4379
27 Mar. (86) . .	1 Sun. .	18	17	30	25 Mar. (84) .	6 Fri. .	9996·9840*		4380
28 Mar. (87) . .	3 Tues. .	0	30	0	15 Mar. (74) .	4 Wed. .	211·2992		4381
27 Mar. (87) . .	4 Wed. .	6	42	30	3 Mar. (63) .	1 Sun. .	86·9826		4382
27 Mar. (86) . .	5 Thur. .	12	55	0	22 Mar. (81) .	0 Sat. .	121·6222		4383
27 Mar. (86) . .	6 Fri. .	19	7	30	11 Mar. (70) .	4 Wed. .	9997·3056*		4384
28 Mar. (87) . .	1 Sun. .	1	20	0	1 Mar. (60) .	2 Mon. .	211·6209		4385
27 Mar. (87) . .	2 Mon. .	7	32	30	19 Mar. (79) .	1 Sun. .	246·2605		4386
27 Mar. (86) . .	3 Tues. .	13	45	0	8 Mar. (67) .	5 Thur. .	121·9439		4387
27 Mar. (86) . .	4 Wed. .	19	57	30	23 Mar. (86) .	4 Wed. .	156·5834		4388
28 Mar. (87) . .	6 Fri. .	2	10	0	16 Mar. (75) .	1 Sun. .	32·2669		4389
27 Mar. (87) . .	0 Sat. .	8	22	30	5 Mar. (65) .	6 Fri. .	246·5821		4390
27 Mar. (86) . .	1 Sun. .	14	35	0	24 Mar. (83) .	5 Thur. .	281·2218		4391
27 Mar. (86) . .	2 Mon. .	20	47	30	13 Mar. (72) .	2 Mon. .	156·9051		4392
28 Mar. (87) . .	4 Wed. .	3	0	0	2 Mar. (61) .	6 Fri. .	32·5885		4393
27 Mar. (87) . .	5 Thur. .	9	12	30	20 Mar. (80) .	5 Thur. .	67·2281		4394
27 Mar. (86) . .	6 Fri. .	15	25	0	10 Mar. (69) .	3 Tues. .	281·5434		4395

* As a mean tithi Chaitra śukla 1 was expunged. The civil day corresponding to it, *i.e.*, the first day of the luni-solar year, was as given in cols. 19, 20.

TABLE

CONCURRENT YEAR.								Mean Intercalated (adhika) lunar month.
Kali.	Śaka.	Chaitrādi Vikrama.	Mēshādi solar year in Bengal.	Kollam.	A.D.	Jovian Saṁvatsara. Southern system.	Jovian Saṁvatsara. Northern system.	
1	2	3	3a	4	5	6	7	8a
4396	1217	1352	701	469-70	1294-95	28 Jaya . .	33 Vikārin .	2 Vaiśākha .
4397	1218	1353	702	470-71	1295-96	29 Manmatha .	34 Śārvarin .	...
4398	1219	1354	703	471-72	*1296-97	30 Durmukha .	35 Plava . .	10 Pausha .
4399	1220	1355	704	472-73	1297-98	31 Hēmalamba .	36 Śubhakṛit .	...
4400	1221	1356	705	473-74	1298-99	32 Vilamba .	37 Śōbhana .	...
4401	1222	1357	706	474-75	1299-1300	33 Vikārin .	38 Krōdhin .	7 Āśvina .
4402	1223	1358	707	475-76	*1300-01	34 Śārvarin .	39 Viśvāvasu .	...
4403	1224	1359	708	476-77	1301-02	35 Plava . .	40 Parābhava .	...
4404	1225	1360	709	477-78	1302-03	36 Śubhakṛit .	41 Plavaṅga .	3 Jyēshṭha .
4405	1226	1361	710	478-79	1303-04	37 Śōbhana .	42 Kīlaka . .	...
4406	1227	1362	711	479-80	*1304-05	38 Krōdhin .	43 Saumya .	12 Phālguna
4407	1228	1363	712	480-81	1305-06	39 Viśvāvasu .	44 Sādhāraṇa .	...
4408	1229	1364	713	481-82	1306-07	40 Parābhava .	45 Virōdhakṛit .	...
4409	1230	1365	714	482-83	1307-08	41 Plavaṅga .	46 Paridhāvin .	8 Kārttika .
4410	1231	1366	715	483-84	*1308-09	42 Kīlaka . .	47 Pramādin .	...
4411	1232	1367	716	484-85	1309-10	43 Saumya .	48 Ānanda .	...
4412	1233	1368	717	485-86	1310-11	44 Sādhāraṇa .	49 Rākshasa .	5 Śrāvaṇa .
4413	1234	1369	718	486-87	1311-12	45 Virōdhakṛit .	50 Anala . .	...
4414	1235	1370	719	487-88	*1312-13	46 Paridhāvin .	51 Piṅgala . .	...
4415	1236	1371	720	488-89	1313-14	47 Pramādin .	52 Kālayukta .	1 Chaitra .
4416	1237	1372	721	489-90	1314-15	48 Ānanda .	53 Siddhārthin .	...
4417	1238	1373	722	490-91	1315-16	49 Rākshasa .	54 Raudra .	10 Pausha .
4418	1239	1374	723	491-92	*1316-17	50 Anala . .	55 Durmati .	...
4419	1240	1375	724	492-93	1317-18	51 Piṅgala .	56 Dundubhi .	...
4420	1241	1376	725	493-94	1318-19	52 Kālayukta .	57 Rudhirōdgārin	7 Āśvina .

LXXVI—*Contd.* 1 Ārya Siddhānta, mean system.

COMMENCEMENT OF THE						
MEAN SOLAR YEAR.			MEAN LUNI-SOLAR YEAR (MEAN SUNRISE OF CIVIL DAY ON WHICH CHAITRA ŚUKLA 1 ENDS).			Kali year.
Day and month, A.D.	Week-day.	Time of mean Mēsha-saṁkrānti.	Day and month, A.D.	Week-day.	*a* (here=*t*, the index of the tithi).	
13	14	17	19	20	25	1
		H. M. S.				
27 Mar. (86) .	0 Sat. .	21 37 40	27 Feb. (58) .	0 Sat. .	157·2268	4396
28 Mar. (87) .	2 Mon. .	3 50 0	18 Mar. (77) .	6 Fri. .	191·8664	4397
27 Mar. (87) .	3 Tues. .	10 2 30	6 Mar. (66) .	3 Tues. .	67·5498	4398
27 Mar. (86) .	4 Wed. .	16 15 0	25 Mar. (84) .	2 Mon. .	102·1894	4399
27 Mar. (86) .	5 Thur. .	22 27 30	15 Mar. (74) .	0 Sat. .	316·5047	4400
28 Mar. (87) .	0 Sat. .	4 40 0	4 Mar. (63) .	4 Wed. .	192·1881	4401
27 Mar. (87) .	1 Sun. .	10 52 30	22 Mar. (82) .	3 Tues. .	226·8277	4402
27 Mar. (86) .	2 Mon. .	17 5 0	11 Mar. (70) .	0 Sat. .	102·5111	4403
27 Mar. (86) .	3 Tues. .	23 17 30	1 Mar. (60) .	5 Thur. .	316·8264	4404
28 Mar. (87) .	5 Thur. .	5 30 0	19 Mar. (78) .	3 Tues. .	12·8341	4405
27 Mar. (87) .	6 Fri. .	11 42 30	8 Mar. (68) .	1 Sun. .	227·1494	4406
27 Mar. (86) .	0 Sat. .	17 55 0	27 Mar. (86) .	0 Sat. .	261·7889	4407
28 Mar. (87) .	2 Mon. .	0 7 30	16 Mar. (75) .	4 Wed. .	137·4728	4408
28 Mar. (87) .	3 Tues. .	6 20 0	5 Mar. (64) .	1 Sun. .	13·1558	4409
27 Mar. (87) .	4 Wed. .	12 32 30	23 Mar. (83) .	0 Sat. .	47·7954	4410
27 Mar. (86) .	5 Thur. .	18 45 0	13 Mar. (72) .	5 Thur. .	262·1106	4411
28 Mar. (87) .	0 Sat. .	0 57 30	2 Mar. (61) .	2 Mon. .	137·7940	4412
28 Mar. (87) .	1 Sun. .	7 10 0	21 Mar. (80) .	1 Sun. .	172·4337	4413
27 Mar. (87) .	2 Mon. .	13 22 30	9 Mar. (69) .	5 Thur. .	48·1170	4414
27 Mar. (86) .	3 Tues. .	19 35 0	27 Feb. (58) .	3 Tues. .	262·4322	4415
28 Mar. (87) .	5 Thur. .	1 47 30	18 Mar. (77) .	2 Mon. .	297·8719	4416
28 Mar. (87) .	6 Fri. .	8 0 0	7 Mar. (66) .	6 Fri. .	172·7553	4417
27 Mar. (87) .	0 Sat. .	14 12 30	25 Mar. (85) .	5 Thur. .	207·3949	4418
27 Mar. (86) .	1 Sun. .	20 25 0	14 Mar. (73) .	2 Mon. .	83·0782	4419
28 Mar. (87) .	3 Tues. .	2 37 30	4 Mar. (63) .	0 Sat. .	297·3935	4420

CONCURRENT YEAR.								Mean Intercalated (adhika) lunar month.
						Jovian Saṁvatsara.		
Kali.	Saka.	Chaitrādi Vikrama.	Mēshādi solar year in Bengal.	Kollam.	A.D.	Southern system.	Northern system.	
1	2	3	3a	4	5	6	7	8a
4421	1242	1377	726	494-95	1319-20	53 Siddhārthin .	58 Raktāksha .	...
4422	1243	1378	727	495-96	*1320-21	54 Raudra .	59 Krōdhana .	...
4423	1244	1379	728	496-97	1321-22	55 Durmati .	60 Kshaya .	3 Jyēshṭha .
4424	1245	1380	729	497-98	1322-23	56 Dundubhi .	1 Prabhava .	...
4425	1246	1381	730	498-99	1323-24	57 Rudhirōdgārin	2 Vibhava .	12 Phālguna
4426	1247	1382	731	499-00	*1324-25	58 Raktāksha .	3 Śukla . .	...
4427	1248	1383	732	500-01	1325-26	59 Krōdhana .	4 Pramōda .	...
4428	1249	1384	733	501-02	1326-27	60 Kshaya .	5 Prajāpati .	8 Kārttika .
4429	1250	1385	734	502-03	1327-28	1 Prabhava .	6 Aṅgiras .	...
4430	1251	1386	735	503-04	*1328-29	2 Vibhava .	7 Śrīmukha .	...
4431	1252	1387	736	504-05	1329-30	3 Śukla . .	8 Bhāva . .	5 Śrāvaṇa .
4432	1253	1388	737	505-06	1330-31	4 Pramōda .	9 Yuvan† .	...
4433	1254	1389	738	506-07	1331-32	5 Prajāpati .	11 *Īśvara* . .	...
4434	1255	1390	739	507-08	*1332-33	6 Aṅgiras .	12 *Bahudhānya* .	1 Chaitra .
4435	1256	1391	740	508-09	1333-34	7 Srīmukha .	13 *Pramādin* .	...
4436	1257	1392	741	509-10	1334-35	8 Bhāva . .	14 *Vikrama* .	10 Pausha .
4437	1258	1393	742	510-11	1335-36	9 Yuvan . .	15 *Vṛisha* . .	...
4438	1259	1394	743	511-12	*1336-37	10 Dhātṛi . .	16 Chitrabhānu .	...
4439	1260	1395	744	512-13	1337-38	11 Īśvara . .	17 Subhānu .	6 Bhādrapada
4440	1261	1396	745	513-14	1338-39	12 Bahudhānya .	18 Tāraṇa . .	...
4441	1262	1397	746	514-15	1339-40	13 Pramāthin .	19 Pārthiva .	...
4442	1263	1398	747	515-16	*1310-41	14 Vikrama .	20 Vyaya . .	3 Jyēshtha .
4443	1264	1399	748	516-17	1341-42	15 Vṛisha .	21 Sarvajit .	...
4444	1265	1400	749	517-18	1342-43	16 Chitrabhānu .	22 Sarvadhārin .	11 Māgha
4445	1266	1401	750	513-19	1313-44	17 Subhānu .	23 Virōdhin	..

† 10 Dhātṛi was suppressed in the north by the mean system, but 11 Īśvara by the true system. The year A.L. 1331-32 was by the latter system called "10 Dhātṛi" in the north.

LXXVI—*Contd.*

1 Ārya Siddhānta, mean system.

COMMENCEMENT OF THE							
MEAN SOLAR YEAR.			MEAN LUNI-SOLAR YEAR (MEAN SUNRISE OF CIVIL DAY ON WHICH CHAITRA ŚUKLA 1 ENDS).			Kali year.
Day and month, A.D.	Week-day.	Time of mean Mēsha-saṁkrānti.	Day and month, A.D.	Week-day.	*a* (here=*t*, the index of the tithi).	
13	14	17	19	20	23	1
		H. M. S.				
28 Mar. (87) . .	4 Wed. .	8 50 0	23 Mar. (82) .	6 Fri. .	332·0331	4421
27 Mar. (87) . .	5 Thur. .	15 2 30	11 Mar. (71) .	3 Tues. .	207·7165	4422
27 Mar. (86) . .	6 Fri. .	21 15 0	28 Feb. (59) .	0 Sat. .	83·3999	4423
28 Mar. (87) . .	1 Sun. .	3 27 30	19 Mar. (78) .	6 Fri. .	118·0395	4424
28 Mar. (87) .	2 Mon. .	9 40 0	9 Mar. (68) .	4 Wed. .	332·3547	4425
27 Mar. (87) . .	3 Tues. .	15 52 30	26 Mar. (86) .	2 Mon. .	28·3624	4426
27 Mar. (86) . .	4 Wed. .	22 5 0	16 Mar. (75) .	0 Sat. .	242·6778	4427
28 Mar. (87) . .	6 Fri. .	4 17 30	5 Mar. (64) .	4 Wed. .	118·3612	4428
28 Mar. (87) . .	0 Sat. .	10 30 0	24 Mar. (83) .	3 Tues. .	153·0008	4429
27 Mar. (87) . .	1 Sun. .	16 42 30	12 Mar. (72) .	0 Sat. .	28·7841	4430
27 Mar. (86) . .	2 Mon. .	22 55 0	2 Mar. (61) .	5 Thur. .	242·9995	4431
28 Mar. (87) . .	4 Wed. .	5 7 30	21 Mar. (80) .	4 Wed. .	277·6391	4432
28 Mar. (87) . .	5 Thur. .	11 20 0	10 Mar. (69) .	1 Sun. .	153·3224	4433
27 Mar. (87) . .	6 Fri. .	17 32 30	27 Feb. (58) .	5 Thur. .	29·0058	4434
27 Mar. (86) . .	0 Sat. .	23 45 0	17 Mar. (76) .	4 Wed. .	63·6455	4435
28 Mar. (87) . .	2 Mon. .	5 57 30	7 Mar. (66) .	2 Mon. .	277·9607	4436
28 Mar. (87) . .	3 Tues. .	12 10 0	25 Mar. (85) .	1 Sun. .	312·6603	4437
27 Mar. (87) . .	4 Wed. .	18 22 30	14 Mar. (74) .	5 Thur. .	188·2837	4438
28 Mar. (87) . .	6 Fri. .	0 35 0	3 Mar. (62) .	2 Mon. .	63·9689	4439
28 Mar. (87) . .	0 Sat. .	6 47 30	22 Mar. (81) .	1 Sun. .	98·6067	4440
28 Mar. (87) . .	1 Sun. .	13 0 0	12 Mar. (71) .	6 Fri. .	312·9231	4441
27 Mar. (87) . .	2 Mon. .	19 12 30	29 Feb. (60) .	3 Tues. .	188·6054	4442
28 Mar. (87) . .	4 Wed. .	1 25 0	19 Mar. (78) .	2 Mon. .	223·2350	4443
28 Mar. (87) . .	5 Thur. .	7 37 30	8 Mar. (67) .	6 Fri. .	98·9284	4444
28 Mar. (87) . .	6 Fri. .	13 50 0	27 Mar. (86) .	5 Thur. .	133·5679	4445

TABLE

CONCURRENT YEAR.								Mean Intercalated (adhika) lunar month.
Kali	Śaka.	Chaitrādi Vikrama.	Mēshādi solar year in Bengal.	Kollam.	A.D.	Jovian Samvatsara. Southern system.	Jovian Samvatsara. Northern system.	
1	2	3	3a	4	5	6	7	8a
4446	1267	1402	751	519-20	*1344-45	18 Tāraṇa	24 Vikṛita	...
4447	1268	1403	752	520-21	1345-46	19 Pārthiva	25 Khara	8 Kārttika
4448	1269	1404	753	521-22	1346-47	20 Vyaya	26 Nandana	...
4449	1270	1405	754	522-23	1347-48	21 Sarvajit	27 Vijaya	...
4450	1271	1406	755	523-24	*1348-49	22 Sarvadhārin	28 Jaya	4 Āshāḍha
4451	1272	1407	756	524-25	1349-50	23 Virōdhin	29 Manmatha	...
4452	1273	1408	757	525-26	1350-51	24 Vikṛita	30 Durmukha	...
4453	1274	1409	758	526-27	1351-52	25 Khara	31 Hēmalamba	1 Chaitra
4454	1275	1410	759	527-28	*1352-53	26 Nandana	32 Vilamba	...
4455	1276	1411	760	528-29	1353-54	27 Vijaya	33 Vikārin	9 Mārgaśira
4456	1277	1412	761	529-30	1354-55	28 Jaya	34 Śārvarin	...
4457	1278	1413	762	530-31	1355-56	29 Manmatha	35 Plava	...
4458	1279	1414	763	531-32	*1356-57	30 Durmukha	36 Śubhakṛit	6 Bhādrapada
4459	1280	1415	764	532-33	1357-58	31 Hēmalamba	37 Śōbhana	...
4460	1281	1416	765	533-34	1358-59	32 Vilamba	38 Krōdhin	...
4461	1282	1417	766	534-35	1359-60	33 Vikārin	39 Viśvāvasu	3 Jyēshṭha
4462	1283	1418	767	535-36	*1360-61	34 Śārvarin	40 Parābhava	...
4463	1284	1419	768	536-37	1361-62	35 Plava	41 Plavaṅga	11 Māgha
4464	1285	1420	769	537-38	1362-63	36 Śubhakṛit	42 Kīlaka	...
4465	1286	1421	770	538-39	1363-64	37 Śōbhana	43 Saumya	...
4466	1287	1422	771	539-40	*1364-65	38 Krōdhin	44 Sādhāraṇa	8 Kārttika
4467	1288	1423	772	540-41	1365-66	39 Viśvāvasu	45 Virōdhakṛit	...
4468	1289	1424	773	541-42	1366-67	40 Parābhava	46 Paridhāvin	...
4469	1290	1425	774	542-43	1367-68	41 Plavaṅga	47 Pramādin	4 Āshāḍha
4470	1291	1426	775	543-44	*1368-69	42 Kīlaka	48 Ānanda	...

LXXVI—*Contd.*

1 Ārya Siddhānta, mean system.

COMMENCEMENT OF THE									
Mean solar year.					Mean luni-solar year (mean sunrise of civil day on which Chaitra śukla 1 ends).			Kali year.
Day and month, A.D.	Week-day.	Time of mean Mēsha-saṁkrānti			Day and month, A.D.	Week-day.	*a* (here=*t*, the index of the tithi).	
13	14	17			19	20	23	1
		H.	M.	S.				
27 Mar. (87)	0 Sat.	20	2	30	15 Mar. (75)	2 Mon.	9·2513	4446
28 Mar. (87)	2 Mon.	2	15	0	5 Mar. (64)	0 Sat.	223·5666	4447
28 Mar (87)	3 Tues.	8	27	30	24 Mar. (83)	6 Fri.	258·2062	4448
28 Mar. (87)	4 Wed.	14	40	0	13 Mar. (72)	3 Tues.	133·8897	4449
27 Mar. (87)	5 Thur.	20	52	30	1 Mar. (61)	0 Sat.	9·5730	4450
28 Mar. (87)	0 Sat.	3	5	0	20 Mar. (79)	6 Fri.	44·2126	4451
28 Mar. (87)	1 Sun.	9	17	30	10 Mar. (69)	4 Wed.	258·5279	4452
28 Mar. (87)	2 Mon.	15	30	0	27 Feb. (58)	1 Sun.	134·2112	4453
27 Mar. (87)	3 Tues.	21	42	30	17 Mar. (77)	0 Sat.	168·8509	4454
28 Mar. (87)	5 Thur.	3	55	0	6 Mar. (65)	4 Wed.	44·5342	4455
28 Mar. (87)	6 Fri.	10	7	30	25 Mar. (84)	3 Tues.	79·1738	4456
28 Mar. (87)	0 Sat.	16	20	0	15 Mar. (74)	1 Sun.	293·4891	4457
27 Mar. (87)	1 Sun.	22	32	30	3 Mar. (68)	5 Thur.	169·1725	4458
28 Mar. (87)	3 Tues.	4	45	0	22 Mar. (81)	4 Wed.	203·8121	4459
28 Mar. (87)	4 Wed.	10	57	30	11 Mar. (70)	1 Sun.	79·4955	4460
28 Mar. (87)	5 Thur.	17	10	0	1 Mar. (60)	6 Fri.	293·8108	4461
27 Mar. (87)	6 Fri.	23	22	30	19 Mar. (79)	5 Thur.	328·4504	4462
28 Mar. (87)	1 Sun.	5	35	0	8 Mar. (67)	2 Mon.	204·1338	4463
28 Mar. (87)	2 Mon.	11	47	30	27 Mar. (86)	1 Sun.	238·7731	4464
28 Mar. (87)	3 Tues.	18	0	0	16 Mar. (75)	5 Thur.	114·4568	4465
28 Mar. (88)	5 Thur.	0	12	30	5 Mar. (65)	3 Tues.	328·7721	4466
28 Mar. (87)	6 Fri.	6	25	0	23 Mar. (82)	1 Sun.	24·7798	4467
28 Mar. (87)	0 Sat.	12	37	30	13 Mar. (72)	6 Fri.	239·0951	4468
28 Mar. (87)	1 Sun.	18	50	0	2 Mar. (61)	3 Tues	114·7785	4469
28 Mar. (88)	3 Tues.	1	2	30	20 Mar. (80)	2 Mon.	149·4181	4470

TABLE

CONCURRENT YEAR.								Mean Intercalated (adhika) lunar month.
						Jovian Saṁvatsara.		
Kali.	Saka.	Chaitrādi Vikrama.	Mēshādi solar year in Bengal.	Kollam.	A.D.	Southern system.	Northern system.	
1	2	3	3a	4	5	6	7	8a
4471	1292	1427	776	544-45	1369-70	43 Saumya	49 Rākshasa	...
4472	1293	1428	777	545-46	1370-71	44 Sādhāraṇa	50 Anala	1 Chaitra
4473	1294	1429	778	546-47	1371-72	45 Virōdhakṛit	51 Piṅgala	...
4474	1295	1430	779	547-48	*1372-73	46 Paridhāvin	52 Kālayukta	9 Mārgaśira
4475	1296	1431	780	548-49	1373-74	47 Pramādin	53 Siddhārthin	...
4476	1297	1432	781	549-50	1374-75	48 Ānanda	54 Raudra	...
4477	1298	1433	782	550-51	1375-76	49 Rākshasa	55 Durmati	6 Bhādrapada
4478	1299	1434	783	551-52	*1376-77	50 Anala	56 Dundubhi	...
4479	1300	1435	784	552-53	1377-78	51 Piṅgala	57 Rudhirōdgārin	...
4480	1301	1436	785	553-54	1378-79	52 Kālayukta	58 Raktāksha	2 Vaiśakha
4481	1302	1437	786	554-55	1379-80	53 Siddhārthin	59 Krōdhana	...
4482	1303	1438	787	555-56	*1380-81	54 Raudra	60 Kshaya	11 Māgha
4483	1304	1439	788	556-57	1381-82	55 Durmati	1 Prabhava	...
4484	1305	1440	789	557-58	1382-83	56 Dundubhi	2 Vibhava	...
4485	1306	1441	790	558-59	1383-84	57 Rudhirōdgārin	3 Śukla	7 Āśvina
4486	1307	1442	791	559-60	*1384-85	58 Raktāksha	4 Pramōda	...
4487	1308	1443	792	560-61	1385-86	59 Krōdhana	5 Prajāpati	...
4488	1309	1444	793	561-62	1386-87	60 Kshaya	6 Aṅgiras	4 Āshādha
4489	1310	1445	794	562-63	1387-88	1 Prabhava	7 Śrīmukha	...
4490	1311	1446	795	563-64	*1388-89	2 Vibhava	8 Bhāva	12 Phālguna
4491	1312	1447	796	564-65	1389-90	3 Śukla	9 Yuvan	...
4492	1313	1448	797	565-66	1390-91	4 Pramōda	10 Dhātṛi	...
4493	1314	1449	798	566-67	1391-92	5 Prajāpati	11 Īśvara	9 Mārgaśira
4494	1315	1450	799	567-68	*1392-93	6 Aṅgiras	12 Bahudhānya	...
4495	1316	1451	800	568-69	1393-94	7 Śrīmukha	13 Pramādin	...

LXXVI—*Contd.*

1 Ārya Siddhānta, mean system.

COMMENCEMENT OF THE									
MEAN SOLAR YEAR.					MEAN LUNI-SOLAR YEAR (MEAN SUNRISE OF CIVIL DAY ON WHICH CHAITRA ŚUKLA 1 ENDS).			Kali year.
Day and month, A.D.	Week-day.	Time of mean Mēsha-saṁkrānti.			Day and month, A.D.	Week-day.	*a* (here=*t*, the index of the tithi).	
13	14	17			19	20	23	1
		H.	M.	S.				
28 Mar. (87) . .	4 Wed. .	7	15	0	9 Mar. (68) .	6 Fri.	25·1015	4471
28 Mar. (87) . .	5 Thur. .	13	27	30	27 Feb. (58) .	4 Wed. .	239·4167	4472
28 Mar. (87) . .	6 Fri. .	19	40	0	18 Mar. (77) .	3 Tues. .	274·0564	4473
28 Mar. (88) . .	1 Sun. .	1	52	30	6 Mar. (66) .	0 Sat. .	149·7397	4474
28 Mar. (87) . .	2 Mon. .	8	5	0	25 Mar. (84) .	6 Fri. .	184·3794	4475
28 Mar. (87) . .	3 Tues. .	14	17	30	14 Mar. (73) .	3 Tues. .	60·0627	4476
28 Mar. (87) . .	4 Wed. .	20	30	0	4 Mar. (63) .	1 Sun. .	274·3779	4477
28 Mar. (88) . .	6 Fri. .	2	42	30	22 Mar. (82) .	0 Sat. .	309·0176	4478
28 Mar. (87) . .	0 Sat. .	8	55	0	11 Mar. (70) .	4 Wed. .	184·7009	4479
28 Mar. (87) . .	1 Sun. .	15	7	30	28 Feb. (59) .	1 Sun. .	60·3844	4480
28 Mar. (87) . .	2 Mon. .	21	20	0	19 Mar. (78) .	0 Sat. .	95·0230	4481
28 Mar. (88) . .	4 Wed. .	3	32	30	8 Mar. (68) .	5 Thur. .	309·3392	4482
28 Mar. (87) . .	5 Thur. .	9	45	0	26 Mar. (85) .	3 Tues. .	5·3469	4483
28 Mar. (87) . .	6 Fri. .	15	57	30	16 Mar. (75) .	1 Sun. .	219·6622	4484
28 Mar. (87) . .	0 Sat. .	22	10	0	5 Mar. (64) .	5 Thur. .	95·3456	4485
28 Mar. (88) . .	2 Mon. .	4	22	30	23 Mar. (83) .	4 Wed. .	129·9852	4486
28 Mar. (87) . .	3 Tues. .	10	35	0	12 Mar. (71) .	1 Sun .	5·6686	4487
28 Mar. (87) . .	4 Wed. .	16	47	30	2 Mar. (61) .	6 Fri. .	219·9839	4488
28 Mar. (87) . .	5 Thur. .	23	0	0	21 Mar. (80) .	5 Thur. .	254·6235	4489
28 Mar. (88) . .	0 Sat. .	5	12	30	9 Mar. (69) .	2 Mon. .	130·3069	4490
28 Mar. (87) . .	1 Sun. .	11	25	0	28 Mar. (87) .	1 Sun. .	164·9464	4491
28 Mar. (87) . .	2 Mon. .	17	37	30	17 Mar. (76) .	5 Thur. .	40·6298	4492
28 Mar. (87) . .	3 Tues. .	23	50	0	7 Mar. (66) .	3 Tues. .	254·9451	4493
28 Mar. (88) . .	5 Thur. .	6	2	30	25 Mar. (85) .	2 Mon. .	289·5848	4494
28 Mar. (87) . .	6 Fri. .	12	15	0	14 Mar. (73) .	6 Fri. .	165·2681	4495

TABLE

CONCURRENT YEAR.								Mean Intercalated (adhika) lunar month.
Kali.	Śaka.	Chaitrādi Vikrama.	Mēshādi solar year in Bengal.	Kollam.	A.D.	Jovian saṁvatsara. Southern system.	Jovian saṁvatsara. Northern system.	
1	2	3	3a	4	5	6	7	8a
4496	1317	1452	801	569-70	1394-95	8 Bhāva . .	14 Vikrama .	6 Bhādrapada
4497	1318	1453	802	570-71	1395-96	9 Yuvan . .	15 Vṛisha . .	...
4498	1319	1454	803	571-72	*1396-97	10 Dhātṛi . .	16 Chitrabhānu .	...
4499	1320	1455	804	572-73	1397-98	11 Īśvara . .	17 Subhānu	2 Vaiśākha .
4500	1321	1456	805	573-74	1398-99	12 Bahudhānya .	18 Tāraṇa . .	...
4501	1322	1457	806	574-75	1399-00	13 Pramāthin .	19 Pārthiva .	11 Māgha .
4502	1323	1458	807	575-76	*1400-01	14 Vikrama .	20 Vyaya . .	...

LXXVI—*Concld.*

COMMENCEMENT OF THE							
MEAN SOLAR YEAR.			MEAN LUNI-SOLAR YEAR (MEAN SUNRISE OF CIVIL DAY ON WHICH CHAITRA ŚUKLA 1 ENDS).				Kali year.
Day and month, A.D.	Week-day.	Time of mean Mēsha-saṁkrānti.	Day and month, A.D.	Week-day.	a (here $=t$, the index of the tithi).		
13	14	17	19	20	23		1
		H. M. S.					
28 Mar. (87) . .	0 Sat. .	18 27 30	3 Mar. (62) .	3 Tues. .	40·9515		4496
29 Mar (88) . .	2 Mon. .	0 40 0	22 Mar. (81) .	2 Mon. .	75·5912		4497
28 Mar. (88) . .	3 Tues. .	6 52 30	11 Mar. (71) .	0 Sat. .	289·9064		4498
28 Mar. (87) . .	4 Wed. .	13 5 0	28 Feb. (59) .	4 Wed. .	165·5893		4499
28 Mar. (87) . .	5 Thur. .	19 17 30	19 Mar. (78) .	3 Tues. .	200·2294		4500
29 Mar (88) . .	0 Sat. .	1 30 0	8 Mar. (67) .	0 Sat. .	75·9127		4501
28 Mar (88) . .	1 Sun. .	7 42 30	26 Mar. (86)	6 Fri. .	110·5523		4502

TABLE LXXVII.

Duration and Collective duration of mean solar months according to the First Arya Siddhānta, with increase of "*a*" at each saṁkrānti.

Mean luni-solar month, ending after the second of the two solar saṁkrāntis connected with it	At the mean solar saṁkrāntis.	Collective duration in time and collective increase of "*a*" from mean Mēsha-saṁkrānti to the several saṁkrāntis.						
		Day.	Week-day.	H.	M.	S.	*a*	
1	2	3					4	
1. Chaitra . .	Mīna-saṁk. (*of previous year*).							
	Mēsha-saṁk.	0	0	0	0	0	0	
2. Vaiśākha . .								
	Vṛishabha-saṁk. .	30	(2)	10	31	2½	307·3526	
3. Jyēshṭha . .								
	Mithuna-saṁk. .	60	(4)	21	2	5	614·7052	The duration of each mean solar month is 30d. 10h. 31m. 2½s.; and this in time the mean moon increases her distance from mean sun, in measurement by 10,000ths of circle, by 307·352623726.
4. Āshāḍha . .								
	Karka-saṁk. .	91	(0)	7	33	7½	922·0579	
5. Śrāvaṇa . .								
	Siṁha-saṁk. .	121	(2)	18	4	10	1229·4105	
6. Bhādrapada . .								
	Kanyā-saṁk. .	152	(5)	4	35	12½	1536·7631	
7. Āśvina . .								
	Tulā-saṁk. . .	182	(0)	15	6	15	1844·1157	
8. Kārttika . .								
	Vṛiśchika-saṁk. .	213	(3)	1	37	17½	2151·4684	
9. Mārgaśira . .								
	Dhanus-saṁk. .	243	(5	12	8	20	2458·8210	
10. Pausha . .								
	Makara-saṁk. .	273	(0)	22	39	22½	2766·1736	
11. Māgha . .								
	Kumbha-saṁk. .	304	(3)	9	10	25	3073·5262	
12. Phālguna . .								
	Mīna-saṁk. .	334	(5)	19	41	27½	3380·8789	
1. Chaitra (*of following year*)	Mēsha-saṁk. (*of following year*).	365	(1)	6	12	30	3688·2315*	

* More fully 3688·231484714.

TABLE LXXVIII.

VALUE OF a $(=t)$ AT BEGINNING OF CENTURIES OF THE KALIYUGA, ACCORDING TO THE FIRST ĀRYA SIDDHĀNTA MEAN SYSTEM.

The value of "a" to be added for beginning of odd years of centuries is given in Table LXXIII above. W.-D.=Week-day.

Century K. Y.	W.-D.	a $(=t)$.
36	1	7715·3525
37	1	6583·1816
38	0	5112·3787
39	0	3980·2078
40	0	2848·0369
41	0	1715·8659
42	0	583·6950
43	0	9451·5240
44	0	8319·3531
45	0	7187·1822
46	6	5716·3793
47	6	4584·2084
48	6	3452·0375

N.B.—These values of "a" agree generally with Professor Jacobi's values (*Epig. Ind. XI, 164*), but the values here stated for the beginnings of centuries 38 to 42 are for mean sunrise on Saturdays, while his are for mean sunrise on the following Sundays.

TABLE LXXIX.

MEAN SUNRISE VALUES OF "a" (DISTANCE OF MEAN MOON FROM MEAN SUN), IN 10,000THS OF CIRCLE, FOR A MONTH PREVIOUS TO THE DAY OF MEAN MĒSHA-SAṀKRĀNTI.

W. D.=Week-day.

Interval of days from mean Mēsha-saṁkrānti day.	W.-D.	a. (mean sunrise value).	Interval of days from mean Mēsha-saṁkrānti day.	W.-D.	a. (mean sunrise value).
31	4	9502·4119	15	6	4920·5219
30	5	9841·0438	14	0	5259·1538
29	6	179·6756	13	1	5597·7856
28	0	518·3075	12	2	5936·4175
27	1	856·9394	11	3	6275·0494
26	2	1195·5713	10	4	6613·6813
25	3	1534·2032	9	5	6952·3131
24	4	1872·8350	8	6	7290·9450
23	5	2211·4669	7	0	7629·5769
22	6	2550·0988	6	1	7968·2088
21	0	2888·7306	5	2	8306·8406
20	1	3227·3625	4	3	8645·4725
19	2	3565·9944	3	4	8984·1044
18	3	3904·6263	2	5	9322·7263
17	4	4243·2581	1	6	9661·3681
16	5	4581·8900	0	0	0

N.B.—The use of this Table is explained in example 1.

TABLE LXXX.

THE SUN'S MEAN LONGITUDE DURING THE HINDU SOLAR YEAR, IN 10,000THS OF CIRCLE, ACCORDING TO THE FIRST ĀRYA SIDDHĀNTA, AT PERIODS OF 24 HOURS EACH, MEASURED FROM THE MOMENT OF MEAN MĒSHA-SAṀKRĀNTI.

The same in degrees, etc., can be calculated by Table XLIV, above.

24-hour period.	Sun's mean longitude.
1	2
At moment of mean Mēsha-saṁkrānti.	*0*
1	27·3779
2	54·7557
3	82·1336
4	109·5114
5	136·8893
6	164·2671
7	191·6450
8	219·0229
9	246·4007
10	273·7786
11	301·1564
12	328·5343
13	355·9121
14	383·2900
15	410·6679
16	438·0457
17	465·4236
18	492·8014
19	520·1793
20	547·5571
21	574·9350
22	602·3129
23	629·6907
24	357·0686
25	684·4464
26	711·8243
27	739·2021
28	766·5800
29	793·9579
30	821·3357
At moment of mean Vrishabha saṁkrānti.	*833·3*
31	848·7136
32	876·0914
33	903·4693
34	930·6471
35	958·2250
36	985·6029
37	012·9807
38	040·3586
39	1067·7364
40	1095·1143
41	122·4921

24-hour period.	Sun's mean longitude.
1	2
42	1149·8700
43	1177·2479
44	1204·5257
45	1232·0036
46	1259·3814
47	1286·7593
48	1314·1371
49	1341·5150
50	1368·8929
51	1396·2707
52	1423·6486
53	1451·0264
54	1478·4043
55	1505·7821
56	1533·1600
57	1560·5379
58	1587·9157
59	1615·2936
60	1642·6714
At moment of mean Mithuna saṁkrānti.	*1666·6̇*
61	1670·0493
62	1697·4271
63	1724·8050
64	1752·1829
65	1779·5607
66	1806·9386
67	1834·3164
68	1861·6943
69	1889·0721
70	1916·4500
71	1943·8279
72	1971·2057
73	1998·5836
74	2025·9614
75	2053·3393
76	2080·7171
77	2108·0950
78	2135·4729
79	2162·8507
80	2190·2286
81	2217·6064
82	2244·9843
83	2272·3621
84	2299·7400
85	2327·1179
86	2354·4957

24-hour period.	Sun's mean longitude.
1	2
87	2381·8736
88	2409·2514
89	2436·6293
90	2464·0071
91	2491·3850
At moment of mean Karka saṁkrānti.	*2500·0*
92	2518·7629
93	2546·1407
94	2573·5186
95	2600·8964
96	2628·2743
97	2655·6521
98	2683·0300
99	2710·4079
00	2737·7857
01	2765·1636
102	2792·5414
103	2819·9193
104	2847·2971
105	2874·6750
106	2902·0529
107	2929·4307
108	2956·8086
109	2984·1864
110	3011·5643
111	3038·9421
112	3066·3200
113	3093·6979
114	3121·0757
115	3148·4536
116	3175·8314
117	3203·2093
118	3230·5872
119	3257·9650
120	3285·3429
121	3312·7207
At moment of mean Siṁha saṁkrānti.	*3333·3̇*
122	3340·0986
123	3367·4764
124	3394·8543
125	3422·2322
126	3449·6100

24-hour period.	Sun's mean longitude.
1	2
127	3476·9879
128	3504·3657
129	3531·7436
130	3559·1214
131	3586·4993
132	3613·8772
133	3641·2550
134	3668·6329
135	3696·0107
136	3723·3886
137	3750·7664
138	3778·1443
139	3805·5222
140	3832·9000
141	3860·2779
142	3887·6557
143	3915·0336
144	3942·4114
145	3969·7893
146	3997·1672
147	4024·5450
148	4051·9229
149	4079·3007
150	4106·6786
151	4134·0564
152	4161·4343
At moment of mean Kanyā saṁkrānti.	*4166·6̇*
153	4188·8122
154	4216·1900
155	4243·5679
156	4270·9457
157	4298·3236
158	4325·7014
159	4353·0793
160	4380·4572
161	4407·8350
162	4435·2129
163	4462·5907
164	4489·9636
165	4517·3464
166	4544·7243
167	4572·1022
168	4599·4800
169	4626·8579
170	4654·2357
171	4681·6136

TABLE LXXX—*Contd.*

24-hour period.	Sun's mean longitude.	24-hour period.	Sun's mean longitude.	24-hour period.	Sun's mean longitude.	24-hour period.	Sun's mean longitude.
1	2	1	2	1	2	1	2
172	4708·9914	220	6023·1286	272	7446·7772	320	8760·9143
173	4736·3693	221	6050·5064	273	7474·1550	321	8788,2922
174	4763·7472	222	6077·8843	*At moment of mean Makara saṁkrānti.*	*7500·0*	322	8815·6700
175	4791·1250	223	6105·2622	274	7501·5329	323	8843·0479
176	4818·5029	224	6132·6400	275	7528·9107	324	8870·4257
177	4845·8807	225	6160·0179	276	7556·2886	325	8897·8936
178	4873·2586	226	6187·3957	277	7583·6664	326	8925·1814
179	4900·6364	227	6214·7736	278	7611·0443	327	8952·5593
180	4928·0143	228	6242·1514	279	7638·4222	328	8979·9372
181	4955·3922	229	6269·0593	280	7665·8000	329	9007·3150
182	4982·7700	230	6296·9072	281	7693·1779	330	9034·6929
At moment mean Tulā saṁkrānti.	*5000·0*	231	6324·2850	282	7720·5557	331	9062·0707
183	5010·1479	232	6351·6629	283	7747·9336	332	9089·4486
184	5037.5257	233	6379·0407	284	7775·3114	333	9116·8264
185	5064·9036	234	6406·[illegible]186	285	7802·6893	334	9144·2043
186	5092·2814	235	6433·7964	286	7830·0672	*At moment of mean Mīna-saṁkrānti.*	*9166·6̇*
187	5119·6593	236	6461·1743	287	7857·4450	335	9171·5822
188	5147·0372	237	6488·5522	288	7884·8229	336	9198·9600
189	5174·4150	238	6515·9300	289	7912·2007	337	9226·3379
190	5201·7929	239	6543·3079	290	7939·5786	338	9253·7157
191	5229·1707	240	6570·6857	291	7966·9564	339	9281·0936
192	5256·5486	241	6598·0636	292	7994·3343	340	9308·4715
193	5283·9264	242	6625·4414	293	8021·7122	341	9335·8493
194	5311·3043	243	6652·8193	294	8049·0900	342	9363·2272
195	5338·6822	*At moment of mean Dhanuḥ saṁkrānti.*	*6666·6̇*	295	8076·4679	343	9390·6050
196	5366·0600	244	6680·1972	296	8103·8457	344	9417·9829
197	5393·4379	245	6707·5750	297	8131·2236	345	9445·3607
198	5420·8157	246	6734·9529	298	8158·6014	346	9472·7386
199	5448·1936	247	6762·3307	299	8185·9793	347	9500·1165
200	5475·5714	248	6789·7086	300	8213·3572	348	9527·4943
201	5502·9493	249	6817·0864	301	8240·7350	349	9554·8722
202	5530·3272	250	6844·4643	302	8268·1129	350	9582·2500
203	5557·7050	251	6871·8422	303	8295·4907	351	9609·6279
204	5585·0829	252	6899·2200	304	8322·8686	352	9637·0057
205	5612·4607	253	6926·5979	*At moment of mean Kumbha saṁkrānti.*	*8333·3̇*	353	9664·3836
206	5639·8386	254	6953·9757	305	8350·2464	354	9691·7615
207	5667·2164	255	6981·3526	306	8377·6243	355	9719·1393
208	5694·5943	256	7008·7314	307	8405·0022	356	9746·5172
209	5721·9722	257	70[illegible]·1093	308	8432·3800	357	9773·8950
210	5749·3500	258	7063·4872	309	8459·7579	358	9801·2729
211	5776·7279	259	7090·8650	310	8487·1357	359	9828·6507
212	5804·1057	260	7118·2429	311	8514·5136	360	9856·0286
213	5831·4836	261	7145·6207	312	8541·8914	361	9883·4065
At moment of mean Vṛiśchika saṁkrānti.	*5833·3̇*	262	7172·9986	313	8569·2693	362	9910·7843
214	5858·8614	263	7200·3764	314	8596·6472	363	9938·1622
215	5886·2393	264	7227·7543	315	8624·0250	364	9965·5400
216	5913·6172	265	7255·1322	316	8651·4029	265	9992·9179
217	5940·9950	266	7282·5100	317	8678·7807	*At moment of mean Mēsha-saṁkrānti of following year*	*10,000·0*
218	5968·3729	267	7309·8879	318	8706·1586		
219	5995·7507	268	7337·2657	319	8733·5364		
		269	7364·6436				
		270	7392·0214				
		271	7419·3993				

TABLE LXXXI.

SUN'S MEAN LONGITUDE. INCREASE IN FRACTIONS OF DAY ACCORDING TO THE FIRST ARYA SIDDHĀNTA.

(For the same in degrees, etc., see above, Table XLIV.)

Increase per hour.		Increase per minute.				Increase per second.			
No.	In 10,000ths of circle.	No.	In 10,000ths of circle.	No.	In 10,000ths of circle.	No.	In 10,000ths of circle.	No.	In 10,000ths of circle.
1	1·1407	1	0·0190	31	0·5894	1	0·0003	31	0·0098
2	2·2815	2	0·0380	32	0·6084	2	0·0006	32	0·0101
3	3·4222	3	0·0570	33	0·6274	3	0·0010	33	0·0105
4	4·5630	4	0·0760	34	0·6464	4	0·0013	34	0·0108
5	5·7037	5	0·0951	35	0·6654	5	0·0016	35	0·0111
6	6·8445	6	0·1141	36	0·6844	6	0·0019	36	0·0114
7	7·9852	7	0·1331	37	0·7035	7	0·0022	37	0·0117
8	9·1260	8	0·1521	38	0·7225	8	0·0025	38	0·0120
9	10·2667	9	0·1711	39	0·7415	9	0·0029	39	0·0124
10	11·4074	10	0·1901	40	0·7605	10	0·0032	40	0·0127
11	12·5482	11	0·2091	41	0·7795	11	0·0035	41	0·0130
12	13·6889	12	0·2281	42	0·7985	12	0·0038	42	0·0133
13	14·8297	13	0·2472	43	0·8175	13	0·0041	43	0·0136
14	15·9704	14	0·2662	44	0·8365	14	0·0044	44	0·0139
15	17·1112	15	0·2852	45	0·8556	15	0·0048	45	0·0143
16	18·2519	16	0·3042	46	0·8746	16	0·0051	46	0·0146
17	19·3926	17	0·3232	47	0·8936	17	0·0054	47	0·0149
18	20·5334	18	0·3422	48	0·9126	18	0·0057	48	0·0152
19	21·6741	19	0·3612	49	0·9316	19	0·0060	49	0·0155
20	22·8149	20	0·3802	50	0·9506	20	0·0063	50	0·0158
21	23·9556	21	0·3993	51	0·9696	21	0·0067	51	0·0162
22	25·0964	22	0·4183	52	0·9886	22	0·0070	52	0·0165
23	26·2371	23	0·4373	53	1·0077	23	0·0073	53	0·0168
		24	0·4563	54	1·0267	24	0·0076	54	0·0171
		25	0·4753	55	1·0457	25	0·0079	55	0·0174
		26	0·4943	56	1·0647	26	0·0082	56	0·0177
		27	0·5133	57	1·0837	27	0·0086	57	0·0181
		28	0·5323	58	1·1027	28	0·0089	58	0·0184
		29	0·5514	59	1·1217	29	0·0092	59	0·0187
		30	0·5704			30	0·0095		

THE BRAHMA-SIDDHANTA OF BRAHMAGUPTA (A.D. 628).

WORKING TABLES FOR COMPUTATION OF ANCIENT DATES BY THE TRUE, OR APPARENT, MOTIONS OF SUN AND MOON.

311. In para. 257 of my article in the *Epigraphia Indica* (Vol. XIV, pp. 241f.) on "*The true longitude of the sun in Hindu astronomy, the Siddhānta-Śirōmaṇi*" and again in a later article (Vol. XV) on *The Siddhānta-Śirōmaṇi*, § 271 I discussed the question of the values assigned in the seventh century A.D. by Brahmagupta to the twenty-four base-sines of angles in the quadrant; and expressed the opinion that when, but not until, definite assurance was obtainable that the values stated in the only available copies of the *Brahma-Siddhānta*[1] were really those fixed by its author, working Tables framed according to its postulates might safely be prepared for the computation of ancient dates.

In response to my appeal Mr. G. R. Kaye (Curator, Board of Education, Simla) has been kind enough to assist me. He tells me that there can be no doubt but that the values given for the several base-sines in the edition of the *Brahma-Siddhānta* printed and published in Benares are correct, and that Brahmagupta certainly made his calculations with a radius (sin. 90°) of 3270′, discarding that of 3438′, which seemingly had been in use in India since the time of the Greeks.[2] Mr. Kaye went fully into the subject in a very learned article, "*Ancient Hindu Spherical Astronomy*," published in the *Journal of the Asiatic Society of Bengal* in 1919 (*New Series, Vol. XV, No. 3*), which contains (*Table 8, p. 187*) a list of the sine-values as determined by the authors of the *Pauliśa-*, *Ārya-*, and *Brahma-Siddhāntas*. He points out that, when properly applied, the equations of the sun's and moon's centres obtained from the sine-values of Brahmagupta agree with those derived from the values assigned by the other authorities.

Accordingly I have prepared the Table of Brahmagupta's sines and resulting base-equations of the sun's centre (*Table LXXXIX below*); and a comparison between these and the equations of the *Siddhānta-Śirōmaṇi* (*Table XLVII above and Prof. Jacobi's Tables, XXIV-B, Epig. Ind., Vol. I*) proves that there is only a very trifling difference whether we use Brahmagupta's, or the older—and later—sine-values. By the *Siddhānta-Śirōmaṇi*, with radius 3438′, the sun's greatest equation, that of 90°, is 2° 10′ 31″, exact. By the *Brahma-Siddhānta*, with radius 3270′, it is 2° 10′ 31″·19. We may therefore safely use Table LXXXIX (*below*)[3] and Table LIX (*above*) for the sun's and moon's equations by the *Brahma-Siddhānta*.

312. The *Brahma-Siddhānta* was composed by Brahmagupta in A.D. 628 and is said to have been extensively used in some parts of India, its principal rival being the *Ārya-Siddhānta* of Āryabhaṭa, known in later years as the *Laghu-Ārya* to distinguish it from the *Mahā-Ārya-Siddhānta* of the tenth century. This last, called also the *Second Ārya-Siddhānta*, seems to have had no great following. The *Rāja-mṛigāṅka*, an astronomical work of A.D. 1042, introduced, according to the information available to the late Sankara Balkrishna Dikshit some important changes into the system of Brahmagupta; but unfortunately no complete copy of it has yet been obtained, and the necessary particulars are not to be found in those fragments

[1] One MS. copy in the India Office, London, and the Benares printed edition.

[2] It would be interesting to learn his reason for the change. Later Indian astronomers reverted to the radius of 3438′, Sin. 90°=radius. With π (ratio of diam. to circumf.) = 3·14159, its accepted modern value, the radius = 3437·74967. According to the *Ārya-* and *Sūrya-Siddhāntas* it is taken as 3438′. Archimedes ratio' was π = 3·14286. The *Sūrya-Siddhānta* alludes to a ratio $\pi = 1 : \sqrt{10}$, which works out to 3·16228. Brahmagupta's radius being 3270′, his ratio must have been $\pi = 3·303$, which is quite different to any of these others.

[3] Or Table XLVII (*above*), col. 9; also Professor Jacobi's Tables XXIV-A, XXIV-B (*Epig. Ind., Vol. I*).

which have come to light. It is not possible therefore to frame any accurate Tables for calculation by the *Rājā-mṛigāṅka*, and we must rest satisfied with the assurance of Mr. S. B. Dikshit[1] that the *Siddhānta-Śirōmaṇi* is the same as the *Rāja-mṛigāṅka* in the matter of calculation of an almanack. Tables for use by the former have already been published by me, comprising the period A.D. 1100-1750 (*above*).

All the authorities appear to arrive at similar or almost similar results in their computation of the lunar tithis, when worked by the true or apparent motions of sun and moon; but, since they differ in their estimate of the position of the sun's apsis at a given date, they necessarily differ somewhat in their estimate of the moment in each year when the true sun reaches long. 0°, the moment, that is, of "true Mēsha-saṁkrānti." This difference leads to differences in the lengths of the true solar months, and consequently to differences in the intercalation and suppression of true lunar months; which differences, again, occasionally cause differences of a whole lunar month in the beginning of the luni-solar year and differences in the names of some of the lunar months therein.

But we are now better able to deal with these matters than before. Dates can be easily computed by the true motions of sun and moon according to the *Sūrya-Siddhānta* for the whole historical period from A.D. 300 to 1900 (*Indian Calendar*)[2]; according to the *Ārya-Siddhānta* from A.D. 900 to 1900 (*above*); according to the *Brahma-Siddhānta* (*the present paper*) from A.D. 600 to 1200; and according to the *Siddhānta-Śirōmaṇi*, *Rāja-mṛigāṅka* and other works of the time of Bhāskarāchārya from A.D. 1100 to 1900 (*above*); these periods comprising the outside limits of use.

And, as regards computation by the mean motions of sun and moon, which system is believed to have been in universal use down to about A.D. 1100, and perhaps in some places to a considerably later date, we now have Tables for work by the *Ārya-Siddhānta* from A.D. 500 to 1400 (*above*), and by the *Brahma-Siddhānta*, from A.D. 500 to 1400 (*below*).

All these Tables are framed on the same system, so as to enable calculation to be made as easily and rapidly as possible.

Elements of the Brahma-Siddhānta.

313. (i) The length of the mean solar sidereal year is 365·2584375 days, or $365^d\ 6^h\ 12^m\ 9^s$ The *Siddhānta-Śirōmaṇi* adhered to this estimate.

(ii) Brahmagupta's sines of angles of the quadrant differ from those of the other authorities. His sine of 90°, the radius, = 3270′ instead of 3438′. His sine of 3° 45′ = 214′ instead of 225′. The 24 base-sines are given in Table LXXXIX below.

(iii) The equations, however, which are based on these sine-values are practically the same as those of the *Siddhānta-Śirōmaṇi* (*compare Table XLVII above, col. 9, and Table LXXXIX below*). Tables LV, LVI, LIX (*above*) may be therefore used as well for the *Brahma-Siddhānta* as for the *Siddhānta-Śirōmaṇi*.

(iv) The greatest equation of the sun's centre, that of 90°, is, in 10,000ths of the circle, 60·425925. The greatest equation of the moon's centre is, in similar measurement, 139·855101852. The sum of the two is 200·284027777.

---

1 *Indian Calendar*, p. 8.

2 Also by the *Indian Chronology* of Dewan Bahadur L. D. Swamikannu Pillai, M.A., whose Tables are framed on different system.

(v) The epoch of the Kaliyuga era was mean sunrise, taken as 6 A.M., on Friday, 18 February, B.C. 3102, that moment being $0^h\ 0^m\ 0^s$ Laṅkā time. This was the moment of mean Mēsha-saṁkrānti, when the mean sun's centre reached long. 0°. True Mēsha-saṁkrānti, when the true sun's centre reached long. 0°, occurred on Tuesday, 15 February, B.C. 3102, at $19^h\ 52^m\ 21^s{\cdot}5$ after mean sunrise at Laṅkā.

(vi) The circumference of the sun's epicycle is 13° 40′, that of the moon 31° 46′. The epicycles are not contracted at any point. In this the *Siddhānta-Śirōmaṇi* concurs (*Jacobi, Epig. Ind., Vol. I, p. 441*).

(vii) The line of apsides of the sun's orbit has a constant forward shift, the perigee-point (on the longitude of which all calculations in this volume are based) moving 0″·144 per ann., or 14″·4 in a century. According to the *Siddhānta-Śirōmaṇi* the movement is more rapid, amounting to 1″·044 per ann. (*Jacobi, op. cit.*).

(viii) The *śōdhya*, or time-interval between true and mean Mēsha-saṁkrāntis, was, in K.Y. 0 or at the epoch of the Kaliyuga era, according to Dr. Schram,[1] 2·171971 or $2^d\ 4^h\ 7^m\ 38^s{\cdot}5$. With this the *Siddhānta-Śirōmaṇi* agrees. But in later years the *śōdhya*, as postulated by the two authorities, differs in value owing to the difference between the two *Siddhāntas* in their estimate of the movement of the sun's apsis. (*See vii above.*)

(ix) The position of the sun's apsis (perigee) at K.Y. 0, the epoch of the Kaliyuga, was 257° 45′ 36″,[2] and his mean anomaly was 102° 14′ 24″, or, in 10,000ths of the circle, 284·0.

(x) The position of the moon's apsis (perigee) at the same moment was 305° 29′ 46″ [3]; and her mean anom. was 54° 30′ 14″, or, in 1,000ths of circle, 151·399691358.

(xi) The sun's mean velocity (he is treated as a planet) and the length of the mean solar year being the same both by the *Brahma-Siddhānta* and the *Siddhānta-Śirōmaṇi*, his mean long. at any moment must be the same by both, and so also the length of the mean solar month. But the two authorities are not in exact accord as to his true long. and the length of the true solar month.

Shift of sun's apsis. The śōdhya. Length of true solar year.

314. The length of the mean solar year being the same, *viz.* $365^d\ 6^h\ 12^m\ 9^s$, by both the *Brahma-Siddhānta* and the *Siddhānta-Śirōmaṇi*, the first portion of § 273 above and accompanying Table A apply as well to the former as to the latter. But for the latter portion of that section and its Table B, the following must be substituted when dealing with the *Brahma-Siddhānta*, the two authorities not being in accord as concerns the matter in question.

315. As stated above, the sun's perigee-point according to the *Brahma-Siddhānta* advances annually 0″·144 along the ecliptic, and in consequence of this shift the true sun's velocity at long. 0″ is a little greater every year than the year before, *i.e.* the true sun reaches long. 0°, or the moment of true Mēsha-saṁkrānti occurs, a little earlier each year. In every year there is a slight increase in the distance and time-difference (our *śōdhya*) between the mean and true suns at that point of the orbit. Dr. Schram has carefully calculated the value of this *śōdhya* at the moment of true Mēsha-saṁkrānti at the beginning of several millenniums, and his results for the period embraced in my general working Table LXXXII are stated in the following Table B.

---

[1] *Indian Chronography*, § 39 D, p. 16.

[2] Jacobi, *Epig. Ind.*, Vol. I, p. 442, § 83, where he gives the place of the apsis (apogee) as 77° 45′ 36″. See also E. Burgess's "*Sūrya-Siddhānta.*"

[3] Moon's apogee given by Jacobi as 125° 29′ 46″.

TABLE B.

VALUE OF ŚŌDHYA BY THE BRAHMA-SIDDHĀNTA.

K.Y. year expired.	A.D.	Exact value of śōdhya at beginning of centuries.				
		days and decimals.	d.	h.	m.	s.
3700	599-600	2·1729145	2	4	8	59·8128
3800	699-700	2·1729400	2	4	9	2·0160
3900	799-800	2·1729655	2	4	9	4·2192
4000	899-900	2·1729910	2	4	9	6·4224
4100	999-1000	2·1730165	2	4	9	8·6256
4200	1099-1100	2·1730420	2	4	9	10·8288
4300	1199-1200	2·1730675	2	4	9	13·0320

One result of this shift of apsis is that, by the *Brahma-Siddhānta*, the true sun reaches the 0° point of long. 0s·022032 earlier every year than the year before, and in consequence the length of the true solar year, or the time needed for the true sun to travel from true Mēsha-saṁkrānti in one year to true Mēsha-saṁkrānti in the next, is (365d 6h 12m 9s—0s·022032) 365d 6h 12m 8s·977968. [The exact moment of true Mēsha-saṁkrānti in each year from A.D. 599 to 1200 is given in the general Table LXXXII below, cols. 13-17. It can be tested by the use of Table A, § 273, referred to above, and Table B here given, using the "longer rule" stated in § 273 or in *Indian Chronography*, p. 61.]

Another result of the shift is that the sun's mean anomaly, or the mean sun's distance from the sun's perigee-point, decreases every year by 0″·144 or 14″·4 in a century. Reckoned in 1,000ths of circle for valuation of our "*c*" (sun's mean anom.) in the Tables, 14″·4 = 0·0$\dot{1}$. The value of "*c*" therefore decreases 0·0$\dot{1}$ in a century, and this decrease has to be taken into account from K.Y. 0, the epoch of the Kaliyuga. This has been done in the preparation of the Tables which follow.

The increase of "a", "b", "c", in centuries, years, days and fractions of days.

316. Following on what has been stated, we learn that Tables LIVA and B, which deal with the periodical increases of "*a*", "*b*" and "*c*" according to the *Siddhānta-Śirōmaṇi*, may safely be used for calculation by the *Brahma-Siddhānta*, with the one reservation as to the increase of "*c*" in a century. "*a*" being the distance of mean moon from mean sun, and the *longitude* of the mean sun not being affected by the shift of apsis, but only his *mean anom.*, or distance from the point of the apsis, it appears that the rate of increase of "*a*" must be same by both authorities.

As to the rate of increase of "*c*" it is, by the *Siddhānta-Śirōmaṇi*, centennially less by 0·080$\dot{5}$ (§ 273 above), and this was taken into account in the preparation of the heading of Table LIVA, where a footnote is appended shewing what the rate of increase would be per century if no such deduction had been made. This rate is, in thousandths of a circle, 997·690008075 in a century of 36525 days, and 0·427795618 in a century of 36526 days. By the *Brahma-Siddhānta*, the centennial decrease in the sun's mean anomaly being 0·0$\dot{1}$, the amount of increase of "*c*" per century is, for a century of 36525 days, 997·678896964, and for a century of 36526 days is

0·416684507. The difference between the two authorities in shorter periods may be ignored except in some extraordinarily close case. If it is ever needed, the increase in "*c*" in one year may be reduced by 0·0001 from the Table quantity.

Otherwise Tables LIV-A and B stand good for calculations by the *Brahma-Siddhānta.*

The values of "a", "b", "c" at the beginning of K.Y. 3700.

317. The general Table LXXXII below begins from the beginning of K.Y. 3700 expired. Table LXXXVI states the value of "*a*", "*b*", "*c*" at that moment, and at the similar moment at the beginning of subsequent centuries. It is necessary therefore to explain how these figures were calculated.

(i) *The value of* "a" (*distance of mean moon from mean sun*) *in K.Y. 3700.* According to Hindu astronomers mean moon and mean sun were in conjunction at the moment of mean Mēsha-saṁkrānti in K.Y. 0, the epoch of the Kaliyuga; or, in other words, at that moment "*a*" = 0. In the 37 succeeding centuries there were 32 common and 5 defective centuries. Taking the century values of "*a*" given in the heading of Table LIV-A and multiplying for 32 common and 5 defective centuries, we arrive at the figure 6567·108945284 as the value of "*a*" at the beginning of the 37th century K.Y., whole revolutions of 10,000 each being omitted. From this figure has to be deducted,—according to the working system of the *Indian Calendar,* which follows Largeteau and Jacobi,—the sum of the greatest equations of sun and moon, *viz.* 200·284027̇ (*above* § *313, iv*). This gives us the value of "*a*" at the beginning of K.Y. 3700 (expired) as 6366·824917506.[1]

Now this value stands for mean sunrise of Sunday, 22 March, A.D. 599, *i.e.* for the sunrise succeeding the moment of occurrence of mean Mēsha-saṁkrānti in K.Y. 3700; but in all my Tables the calculation is for mean sunrise on the actual day of that occurrence, and we have therefore to deduct one day's value of "*a*" (*viz.* 338·631985412—*Table LIV-A above*) from the above estimate. This done, we have, for mean sunrise on Saturday, $a = 6028 \cdot 192932094$.

(ii) *The value of* "b" (*moon's mean anom.*) *at the same moment.* At the *epoch* of the Kaliyuga the moon's mean anom. was, as stated above (§ *313, x*), in 1,000ths of a circle, 151·399691358. Using the century figures of "*b*" in the heading of Table LIV-A, and multiplying for 32 common and 5 defective centuries, it is found that, excluding whole revolutions of 1,000 each, the result is 604·144838202. Adding the value of "*b*" at K.Y. 0, as above, we have for the value of "*b*", at beginning of K.Y. 3700, 755·544529560.[2] But this (*see above, i*) was its value at mean sunrise on Sunday, 22 March, A.D. 599. Deducting one day's value of "*b*" (36·291649786) the fixture for mean sunrise on Saturday, 21 March, amounts to 719·252879774.

(iii) *The value of* "c" (*the sun's mean anom.*) *at the same moment.* The correct increase of "*c*" by the *Brahma-Siddhānta* in centuries of 36525 and 36526 days has been given above in the latter part of § 316. Multiplying those quantities for 32 common and 5 defective centuries, and discarding whole revolutions of 1,000 each, we arrive at the increase, after 37 centuries, of 1·728389044. To this has to be added the value of "*c*" at K.Y. 0 (*above,* § *313, ix*), *viz.* 284·0. The value of "*c*", therefore, at mean sunrise of Sunday, 22 March, A.D.5 99, was 285·728389044[3]. Deducting the "*c*" for one day (2·737787543) we have finally, for mean sunrise on Saturday, 21 March, "*c*" = 282·990601501.

[1] Professor Jacobi differs by about 17 units. He gives the figure 6384·0 (*Epig. Ind., Vol. XI, p. 167, Table IXA*). I can give no explanation of the reason for this; and can only state fully, as in the text, my bases of calculation.

[2] Professor Jacobi's figure for this is 758·1, in my notation, against my 755·5.

[3] This agrees with Professor Jacobi's fixture, which, measured from perigee and in my notation, is 285·7.

The entries, therefore, for the aforesaid Saturday of K.Y. 3700 in Table LXXXVI below are

$$a = 6028{\cdot}1929$$
$$b = 719{\cdot}2529$$
$$c = 282{\cdot}9906.$$

The rest of that Table follows by addition of the proper century values.

Duration of true solar months.

318. It has been mentioned above (§ *313, xi*) that, while the length of the mean solar month must be the same both by the *Brahma-Siddhānta* and the *Siddhānta-Śirōmaṇi*, the lengths of the true solar months according to the two authorities differ because of their different estimate of the shift of the sun's apsis. Thus in K.Y. 4000, the middle year of my general Table LXXXII below, the sun's perigee-point according to the *Siddhānta-Śirōmaṇi* was at long. 258° 55′ 12″, while by the *Brahma-Siddhānta* it was at long. 257° 55′ 12″. Hence the velocity of the true sun (he is always considered as a planet) at the several true solar *saṁkrāntis, i.e.* when the true sun's centre enters the several signs, is not the same by the two authorities quoted. And this has necessitated the preparation of a new Table (*LXXXIII-A below*), giving the lengths of the true solar months and increase of "*a*", "*b*", "*c*" therein individually and collectively according to the *Brahma-Siddhānta*.

There being in K.Y. 4000 a difference of only 4′ 48″ between the positions of the sun's perigee, as estimated by the *Brahma-Siddhānta* and by the *First Ārya-Siddhānta*, the former placing it at 257° 55′ 12″ and the latter at 258°, it was considered sufficiently safe to use Table XLIX (*above*) for the true sun's velocity at different points of his orbit in hours and minutes, and Table L-A for seconds. His true long. at each *saṁkrānti* was computed from his known mean longitude + the equation of the centre, which was calculated in each case. Thus was obtained the length of each month in days, hours, etc. For the increase of "*a*", "*b*", "*c*" during the periods so determined Tables LIV-A and B, which are applicable to the *Brahma-Siddhānta* as well as to the *Siddhānta-Śirōmaṇi*, were used.

Note on work for the nakshatra.

319. In our method of work "*s*" = the true sun's longitude and "*t*" = the *tithi*-index (which shews the true moon's distance from the true sun) at the given moment. $s + t =$ the *nakshatra*-index "*n*", which gives the true moon's place in the heavens, or her apparent longitude. The value of "*t*" is ascertained by the ordinary calculation for a date. The value of "*s*" has to be found.

By the *Ārya-Siddhānta* the formula for finding "*s*", "*c*" being the sun's mean anom. at the given moment, is $s = (c \times 10) + 7226 -$ eqn. c; where the factor 7226, which represents in 10,000ths of circle the long. of sun's perigee *plus* the sun's greatest equation, is a constant.[1]

By the *Sūrya-Siddhānta*, as exemplified in the *Indian Calendar*, the numerical factor is not 7226, but varies in the period A.D. 900 to 1900 from 7206·5077 to 7207·4035, being fixed for rough work at 7207. The variation is due to the postulated shift of the sun's perigee-point.

By the *Siddhānta-Śirōmaṇi* there is, for the same reason, a variation in the numerical factor, *viz.* from 7252·6466 in A.D. 900 to 7259·0910 in A.D. 1700,—roughly from 7253 to 7259.

[1] See *Indian Calendar*, § 156, p. 97; article on the *Siddhānta-Śirōmaṇi*, above, § 273. "*Note on work for the nakshatra*"; article on the *First Ārya-Siddhānta*, above, § 302: and the several examples given in those papers.

By the *Brahma-Siddhānta* the numerical factor varies from $7224{\cdot}5\dot{3}7\dot{0}$ in A.D. 600 to $7225{\cdot}2\dot{0}3\dot{7}$ in A.D. 1200 (the limits of the general Table LXXXII below) For rough work therefore by this authority the formula is $s = (c \times 10) + 7225 -$ eqn. c

For more accurate work the value of "c" should be calculated (by the Tables) with decimals, and instead of multiplying "c" by 10 its value should be changed from thousandths of circle (as in the Table-result) to ten thousandths by moving the decimal point one place to the right and, when the whole number consists of four figures, deleting the last figure on the left[1] the value of "eqn. c" can be obtained from Table LVI with great accuracy: and the numerical factor can be taken from the following summary.

K.Y. century.	A.D. century.	Exact factor in formula.	Roughly.
3700	599-600	$7224{\cdot}5\dot{3}7\dot{0}$	
3800	699-700	$7224{\cdot}6\dot{4}8\dot{1}$	
3900	799-800	$7224{\cdot}7\dot{5}9\dot{2}$	
4000	899-900	$7224{\cdot}8\dot{7}0\dot{3}$	7225
4100	9[illegible]9-1000	$7224{\cdot}9\dot{8}14$	
4200	1099-1100	$7225{\cdot}0\dot{9}2\dot{5}$	
4300	1199-1200	$7225{\cdot}2\dot{0}3\dot{7}$	

Examples.

It is not necessary to give a number of examples of work by the present Tables. The system of calculation being exactly the same as that of the *Indian Calendar* and throughout the present series of articles, the examples already published for computation by other authorities will suffice, *the proper Tables being used*, for work by the *Brahma-Siddhānta*. These Tables are specified in the following pages.

Tables for calculation by the Brahma-Siddhānta.

The system of work for computation of an Indian date will be readily understood by perusal of examples 2 to 11 appended to my paper (*above*) on the *First Ārya-Siddhānta*; but the Tables used are of course not all the same. The following list shews how accurate results by the *Brahma-Siddhānta* are to be obtained in calculation by the movements of true sun and true moon.

Table LXXXII below is the general working Table for the *Brahma-Siddhānta* for the period A.D. 599 to 1200 (K.Y. 3700 to 4300 expired).

For names of months and of nakshatras in different parts of India, see Table LXII above ("*The First Ārya-Siddhānta*").

For collective duration of mean lunar months see Table LXIII-A of the same article, or Table III, Part I, *Indian Calendar*.

Table LXXXIII-A below gives, by the *Brahma-Siddhānta*, the length of the true solar months and their collective dûration, with the corresponding increases of "a", "b", "c".

Table LXXXIII-B states the exact value of "c" and of "equation c" at the several true *saṁkrāntis*, or moments of the true sun's centre reaching the several signs.

[1] Whole revolutions are not necessary for present purposes, and in our system when "a" → 10,000 a whole synodic revolution of the mean moon has been completed.

Table LXXXIII-C shews the value of "*c*" and of "equation *c*" at the beginning of each century of the Kaliyuga.

For the increase of "*a*", "*b*", "*c*" respectively in defective and common centuries, and in common years and Leap-years, see Table LIV-A, heading; but note that by the *Brahma-Siddhānta* the increase of "*c*" in a defective century of 36525 days is 997·678896964 and in a common century of 36526 days is 0·416684507. Tables LIV-A and B contain the necessary figures for days, hours, minutes and seconds.

Table LXXXIV gives the values of "equation *b*," and Table LXXXV those of "equation *c*," for easy calculation by whole numbers, corresponding respectively to Tables VI and VII of the "*Indian Calendar*," which stand for the *Sūrya-Siddhānta*.

For the more detailed values of "equation *b*" and "equation *c*" of moon and sun use Tables LV and LVI above, Vol. XV, as framed for the *Siddhānta-Śirōmaṇi*.

For the indices of *tithis* ("*t*"), *karaṇas*, *yōgas* ("*y*") and *nakshatras* ("*n*") see Table VIII, "*Indian Calendar*," or Table LXVIII (*above*).

For serial numbers of days of a year reckoned from January 1st use Table IX, "*Indian Calendar*," or Table LXIX (*above*).

For conversion of *tithi*-indices and *tithi*-parts into time Table X, "*Indian Calendar*," is to be used, or Table LXX (*above*.)

For finding the week-day according to the European Calendar for any century from A.D. 4 to 2300 see Table LXXI (*above*), or Table XLI-A and B (*pp. 176, 177*, "*Indian Chronography*")

Table LXXXVI gives the values of "*a*", "*b*", "*c*" at the beginning of each century of the Kaliyuga by the *Brahma-Siddhānta*.

Table LXXXVII gives the same for odd years of those centuries.

Table LXXXVIII states the daily sunrise values of "*a*", "*b*", "*c*" for a month previous to the day of Mēsha-saṁkrānti.

Table LXXXIX sets forth the 24 base-sines of angles of the quadrant according to Brahmagupta, and the corresponding equations of the sun's centre.

TABLE LXXXII.

Construction of Table.

The Table is constructed on the lines of Table I of the *Indian Calendar* and is to be used in the same way. The columns are numbered similarly.

Col. 7. The *saṁvatsara*-name,—*i.e.* the name of the Jovian cycle—, of the year is given as determined by my previous calculations (*above, Table XLII*). Entries in italics point to cases where this *saṁvatsara*-name differs from that given to the same year by *Sūrya-Siddhānta* reckoning.

Col. 8. Months noted in roman characters are intercalated (*adhika*) lunar months. Those in italics are suppressed (*kshaya*) months.

Cols. 13, 19. Figures in brackets give the serial number of the day measured from January 1st.

Col. 23. "*a*"=distance, at mean sunrise (taken as 6 A.M.) on the day noted in cols. 19, 20, of mean moon from mean sun, *i.e.*, phase of moon at that moment; stated in 10,000ths of circle and reduced by the sum of the greatest equations of sun and moon, so that calculation of the equations of "*b*" and "*c*" may always be additive.

Col. 24. "*b*"=mean anomaly of moon at the same moment, or mean moon's distance from the perigee-point of her apsis, stated in 1,000ths of circle.

Col. 25. "*c*"=mean anomaly of sun at the same moment, or mean sun's distance from his perigee-point, stated in 1,000ths of circle.

Remarks.

A.D. 629-630, *cols. 19, 20.* A very close case. The moment of true new moon was less than half a minute after mean sunrise at Laṅkā on Wednesday, 1st March. And the first *śukla tithi* of the year ended after mean sunrise on Thursday, 2nd March, which was therefore by rule the first civil day of the luni-solar year. If new moon had taken place more than half a minute earlier the first civil day of the year, "Chaitra śukla 1," would have been 1st March.

A.D. 968-69, *col. 8.* At the Kumbha *saṁkrānti* the true moon was waning. The moment of the next, the Mīna, *saṁkrānti* occurred about $2\frac{1}{2}$ minutes after the moment of true new moon, so that the true moon was waxing at the Mīna *saṁkrānti*. Hence the lunar month Phālguna was intercalated. According to the 19-year sequence we should have expected an intercalation of the lunar month Chaitra next following. The sequence shows similar irregularities when examined by other authorities, but only very rarely.

A.D. 974-75, *cols. 19, 20.* Close case. The 1st true new moon after the Mīna *saṁkrānti* occurred 3 minutes before mean sunrise at Laṅkā on 25th February A.D. 974. That therefore was the day "Chaitra śukla 1."

A.D. 963-64, 982-83, *col. 8.* In both these years an intercalation of the lunar month Śrāvaṇa instead of Āshāḍha would have been more in accordance with the 19-year sequence, seeing that Śrāvaṇa was the intercalated month in A.D. 1001 and 1020; but prior to A.D. 963 at intervals of 19 years there had been eight intercalations of Śrāvaṇa, and towards the close of such a run a change of conditions generally becomes apparent.

A.D. 1001-2, 1020-21, *col. 8.* See the previous note. If in these two years the conditions had made necessary an intercalation of Āshāḍha, the 19-year sequence would have been uninterrupted.

A.D. 1128-29, *col. 8.* By the *Brahma-Siddhānta* the intercalation of Phālguna was clearly demanded. See Remarks preceding Table LX. (*above*), on the same year as worked by the *Siddhānta-Śirōmaṇi*.

TABLE

GENERAL TABLE FOR CALCULATION

Conforming to Table I "Indian Calendar"

(See notes on

CONCURRENT YEAR.								
						Jovian Saṁvatsara.		Intercalated (*adhika*) and suppressed (*kshaya*) true lunar months.
Kali.	Śaka.	Chaitrādi Vikrama.	Mēshādi solar year in Bengal.	Kollam.	A. D.	Southern system.	Northern system.	
1	2	3	3*a*	4	5	6	7	8*a*
3701	522	657	6		599-600	50 Anala . . .		...
3702	523	658	7		*600-01	51 Piṅgala . . .		3 Jyēshṭha .
3703	524	659	8		601-02	52 Kālayukta . . .		...
3704	525	660	9		602-03	53 Siddhārthin .		{ 7 Āśvina 11 *Māgha* (*ksh.*) }
3705	526	661	10		603-04	54 Raudra . . .		1 Chaitra .
3706	527	662	11		*604-05	55 Durmati . . .		...
3707	528	663	12		605-06	56 Dundubhi . . .		5 Śrāvaṇa .
3708	529	664	13		606-07	57 Rudhirōdgārin . .		...
3709	530	665	14		607-08	58 Raktāksha . .		...
3710	531	666	15		*608-09	59 Krōdhana . . .		4 Āshāḍha .
3711	532	667	16		609-10	60 Kshaya . . .		...
3712	533	668	17		610-11	1 Prabhava . . .		...
3713	534	669	18		611-12	2 Vibhava . . .		2 Vaiśākha .
3714	535	670	19		*612-13	3 Śukla . . .		...
3715	536	671	20		613-14	4 Pramōda . . .		6 Bhādrapada
3716	537	672	21		614-15	5 Prajāpati . . .		...
3717	538	673	22		615-16	6 Aṅgiras . . .		...
3718	539	674	23		*616-17	7 Śrīmukha . . .		4 Āshāḍha .
3719	540	675	24		617-18	8 Bhāva . . .		...
3720	541	676	25		618-19	9 Yuvan . . .		...
3721	542	677	26		619-20	10 Dhātri . . .		3 Jyēshṭha
3722	543	678	27		*620-21	11 Īśvara . .		

LXXXII.

BY THE BRAHMA-SIDDHĀNTA.

the columns being similarly numbered.

preceding page.)

COMMENCEMENT OF THE											
SOLAR YEAR.					LUNI-SOLAR YEAR (MEAN SUNRISE OF CIVIL DAY ON WHICH CHAITRA ŚUKLA 1 ENDS).						Kali.
Day and month A. D.	Week-day.	Time of true Mēsha-saṁkrānti.			Day and month A. D.	Week-day.	*a*	*b*	*c*		
13	14	17			19	20	23	24	25		1
		H.	M.	S.							
19 Mar. (78)	5 Thur.	1	6	0	3 Mar. (62)	3 Tues.	9932·8171	66·0032	233·7104		3701
18 Mar. (78)	6 Fri. .	7	18	9	21 Feb. (52)	1 Sun. .	147·1720	949·5390	205·6250		3702
18 Mar. (77)	0 Sat. .	13	30	18	11 Mar. (70)	0 Sat. .	181·8544	885·5324	256·9354		3703
18 Mar. (77)	1 Sun.	19	42	27	28 Feb. (59)	4 Wed.	57·5772	732·7766	226·1121		3704
19 Mar. (78)	3 Tues.	1	54	36	18 Feb. (49)	2 Mon.	271·9320	616·3122	203·5023		3705
18 Mar. (78)	4 Wed.	8	6	45	7 Mar. (67)	0 Sat. .	9967·9825	516·0140	246·5994		3706
18 Mar. (77)	5 Thur.	14	18	54	24 Feb. (55)	4 Wed.	9843·7052	363·2681	215·7762		3707
18 Mar. (77)	6 Fri. .	20	31	3	15 Mar. (74)	3 Tues.	9878·3876	299·1516	267·0865		3708
19 Mar. (78)	1 Sun. .	2	43	12	4 Mar. (63)	0 Sat. .	9754·1105	146·4956	236·2624		3709
18 Mar. (78)	2 Mon. .	8	55	21	22 Feb. (53)	5 Thur.	9968·4653	30·0312	208·1780		3710
18 Mar. (77)	3 Tues.	15	7	30	12 Mar. (71)	4 Wed.	3·1477	966·0247	259·4884		3711
18 Mar. (77)	4 Wed.	21	19	39	2 Mar. (61)	2 Mon.	217·5025	849·5604	231·4029		3712
19 Mar. (78)	6 Fri. .	3	31	48	19 Feb. (50)	6 Fri. .	93·2254	696·8045	200·5797		3713
18 Mar. (78)	0 Sat. .	9	43	57	9 Mar. (69)	5 Thur.	127·9077	632·7980	251·8902		3714
18 Mar. (77)	1 Sun. .	15	56	6	26 Feb. (57)	2 Mon.	3·6306	480·0421	221·0669		3715
18 Mar. (77)	2 Mon. .	22	8	15	16 Mar. (75)	0 Sat. .	9999·6810	379·7440	269·6395		3716
19 Mar. (78)	4 Wed.	4	20	24	6 Mar. (65)	5 Thur.	9914·0358	263·2795	241·5542		3717
18 Mar. (78)	5 Thur.	10	32	33	23 Feb. (54)	2 Mon.	9789·7587	110·5236	210·3710		3718
18 Mar. (77)	6 Fri. .	16	44	42	13 Mar. (72)	1 Sun. .	9824·4420	46·5171	262·0414		3719
18 Mar. (77)	0 Sat. .	22	56	51	3 Mar. (62)	6 Fri. .	38·7959	930·0528	233·9559		3720
19 Mar. (78)	2 Mon.	5	9	0	21 Feb. (52)	4 Wed.	253·1507	813·5885	205·8705		3721
18 Mar. (78)	3 Tues.	11	21	9	11 Mar. (71)	3 Tues.	287·8331	749·5820	257·1816		3722

TABLE

CONCURRENT YEAR.								Intercalated (*adhika*) and suppressed (*kshaya*) true lunar months.
						Jovian Saṁvatsara.		
Kali.	Śaka.	Chaitrādi Vikrama.	Mēshādi solar year in Bengal.	Kollam.	A. D.	Southern system.	Northern system.	
1	2	3	3a	4	5	6	7	8a
3723	544	679	28		621-22	12 Bahudhānya . .		7 Āśvina .
3724	545	680	29		622-23	13 Pramādin . . .		...
3725	546	681	30		623-24	14 Vikrama . . .		...
3726	547	682	31		*624-25	15 Vṛisha . . .		5 Śrāvaṇa .
3727	548	683	32		625-26	16 Chitrabhānu . .		...
3728	549	684	33		626-27	17 Subhānu . . .		...
3729	550	685	34		627-28	18 Tāraṇa . . .		4 Āshāḍha .
3730	551	686	35		*628-29	19 Pārthiva . . .		...
3731	552	687	36		629-30	20 Vyaya . . .		...
3732	553	688	37		630-31	21 Sarvajit . . .		2 Vaiśākha .
3733	554	689	38		631-32	22 Sarvadhārin . .		...
3734	555	690	39		*632-33	23 Virōdhin . . .		6 Bhādrapada
3735	556	691	40		633-34	24 Vikṛita . . .		...
3736	557	692	41		634-35	25 Khara . . .		...
3737	558	693	42		635-36	26 Nandana . . .		4 Āshāḍha .
3738	559	694	43		*636-37	27 Vijaya . . .		...
3739	560	695	44		637-38	28 Jaya		...
3740	561	696	45		638-39	29 Manmatha . . .		3 Jyēshṭha .
3741	562	697	46		639-40	30 Durmukha . . .		...
3742	563	698	47		*640-41	31 Hēmalamba . .		7 Āśvina
3743	564	699	48		641-42	32 Vilamba . . .		...
3744	565	700	49		642-43	33 Vikārin . . .		...
3745	566	701	50		643-44	34 Śārvarin . . .		Śrāvaṇa
3746	567	702	51		*644-45	35 Plava . . .		...
3747	568	703	52		645-46	36 Śubhakṛit . . .		...

LXXXII—*Contd.*

COMMENCEMENT OF THE											Kali.
SOLAR YEAR.					LUNI-SOLAR YEAR (MEAN SUNRISE OF CIVIL DAY ON WHICH CHAITRA ŚUKLA 1 ENDS).						
Day and month A. D.	Week-day.	Time of true Mēsha-saṁkrānti.			Day and month A. D.	Week-day.	*a*	*b*	*c*		
13	14	17			19	20	23	24	25		1
		H.	M.	S.							
18 Mar. (77)	4 Wed.	17	33	18	28 Feb. (59)	0 Sat.	163·5560	596·8261	226·3577		3723
18 Mar. (77)	5 Thur.	23	45	27	18 Mar. (77)	5 Thur.	9859·6063	496·5279	274·9303		3724
19 Mar. (78)	0 Sat.	5	57	36	8 Mar. (67)	3 Tues.	73·9612	380·0635	246·8449		3725
18 Mar. (78)	1 Sun.	12	9	45	25 Feb. (56)	0 Sat.	9949·6840	227·3076	216·0218		3726
18 Mar. (77)	2 Mon.	18	21	54	15 Mar. (74)	6 Fri.	9984·3664	163·3011	267·3321		3727
19 Mar. (78)	4 Wed.	0	34	3	4 Mar. (63)	3 Tues.	9360·0892	10·5451	236·5089		3728
19 Mar. (78)	5 Thur.	6	46	12	22 Feb. (53)	1 Sun.	74·4441	894·0800	208·4235		3729
18 Mar. (78)	6 Fri.	12	58	21	12 Mar. (72)	0 Sat.	109·1265	830·0742	259·7340		3730
18 Mar. (77)	0 Sat.	19	10	30	2 Mar. (61)	5 Thur.††	323·4813	713·6160	231·6485		3731
19 Mar. (78)	2 Mon.	1	22	39	19 Feb. (50)	2 Mon.	199·2041	560·8540	200·8252		3732
19 Mar. (78)	3 Tues.	7	34	47	9 Mar. (68)	0 Sat.	9895·2545	461·5558	249·3979		3733
18 Mar. (78)	4 Wed.	13	46	56	26 Feb. (57)	4 Wed.	9770·9774	307·7999	218·5748		3734
18 Mar. (77)	5 Thur.	19	59	5	16 Mar. (75)	3 Tues.	9805·6597	243·7934	269·8851		3735
19 Mar. (78)	0 Sat.	2	11	14	6 Mar. (65)	1 Sun.	20·0146	127·3290	241·0922		3736
19 Mar. (78)	1 Sun.	8	23	23	23 Feb. (54)	5 Thur.	9895·7375	974·5731	210·9765		3737
18 Mar. (78)	2 Mon.	14	35	32	13 Mar. (73)	4 Wed.	9930·4199	910·5666	262·2870		3738
18 Mar. (77)	3 Tues.	20	47	41	3 Mar. (62)	2 Mon.	144·7746	794·1023	234·2015		3739
19 Mar. (78)	5 Thur.	2	59	50	20 Feb. (51)	6 Fri.	20·4975	641·3463	203·3783		3740
19 Mar. (78)	6 Fri.	9	11	59	11 Mar. (70)	5 Thur.	55·1799	577·3398	254·6887		3741
18 Mar. (78)	0 Sat.	15	24	8	28 Feb. (59)	2 Mon.	9930·9027	424·5838	223·8655		3742
18 Mar. (77)	1 Sun.	21	36	17	18 Mar. (77)	1 Sun.	9965·5851	360·5774	275·1759		3743
19 Mar. (78)	3 Tues.	3	48	26	7 Mar. (66)	5 Thur.	9841·3081	207·8213	244·3527		3744
19 Mar. (78)	4 Wed.	10	0	35	25 Feb. (56)	3 Tues.	55·6628	91·3571	216·2873		3745
18 Mar. (78)	5 Thur.	16	12	44	15 Mar. (75)	2 Mon.	90·3451	27·3506	267·5776		3746
18 Mar. (77)	6 Fri.	22	21	53	4 Mar. (63)	6 Fri.	9966·0680	873·8747	236·7545		3747

†† See "Remarks," above, page 455.

TABLE

CONCURRENT YEAR.								Intercalated (*adhika*) and suppressed (*kshaya*) true lunar months.
						Jovian Saṁvatsara.		
Kali.	Śaka.	Chaitrādi Vikrama.	Mēshādi solar year in Bengal.	Kollam.	A. D.	Southern system.	Northern system.	
1	2	3	3*a*	4	5	6	7	8*a*
3748	569	704	53		646-47	37 Śōbhana		4 Āshāḍha
3749	570	705	54		647-48	38 Krōdhin		...
3750	571	706	55		*648-49	39 Viśvāvasu†		...
3751	572	707	56		649-50	41 *Plavaṅga*		2 Vaiśākha
3752	573	708	57		650-51	42 *Kīlaka*		...
3753	574	709	58		651-52	43 *Saumya*		6 Bhādrapada
3754	575	710	59		*652-53	44 *Sādhāraṇa*		...
3755	576	711	60		653-54	45 *Virōdhakṛit*		...
3756	577	712	61		654-55	46 Paridhāvin		4 Āshāḍha
3757	578	713	62		655-56	47 Pramādin		...
3758	579	714	63		*656-57	48 Ānanda		...
3759	580	715	64		657-58	49 Rākshasa		3 Jyēshṭha
3760	581	716	65		658-59	50 Anala		...
3761	582	717	66		659-60	51 Piṅgala		7 Āśvina
3762	583	718	67		*660-61	52 Kālayukta		...
3763	584	719	68		661-62	53 Siddhārthin		...
3764	585	720	69		662-63	54 Raudra		5 Śrāvaṇa
3765	586	721	70		663-64	55 Durmati		...
3766	587	722	71		*664-65	56 Dundubhi		...
3767	588	723	72		665-66	57 Rudhirōdgārin		4 Āshāḍha
3768	589	724	73		666-67	58 Raktāksha		...
3769	590	725	74		667-68	59 Krōdhana		...
3770	591	726	75		*668-69	60 Kshaya		1 Chaitra
3771	592	727	76		669-70	1 Prabhava		...
3772	593	728	77		670-71	2 Vibhava		5 Śrāvaṇa

† 40 Parābhava was suppressed.

LXXXII—*Contd.*

COMMENCEMENT OF THE										
Solar year.					Luni-solar year (mean sunrise of civil day on which Chaitra śukla 1 ends).					Kali.
Day and month A. D.	Week-day.	Time of true Mēsha-saṁkrānti.			Day and month A. D.	Week-day.	*a*	*b*	*c*	
13	14	17			19	20	23	24	25	1
		H.	M.	S.						
19 Mar. (78)	1 Sun.	4	37	2	22 Feb. (53)	4 Wed.	180·4229	758·1223	208·6691	3748
19 Mar. (78)	2 Mon.	10	49	11	13 Mar. (72)	3 Tues.	215·1052	694·1237	259·9795	3749
18 Mar. (78)	3 Tues.	17	1	20	1 Mar. (61)	0 Sat.	90·8281	541·3679	229·1662	3750
18 Mar. (77)	4 Wed.	23	13	29	18 Feb. (49)	4 Wed.	9966·5509	388·6119	198·3330	3751
19 Mar. (78)	6 Fri.	5	25	38	9 Mar. (68)	3 Tues.	1·2333	324·6053	249·6435	3752
19 Mar. (78)	0 Sat.	11	37	47	26 Feb. (57)	0 Sat.	9876·9561	171·8494	218·8203	3753
18 Mar. (78)	1 Sun.	17	49	56	16 Mar. (76)	6 Fri.	9911·6385	107·8429	270·1306	3754
19 Mar. (78)	3 Tues.	0	2	5	6 Mar. (65)	4 Wed.	125·9934	991·3786	242·0453	3755
19 Mar. (78)	4 Wed.	6	14	14	23 Feb. (54)	1 Sun.	1·7162	838·6227	211·2221	3756
19 Mar. (78)	5 Thur.	12	26	23	14 Mar. (73)	0 Sat.	36·3986	774·6161	262·5325	3757
18 Mar. (78)	6 Fri.	18	38	32	3 Mar. (63)	5 Thur.	250·7534	658·1518	234·4470	3758
19 Mar. (78)	1 Sun.	0	50	41	20 Feb. (51)	2 Mon.	126·5863	505·3958	203·6238	3759
19 Mar. (78)	2 Mon.	7	2	50	10 Mar. (69)	0 Sat.	9822·5266	405·0977	252·1965	3760
19 Mar. (78)	3 Tues.	13	14	59	28 Feb. (59)	5 Thur.	36·8815	288·6334	224·1110	3761
18 Mar. (78)	4 Wed.	19	27	8	17 Mar. (77)	3 Tues.	9732·9319	188·3353	272·6836	3762
19 Mar. (78)	6 Fri.	1	39	17	7 Mar. (66)	1 Sun.	9947·2867	71·8709	244·5982	3763
19 Mar. (78)	0 Sat.	7	51	26	25 Feb. (56)	6 Fri.	161·6415	955·4066	216·5129	3764
19 Mar. (78)	1 Sun.	14	3	35	16 Mar. (75)	5 Thur.	196·2239	891·4001	267·8232	3765
18 Mar. (78)	2 Mon.	20	15	44	4 Mar. (64)	2 Mon.	72·0468	738·6441	237·0000	3766
19 Mar. (78)	4 Wed.	2	27	53	21 Feb. (52)	6 Fri.	9947·7696	585·8882	206·1768	3767
19 Mar. (78)	5 Thur.	8	40	2	12 Mar. (71)	5 Thur.	9982·6410	521·8817	257·4873	3768
19 Mar. (78)	6 Fri.	14	52	11	1 Mar. (60)	2 Mon.	9858·1749	369·1257	226·6640	3769
18 Mar. (78)	0 Sat.	21	4	20	18 Feb. (49)	6 Fri.	9733·8977	216·3699	195·8407	3770
19 Mar. (78)	2 Mon.	3	16	29	8 Mar. (67)	5 Thur.	9768·5801	152·5632	247·1512	3771
19 Mar. (78)	3 Tues.	9	28	38	26 Feb. (57)	3 Tues.	9982·9349	35·8889	219·0059	3772

TABLE

CONCURRENT YEAR.								Intercalated (*adhika*) and suppressed (*kshaya*) true lunar months.
						Jovian Saṁvatsara.		
Kali.	Śaka.	Chaitrādi Vikrama.	Mēshādi solar year in Bengal.	Kollam.	A. D.	Southern system.	Northern system.	
1	2	3	3*a*	4	5	6	7	8*a*
3773	594	729	78		671-72	3 Śukla		...
3774	595	730	79		*672-73	4 Pramōda . . .		...
3775	596	731	80		673-74	5 Prajāpati . . .		4 Āshāḍha .
3776	597	732	81		674-75	6 Aṅgiras . . .		...
3777	598	733	82		675-76	7 Śrīmukha . . .		...
3778	599	734	83		*676-77	8 Bhāva . . .		2 Vaiśākha .
3779	600	735	84		677-78	9 Yuvan . . .		...
3780	601	736	85		678-79	10 Dhātṛi . . .		7 Āśvina .
3781	602	737	86		679-80	11 Īśvara . .		...
3782	603	738	87		*680-81	12 Bahudhānya . .		...
3783	604	739	88		681-82	13 Pramādin . . .		5 Śrāvaṇa
3784	605	740	89		682-83	14 Vikrama . . .		..
3785	606	741	90		683-84	15 Vṛisha . . .		..
3786	607	742	91		*684-85	16 Chitrabhānu . .		3 Jyēshṭha .
3787	608	743	92		685-86	17 Subhānu . . .		...
3788	609	744	93		686-87	18 Tāraṇa . . .		...
3789	610	745	94		687-88	19 Pārthiva . . .		1 Chaitra
3790	611	746	95		*688-89	20 Vijaya . .		...
3791	612	747	96		689-90	21 Sarvajit .		5 Śrāvaṇa
3792	613	748	97		690-91	22 Sarvadhārin .		...
3793	614	749	98		691-92	23 Virōdhin . . .		...
3794	615	750	99		*692-93	24 Vikrita . .		4 Āshāḍha
3795	616	751	100		693-94	25 Khara . . .		...
3796	617	752	101		694-95	26 Nandana . .		...
3797	618	753	102		695-96	27 Vijaya . .		2 Vaiśākha .

LXXXII—*Contd.*

COMMENCEMENT OF THE									Kali.
SOLAR YEAR.			LUNI-SOLAR YEAR (MEAN SUNRISE OF CIVIL DAY ON WHICH CHAITRA ŚUKLA 1 ENDS).						
Day and month A. D.	Week-day.	Time of true Mēsha-saṁkrānti.	Day and month A. D.	Week-day.	*a*	*b*	*c*		
13	14	17	19	20	23	24	25		1
		H. M. S.							
19 Mar. (78)	4 Wed.	15 40 47	17 Mar. (76)	2 Mon.	17·6173	971·8924	270·3762		3773
18 Mar. (78)	5 Thur.	21 52 56	6 Mar. (66)	0 Sat.	231·9621	855·4281	242·2907		3774
19 Mar. (78)	0 Sat.	4 5 5	23 Feb. (54)	4 Wed.	107·6950	702·6722	211·4676		3775
19 Mar. (78)	1 Sun.	10 17 14	14 Mar. (73)	3 Tues.	142·3774	628·6656	262·7781		3776
19 Mar. (78)	2 Mon.	16 29 23	3 Mar. (62)	0 Sat.	18·1001	485·9097	231·9548		3777
18 Mar. (78)	3 Tues.	22 41 31	20 Feb. (51)	4 Wed.	9893·8230	333·1537	201·1315		3778
19 Mar. (78)	5 Thur.	4 53 40	10 Mar. (69)	3 Tues.	9928·5054	269·1472	252·4420		3779
19 Mar. (78)	6 Fri.	11 5 49	27 Feb. (58)	0 Sat.	9804·2283	116·3913	221·6188		3780
19 Mar. (78)	0 Sat.	17 17 58	18 Mar. (77)	6 Fri.	9838·9106	52·4848	272·9292		3781
18 Mar. (78)	1 Sun.	23 30 7	7 Mar. (67)	4 Wed.	53·2655	935·9205	244·8437		3782
19 Mar. (78)	3 Tues.	5 42 16	25 Feb. (56)	2 Mon.	267·6203	819·4561	216·7584		3783
19 Mar. (78)	4 Wed.	11 54 25	16 Mar. (75)	1 Sun.	302·3027	755·4496	268·0688		3784
19 Mar. (78)	5 Thur.	18 6 34	5 Mar. (64)	5 Thur.	178·0255	602·6936	237·5456		3785
19 Mar. (79)	0 Sat.	0 18 43	22 Feb. (53)	2 Mon.	53·7384	449·9378	206·4223		3786
19 Mar. (78)	1 Sun.	6 30 52	12 Mar. (71)	1 Sun.	88·4308	385·9312	257·7328		3787
19 Mar. (78)	2 Mon.	12 43 1	1 Mar. (60)	5 Thur.	9964·1536	233·1752	227·1096		3788
19 Mar. (78)	3 Tues.	18 55 10	18 Feb. (49)	2 Mon.	9839·8765	80·4194	196·0863		3789
19 Mar. (79)	5 Thur.	1 7 19	8 Mar. (68)	1 Sun.	9874·5589	16·4127	247·3967		3790
19 Mar. (78)	6 Fri.	7 19 28	26 Feb. (57)	6 Fri.	88·9137	899·9484	219·3114		3791
19 Mar. (78)	0 Sat.	13 31 37	17 Mar. (76)	5 Thur.	123·5960	835·9419	270·6218		3792
19 Mar. (78)	1 Sun.	19 43 46	6 Mar. (65)	2 Mon.	9999·3189	683·1860	239·7986		3793
19 Mar. (79)	3 Tues.	1 55 55	24 Feb. (55)	0 Sat.	213·6738	566·7217	211·7131		3794
19 Mar. (78)	4 Wed.	8 8 4	13 Mar. (72)	5 Thur.	9909·7241	466·4235	230·1858		3795
19 Mar. (78)	5 Thur.	14 20 13	2 Mar. (61)	2 Mon.	9785·4476	313·6675	229·4626		3796
19 Mar. (73)	6 Fri.	20 32 22	20 Feb. (51)	0 Sat.	9999·8918	197·2032	201·3771		3797

TABLE

CONCURRENT YEAR.								
Kali.	Śaka.	Chaitrādi Vikrama.	Mēshādi solar year in Bengal.	Kollam.	A. D.	Jovian Saṁvatsara. Southern system.	Jovian Saṁvatsara. Northern system.	Intercalated (*adhika*) and suppressed (*kshaya*) true lunar months
1	2	3	3*a*	4	5	6	7	8*a*
3798	619	754	103		*696-97	28 Jaya		...
3799	620	755	104		697-98	29 Manmatha		6 Bhādrapada
3800	621	756	105		698-99	30 Durmukha		...
3801	622	757	106		699-700	31 Hēmalamba		...
3802	623	758	107		*700-01	32 Vilamba		5 Śrāvana
3803	624	759	108		701-02	33 Vikārin		...
3804	625	760	109		702-03	34 Śārvarin		...
3805	626	761	110		703-04	35 Plava		3 Jyēshṭha
3806	627	762	111		*704-05	36 Śubhakṛit		...
3807	628	763	112		705-06	37 Sōbhana		...
3808	629	764	113		706-07	38 Krōdhin		1 Chaitra
3809	630	765	114		707-08	39 Viśvāvasu		...
3810	631	766	115		*708-09	40 Parābhava		5 Śrāvaṇa
3811	632	767	116		709-10	41 Plavaṅga		...
3812	633	768	117		710-11	42 Kīlaka		...
3813	634	769	118		711-12	43 Saumya		4 Āshāḍha
3814	635	770	119		*712-13	44 Sādhāraṇa		...
3815	636	771	120		713-14	45 Virōdhakṛit		...
3816	637	772	121		714-15	46 Paridhāvin		2 Vaiśākha
3817	638	773	122		715-16	47 Pramādin		...
3818	639	774	123		*716-17	48 Ānanda		6 Bhādrapada
3819	640	775	124		717-18	49 Rākshasa		...
3820	641	776	125		718-19	50 Anala		...
3821	642	777	126		719-20	51 Piṅgala		5 Śrāvaṇa
3822	643	778	127		*720-21	52 Kālayukta		...

LXXXII—*Contd.*

COMMENCEMENT OF THE									Kali.
SOLAR YEAR.			LUNI-SOLAR YEAR (MEAN SUNRISE OF CIVIL DAY ON WHICH CHAITRA ŚUKLA 1 ENDS).						
Day and month, A. D.	Week-day.	Time of true Mēsha-saṁkrānti.	Day and month, A. D.	Week-day.	*a*	*b*	*c*		
13	14	17	19	20	23	24	25		1
		H. M. S.							
19 Mar. (79)	1 Sun.	2 44 31	10 Mar. (70)	6 Fri.	34·4841	133·1967	252·6875		3798
19 Mar. (78)	2 Mon.	8 56 40	27 Feb. (58)	3 Tues.	9910·2070	980·4408	221·8643		3799
19 Mar. (78)	3 Tues.	15 8 49	18 Mar. (77)	2 Mon.	9944·8894	916·4343	273·1748		3800
19 Mar. (78)	4 Wed.	21 20 58	8 Mar. (67)	0 Sat.	159·2443	799·9700	245·0671		3801
19 Mar. (79)	6 Fri.	3 33 7	25 Feb. (56)	4 Wed.	34·9671	647·2140	214·2440		3802
19 Mar. (78)	0 Sat.	9 45 16	15 Mar. (74)	3 Tues.	69·6496	583·2074	265·5543		3803
19 Mar. (78)	1 Sun.	15 57 25	4 Mar. (63)	0 Sat.	9945·3723	430·4516	234·7311		3804
19 Mar. (78)	2 Mon.	22 9 34	21 Feb. (52)	4 Wed.	9821·0852	277·6956	203·9079		3805
19 Mar. (79)	4 Wed.	4 21 43	11 Mar. (71)	3 Tues.	9855·7776	213·6890	255·2184		3806
19 Mar. (78)	5 Thur.	10 33 52	1 Mar. (60)	1 Sun.	70·1324	97·2248	227·1329		3807
19 Mar. (78)	6 Fri.	16 46 1	18 Feb. (49)	5 Thur.	9946·0956	944·4986	196·3096		3808
19 Mar. (78)	0 Sat.	22 58 10	9 Mar. (68)	4 Wed.	9980·5376	880·4623	247·6201		3809
19 Mar. (79)	2 Mon.	5 10 19	27 Feb. (58)	2 Mon.	194·8924	773·9979	219·5348		3810
19 Mar. (78)	3 Tues.	11 22 28	17 Mar. (76)	1 Sun.	230·5748	699·9914	270·8451		3811
19 Mar. (78)	4 Wed.	17 34 37	6 Mar. (65)	5 Thur.	105·2977	547·2355	240·0219		3812
19 Mar. (78)	5 Thur.	23 46 46	23 Feb. (54)	2 Mon.	9981·0206	394·4796	209·1987		3813
19 Mar. (79)	0 Sat.	5 58 55	13 Mar. (73)	1 Sun.	15·7029	330·4730	260·5092		3814
19 Mar. (78)	1 Sun.	12 11 4	2 Mar. (61)	5 Thur.	9891·4258	178·7171	229·6859		3815
19 Mar. (78)	2 Mon.	18 23 13	20 Feb. (51)	3 Tues.	105·7806	61·2528	201·6004		3816
20 Mar. (79	4 Wed.	0 35 22	11 Mar. (70)	2 Mon.	140·4629	997·2462	252·9109		3817
19 Mar. (79)	5 Thur.	6 47 31	28 Feb. (59)	6 Fri	16·1858	844·4903	222·0877		3818
19 Mar. (78)	6 Fri.	12 59 40	18 Mar. (77)	5 Thur.	50·8682	780·4838	173·3981		3819
19 Mar. (78)	0 Sat.	19 11 49	8 Mar. (67)	3 Tues.	265·2231	664·0195	245·3126		3820
20 Mar. (79)	2 Mon.	1 23 58	25 Feb. (56)	0 Sat.	140·9458	511·2635	214·4895		3821
19 Mar. (79)	3 Tues.	7 36 7	14 Mar. (74)	5 Thur.	9836·9963	410·9054	263·0622		3822

TABLE

CONCURRENT YEAR.								Intercalated (*adhika*) and suppressed (*kshaya*) true lunar months.
						Jovian Saṁvatsara.		
Kali.	Śaka.	Chaitrādi Vikrama.	Mēshādi solar year in Bengal.	Kollam.	A. D.	Southern system.	Northern system.	
1	2	3	3*a*	4	5	6	7	8*a*
3823	644	779	128		721-22	53 Siddhārthin . . .		...
3824	645	780	129		722-23	54 Raudra . . .		3 Jyēshṭha .
3825	646	781	130		723-24	55 Durmati . . .		...
3826	647	782	131		*724-25	56 Dundubhi . .		7 Āśvina 9 *Mārgaś.* (*ksh*)
3827	648	783	132		725-26	57 Rudhirōdgārin . .		1 Chaitra .
3828	649	784	133		726-27	58 Raktāksha . . .		...
3829	650	785	134		727-28	59 Krōdhana . . .		5 Śrāvaṇa .
3830	651	786	135		*728-29	60 Kshaya . . .		...
3831	652	787	136		729-30	1 Prabhava . . .		...
3832	653	788	137		730-31	2 Vibhava . . .		4 Āshāḍha .
3833	654	789	138		731-32	3 Śukla .		...
3834	655	790	139		*732-33	4 Pramōda . . .		...
3835	656	791	140		733-34	5 Prajāpati . . .		2 Vaiśākha .
3836	657	792	141		734-35	6 Aṅgirasa† . . .		...
3837	658	793	142		735-36	8 *Bhāva*		6 Bhādrapada
3838	659	794	143		*736-37	9 *Yuvan* . . .		...
3839	660	795	144		737-38	10 *Dhātṛi* . . .		...
3840	661	796	145		738-39	11 *Īśvara*		5 Śrāvaṇa .
3841	662	797	146		739-40	12 Bahudhānya . .		...
3842	663	798	147		*740-41	13 Pramādin . .		...
3843	664	799	148		741-42	14 Vikrama . . .		3 Jyēshṭha .
3844	665	800	149		742-43	15 Vṛisha . . .		...
3845	666	801	150		743-44	16 Chitrabhānu . .		7 Āśvina 11 *Māgha* (*ksh*)
3846	667	802	151		*744-45	17 Subhānu . . .		1 Chaitra .
3847	668	803	152		745-46	18 Tāraṇa . . .		...

† 7 Śrīmukha was suppressed.

LXXXII—*Contd.*

COMMENCEMENT OF THE									
SOLAR YEAR.			LUNI-SOLAR YEAR (MEAN SUNRISE OF CIVIL DAY ON WHICH CHAITRA ŚUKLA 1 ENDS).						Kali.
Day and month, A. D.	Week-day.	Time of true Mēsha-saṁkrānti.	Day and month, A. D.	Week-day.	*a*	*b*	*c*		
13	14	17	19	20	23	24	25		1
		H. M. S.							
19 Mar. (78)	4 Wed.	13 48 15	4 Mar. (63)	3 Tues.	51·3511	294·5011	234·9767		3823
19 Mar. (78)	5 Thur.	20 0 24	21 Feb. (52)	0 Sat.	9927·0739	141·7452	204·1534		3824
20 Mar. (79)	0 Sat.	2 12 33	12 Mar. (71)	6 Fri.	9961·7563	77·7385	255·4693		3825
19 Mar. (79)	1 Sun.	8 24 42	1 Mar. (61)	4 Wed.	176·1112	961·2743	227·3785		3826
19 Mar. (78)	2 Mon.	14 36 51	18 Feb. (49)	1 Sun.	51·8342	808·5184	196·5552		3827
19 Mar. (78)	3 Tues.	20 49 0	9 Mar. (68)	0 Sat.	86·5163	744·5118	247·8656		3828
20 Mar. (79)	5 Thur.	3 1 9	26 Feb. (57)	4 Wed.	9962·2392	591·7559	217·0425		3829
19 Mar. (79)	6 Fri.	9 13 18	16 Mar. (76)	3 Tues.	9996·9216	527·7493	268·3529		3830
19 Mar. (78)	0 Sat.	15 25 27	5 Mar. (64)	0 Sat.	9872·6444	374·9934	237·5297		3831
19 Mar. (78)	1 Sun.	21 37 36	22 Feb. (53)	4 Wed.	9748·3673	222·2374	206·7064		3832
20 Mar. (79)	3 Tues.	3 49 45	13 Mar. (72)	3 Tues.	9783·0497	158·2309	258·0169		3833
19 Mar. (79)	4 Wed.	10 1 54	2 Mar. (62)	1 Sun.	9997·4046	41·7666	229·9215		3834
19 Mar. (78)	5 Thur.	16 14 3	20 Feb. (51)	6 Fri.	211·7493	925·3023	201·8460		3835
19 Mar. (78)	6 Fri.	22 26 12	11 Mar. (70)	5 Thur.	246·4417	861·2958	253·1564		3836
20 Mar. (79)	1 Sun.	4 38 21	28 Feb. (59)	2 Mon.	122·1646	708·5398	222·3332		3837
19 Mar. (79)	2 Mon.	10 50 30	18 Mar. (78)	1 Sun.	156·8460	644·5333	274·6437		3838
19 Mar. (78)	3 Tues.	17 2 39	7 Mar. (66)	5 Thur.	32·5698	501·7773	242·8204		3839
19 Mar. (78)	4 Wed.	23 14 48	24 Feb. (55)	2 Mon.	9908·2926	339·0214	211·9973		3840
20 Mar. (79)	6 Fri.	5 26 57	15 Mar. (74)	1 Sun.	9942·9751	275·0149	263·2077		3841
19 Mar. (79)	0 Sat.	11 39 6	3 Mar. (63)	5 Thur.	9818·6978	122·2588	232·4845		3842
19 Mar. (78)	1 Sun.	17 51 15	21 Feb. (52)	3 Tues.	33·0527	5·7947	204·3990		3843
20 Mar. (79)	3 Tues.	0 3 24	12 Mar. (71)	2 Mon.	67·7351	941·7880	255·7105		3844
20 Mar. (79)	4 Wed.	6 15 33	2 Mar. (61)	0 Sat.	282·0900	825·3238	227·6240		3845
19 Mar. (79)	5 Thur.	12 27 42	19 Feb. (50)	4 Wed.	157·8127	672·5678	196·8007		3846
19 Mar. (78)	6 Fri.	18 39 51	9 Mar. (68)	3 Tues.	192·4951	608·5613	248·1112		3847

TABLE

CONCURRENT YEAR.								Intercalated (*adhika*) and suppressed (*kshaya*) true lunar months.
						Jovian Saṁvatsara.		
Kali.	Śaka.	Chaitrādi Vikrama.	Mēshādi solar year in Bengal.	Kollam.	A. D.	Southern system.	Northern system.	
1	2	3	3a	4	5	6	7	8a
3848	669	804	153		746-47	19 Pārthiva . . .		5 Śrāvaṇa .
3849	670	805	154		747-48	20 Vyaya . . .		...
3850	671	806	155		*748-49	21 Sarvajit . . .		...
3851	672	807	156		749-50	22 Sarvadhārin . .		3 Jyēshṭha .
3852	673	808	157		750-51	23 Virōdhin . . .		...
3853	674	809	158		751-52	24 Vikṛita . . .		...
3854	675	810	159		*752-53	25 Khara . . .		2 Vaiśākha .
3855	676	811	160		753-54	26 Nandana . . .		...
3856	677	812	161		754-55	27 Vijaya . . .		6 Bhādrapada
3857	678	813	162		755-56	28 Jaya . . .		...
3858	679	814	163		*756-57	29 Manmatha . . .		...
3859	680	815	164		757-58	30 Durmukha . . .		4 Āshāḍha .
3860	681	816	165		758-59	31 Hēmalamba . .		...
3861	682	817	166		759-60	32 Vilamba . . .		...
3862	683	818	167		*760-61	33 Vikārin . . .		3 Jyēshṭha .
3863	684	819	168		761-62	34 Śārvarin . . .		...
3864	685	820	169		762-63	35 Plava . . .		7 Āśvina .
3865	686	821	170		763-64	36 Śubhakṛit . . .		...
3866	687	822	171		*764-65	37 Śōbhana . . .		...
3867	688	823	172		765-66	38 Krōdhin . . .		5 Śrāvaṇa .
3868	689	824	173		766-67	39 Viśvāvasu . . .		...
3869	690	825	174		767-68	40 Parābhava . . .		...
3870	691	826	175		*768-69	41 Plavaṅga . . .		3 Jyēshṭha .
3871	692	827	176		769-70	42 Kīlaka . . .		...
3872	693	828	177		770-71	43 Saumya . . .		...

LXXXII—*Contd.*

COMMENCEMENT OF THE								Kali.
SOLAR YEAR.			LUNI-SOLAR YEAR (MEAN SUNRISE OF CIVIL DAY ON WHICH CHAITRA ŚUKLA 1 ENDS).					
Day and month, A. D.	Week-day.	Time of true Mēsha-saṁkrānti.	Day and month, A. D.	Week-day.	*a*	*b*	*c*	
13	14	17	19	20	23	24	25	1
		H. M. S.						
20 Mar. (79)	1 Sun. .	0 52 0	26 Feb. (57)	0 Sat. .	68·2180	455·8054	217·2881	3848
20 Mar. (79)	2 Mon.	7 4 9	17 Mar. (76)	6 Fri. .	102·9003	391·7988	268·4984	3849
19 Mar. (79)	3 Tues.	13 16 18	5 Mar. (65)	3 Tues.	9978·6232	239·0429	237·7752	3850
19 Mar. (78)	4 Wed.	19 28 27	22 Feb. (53)	0 Sat. .	9854·3461	86·2869	206·9520	3851
20 Mar. (79)	6 Fri. .	1 40 36	13 Mar. (72)	6 Fri. .	9889·0285	22·2804	258·2625	3852
20 Mar. (79)	0 Sat. .	7 52 45	3 Mar. (62)	4 Wed.	103·3833	905·8161	230·1770	3853
19 Mar. (79)	1 Sun. .	14 4 54	21 Feb. (52)	2 Mon.	317·7384	789·3518	202·0915	3854
19 Mar. (78)	2 Mon.	20 17 3	10 Mar. (69)	0 Sat. .	13·7885	689·0537	250·6642	3855
20 Mar. (79)	4 Wed.	2 29 12	28 Feb. (59)	5 Thur.	228·1433	572·5894	222·5788	3856
20 Mar. (79)	5 Thur.	8 41 21	18 Mar. (77)	3 Tues.	9924·1937	472·2911	271·1514	3857
19 Mar. (79)	6 Fri. .	14 53 30	6 Mar. (66)	0 Sat. .	9799·9166	319·5352	240·3282	3858
19 Mar. (78)	0 Sat. .	21 5 39	24 Feb. (55)	5 Thur.	14·2714	203·0709	212·2428	3859
20 Mar. (79)	2 Mon.	3 17 48	15 Mar. (74)	4 Wed.	48·9538	139·0644	263·5533	3860
20 Mar. (79)	3 Tues.	9 29 57	4 Mar. (63)	1 Sun. .	9924·6766	986·3084	232·7300	3861
19 Mar. (79)	4 Wed.	15 42 6	22 Feb. (53)	6 Fri. .	139·0315	869·8442	204·6445	3862
19 Mar. (78)	5 Thur.	21 54 15	12 Mar. (71)	5 Thur.	173·7138	805·8377	255·9550	3863
20 Mar. (79)	0 Sat. .	4 6 24	1 Mar. (60)	2 Mon.	49·4367	653·0816	225·1318	3864
20 Mar. (79)	1 Sun. .	10 18 33	20 Mar. (79)	1 Sun. .	84·1191	589·0751	276·4422	3865
19 Mar. (79)	2 Mon.	16 30 42	8 Mar. (68)	5 Thur.	9959·8420	436·3192	245·6189	3866
19 Mar. (78)	3 Tues.	22 42 51	25 Feb. (56)	2 Mon.	9835·5647	283·5633	214·7958	3867
20 Mar. (79)	5 Thur.	4 55 0	16 Mar. (75)	1 Sun. .	9870·2472	219·5567	266·1062	3868
20 Mar. (79)	6 Fri. .	11 7 8	6 Mar. (65)	6 Fri. .	84·6020	103·0923	238·0208	3869
19 Mar. (79)	0 Sat. .	17 19 17	23 Feb. (54)	3 Tues.	9960·3248	950·3365	207·1975	3870
19 Mar. (78)	1 Sun. .	23 31 26	13 Mar. (72)	2 Mon.	9995·0072	886·3299	258·5080	3871
20 Mar. (79)	3 Tues.	5 43 35	3 Mar. (62)	0 Sat. .	209·3621	769·8656	230·4226	3872

TABLE

CONCURRENT YEAR.								Intercalated (*adhika*) and suppressed (*kshaya*) true lunar months.
						Jovian Saṁvatsara.		
Kali.	Śaka.	Chaitrādi Vikrama.	Mēshādi solar year in Bengal.	Kollam.	A. D.	Southern system.	Northern system.	
1	2	3	3*a*	4	5	6	7	8*a*
3873	694	829	178		771-72	44 Sādhāraṇa . . .		2 Vaiśākha .
3874	695	830	179		*772-73	45 Virōdhakṛit . . .		...
3875	696	831	180		773-74	46 Paridhāvin . . .		6 Bhādrapada
3876	697	832	181		774-75	47 Pramādin . . .		...
3877	698	833	182		775-76	48 Ānanda . . .		...
3878	699	834	183		*776-77	49 Rākshasa . . .		4 Āshāḍha .
3879	700	835	184		777-78	50 Anala		...
3880	701	836	185		778-79	51 Piṅgala . . .		...
3881	702	837	186		779-80	52 Kālayukta . . .		3 Jyēshṭha
3882	703	838	187		*780-81	53 Siddhārthin . . .		...
3883	704	839	188		781-82	54 Raudra . . .		7 Āśvina .
3884	705	840	189		782-83	55 Durmati . . .		...
3885	706	841	190		783-84	56 Dundubhi . . .		...
3886	707	842	191		*784-85	57 Rudhirōdgārin . .		5 Śrāvaṇa .
3887	708	843	192		785-86	58 Raktāksha . . .		...
3888	709	844	193		786-87	59 Krōdhana . . .		...
3889	710	845	194		787-88	60 Kshaya . . .		3 Jyēshṭha .
3890	711	846	195		*788-89	1 Prabhava . . .		...
3891	712	847	196		789-90	2 Vibhava . . .		...
3892	713	848	197		790-91	3 Śukla		2 Vaiśākha .
3893	714	849	198		791-92	4 Pramōda . . .		...
3894	715	850	199		*792-93	5 Prajāpati . . .		6 Bhādrapada
3895	716	851	200		793-94	6 Aṅgiras . . .		...
3896	717	852	201		794-95	7 Śrīmukha . . .		...
3897	718	853	202		795-96	8 Bhāva . . .		4 Āshāḍha .

LXXXII—*Contd.*

COMMENCEMENT OF THE									
Solar year.			Luni-solar year (mean sunrise of civil day on which Chaitra śukla 1 ends).						Kali.
Day and month, A. D.	Week-day.	Time of true Mēsha-saṁkrānti.	Day and month, A. D.	Week-day.	*a*	*b*	*c*		
13	14	17	19	20	23	24	25		1
		H. M. S.							
20 Mar. (79)	4 Wed. .	11 55 44	20 Feb. (51)	4 Wed. .	75·0849	617·1097	199·5993		3873
19 Mar. (79)	5 Thur. .	18 7 53	10 Mar. (70)	3 Tues. .	119·7672	553·1032	250·9097		3874
20 Mar. (79)	0 Sat. .	0 20 2	27 Feb. (58)	0 Sat. .	9995·4901	400·3472	220·0866		3875
20 Mar. (79)	1 Sun. .	6 32 11	18 Mar. (77)	6 Fri. .	30·1725	336·3306	271·3970		3876
20 Mar. (79)	2 Mon. .	12 44 20	7 Mar. (66)	3 Tues. .	9905·8953	183·5848	240·5738		3877
19 Mar. (79)	3 Tues. .	18 56 29	25 Feb. (56)	1 Sun. .	120·2501	67·1204	212·4883		3878
20 Mar. (79)	5 Thur. .	1 8 38	15 Mar. (74)	0 Sat. .	154·9326	3·1139	263·7988		3879
20 Mar. (79)	6 Fri. .	7 20 47	4 Mar. (63)	4 Wed. .	30·6554	850·3579	232·9756		3880
20 Mar. (79)	0 Sat. .	13 32 56	22 Feb. (53)	2 Mon. .	245·0102	733·8937	204·8901		3881
19 Mar. (79)	1 Sun. .	19 45 5	12 Mar. (72)	1 Sun. .	279·6926	669·8872	256·2005		3882
20 Mar. (79)	3 Tues. .	1 57 14	1 Mar. (60)	5 Thur. .	155·4155	517·1311	225·3773		3883
20 Mar. (79)	4 Wed. .	8 9 23	19 Mar. (78)	3 Tues. .	9851·4659	416·8330	273·9500		3884
20 Mar. (79)	5 Thur. .	14 21 32	8 Mar. (67)	0 Sat. .	9727·1887	264·0770	243·1167		3885
19 Mar. (79)	6 Fri. .	20 33 41	26 Feb. (57)	5 Thur. .	9941·5435	147·6128	215·0413		3886
20 Mar. (79)	1 Sun. .	2 45 50	16 Mar. (75)	4 Wed. .	9976·2260	83·6062	266·3517		3887
20 Mar. (79)	2 Mon. .	8 57 59	6 Mar. (65)	2 Mon. .	190·5807	967·1418	238·2664		3888
20 Mar. (79)	3 Tues. .	15 10 8	23 Feb. (54)	6 Fri. .	66·3036	814·3852	207·4431		3889
19 Mar. (79)	4 Wed. .	21 22 17	13 Mar. (73)	5 Thur. .	100·9860	750·3794	258·7535		3890
20 Mar. (79)	6 Fri. .	3 34 26	2 Mar. (61)	2 Mon. .	9976·7089	597·6235	227·9303		3891
20 Mar. (79)	0 Sat. .	9 46 35	19 Feb. (50)	6 Fri. .	9852·4317	444·8676	197·1071		3892
20 Mar. (79)	1 Sun. .	15 58 44	10 Mar. (69)	5 Thur. .	9887·1140	380·8610	248·4175		3893
19 Mar. (79)	2 Mon. .	22 10 53	27 Feb. (58)	2 Mon. .	9762·8369	228·1051	218·4943		3894
20 Mar. (79)	4 Wed. .	4 23 2	17 Mar. (76)	1 Sun. .	9797·5192	164·0986	268·9047		3895
20 Mar. (79)	5 Thur. .	10 35 11	7 Mar. (66)	6 Fri. .	11·8741	47·6342	240·8194		3896
20 Mar. (79)	6 Fri. .	16 47 20	25 Feb. (56)	4 Wed.	226·2289	931·1699	212·7339		3897

TABLE

CONCURRENT YEAR.								Intercalated (*adhika*) and suppressed (*kshaya*) true lunar months.
Kali.	Śaka.	Chaitrādi Vikrama.	Mēshādi solar year in Bengal.	Kollam.	A. D.	Jovian Saṁvatsara. Southern system.	Jovian Saṁvatsara. Northern system.	
1	2	3	3a	4	5	6	7	8a
3898	719	854	203		*796-97	9 Yuvan . . .		...
3899	720	855	204		797-98	10 Dhātṛi . . .		...
3900	721	856	205		798-99	11 Īśvara . . .		3 Jyēshṭha .
3901	722	857	206		799-800	12 Bahudhānya . .		...
3902	723	858	207		*800-01	13 Pramādin . . .		7 Āśvina .
3903	724	859	208		801-02	14 Vikrama . . .		...
3904	725	860	209		802-03	15 Vṛisha . . .		...
3905	726	861	210		803-04	16 Chitrabhānu . .		5 Śrāvaṇa .
3906	727	862	211		*804-05	17 Subhānu . . .		...
3907	728	863	212		805-06	18 Tāraṇa . . .		...
3908	729	864	213		806-07	19 Pārthiva . . .		3 Jyēshṭha .
3909	730	865	214		807-08	20 Vyaya . . .		...
3910	731	866	215		*808-09	21 Sarvajit . . .		...
3911	732	867	216		809-10	22 Sarvadhārin . .		1 Chaitra .
3912	733	868	217		810-11	23 Virōdhin . . .		...
3913	734	869	218		811-12	24 Vikṛita . . .		5 Śrāvaṇa .
3914	735	870	219		*812-13	25 Khara . . .		...
3915	736	871	220		813-14	26 Nandana . . .		...
3916	737	872	221		814-15	27 Vijaya . . .		4 Āshāḍha .
3917	738	873	222		815-16	28 Jaya . . .		...
3918	739	874	223		*816-17	29 Manmatha . . .		...
3919	740	875	224		817-18	30 Durmukha . . .		3 Jyēshṭha .
3920	741	876	225		818-19	31 Hēmalamba . .		...
3921	742	877	226		819-20	32 Vilamba † . . .		7 Āśvina .
3922	743	878	227		*820-21	34 *Śārvarin* . . .		...

† 33 Vikārin was suppressed.

LXXXII—*Contd.*

COMMENCEMENT OF THE									
SOLAR YEAR.			LUNI-SOLAR YEAR (MEAN SUNRISE OF CIVIL DAY ON WHICH CHAITRA ŚUKLA 1 ENDS).						Kali.
Day and month A. D.	Week-day.	Time of true Mēsha-saṁ-krānti.	Day and month A. D.	Week-day.	*a*	*b*	*c*		
13	14	17	19	20	23	24	25		1
		H. M. S.							
19 Mar. (79)	0 Sat.	22 59 29	15 Mar. (75)	3 Tues.	260·9113	867·1634	264·0442		3898
20 Mar. (79)	2 Mon.	5 11 38	4 Mar. (63)	0 Sat.	136·6341	714·4074	233·2211		3899
20 Mar. (79)	3 Tues.	11 23 47	21 Feb. (52)	4 Wed.	12·3570	561·6515	202·3979		3900
20 Mar. (79)	4 Wed.	17 35 56	12 Mar. (71)	3 Tues.	47·0394	497·6449	253·6621		3901
19 Mar. (79)	5 Thur.	23 48 5	19 Feb. (60)	0 Sat.	9922·7623	344·8890	222·8629		3902
20 Mar. (79)	0 Sat.	6 0 14	19 Mar. (78)	6 Fri.	9957·4347	280·8825	274·1733		3903
20 Mar. (79)	1 Sun.	12 12 23	8 Mar. (67)	3 Tues.	3833·1675	128·1265	243·3500		3904
20 Mar. (79)	2 Mon.	18 24 32	26 Feb. (57)	1 Sun.	47·5223	11·6622	215·2647		3905
20 Mar. (80)	4 Wed.	0 36 41	16 Mar. (76)	0 Sat.	82·2048	947·6557	266·5751		3906
20 Mar. (79)	5 Thur.	6 48 50	6 Mar. (65)	5 Thur.	296·5595	831·1914	238·4897		3907
20 Mar. (79)	6 Fri.	13 0 59	23 Feb. (54)	2 Mon.	172·2824	678·4354	207·6664		3908
20 Mar. (79)	0 Sat.	19 13 8	14 Mar. (73)	1 Sun.	206·9648	614·4289	258·9769		3909
20 Mar. (80)	2 Mon.	1 25 17	2 Mar. (62)	5 Thur.	82·6876	461·6730	228·1537		3910
20 Mar. (79)	3 Tues.	7 37 26	19 Feb. (50)	2 Mon.	9958·4105	308·9171	197·3304		3911
20 Mar. (79)	4 Wed.	13 49 35	10 Mar. (69)	1 Sun.	9993·0928	244·9104	248·6408		3912
20 Mar. (79)	5 Thur.	20 1 44	27 Feb. (58)	5 Thur.	9868·8157	92·1545	217·8177		3913
20 Mar. (80)	0 Sat.	2 13 52	17 Mar. (77)	4 Wed.	9903·4980	28·1481	269·1281		3914
20 Mar. (79)	1 Sun.	8 26 1	7 Mar. (66)	2 Mon.	117·8529	906·6837	251·0427		3915
20 Mar. (79)	2 Mon.	14 38 10	24 Feb. (55)	6 Fri.	9993·5758	758·9278	210·2194		3916
20 Mar. (79)	3 Tues.	20 50 19	15 Mar. (74)	5 Thur.	28·2581	694·9212	264·5299		3917
20 Mar. (80)	5 Thur.	3 2 28	3 Mar. (63)	2 Mon.	9903·9810	542·1653	230·7067		3918
20 Mar. (79)	6 Fri.	9 14 37	21 Feb. (52)	0 Sat.	118·3358	425·7009	202·6212		3919
20 Mar. (79)	0 Sat.	15 26 46	11 Mar. (70)	5 Thur.	9814·3862	325·4028	251·1938		3920
20 Mar. (79)	1 Sun.	21 38 55	1 Mar. (60)	3 Tues.	28·7410	208·9389	223·1084		3921
20 Mar. (80)	3 Tues.	3 51 4	19 Mar. (79)	2 Mon.	63·4234	144·9321	274·3989		3922

TABLE

CONCURRENT YEAR.								Intercalated (*adhika*) and suppressed (*kshaya*) true lunar months.
						Jovian Saṁvatsara.		
Kali.	Śaka.	Chaitrādi Vikrama.	Mēshādi solar year in Bengal.	Kollam.	A. D.	Southern system.	Northern system.	
1	2	3	3*a*	4	5	6	7	8*a*
3923	744	879	228		821-22	35 *Plava*		...
3924	745	880	229		822-23	36 *Śubhakṛit* . . .		5 Śrāvaṇa .
3925	746	881	230		823-24	37 *Śōbhana* . . .		...
3926	747	882	231		*824-25	38 Krōdhin . . .		...
3927	748	883	232	0-1	825-26	39 Viśvāvasu . . .		3 Jyēshṭha .
3928	749	884	233	1-2	826-27	40 Parābhava .		...
3929	750	885	234	2-3	827-28	41 Plavaṅga . . .		...
3930	751	886	235	3-4	*828-29	42 Kīlaka . . .		1 Chaitra .
3931	752	887	236	4-5	829-30	43 Saumya . . .		...
3932	753	888	237	5-6	830-31	44 Sādhāraṇa . . .		5 Śrāvaṇa .
3933	754	889	238	6-7	831-32	45 Virōdhakṛit . . .		...
3934	755	890	239	7-8	*832-33	46 Paridhāvin . . .		...
3935	756	891	240	8-9	833-34	47 Pramādin . . .		4 Āshāḍha .
3936	757	892	241	9-10	834-35	48 Ānanda . . .		...
3937	758	893	242	10-11	835-36	49 Rākshasa . . .		...
3938	759	894	243	11-12	*836-37	50 Anala		2 Vaiśākha .
3939	760	895	244	12-13	837-38	51 Piṅgala . . .		...
3940	761	896	245	13-14	838-39	52 Kālayukta . . .		6 Bhādrapada
3941	762	897	246	14-15	839-40	53 Siddhārthin . . .		...
3942	763	898	247	15-16	*840-41	54 Raudra . . .		...
3943	764	899	248	16-17	841-42	55 Durmati . . .		5 Śrāvaṇa .
3944	765	900	249	17-18	842-43	56 Dundubhi . . .		...
3945	766	901	250	18-19	843-44	57 Rudhirōdgārin . .		...
3946	767	902	251	19-20	*844-45	58 Raktāksha . . .		3 Jyēshṭha .
3947	768	903	252	20-21	845-46	59 Krōdhana . . .		...

LXXXII—*Contd.*

COMMENCEMENT OF THE											
SOLAR YEAR.					LUNI-SOLAR YEAR (MEAN SUNRISE OF CIVIL DAY ON WHICH CHAITRA ŚUKLA 1 ENDS).						Kali.
Day and month A. D.	Week-day.	Time of true Mēsha-saṃkrānti.			Day and month A. D.	Week-day.	*a*	*b*	*c*		
13	14	17			19	20	23	24	25		1
		H.	M.	S.							
20 Mar. (79)	4 Wed. .	10	3	13	8 Mar. (67)	6 Fri. .	9939·1463	992·1760	243·5956		3923
20 Mar. (79)	5 Thur. .	16	15	22	26 Feb. (57)	4 Wed. .	153·5010	875·7118	215·5102		3924
20 Mar. (79)	6 Fri. .	22	27	31	17 Mar. (76)	3 Tues. .	188·1834	811·7052	266·8206		3925
20 Mar. (80)	1 Sun. .	4	39	40	5 Mar. (65)	0 Sat. .	63·9063	658·9493	235·9975		3926
20 Mar. (79)	2 Mon. .	10	51	49	22 Feb. (53)	4 Wed. .	9939·6292	506·1933	205·1642		3927
20 Mar. (79)	3 Tues. .	17	3	58	13 Mar. (72)	3 Tues. .	9974·3115	442·1868	256·4846		3928
20 Mar. (79)	4 Wed. .	23	16	7	2 Mar. (61)	0 Sat. .	9850·0344	289·4309	225·6614		3929
20 Mar. (80)	6 Fri. .	5	28	16	20 Feb. (51)	5 Thur. .	64·6593	172·9666	197·5760		3930
20 Mar. (79)	0 Sat. .	11	40	25	10 Mar. (69)	4 Wed. .	98·8015	108·9590	248·8864		3931
20 Mar. (79)	1 Sun. .	17	52	34	27 Feb. (58)	1 Sun. .	9974·7944	956·2040	218·0632		3932
21 Mar. (80)	3 Tues. .	0	4	43	18 Mar. (77)	0 Sat. .	9·4768	892·1976	269·3736		3933
20 Mar. (80)	4 Wed. .	6	16	52	7 Mar. (67)	5 Thur. .	223·8317	775·7333	241·2883		3934
20 Mar. (79)	5 Thur. .	12	29	1	24 Feb. (55)	2 Mon. .	99·5545	622·9773	210·4650		3935
20 Mar. (79)	6 Fri. .	18	41	10	15 Mar. (74)	1 Sun. .	134·2369	558·9708	261·7754		3936
21 Mar. (80)	1 Sun. .	0	53	19	4 Mar. (63)	5 Thur. .	9·9598	406·2148	230·9522		3937
20 Mar. (80)	2 Mon. .	7	5	28	21 Feb. (52)	2 Mon. .	9885·6826	253·4589	200·1290		3938
20 Mar. (79)	3 Tues. .	13	17	37	11 Mar. (70)	1 Sun. .	9920·3649	189·4523	252·4294		3939
20 Mar. (79)	4 Wed. .	19	29	46	28 Feb. (59)	5 Thur. .	9796·0878	36·6964	220·6162		3940
21 Mar. (80)	6 Fri. .	1	41	55	20 Mar. (79)	5 Thur. .	169·4022	8·9816	274·6644		3941
20 Mar. (80)	0 Sat. .	7	54	4	8 Mar. (68)	2 Mon. .	45·1250	856·2255	243·8412		3942
20 Mar. (79)	1 Sun. .	14	6	13	26 Feb. (57)	0 Sat. .	259·4798	739·7613	215·7558		3943
20 Mar. (79)	2 Mon. .	20	18	22	17 Mar. (76)	6 Fri. .	294·1622	675·7547	267·0662		3944
21 Mar. (80)	4 Wed. .	2	30	31	6 Mar. (65)	3 Tues. .	169·8851	522·9988	236·0990		3945
20 Mar. (80)	5 Thur. .	8	42	40	23 Feb. (54)	0 Sat. .	45·5979	370·2428	205·4197		3946
20 Mar. (79)	6 Fri. .	14	54	49	12 Mar. (71)	5 Thur. .	9741·6583	269·9446	253·9924		3947

TABLE

CONCURRENT YEAR.								Intercalated (*adhika*) and suppressed (*kshaya*) true lunar months.
Kali.	Śaka.	Chaitrādi Vikrama.	Mêshādi solar year in Bengal.	Kollam.	A. D.	Jovian Saṁvatsara. Southern system.	Jovian Saṁvatsara. Northern system.	
1	2	3	3a	4	5	6	7	8a
3948	769	904	253	21-22	846-47	60 Kshaya . . .		...
3949	770	905	254	22-23	847-48	1 Prabhava . . .		1 Chaitra .
3950	771	906	255	23-24	*848-49	2 Vibhava . . .		...
3951	772	907	256	24-25	849-50	3 Śukla		5 Śrāvaṇa .
3952	773	908	257	25-26	850-51	4 Pramōda . . .		...
3953	774	909	258	26-27	851-52	5 Prajāpati . . .		...
3954	775	910	259	27-28	*852-53	6 Aṅgiras . . .		4 Āshāḍha .
3955	776	911	260	28-29	853-54	7 Śrīmukha . . .		...
3956	777	912	261	29-30	854-55	8 Bhāva . . .		...
3957	778	913	262	30-31	855-56	9 Yuvan . . .		2 Vaiśākha
3958	779	914	263	31-32	*856-57	10 Dhātri . . .		...
3959	780	915	264	32-33	857-58	11 Īśvara . . .		6 Bhādrapada
3960	781	916	265	33-34	858-59	12 Bahudhānya .		...
3961	782	917	266	34-35	859-60	13 Pramādin . . .		...
3962	783	918	267	35-36	*860-61	14 Vikrama . . .		5 Śrāvaṇa
3963	784	919	268	36-37	861-62	15 Vrisha . . .		...
3964	785	920	269	37-38	862-63	16 Chitrabhānu . .		...
3965	786	921	270	38-39	863-64	17 Subhānu . . .		3 Jyēshṭha
3966	787	922	271	39-40	*864-65	18 Tāraṇa . . .		...
3967	788	923	272	40-41	865-66	19 Pārthiva . .		{ 7 Āśvina 9 *Mārgaś* : (*ksh*) }
3968	789	924	273	41-42	866-67	20 Vyaya . . .		1 Chaitra
3969	790	925	274	42-43	867-68	21 Sarvajit . . .		...
3970	791	926	275	43-44	*868-69	22 Sarvadhārin . .		5 Śrāvaṇa
3971	792	927	276	44-45	869-70	23 Virōdhin . . .		...
3972	793	928	277	45-46	870-71	24 Vikrita . . .		...

LXXXII—*Contd.*

COMMENCEMENT OF THE								Kali.
SOLAR YEAR.			LUNI-SOLAR YEAR (MEAN SUNRISE OF CIVIL DAY ON WHICH CHAITRA ŚUKLA 1 ENDS).					
Day and month A.D.	Week-day.	Time of true Mēsha-saṁkrānti.	Day and month A. D.	Week-day.	*a*	*b*	*c*	
13	14	17	19	20	23	24	25	1
		H. M. S.						
20 Mar. (79)	0 Sat.	21 6 58	2 Mar. (61)	3 Tues.	9956·0132	153·4804	226·0070	3948
21 Mar. (80)	2 Mon.	3 19 7	19 Feb. (50)	0 Sat.	9832·2167	0·7839	195·0837	3949
20 Mar. (80)	3 Tues.	9 31 16	10 Mar. (70)	0 Sat.	205·0503	973·0095	249·2319	3950
20 Mar. (79)	4 Wed.	15 43 25	27 Feb. (58)	4 Wed.	80·7732	820·2535	218·4088	3951
20 Mar. (79)	5 Thur.	21 55 34	18 Mar. (77)	3 Tues.	115·4556	756·2470	269·6192	3952
21 Mar. (80)	0 Sat.	4 7 43	7 Mar. (66)	0 Sat.	9991·1784	603·4911	238·7960	3953
20 Mar. (80)	1 Sun.	10 19 52	24 Feb. (55)	4 Wed.	9866·9013	450·7353	207·9727	3954
20 Mar. (79)	2 Mon.	16 32 1	14 Mar. (73)	3 Tues.	9900·5837	386·7286	259·2832	3955
20 Mar. (79)	3 Tues.	22 49 10	3 Mar. (62)	0 Sat.	9777·3065	233·9727	228·4600	3956
21 Mar. (80)	5 Thur.	4 56 19	21 Feb. (52)	5 Thur.	9991·6613	117·5084	200·3745	3957
20 Mar. (80)	6 Fri.	11 8 28	11 Mar. (71)	4 Wed.	26·3437	53·5018	251·6849	3958
20 Mar. (79)	0 Sat.	17 20 37	1 Mar. (60)	2 Mon.	240·4285	937·0375	223·5995	3959
20 Mar. (79)	1 Sun.	23 32 45	20 Mar. (79)	1 Sun.	275·3809	873·0310	274·9100	3960
21 Mar. (80)	3 Tues.	5 44 54	9 Mar. (68)	5 Thur.	151·1038	720·2751	244·0867	3961
20 Mar. (80)	4 Wed.	11 57 3	26 Feb. (57)	2 Mon.	26·8266	567·5191	213·2635	3962
20 Mar. (79)	5 Thur.	18 9 12	16 Mar. (75)	1 Sun.	61·5090	503·5126	264·5739	3963
21 Mar. (80)	0 Sat.	0 21 21	5 Mar. (64)	5 Thur.	9937·2318	350·7566	233·5708	3964
21 Mar. (80)	1 Sun.	6 33 30	22 Feb. (53)	2 Mon.	9812·9547	198·0007	202·9275	3965
20 Mar. (80)	2 Mon.	12 45 39	12 Mar. (72)	1 Sun.	9847·6371	132·9941	254·2379	3966
20 Mar. (79)	3 Tues.	18 57 48	2 Mar. (61)	6 Fri.	61·9919	17·5299	226·1525	3967
21 Mar. (80)	5 Thur.	1 9 57	19 Feb. (50)	3 Tues.	9937·7149	864·7741	195·3293	3968
21 Mar. (80)	6 Fri.	7 22 6	11 Mar. (70)	3 Tues.	311·0291	837·0590	249·3775	3969
20 Mar. (80)	0 Sat.	13 34 15	28 Feb. (59)	0 Sat.	186·7519	684·3031	218·5543	3970
20 Mar. (79)	1 Sun.	19 46 24	18 Mar. (77)	6 Fri.	221·4343	620·2265	269·8647	3971
21 Mar. (80)	3 Tues.	1 58 33	7 Mar. (66)	3 Tues.	97·1572	467·5406	239·0416	3972

TABLE

CONCURRENT YEAR.								Intercalated (*adhika*) and suppressed (*Kshaya*) true lunar months.
Kali.	Śaka.	Chaitrādi Vikrama.	Mēshādi solar year in Bengal.	Kollam.	A. D.	Jovian Saṁvatsara. Southern system.	Jovian Saṁvatsara. Northern system.	
1	2	3	3a	4	5	6	7	8a
3973	794	929	278	46-47	871-72	25 Khara . . .		4 Āshāḍha .
3974	795	930	279	47-48	*872-73	26 Nandana . . .		...
3975	796	931	280	48-49	873-74	27 Vijaya . . .		...
3976	797	932	281	49-50	874-75	28 Jaya . . .		2 Vaiśākha .
3977	798	933	282	50-51	875-76	29 Manmatha . .		...
3978	799	934	283	51-52	*876-77	30 Durmukha . .		6 Bhādrapada
3979	800	935	284	52-53	877-78	31 Hēmalamba . .		...
3980	801	936	285	53-54	878-79	32 Vilamba . . .		...
3981	802	937	286	54-55	879-80	33 Vikārin . . .		5 Śrāvaṇa .
3982	803	938	287	55-56	*880-81	34 Śārvarin . . .		...
3983	804	939	288	56-57	881-82	35 Plava . . .		...
3984	805	940	289	57-58	882-83	36 Śubhakṛit . . .		3 Jyēshṭha .
3985	806	941	290	58-59	83-84	37 Śōbhana . .		...
3986	807	942	291	59-60	*884-85	38 Krōdhin . .		{ 7 Āśvina . } { *10 Pausha* (*ksh.*) }
3987	808	943	292	60-61	885-86	39 Viśvāvasu . .		1 Chaitra .
3988	809	944	293	61-62	886-87	40 Parābhava . .		...
3989	810	945	294	62-63	887-88	41 Plavaṅga . . .		5 Śrāvaṇa .
3990	811	946	295	63-64	*888-89	42 Kīlaka . . .		...
3991	812	947	296	64-65	889-90	43 Saumya . . .		...
3992	813	948	297	65-66	890-91	44 Sādhāraṇa . .		3 Jyēshṭha .
3993	814	949	298	66-67	891-92	45 Virōdhakṛit . .		...
3994	815	950	299	67-68	*892-93	46 Paridhāvin . .		...
3995	816	951	300	68-69	893-94	47 Pramādin . . .		2 Vaiśākha .
3996	817	952	331	69-70	894-95	48 Ānanda . . .		...
3997	818	953	302	70-71	895-96	49 Rākshasa . . .		6 Bhādrapada

LXXXII—*Contd.*

COMMENCEMENT OF THE											
Solar year.					Luni-solar year (mean sunrise of civil day on which Chaitra śukla 1 ends).						Kali.
Day and month A. D.	Week-day.	Time of true Mēsha-saṁkrānti.			Day and month A. D.	Week-day.	*a*	*b*	*c*		
13	14	17			19	20	23	24	25		1
		H.	M.	S.							
21 Mar. (80)	4 Wed. .	8	10	42	24 Feb. (55)	0 Sat. .	9972·8801	313·7846	208·2183		3973
20 Mar. (80)	5 Thur. .	14	22	51	14 Mar. (74)	6 Fri. .	7·5624	250·7781	259·5087		3974
20 Mar. (79)	6 Fri. .	20	35	0	3 Mar. (62)	3 Tues. .	9883·2853	98·0222	228·7055		3975
21 Mar. (80)	1 Sun. .	2	47	9	21 Feb. (52)	1 Sun. .	97·6401	981·5579	200·6101		3976
21 Mar. (80)	2 Mon. .	8	59	18	12 Mar. (71)	0 Sat. .	132·3224	917·5514	251·9305		3977
20 Mar. (80)	3 Tues. .	15	11	27	29 Feb. (60)	4 Wed. .	8·0453	764·7954	221·1072		3978
20 Mar. (79)	4 Wed. .	21	23	36	19 Mar. (78)	3 Tues. .	42·7277	700·7889	272·4177		3979
21 Mar. (80)	6 Fri. .	3	35	45	8 Mar. (67)	0 Sat. .	9918·4506	548·0330	241·5146		3980
21 Mar. (80)	0 Sat. .	9	47	54	26 Feb. (57)	5 Thur. .	132·8053	431·5686	213·5091		3981
20 Mar. (80)	1 Sun. .	16	0	3	15 Mar. (75)	3 Tues. .	9828·8558	331·2705	262·0817		3982
20 Mar. (79)	2 Mon. .	22	12	12	5 Mar. (64)	1 Sun. .	43·2106	214·8061	234·0013		3983
21 Mar. (80)	4 Wed. .	4	24	21	22 Feb. (53)	5 Thur. .	9918·9335	62·0502	203·1731		3984
21 Mar. (80)	5 Thur. .	10	36	30	13 Mar. (72)	4 Wed. .	9953·6158	998·0436	254·4835		3985
20 Mar. (80)	6 Fri. .	16	48	39	2 Mar. (62)	2 Mon. .	167·9707	881·5794	226·3980		3986
20 Mar. (79)	0 Sat. .	23	0	48	19 Feb. (50)	6 Fri. .	43·6936	728·9235	195·5748		3987
21 Mar. (80)	2 Mon. .	5	12	57	10 Mar. (69)	5 Thur. .	78·3759	664·8169	246·7168		3988
21 Mar. (80)	3 Tues. .	11	25	6	27 Feb. (58)	2 Mon. .	9954·0987	512·0610	216·0621		3989
20 Mar. (80)	4 Wed. .	17	37	15	17 Mar. (77)	1 Sun. .	9988·7811	448·0544	267·3724		3990
20 Mar. (79)	5 Thur. .	23	49	24	6 Mar. (65)	5 Thur. .	9864·5040	294·2984	236·5493		3991
21 Mar. (80)	0 Sat. .	6	1	33	23 Feb. (54)	2 Mon. .	9740·2868	142·5426	205·7261		3992
21 Mar. (80)	1 Sun. .	12	13	42	14 Mar. (73)	1 Sun. .	9774·9092	78·5360	257·0365		3993
20 Mar. (80)	2 Mon. .	18	25	51	3 Mar. (63)	6 Fri. .	9989·2641	962·0717	228·9510		3994
21 Mar. (80)	4 Wed. .	0	38	0	21 Feb. (52)	4 Wed. .	203·6198	845·6075	200·6968		3995
21 Mar. (80)	5 Thur. .	6	50	9	12 Mar. (71)	3 Tues. .	238·3012	781·6009	252·0073		3996
21 Mar. (80)	6 Fri. .	21	2	18	1 Mar. (60)	0 Sat. .	114·0241	628·8149	221·3528		3997

TABLE

CONCURRENT YEAR.								Intercalated (*adhika*) and suppressed (*kshaya*) true lunar months.
						JOVIAN SAṀVATSARA.		
Kali.	Śaka.	Chaitrādi Vikrama.	Mēshādi solar year in Bengal.	Kollam.	A. D.	Southern system.	Northern system.	
1	2	3	3a	4	5	6	7	8a
3998	819	954	303	71-72	*896-97	50 Anala		...
3999	820	955	304	72-73	897-98	51 Piṅgala		...
4000	821	956	305	73-74	898-99	52 Kālayukta		4 Āshāḍha
4001	822	957	306	74-75	899-900	53 Siddhārthin		...
4002	823	958	307	75-76	*900-01	54 Raudra		...
4003	824	959	308	76-77	901-02	55 Durmati		3 Jyēshṭha
4004	825	960	309	77-78	902-03	56 Dundubhi		...
4005	826	961	310	78-79	903-04	57 Rudhirōdgārin		7 Āśvina
4006	827	962	311	79-80	*904-05	58 Raktāksha †		...
4007	828	963	312	80-81	905-06	59 Krōdhana	60 *Kshaya*	...
4008	829	964	313	81-82	906-07	60 Kshaya	1 *Prabhava*	5 Śrāvaṇa
4009	830	965	314	82-83	907-08	1 Prabhava	2 *Vibhava*	...
4010	831	966	315	83-84	*908-09	2 Vibhava	3 *Śukla*	...
4011	832	967	316	84-85	909-10	3 Śukla	4 Pramōda	3 Jyēshṭha
4012	833	968	317	85-86	910-11	4 Pramōda	5 Prajāpati	...
4013	834	969	318	86-87	911-12	5 Prajāpati	6 Aṅgiras	...
4014	835	970	319	87-88	*912-13	6 Aṅgiras	7 Śrīmukha	2 Vaiśākha
4015	836	971	320	88-89	913-14	7 Śrīmukha	8 Bhāva	...
4016	837	972	321	89-90	914-15	8 Bhāva	9 Yuvan	6 Bhādrapada
4017	838	973	322	90-91	915-16	9 Yuvan	10 Dhātṛi	...
4018	839	974	323	91-92	*916-17	10 Dhātṛi	11 Īśvara	...
4019	840	975	324	92-93	917-18	11 Īśvara	12 Bahudhānya	4 Āshāḍha
4020	841	976	325	93-94	918-19	12 Bahudhānya	13 Pramādin	...
4021	842	977	326	94-95	919-20	13 Pramādin	14 Vikrama	...
4022	843	978	327	95-96	*920-21	14 Vikrama	15 Vṛisha	3 Jyēshṭha

† 59 Krōdhana was suppressed in the North. By Southern reckoning there was no suppression nor has there been any such since.

LXXXII—*Contd.*

COMMENCEMENT OF THE										
SOLAR YEAR.					LUNI-SOLAR YEAR (MEAN SUNRISE OF CIVIL DAY ON WHICH CHAITRA ŚUKLA 1 ENDS).					Kali.
Day and month, A. D.	Week-day.	Time of true Mēsha-saṁkrānti.			Day and month, A. D.	Week-day.	*a*	*b*	*c*	
13	14	17			19	20	23	24	25	1
		H.	M.	S.						
20 Mar. (80)	0 Sat.	19	14	27	19 Mar. (79)	6 Fri.	148·7064	564·8384	272·6632	3998
21 Mar. (80)	2 Mon.	1	26	36	8 Mar. (67)	3 Tues.	24·4293	412·0825	241·8401	3999
21 Mar. (80)	3 Tues.	7	38	45	25 Feb. (56)	0 Sat.	9900·1522	259·3266	211·0169	4000
21 Mar. (80)	4 Wed.	13	50	54	16 Mar. (75)	6 Fri.	9934·8345	195·3200	262·3050	4001
20 Mar. (80)	5 Thur.	20	3	3	4 Mar. (64)	3 Tues.	9810·5573	42·5640	231·4818	4002
21 Mar. (80)	0 Sat.	2	15	12	22 Feb. (53)	1 Sun.	24·9122	926·0097	203·3963	4003
21 Mar. (80)	1 Sun.	8	27	21	13 Mar. (72)	0 Sat.	59·5945	862·0930	254·7067	4004
21 Mar. (80)	2 Mon.	14	29	29	3 Mar. (62)	5 Thur.	273·9494	745·6289	226·6213	4005
20 Mar. (80)	3 Tues.	20	51	38	20 Mar. (80)	3 Tues.	9969·9998	645·3307	275·1940	4006
21 Mar. (80)	5 Thur.	3	3	47	10 Mar. (69)	1 Sun.	184·3546	528·8665	247·1085	4007
21 Mar. (80)	6 Fri.	9	15	56	27 Feb. (58)	5 Thur.	60·0774	376·1105	216·2853	4008
21 Mar. (80)	0 Sat.	15	28	5	17 Mar. (76)	3 Tues.	9756·1279	275·8123	264·8579	4009
20 Mar. (80)	1 Sun.	21	40	14	6 Mar. (66)	1 Sun.	9970·4827	159·3479	236·7726	4010
21 Mar. (80)	3 Tues.	3	52	23	23 Feb. (54)	5 Thur.	9846·2055	6·5921	205·9493	4011
21 Mar. (80)	4 Wed	10	4	32	14 Mar. (73)	4 Wed.	9880·8879	942·5855	257·2597	4012
21 Mar. (80)	5 Thur.	16	16	41	4 Mar. (63)	2 Mon.	95·2428	826·1212	229·1743	4013
20 Mar. (80)	6 Fri.	22	28	50	22 Feb. (53)	0 Sat.	309·5975	709·6569	201·0889	4014
21 Mar. (80)	1 Sun.	4	40	59	11 Mar. (70)	5 Thur.	5·6479	609·3587	249·6615	4015
21 Mar. (80)	2 Mon.	10	53	8	28 Feb. (59)	2 Mon.	9881·3708	456·6028	218·8383	4016
21 Mar. (80)	3 Tues.	17	5	17	19 Mar. (78)	1 Sun.	9916·0531	392·5962	270·1487	4017
20 Mar. (80)	4 Wed.	23	17	26	7 Mar. (67)	5 Thur.	9791·7760	239·8403	239·3256	4018
21 Mar. (80)	6 Fri.	5	29	35	25 Feb. (56)	3 Tues.	6·1309	123·3760	211·2401	4019
21 Mar. (80)	0 Sat.	11	41	44	16 Mar. (75)	2 Mon.	40·8133	59·3695	262·5505	4020
21 Mar. (80)	1 Sun.	17	53	53	5 Mar. (64)	6 Fri.	9916·5360	906·6135	231·6273	4021
21 Mar. (81)	3 Tues.	0	6	2	23 Feb. (54)	4 Wed.	130·8909	790·1493	203·6419	4022

TABLE

CONCURRENT YEAR.								Intercalated (*adhika*) and suppressed (*kshaya*) true lunar months.
						JOVIAN SAṀVATSARA.		
Kali.	Śaka.	Chaitrādi Vikrama.	Mēshādi solar year in Bengal.	Kollam.	A. D.	Southern system.	Northern system.	
1	2	3	3a	4	5	6	7	8a
4023	844	979	328	96-97	921-22	15 Vṛisha . .	16 Chitrabhānu .	...
4024	845	980	329	97-98	922-23	16 Chitrabhānu .	17 Subhānu .	7 Āśvina .
4025	846	981	330	98-99	923-24	17 Subhānu .	18 Tāraṇa .	...
4026	847	982	331	99-100	*924-25	18 Tāraṇa . .	19 Pārthiva .	...
4027	848	983	332	100-01	925-26	19 Pārthiva .	20 Vyaya . .	5 Śrāvaṇa .
4028	849	984	333	101-02	926-27	20 Vyaya . .	21 Sarvajit .	...
4029	850	985	334	102-03	927-28	21 Sarvajit .	22 Sarvadhārin .	...
4030	851	986	335	103-04	*928-29	22 Sarvadhārin .	23 Virōdhin .	3 Jyēshṭha .
4031	852	987	336	104-05	929-30	23 Virōdhin .	24 Vikṛita .	...
4032	853	988	337	105-06	930-31	24 Vikṛita . .	25 Khara . .	...
4033	854	989	338	106-07	931-32	25 Khara . .	26 Nandana .	2 Vaiśākha .
4034	855	990	339	107-08	*932-33	26 Nandana .	27 Vijaya . .	...
4035	856	991	340	108-09	933-34	27 Vijaya . .	28 Jaya . .	6 Bhādrapada
4036	857	992	341	109-10	934-35	28 Jaya . .	29 Manmatha .	...
4037	858	993	342	110-11	935-36	29 Manmatha .	30 Durmukha .	...
4038	859	994	343	111-12	*936-37	30 Durmukha .	31 Hēmalamba .	4 Āshāḍha .
4039	860	995	344	112-13	937-38	31 Hēmalamba .	32 Vilamba .	...
4040	861	996	345	113-14	938-39	32 Vilamba .	33 Vikārin .	...
4041	862	997	346	114-15	939-40	33 Vikārin .	34 Śārvarin .	3 Jyēshtha .
4042	863	998	347	115-16	*940-41	34 Śārvarin .	35 Plava . .	...
4043	864	999	348	116-17	941-42	35 Plava .	36 Śubhakṛit .	7 Āśvina .
4044	865	1000	349	117-18	942-43	36 Śubhakṛit .	37 Śōbhana .	...
4045	866	1001	350	118-19	943-44	37 Śōbhana .	38 Krōdhin .	...
4046	867	1002	351	119-20	*944-45	38 Krōdhin .	39 Viśvāvasu .	5 Śrāvaṇa .
4047	868	1003	352	120-21	945-46	39 Viśvāvasu .	40 Parābhava .	...

LXXXII—*Contd.*

COMMENCEMENT OF THE									Kali.
SOLAR YEAR.			LUNI-SOLAR YEAR (MEAN SUNRISE OF CIVIL DAY ON WHICH CHAITRA ŚUKLA 1 ENDS).						
Day and month, A. D.	Week-day.	Time of true Mēsha-saṁkrānti.	Day and month, A. D.	Week-day.	*a*	*b*	*c*		
13	14	17	19	20	23	24	25		1
		H. M. S.							
21 Mar. (80)	4 Wed. .	6 18 11	13 Mar. (72)	3 Tues. .	165·5733	726·1427	254·9523		4023
21 Mar. (80)	5 Thur. .	12 30 20	2 Mar. (61)	0 Sat. .	41·2961	573·3868	224·1290		4024
21 Mar. (80)	6 Fri. .	18 42 29	21 Mar. (80)	6 Fri. .	75·9785	509·3802	275·4395		4025
21 Mar. (81)	1 Sun. .	0 54 38	9 Mar. (69)	3 Tues. .	9951·7014	356·6243	244·6163		4026
21 Mar. (80)	2 Mon. .	7 6 47	26 Feb. (57)	0 Sat. .	9827·4242	203·8683	213·7931		4027
21 Mar. (80)	3 Tues. .	13 18 56	17 Mar. (76)	6 Fri. .	9862·0966	139·8618	265·1034		4028
21 Mar. (80)	4 Wed. .	19 31 5	7 Mar. (66)	4 Wed. .	76·4614	23·3975	237·0181		4029
21 Mar. (81)	6 Fri. .	1 43 14	24 Feb. (55)	1 Sun. .	9952·1843	870·6416	206·1949		4030
21 Mar. (80)	0 Sat. .	7 55 23	14 Mar (73)	0 Sat. .	9986·8666	806·6351	257·5053		4031
21 Mar. (80)	1 Sun. .	14 7 32	4 Mar. (63)	5 Thur. .	201·2215	690·1707	229·4198		4032
21 Mar. (80)	2 Mon. .	20 19 41	21 Feb. (52)	2 Mon. .	76·9443	537·4148	198·5966		4033
21 Mar. (81)	4 Wed. .	2 31 50	11 Mar. (71)	1 Sun. .	111·6267	473·4083	249·9071		4034
21 Mar. (80)	5 Thur. .	8 43 59	28 Feb. (59)	5 Thur. .	9987·3495	320·6523	219·0839		4035
21 Mar. (80)	6 Fri. .	14 56 8	19 Mar. (78)	4 Wed. .	22·0319	256·6458	270·3942		4036
21 Mar. (80)	0 Sat. .	21 8 17	8 Mar. (67)	1 Sun. .	9897·7548	103·8898	239·5711		4037
21 Mar. (81)	2 Mon. .	3 20 26	26 Feb. (57)	6 Fri. .	112·1097	987·4256	211·4857		4038
21 Mar. (80)	3 Tues. .	9 32 35	16 Mar. (75)	5 Thur. .	146·7920	923·4190	262·7961		4039
21 Mar. (80)	4 Wed. .	15 44 44	5 Mar. (64)	2 Mon. .	22·5148	770·6630	231·9729		4040
21 Mar. (80)	5 Thur. .	21 56 53	23 Feb. (54)	0 Sat. .	236·8697	654·1988	203·8874		4041
21 Mar. (81)	0 Sat. .	4 9 2	12 Mar. (72)	5 Thur. .	9932·9200	553·9006	252·4601		4042
21 Mar. (80)	1 Sun. .	10 21 11	1 Mar. (60)	2 Mon. .	9808·6429	401·1447	221·6368		4043
21 Mar. (80)	2 Mon. .	16 33 20	20 Mar. (79)	1 Sun. .	9843·3253	337·1381	272·9473		4044
21 Mar. (80)	3 Tues. .	22 45 29	9 Mar. (68)	5 Thur. .	9719·0482	184·3821	242·1240		4045
21 Mar. (81)	5 Thur. .	4 57 38	27 Feb. (58)	3 Tues. .	9933·4029	67·9178	214·0386		4046
21 Mar. (80)	6 Fri. .	11 9 47	17 Mar. (76)	2 Mon. .	9968·0854	3·9113	265·3490		4047

TABLE

CONCURRENT YEAR.								Intercalated (*adhika*) and suppressed (*kshaya*) true lunar months.
						JOVIAN SAṀVATSARA.		
Kali.	Śaka.	Chaitrādi Vikrama.	Mēshādi solar year in Bengal.	Kollam.	A. D.	Southern system.	Northern system.	
1	2	3	3a	4	5	6	7	8a
4048	869	1004	353	121-22	946-47	40 Parābhava .	41 Plavaṅga .	...
4049	870	1005	354	122-23	947-48	41 Plavaṅga .	42 Kīlaka . .	3 Jyēshṭha .
4050	871	1006	355	123-24	*948-49	42 Kīlaka . .	43 Saumya .	...
4051	872	1007	356	124-25	949-50	43 Saumya .	44 Sādhāraṇa .	...
4052	873	1008	357	125-26	950-51	44 Sādhāraṇa .	45 Virōdhakṛit .	1 Chaitra .
4053	874	1009	358	126-27	951-52	45 Virōdhakṛit	46 Paridhāvin .	...
4054	875	1010	359	127-28	*952-53	46 Paridhāvin .	47 Pramādin .	5 Śrāvaṇa .
4055	876	1011	360	128-29	953-54	47 Pramādin .	48 Ānanda .	...
4056	877	1012	361	129-30	954-55	48 Ānanda .	49 Rākshasa .	...
4057	878	1013	362	130-31	955-56	49 Rākshasa .	50 Anala . .	4 Āshāḍha .
4058	879	1014	363	131-32	*956-57	50 Anala . .	51 Piṅgala .	...
4059	880	1015	364	132-33	957-58	51 Piṅgala .	52 Kālayukta .	...
4060	881	1016	365	133-34	958-59	52 Kālayukta .	53 Siddhārthin .	3 Jyēshṭha .
4061	882	1017	366	134-35	959-60	53 Siddhārthin .	54 Raudra .	...
4062	883	1018	367	135-36	*960-61	54 Raudra .	55 Durmati .	7 Āśvina .
4063	884	1019	368	136-37	961-62	55 Durmati .	56 Dundubhi .	...
4064	885	1020	369	137-38	962-63	56 Dundubhi .	57 Rudhirōdgārin	...
4065	886	1021	370	138-39	963-64	57 Rudhirōdgārin	58 Raktāksha .	4 Āshāḍha .
4066	887	1022	371	139-40	*964-65	58 Raktāksha .	59 Krōdhana .	...
4067	888	1023	372	140-41	965-66	59 Krōdhana .	60 Kshaya .	...
4068	889	1024	373	141-42	966-67	60 Kshaya .	1 Prabhava .	3 Jyēshṭha
4069	890	1025	374	142-43	967-68	1 Prabhava .	2 Vibhava .	...
4070	891	1026	375	143-44	*968-69	2 Vibhava .	3 Śukla . .	12 Phālguna .
4071	892	1027	376	144-45	969-70	3 Śukla . .	4 Pramōda .	...
4072	893	1028	377	145-46	970-71	4 Pramōda .	5 Prajāpati .	...

LXXXII—*Contd.*

COMMENCEMENT OF THE									Kali.
SOLAR YEAR.			LUNI-SOLAR YEAR (MEAN SUNRISE OF CIVIL DAY ON WHICH CHAITRA ŚUKLA 1 ENDS).						
Day and month, A. D.	Week-day.	Time of true Mēsha-saṁkrānti.	Day and month, A. D.	Week-day.	*a*	*b*	*c*		
13	14	17	19	20	23	24	25		1
		H. M. S.							
21 Mar. (80)	0 Sat. .	17 21 56	7 Mar. (66)	0 Sat. .	182·4402	887·4470	237·2637		4048
21 Mar. (80)	1 Sun. .	23 34 5	24 Feb. (55)	4 Wed. .	58·1630	734·6910	206·4404		4049
21 Mar. (81)	3 Tues. .	5 46 13	14 Mar. (74)	3 Tues. .	92·8454	670·6846	257·7508		4050
21 Mar. (80)	4 Wed. .	11 58 22	3 Mar. (62)	0 Sat. .	9968·5683	517·9286	226·9276		4051
21 Mar. (80)	5 Thur. .	18 10 31	20 Feb. (51)	4 Wed. .	9844·3112	365·1727	196·1044		4052
22 Mar. (81)	0 Sat. .	0 22 40	11 Mar. (70)	3 Tues. .	9878·9735	301·1662	247·4148		4053
21 Mar. (81)	1 Sun. .	6 34 49	28 Feb. (59)	0 Sat. .	9754·6963	148·4102	216·5916		4054
21 Mar. (80)	2 Mon. .	12 46 58	18 Mar. (77)	6 Fri. .	9789·3787	84·4037	267·9020		4055
21 Mar. (80)	3 Tues. .	18 59 7	8 Mar. (67)	4 Wed. .	3·7335	967·9394	239·8167		4056
22 Mar. (81)	5 Thur. .	1 11 16	26 Feb. (57)	2 Mon. .	218·0884	851·4750	211·7312		4057
21 Mar. (81)	6 Fri. .	7 23 25	16 Mar. (76)	1 Sun. .	252·7708	787·4685	263·0416		4058
21 Mar. (80)	0 Sat. .	13 35 34	5 Mar. (64)	5 Thur. .	128·4936	634·7125	232·2184		4059
21 Mar. (80)	1 Sun. .	19 47 43	22 Feb. (53)	2 Mon. .	4·2164	481·9566	201·3952		4060
22 Mar. (81)	3 Tues. .	1 59 52	13 Mar. (72)	1 Sun. .	38·8988	417·9502	252·7056		4061
21 Mar. (81)	4 Wed. .	8 12 1	1 Mar. (61)	5 Thur. .	9914·6217	265·1942	221·8823		4062
21 Mar. (80)	5 Thur. .	14 24 10	20 Mar. (79)	4 Wed. .	9949·3040	201·1877	273·1828		4063
21 Mar. (80)	6 Fri. .	20 36 19	9 Mar. (68)	1 Sun. .	9825·0269	48·5316	242·3696		4064
22 Mar. (81)	1 Sun. .	2 48 28	27 Feb. (58)	6 Fri. .	39·3817	931·9674	214·2842		4065
21 Mar. (81)	2 Mon. .	9 0 37	17 Mar. (77)	5 Thur. .	74·0642	867·9608	265·5946		4066
21 Mar. (80)	3 Tues. .	15 12 46	7 Mar. (66)	3 Tues. .	288·4189	751·4956	237·5093		4067
21 Mar. (80)	4 Wed. .	21 24 55	24 Feb. (55)	0 Sat. .	164·1418	598·7406	206·6860		4068
22 Mar. (81)	6 Fri. .	3 37 4	15 Mar. (74)	6 Fri. .	198·8042	534·7341	257·9964		4069
21 Mar. (81)	0 Sat. .	9 49 13	3 Mar. (63)	3 Tues. .	74·5470	381·9782	227·1731		4070
21 Mar. (80)	1 Sun. .	16 1 22	21 Mar. (80)	1 Sun. .	9770·5974	281·6799	275·7458		4071
21 Mar. (80)	2 Mon. .	22 13 31	11 Mar. (70)	6 Fri. .	9984·9522	616·2156	247·6664		4072

TABLE

CONCURRENT YEAR.								
Kali.	Śaka.	Chaitrādi Vikrama.	Mēshādi solar year in Bengal.	Kollam.	A. D.	Jovian Saṁvatsara. Southern system.	Jovian Saṁvatsara. Northern system.	Intercalated (*adhika*) and suppressed (*kshaya*) true lunar months.
1	2	3	3a	4	5	6	7	8a
4073	894	1029	378	146-47	971-72	5 Prajāpati .	6 Aṅgiras .	5 Śrāvaṇa .
4074	895	1030	379	147-48	*972-73	6 Aṅgiras .	7 Śrīmukha .	...
4075	896	1031	380	148-49	973-74	7 Śrīmukha .	8 Bhāva . .	...
4076	897	1032	381	149-50	974-75	8 Bhāva . .	9 Yuvan . .	4 Āshāḍha .
4077	898	1033	382	150-51	975-76	9 Yuvan . .	10 Dhātṛi . .	...
4078	899	1034	383	151-52	*976-77	10 Dhātṛi . .	11 Īśvara . .	...
4079	900	1035	384	152-53	977-78	11 Īśvara . .	12 Bahudhānya .	2 Vaiśākha .
4080	901	1036	385	153-54	978-79	12 Bahudhānya .	13 Pramāthin .	...
4081	902	1037	386	154-55	979-80	13 Pramāthin .	14 Vikrama .	6 Bhādrapada
4082	903	1038	387	155-56	*980-81	14 Vikrama .	15 Vṛisha . .	...
4083	904	1039	388	156-57	981-82	15 Vṛisha .	16 Chitrabhānu .	...
4084	905	1040	389	157-58	982-83	16 Chitrabhānu .	17 Subhānu	4 Āshāḍha
4085	906	1041	390	158-59	983-84	17 Subhānu .	18 Tāraṇa . .	...
4086	907	1042	391	159-60	*984-85	18 Tāraṇa . .	19 Pārthiva	...
4087	908	1043	392	160-61	985-86	19 Pārthiva .	20 Vyaya . .	3 Jyēshṭha .
4088	909	1044	393	161-62	986-87	20 Vyaya . .	21 Sarvajit .	...
4089	910	1045	394	162-63	987-88	21 Sarvajit .	22 Sarvadhārin .	...
4090	911	1046	395	163-64	*988-89	22 Sarvadhārin .	23 Virōdhin .	1 Chaitra .
4091	912	1047	396	164-65	989-90	23 Virōdhin .	24 Vikṛita† .	...
4092	913	1048	397	165-66	990-91	24 Vikṛita . .	26 *Nandana* .	5 Śrāvaṇa .
4093	914	1049	398	166-67	991-92	25 Khara .	27 *Vijaya* . .	...
4094	915	1050	399	167-68	*992-93	26 Nandana .	28 *Jaya* . .	...
4095	916	1051	400	168-69	993-94	27 Vijaya . .	29 *Manmatha* .	4 Āshāḍha .
4096	917	1052	401	169-70	994-95	28 Jaya . .	30 *Durmukha* .	...
4097	918	1053	402	170-71	995-96	29 Manmatha .	31 Hēmalamba .	...

† 25 Khara was suppressed in the north.

LXXXII—*Contd.*

COMMENCEMENT OF THE									
SOLAR YEAR			LUNI-SOLAR YEAR (MEAN SUNRISE OF CIVIL DAY ON WHICH CHAITRA ŚUKLA 1 ENDS).						Kali.
Day and month, A. D.	Week-day.	Time of true Mēsha-saṁkrānti.	Day and month, A. D.	Week-day.	*a*	*b*	*c*		
13	14	17	19	20	23	24	25		1
		H. M. S.							
22 Mar. (81)	4 Wed. .	4 25 40	28 Feb. (59)	3 Tues. .	9860·6751	12·4597	217·8372		4073
21 Mar. (81)	5 Thur. .	10 37 49	18 Mar. (78)	2 Mon. .	9895·3574	948·4532	268·0475		4074
21 Mar. (80)	6 Fri. .	16 49 58	8 Mar. (67)	0 Sat. .	109·7123	831·9889	240·0622		4075
21 Mar. (80)	0 Sat. .	23 2 7	25 Feb. (56)	4 Wed. .	9985·4352	679·2329	209·2390		4076
22 Mar. (81)	2 Mon. .	5 14 16	16 Mar. (75)	3 Tues. .	20·1175	615·2264	260·5494		4077
21 Mar. (81)	3 Tues. .	11 26 25	4 Mar. (64)	0 Sat. .	9895·8404	462·4704	229·7261		4078
21 Mar. (80)	4 Wed. .	17 38 34	21 Feb. (52)	4 Wed. .	9771·5632	309·7145	198·9029		4079
21 Mar. (80)	5 Thur. .	23 50 43	12 Mar. (71)	3 Tues. .	9806·2456	245·7080	250·2134		4080
22 Mar. (81)	0 Sat. .	6 2 52	2 Mar. (61)	1 Sun. .	20·6004	129·2437	222·1279		4081
21 Mar. (81)	1 Sun. .	12 15 1	20 Mar. (80)	0 Sat. .	55·2828	65·2372	273·4383		4082
21 Mar. (80)	2 Mon. .	18 27 10	9 Mar. (68)	4 Wed. .	9931·0057	912·4811	242·6151		4083
22 Mar. (81)	4 Wed. .	0 39 19	27 Feb. (58)	2 Mon. .	145·3605	796·0169	214·5298		4084
22 Mar. (81)	5 Thur. .	6 51 28	18 Mar. (77)	1 Sun. .	180·0429	732·0103	265·8401		4085
21 Mar. (81)	6 Fri. .	13 3 37	6 Mar. (66)	5 Thur. .	55·7657	579·2544	235·0169		4086
21 Mar. (80)	0 Sat. .	19 15 46	23 Feb. (54)	2 Mon. .	9931·4886	426·4985	204·1937		4087
22 Mar. (81)	2 Mon. .	1 27 55	11 Mar. (73)	1 Sun. .	9966·1709	362·4919	255·5042		4088
22 Mar. (81)	3 Tues. .	7 40 4	3 Mar. (62)	5 Thur. .	9841·8938	209·7360	224·6809		4089
21 Mar. (81)	4 Wed. .	13 52 13	21 Feb. (52)	3 Tues. .	56·2487	93·2717	196·5954		4090
21 Mar. (80)	5 Thur. .	20 4 22	11 Mar. (70)	2 Mon. .	90·8310	29·2651	247·9059		4091
22 Mar. (81)	0 Sat .	2 16 31	28 Feb. (59)	6 Fri. .	9966·6538	876·5093	217·0828		4092
22 Mar. (81)	1 Sun. .	8 28 40	19 Mar. (78)	5 Thur. .	1·3372	812·5027	268·3931		4093
21 Mar. (81)	2 Mon. .	14 40 49	8 Mar. (68)	3 Tues. .	215·6911	696·0384	240·3077		4094
21 Mar. (80)	3 Tues. .	20 52 58	25 Feb. (56)	0 Sat. .	91·4139	543·2825	209·4845		4095
22 Mar. (81)	5 Thur. .	3 5 6	16 Mar. (75)	6 Fri. .	126·0953	479·2759	260·7950		4096
22 Mar. (81)	6 Fri. .	9 17 15	5 Mar (64)	3 Tues. .	1·8192	326·5199	229·9717		4097

TABLE

CONCURRENT YEAR.								Intercalated (*adhika*) and suppressed (*kshaya*) true lunar months.
						JOVIAN SAṀVATSARA.		
Kali.	Saka.	Chaitrādi Vikrama.	Mēshādi solar year in Bengal.	Kollam.	A. D.	Southern system.	Northern system.	
1	2	3	3*a*	4	5	6	7	8*a*
4098	919	1054	403	171-72	*996-97	30 Durmukha	32 Vilamba	2 Vaiśākha
4099	920	1055	404	172-73	997-98	31 Hēmalamba	33 Vikārin	...
4100	921	1056	405	173-74	998-99	32 Vilamba	34 Śārvarin	6 Bhādrapada
4101	922	1057	406	174-75	999-1000	33 Vikārin	35 Plava	...
4102	923	1058	407	175-76	*1000-01	34 Śārvarin	36 Śubhakṛit	...
4103	924	1059	408	176-77	1001-02	35 Plava	37 Śōbhana	5 Śrāvaṇa
4104	925	1060	409	177-78	1002-03	36 Śubhakṛit	38 Krōdhin	...
4105	926	1061	410	178-79	1003-04	37 Śōbhana	39 Viśvāvasu	...
4106	927	1062	411	179-80	*1004-05	38 Krōdhin	40 Parābhava	3 Jyēshṭha
4107	928	1063	412	180-81	1005-06	39 Viśvāvasu	41 Plavaṅga	...
4108	929	1064	413	181-82	1006-07	40 Parābhava	42 Kīlaka	{ 8 Kārttika 9 *Mārgaś:* (*ksh.*) }
4109	930	1065	414	182-83	1007-08	41 Plavaṅga	43 Saumya	1 Chaitra
4110	931	1066	415	183-84	*1008-09	42 Kīlaka	44 Sādhāraṇa	...
4111	932	1067	416	184-85	1009-10	43 Saumya	45 Virōdhakṛit	5 Śrāvaṇa
4112	933	1068	417	185-86	1010-11	44 Sādhāraṇa	46 Paridhāvin	...
4113	934	1069	418	186-87	1011-12	45 Virōdhakṛit	47 Pramādin	...
4114	935	1070	419	187-88	*1012-13	46 Paridhāvin	48 Ānanda	4 Āshāḍha
4115	936	1071	420	188-89	1013-14	47 Pramādin	49 Rākshasa	...
4116	937	1072	421	189-90	1014-15	48 Ānanda	50 Anala	...
4117	938	1073	422	190-91	1015-16	49 Rākshasa	51 Piṅgala	2 Vaiśākha
4118	939	1074	423	191-92	*1016-17	50 Anala	52 Kālayukta	...
4119	940	1075	424	192-93	1017-18	51 Piṅgala	53 Siddhārthin	6 Bhādrapada
4120	941	1076	425	193-94	1018-19	52 Kālayukta	54 Raudra	...
4121	942	1077	426	194-95	1019-20	53 Siddhārthin	55 Durmati	...
4122	943	1078	427	195-96	*1020-21	54 Raudra	56 Dundubhi	5 Śrāvaṇa

LXXXII—*Contd.*

COMMENCEMENT OF THE									Kali.
SOLAR YEAR.				LUNI-SOLAR YEAR (MEAN SUNRISE OF CIVIL DAY ON WHICH CHAITRA ŚUKLA 1 ENDS).					
Day and month, A.D.	Week-day.	Time of true Mēsha-saṁkrānti.		Day and month, A.D.	Week-day.	*a*	*b*	*c*	
13	14	17		19	20	23	24	25	1
		H. M.	S.						
21 Mar. (81)	0 Sat.	15 29	24	22 Feb. (53)	0 Sat.	9877·5419	173·7640	199·1484	4098
21 Mar. (80)	1 Sun.	21 41	33	12 Mar. (71)	6 Fri.	9912·2243	109·7575	251·4589	4099
22 Mar. (81)	3 Tues.	3 53	42	2 Mar. (61)	4 Wed.	126·5792	993·2933	222·3735	4100
22 Mar. (81)	4 Wed.	10 5	51	21 Mar. (80)	3 Tues.	161·2616	929·2867	273·6618	4101
21 Mar. (81)	5 Thur.	16 18	0	9 Mar. (69)	0 Sat.	36·9845	776·5307	242·8385	4102
21 Mar. (80)	6 Fri.	22 30	9	27 Feb. (58)	5 Thur.	251·3393	660·0664	214·7531	4103
22 Mar. (81)	1 Sun.	4 42	18	17 Mar. (76)	3 Tues.	9947·3897	559·7683	263·3257	4104
22 Mar. (81)	2 Mon.	10 54	27	6 Mar. (65)	0 Sat.	9823·1125	407·0122	232·5025	4105
21 Mar. (81)	3 Tues.	17 6	36	24 Feb. (55)	5 Thur.	37·4674	290·5480	204·4171	4106
21 Mar. (80)	4 Wed.	23 18	45	13 Mar. (72)	3 Tues.	9733·5177	190·2498	253·9897	4107
22 Mar. (81)	6 Fri.	5 30	54	3 Mar. (62)	1 Sun.	9947·8726	73·7855	224·9042	4108
22 Mar. (81)	0 Sat.	11 43	3	21 Feb. (52)	6 Fri.	162·2275	957·3273	196·8189	4109
21 Mar. (81)	1 Sun.	17 55	12	11 Mar. (71)	5 Thur.	196·9097	893·3146	248·1293	4110
22 Mar. (81)	3 Tues.	0 7	21	28 Feb. (59)	2 Mon.	72·6326	740·5588	217·3061	4111
22 Mar. (81)	4 Wed.	6 19	30	19 Mar. (78)	1 Sun.	107·3140	676·5522	268·6164	4112
22 Mar. (81)	5 Thur.	12 31	39	8 Mar. (67)	5 Thur.	9983·0379	523·7962	237·7933	4113
21 Mar. (81)	6 Fri.	18 43	48	25 Feb. (56)	2 Mon.	9858·7607	371·0403	206·9701	4114
22 Mar. (81)	1 Sun.	0 55	57	15 Mar. (74)	1 Sun.	9893·4431	307·0338	258·2805	4115
22 Mar. (81)	2 Mon.	7 8	6	4 Mar. (63)	5 Thur.	9769·1660	154·2779	227·4572	4116
22 Mar. (81)	3 Tues.	13 20	15	22 Feb. (53)	3 Tues.	9983·5207	37·8125	199·3718	4117
21 Mar. (81)	4 Wed.	19 32	24	12 Mar. (72)	2 Mon.	18·2031	973·8070	250·6823	4118
22 Mar. (81)	6 Fri.	1 44	33	2 Mar (61)	0 Sat.	232·5580	857·3427	222·5968	4119
22 Mar. (81)	0 Sat.	7 56	42	21 Mar. (80)	6 Fri.	267·2404	793·3362	273·9072	4120
22 Mar. (81)	1 Sun.	14 8	51	10 Mar. (69)	3 Tues.	142·9632	640·5802	243·0840	4121
21 Mar. (81)	2 Mon.	20 21	0	27 Feb. (58)	0 Sat.	18·6860	487·8243	212·2609	4122

TABLE

CONCURRENT YEAR.								Intercalated (*adhika*) and suppressed (*kshaya*) true lunar months.
Kali.	Śaka.	Chaitrādi Vikrama.	Mēshādi solar year in Bengal.	Kollam.	A.D.	Jovian Saṁvatsara. Southern system.	Jovian Saṁvatsara. Northern system.	
1	2	3	3a	4	5	6	7	8 a
4123	944	1079	428	196-97	1021-22	55 Durmati .	57 Rudhirōdgārin	...
4124	945	1080	429	197-98	1022-23	56 Dundubhi .	58 Raktāksha .	...
4125	946	1081	430	198-99	1023-24	57 Rudhirōdgārin	59 Krōdhana .	3 Jyēshṭha .
4126	947	1082	431	199-200	*1024-25	58 Raktāksha .	60 Kshaya .	...
4127	948	1083	432	200-01	1025-26	59 Krōdhana .	1 Prabhava	7 Āśvina 10 *Pausha* (*ksh*)
4128	949	1084	433	201-02	1026-27	60 Kshaya .	2 Vibhava .	1 Chaitra .
4129	950	1085	434	202-03	1027-28	1 Prabhava .	3 Śukla . .	...
4130	951	1086	435	203-04	*1028-29	2 Vibhava .	4 Pramōda .	5 Śrāvaṇa .
4131	952	1087	436	204-05	1029-30	3 Śukla . .	5 Prajāpati .	...
4132	953	1088	437	205-06	1030-31	4 Pramōda .	6 Aṅgiras .	...
4133	954	1089	438	206-07	1031-32	5 Prajāpati .	7 Śrīmukha .	3 Jyēshṭha .
4134	955	1090	439	207-08	*1032-33	6 Aṅgiras .	8 Bhāva . .	...
4135	956	1091	440	208-09	1033-34	7 Śrīmukha .	9 Yuvan . .	...
4136	957	1092	441	209-10	1034-35	8 Bhāva . .	10 Dhātṛi . .	2 Vaiśākha .
4137	958	1093	442	210-11	1035-36	9 Yuvan . .	11 Īśvara . .	...
4138	959	1094	443	211-12	*1036-37	10 Dhātṛi . .	12 Bahudhānya .	6 Bhādrapada
4139	960	1095	444	212-13	1037-38	11 Īśvara . .	13 Pramāthin .	...
4140	961	1096	445	213-14	1038-39	12 Bahudhānya .	14 Vikrama .	...
4141	962	1097	446	214-15	1039-40	13 Pramāthin .	15 Vṛisha . .	4 Āshāḍha .
4142	963	1098	447	215-16	*1040-41	14 Vikrama .	16 Chitrabhānu .	...
4143	964	1099	448	216-17	1041-42	15 Vṛisha . .	17 Subhānu .	...
4144	965	1100	449	217-18	1042-43	16 Chitrabhānu .	18 Tāraṇa . .	3 Jyēshṭha .
4145	966	1101	450	218-19	1043-44	17 Subhānu .	19 Pārthiva .	...
4146	967	1102	451	219-20	*1044-45	18 Tāraṇa . .	20 Vyaya . .	7 Āśvina .
4147	968	1103	452	220-21	1045-46	19 Pārthiva .	21 Sarvajit .	...

LXXXII—*Contd.*

COMMENCEMENT OF THE											Kali.
SOLAR YEAR.					LUNI-SOLAR YEAR (MEAN SUNRISE OF CIVIL DAY ON WHICH CHAITRA ŚUKLA 1 ENDS).						
Day and month, A.D.	Week-day.	Time of true Mēsha-saṁkrānti.			Day and month, A.D.	Week-day.	*a*	*b*	*c*		
13	14	17			19	20	23	24	25		1
		H.	M.	S.							
22 Mar. (81)	4 Wed. .	2	33	9	17 Mar. (76)	6 Fri. .	53·3685	423·8178	263·3090		4123
22 Mar. (81)	5 Thur. .	8	45	18	6 Mar. (65)	3 Tues. .	9929·0902	271·0618	232·7480		4124
22 Mar. (81)	6 Fri. .	14	57	27	23 Feb. (54)	0 Sat. .	9804·8141	118·3068	201·9238		4125
21 Mar. (81)	0 Sat. .	21	9	36	13 Mar. (73)	6 Fri. .	9839·4965	54·2993	253·2353		4126
22 Mar. (81)	2 Mon. .	3	21	45	3 Mar. (62)	4 Wed. .	53·8514	937·8350	225·0498		4127
22 Mar. (81)	3 Tues. .	9	33	54	21 Feb. (52)	2 Mon. .	268·2062	821·3708	197·0643		4128
22 Mar. (81)	4 Wed. .	15	46	3	12 Mar. (71)	1 Sun. .	302·8885	757·3642	248·3748		4129
21 Mar. (81)	5 Thur. .	21	58	12	29 Feb. (60)	5 Thur. .	178·6114	604·6082	217·5517		4130
22 Mar. (81)	0 Sat. .	4	10	21	19 Mar. (78)	4 Wed. .	213·2937	540·6018	268·8620		4131
22 Mar. (81)	1 Sun. .	10	22	30	8 Mar. (67)	1 Sun. .	89·0166	387·8457	238·0388		4132
22 Mar. (81)	2 Mon. .	16	34	39	25 Feb. (56)	5 Thur. .	9964·7395	235·0898	207·2156		4133
21 Mar. (81)	3 Tues. .	22	46	48	15 Mar. (75)	4 Wed. .	9999·4219	171·0833	258·5271		4134
22 Mar. (81)	5 Thur. .	4	58	57	4 Mar. (63)	1 Sun. .	9875·1447	17·3274	227·7028		4135
22 Mar. (81)	6 Fri. .	11	11	6	22 Feb. (53)	6 Fri. .	89·4995	901·8631	199·6173		4136
22 Mar. (81)	0 Sat. .	17	23	5	13 Mar. (72)	5 Thur. .	124·1819	837·8565	250·4278		4137
21 Mar. (81)	1 Sun. .	23	35	24	1 Mar. (61)	2 Mon. .	9999·9048	685·1006	219·6046		4138
22 Mar. (81)	3 Tues. .	5	47	33	20 Mar. (79)	1 Sun. .	34·5871	621·0940	271·4150		4139
22 Mar. (81)	4 Wed. .	11	59	42	9 Mar. (68)	5 Thur. .	9910·3100	468·3381	239·5919		4140
22 Mar. (81)	5 Thur. .	18	11	50	26 Feb. (57)	2 Mon. .	9786·0329	315·5822	209·7686		4141
22 Mar. (82)	0 Sat. .	0	23	59	16 Mar. (76)	1 Sun. .	9820·7152	251·5756	261·0791		4142
22 Mar. (81)	1 Sun. .	6	36	8	6 Mar. (65)	6 Fri. .	35·0700	145·1113	232·9936		4143
22 Mar. (81)	2 Mon. .	12	48	17	23 Feb. (54)	3 Tues. .	9910·7929	982·3553	202·1704		4144
22 Mar. (81)	3 Tues. .	19	0	26	14 Mar. (73)	2 Mon. .	9945·4753	913·3478	253·4808		4145
22 Mar. (82)	5 Thur. .	1	12	35	3 Mar. (63)	0 Sat. .	159·8301	801·8845	225·3953		4146
22 Mar. (81)	6 Fri. .	7	24	44	22 Mar. (81)	6 Fri. .	194·5125	737·8780	276·7058		4147

TABLE

CONCURRENT YEAR								Intercalated (*adhika*) and suppressed (*kshaya*) true lunar months.
						Jovian Saṁvatsara.		
Kali.	Śaka.	Chaitrādi Vikrama.	Mēshādi solar year in Bengal.	Kollam.	A.D.	Southern system.	Northern system.	
1	2	3	3*a*	4	5	6	7	8*a*
4148	969	1104	453	221-22	1046-47	20 Vyaya . .	22 Sarvadhārin .	...
4149	970	1105	454	222-23	1047-48	21 Sarvajit .	23 Virōdhin .	5 Śrāvaṇa .
4150	971	1106	455	223-24	*1048-49	22 Sarvadhārin .	24 Vikṛita . .	...
4151	972	1107	456	224-25	1049-50	23 Virōdhin .	25 Khara . .	...
4152	973	1108	457	225-26	1050-51	24 Vikṛita . .	26 Nandana .	3 Jyēshṭha .
4153	974	1109	458	226-27	1051-52	25 Khara . .	27 Vijaya . .	...
4154	975	1110	459	227-28	*1052-53	26 Nandana .	28 Jaya . .	...
4155	976	1111	460	228-29	1053-54	27 Vijaya . .	29 Manmatha .	2 Vaiśākha .
4156	977	1112	461	229-30	1054-55	28 Jaya . .	30 Durmukha .	...
4157	978	1113	462	230-31	1055-56	29 Manmatha .	31 Hēmalamba .	6 Bhādrapada
4158	979	1114	463	231-32	*1056-57	30 Durmukha .	32 Vilamba .	...
4159	980	1115	464	232-33	1057-58	31 Hēmalamba .	33 Vikārin .	...
4160	981	1116	465	233-34	1058-59	32 Vilamba .	34 Śārvarin .	4 Āshāḍha .
4161	982	1117	466	234-35	1059-60	33 Vikārin .	35 Plava . .	...
4162	983	1118	467	235-36	*1060-61	34 Śārvarin .	36 Śubhakṛit .	...
4163	984	1119	468	236-37	1061-62	35 Plava . .	37 Śōbhana .	3 Jyēshṭha .
4164	985	1120	469	237-38	1062-63	36 Śubhakṛit .	38 Krōdhin .	...
4165	986	1121	470	238-39	1063-64	37 Śōbhana .	39 Viśvāvasu .	7 Āśvina
4166	987	1122	471	239-40	*1064-65	38 Krōdhin .	40 Parābhava .	...
4167	988	1123	472	240-41	1065-66	39 Viśvāvasu .	41 Plavaṅga .	...
4168	989	1124	473	241-42	1066-67	40 Parābhava .	42 Kīlaka . .	5 Śrāvaṇa
4169	990	1125	474	242-43	1067-68	41 Plavaṅga .	43 Saumya .	...
4170	991	1126	475	243-44	*1068-69	42 Kīlaka . .	44 Sādhāraṇa	...
4171	992	1127	476	244-45	1069-70	43 Saumya .	45 Virōdhakrit .	3 Jyēshṭha
4172	993	1128	477	245-46	1070-71	44 Sādhāraṇa .	46 Paridhāvin .	...

LXXXII—*Contd.*

COMMENCEMENT OF THE											
SOLAR YEAR.					LUNI-SOLAR YEAR (MEAN SUNRISE OF CIVIL DAY ON WHICH CHAITRA ŚUKLA 1 ENDS).					Kali.
Day and month, A.D.	Week-day.	Time of true Mēsha-saṁkrānti.			Day and month, A.D.	Week-day.	*a*	*b*	*c*	
13	14	17			19	20	23	24	25	1
		H.	M.	S.						
22 Mar. (81)	0 Sat. .	13	36	53	11 Mar. (70)	3 Tues. .	70·2354	585·1221	245·8826	4148
22 Mar. (81)	1 Sun. .	19	49	2	28 Feb. (59)	0 Sat. .	9945·9581	432·3661	215·0594	4149
22 Mar. (82)	3 Tues. .	2	1	11	18 Mar. (78)	6 Fri. .	9980·6406	368·3596	266·3697	4150
22 Mar. (81)	4 Wed. .	8	13	20	7 Mar. (66)	3 Tues. .	9856·3634	215·6036	235·5466	4151
22 Mar. (81)	5 Thur. .	14	25	29	25 Feb. (56)	1 Sun. .	70·7183	99·1393	207·7536	4152
22 Mar. (81)	6 Fri. .	20	37	38	16 Mar. (75)	0 Sat. .	105·4006	35·1328	258·7716	4153
22 Mar. (82)	1 Sun. .	2	49	47	4 Mar. (64)	4 Wed. .	9981·1235	882·3769	227·9483	4154
22 Mar. (81)	2 Mon. .	9	1	56	22 Feb. (53)	2 Mon. .	195·4783	767·9126	199·8629	4155
22 Mar. (81)	3 Tues. .	15	14	5	13 Mar. (72)	1 Sun. .	230·1606	701·9061	251·1734	4156
22 Mar. (81)	4 Wed. .	21	26	14	2 Mar. (61)	5 Thur. .	105·8835	549·1501	220·3501	4157
22 Mar. (82)	6 Fri. .	3	38	23	20 Mar. (80)	4 Wed. .	140·5659	485·1435	271·6605	4158
22 Mar. (81)	0 Sat. .	9	50	32	9 Mar. (68)	1 Sun. .	16·2888	333·3876	240·8375	4159
22 Mar. (81)	1 Sun. .	16	2	41	26 Feb. (57)	5 Thur. .	9892·0116	179·6317	210·0142	4160
22 Mar. (81)	2 Mon. .	22	14	50	17 Mar. (76)	4 Wed. .	9926·6940	115·6452	261·3246	4161
22 Mar. (82)	4 Wed. .	4	26	59	6 Mar. (66)	2 Mon. .	141·0488	999·1608	233·2391	4162
22 Mar. (81)	5 Thur. .	10	39	8	23 Feb. (54)	6 Fri. .	16·7716	856·4049	202·4159	4163
22 Mar. (81)	6 Fri. .	16	51	17	14 Mar. (73)	5 Thur. .	51·4540	782·3983	253·7264	4164
22 Mar. (81)	0 Sat. .	23	3	26	4 Mar. (63)	3 Tues. .	265·8089	665·9341	225·6409	4165
22 Mar. (82)	2 Mon. .	5	15	35	21 Mar. (81)	1 Sun. .	9961·8593	565·6363	274·2135	4166
22 Mar. (81)	3 Tues. .	11	27	44	10 Mar. (69)	5 Thur. .	9837·5821	412·8799	243·3903	4167
22 Mar. (81)	4 Wed. .	17	39	53	28 Feb. (59)	3 Tues. .	51·9369	296·4157	215·3050	4166
22 Mar. (81)	5 Thur. .	23	52	2	18 Mar. (77)	1 Sun. .	9747·9874	196·1174	263·8775	4169
22 Mar. (82)	0 Sat. .	6	4	11	7 Mar. (67)	6 Fri. .	9962·3421	79·6532	235·7921	4170
22 Mar. (81)	1 Sun. .	12	16	20	25 Feb. (56)	4 Wed. .	176·6970	963·1888	207·7067	4171
22 Mar. (81)	2 Mon. .	18	28	29	16 Mar. (75)	3 Tues. .	211·3794	899·1823	259·0172	4172

TABLE I.

CONCURRENT YEAR.								Intercalated (*adhika*) and suppressed (*kshaya*) true lunar months.
						JOVIAN SAMVATSARA.		
Kali.	Śaka.	Chaitrādi Vikrama	Mēshādi solar year in Bengal	Kollam.	A.D.	Southern system.	Northern system.	
1	2	3	3a	4	5	6	7	8a
4173	994	1129	478	246-47	1071-72	45 Virōdhakrit .	47 Pramādin .	8 Kārttika . / 9 *Mārgas.* (*ksh*)
4174	995	1130	479	247-48	*1072-73	46 Paridhāvin .	48 Ānanda .	2 Vaiśākha .
4175	996	1131	480	248-49	1073-74	47 Pramādin .	49 Rākshasa .	—
4176	997	1132	481	249-50	1074-75	48 Ānanda .	50 Anala* . .	6 Bhādrapada
4177	998	1133	482	250-51	1075-76	49 Rākshasa	52 *Kālayukta* .	—
4178	999	1134	483	251-52	*1076-77	50 Anala . .	53 *Siddhārthin* .	—
4179	1000	1135	484	252-53	1077-78	51 Pingala .	54 *Raudra* . .	4 Āshāḍha
4180	1001	1136	485	253-54	1078-79	52 Kālayukta .	55 *Durmati* .	—
4181	1002	1137	486	254-55	1079-80	53 Siddhārthin .	56 *Dundubhi* .	—
4182	1003	1138	487	255-56	*1080-81	54 Raudra .	57 Rudhirōdgārin	3 Jyēshṭha .
4183	1004	1139	488	256-57	1081-82	55 Durmati .	58 Raktāksha .	—
4184	1005	1140	489	257-58	1082-83	56 Dundubhi .	59 Krōdhana .	7 Āśvina .
4185	1006	1141	490	258-59	1083-84	57 Rudhirōdgārin	60 Kshaya .	—
4186	1007	1142	491	259-60	*1084-85	58 Raktāksha .	1 Prabhava .	—
4187	1008	1143	492	260-61	1085-86	59 Krōdhana .	2 Vibhava .	5 Śrāvaṇa .
4188	1009	1144	493	261-62	1086-87	60 Kshaya .	3 Śukla . .	—
4189	1010	1145	494	262-63	1087-88	1 Prabhava .	4 Pramōda .	—
4190	1011	1146	495	263-64	*1088-89	2 Vibhava .	5 Prajāpati .	3 Jyēshṭha .
4191	1012	1147	496	264-65	1089-90	3 Śukla . .	6 Angiras .	—
4192	1013	1148	497	265-66	1090-91	4 Pramōda .	7 Śrīmukha	8 Kārttika / 10 *Pausha* (*ksh*)
4193	1014	1149	498	266-67	1091-92	5 Prajāpati .	8 Bhāva . .	1 Chaitra .
4194	1015	1150	499	267-68	*1092-93	6 Angiras .	9 Yuvan . .	—
4195	1016	1151	500	268-69	1093-94	7 Śrīmukha .	10 Dhātri . .	6 Bhādrapada
4196	1017	1152	501	269-70	1094-95	8 Bhāva . .	11 Īśvara . .	—
4197	1018	1153	502	270-71	1095-96	9 Yuvan . .	12 Bahudhānya .	—

* 51 Pingala was suppressed in the north.

LXXXII.—*Contd.*

COMMENCEMENT OF THE											
SOLAR YEAR.					LUNI-SOLAR YEAR (MEAN SUNRISE OF CIVIL DAY ON WHICH CHAITRA ŚUKLA 1 ENDS).						Kali.
Day and month, A.D.	Week-day.	Time of true Mēsha-saṁkrānti.			Day and month, A.D.	Week-day.	*a*	*b*	*c*		
13	14	17			19	20	23	24	25		1
		H.	M.	S.							
23 Mar. (82)	4 Wed.	0	40	38	5 Mar. (64)	0 Sat.	87·1023	746·4264	228·1939		4173
22 Mar. (82)	5 Thur.	6	52	47	22 Feb. (53)	4 Wed.	9962·8251	593·6705	197·3706		4174
22 Mar. (81)	6 Fri.	13	4	56	12 Mar. (71)	3 Tues.	9997·5074	530·6639	248·6811		4175
22 Mar. (81)	0 Sat.	19	17	5	1 Mar. (60)	0 Sat.	9873·2303	376·9079	217·8580		4176
23 Mar. (82)	2 Mon.	1	29	14	20 Mar. (79)	6 Fri.	9907·9126	312·9015	269·1683		4177
22 Mar. (82)	3 Tues.	7	41	23	8 Mar. (68)	3 Tues.	9783·6355	160·1454	238·3451		4178
22 Mar. (81)	4 Wed.	13	53	32	26 Feb. (57)	1 Sun.	9997·9904	43·6812	210·2597		4179
22 Mar. (81)	5 Thur.	20	5	41	17 Mar. (76)	0 Sat.	32·6728	979·6747	261·5702		4180
23 Mar. (82)	0 Sat.	2	17	50	7 Mar. (66)	5 Thur.	247·0275	863·2103	235·4847		4181
22 Mar. (82)	1 Sun.	8	29	59	24 Feb. (55)	2 Mon.	122·7504	710·4544	202·6614		4182
22 Mar. (81)	2 Mon.	14	42	8	14 Mar. (73)	1 Sun.	157·4328	646·4478	253·9719		4183
22 Mar. (81)	3 Tues.	20	54	17	3 Mar. (62)	5 Thur.	33·1557	493·6919	223·1487		4184
23 Mar. (82)	5 Thur.	3	6	26	22 Mar. (81)	4 Wed.	67·8380	429·6854	274·4591		4185
22 Mar. (82)	6 Fri.	9	18	35	10 Mar. (70)	1 Sun.	9943·5609	276·9294	245·6358		4186
22 Mar. (81)	0 Sat.	15	30	43	27 Feb. (58)	5 Thur.	9819·2837	124·1735	212·8127		4187
22 Mar. (81)	1 Sun.	21	42	52	18 Mar. (77)	4 Wed.	9853·9661	60·1669	264·1231		4188
23 Mar. (82)	3 Tues.	3	55	1	8 Mar. (67)	2 Mon.	68·3209	943·8027	236·0377		4189
22 Mar. (82)	4 Wed.	10	7	10	26 Feb. (57)	0 Sat.	282·6758	827·2383	207·9522		4190
22 Mar. (81)	5 Thur.	16	19	19	16 Mar. (75)	6 Fri.	317·3582	763·2318	259·2627		4191
22 Mar. (81)	6 Fri.	22	31	28	5 Mar. (64)	3 Tues.	193·0810	610·4759	228·4395		4192
23 Mar. (82)	1 Sun.	4	43	37	22 Feb. (53)	0 Sat.	68·8039	457·7200	197·6162		4193
22 Mar. (82)	2 Mon.	10	55	46	12 Mar. (72)	6 Fri.	103·4862	393·7134	248·9266		4194
22 Mar. (81)	3 Tues.	17	7	55	1 Mar. (60)	3 Tues.	9979·2090	240·9577	218·1035		4195
22 Mar. (81)	4 Wed.	23	20	4	20 Mar. (79)	2 Mon.	13·8914	176·9509	269·4139		4196
23 Mar. (82)	6 Fri.	5	32	13	9 Mar. (68)	6 Fri.	9889·6143	24·1949	238·5907		4197

TABLE

CONCURRENT YEAR.								Intercalated (*adhika*) and suppressed (*kshaya*) true lunar months.
						JOVIAN SAMVATSARA.		
Kali.	Śaka.	Chaitrādi Vikrama.	Mēshādi solar year in Bengal.	Kollam.	A.D.	Southern system.	Northern system.	
1	2	3	3*a*	4	5	6	7	8*a*
4198	1019	1154	503	271-72	*1096-97	10 Dhātri	13 Pramāthin	4 Āshāḍha
4199	1020	1155	504	272-73	1097-98	11 Īśvara	14 Vikrama	...
4200	1021	1156	505	273-74	1098-99	12 Bahudhānya	15 Vṛisha	...
4201	1022	1157	506	274-75	1099-1100	13 Pramāthin	16 Chitrabhānu	3 Jyēshṭha
4202	1023	1158	507	275-76	*1100-01	14 Vikrama	17 Subhānu	...
4203	1024	1159	508	276-77	1101-02	15 Vṛisha	18 Tāraṇa	7 Āśvina
4204	1025	1160	509	277-78	1102-03	16 Chitrabhānu	19 Pārthiva	...
4205	1026	1161	510	278-79	1103-04	17 Subhānu	20 Vyaya	...
4206	1027	1162	511	279-80	*1104-05	18 Tāraṇa	21 Sarvajit	4 Āshāḍha
4207	1028	1163	512	280-81	1105-06	19 Pārthiva	22 Sarvadhārin	...
4208	1029	1164	513	281-82	1106-07	20 Vyaya	23 Virōdhin	...
4209	1030	1165	514	282-83	1107-08	21 Sarvajit	24 Vikṛita	3 Jyēshṭha
4210	1031	1166	515	283-84	*1108-09	22 Sarvadhārin	25 Khara	...
4211	1032	1167	516	284-85	1109-10	23 Virōdhin	26 Nandana	8 Kārttika 10 *Pausha* (*ksh*) 12 Phālguna
4212	1033	1168	517	285-86	1110-11	24 Vikṛita	27 Vijaya	...
4213	1034	1169	518	286-87	1111-12	25 Khara	28 Jaya	...
4214	1035	1170	519	287-88	*1112-13	26 Nandana	29 Manmatha	5 Śrāvaṇa
4215	1036	1171	520	288-89	1113-14	27 Vijaya	30 Durmukha	...
4216	1037	1172	521	289-90	1114-15	28 Jaya	31 Hemalamba	...
4217	1038	1173	522	290-91	1115-16	29 Manmatha	32 Vilamba	4 Āshāḍha
4218	1039	1174	523	291-92	*1116-17	30 Durmukha	33 Vikārin	
4219	1040	1175	524	292-93	1117-18	31 Hēmalamba	34 Śārvarin	...
4220	1041	1176	525	293-94	1118-19	32 Vilamba	35 Plava	2 Vaiśākha
4221	1042	1177	526	294-95	1119-20	33 Vikārin	36 Śubhakṛit	
4222	1043	1178	527	295-96	*1120-21	34 Śārvarin	37 Śobhana	6 Bhādrapada

LXXXII—*Contd.*

COMMENCEMENT OF THE											Kali.
SOLAR YEAR.					LUNI-SOLAR YEAR (MEAN SUNRISE OF CIVIL DAY ON WHICH CHAITRA ŚUKLA 1 ENDS).						
Day and month A. D.	Week-day.	Time of true Mēsha-saṁkrānti.			Day and month A. D.	Week-day.	*a*	*b*	*c*		
13	14	17			19	20	23	24	25		1
		H.	M.	S.							
22 Mar. (82)	0 Sat. .	11	44	22	27 Feb. (58)	4 Wed.	103·9691	907·7307	210·5052		4198
22 Mar. (81)	1 Sun. .	17	56	31	17 Mar. (76)	3 Tues.	138·6515	843·7242	261·8157		4199
23 Mar. (82)	3 Tues.	0	8	40	6 Mar. (65)	0 Sat. .	14·3744	690·9683	230·9925		4200
23 Mar. (82)	4 Wed.	6	20	49	24 Feb. (55)	5 Thur.	228·7291	574·5038	202·8848		4201
22 Mar. (82)	5 Thur.	12	32	58	13 Mar. (73)	3 Tues.	9924·7795	474·2057	251·4575		4202
22 Mar. (81)	6 Fri. .	18	45	7	2 Mar. (61)	0 Sat. .	9800·5024	321·4497	20·6342		4203
23 Mar. (82)	1 Sun. .	0	57	16	21 Mar. (80)	6 Fri. .	9835·1847	257·4432	271·9446		4204
23 Mar. (82)	2 Mon. .	7	9	25	11 Mar. (70)	4 Wed. .	49·5396	140·9788	243·8592		4205
22 Mar. (82)	3 Tues.	13	21	34	28 Feb. (59)	1 Sun. .	9925·2624	988·2229	213·0361		4206
22 Mar. (81)	4 Wed.	19	33	43	18 Mar. (77)	0 Sat. .	9959·9448	924·2154	264·3464		4207
23 Mar. (82)	6 Fri. .	1	45	52	8 Mar. (67)	5 Thur.	174·2996	807·7521	236·2610		4208
23 Mar. (82)	0 Sat. .	7	58	1	25 Feb. (56)	2 Mon. .	50·0225	654·9962	205·4387		4209
22 Mar. (82)	1 Sun. .	14	10	10	15 Mar. (75)	1 Sun. .	84·7048	590·9896	256·7483		4210
22 Mar. (81)	2 Mon. .	20	22	19	4 Mar. (63)	5 Thur.	9960·4277	438·2337	225·9250		4211
23 Mar. (82)	4 Wed.	2	34	28	23 Mar. (82)	4 Wed.	9995·1101	374·2271	277·2354		4212
23 Mar. (82)	5 Thur.	8	46	37	12 Mar. (71)	1 Sun. .	9870·8330	221·4712	246·4122		4213
22 Mar. (82)	6 Fri. .	14	58	46	1 Mar. (61)	6 Fri. .	85·1877	105·0069	218·3269		4214
22 Mar. (81)	0 Sat. .	21	10	55	20 Mar. (79)	5 Thur.	119·8701	41·0004	269·6373		4215
23 Mar. (82)	2 Mon. .	3	23	4	9 Mar. (68)	2 Mon. .	9995·5930	888·3444	238·8140		4216
23 Mar. (82)	3 Tues.	9	35	13	27 Feb. (58)	0 Sat. .	209·9478	771·7891	210·7286		4217
22 Mar. (82)	4 Wed.	15	47	22	17 Mar. (77)	6 Fri. .	244·6302	707·7736	262·0391		4218
22 Mar. (81)	5 Thur.	21	59	31	6 Mar. (65)	3 Tues.	120·3530	555·0176	231·2158		4219
23 Mar. (82)	0 Sat. .	4	11	40	23 Feb. (54)	0 Sat. .	9996·0759	402·2617	200·3925		4220
23 Mar. (82)	1 Sun. .	10	23	49	14 Mar. (73)	6 Fri. .	30·7582	338·2552	251·7030		4221
22 Mar. (82)	2 Mon. .	16	35	58	2 Mar. (62)	3 Tues.	9906·4811	185·4993	220·8798		4222

TABLE

CONCURRENT YEAR.								Intercalated (*adhika*) and suppressed (*kshaya*) true lunar months.
Kali.	Śaka.	Chaitrādi Vikrama.	Mēshādi solar year in Bengal.	Kollam.	A. D.	Jovian Saṁvatsara. Southern system.	Jovian Saṁvatsara. Northern system.	
1	2	3	3a	4	5	6	7	8a
4223	1044	1179	528	296-97	1121-22	35 Plava	38 Krōdhin	...
4224	1045	1180	529	297-98	1122-23	36 Śubhakṛit	39 Viśvāvasu	...
4225	1046	1181	530	298-99	1123-24	37 Śōbhana	40 Parābhava	4 Āshāḍha
4226	1047	1182	531	299-300	*1124-25	38 Krōdhin	41 Plavaṅga	...
4227	1048	1183	532	300-01	1125-26	39 Viśvāvasu	42 Kīlaka	...
4228	1049	1184	533	301-02	1126-27	40 Parābhava	43 Saumya	3 Jyēshṭha
4229	1050	1185	534	302-03	1127-28	41 Plavaṅga	44 Sādhāraṇa	...
4230	1051	1186	535	303-04	*1128-29	42 Kīlaka	45 Virōdhakṛit	12 Phālguna†
4231	1052	1187	536	304-05	1129-30	43 Saumya	46 Paridhāvin	...
4232	1053	1188	537	305-06	1130-31	44 Sādhāraṇa	47 Pramādin	...
4233	1054	1189	538	306-07	1131-32	45 Virōdhakṛit	48 Ānanda	5 Śrāvaṇa
4234	1055	1190	539	307-08	*1132-33	46 Paridhāvin	49 Rākshasa	...
4235	1056	1191	540	308-09	1133-34	47 Pramādin	50 Anala	...
4236	1057	1192	541	309-10	1134-35	48 Ānanda	51 Piṅgala	4 Āshāḍha
4237	1058	1193	542	310-11	1135-36	49 Rākshasa	52 Kālayukta	...
4238	1059	1194	543	311-12	*1136-37	50 Anala	53 Siddhārthin	...
4239	1060	1195	544	312-13	1137-38	51 Piṅgala	54 Raudra	2 Vaiśākha
4240	1061	1196	545	313-14	1138-39	52 Kālayukta	55 Durmati	...
4241	1062	1197	546	314-15	1139-40	53 Siddhārthin	56 Dundubhi	6 Bhādrapada
4242	1063	1198	547	315-16	*1140-41	54 Raudra	57 Rudhirōdgārin	...
4243	1064	1199	548	316-17	1141-42	55 Durmati	58 Raktāksha	...
4244	1065	1200	549	317-18	1142-43	56 Dundubhi	59 Krōdhana	4 Āshāḍha
4245	1066	1201	550	318-19	1143-44	57 Rudhirōdgārin	60 Kshaya	...
4246	1067	1202	551	319-20	*1144-45	58 Raktāksha	1 Prabhava	...
4247	1068	1203	552	320-21	1145-46	59 Krōdhana	2 Vibhava	3 Jyēshṭha

† See "Remarks," p. 455 above.

LXXXII—*Contd.*

COMMENCEMENT OF THE										
SOLAR YEAR.					LUNI-SOLAR YEAR (MEAN SUNRISE OF CIVIL DAY ON WHICH CHAITRA ŚUKLA 1 ENDS).					Kali.
Day and month A. D.	Week-day.	Time of true Mêsha-saṁkrânti.			Day and month A. D.	Week-day.	*a*	*b*	*c*	
13	14	17			19	20	23	24	25	1
		H.	M.	S.						
22 Mar. (81)	3 Tues.	22	48	7	21 Mar. (80)	2 Mon.	9941·1635	121·4928	272·1902	4223
23 Mar. (82)	5 Thur.	5	0	16	11 Mar. (70)	0 Sat.	155·5183	5·0284	244·1047	4224
23 Mar. (82)	6 Fri.	11	12	25	28 Feb. (59)	4 Wed.	31·2411	852·2724	213·2826	4225
22 Mar. (82)	0 Sat.	17	24	34	18 Mar. (78)	3 Tues.	65·9236	788·2659	264·5920	4226
22 Mar. (81)	1 Sun.	23	36	43	8 Mar. (67)	1 Sun.	280·2784	671·8016	236·5066	4227
23 Mar. (82)	3 Tues.	5	48	52	25 Feb. (56)	5 Thur.	156·0012	519·0457	205·6833	4228
23 Mar. (82)	4 Wed.	12	1	1	15 Mar. (74)	3 Tues.	9852·0516	418·7475	254·2560	4229
22 Mar. (82)	5 Thur.	18	13	10	3 Mar. (63)	0 Sat.	9727·7745	265·9915	223·4328	4230
23 Mar. (82)	0 Sat.	0	25	19	22 Mar. (81)	6 Fri.	9762·4568	201·9851	274·7432	4231
23 Mar. (82)	1 Sun.	6	37	27	12 Mar. (71)	4 Wed.	9976·8117	85·5207	246·6577	4232
23 Mar. (82)	2 Mon.	12	49	36	2 Mar. (61)	2 Mon.	191·1665	969·0564	218·5724	4233
22 Mar. (82)	3 Tues.	19	1	45	20 Mar. (80)	1 Sun.	225·8489	905·0499	269·8828	4234
23 Mar. (82)	5 Thur.	1	13	54	9 Mar. (68)	5 Thur.	101·5717	752·2939	239·0596	4235
23 Mar. (82)	6 Fri.	7	26	3	26 Feb. (57)	2 Mon.	9977·2946	599·5380	208·2363	4236
23 Mar. (82)	0 Sat.	13	38	12	17 Mar. (76)	1 Sun.	11·9770	535·5314	259·5468	4237
22 Mar. (82)	1 Sun.	19	50	21	5 Mar. (65)	5 Thur.	9887·6999	382·7755	228·7236	4238
23 Mar. (82)	3 Tues.	2	2	30	22 Feb. (53)	2 Mon.	9763·4226	230·1095	197·9004	4239
23 Mar. (82)	4 Wed.	8	14	39	13 Mar. (72)	1 Sun.	9798·1050	166·0130	249·2108	4240
23 Mar. (82)	5 Thur.	14	26	48	3 Mar. (62)	6 Fri.	12·4599	49·5488	221·1253	4241
22 Mar. (82)	6 Fri.	20	38	57	21 Mar. (81)	5 Thur.	47·1422	985·5422	272·4358	4242
23 Mar. (82)	1 Sun.	2	51	6	11 Mar. (70)	3 Tues.	261·4971	869·0779	244·3503	4243
23 Mar. (82)	2 Mon.	9	3	15	28 Feb. (59)	0 Sat.	137·2199	716·3219	214·5272	4244
23 Mar. (82)	3 Tues.	15	15	24	19 Mar. (78)	6 Fri.	171·9024	652·3154	264·8375	4245
22 Mar. (82)	4 Wed.	21	27	33	7 Mar. (67)	3 Tues.	47·6251	499·5595	234·0143	4246
23 Mar. (82)	6 Fri.	3	39	42	24 Feb. (55)	0 Sat.	9923·3480	346·8035	203·1911	4247

TABLE

CONCURRENT YEAR.								Intercalated (adhika) and suppressed (kshaya) true lunar months.
						Jovian Samvatsara.		
Kali.	Saka.	Chaitrâdi Vikrama.	Mêshâdi solar year in Bengal.	Kollam.	A. D.	Southern system.	Northern system.	
1	2	3	3a	4	5	6	7	8a
4248	1069	1204	553	321-22	1146-47	60 Kshaya . .	3 Śukla . .	...
4249	1070	1205	554	322-23	1147-48	1 Prabhava .	4 Pramôda .	8 Kârttika 9 *Mârgas.* (ksh) 12 Phâlguna
4250	1071	1206	555	323-24	*1148-49	2 Vibhava .	5 Prajâpati .	...
4251	1072	1207	556	324-25	1149-50	3 Śukla . .	6 Angiras .	...
4252	1073	1208	557	325-26	1150-51	4 Pramôda .	7 Śrîmukha .	5 Śrâvaṇa .
4253	1074	1209	558	326-27	1151-52	5 Prajâpati .	8 Bhâva . .	...
4254	1075	1210	559	327-28	*1152-53	6 Angiras .	9 Yuvan . .	...
4255	1076	1211	560	328-29	1153-54	7 Śrîmukha .	10 Dhâtṛi . .	4 Âshâḍha .
4256	1077	1212	561	329-30	1154-55	8 Bhâva . .	11 Îśvara . .	...
4257	1078	1213	562	330-31	1155-56	9 Yuvan . .	12 Bahudhânya .	...
4258	1079	1214	563	331-32	*1156-57	10 Dhâtṛi . .	13 Pramâthin .	2 Vaiśâkha .
4259	1080	1215	564	332-33	1157-58	11 Îśvara . .	14 Vikrama .	...
4260	1081	1216	565	333-34	1158-59	12 Bahudhânya .	15 Vṛisha . .	6 Bhâdrapada
4261	1082	1217	566	334-35	1159-60	13 Pramâthin .	16 Chitrabhânu†	...
4262	1083	1218	567	335-36	*1160-61	14 Vikrama .	18 *Târaṇa* . .	...
4263	1084	1219	568	336-37	1161-62	15 Vṛisha . .	19 *Pârthiva* .	4 Âshâḍha .
4264	1085	1220	569	337-38	1162-63	16 Chitrabhânu .	20 *Vyaya* . .	...
4265	1086	1221	570	338-39	1163-64	17 Subhânu .	21 *Sarvajit* .	...
4266	1087	1222	571	339-40	*1164-65	18 Târaṇa . .	22 *Sarvadhârin* .	3 Jyêshṭha .
4267	1088	1223	572	340-41	1165-66	19 Pârthiva .	23 Virôdhin .	...
4268	1089	1224	573	341-42	1166-67	20 Vyaya . .	24 Vikṛita . .	7 Âśvina 10 *Pausha* (ksh) 12 Phâlguna
4269	1090	1225	574	342-43	1167-68	21 Sarvajit .	25 Khara . .	...
4270	1091	1226	575	343-44	*1168-69	22 Sarvadhârin .	26 Nandana .	...
4271	1092	1227	576	344-45	1169-70	23 Virôdhin .	27 Vijaya . .	5 Śrâvaṇa .
4272	1093	1228	577	345-46	1170-71	24 Vikṛita . .	28 Jaya . .	...

† 17 Subhânu was suppressed in the north.

LXXXII—*Contd.*

COMMENCEMENT OF THE										
SOLAR YEAR.					LUNI-SOLAR YEAR (MEAN SUNRISE OF CIVIL DAY ON WHICH CHAITRA ŚUKLA 1 ENDS).					Kali.
Day and month A. D.	Week-day.	Time of true Mēsha-saṁkrānti.			Day and month A. D.	Week-day.	*a*	*b*	*c*	
13	14	17			19	20	23	24	25	1
		H.	M.	S.						
23 Mar. (82)	0 Sat. .	9	51	51	15 Mar. (74)	6 Fri. .	9958·0304	282·7970	254·5016	4248
23 Mar. (82)	1 Sun. .	16	4	0	4 Mar. (63)	3 Tues.	9833·7532	129·0410	223·6783	4249
22 Mar. (82)	2 Mon. .	22	16	9	22 Mar. (82)	2 Mon. .	9868·4356	66·0346	274·9887	4250
23 Mar. (82)	4 Wed.	4	28	18	12 Mar. (71)	0 Sat. .	82·7905	949·5702	246·9033	4251
23 Mar. (82)	5 Thur.	10	40	27	2 Mar. (61)	5 Thur.	297·1453	833·1059	218·6180	4252
23 Mar. (82)	6 Fri. .	16	52	36	21 Mar. (80)	4 Wed.	331·8276	769·0994	270·1283	4253
22 Mar. (82)	0 Sat. .	23	4	45	9 Mar. (69)	1 Sun. .	207·5505	616·3435	239·3051	4254
23 Mar. (82)	2 Mon. .	5	16	54	26 Feb. (57)	5 Thur.	83·2734	463·5875	208·4819	4255
23 Mar. (82)	3 Tues.	11	29	3	16 Mar. (75)	3 Tues.	9779·3237	363·2894	257·0546	4256
23 Mar. (82)	4 Wed.	17	41	12	6 Mar. (65)	1 Sun. .	9993·6786	246·8250	228·9691	4257
22 Mar. (82)	5 Thur.	23	53	21	23 Feb. (54)	5 Thur.	9869·4024	94·0691	198·1458	4258
23 Mar. (82)	0 Sat. .	6	5	30	13 Mar. (72)	4 Wed.	9904·0838	30·0625	249·4563	4259
23 Mar. (82)	1 Sun. .	12	17	39	3 Mar. (62)	2 Mon.	118·4386	913·5983	221·3709	4260
23 Mar. (82)	2 Mon.	18	29	48	22 Mar. (81)	1 Sun. .	153·1210	849·5918	272·6813	4261
23 Mar. (83)	4 Wed.	0	41	57	10 Mar. (70)	5 Thur.	28·8439	696·8358	241·8581	4262
23 Mar. (82)	5 Thur.	6	54	6	27 Feb. (58)	2 Mon. .	9904·5667	544·0799	211·0349	4263
23 Mar. (82)	6 Fri. .	13	6	15	18 Mar. (77)	1 Sun. .	9939·2491	480·0733	262·3454	4264
23 Mar. (82)	0 Sat. .	19	18	24	7 Mar. (66)	5 Thur.	9814·9719	327·3173	231·5221	4265
23 Mar. (83)	2 Mon.	1	30	33	25 Feb. (56)	3 Tues.	29·3268	210·8530	203·4366	4266
23 Mar. (82)	3 Tues.	7	42	42	15 Mar. (74)	2 Mon.	64·0091	146·8465	255·7471	4267
23 Mar. (82)	4 Wed.	13	54	51	4 Mar. (63)	6 Fri. .	9939·7320	994·0903	223·9239	4268
23 Mar. (82)	5 Thur.	20	7	0	23 Mar. (82)	5 Thur.	9974·4144	930·0840	275·2343	4269
23 Mar. (83)	0 Sat. .	2	19	9	12 Mar. (72)	3 Tues.	188·7692	813·6193	247·1488	4270
23 Mar. (82)	1 Sun.	8	31	18	1 Mar. (60)	0 Sat. .	64·4920	660·3638	216·3257	4271
23 Mar. (82)	2 Mon.	14	43	27	20 Mar. (79)	6 Fri. .	99·1744	596·8573	267·6361	4272

TABLE

CONCURRENT YEAR.								Intercalated (*adhika*) and suppressed (*kshaya*) true lunar months.
						Jovian Saṁvatsara.		
Kali.	Śaka.	Chaitrādi Vikrama.	Mēshādi solar year in Bengal.	Kollam.	A. D.	Southern system.	Northern system.	
1	2	3	3*a*	4	5	6	7	8*a*
4273	1094	1229	578	346-47	1171-72	25 Khara . .	29 Manmatha .	...
4274	1095	1230	579	347-48	*1172-73	26 Nandana .	30 Durmukha .	4 Āshāḍha .
4275	1096	1231	580	348-49	1173-74	27 Vijaya . .	31 Hēmalamba .	...
4276	1097	1232	581	349-50	1174-75	28 Jaya . .	32 Vilamba .	...
4277	1098	1233	582	350-51	1175-76	29 Manmatha .	33 Vikārin .	2 Vaiśākha .
4278	1099	1234	583	351-52	*1176-77	30 Durmukha .	34 Śārvarin .	...
4279	1100	1235	584	352-53	1177-78	31 Hēmalamba .	35 Plava . .	6 Bhādrapada
4280	1101	1236	585	353-54	1178-79	32 Vilamba .	36 Śubhakṛit .	...
4281	1102	1237	586	354-55	1179-80	33 Vikārin .	37 Śōbhana .	...
4282	1103	1238	587	355-56	*1180-81	34 Śārvarin .	38 Krōdhin .	4 Āshāḍha .
4283	1104	1239	588	356-57	1181-82	35 Plava . .	39 Viśvāvasu .	...
4284	1105	1240	589	357-58	1182-83	36 Śubhakṛit .	40 Parābhava .	...
4285	1106	1241	590	358-59	1183-84	37 Śōbhana .	41 Plavaṅga .	2 Vaiśākha .
4286	1107	1242	591	359-60	*1184-85	38 Krōdhin .	42 Kīlaka . .	...
4287	1108	1243	592	360-61	1185-86	39 Viśvāvasu .	43 Saumya .	6 Bhādrapada
4288	1109	1244	593	361-62	1186-87	40 Parābhava .	44 Sādhāraṇa .	...
4289	1110	1245	594	362-63	1187-88	41 Plavaṅga .	45 Virōdhakṛit .	...
4290	1111	1246	595	363-64	*1188-89	42 Kīlaka . .	46 Paridhāvin .	5 Śrāvaṇa .
4291	1112	1247	596	364-65	1189-90	43 Saumya .	47 Pramādin .	...
4292	1113	1248	597	365-66	1190-91	44 Sādhāraṇa .	48 Ānanda .	...
4293	1114	1249	598	366-67	1191-92	45 Virōdhakṛit .	49 Rākshasa .	3 Jyēshṭha .
4294	1115	1250	599	367-68	*1192-93	46 Paridhāvin .	50 Anala . .	...
4295	1116	1251	600	368-69	1193-94	47 Pramādin .	51 Piṅgala .	...
4296	1117	1252	601	369-70	1194-95	48 Ānanda .	52 Kālayukta .	2 Vaiśākha .
4297	1118	1253	602	370-71	1195-96	49 Rākshasa .	53 Siddhārthin .	...

LXXXII—*Contd.*

COMMENCEMENT OF THE											
Solar year.					Luni-solar year (mean sunrise of civil day on which Chaitra śukla 1 ends).						Kali.
Day and month A. D.	Week-day.	Time of true Mēsha-saṁkrānti.			Day and month A. D.	Week-day.	*a*	*b*	*c*		
13	14	17			19	20	23	24	25		1
		H.	M.	S.							
23 Mar. (82)	3 Tues.	20	55	36	9 Mar. (68)	3 Tues.	9974·8973	444·1013	236·8129		**4273**
23 Mar. (83)	5 Thur.	3	7	45	26 Feb. (57)	0 Sat. .	9850·6201	291·3454	205·9896		**4274**
23 Mar. (82)	6 Fri. .	9	19	54	16 Mar. (75)	6 Fri. .	9885·3025	227·3389	257·3001		**4275**
23 Mar. (82)	0 Sat. .	15	32	3	6 Mar. (65)	4 Wed. .	99·6574	110·8745	229·2147		**4276**
23 Mar. (82)	1 Sun. .	21	44	11	23 Feb. (54)	1 Sun. .	9975·3801	958·1187	198·1914		**4277**
23 Mar. (83)	3 Tues.	3	56	20	13 Mar. (73)	0 Sat. .	10·0625	894·1120	249·7018		**4278**
23 Mar. (82)	4 Wed.	10	8	29	3 Mar. (62)	5 Thur.	224·4174	777·6478	221·6164		**4279**
23 Mar. (82)	5 Thur.	16	20	38	22 Mar. (81)	4 Wed.	259·0998	713·6413	272·9269		**4280**
23 Mar. (82)	6 Fri. .	22	32	47	11 Mar. (70)	1 Sun. .	134·8226	560·8853	242·1036		**4281**
23 Mar. (83)	1 Sun.	4	44	56	28 Feb. (59)	5 Thur.	10·5455	408·1294	211·2804		**4282**
23 Mar. (82)	2 Mon.	10	57	5	18 Mar. (77)	4 Wed.	45·2279	344·1228	262·5909		**4283**
23 Mar. (82)	3 Tues.	17	9	14	7 Mar. (66)	1 Sun. .	9920·9507	191·3668	231·7677		**4284**
23 Mar. (82)	4 Wed.	23	21	23	24 Feb. (55)	5 Thur.	9796·6735	38·6109	200·9444		**4285**
23 Mar. (83)	6 Fri. .	5	33	32	15 Mar. (75)	5 Thur.	169·9879	10·8960	254·9926		**4286**
23 Mar. (82)	0 Sat. .	11	45	41	4 Mar. (63)	2 Mon.	45·7108	858·1401	224·1694		**4287**
23 Mar. (82)	1 Sun. .	17	57	50	23 Mar. (82)	1 Sun. .	80·3931	794·1335	275·4799		**4288**
24 Mar. (83)	3 Tues.	0	9	59	13 Mar. (72)	6 Fri. .	294·7480	677·6693	247·3944		**4289**
23 Mar. (83)	4 Wed.	6	22	8	1 Mar. (61)	3 Tues.	170·4708	524·9133	216·5712		**4290**
23 Mar. (82)	5 Thur.	12	34	17	19 Mar. (78)	1 Sun. .	9866·5213	424·6151	265·1438		**4291**
23 Mar. (82)	6 Fri. .	18	46	26	8 Mar. (67)	5 Thur.	9742·2440	271·8592	234·3207		**4292**
24 Mar. (83)	1 Sun. .	0	58	35	26 Feb. (57)	3 Tues.	9956·5989	155·3949	206·2352		**4293**
23 Mar. (83)	2 Mon.	7	10	44	16 Mar. (76)	2 Mon.	9991·2813	91·3884	257·5456		**4294**
23 Mar. (82)	3 Tues.	13	22	53	6 Mar. (65)	0 Sat. .	205·6361	974·9241	229·4602		**4295**
23 Mar. (82)	4 Wed.	19	35	2	23 Feb. (54)	4 Wed.	81·3589	822·1741	198·6370		**4296**
24 Mar. (83)	6 Fri. .	1	47	11	14 Mar. (73)	3 Tues.	116·0413	758·1608	249·9474		**4297**

TABLE

CONCURRENT YEAR.								Intercalated (*adhika*) and suppressed (*kshaya*) true lunar months.
Kali	Śaka.	Chaitrādi Vikrama.	Mēshādi solar year in Bengal.	Kollam.	A. D.	Jovian Saṁvatsara. Southern system.	Jovian Saṁvatsara. Northern system.	
1	2	3	3*a*	4	5	6	7	8*a*
4298	1119	1254	603	371-72	*1196-97	50 Anala . .	54 Raudra .	6 Bhādrapada
4299	1120	1255	604	372-73	1197-98	51 Piṅgala .	55 Durmati .	...
4300	1121	1256	605	373-74	1198-99	52 Kālayukta .	56 Dundubhi .	...
4201	1122	1257	606	374-75	1199-1200	53 Siddhārthin .	57 Rudhirōdgārin	4 Āshādha .
4302	1123	1258	607	375-76	*1200-01	54 Raudra .	58 Raktāksha .	...

LXXXII—*Concld.*

COMMENCEMENT OF THE								
SOLAR YEAR.			LUNI-SOLAR YEAR (MEAN SUNRISE OF CIVIL DAY ON WHICH CHAITRA ŚUKLA 1 ENDS).					Kali.
Day and month A. D.	Week-day.	Time of true Mēsha-saṁkrānti.	Day and month A. D.	Week-day.	*a*	*b*	*c*	
13	14	17	19	20	23	24	25	1
		H. M. S.						
23 Mar. (83)	0 Sat.	7 59 20	2 Mar. (62)	0 Sat	9991·7641	605·4056	219·1242	4298
23 Mar. (82)	1 Sun.	14 11 29	21 Mar. (80)	6 Fri.	26·4465	541·3991	270·4346	4299
23 Mar. (82)	2 Mon.	20 23 38	10 Mar. (69)	3 Tues.	9902·1694	388·6432	239·6115	4300
24 Mar. (83)	4 Wed.	2 35 47	27 Feb. (58)	0 Sat.	9777·8923	235·8872	208·7660	4301
23 Mar. (83)	5 Thur.	8 47 56	17 Mar. (77)	6 Fri.	9812·5747	171·8807	260·0765	4302

TABLE LXXXIII-A.

DUARTION AND COLLECTIVE DURATION OF TRUE SOLAR MONTHS, WITH INCREASE OF "a" "b" "c" AT EACH TRUE SAṀKRĀNTI.

By the Brahma-Siddhānta.

Calculated for the year K. Y. 4500, (expired), A. D. 899—900.

"a" in 10,000 ths of circle; "b" and "c" in 1,000 ths; "saṁ"=solar saṁkrānti.

Luni-solar month (ending at the second of the two solar saṁkrāntis connected with it).	At true solar saṁkrānti.	Collective duration in days, hours, etc., and collective increase of a, b, c from true Mēsha-saṁkrānti to each true saṁkrānti.						At true solar saṁkrānti.	Length of solar month preceding each true saṁkrānti, and increase of a, b, c between each such saṁkrānti.					
		Day.	Week day.	H. M. S.	a	b	c		Day.	Week day.	H. M. S.	a	b	c
1	2	3	4	5	6	7	8	9	10	11	12	13	14	15
	Mīna-saṁ. (*of previous year*).													
1. Chaitra	Mēsha-saṁ.	0	0	0 0 0	0	0	0	Mēsha-saṁ.	0	0	0 0 0	0	0	0
2. Vaiśākha	Vṛishabha-saṁ	30	(2)	22 21 9	474·3381	122·5490	84·6833	Vṛishabha-saṁ.	30	(2)	22 21 9	474·3381	122·5490	84·6833
3. Jyēshṭha	Mithuna-saṁ.	62	(6)	8 15 57	1111·7956	262·5752	170·6856	Mithuna-saṁ.	31	(3)	9 54 48	637·4575	140·0262	86·0023
4. Āshāḍha	Karka-saṁ.	93	(2)	23 12 15	1820·1580	410·2049	257·2610	Karka-saṁ.	31	(3)	14 56 18	708·3624	147·6297	86·5754
5. Śrāvaṇa	Siṁha-saṁ.	125	(6)	10 42 48	2480·1360	552·6492	343·4452	Siṁha-saṁ.	31	(3)	11 30 33	659·9780	142·4443	86·1842
6. Bhādrapada	Kanyā-saṁ.	156	(2)	11 41 2	2991·4178	679·1575	428·4273	Kanyā-saṁ.	31	(3)	0 58 15	511·2818	126·5083	84·9821
7. Āśvina	Tulā-saṁ.	186	(4)	22 35 29	3304·2747	784·4003	511·8051	Tulā saṁ.	30	(2)	10 54 27	312·8569	105·2428	83·3778
8. Kārttika	Vṛiśchika-saṁ.	216	(6)	20 28 50	3433·4472	869·9574	593·6979	Vṛiśchika-saṁ.	29	(1)	21 53 21	129·1725	85·5571	81·8928
9. Mārgaśīra	Dhanus-saṁ.	246	(1)	8 0 47	3416·4906	939·8537	674·4092	Dhanus-saṁ.	29	(1)	11 31 57	9983·0434	69·8963	80·7113
10. Pausha	Makara-saṁ.	275	(2)	16 6 58	3351·2241	4·5725	754·7299	Makara-saṁ.	29	(1)	8 6 11	9934·7335	64·7188	80·3207
11. Māgha	Kumbha-saṁ.	305	(4)	2 49 9	3322·5644	73·2145	835·3466	Kumbha-saṁ.	29	(1)	10 42 11	9971·3403	68·6420	80·6167
12. Phālguna	Mīna-saṁ.	334	(5)	22 4 25	3414·5580	154·7871	916·9387	Mīna-saṁ.	29	(1)	19 15 16	91·9936	81·5726	81·5921
1. Chaitra (*of following year*).	Mēsha-saṁ. (*of following year*)	365	[illegible]	6 12 9	3688·2056	255·8315	1000·0	Mēsha-saṁ. (*of following year*).	30	(2)	8 7 44	273·6476	101·0407	83·0608

TABLE LXXXIII-B.

VALUE OF "c" AND OF "EQUATION c" AT THE SEVERAL TRUE SAMKRÁNTIS.

Correct for K. Y. 4000, A. D. 899-900.

"c" in 1,000ths of circle, "equation c" in 10,000ths.

Saṁkrānti.	c	"Equation c."
Mēsha-saṁ. .	277·6064	0·9037
Vṛishabha-saṁ. .	362·2899	14·4355
Mithuna-saṁ. .	448·2921	41·1356
Karka-saṁ. .	534·8676	73·5542
Siṁha-saṁ. .	621·0519	102·0578
Kanyā-saṁ. .	706·0241	118·5381
Tulā-saṁ. .	789·4020	118·9561
Vṛiśchika-saṁ. .	871·2948	104·1144
Dhanus-saṁ. .	952·0062	78·3666
Makara-saṁ. .	32·3264	48·2336
Kumbha-saṁ. .	112·9432	21·0624
Mīna-saṁ .	194·5355	3·6494

TABLE LXXXIII-C.

EXACT VALUE OF "c" AND OF "EQUATION c" AT THE MOMENT OF TRUE MĒSHA-SAMKRÁNTI AT BEGINNING OF EACH CENTURY K. Y.

"c" in 1,000ths of circle. "Equation c" in 10,000ths.

K. Y.	A. D.	c	"Eqn. c."
3700	599-600	277·6399	0·93
3800	699-700	277·6287	0·93
3900	799-800	277·6175	0·93
4000	899-900	277·6064	0·93
4100	999-1000	277·5952	0·93
4200	1099-1100	277·5840	0·93
4300	1199-1200	277·5728	0·93

TABLES LXXXIV, LXXXV.

"EQUATION *b*" AND "EQUATION *c*" IN WHOLE NUMBERS BY THE BRAHMA-SIDDHĀNTA AND SIDDHĀNTA-ŚIRŌMAṆI.

Corresponding to Tables VI, VII, "Indian Calendar."

For close detail Tables LV, LVI, (*Vol. XV above*) are to be used.

"Arg."=moon's (*b*) or sun's (*c*) mean anom. in 1,000ths of circle.

TABLE LXXXIV.

LUNAR "EQUATION *b*."

Arg.	Eqn.	Arg.	Arg.	Eqn.	Arg.
0	140	500	500	140	1000
10	149	490	510	131	990
20	158	480	520	122	980
30	166	470	530	114	970
40	174	460	540	105	960
50	183	450	550	97	950
60	191	440	560	88	940
70	199	430	570	80	930
80	207	420	580	73	920
90	214	410	590	65	910
100	222	400	600	58	900
110	229	390	610	51	890
120	235	380	620	44	880
130	241	370	630	38	870
140	247	360	640	32	860
150	253	350	650	27	850
160	258	340	660	22	840
170	262	330	670	17	830
180	266	320	680	13	820
190	270	310	690	10	810
200	273	300	700	7	800
210	275	290	710	4	790
220	277	280	720	2	780
230	279	270	730	1	770
240	279	260	740	0	760
250	289	250	750	0	750

TABLE LXXXV.

SOLAR "EQUATION *c*."

Arg.	Eqn.	Arg.	Arg.	Eqn.	Arg.
0	60	500	500	60	1000
10	56	490	510	64	990
20	53	480	520	68	980
30	49	470	530	72	970
40	46	460	540	75	960
50	42	450	550	79	950
60	38	440	560	82	940
70	34	430	570	86	930
80	31	420	580	89	920
90	28	410	590	93	910
100	25	400	600	96	900
110	22	390	610	99	890
120	19	380	620	102	880
130	16	370	630	104	870
140	14	360	640	107	860
150	12	350	650	109	850
160	9	340	660	111	840
170	7	330	670	113	830
180	6	320	680	115	820
190	4	310	690	117	810
200	3	300	700	118	800
210	2	290	710	119	790
220	1	280	720	120	780
230	0	270	730	120	770
240	0	260	740	121	760
250	0	250	750	121	750

AUXILIARY TABLE.

Difference in Equation.	Last figure of argument								
	9	8	7	6	5	4	3	2	1
	Add or substract								
9	8	7	6	5	4 or 5	4	3	2	1
8	7	6	6	5	4	3	2	2	1
7	6	6	5	4	3 or 4	3	2	1	1
6	5	5	4	4	3	2	2	1	1
5	4 or 5	4	3 or 4	3	2 or 3	2	1 or 2	1	0 or 1
4	4	3	3	2	2	2	1	1	0
3	3	2	2	2	1 or 2	1	1	1	0
2	2	2	1	1	1	1	1	0	0
1	1	1	1	1	0 or 1	0	0	0	0

TABLE LXXXVI.

VALUE OF "*a*", "*b*", "*c*" AT BEGINNING OF CENTURIES OF THE KALIYUGA, BY THE BRAHMA-SIDDHĀNTA.

K.Y. Century.	Beginning in A.D.	Week-day.	*a*	*b*	*c*
37	599	0	6028·1929	719·2529	282·9906
38	699	6	4900·0921	308·0530	283·3962
39	799	6	3433·3593	860·5614	281·0640
40	899	6	2305·2584	449·3615	281·4695
41	999	6	1177·1576	38·1616	281·8751
42	1099	6	49·0567	626·9616	282·2807
43	1199	0	8920·9559	215·7617	282·6863

TABLE LXXXVII.

INCREASE OF *a*, *b*, *c* FOR YEARS OF KALIYUGA CENTURY.

* = *year of 366 days.*

Year.	Week-day.	*a*	*b*	*c*	Year.	Week-day.	*a*	*b*	*c*
0	0	0	0	0	30	3	729·2961	683·8984	0·6759
1	1	3600·6747	246·4522	999·2925	31	4	4329·9708	930·3505	999·9683
*2	2	7201·3494	492·9043	998·5849	32	5	7930·6455	176·8027	999·2608
3	4	1140·6560	775·6482	0·6151	*33	6	1531·3202	423·2549	998·5533
4	5	4741·3307	22·1003	999·9076	34	1	5470·6268	705·9987	0·5835
5	6	8342·0054	268·5525	999·2001	35	2	9071·3015	952·4509	999·8759
*6	0	1942·6800	515·0047	998·4925	36	3	2671·9762	198·9030	999·1684
7	2	5881·9867	797·7485	0·5227	*37	4	6272·6509	445·3552	998·4609
8	3	9482·6614	44·2007	999·8152	38	6	211·9575	728·0990	0·4911
9	4	3083·3360	290·6528	999·1077	39	0	3812·6322	974·5512	999·7836
*10	5	6684·0107	537·1050	998·4001	40	1	7413·3069	221·0034	999·0760
11	0	623·3174	819·8488	0·4303	*41	2	1013·9815	467·4555	998·3685
12	1	4223·9921	66·3010	999·7228	42	4	4953·2882	750·1994	0·3987
*13	2	7824·6667	312·7532	999·0153	43	5	8553·9629	996·6515	999·6912
14	4	1763·9734	595·4970	1·0455	*44	6	2154·6376	243·1037	998·9836
15	5	5364·6481	841·9492	0·3379	45	1	6093·9442	525·8475	1·0158
16	6	8965·3227	88·4013	999·6304	46	2	9694·6189	772·2997	0·3063
*17	0	2565·9974	334·8535	998·9229	47	3	3295·2936	18·7519	999·5988
18	2	6505·3041	617·5973	0·9531	*48	4	6895·9682	265·2040	998·8912
19	3	105·9788	864·0495	0·2455	49	6	835·2749	547·9479	0·9214
20	4	3706·6534	110·5017	999·5380	50	0	4435·9496	794·4000	0·2139
*21	5	7307·3281	356·9539	998·8305	51	1	8036·6243	40·8522	999·5064
22	0	1246·6348	639·6977	0·8607	*52	2	1637·2989	287·3044	998·7988
23	1	4847·3094	886·1499	0·1531	53	4	5576·6056	570·0482	0·8290
24	2	8447·9841	132·6020	999·4456	54	5	9177·2803	816·5004	0·1215
*25	3	2048·6588	379·0542	993·7381	55	6	2777·9549	62·9526	999·4140
26	5	5987·9655	661·7980	0·7683	*56	0	6378·6296	309·4047	998·7064
27	6	9588·6401	908·2502	0·0607	57	2	317·9363	592·1485	0·7366
28	0	3189·3148	154·7024	999·3532	58	3	3918·6110	838·6007	0·0291
*29	1	6789·9895	401·1545	998·6457	59	4	7519·2856	85·0529	999·3216

TABLE LXXXVII—*Contd.*

Year.	Week-day	a	b	c
*60	5	1119·9603	331·5051	998·6140
61	0	5059·2670	614·2489	0·6442
62	1	8659·9416	860·7011	999·9367
63	2	2260·6163	107·1532	999·2292
*64	3	5861·2910	353·6054	998·5216
65	5	9800·5977	636·3492	0·5518
66	6	3401·2723	882·8014	999·8443
67	0	7001·9470	129·2536	999·1368
*68	1	602·6217	375·7057	998·4292
69	3	4541·9283	658·4496	0·4594
70	4	8142·6030	904·9017	999·7519
*71	5	1743·2777	151·3539	999·0444
72	0	5682·5844	434·0977	1·0746
73	1	9283·2590	680·5499	0·3670
74	2	2883·9337	927·0021	999·6595
*75	3	6484·6084	173·4542	998·9520
76	5	423·9159	456·1981	0·9822
77	6	4024·5897	702·6502	0·2746
78	0	7625·2644	949·1024	999·5671
*79	1	1225·9391	195·5546	998·8596
80	3	5165·2457	478·2984	0·8898
81	4	8765·9204	724·7506	0·1822
82	5	2366·5951	971·2027	999·4747
*83	6	5967·2698	217·6549	998·7672
84	1	9906·5764	500·3987	0·7974
85	2	3507·2511	746·8509	0·0898
86	3	7107·9258	993·3031	999·3823
*87	4	708·6004	239·7552	998·6748
88	6	4647·9071	522·4991	0·7050
89	0	8248·5818	768·9512	999·9974
90	1	1849·2565	15·4034	999·2899
*91	2	5449·9311	261·8556	998·5824
92	4	9389·2378	544·5994	0·6126
93	5	2989·9125	791·0516	999·9050
94	6	6590·5871	37·5038	999·1975
*95	0	191·2618	283·9559	998·4900
96	2	4130·5685	566·6997	0·5202
97	3	7731·2431	813·1519	999·8126
98	4	1331·9178	59·6041	999·1051
*99	5	4932·5925	306·0563	998·3976
100	0	8871·8992	588·8001	0·4278

TABLE LXXXVIII.

Values of "*a*", "*b*", "*c*" per day from Mina 1 to Mesha 2, the day of mean Mêsha-saṁkrânti.

No. of days interval from 0 Mêsha.	Month and day.	Week-day.	a	b	c
1	2	3	4	5	6
29	Mina 1	4	9502·4085	874·9589	915·1286
28	,, 2	5	9841·0404	911·2506	917·8664
27	,, 3	6	179·6724	947·5422	920·6042
26	,, 4	0	518·3044	983·8339	923·3419
25	,, 5	1	856·9364	20·1255	926·0797
24	,, 6	2	1195·5684	56·4172	928·8175
23	,, 7	3	1534·2004	92·7088	931·5553
22	,, 8	4	1872·8324	129·0005	934·2931
21	,, 9	5	2211·4643	165·2921	937·0309
20	,, 10	6	2550·0963	201·5838	939·7687
19	,, 11	0	2888·7283	237·8754	942·5065
18	,, 12	1	3227·3603	274·1671	945·2442
17	,, 13	2	3565·9923	310·4587	947·9820
16	,, 14	3	3904·6243	346·7504	950·7198
15	,, 15	4	4243·2563	383·0420	953·4576
14	,, 16	5	4581·8882	419·3336	956·1954
13	,, 17	6	4920·5202	455·6253	958·9332
12	,, 18	0	5259·1522	491·9169	961·6710
11	,, 19	1	5597·7842	528·2086	964·4088
10	,, 20	2	5936·4162	564·5002	967·1465
9	,, 21	3	6275·0482	600·7919	969·8843
8	,, 22	4	6613·6801	637·0835	972·6221
7	,, 23	5	6952·3121	673·3752	975·3599
6	,, 24	6	7290·9441	709·6668	978·0977
5	,, 25	0	7629·5761	745·9585	980·8355
4	,, 26	1	7968·2081	782·2501	983·5733
3	,, 27	2	8306·8401	818·5418	986·3111
2	,, 28	3	8645·4721	854·8334	989·0488
1	,, 29	4	8984·1040	891·1251	991·7866
	Mêsha 0	5	9322·7360	927·4167	994·5244
	,, 1	6	9661·3680	963·7084	997·2622
	,, 2	0	0	0	0

TABLE LXXXIX.

SUN'S EQUATION OF THE CENTRE AND SINE-VALUES ACCORDING TO THE BRAHMA-SIDDHĀNTA.

Serial No. of sine.	Sun's mean anom.				Sine of anom. angle. Value in minutes.	Sine of anom. angle. Diff.	Equation. Equation.			Equation. Difference per minute of anom.	Sun's mean anom.				Serial No. of sine.
1	2				3	4	5			6	7				1
	°	′	°	′	′	′	°	′	″	″	°	′	°	′	
0	0	0	180	0	0	214	0	0	0	2·27̇	180	0	360	0	0
1	3	45	176	15	214	213	0	8	32·50	2·2760	183	45	356	15	1
2	7	30	172	30	427	211	0	17	2·61	2·2458	187	30	352	30	2
3	11	15	168	45	638	208	0	25	27·92	2·2128	191	15	348	45	3
4	15	0	165	0	846	205	0	33	46·05	2·1822	195	0	345	0	4
5	18	45	161	15	1051	200	0	41	57·02	2·1287	198	45	341	15	5
6	22	30	157	30	1251	195	0	49	55·97	2·0755	202	30	337	30	6
7	26	15	153	45	1446	189	0	57	42·97	2·0117	206	15	333	45	7
8	30	0	150	0	1635	182	1	5	15·60	1·9372	210	0	330	0	8
9	33	45	146	15	1817	174	1	12	31·46	1·8520	213	45	326	15	9
10	37	30	142	30	1991	165	1	19	28·17	1·7562	217	30	322	30	10
11	41	15	138	45	2156	156	1	26	3·32	1·6604	221	15	318	45	11
12	45	0	135	0	2312	147	1	32	16·92	1·5646	225	0	315	0	12
13	48	45	131	15	2459	135	1	38	8·96	1·4369	228	45	311	15	13
14	52	30	127	30	2594	125	1	43	32·27	1·3305	232	30	307	30	14
15	56	15	123	45	2719	113	1	48	31·62	1·2028	236	15	303	45	15
16	60	0	120	0	2832	101	1	53	2·24	1·0750	240	0	300	0	16
17	63	45	116	15	2933	88	1	57	4·12	0·9367	243	45	296	15	17
18	67	30	112	30	3021	75	2	0	34·87	0·7982	247	30	292	30	18
19	71	15	108	45	3096	63	2	3	34·49	0·6706	251	15	288	45	19
20	75	0	105	0	3159	48	2	6	5·36	0·5184	255	0	285	0	20
21	78	45	101	15	3207	35	2	8	1·99	0·3651	258	45	281	15	21
22	82	30	97	30	3242	21	2	9	24·14	0·2235	262	30	277	30	22
23	86	15	93	45	3263	7	2	10	14·43	0·0745	266	15	273	45	23
24	90	0	90	0	3270		2	10	31·19		270	0	270	0	24

THE BRAHMA-SIDDHĀNTA OF BRAHMAGUPTA, A.D. 628.

WORKING TABLES FOR COMPUTATION OF ANCIENT DATES BY THE MEAN MOTIONS OF SUN AND MOON.

321. The Tables published in my last article enabled the dates of ancient Indian inscriptions and records to be verified according to the requirements of the *Brahma-Siddhānta* with, as basis of calculation, the "true" or apparent motions of sun and moon. This mode of reckoning appears to have been introduced in the 11th century A.D. But the *Brahma-Siddhānta* was composed in A.D. 628 and for at least four centuries after its appearance details for the calendar were almost certainly based on mean planetary motions; while it is believed that this mean system continued to guide the preparation of *pañchāngs* (almanacs) till a much later date—perhaps for several centuries in some parts of the country.

For the correct verification, therefore, of early dates it is necessary for historians to be provided with a set of Tables based on mean planetary motions and the postulates of the *Brahma-Siddhānta* in addition to those based on mean motions and the postulates of the *Ārya-Siddhānta*. The latter were provided in a previous article in this volume. The former are presented herewith. They cover a period of 800 years, from K.Y. 3700 to 4500, or from A.D. 599 to 1400.

The system of work is the same as in all my previous Tables, that is to say, it is the system of Largeteau as adopted by Professor H. Jacobi in the *Indian Antiquary*, Vol. VIII, and in the *Epigraphia Indica*, Vol. XI. Full examples shewing the method of work, which is very simple, are given in my former articles; others, specially concerning the system of mean reckoning on *Brahma-Siddhānta* principles, are given below.

In case of doubt as to which of the Tables already published should be used in the present case attention is directed to the accompanying § 330.

322. In examining the dates of records in earlier years it is necessary to remember that the modes of reckoning adopted were not always the same as those used in more recent years. As to eras, reference to articles 6-12 of my former work, *Indian Chronography*, is recommended. For other matters the late Dr. J. F. Fleet's remarks in the *Journal of the Royal Asiatic Society* for 1912, pp. 704-5, will be found very valuable.

Especially let it be borne in mind that the lunar month reckoning in early years was probably carried out on the *pūrṇimānta* system. According to the late Professor Kielhorn the earliest known date certainly in *amānta* reckoning belonged to the year A.D. 794. It is contained in the Paiṭhān plates of the Rāshṭrakūṭa king Govinda III (*Epig. Ind.*, III, 105; *Ind. Ant.*, XVII, p. 142, No. 9). As regards these two systems, the *amānta* and *pūrṇimānta* names of lunar months, see *Indian Calendar*, §§ 13. 45 (with Table on p. 26), 47, 51, and the late Sankara Balkrishna Dikshit's footnote on p. 31; also *Indian Chronography*, §§ 75, 76, p. 31.

Elements of the Brahma-Siddhānta mean reckoning.

323. The principal elements are fully stated in my former article on this authority (*above, p. 448*, § 313). For calculation on the mean system the following notes are necessary.

(i) The length of the mean sidereal solar year is 365d 6h 12m 9s, a fixture afterwards adopted by Bhāskarāchārya in his *Siddhānta-Śiromaṇi*, A.D. 1150.

(ii) The advance of "*a*" (distance of mean moon from mean sun)—which finally fixes the index of the *tithi* ($\frac{1}{30}$th of a mean lunation) in measurement by 10,000ths of circle—in every civil day of 24 hours and in hours, minutes and seconds, is given for the *Siddhānta-Śirōmaṇi* in Tables LIV-A and B above, pp. 148, 152. These Tables are applicable to the *Brahma-Siddhānta*.

(iii) For the sun's mean motion per day, hour, minute, etc., see Tables XLIII and XLIV above pp. 59, 60.

(iv) The advance of *a* in one mean solar month is, in 10,000ths of circle, 307·349156595.

(v) Each solar month consists of $30^{d}\ 10^{h}\ 31^{m}\ 0^{s}{\cdot}75$. Table XCI below shews the interval of days, hours, etc., between the moment of mean Mēsha-*saṁkrānti*, when the mean sun is at celestial long. 0° (*Table XC, cols. 13-17*), and the moment of each subsequent *saṁkrānti* when the mean sun enters each of the twelve signs; and so enables the day and time when each mean solar month begins to be ascertained. The same Table gives the advance of "*a*" from its value at the moment of mean Mēsha-*saṁkrānti* to the same at each subsequent *saṁkrānti*.

(vi) The interval between the moments of true and mean Mēsha-*saṁkrānti*, i.e. between the moments of the astronomical beginning respectively of the true and mean solar year, which interval we call the *śōdhya*, varies slightly year by year in consequence of the postulated shift of the sun's apsis (§ 313, VII, *above* p. 449). The exact intervals, century by century from K.Y. 3700 to 4300, were given above in § 315. The Table is here repeated and extended so as to embrace the whole period of the general Table XC below. The quantities were computed by Dr. Robert Schram.

TABLE

VALUE OF *śōdhya* BY THE *Brahma-Siddhānta*.

Kaliyuga.	A.D.	Śōdhya at beginning of centuries.				
		D.	H.	M.	S.	Days and decimals.
3700	599-600	2	4	8	59·8128	2·1729145
3800	699-700	2	4	9	2·0160	2·1729400
3900	799-800	2	4	9	4·2192	2·1729655
4000	899-900	2	4	9	6·4224	2·1729910
4100	999-1000	2	4	9	8·6256	2·1730165
4200	1099-1100	2	4	9	10·8288	2·1730420
4300	1199-1200	2	4	9	13·0320	2·1730675
4400	1299-1300	2	4	9	15·2352	2·1730930
4500	1399-1400	2	4	9	17·4384	2·1731185

The moment of mean Mēsha-saṁkrānti, or the beginning of the mean solar year.

324. The general Table which follows states (*Table XC, cols. 13-17*) the moment of beginning of each mean solar year according to the *Brahma-Siddhānta*. The first entry is for the expired year 3700 of the Kaliyuga (A.D. 599-600), in which year the astronomical beginning is fixed as at 5^h 15^m after mean sunrise on Saturday, 21 March, A.D. 599. It is incumbent on me to prove the correctness of this fixture. Subsequent entries are based on it by the addition to it year by year of 365^d 6^h 12^m 9^s. Proof may be offered in three ways :—(A) by comparison with the date and time already found for the beginning of the true solar year K.Y. 3700, utilizing Dr. Schram's determination of the interval between the two occurrences; (B) by comparison with the date and time fixed for the beginning of the same mean solar year according to the *First Ārya-Siddhānta*, allowing for the time-difference between the two authorities caused by their different estimate as to the length of the mean solar year, viz. 21^s; (C) by direct computation from the moment of mean Mēsha-*saṁkrānti*, at the beginning of the Kaliyuga era, 3,700 years earlier, which, according to the *Brahma-Siddhānta* (§ 313, v, *above, p. 449*), was exactly at mean sunrise, or 0^h 0^m 0^s Laṅkā time, on Friday, 18 Febr. (B.C. 3102).

A

			h.	*m.*	*s.*
Moment of true Mēsha-*saṁkrānti* in K. Y. 3700 (A.D. 599) (*Table LXXXII*, above.)	(5) Thur.,	19 Mar.	1	6	0·1872
Śōdhya as above (§ 323, *Table*) . .	+ (2)	2	4	8	59·8128
Moment of mean Mēsha-*saṁkrānti* .	(0) Sat.,	21 Mar.	5	15	0

B

[See *Indian Calendar*, Table I, cols. 13-17, for A.D. 599-600.]

			h.	*m.*	*s.*
True Mēsha-*saṁkrānti* by *Ārya-Siddhānta*	(5) Thur.,	19 Mar.	23	17	30
Ārya-Siddhānta śōdhya . . .	+ (2)	2	3	32	30
Mean Mēsha-*saṁkrānti* by *Ārya-Siddhānta*	(1) Sun.,	22 Mar.	2	50	0
Less Time-difference in 3,700 years[1] .			−21	35	0
Mean Mēsha-*saṁkrānti* by *Brahma-Siddhānta*	(0) Sat.,	21 Mar	5	15	0

The epoch of the Kaliyuga was, as stated above 0^h 0^m 0^s Laṅkā time, or exactly at mean sunrise on Friday 18 Feb. B.C. 3102. The length of the mean solar year being 365^d 6^h 12^m 9^s, the beginning of the next mean solar year took place 6^h 12^m 9^s after mean sunrise; and after the expiration of a century from the epoch the mean solar year began at 20^h 15^m 0^s after mean sunrise; so that after 37 centuries had passed the mean solar year K.Y. 3700 began at 5^h 15^m 0^s after mean sunrise.

When this latter calculation is carried out century by century, the figures shew that centuries 6, 12, 19, 25 and 32, five in all, were defective centuries consisting each of 36,525 days, the remainder being common centuries of 36,526 days. Since 36,526 divided by 7 leaves no

[1] See Table, § 273, in Article on the *Siddhānta-Śirōmaṇi* (above, p. 133), which is equally applicable to the *Brahma-Siddhānta*; or refer to *Indian Chronography*, p. 61. The time-difference in 3,000 years is 17^h 30^m, in 00 years 4^h 5^m, total 21^h 35^m.

remainder and 36,525 divided by 7 leaves remainder 6, the results shew that whereas century 0 began on a Friday, century 37 began on a Saturday.

Table XC therefore, as regards the moment of mean Mēsha-*saṁkrānti* in K.Y. 3700 expired, A.D. 599-600, is proved to be correct.

The beginning of the mean luni-solar year. Amānta system.

325. In § 317 of my article on the *Brahma-Siddhānta* as calculated by the true motions of the sun and moon (*above, p. 451*) it will be seen that the value of "*a*" at mean sunrise of Sunday, 22 March, A.D. 599 (K.Y. 3700) was proved to be, in measurement by 10,000ths of a circle, 6567·108945284. The mean solar century, however, began on the previous day, Saturday, 21 March. Deducting one day's value of *a*, viz. 338·631985412, from the above, we find that at mean sunrise of that Saturday the value of *a*, or the mean moon's distance from mean sun, was 6228·476959872. This was its value at the beginning of the 37th century K.Y. Hence the first entry in Table XCII below which gives the values at mean sunrise on the day on which each century began. The remaining figures in that Table were obtained by the addition to this value of the increase of "*a*" in a century. [See § 316 of the my article on the *Brahma-Siddhānta* "*true*" *System*, above, p. 450. The increase of *a* in a century of 36,525 days is 997·678896964, and in a common century of 36,526 days is 0·416684507.] Centuries 38 and 44 were defective centuries; the rest were common ones. For the beginnings of the odd years of centuries Table LXXXVII above, p. 509 was used, the value of "*a*" there given being added to that for the century.

Thus was determined the value of "*a*" at mean sunrise of the day on which each mean *solar* year begins (*see Example 1 below*). From this is found the value of "*a*" at mean sunrise of the day on which the mean *luni-solar* year begins.

326. The first day of the luni-solar year is, according to the general rule, the civil day on which expired the first *tithi* of the bright half (*śukla*) of the *amānta* lunar month Chaitra, i.e. the *tithi* which begins at the moment of the first new moon after the Mina-*saṁkrānti*, or at the moment of the new moon when that *amānta* lunar month begins within the limits of which the Mēsha-*saṁkrānti* occurs. Having already established the value of "*a*" on the day in any year on which mean Mēsha-*saṁkrānti* occurred, we have to subtract from that value the increase of "*a*" in whole days between the two dates, the day on which the luni-solar year began being the earlier. The first 30 days' entries in Table LIVA (*above*, p. 148) enable this to be done. We select in that Table the "*a*" in col. 3 the value of which is next lower than the "*a*" of mean Mēsha-*saṁkrānti*, and the Table then shews in col. 1 the number of intervening days, and therefrom the European day and month, and, by subtraction, also (col. 2), the week-day. Deducting the selected "*a*" from the "*a*" of mean Mēsha-*saṁkrānti*, we have the "*a*" of mean sunrise of the day, Chaitra *śukla* 1, on which the luni-solar year begins.

Thus,—mean Mēsha-*saṁkrānti* of the year K.Y. 3700, A.D. 599-600, was shewn above to have occurred on (0) Saturday, 21 March A.D. 599, at mean sunrise on which day the mean moon's *tithi*-index *a* was 6228·4770. In Table LIVA, amongst the values of "*a*" in the first 30 days, it is seen that the next lower value is 6095·3757. 6228·4770 − 6095·3757 = 133·1013[1]. *Col. 1* shews that the interval of days was 18, and col. 2 shews the week-day 4. Mean Mēsha-*saṁkrānti* occurred on (0) Saturday. 0 (or 7) − 4 = 3 Tuesday. It is therefore found that the day Chaitra *śukla* 1, the first civil day of the mean luni-solar year, was (3) Tuesday, 3 March A.D. 599, and that the value of "*a*" at mean sunrise on that day was 133·1013, shewing the currency of the *tithi śukla* 1. This is the entry in Table XC below.

It comes to the same thing if the "*a*" of Table XCIII below is added to the "*a*" of mean Mēsha-*saṁkrānti*, the Table being prepared for that purpose. The "*a*" of mean Mēsha-

[1] All values of *a* below 333·3 prove the *tithi* to have been the first of the *amānta* lunar month, or the first *tithi* of the first (*śukla*) fortnight.

saṁkrānti was 6228·4770. We select such a value of "*a*" in col. 3 of that Table as, added to the former, makes a value between 0 and 333·3̇, the limits of the *tithi śukla* 1; and note the interval of days, and the week-day resulting by addition of the given week-day (col. 2) to the week-day of mean Mēsha-*saṁkrānti*. Here the selected value of "*a*" is 3904·6243, since 6228·4770 + 3904·6243 = 133·1013. The interval of days is 18 (col. 1). The week-day corresponding to the day Chaitra *śukla* 1 is (0 + 3 =) 3. The result is the same as obtained by the former process.

All the entries in the general Table XC, cols. 19-23, can be proved in this way.

To find the exact phase of the mean moon, i.e. the mean *tithi*-index "*a*", on any day of any year, or at any particular moment of any day, it is only necessary to add to the value of "*a*" given in col. 23 of Table XC for the first day of the luni-solar year the amount of increase of "*a*" during the intervening whole days, hours, etc., given in Tables LIV A and B above, pp. 148, 152.

The pūrṇimānta system of lunar months.

327. The *amānta* lunar month begins at the moment of new moon, the *pūrṇimānta* month at the moment of full moon a fortnight earlier; so that the fortnight (*śukla*) between new moon and full moon bears the same month-name by both systems, while the fortnight (*krishna*) between full moon and new moon bears, in the *pūrṇimānta* system, the name of the lunar month next after that which it bears in the *amānta* system. The *śukla* fortnight of the first lunar month, for instance, belongs to Chaitra by both systems. The following *krishna* fortnight, however, belongs to Chaitra by the *amānta* system, but to Vaiśākha by the *pūrṇimānta* system.

This should always be borne in mind when examining dates of inscriptions, especially in earlier years. For references to already published explanations see § 322 above, and for a Table of corresponding fortnights and lunar months see *Indian Calendar*, Table II, Part I.

The mean moon's nakshatra.

328. The note on this subject already given (§ 308, p. 362) in dealing with calculation by the *First Ārya-Siddhānta mean system* applies equally to the *Brahma-Siddhānta* mean system. It is unnecessary to repeat it.

Tables LXXX and LXXXI, (pp. 444, 446), fixing the sun's mean longitude for every day of the mean solar year according to the *First Ārya-Siddhānta*, may safely be used for general calculation by the *Brahma-Siddhānta*, since the difference between the two authorities in their estimates of the length of the year only amounts to 21 seconds.[1] But in any exceptionally close case the exact value, at mean sunrise of any day in the year of "*s*", or the sun's mean longitude, can be found by multiplying the sun's mean motion in one day (*Table XLIII*, p. 59), by the number of days' interval between the day on which mean Mēsha-*saṁkrānti* occurred and the given day. The sun's mean motion in one day by the *Brahma-Siddhānta* is 59ᵐ 8ˢ·172655, or in 10,000ths of circle 27·377875426.

The *Rule for work* is as follows. (i) Find, as above, value of "*a*" at mean sunrise of given day. (ii) Note number of whole days intervening between the day of mean Mēsha-*saṁkrānti* (*Table XC below, col. 13, figure in brackets*) and the given day. Turn to Table LXXX and note the increase of sun's mean long., "*s*", during that interval. Deduct from this, by Table LXXXI, the increase of long. during the hours and minutes stated in col. 17 of Table XC. The result is the sun's mean long., "*s*", at mean sunrise of given day. (iii) Add *s* to *a*. This = "*n*", the required index of the mean *nakshatra*, or the mean moon's place in the heavens at that moment. Table LXVIII above, p. 350 or Table VIII, *Indian Calendar*, will shew in which *nakshatra* the mean moon stood at the time.

[1] In measurement by 10,000ths of circle the total difference in 365 days is 0·00665, by which amount the *Brahma-Siddhānta* is the greater.

The 19-year intercalation cycle.

329. [See *Indian Calendar*, § 50, p. 29, and notes in previous articles above on the working of the cycle by different systems.] The sequence in the present case works perfectly regularly except in four instances. In every case except these, after four successive intercalations of the same lunar month at intervals of 19 years each, the intercalated month gives way to the month next preceding it. The exceptions are—a run of five mean intercalary Bhādrapadas between A.D. 746 and 822, five Āśvinas between 952 and 1009, five Kārttikas between 1120 and 1196, and five Paushas between 1231 and 1307.

Working Tables.

330. For general guidance the following Tables, as given for work by the *Ārya-Siddhānta* (*above*), should be used, or the similar Tables published in the *Indian Calendar*.

Table LXII, or *Ind. Cal.*, Table II, Parts I and II, for names of months and *nakshatras*.

Table LXIIIA, or *Ind. Cal.*, Table III, Part I, for collective duration of mean lunar months.

Table LXVIII, or *Ind. Cal.*, Table VIII, for indices of *tithis, karaṇas, nakshatras* and *yōgas*.

Table LXIX, or *Ind. Cal.*, Table IX, for the serial number of days of the year and their names and numbers in European reckoning.

Table LXX, or *Ind. Cal.*, Table X, for conversion of the indices of *tithis, nakshatras* and *yōgas* into time.

Table LXXI, the European Calendar for 23 centuries. [Table XIII, *Indian Calendar*, may also be used, but the former is easier.]

Table XCI below gives the collective duration of mean solar months, measured from the moment of mean Mēsha-*saṁkrānti*, the astronomical beginning of the mean solar year; also the increase of "*a*", the mean *tithi*-index, during the interval.

Table XCII shews the value of "*a*" at the beginning of each mean solar century of the Kaliyuga, that is to say, its value at mean sunrise of the day on which each such solar century began.

For odd years of such centuries Table LXXXVII (*above*, p. 509) is to be used in conjunction with Table XCII, addition of the two given values of "*a*" yielding the value of "*a*" at mean sunrise of the day on which each mean year of the Kaliyuga solar century began.

For increase of "*a*" in subsequent days, hours, etc., in any K.Y. year, or any moment of any day Tables LIVA and B (*above*) are to be used.

The use of Table XCIII is explained in § 326 above.

Table XCIV-A to F enables the units and decimals of units of results obtained from our system of reckoning in measurement by 10,000ths of a circle, to be converted readily into time, if required. The same can be converted into space-measurement in degrees, etc., by Table XLV-B above.

EXAMPLES.

[*N.B.*—Work may always be done in whole numbers, resorting to decimals only in close cases.]

Example 1. To find the mean tithi-index, or phase of moon, at mean sunrise of the day on which mean Mēsha-saṁkrānti occurred in any year.

This is a necessary operation for finding the *tithi*-index "*a*" at the *moment* of mean **Mēsha-*saṁkrānti***, which is obtained by addition of the "*a*" of subsequent hours, minutes, etc., to the *a* of mean sunrise. [The intercalation of lunar months is decided by the value of "*a*" at the moment of mean Mēsha-*saṁkrānti*.] Two cases are considered, A and B.

A. Take the year Kaliyuga 3851 expired. This was Śaka expired 672. It began (*Table XC, cols. 13-17*) astronomically at $5^h\ 49^m\ 39^s$ after mean sunrise on Sunday, 22 March A.D. 750. We want to know the mean moon's phase, as shewn by the *tithi*-index "*a*", at mean sunrise of that day. ["*w.-d.*" = week-day.]

	w.-d.	*a.*
(*Table XCII.*) At beginning of K.Y. Century 38, mean sunrise	(0)	5100·3761
(*Table LXXXVII.*) At beginning of K.Y. year 51, mean sunrise	(1)	8036·6243
At mean sunrise on the Sunday in question "*a*" = . .	(1)	3137·0004

The moon was then (*Table VIII or LXVIII*, p. 350 above, *col. 3*) about 10 days old.

B. The year K.Y. 3849, Śaka 670 both expired. This began (*Table XC*) at $17^h\ 25^m\ 21^s$ after mean sunrise on Thursday, 21 March A.D. 748. The first result shews the "*a*" for mean sunrise on Friday, 22 March, and the "*a*" for one day has to be deducted. This is due to the fact that Table LXXXVII has to serve for all K.Y. centuries, common or defective. The correction required is never more than that for one day.

(*Table XCII.*) At beginning of K.Y. Century 38, mean sunrise	(0)	5100·3761
(*Table LXXXVII.*) Add for odd K.Y. year 49, mean sunrise	(6)	835·2749
At mean sunrise on Friday, 22 Mar.	(6)	5935·6510
Deduct one day's value of "*a*" (*Table LIV-A*) . . .	–(1)	–338·6320
At mean sunrise on Thursday, 21 Mar.	(5)	5597·0190

Example 2. To find the civil day corresponding to Chaitra śukla 1, or the first civil day of the luni-solar year; and the value of "a" *(place of mean moon) at mean sunrise thereon.*

The civil day corresponding to mean Chaitra *śukla* 1 is that on which the mean *tithi* "*śukla* 1" expired. The *tithi*-index (a=) 333·$\dot{3}$ marks the last instant of the first *śukla tithi*, so that we have to find a day on which at mean sunrise the *tithi*-index "*a*" was between 0 and 333·$\dot{3}$. The *amānta* lunar month called "Chaitra" begins with the first new moon after the Mīna-*saṁkrānti*, and the civil day called "Chaitra *śukla* 1" is necessarily earlier than the day on which mean Mēsha-*saṁkrānti* occurred. We have to find the number of days' interval between these two days. There are two ways of ascertaining these points, one by using Table XCIII (p. 591 below) and adding its figures, one by using Table LIVA (p. 148 above) and subtracting its figures.

(i) Take the year in Example 1, A, above. The value of "*a*" at mean sunrise of Sunday, 22 March A.D. 750, was found to be 3137·0004. We turn to Table XCIII and select in col. 3 such a value of "*a*" as, added to 3137·0004, will result in a total value of "*a*" between 0 and 333·$\dot{3}$. This is found to be 6952·3121, the sum of the two (always disregarding quantities over 10,000) being 89·3125. The interval of whole days from mean Mēsha-*saṁkrānti* day was 9 (*col. 1*). Adding the number of the week-day (*col. 2*), viz. 5, to the week-day of mean Mēsha-*saṁkrānti*, viz. 1 Sunday, we have the week-day 6 Friday. Mean Mēsha-*saṁkrānti* occurred on Sunday, 22 March; and, therefore, it has been determined that the day Chaitra *śukla* 1, the first day of the luni-solar year, was Friday, 13 March A.D. 750, on which day, "*a*" being 89·3125, Chaitra *śukla* 1 was the current *tithi* at mean sunrise.

Similarly in Example 1, B. At mean sunrise of (5) Thursday, 21 March A.D. 748, "*a*" was 5597·0190. Add (*Table XCIII col. 3*) 4581·8882. Result 178·9072. The interval of days was (*col. 1*) 16. The week-day number was 5. The week-day of 21 March was 5 (Thursday). Hence the week-day 16 days earlier was 5+5=3 Tuesday. So the beginning of the mean luni-solar year was on Tuesday, 5 March A.D. 748, on which date at mean sunrise the mean *tithi* *śukla* 1 was current, the value of "*a*" at that moment being 178·9072.

The entries in Table XC against these years correspond to these results.

(ii) The same results are obtained by using Table LIVA *above*, and deducting the figures for the interval of whole days between the two occurrences. We note that value of "*a*" in the first 30 days of that Table which is next lower than the value of "*a*" already found for the day of mean Mēsha-*saṁkrānti*, and deduct the former from the latter. The number of intervening days (*col. 1*) and the number of week-days (*col. 2*) stand against the selected entry. This week-day number is deducted, of course, from the week-day of mean Mēsha-*saṁkrānti*. Thus—

A. For K.Y. 3851, A.D. 750.

	w.-d.	*a.*
(*Example 1, A.*) For mean sunrise on Sunday, 22 March A.D. 750.	(1)	3137·0004
(*Table LIVA.*) Next lower value of "*a*," and week-day	—(2)	—3047·6879
At mean sunrise of the day Chaitra *śukla* 1 . . .	(6)	89·3125

The interval of days (*col. 1*) was nine. 6=Friday. Hence the day corresponding to Chaitra *śukla* 1 was Friday, 13 March, and at mean sunrise the mean *tithi* Chaitra *śukla* 1 was current, the value of "*a*" being 89·3125.

B. For K.Y. 3849, A.D. 748.

(*Example 1, B.*) At mean sunrise on Thursday, 21 March, A.D. 748.	(5)	5597·0190
(*Table LIVA.*) Next lower value of *a*, and week-day .	—(2)	—5418·1118
At mean sunrise of the day Chaitra *śukla* 1 . . .	(3)	178·9072

The interval of days was 16. 3=Tuesday. Hence the day corresponding to Chaitra *śukla* 1 was Tuesday, 5 March A.D. 748, and at mean sunrise the value of *a* was 178·9072.

These results are the same as those found by the former process. The examples enable any worker to prove the correctness of all my entries in cols. 19-23 of the general Table XC below.

Example 3. To find if a lunar month was or was not intercalated in the given year.

It will be enough, for this problem, to refer to Example 3 of my article (*above*) on the *Arya-Siddhānta—mean system*. The work here is precisely similar; but for the values of "*a*" for hours and minutes Table LIVB should be used, and Table XCI for the advance of "*a*" during the mean solar months, etc.

Example 4. To find the mean tithi-index "**a**", *shewing phase of moon, at mean sunrise of any day in the year; or at any moment of any day.*

Table XC (cols. 19-23) gives the civil day corresponding to mean Chaitra *śukla* 1 (the initial day of the mean luni-solar year), its serial number (in brackets) from January 1st of the equivalent A.D. year, and the mean *tithi*-index *a* at mean sunrise. Calculate by Table III, *Indian Calendar*, or by Table LXIIIA (*above*,) the interval of whole days from that day to the given day, and, if necessary, the excess of hours, minutes, etc., to the given moment on that day. Add the increment of "*a*" for the interval of whole days from Table LIV-A and for fractions of days from Table LIV-B to the "*a*", as above, of the initial day; as also the number of days' interval and the corresponding week-day

E.g. Required the *tithi*-index at mean sunrise of the day called "Āshāḍha *śukla* 4" in Saka 547 expired, or A.D. 625-26, and the corresponding A.D. day and week-day.

In this year there was no intercalated month. The interval from the day "Chaitra *śukla* 1" day to the day "Āshāḍha *śukla* 4" is approximately (Table LXIII-A above, p. 335) 93 days. We try this—

	d.	*w.-d.*	*a.*
Table XC. Chaitra *śukla* 1, mean sunrise . .	(74)	(6)	184·6506
Table LIVA for 93 days	+(93)	(2)	1492·7746
	(167)	(1)	1677·4252
This value of "*a*" (*Table LXVIII*) shews that the 6th *śukla tithi* was current at mean sunrise. ∴ Deduct for 2 days . . .	−(2)	−(2)	−677·2640
At mean sunrise on Āshāḍha *śukla* 4 . . .	(165)	(6)	1000·1612

Table LXVIII or VIII *Indian Calendar*, shews the currency of the 4th *śukla tithi*, at that mean sunrise, since its first point is when $a=1{,}000$. Day 165 was (*Table IX, Indian Calendar, or LXIX, above*) 14th June A.D. 625. 6=Friday. We learn, however, that the 4th mean *tithi* had begun only about $\frac{1}{4}$ of a minute before the moment of mean sunrise; so that if the basis of calculation had been the moment of true sunrise (a little earlier than mean sunrise) the corresponding day might have been Thursday, 13 June.

Example 5. To find the nakshatra, or place in the heavens of the mean moon, at mean sunrise of any day or of any later moment in the day.

Take the case in the last example. It is required to find the value of "*n*", the *nakshatra*-index, at mean sunrise of the day called, in the mean system, "Ashāḍha *śukla* 4" in the given year, A.D. 625.

The mean *tithi*-index, "*a*", at that mean sunrise was found to be 1000·1612. Since $s+a=n$ (§ *327 above*), we have to ascertain the value of "*s*", the sun's mean longitude at that moment.

The day, 14 June, was the 165th day after Jan. 1 in that year. Mean Mēsha-*saṁkrānti* had taken place on (Table XC, *cols. 13-17*) the 79th day at $22^h\ 30^m\ 54^s$ after mean sunrise. The day 14 June was (165−79) 86 days later. We proceed as follows:—

		s.
Table LXXX, *p. 444*. Interval of 86 days		2354·4957
Less (Table LXXXI) for 22^h	25·0964	
30^m	0·5704	
54^s	0·0171	
	25·6839	−25·6839
At mean sunrise on the day Āshāḍha *śukla* 4 sun's mean long., "*s*" =		2328·8118
Add "*a*" as already found for that moment		1000·1612
At mean sunrise on that day "*n*"=		3328·9730

This last is the required *nakshatra*-index. Reference to Table VIII, *Indian Calendar*, or Table LXVIII above shews that the moon was then in the *nakshatra* Aślēsha by the equal space system of division of the ecliptic, which ended when "n" $=3333\cdot\dot{3}$; but that by the system of Garga or the *Brahma-Siddhānta* (our present authority) she was in Maghā, of which the ending points are respectively 3518·5 and 3477·1. Converted into degrees (Table VIII-B, *Indian Calendar, or Table XLV-B, above*) the moon at that mean sunrise stood at about 119°51′.

For the value of "n" at any later hour of the given day the index-value for the time since mean sunrise must be added (*Table LXXXI*) to the "n" of mean sunrise. At about 3 hours 50 min. after mean sunrise, for instance, the mean moon entered Maghā by the equal-space system; for the beginning point of that *nakshatra* is $3333\cdot\dot{3}$. The increase of "n" in 3 hours 50 min. is 4·3728, and $3328\cdot9730+4\cdot3728=3333\cdot3458$.

Example 6. To find the yōga, "y", at the same moment as in Example 5.

The formula for finding the *yōga*-index is either $s+n=$"y", the *yōga*-index; or, in cases where it is not necessary to calculate n (the *nakshatra*), 2 "s"$+a=$"y". Here, at mean sunrise on 14 June A.D. 625, we have found "s"$=2328\cdot8118$ and "n"$=3328\cdot9730$. The *yōga*-index, "y", therefore, $=5657\cdot7848$; and reference to Table VIII, *Indian Calendar*, cols. 12-13, or Table LXVIII (*above*, cols. 6, 8, 9, 10), shews that the mean moon was at that moment in the *yōga* Siddhi. Again $2\ s=4657\cdot6236$, and this $+\ a$, which was found to be $1000\cdot1612=5657\cdot7848$, the same as before.

TABLE XC.

Remarks.

K.Y. 3736 expired, A.D. 635-36. A very close case in the matter of intercalation of lunar month. Mean new moon occurred about 2^m after the moment of the Karka-*saṁkrānti* (mean sun at long. 90°), and, therefore, at that moment the mean moon was waning, while she was waxing at the next, Siṁha-*saṁkrānti* (mean sun at 120°). Accordingly the intercalated month was Śrāvaṇa.

K.Y. 3923 expired, A.D. 822-23. According to the 19-year sequence of intercalations the same month is generally intercalated four times running, *i.e.* at intervals of 19 years each. Here, however, is an instance of a fifth intercalation of the same month [See § 329 of text above.]

K.Y. 4110 expired, A.D. 1009-10. A similar case. Āśvina intercalated for the fifth time.

K.Y. 4297 expired, A.D. 1196-97. Another. Kārttika intercalated for the fifth time.

K.Y. 4408 expired, A.D. 1307-08. Another. Pausha intercalated for the fifth time. This was a very close case. The moment of mean new moon was about 1 minute after the mean sun reached the Dhanus-*saṁkrānti* (mean sun at long. 240°), but she was actually waning at the moment of the *saṁkrānti* and was waxing at the next, Makara, *saṁkrānti*. Consequently the lunar month Pausha was intercalated.

TABLE

MEAN SYSTEM TABLE.

Numbers of columns conform

(*Cols. 1 to 4.*)—The years herein stated are the *current* years corresponding

(*Cols. 6 and 7.*)—*Saṁvatsara-names of mean solar years in italics shew cases*

CONCURRENT YEAR.								
Kali.	Saka.	Chaitrādi Vikrama.	Mēshādi solar year in Bengal.	Kollam.	A.D.	Jovian Saṁvatsara. Southern system.	Jovian Saṁvatsara. Northern system.	Mean intercalated (*adhika*) lunar month.
1	2	3	3a	4	5	6	7	8a
3701	522	657	6		599-600	50 Anala . .		...
3702	523	658	7		*600-01	51 Piṅgala . .		2 Vaiśākha .
3703	524	659	8		601-02	52 Kālayukta . .		...
3704	525	660	9		602-03	53 Siddhārthin . .		10 Pausha .
3705	526	661	10		603-04	54 Raudra . .		...
3706	527	662	11		*604-05	55 Durmati . .		...
3707	528	663	12		605-06	56 Dundubhi . .		7 Āśvina .
3708	529	664	13		606-07	57 Rudhirōdgārin .		...
3709	530	665	14		607-08	58 Raktāksha . .		...
3710	531	666	15		*608-09	59 Krōdhana . .		3 Jyēshṭha .
3711	532	667	16		609-10	60 Kshaya . .		...
3712	533	668	17		610-11	1 Prabhava . .		12 Phālguna .
3713	534	669	18		611-12	2 Vibhava . .		...
3714	535	670	19		*612-13	3 Śukla . .		...
3715	536	671	20		613-14	4 Pramōda . .		8 Kārttika .
3716	537	672	21		614-15	5 Prajāpati . .		...
3717	538	673	22		615-16	6 Aṅgiras . .		...
3718	539	674	23		*616-17	7 Srīmukha . .		5 Srāvaṇa .
3719	540	675	24		617-18	8 Bhāva . .		...
3720	541	676	25		618-19	9 Yuvan . .		...

XC.

BRAHMA-SIDDHĀNTA.

to Table I, "Indian Calendar."

to the A.D. years in col. 5; as in Table I, "Indian Calendar."

where differences exist from Sūrya-Siddhānta nomenclature in true solar years.

COMMENCEMENT OF THE									Kali.
MEAN SOLAR YEAR.					MEAN LUNI-SOLAR YEAR (MEAN SUNRISE OF THE CIVIL DAY ON WHICH CHAITRA ŚUKLA 1 ENDS).				
Day and month, A.D.	Week-day.	Time of mean Mēsha-saṁkrānti.			Day and month, A.D.	Week-day.	a (here = t, the index of the *tithi*).		
13	14	17			19	20	23		1
		H.	M.	S.					
21 Mar. (80) . .	0 Sat. .	5	15	0	3 Mar. (62) .	3 Tues. .	133·1013		3701
20 Mar. (80) . .	1 Sun. .	11	27	9	20 Feb. (51) .	0 Sat. .	8·8241		3702
20 Mar. (79) . .	2 Mon. .	17	39	18	10 Mar. (69) .	6 Fri. .	43·5065		3703
20 Mar. (79) . .	3 Tues. .	23	51	27	28 Feb. (59) .	4 Wed. .	257·8614		3704
21 Mar. (80) . .	5 Thur. .	6	3	36	19 Mar. (78) .	3 Tues. .	292·5437		3705
20 Mar. (80) . .	6 Fri. .	12	15	45	7 Mar. (67) .	0 Sat. .	168·2666		3706
20 Mar. (79) . .	0 Sat. .	18	27	54	24 Feb. (55) .	4 Wed. .	43·3394		3707
21 Mar. (80) . .	2 Mon. .	0	40	3	15 Mar. (74) .	3 Tues. .	78·6718		3708
21 Mar. (80) . .	3 Tues. .	6	52	12	5 Mar. (64) .	1 Sun. .	293·0266		3709
20 Mar. (80) . .	4 Wed. .	13	4	21	22 Feb. (53) .	5 Thur. .	168·7494		3710
20 Mar. (79) . .	5 Thur. .	19	16	30	12 Mar. (71) .	4 Wed. .	203·4218		3711
21 Mar. (80) . .	0 Sat. .	1	28	39	1 Mar. (60) .	1 Sun. .	79·1547		3712
21 Mar. (80) . .	1 Sun. .	7	40	48	20 Mar. (79) .	0 Sat. .	113·8371		3713
20 Mar. (80) . .	2 Mon. .	13	52	57	9 Mar. (69) .	5 Thur. .	328·1918		3714
20 Mar. (79) . .	3 Tues. .	20	5	6	26 Feb. (57) .	2 Mon. .	203·9147		3715
21 Mar. (80) . .	5 Thur. .	2	17	15	17 Mar. (76) .	1 Sun. .	238·5972		3716
21 Mar. (80) . .	6 Fri. .	8	29	24	6 Mar. (65) .	5 Thur. .	114·3199		3717
20 Mar. (80) . .	0 Sat. .	14	41	33	24 Feb. (55) .	3 Tues. .	328·6747		3718
20 Mar. (79) . .	1 Sun. .	20	53	42	13 Mar. (72) .	1 Sun. .	24·7252		3719
21 Mar. (80) . .	3 Tues. .	3	5	51	3 Mar. (62) .	6 Fri. .	239·0801		3720

TABLE

CONCURRENT YEAR.								Mean intercalated (adhika) lunar month.
Kali.	Śaka.	Chaitrâdi Vikrama.	Mêshâdi solar year in Bengal.	Kollam.	A.D.	Jovian Samvatsara. Southern system.	Jovian Samvatsara. Northern system.	
1	2	3	3a	4	5	6	7	8a
3721	542	677	26		619-20	10 Dhâtṛi . . .		1 Chaitra .
3722	543	678	27		*620-21	11 Îśvara . . .		...
3723	544	679	28		621-22	12 Bahudhânya . .		10 Pausha .
3724	545	680	29		622-23	13 Pramâthin . . .		...
3725	546	681	30		623-24	14 Vikrama . . .		...
3726	547	682	31		*624-25	15 Vṛisha . . .		6 Bhâdrapada .
3727	548	683	32		625-26	16 Chitrabhânu . .		...
3728	549	684	33		626-27	17 Subhânu . . .		...
3729	550	685	34		627-28	18 Târaṇa . . .		3 Jyêshṭha .
3730	551	686	35		*628-29	19 Pârthiva . . .		...
3731	552	687	36		629-30	20 Vyaya . . .		11 Mâgha .
3732	553	688	37		630-31	21 Sarvajit . . .		...
3733	554	689	38		631-32	22 Sarvadhârin . .		...
3734	555	690	39		*632-33	23 Virôdhin . . .		8 Kârttika .
3735	556	691	40		633-34	24 Vikṛita . . .		...
3736	557	692	41		634-35	25 Khara . . .		...
3737	558	693	42		635-36	26 Nandana . . .		5 Śrâvaṇa § .
3738	559	694	43		*636-37	27 Vijaya . . .		...
3739	560	695	44		637-38	28 Jaya . . .		...
3740	561	696	45		638-39	29 Manmatha . .		1 Chaitra .
3741	562	697	46		639-40	30 Durmukha . .		...
3742	563	698	47		*640-41	31 Hêmalamba . .		10 Pausha .
3743	564	699	48		641-42	32 Vilamba . . .		...
3744	565	700	49		642-43	33 Vikârin . . .		...
3745	566	701	50		643-44	34 Śârvarin . . .		6 Bhâdrapada .

§ See "Remarks," p. 523 above.

XC—*contd.*

COMMENCEMENT OF THE									Kali.
MEAN SOLAR YEAR.					MEAN LUNI-SOLAR YEAR (MEAN SUNRISE OF THE CIVIL DAY ON WHICH CHAITRA ŚUKLA 1 ENDS).				
Day and month, A.D.	Week-day.	Time of mean Mêsha-saṁkrânti.			Day and month, A.D.	Week-day.	*a* (here = *t*, the index of the *tithi*).		
13	14	17			19	20	23		1
		H.	M.	S.					
21 Mar. (80)	4 Wed.	9	18	0	20 Feb. (51)	3 Tues.	114·8028		3721
20 Mar. (80)	5 Thur.	15	30	9	10 Mar. (70)	2 Mon.	149·4852		3722
20 Mar. (79)	6 Fri.	21	42	18	27 Feb. (58)	6 Fri.	25·2081		3723
21 Mar. (80)	1 Sun.	3	54	27	18 Mar. (77)	5 Thur.	59·8904		3724
21 Mar. (80)	2 Mon.	10	6	36	8 Mar. (67)	3 Tues.	274·2453		3725
20 Mar. (80)	3 Tues.	16	18	45	25 Feb. (56)	0 Sat.	149·9682		3726
20 Mar. (79)	4 Wed.	22	30	54	15 Mar. (74)	6 Fri.	184·6506		3727
21 Mar. (80)	6 Fri.	4	43	3	4 Mar. (63)	3 Tues.	60·3734		3728
21 Mar. (80)	0 Sat.	10	55	12	22 Feb. (53)	1 Sun.	274·7282		3729
20 Mar. (80)	1 Sun.	17	7	21	12 Mar. (72)	0 Sat.	309·4106		3730
20 Mar. (79)	2 Mon.	23	19	30	1 Mar. (60)	4 Wed.	185·1334		3731
21 Mar. (80)	4 Wed.	5	31	39	20 Mar. (79)	3 Tues.	219·8158		3732
21 Mar. (80)	5 Thur.	11	43	48	9 Mar. (68)	0 Sat.	95·5387		3733
20 Mar. (80)	6 Fri.	17	55	57	27 Feb. (58)	5 Thur.	309·8935		3734
21 Mar. (80)	1 Sun.	0	8	6	16 Mar. (75)	3 Tues.	5·9439		3735
21 Mar. (80)	2 Mon.	6	20	15	6 Mar. (65)	1 Sun.	220·2987		3736
21 Mar. (80)	3 Tues.	12	32	24	23 Feb. (54)	5 Thur.	96·0216		3737
20 Mar. (80)	4 Wed.	18	44	33	13 Mar. (73)	4 Wed.	130·7040		3738
21 Mar. (80)	6 Fri.	0	56	42	2 Mar. (61)	1 Sun.	6·4268		3739
21 Mar. (80)	0 Sat.	7	8	51	20 Feb. (51)	6 Fri.	220·7816		3740
21 Mar. (80)	1 Sun.	13	21	0	11 Mar. (70)	5 Thur.	255·4640		3741
20 Mar. (80)	2 Mon.	19	33	9	28 Feb. (59)	2 Mon.	131·1868		3742
21 Mar. (80)	4 Wed.	1	45	18	18 Mar. (77)	1 Sun.	165·8692		3743
21 Mar. (80)	5 Thur.	7	57	27	7 Mar. (66)	5 Thur.	41·5921		3744
21 Mar. (80)	6 Fri.	14	9	36	25 Feb. (56)	3 Tues.	255·9470		3745

TABLE

CONCURRENT YEAR.								
Kali.	Śaka.	Chaitrādi Vikrama.	Mēshādi solar year in Bengal.	Kollam.	A.D.	Jovian Saṁvatsara. Southern system.	Jovian Saṁvatsara. Northern system.	Mean intercalated (*adhika*) lunar month.
1	2	3	3a	4	5	6	7	8a
3746	567	702	51		*644-45	35 Plava . . .		...
3747	568	703	52		645-46	36 Śubhakṛit . . .		...
3748	569	704	53		646-47	37 Śōbhana . . .		3 Jyēshṭha .
3749	570	705	54		647-48	38 Krōdhin . . .		...
3750	571	706	55		*648-49	39 †Viśvāvasu . .		11 Māgha .
3751	572	707	56		649-50	41 *Plavaṅga* . . .		...
3752	573	708	57		650-51	42 *Kīlaka* . . .		...
3753	574	709	58		651-52	43 *Saumya* . . .		8 Kārttika .
3754	575	710	59		*652-53	44 *Sādhāraṇa* . .		...
3755	576	711	60		653-54	45 *Virōdhakṛit* . .		...
3756	577	712	61		654-55	46 Paridhāvin . .		4 Āshāḍha .
3757	578	713	62		655-56	47 Pramādin . . .		...
3758	579	714	63		*656-57	48 Ānanda . . .		...
3759	580	715	64		657-58	49 Rākshasa . . .		1 Chaitra .
3760	581	716	65		658-59	50 Anala . . .		...
3761	582	717	66		659-60	51 Piṅgala . . .		9 Mārgaśira .
3762	583	718	67		*660-61	52 Kālayukta . . .		...
3763	584	719	68		661-62	53 Siddhārthin . .		...
3764	585	720	69		662-63	54 Raudra . . .		6 Bhādrapada .
3765	586	721	70		663-64	55 Durmati . . .		...
3766	587	722	71		*664-65	56 Dundubhi . . .		...
3767	588	723	72		665-66	57 Rudhirōdgārin . .		2 Vaiśākha .
3768	589	724	73		666-67	58 Raktaksha . . .		...
3769	590	725	74		667-68	59 Krōdhana . . .		11 Māgha .
3770	591	726	75		*668-69	60 Kshaya . . .		...

† 40 Parābhava was suppressed, both in mean and true reckoning.

XC—*contd.*

COMMENCEMENT OF THE							Kali.
MEAN SOLAR YEAR.			MEAN LUNI-SOLAR YEAR (MEAN SUNRISE OF THE CIVIL DAY ON WHICH CHAITRA ŚUKLA 1 ENDS).				
Day and month, A.D.	Week-day.	Time of mean Mēsha-saṁkrānti.	Day and month, A.D.	Week-day.	*a* (here = *t*, the index of the *tithi*).		
13	14	17	19	20	23		1
		H. M. S.					
20 Mar. (80)	0 Sat.	20 21 45	15 Mar. (75)	2 Mon.	290·6293		3746
21 Mar. (80)	2 Mon.	2 33 54	4 Mar. (63)	6 Fri.	166·3522		3747
21 Mar. (80)	3 Tues.	8 46 3	21 Feb. (52)	3 Tues.	42·0750		3748
21 Mar. (80)	4 Wed.	14 58 12	12 Mar. (71)	2 Mon.	76·7573		3749
20 Mar. (80)	5 Thur.	21 10 21	1 Mar. (61)	0 Sat.	291·1122		3750
21 Mar. (80)	0 Sat.	3 22 30	20 Mar. (79)	6 Fri.	325·7946		3751
21 Mar. (80)	1 Sun.	9 34 39	9 Mar. (68)	3 Tues.	201·5175		3752
21 Mar. (80)	2 Mon.	15 46 48	26 Feb. (57)	0 Sat.	77·2402		3753
20 Mar. (80)	3 Tues.	21 58 57	16 Mar. (76)	0 Sat.	111·9227		3754
21 Mar. (80)	5 Thur.	4 11 6	6 Mar. (65)	4 Wed.	326·2775		3755
21 Mar. (80)	6 Fri.	10 23 15	23 Feb. (54)	1 Sun.	202·0003		3756
21 Mar. (80)	0 Sat.	16 35 24	14 Mar. (73)	0 Sat.	236·6827		3757
20 Mar. (80)	1 Sun.	22 47 33	2 Mar. (62)	4 Wed.	112·4056		3758
21 Mar. (80)	3 Tues.	4 59 42	20 Feb. (51)	2 Mon.	326·7604		3759
21 Mar. (80)	4 Wed.	11 11 51	10 Mar. (69)	0 Sat.	22·8108		3760
21 Mar. (80)	5 Thur.	17 24 0	28 Feb. (59)	5 Thur.	237·1656		3761
20 Mar. (80)	6 Fri.	23 36 9	18 Mar. (78)	4 Wed.	271·8480		3762
21 Mar. (80)	1 Sun.	5 48 18	7 Mar. (66)	1 Sun.	147·5708		3763
21 Mar. (80)	2 Mon.	12 0 27	24 Feb. (55)	5 Thur.	23·2937		3764
21 Mar. (80)	3 Tues.	18 12 36	15 Mar. (74)	4 Wed.	57·9761		3765
21 Mar. (81)	5 Thur.	0 24 45	4 Mar. (64)	2 Mon.	272·3310		3766
21 Mar. (80)	6 Fri.	6 36 54	21 Feb. (52)	6 Fri.	148·0537		3767
21 Mar. (80)	0 Sat.	12 49 3	12 Mar. (71)	5 Thur.	182·7361		3768
21 Mar. (80)	1 Sun.	19 1 12	1 Mar. (60)	2 Mon.	58·4590		3769
21 Mar. (81)	3 Tues.	1 13 21	19 Mar. (79)	1 Sun.	93·1413		3770

TABLE

CONCURRENT YEAR.								Mean intercalated (*adhika*) lunar month.
Kali.	Śaka.	Chaitrādi Vikrama.	Mēshādi solar year in Bengal.	Kollam.	A.D.	Jovian Saṁvatsara. Southern system.	Jovian Saṁvatsara. Northern system.	
1	2	3	3a	4	5	6	7	8a
3771	592	727	76		669-70	1 Prabhava . . .		...
3772	593	728	77		670-71	2 Vibhava . . .		7 Āśvina .
3773	594	729	78		671-72	3 Sukla . . .		...
3774	595	730	79		*672-73	4 Pramōda . . .		...
3775	596	731	80		673-74	5 Prajāpati . . .		4 Āshāḍha .
3776	597	732	81		674-75	6 Aṅgiras . . .		...
3777	598	733	82		675-76	7 Srīmukha . . .		...
3778	599	734	83		*676-77	8 Bhāva . . .		1 Chaitra
3779	600	735	84		677-78	9 Yuvan . . .		...
3780	601	736	85		678-79	10 Dhātṛi . .		9 Mārgaśira
3781	602	737	86		679-80	11 Īśvara . . .		...
3782	603	738	87		*680-81	12 Bahudhānya . .		...
3783	604	739	88		681-82	13 Pramāthin . . .		6 Bhādrapada .
3784	605	740	89		682-83	14 Vikrama . . .		...
3785	606	741	90		683-84	15 Vṛisha . . .		...
3786	607	742	91		*684-85	16 Chitrabhānu . .		2 Vaiśākha .
3787	608	743	92		685-86	17 Subhānu . . .		...
3788	609	744	93		686-87	18 Tāraṇa . . .		11 Māgha .
3789	610	745	94		687-88	19 Pārthiva . . .		...
3790	611	746	95		*688-89	20 Vyaya . . .		...
3791	612	747	96		689-90	21 Sarvajit . . .		7 Āśvina
3792	613	748	97		690-91	22 Sarvadhārin .		...
3793	614	749	98		691-92	23 Virōdhin . . .		...
3794	615	750	99		*692-93	24 Vikṛita . . .		4 Āshāḍha .
3795	616	751	100		693-94	25 Khara . .		...

XC—*contd.*

COMMENCEMENT OF THE									
MEAN SOLAR YEAR.					MEAN LUNI-SOLAR YEAR (MEAN SUNRISE OF THE CIVIL DAY ON WHICH CHAITRA ŚUKLA 1 ENDS).			Kali.
Day and month, A.D.	Week-day.	Time of mean Mēsha-saṁkrānti.			Day and month, A.D.	Week-day.	*a* (here = *t*, the index of the *tithi*).	
13	14	17			19	20	23	1
		H.	M.	S.				
21 Mar. (80) . .	4 Wed. .	7	25	30	9 Mar. (68) .	6 Fri. .	307·4962	3771
21 Mar. (80) . .	5 Thur. .	13	37	39	26 Feb. (57) .	3 Tues. .	183·2190	3772
21 Mar. (80) . .	6 Fri. .	19	49	48	17 Mar. (76) .	2 Mon. .	217·9015	3773
21 Mar. (81) . .	1 Sun. .	2	1	57	5 Mar. (65) .	6 Fri. .	93·6242	3774
21 Mar. (80) . .	2 Mon. .	8	14	6	23 Feb. (54) .	4 Wed. .	307·9791	3775
21 Mar. (80) . .	3 Tues. .	14	26	15	13 Mar. (72) .	2 Mon. .	4·0295	3776
21 Mar. (80) . .	4 Wed. .	20	38	24	3 Mar. (62) .	0 Sat. .	218·3843	3777
21 Mar. (81) . .	6 Fri. .	2	50	33	20 Feb. (51) .	4 Wed. .	94·1071	3778
21 Mar. (80) . .	0 Sat. .	9	2	42	10 Mar. (69) .	3 Tues. .	128·7896	3779
21 Mar. (80) . .	1 Sun. .	15	14	51	27 Feb. (58) .	0 Sat. .	4·5124	3780
21 Mar. (80) . .	2 Mon. .	21	27	0	18 Mar. (77) .	6 Fri. .	39·1947	3781
21 Mar. (81) . .	4 Wed. .	3	39	9	7 Mar. (67) .	4 Wed. .	253·5496	3782
21 Mar. (80) . .	5 Thur. .	9	51	18	24 Feb. (55) .	1 Sun. .	129·2725	3783
21 Mar. (80) . .	6 Fri. .	16	3	27	15 Mar. (74) .	0 Sat. .	163·9549	3784
21 Mar. (80) . .	0 Sat. .	22	15	36	4 Mar. (63) .	4 Wed. .	39·6776	3785
21 Mar. (81) . .	2 Mon. .	4	27	45	22 Feb. (53) .	2 Mon. .	254·0325	3786
21 Mar. (80) . .	3 Tues. .	10	39	54	12 Mar. (71) .	1 Sun. .	288·7149	3787
21 Mar. (80) . .	4 Wed. .	16	52	3	1 Mar. (60) .	5 Thur. .	164·4377	3788
21 Mar. (80) . .	5 Thur. .	23	4	12	20 Mar. (79) .	4 Wed. .	199·1200	3789
21 Mar. (81) . .	0 Sat. .	5	16	21	8 Mar. (68) .	1 Sun. .	74·8430	3790
21 Mar. (80) . .	1 Sun. .	11	28	30	26 Feb. (57) .	6 Fri. .	289·1978	3791
21 Mar. (80) . .	2 Mon. .	17	40	39	17 Mar. (76) .	5 Thur. .	323·8802	3792
21 Mar. (80) . .	3 Tues. .	23	52	48	6 Mar. (65) .	2 Mon. .	199·6030	3793
21 Mar. (81) . .	5 Thur. .	6	4	57	23 Feb. (54) .	6 Fri. .	75·3259	3794
21 Mar. (80) . .	6 Fri. .	12	17	6	13 Mar. (72) .	5 Thur. .	110·0082	3795

TABLE

CONCURRENT YEAR.								Mean intercalated (*adhika*) lunar month.
Kali.	Saka.	Chaitrâdi Vikrama.	Mêshâdi solar year in Bengal.	Kollam.	A.D.	Jovian Samvatsara. Southern system.	Jovian Samvatsara. Northern system.	
1	2	3	3a	4	5	6	7	8a
3796	617	752	101		694-95	26 Nandana . . .		12 Phâlguna .
3797	618	753	102		695-96	27 Vijaya . . .		...
3798	619	754	103		*696-97	28 Jaya		...
3799	620	755	104		697-98	29 Manmatha . . .		9 Mârgaśira .
3800	621	756	105		698-99	30 Durmukha . . .		...
3801	622	757	106		699-700	31 Hêmalamba . .		...
3802	623	758	107		*700-01	32 Vilamba . . .		5 Śrâvaṇa .
3803	624	759	108		701-02	33 Vikârin . . .		...
3804	625	760	109		702-03	34 Śârvarin . . .		...
3805	626	761	110		703-04	35 Plava		2 Vaiśâkha .
3806	627	762	111		*704-05	36 Śubhakṛit . . .		...
3807	628	763	112		705-06	37 Śôbhana . . .		10 Pausha .
3808	629	764	113		706-07	38 Krôdhin . . .		...
3809	630	765	114		707-08	39 Viśvâvasu . . .		...
3810	631	766	115		*708-09	40 Parâbhava . . .		7 Aśvina .
3811	632	767	116		709-10	41 Plavaṅga . . .		...
3812	633	768	117		710-11	42 Kîlaka . . .		...
3813	634	769	118		711-12	43 Saumya . . .		4 Ashâḍha .
3814	635	770	119		*712-13	44 Sâdhâraṇa . . .		...
3815	636	771	120		713-14	45 Virôdhakṛit . .		12 Phâlguna .
3816	637	772	121		714-15	46 Paridhâvin . . .		...
3817	638	773	122		715-16	47 Pramâdin . . .		...
3818	639	774	123		*716-17	48 Ânanda . . .		9 Mârgaśira .
3819	640	775	124		717-18	49 Râkshasa . . .		...
3820	641	776	125		718-19	50 Anala . . .		...

XC—*contd.*

COMMENCEMENT OF THE						
MEAN SOLAR YEAR.			MEAN LUNI-SOLAR YEAR (MEAN SUNRISE OF THE CIVIL DAY ON WHICH CHAITRA ŚUKLA 1 ENDS).			Kali.
Day and month, A.D.	Week-day.	Time of mean Mēsha-saṁkrānti.	Day and month, A.D.	Week-day.	a (here = t, the index of the *tithi*).	
13	14	17	19	20	23	1
		H. M. S.				
21 Mar. (80) . .	0 Sat. .	18 29 15	3 Mar. (62) .	3 Tues. .	324·3631	3796
22 Mar. (81) . .	2 Mon. .	0 41 24	21 Mar. (80) .	1 Sun. .	20·4135	3797
21 Mar. (81) . .	3 Tues. .	6 53 33	10 Mar. (70) .	6 Fri. .	234·7683	3798
21 Mar. (80) . .	4 Wed. .	13 5 42	27 Feb. (58) .	3 Tues. .	110·4911	3799
21 Mar. (80) . .	5 Thur. .	19 17 51	18 Mar. (77) .	2 Mon. .	145·1735	3800
22 Mar. (81) . .	0 Sat. .	1 30 0	7 Mar. (66) .	6 Fri. .	20·8963	3801
21 Mar. (81) . .	1 Sun. .	7 42 9	25 Feb. (56) .	4 Wed. .	235·2512	3802
21 Mar. (80) . .	2 Mon. .	13 54 18	15 Mar. (74) .	3 Tues. .	269·9336	3803
21 Mar. (80) . .	3 Tues. .	20 6 27	4 Mar. (63) .	0 Sat. .	145·6564	3804
22 Mar. (81) . .	5 Thur. .	2 18 36	21 Feb. (52) .	4 Wed. .	21·3792	3805
21 Mar. (80 . .	6 Fri. .	8 30 45	11 Mar. (71) .	3 Tues. .	56·0616	3806
21 Mar. (80) . .	0 Sat. .	14 42 54	1 Mar. (60) .	1 Sun. .	270·4164	3807
21 Mar. (80) . .	1 Sun. .	20 55 3	20 Mar. (79) .	0 Sat. .	305·0988	3808
22 Mar. (81) . .	3 Tues. .	3 7 12	9 Mar. (68) .	4 Wed. .	180·8217	3809
21 Mar. (81) . .	4 Wed. .	9 19 21	26 Feb. (57) .	1 Sun. .	56·5444	3810
21 Mar. (80) . .	5 Thur. .	15 31 30	16 Mar. (75) .	0 Sat. .	91·2269	3811
21 Mar. (80) . .	6 Fri. .	21 43 39	6 Mar. (65) .	5 Thur. .	305·5817	3812
22 Mar. (81) . .	1 Sun. .	3 55 48	23 Feb. (54) .	2 Mon. .	181·3046	3813
21 Mar. (81) . .	2 Mon. .	10 7 57	13 Mar. (73) .	1 Sun. .	215·9869	3814
21 Mar. (80) . .	3 Tues. .	16 20 6	2 Mar. (61) .	5 Thur. .	91·7098	3815
21 Mar. (80) . .	4 Wed. .	22 32 15	21 Mar. (80) .	4 Wed. .	126·3922	3816
22 Mar. (81) . .	6 Fri. .	4 44 24	10 Mar. (69) .	1 Sun. .	2·1150	3817
21 Mar. (81) . .	0 Sat. .	10 56 33	28 Feb. (59) .	6 Fri. .	216·4698	3818
21 Mar. (80) . .	1 Sun. .	17 8 42	18 Mar. (77) .	5 Thur. .	251·1632	3819
21 Mar. (80) . .	2 Mon. .	23 20 51	7 Mar. (66) .	2 Mon. .	126·8751	3820

TABLE

CONCURRENT YEAR.								
Kali.	Śaka.	Chaitrādi Vikrama.	Mēshādi solar year in Bengal.	Kollam.	A.D.	Jovian Saṁvatsara. Southern system.	Jovian Saṁvatsara. Northern system.	Mean intercalated (*adhika*) lunar month.
1	2	3	3a	4	5	6	7	8a
3821	642	777	126		719-20	51 Piṅgala . . .		5 Śrāvaṇa .
3822	643	778	127		*720-21	52 Kālayukta . . .		...
3823	644	779	128		721-22	53 Siddhārthin . .		...
3824	645	780	129		722-23	54 Raudra . . .		2 Vaiśākha .
3825	646	781	130		723-24	55 Durmati . . .		...
3826	647	782	131		*724-25	56 Dundubhi . . .		10 Pausha .
3827	648	783	132		725-26	57 Rudhirōdgārin . .		...
3828	649	784	133		726-27	58 Raktāksha . . .		...
3829	650	785	134		727-28	59 Krōdhana . . .		7 Āśvina .
3830	651	786	135		*728-29	60 Kshaya . . .		...
3831	652	787	136		729-30	1 Prabhava . . .		...
3832	653	788	137		730-31	2 Vibhava . . .		3 Jyēshtha .
3833	654	789	138		731-32	3 Sukla . . .		...
3834	655	790	139		*732-33	4 Pramōda . . .		12 Phālguna .
3835	656	791	140		733-34	5 Prajāpati† . . .		...
3836	657	792	141		734-35	7 *Śrīmukha* . . .		...
3837	658	793	142		735-36	8 *Bhāva* . . .		8 Kārttika .
3838	659	794	143		*736-37	9 *Yuvan* . . .		...
3839	660	795	144		737-38	10 *Dhātri* . . .		...
3840	661	796	145		738-39	11 *Īśvara* . . .		5 Srāvana .
3841	662	797	146		739-40	12 Bahudhānya . .		...
3842	663	798	147		*740-41	13 Pramāthin . . .		...
3843	664	799	148		741-42	14 Vikrama . . .		1 Chaitra .
3844	665	800	149		742-43	15 Vrisha . . .		...
3845	666	801	150		743-44	16 Chitrabhānu . .		10 Pausha .

† No. 6 Aṅgiras was suppressed according to the mean system. By the *Brahma-Siddhanta* "true" system K.Y. 3836, A.D. 734-735, was called Aṅgiras, 7 Srimukha being suppressed. K.Y. 3837, A.D. 735-36, was 8 Bhāva by both systems.

XC—*contd.*

COMMENCEMENT OF THE							
MEAN SOLAR YEAR.			MEAN LUNI-SOLAR YEAR (MEAN SUNRISE OF THE CIVIL DAY ON WHICH CHAITRA ŚUKLA 1 ENDS).				Kali.
Day and month, A.D.	Week-day.	Time of mean Mēsha-saṁkrānti.		Day and month, A.D.	Week-day.	*a* (here = *t*, the index of the *tithi*).	
13	14	17		19	20	23	1
		H. M. S.					
22 Mar. (81) . .	4 Wed. .	5 33 0		24 Feb. (55) .	6 Fri. .	2·5979	3821
21 Mar. (81) . .	5 Thur. .	11 45 9		14 Mar. (74) .	5 Thur. .	37·2803	3822
21 Mar. (80) . .	6 Fri. .	17 57 18		4 Mar. (63) .	3 Tues. .	251·6352	3823
22 Mar. (81) . .	1 Sun. .	0 9 27		21 Feb. (52) .	0 Sat. .	127·3579	3824
22 Mar. (81) . .	2 Mon. .	6 21 36		12 Mar. (71) .	6 Fri. .	162·0403	3825
21 Mar. (81) . .	3 Tues. .	12 33 45		29 Feb. (60) .	3 Tues. .	37·7632	3826
21 Mar. (80) . .	4 Wed. .	18 45 54		19 Mar. (78) .	2 Mon. .	72·4457	3827
22 Mar. (81) . .	6 Fri. .	0 58 3		9 Mar. (68) .	0 Sat. .	286·8004	3828
22 Mar. (81) . .	0 Sat. .	7 10 12		26 Feb. (57) .	4 Wed. .	162·5233	3829
21 Mar. (81) . .	1 Sun. .	13 22 21		16 Mar. (76) .	3 Tues. .	197·2057	3830
21 Mar. (80) . .	2 Mon. .	19 34 30		5 Mar. (64) .	0 Sat. .	72·9284	3831
22 Mar. (81, . .	4 Wed. .	1 46 39		23 Feb. (54) .	5 Thur. .	287·2833	3832
22 Mar. (81) . .	5 Thur. .	7 58 48		14 Mar. (73) .	4 Wed. .	321·9657	3833
21 Mar. (81) . .	6 Fri. .	14 10 57		2 Mar. (62) .	1 Sun. .	197·6886	3834
21 Mar. (80) . .	0 Sat. .	20 23 6		21 Mar. (80) .	0 Sat. .	232·3709	3835
22 Mar. (81) . .	2 Mon. .	2 35 15		10 Mar. (69) .	4 Wed. .	108·0938	3836
22 Mar. (81) . .	3 Tues. .	8 47 24		28 Feb. (59) .	2 Mon. .	322·4486	3837
21 Mar. (81) . .	4 Wed. .	14 59 33		17 Mar. (77) .	0 Sat. .	18·4990	3838
21 Mar. (80) . .	5 Thur. .	21 11 42		7 Mar. (66) .	5 Thur. .	232·8538	3839
22 Mar. (81) . .	0 Sat. .	3 23 51		24 Feb. (55) .	2 Mon. .	108·5767	3840
22 Mar. (81) . .	1 Sun. .	9 36 0		15 Mar. (74) .	1 Sun. .	143·2591	3841
21 Mar. (81) . .	2 Mon. .	15 48 9		3 Mar. (63) .	5 Thur. .	18·9819	3842
21 Mar. (80) . .	3 Tues. .	22 0 18		21 Feb. (52) .	3 Tues. .	233·3367	3843
22 Mar. (81) . .	5 Thur. .	4 12 27		12 Mar. (71) .	2 Mon. .	268·0191	3844
22 Mar. (81) . .	6 Fri. .	10 24 36		1 Mar. (60) .	6 Fri. .	143·7420	3845

TABLE

CONCURRENT YEAR.						Jovian Saṁvatsara.		Most intercalated (*adhika*) lunar month.
Kali.	Śaka.	Chaitrādi Vikrama.	Mēshādi solar year in Bengal.	Kollam.	A.D.	Southern system.	Northern system.	
1	2	3	3*a*	4	5	6	7	8*a*
3846	667	802	151		*744-45	17 Subhānu . . .		...
3847	668	803	152		745-46	18 Tāraṇa . . .		...
3848	669	804	153		746-47	19 Pārthiva . . .		6 Bhādrapada .
3849	670	805	154		747-48	20 Vyaya . . .		...
3850	671	806	155		*748-49	21 Sarvajit . . .		...
3851	672	807	156		749-50	22 Sarvadhārin . .		3 Jyēshṭha .
3852	673	808	157		750-51	23 Virōdhin . . .		...
3853	674	809	158		751-52	24 Vikṛita . . .		12 Phālguna .
3854	675	810	159		*752-53	25 Khara . . .		...
3855	676	811	160		753-54	26 Nandana . . .		...
3856	677	812	161		754-55	27 Vijaya . . .		8 Kārttika .
3857	678	813	162		755-56	28 Jaya . . .		...
3858	679	814	163		*756-57	29 Manmatha . . .		...
3859	680	815	164		757-58	30 Durmukha . . .		5 Srāvaṇa .
3860	681	816	165		758-59	31 Hēmalamba . .		...
3861	682	817	166		759-60	32 Vilamba . . .		...
3862	683	818	167		*760-61	33 Vikārin . . .		1 Chaitra .
3863	684	819	168		761-62	34 Sārvarin . . .		...
3864	685	820	169		762-63	35 Plava . . .		10 Pausha .
3865	686	821	170		763-64	36 Subhakṛit . . .		...
3866	687	822	171		*764-65	37 Sōbhana . . .		...
3867	688	823	172		765-66	38 Krōdhin . . .		6 Bhādrapada .
3868	689	824	173		766-67	39 Viśvāvasu . . .		...
3869	690	825	174		767-68	40 Parābhava . . .		...
3870	691	826	175		*768-69	41 Plavaṅga . . .		3 Jyēshṭha

XC—*Contd.*

COMMENCEMENT OF THE									Kali.
MEAN SOLAR YEAR.					MEAN LUNI-SOLAR YEAR (MEAN SUNRISE OF THE CIVIL DAY ON WHICH CHAITRA ŚUKLA 1 ENDS).				
Day and month, A.D	Week-day.	Time of mean Mēsha-saṁkrānti.			Day and month, A.D.	Week-day.	*a* (here = *t*, the index of the *tithi*).		
13	14	17			19	20	23		1
		H.	M.	S.					
21 Mar. (81)	0 Sat.	16	36	45	19 Mar. (79)	5 Thur.	178·4243		3846
21 Mar. (80)	1 Sun.	22	48	54	8 Mar. (67)	2 Mon.	54·1472		3847
22 Mar. (81)	3 Tues.	5	1	3	26 Feb. (57)	0 Sat.	268·5021		3848
22 Mar. (81)	4 Wed.	11	13	12	17 Mar. (76)	6 Fri.	303·1844		3849
21 Mar. (81)	5 Thur.	17	25	21	5 Mar. (65)	3 Tues.	178·9072		3850
21 Mar. (80)	6 Fri.	23	37	30	22 Feb. (53)	0 Sat.	54·6301		3851
22 Mar. (81)	1 Sun.	5	49	39	13 Mar. (72)	6 Fri.	89·3125		3852
22 Mar. (81)	2 Mon.	12	1	48	3 Mar. (62)	4 Wed.	303·6673		3853
21 Mar. (81)	3 Tues.	18	13	57	20 Mar. (80)	2 Mon.	9999·7177§		3854
22 Mar. (81)	5 Thur.	0	26	6	10 Mar. (69)	0 Sat.	214·0726		3855
22 Mar. (81)	6 Fri.	6	38	15	27 Feb. (58)	4 Wed.	89·7953		3856
22 Mar. (81)	0 Sat.	12	50	24	18 Mar. (77)	3 Tues.	124·4778		3857
21 Mar. (81)	1 Sun.	19	2	33	6 Mar. (66)	0 Sat.	0·2006		3858
22 Mar. (81)	3 Tues	1	14	42	24 Feb. (55)	5 Thur.	214·5555		3859
22 Mar. (81)	4 Wed.	7	26	51	15 Mar. (74)	4 Wed.	249·2378		3860
22 Mar. (81)	5 Thur.	13	39	0	4 Mar. (63)	1 Sun.	124·9607		3861
21 Mar. (81)	6 Fri.	19	51	9	21 Feb. (52)	5 Thur.	0·6835		3862
22 Mar. (81)	1 Sun.	2	3	18	11 Mar. (70)	4 Wed.	35·3658		3863
22 Mar. (81)	2 Mon.	8	15	27	1 Mar. (60)	2 Mon.	249·7207		3864
22 Mar. (81)	3 Tues.	14	27	36	20 Mar. (79)	1 Sun.	284·4031		3865
21 Mar. (81)	4 Wed.	20	39	45	8 Mar. (68)	5 Thur.	160·1261		3866
22 Mar. (81)	6 Fri.	2	51	54	25 Feb. (56)	2 Mon.	35·8488		3867
22 Mar. (81)	0 Sat.	9	4	3	16 Mar. (75)	1 Sun.	70·5312		3868
22 Mar. (81)	1 Sun.	15	16	12	6 Mar. (65)	6 Fri.	284·8860		3869
21 Mar. (81)	2 Mon.	21	28	21	23 Feb. (54)	3 Tues.	160·6088		3870

§Chaitra *śukla* 1 was suppressed.

TABLE

CONCURRENT YEAR.								Mean intercalated (*adhika*) lunar month.
Kali.	Śaka.	Chaitrâdi Vikrama.	Mêshâdi solar year in Bengal.	Kollam.	A.D.	Jovian Saṁvatsara. Southern system.	Jovian Saṁvatsara. Northern system.	
1	2	3	3a	4	5	6	7	8a
3871	692	827	176		769-70	42 Kîlaka . . .		...
3872	693	828	177		770-71	43 Saumya . . .		11 Mâgha .
3873	694	829	178		771-72	44 Sâdhâraṇa . . .		...
3874	695	830	179		*772-73	45 Virôdhakṛit . .		...
3875	696	831	180		773-74	46 Paridhâvin . . .		8 Kârttika .
3876	697	832	181		774-75	47 Pramâdin . . .		...
3877	698	833	182		775-76	48 Ânanda . . .		...
3878	699	834	183		*776-77	49 Râkshasa . . .		4 Âshâḍha .
3879	700	835	184		777-78	50 Anala . . .		
3880	701	836	185		778-79	51 Pingala . . .		...
3881	702	837	186		779-80	52 Kâlayukta . . .		1 Chaitra .
3882	703	838	187		*780-81	53 Siddhârthin . . .		...
3883	704	839	188		781-82	54 Raudra . . .		9 Mârgaśira .
3884	705	840	189		782-83	55 Durmati . . .		...
3885	706	841	190		783-84	56 Dundubhi . . .		...
3886	707	842	191		*784-85	57 Rudhirôdgârin . .		6 Bhâdrapada .
3887	708	843	192		785-86	58 Raktâksha . . .		...
3888	709	844	193		786-87	59 Krôdhana . . .		...
3889	710	845	194		787-88	60 Kshaya . . .		3 Jyêshṭha .
3890	711	846	195		*788-89	1 Prabhava . . .		...
3891	712	847	196		789-90	2 Vibhava . . .		11 Mâgha
3892	713	848	197		790-91	3 Śukla . . .		...
3893	714	849	198		791-92	4 Pramôda . . .		...
3894	715	850	199		*792-93	5 Prajâpati . . .		8 Kârttika .
3896	716	851	200		793-94	6 Angiras . . .		...

XC—*Contd.*

COMMENCEMENT OF THE							
MEAN SOLAR YEAR.			MEAN LUNI-SOLAR YEAR (MEAN SUNRISE OF THE CIVIL DAY ON WHICH CHAITRA ŚUKLA 1 ENDS).			Kali.
Day and month, A.D.	Week-day.	Time of mean Mēsha-saṁkrānti.	Day and month, A.D.	Week-day.	*a* (here = *t*, the index of the *tithi*).	
13	14	17	19	20	23	1
		H. M. S.				
22 Mar. (81) . .	4 Wed. .	3 40 30	13 Mar. (72) .	2 Mon. .	195·2912	3871
22 Mar. (81) . .	5 Thur. .	9 52 39	2 Mar. (61) .	6 Fri. .	71·0141	3872
22 Mar. (81) . .	6 Fri. .	16 4 48	21 Mar. (80) .	5 Thur. .	105·6965	3873
21 Mar. (81) . .	0 Sat. .	22 16 57	10 Mar. (70) .	3 Tues. .	320·0513	3874
22 Mar. (81) . .	2 Mon. .	4 29 6	27 Feb. (58) .	0 Sat. .	195·7741	3875
22 Mar. (81) . .	3 Tues. .	10 41 15	18 Mar. (77) .	6 Fri. .	230·4566	3876
22 Mar. (81) . .	4 Wed. .	16 53 24	7 Mar. (66) .	3 Tues. .	106·1793	3877
21 Mar. (81) . .	5 Thur. .	23 5 33	25 Feb. (56) .	1 Sun. .	320·5342	3878
22 Mar. (81) . .	0 Sat. .	5 17 42	14 Mar. (73) .	6 Fri. .	16·5846	3879
22 Mar. (81) . .	1 Sun. .	11 29 51	4 Mar. (63) .	4 Wed. .	230·9395	3880
22 Mar. (81) . .	2 Mon. .	17 42 0	21 Feb. (52) .	1 Sun. .	106·6622	3881
21 Mar. (81) . .	3 Tues. .	23 54 9	11 Mar. (71) .	0 Sat. .	141·3446	3882
22 Mar. (81) . .	5 Thur. .	6 6 18	28 Feb. (59) .	4 Wed. .	17·0675	3883
22 Mar. (81) . .	6 Fri. .	12 18 27	19 Mar. (78) .	3 Tues. .	51·7499	3884
22 Mar. (81) . .	0 Sat. .	18 30 36	9 Mar. (68) .	1 Sun. .	266·1047	3885
22 Mar. (82) . .	2 Mon. .	0 42 45	26 Feb. (57) .	5 Thur. .	141·8276	3886
22 Mar. (81) . .	3 Tues. .	6 54 54	16 Mar. (75) .	4 Wed. .	176·5100	3887
22 Mar. (81) . .	4 Wed. .	13 7 3	5 Mar. (64) .	1 Sun. .	52·2327	3888
22 Mar. (81) . .	5 Thur. .	19 19 12	23 Feb. (54) .	6 Fri. .	266·5876	3889
22 Mar. (82) . .	0 Sat. .	1 31 21	13 Mar. (73) .	5 Thur. .	301·2700	3890
22 Mar. (81) . .	1 Sun. .	7 43 30	2 Mar. (61) .	2 Mon. .	176·9929	3891
22 Mar. (81) . .	2 Mon. .	13 55 39	21 Mar. (80) .	1 Sun. .	211·6752	3892
22 Mar. (81) . .	3 Tues. .	20 7 48	10 Mar. (69) .	5 Thur. .	87·3981	3893
22 Mar. (82) . .	5 Thur. .	2 19 57	28 Feb. (59) .	3 Tues. .	301·7530	3894
22 Mar. (81) . .	6 Fri. .	8 32 6	17 Mar. (76) .	1 Sun. .	9997·8033 §	3895

§ Chaitra *śukla* 1 was suppressed.

TABLE

CONCURRENT YEAR.								Mean intercalated (*adhika*) lunar month.
						Jovian Saṁvatsara.		
Kali.	Śaka.	Chaitrādi Vikrama.	Mēshādi solar year in Bengal.	Kollam.	A.D.	Southern system.	Northern system.	
1	2	3	3a	4	5	6	7	8a
3896	717	852	201		794-95	7 Śrīmukha . . .		...
3897	718	853	202		795-96	8 Bhāva . . .		4 Āshāḍha .
3898	719	854	203		*796-97	9 Yuvan . . .		...
3899	720	855	204		797-98	10 Dhātṛi . . .		...
3900	721	856	205		798-99	11 Īśvara . . .		1 Chaitra .
3901	722	857	206		799-800	12 Bahudhānya . .		...
3902	723	858	207		*800-01	13 Pramāthin . . .		9 Mārgaśira .
3903	724	859	208		801-02	14 Vikrama . . .		...
3904	725	860	209		802-03	15 Vṛisha . . .		...
3905	726	861	210		803-04	16 Chitrabhānu . .		6 Bhādrapada .
3906	727	862	211		*804-05	17 Subhānu . . .		...
3907	728	863	212		805-06	18 Tāraṇa . . .		...
3908	729	864	213		806-07	19 Pārthiva . . .		2 Vaiśākha .
3909	730	865	214		807-08	20 Vyaya . . .		...
3910	731	866	215		*808-09	21 Sarvajit . . .		11 Māgha .
3911	732	867	216		809-10	22 Sarvadhārin . .		...
3912	733	868	217		810-11	23 Virōdhin . . .		...
3913	734	869	218		811-12	24 Vikṛita . . .		7 Āśvina .
3914	735	870	219		*812-13	25 Khara . . .		...
3915	736	871	220		813-14	26 Nandana . . .		...
3916	737	872	221		814-15	27 Vijaya . . .		4 Āshāḍha .
3917	738	873	222		815-16	28 Jaya . . .		...
3918	739	874	223		*816-17	29 Manmatha . . .		12 Phālguna .
3919	740	875	224		817-18	30 Durmukha . . .		...
3920	741	876	225		818-19	31 Hēmalamba† . .		...

† 32 Vilamba was suppressed by mean reckoning. By *Brahma-Siddhānta* "true" reckoning the year K. Y. 3921, A.D. 819-20, was 32 "Vilamba," and 33 Vikārin was suppressed.

XC—*Contd.*

COMMENCEMENT OF THE							
MEAN SOLAR YEAR.			MEAN LUNI-SOLAR YEAR (MEAN SUNRISE OF THE CIVIL DAY ON WHICH CHAITRA ŚUKLA 1 ENDS).			Kali.
Day and month, A.D.	Week-day.	Time of mean Mēsha-saṁkrānti.	Day and month, A.D.	Week-day.	*a* (here = *t*, the index of the *tithi*).	
13	14	17	19	20	23	1
		H. M. S.				
22 Mar. (81) . .	0 Sat. .	14 44 15	7 Mar. (66) .	6 Fri. .	212·1581	3896
22 Mar. (81) . .	1 Sun. .	20 56 24	24 Feb. (55) .	3 Tues. .	87·8810	3897
22 Mar. (82) . .	3 Tues. .	3 8 33	14 Mar. (74) .	2 Mon. .	122·5633	3898
22 Mar. (81) . .	4 Wed. .	9 20 42	3 Mar. (62) .	6 Fri. .	9998·2862§	3899
22 Mar. (81) . .	5 Thur. .	15 32 51	21 Feb. (52) .	4 Wed. .	212·6410	3900
22 Mar. (81) . .	6 Fri. .	21 45 0	12 Mar. (71) .	3 Tues. .	247·3234	3901
22 Mar. (82) . .	1 Sun. .	3 57 9	29 Feb. (60) .	0 Sat. .	123·0463	3902
22 Mar. (81) . .	2 Mon. .	10 9 18	19 Mar. (78) .	6 Fri. .	157·7287	3903
22 Mar. (81) . .	3 Tues. .	16 21 27	8 Mar. (67) .	3 Tues. .	33·4515	3904
22 Mar. (81) . .	4 Wed. .	22 33 36	26 Feb. (57) .	1 Sun. .	247·8064	3905
22 Mar. (82) . .	6 Fri. .	4 45 45	16 Mar. (76) .	0 Sat. .	282·4888	3906
22 Mar. (81) . .	0 Sat. .	10 57 54	5 Mar. (64) .	4 Wed. .	158·2115	3907
22 Mar. (81) . .	1 Sun. .	17 10 3	22 Feb. (53) .	1 Sun. .	33·9344	3908
22 Mar. (81) . .	2 Mon. .	23 22 12	13 Mar. (72) .	0 Sat. .	68·6108	3909
22 Mar. (82) . .	4 Wed. .	5 34 21	2 Mar. (62) .	5 Thur. .	282·9716	3910
22 Mar. (81) . .	5 Thur. .	11 46 30	21 Mar. (80) .	4 Wed. .	317·6540	3911
22 Mar. (81) . .	6 Fri. .	17 58 39	10 Mar. (69) .	1 Sun. .	193·3769	3912
23 Mar. (82) . .	1 Sun. .	0 10 48	27 Feb. (58) .	5 Thur. .	69·0998	3913
22 Mar. (82) . .	2 Mon. .	6 22 57	17 Mar. (77) .	4 Wed. .	103·7821	3914
22 Mar. (81) . .	3 Tues. .	12 35 6	7 Mar. (66) .	2 Mon. .	318·1369	3915
22 Mar. (81) . .	4 Wed. .	18 47 15	24 Feb. (55) .	6 Fri. .	193·8598	3916
23 Mar. (82) . .	6 Fri. .	0 59 24	15 Mar. (74) .	5 Thur. .	228·5421	3917
22 Mar. (82) . .	0 Sat. .	7 11 33	3 Mar. (63) .	2 Mon. .	104·2650	3918
22 Mar. (81) . .	1 Sun. .	13 23 42	22 Mar. (81) .	1 Sun. .	138·9474	3919
22 Mar. (81) . .	2 Mon. .	19 35 51	11 Mar. (70) .	5 Thur. .	14·6703	3920

§ Chaitra *śukla* 1 was suppressed.

TABLE

CONCURRENT YEAR.								Mean intercalated (*adhika*) lunar month.
						JOVIAN SAṀVATSARA.		
Kali.	Śaka.	Chaitrādi Vikrama.	Mēshādi solar year in Bengal.	Kollam.	A.D.	Southern system.	Northern system.	
1	2	3	3*a*	4	5	6	7	8*a*
3921	742	877	226		819-20	33 *Vikārin* . . .		9 Mārgaśira .
3922	743	878	227		*820-21	34 *Śārvarin* . . .		...
3923	744	879	228		821-22	35 *Plava* . . .		...
3924	745	880	229		822-23	36 *Śubhakṛit* . . .		6 Bhādrapada‡
3925	746	881	230		823-24	37 *Śōbhana* . . .		...
3926	747	882	231		*824-25	38 Krōdhin . . .		...
3927	748	883	232	0-1	825-26	39 Viśvāvasu . . .		2 Vaiśākha .
3928	749	884	233	1-2	826-27	40 Parābhava . . .		...
3929	750	885	234	2-3	827-28	41 Plavaṅga . . .		11 Māgha .
3930	751	886	235	3-4	*828-29	42 Kīlaka . . .		...
3931	752	887	236	4-5	829-30	43 Saumya . . .		...
3932	753	888	237	5-6	830-31	44 Sādhāraṇa . . .		7 Āśvina .
3933	754	889	238	6-7	831-32	45 Virōdhakṛit . .		...
3934	755	890	239	7-8	*832-33	46 Paridhāvin . . .		...
3935	756	891	240	8-9	833-34	47 Pramādin . . .		4 Āshāḍha .
3936	757	892	241	9-10	834-35	48 Ānanda . . .		...
3937	758	893	242	10-11	835-36	49 Rākshasa . . .		12 Phālguna .
3938	759	894	243	11-12	*836-37	50 Anala . . .		...
3939	760	895	244	12-13	837-38	51 Piṅgala . . .		...
3940	761	896	245	13-14	838-39	52 Kālayukta . . .		9 Mārgaśira .
3941	762	897	246	14-15	839-40	53 Siddhārthin . . .		...
3942	763	898	247	15-16	*840-41	54 Raudra . . .		...
3943	764	899	248	16-17	841-42	55 Durmati . . .		5 Śrāvaṇa .
3944	765	900	249	17-18	842-43	56 Dundubhi . . .		...
3945	766	901	250	18-19	843-44	57 Rudhirōdgārin . .		...

‡ See "Remarks," p. 215 above.

XC—*Contd.*

COMMENCEMENT OF THE							
MEAN SOLAR YEAR.			MEAN LUNI-SOLAR YEAR (MEAN SUNRISE OF THE CIVIL DAY ON WHICH CHAITRA ŚUKLA 1 ENDS).			Kali.
Day and month, A.D.	Week-day.	Time of mean Mēsha-saṁkrānti.	Day and month, A.D.	Week-day.	*a* (here = *t*, the index of the *tithi*).	
13	14	17	19	20	23	1
		H. M. S.				
23 Mar. (82)	4 Wed.	1 48 0	1 Mar. (60)	3 Tues.	229·0250	3921
22 Mar. (82)	5 Thur.	8 0 9	19 Mar. (79)	2 Mon.	263·7074	3922
22 Mar. (81)	6 Fri.	14 12 18	8 Mar. (67)	6 Fri.	139·4313	3923
22 Mar. (81)	0 Sat.	20 24 27	25 Feb. (56)	3 Tues.	15·1531	3924
23 Mar. (82)	2 Mon.	2 36 36	16 Mar. (75)	2 Mon.	49·8355	3925
22 Mar. (82)	3 Tues.	8 48 45	5 Mar. (65)	0 Sat.	264·1904	3926
22 Mar. (81)	4 Wed.	15 0 54	22 Feb. (53)	4 Wed.	139·9132	3927
22 Mar. (81)	5 Thur.	21 13 3	13 Mar. (72)	3 Tues.	174·5955	3928
23 Mar. (82)	0 Sat.	3 25 12	2 Mar. (61)	0 Sat.	50·3184	3929
22 Mar. (82)	1 Sun.	9 37 21	20 Mar. (80)	6 Fri.	85·0009	3930
22 Mar. (81)	2 Mon.	15 49 30	10 Mar. (69)	4 Wed.	299·3556	3931
22 Mar. (81)	3 Tues.	22 1 39	27 Feb. (58)	1 Sun.	175·0784	3932
23 Mar. (82)	5 Thur.	4 13 48	18 Mar. (77)	0 Sat.	209·7609	3933
22 Mar. (82)	6 Fri.	16 25 57	6 Mar. (66)	4 Wed.	85·4837	3934
22 Mar. (81)	0 Sat.	16 38 6	24 Feb. (55)	2 Mon.	299·8385	3935
22 Mar. (81)	1 Sun.	22 50 15	14 Mar. (73)	0 Sat.	9995·8889 §	3936
23 Mar. (82)	3 Tues.	5 2 24	4 Mar. (63)	5 Thur.	210·2438	3937
22 Mar. (82)	4 Wed.	11 14 33	22 Mar. (82)	4 Wed.	244·9262	3938
22 Mar. (81)	5 Thur.	17 26 42	11 Mar. (70)	1 Sun.	120·6490	3939
22 Mar. (81)	6 Fri.	23 38 51	28 Feb. (59)	5 Thur.	9996·3718 §	3940
23 Mar. (82)	1 Sun.	5 51 0	19 Mar. (78)	4 Wed.	31·0542	3941
22 Mar. (82)	2 Mon.	12 3 9	8 Mar. (68)	2 Mon.	245·4090	3942
22 Mar. (81)	3 Tues.	18 15 18	25 Feb. (56)	6 Fri.	121·1319	3943
23 Mar. (82)	5 Thur.	0 27 27	16 Mar. (75)	5 Thur.	155·8143	3944
23 Mar. (82)	6 Fri.	6 39 36	5 Mar. (64)	2 Mon	31·5372	3945

§ Chaitra *śukla* 1 was suppressed.

TABLE

CONCURRENT YEAR.								Mean intercalated (*adhika*) lunar month.
Kali.	Śaka.	Chaitrādi Vikrama.	Mēshādi solar year in Bengal.	Kollam.	A.D.	Jovian Saṁvatsara. Southern system.	Jovian Saṁvatsara. Northern system.	
1	2	3	3a	4	5	6	7	8a
3946	767	902	251	19-20	*844-45	58 Raktāksha . . .		2 Vaiśākha .
3947	768	903	252	20-21	845-46	59 Krōdhana . . .		...
3948	769	904	253	21-22	846-47	60 Kshaya . . .		10 Pausha .
3949	770	905	254	22-23	847-48	1 Prabhava . . .		...
3950	771	906	255	23-24	*848-49	2 Vibhava . . .		...
3951	772	907	256	24-25	849-50	3 Śukla . . .		7 Aśvina .
3952	773	908	257	25-26	850-51	4 Pramōda . . .		...
3953	774	909	258	26-27	851-52	5 Prajāpati . . .		...
3954	775	910	259	27-28	*852-53	6 Aṅgiras . . .		3 Jyēshṭha .
3955	776	911	260	28-29	853-54	7 Śrīmukha . . .		...
3956	777	912	261	29-30	854-55	8 Bhāva . . .		12 Phālguna .
3957	778	913	262	30-31	855-56	9 Yuvan . . .		...
3958	779	914	263	31-32	*856-57	10 Dhātṛi . . .		...
3959	780	915	264	32-33	857-58	11 Īśvara . . .		8 Kārttika .
3960	781	916	265	33-34	858-59	12 Bahudhānya . . .		...
3961	782	917	266	34-35	859-60	13 Pramāthin . . .		...
3962	783	918	267	35-36	*860-61	14 Vikrama . . .		5 Śrāvaṇa .
3963	784	919	268	36-37	861-62	15 Vṛisha . . .		...
3964	785	920	269	37-38	862-63	16 Chitrabhānu . . .		...
3965	786	921	270	38-39	863-64	17 Subhānu . . .		2 Vaiśākha .
3966	787	922	271	39-40	*864-65	18 Tāraṇa . . .		...
3967	788	923	272	40-41	865-66	19 Pārthiva . . .		10 Pausha .
3968	789	924	273	41-42	866-67	20 Vyaya . . .		...
3969	790	925	274	42-43	867-68	21 Sarvajit . . .		...
3970	791	926	275	43-44	*868-69	22 Sarvadhārin . . .		7 Aśvina .

XC—*contd.*

COMMENCEMENT OF THE							
MEAN SOLAR YEAR.			MEAN LUNI-SOLAR YEAR (MEAN SUNRISE OF THE CIVIL DAY ON WHICH CHAITRA ŚUKLA 1 ENDS).			Kali
Day and month, A.D.	Week-day.	Time of mean Mēsha-saṁkrānti.	Day and month, A.D.	Week-day.	*a* (here = *t*, the index of the *tithi*).	
13	14	17	19	20	23	1
		H. M. S.				
22 Mar. (82)	0 Sat.	12 51 45	23 Feb. (54)	0 Sat.	245·8919	3946
22 Mar. (81)	1 Sun.	19 3 54	13 Mar. (76)	6 Fri.	280·5743	3947
23 Mar. (82)	3 Tues.	1 16 3	2 Mar. (61)	3 Tues.	156·2972	3948
23 Mar. (82)	4 Wed.	7 28 12	21 Mar. (80)	2 Mon.	190·9796	3949
22 Mar. (82)	5 Thur.	13 40 21	9 Mar. (69)	6 Fri.	66·7024	3950
22 Mar. (81)	6 Fri.	19 52 30	27 Feb. (58)	4 Wed.	281·0572	3951
23 Mar. (82)	1 Sun.	2 4 39	18 Mar. (77)	3 Tues.	315·7397	3952
23 Mar. (82)	2 Mon.	8 16 48	7 Mar. (66)	0 Sat.	191·4624	3953
22 Mar. (82)	3 Tues.	14 28 57	24 Feb. (55)	4 Wed.	67·1853	3954
22 Mar. (81)	4 Wed.	20 41 6	14 Mar. (73)	3 Tues.	101·8677	3957
23 Mar. (82)	6 Fri.	2 53 15	4 Mar. (63)	1 Sun.	316·2225	3956
23 Mar. (82)	0 Sat.	9 5 24	22 Mar. (81)	6 Fri.	12·2729	3957
22 Mar. (82)	1 Sun.	15 17 33	11 Mar. (71)	4 Wed.	226·6278	3958
22 Mar. (81)	2 Mon.	21 29 42	28 Feb. (59)	1 Sun.	102·3506	3959
23 Mar. (82)	4 Wed.	3 41 51	19 Mar. (78)	0 Sat.	137·0329	3960
23 Mar. (82)	5 Thur.	9 54 0	8 Mar. (67)	4 Wed.	12·7558	3961
22 Mar. (82)	6 Fri.	16 6 9	26 Feb. (57)	2 Mon.	227·1107	3962
22 Mar. (81)	0 Sat.	22 18 18	16 Mar. (75)	1 Sun.	261·7930	3963
23 Mar. (82)	2 Mon.	4 30 27	5 Mar. (64)	5 Thur.	137·5159	3964
23 Mar. (82)	3 Tues.	10 42 36	22 Feb. (53)	2 Mon.	13·2387	3965
22 Mar. (82)	4 Wed.	16 54 45	12 Mar. (72)	1 Sun.	47·9211	3966
22 Mar. (81)	5 Thur.	23 6 54	2 Mar. (61)	6 Fri.	262·2759	3967
23 Mar. (82)	0 Sat.	5 19 3	21 Mar. (80)	5 Thur.	296·9584	3968
23 Mar. (82)	1 Sun.	11 31 12	10 Mar. (69)	2 Mon.	172·6812	3969
22 Mar. (82)	2 Mon.	17 43 21	27 Feb. (58)	6 Fri.	48·4039	3970

TABLE

CONCURRENT YEAR.								
						Jovian Saṁvatsara.		Mean intercalated (*adhika*) lunar month.
Kali.	Śaka.	Chaitrâdi Vikrama.	Mêshâdi solar year in Bengal.	Kollam.	A.D.	Southern system.	Northern system.	
1	2	3	3*a*	4	5	6	7	8*a*
3971	792	927	276	44-45	869-70	23 Virōdhin		...
3972	793	928	277	45-46	870-71	24 Vikṛita		...
3973	794	929	278	46-47	871-72	25 Khara		3 Jyēshṭha
3974	795	930	279	47-48	*872-73	26 Nandana		...
3975	796	931	280	48-49	873-74	27 Vijaya		12 Phâlguna
3976	797	932	281	49-50	874-75	28 Jaya		...
3977	798	933	282	50-51	875-76	29 Manmatha		...
3978	799	934	283	51-52	*876-77	30 Durmukha		8 Kârttika
3979	800	935	284	52-53	877-78	31 Hēmalamba		...
3980	801	936	285	53-54	878-79	32 Vilamba		...
3981	802	937	286	54-55	879-80	33 Vikârin		5 Śrâvaṇa
3982	803	938	287	55-56	*880-81	34 Śârvarin		...
3983	804	939	288	56-57	881-82	35 Plava		...
3984	805	940	289	57-58	882-83	36 Śubhakṛit		1 Chaitra
3985	806	941	290	58-59	883-84	37 Śōbhana		...
3986	807	942	291	59-60	*884-85	38 Krōdhin		10 Pausha
3987	808	943	292	60-61	885-86	39 Viśvâvasu		...
3988	809	944	293	61-62	886-87	40 Parâbhava		...
3989	810	945	294	62-63	887-88	41 Plavaṅga		6 Bhâdrapada
3990	811	946	295	63-64	*888-89	42 Kîlaka		...
3991	812	947	296	64-65	889-90	43 Saumya		...
3992	813	948	297	65-66	890-91	44 Sâdhâraṇa		3 Jyēshṭha
3993	814	949	298	66-67	891-92	45 Virōdhakṛit		...
3994	815	950	299	67-68	*892-93	46 Paridhâvin		11 Mâgha
3995	816	951	300	68-69	893-94	47 Pramâdin		...

XC—*contd.*

COMMENCEMENT OF THE						
MEAN SOLAR YEAR.			MEAN LUNI-SOLAR YEAR (MEAN SUNRISE OF THE CIVIL DAY ON WHICH CHAITRA ŚUKLA 1 ENDS).			Kali.
Day and month, A.D.	Week-day.	Time of mean Mēsha-saṁkrānti.	Day and month, A.D.	Week-day.	*a* (here = *t*, the index of the *tithi*).	
13	14	17	19	20	23	1
		H. M. S.				
22 Mar. (81)	3 Tues.	23 55 30	17 Mar. (76)	5 Thur.	83·0864	3971
23 Mar. (82)	5 Thur.	6 7 39	7 Mar. (66)	3 Tues.	297·4412	3972
23 Mar. (82)	6 Fri.	12 19 48	24 Feb. (55)	0 Sat.	173·1641	3973
22 Mar. (82)	0 Sat.	18 31 57	14 Mar. (74)	6 Fri.	207·8464	3974
23 Mar. (82)	2 Mon.	0 44 6	3 Mar. (62)	3 Tues.	83·5693	3975
23 Mar. (82)	3 Tues.	6 56 15	22 Mar. (81)	2 Mon.	118·2517	3976
23 Mar. (82)	4 Wed.	13 8 24	12 Mar. (71)	0 Sat.	332·6065	3977
22 Mar. (82)	5 Thur.	19 20 33	29 Feb. (60)	4 Wed.	208·3293	3978
23 Mar. (82)	0 Sat.	1 32 42	19 Mar. (78)	3 Tues.	243·0118	3979
23 Mar. (82)	1 Sun.	7 44 51	8 Mar. (67)	0 Sat.	118·7346	3980
23 Mar. (82)	2 Mon.	13 57 0	26 Feb. (57)	5 Thur.	333·0894	3981
22 Mar. (82)	3 Tues.	20 9 9	15 Mar. (75)	3 Tues.	29·1398	3982
23 Mar. (82)	5 Thur.	2 21 18	5 Mar. (64)	1 Sun.	243·4947	3983
23 Mar. (82)	6 Fri.	8 33 27	22 Feb. (53)	5 Thur.	119·2175	3984
23 Mar. (82)	0 Sat.	14 45 36	13 Mar. (72)	4 Wed.	153·8998	3985
22 Mar. (82)	1 Sun.	20 57 45	1 Mar. (61)	1 Sun.	29·6227	3986
23 Mar. (82)	3 Tues.	3 9 54	20 Mar. (79)	0 Sat.	64·3052	3987
23 Mar. (82)	4 Wed.	9 22 3	10 Mar. (69)	5 Thur.	278·6599	3988
23 Mar. (82)	5 Thur.	15 34 12	27 Feb. (58)	2 Mon.	154·3828	3989
22 Mar. (82)	6 Fri.	21 46 21	17 Mar. (77)	1 Sun.	189·0652	3990
23 Mar. (82)	1 Sun.	3 58 30	6 Mar. (65)	5 Thur.	64·7881	3991
23 Mar. (82)	2 Mon.	10 10 39	24 Feb. (55)	3 Tues.	279·1428	3992
23 Mar. (82)	3 Tues.	16 22 48	15 Mar. (74)	2 Mon.	313·8252	3993
22 Mar. (82)	4 Wed.	22 34 57	3 Mar. (63)	6 Fri.	189·5481	3994
23 Mar. (82)	6 Fri.	4 47 6	22 Mar. (81)	5 Thur.	224·2204	3995

TABLE

CONCURRENT YEAR.								Mean intercalated (*adhika*) lunar month.
Kali.	Śaka.	Chaitrādi Vikrama.	Mēshādi solar year in Bengal.	Kollam.	A.D.	JOVIAN SAṀVATSARA. Southern system.	JOVIAN SAṀVATSARA. Northern system.	
1	2	3	3a	4	5	6	7	8a
3996	817	952	301	69-70	894-95	48 Ānanda . . .		...
3997	818	953	302	70-71	895-96	49 Rākshasa . . .		8 Kārttika .
3998	819	954	303	71-72	*896-97	50 Anala . .		...
3999	820	955	304	72-73	897-98	51 Piṅgala . .		...
4000	821	956	305	73-74	898-99	52 Kālayukta . . .		5 Śrāvaṇa .
4001	822	957	306	74-75	899-900	53 Siddhārthin . .		...
4002	823	958	307	75-76	*900-01	54 Raudra . . .		...
4003	824	959	308	76-77	901-02	55 Durmati . . .		1 Chaitra .
4004	825	960	309	77-78	902-03	56 Dundubhi . . .		...
4005	826	961	310	78-79	903-04	57 Rudhirōdgārin† . .		10 Pausha .
4006	827	962	311	79-80	*904-05	58 Raktāksha .	59 *Krōdhana* .	...
4007	828	963	312	80-81	905-06	59 Krōdhana .	60 *Kshaya* . .	...
4008	829	964	313	81-82	906-07	60 Kshaya . .	1 *Prabhava* .	6 Bhādrapada .
4009	830	965	314	82-83	907-08	1 Prabhava .	2 *Vibhava* .	...
4010	831	966	315	83-84	*908-09	2 Vibhava .	3 *Śukla* . .	...
4011	832	967	316	84-85	909-10	3 Śukla . .	4 Pramōda .	3 Jyēshtha .
4012	833	968	317	85-86	910-11	4 Pramōda .	5 Prajāpati .	...
4013	834	969	318	86-87	911-12	5 Prajāpati .	6 Aṅgiras . .	11 Māgha .
4014	835	970	319	87-88	*912-13	6 Aṅgiras . .	7 Śrīmukha .	...
4015	836	971	320	88-89	913-14	7 Śrīmukha .	8 Bhāva . .	...
4016	837	972	321	89-90	914-15	8 Bhāva . .	9 Yuvan . .	8 Kārttika
4017	838	973	322	90-91	915-16	9 Yuvan . .	10 Dhātṛi . .	...
4018	839	974	323	91-92	*916-17	10 Dhātṛi . .	11 Īśvara . .	...
4019	840	975	324	92-93	917-18	11 Īśvara . .	12 Bahudhānya .	4 Ashāḍha .
4020	841	976	325	93-94	918-19	12 Bahudhānya .	13 Pramāthin .	...

† 58 Raktāksha was suppressed in the north. By southern reckoning there was no suppression, and there has been none since. By *Brahma-Siddhānta* "true" reckoning K.Y. 4006, A.D. 904-05, was 58 Raktaksha, 59 Krōdhana being suppressed in the north.

XC—*contd.*

COMMENCEMENT OF THE							
MEAN SOLAR YEAR.			MEAN LUNI-SOLAR YEAR (MEAN SUNRISE OF THE CIVIL DAY ON WHICH CHAITRA ŚUKLA 1 ENDS).			Kali.
Day and month, A.D.	Week-day.	Time of mean Mēsha-saṁkrānti.	Day and month, A.D.	Week-day.	*a* (here = *t*, the index of the *tithi*).	
13	14	17	19	20	23	1
		H. M. S.				
23 Mar. (82)	0 Sat.	10 59 15	11 Mar. (70)	2 Mon.	99·9533	3996
23 Mar. (82)	1 Sun.	17 11 24	1 Mar. (60)	0 Sat.	314·3081	3997
22 Mar. (82)	2 Mon.	23 23 33	18 Mar. (78)	5 Thur.	10·3584	3998
23 Mar. (82)	4 Wed.	5 35 42	8 Mar. (67)	3 Tues.	224·7133	3999
23 Mar. (82)	5 Thur.	11 47 51	25 Feb. (56)	0 Sat.	100·4362	4000
23 Mar. (82)	6 Fri.	18 0 0	16 Mar. (75)	6 Fri.	135·1186	4001
23 Mar. (83)	1 Sun.	0 12 9	4 Mar. (64)	3 Tues.	10·8415	4002
23 Mar. (82)	2 Mon.	6 24 18	22 Feb. (53)	1 Sun.	225·4963	4003
23 Mar. (82)	3 Tues.	12 36 27	13 Mar. (72)	0 Sat.	259·8786	4004
23 Mar. (82)	4 Wed.	18 48 36	2 Mar. (61)	4 Wed.	135·6015	4005
23 Mar. (83)	6 Fri.	1 0 45	20 Mar. (80)	3 Tues.	170·2839	4006
23 Mar. (82)	0 Sat.	7 12 54	9 Mar. (68)	0 Sat.	46·0067	4007
23 Mar. (82)	1 Sun.	13 25 3	27 Feb. (58)	5 Thur.	260·3616	4008
23 Mar. (82)	2 Mon.	19 37 12	18 Mar. (77)	4 Wed.	295·0440	4009
23 Mar. (83)	4 Wed.	1 49 21	6 Mar. (66)	1 Sun.	170·7668	4010
23 Mar. (82)	5 Thur.	8 1 30	23 Feb. (54)	5 Thur.	46·4896	4011
23 Mar. (82)	6 Fri.	14 13 39	14 Mar. (73)	4 Wed.	81·1720	4012
23 Mar. (82)	0 Sat.	20 25 48	4 Mar. (63)	2 Mon.	295·5269	4013
23 Mar. (83)	2 Mon.	2 37 57	22 Mar. (82)	1 Sun.	330·2092	4014
23 Mar. (82)	3 Tues.	8 50 6	11 Mar. (70)	5 Thur.	205·[illegible]	4015
23 Mar. (82)	4 Wed.	15 2 15	28 Feb. (59)	2 Mon.	81·6549	4016
23 Mar. (82)	5 Thur.	21 14 24	19 Mar. (78)	1 Sun.	116·3373	4017
23 Mar. (83)	0 Sat.	3 26 33	8 Mar. (68)	6 Fri.	330·6921	4018
23 Mar. (82)	1 Sun.	9 38 42	25 Feb. (56)	3 Tues.	206·4150	4019
23 Mar. (82)	2 Mon.	15 50 51	16 Mar. (75)	2 Mon.	241·[illegible]	4020

TABLE

CONCURRENT YEAR.								Mean intercalated (*adhika*) lunar month.
Kali.	Śaka.	Chaitrādi Vikrama.	Mēshādi solar year in Bengal.	Kollam.	A.D.	Jovian Saṁvatsara. Southern system.	Jovian Saṁvatsara. Northern system.	
1	2	3	3a	4	5	6	7	8a
4021	842	977	326	94-95	919-20	13 Pramāthin .	14 Vikrama .	...
4022	843	978	327	95-96	*920-21	14 Vikrama .	15 Vṛisha .	1 Chaitra .
4023	844	979	328	96-97	921-22	15 Vṛisha .	16 Chitrabhānu .	...
4024	845	980	329	97-98	922-23	16 Chitrabhānu .	17 Subhānu .	9 Mārgaśira .
4025	846	981	330	98-99	923-24	17 Subhānu .	18 Tāraṇa .	...
4026	847	982	331	99-100	*924-25	18 Tāraṇa .	19 Pārthiva .	...
4027	848	983	332	100-01	925-26	19 Pārthiva .	20 Vyaya .	6 Bhādrapada .
4028	849	984	333	101-02	926-27	20 Vyaya .	21 Sarvajit .	...
4029	850	985	334	102-03	927-28	21 Sarvajit .	22 Sarvadhārin .	...
4030	851	986	335	103-04	*928-29	22 Sarvadhārin .	23 Virōdhin .	2 Vaiśākha .
4031	852	987	336	104-05	929-30	23 Virōdhin .	24 Vikṛita .	...
4032	853	988	337	105-06	930-31	24 Vikṛita .	25 Khara .	11 Māgha .
4033	854	989	338	106-07	931-32	25 Khara .	26 Nandana .	...
4034	855	990	339	107-08	*932-33	26 Nandana .	27 Vijaya .	...
4035	856	991	340	108-09	933-34	27 Vijaya .	28 Jaya .	7 Āśvina .
4036	857	992	341	109-10	934-35	28 Jaya .	29 Manmatha .	...
4037	858	993	342	110-11	935-36	29 Manmatha .	30 Durmukha .	...
4038	859	994	343	111-12	*936-37	30 Durmukha .	31 Hēmalamba .	4 Āshāḍha
4039	860	995	344	112-13	937-38	31 Hēmalamba .	32 Vilamba .	...
4040	861	996	345	113-14	938-39	32 Vilamba .	33 Vikārin .	...
4041	862	997	346	114-15	939-40	33 Vikārin .	34 Śārvarin .	1 Chaitra
4042	863	998	347	115-16	*940-41	34 Śārvarin .	35 Plava .	...
4043	864	999	348	116-17	941-42	35 Plava .	36 Śubhakṛit .	9 Mārgaśira .
4044	865	1000	349	117-18	942-43	36 Śubhakṛit .	37 Śōbhana .	...
4045	866	1001	350	118-19	943-44	37 Śōbhana .	38 Krōdhin .	...

XC—*contd.*

COMMENCEMENT OF THE							
MEAN SOLAR YEAR.			MEAN LUNI-SOLAR YEAR (MEAN SUNRISE OF THE CIVIL DAY ON WHICH CHAITRA ŚUKLA 1 ENDS).			Kali.
Day and month, A.D.	Week-day.	Time of mean Mēsha-saṁkrānti.	Day and month, A.D.	Week-day.	*a* (here = *t*, the index of the *tithi*).	
13	14	17	19	20	23	1
		H. M. S.				
23 Mar. (82)	3 Tues.	22 3 0	5 Mar. (64)	6 Fri.	116·8202	4021
23 Mar. (83)	5 Thur.	4 15 9	23 Feb. (54)	4 Wed.	331·1750	4022
23 Mar. (82)	6 Fri.	10 27 18	12 Mar. (71)	2 Mon.	27·2254	4023
23 Mar. (82)	0 Sat.	16 39 27	2 Mar. (61)	0 Sat.	241·5802	4024
23 Mar. (82)	1 Sun.	22 51 36	21 Mar. (80)	6 Fri.	276·2626	4025
23 Mar. (83)	3 Tues.	5 3 45	9 Mar. (69)	3 Tues.	151·9855	4026
23 Mar. (82)	4 Wed.	11 15 54	26 Feb. (57)	0 Sat.	27·7084	4027
23 Mar. (82)	5 Thur.	17 28 3	17 Mar. (76)	6 Fri.	62·3907	4028
23 Mar. (82)	6 Fri.	23 40 12	7 Mar. (66)	4 Wed.	276·7455	4029
23 Mar. (83)	1 Sun.	5 52 21	24 Feb. (55)	1 Sun.	152·4684	4030
23 Mar. (82)	2 Mon.	12 4 30	14 Mar. (73)	0 Sat.	187·1507	4031
23 Mar. (82)	3 Tues.	18 16 39	3 Mar. (62)	4 Wed.	62·8736	4032
24 Mar. (83)	5 Thur.	0 28 48	22 Mar. (81)	3 Tues.	97·5560	4033
23 Mar. (83)	6 Fri.	6 40 57	11 Mar. (71)	1 Sun.	311·9109	4034
23 Mar. (82)	0 Sat.	12 53 6	28 Feb. (59)	5 Thur.	187·6336	4035
23 Mar. (82)	1 Sun.	19 5 15	19 Mar. (78)	4 Wed.	222·3161	4036
24 Mar. (83)	3 Tues.	1 17 24	8 Mar. (67)	1 Sun.	98·0389	4037
23 Mar. (83)	4 Wed.	7 29 33	26 Feb. (57)	6 Fri.	312·3938	4038
23 Mar. (82)	5 Thur.	13 41 42	15 Mar. (74)	4 Wed.	8·4441	4039
23 Mar. (82)	6 Fri.	19 53 51	5 Mar. (64)	2 Mon.	222·7990	4040
24 Mar. (83)	1 Sun.	2 6 0	22 Feb. (53)	6 Fri.	98·5218	4041
23 Mar. (83)	2 Mon.	8 18 9	12 Mar. (72)	5 Thur.	133·2042	4042
23 Mar. (82)	3 Tues.	14 30 18	1 Mar. (60)	2 Mon	8·9270	4043
23 Mar. (82)	4 Wed.	20 42 27	20 Mar. (79)	1 Sun.	43·6094	4044
24 Mar. (83)	6 Fri.	2 54 36	10 Mar. (69)	6 Fri.	257·9643	4045

TABLE

CONCURRENT YEAR.								Mean intercalated (*adhika*) lunar month.
Kali.	Śaka.	Chaitrādi Vikrama.	Mēshādi solar year in Bengal.	Kollam.	A.D.	Jovian Saṁvatsara. Southern system.	Jovian Saṁvatsara. Northern system.	
1	2	3	3*a*	4	5	6	7	8*a*
4046	867	1002	351	119-20	*944-45	38 Krōdhin	39 Viśvāvasu	6 Bhādrapada
4047	868	1003	352	120-21	945-46	39 Viśvāvasu	40 Parābhava	...
4048	869	1004	353	121-22	946-47	40 Parābhava	41 Plavaṅga	...
4049	870	1005	354	122-23	947-48	41 Plavaṅga	42 Kīlaka	2 Vaiśākha
4050	871	1006	355	123-24	*948-49	42 Kīlaka	43 Saumya	...
4051	872	1007	356	124-25	949-50	43 Saumya	44 Sādhāraṇa	11 Māgha
4052	873	1008	357	125-26	950-51	44 Sādhāraṇa	45 Virōdhakṛit	...
4053	874	1009	358	126-27	951-52	45 Virōdhakṛit	46 Paridhāvin	...
4054	875	1010	359	127-28	*952-53	46 Paridhāvin	47 Pramādin	7 Āśvina
4055	876	1011	360	128-29	953-54	47 Pramādin	48 Ānanda	...
4056	877	1012	361	129-30	954-55	48 Ānanda	49 Rākshasa	...
4057	878	1013	362	130-31	955-56	49 Rākshasa	50 Anala	4 Āshāḍha
4058	879	1014	363	131-32	*956-57	50 Anala	51 Piṅgala	...
4059	880	1015	364	132-33	957-58	51 Piṅgala	52 Kālayukta	12 Phālguna
4060	881	1016	365	133-34	958-59	52 Kālayukta	53 Siddhārthin	...
4061	882	1017	366	134-35	959-60	53 Siddhārthin	54 Raudra	...
4062	883	1018	367	135-36	*960-61	54 Raudra	55 Durmati	9 Mārgaśira
4063	884	1019	368	136-37	961-62	55 Durmati	56 Dundubhi	...
4064	885	1020	369	137-38	962-63	56 Dundubhi	57 Rudhirōdgārin	...
4065	886	1021	370	138-39	963-64	57 Rudhirōdgārin	58 Raktāksha	5 Śrāvaṇa
4066	887	1022	371	139-40	*964-65	58 Raktāksha	59 Krōdhana	...
4067	888	1023	372	140-41	965-66	59 Krōdhana	60 Kshaya	...
4068	889	1024	373	141-42	966-67	60 Kshaya	1 Prabhava	2 Vaiśākha
4069	890	1025	374	142-43	967-68	1 Prabhava	2 Vibhava	...
4070	891	1026	375	143-44	*968-69	2 Vibhava	3 Śukla	10 Pausha

XC—*contd.*

COMMENCEMENT OF THE									
MEAN SOLAR YEAR.					MEAN LUNI-SOLAR YEAR (MEAN SUNRISE OF THE CIVIL DAY ON WHICH CHAITRA ŚUKLA 1 ENDS).			Kali.
Day and month, A.D.	Week-day.	Time of mean Mēsha-saṁkrānti.			Day and month, A.D.	Week-day.	*a* (here = *t*, the index of the *tithi*).	
13	14	17			19	20	23	1
		H.	M.	S.				
23 Mar. (83)	0 Sat.	9	6	45	27 Feb. (58)	3 Tues.	133·6871	4046
23 Mar. (82)	1 Sun.	15	18	54	17 Mar. (76)	2 Mon.	168·3695	4047
23 Mar. (82)	2 Mon.	21	31	3	6 Mar. (65)	6 Fri.	44·0923	4048
24 Mar. (83)	4 Wed.	3	43	12	24 Feb. (55)	4 Wed.	258·4471	4049
23 Mar. (83)	5 Thur.	9	55	21	14 Mar. (74)	[illegible] Tues.	293·1295	4050
23 Mar. (82)	6 Fri.	16	7	30	3 Mar. (62)	0 Sat.	168·8524	4051
23 Mar. (82)	0 Sat.	22	19	39	22 Mar. (81)	6 Fri.	203·5348	4052
24 Mar. (83)	2 Mon.	4	31	48	11 Mar. (70)	3 Tues.	79·2576	4053
23 Mar. (83)	3 Tues.	10	43	57	29 Feb. (60)	1 Sun.	293·6125	4054
23 Mar. (82)	4 Wed.	16	56	6	19 Mar. (78)	0 Sat.	228·2949	4055
23 Mar. (82)	5 Thur.	23	8	15	8 Mar. (67)	4 Wed.	201·0176	4056
24 Mar. (83)	0 Sat.	5	20	24	25 Feb. (56)	1 Sun.	79·7405	4057
23 Mar. (83)	1 Sun.	11	32	33	15 Mar. (75)	0 Sat.	114·4229	4058
23 Mar. (82)	2 Mon.	17	44	42	5 Mar. (64)	5 Thur.	328·7778	4059
23 Mar. (82)	3 Tues.	23	56	51	23 Mar. (82)	1 Sun.	24·8281	4060
24 Mar. (83)	5 Thur.	6	9	0	13 Mar. (72)	1 Sun.	239·1830	4061
23 Mar. (83)	6 Fri.	12	21	9	1 Mar. (61)	5 Thur.	114·2058	4062
23 Mar. (82)	0 Sat.	18	33	18	20 Mar. (79)	4 Wed.	149·5881	4063
24 Mar. (83)	2 Mon.	0	45	27	9 Mar. (68)	1 Sun.	25·3110	4064
24 Mar. (83)	3 Tues.	6	57	36	27 Feb. (58)	6 Fri.	239·6659	4065
23 Mar. (83)	4 Wed.	13	9	45	17 Mar. (77)	5 Thur.	274·3483	4066
23 Mar. (82)	5 Thur.	19	21	54	6 Mar. (65)	2 Mon.	150·0710	4067
24 Mar. (83)	0 Sat.	1	34	3	23 Feb. (54)	6 Fri.	25·7939	4068
24 Mar. (83)	1 Sun.	7	46	12	14 Mar. (73)	5 Thur.	60·4763	4069
23 Mar. (83)	2 Mon.	13	58	21	3 Mar. (63)	3 Tues.	274·8311	4070

TABLE

CONCURRENT YEAR.								
Kali.	Śaka.	Chaitrādi Vikrama.	Mēshādi solar year in Bengal.	Kollam.	A.D.	Jovian Saṁvatsara. Southern system.	Jovian Saṁvatsara. Northern system.	Mean intercalated (*adhika*) lunar month.
1	2	3	3a	4	5	6	7	8a
4071	892	1027	376	144-45	969-70	3 Sukla . .	4 Pramōda .	...
4072	893	1028	377	145-46	970-71	4 Pramōda .	5 Prajāpati .	...
4073	894	1029	378	146-47	971-72	5 Prajāpati .	6 Aṅgiras . .	7 Āśvina .
4074	895	1030	379	147-48	*972-73	6 Aṅgiras . .	7 Śrīmukha .	...
4075	896	1031	380	148-49	973-74	7 Śrīmukha .	8 Bhāva . .	...
4076	897	1032	381	149-50	974-75	8 Bhāva . .	9 Yuvan . .	4 Āshāḍha .
4077	898	1033	382	150-51	975-76	9 Yuvan . .	10 Dhātṛi . .	...
4078	899	1034	383	151-52	*976-77	10 Dhātṛi . .	11 Īśvara . .	12 Phālguna .
4079	900	1035	384	152-53	977-78	11 Īśvara . .	12 Bahudhānya .	...
4080	901	1036	385	153-54	978-79	12 Bahudhānya .	13 Pramāthin .	...
4081	902	1037	386	154-55	979-80	13 Pramāthin .	14 Vikrama .	9 Mārgaśira .
4082	903	1038	387	155-56	*980-81	14 Vikrama .	15 Vṛisha . .	...
4083	904	1039	388	156-57	981-82	15 Vṛisha . .	16 Chitrabhānu .	...
4084	905	1040	389	157-58	982-83	16 Chitrabhānu .	17 Subhānu .	5 Śrāvaṇa .
4085	906	1041	390	158-59	983-84	17 Subhānu .	18 Tāraṇa . .	...
4086	907	1042	391	159-60	*984-85	18 Tāraṇa . .	19 Pārthiva .	...
4087	908	1043	392	160-61	985-86	19 Pārthiva .	20 Vyaya . .	2 Vaiśākha .
4088	909	1044	393	161-62	986-87	20 Vyaya . .	21 Sarvajit . .	
4089	910	1045	394	162-63	987-88	21 Sarvajit . .	22 Sarvadhārin .	10 Pausha
4090	911	1046	395	163-64	*988-89	22 Sarvadhārin .	23 Virōdhin .	
4091	912	1047	396	164-65	989-90	23 Virōdhin .	24 Vikṛita † .	
4092	913	1048	397	165-66	990-91	24 Vikṛita .	26 *Nandana* .	7 Āśvina .
4093	914	1049	398	166-67	991-92	25 Khara . .	27 *Vijaya* .	
4094	915	1050	399	167-68	*992-93	26 Nandana . .	28 *Jaya* . .	
4095	916	1051	400	168-69	993-94	27 Vijaya . .	29 *Manmatha* .	3 Jyēshṭha .

† 25 Khara was suppressed in the north by the *Brahma-Siddhānta* system, whether calculated by "true" or mean reckoning.

XC—*contd.*

COMMENCEMENT OF THE									
MEAN SOLAR YEAR.					MEAN LUNI-SOLAR YEAR (MEAN SUNRISE OF THE CIVIL DAY ON WHICH CHAITRA ŚUKLA 1 ENDS).			Kali.
Day and month, A.D.	Week-day.	Time of mean Mēsha-saṁkrānti.			Day and month, A.D.	Week-day.	*a* (here = *t*, the index of the *tithi*).	
13	14	17			19	20	23	1
		H.	M.	S.				
23 Mar. (82)	3 Tues.	20	10	30	22 Mar. (81)	2 Mon.	309·5135	4071
24 Mar. (83)	5 Thur.	2	22	39	11 Mar. (70)	6 Fri.	185·2364	4072
24 Mar. (83)	6 Fri.	8	34	48	28 Feb. (59)	3 Tues.	60·9593	4073
23 Mar. (83)	0 Sat.	14	46	57	18 Mar. (78)	2 Mon.	95·6416	4074
23 Mar. (82)	1 Sun.	20	59	6	8 Mar. (67)	0 Sat.	309·9964	4075
24 Mar. (83)	3 Tues.	3	11	15	25 Feb. (56)	4 Wed.	185·7193	4076
24 Mar. (83)	4 Wed.	9	23	24	16 Mar. (75)	3 Tues.	220·4016	4077
23 Mar. (83)	5 Thur.	15	35	33	4 Mar. (64)	0 Sat.	96·1245	4078
23 Mar. (82)	6 Fri.	21	47	42	23 Mar. (82)	6 Fri.	130·8069	4079
24 Mar. (83)	1 Sun.	3	59	51	12 Mar. (71)	3 Tues.	6·5298	4080
24 Mar. (83)	2 Mon.	10	12	0	2 Mar. (61)	1 Sun.	220·8845	4081
23 Mar. (83)	3 Tues.	16	24	9	20 Mar. (80)	0 Sat.	255·5669	4082
23 Mar. (82)	4 Wed.	22	36	18	9 Mar. (68)	4 Wed.	131·2898	4083
24 Mar. (83)	6 Fri.	4	48	27	26 Feb. (57)	1 Sun.	7·0127	4084
24 Mar. (83)	0 Sat.	11	0	36	17 Mar. (76)	0 Sat.	41·6950	4085
23 Mar. (83)	1 Sun.	17	12	45	6 Mar. (66)	5 Thur.	256·0499	4086
23 Mar. (82)	2 Mon.	23	24	54	23 Feb. (54)	2 Mon.	131·7727	4087
24 Mar. (83)	4 Wed.	5	37	3	14 Mar. (73)	1 Sun.	166·4550	4088
24 Mar. (83)	5 Thur.	11	49	12	3 Mar. (62)	5 Thur.	42·1779	4089
23 Mar. (83)	6 Fri.	18	1	21	21 Mar. (81)	4 Wed.	76·8603	4090
24 Mar. (83)	1 Sun.	0	13	30	11 Mar. (70)	2 Mon.	291·2152	4091
24 Mar. (83)	2 Mon.	6	25	39	28 Feb. (59)	6 Fri.	166·9398	4092
24 Mar. (83)	3 Tues.	12	37	48	19 Mar. (78)	5 Thur.	201·6204	4093
23 Mar. (83)	4 Wed.	18	49	57	7 Mar. (67)	2 Mon.	77·3432	4094
24 Mar. (83)	6 Fri.	1	2	6	25 Feb. (56)	0 Sat.	291·6980	4095

TABLE

CONCURRENT YEAR.								Mean intercalated (*adhika*) lunar month.
						Jovian Saṁvatsara.		
Kali.	Śaka.	Chaitrādi Vikrama.	Mēshādi solar year in Bengal.	Kollam.	A.D.	Southern system.	Northern system.	
1	2	3	3a	4	5	6	7	8a
4096	917	1052	401	169-70	994-95	28 Jaya . .	30 *Durmukha* .	
4097	918	1053	402	170-71	995-96	29 Manmatha .	31 Hēmalamba .	12 Phālguna .
4098	919	1054	403	171-72	*996-97	30 Durmukha .	32 Vilamba .	...
4099	920	1055	404	172-73	997-98	31 Hēmalamba .	33 Vikārin . .	...
4100	921	1056	405	173-74	998-99	32 Vilamba .	34 Śārvarin . .	8 Kārttika .
4101	922	1057	406	174-75	999-1000	33 Vikārin . .	35 Plava . .	...
4102	923	1058	407	175-76	*1000-01	34 Śārvarin .	36 Śubhakṛit .	...
4103	924	1059	408	176-77	1001-02	35 Plava . .	37 Śōbhana .	5 Śrāvaṇa
4104	925	1060	409	177-78	1002-03	36 Subhakṛit .	38 Krōdhin .	...
4105	926	1061	410	178-79	1003-04	37 Sōbhana .	39 Viśvāvasu .	...
4106	927	1062	411	179-80	*1004-05	38 Krōdhin .	40 Parābhava .	1 Chaitra .
4107	928	1063	412	180-81	1005-06	39 Viśvāvasu .	41 Plavaṅga	...
4108	929	1064	413	181-82	1006-07	40 Parābhava .	42 Kīlaka . .	10 Pausha .
4109	930	1065	414	182-83	1007-08	41 Plavaṅga .	43 Saumya .	...
4110	931	1066	415	183-84	*1008-09	42 Kīlaka . .	44 Sādhāraṇa .	
4111	932	1067	416	184-85	1009-10	43 Saumya .	45 Virōdhakṛit .	7 Āśvina† .
4112	933	1068	417	185-86	1010-11	44 Sādhāraṇa .	46 Paridhāvin .	
4113	934	1069	418	186-87	1011-12	45 Virōdhakṛit .	47 Pramādin .	...
4114	935	1070	419	187-88	*1012-13	46 Paridhāvin .	48 Ānanda . .	3 Jyēshṭha .
4115	936	1071	420	188-89	1013-14	47 Pramādin .	49 Rākshasa .	...
4116	937	1072	421	189-90	1014-15	48 Ānanda . .	50 Anala . .	12 Phālguna .
4117	938	1073	422	190-91	1015-16	49 Rākshasa .	51 Piṅgala . .	...
4118	939	1074	423	191-92	*1016-17	50 Anala .	52 Kālayukta .	...
4119	940	1075	424	192-93	1017-18	51 Piṅgala .	53 Siddhārthin .	8 Kārttika .
4120	941	1076	425	193-94	1018-19	52 Kālayukta .	54 Raudra . .	

† See "Remarks," p. 523 above.

XC—*contd.*

COMMENCEMENT OF THE									
MEAN SOLAR YEAR.					MEAN LUNI-SOLAR YEAR (MEAN SUNRISE OF THE CIVIL DAY ON WHICH CHAITRA ŚUKLA 1 ENDS).			Kali.
Day and month, A.D.	Week-day.	Time of mean Mēsha-saṁkrānti.			Day and month, A.D.	Week-day.	*a* (here = *t*, the index of the *tithi*).	
13	14	17			19	20	23	1
		H.	M.	S.				
24 Mar. (83)	0 Sat.	7	14	15	16 Mar. (75)	6 Fri.	326·3804	4096
24 Mar. (83)	1 Sun.	13	26	24	5 Mar. (64)	3 Tues.	202·1033	4097
23 Mar. (83)	2 Mon.	19	38	33	23 Mar. (83)	2 Mon.	236·7856	4098
24 Mar. (83)	4 Wed.	1	50	42	12 Mar. (71)	6 Fri.	112·5085	4099
24 Mar. (83)	5 Thur.	8	2	51	2 Mar. (61)	4 Wed.	326·8633	4100
24 Mar. (83)	6 Fri.	14	15	0	20 Mar. (79)	2 Mon.	22·9136	4101
23 Mar. (83)	0 Sat.	20	27	9	9 Mar. (69)	0 Sat.	237·2685	4102
24 Mar. (83)	2 Mon.	2	39	18	26 Feb. (57)	4 Wed.	112·9914	4103
24 Mar. (83)	3 Tues.	8	51	27	17 Mar. (76)	3 Tues.	147·6737	4104
24 Mar. (83)	4 Wed.	15	3	36	6 Mar. (65)	0 Sat.	23·3966	4105
23 Mar. (83)	5 Thur.	21	15	45	24 Feb. (55)	5 Thur.	237·7514	4106
24 Mar. (83)	0 Sat.	3	27	54	14 Mar. (73)	4 Wed.	272·4338	4107
24 Mar. (83)	1 Sun.	9	40	3	3 Mar. (62)	1 Sun.	148·1566	4108
24 Mar. (83)	2 Mon.	15	52	12	22 Mar. (81)	0 Sat.	182·8390	4109
23 Mar. (83)	3 Tues.	22	4	21	10 Mar. (70)	4 Wed.	58·5618	4110
24 Mar. (83)	5 Thur.	4	16	30	28 Feb. (59)	2 Mon.	272·9167	4111
24 Mar. (83)	6 Fri.	10	28	39	19 Mar. (78)	1 Sun.	307·5991	4112
24 Mar. (83)	0 Sat.	16	40	48	8 Mar. (67)	5 Thur.	183·3219	4113
23 Mar. (83)	1 Sun.	22	52	57	25 Feb. (56)	2 Mon.	59·0447	4114
24 Mar. (83)	3 Tues.	5	5	6	15 Mar. (74)	1 Sun.	93·7270	4115
24 Mar. (83)	4 Wed.	11	17	15	5 Mar. (64)	6 Fri.	308·0820	4116
24 Mar. (83)	5 Thur.	17	29	24	23 Mar. (82)	4 Wed.	4·1323	4117
23 Mar. (83)	6 Fri.	23	41	33	12 Mar. (72)	2 Mon.	218·4872	4118
24 Mar. (83)	1 Sun.	5	53	42	1 Mar. (60)	6 Fri.	94·2100	4119
24 Mar. (83)	2 Mon.	12	5	51	20 Mar. (79)	5 Thur.	128·8924	4120

TABLE

CONCURRENT YEAR.								Mean intercalated (*adhika*) lunar month.
Kali.	Śaka.	Chaitrādi Vikrama.	Mēshādi solar year in Bengal.	Kollam.	A.D.	Jovian Saṁvatsara. Southern system.	Jovian Saṁvatsara. Northern system.	
1	2	3	3a	4	5	6	7	8a
4121	942	1077	426	194-95	1019-20	53 Siddhārthin .	55 Durmati . .	...
4122	943	1078	427	195-96	*1020-21	54 Raudra .	56 Dundubhi .	5 Śrāvaṇa .
4123	944	1079	428	196-97	1021-22	55 Durmati . .	57 Rudhirōdgārin .	...
4124	945	1080	429	197-98	1022-23	56 Dundubhi .	58 Raktāksha .	...
4125	946	1081	430	198-99	1023-24	57 Rudhirōdgārin .	59 Krōdhana .	1 Chaitra .
4126	947	1082	431	199-200	*1024-25	58 Raktāksha .	60 Kshaya . .	...
4127	948	1083	432	200-01	1025-26	59 Krōdhana .	1 Prabhava .	10 Pausha .
4128	949	1084	433	201-02	1026-27	60 Kshaya . .	2 Vibhava . .	...
4129	950	1085	434	202-03	1027-28	1 Prabhava .	3 Śukla . .	...
4130	951	1086	435	203-04	*1028-29	2 Vibhava .	4 Pramōda .	6 Bhādrapada .
4131	952	1087	436	204-05	1029-30	3 Śukla . .	5 Prajāpati .	...
4132	953	1088	437	205-06	1030-31	4 Pramōda .	6 Aṅgiras . .	...
4133	954	1089	438	206-07	1031-32	5 Prajāpati .	7 Śrīmukha .	3 Jyēshṭha .
4134	955	1090	439	207-08	*1032-33	6 Aṅgiras . .	8 Bhāva . .	...
4135	956	1091	440	208-09	1033-34	7 Śrīmukha .	9 Yuvan . .	11 Māgha .
4136	957	1092	441	209-10	1034-35	8 Bhāva . .	10 Dhātṛi . .	...
4137	958	1093	442	210-11	1035-36	9 Yuvan . .	11 Īśvara . .	...
4138	959	1094	443	211-12	*1036-37	10 Dhātṛi . .	12 Bahudhānya .	8 Kārttika .
4139	960	1095	444	212-13	1037-38	11 Īśvara . .	13 Pramāthin .	...
4140	961	1096	445	213-14	1038-39	12 Bahudhānya .	14 Vikrama .	...
4141	962	1097	446	214-15	1039-40	13 Pramāthin .	15 Vṛisha . .	4 Āshāḍha .
4142	963	1098	447	215-16	*1040-41	14 Vikrama . .	16 Chitrabhānu .	...
4143	964	1099	448	216-17	1041-42	15 Vṛisha . .	17 Subhānu .	... *
4144	965	1100	449	217-18	1042-43	16 Chitrabhānu .	18 Tāraṇa . .	1 Chaitra .
4145	966	1101	450	218-19	1043-44	17 Subhānu .	19 Pārthiva .	...

XC—*contd.*

COMMENCEMENT OF THE							
MEAN SOLAR YEAR.			MEAN LUNI-SOLAR YEAR (MEAN SUNRISE OF THE CIVIL DAY ON WHICH CHAITRA ŚUKLA 1 ENDS).			Kali.	
Day and month, A.D.	Week-day.	Time of mean Mēsha-saṁkrānti.		Day and month, A.D.	Week-day.	*a* (here = *t*, the index of the *tithi*).	
13	14	17		19	20	23	1
		H. M. S.					
24 Mar. (83)	3 Tues.	18 18 0		9 Mar. (68)	2 Mon.	4·6131	4121
24 Mar. (84)	5 Thur.	0 30 9		27 Feb. (58)	0 Sat.	218·9701	4122
24 Mar. (83)	6 Fri.	6 42 18		17 Mar. (76)	6 Fri.	253·6525	4123
24 Mar. (83)	0 Sat.	12 54 27		6 Mar. (65)	3 Tues.	129·3753	4124
24 Mar. (83)	1 Sun.	19 6 36		23 Feb. (54)	0 Sat.	5·0981	4125
24 Mar. (84)	3 Tues.	1 18 45		13 Mar. (73)	6 Fri.	39·7806	4126
24 Mar. (83)	4 Wed.	7 30 54		3 Mar. (62)	4 Wed.	254·1354	4127
24 Mar. (83)	5 Thur.	13 43 3		22 Mar. (81)	3 Tues.	288·8177	4128
24 Mar. (83)	6 Fri.	19 55 12		11 Mar. (70)	0 Sat.	164·5406	4129
24 Mar. (84)	1 Sun.	2 7 21		28 Feb. (59)	4 Wed.	40·2635	4130
24 Mar. (83)	2 Mon.	8 19 30		18 Mar. (77)	3 Tues.	74·9458	4131
24 Mar. (83)	3 Tues.	14 31 39		8 Mar. (67)	1 Sun.	289·3006	4132
24 Mar. (83)	4 Wed.	20 43 48		25 Feb. (56)	5 Thur.	165·0235	4133
24 Mar. (84)	6 Fri.	2 55 57		15 Mar. (75)	4 Wed.	199·7059	4134
24 Mar. (83)	0 Sat.	9 8 6		4 Mar. (63)	1 Sun.	75·4287	4135
24 Mar. (83)	1 Sun.	15 20 15		23 Mar. (82)	0 Sat.	110·1111	4136
24 Mar. (83)	2 Mon.	21 32 24		13 Mar. (72)	5 Thur.	324·4660	4137
24 Mar. (84)	4 Wed.	3 44 33		1 Mar. (61)	2 Mon.	200·1888	4138
24 Mar. (83)	5 Thur.	9 56 42		20 Mar. (79)	1 Sun.	234·8712	4139
24 Mar. (83)	6 Fri.	16 8 51		9 Mar. (68)	5 Thur.	110·5940	4140
24 Mar. (83)	0 Sat.	22 21 0		27 Feb. (58)	3 Tues.	324·9489	4141
24 Mar. (84)	2 Mon.	4 33 9		16 Mar. (76)	1 Sun.	20·9992	4142
24 Mar. (83)	3 Tues.	10 45 18		6 Mar. (65)	6 Fri.	235·3541	4143
24 Mar. (83)	4 Wed.	16 57 27		23 Feb. (54)	3 Tues.	111·0793	4144
24 Mar. (83)	5 Thur.	23 9 36		14 Mar. (73)	2 Mon.	145·759[illegible]	4145

TABLE

CONCURRENT YEAR.								Mean intercalated (*adhika*) lunar month.
Kali.	Śaka.	Chaitrādi Vikrama.	Mēshādi solar year in Bengal.	Kollam.	A.D.	Jovian Samvatsara. Southern system.	Jovian Samvatsara. Northern system.	
1	2	3	3a	4	5	6	7	8a
4146	967	1102	451	219-20	*1044-45	18 Tāraṇa . .	20 Vyaya . .	9 Mārgaśira
4147	968	1103	452	220-21	1045-46	19 Pārthiva .	21 Sarvajit .	...
4148	969	1104	453	221-22	1046-47	20 Vyaya . .	22 Sarvadhārin .	...
4149	970	1105	454	222-23	1047-48	21 Sarvajit . .	23 Virōdhin .	6 Bhādrapada .
4150	971	1106	455	223-24	*1048-49	22 Sarvadhārin .	24 Vikṛita . .	...
4151	972	1107	456	224-25	1049-50	23 Virōdhin .	25 Khara . .	
4152	973	1108	457	225-26	1050-51	24 Vikṛita . .	26 Nandana .	3 Jyēshṭha .
4153	974	1109	458	226-27	1051-52	25 Khara . .	27 Vijaya . .	...
4154	975	1110	459	227-28	*1052-53	26 Nandana .	28 Jaya . .	11 Māgha .
4155	976	1111	460	228-29	1053-54	27 Vijaya . .	29 Manmatha .	
4156	977	1112	461	229-30	1054-55	28 Jaya . .	30 Durmukha .	
4157	978	1113	462	230-31	1055-56	29 Manmatha .	31 Hēmalamba .	8 Kārttika .
4158	979	1114	463	231-32	*1056-57	30 Durmukha .	32 Vilamba .	
4159	980	1115	464	232-33	1057-58	31 Hēmalamba .	33 Vikārin . .	...
4160	981	1116	465	233-34	1058-59	32 Vilamba .	34 Śārvarin .	4 Āshāḍha .
4161	982	1117	466	234-35	1059-60	33 Vikārin . .	35 Plava . .	...
4162	983	1118	467	235-36	*1060-61	34 Śārvarin .	36 Śubhakṛit .	...
4163	984	1119	468	236-37	1061-62	35 Plava . .	37 Śōbhana .	1 Chaitra .
4164	985	1120	469	237-38	1062-63	36 Śubhakṛit .	38 Krōdhin .	
4165	986	1121	470	238-39	1063-64	37 Śōbhana .	39 Viśvāvasu .	9 Mārgaśira .
4166	987	1122	471	239-40	*1064-65	38 Krōdhin .	40 Parābhava .	
4167	988	1123	472	240-41	1065-66	39 Viśvāvasu .	41 Plavanga .	
4168	989	1124	473	241-42	1066-67	40 Parābhava .	42 Kīlaka .	6 Bhādrapada .
4169	990	1125	474	242-43	1067-68	41 Plavanga .	43 Saumya .	...
4170	991	1126	475	243-44	*1068-69	42 Kīlaka . .	44 Sādhāraṇa .	...

XC—*contd.*

COMMENCEMENT OF THE							
MEAN SOLAR YEAR.			MEAN LUNI-SOLAR YEAR (MEAN SUNRISE OF THE CIVIL DAY ON WHICH CHAITRA ŚUKLA 1 ENDS).			Kali.
Day and month, A.D.	Week-day.	Time of mean Mēsha-saṁkrānti.	Day and month, A.D.	Week-day.	*a* (here = *i*, the index of the *tithi*).	
13	14	17	19	20	23	1
		H. M. S.				
24 Mar. (84)	0 Sat.	5 21 45	2 Mar. (62)	6 Fri.	21·4821	4146
24 Mar. (83)	1 Sun.	11 33 54	21 Mar. (80)	5 Thur.	56·1645	4147
24 Mar. (83)	2 Mon.	17 46 3	11 Mar. (70)	3 Tues.	270·5194	4148
24 Mar. (83)	3 Tues.	23 58 12	28 Feb. (59)	0 Sat.	146·2422	4149
24 Mar. (84)	5 Thur.	6 10 21	18 Mar. (78)	6 Fri.	180·9246	4150
24 Mar. (83)	6 Fri.	12 22 30	7 Mar. (66)	3 Tues.	56·6475	4151
24 Mar. (83)	0 Sat.	18 34 39	25 Feb. (56)	1 Sun.	271·0023	4152
25 Mar. (84)	2 Mon.	0 46 48	16 Mar. (75)	0 Sat.	305·6846	4153
24 Mar. (84)	3 Tues.	6 58 57	4 Mar. (64)	4 Wed.	181·4075	4154
24 Mar. (83)	4 Wed.	13 11 6	23 Mar. (82)	3 Tues.	216·0899	4155
24 Mar. (83)	5 Thur.	19 23 15	12 Mar. (71)	0 Sat.	91·8127	4156
25 Mar. (84)	0 Sat.	1 35 24	2 Mar. (61)	5 Thur.	306·1675	4157
24 Mar. (84)	1 Sun.	7 47 33	19 Mar. (79)	3 Tues.	2·2180	4158
24 Mar. (83)	2 Mon.	13 59 42	9 Mar. (68)	1 Sun.	216·5728	4159
24 Mar. (83)	3 Tues.	20 11 51	26 Feb. (57)	5 Thur.	92·2956	4160
25 Mar. (84)	5 Thur.	2 24 0	17 Mar. (76)	4 Wed.	126·9780	4161
24 Mar. (84)	6 Fri.	8 36 9	5 Mar. (65)	1 Sun	2·7009	4162
24 Mar. (83)	0 Sat.	14 48 18	23 Feb. (54)	6 Fri.	217·0556	4163
24 Mar. (83)	1 Sun.	21 0 27	14 Mar. (73)	5 Thur.	251·7380	4164
25 Mar. (84)	3 Tues.	3 12 36	3 Mar. (62)	2 Mon.	127·4609	4165
24 Mar. (84)	4 Wed.	9 24 45	21 Mar. (81)	1 Sun.	162·1433	4166
24 Mar. (83)	5 Thur.	15 36 54	10 Mar. (69)	5 Thur.	37·8661	4167
24 Mar. (83)	6 Fri.	21 49 3	28 Feb. (59)	3 Tues.	252·2210	4168
25 Mar. (84)	1 Sun.	4 1 12	19 Mar. (78)	2 Mon.	286·9051	4169
24 Mar. (84)	2 Mon.	10 13 21	7 Mar. (67)	6 Fri.	162·6262	4170

TABLE

CONCURRENT YEAR.								Mean intercalated (*adhika*) lunar month.
Kali.	Saka.	Chaitrādi Vikrama.	Mēshādi solar year in Bengal.	Kollam.	A.D.	Jovian Samvatsara. Southern system.	Jovian Samvatsara. Northern system.	
1	2	3	3*a*	4	5	6	7	8*a*
4171	992	1127	476	244-45	1069-70	43 Saumya	45 Virōdhakṛit	2 Vaiśākha
4172	993	1128	477	245-46	1070-71	44 Sādhāraṇa	46 Paridhāvin	...
4173	994	1129	478	246-47	1071-72	45 Virōdhakṛit	47 Pramādin	11 Māgha
4174	995	1130	479	247-48	*1072-73	46 Paridhāvin	48 Ānanda	...
4175	996	1131	480	248-49	1073-74	47 Pramādin	49 Rākshasa	...
4176	997	1132	481	249-50	1074-75	48 Ānanda	50 Anala †	7 Āśvina
4177	998	1133	482	250-51	1075-76	49 Rākshasa	52 *Kālayukta*	...
4178	999	1134	483	251-52	*1076-77	50 Anala	53 *Siddhārthin*	...
4179	1000	1135	484	252-53	1077-78	51 Piṅgala	54 *Raudra*	4 Āshāḍha
4180	1001	1136	485	253-54	1078-79	52 Kālayukta	55 *Durmati*	...
4181	1002	1137	486	254-55	1079-80	53 Siddhārthin	56 *Dundubhi*	12 Phālguna
4182	1003	1138	487	255-56	*1080-81	54 Raudra	57 Rudhirōdgārin	...
4183	1004	1139	488	256-57	1081-82	55 Durmati	58 Raktāksha	...
4184	1005	1140	489	257-58	1082-83	56 Dundubhi	59 Krōdhana	9 Mārgaśira
4185	1006	1141	490	258-59	1083-84	57 Rudhirōdgārin	60 Kshaya	...
4186	1007	1142	491	259-60	*1084-85	58 Raktāksha	1 Prabhava	...
4187	1008	1143	492	260-61	1085-86	59 Krōdhana	2 Vibhava	6 Bhādrapada
4188	1009	1144	493	261-62	1086-87	60 Kshaya	3 Śukla	...
4189	1010	1145	494	262-63	1087-88	1 Prabhava	4 Pramōda	...
4190	1011	1146	495	263-64	*1088-89	2 Vibhava	5 Prajāpati	2 Vaiśākha
4191	1012	1147	496	264-65	1089-90	3 Śukla	6 Aṅgiras	...
4192	1013	1148	497	265-66	1090-91	4 Pramōda	7 Śrīmukha	11 Māgha
4193	1014	1149	498	266-67	1091-92	5 Prajāpati	8 Bhāva	...
4194	1015	1150	499	267-68	*1092-93	6 Aṅgiras	9 Yuvan	...
4195	1016	1151	500	268-69	1093-94	7 Śrīmukha	10 Dhātṛi	7 Āśvina

† 51 Piṅgala was suppressed in the north, according to both "true" and mean systems, in *Brahma-Siddhānta* reckoning

XC—*contd.*

COMMENCEMENT OF THE									
MEAN SOLAR YEAR.					MEAN LUNI-SOLAR YEAR (MEAN SUNRISE OF THE CIVIL DAY ON WHICH CHAITRA ŚUKLA 1 ENDS).			Kali.
Day and month, A.D.	Week-day.	Time of mean Mēsha-saṁkrānti.			Day and month, A.D.	Week-day.	*a* (here = *t*, the index of the *tithi*).	
13	14	17			19	20	23	1
		H.	M.	S.				
24 Mar. (83) . .	3 Tues. .	16	25	30	24 Feb. (55) .	3 Tues. .	38·3490	4171
24 Mar. (83) . .	4 Wed. .	22	37	39	15 Mar. (74) .	2 Mon. .	73·0314	4172
25 Mar. (84) . .	6 Fri. .	4	49	48	5 Mar. (64) .	0 Sat. .	287·3863	4173
24 Mar. (84) . .	0 Sat. .	11	1	57	23 Mar. (83) .	6 Fri. .	322·0686	4174
24 Mar. (83) .	1 Sun. .	17	14	6	12 Mar. (71) .	3 Tues. .	197·7915	4175
24 Mar. (83) . .	2 Mon. .	23	26	15	1 Mar. (60) .	0 Sat. .	73·5143	4176
25 Mar. (84) . .	4 Wed. .	5	38	24	20 Mar. (79) .	6 Fri. .	108·1967	4177
24 Mar. (84) . .	5 Thur. .	11	50	33	9 Mar. (69) .	4 Wed. .	322·5515	4178
24 Mar. (83) . .	6 Fri. .	18	2	42	26 Feb. (57) .	1 Sun. .	198·2744	4179
25 Mar. (84) .	1 Sun. .	0	14	51	17 Mar (76) .	0 Sat. .	232·9568	4180
25 Mar. (84) .	2 Mon. .	6	27	0	6 Mar. (65) .	4 Wed. .	108·6796	4181
24 Mar. (84) . .	3 Tues. .	12	39	9	24 Mar. (84) .	3 Tues. .	143·3620	4182
24 Mar. (83) . .	4 Wed. .	18	51	18	13 Mar. (72) .	0 Sat. .	19·0848	4183
25 Mar. (84) . .	6 Fri. .	1	3	27	3 Mar. (62) .	5 Thur. .	233·4397	4184
25 Mar. (84) . .	0 Sat. .	7	15	36	22 Mar. (81) .	4 Wed. .	268·1220	4185
24 Mar. (84) . .	1 Sun. .	13	27	45	10 Mar. (70) .	1 Sun. .	143·8449	4186
24 Mar. (83) . .	2 Mon. .	19	39	54	27 Feb. (58) .	5 Thur. .	19·5678	4187
25 Mar. (84) . .	4 Wed. .	1	52	3	18 Mar. (77) .	4 Wed. .	54·2501	4188
25 Mar. (84) . .	5 Thur. .	8	4	12	8 Mar. (67) .	2 Mon. .	268·6050	4189
24 Mar. (84) . .	6 Fri. .	14	16	21	25 Feb. (56) .	6 Fri. .	144·3278	4190
24 Mar. (83) . .	0 Sat. .	20	28	30	15 Mar. (74) .	5 Thur. .	179·0102	4191
25 Mar. (84) . .	2 Mon. .	2	40	39	4 Mar. (63) .	2 Mon. .	54·7330	4192
25 Mar. (84) . .	3 Tues. .	8	52	48	23 Mar. (82) .	1 Sun. .	89·4154	4193
24 Mar. (84) . .	4 Wed. .	15	4	57	12 Mar. (72) .	6 Fri. .	203·776[illegible]	4194
24 Mar. (83) . .	5 Thur. .	21	17	6	1 Mar. (60) .	3 Tues. .	179·4930	4195

TABLE

CONCURRENT YEAR.								Mean intercalated (*adhika*) lunar month.
						Jovian Samvatsara.		
Kali.	Saka.	Chaitrâdi Vikrama.	Mêshâdi solar year in Bengal.	Kollam.	A.D.	Southern system.	Northern system.	
1	2	3	3a	4	5	6	7	8a
4196	1017	1152	501	269-70	1094-95	8* Bhāva	11 Īśvara	...
4197	1018	1153	502	270-71	1095-96	9 Yuvan	12 Bahudhānya	...
4198	1019	1154	503	271-72	*1096-97	10 Dhātṛi	13 Pramādin	4 Āshāḍha
4199	1020	1155	504	272-73	1097-98	11 Īśvara	14 Vikrama	...
4200	1021	1156	505	273-74	1098-99	12 Bahudhānya	15 Vṛisha	12 Phālguna
4201	1022	1157	506	274-75	1099-1100	13 Pramādin	16 Chitrabhānu	...
4202	1023	1158	507	275-76	*1100-01	14 Vikrama	17 Subhānu	...
4203	1024	1159	508	276-77	1101-02	15 Vṛisha	18 Tāraṇa	9 Mārgaśira
4204	1025	1160	509	277-78	1102-03	16 Chitrabhānu	19 Pārthiva	...
4205	1026	1161	510	278-79	1103-04	17 Subhānu	20 Vyaya	...
4206	1027	1162	511	279-80	*1104-05	18 Tāraṇa	21 Sarvajit	5 Śrāvaṇa
4207	1028	1163	512	280-81	1105-06	19 Pārthiva	22 Sarvadhārin	...
4208	1029	1164	513	281-82	1106-07	20 Vyaya	23 Virōdhin	...
4209	1030	1165	514	282-83	1107-08	21 Sarvajit	24 Vikṛita	2 Vaiśākha
4210	1031	1166	515	283-84	*1108-09	22 Sarvadhārin	25 Khara	...
4211	1032	1167	516	284-85	1109-10	23 Virōdhin	26 Nandana	10 Pausha
4212	1033	1168	517	285-86	1110-11	24 Vikṛita	27 Vijaya	...
4213	1034	1169	518	286-87	1111-12	25 Khara	28 Jaya	...
4214	1035	1170	519	287-88	*1112-13	26 Nandana	29 Manmatha	7 Āśvina
4215	1036	1171	520	288-89	1113-14	27 Vijaya	30 Durmukha	...
4216	1037	1172	521	289-90	1114-15	28 Jaya	31 Hēmalamba	...
4217	1038	1173	522	290-91	1115-16	29 Manmatha	32 Vilamba	3 Jyēshṭha
4218	1039	1174	523	291-92	*1116-17	30 Durmukha	33 Vikārin	...
4219	1040	1175	524	292-93	1117-18	31 Hēmalamba	34 Śārvarin	12 Phālguna
4220	1041	1176	525	293-94	1118-19	32 Vilamba	35 Plava	

XC—*contd.*

COMMENCEMENT OF THE									
MEAN SOLAR YEAR.					MEAN LUNI-SOLAR YEAR (MEAN SUNRISE OF THE CIVIL DAY ON WHICH CHAITRA ŚUKLA 1 ENDS).			Kali.
Day and month, A.D.	Week-day	Time of mean Mēsha-saṁkrānti			Day and month, A.D.	Week-day.	*a* (here = *t*, the index of the *tithi*).	
13	14	17			19	20	23	1
		H.	M.	S.				
25 Mar. (84)	0 Sat.	3	29	15	20 Mar. (79)	2 Mon.	214·1755	4196
25 Mar. (84)	1 Sun.	9	41	24	9 Mar. (68)	6 Fri.	89·8983	4197
24 Mar. (84)	2 Mon.	15	53	33	27 Feb. (58)	4 Wed.	304·2531	4198
24 Mar. (83)	3 Tues.	22	5	42	16 Mar. (75)	2 Mon.	0·3035	4199
25 Mar. (84)	5 Thur.	4	17	51	6 Mar. (65)	0 Sat.	214·6584	4200
25 Mar. (84)	6 Fri.	10	30	0	25 Mar. (84)	6 Fri.	249·3408	4201
24 Mar. (84)	0 Sat.	16	42	9	13 Mar. (73)	3 Tues.	125·0637	4202
24 Mar. (83)	1 Sun.	22	54	18	2 Mar. (61)	0 Sat.	0·7865	4203
25 Mar. (84)	3 Tues.	5	6	27	21 Mar. (80)	6 Fri.	35·4689	4204
25 Mar. (84)	4 Wed.	11	18	36	11 Mar. (70)	4 Wed.	249·8237	4205
24 Mar. (84)	5 Thur.	17	30	45	28 Feb. (59)	1 Sun.	125·5466	4206
24 Mar. (83)	6 Fri.	23	42	54	18 Mar. (77)	0 Sat.	160·2289	4207
25 Mar. (84)	1 Sun.	5	55	3	7 Mar. (66)	4 Wed.	35·9518	4208
25 Mar. (84)	2 Mon.	12	7	12	25 Feb. (56)	2 Mon.	250·3066	4209
24 Mar. (84)	3 Tues.	18	19	21	15 Mar. (75)	1 Sun.	284·9889	4210
25 Mar. (84)	5 Thur.	0	31	30	4 Mar. (63)	5 Thur.	160·7118	4211
25 Mar. (84)	6 Fri.	6	43	39	23 Mar. (82)	4 Wed.	195·3942	4212
25 Mar. (84)	0 Sat.	12	55	48	12 Mar. (71)	1 Sun.	71·1171	4213
24 Mar. (84)	1 Sun.	19	7	57	1 Mar. (61)	6 Fri.	285·4718	4214
25 Mar. (84)	3 Tues.	1	20	6	20 Mar. (79)	5 Thur.	320·1543	4215
25 Mar. (84)	4 Wed.	7	32	15	9 Mar. (68)	2 Mon.	195·8771	4216
25 Mar. (84)	5 Thur.	13	44	24	26 Feb. (57)	6 Fri.	71·5999	4217
24 Mar. (84)	6 Fri.	19	56	33	16 Mar. (76)	5 Thur.	106·2823	4218
25 Mar. (84)	1 Sun.	2	8	42	6 Mar. (65)	[illegible] Tues.	320·6372	4219
25 Mar. (84)	2 Mon.	8	20	51	24 Mar. (83)	1 Sun.	16·6876	4220

TABLE

CONCURRENT YEAR.								Mean intercalated (*adhika*) lunar month.
						JOVIAN SAMVATSARA.		
Kali.	Śaka.	Chaitrādi Vikrama.	Mēshādi solar year in Bengal.	Kollam.	A.D.	Southern system.	Northern system.	
1	2	3	3*a*	4	5	6	7	8*a*
4221	1042	1177	526	294-95	1119-20	33 Vikārin . .	36 Śubhakṛit .	...
4222	1043	1178	527	295-96	*1120-21	34 Śārvarin .	37 Śōbhana . .	8 Kārttika .
4223	1044	1179	528	296-97	1121-22	35 Plava . .	38 Krōdhin . .	...
4224	1045	1180	529	297-98	1122-23	36 Śubhakṛit .	39 Viśvāvasu .	...
4225	1046	1181	530	298-99	1123-24	37 Śōbhana . .	40 Parābhava .	5 Srāvaṇa .
4226	1047	1182	531	299-300	*1124-25	38 Krōdhin . .	41 Plavaṅga .	...
4227	1048	1183	532	300-01	1125-26	39 Viśvāvasu .	42 Kīlaka . .	...
4228	1049	1184	533	301-02	1126-27	40 Parābhava .	43 Saumya . .	2 Vaiśākha
4229	1050	1185	534	302-03	1127-28	41 Plavaṅga .	44 Sādhāraṇa .	...
4230	1051	1186	535	303-04	*1128-29	42 Kīlaka . .	45 Virōdhakṛit .	10 Pausha
4231	1052	1187	536	304-05	1129-30	43 Saumya . .	46 Paridhāvin .	...
4232	1053	1188	537	305-06	1130-31	44 Sādhāraṇa .	47 Pramādin .	...
4233	1054	1189	538	306-07	1131-32	45 Virōdhakṛit .	48 Ānanda . .	7 Āśvina
4234	1055	1190	539	307-08	*1132-33	46 Paridhāvin .	49 Rākshasa .	...
4235	1056	1191	540	308-09	1133-34	47 Pramadin .	50 Anala . .	...
4236	1057	1192	541	309-10	1134-35	48 Ānanda . .	51 Piṅgala . .	3 Jyēshṭha
4237	1058	1193	542	310-11	1135-36	49 Rākshasa .	52 Kālayukta .	...
4238	1059	1194	543	311-12	*1136-37	50 Anala . .	53 Siddhārthin .	12 Phālguna
4239	1060	1195	544	312-13	1137-38	51 Piṅgala . .	54 Raudra . .	...
4240	1061	1196	545	313-14	1138-39	52 Kālayukta .	55 Durmati . .	...
4241	1062	1197	546	314-15	1139-40	53 Siddhārthin .	56 Dundubhi .	8 Kārttika .
4242	1063	1198	547	315-16	*1140-41	54 Raudra . .	57 Rudhirōdgārin .	...
4243	1064	1199	548	316-17	1141-42	55 Durmati . .	58 Raktāksha .	...
4244	1065	1200	549	317-18	1142-43	56 Dundubhi .	59 Krōdhana .	5 Srāvaṇa
4245	1066	1201	550	318-19	1143-44	57 Rudhirōdgārin .	60 Kshaya . .	...

XC—*contd.*

COMMENCEMENT OF THE									
MEAN SOLAR YEAR.					MEAN LUNI-SOLAR YEAR (MEAN SUNRISE OF THE CIVIL DAY ON WHICH CHAITRA ŚUKLA 1 ENDS).			Kali.
Day and month, A.D.	Week-day.	Time of mean Mēsha-saṁkrānti.			Day and month, A.D.	Week-day.	*a* (here = *t*, the index of the *tithi*).	
13	14	17			19	20	23	1
		H.	M.	S.				
25 Mar. (84) . .	3 Tues. .	14	33	0	14 Mar. (73) .	6 Fri. .	231·0424	4221
24 Mar. (84) . .	4 Wed. .	20	45	9	2 Mar. (62) .	3 Tues. .	106·7652	4222
25 Mar. (84) . .	6 Fri. .	2	57	18	21 Mar. (80) .	2 Mon. .	141·4477	4223
25 Mar. (84) . .	0 Sat. .	9	9	27	10 Mar. (69) .	6 Fri. .	17·1704	4224
25 Mar. (84) . .	1 Sun. .	15	21	36	28 Feb. (59) .	4 Wed. .	231·5253	4225
24 Mar. (84) . .	2 Mon. .	21	33	45	18 Mar. (78) .	3 Tues. .	266·2077	4226
25 Mar. (84) . .	4 Wed. .	3	45	54	7 Mar. (66) .	0 Sat. .	141·9306	4227
25 Mar. (84) . .	5 Thurs. .	9	58	3	24 Feb. (55) .	4 Wed. .	17·6533	4228
25 Mar. (84) . .	6 Fri. .	16	10	12	15 Mar. (74) .	3 Tues. .	52·3357	4229
24 Mar. (84) . .	0 Sat. .	22	22	21	4 Mar. (64) .	1 Sun. .	266·6906	4230
25 Mar. (84) . .	2 Mon. .	4	34	30	23 Mar. (82) .	0 Sat. .	301·3729	4231
25 Mar. (84) . .	3 Tues. .	10	46	39	12 Mar. (71) .	4 Wed. .	177·0958	4232
25 Mar. (84) . .	4 Wed. .	16	58	48	1 Mar. (60) .	1 Sun. .	52·8186	4233
24 Mar. (84) . .	5 Thurs. .	23	10	57	19 Mar. (79) .	0 Sat. .	87·5011	4234
25 Mar. (84) . .	0 Sat. .	5	23	6	9 Mar. (68) .	5 Thurs. .	301·8558	4235
25 Mar. (84) . .	1 Sun. .	11	35	15	26 Feb. (57) .	2 Mon. .	177·5787	4236
25 Mar. (84) . .	2 Mon. .	17	47	24	17 Mar. (76) .	1 Sun. .	212·2611	4237
24 Mar. (84) . .	3 Tues. .	23	59	33	5 Mar. (65) .	5 Thurs. .	87·9840	4238
25 Mar. (84) . .	5 Thurs. .	6	11	42	24 Mar. (83) .	4 Wed. .	122·6663	4239
25 Mar. (84) . .	6 Fri. .	12	23	51	13 Mar. (72) .	1 Sun. .	9998·3892 §	4240
25 Mar. (84) . .	0 Sat. .	18	36	0	3 Mar. (62) .	6 Fri. .	212·7440	4241
25 Mar. (85) . .	2 Mon. .	0	48	9	21 Mar. (81) .	5 Thurs. .	247·4264	4242
25 Mar. (84) . .	3 Tues. .	7	0	18	10 Mar. (69) .	2 Mon. .	123·0492	4243
25 Mar. (84) . .	4 Wed. .	13	12	27	27 Feb. (58) .	6 Fri. .	9998·8721 §	4244
25 Mar. (84) . .	5 Thurs. .	19	24	36	18 Mar. (77) .	5 Thurs. .	33·5545	4245

§ Chaitra śukla 1 was suppressed.

TABLE

CONCURRENT YE.								Mean intercalated (*adhika*) lunar month.
						Jovian Samvatsara.		
Kali.	Śaka.	Chaitrādi Vikrama.	Mēshādi solar year in Bengal.	Kollam.	A.D.	Southern system.	Northern system.	
1	2	3	3*a*	4	5	6	7	8*a*
4246	1067	1202	551	319-20	*1144-45	58 Raktāksha .	1 Prabhava .	...
4247	1068	1203	552	320-21	1145-46	59 Krōdhana .	2 Vibhava . .	1 Chaitra .
4248	1069	1204	553	321-22	1146-47	60 Kshaya . .	3 Śukla . .	...
4249	1070	1205	554	322-23	1147-48	1 Prabhava .	4 Pramōda . .	10 Pausha .
4250	1071	1206	555	323-24	*1148-49	2 Vibhava . .	5 Prajāpati .	...
4251	1072	1207	556	324-25	1149-50	3 Śukla . .	6 Aṅgiras . .	...
4252	1073	1208	557	325-26	1150-51	4 Pramōda .	7 Śrīmukha .	6 Bhādrapada .
4253	1074	1209	558	326-27	1151-52	5 Prajāpati .	8 Bhāva . .	...
4254	1075	1210	559	327-28	*1152-53	6 Aṅgiras . .	9 Yuvan . .	...
4255	1076	1211	560	328-29	1153-54	7 Śrīmukha .	10 Dhātṛi . .	3 Jyēshtha .
4256	1077	1212	561	329-30	1154-55	8 Bhāva . .	11 Īśvara . .	...
4257	1078	1213	562	330-31	1155-56	9 Yuvan . .	12 Bahudhānya .	11 Māgha .
4258	1079	1214	563	331-32	*1156-57	10 Dhātṛi . .	13 Pramādin .	...
4259	1080	1215	564	332-33	1157-58	11 Īśvara . .	14 Vikrama .	...
4260	1081	1216	565	333-34	1158-59	12 Bahudhānya .	15 Vṛisha . .	8 Kārttika .
4261	1082	1217	566	334-35	1159-60	13 Pramādin .	16 Chitrabhānu † .	...
4262	1083	1218	567	335-36	*1160-61	14 Vikrama .	18 *Tāraṇa* . .	
4263	1084	1219	568	336-37	1161-62	15 Vṛisha . .	19 *Pārthiva* .	5 Śrāvaṇa .
4264	1085	1220	569	337-38	1162-63	16 Chitrabhānu .	20 *Vyaya* . .	...
4265	1086	1221	570	338-39	1163-64	17 Subhānu .	21 *Sarvajit* . .	
4266	1087	1222	571	339-40	*1164-65	18 Tāraṇa . .	22 *Sarvadhārin* .	1 Chaitra .
4267	1088	1223	572	340-41	1165-66	19 Pārthiva .	23 Virōdhin .	...
4268	1089	1224	573	341-42	1166-67	20 Vyaya . .	24 Vikṛita . .	10 Pausha .
4269	1090	1225	574	342-43	1167-68	21 Sarvajit . .	25 Khara . .	
4270	1091	1226	575	343-44	*1168-69	22 Sarvadhārin .	26 Nandana .	...

† 17 Subhānu was suppressed in the north by the *Brahma-Siddhānta*, both in true and mean reckoning.

XC—*contd.*

COMMENCEMENT OF THE							
MEAN SOLAR YEAR.			MEAN LUNI-SOLAR YEAR (MEAN SUNRISE OF THE CIVIL DAY ON WHICH CHAITRA ŚUKLA 1 ENDS).			Kali.
Day and month, A.D.	Week-day.	Time of mean Mēsha-saṁkrānti.	Day and month, A.D.	Week-day.	*a* (here = *t*, the index of the *tithi*).	
13	14	17	19	20	23	1
		H. M. S.				
25 Mar. (85) . .	0 Sat. .	1 36 45	7 Mar. (67) .	3 Tues. .	247·9093	4246
25 Mar. (84) . .	1 Sun. .	7 48 54	24 Feb. (55) .	0 Sat. .	123·6321	4247
25 Mar. (84) . .	2 Mon. .	14 1 3	15 Mar. (74) .	6 Fri. .	158·3145	4248
25 Mar. (84) . .	3 Tues. .	20 13 12	4 Mar. (63) .	3 Tues. .	34·0373	4249
25 Mar. (85) . .	5 Thurs. .	2 25 21	22 Mar. (82) .	2 Mon. .	68·7197	4250
25 Mar. (84) . .	6 Fri. .	8 37 30	12 Mar. (71) .	0 Sat. .	283·0746	4251
25 Mar. (84) . .	0 Sat. .	14 49 39	1 Mar. (60) .	4 Wed. .	158·7974	4252
25 Mar. (84) . .	1 Sun. .	21 1 48	20 Mar. (79) .	3 Tues. .	193·4798	4253
25 Mar. (85) . .	3 Tues. .	3 13 57	8 Mar. (68) .	0 Sat. .	69·2026	4254
25 Mar. (84) . .	4 Wed. .	9 26 6	26 Feb. (57) .	5 Thur. .	283·5575	4255
25 Mar. (84) . .	5 Thur. .	15 38 15	17 Mar. (76) .	4 Wed. .	318·2398	4256
25 Mar. (84) . .	6 Fri. .	21 50 24	6 Mar. (65) .	1 Sun. .	193·9627	4257
25 Mar. (85) . .	1 Sun. .	4 2 33	24 Mar. (84) .	0 Sat. .	228·6451	4258
25 Mar. (84) . .	2 Mon. .	10 14 42	13 Mar. (72) .	4 Wed. .	104·3680	4259
25 Mar. (84) . .	3 Tues. .	16 26 51	3 Mar. (62) .	2 Mon. .	318·7227	4260
25 Mar. (84) . .	4 Wed. .	22 39 0	21 Mar. (80) .	0 Sat. .	14·7731	4261
25 Mar. (85) . .	6 Fri. .	4 51 9	10 Mar. (70) .	5 Thur. .	229·1280	4262
25 Mar. (84) . .	0 Sat. .	11 3 18	27 Feb. (58) .	2 Mon. .	104·8508	4263
25 Mar. (84) . .	1 Sun. .	17 15 27	18 Mar. (77) .	1 Sun. .	139·5332	4264
25 Mar. (84) . .	2 Mon. .	23 27 36	7 Mar. (66) .	5 Thur. .	15·2561	4265
25 Mar. (85) . .	4 Wed. .	5 39 45	25 Feb. (56) .	3 Tues. .	229·6109	4266
25 Mar. (84) . .	5 Thur. .	11 51 54	15 Mar. (74) .	2 Mon. .	264·2932	4267
25 Mar. (84) . .	6 Fri. .	18 4 3	4 Mar. (63) .	6 Fri. .	140·0161	4268
26 Mar. (85) . .	1 Sun. .	0 16 12	23 Mar. (82) .	5 Thur. .	174·6985	4269
25 Mar. (85) . .	2 Mon. .	6 28 21	11 Mar. (71) .	2 Mon. .	50·4213	4270

TABLE

CONCURRENT YEAR.								Mean intercalated (*adhika*) lunar month.
						Jovian Saṁvatsara.		
Kali.	Saka.	Chaitrādi Vikrama.	Mēshādi solar year in Bengal.	Kollam.	A.D.	Southern system.	Northern system.	
1	2	3	3a	4	5	6	7	8a
4271	1092	1227	576	344-45	1169-70	23 Virōdhin .	27 Vijaya .	6 Bhādrapada .
4272	1093	1228	577	345-46	1170-71	24 Vikṛita .	28 Jaya .	...
4273	1094	1229	578	346-47	1171-72	25 Khara .	29 Manmatha .	...
4274	1095	1230	579	347-48	*1172-73	26 Nandana .	30 Durmukha .	3 Jyēshṭha .
4275	1096	1231	580	348-49	1173-74	27 Vijaya .	31 Hēmalamba .	...
4276	1097	1232	581	349-50	1174-75	28 Jaya .	32 Vilamba .	11 Māgha .
4277	1098	1233	582	350-51	1175-76	29 Manmatha .	33 Vikārin .	...
4278	1099	1234	583	351-52	*1176-77	30 Durmukha .	34 Śārvarin .	...
4279	1100	1235	584	352-53	1177-78	31 Hēmalamba .	35 Plava .	8 Kārttika .
4280	1101	1236	585	353-54	1178-79	32 Vilamba .	36 Śubhakṛit .	...
4281	1102	1237	586	354-55	1179-80	33 Vikārin .	37 Śōbhana .	...
4282	1103	1238	587	355-56	*1180-81	34 Śārvarin .	38 Krōdhin .	4 Āshāḍha .
4283	1104	1239	588	356-57	1181-82	35 Plava .	39 Viśvāvasu .	...
4284	1105	1240	589	357-58	1182-83	36 Śubhakṛit .	40 Parābhava .	...
4285	1106	1241	590	358-59	1183-84	37 Śōbhana .	41 Plavaṅga .	1 Chaitra .
4286	1107	1242	591	359-60	*1184-85	38 Krōdhin .	42 Kīlaka .	...
4287	1108	1243	592	360-61	1185-86	39 Viśvāvasu .	43 Saumya .	9 Mārgaśira .
4288	1109	1244	593	361-62	1186-87	40 Parābhava .	44 Sādhāraṇa .	...
4289	1110	1245	594	362-63	1187-88	41 Plavaṅga .	45 Virōdhakṛit .	...
4290	1111	1246	595	363-64	*1188-89	42 Kīlaka .	46 Paridhāvin .	6 Bhādrapada .
4291	1112	1247	596	364-65	1189-90	43 Saumya .	47 Pramādin .	...
4292	1113	1248	597	365-66	1190-91	44 Sādhāraṇa .	48 Ānanda .	...
4293	1114	1249	598	366-67	1191-92	45 Virōdhakṛit .	49 Rākshasa .	2 Vaiśākha .
4294	1115	1250	599	367-68	*1192-93	46 Paridhāvin .	50 Anala .	...
4295	1116	1251	600	368-69	1193-94	47 Pramādin .	51 Piṅgala .	11 Māgha .

XC—*contd.*

COMMENCEMENT OF THE							Kali.
MEAN SOLAR YEAR.			MEAN LUNI-SOLAR YEAR (MEAN SUNRISE OF THE CIVIL DAY ON WHICH CHAITRA ŚUKLA 1 ENDS).				
Day and month, A.D.	Week-day.	Time of mean Mēsha-saṁkrānti.	Day and month, A.D.	Week-day.	*a* (here = *t*, the index of the *tithi*).		
13	14	17	19	20	23		1
		H. M. S.					
25 Mar. (84) . .	3 Tues. .	12 40 30	1 Mar. (60) .	0 Sat. .	264·7762		4271
25 Mar. (84) . .	4 Wed. .	18 52 39	20 Mar. (79) .	6 Fri. .	299·4586		4272
26 Mar. (85) . .	6 Fri. .	1 4 48	9 Mar. (68) .	3 Tues. .	175·1815		4273
25 Mar. (85) . .	0 Sat. .	7 16 57	26 Feb. (57) .	0 Sat. .	50·9042		4274
25 Mar. (84) . .	1 Sun. .	13 29 6	16 Mar. (75) .	6 Fri. .	85·5866		4275
25 Mar. (84) . .	2 Mon. .	19 41 15	6 Mar. (65) .	4 Wed. .	299·9415		4276
26 Mar. (85) . .	4 Wed. .	1 53 24	24 Mar. (83) .	2 Mon. .	9995·9918 §		4277
25 Mar. (85) . .	5 Thur. .	8 5 33	13 Mar. (73) .	0 Sat. .	210·3467		4278
25 Mar. (84) . .	6 Fri. .	14 17 42	2 Mar. (61) .	4 Wed. .	86·0695		4279
25 Mar. (84) . .	0 Sat. .	20 29 51	21 Mar. (80) .	3 Tues. .	120·751[illegible]		4280
26 Mar. (85) . .	2 Mon. .	2 42 0	10 Mar. (69) .	0 Sat. .	9996·4747 §		4281
25 Mar. (85) . .	3 Tues. .	8 54 9	28 Feb. (59) .	5 Thur. .	210·8296		4282
25 Mar. (84) . .	4 Wed. .	15 6 18	18 Mar. (77) .	4 Wed. .	245·5120		4283
25 Mar. (84) . .	5 Thur. .	21 18 27	7 Mar. (66) .	1 Sun. .	121·2349		4284
26 Mar. (85) . .	0 Sat. .	3 30 36	24 Feb. (55) .	5 Thur. .	9996·9576 §		4285
25 Mar. (85) . .	1 Sun. .	9 42 45	14 Mar. (74) .	4 Wed. .	31·6400		4286
25 Mar. (84) . .	2 Mon. .	15 54 54	4 Mar. (63) .	2 Mon. .	245·9949		4287
25 Mar. (84) . .	3 Tues. .	22 7 3	23 Mar. (82) .	1 Sun. .	280·6772		4288
26 Mar. (85) . .	5 Thur. .	4 19 12	12 Mar. (71) .	5 Thur. .	156·4061		4289
25 Mar. (85) . .	6 Fri. .	10 31 21	29 Feb. (60) .	2 Mon. .	32·1230		4290
25 Mar. (84) . .	0 Sat. .	16 43 30	19 Mar. (78) .	1 Sun. .	66·8054		4291
25 Mar. (84) . .	1 Sun. .	22 55 39	9 Mar. (68) .	6 Fri. .	281·1602		4292
26 Mar. (85) . .	3 Tues. .	5 7 48	26 Feb. (57) .	3 Tues. .	156·8830		4293
25 Mar. (85) . .	4 Wed. .	11 19 57	16 Mar. (76) .	2 Mon. .	191·5654		4294
25 Mar. (84) . .	5 Thur. .	17 32 6	5 Mar. (64) .	6 Fri. .	67·2882		4295

§ Chaitra śukla 1 was suppressed.

TABLE

CONCURRENT YEAR.								Mean intercalated (*adhika*) lunar month.
Kali.	Śaka.	Chaitrādi Vikrama.	Mēshādi solar year in Bengal.	Kollam.	A.D.	Jovian Saṁvatsara. Southern system.	Jovian Saṁvatsara. Northern system.	
1	2	3	3*a*	4	5	6	7	8*a*
4296	1117	1252	601	369-70	1194-95	48 Ānanda	52 Kālayukta	...
4297	1118	1253	602	370-71	1195-96	49 Rākshasa	53 Siddhārthin	...
4298	1119	1254	603	371-72	*1196-97	50 Anala	54 Raudra	8 Kārttika ‡
4299	1120	1255	604	372-73	1197-98	51 Piṅgala	55 Durmati	...
4300	1121	1256	605	373-74	1198-99	52 Kālayukta	56 Dundubhi	...
4301	1122	1257	606	374-75	1199-1200	53 Siddhārthin	57 Rudhirōdgārin	4 Āshāḍha
4302	1123	1258	607	375-76	*1200-01	54 Raudra	58 Raktāksha	
4303	1124	1259	608	376-77	1201-02	55 Durmati	59 Krōdhana	...
4304	1125	1260	609	377-78	1202-03	56 Dundubhi	60 Kshaya	1 Chaitra
4305	1126	1261	610	378-79	1203-04	57 Rudhirōdgārin	1 Prabhava	...
4306	1127	1262	611	379-80	*1204-05	58 Raktāksha	2 Vibhava	9 Mārgaśira
4307	1128	1263	612	380-81	1205-06	59 Krōdhana	3 Sukla	...
4308	1129	1264	613	381-82	1206-07	60 Kshaya	4 Pramōda	...
4309	1130	1265	614	382-83	1207-08	1 Prabhava	5 Prajāpati	6 Bhādrapada
4310	1131	1266	615	383-84	*1208-09	2 Vibhava	6 Aṅgiras	...
4311	1132	1267	616	384-85	1209-10	3 Śukla	7 Śrīmukha	...
4312	1133	1268	617	385-86	1210-11	4 Pramōda	8 Bhāva	2 Vaiśākha
4313	1134	1269	618	386-87	1211-12	5 Prajāpati	9 Yuvan	...
4314	1135	1270	619	387-88	*1212-13	6 Aṅgiras	10 Dhātṛi	11 Māgha
4315	1136	1271	620	388-89	1213-14	7 Śrīmukha	11 Īśvara	...
4316	1137	1272	621	389-90	1214-15	8 Bhāva	12 Bahudhānya	...
4317	1138	1273	622	390-91	1215-16	9 Yuvan	13 Pramādin	7 Āśvina
4318	1139	1274	623	391-92	*1216-17	10 Dhātri	14 Vikrama	
4319	1140	1275	624	392-93	1217-18	11 Īśvara	15 Vṛisha	...
4320	1141	1276	625	393-94	1218-19	12 Bahudhānya	16 Chitrabhān	4 Āshaḍha

‡ See "Remarks." p. 523 above.

XC—*contd.*

COMMENCEMENT OF THE						Kali.
MEAN SOLAR YEAR.			MEAN LUNI-SOLAR YEAR (MEAN SUNRISE OF THE CIVIL DAY ON WHICH CHAITRA ŚUKLA 1 ENDS).			
Day and month, A.D.	Week-day.	Time of mean Mēsha-saṁkrānti.	Day and month, A.D.	Week-day.	*a* (here = *t*, the index of the *tithi*).	
13	14	17	19	20	23	1
		H. M. S.				
25 Mar. (84)	6 Fri.	23 44 15	24 Mar. (83)	5 Thur.	101·9706	4296
26 Mar. (85)	1 Sun.	5 56 24	14 Mar. (73)	3 Tues.	316·3255	4297
25 Mar. (85)	2 Mon.	12 8 33	2 Mar. (62)	0 Sat.	192·0482	4298
25 Mar. (84)	3 Tues.	18 20 42	21 Mar. (80)	6 Fri.	226·7307	4299
26 Mar. (85)	5 Thur.	0 32 51	10 Mar. (69)	3 Tues.	102·4535	4300
26 Mar. (85)	6 Fri.	6 45 0	28 Feb. (59)	1 Sun.	316·8083	4301
25 Mar. (85)	0 Sat.	12 57 9	17 Mar. (77)	6 Fri.	12·8587	4302
25 Mar. (84)	1 Sun.	19 9 18	7 Mar. (66)	4 Wed.	227·2136	4303
26 Mar. (85)	3 Tues.	1 21 27	24 Feb. (55)	1 Sun.	102·9363	4304
26 Mar. (85)	4 Wed.	7 33 36	15 Mar. (74)	0 Sat.	137·6188	4305
25 Mar. (85)	5 Thur.	13 45 45	3 Mar. (63)	4 Wed.	13·3416	4306
25 Mar. (84)	6 Fri.	19 57 54	22 Mar. (81)	3 Tues.	48·0239	4307
26 Mar. (85)	1 Sun.	2 10 3	12 Mar. (71)	1 Sun.	262·3788	4308
26 Mar. (85)	2 Mon.	8 22 12	1 Mar. (60)	5 Thur.	138·1017	4309
25 Mar. (85)	3 Tues.	14 34 21	19 Mar. (79)	4 Wed.	172·7840	4310
25 Mar. (84)	4 Wed.	20 46 30	8 Mar. (67)	1 Sun.	48·5069	4311
26 Mar. (85)	6 Fri.	2 58 39	26 Feb. (57)	6 Fri.	262·8617	4312
26 Mar. (85)	0 Sat.	9 10 48	17 Mar. (76)	5 Thur.	297·5441	4313
25 Mar. (85)	1 Sun.	15 22 57	5 Mar. (65)	2 Mon.	173·2669	4314
25 Mar. (84)	2 Mon.	21 35 6	24 Mar. (83)	1 Sun.	207·9493	4315
26 Mar. (85)	4 Wed.	3 47 15	13 Mar. (72)	5 Thur.	83·6722	4316
26 Mar. (85)	5 Thur.	9 59 24	3 Mar. (62)	3 Tues.	298·0269	4317
25 Mar. (85)	6 Fri.	16 11 33	21 Mar. (81)	2 Mon.	332·7094	4318
25 Mar. (84)	0 Sat.	22 23 42	10 Mar. (69)	6 Fri.	208·4322	4319
26 Mar. (85)	2 Mon.	4 35 51	27 Feb. (58)	3 Tues.	84·1551	4320

TABLE

CONCURRENT YEAR.								Mean intercalated (*adhika*) lunar month.
Kali.	Saka.	Chaitrâdi Vikrama.	Mêshâdi solar year in Bengal.	Kollam.	A.D.	Jovian Saṁvatsara. Southern system.	Jovian Saṁvatsara. Northern system.	
1	2	3	3a	4	5	6	7	8a
4321	1142	1277	626	394-95	1219-20	13 Pramādin .	17 Subhānu .	...
4322	1143	1278	627	395-96	*1220-21	14 Vikrama .	18 Tāraṇa . .	12 Phālguna .
4323	1144	1279	628	396-97	1221-22	15 Vṛisha . .	19 Pārthiva .	...
4324	1145	1280	629	397-98	1222-23	16 Chitrabhānu .	20 Vyaya . .	...
4325	1146	1281	630	398-99	1223-24	17 Subhānu .	21 Sarvajit . .	9 Mārgaśira .
4326	1147	1282	631	399-400	*1224-25	18 Tāraṇa . .	22 Sarvadhārin .	...
4327	1148	1283	632	400-01	1225-26	19 Pārthiva .	23 Virōdhin .	...
4328	1149	1284	633	401-02	1226-27	20 Vyaya . .	24 Vikṛita . .	5 Śrāvaṇa .
4329	1150	1285	634	402-03	1227-28	21 Sarvajit . .	25 Khara . .	...
4330	1151	1286	635	403-04	*1228-29	22 Sarvadhārin .	26 Nandana .	...
4331	1152	1287	636	404-05	1229-30	23 Virōdhin .	27 Vijaya . .	2 Vaiśākha .
4332	1153	1288	637	405-06	1230-31	24 Vikṛita . .	28 Jaya . .	...
4333	1154	1289	638	406-07	1231-32	25 Khara . .	29 Manmatha .	10 Pausha .
4334	1155	1290	639	407-08	*1232-33	26 Nandana . .	30 Durmukha .	...
4335	1156	1291	640	408-09	1233-34	27 Vijaya . .	31 Hēmalamba .	...
4336	1157	1292	641	409-10	1234-35	28 Jaya . .	32 Vilamba .	7 Āśvina .
4337	1158	1293	642	410-11	1235-36	29 Manmatha .	33 Vikārin . .	...
4338	1159	1294	643	411-12	*1236-37	30 Durmukha .	34 Śārvarin . .	...
4339	1160	1295	644	412-13	1237-38	31 Hēmalamba .	35 Plava . .	4 Āshāḍha .
4340	1161	1296	645	413-14	1238-39	32 Vilamba . .	36 Śubhakṛit .	...
4341	1162	1297	646	414-15	1239-40	33 Vikārin . .	37 Śōbhana . .	12 Phālguna
4342	1163	1298	647	415-16	*1240-41	34 Śārvarin . .	38 Krōdhin .	...
4343	1164	1299	648	416-17	1241-42	35 Plava . .	39 Viśvāvasu .	...
4344	1165	1300	649	417-18	1242-43	36 Subhakṛit .	40 Parābhava .	9 Mārgaśira .
4345	1166	1301	650	418-19	1243-44	37 Sōbhana . .	41 Plavaṅga .	...

XC—*contd.*

COMMENCEMENT OF THE						Kali.
MEAN SOLAR YEAR.			MEAN LUNI-SOLAR YEAR (MEAN SUNRISE OF THE CIVIL DAY ON WHICH CHAITRA ŚUKLA 1 ENDS).			
Day and month, A.D.	Week-day.	Time of mean Mêsha-saṁkrānti.	Day and month, A.D.	Week-day.	*a* (here = *t*, the index of the *tithi*).	
13	14	17	19	20	23	1
		H. M. S.				
26 Mar. (85)	3 Tues.	10 48 0	18 Mar. (77)	2 Mon.	118·8374	4321
25 Mar. (85)	4 Wed.	17 0 9	7 Mar. (67)	0 Sat.	333·1923	4322
25 Mar. (84)	5 Thur.	23 12 18	25 Mar. (84)	5 Thur.	29·2427	4323
26 Mar. (85)	0 Sat.	5 24 27	15 Mar. (74)	3 Tues.	243·5975	4324
26 Mar. (85)	1 Sun.	11 36 36	4 Mar. (63)	0 Sat.	119·3203	4325
25 Mar. (85)	2 Mon.	17 48 45	22 Mar. (82)	6 Fri.	154·0027	4326
26 Mar. (85)	4 Wed.	0 0 54	11 Mar. (70)	3 Tues.	29·7256	4327
26 Mar. (85)	5 Thur.	6 13 3	1 Mar. (60)	1 Sun.	244·0804	4328
26 Mar. (85)	6 Fri.	12 25 12	20 Mar. (79)	0 Sat.	278·7628	4329
25 Mar. (85)	0 Sat.	18 37 21	8 Mar. (68)	4 Wed.	154·4857	4330
26 Mar. (85)	2 Mon.	0 49 30	25 Feb. (56)	1 Sun.	30·2084	4331
26 Mar. (85)	3 Tues.	7 1 39	16 Mar. (75)	0 Sat.	64·8908	4332
26 Mar. (85)	4 Wed.	13 13 48	6 Mar. (65)	5 Thur.	279·2457	4333
25 Mar. (85)	5 Thur.	19 25 57	24 Mar. (84)	4 Wed.	313·9281	4334
26 Mar. (85)	0 Sat.	1 38 6	13 Mar. (72)	1 Sun.	189·6509	4335
26 Mar. (85)	1 Sun.	7 50 15	2 Mar. (61)	5 Thur.	65·3738	4336
26 Mar. (85)	2 Mon.	14 2 24	21 Mar. (80)	4 Wed.	100·0562	4337
25 Mar. (85)	3 Tues.	20 14 33	10 Mar. (70)	2 Mon.	314·4110	4338
26 Mar. (85)	5 Thur.	2 26 42	27 Feb. (58)	6 Fri.	190·1338	4339
26 Mar. (85)	6 Fri.	8 38 51	18 Mar. (77)	5 Thur.	224·8162	4340
26 Mar. (85)	0 Sat.	14 51 0	7 Mar. (66)	2 Mon.	100·5391	4341
25 Mar. (85)	1 Sun.	21 3 9	25 Mar. (85)	1 Sun.	135·2214	4342
26 Mar. (85)	3 Tues.	3 15 18	14 Mar. (73)	5 Thur.	10·9443	4343
26 Mar. (85)	4 Wed.	9 27 27	4 Mar. (63)	3 Tues.	225·2991	4344
26 Mar. (85)	5 Thur.	15 39 36	23 Mar. (82)	2 Mon.	259·9815	4345

TABLE

CONCURRENT YEAR.									Mean intercalated (*adhika*) lunar month.
Kali.	Saka.	Chaitrādi Vikrama.	Mēshādi solar year in Bengal.	Kollam.	A.D.	Jovian Saṁvatsara. Southern system.	Jovian Saṁvatsara. Northern system.		
1	2	3	3a	4	5	6	7		8a
4346	1167	1302	651	419-20	*1244-45	38 Krōdhin .	42 Kīlaka† . .		...
4347	1168	1303	652	420-21	1245-46	39 Viśvāvasu .	44 *Sādhāraṇa* .		5 Śrāvaṇa .
4348	1169	1304	653	421-22	1246-47	40 Parābhava .	45 *Virōdhakṛit* .		...
4349	1170	1305	654	422-23	1247-48	41 Plavaṅga .	46 *Paridhāvin* .		...
4350	1171	1306	655	423-24	*1248-49	42 Kīlaka . .	47 *Pramādin* .		2 Vaiśākha .
4351	1172	1307	656	424-25	1249-50	43 Saumya . .	48 *Ānanda* .		...
4352	1173	1308	657	425-26	1250-51	44 Sādhāraṇa .	49 *Rākshasa* .		10 Pausha .
4353	1174	1309	658	426-27	1251-52	45 Virōdhakṛit .	50 Anala . .		...
4354	1175	1310	659	427-28	*1252-53	46 Paridhāvin .	51 Piṅgala .		...
4355	1176	1311	660	428-29	1253-54	47 Pramādin .	52 Kālayukta .		7 Āśvina .
4356	1177	1312	661	429-30	1254-55	48 Ānanda .	53 Siddhārthin .		...
4357	1178	1313	662	430-31	1255-56	49 Rākshasa .	54 Raudra . .		...
4358	1179	1314	663	431-32	*1256-57	50 Anala . .	55 Durmati .		3 Jyēshṭha .
4359	1180	1315	664	432-33	1257-58	51 Piṅgala . .	56 Dundubhi .		...
4360	1181	1316	665	433-34	1258-59	52 Kālayukta .	57 Rudhirōdgārin		12 Phālguna .
4361	1182	1317	666	434-35	1259-60	53 Siddhārthin .	58 Raktāksha .		...
4362	1183	1318	667	435-36	*1260-61	54 Raudra . .	59 Krōdhana .		...
4363	1184	1319	668	436-37	1261-62	55 Durmati	60 Kshaya . .		8 Kārttika .
4364	1185	1320	669	437-38	1262-63	56 Dundubhi .	1 Prabhava .		...
4365	1186	1321	670	438-39	1263-64	57 Rudhirōdgārin	2 Vibhava .		...
4366	1187	1322	671	439-40	*1264-65	58 Raktāksha .	3 Śukla . .		5 Śrāvaṇa .
4367	1188	1323	672	440-41	1265-66	59 Krōdhana .	4 Pramōda .		...
4368	1189	1324	673	441-42	1266-67	60 Kshaya .	5 Prajāpati .		...
4369	1190	1325	674	442-43	1267-68	1 Prabhava .	6 Aṅgiras . .		1 Chaitra .
4370	1191	1326	675	443-44	*1268-69	2 Vibhava .	7 Śrīmukha .		...

† 43 Saumya was suppressed in the north by the mean system. By the "true" system K.Y. 4346 (expired), A.D. 1245-46, was called "Saumya," 44 Sādhāraṇa being suppressed. The next year was 45 Virōdhakrit by both system of reckoning.

XC—*contd.*

COMMENCEMENT OF THE									
MEAN SOLAR YEAR.					MEAN LUNI-SOLAR YEAR (MEAN SUNRISE OF THE CIVIL DAY ON WHICH CHAITRA ŚUKLA 1 ENDS).			Kali.
Day and month, A.D.	Week-day.	Time of mean Mēsha-saṁkrānti.			Day and month, A.D.	Week-day.	*a* (here = *t*, the index of the *tithi*).	
13	14	17			19	20	23	1
		H.	M.	S.				
25 Mar. (85) . .	6 Fri. .	21	51	45	11 Mar. (71) .	6 Fri. .	135·7043	4346
26 Mar. (85) . .	1 Sun. .	4	3	54	28 Feb. (59) .	3 Tues.	11·4272	4347
26 Mar. (85) . .	2 Mon. .	10	16	3	19 Mar. (78) .	2 Mon. .	46·1096	4348
26 Mar. (85) . .	3 Tues. .	16	28	12	9 Mar. (68) .	0 Sat. .	260·4644	4349
25 Mar. (85) . .	4 Wed. .	22	40	21	26 Feb. (57) .	4 Wed. .	136·1872	4350
26 Mar. (85) . .	6 Fri. .	4	52	30	16 Mar. (75) .	3 Tues. .	170·8696	4351
26 Mar. (85) . .	0 Sat. .	11	4	39	5 Mar. (64) .	0 Sat. .	46·5925	4352
26 Mar. (85) . .	1 Sun. .	17	16	48	24 Mar. (83) .	6 Fri. .	81·2748	4353
25 Mar. (85) . .	2 Mon. .	23	28	57	13 Mar. (73) .	4 Wed. .	295·6297	4354
26 Mar. (85) . .	4 Wed. .	5	41	6	2 Mar. (61) .	1 Sun. .	171·3526	4355
26 Mar. (85) . .	5 Thur. .	11	53	15	21 Mar. (80) .	0 Sat. .	206·0349	4356
26 Mar. (85) . .	6 Fri. .	18	5	24	10 Mar. (69) .	4 Wed. .	81·7577	4357
26 Mar. (86) . .	1 Sun. .	0	17	33	28 Feb. (59) .	2 Mon. .	296·1126	4358
26 Mar. (85) . .	2 Mon. .	6	29	42	18 Mar. (77) .	1 Sun. .	330·7950	4359
26 Mar. (85) . .	3 Tues. .	12	41	51	7 Mar. (66) .	5 Thur. .	206·5178	4360
26 Mar. (85) . .	4 Wed. .	18	54	0	26 Mar. (85) .	4 Wed. .	241·2002	4361
26 Mar. (86) . .	6 Fri. .	1	6	9	14 Mar. (74) .	1 Sun. .	116·9231	4362
26 Mar. (85) . .	0 Sat. .	7	18	18	4 Mar. (63) .	6 Fri. .	331·2778	4363
26 Mar. (85) . .	1 Sun. .	13	30	27	22 Mar. (81) .	4 Wed. .	27·3283	4364
26 Mar. (85) . .	2 Mon. .	19	42	36	12 Mar. (71) .	2 Mon. .	241·6831	4365
26 Mar. (86) . .	4 Wed. .	1	54	45	29 Feb. (60) .	6 Fri. .	117·4060	4366
26 Mar. (85) . .	5 Thur. .	8	6	54	19 Mar. (78) .	5 Thur. .	152·0883	4367
26 Mar. (85) . .	6 Fri. .	14	19	3	8 Mar. (67) .	2 Mon. .	27·8112	4368
26 Mar. (85) . .	0 Sat. .	20	31	12	26 Feb. (57) .	0 Sat. .	242·1660	4369
26 Mar. (86) . .	2 Mon. .	2	43	21	16 Mar. (76) .	6 Fri. .	276·8483	4370

TABLE

CONCURRENT YEAR.								Mean intercalated (adhika) lunar month.
Kali.	Saka.	Chaitrādi Vikrama.	Mēshādi solar year in Bengal.	Kollam.	A.D.	Jovian Saṁvatsara. Southern system.	Jovian Saṁvatsara. Northern system.	
1	2	3	3a	4	5	6	7	8a
4371	1192	1327	676	444-45	1269-70	3 Śukla . .	8 Bhāva . .	10 Pausha .
4372	1193	1328	677	445-46	1270-71	4 Pramōda .	9 Yuvan . .	...
4373	1194	1329	678	446-47	1271-72	5 Prajāpati .	10 Dhātṛi . .	...
4374	1195	1330	679	447-48	*1272-73	6 Aṅgiras . .	11 Īśvara . .	7 Āśvina .
4375	1196	1331	680	448-49	1273-74	7 Śrīmukha .	12 Bahudhānya .	...
4376	1197	1332	681	449-50	1274-75	8 Bhāva . .	13 Pramāthin .	...
4377	1198	1333	682	450-51	1275-76	9 Yuvan . .	14 Vikrama .	3 Jyēshṭha .
4378	1199	1334	683	451-52	*1276-77	10 Dhātṛi . .	15 Vṛisha . .	...
4379	1200	1335	684	452-53	1277-78	11 Īśvara . .	16 Chitrabhānu .	12 Phālguna .
4380	1201	1336	685	453-54	1278-79	12 Bahudhānya .	17 Subhānu .	...
4381	1202	1337	686	454-55	1279-80	13 Pramāthin .	18 Tāraṇa . .	...
4382	1203	1338	687	455-56	*1280-81	14 Vikrama .	19 Pārthiva .	8 Kārttika .
4383	1204	1339	688	456-57	1281-82	15 Vṛisha . .	20 Vyaya . .	
4384	1205	1340	689	457-58	1282-83	16 Chitrabhānu .	21 Sarvajit . .	...
4385	1206	1341	690	458-59	1283-84	17 Subhānu .	22 Sarvadhārin .	5 Śrāvaṇa .
4386	1207	1342	691	459-60	*1284-85	18 Tāraṇa . .	23 Virōdhin .	...
4387	1208	1343	692	460-61	1285-86	19 Pārthiva .	24 Vikṛita . .	...
4388	1209	1344	693	461-62	1286-87	20 Vyaya . .	25 Khara . .	1 Chaitra .
4389	1210	1345	694	462-63	1287-88	21 Sarvajit . .	26 Nandana .	...
4390	1211	1346	695	463-64	*1288-89	22 Sarvadhārin .	27 Vijaya . .	10 Pausha .
4391	1212	1347	696	464-65	1289-90	23 Virōdhin .	28 Jaya . .	...
4392	1213	1348	697	465-66	1290-91	24 Vikṛita . .	29 Manmatha .	...
4393	1214	1349	698	466-67	1291-92	25 Khara . .	30 Durmukha .	6 Bhādrapada .
4394	1215	1350	699	467-68	*1292-93	26 Nandana .	31 Hēmalamba .	...
4395	1216	1351	700	468-69	1293-94	27 Vijaya . .	32 Vilamba .	...

XC—*contd.*

COMMENCEMENT OF THE									Kali.
MEAN SOLAR YEAR.					MEAN LUNI-SOLAR YEAR (MEAN SUNRISE OF THE CIVIL DAY ON WHICH CHAITRA ŚUKLA 1 ENDS).				
Day and month, A.D.	Week-day.	Time of mean Mēsha-saṁkrānti.			Day and month, A.D.	Week day.	*a* (here = *t*, the index of the *tithi*).		
13	14	17			19	20	23		1
		H.	M.	S.					
26 Mar. (85) . .	3 Tues. .	8	55	30	5 Mar. (64) .	3 Tues.	152·5712		4371
26 Mar. (85) . .	4 Wed. .	15	7	39	24 Mar. (83) .	2 Mon. .	187·2536		4372
26 Mar. (85) . .	5 Thur. .	21	19	48	13 Mar. (72) .	6 Fri. .	62·9765		4373
26 Mar. (86) . .	0 Sat. .	3	31	57	2 Mar. (62) .	4 Wed. .	277·3313		4374
26 Mar. (85) . .	1 Sun. .	9	44	6	21 Mar. (80) .	3 Tues. .	312·0137		4375
26 Mar. (85) . .	2 Mon. .	15	56	15	10 Mar. (69) .	0 Sat. .	187·7365		4376
26 Mar. (85) . .	3 Tues. .	22	8	24	27 Feb. (58) .	4 Wed. .	63·4593		4377
26 Mar. (86) . .	5 Thur. .	4	20	33	17 Mar. (77) .	3 Tues. .	98·1417		4378
26 Mar. (85) . .	6 Fri. .	10	32	42	7 Mar. (66) .	1 Sun. .	312·4966		4379
26 Mar. (85) . .	0 Sat. .	16	44	51	25 Mar. (84) .	6 Fri. .	8·5470		4380
26 Mar. (85) . .	1 Sun. .	22	57	0	15 Mar. (74) .	4 Wed. .	222·9018		4381
26 Mar. (86) . .	3 Tues. .	5	9	9	3 Mar. (63) .	1 Sun. .	98·6246		4382
26 Mar. (85) . .	4 Wed. .	11	21	18	22 Mar. (81) .	0 Sat. .	133·3071		4383
26 Mar. (85) . .	5 Thur. .	17	33	27	11 Mar. (70) .	4 Wed. .	9·0299		4384
26 Mar. (85) . .	6 Fri. .	23	45	36	1 Mar. (60) .	2 Mon. .	223·3847		4385
26 Mar. (86) . .	1 Sun. .	5	57	45	19 Mar. (79) .	1 Sun. .	258·0671		4386
26 Mar. (85) . .	2 Mon. .	12	9	54	8 Mar. (67) .	5 Thur. .	133·7900		4387
26 Mar. (85) . .	3 Tues. .	18	22	3	25 Feb. (56) .	2 Mon. .	9·5127		4388
27 Mar. (86) . .	5 Thur. .	0	34	12	16 Mar. (75) .	1 Sun. .	44·1952		4389
26 Mar. (86) . .	6 Fri. .	6	46	21	5 Mar. (65) .	6 Fri. .	258·5500		4390
26 Mar. (85) . .	0 Sat. .	12	58	30	24 Mar. (83) .	5 Thur. .	293·2324		4391
26 Mar. (85) . .	1 Sun. .	19	10	39	13 Mar. (72) .	2 Mon. .	168·9552		4392
27 Mar. (86) . .	3 Tues. .	1	22	48	2 Mar. (61) .	6 Fri. .	44·6781		4393
26 Mar. (86) . .	4 Wed. .	7	34	57	20 Mar. (80) .	5 Thur. .	79·3605		4394
26 Mar. (85) . .	5 Thur. .	13	47	6	10 Mar. (69) .	3 Tues. .	293·7152		4395

TABLE

CONCURRENT YEAR.								
						Jovian Saṁvatsara.		Mean intercalated (*adhika*) lunar month.
Kali.	Saka.	Chaitrādi Vikrama.	Mēshādi solar year in Bengal.	Kollam.	A.D.	Southern system.	Northern system.	
1	2	3	3*a*	4	5	6	7	8*a*
4396	1217	1352	701	469-70	1294-95	28 Jaya	33 Vikārin	3 Jyēshṭha
4397	1218	1353	702	470-71	1295-96	29 Manmatha	34 Śārvarin	...
4398	1219	1354	703	471-72	*1296-97	30 Durmukha	35 Plava	11 Māgha
4399	1220	1355	704	472-73	1297-98	31 Hēmalamba	36 Subhakṛit	...
4400	1221	1356	705	473-74	1298-99	32 Vilamba	37 Śōbhana	...
4401	1222	1357	706	474-75	1299-1300	33 Vikārin	38 Krōdhin	8 Kārttika
4402	1223	1358	707	475-76	*1300-01	34 Śārvarin	39 Viśvāvasu	...
4403	1224	1359	708	476-77	1301-02	35 Plava	40 Parābhava	...
4404	1225	1360	709	477-78	1302-03	36 Śubhakṛit	41 Plavaṅga	4 Āshāḍha
4405	1226	1361	710	478-79	1303-04	37 Śōbhana	42 Kīlaka	...
4406	1227	1362	711	479-80	*1304-05	38 Krōdhin	43 Saumya	...
4407	1228	1363	712	480-81	1305-06	39 Viśvāvasu	44 Sādhāraṇa	1 Chaitra
4408	1229	1364	713	481-82	1306-07	40 Parābhava	45 Virōdhakṛit	...
4409	1230	1365	714	482-83	1307-08	41 Plavaṅga	46 Paridhāvin	10 Pausha ‡
4410	1231	1366	715	483-84	*1308-09	42 Kīlaka	47 Pramādin	...
4411	1232	1367	716	484-85	1309-10	43 Saumya	48 Ānanda	...
4412	1233	1368	717	485-86	1310-11	44 Sādhāraṇa	49 Rākshasa	6 Bhādrapada
4413	1234	1369	718	486-87	1311-12	45 Virōdhakṛit	50 Anala	...
4414	1235	1370	719	487-88	*1312-13	46 Paridhāvin	51 Piṅgala	...
4415	1236	1371	720	488-89	1313-14	47 Pramādin	52 Kālayukta	3 Jyēshṭha
4416	1237	1372	721	489-90	1314-15	48 Ānanda	53 Siddhārthin	...
4417	1238	1373	722	490-91	1315-16	49 Rākshasa	54 Raudra	11 Māgha
4418	1239	1374	723	491-92	*1316-17	50 Anala	55 Durmati	...
4419	1240	1375	724	492-93	1317-18	51 Piṅgala	56 Dundubhi	...
4420	1241	1376	725	493-94	1318-19	52 Kālayukta	57 Rudhirōdgārin	8 Kārttika

‡ See "Remarks," p. 523, preceding this Table.

XC—*contd.*

COMMENCEMENT OF THE							
MEAN SOLAR YEAR.				MEAN LUNI-SOLAR YEAR (MEAN SUNRISE OF THE CIVIL DAY ON WHICH CHAITRA ŚUKLA 1 ENDS).			Kali.
Day and month, A.D.	Week-day.	Time of mean Mēsha-saṁkrānti.		Day and month, A.D.	Week-day.	*a* (here = *t*, the index of the *tithi*).	
13	14	17		19	20	23	1
		H.	M. S.				
26 Mar. (85)	6 Fri.	19	59 15	27 Feb. (58)	0 Sat.	169·4381	4396
27 Mar. (86)	1 Sun.	2	11 24	18 Mar. (77)	6 Fri.	204·1205	4397
26 Mar. (86)	2 Mon.	8	23 33	6 Mar. (66)	3 Tues.	79·8433	4398
26 Mar. (85)	3 Tues.	14	35 42	25 Mar. (84)	2 Mon.	114·5257	4399
26 Mar. (85)	4 Wed.	20	47 51	15 Mar. (74)	0 Sat.	328·8806	4400
27 Mar. (86)	6 Fri.	3	0 0	4 Mar. (63)	4 Wed.	204·6034	4401
26 Mar. (86)	0 Sat.	9	12 9	22 Mar. (82)	3 Tues.	239·2859	4402
26 Mar. (85)	1 Sun.	15	24 18	11 Mar. (70)	0 Sat.	115·0087	4403
26 Mar. (85)	2 Mon.	21	36 27	1 Mar. (60)	5 Thur.	329·3635	4404
27 Mar. (86)	4 Wed.	3	48 36	19 Mar. (78)	3 Tues.	25·4139	4405
26 Mar. (86)	5 Thur.	10	0 45	8 Mar. (68)	1 Sun.	239·7688	4406
26 Mar. (85)	6 Fri.	16	12 54	25 Feb. (56)	5 Thur.	115·4915	4407
26 Mar. (85)	0 Sat.	22	25 3	16 Mar. (75)	4 Wed.	150·1739	4408
27 Mar. (86)	2 Mon.	4	37 12	5 Mar. (64)	1 Sun.	25·8968	4409
26 Mar. (86)	3 Tues.	10	49 21	23 Mar. (83)	0 Sat.	60·5791	4410
26 Mar. (85)	4 Wed.	17	1 30	13 Mar. (72)	5 Thur.	274·9340	4411
26 Mar. (85)	5 Thur.	23	13 39	2 Mar. (61)	2 Mon.	150·6569	4412
27 Mar. (86)	0 Sat.	5	25 48	21 Mar. (80)	1 Sun.	185·3395	4413
26 Mar. (86)	1 Sun.	11	37 57	9 Mar. (69)	5 Thur.	61·0621	4414
26 Mar. (85)	2 Mon.	17	50 6	27 Feb. (58)	3 Tues.	275·4169	4415
27 Mar. (86)	4 Wed.	0	2 15	18 Mar. (77)	2 Mon.	310·0993	4416
27 Mar. (86)	5 Thur.	6	14 24	7 Mar. (66)	6 Fri.	185·8221	4417
26 Mar. (86)	6 Fri.	12	26 33	25 Mar. (85)	5 Thur.	220·5045	4418
26 Mar. (85)	0 Sat.	18	38 42	14 Mar. (73)	2 Mon.	96·2274	4419
27 Mar. (86)	2 Mon.	0	50 51	4 Mar. (63)	0 Sat.	310·5822	4420

TABLE

CONCURRENT YEAR.								Mean intercalated (*adhika*) lunar month.
Kali.	Saka.	Chaitrâdi Vikrama.	Mêshâdi solar year in Bengal.	Kollam.	A.D.	Jovian Samvatsara. Southern system.	Jovian Samvatsara. Northern system.	
1	2	3	3*a*	4	5	6	7	8*a*
4421	1242	1377	726	494-95	1319-20	53 Siddhârthin .	58 Raktâksha .	...
4422	1243	1378	727	495-96	*1320-21	54 Raudra . .	59 Krôdhana .	...
4423	1244	1379	728	496-97	1321-22	55 Durmati .	60 Kshaya . .	4 Âshâḍha .
4424	1245	1380	729	497-98	1322-23	56 Dundubhi .	1 Prabhava .	...
4425	1246	1381	730	498-99	1323-24	57 Rudhirôdgârin .	2 Vibhava .	...
4426	1247	1382	731	499-500	*1324-25	58 Raktâksha .	3 Śukla . .	1 Chaitra .
4427	1248	1383	732	500-01	1325-26	59 Krôdhana .	4 Pramôda .	...
4428	1249	1384	733	501-02	1326-27	60 Kshaya . .	5 Prajâpati .	9 Mârgaśira .
4429	1250	1385	734	502-03	1327-28	1 Prabhava .	6 Aṅgiras . .	...
4430	1251	1386	735	503-04	*1328-29	2 Vibhava .	7 Śrîmukha .	...
4431	1252	1387	736	504-05	1329-30	3 Śukla . .	8 Bhâva† . .	6 Bhâdrapada .
4432	1253	1388	737	505-06	1330-31	4 Pramôda .	10 *Dhâtṛi* . .	...
4433	1254	1389	738	506-07	1331-32	5 Prajâpati .	11 *Îśvara* . .	...
4434	1255	1390	739	507-08	*1332-33	6 Aṅgiras . .	12 *Bahudhânya* .	2 Vaiśâkha .
4435	1256	1391	740	508-09	1333-34	7 Śrîmukha .	13 *Pramâthin* .	...
4436	1257	1392	741	509-10	1334-35	8 Bhâva . .	14 *Vikrama* .	11 Mâgha .
4437	1258	1393	742	510-11	1335-36	9 Yuvan . .	15 *Vṛisha* . .	...
4438	1259	1394	743	511-12	*1336-37	10 Dhâtṛi . .	16 Chitrabhânu .	...
4439	1260	1395	744	512-13	1337-38	11 Îśvara . .	17 Subhânu .	7 Âśvina .
4440	1261	1396	745	513-14	1338-39	12 Bahudhânya .	18 Târaṇa . .	...
4441	1262	1397	746	514-15	1339-40	13 Pramâthin .	19 Pârthiva . .	...
4442	1263	1398	747	515-16	*1340-41	14 Vikrama .	20 Vyaya . .	4 Âshâḍha .
4443	1264	1399	748	516-17	1341-42	15 Vṛisha . .	21 Sarvajit . .	...
4444	1265	1400	749	517-18	1342-43	16 Chitrabhânu .	22 Sarvadhârin .	12 Phâlguna .
4445	1266	1401	750	518-19	1343-44	17 Subhânu .	23 Virôdhin .	...

† 9 Yuvan was suppressed in the north by the mean system. By the "true" system K.Y. 4431 (expired), A. D. 1330-31, was called "Yuvan," and 10 Dhâtṛi was suppressed. The next year was 11 Îśvara by both systems.

XC—*contd*

COMMENCEMENT OF THE							
MEAN SOLAR YEAR.			MEAN LUNI-SOLAR YEAR (MEAN SUNRISE OF THE CIVIL DAY ON WHICH CHAITRA ŚUKLA 1 ENDS).			Kali.
Day and month, A.D.	Week-day.	Time of mean Mēsha-saṁkrānti.		Day and month, A.D.	Week-day.	*a* (here = *t*, the index of the *tithi*).	
13	14	17		19	20	23	1
		H. M.	S.				
27 Mar. (86) . .	3 Tues. .	7 3	0	22 Mar. (81) .	5 Thur. .	6·6326	4421
26 Mar. (86) . .	4 Wed. .	13 15	9	11 Mar. (71) .	3 Tues. .	220·9874	4422
26 Mar. (85) . .	5 Thur. .	19 27	18	28 Feb. (59) .	0 Sat. .	96·7103	4423
27 Mar. (86) . .	0 Sat. .	1 39	27	19 Mar. (78) .	6 Fri. .	131·3926	4424
27 Mar. (86) . .	1 Sun. .	7 51	36	8 Mar. (67) .	3 Tues. .	7·1155	4425
26 Mar. (86) . .	2 Mon. .	14 3	45	26 Feb. (57) .	1 Sun. .	221·4703	4426
26 Mar. (85) . .	3 Tues. .	20 15	54	16 Mar. (75) .	0 Sat. .	256·1527	4427
27 Mar. (86) . .	5 Thur. .	2 28	3	5 Mar. (64) .	4 Wed. .	131·8755	4428
27 Mar. (86) . .	6 Fri. .	8 40	12	24 Mar. (83) .	3 Tues. .	166·5579	4429
26 Mar. (86) . .	0 Sat. .	14 52	21	12 Mar. (72) .	0 Sat. .	42·2808	4430
26 Mar. (85) . .	1 Sun. .	21 4	30	2 Mar. (61) .	5 Thur. .	256·6356	4431
27 Mar. (86) . .	3 Tues. .	3 16	39	21 Mar. (80) .	4 Wed. .	291·3180	4432
27 Mar. (86) . .	4 Wed. .	9 28	48	10 Mar. (69) .	1 Sun. .	167·0409	4433
26 Mar. (86) . .	5 Thur. .	15 40	57	27 Feb. (58) .	5 Thur. .	42·7637	4434
26 Mar. (85) . .	6 Fri. .	21 53	6	17 Mar. (76) .	4 Wed. .	77·4460	4435
27 Mar. (86) . .	1 Sun. .	4 5	15	7 Mar. (66) .	2 Mon. .	291·8009	4436
27 Mar. (86) . .	2 Mon. .	10 17	24	25 Mar. (85) .	1 Sun. .	326·4833	4437
26 Mar. (86) . .	3 Tues. .	16 29	33	14 Mar. (74) .	5 Thur. .	202·2062	4438
26 Mar. (85) . .	4 Wed. .	22 41	42	3 Mar. (62) .	2 Mon. .	77·9289	4439
27 Mar. (86) . .	6 Fri. .	4 53	51	22 Mar. (81) .	1 Sun. .	112·6114	4440
27 Mar. (86) . .	0 Sat. .	11 6	0	12 Mar. (71) .	6 Fri. .	326·9662	4441
26 Mar. (86) . .	1 Sun. .	17 18	9	29 Feb. (60) .	3 Tues. .	202·6890	4442
26 Mar. (85) . .	2 Mon. .	23 30	18	19 Mar. (78) .	2 Mon. .	237·3714	4443
27 Mar. (86) . .	4 Wed. .	5 42	27	8 Mar. (67) .	6 Fri. .	113·0943	4444
27 Mar. (86) . .	5 Thur. .	11 54	36	27 Mar. (85) .	5 Thur. .	147·7767	4445

TABLE

CONCURRENT YEAR.								
Kali.	Śaka.	Chaitrâdi Vikrama.	Mêshâdi solar year in Bengal.	Kollam.	A.D.	Jovian Saṁvatsara. Southern system.	Jovian Saṁvatsara. Northern system.	Mean intercalated (*adhika*) lunar month.
1	2	3	3*a*	4	5	6	7	8*a*
4446	1267	1402	751	519-20	*1344-45	18 Târaṇa . .	24 Vikṛita . .	...
4447	1268	1403	752	520-21	1345-46	19 Pârthiva . .	25 Khara . .	9 Mârgaśira .
4448	1269	1404	753	521-22	1346-47	20 Vyaya . .	26 Nandana . .	...
4449	1270	1405	754	522-23	1347-48	21 Sarvajit . .	27 Vijaya . .	...
4450	1271	1406	755	523-24	*1348-49	22 Sarvadhârin .	28 Jaya . .	6 Bhâdrapada .
4451	1272	1407	756	524-25	1349-50	23 Virôdhin .	29 Manmatha .	...
4452	1273	1408	757	525-26	1350-51	24 Vikṛita . .	30 Durmukha .	...
4453	1274	1409	758	526-27	1351-52	25 Khara . .	31 Hêmalamba .	2 Vaiśâkha .
4454	1275	1410	759	527-28	*1352-53	26 Nandana . .	32 Vilamba . .	...
4455	1276	1411	760	528-29	1353-54	27 Vijaya . .	33 Vikârin . .	11 Mâgha .
4456	1277	1412	761	529-30	1354-55	28 Jaya . .	34 Śârvarin . .	...
4457	1278	1413	762	530-31	1355-56	29 Manmatha .	35 Plava . .	...
4458	1279	1414	763	531-32	*1356-57	30 Durmukha .	36 Śubhakṛit .	7 Âśvina .
4459	1280	1415	764	532-33	1357-58	31 Hêmalamba .	37 Śôbhana . .	...
4460	1281	1416	765	533-34	1358-59	32 Vilamba . .	38 Krôdhin . .	...
4461	1282	1417	766	534-35	1359-60	33 Vikârin . .	39 Viśvâvasu .	4 Âshâḍha .
4462	1283	1418	767	535-36	*1360-61	34 Śârvarin . .	40 Parâbhava .	...
4463	1284	1419	768	536-37	1361-62	35 Plava . .	41 Plavaṅga .	12 Phâlguna .
4464	1285	1420	769	537-38	1362-63	36 Śubhakṛit .	42 Kîlaka . .	...
4465	1286	1421	770	538-39	1363-64	37 Śôbhana . .	43 Saumya . .	...
4466	1287	1422	771	539-40	*1364-65	38 Krôdhin . .	44 Sâdhâraṇa .	9 Mârgaśira .
4467	1288	1423	772	540-41	1365-66	39 Viśvâvasu .	45 Virôdhakṛit .	...
4468	1289	1424	773	541-42	1366-67	40 Parâbhava .	46 Paridhâvin .	...
4469	1290	1425	774	542-43	1367-68	41 Plavaṅga .	47 Pramâdin .	5 Srâvaṇa .
4470	1291	1426	775	543-44	*1368-69	42 Kîlaka . .	48 Ânanda . .	...

XC—*contd.*

COMMENCEMENT OF THE								
MEAN SOLAR YEAR.					MEAN LUNI-SOLAR YEAR (MEAN SUNRISE OF THE CIVIL DAY ON WHICH CHAITRA SUKLA 1 ENDS).			Kali.
Day and month, A.D.	Week-day.	Time of mean Mêsha-saṁkrānti.			Day and month, A.D.	Week-day.	*a* (here = *t*, the index of the *tithi*).	
13	14	17			19	20	23	1
		H.	M.	S.				
26 Mar. (86)	6 Fri.	18	6	45	15 Mar. (75)	2 Mon.	23·4995	4446
27 Mar. (86)	1 Sun.	0	18	54	5 Mar. (64)	0 Sat.	237·8543	4447
27 Mar. (86)	2 Mon	6	31	3	24 Mar. (83)	6 Fri.	272·5367	4448
27 Mar. (86)	3 Tues.	12	43	12	13 Mar. (72)	3 Tues.	148·2595	4449
26 Mar. (86)	4 Wed.	18	55	21	1 Mar. (61)	0 Sat.	23·9824	4450
27 Mar. (86)	6 Fri.	1	7	30	20 Mar. (79)	6 Fri.	58·6648	4451
27 Mar. (86)	0 Sat.	7	19	39	10 Mar. (69)	4 Wed.	273·0197	4452
27 Mar. (86)	1 Sun.	13	31	48	27 Feb. (58)	1 Sun.	148·7424	4453
26 Mar. (86)	2 Mon.	19	43	57	17 Mar. (77)	0 Sat.	183·4248	4454
27 Mar. (86)	4 Wed.	1	56	6	6 Mar. (65)	4 Wed.	59·1477	4455
27 Mar. (86)	5 Thur.	8	8	15	25 Mar. (84)	3 Tues.	93·8300	4456
27 Mar. (86)	6 Fri.	14	20	24	15 Mar. (74)	1 Sun.	308·1849	4457
26 Mar. (86)	0 Sat.	20	32	33	3 Mar. (63)	5 Thur.	183·9077	4458
27 Mar. (86)	2 Mon.	2	44	42	22 Mar. (81)	4 Wed.	218·5902	4459
27 Mar. (86)	3 Tues.	8	56	51	11 Mar. (70)	1 Sun.	94·3129	4460
27 Mar. (86)	4 Wed.	15	9	0	1 Mar. (60)	6 Fri.	308·6678	4461
26 Mar. (86)	5 Thur.	21	21	9	18 Mar. (78)	4 Wed.	4·7182	4462
27 Mar. (86)	0 Sat.	3	33	18	8 Mar. (67)	2 Mon.	219·0730	4463
27 Mar. (86)	1 Sun.	9	45	27	27 Mar. (86)	1 Sun.	253·7554	4464
27 Mar. (86)	2 Mon.	15	57	36	16 Mar. (75)	5 Thur.	129·4783	4465
26 Mar. (86)	3 Tues.	22	9	45	4 Mar. (64)	2 Mon.	5·2011	4466
27 Mar. (86)	5 Thur.	4	21	54	23 Mar. (82)	1 Sun.	39·8835	4467
27 Mar. (86)	6 Fri.	10	34	3	13 Mar. (72)	6 Fri.	254·2383	4468
27 Mar. (86)	0 Sat.	16	46	12	2 Mar. (61)	3 Tues.	129·9612	4469
26 Mar. (86)	1 Sun.	22	58	21	20 Mar. (80)	2 Mon.	164·6435	4470

TABLE

CONCURRENT YEAR.								Mean intercalated (*adhika*) lunar month.
Kali.	Śaka.	Chaitrādi Vikrama.	Mēshādi solar year in Bengal.	Kollam.	A.D.	Jovian Saṁvatsara. Southern system.	Jovian Saṁvatsara. Northern system.	
1	2	3	3a	4	5	6	7	8a
4471	1292	1427	776	544-45	1369-70	43 Saumya .	49 Rākshasa .	...
4472	1293	1428	777	545-46	1370-71	44 Sādhāraṇa	50 Anala . .	2 Vaiśākha .
4473	1294	1429	778	546-47	1371-72	45 Virōdhakṛit .	51 Piṅgala .	...
4474	1295	1430	779	547-48	*1372-73	46 Paridhāvin .	52 Kālayukta .	10 Pausha .
4475	1296	1431	780	548-49	1373-74	47 Pramādin .	53 Siddhārthin .	...
4476	1297	1432	781	549-50	1374-75	48 Ānanda . .	54 Raudra . .	...
4477	1298	1433	782	550-51	1375-76	49 Rākshasa .	55 Durmati .	7 Āśvina .
4478	1299	1434	783	551-52	*1376-77	50 Anala . .	56 Dundubhi .	...
4479	1300	1435	784	552-53	1377-78	51 Piṅgala . .	57 Rudhirōdgārin	...
4480	1301	1436	785	553-54	1378-79	52 Kālayukta .	58 Raktāksha .	3 Jyēshṭha .
4481	1302	1437	786	554-55	1379-80	53 Siddhārthin .	59 Krōdhana .	...
4482	1303	1438	787	555-56	*1380-81	54 Raudra . .	60 Kshaya . .	12 Phālguna .
4483	1304	1439	788	556-57	1381-82	55 Durmati .	1 Prabhava .	...
4484	1305	1440	789	557-58	1382-83	56 Dundubhi .	2 Vibhava .	...
4485	1306	1441	790	558-59	1383-84	57 Rudhirōdgārin .	3 Śukla . .	9 Mārgaśira .
4486	1307	1442	791	559-60	*1384-85	58 Raktāksha .	4 Pramōda .	...
4487	1308	1443	792	560-61	1385-86	59 Krōdhana .	5 Prajāpati .	...
4488	1309	1444	793	561-62	1386-87	60 Kshaya . .	6 Aṅgiras . .	5 Śrāvaṇa .
4489	1310	1445	794	562-63	1387-88	1 Prabhava .	7 Śrīmukha .	...
4490	1311	1446	795	563-64	*1388-89	2 Vibhava .	8 Bhāva . .	...
4491	1312	1447	796	564-65	1389-90	3 Śukla . .	9 Yuvan . .	2 Vaiśākha .
4492	1313	1448	797	565-66	1390-91	4 Pramōda .	10 Dhātṛi . .	...
4493	1314	1449	798	566-67	1391-92	5 Prajāpati .	11 Īśvara . .	10 Pausha .
4494	1315	1450	799	567-68	*1392-93	6 Aṅgiras . .	12 Bahudhānya .	...
4495	1316	1451	800	568-69	1393-94	7 Śrīmukha .	13 Pramāthin .	...

XC—*contd.*

COMMENCEMENT OF THE									Kali.
MEAN SOLAR YEAR.					MEAN LUNI-SOLAR YEAR (MEAN SUNRISE OF THE CIVIL DAY ON WHICH CHAITRA ŚUKLA 1 ENDS).				
Day and month, A.D.	Week-day.	Time of mean Mêsha-saṁkrânti.			Day and month, A.D.	Week-day.	*a* (here = *t*, the index of the *tithi*).		
13	14	17			19	20	23		1
		H.	M.	S.					
27 Mar. (86)	3 Tues.	5	10	30	9 Mar. (68)	6 Fri.	40·3664		4471
27 Mar. (86)	4 Wed.	11	22	39	27 Feb. (58)	4 Wed.	254·7212		4472
27 Mar. (86)	5 Thur.	17	34	48	18 Mar. (77)	3 Tues.	289·4036		4473
26 Mar. (86)	6 Fri.	23	46	57	6 Mar. (66)	0 Sat.	165·1264		4474
27 Mar. (86)	1 Sun.	5	59	6	25 Mar. (84)	6 Fri.	199·8088		4475
27 Mar. (86)	2 Mon.	12	11	15	14 Mar. (73)	3 Tues.	75·5317		4476
27 Mar. (86)	3 Tues.	18	23	24	4 Mar. (63)	1 Sun.	289·8864		4477
27 Mar. (87)	5 Thur.	0	35	33	22 Mar. (82)	0 Sat.	324·5689		4478
27 Mar. (86)	6 Fri.	6	47	42	11 Mar. (70)	4 Wed.	200·2917		4479
27 Mar. (86)	0 Sat.	12	59	51	28 Feb. (59)	1 Sun.	76·0146		4480
27 Mar. (86)	1 Sun.	19	12	0	19 Mar. (78)	0 Sat.	110·6969		4481
27 Mar. (87)	3 Tues.	1	24	9	8 Mar. (68)	5 Thur.	325·0518		4482
27 Mar. (86)	4 Wed.	7	36	18	26 Mar. (85)	3 Tues.	21·1022		4483
27 Mar. (86)	5 Thur.	13	48	27	16 Mar. (75)	1 Sun.	235·4571		4484
27 Mar. (86)	6 Fri.	20	0	36	5 Mar. (64)	5 Thur.	111·1798		4485
27 Mar. (87)	1 Sun.	2	12	45	23 Mar. (83)	4 Wed.	145·8623		4486
27 Mar. (86)	2 Mon.	8	24	54	12 Mar. (71)	1 Sun.	21·5851		4487
27 Mar. (86)	3 Tues.	14	37	3	2 Mar. (61)	6 Fri.	235·9399		4488
27 Mar. (86)	4 Wed.	20	49	12	21 Mar. (80)	5 Thur.	270·6223		4489
27 Mar. (87)	6 Fri.	3	1	21	9 Mar. (69)	2 Mon.	146·3452		4490
27 Mar. (86)	0 Sat.	9	13	30	26 Feb. (57)	6 Fri.	22·0680		4491
27 Mar. (86)	1 Sun.	15	25	39	17 Mar. (76)	5 Thur.	56·7503		4492
27 Mar. (86)	2 Mon.	21	37	48	7 Mar. (66)	3 Tues.	271·1052		4493
27 Mar. (87)	4 Wed.	3	49	57	25 Mar. (85)	2 Mon.	305·7876		4494
27 Mar. (86)	5 Thur.	10	2	6	14 Mar. (73)	6 Fri.	181·5104		4495

TABLE

CONCURRENT YEAR.								Mean intercalated (*adhika*) lunar month.
						Jovian Saṁvatsara.		
Kali.	Śaka.	Chaitrādi Vikrama.	Mēshādi solar year in Bengal.	Kollam.	A.D.	Southern system.	Northern system.	
1	2	3	3a	4	5	6	7	8a
4496	1317	1452	801	569-70	1394-95	8 Bhāva . .	14 Vikrama .	7 Āśvina .
4497	1318	1453	802	570-71	1395-96	9 Yuvan . .	15 Vṛisha . .	...
4498	1319	1454	803	571-72	*1396-97	10 Dhātṛi . .	16 Chitrabhānu .	...
4499	1320	1455	804	572-73	1397-98	11 Īśvara . .	17 Subhānu .	3 Jyēshṭha .
4500	1321	1456	805	573-74	1398-99	12 Bahudhānya .	18 Tāraṇa . .	...
4501	1322	1457	806	574-75	1399-1400	13 Pramāthin .	19 Pārthiva .	12 Phālguna .
4502	1323	1458	807	575-76	*1400-01	14 Vikrama .	20 Vyaya . .	...

XC—*concld.*

COMMENCEMENT OF THE									Kali
MEAN SOLAR YEAR.					MEAN LUNI-SOLAR YEAR (MEAN SUNRISE OF THE CIVIL DAY ON WHICH CHAITRA ŚUKLA 1 ENDS).				
Day and month, A.D.	Week-day	Time of mean Mēsha-saṁkrānti.			Day and month, A.D.	Week-day.	*a* (here = *t*, the index of the *tithi*).		
13	14	17			19	20	23		1
		H.	M.	S.					
27 Mar. (86) . .	6 Fri. .	16	14	15	3 Mar. (62) .	3 Tues. .	57·2333		4496
27 Mar. (86) . .	0 Sat. .	22	26	24	22 Mar. (81) .	2 Mon. .	91·9157		4497
27 Mar. (87) . .	2 Mon. .	4	38	33	11 Mar. (71) .	0 Sat. .	306·2704		4498
27 Mar. (86) . .	3 Tues. .	10	50	42	28 Feb. (59) .	4 Wed. .	181·9933		4499
27 Mar. (86) . .	4 Wed. .	17	2	51	19 Mar. (78) .	3 Tues. .	216·6757		4500
27 Mar. (86) . .	5 Thur. .	23	15	0	8 Mar. (67) .	0 Sat. .	92·3986		4501
27 Mar. (87) . .	0 Sat. .	5	27	9	26 Mar. (86) .	6 Fri. .	127·0810		4502

TABLE XCI.

Duration and Collective Duration of mean solar months according to the Brahma-Siddhānta, with increase of "*a*" at each saṁkrānti.

Mean luni-solar month, ending after the second of the two solar saṁkrāntis connected with it.	At the mean solar saṁkrāntis.	Collective duration in time, and collective increase of "*a*" from mean Mēsha-saṁkrānti to the several saṁkrāntis.						
		Day.	Week-day.	H.	M.	S.	*a*	
1	2	3					4	
	Mīna-saṁk. (*of previous year*).							
1 Chaitra	Mēsha-saṁk.	0	0	0	0	0	0	
2 Vaiśākha	Vṛishabha-saṁk.	30	(2)	10	31	0¾	307·3492	The duration of each mean solar month is 30ᵈ 10ʰ 31ᵐ 0¾ˢ; and in this time the mean moon's increase of distance from mean sun (our *a*), in measurement by 10,000ths of circle, is 307·349156595.
3 Jyēshṭha	Mithuna-saṁk.	60	(4)	21	2	1½	614·6983	
4 Āshāḍha	Karka-saṁk.	91	(0)	7	33	2¼	922·0475	
5 Śrāvaṇa	Siṁha-saṁk.	121	(2)	18	4	3	1229·3966	
6 Bhādrapada	Kanyā-saṁk.	152	(5)	4	35	3¾	1536·7458	
7 Āśvina	Tulā-saṁk.	182	(0)	15	6	4½	1844·0949	
8 Kārttika	Vṛiśchika-saṁk.	213	(3)	1	37	5¼	2151·4441	
9 Mārgaśira	Dhanus-saṁk.	243	(5)	12	8	6	2458·7933	
10 Pausha	Makara-saṁk.	273	(0)	22	39	6¾	2766·1424	
11 Māgha	Kumbha-saṁk.	304	(3)	9	10	7½	3073·4916	
12 Phālguna	Mīna-saṁk.	334	(5)	19	41	8¼	3380·8407	
1 Chaitra (*of following year*).	Mēsha-saṁk. (*of following year*).	365	(1)	6	12	9	3688·1899	

A saṁkrānti occurs at the moment when the mean sun enters a zodiacal sign.

TABLE XCII.

Century-Table.

Value of "a" (="t") at beginning of centuries K.Y., i.e. at mean sunrise on day of occurrence of mean Mēsha-saṁkrānti (mean sun at 0°) in first year of century. [Centuries 38, 44, were defective; the rest common.]

Beginning of K.Y. century.	Beginning in A.D.	Week-day.	a (= t).
37	599	(0)	6228·4770
38	699	(0)	5100·3761
39	799	(6)	3633·6433
40	899	(6)	2505·5425
41	999	(6)	1377·4416
42	1099	(6)	249·3408
43	1199	(6)	9121·2399
44	1299	(6)	7993·1391
45	1399	(5)	6526·4063

For odd years of centuries use the *Siddhānta-Śirōmaṇi* Table LVII-B *above*.

TABLE XCIII.

Mean sunrise values of "a" (distance of mean moon from mean sun) in 10,000ths of circle for a month previous to the day on which mean Mēsha-saṁkrānti occurred.

Interval of days from mean Mēsha-saṁkrānti day.	Week-day.	a (mean sunrise value).	Interval of days from mean Mēsha-saṁkrānti day.	Week-day.	a (mean sunrise value).
1	2	3	1	2	3
31	(4)	9502·4085	15	(6)	4920·5202
30	(5)	9841·0404	14	(0)	5259·1522
29	(6)	179·6724	13	(1)	5597·7842
28	(0)	518·3044	12	(2)	5936·4162
27	(1)	856·9364	11	(3)	6275·0482
26	(2)	1195·5684	10	(4)	6613·6801
25	(3)	1534·2004	9	(5)	6952·3121
24	(4)	1872·8324	8	(6)	7290·9441
23	(5)	2211·4643	7	(0)	7629·5761
22	(6)	2550·0963	6	(1)	7968·2081
21	(0)	2888·7283	5	(2)	8306·8401
20	(1)	3227·3603	4	(3)	8645·4721
19	(2)	3565·9923	3	(4)	8984·1040
18	(3)	3904·6243	2	(5)	9322·7360
17	(4)	4243·2563	1	(6)	9661·3680
16	(5)	4581·8882	0	(0)	0·0

The use of this Table is explained in Example 2 of this article, and in Example 1 of article on the *First Ārya-Siddhānta, mean system above.*

TABLE XCIV.

TIME-EQUIVALENTS OF THE TITHI ("*a*" OR "*t*"), NAKSHATRA ("*n*"), AND YŌGA ("*y*") UNITS.

In very close cases it is sometimes necessary to calculate the exact moment of the beginning and ending of *tithis, nakshatras* and *yōgas,* with greater accuracy than can be obtained by the use of Table X, *Indian Calendar,* or Table LXX (above, where the time-equivalent of the unit, respectively, is given only in hours and minutes). My general working Tables given in this volume for the Hindu astronomical *Siddhāntas* yield results, stated in measurement by 10,000ths of the circle, with an accuracy extending to four places of decimals, and the following Table enables the result to be translated into time down to a fraction of a second. It may be used for all astronomical authorities.

The tithi-index unit.

The *tithi*-unit is $\frac{1}{10,000}$th of a mean lunation. The mean lunation, according to the *Ārya-* and *Sūrya-Siddhāntas,* occupies $29^d\ 12^h\ 44^m\ 2^s{\cdot}79$. The unit, or 10,000th part of this, is $4^m{\cdot}2524046$, or $4^m\ 15^s{\cdot}144279$.

The nakshatra-index unit.

The moon's *nakshatra*, or her position in the heavens, mean or true, is found by adding the *tithi*-index, "*a*" or "*t*", to the index of the sun's longitude, "*s*", mean or true. Both these values are found in the ordinary course of calculation for a date.

The mean *nakshatra*-value $n = 10,000$ is reached in $27^d\ 7^h\ 43^m\ 12^s{\cdot}3$. In this period the sun's mean motion amounts, in 10,000ths of circle measurement, to 748·0087 (*Table XLIV above* and the moon's mean distance from mean sun increases (*Table LIV A, B*) to 9251·9913. Total 10,000.

$27^d\ 7^h\ 43^m\ 12^s{\cdot}3 = 39343^m{\cdot}205$, and this divided by 10,000 fixes the time-equivalent of the *nakshatra*-unit as $3^m{\cdot}9343205$, or $3^m\ 56^s{\cdot}05923$.

The yōga-index unit.

Similarly the *yōga-chakra* is estimated by the *Sūrya-Siddhānta* (*Indian Calendar, p. 62, § 113*) as occupying 36605·116 minutes of time, or $25^d\ 10^h\ 5^m\ 6^s{\cdot}96$.[1] The *yōga*-unit therefore is $3^m{\cdot}6605116$, or $3^m\ 39^s{\cdot}6307$.

---

[1] The *yōga* formula is "*y*" = "*s*" (sun's long.) + "*n*" (moon's *nakshatra*), and, since $n = s + a$, $y = 2s + a$. In the period noted it will be found by calculation, using Table XLIV above, that the mean sun "*s*" arrives, in 10,000ths of circle measurement, at long. 695·9511; and by using Table LXIV that in the same period the mean moon has increased her distance from mean sun "(*a*)" by 8608·0964. Twice "*s*" = 1391·9022, and this + 8608·0964 (the value of "*a*") = 9999·9988, practically 10,000 exactly. Table LXIV was prepared according to the *First Ārya-Siddhānta.* Using *Siddhānta-Śirōmaṇi* and *Brahma-Siddhānta* estimates (*Table LIV*) the total amounts to 10,000·0015, I have as yet no similar Table according to *Sūrya-Siddhānta* requirements; but from what has been said it may be assumed that its estimate of the time occupied by one *yōga-chakra* (= 10,000) is correct.

TABLE XCIV-A.

Time-equivalents.

Tithi-index units.

(" Arg."$= a$ or t.)

Arg.	H.	M.	S.	Arg.	H.	M.	S.	Arg.	H.	M.	S.	Arg.	H.	M.	S.
1	0	4	15·14	30	2	7	34·33	59	4	10	53·51	88	6	14	12·70
2	0	8	30·29	31	2	11	49·47	60	4	15	8·7	89	6	18	27·84
3	0	12	45·43	32	2	16	4·62	61	4	19	23·80	90	6	22	42·99
4	0	17	0·58	33	2	20	19·76	62	4	23	38·95	91	6	26	58·13
5	0	21	15·72	34	2	24	34·91	63	4	27	54·09	92	6	31	13·27
6	0	25	30·87	35	2	28	50·05	64	4	32	9·23	93	6	35	28·42
7	0	29	46·01	36	2	33	5·19	65	4	36	24·38	94	6	39	43·56
8	0	34	1·15	37	2	37	20·34	66	4	40	39·52	95	6	43	58·71
9	0	38	16·30	38	2	41	35·48	67	4	44	54·67	96	6	48	13·85
10	0	42	31·44	39	2	45	50·63	68	4	49	9·81	97	6	52	29·00
11	0	46	46·59	40	2	50	5·77	69	4	53	24·96	98	6	56	44·14
12	0	51	1·73	41	2	54	20·92	70	4	57	40·10	99	7	0	59·28
13	0	55	16·88	42	2	58	36·06	71	5	1	55·24	100	7	5	14·43
14	0	59	32·02	43	3	2	51·20	72	5	6	10·39	200	14	10	28·86
15	1	3	47·16	44	3	7	6·35	73	5	10	25·53	300	21	15	43·28
16	1	8	2·31	45	3	11	21·49	74	5	14	40·68	400	28	20	57·71
17	1	12	17·45	46	3	15	36·64	75	5	18	55·82	500	35	26	12·14
18	1	16	32·60	47	3	19	51·78	76	5	23	10·97	600	42	31	26·57
19	1	20	47·74	48	3	24	6·93	77	5	27	26·11	700	49	36	41·00
20	1	25	2·29	49	3	28	22·07	78	5	31	41·25	800	56	41	55·42
21	1	29	18·03	50	3	32	37·21	79	5	35	56·40	900	63	47	9·85
22	1	33	33·17	51	3	36	52·36	80	5	40	11·54	1000	70	52	24·28
23	1	37	48·32	52	3	41	7·50	81	5	44	26·69				
24	1	42	3·46	53	3	45	22·65	82	5	48	41·83				
25	1	46	18·61	54	3	49	37·79	83	5	52	56·98				
26	1	50	33·75	55	3	53	52·94	84	5	57	12·20				
27	1	54	48·90	56	3	58	8·08	85	6	1	27·26				
28	1	59	4·04	57	4	2	23·22	86	6	5	42·41				
29	2	3	19·18	58	4	6	38·37	87	6	9	57·55				

TABLE XCIV-B.

Time-equivalents.

Decimals of tithi-index units.

First 2 decimals.	M.	S.	First 2 decimals.	M.	S.	First 2 decimals.	M.	S.
·01	0	2·55	·34	1	26·75	·67	2	50·95
·02	0	5·10	·35	1	29·30	·68	2	53·50
·03	0	7·65	·36	1	31·85	·69	2	56·05
·04	0	10·21	·37	1	34·40	·70	2	58·60
·05	0	12·76	·38	1	36·95	·71	3	1·15
·06	0	15·31	·39	1	39·51	·72	3	3·70
·07	0	17·86	·40	1	42·06	·73	3	6·26
·08	0	20·41	·41	1	44·61	·74	3	8·81
·09	0	22·96	·42	1	47·16	·75	3	11·36
·10	0	25·51	·43	1	49·71	·76	3	13·91
·11	0	28·07	·44	1	52·26	·77	3	16·46
·12	0	30·62	·45	1	54·81	·78	3	19·01
·13	0	33·17	·46	1	57·37	·79	3	21·56
·14	0	35·72	·47	1	59·92	·80	3	24·12
·15	0	38·27	·48	2	2·47	·81	3	26·67
·16	0	40·82	·49	2	5·02	·82	3	29·22
·17	0	43·37	·50	2	7·57	·83	3	31·78
·18	0	45·93	·51	2	10·12	·84	3	34·32
·19	0	48·48	·52	2	12·68	·85	3	36·87
·20	0	51·03	·53	2	15·23	·86	3	39·42
·21	0	53·58	·54	2	17·78	·87	3	41·98
·22	0	56·13	·55	2	20·33	·88	3	44·53
·23	0	58·68	·56	2	22·88	·89	3	47·08
·24	0	61·23	·57	2	25·43	·90	3	49·63
·25	1	3·79	·58	2	27·98	·91	3	52·18
·26	1	6·34	·59	2	30·54	·92	3	54·73
·27	1	8·89	·60	2	33·09	·93	3	57·28
·28	1	11·44	·61	2	35·64	·94	3	59·84
·29	1	13·99	·62	2	38·19	·95	4	2·39
·30	1	16·54	·63	2	40·74	·96	4	4·94
·31	1	19·09	·64	2	43·29	·97	4	7·49
·32	1	21·65	·65	2	45·84	·98	4	10·04
·33	1	24·20	·66	2	48·40	·99	4	12·59

3rd and 4th decimals.	S.	3rd and 4th decimals.	S.	3rd and 4th decimals.	S.
·0001	0·03	·0034	0·87	·0067	1·71
·0002	0·05	·0035	0·89	·0068	1·73
·0003	0·08	·0036	0·92	·0069	1·76
·0004	0·10	·0037	0·94	·0070	1·79
·0005	0·13	·0038	0·97	·0071	1·81
·0006	0·15	·0039	1·00	·0072	1·84
·0007	0·18	·0040	1·02	·0073	1·86
·0008	0·20	·0041	1·05	·0074	1·89
·0009	0·23	·0042	1·07	·0075	1·91
·0010	0·26	·0043	1·10	·0076	1·94
·0011	0·28	·0044	1·12	·0077	1·96
·0012	0·31	·0045	1·15	·0078	1·99
·0013	0·33	·0046	1·17	·0079	2·02
·0014	0·36	·0047	1·20	·0080	2·04
·0015	0·38	·0048	1·22	·0081	2·07
·0016	0·41	·0049	1·25	·0082	2·09
·0017	0·43	·0050	1·28	·0083	2·12
·0018	0·46	·0051	1·30	·0084	2·14
·0019	0·48	·0052	1·33	·0085	2·17
·0020	0·51	·0053	1·35	·0086	2·19
·0021	0·54	·0054	1·38	·0087	2·22
·0022	0·56	·0055	1·40	·0088	2·25
·0023	0·59	·0056	1·43	·0089	2·27
·0024	0·61	·0057	1·45	·0090	2·30
·0025	0·64	·0058	1·48	·0091	2·32
·0026	0·66	·0059	1·51	·0092	2·35
·0027	0·69	·0060	1·53	·0093	2·37
·0028	0·71	·0061	1·56	·0094	2·40
·0029	0·74	·0062	1·58	·0095	2·42
·0030	0·77	·0063	1·61	·0096	2·45
·0031	0·79	·0064	1·63	·0097	2·47
·0032	0·82	·0065	1·66	·0098	2·50
·0033	0·84	·0066	1·68	·0099	2·52

TABLE XCIV-C.

Time-equivalents.

Nakshatra-index units.

Arg.	H.	M.	S.	Arg.	H.	M.	S.	Arg.	H.	M.	S.	Arg.	H.	M.	S.
1	0	3	56·06	31	2	1	57·84	61	3	59	59·61	91	5	58	1·39
2	0	7	52·12	32	2	5	53·90	62	4	3	55·67	92	6	1	57·45
3	0	11	48·18	33	2	9	49·95	63	4	7	51·73	93	6	5	53·51
4	0	15	44·24	34	2	13	46·01	64	4	11	47·79	94	6	9	49·57
5	0	19	40·30	35	2	17	42·07	65	4	15	43·85	95	6	13	45·63
6	0	23	36·36	36	2	21	38·13	66	4	19	39·91	96	6	17	41·69
7	0	27	32·41	37	2	25	34·19	67	4	23	35·97	97	6	21	37·75
8	0	31	28·47	38	2	29	30·25	68	4	27	32·03	98	6	25	33·80
9	0	35	24·53	39	2	33	26·31	69	4	31	28·09	99	6	29	29·86
10	0	39	20·59	40	2	37	22·37	70	4	35	24·15	100	6	33	25·92
11	0	43	16·65	41	2	41	18·43	71	4	39	20·21	200	13	6	51·85
12	0	47	12·71	42	2	45	14·49	72	4	43	16·26	300	19	40	17·78
13	0	51	8·77	43	2	49	10·55	73	4	47	12·32				
14	0	55	4·83	44	2	53	6·61	74	4	51	8·38				
15	0	59	0·89	45	2	57	2·67	75	4	55	4·44				
16	1	2	56·95	46	3	0	58·72	76	4	59	0·50				
17	1	6	53·01	47	3	4	54·78	77	5	2	56·56				
18	1	10	49·07	48	3	8	50·84	78	5	6	52·62				
19	1	14	45·13	49	3	12	46·90	79	5	10	48·68				
20	1	18	41·18	50	3	16	42·96	80	5	14	44·74				
21	1	22	37·24	51	3	20	39·02	81	5	18	40·80				
22	1	26	33·30	52	3	24	35·08	82	5	22	36·86				
23	1	30	29·36	53	3	28	31·14	83	5	26	32·92				
24	1	34	25·42	54	3	32	27·20	84	5	30	28·98				
25	1	38	21·48	55	3	36	23·26	85	5	34	25·03				
26	1	42	17·54	56	3	40	19·32	86	5	38	21·09				
27	1	46	13·60	57	3	44	15·38	87	5	42	17·15				
28	1	50	9·66	58	3	48	11·44	88	5	46	13·21				
29	1	54	5·72	59	3	52	7·49	89	5	50	9·27				
30	1	58	1·78	60	3	56	3·55	90	5	54	5·33				

TABLE XCIV-D.

Time-equivalents.

Decimals of Nakshatra-index units.

First 2 decimals.	M.	S.	First 2 decimals.	M.	S.	First 2 decimals.	M.	S.
·01	0	2·36	·34	1	20·26	·67	2	38·16
·02	0	4·72	·35	1	22·62	·68	2	40·52
·03	0	7·08	·36	1	24·98	·69	2	42·88
·04	0	9·44	·37	1	27·34	·70	2	45·24
·05	0	11·80	·38	1	29·70	·71	2	47·60
·06	0	14·16	·39	1	32·06	·72	2	49·96
07	0	16·52	·40	1	34·42	·73	2	52·32
·08	0	18·88	·41	1	36·78	·74	2	54·68
·09	0	21·25	·42	1	39·14	·75	2	57·04
·10	0	23·61	·43	1	41·51	·76	2	59·40
·11	0	25·97	·44	1	43·87	·77	3	1·77
·12	0	28·33	·45	1	46·23	·78	3	4·13
·13	0	30·69	·46	1	48·59	·79	3	6·49
·14	0	33·05	·47	1	50·95	·80	3	8·85
·15	0	35·41	·48	1	53·31	·81	3	11·21
·16	0	37·77	·49	1	55·67	·82	3	13·57
·17	0	40·13	·50	1	58·03	·83	3	15·93
·18	0	42·49	·51	2	0·39	·84	3	18·29
·19	0	44·85	·52	2	2·75	·85	3	20·65
·20	0	47·21	·53	2	5·11	·86	3	23·01
·21	0	49·57	·54	2	7·47	·87	3	25·37
·22	0	51·93	·55	2	9·83	·88	3	27·73
·23	0	54·29	·56	2	12·19	·89	3	30·09
·24	0	56·65	·57	2	14·55	·90	3	32·45
·25	0	59·01	·58	2	16·91	·91	3	34·81
·26	1	1·38	·59	2	19·28	·92	3	37·17
·27	1	3·74	·60	2	21·64	·93	3	39·54
·28	1	6·10	·61	2	24·00	·94	3	41·90
·29	1	8·46	·62	2	26·36	·95	3	44·26
·30	1	10·82	·63	2	28·72	·96	3	46·62
·31	1	13·18	·64	2	31·08	·97	3	48·98
·32	1	15·54	·65	2	33·44	·98	3	51·34
·33	1	17·90	·66	2	35·80	·99	3	53·70

3rd and 4th decimals.	S.	3rd and 4th decimals.	S.	3rd and 4th decimals.	S.
·0001	0·02	·0034	0·80	·0067	1·58
·0002	0·05	·0035	0·83	·0068	1·61
·0003	0·07	·0036	0·85	·0069	1·63
·0004	0·09	·0037	0·87	·0070	1·65
·0005	0·12	·0038	0·90	·0071	1·68
·0006	0·14	·0039	0·92	·0072	1·70
·0007	0·17	·0040	0·94	·0073	1·72
·0008	0·19	·0041	0·97	·0074	1·75
·0009	0·21	·0042	0·99	·0075	1·77
·0010	0·24	·0043	1·02	·0076	1·79
·0011	0·26	·0044	1·04	·0077	1·82
·0012	0·28	·0045	1·06	·0078	1·84
·0013	0·31	·0046	1·09	·0079	1·86
·0014	0·33	·0047	1·11	·0080	1·89
·0015	0·35	·0048	1·13	·0081	1·91
·0016	0·38	·0049	1·16	·0082	1·94
·0017	0·40	·0050	1·18	·0083	1·96
·0018	0·42	·0051	1·20	·0084	1·98
·0019	0·45	·0052	1·23	·0085	2·01
·0020	0·47	·0053	1·25	·0086	2·03
·0021	0·50	·0054	1·27	·0087	2·05
·0022	0·52	·0055	1·30	·0088	2·08
·0023	0·54	·0056	1·32	·0089	2·10
·0024	0·57	·0057	1·35	·0090	2·12
·0025	0·59	·0058	1·37	·0091	2·15
·0026	0·61	·0059	1·39	·0092	2·17
·0027	0·64	·0060	1·42	·0093	2·20
·0028	0·66	·0061	1·44	·0094	2·22
·0029	0·68	·0062	1·46	·0095	2·24
·0030	0·71	·0063	1·49	·0096	2·27
·0031	0·73	·0064	1·51	·0097	2·29
·0032	0·76	·0065	1·53	·0098	2·31
·0033	0·78	·0066	1·56	·0099	2·34

TABLE XCIV-E.

TIME-EQUIVALENTS.

YOGA-INDEX UNITS.

Arg.	H.	M.	S.	Arg.	H.	M.	S.	Arg.	H.	M.	S.	Arg.	H.	M.	S.
1	0	3	39·63	31	1	53	28·55	61	3	43	17·47	91	5	33	6·39
2	0	7	19·26	32	1	57	8·18	62	3	46	57·10	92	5	36	46·02
3	0	10	58·89	33	2	0	47·81	63	3	50	36·73	93	5	40	25·65
4	0	14	38·52	34	2	4	27·44	64	3	54	16·36	94	5	44	5·29
5	0	18	18·15	35	2	8	7·07	65	3	57	56·00	95	5	47	44·92
6	0	21	57·78	36	2	11	46·71	66	4	1	35·63	96	5	51	24·55
7	0	25	37·41	37	2	15	26·34	67	4	5	15·26	97	5	55	4·18
8	0	29	17·05	38	2	19	5·97	68	4	8	54·89	98	5	58	43·81
9	0	32	56·68	39	2	22	45·60	69	4	12	34·52	99	6	2	23·44
10	0	36	36·31	40	2	26	25·23	70	4	16	14·15	100	6	6	3·07
11	0	40	15·94	41	2	30	4·86	71	4	19	53·78	200	12	12	6·14
12	0	43	55·57	42	2	33	44·49	72	4	23	33·41	300	18	18	9·21
13	0	47	35·20	43	2	37	24·12	73	4	27	13·04				
14	0	51	14·83	44	2	41	3·75	74	4	30	52·67				
15	0	54	54·46	45	2	44	43·38	75	4	34	32·30				
16	0	58	34·09	46	2	48	23·01	76	4	38	11·93				
17	1	2	13·72	47	2	52	2·64	77	4	41	51·56				
18	1	5	53·35	48	2	55	42·27	78	4	45	31·19				
19	1	9	32·98	49	2	59	21·90	79	4	49	10·83				
20	1	13	12·61	50	3	3	1·53	80	4	52	50·46				
21	1	16	52·24	51	3	6	41·17	81	4	56	30·09				
22	1	20	31·88	52	3	10	20·80	82	5	0	9·72				
23	1	24	11·51	53	3	14	0·43	83	5	3	49·35				
24	1	27	51·14	54	3	17	40·06	84	5	7	28·98				
25	1	31	30·77	55	3	21	19·69	85	5	11	8·61				
26	1	35	10·40	56	3	24	59·32	86	5	14	48·24				
27	1	38	50·03	57	3	28	38·95	87	5	18	27·87				
28	1	42	29·66	58	3	32	18·58	88	5	22	7·50				
29	1	46	9·29	59	3	35	58·21	89	5	25	47·13				
30	1	49	48·92	60	3	39	37·84	90	5	29	26·76				

TABLE XCIV-F.

Time-equivalents.

Decimals of yoga-index units.

First 2 decimals.	M.	S.	First 2 decimals.	M.	S.	First 2 decimals.	M.	S.
·01	0	2·20	·34	1	14·67	·67	2	27·15
·02	0	4·39	·35	1	16·87	·68	2	29·35
·03	0	6·59	·36	1	19·07	·69	2	31·55
·04	0	8·79	·37	1	21·26	·70	2	33·74
·05	0	10·98	·38	1	23·46	·71	2	35·94
·06	0	13·18	·39	1	25·66	·72	2	38·13
·07	0	15·37	·40	1	27·85	·73	2	40·33
·08	0	17·57	·41	1	30·05	·74	2	42·53
·09	0	19·77	·42	1	32·24	·75	2	44·72
·10	0	21·96	·43	1	34·44	·76	2	46·92
·11	0	24·16	·44	1	36·64	·77	2	49·12
·12	0	26·36	·45	1	38·83	·78	2	51·31
·13	0	28·55	·46	1	41·03	·79	2	53·51
·14	0	30·75	·47	1	43·23	·80	2	55·70
·15	0	32·94	·48	1	45·42	·81	2	57·90
·16	0	35·14	·49	1	47·62	·82	3	0·10
·17	0	37·34	·50	1	49·82	·83	3	2·29
·18	0	39·53	·51	1	52·01	·84	3	4·49
·19	0	41·73	·52	1	54·21	·85	3	6·69
·20	0	43·93	·53	1	56·40	·86	3	8·88
·21	0	46·12	·54	1	58·60	·87	3	11·08
·22	0	48·32	·55	2	0·80	·88	3	13·28
·23	0	50·52	·56	2	2·99	·89	3	15·47
·24	0	52·71	·57	2	5·19	·90	3	17·67
·25	0	54·91	·58	2	7·39	·91	3	19·86
·26	0	57·10	·59	2	9·58	·92	3	22·06
·27	0	59·30	·60	2	11·78	·93	3	24·26
·28	1	1·50	·61	2	13·97	·94	3	26·45
·29	1	3·69	·62	2	16·17	·95	3	28·65
·30	1	5·89	·63	2	18·37	·96	3	30·85
·31	1	8·09	·64	2	20·56	·97	3	33·04
·32	1	10·28	·65	2	22·76	·98	3	35·24
·33	1	12·48	·66	2	24·96	·99	3	37·43

3rd and 4th decimals.	S.	3rd and 4th decimals.	S.	3rd and 4th decimals.	S.
·0001	0·02	·0034	0·75	·0067	1·47
·0002	0·04	·0035	0·77	·0068	1·49
·0003	0·07	·0036	0·79	·0069	1·52
·0004	0·09	·0037	0·81	·0070	1·54
·0005	0·11	·0038	0·83	·0071	1·56
·0006	0·13	·0039	0·86	·0072	1·58
·0007	0·15	·0040	0·88	·0073	1·60
·0008	0·18	·0041	0·90	·0074	1·63
·0009	0·20	·0042	0·92	·0075	1·65
·0010	0·22	·0043	0·94	·0076	1·67
·0011	0·24	·0044	0·97	·0077	1·69
·0012	0·26	·0045	0·99	·0078	1·71
·0013	0·29	·0046	1·01	·0079	1·74
·0014	0·31	·0047	1·03	·0080	1·76
·0015	0·33	·0048	1·05	·0081	1·78
·0016	0·35	·0049	1·08	·0082	1·80
·0017	0·37	·0050	1·10	·0083	1·82
·0018	0·40	·0051	1·12	·0084	1·84
·0019	0·42	·0052	1·14	·0085	1·87
·0020	0·44	·0053	1·16	·0086	1·89
·0021	0·46	·0054	1·19	·0087	1·91
·0022	0·48	·0055	1·21	·0088	1·93
·0023	0·51	·0056	1·23	·0089	1·95
·0024	0·53	·0057	1·25	·0090	1·98
·0025	0·55	·0058	1·27	·0091	2·00
·0026	0·57	·0059	1·30	·0092	2·02
·0027	0·59	·0060	1·32	·0093	2·04
·0028	0·61	·0061	1·34	·0094	2·06
·0029	0·64	·0062	1·36	·0095	2·09
·0030	0·66	·0063	1·38	·0096	2·11
·0031	0·68	·0064	1·41	·0097	2·13
·0032	0·70	·0065	1·43	·0098	2·15
·0033	0·72	·0066	1·45	·0099	2·17

TABLES FOR FINDING THE MEAN PLACE OF THE PLANET SATURN.

By J. F. Fleet, I.C.S. (Retd.), Ph.D., C.I.E.

Extracted and reprinted from the Journal of the Royal Asiatic Society, October 1915, by permission of the Council of the Society.

In examining the astrological details of a date in Śaka 380 (J. R. A. S., 1915, p. 482), I had to work out the bases for tables, and to make parts of the tables themselves, for finding the mean place of the planet Saturn, that is, his mean longitude, according to the *First Ārya-Siddhānta* and the *Original* and *Present Sūrya-Siddhāntas*. It has seemed useful to complete the tables and publish them, with examples of the use of them, so that they may be available for any future work of the same kind.[1] At the same time, I seek to give them an interest by attaching some general remarks and showing the bases from which they have been made.

General Remarks.

The starting-point of my tables is the beginning of the Kaliyuga era in B.C. 3102, when, according to the Hindū astronomy, there was the latest recurrence of a conjunction of all the planets (including the sun and the moon), by their mean longitudes, at the initial point of the Hindū sphere, namely, the point 0° of the sidereal sign Mēsha (Aries).[2] According to the *First Ārya-Siddhānta* this conjunction was at mean sunrise, 6.0 a.m., for the prime meridian of Laṅkā-Ujjain, on 18 February in the said year. According to the two *Sūrya-Siddhāntas* it was at the preceding midnight.

The years in my tables are the mean sidereal solar years of the Kaliyuga: and, as a first step in using the tables, for any given year of the Śaka or any other Hindū era, or of our era, we must take the corresponding year of the Kaliyuga.[3] Each year is the period in which the sun by mean motion travels round the circle of the heavens from the point 0° of the sign Mēsha back to the same point. The length of this year differs slightly according to each of the three authorities, as a result of the difference in the number of days assigned by them (see farther on, under the Bases) to the *exeligmos* or calculative period of 4,320,000 years which constitutes the Yuga, Mahāyuga, or Chaturyuga, the cycle of Four Ages. The lengths of the years are as follows: –

	days.		*d.*	*h.*	*m.*	*s.*
First Ārya-S.	365·258680$\dot{5}$	=	365	6	12	30
Original Sūrya-S. . . .	365·25875	=	365	6	12	36
Present Sūrya-S.	36·2587564$\dot{8}\dot{1}$	=	365	6	12	36·56

The days are mean natural or civil days, each of exactly twenty-four hours. For calculative purposes they run from mean sunrise to mean sunrise according to the *First Ārya-Siddhānta*, and from the preceding midnight to midnight according to the two *Sūrya-Siddhāntas*. But for ordinary use the Hindū day runs from true sunrise to true sunrise according to both the schools.

The revolution of Saturn is his journey round the heavens, through the twelve signs of the zodiac and the twenty-seven *nakshatras* or "lunar mansions", from the point 0° of the sign

[1] Tables by Professor Jacobi (on quite different lines) for finding both the mean and the true places of all the planets according to the *Present Sūrya-Siddhānta*, have been published in the *Epigraphia Indica*, Vol. 12, p. 79 ff. I had not seen these when my paper in question was written. Professor Jacobi's process is a shorter one, as a result of much work done by him in making his tables. But his tables do not make mine unnecessary, even for the *Present Sūrya-Siddhānta*; in the first place, because we want, for any time before about A.D. 1000, a much earlier guide than that work; and secondly, because they do not give the very close results which are to be got from my tables.

[2] On this matter see my paper on the Kaliyuga in J. R. A. S., 1911, p. 493.

[3] We might, of course, lay down as an additive constant the place of Saturn, according to each of the three authorities, for the beginning of the Śaka era in A.D. 78, or for any other chosen time, and then work for only the remaining years. But in my opinion little, if anything, is really gained by that method.

Mēsha back to the same point. His revolution and longitude are, of course, geocentric; the earth being regarded as the centre of the universe in the Hindū astronomy.

From Table I, which gives Saturn's mean yearly motion, we get, as the first step in any working, the number of revolutions completed by him, and, over and above that, his mean place or longitude in signs, degrees, minutes, and seconds, reckoned from the point 0° of Mēsha, at the moment of the mean Mēsha-saṁkrānti, or entrance of the sun into Mēsha, of the given year; that is, at the moment of the mean vernal equinox, which is the astronomical beginning of the year. The date and time of that moment may be ascertained from Sewell and Dīkshit's *Indian Calendar*, Table I, taken with the intervals between the true and mean Mēsha-saṁkrāntis given on p. 12, and Sewell's *Indian Chronography*, tables 17 and 38, A, and p. 57. It is not always necessary to reduce Saturn's place at that moment to his place at mean sunrise on that same day, as I have done in Example 1 below (p. 616): but it is generally useful to do so; especially if we are likely to work for more days than one in one and the same year.

In using Table I, the seconds in the first nine years may be turned into even numbers by rejecting anything up to ·5 and taking anything over ·5 as 1 to be added to the integral number.

Table II, which gives Saturn's mean daily motion and supplies what is wanted for finding his mean place or longitude at any subsequent time in the same year, is in two parts: A, for general use, with the seconds treated on those same lines; and B, for closer work, with the actual seconds to three places of decimals, determined by rejecting anything up to ·0005 and treating anything over that as 1 to be added to the third figure.

Results worked from Table I, with the seconds treated as indicated above, and Table II, Part A, will be close enough for all general purposes. But, if it is ever necessary,—as, for instance if a resulting place is very near to the beginning of a sign or a *nakshatra*, when a few seconds of arc may make a difference in the sign or the *nakshatra*; or if a resulting time is very near to sunrise, when a few minutes of time or seconds of arc may make a difference in the day,—to get a still closer result, then we must work with the decimals given in Table I and Table II, Part B, and must also use actual minutes and seconds, instead of even minutes, in the time of the Mēsha saṁkrānti: in short, we must then work with exactness all through.

Means may perhaps be added hereafter for finding the true place of Saturn, that is, his true or apparent longitude. But that does not seem necessary at present: there are various indications that the mean places are the right ones to take for the planets down to at any rate about A.D. 1000. And certainly, if a statement about any planet is found to be correct for its mean place though not for its true place, we need not condemn the statement on that account.

In addition to the details given in the next section, which explains the bases of my tables the following may be noted here:—

The period of Saturn, the time in which he makes one revolution, works out according to the three authorities as follows:—

First Ārya-S.	10766·0646543489...	days.
Original Sūrya-S.	10766·0667012363...	,,
Present Sūrya-S.	10765·7730746138...	,,

In terms of the mean Julian year of 365·25 days, these figures represent—

First Ārya-S.	29·4758785882...	years
Original Sūrya-S	29·4758841922...	,,
Present Sūrya-S.	29·4750802864...	,,

These cannot be expressed exactly in years, months, and days, because our months have not a uniform number of days. But, with the month taken at 365·25 ÷ 12 = 30·4375 days, they represent (say)—

First Ārya-S. . . .	29y.	5m.	21·62715d.
Original Sūrya-S. . . .	29y.	5m.	21·62920d.
Present Sūrya . . .	29y.	5m.	21·33557d.

The periods given above are geocentric, as has already been said. Modern science gives the period of Saturn's sidereal revolution round the sun as—

10759·2198 days,[1] = 29·457... years.

Slightly better Hindū approximations were got by Lalla and the person who devised the corrections for the *Present Sūrya-Siddhānta* : see pp. 603, 605, below. Of these, Lalla's result was the nearer, but only by a little more than three minutes : this is due to his *exeligmos* being shorter by 328 days.

Bases of the Tables.

First Ārya-Siddhānta.

By this name is meant the *Āryabhaṭīya*, which was written by Aryabhaṭa at Kusumapura, *i.e.*, Pāṭaliputra, Paṭna, in or soon after A.D. 499.[2] The text, with the commentary by Paramādiśvara, has been edited by Professor Kern (Leiden, 1874). Its elements in this matter are :—

146,564 revolutions of Saturn in the Yuga of 4,320,000 years comprising 1,577,917,500 days.

The mean yearly motion is—

$$\frac{146564 \times 360^\circ}{4320000} = 12^\circ{\cdot}213\dot{6} = 12^\circ\ 12'\ 49{\cdot}2''$$

The mean daily motion is—

$$\frac{146564 \times 360^\circ \times 60'}{1577917500} = 2'{\cdot}0063041318\ldots$$

$$= 2'\ 0''{\cdot}3782479122\ldots$$

Saturn's period of revolution has been given on p. 601 above. A sign being one-twelfth of a revolution, and a *nakshatra* being one-twenty-seventh of the same,[3] it follows that he spends in one sign 897·1720545290 days, = 2·4563232156 Julian years, or (say)—2y. 5m. 14·48455d.; and in one *nakshatra* 398·7431353462 days, or (say)—398d. 17h 50·11490m.

Lalla, who was the exponent of Āryabhaṭa and seems to have written in the period A.D. 600-650, introduced certain *bījas* or corrections for the mean motions of all the planets, to be applied to the *First Ārya-Siddānta* with effect from the year Śaka 420 expired, so as to bring their calculated places into agreement with their places as determined by observation.[4] In the case of Saturn he added $\frac{20'}{250} = 4''{\cdot}8$, by which he raised the mean yearly motion from 12° 12′ 49″·2 to 12° 12′ 54″. Since one revolution in 4,320,000 years would represent 0″·3 mean yearly

[1] Lockyer, *Elementary Lessons in Astronomy* (1907), p. 350.

[2] See my paper in J. R. A. S., 1911, p. 110.

[3] That is, according to the equal-space system, by which each *nakshatra* measures 13° 20′.

[4] See his *Sishyadhīvriddhida*, ed. Sudhakara Dvivedi, Benares, 1886, p. 10 verses 59, 60 ; p. 56 verses 18, 19.

motion, and 4·8 divided by 0·3 = 16, this *bīja* had the effect of increasing the revolutions of Saturn in such a period from 146,564 to 146,580[1]; and (since the number of days in the *exeligmos* remained the same) of increasing also the mean daily motion, and of shortening the period of revolution. Thus, according to Lalla,—

The mean yearly motion became—

$$\frac{146580 \times 360^\circ}{4320000} = 12^\circ \cdot 215 = 12^\circ\ 12'\ 54''$$

The mean daily motion became—

$$\frac{146580 \times 360^\circ \times 60'}{1577917500} = 2' \cdot 0065231547\ldots$$
$$= 2'\ 0'' \cdot 3913892836\ldots$$

And Saturn's period of revolution became—

$$\frac{1577917500}{146580} = 10764 \cdot 8894801473\ldots \text{ days}$$
$$= 29 \cdot 4726611366\ldots \text{ Julian years}$$
$$\text{or (say) } 29\text{y} \quad 5\text{m.} \quad 20 \cdot 45198\text{d.}$$

The place of Saturn according to Lalla is got by adding 4″·8 for each year after Śaka 420 expired, = Kaliyuga 3599 expired, to his place as found according to the *First Ārya-Siddhānta*

Original Sūrya-Siddhānta.

This work is only known from Varāhamihira's statements about it in his *Pañchasiddhāntikā*, which was written about A.D. 550.[2] The Siddhānta itself (its author is not known) seems to date from much about the same time with the *First Ārya-Siddhānta*, but is perhaps rather earlier than that work The *Pañchasiddhāntikā* has been edited by Dr. Thibaut and the Mahāmahōpādhyāya Sudhākara Dvivedi, with a Sanskrit commentary by the editors and an English translation (Benares, 1889). Here the elements are :—

146,564 revolutions of Saturn in 4,320,000 years comprising 1,577,917,800 days.[3]

The number of revolutions being the same, the mean yearly motion is also exactly the same as by the *First Ārya-Siddhānta*; viz.—

$$\frac{146564 \times 360^\circ}{4320000} = 12^\circ \cdot 2\dot{1}3\dot{6} = 12^\circ\ 12'\ 49'' \cdot 2:$$

and so the place of Saturn according to this work at the beginning of a year differs from his place according to the *First Ārya-Siddhānta* only in proportion to the time by which the mean Mēsha-saṁkrānti of this work differs from that of the mean Mēsha-saṁkrānti of the *First Ārya-Siddhānta*.

The number of days being more by 300, the mean daily motion is slightly less, viz.—

$$\frac{146564 \times 360^\circ \times 60'}{1577917800} = 2' \cdot 0063037504\ldots$$
$$= 2'\ 0'' \cdot 3782250252\ldots$$

---

[1] Lalla, however, did not put his corrections in this shape.

[2] There is a very useful paper on the *Original Sūrya-Siddhānta*, by Sh. B. Dikshit, in the *Indian Antiquary*, Vol. 19 (1890), p. 45. It seems likely that the text of the work might be found in Burma or Arakan, as it has been followed there down to quite recent times: see, *e.g.*, Sir Alfred Irwin's *Burmese and Arakanese Calendars* (1909), p. 3, and his "Elements of the Burmese Calendar from A.D. 638 to 1752" in *Ind. Ant.*, 1910, p. 289.

[3] The actual *exeligmos* or calculative period of this work is one of 180,000 years comprising 65,746,575 days, and the numbers of the revolutions of the planets are not stated in actual words. The editors have worked out the numbers of the revolutions for the longer *exeligmos* from the details given in *Pañchasiddhāntikā*, Chapter 16; trans., p. 91 comment., p. 88; introd., p. 19.

Saturn's period has been given on p. 612 above. It follows that he spends—

in one sign 897·1722251030 days, =
2·4563236826.. Julian years, or (say)—
2y. 5m. 14·48473d.; and—
in one *nakshatra* 398·7432111569 .. days, or (say)—
398d. 17h. 50·22407m.

Present Sūrya-Siddhānta.

This work is well known from the translation by E. Burgess, with Whitney's invaluable notes, published in the Journal of the American Oriental Society, Vol. 6 (1860), pp. 141-498.[1] Its text, with the commentary by Raṅganātha, has been given by F. E. Hall and Pandit Bapu Deva Sastri in the *Bibliotheca Indica* series (Calcutta, 1859) and by Pandit Hari Shankar (Benares, 1881). It is not known when and by whom the work was written. But, as was pointed out by Whitney (*loc. cit.*, p. 424), its general system is older than that of Bhāskarāchārya's *Siddhānta Śirōmaṇi* (written A.D. 1150). And Sh. B. Dikshit has said that it superseded the *Original Surya-Siddhānta* probably not later than A.D. 1000.[2] Bhaṭṭōtpala, writing his commentary on the *Bṛihat-Saṁhitā*, Chapter 2, at some time about A.D. 966 does not seem to quote there any of the elements in which the *Present* differs from the *Original Sūrya-Siddhānta*. According to this work, the elements in our present matter are:—

146,568 revolutions of Saturn in 4,320,000 years comprising 1,577,917,828 days: which figures increase the yearly and daily motion and shorten the period of revolution.

The mean yearly motion is—

$$\frac{146568 \times 360^\circ}{4320000} = 12^\circ{\cdot}214 = 12^\circ\ 12'\ 50''{\cdot}4$$

The mean daily motion is—

$$\frac{146568 \times 360^\circ \times 60'}{1577917828} = 2'{\cdot}0063584705\ldots$$
$$= 2'\ 0''{\cdot}3815082314\ldots$$

Saturn's period has been given on p. 601 above. It follows that he spends—

in one sign 897·1477562178... days, =
2·4562566905.. Julian years, or (say)—
2y. 5m. 14·46026d.; and—
in one *nakshatra* 398·7323360968... days, or (say)—
398d. 17h. 34·56398m.

The elements of the *Present Sūrya-Siddhānta*, that is, its number of days for the 4,320,000 years and its numbers of the revolutions of the planets in that period, may be regarded as the results of *bījas* or corrections applied to the *Original Sūrya-Siddhānta*. To the *Present Sūrya-Siddhānta* itself certain *bījas* were applied in the fifteenth century, with effect from the beginning of the Kaliyuga; and by one of them the number of revolutions of Saturn was raised to 146,580 in the *exeligmos* of the same number of years and days.[3]

---

[1] There is also a translation, with a few notes, by Pandit Bapu Deva Sastri (Calcutta, 1861).

[2] *Indian Calendar*, p. 8.

[3] For a useful note on these *bījas*, see Sh. B. Dikshit's *Bhāratīya-Jyōtiḥśāstra* or "*History of Indian Astronomy*," p. 184. Who devised these corrections is not known: but they are stated in the shape of the resulting numbers of the revolutions, in the *Makaranda*, a work composed by an author of that same name, a resident of Benares, who is believed to have written it in A.D. 1478. It seems to be only by a coincidence that the number of revolutions thus assigned to Saturn, *viz.* 146,580, is the same with that which results from the correction for Saturn applied by Lalla to the *First Arya-Siddhānta*.

This further raised—

the mean yearly motion to $12''{\cdot}215 = 12''$ $12'''$ $54''''$, and

the mean daily motion to $2'$ $0''{\cdot}3913642560$ .. ;

and reduced—

the period of revolution to $10764{\cdot}8917178332$... days.

The place of Saturn according to this *bîja* is got by adding $3''{\cdot}6$ for each year from the beginning of the Kaliyuga, to his place as found according to the *Present Sūrya-Siddhânta*.

EXAMPLES.

The place of Saturn means here his place by mean motion; that is, his mean longitude.

The times are for mean sunrise, 6.0 A.M., at Ujjain, the Hindū Greenwich.

The *nakshatras* are taken according to the equal-space system, by which each of them measures $13°$ $20'$.[1]

1. What was the place of Saturn, according to the *First Ārya-Siddhānta*, at mean sunrise on 25 August, A.D. 458, on which day there began the *tithi* Āśvina śukla 1, Śaka 380 expired?

Śaka 380 expired being the Kaliyuga year 3559 expired, we proceed as follows; omitting the revolutions as not being wanted for present purposes, but bearing in mind that every twelve signs add one more revolution, and that we have to take into account here only the excess over the revolutions:—

By Table I, col. A :—

	Signs.	°	′	″
years: 3000	9	11	0	0
500	11	16	50	0
50	8	10	41	0
9	3	19	55	23
Place of Saturn at mean Mesha-saṁkrānti, Śaka 380 expired, viz. on 20 March, A.D. 458, at 15^h 27^m [2]	8	28	26	23

We reduce this for mean sunrise on that same day by deducting his motion for 15^h 27^m or say 15^h 24^m, at 1 hour$=5''$ and 12 minutes$=1''$, $=77''$, $=1'$ $17''$:—

	Signs.	°	′	″
from	8	28	26	23
deduct for 15^h 24^m			1	17
Place of Saturn at mean sunrise on 20 March, A.D. 458	8	28	25	6

Since 20 March is the day 79 of the year A.D. 458,[3] and 25 August is the day 237, we proceed for $237-79=158$ days, which will take us from any particular moment (in this case mean sunrise) on 20 March to the same moment on 25 August :—

	Signs.	°	′	″
Therefore to	8	28	25	6
add for days (Table II, Part A) :—				
days: 100		3	20	28
50		1	40	19
8			16	2
Place of Saturn at mean sunrise on 25 August, A.D. 458	9	3	42	6

[1] For the necessary details of the *nakshatras*, according to both this system and the two systems of unequal spaces, see Sewell's *Indian Chronology*, Table 22.

[2] See *Indian Calendar*, Table I. [R. S.]

[3] *Ind. Cal.*, Table IX, or above, Table LXIX. [R. S.]

Accordingly, at mean sunrise on the given day, Saturn had completed nine signs of his current revolution; and was at the point 3° 42′ 6″ of the tenth sign Makara (Capricornus).

Also, since 9ˢ 3° = 273°, and the *nakshatra* Uttara-Ashāḍhā begins at 266° 40′ and ends at 280°, he was at the point 273° 42′ 6″ − 266° 40′ = 7° 2′ 6″ of that *nakshatra*.

2. When, according to the *First Ārya-Siddhānta*, did Saturn enter the *nakshatra* Uttara-Ashāḍhā, in which, as we have found above, he was on 25 August, A.D. 458, in Śaka 380 expired?

It is seen almost at a glance that this must have been before the beginning of Śaka 380 expired, *i.e.*, in the preceding Śaka year. Accordingly, we proceed as follows:—

From Example 1.—

	Signs.	°	′	″
Place of Saturn at mean Mēsha-saṁkrānti, Śaka 380 expired	8	28	26	23
Deduct mean yearly motion for one year (Table I, col. A)		12	12	49
Place of Saturn at mean Mēsha-saṁkrānti, Śaka 379 expired, on 20 March, A.D. 457, at 9ʰ 14ᵐ [1]	8	16	13	34
Deduct for 9ʰ 14ᵐ, or say 9ʰ 12ᵐ, at 1ʰ = 5′ and 12ᵐ = 1″				46
Place of Saturn at mean sunrise on 20 March, A.D. 457	8	16	12	48

Since 8ˢ 16° = 256°, and Uttara-Ashāḍhā begins at 266° 40′,—

	°	′	″
from	266	40	0
deduct place at mean sunrise on 20 March, A.D. 457	256	12	48
remainder	10	27	12

This remainder is the distance which Saturn then had to go to enter Uttara-Ashāḍhā. It amounts to 627′ 12″, which, at 2′ per day, represents roughly (but appreciably less than) 313½ days. We try for 312 days:—

	°	′	″
distance to go	10	27	12
deduct for days (Table II, Part A):—			
days: 300	10	1	53
10		20	4
2		4	1
	10	25	58
remainder still to go		1	14

This remainder being less than the mean motion for one day, viz. 2′ we see that we have got the right day.

Now, 20 March being the day 79[2] of the year A.D. 457, we have 79 + 312 = 391 − 365 = 26, which takes us from any particular moment (in this case, mean sunrise) on 20 March, A.D. 457, to the same moment on 26 January, A.D. 458. Accordingly, we have:—

	°	′	″
Place of Saturn at mean sunrise on 20 March, A.D. 457 .	256	12	48
add for 312 days, as above	10	25	58
Place of Saturn at mean sunrise on 26 January, A.D. 458 .	266	38	46

[1] See *Indian Calendar*, Table I. [R. S.]

[2] Table IX, *Indian Calendar*, or Table LXIX above. [R. S.]

Saturn then still had to go 1′ 14″, or say 1′ 15″, to enter Uttara-Ashāḍhā : and at 5″ per hour this represents 75÷5=15 hours.

Accordingly, he entered Uttara-Ashāḍhā at 15 hours after mean sunrise on 26 January, A.D. 458.

3. In the same period, and again according to the *First Ārya-Siddhānta*, on what day did Saturn leave Uttara-Ashāḍhā and enter the next *nakshatra* Śravana ?

This can be got from what we have worked under Example 2, thus :—

We have found there that Saturn entered Uttara-Ashāḍhā at 15 hours after mean sunrise on 26 January, A.D. 458.

His time in each *nakshatra* (see p. 602 above) is $398^d\ 17^h\ 50{\cdot}11490^m$.

		d.	h.	m.
To the day and time in January, A.D. 458		26	15	0
add for one *nakshatra*		398	17	50
		425	8	50
deduct days—				
in A.D. 458	365			
in Jan., A.D. 459	31			
in Feb., ,, ...	28 =	424		
remainder		1	8	50

That is, he left Uttara-Ashāḍhā and entered Śravaṇa at $8^h\ 50^m$ after mean sunrise on the day 1 after 28 February, that is, on 1 March, A.D. 459.

Remark.—By actual working from the mean Mēsha-saṁkrānti in A.D. 458, we should find the time to be 9 hours. The difference, 10 minutes, = less than 1″ of longitude, is due to the way in which we have worked, and is negligible for present purposes : we only wanted to fix the day ; and the time is so far from sunrise as to leave no doubt as to that. But this process of carrying on—(and so, also, that of carrying back, used under Example 2 by deducting for a year instead of making a separate calculation)—must be used cautiously.

I. Mean Yearly Motion.

"Rev."—complete revolutions . . . . "Sign."—sign of the zodiac.

	A.					B.				
	First Ārya- and Original Sūrya-Siddhāntas.					Present Sūrya-Siddhānta.				
years.	Rev.	Sign.	°	′	″	Rev.	Sign.	°	′	″
1			12	12	49·2			12	12	50·4
2			24	25	38·4			24	25	40·8
3		1	6	38	27·6		1	6	38	31·2
4		1	18	51	16·8		1	18	51	21·6
5		2	1	4	6·0		2	1	4	12·0
6		2	13	16	55·2		2	13	17	2·4
7		2	25	29	44·4		2	25	29	52·8
8		3	7	42	33·6		3	7	42	43·2
9		3	19	55	22·8		3	19	55	33·6
10		4	2	8	12·0		4	2	8	24·0
20		8	4	16	24·0		8	4	16	48·0
30	1	0	6	24	36·0	1	0	6	25	12·0
40	1	4	8	32	48·0	1	4	8	33	36·0
50	1	8	10	41	0·0	1	8	10	42	0·0
60	2	0	12	49	12·0	2	0	12	50	24·0
70	2	4	14	57	24·0	2	4	14	58	48·0
80	2	8	17	5	36·0	2	8	17	7	12·0
90	3	0	19	13	48·0	3	0	19	15	36·0
100	3	4	21	22	0·0	3	4	21	24	0·0
200	6	9	12	44	0·0	6	9	12	48	0·0
300	10	2	4	6	0·0	10	2	4	12	0·0
400	13	6	25	28	0·0	13	6	25	36	0·0
500	16	11	16	50	0·0	16	11	17	0	0·0
600	20	4	8	12	0·0	20	4	8	24	0·0
700	23	8	29	34	0·0	23	8	29	48	0·0
800	27	1	20	56	0·0	27	1	21	12	0·0
900	30	6	12	18	0·0	30	6	12	36	0·0
1000	33	11	3	40	0·0	33	11	4	0	0·0
2000	67	10	7	20	0·0	67	10	8	0	0·0
3000	101	9	11	0	0·0	101	9	12	0	0·0
4000	135	8	14	40	0·0	135	8	16	0	0·0
5000	169	7	18	20	0·0	169	7	20	0	0·0

II. Mean daily motion.

A. For all the three Siddhāntas: with even seconds.

For parts of a day, 1 hour = 5″; 12 minutes = 1″.

days.	°	′	″	days.	°	′	″	days.	°	′	″	days.	°	′	″
1		2	0	6		12	2	20		40	8	70	2	20	26 [a]
2		4	1	7		14	3	30	1	0	11	80	2	40	30 [b]
3		6	1	8		16	3	40	1	20	15	90	3	0	4
4		8	2	9		18	3	50	1	40	19	100	3	20	38
5		10	2	10		20	4	60	2	0	23	200	6	41	16
												300	10	1	3 [c]

a For the *Present Sūrya-Siddhānta*, the seconds here are 27.
b For the *Present Sūrya Siddhānta*, the seconds here are 31.
c For the *Present Sūrya Siddhānta*, the seconds here are 54.

B. For the separate Siddhāntas: with actual seconds.

	First Ārya.			Original Sūrya.			Present Sūrya.		
days.	°	′	″	°	′	″	°	′	°
1		2	0·378		2	0·378		2	0·382
2		4	0·756		4	0·756		4	0·763
3		6	1·135		6	1·135		6	1·145
4		8	1·513		8	1·513		8	1·526
5		10	1·891		10	1·891		10	1·908
6		12	2·269		12	2·269		12	2·289
7		14	2·648		14	2·648		14	2·671
8		16	3·026		16	3·026		16	3·052
9		18	3·404		18	3·404		18	3·434
10		20	3·782		20	3·782		20	3·815
20		40	7·565		40	7·565		40	7·630
30	1	0	11·347	1	0	11·347	1	0	11·445
40	1	20	15·130	1	20	15·129	1	20	15·260
50	1	40	18·912	1	40	18·911	1	40	19·075
60	2	0	22·695	2	0	22·694	2	0	22·890
70	2	20	26·477	2	20	26·476	2	20	26·706
80	2	40	30·260	2	40	30·258	2	40	30·521
90	3	0	34·042	3	0	34·040	3	0	34·336
100	3	20	37·825	3	20	37·823	3	20	38·151
200	6	41	15·650	6	41	15·645	6	41	16·302
300	10	1	53·474	10	1	53·458	10	1	54·452

SGPI—S17—X-8-16—27-8-24—800.